TABLE A-3 PHYSICAL CONSTANTS

Avogadro's number	6.022169
Boltzmann constant	1.380622×10^{-23} J/K
Planck's constant	6.626196×10^{-34} J·s
Speed of light	2.9979250×10^8 m/s
Electronic charge	$1.6021917 \times 10^{-19}$ C
Bohr magneton	9.274096×10^{-24} A·m^2
Permeability of free space	4×10^{-7} N/A^2
Standard gravitational acceleration	9.80665 m/s^2 = 32.174 ft/s^2

TABLE A-4 UNIVERSAL GAS CONSTANT

8314.29 J/kgmole·K
8.31429 kJ/kgmole·K
0.0820560 atm·m^3/kgmole·K
1.98583 kcal/kgmole·K
1545.31 ft·lbf/lbmole·°R
1.98583 Btu/lbmole·°R
0.730225 atm·ft^3/lbmole·°R

Engineering Thermodynamics

Jui Sheng Hsieh

New Jersey Institute of Technology

PRENTICE HALL, Englewood Cliffs, NJ 07632

Library of Congress Cataloging-in-Publication Data

Hsieh, Jui Sheng
    Engineering thermodynamics / Jui Sheng Hsieh.
        p. cm.
    Includes bibliographical references and index.
    ISBN 0-13-275702-8
    1. Thermodynamics. I. Title.
TJ265.H76 1993
621.402'1--dc20   91-38933
                        CIP

Acquisitions editor: **DOUG HUMPHREY**
Editorial/production supervision and
  interior design: **RICHARD DeLORENZO**
Copy editor: **PETER ZURITA**
Cover design: **WANDA LUBELSKA**
Prepress buyer: **LINDA BEHRENS**
Manufacturing buyer: **DAVID DICKEY**
Editorial assistant: **JAIME ZAMPINO**
Supplements editor: **ALICE DWORKIN**

© 1993 by Prentice-Hall, Inc.
A Simon & Schuster Company
Englewood Cliffs, New Jersey 07632

Printed in the United States of America

10 9 8 7 6 5 4 3 2 1

ISBN 0-13-275702-8

Prentice-Hall International (UK) Limited, London
Prentice-Hall of Austria Pty. Limited, Sydney
Prentice-Hall Canada Inc., Toronto
Prentice-Hall Hispanoamericana, S.A., Mexico
Prentice-Hall of India Private Limited, New Delhi
Prentice-Hall of Japan, Inc., Tokyo
Simon & Schuster Asia Pte. Ltd., Singapore
Editora Prentice-Hall do Brasil, Ltda., Rio de Janeiro

Contents

Contents

Contents

Preface

Thermodynamics is the science of energy. This book provides a rigorous and comprehensive treatment of the basic principles and engineering applications of thermodynamics. The presentation of the subject follows the traditional classical, or macroscopic, approach. It is intended to fit undergraduate engineering curricula and contains enough material for a two-semester thermodynamics course. With proper selection of materials it can be used for a single course given in one semester.

Throughout the preparation of this book the student has been foremost in the author's mind. Sufficient detail is given in the presentation of the subject matter. Important derivations and calculations are not left to the student but are contained in the main body of the text. A large number of completely solved examples are provided to illustrate the theories and applications. There are many end-of-chapter problems which model practical engineering situations, with accompanying schematics to enhance the student's understanding of the problem.

Chapter 1 presents basic definitions, concepts, and schematics of some typical engineering applications. The first law of thermodynamics is introduced in Chapter 2 along with the formulation of various work modes. Thermodynamic properties of pure substances are discussed in Chapter 3 with emphasis on the use of tabulated property data in energy analyses. The concept of ideal gas and its use as a simple model of the actual behavior of a pure substance is introduced in Chapter 4. The conservation-of-mass and conservation-of-energy equations are presented in Chapter 5, including steady-flow and uniform-flow typical processes of application.

Chapters 6 and 7 are devoted to a thorough treatment of the second law of thermodynamics and its consequences. The property "entropy" is developed from the macroscopic viewpoint, with a microscopic interpretation added as an aid to understanding the nature of this property. The concepts of availability and irreversibility are developed in Chapter 8, laying the foundation for the study of second-law analysis and second-law effectiveness.

Chapters 9 through 12 illustrate the engineering applications concerning air-conditioning, gas power, vapor power, and refrigeration. Innovative energy systems, such as combined cycles, cogeneration systems, and low-temperature Rankine cycles, are included in the study. Also included is the subject of cryogenics, with an emphasis on gas liquefaction.

Thermodynamic relations for simple compressible systems are presented in Chapter 13, including general equations for specific heats, internal energy, enthalpy, entropy, Helmholtz function, Gibbs function, and Maxwell relations. In addition, general equations for simple paramagnetic systems are included in this chapter to lay the basis for the study of magnetic cooling in cryogenics for attaining extremely low temperatures, approaching the absolute zero as a limit. Analytical and graphical equations of state for real gases and real-gas mixtures are treated in Chapter 14, with detailed numerical illustrations on the use of these equations along with the general equations developed in Chapter 13 to evaluate various thermodynamic properties and heat and work interactions.

Chemical reactions, with emphasis on the first-law analysis of combustion processes are given in Chapter 15. Whereas second-law analyses of reactive systems are presented in Chapter 16 when the absolute entropy and Gibbs function of formation are studied. In addition to availability analysis of reactive systems, Chapter 16 covers the topics of stability, phase and reaction equilibrium, equilibrium constant, and the third law of thermodynamics. A detailed study of absorption refrigeration analysis is presented as an illustration of phase equilibrium of binary vapor-liquid mixtures.

The presentation of the second-law analysis and second-law effectiveness in the main body of the text and in the end-of-chapter problems are arranged in such a way that give the instructor the choice of covering the first-law and second-law analyses together or separately. The structure of the book can be easily adopted to the case where a brief coverage of the second-law analysis in one thermodynamics course and a thorough coverage of this material in another thermodynamics course are called for.

A simplified, but comprehensive, summary is included at the end of each chapter. These summaries can be used for review classes by the instructor. They can serve as the last-minute quick review materials by the student before taking an examination. They can also be used by the student as formula sheets during exams if the instructor prefers closed-book tests and yet wants to make the important equations available to the students.

A bibliography at the end of the book gives a selected group of references that can be helpful to the students for their current and future study of thermodynamics. It should be noted that the number in square brackets in the text refers to the number in the Bibliography.

Most countries in the world use the metric system of units. Old English units, however, are still widely used in some industries and everyday life in the United States. This book uses both SI and English units, with an emphasis on SI. There are more examples and chapter-end-problems which use SI than English units, the ratio of examples written in English units to that in SI being about 3 to 5. This book can be covered using combined SI and English units or SI units alone, depending on the preference of the instructor. Tables and charts of properties are provided in both sets of units. A table of unit conversion is included.

It is a great pleasure to acknowledge my indebtedness to Dr. E. M. Sparrow of University of Minnesota for his detailed review and invaluable suggestions on the manuscript. It is with deep appreciation that I express thanks to the numerous and valuable comments, suggestions, criticism, and praise of the following academic reviewers: Dr. P. S. Ayyaswamy of University of Pennsylvania, Dr. J. E. Drummond of University of Akron, Dr. S. Goplen of North Dakota State University, Dr. G. S. Jakubowski of Memphis State University, Dr. J. E. Peters of University of Illinois-Chicago, Dr. C. S. Reddy of Union College, and Dr. J. W. Sheffield of University of Missouri-Rolla. Thanks are also due to Dr. R. P. Kirchner of New Jersey Institute of Technology for using the manuscript in his thermodynamics class at NJIT. Finally, I wish to express my thanks and appreciation to my wife Mary, my son Lawrence, and my daughters Esther and Vivian for their encouragement, support, and typing efforts throughout the preparation of this text.

J. S. Hsieh

Symbols

a	Acceleration	KE	Kinetic energy
a, A	Specific Helmholtz function, Helmholtz function	l, L	Length
		L	Latent heat
A	Area	LHV	Lower heating value
AF	Air-fuel ratio	m	Mass
c	Specific heat	M	Molar mass or molecular weight
c_p	Constant-pressure specific heat	mep	Mean effective pressure
c_v	Constant-volume specific heat	$\mathbf{M}$	Magnetization or magnetic moment per unit volume
C_c	Curie constant		
COP	Coefficient of performance	n	Number of moles
COP_{ref}	Coefficient of performance of a refrigerator	n	Polytropic exponent
		p	Pressure
$\text{COP}_{H.P.}$	Coefficient of performance of a heat pump	p_c	Critical pressure
		p_i	Partial pressure of component i
$C_{\mathbf{H}}$	Heat capacity at constant magnetic field	p_r	Reduced pressure p/p_c
$C_{\mathbf{M}}$	Heat capacity at constant magnetic moment	p_r	Relative pressure as used in gas tables
		PE	Potential energy
d	Differential change in a property	$\mathbf{P}$	Electric polarization or electric dipole moment per unit volume
đ	Differential change in a path function		
e	Base of natural logarithm	q, Q	Heat transfer per unit mass, heat transfer
e, E	Specific total energy, total energy	Q_{av}	Available energy
$\mathbf{E}$	Electric field strength	Q_{unav}	Unavailable energy
$\mathscr{E}$	Electrical potential	r	Compression ratio
f	Functional relation	r_c	Cutoff ratio
F	Degree of freedom	r_p	Pressure ratio
F	Force	R	Gas constant
g	gravitational acceleration	$\mathscr{R}$	Universal gas constant
g, G	Specific Gibbs function, Gibbs function	s, S	Specific entropy, entropy
g_c	$g_c = 32.174 \text{ ft} \cdot \text{lbm/lbf} \cdot \text{s}^2$	ΔS_R	Entropy change of reaction
Δg_f°	Gibbs function of formation at standard state	S_{prod}	Entropy production
		t	Time
ΔG_R	Gibbs-function change of reaction	T	Temperature
h	Vertical height	T_c	Critical temperature
h, H	Specific enthalpy, enthalpy	T_{db}	Dry-bulb temperature
Δh_f°	Enthalpy of formation at standard state	T_{dp}	Dew-point temperature
ΔH_R°	Enthalpy of reaction at standard state	T_{wb}	Wet-bulb temperature
		T_r	Reduced temperature T/T_c
HHV	Higher heating value	u, U	Specific internal energy, internal energy
$\mathbf{H}$	Magnetic field streagth		
i	Electric current	ΔU_R°	Internal energy of reaction at standard state
i, I	Specific irreversibility, irreversibility		
k	Boltzmann constant	v, V	Specific volume, volume
k	Specific heat ratio, c_p/c_v	v_c	Critical volume
K_p	Diffuser pressure coefficient	v_i	Partial volume of component i
K_p	Equilibrium constant	v_r	Reduced volume v/v_c

v_r	Relative volume as used in gas tables
V	Velocity
w, W	Work per unit mass, Work
x	Mole fraction
x	Quality
y	Mass fraction
z	Elevation
Z	Compressibility factor
Z	Electric charge
Z_c	Critical compressibility factor

Greek Letters

α	Coefficient of thermal expansion
γ	Specific weight
γ	Surface (or interfacial) tension
δ	Virtual variation
Δ	Finite change = final minus initial
ϵ	Second-law effectiveness
ϵ	Strain
ϵ_0	Permittivity of free space
η	Efficiency
η_{th}	Thermal efficiency
η_T	Turbine efficiency
θ	Angle
κ	Isothermal compressibility
κ_s	Adiabatic compressibility
μ	Chemical potential
μ_J	Joule-Thomson coefficient
μ_0	Permeability of free space
ν	Stoichiometric coefficient
π	$\pi = 3.14159$
ρ	Density
σ	Stress
Σ	Summation
τ	Torque
ϕ	$\phi = \displaystyle\int_0^T c_p \frac{dT}{T}$ as defined in gas tables
ϕ	Relative humidity
φ, Φ	Closed system specific availability, availability
ψ, Ψ	Open system specific stream availability, stream availability
ω	Specific humidity
Ω	Thermodynamic probability

Subscripts

| a | Air, dry air |
| a, act | Actual |

abs	Absolute
atm	Atmosphere
av	Average
c	Compressor
c	Critical point property
cv	Control volume
f	Final state
f	Saturated liquid
fg	Difference in property between saturated vapor and saturated liquid
g	Saturated vapor
H	High-temperature reservoir
i	Initial state
i	ith component in a mixture
i	Saturated solid
in	Input
irr	Irreversible
int rev	Internally reversible
L	Low-temperature reservoir
max	Maximum
min	Minimum
N	Nozzle
out	Output
P	Pump
prop	Propulsive
R	Chemical reaction
R	Energy reservoir
reg	Regenerator
reh	Reheater
rev	Reversible
s	Isentropic
sat	Saturated
surr	Surroundings
sys	System
st gen	Steam generator
th	Thermal
T	Turbine
v	Water vapor
0	Dead state
0	Standard state
1	State 1
1	Component 1 in a mixture

Superscripts

•	Quantity per unit time
°	Property at standard state
°	Property at unit pressure
*	Ideal gas state

Introduction

1

1-1 THE NATURE OF THERMODYNAMICS

Thermodynamics is the basic science that deals with energy, matter, and their transformations and interactions. It is based on two general laws of nature, the first and second laws of thermodynamics. The first law is essentially the law of conservation of energy to account for the balance of thermal and other forms of energy taking part in a transformation. The second law places limitations on certain kinds of energy transformation. Based on these laws, engineers design and build various useful devices including stationary and vehicular heat engines, refrigeration and air-conditioning machines, and chemical processing plants.

The science of *classical thermodynamics* was developed without an inquiry into the structure of matter. It is concerned only with the average characteristics of large aggregations of molecules, not with the characteristics of individual molecules. In other words, classical thermodynamics takes the macroscopic point of view and deals with macroscopic phenomena. On the other hand, *statistical thermodynamics* considers the microscopic structure of matter and adopts the laws of mechanics on the statistical analysis of the individual particles. This text is based on the classical approach.

1-2 ENGINEERING APPLICATIONS OF THERMODYNAMICS

Engineering thermodynamics is a branch of thermodynamics in which emphasis is placed on the engineering analysis and design of processes, devices, and systems involving the beneficial utilization of energy and material. It covers a wide variety of applications, from the design of steam power stations and gas-liquefaction plants to the analysis of rocket engines. In order to give the students some familiarity with the processes, the equipment, and the technical terms involved in a thermodynamic analysis, we offer now a bird's-eye view of a number of engineering applications. Bear in mind, however, that what we mention here is only a few of the types of systems that can be analyzed thermodynamically.

Figure 1-1 shows a schematic diagram of a simple steam power plant. Steam at a high pressure and temperature leaves the steam generator and enters the turbine, where it expands to a lower pressure and temperature and does work to drive the electric generator, resulting in the output of electric power. The lower-pressure and lower-temper-

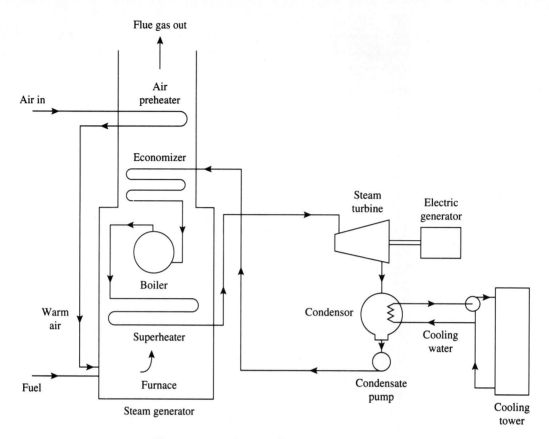

Figure 1-1 Schematic of a steam power plant.

ature exhaust steam from the turbine then enters the condenser and condenses to liquid by transferring heat to the cooling water, which in turn transfers the waste heat to a river, a lake, or a cooling tower. The liquid condensate from the condenser is pumped into the steam generator to be vaporized and heated to a high temperature, thus completing a thermodynamic cycle.

Some details of the steam generator are also shown in Fig. 1-1. The economizer is a heat exchanger, where heat is transferred from the products of combustion to the condensate coming from the condensate pump, thus raising the temperature of the liquid water without evaporation. The evaporation of water occurs in the boiler section. The vapor formed in the boiler flows into the superheater, where additional heat is transferred from the hot products of combustion to increase the temperature of the vapor to a high value before entering the steam turbine. The air preheater shown in Fig. 1-1 is used to warm up the incoming outside air before entering the furnace for efficient burning of the fuel.

Although the basic components of a steam power plant are those shown in the simple drawing of Fig. 1-1, actual steam power generation systems are more complex. To help gain a general feel for what the actual equipment looks like, we include a sectional drawing of a fossil-fuel steam power station (Fig. 1-2) and a cutaway view of a steam generator (Fig. 1-3). In these figures, the names of the essential elements are indicated.

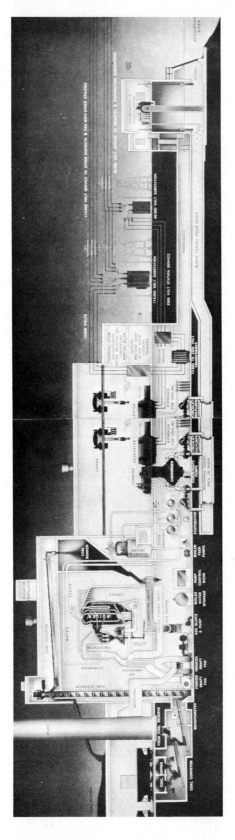

Figure 1-2 A steam power station. (Illustration of Jennison Station. Adapted by permission of New York State Electric and Gas Corp. and Gilbert/Commonwealth, Inc.)

3

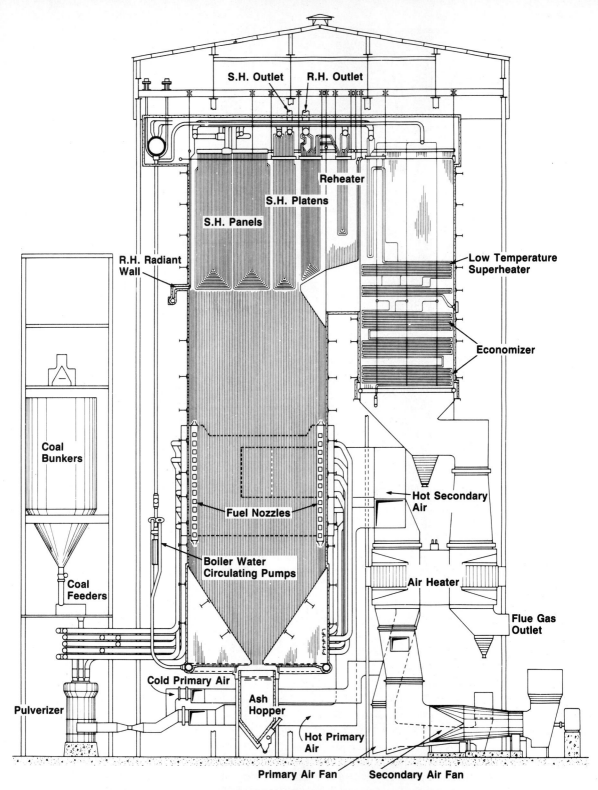

Figure 1-3 A steam generator. (A 500-MW pulverized coal boiler, rated at 4,100,000 lb of steam per hour with superheater outlet conditions of 2620 psig and 1005°F. Courtesy of Combustion Engineering, Inc.)

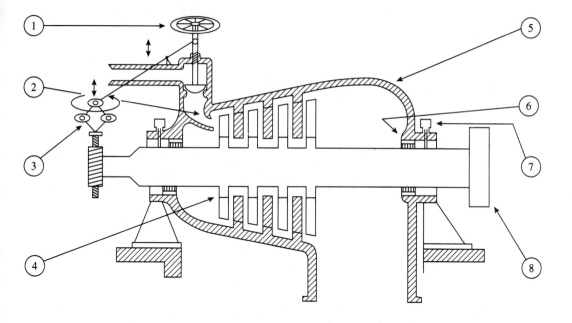

① A valve to control steam flow = speed or load
② Nozzle(s) to organize steam flow to buckets
③ Overspeed/governor device to control speed
④ Rotor/buckets combination to convert kinetic energy to shaft rotation
⑤ Shell to contain pressure, mount stationary vanes and provide attachments to standards bearings
⑥ Sealing devices for shaft
⑦ Lubrication for bearings
⑧ Coupling to transmit power

Figure 1-4   Schematic of a steam turbine. (Courtesy of General Electric Co.)

Figure 1-4 shows a schematic diagram of a simple steam turbine with the names and functions of the various elements illustrated. It is clear that the main function of the steam turbine is the expansion of the high-pressure steam with the resulting work done against the series of curved blades attached on the rotating shaft. Figure 1-5 shows a cutaway view of an actual steam turbine. High-pressure steam is first admitted to the high-pressure turbine to be expanded to an intermediate pressure and is then discharged to the reheater to be reheated back to a high temperature before being led into the intermediate-pressure turbine for further expansion. It is finally admitted to the tandem-compound low-pressure turbines for final expansion to the condenser pressure.

Figure 1-6 shows a schematic cutaway diagram of a steam condenser. It is a shell-and-tube heat exchanger with cooling water flowing through many small tubes and steam flowing around the tubes. Heat is transferred from the steam to the cooling water, causing the steam to condense. Condensate is collected and stored in the hotwell, ready to be pumped out. Noncondensable gases, such as air, CO_2, and H_2, are removed from the condenser by an air-removal pump (not shown). The water vapor entrained with the noncondensable gases are recovered later by condensation.

Figure 1-7 shows a cutaway view of a turbojet engine. The incoming air is slowed down in the inlet diffuser and then compressed by the compressor before entering the combustion chamber, where fuel is added and burned. The high-pressure and high-temperature products of combustion are subsequently expanded in the turbine to a lower

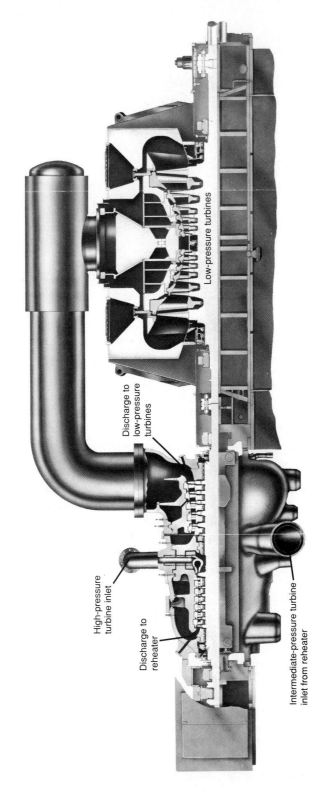

High-pressure turbine inlet

Discharge to reheater

Discharge to low-pressure turbines

Low-pressure turbines

Intermediate-pressure turbine inlet from reheater

Figure 1-5 A tandem-compound two-flow reheat steam turbine. (Courtesy of General Electric Co.)

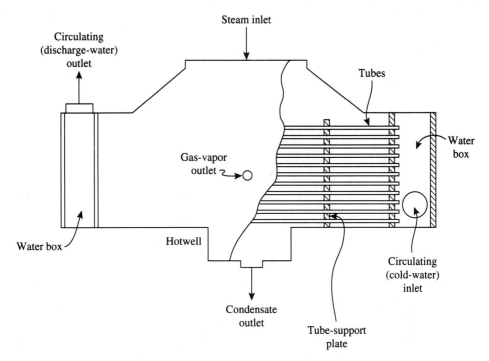

Figure 1-6 A steam condenser.

pressure so as to generate power to drive the compressor. The gases exhausted from the turbine are further expanded in the exit nozzle, resulting in a high-velocity jet rushing out from the engine, thus creating a forward thrust according to Newton's third law.

Figure 1-8 shows a schematic of a window-type air conditioner. It incorporates a small self-contained mechanical refrigeration plant that has four basic components, namely, a compressor, a condenser, an expansion valve (or a capillary tube), and an evaporator. Vapor refrigerant is first compressed to a higher pressure and subsequently condensed to liquid by transferring heat to the outdoor air. The liquid refrigerant is then expanded through the expansion valve to a lower pressure and is admitted to the evaporator to be vaporized by absorbing heat from the indoor air.

Figure 1-9 shows a schematic diagram of a 100-MW solar central-receiver power plant where a large number of individually guided two-axis tracking mirrors or heliostats reflect and focus solar rays onto a single central receiver located on the top of a slim tower of some 60-story height. The cherry-red glare of concentrated sunlight of the order of 1000 times its ground-level normal intensity heats up the collector fluid (liquid sodium in this case) circulating inside the receiver to a temperature of 600°C (1100°F). The hot sodium is directed down the tower and into the storage tank. The stored energy in the hot sodium is then utilized via a heat exchanger to provide steam for a steam-powered turbine-generator unit. Some of the collected energy is held temporarily in the storage to bridge over cloud periods or to extend the benefit of the solar unit into the early evening. Such a solar plant would occupy an open area of over 1 square mile (2.6 square kilometers).

Figure 1-10 shows a schematic diagram of a thermoelectric generator with three *p-n* units arranged in series electrically and in parallel thermally. Each *p-n* unit consists of a *p*-type semiconductor and an *n*-type semiconductor, connected by electrodes as

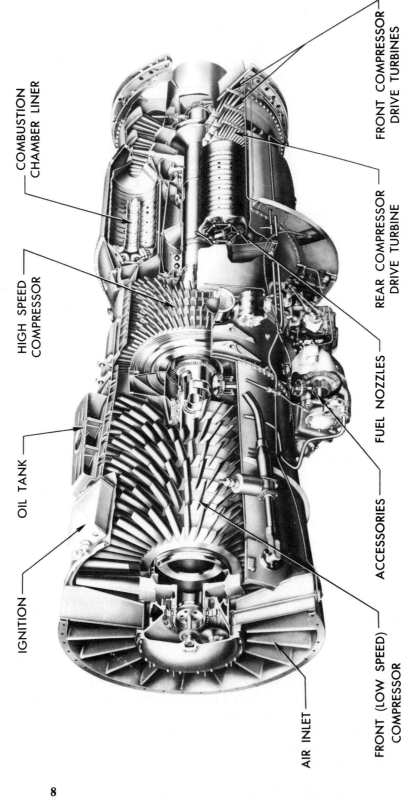

Figure 1-7 A turbojet engine. (Pratt & Whitney J75 (JT4) turbojet engine, used for aircrafts such as Boeing 707 and McDonnell Douglas DC-8. Courtesy of United Technologies—Pratt & Whitney Co.)

COMBUSTION CHAMBER LINER

HIGH SPEED COMPRESSOR

OIL TANK

IGNITION

FRONT (LOW SPEED) COMPRESSOR

AIR INLET

ACCESSORIES

FUEL NOZZLES

REAR COMPRESSOR DRIVE TURBINE

FRONT COMPRESSOR DRIVE TURBINES

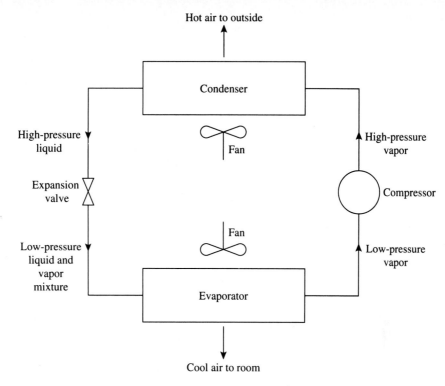

Figure 1-8 Schematic of an air conditioner (window type).

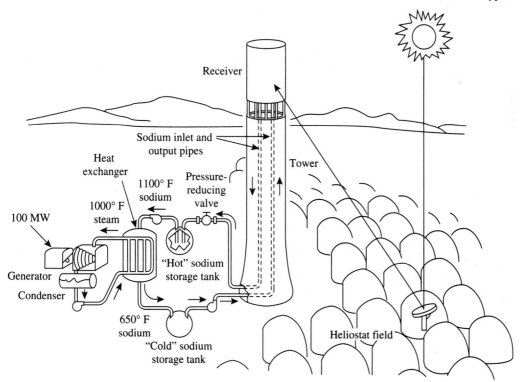

Figure 1-9 A solar central-receiver power plant.

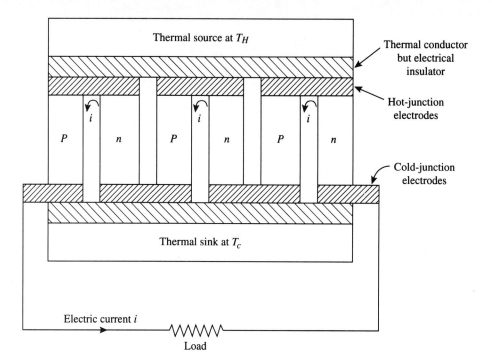

Figure 1-10 A thermoelectric generator.

shown in the figure. As heat is added from the high-temperature source and heat is rejected to the low-temperature sink, electrons move in the *n*-type semiconductor from the high-temperature end to the low-temperature end, and subsequently return to the *n*-type semiconductor after passing through the external circuit and the *p*-type semiconductor, resulting in the flow of a positive current *i* as indicated in the figure.

1-3 THERMODYNAMIC SYSTEM, PROPERTY, STATE, AND PROCESS

In a thermodynamic analysis, a collection of matter or a region of space within certain prescribed boundaries is specified in order to focus study. The boundaries can be either real or imaginary. The collection of matter or region of space thus specified is called a thermodynamic *system,* and everything external to the system that has relation with the system is called the *surroundings* or the *environment.* A system of fixed content is referred to as *closed,* and one whose content can be varied by passage of matter across its boundaries is referred to as *open.* As a closed system is for a fixed quantity of mass, it is alternatively called a *control mass.* Similarly, as an open system is for a region of space within a boundary that matter may cross, it is alternatively called a *control volume.* The boundary of a control volume is referred to as the *control surface.*

Figure 1-11 depicts two examples of thermodynamic systems. In part (a) of the figure, a gas is contained in a cylinder fitted by a weighted piston. The gas is considered as the closed system bounded by the cylinder walls and the piston. As the piston moves up and down, the system boundary varies. In part (b), a fluid flows through the pipe. The space enclosed by a portion of the pipe wall and two imaginary cross-sectional surfaces form the control volume.

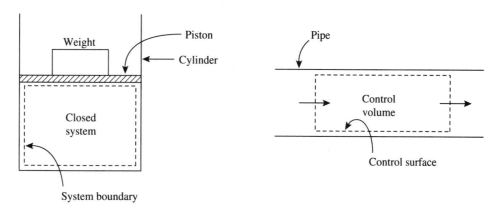

Figure 1-11 Thermodynamic systems. (a) A closed system or control mass. (b) An open system or control volume.

In a closed system, even though there can be no mass transfer across the boundary, energy transfer between the system and its surroundings is allowed. When both mass and energy are prohibited to cross the system boundary, this system is referred to as *isolated*. Notice that any system, closed or open, and its surroundings together constitute an isolated system.

A quantity of matter that is homogeneous in chemical composition and in physical structure is called a *phase*. All substances can exist in solid, liquid, and gaseous phases; solids can exist in different forms. A system consisting of a single phase is called a homogeneous system; a system consisting of more than one phase is called a heterogeneous system.

The *state* of a system at any instant is its condition of existence at that instant. A *property* of a system is any quantity or characteristic that depends upon the state of the system. Thermodynamic properties can be classified as intensive properties and extensive properties. An *intensive property* of a system is independent of the extent of mass of the system. Pressure, temperature, and density are familiar examples of intensive properties. An *extensive property* of a system is one whose value for the entire system is equal to the sum of its values for all parts of the system. Volume, energy, and mass are examples of extensive properties. If the value of any extensive property is divided by the mass of the system, the resulting property is intensive and is called a specific property. For example, specific volume is the ratio of volume to mass and is an intensive property.

When any property of a system changes, there is a change of state, and the system is said to undergo a *process*. The *path* of a process is the series of states through which the system passes during the process. A process whose initial and final states are identical is called a *cycle*.

A property of a system depends only on the state of the system and not on how that state was attained. The uniqueness in the value of a property at a state introduces the name state or *point function* for a property. In contrast, quantities that depend on the path of the process by which a system changes between two states are called *path functions*. Because a property is a point function, its differential must be an exact or perfect differential in mathematical terms. The line integral of the differential of a property is independent of the path or curve connecting the end states, and in the special case of a complete cycle, this integral vanishes.

The preceding statement can be made more clear with the help of a property diagram such as the one shown in Fig. 1-12, in which we assume that two properties x and y are sufficient to determine the state of the system under consideration. At state A, $x = x_A$ and $y = y_A$; and at state B, $x = x_B$ and $y = y_B$. The changes in properties x and y between the two states are given by

$$\int_A^B dx = \Delta x_{AB} = x_B - x_A \qquad \text{and} \qquad \int_A^B dy = \Delta y_{AB} = y_B - y_A$$

regardless of whether path I or path II is used to connect the two end states A and B. Moreover, when the system changes from state A to state B by path I and comes back from B to A by path II, the system thus undergoes a cycle, for which we have

$$\oint dx = 0 \qquad \text{and} \qquad \oint dy = 0$$

where the symbol $\oint$ indicates integration around a cycle.

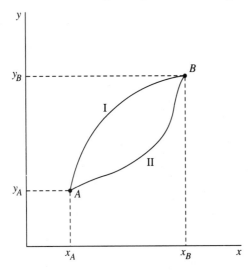

Figure 1-12 A property diagram showing two paths I and II between two states A and B.

It should be noted that in Fig. 1-12, the areas underneath paths I and II are different between the same end states A and B. The quantity represented by the area underneath a process curve is therefore a path function. It is not a thermodynamic property. The cyclic integration of a path function is not zero. We will see in Chapter 2 that work and heat interactions between a system and its surroundings are path functions and not thermodynamic properties.

1-4 EQUILIBRIUM STATE AND QUASIEQUILIBRIUM PROCESS

A system is in a state of equilibrium if a change of state cannot occur while the system is not subject to interactions with the environment. When a system is in equilibrium, there must be no unbalanced potential that tends to promote a change of state. The unbalanced potential can be mechanical, thermal, chemical, or electrical, or any combination of them. When there is no unbalanced force within a system, it is said to be in a state of *mechanical equilibrium*. If the change in pressure with elevation due to gravitational force is neglected, a system of fluid in a state of mechanical equilibrium should have uniform pressure. When there is no temperature gradient within a system, it is said to be in a state of *thermal equilibrium*. When a system has no tendency to undergo either a chemical reaction or a process such as diffusion or solution, the system is said to be in a state of *chemical equilibrium*. When there is no electrical potential gradient within a system, it is said to be in a state of *electrical equilibrium*. If all the conditions for mechanical, thermal, chemical, and electrical equilibrium are satisfied, the system is said to be in a state of *thermodynamic equilibrium,* or simply in an *equilibrium state*. When any of the conditions for a complete thermodynamic equilibrium is not satisfied, the system is said to be in a *nonequilibrium state*. Classical thermodynamics deals only with systems in equilibrium states.

When a system is in thermodynamic equilibrium, its state can be represented by a single point on a set of thermodynamic coordinates. If it is not in thermodynamic equilibrium, some of its macroscopic properties vary from one part to another and are not single-valued and unique so that it is impossible to locate a single point on a state diagram.

During a thermodynamic process, some unbalanced potential must exist either within the system or between it and the surroundings to promote the change of state. If the unbalanced potential is infinitesimal so that the system is infinitesimally close to a state of equilibrium at all times, such a process is called *quasiequilibrium* (or *quasistatic*). A quasiequilibrium process can be considered practically as a series of equilibrium states and its path can be represented graphically as a continuous line on a state diagram.

In contrast, any process taking place due to finite unbalanced potentials is *nonquasiequilibrium*. Such a process cannot be described by means of macroscopic coordinates that are characteristic qualities of the entire system. Thus, the complete path of a nonquasiequilibrium process is indeterminate and cannot be represented on a state diagram, although it is possible to indicate the general direction of change by locating the terminal equilibrium states.

1-5 DIMENSIONS AND UNITS

A *dimension* is a name given to a physical quantity. For example, length is the dimension used to describe the distance between two points. In thermodynamic analysis, length, mass (or force), time, and temperature are generally chosen as the primary dimensions. Other physical quantities are given dimensions in terms of the primary ones. For example, velocity has the dimension of length divided by time.

A *unit* is a definite measure of a dimension. For example, meter, foot, and mile are different units for the dimension of length. The *SI* (*Système Internationale*), or *International System* of units, is the most widely used system throughout the world. In the United States, however, in addition to the SI units, the English units are still in use and are expected to be used for many years to come. This textbook uses both systems of units.

The SI system is based on a decimal relationship between the various units for each dimension and is therefore convenient to use. Table A-1 in Appendix 2 gives the basic and derived units of the SI system, together with the names and symbols of decimal multipliers.

Because in the SI system, mass, length, and time are chosen as primary dimensions, the unit of force is derived from Newton's second law, which states that the force F acting on a body is proportional to the product of the mass m and the acceleration a in the direction of the force, or in SI units

$$F = ma \qquad (1\text{-}1)$$

where m is in kilograms (kg), a in meters/seconds^2 (m/s^2), and F in newtons (N) according to the definition

$$1 \text{ N} = 1 \text{ kg} \times 1 \text{ m/s}^2 \qquad (1\text{-}2)$$

meaning that a force of 1 newton is required to accelerate a mass of 1 kilogram at the rate of 1 meter per second per second.

In English units, both force and mass plus length and time are chosen as the primary dimensions. Force and mass are assigned the units of pound-force (lbf) and pound-mass (lbm), respectively; and length and time are assigned the units of foot (ft) and second (s), respectively. With these units, Newton's second law is written as

$$F = \frac{1}{g_c} ma \qquad (1\text{-}3)$$

where g_c is a conversion constant. Because the weight w of a body is defined as the force exerted on the body as a result of the acceleration of gravity, based on the preceding Newton's second law equation, we have

$$w = \frac{1}{g_c} mg$$

where g is the acceleration of gravity. In the English system, the units of force and mass are arbitrarily chosen so that the weight of a body at sea level on the earth's surface will have the same numerical value as its mass. At sea level, the acceleration of gravity is 32.174 ft/s^2. It follows that at any location

$$g_c = 32.174 \text{ ft·lbm/lbf·s}^2 \qquad (1\text{-}4)$$

Notice that the conversion constant g_c has both a numerical value and dimensions. It is imperative to carry the conversion constant g_c along in all calculations using the English units in which the conversion between mass and force is involved.

Example 1-1

A body having a mass of 7 lbm is accelerated upward at 60 ft/s^2. The local acceleration of gravity is 32.174 ft/s^2. Determine the required upward force in lbf.

Solution The given data are $m = 7$ lbm, $a = 60$ ft/s^2 upward, and $g = 32.174$ ft/s^2 downward.

Let F represent the required upward force. The net upward force on the body is $F - mg$. Applying Newton's second law, we have

$$F - mg = ma$$

or

$$F = m(a + g)$$

$$= (7 \text{ lbm})(60 + 32.174 \text{ ft/s}^2) \frac{1}{32.174 \text{ ft·lbm/lbf·s}^2}$$

$$= 20.05 \text{ lbf}$$

1-6 DENSITY, SPECIFIC WEIGHT, AND PRESSURE

Density ρ is defined as the mass m of a substance divided by its volume V, or

$$\rho = \frac{\text{mass}}{\text{volume}} = \frac{m}{V}$$

Specific volume v is the reciprocal of density, or

$$v = \frac{1}{\rho} = \frac{\text{volume}}{\text{mass}} = \frac{V}{m}$$

Specific weight γ is defined as the weight w of a substance divided by its volume, or

$$\gamma = \frac{\text{weight}}{\text{volume}} = \frac{w}{V}$$

The relation between specific weight and density can be obtained by applying Newton's second law for the weight of a substance to get $w = mg$. Dividing both sides of this equation by the volume of the substance, we have

$$\gamma = \rho g \qquad (1\text{-}5)$$

where g is the acceleration of gravity. In English units, this equation can be written as

$$\gamma = \frac{1}{g_c} \rho g$$

where g_c is the conversion constant (32.174 ft·lbm/lbf·s^2).

The basic unit of density is kg/m^3 in SI units and lbm/ft^3 in English units; the basic unit of specific volume is m^3/kg in SI units and ft^3/lbm in English units; and the basic unit of specific weight is N/m^3 in SI units and lbf/ft^3 in English units.

In thermodynamics, a specific property, such as specific volume, is frequently expressed on a mole basis. A *mole* is a quantity of a substance having a mass numerically equal to its molecular mass. For example, 1 kg-mole of diatomic oxygen (O_2) is a mass of 32 kg and 1 lb-mole of O_2 is a mass of 32 lbm. Thus, specific volume can have the unit of m^3/kgmole in the SI system and ft^3/lbmole in the English system. A specific property expressed on a mole basis is sometimes called a molar specific property. In this book, we will not introduce a separate set of symbols for molar specific properties and will rely on the fact that the nature of the quantity is usually clear from the content or from the unit used.

Analogous to normal stress in solids, *pressure* of fluids is defined as the normal force per unit area acting on some real or imaginary surface. In fluid systems (either liquids or gases), the pressure on the surface of the container is due to the cumulative effect of individual molecules striking the walls of the container. For a fluid in equilibrium, the pressure at a point or, more correctly speaking, around the vicinity of an elemental volume is the same in all directions. The actual pressure at a given position in a system is called its *absolute pressure*. Most pressure-measuring instruments measure the difference between the pressure of a fluid and the pressure of the atmosphere and give readings with the atmospheric pressure as the reference point. Such a pressure reading is called the *gage pressure* of the fluid. The absolute pressure p_{abs} is related to the gage pressure p_{gage} and the atmospheric pressure p_{atm} as follows:

$$p_{abs} = p_{atm} + p_{gage} \qquad (1\text{-}6)$$

If a fluid exists at a pressure lower than the atmospheric pressure, its gage pressure is negative. It is convenient to apply the term *vacuum* to the magnitude of the gage pressure and make a vacuum measurement a positive value, so that a vacuum p_{vacuum} is related to the absolute pressure p_{abs} and the atmospheric pressure p_{atm} in the following manner:

$$p_{abs} = p_{atm} - p_{vacuum} \qquad (1\text{-}7)$$

The relationships among absolute pressure, atmospheric pressure, gage pressure, and vacuum are shown graphically in Fig. 1-13. In thermodynamic calculations, the use of absolute pressure is required. Equation (1-6) or (1-7) can be used to do the conversion when gage or vacuum pressure is given.

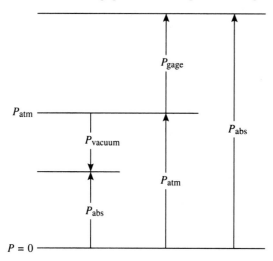

Figure 1-13 Relationships among absolute, atmospheric, gage, and vacuum pressures.

The dimension of pressure is force/area. In SI units, the basic unit of pressure is the pascal (Pa) as defined by

$$1 \text{ Pa} = 1 \text{ N/m}^2$$

In this book, the term pascal without any modifying word always refers to absolute pressure. For a gage pressure, the unit will be so indicated. In English units, the basic unit of pressure is lbf/ft² (psf). It is customary in English units to affiliate in the units the word *absolute* for absolute pressure and the word *gage* for gage pressure. For example, lbf/in.² abs (psia) is used for absolute pressure and lbf/in.² gage (psig) is used for gage pressure. Notice that a vacuum reading is a negative gage pressure, as, for example, a pressure of 50 kPa vacuum is equivalent to a gage pressure of −50 kPa.

Two additional units of pressure are used extensively. One of these is the bar, which is defined as 10^5 N/m² or 10^5 Pa. The other widely used pressure unit is the atmosphere (atm). Furthermore, there is another frequently used pressure unit that is expressed in terms of the height of a column of mercury (Hg) that the pressure will support at a specified temperature. The relation between the standard atmosphere and other pressure units are as follows:

$$1 \text{ standard atmosphere (atm)} = 14.696 \text{ lbf/in.}^2 \text{ (psi)}$$
$$= 1.01325 \times 10^5 \text{ N/m}^2$$
$$= 1.01325 \text{ bar}$$
$$= 760 \text{ mm Hg at } 0°C$$

Many devices are available for measuring pressure. One relatively simple device is the Bourdon-tube pressure gage, as shown schematically in Fig. 1-14. When fluid under pressure is applied to the inside of the tube, the elliptical tube section tends to become circular and the C-shaped tube straightens, causing a motion of the pointer. The movement of the pointer on a dial, after necessary calibration, gives the difference in pressure between the fluid inside the tube and the surrounding fluid outside the tube. As the surrounding is at atmospheric pressure, the reading on the dial of this instrument is for gage pressures. Vacuum pressures can be also calibrated with this device.

Another common pressure-measuring device is a U-tube manometer. Figure 1-15 shows an open U-tube, partially filled with a manometer liquid (mercury or water) having a density ρ, with one end open to the atmosphere and the other end connected to a closed tank containing a gas at a uniform absolute pressure p_{abs}. The difference in pressure between the gas and the atmosphere is able to support a column of manometer liq-

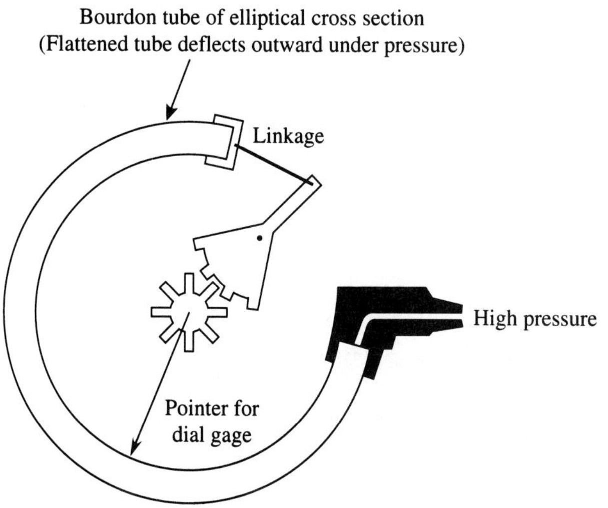

Bourdon tube of elliptical cross section
(Flattened tube deflects outward under pressure)

Linkage

High pressure

Pointer for
dial gage

Figure 1-14   Bourdon-tube pressure gage.

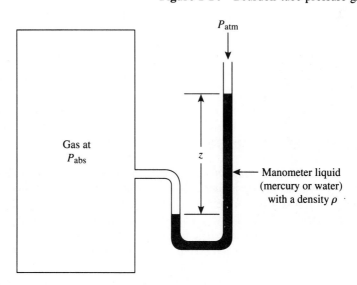

Figure 1-15   U-tube manometer.

uid of height z. The gage pressure of the gas in the tank can be written by the expression:

$$p_{gage} \text{ in N/m}^2 = p_{abs} - p_{atm} = (\rho \text{ kg/m}^3)(z \text{ m})(g \text{ m/s}^2) \qquad (1\text{-}8a)$$

or

$$p_{gage} \text{ in lbf/ft}^2 = p_{abs} - p_{atm}$$

$$= (\rho \text{ lbm/ft}^3)(z \text{ ft}) \frac{g \text{ ft/s}^2}{g_c \text{ ft·lbm/lbf·s}^2} \qquad (1\text{-}8b)$$

Example 1-2

(a) Specific gravity of a liquid is defined as the ratio of its density to that of water at a specified temperature. The specific gravity of mercury based on water at 0°C is 13.6. Determine the specific weight of mercury in N/m³ at 0°C.

(b) A gas has a vacuum pressure of 300 mm Hg when the barometer stands at 750 mm Hg, both based on 0°C. Determine the absolute pressure of the gas in pascals.

Solution

(a) At 0°C, the density of water is $\rho_{water} = 1000$ kg/m³. The density of mercury ρ_{Hg} at 0°C is then

$$\rho_{Hg} = (\text{sp. gravity of Hg}) \rho_{water}$$

$$= 13.6 \times 1000 = 13600 \text{ kg/m}^3$$

The specific weight of mercury γ_{Hg} at 0°C is

$$\gamma_{Hg} = \rho_{Hg}g = (13,600 \text{ kg/m}^3)(9.807 \text{ m/s}^2) = 133,400 \text{ N/m}^3$$

(b) The absolute pressure of the gas is given by

$$p_{abs} = p_{atm} - p_{vacuum} = 750 - 300 = 450 \text{ mm Hg}$$

$$= 0.450 \text{ m Hg} = (133,400 \text{ N/m}^3)(0.450 \text{ m})$$

$$= 60,000 \text{ N/m}^2 = 60,000 \text{ Pa}$$

The value of p_{abs} in pascal can be also calculated as follows:

$$p_{abs} = 450 \text{ mm Hg}$$

$$= (450 \text{ mm Hg})\left(\frac{1.01325 \times 10^5 \text{ Pa}}{760 \text{ mm Hg}}\right)$$

$$= 60,000 \text{ Pa}$$

1-7 TEMPERATURE AND THERMOMETRY

When two systems that initially are each in thermal equilibrium are brought in thermal contact while isolated from all other surroundings, each of the two systems will in general undergo a certain change of state until mutual thermal equilibrium between the two is reached. Two systems in mutual thermal equilibrium have a common characteristic. This characteristic is solely a function of the states of the systems, and therefore is a property of each system. This property is the *temperature*.

When two systems are each in thermal equilibrium with a third system, they are in thermal equilibrium with each other. This statement is known as the *zeroth law* of thermodynamics. In terms of temperature, the zeroth law states that when two systems are each equal in temperature to a third system, they are equal in temperature to each other. This obvious fact forms the basis of temperature measurement.

In order to establish the means of temperature measurement, we choose a reference system for which there is some physical property that changes with temperature. This reference system is called a thermometer and the selected physical property whose changes are used as indication of changes in temperature is called the *thermometric property*. A small quantity of mercury enclosed in a glass capillary is called a mercury-in-glass thermometer, in which the volume of the mercury is used as the thermometric property. When two wires of different materials are joined at their ends with different temperatures existing at the two junctions, an electromotive force is generated that can be used as a measure of the temperature of one junction if the temperature at the other junction is known; such a thermometer is known as a *thermocouple*. Gases, such as oxygen, nitrogen, or helium, at very low pressures are known to have a special relationship among their pressure, specific volume, and temperature. The temperature of such a gas is a unique function of volume when its pressure is held constant, or a unique function of pressure when its volume is held constant. Thus, if the pressure of the gas is held constant while its volume is used as the thermometric property, a *constant-pressure gas thermometer* is formed. On the other hand, if the volume of the gas is held constant while its pressure is used as the thermometric property, a *constant-volume gas thermometer* can be constructed.

To establish quantitative measurements of temperature in terms of numbers, easily reproducible states (called fixed points) of an arbitrary standard system are assigned arbitrary numbers. Formerly, two standard fixed points, the ice point and the steam point, were used. The ice point is defined as the equilibrium temperature of pure ice and air-saturated water, and the steam point as the equilibrium temperature of pure water and water vapor, both at 1 atmosphere pressure. However, in practice it is difficult to realize the condition of equilibrium between pure ice and air-saturated water as required by the definition of the ice point. As ice melts, it forms a layer of pure water around itself that prevents the direct contact of pure ice and air-saturated water. Due to this difficulty in obtaining the ice point with accuracy, the two-fixed-point method was abandoned in 1954 by international agreement. The new method adopts a single standard fixed point, namely, the triple point of water, which is the equilibrium state in which ice, liquid water, and water vapor coexist. A value of 273.16 degrees Kelvin (K) on the Kelvin scale is assigned for the temperature at this fixed point. The zero of the Kelvin scale is the absolute zero of temperature.

A temperature scale based on the single-fixed-point method mentioned before can be established through the use of a gas thermometer. Figure 1-16 shows schematically a constant-volume gas thermometer. It consists of a quantity of gas contained in a bulb with a capillary tube leading to a mercury monometer. The volume of the gas is kept constant by raising or lowering the mercury reservoir until the mercury in the left-hand side of the U-tube stands at the level indicated by a fixed mark. The height z of the mercury column gives the gas pressure, which is an indication of the gas temperature.

Let P_{TP} be the absolute pressure of the gas when the bulb of a constant-volume gas thermometer is immersed in water at its triple-point temperature, which has an assigned value of 273.16 K; and p be the absolute pressure of the gas when the bulb is immersed in any system whose temperature T is to be measured. Experiment shows that the value of the ratio P/P_{TP} depends on the kind and the amount of gas in the bulb. However,

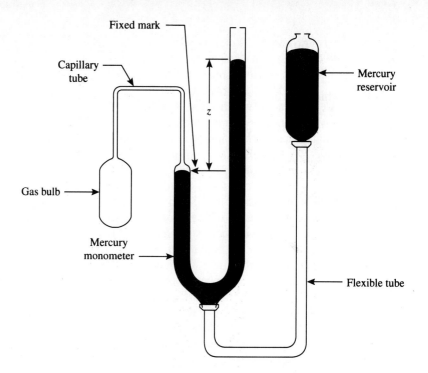

Figure 1-16 Schematic diagram of a constant-volume gas thermometer.

when the procedure is repeated with less and less gas in the bulb, the limiting value of the ratio p/p_{TP} and therefore the value of $T/273.16$, when extrapolated to $p_{TP} = 0$, becomes fixed irrespective of the kind of gas used. This is illustrated in Fig. 1-17, which shows the uniqueness of the indicated temperature when $p_{TP} \rightarrow 0$ as given by a gas thermometer using different gases for different series of measurements. Thus, in the limiting condition, a unique value of temperature T is defined by the equation

$$T = 273.16 \lim_{p_{TP} \rightarrow 0} \left(\frac{p}{p_{TP}} \right) \tag{1-9}$$

The temperature so defined is called the *ideal gas temperature* in degrees Kelvin.

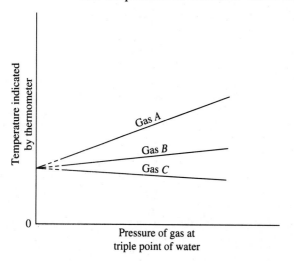

Figure 1-17 Results of measurement from a constant-volume gas thermometer.

A second absolute temperature scale in use is the Rankine scale, defined by setting

$$T \ (°R) = 1.8T \ (K) \tag{1-10}$$

which corresponds to the assignment of 491.69 degrees Rankine (°R) to the triple point of water.

Associated with the Kelvin scale is the Celsius scale on which the triple point of water is, by definition, 0.01 degrees Celsius (°C). The relation between the Kelvin and Celsius scales is

$$T \ (K) = T \ (°C) + 273.15 \tag{1-11}$$

Associated with the Rankine scale is the Fahrenheit scale. The relation between the Rankine and Fahrenheit scales is

$$T \ (°R) = T \ (°F) + 459.67 \tag{1-12}$$

and that between the Fahrenheit and Celsius scales is

$$T \ (°F) = 1.8T \ (°C) + 32 \tag{1-13}$$

These four temperature scales are compared in Fig. 1-18.

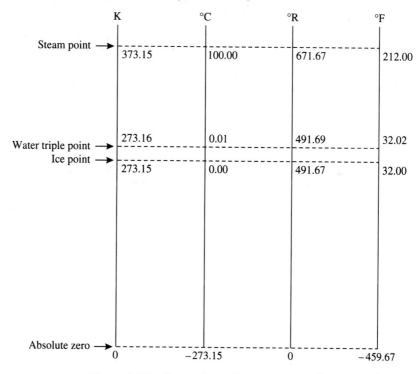

Figure 1-18   Comparison of temperature scales.

Example 1-3

(a) The temperature of a room initially is 95°F. What is this room temperature in °R, °C, and K?

(b) During a cooling process, the room temperature drops by 20°F. Express this drop in temperature in °R, °C, and K.

(c) What is the final room temperature after cooling in °F, °R, °C, and K?

Solution The given data are T_1 (°F) = 95°F and ΔT (°F) = T_2 (°F) − T_1 (°F) = −20°F.

(a) The initial room temperature expressed in °R, °C, and K, respectively, are

$$T_1 \text{ (°R)} = T_1 \text{ (°F)} + 459.67 = 95 + 459.67 = 554.67°R$$

$$T_1 \text{ (°C)} = \frac{T_1 \text{ (°F)} - 32}{1.8} = \frac{95 - 32}{1.8} = 35°C$$

$$T_1 \text{ (K)} = T_1 \text{ (°C)} + 273.15 = 35 + 273.15 = 308.15 \text{ K}$$

$$= \frac{T_1 \text{ (°R)}}{1.8} = \frac{554.67}{1.8} = 308.15 \text{ K}$$

(b) The temperature changes are identical in the Rankine and Fahrenheit scales. Thus,

$$\Delta T \text{ (°R)} = \Delta T \text{ (°F)} = -20°R$$

The temperature change in the Kelvin scale is

$$\Delta T \text{ (K)} = \frac{\Delta T \text{ (°R)}}{1.8} = \frac{-20}{1.8} = -11.11 \text{ K}$$

The temperature changes are identical in the Celsius and Kelvin scales. Thus,

$$\Delta T \text{ (°C)} = \Delta T \text{ (K)} = -11.11°C$$

(c) The final room temperature expressed in °F, °R, °C, and K, respectively, are

$$T_2 \text{ (°F)} = T_1 \text{ (°F)} - 20 = 95 - 20 = 75°F$$

$$T_2 \text{ (°R)} = T_2 \text{ (°F)} + 459.67 = 75 + 459.67 = 534.67°R$$

$$= T_1 \text{ (°R)} - 20 = 554.67 - 20 = 534.67°R$$

$$T_2 \text{ (°C)} = \frac{T_2 \text{ (°F)} - 32}{1.8} = \frac{75 - 32}{1.8} = 23.89°C$$

$$T_2 \text{ (K)} = \frac{T_2 \text{ (°R)}}{1.8} = \frac{534.67}{1.8} = 297.04 \text{ K}$$

$$= T_2 \text{ (°C)} + 273.15 = 23.89 + 273.15 = 297.04 \text{ K}$$

1-8 ENERGY

Energy is the ability to force a change or to produce an effect. All matter possesses energy arising from the motions and the configurations of its internal particles. The microscopic energy of matter includes such energy modes as the molecular translation, rotation, vibration and binding, the electron orbiting and spinning, the nuclear spinning and binding, and the intermolecular potential. In the study of macroscopic thermodynamics, the actual nature of these energy modes on a microscopic level is unimportant, because we are interested solely in the gross behavior of a large number of particles. We will group these "hidden microscopic modes" together and called it the *internal energy*. The internal energy of matter is a function of its state and is therefore a thermodynamic property. It is denoted by the symbol U.

When a system of matter as a whole has a velocity, it possesses a bulk *kinetic energy*. When the system as a whole is at an elevation with respect to some reference level, it possesses a bulk *potential energy*. The bulk kinetic energy (KE) and potential energy (PE) are associated with a selected coordinate frame and can be specified by the macroscopic parameters of mass, velocity, and elevation. Together with the internal energy, they make up the total energy E stored in the system in a given state, or

$$E = U + KE + PE \tag{1-14}$$

The energy of a system can be altered by the transfer of energy in the form of heat or work through its boundary from the sorroundings. Heat is a form of energy in transit due to a temperature difference. Work is defined as the energy expended by a force acting through a distance. It is also a form of energy in transit. Neither heat nor work is a property of a system. We will study these transitional forms of energy in more detail in the next chapter.

The dimension of energy, either stored or in transit, is the product of force and length. The basic unit of energy in SI units is

$$1 \text{ joule (J)} = 1 \text{ newton·meter (N·m)} \tag{1-15}$$

The basic unit of energy in English system is the foot-pound force (ft·lbf). Another energy unit in English system is the Btu, which is related to the foot-pound force by the following:

$$1 \text{ Btu} = 778.169 \text{ ft·lbf} \tag{1-16}$$

We now turn to the derivation of the expressions for bulk kinetic and potential energies. Kinetic energy is evaluated by determining the work required to accelerate a body from rest to a certain velocity. From Newton's second law, Eq. (1-1),

$$F = ma = m\frac{d\mathbf{V}}{dt} = m\frac{dx}{dt}\frac{d\mathbf{V}}{dx} = m\mathbf{V}\frac{d\mathbf{V}}{dx}$$

where $\mathbf{V}$ is velocity, t is time, and x is displacement. It follows that

$$d\,KE = F\,dx = m\mathbf{V}\,d\mathbf{V}$$

Integrating,

$$\int_0^{KE} d\,KE = \int_0^{\mathbf{V}} m\mathbf{V}\,d\mathbf{V}$$

so that

$$KE = \tfrac{1}{2}m\mathbf{V}^2 \tag{1-17}$$

where in SI units, m is in kg, $\mathbf{V}$ in m/s, and KE in J. In the English system, the equation can be rewritten as

$$KE = \frac{1}{2}\frac{m}{g_c}\mathbf{V}^2$$

where m is in lbm, $\mathbf{V}$ in ft/s, $g_c = 32.174$ ft·lbm/lbf·s^2, and KE in ft·lbf.

Bulk potential energy is evaluated by determining the work required to lift a body at constant velocity from the elevation of the reference level to a higher level in the gravitational field. That is,

$$d\,PE = mg\,dz$$

Integrating,

$$\int_0^{PE} d\,PE = \int_0^z mg\,dz$$

so that

$$PE = mgz \qquad\qquad (1\text{-}18)$$

where g is the gravitational acceleration, and z is the height above the reference level. In SI units, m is in kg, z in m, g in m/s^2, and PE in J. In the English system, the equation can be rewritten as

$$PE = \frac{m}{g_c}gz$$

where m is in lbm, z in ft, $g_c = 32.174$ ft·lbm/lbf·s^2, g in ft/s^2, and PE in ft·lbf.

Example 1-4

(a) An automobile weighing 3000 lbf travels at a speed of 55 mph before reaching a hill. Without additional engine power, it goes uphill to a standstill at an elevation 70 ft above the bottom of the hill (see Fig. 1-19). Determine the energy dissipated during the ascent. The acceleration of gravity is 32.174 ft/s^2.

(b) Convert the given data to SI units and repeat the calculations.

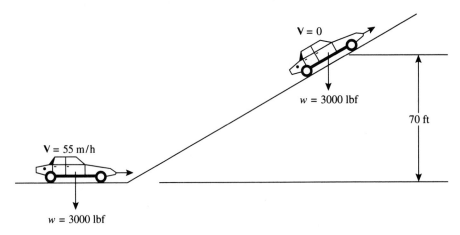

Figure 1-19   Sketch for Example 1-4.

Solution   Let the subscripts 1 and 2 denote the initial and final locations, respectively, of the automobile.

(a) The given data in English units are

$$w = 3000 \text{ lbf}$$

$$V_1 = 55 \text{ mph} = (55 \text{ miles/hour})(5280 \text{ ft/mile})/(3600 \text{ s/hour})$$

$$= 80.67 \text{ ft/s}$$

$$V_2 = 0, \qquad z_2 - z_1 = 70 \text{ ft}, \qquad \text{and } g = 32.174 \text{ ft/s}^2$$

The mass of the automobile is

$$m = w\frac{g_c}{g} = (3000 \text{ lbf})\frac{32.174 \text{ ft·lbm/lbf·s}^2}{32.174 \text{ ft/s}^2} = 3000 \text{ lbm}$$

$$\text{Decrease in KE} = \text{KE}_1 - \text{KE}_2 = \tfrac{1}{2}mV_1^2 - 0$$

$$= \tfrac{1}{2}(3000 \text{ lbm})(80.67 \text{ ft/s})^2 \frac{1}{32.174 \text{ ft·lbm/lbf·s}^2}$$

$$= 303,400 \text{ ft·lbf}$$

$$\text{Increase in PE} = \text{PE}_2 - \text{PE}_1 = mg(z_2 - z_1)$$

$$= (3000 \text{ lbm}) \frac{32.174 \text{ ft/s}^2}{32.174 \text{ ft·lbm/lbf·s}^2} (70 \text{ ft})$$

$$= 210,000 \text{ ft·lbf}$$

Thus,

$$\text{Energy dissipated} = \text{decrease in KE} - \text{increase in PE}$$

$$= 303,400 - 210,000 = 93,400 \text{ ft·lbf}$$

(b) The given data in SI units are

$$m = (3000 \text{ lbm})(0.453592 \text{ kg/lbm}) = 1360.78 \text{ kg}$$

$$V_1 = (55 \text{ miles/hour})(1.60934 \text{ km/mile}) = 88.5137 \text{ km/hour}$$

$$= (88.5137 \text{ km/hour})(1000 \text{ m/km})/(3600 \text{ s/hour})$$

$$= 24.587 \text{ m/s}$$

$$V_2 = 0, \qquad (z_2 - z_1) = (70 \text{ ft})(0.3048 \text{ m/ft}) = 21.336 \text{ m},$$

and $g = (32.174 \text{ ft/s}^2)(0.3048 \text{ m/ft}) = 9.807 \text{ m/s}^2$

$$\text{Decrease in KE} = \text{KE}_1 - \text{KE}_2 = \tfrac{1}{2}mV_1^2 - 0$$

$$= \tfrac{1}{2}(1360.78 \text{ kg})(24.587 \text{ m/s})^2$$

$$= 411,300 \text{ J} = 411.3 \text{ kJ}$$

$$\text{Increase in PE} = \text{PE}_2 - \text{PE}_1 = mg(z_2 - z_1)$$

$$= (1360.78 \text{ kg})(9.807 \text{ m/s}^2)(21.336 \text{ m})$$

$$= 284,700 \text{ J} = 284.7 \text{ kJ}$$

Thus,

$$\text{Energy dissipated} = \text{decrease in KE} - \text{increase in PE}$$

$$= 411.3 - 284.7 = 126.6 \text{ kJ}$$

1-9 SUMMARY

Thermodynamics is the basic science that deals with energy and matter, and their transformations and interactions. Engineering thermodynamics is a branch of thermodynamics in which emphasis is placed on the engineering analysis and design of processes, devices, and systems involving the beneficial utilization of energy and material.

A thermodynamic system is defined as a collection of matter or a region of space within certain prescribed boundaries. Everything external to the system that has relation with the system is called the surroundings or environment. A system of fixed content is called a closed system (or a control mass). A system whose content can be varied by passage of matter across its boundaries is called an open system (or a control volume).

The state of a system at any instant is its condition of existence at that instant. A property of a system is any quantity or characteristic that depends upon the state of the system. An intensive property (such as pressure and temperature) of a system is independent of the extent of mass of the system. An extensive property (such as total volume and energy) of a system is one whose value for the entire system is equal to the sum of its values for all parts of the system. If the value of any extensive property is divided by the mass of the system, the resulting property is intensive and is called a specific property.

When any property of a system changes, there is a change of state, and the system is said to undergo a process. The path of a process is the series of states through which the system passes during the process. A process whose initial and final states are identical is called a cycle.

A system is in thermodynamic equilibrium if it is incapable of making spontaneous state change without an interaction with its environment. A quasiequilibrium (or quasistatic) process is one in which departures from equilibrium during the process are infinitesimal.

The pressure is the normal force exerted per unit area on the surface of a system. The absolute pressure p_{abs} is related to the atmospheric pressure p_{atm} and the gage pressure p_{gage}, or the vacuum pressure p_{vacuum} as follows:

$$p_{abs} = p_{atm} + p_{gage}$$

$$p_{abs} = p_{atm} - p_{vacuum}$$

The temperature of a substance can be visualized as a measure of the mean molecular velocity of the substance. The temperature scales in SI and English systems, respectively, are the Celsius scale (°C) and the Fahrenheit scale (°F). The absolute temperature scales in SI and English systems, respectively, are the Kelvin scale (K) and the Rankine scale (°R). The temperature scales are related by

$$T\ (\text{K}) = T\ (°\text{C}) + 273.15$$

$$T\ (°\text{R}) = T\ (°\text{F}) + 459.67$$

$$T\ (°\text{R}) = 1.8T\ (\text{K})$$

$$T\ (°\text{F}) = 1.8T\ (°\text{C}) + 32$$

Energy is the ability to force a change or to produce an effect. When a system as a whole has a velocity, it possesses a bulk kinetic energy. When a system as a whole is at an elevation with respect to some reference level, it possesses a bulk potential energy. The internal energy of a substance is related to the kinetic and potential energies of the particles of the substance. In the absence of electricity, magnetism, and surface tension, the total energy E of a system is the sum of internal energy U, kinetic energy $m\mathbf{V}^2/2$, and potential energy mgz, or

$$E = U + m\mathbf{V}^2/2 + mgz$$

PROBLEMS

1-1. The gravitational acceleration g can be expressed as a function of elevation z above sea level in the form

$$g = A - Bz$$

where $A = 32.17$ ft/s^2, and $B = 3.32 \times 10^{-6}$ for z in feet. Find the weight of a mass that weighs 1 lbf at sea level when it is at 20,000 ft.

1-2. At a certain latitude, the gravitational acceleration g can be expressed as a function of elevation z above sea level as follows:

$$g = 9.807 - 3.32 \times 10^{-6}z$$

where g is in m/s^2, and z in meters. Find the height, in kilometers, above sea level where the weight of an object will have decreased by (a) 1 percent and (b) 5 percent.

1-3. Determine the weight of a 10-pound mass at standard gravity in (a) lbf, (b) dynes, and (c) newtons.

1-4. One system of units takes length in feet (ft), time in seconds (s), and force in pound-force (lbf) as the base units. The derived unit for mass is called the slug. Based on Newton's second law, determine the relation between the slug and the length, time, and force units mentioned before. What is the equivalence between the slug and the pound-mass (lbm)?

1-5. A spring scale (such as a bathroom scale) measures the weight of a body, which is the local gravitational force applied on the body. A beam scale compares masses on the two sides of the beam; thus, its reading is not affected by the variations in gravitational acceleration. A 75-kg astronaut weighs 735 N on earth. How much will the astronaut weigh (a) on a spring scale and (b) on a beam scale when the measurements are made on the moon, where the local gravitational acceleration is 1.67 m/s^2?

1-6. A piece of metal is weighed in a satellite, where the gravitational acceleration is 3 m/s^2. The observed weight is 12 N. Determine the mass of the metal in kilograms. How much will the metal weigh on earth?

1-7. An astronaut weighs 185 lbf on earth. What will be the astronaut's weight on the moon? The acceleration due to the moon's gravity is 1/6 that of the earth's.

1-8. A 1.1-ft^3 rigid block of material has a weight of 43.7 lbf at earth's sea level. The gravitational acceleration on the surface of Mars is 12.95 ft/s^2. For this block of material, determine (a) the mass on earth and on Mars, (b) the density on earth and on Mars, (c) the specific volume on earth and on Mars, (d) the weight on Mars, and (e) the specific weight on earth and on Mars.

1-9. On the surface of the moon, where the local acceleration of gravity is 5.47 ft/s^2, 10 lbm of a gas occupy a volume of 40 ft^3. For the gas, determine (a) the specific volume, (b) the density, and (c) the specific weight.

1-10. An astronaut circling the earth in a spaceship weighs 275 N at a location where the gravitational acceleration is 3.35 m/s^2. Determine the mass of the astronaut and his weight at sea level on the earth.

1-11. Acceleration during takeoff of space ships is sometimes measured in g's, or multiples of the standard acceleration of gravity. Determine the net upward force that a 70-kg astronaut experiences if the acceleration on lift-off is 10 g's.

1-12. A pressure cooker has a well-sealed lid and a petcock tightly against a small opening on top of the lid (see Fig. P1-12). Steam in the cooker can escape only through the small opening when the

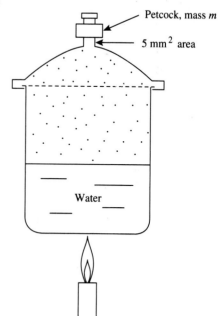

Figure P1-12

pressure force inside the cooker is large enough to overcome the weight of the petcock. Determine the mass of the petcock of a pressure cooker with an operating pressure of 100 kPa gage and an opening cross-sectional area of 5 mm². The atmospheric pressure is 101 kPa.

1-13. A 10-m diameter balloon is filled with helium gas having a density of 0.29 kg/m³. The balloon carrying two 75-kg people is released to float in air, which has a density $\rho_{air} = 1.16$ kg/m³. The buoyancy force F_b that pushes the balloon upward can be expressed as

$$F_b = \rho_{air} g V_{balloon}$$

where g is the local gravitational acceleration, and $V_{balloon}$ is the volume of the balloon. Determine the acceleration of the balloon when it is first released, neglecting the weight of the ropes and the cage, as shown in Fig. P1-13.

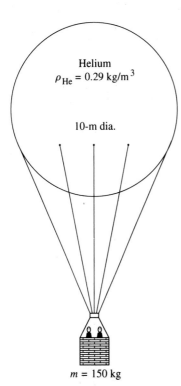

Helium
$\rho_{He} = 0.29$ kg/m³

10-m dia.

$m = 150$ kg

Figure P1-13

1-14. The mass of the earth is approximately 6×10^{24} kg. What is the average density of the earth, assuming the earth to be a sphere of radius 6.4×10^6 m?

1-15. What is the temperature for which thermometers using the Fahrenheit scale and the Celsius scale give the same reading?

1-16. The temperature of a system is 22°C. Express this temperature in °F, °R, and K.

1-17. The temperature of a system rises by 50°F during a heating process. Express this rise in °R, °C, and K.

1-18. The normal human body temperature is 37°C. What is it on the Fahrenheit scale?

1-19. A thermocouple generates a voltage $\mathscr{E}$ (in mV) in a circuit when the cold junction is maintained at 0°C and the hot junction is used as a probe to measure temperature T (in °C). See Fig. P1-19. The voltage–temperature relationship is $\mathscr{E} = aT + bT^2$, where $a = 0.26$ mV/°C, and $b = 5 \times 10^{-4}$ mV/°C². Determine the temperature for the millivolt reading of (a) 10 mV and (b) 50 mV.

1-20. Convert the following readings of pressure to psia, assuming the barometer reads 29.92 inches mercury (in. Hg): (a) 32 psig, (b) 76 in. Hg gage, (c) 16 in. Hg vacuum, and (d) 3.3 ft water gage. Note that 1 in. Hg = 0.491 psi and 1 ft water = 0.433 psi.

1-21. A mercury barometer, as shown in Fig. P1-21, is a manometer for measuring atmospheric pressure. What will be the atmospheric pressure if the mercury height h is 30 in.? The density of mercury is 810 lbm/ft³.

1-22. A standard atmosphere will support a column of mercury 0.760 meter high at 0°C. Calculate the pressure exerted by this atmosphere in Pa and psi. The density of mercury at 0°C is 13.6 g/cm³.

1-23. Convert the pressure units (a) feet of water and (b) inches of mercury into pound-force per square inch (psi), taking the density of water as 62.4 lbm/ft³ and the specific gravity of mercury as 13.6.

1-24. Figure P1-24 shows an open manometer for the measurement of the pressure of a pipeline containing oil with a specific gravity of 0.90. Water with a density value of 1000 kg/m³ is in the U-tube. For the monometer readings indicated in the figure, determine the pressure at the centerline of the pipe. The pressure of the atmosphere is 100 kPa.

1-25. Figure P1-25 gives the readings of a monometer measurement of the gas pressure in a pipeline in a location where the atmospheric pressure is 14.7 psia and the gravitational acceleration $g = 29.4$ ft/s². The density of the monometer liquid is 0.0136 kg/cm³. Determine the absolute gas pressure in the pipeline.

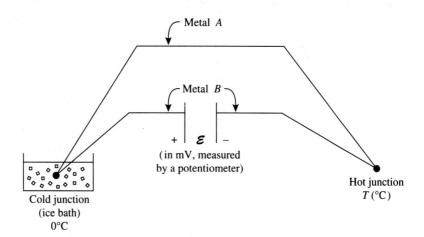

Metal *A*

Metal *B*

+ | $\mathcal{E}$ | −

(in mV, measured
by a potentiometer)

Hot junction
T (°C)

Cold junction
(ice bath)
0°C

Figure P1-19

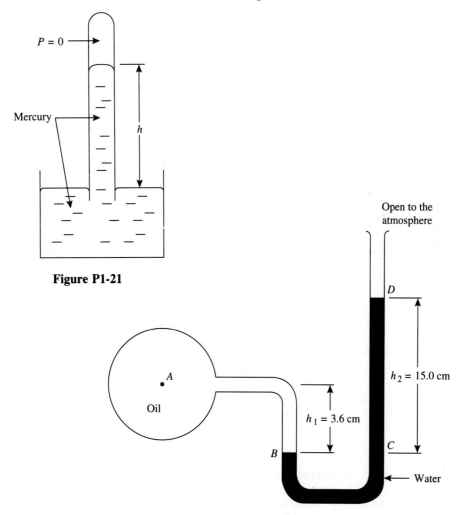

P = 0

Mercury

h

Figure P1-21

Open to the
atmosphere

D

$h_2 = 15.0$ cm

Oil

A

$h_1 = 3.6$ cm

B

C

Water

Figure P1-24

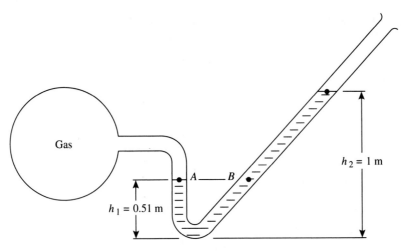

Figure P1-25

1-26. A person drives up a mountain in a car with a barometer. At the foothills of the mountain, the barometer indicates that the local atmospheric pressure is 75 cm Hg absolute. At some location up the mountain, the barometer indicates a local atmospheric pressure of 70 cm Hg absolute. Assuming that the average density of the atmospheric air is 1.2 kg/m³, estimate the change in altitude of this trip.

1-27. The standard atmospheric pressure at sea level is 14.7 psia. (a) What is the pressure at a mountaintop having an elevation of 30,000 ft above sea level? The average air density between sea level and the mountaintop can be assumed to be a constant value of 0.05 lbm/ft³. (b) What is the pressure at the bottom of a sea having a depth of 1300 ft? Take the water density to be 62.4 lbm/ft³. Neglect variations in gravitational acceleration.

1-28. The pilot of an airplane reports a barometric reading of 680 mm Hg while the ground control reports a barometric reading of 750 mm Hg. Estimate the altitude of the airplane from ground level if the average air density is 1.20 kg/m³ and g = 9.81 m/s².

1-29. A skin diver descends 100 m to a sunken ship. Determine the pressure exerted by the seawater on the body of the diver, assuming the density of seawater to be 1020 kg/m³. The atmospheric pressure is 101 kPa.

1-30. A skin diver descends 100 m to a sunken ship and finds a tank with a pressure gage attached. The gage reading is 90 kPa. What is the absolute pressure of the gas in the tank? The density of seawater is 1020 kg/m³ and the atmospheric pressure is 101 kPa.

1-31. A submarine is cruising at a depth of 280 m in seawater. The inside of the submarine is pressurized to atmospheric pressure. The average local gravity is 9.70 m/s² and the average density of the seawater is 1.03×10^3 kg/m³. Determine the pressure difference across the hull in kPa.

1-32. A 5-lbm object is moving at a speed of 10 ft/s at an elevation 30 ft above the sea-level ground. Determine the weight of the object and its kinetic and potential energies.

1-33. A 35-lbm body has a potential energy of 400 ft · lbf relative to datum plane A and a potential energy of −300 ft · lbf relative to datum plane B. The local acceleration of gravity is 31.6 ft/s². What is the relative location of plane A with respect to plane B?

1-34. An automobile having a mass of 3000 lbm travels at a speed of 60 miles per hour. (a) What is the mass of the automobile in kg? (b) What is the speed in km/h? (c) What is the kinetic energy of the automobile in kJ?

1-35. An automobile weighing 2000 lbf is 8 ft above the floor on a hoist in a repair shop. Determine the potential energy of the automobile in ft·lbf and N·m.

1-36. An airplane weighing 5 tons at ground is flying at a height of 30,000 ft at 500 miles per hour. Calculate the kinetic and potential energies of the airplane. Give your answer in both English and SI units.

1-37. An airplane having a mass of 2000 kg is flying at an altitude of 400 m with a velocity of 50 m/s. Assume that the gravitational acceleration g is a constant value of 9.8 m/s². (a) What is the final velocity if the kinetic energy is increased by 2400 kJ while flying at the same altitude? (b) What is the altitude if the potential energy is increased by 2400 kJ while flying at the same speed?

1-38. One method of meeting the extra electric power demand at peak periods is to pump some water from a lake to a water reservoir at a higher elevation at times of low power demand and to use the potential energy of the stored water to generate electricity through a hydraulic turbine at times of high power demand. See Fig. P1-38. For an energy storage capacity of 5×10^6 kW · h, determine the minimum amount of water that needs to be stored at an elevation of 100 m above the ground level.

1-39. The molecular mass of water (H_2O) is 18. Determine the number of moles for 18 lbm of water in (a) pound-mole (lbmole), (b) gram-mole (gmole), and (c) kilogram-mole (kgmole).

1-40. For each of the following substances: N_2, CO_2, and NH_3, determine (a) the number of kgmole in 50 kg of the substance, and (b) the mass, in lbm, of 20 lbmole of the substance.

1-41. On a certain location on earth and a certain time of the year, the sun is shining for 10 hours a day and has 200 Btu/ft²·h of average total radiation falling on a horizontal earth surface. How many ft² of earth surface would be necessary to capture the solar energy equivalent to a person's average daily consumption of food energy of 3000 kcal?

1-42. An automobile consumes 10 liters of gasoline while running at a steady speed of 80 km/h for 1 hour and 10 minutes. Determine the fuel economy in km/liter and mi/gallon.

1-43. A 2000-kg automobile is to climb a 100-m long uphill road with a 30° slope in 10 seconds. See Fig. P1-43. Neglecting friction, air drag, and rolling resistance, determine the power required for the following cases: (a) at a constant velocity, (b) from rest to a final velocity of 30 m/s, and (c) from 35 m/s to a final velocity of 5 m/s.

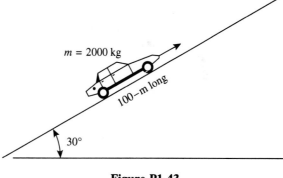

Figure P1-43

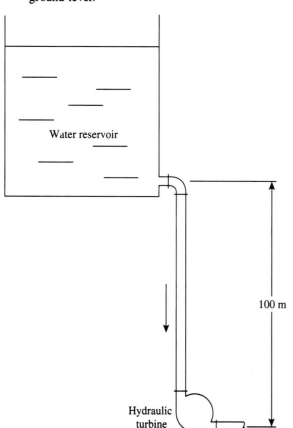

Figure P1-38

2

The First Law of Thermodynamics

The first law of thermodynamics is a law of nature governing the transformation of energy quantities. In this chapter, we first study the various forms of energy taking part in transformations during common thermodynamic processes. We then turn our attention to first-law formulations, particularly for closed systems containing compressible substances.

2-1 HEAT: A FORM OF ENERGY TRANSFER

When two systems at different temperatures are placed in thermal contact with each other and isolated from all other systems, a change in the temperature or other thermodynamic properties in both systems will always occur due to energy transfer. The energy being transferred from one system to another solely by reason of a temperature difference between the two systems is called *heat*. According to this definition, heat can be identified only as it crosses the boundary of a system, and therefore is a form of energy in transit. It is not an energy contained in a system and not a property of a system. Upon entering a system, heat is converted into potential or kinetic energy of the molecules, the atoms, or the subatomic particles that form the system.

As a convention of sign, heat transferred to a system is considered positive and heat transferred from a system is considered negative. The symbol Q is used to represent heat transfer to a system, and q is used to represent heat transfer per unit mass or per mole of the system.

A process that involves no heat interaction between the system and its surroundings is called an *adiabatic* process.

The amount of heat transferred to or from a system depends on the path of the process by which the system changes between two given end states. Heat is therefore a path function. The differential of a path function is inexact and will be denoted by the symbol đ to distinguish from the symbol d for exact differentials. Thus, we will use $đQ$, instead of dQ, to express the differential of a heat interaction.

2-2 WORK: A FORM OF ENERGY TRANSFER

In mechanics, work is defined as the product of a force and the displacement when both are measured in the same direction. In thermodynamics, work is an interaction between a system and its surroundings by reasons other than a temperature difference. The system under consideration may be, for instance, a compressible fluid, a paramagnetic solid, or an electric battery. However, a work interaction between a system and its surroundings may not always involve a recognizable force and a displacement. We therefore adopt the following general definition: *work* is done by a system if the sole effect on the surroundings can be reduced to the lifting of a weight. The magnitude of work is the product of the weight and the distance it could be lifted. It is always possible to reduce the external effect of any work interaction to the lifting of a weight through some mechanism whether it is real or by reasonable imagination.

To illustrate the suitability of the definition we adopted for work interaction, let us consider an electric battery as a system (see Fig. 2-1). The battery supplies electric current to drive an electric motor, which in turn drives a pulley–rope mechanism to lift a weight. Assume that all electrical and frictional losses and air resistance are absent. The electric current crossing the system boundary constitutes a work interaction with the sole effect external to the system being the lifting of the weight.

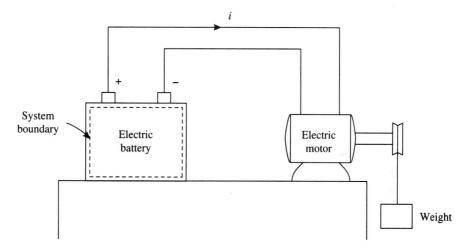

Figure 2-1 An illustration for work interaction.

Like heat, work is an interaction across the boundary of a system and therefore is also a form of energy in transit. Unlike heat, however, work is not an energy in transit because of a difference in temperature. Neither heat nor work is an energy contained in a system and neither is a property of a system.

The amount of work done by or on a system depends on the path of the process by which the system changes between two given end states. Work is therefore a path function.

As a convention of sign, work done by a system is considered positive and work done on a system is considered negative. The symbol W is used to represent work done by a system, and w is used to represent work per unit mass or per mole of the system.

Because work is a path function, we will use đW, instead of dW, to express the differential of a work interaction.

The time rate at which work is done by or on a system is defined as *power*.

2-3 EXPANSION OR COMPRESSION WORK

In this section, we seek to derive the expression for work in relation to volume changes. Consider as a system a quantity of a compressible fluid contained in a cylinder provided with a movable piston, as shown in Fig. 2-2. As the piston moves to the right a small distance d*l* under the action of fluid pressure, work is done by the fluid of the amount

$$đW = F\,dl$$

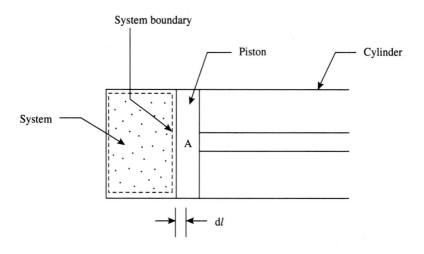

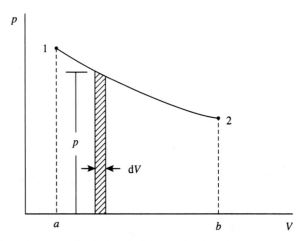

Figure 2-2 Work done on the moving boundary of a closed system in a quasiequilibrium process.

where F is the total force on the piston due to fluid pressure. The force F will probably vary as the piston moves, but for any position of the piston, it can be written as

$$F = pA$$

where p is the fluid pressure against the piston, and A the piston area. If the process undergone by the fluid is quasiequilibrium, p is then the uniform pressure of the fluid that is unique throughout the fluid system. Thus, we have

$$đW = pA \, dl$$

But the product $A \, dl$ is a differential volume dV, so, finally,

$$đW = p \, dV \tag{2-1}$$

It is often convenient to write the equation for a unit mass of the system. Thus,

$$đw = p \, dv \tag{2-1a}$$

where v is the specific volume.

Equation (2-1) represents the work done by the fluid during an infinitesimal quasiequilibrium process. However, if there are dissipative effects, as when the piston moves with friction, part of the work done by the fluid would become energies of the fluid and the piston–cylinder machine parts. Thus, less work would be delivered out through the piston. In other words, Eq. (2-1) applies to quasiequilibrium processes only.

The total work done by the fluid as the fluid expands under quasiequilibrium conditions from state 1 to state 2 is

$$W_{12} = \int_1^2 p \, dV \tag{2-2}$$

This work is represented by the area $1–2–b–a–1$ under the curve $1–2$ on a pressure–volume, or simply $p–V$, diagram, as shown in Fig. 2-2. If the process had proceeded from 2 to 1, work of the amount

$$W_{21} = \int_2^1 p \, dV = \text{area } 2–1–a–b–2$$

would be done on the fluid.

It is possible to connect two states of a system by many different quasiequilibrium processes. The work done by or on the system depends on the path of the process, indicating that work is a path function.

Although the equation $đW = p \, dV$ was derived for the case of fluid expansion against a piston, it is correct for a closed system of any shape under a uniform pressure over the moving boundary.

It must be emphasized that only during a quasiequilibrium process, the pressure p in Eqs. (2-1) and (2-2) is the pressure of the entire system undergoing the process and the area under the process curve on a $p–V$ diagram represents the work done. For a nonquasiequilibrium process, the pressure p in Eqs. (2-1) and (2-2) is just the average pressure against the moving boundary and not the pressure of the entire system. Because the process curve is not defined in a nonquasiequilibrium process, it is, of course, meaningless to talk about the area under a process curve.

Consider now a gas system contained in a cylinder–piston arrangement that undergoes a cyclic process 1–2–3–4–5–1, as depicted in the p–V diagram of Fig. 2-3. In process 1–2–3, work is done by the gas on the piston, as represented by the area 1–2–3–b–a, whereas in process 4–5, work is done by the piston on the gas, as represented by the area 4–5–a–b. There is no boundary work involved in processes 3–4 and 5–1, that is, the volume of the system does not change. The net work done by the gas on the piston is represented by the cross-hatched area enclosed by the cyclic process. Notice that in a cyclic process, there is no change in all properties of the system, but there is, in general, a net work.

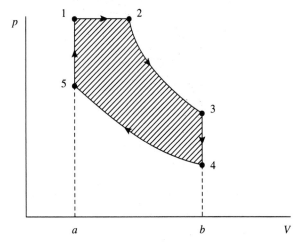

Figure 2-3 Net work of a cycle.

The evaluation of the integral of Eq. (2-2) requires a knowledge of the functional relationship between pressure and volume during a quasiequilibrium process. One common quasiequilibrium process, known as a *polytropic* process, has the following p–V relation:

$$pV^n = \text{constant} \tag{2-3}$$

where n is a constant that can be any value from $-\infty$ to $+\infty$ depending on the particular process. Performing the integration of Eq. (2-2) with this p–V relation, we have

$$W = \int_1^2 p\, dV = \text{constant} \int_1^2 \frac{dV}{V^n}$$

$$= \text{constant}\left(\frac{V_2^{-n+1} - V_1^{-n+1}}{-n + 1}\right)$$

Substituting $p_1 V_1^n = p_2 V_2^n = \text{constant}$, we have

$$W = \frac{p_2 V_2 - p_1 V_1}{1 - n} \tag{2-4}$$

Notice that Eq. (2-4) is valid for any value of n except $n = 1$. For the case when $n = 1$, we have

$$pV = \text{constant} = p_1 V_1 = p_2 V_2$$

so that

$$W = \int_1^2 p \, dV = p_1 V_1 \int_1^2 \frac{dV}{V}$$

or

$$W = p_1 V_1 \ln \frac{V_2}{V_1} \qquad (2\text{-}5)$$

For the special case of $n = 0$, $pV^0 = p = $ constant, the process occurs at constant pressure so that

$$W = p_1(V_2 - V_1)$$

When $n = \infty$, the p–V relation becomes $p^{1/n}V = V = $ constant, the process occurs at constant volume so that

$$W = 0$$

Example 2-1

A cylinder with a movable piston contains a gas at an initial pressure of 300 kPa and an initial volume of 0.2 m³. The gas expands in a quasiequilibrium process until its volume is doubled. Determine the work done by the gas if the process follows the relation (a) $p = $ constant, (b) $pV = $ constant, (c) $pV^2 = $ constant, and (d) $p/V = $ constant.

Solution

$$p_1 = 300 \text{ kpa}, \qquad V_1 = 0.2 \text{ m}^3, \qquad V_2 = 2V_1 = 0.4 \text{ m}^3$$

(a) $p = $ constant $= p_1 = p_2$

$$W = \int_1^2 p \, dV = p_1(V_2 - V_1)$$

$$= (300{,}000 \text{ N/m}^2)(0.4 \text{ m}^3 - 0.2 \text{ m}^3)$$

$$= 60{,}000 \text{ N·m} = 60{,}000 \text{ J} = 60 \text{ kJ}$$

(b) $pV = $ constant $= p_1 V_1 = p_2 V_2$
From Eq. (2-5),

$$W = p_1 V_1 \ln \frac{V_2}{V_1}$$

$$= (300{,}000 \text{ N/m}^2)(0.2 \text{ m}^3) \ln \frac{0.4}{0.2}$$

$$= 41{,}590 \text{ N·m} = 41{,}590 \text{ J} = 41.59 \text{ kJ}$$

(c) $pV^2 = $ constant $= p_1 V_1^2 = p_2 V_2^2$
Thus,

$$p_2 = p_1 \left(\frac{V_1}{V_2}\right)^2 = 300\left(\frac{0.2}{0.4}\right)^2 = 75 \text{ kPa}$$

From Eq. (2-4) with $n = 2$,

$$W = \frac{p_2 V_2 - p_1 V_1}{1 - n}$$

$$= \frac{(75,000 \text{ N/m}^2)(0.4 \text{ m}^3) - (300,000 \text{ N/m}^2)(0.2 \text{ m}^3)}{1 - 2}$$

$$= 30,000 \text{ N·m} = 30,000 \text{ J} = 30 \text{ kJ}$$

(d) $\dfrac{p}{V} = pV^{-1} = \text{constant} = \dfrac{p_1}{V_1} = \dfrac{p_2}{V_2}$

Thus,

$$p_2 = p_1 \left(\frac{V_2}{V_1}\right) = 300\left(\frac{0.4}{0.2}\right) = 600 \text{ kPa}$$

From Eq. (2-4) with $n = -1$,

$$W = \frac{p_2 V_2 - p_1 V_1}{1 - n}$$

$$= \frac{600,000(0.4) - (300,000(0.2)}{1 - (-1)}$$

$$= 90,000 \text{ N·m} = 90,000 \text{ J} = 90 \text{ kJ}$$

or by integration,

$$W = \int_1^2 p \, dV = \text{constant} \int_1^2 V \, dV = \frac{p_1}{V_1}\left(\frac{V_2^2 - V_1^2}{2}\right)$$

$$= \frac{300,000 \text{ N/m}^2}{0.2 \text{ m}^3}\left(\frac{0.4^2 \text{ m}^6 - 0.2^2 \text{ m}^6}{2}\right) = 90,000 \text{ Nm}$$

$$= 90 \text{ kJ}$$

Figure 2-4 shows the processes of this example in a p–V diagram.

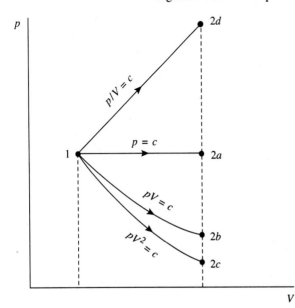

V **Figure 2-4** p–V diagram for Example 2-1.

The First Law of Thermodynamics   Chap. 2

2-4 OTHER QUASIEQUILIBRIUM MODES OF WORK

We derived in Sec. 2-3 the work done by a compressible system on the moving boundary due to volume change ($p\,dV$ work). There are a number of other quasiequilibrium work modes that occur frequently in thermodynamic analyses. The objective of this section is to introduce several systems with work modes other than $p\,dV$.

Elastic work

First, consider the work done in stretching an elastic thin solid rod or wire of cross-sectional area A and natural length L, as illustrated in Fig. 2-5. If the stretching force F acts through an elongation dL, the work done is

$$đW = -F\,dL \qquad (2\text{-}6)$$

where the negative sign indicates that work input is required to increase the elongation of a rod, in accordance with our sign convention.

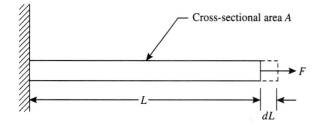

Figure 2-5   Elastic thin rod or wire under tension.

It is appropriate in the study of elastic solids to express work in terms of the *stress* σ and the *strain* ϵ, which are defined as

$$\sigma = \frac{F}{A} \qquad (2\text{-}7)$$

$$d\epsilon = \frac{dL}{L} \qquad (2\text{-}8)$$

In Eq. (2-7), tensile stress will be taken as positive and compressive stress as negative. We are to neglect the lateral strain when the rod is under axial stress. Within the elastic limit, the process of loading and unloading can be performed in a quasiequilibrium condition. Upon substituting Eqs. (2-7) and (2-8) into Eq. (2-6), we get the work of elastic stretching,

$$đW = -AL\sigma\,d\epsilon$$

or

$$đW = -V_0\sigma\,d\epsilon \qquad (2\text{-}9)$$

where $V_0 = AL$ = volume of the rod at the unstrained state. It is convenient to introduce specific quantities referred to a unit of volume of the system. Thus, the work per unit unstrained volume (for which the symbol w is used) is

$$đw = -\sigma\,d\epsilon \qquad (2\text{-}9a)$$

where the minus sign indicates that work input is required to increase the strain of an elastic solid.

Surface-tension work

We now seek to study the work done in stretching a liquid surface film. A liquid and a gas that coexist in equilibrium are separated by a finite but extremely thin layer of phase boundary of a magnitude comparable with molecular dimensions across which the transition between the two phases takes place. A molecule in the interior of the liquid phase experiences, in general, no resultant intermolecular attractions from its neighbors. However, a molecule in the phase boundary is attracted more strongly by the closely oriented liquid molecules on one side than by the very distant gas molecules on the other side. Work must be done against the resulting unbalanced attraction force to bring a molecule from the interior of the liquid to the surface. A molecule in the phase boundary can thus be considered as having a greater potential energy than a molecule in the interior of the liquid. The total additional potential energy is obviously proportional to the surface area between the phases. The surface tends to assume a shape of minimum area to conform with the conditions of stable equilibrium at minimum potential energy. In other words, the net inward attraction from the liquid causes the surface to diminish in area until the surface is the smallest possible for a given volume, subject to the external conditions or forces acting on the system. For a given volume, a sphere has the minimum surface. This is the reason why liquid droplets in a gas and gas bubbles in a liquid are spherical.

In ordinary thermodynamic calculations, surface work effects are usually small compared to such bulk contributions as volume changes. A system chosen to demonstrate surface work must have a geometry that will emphasize surface effects. In addition, the disturbing effect of gravity should be eliminated. Let us consider as a system a soap film streched across a rectangular wire frame, one side of which is movable, as illustrated in Fig. 2-6. The soap film consists of two surface films with a small amount of liquid in between. During a change in the film area, there are mass transfers between the interior of the liquid and the surface film. Because the surface film tends to assume a minimum area, it tends to pull the movable wire to the left, and a force F as indicated is required to keep the wire in position. When the film is stretched slowly to the right a

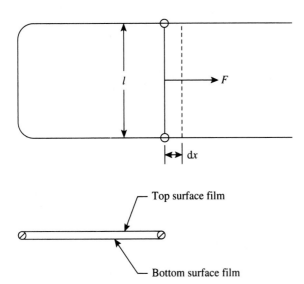

Figure 2-6 Surface films stretched on a wire frame.

distance dx, the film area A will increase by an amount $dA = 2l\,dx$. Thus, the work required is

$$dW = -F\,dx = -\frac{F}{2 \times l}\,dA$$

Here we neglect the small work done due to volume change. The work required per unit increase in surface area is defined as the *interfacial tension* γ, which is also the force acting perpendicularly to a line of unit length in the surface. That is,

$$\gamma = \frac{F\,dx}{2 \times l \times dx} = \frac{F}{2 \times l}$$

Hence, the work required becomes

$$dW = -\gamma\,dA \tag{2-10}$$

where, in accordance with our sign convention, the negative sign indicates that work input is required to increase the area of an interface.

Electrical work

We now consider the electric work in displacing an electric charge across the boundary of a system in the presence of an electric potential. In Fig. 2-7, a storage battery is connected to an external circuit through which an electric current flows. The battery is considered as the system. For a potential difference $\mathscr{E}$, a current i performs work at the rate

$$\dot{W} = \frac{dW}{dt} = -\mathscr{E}i$$

But

$$i = \frac{dZ}{dt}$$

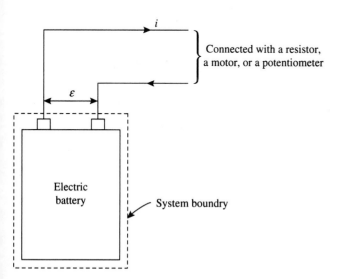

Connected with a resistor, a motor, or a potentiometer

Electric battery

System boundry

Figure 2-7 Electric work mode.

We have then

$$\text{đ}W = -\mathscr{E}\,dZ \tag{2-11}$$

where Z is the electric charge. The minus sign is required to conform with our sign convention for work interactions. Work is done by the battery when it is discharging, and work is done on it when it is being charged. Because dZ is negative in a discharging process, $\text{đ}W$ is thus a positive value, corresponding to the work done by the system. It should be noted that if the potential difference across the terminals is infinitely close to the electromotive force (emf) of the battery cell, charging and discharging processes can be made in quasiequilibrium conditions.

Torsion work

When a solid bar is constrained to a fixed position at one end and a tangential force F is applied at some radial length r, the torque τ applied to the bar is given by

$$\tau = Fr$$

When the torque acts through a differential angular displacement $d\theta$, the differential torsion work done is given by

$$\text{đ}W = -\tau\,d\theta \tag{2-12}$$

where the angular displacement θ is in rad, noting that

$$360° = (2\pi)\ \text{rad}$$

The minus sign in Eq. (2-12) is required to conform with our sign convention for work interactions. Work is done on the bar when the direction of action of the torque is the same as the direction of rotation of the bar.

When the rotation of a shaft is under a constant torque, the expression for torsional work becomes

$$W = -\tau\,(\Delta\theta)$$

Work of magnetization and polarization

There are two common thermodynamic systems with quasiequilibrium work modes involving external force fields, namely, a paramagnetic system in a magnetic field and a dielectric system in an electric field. The paramagnetic system is of prime importance in low-temperature thermodynamics. We will study this system in some detail in Chapter 12. For now, it suffices to note that the work required in the magnetization of a magnetic material is given by the expression

$$\text{đ}W = -\mu_0\,\mathbf{H}\,d(V\mathbf{M}) \tag{2-13}$$

where $\mathbf{H}$ is the magnetic field or intensity, $\mathbf{M}$ the magnetization or magnetic moment per unit volume, V the volume, and μ_0 the permeability of free space, which has a value of $4\pi \times 10^{-7}$ weber/ampere-meter (Wb/A·m).

In contrast with an electric conductor, which has a sufficiently large number of free electrons, a dielectric or electric insulator has none or only a relatively small number of free electrons. The major effect of an electric field on a dielectric is the polarization of the electric dipoles. Work is done by the electric field on the dielectric material during the polarization process. This work is given by the expression

$$đW = -\mathbf{E}\, d(V\mathbf{P}) \tag{2-14}$$

where $\mathbf{E}$ is the electric field or intensity, $\mathbf{P}$ the electric polarization or dipole moment per unit volume, and V the volume.

Summary of quasiequilibrium work modes

In this and the previous section, we have studied a few different systems that can perform quasiequilibrium work. The expressions for work always take the form

$$đ(\text{work}) = (\text{intensive property}) \cdot d(\text{extensive property})$$

Even though the factors within a given work expression may not bring to mind recognizable force and displacement, it is convenient to refer to all types of work as a product of a force and a displacement, and to write

$$đ(\text{work}) = (\text{generalized force}) \cdot d(\text{generalized displacement})$$

For a system that can perform quasiequilibrium work in different independent modes, we can write

$$đ(\text{work})_{\text{total}} = \Sigma(\text{generalized force}) \cdot d(\text{generalized displacement})$$

If a system involves all the quasiequilibrium work modes we have studied, then the total work is given by

$$đW_{\text{total}} = p\, dV - F\, dL - \gamma\, dA - \mathscr{E}\, dZ - \tau\, d\theta - \mu_0 \mathbf{H}\, d(V\mathbf{M}) - \mathbf{E}\, d(V\mathbf{P})$$

Other quasiequilibrium work modes may be included in the preceding equation. Non-quasiequilibrium work, such as work done due to frictional effect, however, should not be included.

Example 2-2

An atomizer slowly shoots out minute water droplets of an average radius of 10^{-4} cm into air. Estimate the work required when 1 kg of water is atomized at a constant temperature of 25°C. The interfacial tension of water in contact with air at 25°C is 7.2×10^{-4} N/cm. The specific volume of water at 25°C is 1.0029 cm³/g.

Solution The number of droplets formed by 1 kg of water is

$$\frac{(1.0029 \text{ cm}^3/\text{g})(10^3 \text{g/kg})}{\frac{4}{3}\pi (10^{-4})^3 \text{ cm}^3} = 2.394 \times 10^{14}$$

The total surface area of all these droplets is

$$A = 4\pi (10^{-4})^2(2.394 \times 10^{14}) = 3.009 \times 10^7 \text{ cm}^2/\text{kg}$$

From Eq. (2-11), $đW = -\gamma\, dA$, we then have

$$W = -\gamma(A - A_0)$$

where A_0 is the initial surface area of 1 kg of water before being atomized. Because $A \gg A_0$, we can write

$$W = -\gamma A$$

Therefore,

$$W = -(7.2 \times 10^{-4} \text{ N/cm})(3.009 \times 10^7 \text{ cm}^2/\text{kg})$$

$$= -2.166 \times 10^4 \text{ N·cm/kg}$$

$$= -2.166 \times 10^2 \text{ J/kg}$$

where the negative sign indicates that work input is required to form the droplets.

Example 2-3

Liquid nitrobenzene ($C_6H_5NO_2$) is a dielectric material. In a certain range of temperatures, the electric polarization or dipole moment per unit volume, $\mathbf{P}$, is related to the electric field, $\mathbf{E}$, as follows:

$$\mathbf{P} = \epsilon_0(23.8 - 4.16 \times 10^{-4}\mathbf{E})\mathbf{E}$$

where $\mathbf{P}$ is in coulomb/meter^2 (C/m^2), $\mathbf{E}$ is in volt/meter (V/m), which is newton/coulomb (N/C), and ϵ_0 is the permittivity of free space and has a value of 8.854×10^{-12} coulomb^2/newton-meter^2 (C^2/N·m^2). Notice that the constant 4.16×10^{-4} has the dimension of $\mathbf{E}^{-1}$. Calculate the work done per cm^3 of nitrobenzene when the electric field is increased from zero to 10^4 volts/meter.

Solution   By differentiating the given $\mathbf{P}$–$\mathbf{E}$ relation,

$$d\mathbf{P} = \epsilon_0(23.8 - 4.16 \times 10^{-4} \times 2\mathbf{E}) \, d\mathbf{E}$$

Inserting this equation in Eq. (2-14), we have

$$W = -\int \mathbf{E} \, d(V\mathbf{P}) = -V \int \mathbf{E} \, d\mathbf{P}$$

$$= -V\epsilon_0 \int_{\mathbf{E}_1}^{\mathbf{E}_2} (23.8\mathbf{E} - 8.32 \times 10^{-4} \, \mathbf{E}^2) \, d\mathbf{E}$$

$$= -V\epsilon_0 \left[\frac{23.8}{2}(\mathbf{E}_2^2 - \mathbf{E}_1^2) - \frac{8.32 \times 10^{-4}}{3}(\mathbf{E}_2^3 - \mathbf{E}_1^3) \right]$$

With $V = 1$ cm^3 $= 10^{-6}$ m^3, $\mathbf{E}_1 = 0$, and $\mathbf{E}_2 = 10^4$ N/C, we obtain

$$W = -(10^{-6} \text{ m}^3)(8.854 \times 10^{-12} \text{ C}^2/\text{N·m}^2)\left[\frac{23.8}{2}(10^4 \text{ N/C})^2 \right.$$

$$\left. - \frac{1}{3}(8.32 \times 10^{-4} \text{ C/N})(10^4 \text{ N/C})^3 \right]$$

$$= -8.08 \times 10^{-9} \text{ N·m} = -8.08 \times 10^{-9} \text{ J}$$

where the minus sign indicates that work is done on the dielectric material.

Example 2-4

An electronic device is operated by a battery at a constant voltage of 120 volts. The current i drawn by the device varies with time t according to the relation

$$i = 15e^{-t/60}$$

where i is in amperes, and t is in seconds. Determine (a) the work performed on the device during the first 5 minutes of operation, and (b) the power delivered to the device at $t = 0$ and $t = 5$ min.

Solution The work done on the device is given by Eq. (2-12),

$$dW_{in} = \mathscr{E}i \, dt = \mathscr{E}(15e^{-t/60}) \, dt$$

(a) The total work done on the device between $t_1 = 0$ and $t_2 = 300$ s is then

$$W_{in} = 15 \, \mathscr{E} \int_{t_1}^{t_2} e^{-t/60} \, dt$$

$$= 15(-60)\mathscr{E} \int_{t_1}^{t_2} e^{-t/60} \, d(-t/60)$$

$$= -900\mathscr{E}(e^{-t_2/60} - e^{-t_1/60})$$

$$= -900(120)(e^{-300/60} - e^0)$$

$$= 107,300 \text{ J} = 107.3 \text{ kJ}$$

(b) When $t = 0$, $i = 15e^{-t/60} = 15$ amperes. The power input to the device is given by Eq. (2-11),

$$\dot{W}_{in} = \mathscr{E}i = (120 \text{ volts})(15 \text{ amperes}) = 1800 \text{ watts}$$

When $t = 300$ s, $i = 15e^{-300/60} = 0.1011$ ampere. The power input to the device is

$$\dot{W}_{in} = (120 \text{ volts})(0.1011 \text{ ampere}) = 12.1 \text{ watts}$$

2-5 THE FIRST LAW OF THERMODYNAMICS

The *first law of thermodynamics* is essentially the law of conservation of energy applied to thermodynamic systems. It expresses the fact that work and heat are different aspects of the same physical quantity. This is a fundamental law of nature and cannot be deduced or established from other fundamental laws. Its establishment is based on empirical facts. Many experiments have been conducted and all such experiments have firmly supported it.

As an illustration, consider as a system the gas contained in a rigid tank provided with a thermometer and a paddle-wheel pulley-weight assembly, as shown in Fig. 2-8(a). The tank is initially insulated with the weight at the top position and has a thermometer indicating a temperature T_a. As the weight falls, work is done by the paddle wheel on the gas, resulting in a raise of gas temperature. When the weight falls to the bottom position, the thermometer indicates a temperature T_b. By neglecting friction of the pulley assembly and the air resistance on the falling weight, the work done on the gas is given by $mg \, L$. Now let a portion of the tank insulation be removed and thermal contact initiated with a cold body, as shown in Fig. 2-8(b). As heat flows out, the thermometer reading drops. When the initial temperature T_a is reached, the thermal contact between the gas and the cold body is cut off and the total heat flow is evaluated. The gas in the container has gone through a cycle in the two parts of the experiment. The conclusion of such an experiment is always that the work done on the gas in the first part is equal to the heat transferred out in the second part, when both work and heat are expressed in the same units.

Many experiments for different systems and different experiment arrangements have led to the formulation of the first law of thermodynamics, which in equation form is written as

$$\oint dW = \oint dQ \tag{2-15}$$

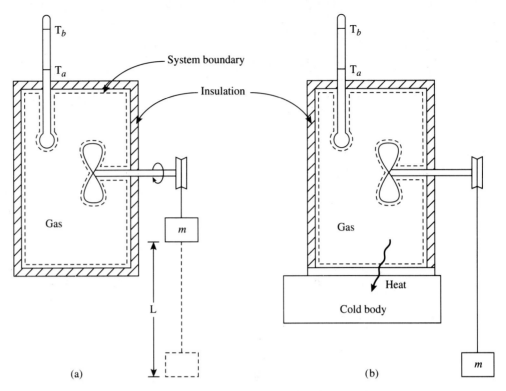

Figure 2-8 A closed system undergoing a cycle.

where the symbol $\oint$ indicates integration around a cycle, and both work W and heat Q are measured in the same units.

Equation (2-15) can be rewritten as

$$\oint (đQ - đW) = 0$$

It follows that because the cyclic integral of the quantity $đQ - đW$ is always zero, this quantity $(đQ - đW)$ must be the differential of a property of the system. To define this property more clearly, consider a system that changes from state 1 to state 2 by process A and returns to state 1 by either process B or process C, as depicted on the property diagram of Fig. 2-9. For the cycle composed of processes A and B, we have

$$\int_{1,A}^{2} (đQ - đW) + \int_{2,B}^{1} (đQ - đW) = 0$$

and for the cycle composed of processes A and C, we have

$$\int_{1,A}^{2} (đQ - đW) + \int_{2,C}^{1} (đQ - đW) = 0$$

Comparing the preceding two equations reveals that

$$\int_{2,B}^{1} (đQ - đW) = \int_{2,C}^{1} (đQ - đW)$$

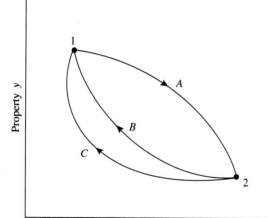

Property y

Property x

Figure 2-9 Energy is a point function.

Because B and C are arbitrary processes between states 2 and 1, it follows that the quantity $\int (đQ - đW)$ is the same for all processes between the two states. In other words, $\int (đQ - đW)$ depends only on the end states and not on the path followed between the two end states. Therefore, $\int (đQ - đW)$ is a point function and is a property of the system. This property is called the *total energy* (or simply the *energy*) and is denoted by the symbol E. Thus, we can write

$$\Delta E = \int (đQ - đW) = Q - W$$

or

$$Q = \Delta E + W \qquad (2\text{-}16)$$

Equation (2-16) is a useful expression of the first law for a system of fixed mass. In words, it reads that the heat added to a system during a process is equal to the gain in total energy of the system plus work done by the system. Note that heat added to a system is considered positive and heat rejected from a system is considered negative; and work done by a system is considered positive and work done on a system is considered negative. The symbol Δ will always be used to mean "final minus initial." Thus, in Eq. (2-16)

$$\Delta E = E_{\text{final}} - E_{\text{initial}}$$

It should be noted that Eq. (2-16) expresses the conservation of energy in a thermodynamic process. It is applicable no matter if the process is in quasiequilibrium or nonquasiequilibrium.

Because E is a property, its differential is exact and will be denoted as dE. Thus, for an infinitesimal state change, the first law is written as

$$đQ = dE + đW \qquad (2\text{-}17)$$

The property E represents the total energy of a system in a given state. It includes the kinetic energy KE, the potential energy PE, and the internal energy U of the system. Expressed mathematically,

$$E = U + \text{KE} + \text{PE}$$

which is the same equation as Eq. (1-14). Substituting Eqs. (1-17) and (1-18) into this equation results in

$$E = U + \frac{m\mathbf{V}^2}{2} + mgz \qquad (2\text{-}18)$$

For the unit mass of a system, this is written as

$$e = u + \frac{\mathbf{V}^2}{2} + gz \qquad (2\text{-}18a)$$

where e is the specific total energy, u is the specific internal energy, $\mathbf{V}$ is the velocity of the system, z is the elevation of the center of gravity of the system from an arbitrary datum plane, and g is the acceleration due to gravity.

Internal energy U is an extensive property of a system and represents all the energy associated with the motion and configuration of the molecules and a number of other microscopic forms. In the absence of bulk KE and PE changes, the first law for a system of fixed mass can be written in integrated form as

$$Q = \Delta U + W \qquad (2\text{-}19)$$

and in differential form

$$đQ = dU + đW \qquad (2\text{-}20)$$

By applying the first law of thermodynamics to a cyclic process, the result can be written as

$$\text{net } W_{\text{out}} = \text{net } Q_{\text{in}} \qquad (2\text{-}21)$$

Even though energy is conserved, the desirability of different forms of energy is not the same. For example, in conventional heat engines, we intend to transform heat into work. Work is therefore the desired output and heat is the required input. The extent of this transformation is of prime interest. A measure of accomplishment of a heat engine is expressed in terms of the *thermal efficiency* η_{th} as defined:

$$\eta_{\text{th}} = \frac{\text{net work output}}{\text{heat input}} \qquad (2\text{-}22)$$

The first law allows this efficiency to be 100%, but we will learn later that the second law of thermodynamics prohibits a 100% conversion from heat to work, making $\eta_{\text{th}} < 100\%$ no matter how idealized the processes are assumed to be.

Example 2-5

An ideal rubber is defined as a rubber strip for which internal energy is a function of temperature only and is independent of extension. For an ideal rubber strip, the relation among the applied force F, the temperature T, and the length L is given by the expression

$$F = 0.0133T\left(\frac{L}{L_0} - \frac{L_0^2}{L^2}\right)$$

where F is in N, T in K, L in m, and $L_0 = L$ at zero tension. Determine the work done and the heat transfer when the rubber strip is stretched slowly from $L_0 = 1$ m to $L = 2L_0 = 2$ m at a constant temperature of $T = 300$ K.

The First Law of Thermodynamics   Chap. 2

Solution   From Eq. (2-6), we have

$$\mathrm{d}W = -F\,\mathrm{d}L = -0.0133T\left(\frac{L}{L_0} - \frac{L_0^2}{L^2}\right)\mathrm{d}L$$

Integrating at constant temperature gives

$$W = -0.0133T\left[\frac{1}{2L_0}(L^2 - L_0^2) - L_0^2\left(\frac{1}{L_0} - \frac{1}{L}\right)\right]$$

$$= -(0.0133)(300)[\tfrac{1}{2}(4 - 1) - (1 - \tfrac{1}{2})] = -3.99\text{ J}$$

where the minus sign indicates that work is done on the rubber strip.

Because for an ideal rubber the internal energy is a function of temperature only, $\Delta U = 0$ for the isothermal stretching process. By the first law, we have then

$$Q = W = -3.99\text{ J}$$

where the minus sign indicates that heat is rejected from the rubber strip.

Example 2-6

A $0.2-m^3$ insulated rigid tank contains air with an initial density of 1.2 kg/m^3. A paddle wheel inside the tank transfers energy to the air at a constant rate of 4 W for 20 min. Determine the increase in internal energy of the air in kJ and kJ/kg.

Solution   Figure 2-10 is a schematic diagram of a paddle wheel doing work on the air. Because the paddle wheel does work on the air by frictional effect, this work interaction is nonquasiequilibrium. The first law of thermodynamics is nevertheless applicable. In the absence of bulk KE and PE changes, Eq. (2-19) will be used for the system (i.e., the air). As the tank is insulated, $Q = 0$, the first-law equation reduces to

$$\overset{0}{\cancel{Q}} = \Delta U + W$$

or

$$\Delta U = U_2 - U_1 = -W$$

where subscript 1 denotes the initial state of the air before the turning of the paddle wheel, and subscript 2 denotes the final state of the air after the 20-min working time of the paddle wheel.

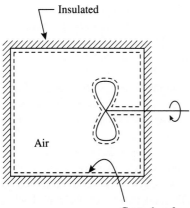

Control surface   **Figure 2-10**   Paddle-wheel work.

The given data are $V = 0.2$ m³ (neglecting the volume occupied by the paddle wheel), $\rho_1 = 1.2$ kg/m³, and $\dot{W} = -4$W $= -4$ J/s for $t = 20$ min $= 1200$ s. The minus sign for the paddle-wheel power expression is used because work is done on the system (air). The mass of air in the tank is calculated by the equation

$$m = \rho_1 V = (1.2 \text{ kg/m}^3)(0.2 \text{ m}^3) = 0.24 \text{ kg}$$

The paddle-wheel work is evaluted by the expression

$$W = \int_{t=0}^{t=1200\,s} \dot{W}\, dt = (-4 \text{ J/s})(1200 \text{ s}) = -4800 \text{ J} = -4.8 \text{ kJ}$$

Therefore,

$$\Delta U = U_2 - U_1 = -(-4.8) = 4.8 \text{ kJ}$$

and

$$\Delta u = u_2 - u_1 = \Delta U/m = (4.8 \text{ kJ})/(0.24 \text{ kg}) = 20 \text{ kJ/kg}$$

2-6 THE STATE POSTULATE AND SIMPLE SYSTEMS

Experimental observations have shown that there is a uniqueness about the number of independent properties required to specify an equilibrium state of a system. The number of independent properties required obviously depends on the number of different energy interactions. Besides heat interaction, there may be a number of different modes of work interactions. The relation between the number of independent properties and the number of different energy interactions may be summarized in a postulate, known as the *state postulate*, which is as follows: The number of independent properties required to specify the equilibrium state of a system is equal to the number of possible quasiequilibrium work modes plus one. The "plus one" accounts for heat interaction.

The basic laws of thermodynamics are quite general and can be applied to systems of considerable complexity, involving mechanical, chemical, electrical, magnetic, and other effects. In Sections 2-3 and 2-4, we introduced a few common quasiequilibrium work modes. A thermodynamic system can involve any one or more of those work modes. Fortunately, the majority of systems encountered in thermodynamic analyses are of restricted nature. For instance, there are a vast number of thermodynamic systems composed of so-called pure substances that are homogeneous and invariable in chemical composition. In addition, in many cases, the effects of motion, capillarity, solid distortion, and external force fields (electric, magnetic, and gravitational) are absent or can be neglected. A system composed of such substances can perform work in a quasiequilibrium process by volume change only ($p\, dV$ work). The term *simple compressible system* will be applied to such a system. The word "simple" is used to indicate that the system has only one quasiequilibrium work mode. There are other simple systems, such as a *simple elastic system*, for which the only quasiequilibrium work mode is elastic elongation, and a *simple magnetic system*, for which the only quasiequilibrium work mode is magnetization.

According to the state postulate, two independent intensive thermodynamic properties are required to specify an equilibrium state of a simple system. This is known as the two-property rule. The significance of the term independent property should be emphasized, because only when the properties are independent do they suffice to specify the state. For example, as we will show in Chapter 3, pressure and temperature are not independent properties for a pure substance in the form of a mixture of liquid and vapor because an indefinite number of states could exist at a given pressure and temperature.

2-7 SIMPLE COMPRESSIBLE SYSTEMS

The sum of the internal energy and the pressure–volume product consists entirely of properties and is itself a property. The frequent occurrence of this new property in the analysis of simple compressible systems warrants giving it a name and a symbol. We call it *enthalpy*. It is represented by the symbol H. Thus,

$$H = U + pV \tag{2-23}$$

For a unit mass or unit mole, we write

$$h = u + pv \tag{2-23a}$$

For a simple compressible closed system, the only quasiequilibrium work mode is

$$đW = p\,dV$$

By substituting this equation into the first law,

$$đQ = dU + đW$$

results in the useful expression

$$đQ = dU + p\,dV \tag{2-24}$$

or for a unit mass or unit mole

$$đq = du + p\,dv \tag{2-24a}$$

With the enthalpy as defined by Eq. (2-23), we have

$$dH = dU + p\,dV + V\,dp$$

Substituting Eq. (2-24) into this equation yields

$$đQ = dH - V\,dp \tag{2-25}$$

or

$$đq = dh - v\,dp \tag{2-25a}$$

Inasmuch as the $p\,dV$ work is for quasiequilibrium processes only, we must recognize that Eqs. (2-24) and (2-25) are two different forms of the first law for quasiequilibrium processes of simple compressible systems with fixed contents.

Example 2-7

A gas contained in a cylinder–piston arrangement is compressed in a quasiequilibrium process from an initial state of 10 ft³ and 15 psia to a final state of 5 ft³ and 15 psia. There is a heat rejection of 35 Btu from the gas to its surroundings during the constant-pressure process. Changes in bulk KE and PE are zero. Determine the work done and the internal energy and enthalpy changes for the gas.

Solution Figure 2-11 shows a schematic and the $p–V$ diagram for the quasiequilibrium process. The given data are $p_1 = p_2 = 15$ psia, $V_1 = 10$ ft³, $V_2 = 5$ ft³, and $Q = -35$ Btu.

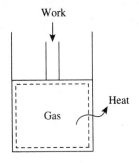

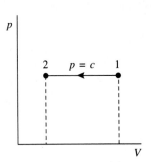

Figure 2-11 Schematic and p–V diagrams for Example 2-7.

The work done is calculated as

$$W = \int_1^2 p \, dV = p_1(V_2 - V_1)$$

$$= (15 \times 144 \text{ lbf/ft}^2)(5 \text{ ft}^3 - 10 \text{ ft}^3)$$

$$= -10,800 \text{ ft-lbf} = -\frac{10,800}{778} = -13.9 \text{ Btu}$$

where the negative sign indicates that work is done on the gas. The amount of work is represented by the area under the process curve 1–2 in the p–V plot.

By the first law,

$$Q = \Delta U + W$$

so that

$$\Delta U = U_2 - U_1 = Q - W = -35 - (-13.9) = -21.1 \text{ Btu}$$

where the negative sign indicates that the internal energy of the gas decreases 21.1 Btu in the process.

The change in enthalpy of the gas is

$$\Delta H = H_2 - H_1 = \Delta U + \Delta(pV) = (U_2 - U_1) + (p_2 V_2 - p_1 V_1)$$

$$= (U_2 - U_1) + p_1(V_2 - V_1) = -21.1 + (-13.9) = -35.0 \text{ Btu}$$

where the negative sign indicates that the enthalpy of the gas decreases 35.0 Btu in the process. Notice that for this example, we have $Q = \Delta H$. For a closed system, this is true only for a constant-pressure process.

2-8 SPECIFIC HEATS

Specific heat was originally defined as the amount of heat required, under certain conditions, to change the temperature of a unit mass of a substance by 1 degree. Mathematically, it is expressed as

$$c_x = \left(\frac{dq}{dT} \right)_x$$

where the subscript x denotes the particular process under which the heat transfer takes place. For a compressible substance, the greatest practical interest lies in constant-

The First Law of Thermodynamics   Chap. 2

volume and constant-pressure changes, for which the specific heats are defined by

$$c_v = \left(\frac{đq}{dT}\right)_v$$

$$c_p = \left(\frac{đq}{dT}\right)_p$$

The specific heats as defined in the preceding expressions are useful in the calculation of heat transfer between a system and its surroundings under special conditions. Because heat is an energy in transit and not a thermodynamic property, a quantity defined in terms of heat should be, in general, not a thermodynamic property. Under a more general definition, however, specific heat is a thermodynamic property that has much more important roles in thermodynamics than simply the role in the calculation of heat interaction.

We now redefine the property specific heat for simple compressible substances. According to the state postulate, the state of a simple compressible substance is determined by values of two independent intensive thermodynamic properties. As a result, the internal energy of a simple compressible homogeneous substance can be considered a function of its temperature T and specific volume v. The choice of T and v as the independent variables is entirely arbitrary, but this choice serves our present purpose. Thus, we write

$$u = u(T, v)$$

The total differential of u is then

$$du = \left(\frac{\partial u}{\partial T}\right)_v dT + \left(\frac{\partial u}{\partial v}\right)_T dv$$

Substituting this equation into Eq. (2-24a), we have

$$đq = du + p\,dv = \left(\frac{\partial u}{\partial T}\right)_v dT + \left[\left(\frac{\partial u}{\partial v}\right)_T + p\right] dv$$

For a constant-volume process, the preceding equation becomes

$$\left(\frac{đq}{dT}\right)_v = \left(\frac{\partial u}{\partial T}\right)_v$$

In light of the original definition of c_v, we now redefine the constant-volume specific heat as follows:

$$c_v = \left(\frac{\partial u}{\partial T}\right)_v \qquad\qquad (2\text{-}26)$$

In a similar manner, let temperature T and pressure p be the independent variables and enthalpy h the dependent variable, or

$$h = h(T, p)$$

The total differential of h is then

$$dh = \left(\frac{\partial h}{\partial T}\right)_p dT + \left(\frac{\partial h}{\partial p}\right)_T dp$$

Substituting this equation into Eq. (2-25a), we have

$$dq = dh - v\,dp = \left(\frac{\partial h}{\partial T}\right)_p dT + \left[\left(\frac{\partial h}{\partial p}\right)_T - v\right] dp$$

For a constant-pressure process, the preceding equation becomes

$$\left(\frac{dq}{dT}\right)_p = \left(\frac{\partial h}{\partial T}\right)_p$$

In light of the original definition of c_p, we now redefine the constant-pressure specific heat as follows:

$$c_p = \left(\frac{\partial h}{\partial T}\right)_p \tag{2-27}$$

Notice that the constant-volume and constant-pressure specific heats as defined by Eqs. (2-26) and (2-27) are thermodynamic properties because the only terms appearing in the definitions are properties.

Because the specific internal energy u and enthalpy h can be expressed either on a unit mass or unit mole basis, the value of c_v and c_p can be expressed also on a unit mass or unit mole basis. The commonly used units of specific heats are kJ/kg·K (kJ/kg·°C) or kJ/kgmole·K (kJ/kgmole·°C) in the SI system, and Btu/1bm·°R (Btu/1bm·°F) or Btu/lbmole·°R (Btu/lbmole·°F) in the English system. Notice that the temperature quantity in the denominator of the specific-heat units involves a change in temperature, and not a value of the temperature itself. Consequently, the specific heats expressed in per °C and per K in metric units are the same. The same is true for per °F and per °R in English units.

Specific heats of a substance vary strongly with temperature and to a lesser degree with pressure. For liquids and solids, c_p and c_v are almost equal at low temperatures. As the temperature increases, c_p continues to rise, but c_v approaches a fixed value on a mole basis. This is demonstrated in Fig. 2-12 for the case of copper. For gases, there is an appreciable difference between c_p and c_v, this difference being equal to a fixed value

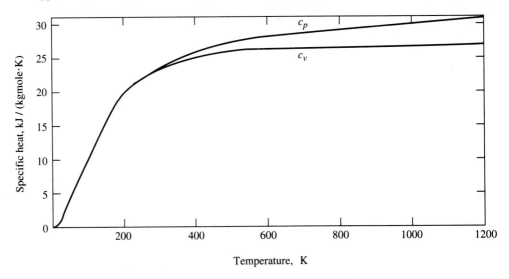

Figure 2-12 The variation in c_p and c_v with temperature for copper.

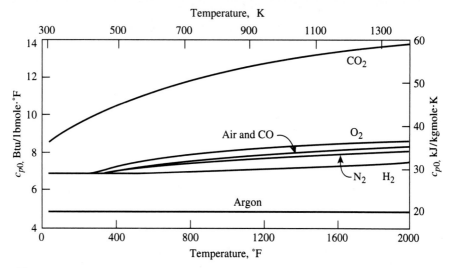

Figure 2-13 The variation in c_{p0} with temperature for several gases. (Based on data from NBS Circular 564, 1955.)

on a mole basis for the so-called ideal gases. Figure 2-13 shows the effect of temperature on c_p for several gases as the pressure approaches zero. The symbol c_{p0} refers to constant-pressure specific heat at zero pressure.

Example 2-8

Ten pounds of a gas are contained in an insulated rigid tank. Mechanical work of 70,000 ft · lbf is done on the gas by rotating a paddle wheel of negligible mass within the tank. Additionally, 100 Btu of electric energy is supplied to the gas through a resistance heater of negligible mass located inside the tank. The temperature of the gas increases 100°F during the process. Determine the average value of c_v for the gas.

Solution Figure 2-14 shows a schematic and the p–v diagram for the nonquasiequilibrium process of this example. The tank is rigid, so that the boundary work $p\,dV = 0$. There are, however, paddle-wheel and electric work interactions. Thus,

$$W = \text{total work interaction}$$

$$= \text{paddle-wheel work } W_{pw} + \text{electric work } W_{ele}$$

$$= -\frac{70,000 \text{ ft·lbf}}{778 \text{ ft·lbf/Btu}} + (-100 \text{ Btu}) = -190 \text{ Btu}$$

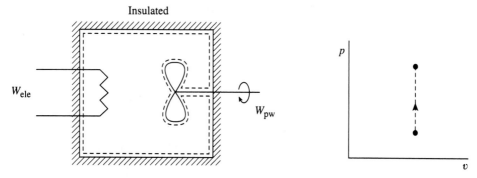

Figure 2-14 Schematic and p–v diagrams for Example 2-8.

Because the tank is insulated, $Q = 0$, the first law gives

$$\cancelto{0}{Q} = \Delta U + W = \Delta U + (-190)$$

so that

$$\Delta U = 190 \text{ Btu}$$

The average value of the constant-volume specific heat of the gas is then

$$c_v = \left(\frac{\Delta u}{\Delta T}\right)_v = \frac{1}{m}\left(\frac{\Delta U}{\Delta T}\right)_v = \frac{190 \text{ Btu}}{(10 \text{ lbm})(100°\text{F})}$$

$$= 0.190 \text{ Btu/lbm·°F}$$

2-9 SUMMARY

Heat is the energy transfer from one system to another solely by reason of a temperature difference between the two systems. Work is the energy transfer between a system and its surroundings if the sole effect on the surroundings can be reduced to the lifting of a weight. Neither heat nor work is an energy contained in a system and neither is a property of a system. As conventions of sign, heat transferred to a system is considered positive and work done by a system is considered positive.

The quasiequilibrium work modes introduced in this chapter are as follows:

Expansion work	$đW = p\, dV$
Elastic work	$đW = -F\, dL$
	$đw = -\sigma\, d\epsilon$
Surface-tension work	$đW = -\gamma\, dA$
Electrical work	$đW = -\mathscr{E}\, dZ$
Magnetization work	$đW = -\mu_0 \mathbf{H}\, d(V\mathbf{M})$
Polarization work	$đW = -\mathbf{E}\, d(V\mathbf{P})$

The first law of thermodynamics is essentially the law of conservation of energy applied to thermodynamic systems. For an infinitesimal state change, the first law is written as

$$đQ = dE + đW$$

or

$$đq = de + đw$$

In the absence of bulk KE and PE changes, the first law for a system of fixed mass is written as

$$đQ = dU + đW$$

or

$$đq = du + đw$$

A measure of accomplishment of a heat engine is expressed in terms of the thermal efficiency as defined by

$$\eta_{th} = \frac{\text{net work output}}{\text{heat input}}$$

A system that has only one quasiequilibrium work mode is referred to as a simple system. A simple compressible system has only the expansion work mode ($p\, dV$).

Enthalpy is a thermodynamic property as defined by

$$H = U + pV \qquad \text{or} \qquad h = u + pv$$

The first law for quasiequilibrium processes of a simple compressible closed system can be written as

$$đQ = dU + p\, dV \qquad \text{or} \qquad đq = du + p\, dv$$

and

$$đQ = dH - V\, dp \qquad \text{or} \qquad đq = dh - v\, dp$$

For a compressible substance, the specific heats at constant pressure (c_p) and constant volume (c_v) are defined by

$$c_p = \left(\frac{\partial h}{\partial T}\right)_p$$

$$c_v = \left(\frac{\partial u}{\partial T}\right)_v$$

PROBLEMS

2-1. A gas originally at a temperature T_1, a pressure p_1, and a volume V_1 expands slowly according to the relation $p/V = A$, where A is a constant, to a volume equal to two times its original volume. (a) Draw the process in the p–V plane. (b) Find the work done by the gas in terms of p_1 and V_1.

2-2. A quantity of a substance in a closed system is made to undergo a quasiequilibrium process, starting from an initial volume of 1 m³ and an initial pressure of 100 kPa. The final volume is 2 m³. Calculate the work done by the substance if pressure times volume remains constant.

2-3. A quantity of a fluid with an initial volume of 1m³ and an initial pressure of 500 kPa is made to undergo a quasiequilibrium process within a piston–cylinder system. The final volume is 2 m³. Calculate the work done by the fluid according to the condition that pressure is proportional to the square of volume.

2-4. Calculate the work of the following quasiequilibrium processes for a unit mass of a fluid in a closed system from the state of 100 psia and 40 ft³/1bm: (a) To a final state of 246 psia and 20 ft³/lbm, according to $pv^n = $ constant. (b) To a final volume of 20 ft³/lbm, according to $pv = $ constant. (c) At constant pressure until the final volume is 80 ft³/lbm. (d) At constant volume until the pressure is 50 psia. Sketch the processes on a p–v diagram.

2-5. Three cubic feet of a gas initially at 20 psia are compressed by a frictionless process in a closed system to a volume of 0.6 ft³. If the process takes place in such a way that the product $pV^{1.35}$ remains constant, how much work is done on the gas?

2-6. A certain fluid in a closed system initially at a pressure of 200 psia and having a volume of 3 ft³ undergoes a frictionless process during which the product $pV^{1.3}$ remains constant. If the final pressure is 40 psia, how much work does the system do?

2-7. A cylinder with a movable piston contains a gas at an initial pressure of 600 kPa and an initial volume of 0.10 m³. The gas expands quasistatically to a final volume of 0.5 m³. Determine the work done by the gas if the process follows the relation (a) $p = $ constant, (b) $pV = $ constant, (c) $pV^{1.4} = $ constant, and (d) $p = -300V + 630$, where p is in kPa, and V in m³.

2-8. A horizontal cylinder fitted with a sliding piston contains 0.3 m³ of a gas at a pressure of 1 atm. The piston is restrained by a liner spring. See Fig. P2-8. In the initial state, the gas pressure inside the cylinder just balances the atmospheric pressure of 1 atm on the outside of the piston and the spring exerts no force on the piston. The gas is then heated slowly until its volume and pressure become 0.15 m³ and 2 atm, respectively. (a) Write

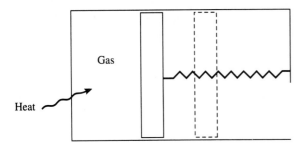

Figure P2-8

the equation for the relation beween the pressure and volume of the gas. (b) Calculate the work done by the gas. (c) Of the total work done by the gas, how much is done against the atmosphere? How much against the spring?

2-9. A cylinder having a piston being restrained by a spring contains 1 ft³ of air at a pressure of 15 psia, which just balances the atmospheric pressure of 15 psia on the other side of the piston. In this initial state, the spring exerts no force on the piston. The gas is then heated slowly until the volume becomes 1.5 ft³. The final pressure of the gas is 35 psia, and during the process, the spring exerts a force that is proportional to the displacement of the piston from the initial position. (a) Write the equation for the relation between the pressure and volume of the gas. (b) Show the process on a pressure–volume diagram. (c) Calculate the work done by the gas. (d) Of the total work, how much is done against the atmosphere? How much against the spring?

2-10. A gas is compressed at a constant temperature of 27°C from a specific volume of 0.12 m³/kg to 0.04 m³/kg. The p–v–T relationship for the gas is given by

$$\frac{pv}{RT} = 1 + \frac{b}{v}$$

where p is pressure in Pa (N/m²), v is specific volume in m³/kg, T is temperature in K, $R = 138.6$ J/kg·K, and $b = 0.012$ m³/kg. Calculate the work done on the gas in J/kg.

2-11. A gas is expanded slowly at a constant temperature of 17°C from 1 m³ to 3 m³. The p–v–T relationship for the gas is given by

$$\left(p + \frac{a}{v^2}\right)v = \mathcal{R}T$$

where p is in N/m², v in m³/kgmole, T in K, and

$$a = 15 \times 10^5 \text{ N·m}^4/(\text{kgmole})^2$$

$$\mathcal{R} = 8314 \text{ N·m/kgmole·K}$$

Compute the work done in joules by 0.3 kilogram-moles of the gas.

2-12. An elastic sphere initially has a diameter of 1 m and contains a gas at a pressure of 1 atm. Due to heat transfer, the diameter of the sphere increases slowly to 1.1 m. See Fig. P2-12. During the heating process, the gas pressure inside the sphere is proportional to the sphere diameter. Calculate the work done by the gas.

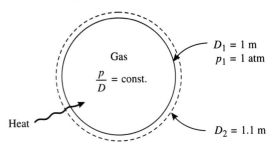

Figure P2-12

2-13. An elastic sphere initially has a diameter of 10 in. and contains air at a pressure of 20 psia. Due to heating, the diameter of the sphere increases slowly to 12 in. During the heating process, the air pressure inside the sphere is proportional to the sphere diameter. Calculate the work done by the air.

2-14. A rectangular chamber of 1 ft width contains water with a free surface. One end of the chamber is fitted with a sliding piston, as shown in Fig. P2-14. In the initial condition, the chamber length x is 2 ft and the water height y is 1.5 ft. Calculate the work done by the water on the piston when the chamber length is increased slowly from 2 ft to 2.5 ft. The density of water is 62.4 lbm/ft³. The atmospheric pressure is 14.7 psia.

2-15. A cube of solid CO_2 2 in. on a side vaporizes at atmospheric pressure of 14.7 psia. See Fig. P2-15. Determine the boundary work done by the CO_2. The density of solid CO_2 is 97.56 lbm/ft³ and that of vapor CO_2 is 0.174 lbm/ft³.

2-16. A cube of solid water (ice) 2 in. on a side melts into liquid water at 14.7 psia and 32°F. See Fig. P2-16. Determine the boundary work done on the water. At 14.7 psia and 32°F, the density of ice is

The First Law of Thermodynamics　Chap. 2

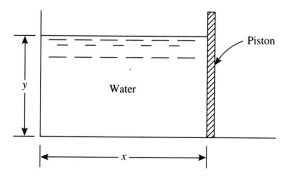

Figure P2-14

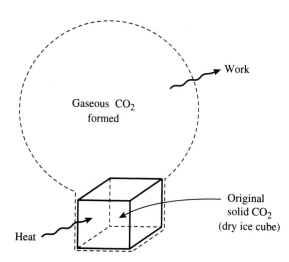

Figure P2-15

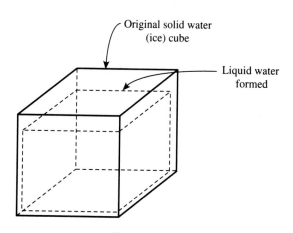

Figure P2-16

57.2 lbm/ft^3 and that of liquid water is 62.4 lbm/ft^3.

2-17. The stress–elongation relationship for a rubber substance is given by the expression.

$$\sigma = (A_0/A)KT(\lambda - 1/\lambda^2)$$

where σ is the stress, K is a constant, λ is the extension ratio $= L/L_0$, L_0 is the no-load length, and A_0 is the no-load area. (a) Using the equation for work in the form (Eq. 2-9)

$$đW = -V_0\sigma\, dє$$

show that

$$\frac{đW}{V_0} = -KT(\lambda - 1/\lambda^2)\, d\lambda$$

for isothermal quasiequilibrium expansion or compression. (b) For a sample of rubber, $L_0 = 1$ ft, $A_0 = 0.1$ in.^{2}, and $K = 0.15$ lbf/in.$^2 \cdot$ °R at $T = 500$°R, calculate W for elongation of the sample from $L = L_0$ to $L = 2L_0$.

2-18. For a certain rubber band, the restoring force F is related to its initial length L_0 and the displacement x as follows:

$$F = 0.80\left[\frac{x}{L_0} + \left(\frac{x}{L_0}\right)^2\right]$$

where F is in lbf, x and L_0 in in. See Fig. P2-18. Determine the work required to stretch the rubber band from an initial length of 2.0 in. to a final length of 3.0 in.

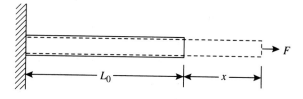

Figure P2-18

2-19. The tensile force in a steel wire 1 m long and 1 mm diameter is increased slowly at a constant temperature from zero to 100 N. Calculate work done, assuming $Y = 2.00 \times 10^8$ kN/m^2, where Y is Young's isothermal modulus of elasticity, as defined by the expression

$$Y = \left(\frac{\partial\sigma}{\partial є}\right)_T$$

where σ is stress, and $є$ is strain.

2-20. Derive the expression for the work required to blow up slowly a soap bubble of radius R into air that is at atmospheric pressure p_{atm} if γ is the surface tension of the soap solution in contact with air.

2-21. An empirical equation for the temperature dependence of surface tension γ of a pure liquid in equilibrium with its vapor has the form

$$\gamma = \gamma_0\left(1 - \frac{T}{T_c'}\right)^n$$

where γ is the surface tension at $T\,°C$, γ_0 is the surface tension at $0°C$, T_c' is a temperature in $°C$ determined empirically for the liquid, and n is a constant. For water, $\gamma_0 = 7.55 \times 10^{-4}$ N/cm, $T_c' = 368°C$, and $n = 1.2$. Liquid water is atomized slowly at constant temperature of $100°C$ into droplets of an average radius of 10^{-4} cm in equilibrium with its vapor. Estimate the work required per kg of water atomized.

2-22. A spherical air bubble with an initial radius of 10^{-4} in. is suspended in a pool of water. The interfacial tension γ is a constant value of 4×10^4 lbf/in. and the water pressure is a constant value of 18 psia. When the air bubble grows to a final radius of 10^{-3} in., what is the total work done by the air? Of this total work, how much goes to the forming of the additional surface area of the bubble, and how much goes to the pushing of the water?

2-23. For a dielectric material, the polarization is related to the electric field by the following equation:

$$\mathbf{P} = \epsilon_0(22.4\mathbf{E} - 4.9 \times 10^{-4}\mathbf{E}^2)$$

where $\mathbf{P}$ = polarization in C/m^2
$\mathbf{E}$ = electric field strength in V/m
$\epsilon_0 = 8.854 \times 10^{-12} C^2 \cdot s^2/kg \cdot m^3$

Compute the work done (in J) on 1 cm³ of this material when the electric field is increased slowly from zero to 10^4 V/m.

2-24. An electric motor draws a constant current of 10 amperes at a constant voltage of 110 volts while rotating at a constant speed of 1000 rpm (revolutions per minute), thus developing a constant torque of 10 N · m at the output shaft. Determine the electric power required, the shaft power developed, and the net power input to the motor, all in kW.

2-25. A dc motor draws a constant current of 50 amperes at a constant voltage of 24 volts while rotating at a constant speed of 1500 revolutions per minute,

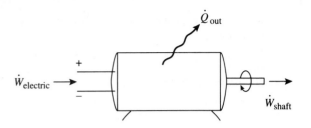

Figure P2-25

thus developing a constant torque of 6.8 N · m at the output shaft. See Fig. P2-25. Determine the rate of heat transfer from the motor in steady operation.

2-26. An electric potential of 115 V is applied on a resistor that draws a constant current of 9 A for 5 min. Determine the total amount of electrical work, in kJ, and the instantaneous power, in kW.

2-27. An electric potential of 220 V is applied to an electric heater that draws 6 A for 10 h. Determine the total amount of electrical work, in kW·h.

2-28. A baseboard resistance heater is used to supply heat to a room that has a heat loss of 10,000 kJ/h. Even though the heater operates continuously, the room air temperature remains constant. Determine the power rating of the heater, in kW.

2-29. An electronic device is operated by a battery at a constant voltage of 120 volts. The current i drawn by the device varies with time t according to the relation

$$i = 10e^{-t/60}$$

where i is in amperes, and t is in seconds. Determine (a) the work performed on the device during the first 2 min, and (b) the power delivered to the device at $t = 0$ and $t = 2$ min.

2-30. An electrical device draws a current from a 12-volt battery. The current varies linearly with time, beginning with 1 A and ending with 0.8 A after 5 min. Determine the total work done on the device.

2-31. An automobile battery is charged from the engine–generator across 12.5-volt dc terminals with 8-ampere current. Determine the power input to the battery in watts.

2-32. A torque of 68 N · m is associated with a shaft rotating at 1000 revolutions per minute. Determine the power transmitted, in kW.

2-33. The drive shaft of an automobile transmits 80 kW of power from the engine to the rear wheels. The rotating speed of the shaft is 2500 revolutions per

The First Law of Thermodynamics   Chap. 2

minute. Determine the torque developed by the engine, in N·m.

2-34. When a garage door is lowered, it stretches a spring 1.3 m. The spring constant is 675 N/m. Determine the work done by the door on the spring.

2-35. A dynamometer is used to measure the power delivered by an internal combustion engine. The output torque τ (in N·m) of the engine is found to vary with the rpm (revolutions per minute) of the rotating shaft according to the relation

$$\tau = 400 \sin\left(\frac{\pi}{2}\frac{\text{rpm}}{3000}\right)$$

Calculate the power delivered (in kW) by the engine at speeds of 1000 and 2000 rpm. If the engine accelerates linearly from rest to 1000 rpm in 1 min, calculate the work delivered by the engine, in kJ.

2-36. For a linear spring, the displacement x is proportional to the force F applied. That is, $F = kx$, where k is the spring constant. A linear spring is compressed from its unstressed length L_0 to 0.40 m by a force of -100 N. It is later maintained at a length of 0.70 m by a force of $+200$ N. Determine the unstressed length L_0 (in m), the spring constant k (in N/m), and the work required to change the length from 0.40 to 0.70 m (in N·m).

2-37. A rubber band is stretched adiabatically and horizontally from an initial length of 3 in. to a final length of 4 in. The band has a mass of 0.001 lbm and a cross-sectional area of 0.008 in². Assume that the rubber band obeys Hooke's law of elasticity:

$$\sigma = Y\epsilon$$

where σ and ϵ are the stress and strain, respectively, and Y is the elastic modulus and has a value of 1000 lbf/in². (a) Determine the internal energy change of the band due to stretching. (b) When the rubber band is suddenly released horizontally, it flies away. Determine its initial flying velocity, neglecting any heat transfer and air friction.

2-38. A large-diameter blowpipe is used to blow a soap bubble. See Fig. P2-38. When the opposite end of the blowpipe is uncovered, the surface tension of the soap bubble causes the bubble to collapse, thus forcing the air within the soap bubble to flow through the blowpipe and discharge into the atmosphere. Determine the velocity of the air in the blowpipe when a 12-cm soap bubble collapses. For

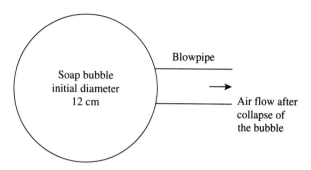

Figure P2-38

the soap bubble, the surface tension $\gamma = 0.04$ N/m. The air density $\rho = 1.20$ kg/m³.

2-39. A fluid in a closed system expands slowly such that the pressure p varies linearly with the volume V, and the internal energy U varies with the pV product according to the relation

$$U = 32 + 0.004pV$$

where U is in Btu, p in lbf/ft² abs, and V in ft³. When the fluid changes from an initial state of 25 psia, 1 ft³, to a final state of 60 psia, 2 ft³, what are the work and the heat transfer, in Btu?

2-40. A cylinder with a moving piston contains ethane gas initially at 35 kPa with a volume of 0.12 m³. The gas is compressed slowly without friction until its volume is halved in such a manner that its pressure and volume are linearly related, that is,

$$p = a + bv$$

The final pressure is 80 kPa, and the change in internal energy of the ethane is 3.22 kJ. Determine the work done and the heat transferred.

2-41. A cylinder with a moving piston contains 0.5 lbm of a fluid initially having a volume of 2 ft³ at 20 psia and 50°F. The fluid is compressed slowly to 120 psia in such a way that

$$p = a - 100V$$

where p is in psia, V in ft³, and a is a constant. Its final temperature is 20°F. The internal energy of the fluid is given by

$$u = 0.20T$$

where u is in Btu/lbm, and T in °F. Write the equation for the relation between p and V and draw the process on p–V coordinates. Calculate (a) the work required in the compression process, (b) the heat transferred, and (c) the enthalpy change.

2-42. A fluid undergoes a frictionless process in a closed system from $V_1 = 7$ ft^3 to $V_2 = 2$ ft^3 in accordance with

$$V = 35/(p - 8)$$

where V is in ft^3 and p in psia. During this process, the fluid rejects 20 Btu of heat. What is the change in enthalpy?

2-43. (a) A gas expands at constant pressure from 700 kPa, 0.03 m^3, to 0.06 m^3, while receiving 85 kJ of heat. There is no work other than that done on a piston. Find the change in internal energy of the gas. (b) The same gas expands from 0.03 m^3 to 0.06 m^3, exerting a constant pressure of 700 kPa against a piston, while a stirring device does 10 kJ of work on the gas. Find the change of internal energy, the work, and the heat transfer for the process.

2-44. A fluid is compressed quasistatically in a closed system in such a way that $pv = 14400$ (where p is in psfa, and v in ft^3/lbm) from an initial pressure of 100 psia to a final pressure of 200 psia. During the compression, the internal energy of the fluid changes from 60 to 65 Btu/lbm. What is the heat transfer?

2-45. Air is contained in a cylinder–piston device. Initially, $p = 700$ kPa, $v = 0.10$ m^3/kg, and $u = 160$ kJ/kg. The air expands slowly with the pressure remaining constant until the volume becomes 0.20 m^3/kg and the internal energy becomes 320 kJ/kg. How much work is done by the gas and how much heat is transferred?

2-46. The constant-volume specific heat for a material is expressed as

$$c_v = 0.1545 + 0.375 \times 10^{-8}T^2$$

where T is in °R, and c_v in Btu/lbm · °F. How much heat is required to raise the temperature of 1 pound of the material from 0°F to 1000°F at constant volume? Determine the average value of c_v for the temperature range specified.

2-47. One pound of a certain substance undergoes a process during which its temperature rises from 50°F to 200°F. If the substance absorbs 47 Btu of heat while its temperature changes from 50°F to 150°F and absorbs 28 Btu while its temperature changes from 150°F to 200°F, and if the relation between its specific heat for the process and its temperature is known to be linear, what is its specific heat at 200°F?

2-48. The constant-pressure specific heat for oxygen at low pressure can be expressed as

$$c_p = 8.314(3.626 - 1.878 \times 10^{-3}T$$
$$+ 7.056 \times 10^{-6}T^2 - 6.764 \times 10^{-9}T^3$$
$$+ 2.156 \times 10^{-12}T^4)$$

where c_p is in kJ/kgmole·K, and T is in K. How much heat is required to raise the temperature of 32 kg of oxygen from 300 K to 700 K at a constant low pressure?

2-49. The constant-pressure specific heat for a substance is expressed as

$$c_p = 0.0587 + 0.291 \times 10^{-3}T$$

where T is expressed in °R, and c_p in Btu/lbm·°F. How much heat is required to raise the temperature of 5 lbm of the substance from 40°F to 540°F?

2-50. The constant-pressure specific heat for air is given by the equation

$$c_p = 0.219 + 0.342 \times 10^{-4}T - 0.293 \times 10^{-8}T^2$$

where T is in °R, and c_p in Btu/lbm·°R. (a) How much heat is required to raise the temperature of 3 lbm of air from 40°F to 1540°F, with the pressure remaining constant? (b) What is the average c_p for air between 40°F and 1540°F?

2-51. The equation of c_p for a certain system can be written as

$$c_p = 0.5 + 10/(T + 100)$$

where c_p is in cal/°C, and T is the temperature of the system in °C. The system is heated under a constant pressure of 1 atm until its volume increases from 2000 cm^3 to 2400 cm^3 and its temperature increases from 0°C to 100°C. (a) How much heat is added to the system? (b) How much does the internal energy of the system increase?

2-52. Write a computer program to solve the following problem. A gas is expanded quasistatically at a constant temperature of $T = 290$ K from $V_1 = 1$ m^3 to $V_2 = 3$ m^3. The p–v–T relationship for the gas is given by

$$\left(p + \frac{a}{v^2}\right)v = \mathfrak{R}T$$

where p is in N/m^2, v in m^3/kgmole, T in K, and

$$a = 15 \times 10^5 \text{ N·m}^4/\text{kgmole}^2$$

$$\mathfrak{R} = 8314 \text{ N·m/kgmole·K}$$

Compute the work done in J by 0.3 kilogram-moles of the gas.

2-53. Solve Problem 2-52 if the final volume V_2 is (a) 2 m^3, (b) 4 m^3, (c) 6 m^3, (d) 8 m^3, and (e) 10 m^3.

3

Thermodynamic Properties of Pure Substances

All substances can exist in solid, liquid, and gaseous phases. The study of the properties of different phases of a substance and the understanding of the phenomenon of phase transition are of prime importance in thermodynamics. In this chapter, we study the phase behaviour of a common group of subatances, namely, pure substances.

3-1 PURE SUBSTANCE AND PHASE TRANSITION

A substance that is homogeneous and invariable in chemical composition, irrespective of the phase or phases in which it exists, is called a *pure substance*. In other words, a pure substance is a one-component system. For instance, pure oxygen or pure nitrogen in any phase or any combination of phases is a pure substance. A chemically stable mixture, such as air, which is essentially a mixture of oxygen and nitrogen, can be treated as a pure substance as long as it remains in one phase. A system of gaseous air and liquid air, however, is not a pure substance, because the two phases have different composition.

We now proceed to examine in a descriptive way the properties of pure substances and the phases in which a pure substance can exist. In this chapter, we are interested in systems whose volume changes are of most importance. The effects of motion, capillarity, solid distortion, and external force fields (electric, magnetic, and gravitational) are not significant.

In order to become familiar with the technical terms in a phase-transition process, let us consider the heating of an amount of water (H_2O) contained in a cylinder fitted with a piston under 1 atmospheric pressure. A temperature–volume diagram for this process is depicted in Fig. 3-1. At state 1, the water is solid (ice) at a pressure of 1 atm and a temperature of less than $0°C = 32°F$. When the ice is heated slowly, its volume increases slightly while its temperature gradually rises until state 2 at $0°C = 32°F$ is reached. Upon further heating, the ice begins to melt, but there is no change in temperature until all the ice has been melted and appears as liquid water at state 3. During the melting process 2–3, the total volume decreases. The phenomenon of contracting on melting or expanding on freezing is a special characteristic of water. When the all-liquid water at state 3 is further heated, its volume continues to decrease, reaching a minimum value when the temperature is increased to about $4°C = 39°F$ at state a. Upon

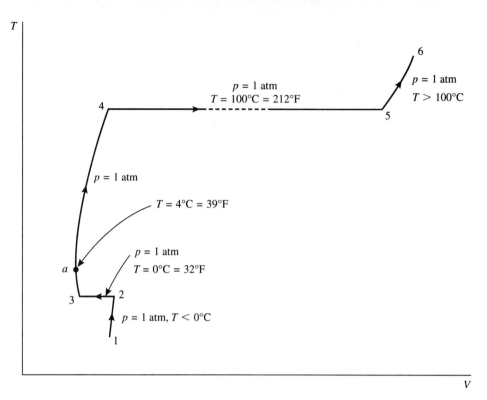

Figure 3-1 *T–V* diagram for heating of water at 1 atmospheric pressure.

further heating after state *a*, the liquid water volume increases slightly while its temperature increases greatly until a temperature of 100°C = 212°F at state 4 is reached. At state 4, the liquid water begins to vaporize. Both pressure and temperature remain constant during the entire vaporization process 4–5 while its volume increases tremendously. At state 5, the water is all in the vapor form at 1 atm and 100°C = 212°F. Further heating after state 5 results in continuous rise in temperature and volume, say, to state 6 as depicted in Fig. 3-1.

In the heating process described before, the water changes from solid phase (states 1 to 2) to liquid phase (states 3 to 4) and then to vapor phase (states 5 to 6). Between states 2 and 3, it consists of a mixture of solid and liquid phases, and between states 4 and 5, it consists of a mixture of liquid and vapor. States 2, 3, 4, and 5 are known as saturation states. A *saturation state* is a state from which a change of phase may occur without a change of pressure or temperature. Thus, state 2 is a saturated solid state, states 3 and 4 are saturated liquid states, and state 5 is a saturated vapor state. Notice that saturated liquid state 3 is with respect to solidification, whereas saturated liquid state 4 is with respect to vaporization. In a saturation state, its temperature is called the saturation temperature corresponding to its pressure, and its pressure is called the saturation pressure corresponding to its temperature. The liquid states other than states 3 and 4 shown in Fig. 3-1 are not saturation states because the temperature of those states are lower than the saturation temperature (T_4) corresponding to the pressure; they are called *subcooled liquid* states. Similarly, any solid other than at state 2 shown in Fig. 3-1 is not a saturated solid because its temperature is lower than the saturation temperature (T_2) corresponding to the pressure; it is called a *subcooled solid*.

When the temperature in a vapor phase, say, at state 6 in Fig. 3-1, is higher than the saturation temperature (T_5) corresponding to the pressure, it is called a *superheated vapor*. Notice that the temperature readings indicated in Fig. 3-1 are for heating at a constant pressure of 1 atmosphere. For heating at other constant pressures, the process curves are different but follow the same general pattern shown.

As illustrated in Fig. 3-1, water has the peculiar characteristic of expanding on freezing. All other known materials contract on freezing. A constant-pressure heating process of any material other than water will show a *T–V* variation such as depicted in Fig. 3-2, with the phase changes labeled the same as in Fig. 3-1.

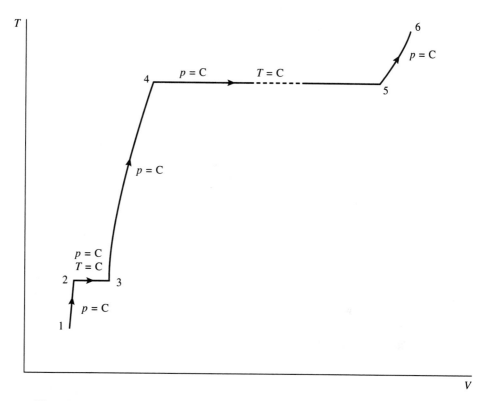

Figure 3-2 *T–V* diagram at a constant pressure for a substance that contracts on freezing.

Experimental data reveal that in the saturation states of a pure substance, the pressure, temperature, and specific volume always vary in some unique pattern. Thus, if the vaporization process 4–5 in Figs. 3-1 and 3-2 is plotted for a number of different pressures, a series of saturated liquid states 4 and saturated vapor states 5 would appear as shown in the *T–v* diagram of Fig. 3-3 and in the *p–v* diagram of Fig. 3-4. The locus of saturated liquid states is called the *saturated liquid line,* and the locus of saturated vapor states is called the *saturated vapor line.* A point to the left of the saturated liquid line represents a *subcooled* or *compressed liquid* state, and a point to the right of the saturated vapor line represents a superheated vapor state. A point lying between the saturated liquid line and saturated vapor line represents a mixture where saturated liquid and saturated vapor coexist in equilibrium.

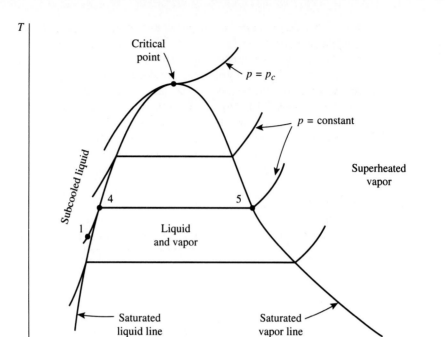

Figure 3-3 Saturated liquid and saturated vapor lines on a T–v diagram, with a few constant-pressure lines.

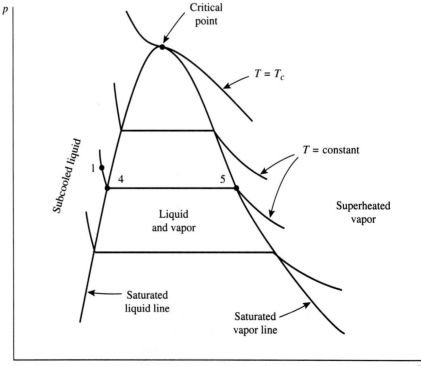

Figure 3-4 Saturated liquid and saturated vapor lines on a p–v diagram, with a few constant-temperature lines.

The reason that a point to the left of the saturated liquid line in Figs. 3-3 and 3-4 can be referred to as either a subcooled liquid or a compressed liquid is made clear in the following illustration. Consider the liquid state 1 as indicated in both Figs. 3-3 and 3-4. In Fig. 3-3, it is seen that T_1 is lower than the saturation temperature (T_4) corresponding to its pressure ($p_1 = p_4$), thus initiating the name subcooled liquid to state 1. On the other hand, in Fig. 3-4, it is seen that p_1 is higher than the saturation pressure (p_4) corresponding to its temperature ($T_1 = T_4$), thus initiating the name compressed liquid to state 1.

The saturated liquid line and saturated vapor line meet at a point, called the *critical point*, which marks the termination of any distinction between liquid and vapor phases. In the neighborhood of the critical point, the properties of the liquid phase and the vapor phase approach each other, and at the critical point, the properties of both phases become identical. In a pure substance, the critical point is the state of highest pressure and temperature at which distinct liquid and vapor phases can coexist. Beyond the critical point, the liquid phase and vapor phase merge into each other. The pressure, temperature, and volume at the critical point are called the *critical pressure, critical temperature,* and *critical volume,* respectively. The critical point data for substances of interest are given in Table A-5 in the Appendix.

As mentioned before, at the critical point, the liquid and vapor phases are indistinguishable from each other. To demonstrate this phase indistinguishability, let us consider a rigid transparent vessel filled with a liquid–vapor mixture of a pure substance in the proportion as depicted by state a in Fig. 3-5. Upon heating, the mixture goes through a constant-volume process with rising pressure and temperature. As the liquid vaporizes, the meniscus separating the two phases falls. When the fluid reaches state g, shown in the p–v diagram of Fig. 3-5, it all becomes saturated vapor. Further heating results in superheating of the vapor. Consider another case in which the vessel is initially filled with a mixture in the proportion as depicted by state b. When heat is added to the mixture, the meniscus separating the two phases rises, indicating the condensation of vapor to liquid. When the fluid reaches state f, it becomes all saturated liquid. The fluid remains in the liquid phase during further heating. Finally, consider the case in which the vessel is initially filled with a mixture in a proportion as depicted by state

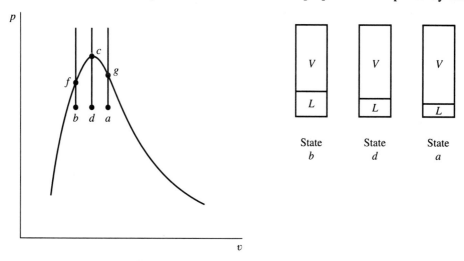

Figure 3-5 Critical-point experiments (V = vapor, L = liquid; not to scale).

d such that its specific volume is equal to that of the critical point. As the mixture is heated, the liquid becomes less dense, and the vapor becomes more dense, with the location of the meniscus stationary. When the critical point *c* is reached, the meniscus vanishes, resulting in a loss of any phase identity. During further heating, the fluid remains in a single phase, which can be considered as either a liquid or a vapor.

3-2 *p–v–T* SURFACE FOR A SUBSTANCE THAT CONTRACTS ON FREEZING

In the last section, we defined the saturated liquid line and saturated vapor line as the loci of saturated liquid states (state 4 in Fig. 3-2) and saturated vapor states (state 5 in Fig. 3-2), respectively. This saturated liquid line is with respect to vaporization. There is another saturated liquid line defined as the locus of the saturated liquid states (state 3 in Fig. 3-2), which is with respect to solidification. Similarly, the locus of the saturated solid states (state 2 in Fig. 3-2) is called the saturated solid line. All the saturation lines are shown in a three-dimensional pressure–volume–temperature (*p–v–T*) surface of

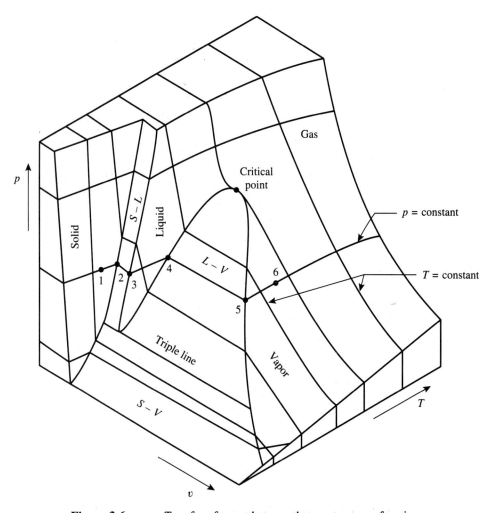

Figure 3-6   *p–v–T* surface for a substance that contracts on freezing.

Fig. 3-6 for a substance that contracts on freezing. The projections of this $p-v-T$ surface on the $p-v$ and $p-T$ planes are shown in Figs. 3-7 and 3-8, respectively. The numbers 1 to 6 indicated in Figs. 3-6 and 3-7 correspond to the same numbers indicated in Figs. 3-2.

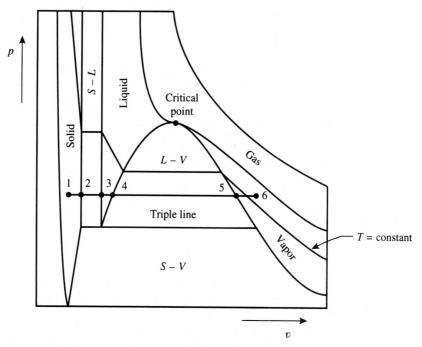

Figure 3-7 $p-v$ diagram for a substance that contracts on freezing.

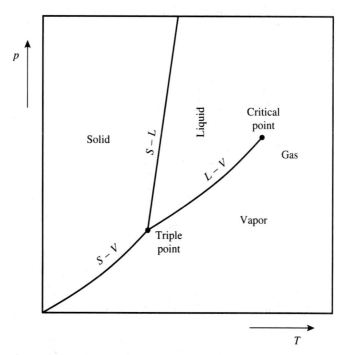

Figure 3-8 $p-T$ diagram for a substance that contracts on freezing.

There are three regions, labeled solid, liquid, and gas (or vapor), in Figs. 3-6 to 3-8 where the substance exists only in a single phase. Between the single-phase regions lie the two-phase regions, where two phases coexist in equilibrium. They are labeled $S-L$, $L-V$, and $S-V$ for solid–liquid, liquid–vapor, and solid–vapor coexistent regions, respectively. In the two-phase regions, lines of constant pressure are also lines of constant temperature. Thus, the two-phase portions of the $p-v-T$ surface are made up of straight lines parallel to the volume axis; such surfaces are called ruled surfaces. Therefore, the two-phase surfaces appear as lines when projected on the $p-T$ plane.

There are states of fixed pressure and temperature where solid, liquid, and vapor phases can coexist in equilibrium. These states are called the triple phase, or, more commonly, the *triple point*. On a $p-v-T$ surface, the triple point is a line of constant pressure and temperature. On some diagrams, as will be seen later, it appears as an area of constant pressure and temperature. The triple point appears as a point only on a $p-T$ diagram. Triple-point data for a number of substances are given in Table 3-1.

At pressures lower than the triple-point pressure, there is a region where solid and vapor coexist in equilibrium. A phase transition from vapor directly to solid is called *sublimation*.

TABLE 3-1 TRIPLE-POINT DATA

Substance	Formula	Temperature (K)	Pressure (mm Hg)
Acetylene	C_2H_2	192.4	962
Ammonia	NH_3	195.42	45.58
Argon	A	83.78	515.7
Carbon dioxide	CO_2	216.55	3885.1
Carbon monoxide	CO	68.14	115.14
Ethane	C_2H_6	89.88	0.006
Ethylene	C_2H_4	104.00	0.9
Hydrogen	H_2	13.84	52.8
Hydrogen sulfide	H_2S	187.66	173.9
Krypton	Kr	115.6	538
Methane	CH_4	90.67	87.7
Neon	Ne	24.57	324
Nitric oxide	NO	109.50	164.4
Nitrogen	N_2	63.15	94.01
Oxygen	O_2	54.35	1.14
Sulfur dioxide	SO_2	197.69	1.256
Water	H_2O	273.16	4.587
Xenon	Xe	161.3	611

Our discussion so far has been in reference to $p-v-T$ surfaces. However, properties other than pressure, volume, or temperature can also be used as coordinates to construct a thermodynamic surface. Figure 3-9 shows a $p-v-u$ surface for a normal type of substance that contracts on freezing. This figure is used here to illustrate the characteristic at the triple point. It is to be noted that the triple point becomes a triangular area of constant pressure and temperature on the $p-v-u$ surface. The three corners of the triangle represent the states of vapor, liquid, and solid phases, and any point in the triangle, such as point M in Fig. 3-9, represents a mixture of the three phases coexisting at the triple-point pressure and temperature. Let mass fractions of vapor, liquid, and

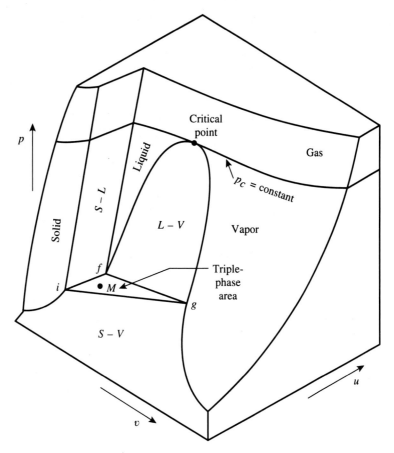

Figure 3-9 *p–v–u* surface for a substance that contracts on freezing.

solid in the mixture be denoted by x, y, and z, respectively. That is, symbolically

$$x = \frac{m_{\text{vapor}}}{m_{\text{total}}} \qquad y = \frac{m_{\text{liquid}}}{m_{\text{total}}} \qquad z = \frac{m_{\text{solid}}}{m_{\text{total}}}$$

where

$$m_{\text{total}} = m_{\text{vapor}} + m_{\text{liquid}} + m_{\text{solid}}$$

Consequently,

$$x + y + z = 1$$

Any specific property, such as specific volume, of the three-phase mixture can be determined from an equation of the form

$$v = xv_g + yv_f + zv_i \qquad (3\text{-}1)$$

where v_g, v_f, and v_i denote the specific volumes of saturated vapor, liquid, and solid, respectively, at the triple-phase conditions. Notice that the specific volume v of a three-phase mixture as given by Eq. (3-1) represents the weighted-average v of the three phases present. The actual state of the mixture is the saturated states of the three separate phases that are coexisting.

Water is the only known substance that expands on freezing. The phase transitions during a constant-pressure heating process are similar to those shown in the *T–v* diagram of Fig. 3-1, which is for 1 atmospheric pressure. The three-dimensional *p–v–T* surface for water is shown in Fig. 3-10, which is similar to Fig. 3-6 except for the difference in the solid–liquid coexist region due to the peculiar characteristic of water prevailing in the solidification process. The projections of this *p–v–T* surface on the *p–v* and *p–T* planes are shown in Figs. 3-11 and 3-12, respectively. The numbers 1 to 6 indicated in Fig. 3-10 correspond to the same numbers indicated in Fig. 3-1 but for a general constant-pressure process not necessarily at 1 atmosphere.

It should be mentioned that a pure substance can exist in different solid phases. At extremely high pressures, several different crystalline solid phases of water have been observed. These are illustrated in Fig. 3-13. In this figure, different two-phase and triple-phase regions are clearly shown. Notice that besides the usual solid–liquid–vapor triple point, there are other triple points involving two or three solid phases.

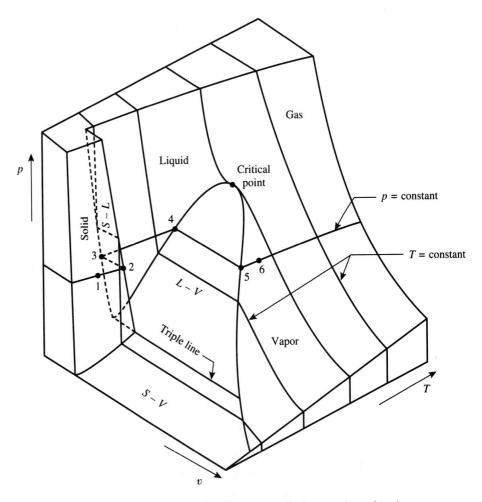

Figure 3-10   *p–v–T* surface for water, which expands on freezing.

Thermodynamic Properties of Pure Substances   Chap. 3

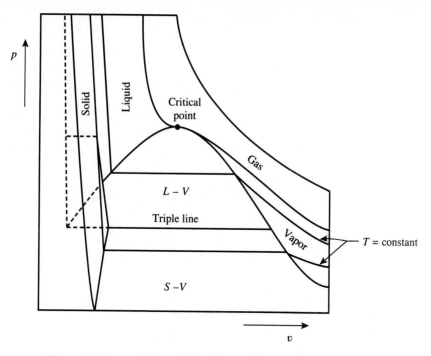

Figure 3-11 $p-v$ diagram for water, which expands on freezing.

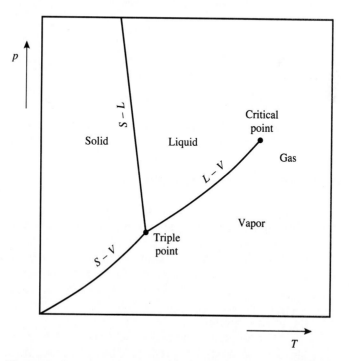

Figure 3-12 $p-T$ diagram for water, which expands on freezing.

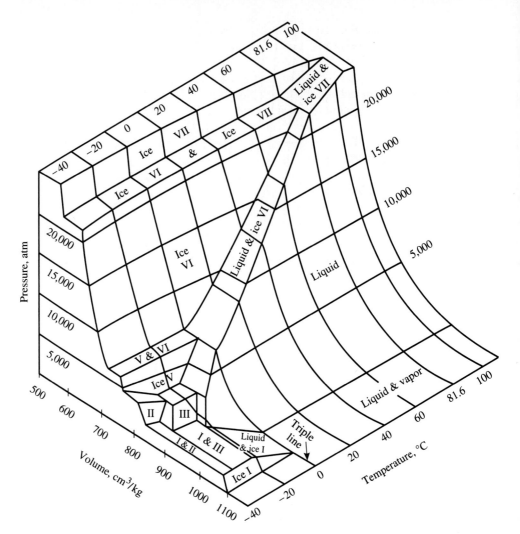

Figure 3-13 Pressure–volume–temperature surface showing different solid phases of water. (Based on measurements of Bridgman.)

3-4 TABLES OF THERMODYNAMIC PROPERTIES

In the analysis and solution of thermodynamic problems so far in this book, we have been using these fundamental properties: pressure p, temperature T, specific volume v, specific internal energy u, and specific enthalpy h. When we study the second law of thermodynamics, we will define another fundamental property, namely, entropy. It is convenient to include entropy when we present tables and charts of thermodynamic properties. The symbol s is used for specific entropy, for which the common units are kJ/kg·K (or kJ/kgmole·K) in the SI system and Btu/lbm·°R (or Btu/lbmole·°R) in the English system.

For all pure compressible substances, the format of thermodynamic property tables is alike. Complete property tables of several substances, such as steam (H_2O) and refrigerant 12, are given in Appendix 3. We will use steam (H_2O) tables in our illustrations. Four steam tables are usually presented: one for saturated liquid and vapor, one for superheated vapor, one for compressed liquid, and one for saturated solid and vapor. For convenience of interpolation, two saturated liquid–vapor tables are usually given, one with temperature as the independent variable and the other with pressure as the independent variable. Table 3-2(a) gives the table headings of the properties of saturated liquid and vapor as a function of intergral values of temperature in SI units; and Table 3-2(b) gives the same table, but in terms of integral values of pressure in English units. The symbols used in these tables are as shown in Fig. 3-14, where f and g denote saturated liquid and saturated vapor, respectively, both at the same p and T. State X in Fig. 3-14 represents a mixture of saturated liquid and saturated vapor coexisting in equilibrium at the given p and T with

$$m = m_f + m_g \quad \text{and} \quad V = V_f + V_g$$

where m and V are the mass and volume of the mixture, respectively, m_f and V_f are the mass and volume of the liquid phase, respectively, and m_g and V_g are the mass and volume of the vapor phase, respectively.

The *quality* x of the mixture at state X is defined as the fraction of mass present in the vapor phase, or

$$x = \frac{m_g}{m} = \frac{m_g}{m_f + m_g} \tag{3-2}$$

and

$$1 - x = \frac{m_f}{m} = \frac{m_f}{m_f + m_g}$$

The specific volume v of the mixture is then given by

$$v = \frac{V}{m} = \frac{V_f + V_g}{m_f + m_g} = \frac{m_f v_f + m_g v_g}{m_f + m_g} = (1 - x)v_f + xv_g$$

or

$$v = v_f + x(v_g - v_f) = v_f + xv_{fg} \tag{3-3}$$

where

$$v_{fg} = v_g - v_f$$

Equation (3-3) can also be written as

$$v = v_g - (1 - x)v_{fg} \tag{3-3a}$$

Expressions analogous to the preceding equations can be derived for specific internal energy u, specific enthalpy h, and specific entropy s for a mixture of liquid and vapor.

Accordingly, we have

$$u = u_f + xu_{fg} = u_g - (1 - x)u_{fg} \tag{3-4}$$

$$h = h_f + xh_{fg} = h_g - (1 - x)h_{fg} \tag{3-5}$$

$$s = s_f + xs_{fg} = s_g - (1 - x)s_{fg} \tag{3-6}$$

Sec. 3-4   Tables of Thermodynamic Properties

75

TABLE 3-2 TABLE HEADINGS FOR PROPERTIES OF SATURATED LIQUID AND VAPOR

(a) TEMPERATURE TABLE, SI UNITS

Temp., T (°C)	Press., P (kPa)	Specific volume (m³/kg)		Internal energy (kJ/kg)			Enthalpy (kJ/kg)			Entropy (kJ/kg·K)		
		Sat. liquid, v_f	Sat. vapor, v_g	Sat. liquid, u_f	Evap., u_{fg}	Sat. vapor, u_g	Sat. liquid, h_f	Evap., h_{fg}	Sat. vapor, h_g	Sat. liquid, s_f	Evap., s_{fg}	Sat. vapor, s_g

(b) PRESSURE TABLE, ENGLISH UNITS

Press., P (psia)	Temp., T (°F)	Specific volume (ft³/lbm)		Internal energy (Btu/lbm)			Enthalpy (Btu/lbm)			Entropy (Btu/lbm·°R)		
		Sat. liquid, v_f	Sat. vapor, v_g	Sat. liquid, u_f	Evap., u_{fg}	Sat. vapor, u_g	Sat. liquid, h_f	Evap., h_{fg}	Sat. vapor, h_g	Sat. liquid, s_f	Evap., s_{fg}	Sat. vapor, s_g

TABLE 3-3 TABLE HEADINGS FOR PROPERTIES OF SUPERHEATED VAPOR (TWO POSSIBLE FORMATS)

Temp., T	Pressure, p				Pressure, p			
	v	u	h	s	v	u	h	s

Pressure, p	Temp., T	Temp., T
	v u h s	

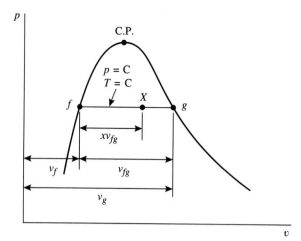

Figure 3-14 p–v diagram showing the saturated liquid state f, saturated vapor state g, and a mixture state X with quality x.

where

$$u_{fg} = u_g - u_f \qquad h_{fg} = h_g - h_f \qquad \text{and} \qquad s_{fg} = s_g - s_f$$

It should be noticed that the specific properties v, u, h, and s of a mixture as given by Eqs. (3-3) to (3-6) are the weighted-average values of the two phases present. The true properties of the mixture are those of the saturated liquid and vapor phases that are coexisting.

The table format for solid–vapor sublimation region is the same as Table 3-2(a), with the subscript i denoting the solid phase and g denoting the vapor phase.

It is fitting at this point to note that heat must be added to or rejected from a substance to cause it to change phase in a quasiequilibrium process. Application of the first law to a phase change for a pure substance shows that the heat required is equal to the difference in enthalpy between the final and the initial phases at the same pressure and temperature. For example, the heat required to vaporize a saturated liquid to saturated vapor is given by

$$q_{fg} = u_{fg} + \int_f^g p \, dv = (u_g - u_f) + (p_g v_g - p_f v_f) = h_g - h_f$$

The amount of heat required to cause a phase change is what was called the *latent heat*. A more appropriate name is the enthalpy of phase change because it is a property of the system. Thus, we can write

$$\text{Enthalpy (or latent heat) of vaporization} = h_{fg} = h_g - h_f$$

$$\text{Enthalpy (or latent heat) of fusion} = h_{if} = h_f - h_i$$

$$\text{Enthalpy (or latent heat) of sublimation} = h_{ig} = h_g - h_i$$

wherein the subscripts g, f, and i denote saturated vapor, saturated liquid, and saturated solid phases, respectively.

For a single-phase region, such as the superheated vapor region or compressed liquid region, two intensive properties are required to determine an equilibrium state. Pressure and temperature are usually used as the coordinates to build a single–phase property table. Table 3-3 illustrates two possible formats for a superheated vapor table. A compressed liquid table is also built according to the same format. It must be recognized that the properties of compressed liquids are primarily a function of temperature, with pressure exerting only a minor effect. A compressed liquid table usually does not extend the pressure entry to low values. Below the lowest pressure entry, the properties of a compressed liquid can be taken as those of the saturated liquid at the temperature of the compressed liquid.

In the tabulation of thermodynamic properties, the reference point of a property with zero tabulated value is selected arbitrarily. Steam tables assign $u = 0$ and $s = 0$ to saturated liquid at the triple-point temperature of water, which is $0.01°C = 32.018°F$. Tables for refrigerants, such as refrigerant 12 (Freon 12), assign $h = 0$ and $s = 0$ to saturated liquid at $-40°C$ in SI-unit tables and $-40°F$ in English-unit tables, noting that $-40°C = -40°F$. The arbitrary assignment of zero point to properties does not really matter to us as the user, because we are more concerned with changes in the properties, and any assigned reference value would cancel out in our calculations.

In solving numerical problems, interpolation of table data is often required. For the purpose of learning, we will assume that linear interpolation is valid even in the condensed tables in Appendix 3. For professional use in more accurate calculations, one should always employ the more complete original tables, such as those given in the reference for the condensed tables in Appendix 3.

Example 3-1

A rigid uninsulated tank contains 1 kg of H_2O at a pressure of 5.0 MPa and at a temperature of 500°C (state 1). At two different times later, it is observed that the pressure has dropped to (a) 4.0 MPa (state 2) and (b) 1.0 MPa (state 3). Determine the temperature, specific internal energy, specific enthalpy, and specific entropy of the fluid at the three different states.

Solution Referring to Fig. 3-15, at the initial pressure of $p_1 = 5.0$ MPa, we have from the saturated steam table (A-19M):*

$$T_a = T_{sat} = 263.99°C$$

$$v_a = v_g = 0.03944 \text{ m}^3/\text{kg}$$

Because T_a is less than the given temperature of $T_1 = 500°C$, we conclude that state 1 with $p_1 = 5.0$ MPa and $T_1 = 500°C$ must be a superheated steam. Then from the superheated steam table (A-20M), we obtain

$$v_1 = 0.06857 \text{ m}^3/\text{kg}$$

$$u_1 = 3091.0 \text{ kJ/kg}$$

$$h_1 = 3433.8 \text{ kJ/kg}$$

$$s_1 = 6.9759 \text{ kJ/kg·K}$$

(a) From the saturated steam table (A-19M), at $p_2 = 4.0$ MPa, we have

$$T_b = T_{sat} = 250.40°C$$

$$v_b = v_g = 0.04978 \text{ m}^3/\text{kg}$$

---

*In this book, all steam (H_2O) property data are based on the original complete tables of Ref. [21] in the Bibliography.

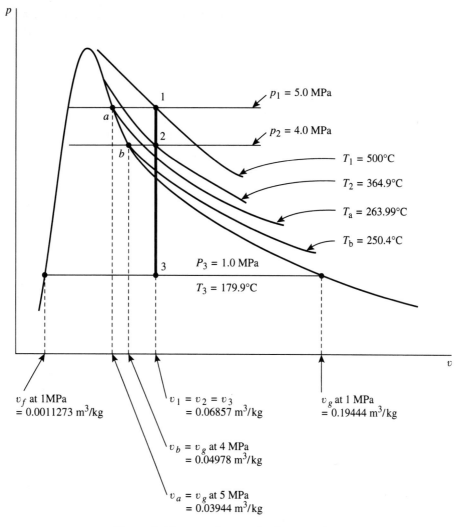

Figure 3-15 p–v diagram for Example 3-1.

Because $v_2 = v_1 = 0.06857$ m^3/kg is greater than $v_b = 0.04978$ m^3/kg, state 2 is a superheated steam. From the superheated steam table (A-20M), by interpolation, we have

$$T_2 = 364.9°C$$

$$u_2 = 2854.9 \text{ kJ/kg}$$

$$h_2 = 3129.2 \text{ kJ/kg}$$

$$s_2 = 6.6402 \text{ kJ/kg·K}$$

(b) From the saturated steam table (A-19M), at $p_3 = 1.0$ MPa, we have

$$T_{\text{sat}} = 179.91°C$$

$$v_f = 0.0011273 \text{ m}^3/\text{kg}$$

$$v_g = 0.19444 \text{ m}^3/\text{kg}$$

Because $v_3 = v_1 = 0.06857$ m³/kg is greater than v_f and smaller than v_g at the given pressure of $p_3 = 1.0$ MPa, state 3 is in the two-phase region with $T_3 = T_{sat}$ (at 1.0 MPa) = 179.91°C. We now calculate the quality x_3 of the liquid–vapor mixture at state 3:

$$v_3 = v_1 = 0.06857 \text{ m}^3/\text{kg} = v_f + x_3(v_g - v_f) \qquad \text{(at 1.0 MPa)}$$

$$= 0.0011273 + x_3(0.19444 - 0.0011273)$$

so that

$$x_3 = 0.349 = 34.9\%$$

Therefore, the following values of u_3, h_3, and s_3 are calculated with data from the saturated steam table (A-19M) at 1.0 MPa:

$$u_3 = u_f + x_3 u_{fg} = 761.68 + 0.349(1822.0) = 1397.6 \text{ kJ/kg}$$

$$h_3 = h_f + x_3 h_{fg} = 762.81 + 0.349(2015.3) = 1466.1 \text{ kJ/kg}$$

$$s_3 = s_f + x_3 s_{fg} = 2.1387 + 0.349(4.4478) = 3.691 \text{ kJ/kg·K}$$

Example 3-2

One pound of H_2O contained in a cylinder with a movable piston is compressed at constant temperature from $T_1 = 400°F$ and $p_1 = 60$ psia to (a) $p_2 = 120$ psia, (b) $v_3 = 1.20$ ft³/lbm, (c) $p_4 = 300$ psia, and (d) $p_5 = 1000$ psia. Determine the values of v, u, h, and s at the initial and the various final states.

Solution   The various state points are depicted in Fig. 3-16 along the isothermal line 1–2–3–4–5. From the saturation steam table (Table A-18E), we get

$$p_a = p_{sat} \text{ at } 400°F = 247.1 \text{ psia}$$

which is higher than the given $p_1 = 60$ psia. Thus, we conclude that point 1 is in the superheated region. From superheated steam table (A-20E), at $T_1 = 400°F$ and $p_1 = 60$ psia, we have

$$v_1 = 8.353 \text{ ft}^3/\text{lbm}, \qquad u_1 = 1140.8 \text{ Btu/lbm},$$

$$h_1 = 1233.5 \text{ Btu/lbm}, \qquad s_1 = 1.7134 \text{ Btu/lbm·°R}$$

(a) The value of $p_2 = 120$ psia is still lower than the saturation pressure of $p_a = 247.1$ psia at the given temperature of $T_2 = 400°F$, so that point 2 is still in the superheated region. From the superheated steam table (A-20E), at $T_2 = 400°F$ and

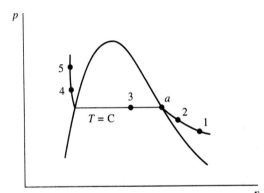

p

5

4

3   a

2

1

$T = C$

v **Figure 3-16** p–v diagram for Example 3-2.

$p_2 = 120$ psia, we obtain

$$v_2 = 4.079 \text{ ft}^3/\text{lbm}, \qquad u_2 = 1133.8 \text{ Btu/lbm},$$

$$h_2 = 1224.4 \text{ Btu/lbm}, \qquad s_2 = 1.6288 \text{ Btu/lbm·°R}$$

(b) At $T_3 = 400°F$, the saturated steam table (A-18E) gives

$$v_f = 0.018638 \text{ ft}^3/\text{lbm} \qquad \text{and} \qquad v_g = 1.8661 \text{ ft}^3/\text{lbm}$$

With a specific volume $v_3 = 1.20 \text{ ft}^3/\text{lbm}$, it is clear that state 3 is in the two-phase region. The quality x_3 is calculated by the equation

$$v_3 = [v_f + x_3(v_g - v_f)] \text{ at } 400°F$$

or

$$1.20 = 0.018638 + x_3(1.8661 - 0.018638)$$

so that

$$x_3 = 0.6395$$

Therefore, at $T_3 = 400°F$, we obtain the following values:

$$u_3 = u_f + x_3 u_{fg} = 374.27 + 0.6395(742.4) = 849.0 \text{ Btu/lbm}$$

$$h_3 = h_f + x_3 h_{fg} = 375.12 + 0.6395(826.8) = 903.9 \text{ Btu/lbm}$$

$$s_3 = s_f + x_3 s_{fg} = 0.56672 + 0.6395(0.9617) = 1.1817 \text{ Btu/lbm·°R}$$

(c) Because $p_4 = 300$ psia is greater than p_{sat} at $400°F(= 247.1$ psia), point 4 is in the compressed liquid region. As the compressed liquid table starts at a minimum pressure of 500 psia, we use the saturation table (A-18E) to get approximate values for the compressed liquid. Thus, at $T_4 = 400°F$, we get

$$v_4 = v_f = 0.018638 \text{ ft}^3/\text{lbm}$$

$$u_4 = u_f = 374.27 \text{ Btu/lbm}$$

$$h_4 = h_f = 375.12 \text{ Btu/lbm}$$

$$s_4 = s_f = 0.56672 \text{ Btu/lbm·°R}$$

(d) At $T_5 = 400°F$ and $p_5 = 1000$ psia, the compressed liquid table (A-21E) gives

$$v_5 = 0.018550 \text{ ft}^3/\text{lbm}, \qquad u_5 = 372.55 \text{ Btu/lbm},$$

$$h_5 = 375.98 \text{ Btu/lbm}, \qquad s_5 = 0.56472 \text{ Btu/lbm·°R}$$

3-5 USE OF TABULAR DATA IN CLOSED-SYSTEM ENERGY ANALYSIS

In the previous section, we introduced the general format of typical thermodynamic property tables and briefly described the use of these tables to acquire data for problem solution. We now illustrate the use the table data in energy analysis for closed systems through several numerical examples.

Example 3-3

The radiator of a heating system has a volume of 0.02 m³. After it is filled with dry saturated steam at 200 kPa, the radiator valves are then closed. As a result of heat transfer to the room, the steam pressure drops to 100 kPa. Determine the amount of heat transfer to the room.

Solution Referring to Fig. 3-17, from the saturated steam table (A-19M) at $p_1 = 200$ kPa $= 0.2$ MPa, we obtain

$$v_1 = v_g = 0.8857 \text{ m}^3/\text{kg} \qquad \text{and} \qquad u_1 = u_g = 2529.5 \text{ kJ/kg}$$

The mass of steam in the radiator is

$$m = \frac{V}{v_1} = \frac{0.02 \text{ m}^3}{0.8857 \text{ m}^3/\text{kg}} = 0.0226 \text{ kg}$$

After a constant-volume heat rejection to the room, the steam in the radiator becomes a liquid–vapor mixture with a quality x_2; thus,

$$v_2 = v_1 = (v_f + x_2 v_{fg}) \text{ at } 100 \text{ kPa}$$

or

$$0.8857 = 0.001043 + x_2(1.6940 - 0.001043)$$

so that

$$x_2 = 0.5226$$

Therefore,

$$u_2 = (u_f + x_2 u_{fg}) \text{ at } 100 \text{ kPa}$$

$$= 417.36 + 0.5226(2088.7) = 1508.9 \text{ kJ/kg}$$

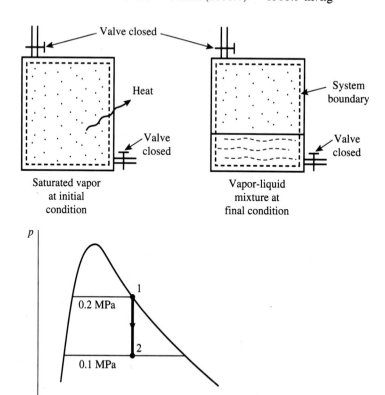

Figure 3-17   Schematic and p–v diagrams for Example 3-3.

Thermodynamic Properties of Pure Substances   Chap. 3

Because for the constant-volume process, $W = \int_1^2 p \, dV = 0$, we have from the first law of thermodynamics,

$$Q = \Delta U + \overset{0}{\cancel{W}} = \Delta U = m(u_2 - u_1)$$

$$= (0.0226 \text{ kg})(1508.9 \text{ kJ/kg} - 2529.5 \text{ kJ/kg}) = -23.1 \text{ kJ}$$

meaning that the heat transfer from the steam to the room is 23.1 kJ.

Example 3-4

A cylinder fitted with a piston contains 5 pounds of saturated liquid water at 300 psia. Heat is slowly transferred to the system in a constant-pressure process until the temperature reaches 600°F. Determine the work done and the heat transfer for this process.

Solution The constant-pressure heating process is depicted in Fig. 3-18. From the saturated steam table (A-19E), at 300 psia, we have

$$T_1 = T_{\text{sat}} = 417.43°\text{F} \qquad v_1 = v_f = 0.018896 \text{ ft}^3/\text{lbm},$$

$$u_1 = u_f = 393.0 \text{ Btu/lbm}, \qquad h_1 = h_f = 394.1 \text{ Btu/lbm}$$

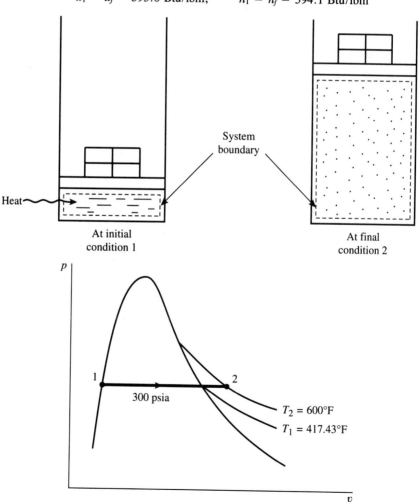

Figure 3-18 Schematic and p–v diagrams for Example 3-4.

Because $T_2 = 600°F$ is greater than the saturation temperature at 300 psia, state 2 is a superheated vapor. From the superheated steam table (A-20E), at $p_2 = 300$ psia and $T_2 = 600°F$, we have

$$v_2 = 2.004 \text{ ft}^3/\text{lbm}, \qquad u_2 = 1203.2 \text{ Btu/lbm}, \qquad h_2 = 1314.5 \text{ Btu/lbm}$$

The work done by the system is calculated as

$$W = m \int_1^2 p \, dv = mp_1(v_2 - v_1)$$

$$= (5 \text{ lbm})(300 \times 144 \text{ lbf/ft}^2)(2.004 \text{ ft}^3/\text{lbm} - 0.018896 \text{ ft}^3/\text{lbm})$$

$$= 428{,}780 \text{ ft·lbf} = \frac{428{,}780 \text{ ft·lbf}}{778 \text{ ft·lbf/Btu}} = 551 \text{ Btu}$$

By the first law, the heat added to the system is calculated as

$$Q = \Delta U + W = m(u_2 - u_1) + W$$

$$= (5 \text{ lbm})(1203.2 \text{ Btu/lbm} - 393.0 \text{ Btu/lbm}) + 551 \text{ Btu}$$

$$= 4602 \text{ Btu}$$

The heat added during the constant-pressure can also be calculated by the equation

$$Q = \Delta U + W = m(u_2 - u_1) + m(p_2v_2 - p_1v_1)$$

$$= m[(u_2 + p_2v_2) - (u_1 + p_1v_1)]$$

$$= m(h_2 - h_1) = (5 \text{ lbm})(1314.5 \text{ Btu/lbm} - 394.1 \text{ Btu/lbm})$$

$$= 4602 \text{ Btu}$$

Example 3-5

Refrigerant-12 (Freon-12) contained in a piston–cylinder device initially has a volume of 1 ft³ at 40 psia and 60°F (state 1). Heat is slowly added to the gas in a constant-volume process until the pressure reaches 50 psia (state 2). Heat is then slowly removed from the gas in a constant-pressure process until the temperature reaches 100°F (state 3). Determine the amounts of heat transfer in the two processes.

Solution By referring to Fig. 3-19, $p_1 = 40$ psia, $T_1 = 60°F$, $V_1 = 1$ ft³, $p_2 = 50$ psia, and $T_3 = 100°F$. All three states are in the superheated vapor region. From the superheated Freon-12 table (Table A-23E), at $p_1 = 40$ psia and $T_1 = 60°F$, we get

$$v_1 = 1.0789 \text{ ft}^3/\text{lbm} \qquad \text{and} \qquad h_1 = 85.206 \text{ Btu/lbm}$$

so that

$$u_1 = h_1 - p_1v_1$$

$$= (85.206 \text{ Btu/lbm}) - (40 \times 144 \text{ lbf/ft}^2)(1.0789 \text{ ft}^3/\text{lbm})/(778 \text{ ft·lbf/Btu})$$

$$= 77.2 \text{ Btu/lbm}$$

At $p_2 = 50$ psia and $v_2 = v_1 = 1.0789$ ft³/lbm, from the superheated Freon-12 table by interpolation, we get

$$T_2 = 173.3°F \qquad \text{and} \qquad h_2 = 102.6 \text{ Btu/lbm}$$

Thermodynamic Properties of Pure Substances   Chap. 3

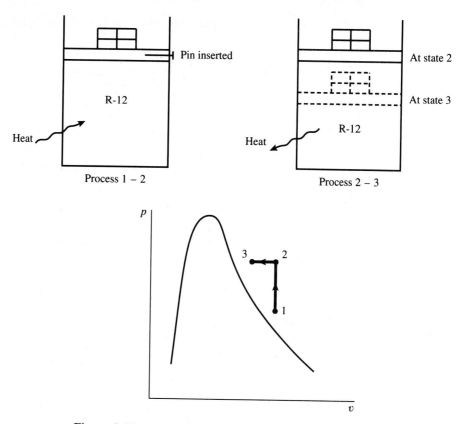

Figure 3-19 Schematic and $p-v$ diagrams for Example 3-5.

Accordingly,

$$u_2 = h_2 - p_2 v_2$$

$$= (102.6 \text{ Btu/lbm}) - (50 \times 144 \text{ lbf/ft}^2)(1.0789 \text{ ft}^3/\text{lbm})/(778 \text{ ft·lbf/Btu})$$

$$= 92.6 \text{ Btu/lbm}$$

At $p_3 = p_2 = 50$ psia and $T_3 = 100°F$, the superheated Freon-12 table gives

$$h_3 = 90.953 \text{ Btu/lbm}$$

The mass of Freon-12 is

$$m = \frac{V_1}{v_1} = \frac{1 \text{ ft}^3}{1.0789 \text{ ft}^3/\text{lbm}} = 0.927 \text{ lbm}$$

For the constant-volume process 12,

$$W_{12} = m \int_1^2 p \, dv = 0$$

so that the first law gives

$$Q_{12} = \Delta U_{12} + W_{12} = \Delta U_{12} = m(u_2 - u_1)$$

$$= (0.927 \text{ lbm})(92.6 \text{ Btu/lbm} - 77.2 \text{ Btu/lbm}) = 14.3 \text{ Btu}$$

meaning that 14.3 Btu of heat is added to the Freon in process 1–2. For the constant-pressure process 2–3,

$$W_{23} = m \int_{2}^{3} p \, dv = mp_2(v_3 - v_2) = m(p_3 v_3 - p_2 v_2)$$

so that

$$Q_{23} = \Delta U_{23} + W_{23} = m(u_3 - u_2) + m(p_3 v_3 - p_2 v_2) = m(h_3 - h_2)$$

$$= (0.927 \text{ lbm})(90.95 \text{ Btu/lbm} - 102.6 \text{ Btu/lbm})$$

$$= -10.8 \text{ Btu}$$

meaning that 10.8 Btu of heat is removed from the Freon in process 2–3.

3-6 PROPERTIES OF INCOMPRESSIBLE SUBSTANCES

A substance whose density (or specific volume) remains constant regardless of changes in other properties is said to be *incompressible*. Many solids and liquids can be considered to be essentially incompressible. The incompressibility of solids and liquids can be visualized when one notices that on the p–v–T equilibrium surfaces of Figs. 3-6 and 3-10, the specific-volume variations are so small in the solid and liquid regions for wide ranges of pressure and temperature changes.

The state of an incompressible substance can be represented by

$$\rho = \text{constant} \qquad \text{or} \qquad v = \text{constant} \qquad (3\text{-}7)$$

The internal energy u of a simple compressible homogenous substance can be considered a function of its temperature T and specific volume v, or

$$u = u(T, v)$$

The total differential of u is

$$du = \left(\frac{\partial u}{\partial T}\right)_v dT + \left(\frac{\partial u}{\partial v}\right)_T dv$$

Since $c_v = (\partial u / \partial T)_v$ by definition, the preceding equation becomes

$$du = c_v \, dT + \left(\frac{\partial u}{\partial v}\right)_T dv$$

This is a general expression for a simple compressible substance. If the substance is incompressible, we have then $dv = 0$ and

$$du = c_v \, dT \qquad (3\text{-}8)$$

This is, the internal energy of an incompressible substance is a function of temperature only. From the definition of c_v, we conclude that for an incompressible substance, c_v is also a function of temperature only.

Additionally, because by definition $h = u + pv$, so that

$$dh = du + p \, dv + v \, dp$$

For an incompressible substance the preceding expression reduces to

$$dh = c_v \, dT + v \, dp \qquad (3\text{-}9)$$

It follows that for an incompressible substance, we must have

$$\left(\frac{\partial h}{\partial T}\right)_p = c_v$$

But, by definition,

$$c_p = \left(\frac{\partial h}{\partial T}\right)_p$$

Consequently, for an incompressible substance, the specific heats c_p and c_v are equal and the symbol c is usually used for both, or written in equation form,

$$c_p = c_v = c \tag{3-10}$$

As a result, Eq. (3-8) can be simply written as

$$du = c\, dT \tag{3-11}$$

where the specific heat c is a function of temperature only. Intergrating the preceding equation between states 1 and 2 for an incompressible substance yields

$$u_2 - u_1 = \int_1^2 c\, dT \tag{3-12}$$

On the basis of Eq. (3-9), we realize that the enthalpy of an incompressible substance is not just a function of temperature alone. Integrating Eq. (3-9) between states 1 and 2, we have

$$h_2 - h_1 = \int_1^2 c\, dT + v(p_2 - p_1) \tag{3-13}$$

Or from the definition of enthalpy $h = u + pv$, we can write directly

$$h_2 - h_1 = u_2 - u_1 + v(p_2 - p_1) \tag{3-14}$$

Example 3-6

An insulated tank contains 50 kg of water at 20°C and 1 atm. At 20-kg of copper at 60°C is dropped into the tank. Determine the final equilibrium temperature of the copper-and-water system at 1 atm pressure.

Solution Both copper and water can be considered as incompressible. Assume the specific heat of copper to be a constant value of 0.386 kJ/kg·°C and that of water to be a constant value of 4.180 kJ/kg·°C. Because the tank is insulated, an energy balance for the copper-and-water system gives

$$m_{copper} C_{copper}(T_{copper,\ initial} - T_{final}) = m_{water} C_{water}(T_{final} - T_{water,\ initial})$$

so that the final equilibrium temperature is

$$\begin{aligned} T_{final} &= \frac{m_{copper} C_{copper} T_{copper,\ initial} + m_{water} C_{water} T_{water,\ initial}}{m_{copper} C_{copper} + m_{water} C_{water}} \\[2mm] &= \frac{(20)(0.386)(60) + (50)(4.180)(20)}{(20)(0.386) + (50)(4.180)} \\[2mm] &= 21.4°C \end{aligned}$$

Sec. 3-6   Properties of Incompressible Substances

3-7 SUMMARY

A substance that is homogeneous and invariable in chemical composition, irrespective of the phase or phases in which it exists, is called a pure substance. Pure substances can exist in vapor, liquid, and solid phases. Two phases of a pure substance can coexist in equilibrium only if their pressure and temperature bear a certain fixed relationship to each other. The corresponding values of pressure and temperature under which two phases can coexist in equilibrium are called the saturation pressure and the saturation temperature. Any phase (vapor, liquid, or solid) of a pure substance existing under corresponding values of saturation pressure and temperature is called a saturated phase (saturated vapor, liquid, or solid).

When the temperature in a vapor phase is higher than the saturation temperature corresponding to its pressure, it is called a superheated vapor. When the temperature of a liquid is lower than the saturation temperature corresponding to its pressure, or when the pressure of a liquid is higher than the saturation pressure corresponding to its temperature, it is called a subcooled liquid or a compressed liquid.

The critical state (or critical point) is the limiting condition of pressure and temperature under which separate liquid and vapor phases of a pure substance can be distinguished. The pressure, temperature, and volume at the critical point are called the critical pressure, critical temperature, and critical volume, respectively.

The condition of pressure and temperature under which vapor, liquid, and solid phases of a pure substance can coexist in equilibrium is called the triple phase (or triple point). Any specific property, such as specific volume v, of a triple-phase mixture can be determined from an equation of the form

$$v = xv_g + yv_f + zv_i$$

where x, y, and z are mass fractions of vapor, liquid, and solid, respectively. The subscripts g, f, and i denote saturated vapor, saturated liquid, and saturated solid, respectively, at the triple-phase conditions.

Any specific property, such as specific volume v, of a liquid–vapor two-phase mixture can be determined from an equation of the form

$$v = v_f + x(v_g - v_f)$$

where x is the quality of the two-phase mixture as defined by

$$x = \frac{\text{mass of vapor}}{\text{total mass of the mixture}}$$

Steam tables are used in this chapter to demonstrate the use of thermodynamic property tables. Property tables of other substances are used in the same manner. Readers should get familiar with the property tables.

PROBLEMS

3-1. Complete the following table of properties of H_2O.

	Pressure (MPa)	Temp. (°C)	Specific volume (m³/kg)	Specific internal energy (kJ/Kg)	Specific enthalpy (kJ/kg)	Specific entropy (kJ/kg·K)	Quality (if appropriate)
(a)		150	0.3928				
(b)	1.50		0.001154				
(c)		200	0.09				
(d)	0.50	300					
(e)		250	0.19234				
(f)		200	0.001143				
(g)	50.0		0.001035				
(h)	4.0		0.08643				
(i)	10.0		0.01				
(j)	2.5		0.07998				
(k)		260	0.001276				
(l)	1.50	160					

3-2. Complete the following table of properties of H_2O.

	Pressure (psia)	Temp. (°F)	Specific volume (ft³/lbm)	Specific internal energy (Btu/lbm)	Specific enthalpy (Btu/lbm)	Specific entropy (Btu/lbm·°R)	Quality (if appropriate)
(a)	30		13.748				
(b)		90	0.016099				
(c)	20			950			
(d)		200	25				
(e)	120	400					
(f)	300				1368.3		
(g)		800	0.05932				
(h)	2000	100					
(i)	100	150					

3-3. Complete the following table of properties of Refrigerant-12.

	Pressure (psia)	Temp. (°F)	Specific volume (ft³/lbm)	Specific enthalpy (Btu/lbm)	Specific entropy (Btu/lbm·°R)	Quality (if appropriate)
(a)		−50	4.9742			
(b)	29.335		0.011160			
(c)		30	0.8			
(d)	19.189			20.0		
(e)		60	0.011913			
(f)	114.49				0.16353	
(g)	15	40				
(h)	30		1.5306			
(i)		100	1.1812			
(j)	25			85.965		
(k)	50	20				

3-4. Complete the following table of properties of Refrigerant-12.

	Temp. (°C)	Pressure (MPa)	Specific volume (m³/kg)	Specific enthalpy (kJ/kg)	Specific entropy (kJ/kg·K)	Quality (if appropriate)
(a)	0		0.055389			
(b)	−10		0.06			
(c)	5	0.5				
(d)	20	0.5				
(e)	100		0.018812			
(f)	−40			0		

Thermodynamic Properties of Pure Substances   Chap. 3

3-5. Complete the following table of properties of ammonia.

	Temp. (°C)	Pressure (kPa)	Specific volume (m³/kg)	Specific enthalpy (kJ/kg)	Specific entropy (kJ/kg·K)	Quality (if appropriate)
(a)	−20		0.6237			
(b)		614.95		227.6		
(c)	−32		0.50			
(d)	50	300				
(e)	0	1000				

3-6. Complete the following table of properties of ammonia.

	Temp. (°F)	Pressure (psia)	Specific volume (ft³/lbm)	Specific enthalpy (Btu/lbm)	Specific entropy (Btu/lbm·°R)	Quality (if appropriate)
(a)	−5		0.02406			
(b)	50			625.2		
(c)	70		2.0			
(d)	10	50				
(e)	60	80				
(f)		40	8.096			

3-7. Determine the properties as required in the following table.

	Substance	Given	Find
(a)	Water	$p = 10$ MPa, $T = 100°C$	$u = ?$
(b)	Water	$p = 1$ MPa, $T = 150°C$	$u = ?$
(c)	Refrigerant-12	$T = 20°C$, $v = 0.028$ m³/kg	$h = ?$
(d)	Ammonia	$T = 0°C$, $x = 0.2$	$v = ?$

3-8. Determine the specific volume, internal energy, enthalpy, and entropy of compressed liquid water at 100°C and 10 MPa using (a) the compressed liquid table, and (b) the saturated liquid table for approximation. Give the percent error for the values obtained in (b) as compared to (a).

3-9. A 20-cm diameter cooking pan has a 0.5-kg lid. See Fig. P3-9. It is filled with water at a location where the atmospheric pressure is 101.3 kPa. At what temperature will the water start to boil when it is heated?

Atmosphere 101.3 kPa

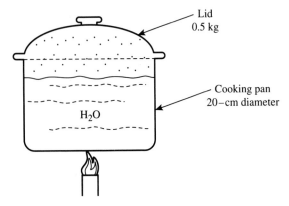

Lid
0.5 kg

Cooking pan
20-cm diameter

H_2O

Figure P3-9

3-10. A 15-ft³ tank contains 5 pounds of a liquid and vapor water mixture at 100 psia. Determine the volume and mass of vapor in the tank.

3-11. A rigid uninsulated tank contains 1 kg of H_2O at a pressure of 200 kPa and a temperature of 200°C. After some time, it is observed that the temperature has dropped to 100°C. Determine the internal energy, enthalpy, and entropy of the fluid at the initial and final conditions.

3-12. A rigid container is filled with steam at 700 kPa and 200°C. (a) At what temperature will the steam start to condense when the container is cooled? (b) To what temperature must the container be cooled to condense 50% of the steam mass?

3-13. A 0.5-m³ rigid tank contains 10 kg of Refrigerant-12 at −20°C. Determine the pressure, the volume occupied by the liquid, and the total internal energy.

3-14. A 0.5-m³ rigid tank contains a liquid–vapor mixture of ammonia at a temperature of 30°C. The liquid occupies 10% of the volume. Determine the quality and the total mass of the fluid in the tank.

3-15. A 0.5-m³ rigid tank contains a liquid–vapor mixture of Refrigerant-12 at 200 kPa. The liquid occupies 10% of the volume. Determine the quality and the total mass of the refrigerant in the tank.

3-16. A liquid–vapor mixture of Refrigerant-12 at 35°C is stored in a 0.1-m³ rigid container, with liquid and vapor initially occupying equal volumes. Additional Refrigerant-12 is charged into the container until the final mass is 80 kg. If the temperature remains at 35°C, how much mass enters the container and what is the final volume occupied by the liquid?

3-17. A vapor–liquid mixture of Refrigerant-12 is stored in a sealed rigid container at 200°F. When the container is heated, the refrigerant passes exactly through its critical state. What are the initial ratio of mass of vapor to liquid and the initial ratio of volume of vapor to liquid?

3-18. A vapor–liquid mixture of water is stored in a 0.5-ft³ sealed rigid container at 1 atm. When the container is heated, the water passes exactly through its critical state. Determine the total mass of water in the container, the initial mass of vapor, and the initial volume of vapor.

3-19. Two cubic meters of steam at 2.0 MPa and 300°C are contained in a cylinder and piston arrangement (state 1). Heat is transferred out from the steam at constant temperature until its volume has been reduced to one-tenth of its original (state 2). The steam is then heated at constant volume to 20 MPa (state 3). Determine (a) p_2 and h_2, (b) T_3 and h_3, and (c) the mass of steam in the cylinder.

3-20. A triple-phase mixture of H_2O consists of 40% solid, 50% liquid, and 10% vapor by mass. Determine the specific internal energy, enthalpy, and volume of the mixture. Show the mixture state on a u–v diagram and a p–h diagram.

3-21. Two insulated tanks are connected by an insulated pipe with a valve. See Fig. P3-21. Tank A has a volume of 0.6 m³ and contains water at 200 kPa, 200°C. Tank B has a volume of 0.3 m³ and contains water at 500 kPa, 90% quality. When the valve is opened, what is the pressure of the final equilibrium state?

3-22. A rigid tank has a volume of 2 ft³ and contains saturated water vapor at a temperature of 230°F. Because of heat transfer, the temperature inside the tank drops to 212°F. Calculate the total mass of steam in the tank, the quality of the steam at the final state, the volume and mass of liquid in the final state, and the volume and mass of vapor in the final state. Also calculate the heat transfer.

Thermodynamic Properties of Pure Substances   Chap. 3

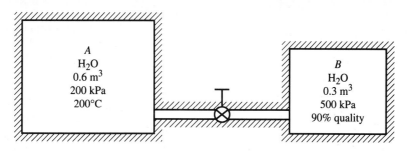

Figure P3-21

3-23. A 0.02-m³ radiator of a steam heating system is filled with saturated steam at a pressure of 150 kPa. When all valves to the radiator are closed, how much heat is transferred to the room when the pressure of the steam falls to 75 kPa?

3-24. A 2-ft³ radiator of a heating system contains saturated steam at 20 psia. When all the valves on the radiator are closed, as a result of heat transfer to the room, the pressure drops to 15 psia. Calculate the total mass of steam in the radiator, the quality of the steam at the final state, the volume and mass of liquid in the final state, and the volume and mass of vapor in the final state. Also calculate the heat transfer.

3-25. A cylinder–piston device holds a mixture of 1 kg liquid water and 1 kg of water vapor in equilibrium at 700 kPa. Heat is transferred to the system in a constant-pressure process until the temperature reaches 350°C. Determine the work done by the system in kJ.

3-26. Two pounds of steam in a closed system initially at 300°F and having a quality of 90% are heated at constant pressure to 600°F. How much heat is added to the steam?

3-27. A cylinder fitted with a piston contains 0.5 kg of steam with an initial specific volume of 0.2 m³/kg at a pressure of 5 bars. Heat is transferred slowly to the steam at a constant pressure of 5 bars until the temperature reaches 320°C. Determine the heat transferred and the work done for the process. Draw the p–v diagram for the process relative to the saturation line.

3-28. A rigid tank of 5 ft³ volume contains a liquid–vapor mixture of H_2O initially at 70 psia and 40% quality. What is the mass of vapor initially in the tank? If the pressure is raised to 100 psia by the transfer of heat, what will be the final mass of the vapor and what is the amount of heat transfer?

3-29. Steam initially at 400°F and 40 psia is cooled at constant volume to 200°F. What is the quality when the temperature reaches 200°F? How much heat is transferred, per pound of steam, in the process between 400°F and 200°F?

3-30. A steam boiler having a total volume of 4 m³ initially contains 3 m³ of liquid water and 1 m³ of vapor water in equilibrium at 0.1 MPa. The boiler is fired up. If both the inlet and discharge valves are kept closed, the relief valve will lift when the pressure reaches 5 MPa. How much heat is transferred to the steam before the relief valve lifts?

3-31. A steam boiler having a total volume of 80 ft³ initially contains 60 ft³ of liquid water and 20 ft³ of water vapor in equilibrium at 14.7 psia. The boiler is fired up and heat is transferred to the steam. If both the inlet and discharge valves are kept closed, the relief valve will lift when the pressure reaches 800 psia. How much heat is transferred to the steam before the relief valve lifts?

3-32. A cylinder fitted with a piston contains 3 lbm of steam at 150 psia and 90% quality. The piston is permitted to move in such a way that when the contained volume is six times the initial volume, the pressure has dropped to 15 psia. During the process, there is a heat rejection of 10 Btu from the steam. Calculate the work done by the steam.

3-33. Twenty kilograms of Refrigerant-12 in a piston–cylinder device initially at a temperature of −30°C and a quality of 80% undergo a frictionless process at constant pressure to a temperature of 40°C. Determine the work done and heat transfer, in kJ.

3-34. A piston–cylinder device contains Refrigerant-12 vapor initially at 0.8 MPa and 50°C. The fluid undergoes a constant-pressure process ending at the saturated vapor state. Determine the work and heat trasfer per kilogram of the fluid.

3-35. Refrigerant-12 undergoes a process in a piston–cylinder assembly from 0.2 MPa, 10°C to 1 MPa, 60°C, according to the relation

$$pv^n = \text{constant}$$

where n is a constant. Determine the value of the constant n and calculate the work in kJ/kg.

3-36. Refrigerant-12 initially at 1 MPa and 100°C expands slowly in a cylinder–piston device to a final pressure of 0.50 MPa according to the relation

$$pv = \text{constant}$$

Determine the work done per kilogram of refrigerant.

3-37. One pound of steam initially at 300 psia and 560°F undergoes a quasiequilibrium polytropic process in a closed system. If in the final state the steam is at 100 psia and 360°F, how much heat does it absorb during the process?

3-38. Three kilograms of H_2O at 120 bars with a specific volume of 0.00535 m^3/kg are contained behind a piston in a cylinder. The volume of the H_2O is doubled and the temperature is maintained constant. Find the state of the H_2O at the final condition. What is the work done involved in the process?

3-39. A 2.5-ft^3 tank contains liquid water and vapor water in equilibrium at 100°F in such a liquid-to-vapor proportion that when enough heat is added, it can become saturated vapor at 1 atm. Determine the initial quality of the two-phase mixture and the amount of heat that must be added to produce saturated vapor at 1 atm.

3-40. A rigid tank of 1 m^3 total volume contains 0.05 m^3 of saturated liquid water and 0.95 m^3 of saturated vapor water at 0.1 MPa. Heat is added to the tank to vaporize the water. Determine the amount of heat required to vaporize the liquid and the pressure at which vaporization takes place.

3-41. A piston–cylinder device contains 0.025 kg of saturated water vapor at 300 kPa. See Fig. P3-41. Electric work of the amount of 7.2 kJ is done on the steam and heat of the amount of 3.7 kJ is transferred out from the steam while the steam undergoes a constant-pressure process. Determine the final temperature of the steam.

3-42. Two pounds of steam at 200 psia and 90% quality are heated slowly in a closed system until the temperature reaches 500°F. Determine the amount of heat addition if the process is at (a) constant pressure and (b) constant volume.

3-43. Five pounds of water initially at 220°F with a quality of 50% are heated at constant pressure to a second state, and then cooled at constant volume to a third state. The third state is a saturated vapor at 120°F. Draw the processes on a p–v diagram and determine the total work done during the combined process.

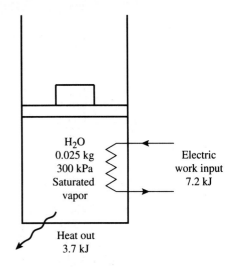

H₂O
0.025 kg
300 kPa
Saturated
vapor

Electric
work input
7.2 kJ

Heat out
3.7 kJ

Figure P3-41

3-44. A piston–cylinder device contains steam initially at 3 MPa and 400°C with a total volume of 2.0 m^3 (state 1). The steam is cooled at constant volume to 200°C (state 2). This is followed by a constant-temperature heat rejection process to bring the system to the saturated liquid state (state 3). Sketch the processes on a p–v diagram and determine the total heat rejection.

3-45. A rigid tank contains Refrigerant-12 at 700 kPa and 40°C. Heat is transferred to the surroundings until the temperature drops to 20°C. Determine the initial and final values of specific volume and internal energy, per kg of the fluid. Also determine the heat transfer, in kJ/kg.

3-46. Refrigerant-12 contained in a piston–cylinder device initially has a volume of 0.1 m^3 at 0.25 MPa and 40°C (state 1). See Fig. P3-46. Heat is slowly added to the fluid in a constant-volume process until the pressure reaches 0.30 MPa (state 2). Heat is then slowly removed from the fluid in a constant-pressure process until the temperature reaches 50°C (state 3). Determine the amounts of heat transfer in the two processes.

3-47. A 0.014-m^3 tea kettle initially contains 2.3 kg of liquid–vapor water mixture at the atmospheric pressure of 100 kPa. See Fig. P3-47. The "pop-off" valve of the kettle keeps the water vapor in the kettle until its pressure reaches 35 kPa gage. At this internal pressure, the valve opens and allows the vapor to escape into the atmosphere so as to maintain the internal pressure constant. A young housewife puts this kettle on a stove and turns on the fire, hoping to humidify the room air.

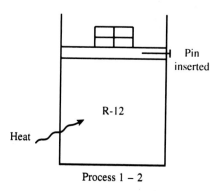

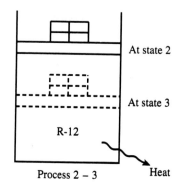

Process 1 – 2   Process 2 – 3

Figure P3-46

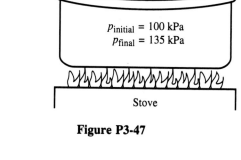

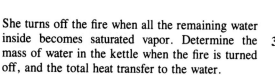

Figure P3-47

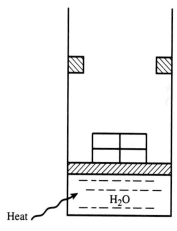

Figure P3-48

She turns off the fire when all the remaining water inside becomes saturated vapor. Determine the mass of water in the kettle when the fire is turned off, and the total heat transfer to the water.

3-48. A cylindrical piston device contains a 4 pound mass of water initially at $p_1 = 100$ psia and $T_1 = 60°F$. Heat is added to the water, making it go through a constant-pressure process until $V_2 = 12.88$ ft^3, when the piston is stopped by the stops, as shown in Fig. P3-48. Additional heat is added to the steam until it becomes a saturated vapor (state 3). Determine the heat and work interactions for processes 1–2 and 2–3.

3-49. A cylinder–piston arrangement contains 0.1 lbm of Refrigerant-12 initially at 100 psia, 180°F. The refrigerant is first expanded adiabatically to 30 psia, 120°F, and then compressed under constant pressure to half its initial volume. Determine the work and the heat quantities of the two processes and the final temperature of the refrigerant.

3-50. Dry saturated steam at 690 kPa is heated in a closed system at constant volume (process 1–2) until its pressure is 1200 kPa. It is then expanded adiabatically to 690 kPa, 370°C (process 2–3). Finally, it is cooled at constant pressure to the saturation temperature. All processes are quasistatic. Draw the p–v diagram for the processes and calculate the net work done, per kg of steam.

3-51. A mixture of H$_2$O at the triple-point condition (0.01°C) containing 20% solid, 30% liquid, and 50% vapor by mass enters a tube at the rate of 5 kg/s. Heat is added to the mixture at constant pressure and the H$_2$O emerges out from the tube as all saturated vapor. Calculate the rate of heat supplied.

For saturated solid ice, the enthalpy $h_i = -333.4$ kJ/kg.

3-52. A food processor contains 2.08 lbm of water initially at 60°F, 14.7 psia. See Fig. P3-52. The mixing blade of the processor is driven by a 0.25-hp electric motor. Determine the water temperature when the motor has been turned on for 10 min, using (a) data taken from a steam table, and (b) constant specific heat $c = 1.0$ Btu/lbm · °F for incompressible liquid water. Assume that the machine is insulated and that the water pressure is kept constant.

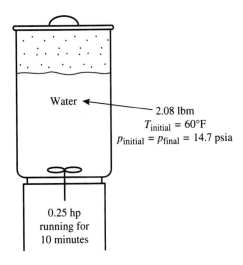

Water ← 2.08 lbm
$T_{\text{initial}} = 60°F$
$p_{\text{initial}} = p_{\text{final}} = 14.7$ psia

0.25 hp
running for
10 minutes

Figure P3-52

3-53. A 0.05-m³ insulated tank contains water at 20°C and 101 kPa. A 4-kW resistance heater is placed in the water and is allowed to operate for 20 min. The heater element has a mass of 2 kg and a specific heat of 0.75 kJ/kg·°C. Determine the final temperature of the water and the heater element, ne-

glecting the heat capacity of the tank material itself.

3-54. Four-kg of liquid water at 50°C is poured into a 13-kg copper tank that is initially at 27°C. The tank is perfectly insulated. What is the equilibrium temperature of the water-and-tank system?

3-55. Copper changes from an initial state of 100 kPa, 20°C, to a final state of 3000 kPa, 200°C. Determine the changes in internal energy and enthalpy per kilogram of copper, assuming a constant density of 8930 kg/m³ for copper.

3-56. A block of iron at 85°C is dropped into an insulated tank containing 100 kg of water at 20°C. See Fig. P3-56. A paddle wheel driven by a 200-W motor is activated to stir the water for 20 min, resulting in a final equilibrium temperature of 24°C. Neglecting any energy stored in the paddle wheel, determine the mass of the iron block.

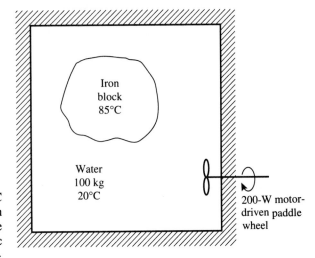

Iron
block
85°C

Water
100 kg
20°C

200-W motor-
driven paddle
wheel

Figure P3-56

Thermodynamic Properties of Pure Substances   Chap. 3

4

Ideal Gases

When a homogeneous substance is in an equilibrium state, its thermodynamic properties must have unique values and are interrelated in a certain pattern. One class of substances that can be modeled by a simple property relationship is the so-called ideal gas. In this chapter, we study the properties of ideal gases.

4-1 IDEAL-GAS EQUATION OF STATE

An equation that expresses the relationship among properties of a substance is called an *equation of state*. For a simple compressible substance, the absolute pressure p, specific volume v, and absolute temperature T are the state parameters that are linked by a functional relationship in the form

$$f(p, v, T) = 0$$

On the basis of experimental observations, it has been established that the p–v–T behavior of gases at low densities can be approximated by the following equation, known as the *ideal-gas equation of state:*

$$pv = \mathcal{R}T \tag{4-1}$$

where $\mathcal{R}$ is the *universal gas constant*, which is the same for all gases. For Eq. (4-1) in SI units, p is in N/m^2 (Pa), v in $m^3/kgmole$, T in K, and $\mathcal{R} = 8314$ J/kgmole·K. In English units, p is in lbf/ft² abs (psfa), v in ft³/lbmole, T in °R, and $\mathcal{R} = 1545$ ft·lbf/lbmole·°R. Values of the universal gas constant for various different units are given in Table A-4.

Dividing Eq. (4-1) by the molecular mass M, we have the ideal-gas equation of state on a unit-mass basis:

$$pv = RT \tag{4-2}$$

where R is the *gas constant* for a particular gas as defined by

$$R = \frac{\mathcal{R}}{M} \tag{4-3}$$

Because the gas constant R depends upon the molecular mass of a substance, its value is different for different substances. In Eq. (4-2), the units are p in N/m^2 (Pa), v in m^3/kg, T in K, and R in J/kg·K in the SI system; and p in lbf/ft^2 abs (psfa), v in ft^3/lbm, T in °R, and R in ft·lbf/lbm·°R in the English system.

For a gas of n moles having a total volume of V, the ideal-gas equation of state is written as

$$pV = n\mathscr{R}T \qquad (4\text{-}4)$$

Whereas for a gas of mass m and total volume V, the ideal-gas equation of state is written as

$$pV = mRT \qquad (4\text{-}5)$$

An *ideal gas* is a gas that obeys the ideal-gas equation of state. The ideal-gas equation holds exactly for a real gas only as the pressure approaches zero. It holds approximately for a real gas at higher pressures when the gas is at relatively high temperatures. The deviation of a real gas from the ideal-gas behavior at a given pressure and temperature can be accounted for by the introduction of a correction factor, called the *compressibility factor Z*:

$$Z = \frac{pv}{RT}$$

or

$$pv = ZRT \qquad (4\text{-}6)$$

where Z is, in general, a function of the pressure and temperature of the gas. For an ideal gas, $Z = 1$ for all pressures and temperatures. Thus, the value of Z expresses the extent of deviation of the gas from ideal-gas behavior.

Figure 4-1 shows the compressibility diagram for nitrogen, illustrating how the compressibility factor varies with pressure along different constant-temperature lines. Notice that Z approaches unity as p approaches zero at any temperature. At higher pressures, the values of Z are different at different temperatures. At room temperature of 300 K (and above), the compressibility factor is near unity up to about 10 MPa. This means that nitrogen can be considered as an ideal gas over these ranges of pressure and temperature.

The variation of Z with respect to pressure and temperature for other gases follows a pattern similar to the one shown in Fig. 4-1 for nitrogen. They differ, however, in quantitative details, as demonstrated in Fig. 4-2 for the case of carbon dioxide as compared to nitrogen at room temperature of 300 K. Carbon dioxide behaves as an ideal gas at room temperatures only when the pressure is very low, say, 0.1 MPa.

In common engineering applications, diatomic and monatomic gases, such as nitrogen, oxygen (also air), hydrogen, carbon monoxide, helium, and argon, can be treated as ideal gases at room temperature (and above) up to 1 to 2 MPa. Use of the ideal-gas equation for heavier gases such as carbon dioxide will result in a higher degree of inaccuracy. Never consider dense gases such as water vapor in steam power plants and refrigerant vapor in refrigerators as ideal gases.

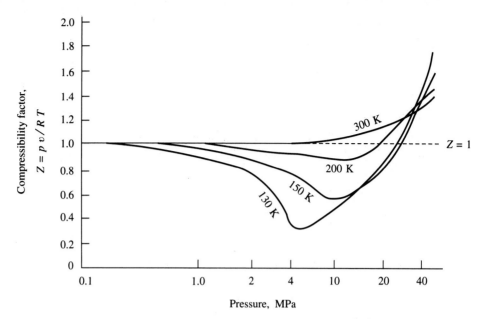

Figure 4-1 Compressibility diagram for nitrogen.

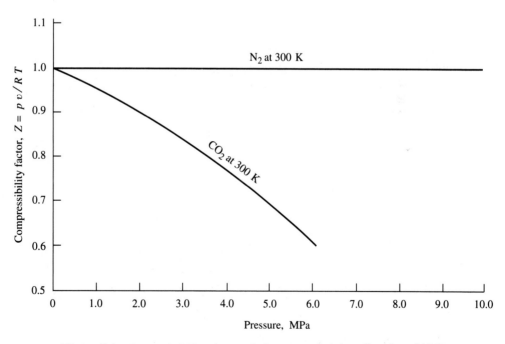

Figure 4-2 Compressibility factor of nitrogen and carbon dioxide at 300 K.

Example 4-1

A rigid tank contains 50 lbm of an ideal gas with a molecular mass of 28 at 80 psia and 100°F. A leak occurs in the tank that is not detected until the pressure has dropped to 50 psia and the temperature has dropped to 70°F. Determine the mass of gas remaining in the tank and the mass of gas that has leaked out.

Solution The given data are $m_1 = 50$ lbm, $p_1 = 80$ psia, $T_1 = 100°F = 100 + 460 = 560°R$, $V_2 = V_1 = V$, $p_2 = 50$ psia, $T_2 = 70°F = 70 + 460 = 530°R$, and $M = 28$ lbm/lbmole.

The gas constant is

$$R = \frac{\mathcal{R}}{M} = \frac{1545 \text{ ft·lbf/lbmole·°R}}{28 \text{ lbm/lbmole}} = 55.2 \text{ ft·lbf/lbm·°R}$$

For the initial conditions, the ideal-gas equation gives

$$V = \frac{m_1 R T_1}{p_1} = \frac{(50 \text{ lbm})(55.2 \text{ ft·lbf/lbm·°R})(560°R)}{80 \times 144 \text{ lbf/ft}^2 \text{ abs}}$$

$$= 134.2 \text{ ft}^3$$

For the final conditions, the ideal-gas equation gives

$$m_2 = \frac{p_2 V}{R T_2} = \frac{(50 \times 144 \text{ lbf/ft}^2 \text{ abs})(134.2 \text{ ft}^3)}{(55.2 \text{ ft·lbf/lbm·°R})(530°R)}$$

$$= 33.0 \text{ lbm}$$

which is the mass of gas remaining in the tank.

The mass of gas that has leaked out is given by

$$m_1 - m_2 = 50 - 33 = 17 \text{ lbm}$$

4-2 INTERNAL ENERGY, ENTHALPY, AND SPECIFIC HEATS OF IDEAL GASES

Internal energy of a simple compressible substance is, in general, a function of two independent intensive properties, say, $u = u(T, v)$. It was found, however, by both microscopic reasoning and macroscopic measurements that the internal energy of a gas is not a function of its specific volume as long as the gas is at a low density so that the ideal-gas equation of state is followed. This important fact is commonly called Joule's law and can be written mathematically as $(\partial u / \partial v)_T = 0$ for an ideal gas. It implies that the internal energy of an ideal gas is a function of temperature only. That is, for an ideal gas,

$$u = u(T) \tag{4-7}$$

This conclusion will be demonstrated mathematically in Sec. 13-4. For now, let us see how Joule conducted his original experiment in 1843 to establish this conclusion.

In his classical experiment, Joule submerged two tanks connected by a pipe and a valve in a water bath, as shown schematically in Fig. 4-3. Initially, tank A contained air at a high pressure (22 atm) and tank B was evacuated. When thermal equilibrium was achieved between the tanks and the water bath, the valve was opened to allow the pressures in the two tanks to equalize. During the expansion process of the air, no temperature change in the water bath was detected. That led the observer to make the as-

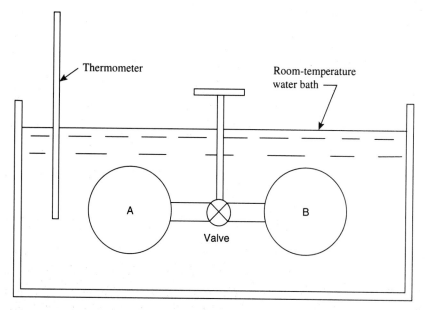

Figure 4-3 Schematic of the apparatus for Joule's experiment regarding internal energy of an ideal gas.

sumption that no heat was transferred to or from the air. As there was also no work done, from the first law of thermodynamics, it was concluded that there was no change in the internal energy of the gas even though its specific volume and pressure changed. Therefore, it was asserted that the internal energy of an ideal gas is a function of temperature only and not a function of pressure or specific volume, thus verifying the truth of Eq. (4-7). It should be emphasized that this equation is for ideal gases only. Because air at high pressures does not behave as an ideal gas, temperature changes were detected in Joule's experiments when very accurate measurements were made.

Because for an ideal gas,

$$h = u + pv = u + RT \tag{4-8}$$

and u is a function of temperature only, it follows that the enthalpy h of an ideal gas is also a function of temperature only. That is, for an ideal gas,

$$h = h(T) \tag{4-9}$$

According to the definitions for constant-volume and constant-pressure specific heats,

$$c_v = \left(\frac{\partial u}{\partial T}\right)_v \qquad \text{and} \qquad c_p = \left(\frac{\partial h}{\partial T}\right)_p$$

we see that c_v and c_p for an ideal gas are also functions of temperature only. Thus, for an ideal gas, the preceding equations for c_v and c_p can be expressed in terms of ordinary derivatives,

$$c_v = \frac{du}{dT} \qquad \text{and} \qquad c_p = \frac{dh}{dT}$$

We accordingly write for any process of an ideal gas,

$$du = c_v\, dT \tag{4-10}$$

$$dh = c_p\, dT \tag{4-11}$$

Carefully notice that the preceding two equations apply to all processes, not just for constant-volume and constant-pressure processes.

Differentiating Eq. (4-8), we have

$$dh = du + R\, dT$$

Substituting Eqs. (4-10) and (4-11) into the preceding equation yields

$$c_p\, dT = c_v\, dT + R\, dT$$

or

$$c_p - c_v = R \tag{4-12}$$

This equation allows the calculation of either c_p or c_v when the other one is known. When the specific heats are given on a mole basis, the universal gas constant $\mathcal{R}$ is used to replace the individual gas constant R in Eq. (4-12).

The ratio of c_p and c_v often appears in thermodynamic analyses and calculations. This ratio is given the symbol k, or

$$k = \frac{c_p}{c_v} \tag{4-13}$$

From Eq. (4-12), we must conclude that $c_p > c_v$ and $k > 1$.

4-3 VARIATION OF SPECIFIC HEATS WITH TEMPERATURE

Figure 2-8 in Chapter 2 illustrates the temperature effect upon the zero-pressure or ideal-gas specific heats of a few common gases. It is seen from this figure that a monatomic gas (such as argon) shows little or no variation of specific heat over a wide range of temperatures. As the number of atoms in the molecules of a gas increases, the effect of temperature on specific heat becomes more prominent. We can gain some insight into this behavior from a microscopic consideration (though in a simplified fashion) on the various modes of molecular motion—translational, rotational, and vibrational. A monatomic gas possesses only a translational energy mode, which is proportional to its temperature. As a result, the specific heat of a monatomic gas is not temperature-dependent. On the other hand, when heat flows into a polyatomic gas, only a portion of it is available for increasing the average molecular translational energy, the remainder going into an increase in molecular rotational and vibrational energies. Because temperature is proportional to the mean translational energy, the rise in temperature for a given heat inflow is smaller in a polyatomic than in a monatomic gas, resulting in a larger specific heat in a polyatomic gas. Because translational and rotational energies increase linearly with temperature, the increase in specific heat with temperature in a polyatomic gas is due primarily to the increase in vibrational energy. Diatomic gases (such N_2 and O_2) have less vibrational modes as compared to triatomic gases (such as CO_2 and H_2O). Thus, the increase in specific heat with the increase in temperature is less prominent in diatomic gases as compared to triatomic gases.

Values of the constant-pressure specific heat c_p and constant-volume specific heat c_v, together with their ratio $k = c_p/c_v$, as a function of temperature for a few common gases at very low pressures ($p \rightarrow 0$) are given in Appendix Table A-6. Empirical equations for c_p as a function of temperature at zero pressure for various gases are given in Appendix Table A-7. Empirical equations for c_v can be obtained by applying the relation $c_p - c_v = R$.

With the temperature functions of specific heats established, one can use Eqs. (4-10) and (4-11) to evaluate the changes in internal energy and enthalpy of an ideal gas between two equilibrium states. Thus,

$$\Delta u = u_2 - u_1 = \int_1^2 c_v \, dT \tag{4-14}$$

$$\Delta h = h_2 - h_1 = \int_1^2 c_p \, dT \tag{4-15}$$

Direct integration of these equations is possible if analytical equations of specific heats as functions of temperature are available. If the specific heats in the range of temperature considered are constants, or average values of specific heats can be used, we then have

$$\Delta u = u_2 - u_1 = c_v(T_2 - T_1) \tag{4-16}$$

$$\Delta h = h_2 - h_1 = c_p(T_2 - T_1) \tag{4-17}$$

By assigning a zero internal energy and enthalpy at a reference temperature T_0, the integration of Eqs. (4-10) and (4-11) gives

$$\int_0^u du = u = \int_{T_0}^T c_v \, dT$$

and

$$\int_0^h dh = h = \int_{T_0}^T c_p \, dT$$

Accordingly, by the use of accurate specific-heat data to perform the previous integrations, a table for the values of internal energy and enthalpy with temperature as the single entry can be established for any gas having ideal-gas behavior. Appendix Table A-9 is such a table for air, and Appendix Tables A-10 to A-16 are similar tables for N_2, O_2, H_2O, CO_2, H_2, and CO. All tables are based on $T_0 = 0$ K $= 0°R$.

Example 4-2

The temperature of 1 kilogram of oxygen is increased from 400 K to 800 K at a low pressure. Determine the changes in enthalpy and internal energy by the use of (a) data from the ideal-gas property table for oxygen, (b) arithmetic-mean specific heats for oxygen between the given temperature limits, and (c) empirical specific-heat equation for oxygen as given by

$$c_p = \mathscr{R}(a + bT + cT^2 + dT^3 + eT^4)$$

where c_p is in kJ/kgmole·K, $\mathscr{R} = 8.314$ kJ/kgmole·K, $a = 3.626$, $b = -1.878 \times 10^{-3}$, $c = 7.056 \times 10^{-6}$, $d = -6.764 \times 10^{-9}$, and $e = 2.156 \times 10^{-12}$.

Solution

$$T_1 = 400 \text{ K}, \quad \text{and} \quad T_2 = 800 \text{ K}, \quad \text{and} \quad M = 32.00 \text{ kg/kgmole}$$

(a) From the ideal-gas property table for oxygen (Table A-11)*, we obtain

$$h_1 = 11,709 \text{ kJ/kgmole}, \quad u_1 = 8,383 \text{ kJ/kgmole} \quad \text{at } T_1 = 400 \text{ K}$$

$$h_2 = 24,519 \text{ kJ/kgmole}, \quad u_2 = 17,867 \text{ kJ/kgmole} \quad \text{at } T_2 = 800 \text{ K}$$

Thus,

$$\Delta h = h_2 - h_1 = 24,519 - 11,709$$

$$= 12,810 \text{ kJ/kgmole} = \frac{12,810}{32} = 400.3 \text{ kJ/kg}$$

$$\Delta u = u_2 - u_1 = 17,867 - 8,383$$

$$= 9,484 \text{ kJ/kgmole} = \frac{9,484}{32} = 296.4 \text{ kJ/kg}$$

(b) From Table A-6M for oxygen between 400 K and 800 K, the arithmetic-mean specific heats are

$$c_p = \tfrac{1}{2}(0.941 + 1.054) = 0.998 \text{ kJ/kg·K}$$

$$c_v = \tfrac{1}{2}(0.681 + 0.794) = 0.738 \text{ kJ/kg·K}$$

Thus,

$$\Delta h = c_p(T_2 - T_1) = 0.998(800 - 400) = 399.2 \text{ kJ/kg}$$

$$\Delta u = c_v(T_2 - T_1) = 0.738(800 - 400) = 295.2 \text{ kJ/kg}$$

(c) From the given c_p equation, we have

$$\Delta h = \int_{T_1}^{T_2} c_p \, dT = \mathcal{R} \int_{T_1}^{T_2} (a + bT + cT^2 + dT^3 + eT^4) \, dT$$

$$= \mathcal{R}\left[a(T_2 - T_1) + \frac{b}{2}(T_2^2 - T_1^2) + \frac{c}{3}(T_2^3 - T_1^3) + \frac{d}{4}(T_2^4 - T_1^4) + \frac{e}{5}(T_2^5 - T_1^5) \right]$$

Substituting numerical values, we obtain

$$\Delta h = 12,811 \text{ kJ/kgmole} = \frac{12,811}{32} = 400.3 \text{ kJ/kg}$$

Now

$$c_v = c_p - \mathcal{R} = \mathcal{R}[(a - 1) + bT + cT^2 + dT^3 + eT^4]$$

so that

$$\Delta u = \int_{T_1}^{T_2} c_v \, dT$$

$$= \mathcal{R}\left[(a - 1)(T_2 - T_1) + \frac{b}{2}(T_2^2 - T_1^2) \right.$$

$$\left. + \frac{c}{3}(T_2^3 - T_1^3) + \frac{d}{4}(T_2^4 - T_1^4) + \frac{e}{5}(T_2^5 - T_1^5) \right]$$

$$= 9,486 \text{ kJ/kgmole} = \frac{9,486}{32} = 296.4 \text{ kJ/kg}$$

*In this book, all table values of ideal-gas properties for air, N_2, O_2, etc., are based on the original complete tables of Ref. [20] in the Bibliography.

Notice that the value of Δu can be obtained simply by the relation

$$\Delta u = \Delta h - \mathcal{R}(\Delta T) = 12,811 - 8.314(800 - 400)$$
$$= 9,485 \text{ kJ/kgmole} = 296.4 \text{ kJ/kg}$$

4-4 POLYTROPIC PROCESSES FOR IDEAL GASES

In Section 2-3, we introduced a common thermodynamic process, known as the quasiequilibrium polytropic process, which has the following $p-v$ relation:

$$pv^n = \text{constant}$$

If the working substance is an ideal gas, we have the additional relation

$$pv = RT$$

Eliminating pressure p between these two equations gives

$$\left(\frac{RT}{v}\right)v^n = \text{constant}$$

or

$$Tv^{n-1} = \text{constant} \tag{4-18}$$

On the other hand, eliminating volume v between the same two equations gives

$$p^{1/n}\left(\frac{RT}{p}\right) = \text{constant}$$

or

$$\frac{T}{p^{(n-1)/n}} = \text{constant} \tag{4-19}$$

Writing Eqs. (4-18) and (4-19) between the end states 1 and 2 of a polytropic process for an ideal gas yields

$$\frac{T_2}{T_1} = \left(\frac{p_2}{p_1}\right)^{(n-1)/n} = \left(\frac{v_1}{v_2}\right)^{n-1} \tag{4-20}$$

When the polytropic constant n takes on certain particular values, several important familiar thermodynamic processes can be identified. The value of n for these familiar processes for ideal gases are summarized in what follows.

Constant-pressure (isobaric) process: $n = 0$
Constant-temperature (isothermal) process: $n = 1$
A process with $n = k = c_p/c_v$ (when k is a constant and $1 < k < \infty$)
Constant-volume (isovolumic) process: $n = \infty$

When $n = 0$, $pv^0 = p = \text{constant}$, which represents a constant-pressure process, and when $n = \infty$, $p^{1/n}v = p^0 v = v = \text{constant}$, which represents a constant-volume process. When $n = 1$ for an ideal gas, $pv = RT = \text{constant}$, which represents a constant-temperature process. In Chapter 6, we will define a new property called entropy s, and

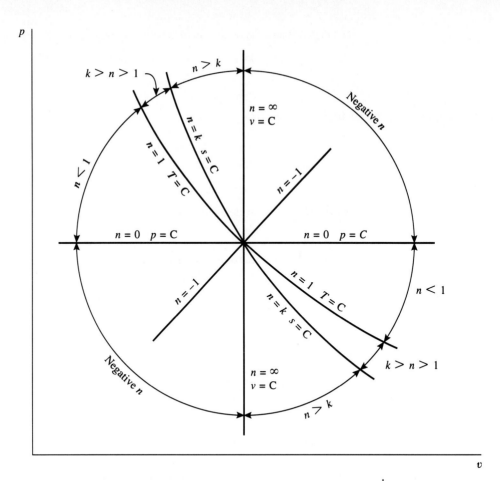

Figure 4-4 Polytropic processes on a p–v diagram for ideal gases.

in Chapter 7, we will establish that when $n = k = c_p/c_v =$ constant, the process $pv^k =$ constant represents a constant s process for an ideal gas with constant specific heats. These processes are illustrated in Fig. 4-4 on a p–v diagram. A process with $n = -1$, such that $pv^{-1} = p/v =$ constant is also plotted in Fig. 4-4. In addition, different regions of positive and negative values of n are also indicated in Fig. 4-4 for reference.

Example 4-3

A cylinder with a movable piston contains an ideal gas at an initial pressure of 300 kPa and an initial temperature of 373 K. The gas expands in a quasiequilibrium process until its volume is doubled. Determine the final pressure and temperature of the gas if the process follows the polytropic relation for (a) $p =$ constant, (b) $pV =$ constant, (c) $pV^2 =$ constant, and (d) $p/V =$ constant.

Solution The given data are $p_1 = 300$ kPa, $T_1 = 373$ K, $V_2/V_1 = 2$. All calculations of this problem are based on the quasiequilibrium polytropic relation

$$\frac{T_2}{T_1} = \left(\frac{p_2}{p_1}\right)^{(n-1)/n} = \left(\frac{V_1}{V_2}\right)^{n-1}$$

(a) $p = $ constant, $n = 0$.

$$p_2 = p_1 = 300 \text{ kPa}$$

$$T_2 = T_1\left(\frac{V_1}{V_2}\right)^{n-1} = 373(\tfrac{1}{2})^{-1} = 746 \text{ K}$$

(b) $pV = $ constant $= RT$, $n = 1$.

$$p_2 = p_1\left(\frac{V_1}{V_2}\right)^{n} = 300(\tfrac{1}{2})^{1} = 150 \text{ kPa}$$

$$T_2 = T_1 = 373 \text{ K}$$

(c) $pV^2 = $ constant, $n = 2$.

$$p_2 = p_1\left(\frac{V_1}{V_2}\right)^{n} = 300(\tfrac{1}{2})^{2} = 75 \text{ kPa}$$

$$T_2 = T_1\left(\frac{V_1}{V_2}\right)^{n-1} = 373(\tfrac{1}{2})^{2-1} = 186.5 \text{ K}$$

(d) $p/V = pV^{-1} = $ constant, $n = -1$.

$$p_2 = p_1\left(\frac{V_1}{V_2}\right)^{n} = 300(\tfrac{1}{2})^{-1} = 600 \text{ kPa}$$

$$T_2 = T_1\left(\frac{V_1}{V_2}\right)^{n-1} = 373(\tfrac{1}{2})^{-1-1} = 1492 \text{ K}$$

The processes are shown on the p–v diagram of Fig. 4-5.

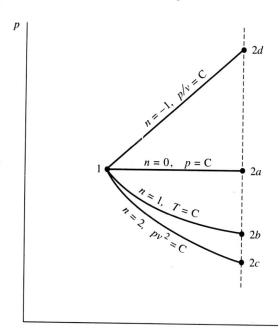

v **Figure 4-5** p–v diagram for Example 4-3.

4-5 ENERGY ANALYSIS INVOLVING IDEAL GASES

In the previous sections of this chapter, we have introduced some very important equations for ideal gases. We now illustrate the use of these equations in energy analyses for closed systems through the presentation of several numerical examples.

Example 4-4

Five pounds of a certain ideal gas with $k = 1.30$ and $R = 35$ ft·lbf/1bm·°R is compressed in a quasiequilibrium polytropic process in a closed system from 15 psia and 80°F to 60 psia and 315°F. Determine the polytropic constant n and the work and heat interactions with the surroundings.

Solution

$$m = 5 \text{ lbm}, \; p_1 = 15 \text{ psia}, \; T_1 = 80°F = 80 + 460 = 540°R,$$

$$p_2 = 60 \text{ psia}, \; T_2 = 315°F = 315 + 460 = 775°R$$

For the polytropic process, we have

$$\frac{T_2}{T_1} = \left(\frac{p_2}{p_1}\right)^{(n-1)/n}$$

It follows that

$$\ln\left(\frac{T_2}{T_1}\right) = \frac{n-1}{n}\ln\left(\frac{p_2}{p_1}\right)$$

or

$$\frac{n-1}{n} = \frac{\ln(T_2/T_1)}{\ln(p_2/p_1)} = \frac{\ln(775/540)}{\ln(60/15)} = 0.2606$$

Therefore,

$$n = \frac{1}{1 - 0.2606} = 1.35$$

The work interaction is calculated by the equation

$$W = \int_1^2 p \, dV = \frac{p_2 V_2 - p_1 V_1}{1 - n} = \frac{mR(T_2 - T_1)}{1 - n}$$

$$= \frac{(5 \text{ lbm})(35 \text{ ft·lbf/lbm·°R})(775°R - 540°R)}{1 - 1.35}$$

$$= -117,500 \text{ ft·lbf} = -\frac{117,500}{778} = -151.0 \text{ Btu}$$

where the negative sign indicates that work is done on the gas.

Because $k = c_p/c_v$ and $c_p - c_v = R$, we have $c_v + R = kc_v$, or

$$c_v = \frac{R}{k-1} = \frac{35/778}{1.30 - 1} = 0.150 \text{ Btu/lbm·°F}$$

Accordingly,

$$\Delta U = mc_v(T_2 - T_1)$$

$$= (5 \text{ lbm})(0.150 \text{ Btu/lbm·°F})(315°F - 80°F)$$

$$= 176.3 \text{ Btu}$$

and

$$Q = \Delta U + W = 176.3 - 151.0 = 25.3 \text{ Btu}$$

where the positive value indicates that heat is added to the gas.

Example 4-5

A rigid tank with a volume of 0.035 m³ contains oxygen initially at 350 K and 120 kPa. During a process, paddle-wheel work on the gas amounts to 1.5 kJ and there is a heat loss of 3.5 kJ to the surroundings. Taking data from the oxygen property table, determine the final temperature and pressure of the gas.

Solution Figure 4-6 shows a schematic and the p–v diagram for the nonquasiequilibrium process of this example. The given data are $V = 0.035$ m³ = constant, $T_1 = 350$ K, $p_1 = 120$ kPa, $W = -1.5$ kJ, and $Q = -3.5$ kJ.

The gas constant of oxygen is calculated by

$$R = \frac{\mathcal{R}}{M} = \frac{8.314 \text{ kJ/kgmole·K}}{32 \text{ kg/kgmole}} = 0.2598 \text{ kJ/kg·K}$$

The mass of oxygen in the tank is

$$m = \frac{p_1 V_1}{R T_1} = \frac{(120 \text{ kN/m}^2)(0.035 \text{ m}^3)}{(0.2598 \text{ kJ/kg·K})(350 \text{ K})} = 0.04619 \text{ kg}$$

The number of moles of oxygen in the tank is

$$n = \frac{m}{M} = \frac{0.04619 \text{ kg}}{32 \text{ kg/kgmole}} = 0.001443 \text{ kgmole}$$

The number of moles of oxygen can also be calculated by

$$n = \frac{p_1 V_1}{\mathcal{R} T_1} = \frac{(120 \text{ kN/m}^2)(0.035 \text{ m}^3)}{(8.314 \text{ kJ/kgmole·K})(350 \text{ K})} = 0.001443 \text{ kgmole}$$

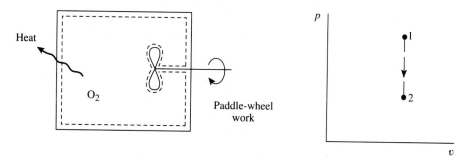

Figure 4-6 Schematic and p–v diagrams for Example 4-5.

From the ideal-gas property table for oxygen (Table A-11M), at $T_1 = 350$ K, we get $u_1 = 7304.1$ kJ/kgmole. The first law reads

$$Q = \Delta U + W$$

or

$$\Delta U = n(u_2 - u_1) = Q - W$$

By substituting numerical values,

$$(0.001443 \text{ kgmole})(u_2 \text{ kJ/kgmole} - 7304.1 \text{ kJ/kgmole}) = -3.5 \text{ kJ} - (-1.5 \text{ kJ})$$

so that

$$u_2 = 5918.1 \text{ kJ/kgmole}$$

From Table A-11 M, at $u_2 = 5918.1$ kJ/kgmole, we get $T_2 = 284.5$ K.

The final pressure p_2 of oxygen can be calculated by either of the following two equations:

$$p_2 = \frac{n\mathscr{R}T_2}{V_2}$$

$$= \frac{(0.001443 \text{ kgmole})(8.314 \text{ kJ/kgmole·K})(284.5 \text{ K})}{0.035 \text{ m}^3}$$

$$= 97.5 \text{ kPa}$$

or

$$p_2 = \frac{mRT_2}{V_2}$$

$$= \frac{(0.04619 \text{ kg})(0.2598 \text{ kJ/kg·K})(284.5 \text{ K})}{0.035 \text{ m}^3} = 97.5 \text{ kPa}$$

Example 4-6

An ideal gas with $c_p = 1.044$ kJ/kg·K and $c_v = 0.745$ kJ/kg·K contained in a cylinder–piston assembly initially has a pressure of 150 kPa, a temperature of 30°C, and a volume of 0.22 m³. It is heated slowly at constant volume (process 1–2) until the pressure is doubled. See Fig. 4-7(a). It is then expanded slowly at constant pressure (process 2–3) until the volume is doubled. Determine the work done and heat added in the combined process.

Solution The two processes are depicted in Fig. 4-7 in a schematic and on p–V coordinates. The given data are $p_1 = 150$ kPa, $T_1 = 30°C = 30 + 273 = 303$ K, $V_1 = 0.22$ m³, $V_2 = V_1$, $p_2 = 2p_1$, $p_2 = p_3$, and $V_3 = 2V_2$. The gas constant R is calculated as

$$R = c_p - c_v = 1.044 - 0.745 = 0.299 \text{ kJ/kg·K} = 299 \text{ J/kg·K}$$

The mass of the gas is

$$m = \frac{p_1 V_1}{RT_1} = \frac{(150 \times 10^3 \text{ N/m}^2)(0.22 \text{ m}^3)}{(299 \text{ N·m/kg·K})(303 \text{ K})} = 0.364 \text{ kg}$$

For the constant-volume process 1–2, we obtain

$$T_2 = T_1\left(\frac{p_2}{p_1}\right) = 303(2) = 606 \text{ K}$$

For the constant-pressure process 2–3, we obtain

$$T_3 = T_2\left(\frac{V_3}{V_2}\right) = 606(2) = 1212 \text{ K}$$

For process 1–2,

$$W_{12} = \int_1^2 p \, dV = 0$$

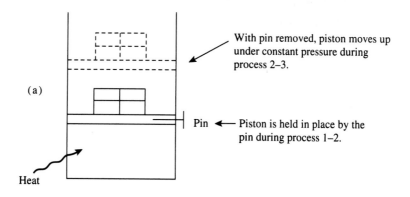

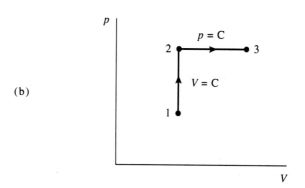

(a)

(b)

With pin removed, piston moves up under constant pressure during process 2–3.

Pin ← Piston is held in place by the pin during process 1–2.

Heat

p

$p = C$

2   3

$V = C$

1

V

Figure 4-7 Schematic and p–v diagrams for Example 4-6.

and

$$Q_{12} = \Delta U_{12} + W_{12} = \Delta U_{12} = mc_v(T_2 - T_1)$$
$$= (0.364 \text{ kg})(0.745 \text{ kJ/kg·K})(606 \text{ K} - 303 \text{ K}) = 82.2 \text{ kJ}$$

For process 2–3,

$$W_{23} = \int_2^3 p \, dV = p_2(V_3 - V_2) = p_3 V_3 - p_2 V_2 = mR(T_3 - T_2)$$
$$= (0.364 \text{ kg})(0.299 \text{ kJ/kg·K})(1212 \text{ K} - 606 \text{ K})$$
$$= 66.0 \text{ kJ}$$

or

$$W_{23} = p_2(V_3 - V_2) = (2p_1)(2V_2 - V_2)$$
$$= (2p_1)(V_2) = (2p_1)(V_1)$$
$$= 2(150 \text{ kN/m}^2)(0.22 \text{ m}^3) = 66.0 \text{ kJ}$$
$$Q_{23} = \Delta U_{23} + W_{23} = mc_v(T_3 - T_2) + W_{23}$$
$$= (0.364 \text{ kg})(0.745 \text{ kJ/kg·K})(1212 \text{ K} - 606 \text{ K}) + (66.0 \text{ kJ})$$
$$= 230.3 \text{ kJ}$$

Because process 2–3 is a constant-pressure process, Q_{23} can also be calculated by the following:

$$Q_{23} = \Delta U_{23} + W_{23} = (U_3 - U_2) + (p_3 V_3 - p_2 V_2)$$

$$= H_3 - H_2 = m(h_3 - h_2) = mc_p(T_3 - T_2)$$

$$= (0.364 \text{ kg})(1.044 \text{ kJ/kg·K})(1212 \text{ K} - 606 \text{ K}) = 230.3 \text{ kJ}$$

The total work done and heat added in the combined process are

$$W_{123} = W_{12} + W_{23} = 0 + 66.0 = 66.0 \text{ kJ}$$

$$Q_{123} = Q_{12} + Q_{23} = 82.2 + 230.3 = 312.5 \text{ kJ}$$

Example 4-7

An ideal gas with $c_p = 0.220$ Btu/lbm·°R and $R = 48.3$ ft·lbf/lbm·°R undergoes a quasiequilibrium three-process cycle in a closed system. From the initial state of $p_1 = 20$ psia and $T_1 = 110°F$, the gas is compressed isothermally to one-fifth of its initial volume (process 1–2); it is then heated at constant volume to state 3 (process 2–3); and, finally, it is expanded in a polytropic process with $n = 1.45$ back to its initial state (process 3–1). Determine (a) the pressures, temperatures, and specific volumes around the cycle, (b) the work and heat interactions for each process per lbm of the gas, and (c) the cycle thermal efficiency.

Solution The three-process cycle is depicted in Fig. 4-8 on p–v coordinates. The given data are $p_1 = 20$ psia. $T_1 = 110°F = 570°R$, $v_2 = (1/5)v_1$, $T_2 = T_1$, $v_3 = v_2$, $p_3 v_3^{1.45} = p_1 v_1^{1.45}$, $c_p = 0.220$ Btu/lbm·°R, and $R = 48.3$ ft·lbf/lbm·°R. Accordingly, we have

$$c_v = c_p - R = 0.220 - \frac{48.3}{778} = 0.158 \text{ Btu/lbm·°R}$$

(a) The values of p, T, and v for states 1, 2, and 3 are calculated as follows:

$$v_1 = \frac{RT_1}{p_1} = \frac{(48.3 \text{ ft·lbf/lbm·°R})(570°R)}{20 \times 144 \text{ lbf/ft}^2 \text{ abs}} = 9.559 \text{ ft}^3/\text{lbm}$$

$$v_2 = \frac{1}{5}v_1 = \frac{1}{5}(9.559) = 1.912 \text{ ft}^3/\text{lbm}$$

$$p_2 = p_1\left(\frac{v_1}{v_2}\right) = 20(5) = 100 \text{ psia}$$

$$p_3 = p_1\left(\frac{v_1}{v_3}\right)^n = p_1\left(\frac{v_1}{v_2}\right)^n = 20(5)^{1.45} = 206.3 \text{ psia}$$

$$T_3 = T_1\left(\frac{v_1}{v_3}\right)^{n-1} = T_1\left(\frac{v_1}{v_2}\right)^{n-1} = 570(5)^{1.45-1}$$

$$= 1176°R = 1176 - 460 = 716°F$$

$$v_3 = v_2 = 1.912 \text{ ft}^3/\text{lbm}$$

or

$$v_3 = \frac{RT_3}{p_3} = \frac{48.3(1176)}{206.3 \times 144} = 1.912 \text{ ft}^3/\text{lbm}$$

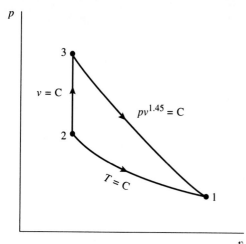

v **Figure 4-8** p–v diagram for Example 4-7.

(b) The work and heat interactions for the three processes are calculated as follows:
For process 1–2, $pv = RT = RT_1 = $ constant.

$$w_{12} = \int_1^2 p\,dv = RT_1 \int_1^2 \frac{dv}{v} = RT_1 \ln \frac{v_2}{v_1}$$

$$= (48.3 \text{ ft·lbf/lbm·°R})(570°R) \ln (1/5)$$

$$= -44{,}309 \text{ ft·lbf/lbm}$$

$$= -\frac{44309}{778} = -56.95 \text{ Btu/lbm}$$

$$q_{12} = \Delta u_{12} + w_{12} = w_{12} = -56.95 \text{ Btu/lbm}$$

For process 2–3, $v = $ constant.

$$w_{23} = \int_2^3 p\,dv = 0$$

$$q_{23} = \Delta u_{23} + w_{23} = \Delta u_{23} = c_v(T_3 - T_2)$$

$$= (0.158 \text{ Btu/lbm·°F})(716°F - 110°F) = 95.75 \text{ Btu/lbm}$$

For process 3–1, $pv^{1.45} = $ constant.

$$w_{31} = \int_3^1 p\,dv = \frac{p_1v_1 - p_3v_3}{1 - n} = \frac{R(T_1 - T_3)}{1 - n}$$

$$= \frac{(48.3 \text{ ft·lbf/lbm·°R})(570°R - 1176°R)}{1 - 1.45}$$

$$= 65{,}044 \text{ ft·lbf/lbm}$$

$$= \frac{65{,}044}{778} = 83.60 \text{ Btu/lbm}$$

$$q_{31} = \Delta u_{31} + w_{31} = c_v(T_1 - T_3) + w_{31}$$

$$= (0.158 \text{ Btu/lbm·°F})(110°F - 716°F) + 83.60 \text{ Btu/lbm}$$

$$= -12.15 \text{ Btu/lbm}$$

(c) The thermal efficiency is

$$\eta_{th} = \frac{\text{net } w_{out}}{q_{in}} = \frac{w_{12} + w_{23} + w_{31}}{q_{23}}$$

$$= \frac{-56.95 + 0 + 83.60}{95.75} = \frac{26.65}{95.75} = 0.278 = 27.8\%$$

or

$$\eta_{th} = \frac{\text{net } q_{in}}{q_{in}} = \frac{q_{12} + q_{23} + q_{31}}{q_{23}}$$

$$= \frac{-56.95 + 95.75 - 12.15}{95.75} = \frac{26.65}{95.75}$$

$$= 0.278 = 27.8\%$$

4-6 SUMMARY

An ideal gas is a gas that obeys the ideal-gas equation of state as given by

$$pv = RT$$

where R is the gas constant as defined by

$$R = \frac{\mathcal{R}}{M}$$

where $\mathcal{R}$ is the universal gas constant, and M is the molar mass. $\mathcal{R}$ has the same value for all gases, and R has a different value for different gases.

The ideal-gas equation holds for a real gas only as the pressure approaches zero. It holds approximately for a real gas at higher pressures when the gas is at relatively high temperatures. The deviation of a real gas from the ideal-gas behavior at a given pressure and temperature can be determined by examining the compressibility factor Z, which is defined as

$$Z = pv/RT$$

For an ideal gas, $Z = 1$ for all pressures and temperatures. Thus, the value of Z expresses the extent of deviation of the gas from ideal-gas behavior.

In common engineering applications, diatomic and monatomic gases, such as nitrogen, oxygen (also air), hydrogen, carbon monoxide, helium, and argon, can be treated as ideal gases at room temperatures (and above) up to 1 to 2 MPa. Use of the ideal-gas equation for heavier gases such as carbon dioxide will result in a higher degree of inaccuracy. Never consider dense gases such as water vapor in steam power plants and refrigerant vapor in refrigerators as ideal gases.

The internal energy and enthalpy of an ideal gas are functions of temperature only. Thus, for any process of an ideal gas, we have

$$du = c_v \, dT$$

$$dh = c_p \, dT$$

The specific heats of an ideal gas are also functions of temperature only, and

$$c_p - c_v = R$$

for all conditions of an ideal gas.

The ideal-gas property tables given in the Appendix for air, N_2, O_2, H_2O, CO_2, H_2, and CO are based on $u = 0$ and $h = 0$ at the absolute zero temperature. In these tables, the variations of specific heats with temperature have been taken into account.

If the specific heats in the range of temperature considered are constant, or average values of specific heats can be used, we have

$$\Delta u = c_v \, \Delta T \qquad \text{and} \qquad \Delta h = c_p \, \Delta T$$

A quasiequilibrium polytropic process has the p–v relation,

$$pv^n = \text{constant}$$

where n is a constant. If the working substance is an ideal gas, between states 1 and 2, we then have

$$\frac{T_2}{T_1} = \left(\frac{p_2}{p_1}\right)^{(n-1)/n} = \left(\frac{v_1}{v_2}\right)^{n-1}$$

When $n = 0$, it is a constant-pressure process; when $n = 1$, it is a constant-temperature process; and when $n = \infty$, it is a constant-volume process.

PROBLEMS

4-1. A 3-m^3 tank contains carbon dioxide at 0.5 MPa and 30°C. Gas leaks out of the tank. At the time of detection of the leak, there is 22 kg of the gas remaining in the tank at a temperature of 20°C. Determine the original mass of carbon dioxide in the tank and the final pressure of the gas after the leakage.

4-2. One-tenth pound of nitrogen with $c_p = 0.248$ Btu/lbm·°R is contained in a piston–cylinder assembly that initially has a volume of 1.0 ft³. Heat transfer is allowed to take place until the volume is 90% of its initial value while the pressure is kept constant at 20 psia. Find the heat transfer and the final temperature.

4-3. One pound of air ($R = 53.3$ ft·lbf/lbm·°R) is compressed isothermally in a closed system from 15 psia, 80°F, to 90 psia. Determine the heat transfer.

4-4. A 20-cm^3 small cylinder containing helium at 500 kPa and 50°C is placed inside a 2000-cm^3 insulated tank. See Fig. P4-4. The tank is initially evacuated. The helium gas leaks out and fills the entire big tank. When equilibrium is reached, what will be the pressure and temperature of the helium?

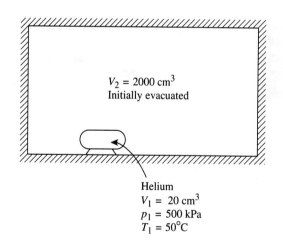

$V_2 = 2000$ cm³
Initially evacuated

Helium
$V_1 = 20$ cm³
$p_1 = 500$ kPa
$T_1 = 50°C$

Figure P4-4

4-5. The buoyant or lift force on a submerged body is equal to the weight of the displaced fluid. A 10-ft-diameter balloon is filled with helium at 3 psig and 70°F. This balloon is floating in standard atmospheric air at 14.7 psia and 70°F. Neglecting the weight of the balloon material, calculate the net lifting force of the balloon.

4-6. An ideal gas contained in a cylinder–piston device initially has a pressure of 0.12 MPa, a temperature of 30°C, with a total volume of 0.2 m³. Electric current flows through a resistor within the cylinder until 0.02 kW·h of electric energy has entered the gas. If the process occurs at constant pressure and the final volume is 0.3 m³, determine the heat transfer to/from the surrounding. For the gas, c_v = 0.725 kJ/kg·K and R = 0.287 kJ/kg·K.

4-7. One kilogram of air is contained in a cylinder–piston device initially at 500 kPa and 300 K. Electric work in the amount of 75 kJ/kg is performed on the air while it undergoes a constant-temperature process to a final volume of two times its initial volume. Determine the amount of heat transfer in kJ/kg.

4-8. A cylinder has an inside cross-sectional area of 10 ft² and a length of 30 ft, with its top closed and its bottom open. See Fig. P4-8. It is partially submerged in water with its axis vertical until 250 ft³ of air is trapped in the cylinder and the water level inside the cylinder is 15 ft below that outside. The air is initially at 150°F. Neglecting any effects of water vapor in the air, determine the heat transfer so that the air temperature in the cylinder is reduced to 100°F. The water level outside the cylinder remains 10 ft below the top of the cylinder.

4-9. The elevator shaft of a building is 400 ft high. The lower end of the shaft is open to the outdoors. See Fig. P4-9. When the air inside the shaft is 80°F and the outdoor temperature is 0°F, what is the pressure difference between the air at the top of the shaft and the outdoor air? Barometric pressure is 14.7 psia.

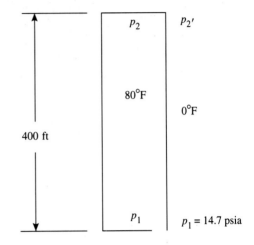

Figure P4-9

4-10. A rigid tank and a frictionless piston–cylinder device contain 10 kilograms each of an ideal gas with a molecular mass of 25, initially, at the same

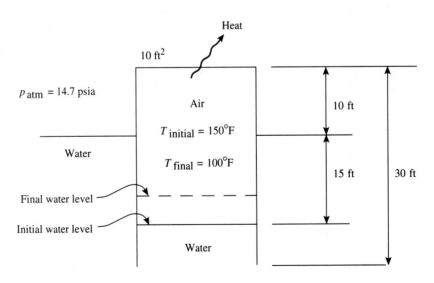

Figure P4-8

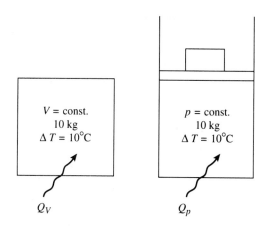

Figure P4-10

pressure, temperature, and volume. See Fig. P4-10. Heat is added to both systems separately in order to increase the temperature of each system by 10°C. During the heating process, the gas in the piston–cylinder device undergoes a constant-pressure process. Determine the difference in the amounts of heat supplied to the two systems.

4-11. Determine the specific internal energy, specific enthalpy, and specific volume, per kilogram of hydrogen at a pre sure of 50 kPa and a temperature of 400 K.

4-12. Determine the specific internal energy, specific enthalpy, and specific volume, per kilogram of oxygen at 100 kPa and 400 K.

4-13. Determine the specific internal energy, specific enthalpy, and specific volume, per pound of carbon dioxide at 14.7 psia and 600°R.

4-14. Determine the specific internal energy, specific enthalpy, and specific volume, per pound of carbon monoxide at 100 psia and 1000°R.

4-15. To reduce the volume of an ideal gas to one-half of its initial volume at a constant temperature of $T_a = 350°C$ requires a certain amount of work. At what constant temperature T_b will the same amount of work reduce the volume of the gas to one-quarter of the same initial volume?

4-16. A balloon having a volume of 30 ft³ is filled with air ($c_v = 0.171$ Btu/lbm·°R) at 70°F and 14.7 psia. After exposure to solar radiation, the volume of the balloon increases slowly to 35 ft³ while the pressure remains constant. Calculate the work done and the heat transferred.

4-17. A 10-m radius spherical balloon is made from an inelastic material. It is to be filled with helium

from a storage tank that contains helium initially at 15 MPa and 25°C. The balloon is initially flat, and the atmospheric pressure is 101 kPa. (a) How much work is done against the atmosphere as the balloon is inflated. Assume that the pressure in the balloon is essentially equal to the atmospheric pressure during the inflation process. (b) What is the required volume of the storage tank, if the final pressure in the tank is the same as that in the balloon and the temperature of the helium in the tank remains constant at 25°C?

4-18. Carbon monoxide at a low pressure changes temperature from 300 K to 600 K. Determine the changes in specific internal energy and enthalpy for the process.

4-19. Argon at a low pressure changes temperature from 500°R to 1000°R. Determine the changes in specific internal energy and enthalpy, per pound of argon.

4-20. A rigid tank contains nitrogen at 27°C and a low pressure. It is heated until the temperature is doubled. Determine the heat transfer, in kJ/kg, by using (a) average specific-heat data, and (b) data from the ideal-gas nitrogen table.

4-21. One-tenth cubic meter of nitrogen at 0.1 MPa and 25°C undergoes a quasiequilibrium constant-pressure process in an adiabatic piston–cylinder device. See Fig. P4-21. During the process there are 20,000 N·m of work input from a paddle wheel of negligible mass within the cylinder. Using data

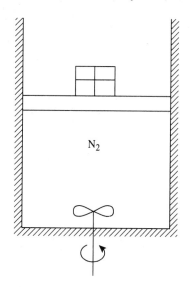

Figure P4-21

taken from the ideal-gas nitrogen table, determine the final temperature of nitrogen in the cylinder.

4-22. An insulated rigid tank contains 0.5 kg of a gas. Paddle-wheel work of the amount of 10 kJ is done on the gas, and electric work of the amount of 15 kJ is also added to the gas through the use of a resistance heater. The temperature of the gas increases by 50°C during the process. What is the average value of c_v for the gas, in kJ/kg·°C?

4-23. The temperature of 1 kilogram of nitrogen is increased from 400 K to 800 K at a low pressure. Determine the changes in enthalpy and internal energy by the use of (a) data from the ideal-gas property table for nitrogen (Table A-10M), (b) arithmetic-mean specific heat for nitrogen between the given temperature limits (Table A-6M), and (c) the empirical specific-heat equation for nitrogen as given by

$$c_p = \mathcal{R}(a + bT + cT^2 + dT^3 + eT^4)$$

where c_p is in kJ/kgmole·K, $\mathcal{R} = 8.314$ kJ/kgmole·K, $a = 3.675$, $b = -1.208 \times 10^{-3}$, $c = 2.324 \times 10^{-6}$, $d = -0.632 \times 10^{-9}$, and $e = -0.226 \times 10^{-12}$.

4-24. For CO_2 gas, the specific heat at constant pressure (c_p) as a function of temperature (T) is as follows:

$$c_p = 16.2 - 6.53 \times 10^3 T^{-1} + 1.41 \times 10^6 T^{-2}$$

where c_p is in Btu/lbmole·°R, and T in °R. Calculate the change in enthalpy (per lbmole) when the temperature is increased from 500°R to 1200°R. (a) Use $\Delta h = \int c_p \, dT$ and (b) $\Delta h = c_{p,\text{at mean temp}} (\Delta T)$.

4-25. The c_p equation for H_2 gas can be expressed as follows:

$$c_p = 56.505 - 2.2223 \times 10^4 T^{-0.75} + 1.165$$
$$\times 10^5 T^{-1} - 5.607 \times 10^5 T^{-1.5}$$

where c_p is in kJ/kgmole·K, and T in K. Calculate the change in enthalpy per kgmole of H_2 when the temperature is increased from 300 K to 1000 K. (a) Use $\Delta h = \int c_p \, dT$. (b) Use $\Delta h = c_{p,\text{mean}}(\Delta T)$, where $c_{p,\text{mean}} = \frac{1}{2}(c_p$ at 300 K $+ c_p$ at 1000 K). (c) Use $\Delta h = c_{p,\text{mean}}(\Delta T)$, where $c_{p,\text{mean}} = c_p$ at 650 K.

4-26. The c_p equation for O_2 gas can be expressed as follows:

$$c_p = 11.515 - 172T^{-1/2} + 1530T^{-1}$$

where c_p is in Btu/lbmole·°R, and T in °R. Calculate the change in enthalpy per lbmole of O_2 when

the temperature is increased from $T_1 = 600°R$ to $T_2 = 1200°R$. (a) Use $\Delta h = \int c_p \, dT$. (b) Use $\Delta h = c_{p,\text{mean}}(\Delta T)$, where $C_{p,\text{mean}} = \frac{1}{2}(c_p$ at $T_1 + c_p$ at T_2). (c) Use $\Delta h = c_{p,\text{mean}}(\Delta T)$, where $c_{p,\text{mean}} = $ mean c_p at $\frac{1}{2}(T_1 + T_2)$.

4-27. For an ideal gas, the constant-pressure specific heat c_p as a function of temperature is given by the relation

$$c_p = 9.47 - 3.47 \times 10^3 T^{-1} + 1.16 \times 10^6 T^{-2}$$

where c_p is in Btu/lbmole·°R, and T in °R. The molecular mass of the gas is 28. (a) How much heat is required to raise the temperature of 1 lbm of the gas from 500°R to 5000°R slowly at constant pressure? (b) Repeat part (a) but at constant volume. (c) What are the values of k at each end of the temperature range? (d) What are the mean values of c_p and c_v for the temperature range?

4-28. Helium is compressed in an adiabatic cylinder from 70°F, 10 ft³, 15 psia, to 1 ft³ according to the relation $pv^k = $ constant. Find the change of enthalpy, the work done, and the final temperature. For helium, $c_p = 1.25$ Btu/lbm·°F and $c_v = 0.753$ Btu/lbm·°F.

4-29. Air is compressed in a cylinder cooled by a water jacket. The indicator diagram shows that pressure and volume are related by $pv^{1.30} = $ constant. The process begins at 80°F, 14 psia, and the pressure rises to 80 psia. If the process is frictionless, how much heat is transferred per lbm of air?

4-30. Air with $c_v = 0.723$ kJ/kg·K is compressed in a closed system according to $pv^{1.30} = $ constant from an initial temperature of 17°C and pressure of 1 bar, to a final pressure of 5 bars. Determine (a) the final temperature, (b) the work, and (c) the heat transfer.

4-31. Helium gas contained in a cylinder fitted with a piston expands slowly according to the relation $pv^{1.30} = $ constant. The initial volume is 3 ft³, the initial pressure is 70 psia, and the initial temperature is 500°R. The gas expands to 20 psia. Determine the work and the heat transfer for the process.

4-32. One-tenth kilogram of an ideal gas with $M = 28.97$ kg/kgmole and $c_p = 1.0035$ kJ/kg·K contained in a cylinder fitted with a piston is compressed in a quasistatic process. During the compression process, the relation between pressure and volume is $pv^{1.25} = $ constant. The initial pressure is 100 kPa and the initial temperature is 20°C. The final volume is 1/8 of the initial volume. Determine (a) the final pressure and temperature of the gas, and (b) the work and the heat transfer.

4-33. Two-tenths pound of air is compressed in a cylinder in a quasiequilibrium process. During the process, the relation between pressure and volume is $pv^{1.25} = $ constant. The initial pressure is 20 psia and the initial temperature is 60°F. The final volume is 1/8 of the initial volume. Determine the work and the heat transfer.

4-34. Two pounds of an ideal gas for which $R = 26$ ft·lbf/lbm·°R and $k = 1.1$ undergo a frictionless polytropic process in a closed system from 15 psia, 100°F, to 75 psia, 3.72 ft³. Determine (a) the value of the polytropic exponent n, and (b) the work and the heat transfer.

4-35. One kilogram of an ideal gas for which $R = 0.296$ kJ/kg·K and $k = 1.4$ undergoes a quasistatic polytropic process from $p_1 = 101.3$ kPa and $T_1 = 50$°C to $p_2 = 3p_1$ and $v_2 = 0.5$ m³. Determine (a) the final termperature T_2, (b) the polytropic exponent n, (c) the work in kJ, and (d) the heat in kJ.

4-36. Helium with $c_p = 1.24$ Btu/lbm·°R and having an initial volume of 2 ft³ expands slowly in a closed system according to the relation $pv^{1.5} = $ contant from 70 psia, 400°R, to 30 psia. Determine the work done and the heat transfer.

4-37. Ammonia vapor (NH_3) is compressed in a piston–cylinder machine from 15 psia, 3 ft³, to 150 psia, 0.4 ft³, according to the relation $pv^n = $ constant. Assuming the vapor to be an ideal gas with $R = 90.72$ ft·lbf/lbm·°R and $c_v = 0.384$ Btu/lbm·°R, calculate the work done and the heat transfer.

4-38. One pound of hydrogen initially at a pressure of 100 psia with a specific volume of 16 ft³/lbm is contained in a cylinder with a movable piston. 1000 Btu of heat is added to the gas in an irreversible process, and after equilibrium is established, the gas is at a pressure of 15 psia and has a specific volume of 200 ft³/lbm. Assuming hydrogen to be an ideal gas, find (a) the work done by the irreversible process, and (b) the work necessary to return the gas to its original state by means of a quasiequilibrium constant-pressure process, followed by a quasiequilibrium constant-volume process. Use $c_p = 3.42$ Btu/lbm·°R for hydrogen.

4-39. A 5 m × 5 m × 6 m dormitory room is occupied by a summer-school student who has a 200-W fan in the room. See Fig. P4-39. One morning, when the room temperature is 15°C, she turns on the fan and leaves the room, hoping that the room will be

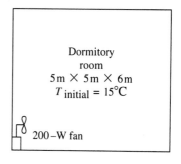

Dormitory
room
5m × 5m × 6m
$T_{\text{initial}} = 15$°C

200–W fan

Figure P4-39

cool when she comes back in the evening. Assuming the room to be well insulated and well sealed, determine the temperature in the room when she comes back 8 hours later. Barometric pressure is 101 kPa.

4-40. A 0.5-m³ rigid tank containing air at 300 kPa and 20°C is connected by a pipe with a valve to a piston–cylinder device that initially contains no air under the piston. See Fig. P4-40. The weight of the piston is such that a pressure of 200 kPa is required to raise the piston. When the valve is opened, air flows slowly into the cylinder until the tank pressure drops to 200 kPa. During this process, there is heat transfer from the surroundings to the cylinder to maintain the tank temperature at 20°C. Determine the work and heat transfer for the process.

4-41. A rigid insulated tank has a free-moving uninsulated piston in it. See Fig. P4-41. One side of the piston contains 1 kilogram of air initially at 0.5 MPa, 350 K; the other side of the piston contains 3 kilograms of carbon dioxide initially at 0.2 MPa, 450 K. Assuming air and carbon dioxide to behave as ideal gases with constant specific heats, determine the final equilibrium temperature and pressure. Neglect any energy storage capacity of the piston.

4-42. A cycle involving 1 lbm of a gas is made up of the following quasiequilibrium processes:

Process 1-2: The gas is compressed according to $pv^{1.4} = $ constant from $p_1 = 14.7$ psia, $v_1 = 14$ ft³/lbm, to $p_2 = 2p_1$.

Process 2-3: The gas is heated at constant pressure until $v_3 = v_1$.

Process 3-1: The gas is cooled at constant volume.

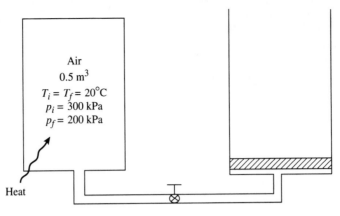

Air
0.5 m^3
$T_i = T_f = 20°C$
$p_i = 300$ kPa
$p_f = 200$ kPa

Heat

Figure P4-40

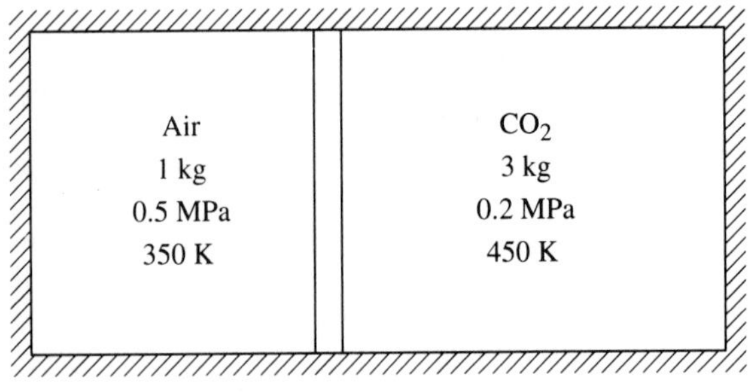

Air	CO$_2$
1 kg	3 kg
0.5 MPa	0.2 MPa
350 K	450 K

Figure P4-41

Draw the p–v diagram and calculate the net work of the cycle. If the machine operating on this cycle handles 5 ft^3/s of the gas at state 1, what is the power of the machine?

4-43. A piston–cylinder arrangement contains an ideal gas with a molar mass of 50 initially at 20°C and 0.24 m^3/kg (state 1). It undergoes an isothermal process to 0.12 m^3/kg (state 2). It then expands at constant pressure to 0.36 m^3/kg (state 3). Finally, it returns to its initial state along a path that is a straight line on p–v diagram. Draw the cyclic proc-

ess on a p–v diagram and determine the values of p_2 (in kPa) and the net work of the cycle (in kJ/kg).

4-44. One kilogram of an ideal gas with $c_v = 0.718$ kJ/kg·K and $R = 0.287$ kJ/kg·K initially at $p_1 = 100$ kPa and $T_1 = 300$ K is heated in a closed system until $p_2 = 2p_1$ and $v_2 = 2v_1$ according to the relation $p/v =$ constant. Heat is then removed at constant volume until $p_3 = p_1$. Finally, heat is removed at constant pressure until the initial state is reached. Determine the work and heat interactions for the three processes and the net work produced in the cycle. Draw the processes on the p–v diagram.

4-45. A closed system consists of 2 kg of carbon dioxide initially at 0.1 MPa and 300 K (state 1). It is heated at constant volume to 0.4 MPa (state 2). It is then expanded back to 0.1 MPa (state 3) according to the relation $pv^{1.28} =$ constant. It is finally compressed at constant pressure back to 300 K (state 1 again). Taking data from the ideal-gas carbon dioxide table, determine the quantities of work and heat (in kJ) for the three processes and calculate the thermal efficiency of the cycle.

Ideal Gases   Chap. 4

5

First Law Analysis for Control Volumes

Although the development of classical thermodynamics is mainly based on the concept of a closed system, a major portion of its application involves bulk mass flow through an open system. In the analysis of an open system, the study is no longer focused on a fixed mass, but rather on a region of space called a control volume. This chapter is concerned with the development and analysis of the control volume form of the first law of thermodynamics.

5-1 CONSERVATION OF MASS FOR CONTROL VOLUMES

A *control volume* is a region of space specified in order to focus study. The boundary of a control volume is called a *control surface*. This surface is always a closed envelope, real or imaginary, enclosing the control volume. The selection of a control volume is arbitrary as long as it makes the analysis convenient to accomplish. For example, if we wish to analyze the flow of air through a nozzle, the best choice for a control volume would be the region within the nozzle enclosed by the nozzle walls and an imaginary cross-section at each end. A control volume may have a moving control surface. For example, for the charging of air to a cylinder with a movable piston, the underside of the piston would be a part of a control volume.

Because this book is not concerned with any nuclear reaction, the subject of the equivalence of mass and energy is not involved in our analyses. Accordingly, the principle of conservation of mass can be simply stated in words for a control volume (cv) as

(Rate of mass flow entering at the boundary of the control volume)
— (rate of mass flow leaving at the boundary of the control volume)
= (time rate of change of the mass within the control volume)

Converting this word expression into a mathematical expression, we write

$$\sum_{\text{inlet}} \dot{m} - \sum_{\text{exit}} \dot{m} = \frac{dm_{\text{cv}}}{dt} \tag{5-1}$$

where $\sum_{\text{inlet}} \dot{m}$ represents the summation of mass flow per unit time into the control volume, $\sum_{\text{exit}} \dot{m}$ the summation of mass flow per unit time out of the control volume, and

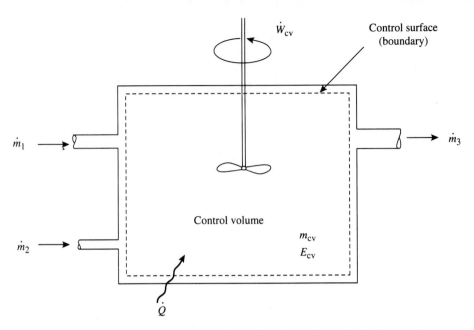

Figure 5-1 A control volume for a mixing chamber with two inlet flows and one exit flow plus work and heat interactions.

dm_{cv}/dt the mass accumulation per unit time (t) within the control volume. Equation (5-1) for the conservation of mass is commonly known as the *continuity equation*.

Figure 5-1 illustrates a control volume for a mixing chamber with two inlet flows and one exit flow. For this control volume, the continuity equation is written as

$$(\dot{m}_1 + \dot{m}_2) - \dot{m}_3 = \frac{dm_{cv}}{dt}$$

For a single-inlet and single-exit open system, such as shown schematically in Fig. 5-2, Eq. (5-1) reduces to

$$\dot{m}_1 - \dot{m}_2 = \frac{dm_{cv}}{dt}$$

The mass-flow rate of fluid entering or leaving a control volume can be expressed in terms of the properties associated with the material at the crossing of the control surface. Assume that the fluid flows across the control surface via a channel of constant cross-sectional area A with an average velocity $\mathbf{V}$ (see Fig. 5-3). The volume-flow rate of fluid crossing the control surface is given by

$$\dot{V} = A\mathbf{V}$$

The mass-flow rate of fluid crossing the control surface is then

$$\dot{m} = \frac{\dot{V}}{v} = \frac{A\mathbf{V}}{v} = \rho A \mathbf{V} \tag{5-2}$$

where v is the specific volume, and ρ is the density of the fluid at the crossing section.

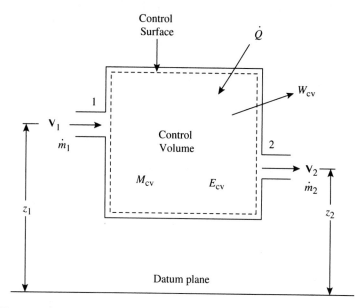

Control
Surface

$\dot{Q}$

W_{cv}

1

$\mathbf{V}_1$

$\dot{m}_1$

Control
Volume

2

M_{cv} E_{cv}

$\mathbf{V}_2$

$\dot{m}_2$

z_1

z_2

Datum plane

Figure 5-2   Schematic for a single-inlet and single-exit control volume.

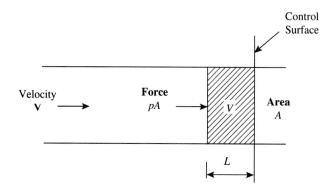

Control
Surface

Velocity
V

Force
pA

V

Area
A

L

Figure 5-3   Fluid flow through a channel at the crossing of a control surface.

Example 5-1

Air flows through a pipe with a variable cross-section. At the pipe inlet, the pressure is 500 kPa, the temperature is 300 K, the area is 30 cm², and the velocity is 50 m/s. At the pipe exit, the pressure is 400 kPa, the temperature is 320 K, and the area is 20 cm². Determine the mass-flow rate and the exit velocity.

Solution   Air at the given conditions of this problem can be considered as an ideal gas. Thus, the specific volume of air at the inlet condition is

$$v_1 = \frac{RT_1}{p_1} = \frac{8.314 \text{ kJ/kgmole·K}}{29 \text{ kg/kgmole}}\left(\frac{300 \text{ K}}{500 \text{ kN/m}^2}\right) = 0.1720 \text{ m}^3/\text{kg}$$

The mass-flow rate of air is then

$$\dot{m} = \frac{A_1 \mathbf{V}_1}{v_1} = \frac{(0.0030 \text{ m}^2)(50 \text{ m/s})}{0.1720 \text{ m}^3/\text{kg}} = 0.8721 \text{ kg/s}$$

The mass-flow rate at the exit section must be the same as that at the inlet section. Thus,

$$\dot{m} = \frac{A_2 V_2}{v_2} \qquad \text{or} \qquad V_2 = \frac{\dot{m} v_2}{A_2}$$

where the specific volume of air at the exit section is

$$v_2 = \frac{RT_2}{p_2} = \frac{8.314}{29}\left(\frac{320}{400}\right) = 0.2294 \text{ m}^3/\text{kg}$$

Therefore, the air velocity at the exit is

$$V_2 = \frac{(0.8721 \text{ kg/s})(0.2294 \text{ m}^3/\text{kg})}{0.0020 \text{ m}^2} = 100.0 \text{ m/s}$$

The exit velocity can also be calculated as follows.

$$\dot{m} = \frac{A_1 V_1}{v_1} = \frac{A_2 V_2}{v_2}$$

or

$$\dot{m} = \frac{A_1 V_1 p_1}{RT_1} = \frac{A_2 V_2 p_2}{RT_2}$$

Thus,

$$V_2 = V_1\left(\frac{A_1}{A_2}\right)\left(\frac{p_1}{p_2}\right)\left(\frac{T_2}{T_1}\right)$$

$$= (50)\left(\frac{30}{20}\right)\left(\frac{500}{400}\right)\left(\frac{320}{300}\right) = 100 \text{ m/s}$$

5-2 CONSERVATION OF ENERGY FOR CONTROL VOLUMES

The first law of thermodynamics, or the law of conservation of energy, as applied to open systems (control volumes) can be stated in words as

(Heat added to the system) − (work done by the system)

+ (total energy of matter entering the system)

− (total energy of matter leaving the system)

= increase in total energy of the system

where the term total energy is the sum of the internal energy and the bulk kinetic and potential energies as defined in Eq. (2-18). When a fluid flows across a control surface, in addition to the transportation of stored energy, work is required to push the fluid into or out of the control volume. Consider the process by which a fluid element of volume V with cross-sectional area A and length L is pushed across a control surface, as depicted in Fig. 5-3. The force acting on this element is pA, where p is the pressure. This force acts through a distance L, and hence the work done in pushing the element across the control surface is $pAL = pV$. This work is called *flow work*. When unit mass of a fluid with specific volume v crosses a control surface, the flow work is pv. Thus, for the control volume as depicted in the schematic diagram of Fig. 5-2 with a single inlet and a single exit, the flow work per unit mass at section 1 is $p_1 v_1$, which is the work done on the system by the fluid outside the control volume. Similarly, the flow work per unit mass at section 2 is $p_2 v_2$, which is the work done by the system on the fluid outside the control volume.

Obviously, in addition to flow works, a control volume may involve other forms of work, such as work done on a rotating shaft or a moving boundary. We will denote these other forms of work by the symbol W_{cv} and call it shaft work or simply work. The first law equation for a control volume can be written on a rate basis as

$$\dot{Q} - \dot{W}_{cv} + \sum_{\text{inlet}} \dot{m}(e + pv) - \sum_{\text{exit}} \dot{m}(e + pv) = \frac{dE_{cv}}{dt} \tag{5-3}$$

where E_{cv} is the total energy of the control volume, and each inlet and exit stream to be represented by a term $\dot{m}(e + pv)$.

If a fluid stream crossing the control surface has an average velocity $\mathbf{V}$ and the vertical distance of the stream from a horizontal datum plane is z, then the specific total energy of fluid crossing the control surface is given by

$$e = u + \frac{\mathbf{V}^2}{2} + gz \tag{5-4}$$

It is imperative to note that in the present study, we are concerned with simple compressible systems without any electricity, magnetism, or surface tension effects, and Eq. (5-4) is written accordingly.

Substituting Eq. (5-4) into Eq. (5-3) and replacing $(u + pv)$ by h, we obtain the following more convenient form of the first-law equation for a control volume:

$$\dot{Q} - \dot{W}_{cv} + \sum_{\text{inlet}} \dot{m}\left(h + \frac{\mathbf{V}^2}{2} + gz\right) - \sum_{\text{exit}} \dot{m}\left(h + \frac{\mathbf{V}^2}{2} + gz\right) = \frac{dE_{cv}}{dt} \tag{5-5}$$

As illustrations for the preceding equation, for the mixing chamber as depicted in Fig. 5-1, it becomes

$$\dot{Q} - \dot{W}_{cv} + \dot{m}_1\left(h_1 + \frac{\mathbf{V}_1^2}{2} + gz_1\right) + \dot{m}_2\left(h_2 + \frac{\mathbf{V}_2^2}{2} + gz_2\right)$$
$$- \dot{m}_3\left(h_3 + \frac{\mathbf{V}_3^2}{2} + gz_3\right) = \frac{dE_{cv}}{dt}$$

and for the single-inlet and single-exit control volume shown in Fig. 5-2, it becomes

$$\dot{Q} - \dot{W}_{cv} + \dot{m}_1\left(h_1 + \frac{\mathbf{V}_1^2}{2} + gz_1\right) - \dot{m}_2\left(h_2 + \frac{\mathbf{V}_2^2}{2} + gz_2\right) = \frac{dE_{cv}}{dt}$$

It is instructive to mention that the control-volume (open-system) energy-conservation expression can be obtained by transforming the proper control-mass (closed-system) expression established in Sec. 2-4. This book, however, prefers the straightforward writing of the control-volume expression without referring back to the control-mass expression, because it is easier and more suitable to beginning students of thermodynamics. For those who would like to see a proper conversion of equations for the two systems, please read [32, Chapter 5] in the Bibliography at the end of the book.

5-3 STEADY-FLOW, STEADY-STATE PROCESSES

A simple but important special case of open-system processes is that of *steady flow* and *steady state*. The flow is steady when no mass is accumulating within the control volume, and the state is steady when the properties at each point within the control volume are unchanging in time. In other words, in a steady-flow, steady-state system, the rate of mass flow into the control volume equals the rate of mass flow out of it, and the energy stored within the control volume remains constant. Expressed mathematically,

$$\frac{dm_{cv}}{dt} = 0$$

$$\frac{dE_{cv}}{dt} = 0$$

Accordingly, for a steady-flow, steady-state system, the conservation of mass, Eq. (5-1), and the conservation of energy, Eq. (5-5), simplify to

$$\sum_{\text{inlet}} \dot{m} - \sum_{\text{exit}} \dot{m} = 0 \tag{5-6}$$

$$\dot{Q} - \dot{W}_{cv} + \sum_{\text{inlet}} \dot{m}\left(h + \frac{\mathbf{V}^2}{2} + gz \right) - \sum_{\text{exit}} \dot{m}\left(h + \frac{\mathbf{V}^2}{2} + gz \right) = 0 \tag{5-7}$$

For a steady-flow, steady-state system with a single inlet at section 1 and a single exit at section 2, as depicted in Fig. 5-2, the conservation of mass is written as

$$\dot{m}_1 = \dot{m}_2 = \dot{m} \tag{5-8}$$

and the conservation of energy is written as

$$\dot{Q} - \dot{W}_{cv} + \dot{m}\left(h_1 + \frac{\mathbf{V}_1^2}{2} + gz_1 \right) - \dot{m}\left(h_2 + \frac{\mathbf{V}_2^2}{2} + gz_2 \right) = 0 \tag{5-9}$$

Dividing this equation by the constant mass-flow rate leads to

$$q - w_{cv} + \left(h_1 + \frac{\mathbf{V}_1^2}{2} + gz_1 \right) - \left(h_2 + \frac{\mathbf{V}_2^2}{2} + gz_2 \right) = 0 \tag{5-10}$$

where q and w_{cv} are, respectively, heat added to the control volume and shaft work done by the control volume per unit mass flowing in and out the control volume.

The steady-flow, steady-state energy equation given in this section is extremely useful in mechanical engineering thermodynamics. Except for the short-term transient startup or shutdown, the long-term operation of many mechanical devices can be analyzed by these equations. In the next few sections, we will present a number of applications to illustrate the use of this equation.

5-4 NOZZLES AND DIFFUSERS

When a fluid flows through a passage of varying cross-section, its velocity varies along the passage. If the velocity increases with a corresponding decrease in pressure, the passage is called a *nozzle*. If the velocity decreases with a corresponding increase in pressure, the passage is called a *diffuser*. Figures 5-4 and 5-5 show schematically the

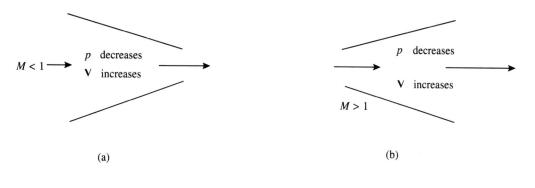

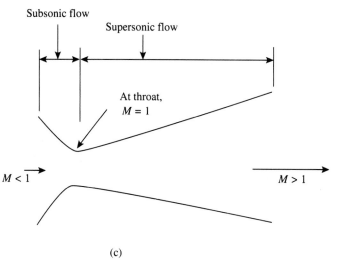

Figure 5-4 Nozzles: (a) converging (subsonic), (b) diverging (supersonic), and (c) converging–diverging.

types of nozzles and diffusers, respectively, for compressible fluids. The parameter M indicated in these figures is the *Mach number,* which is defined as the ratio of the velocity of the fluid at a point in the flow passage to the velocity of sound at the same point. Flow with $M < 1$ is called subsonic and flow with $M > 1$ is called supersonic. This book will not address the thermodynamic aspects of compressible fluid flow that result in the flow pattern shown in Figs. 5-4 and 5-5. For those interested in this subject, please see [30] in the Bibliography at the end of the book. We will, however, offer the following brief introduction on the relationship of velocity and flow area for a nozzle to help gain some understanding of the phenomenon.

Equation (5-2) gives the mass-flow rate $\dot{m}$ in terms of flow area A, average velocity V, and specific volume v of the fluid. At low flow velocities, the specific-volume change is small; thus, in order to maintain a constant mass flow, the flow area must decrease as velocity increases. The specific volume increases gradually as velocity continues to increase; thus, lesser decrease in flow area is needed to maintain the same mass flow. At very high flow velocities, the rate of increase of specific volume is greater than the rate of increase of velocity, making it necessary to increase the flow area to maintain constant mass-flow rate.

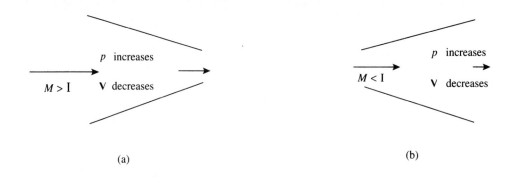

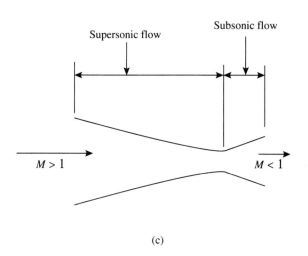

Figure 5-5 Diffusers: (a) converging (supersonic), (b) diverging (subsonic), and (c) converging–diverging.

Because both nozzle and diffuser are merely fluid passages, no shaft work is involved in these devices. In addition, as these passages are usually short in length, the change in potential energy, if any, would be negligible. Denoting the inlet and exit sections of a nozzle or a diffuser by the subscripts 1 and 2, respectively, the steady-flow, steady-state energy equation, Eq. (5-10), on a unit-mass basis becomes

$$q - \overset{0}{\cancel{w}_{cv}} + (h_1 - h_2) + \tfrac{1}{2}(\mathbf{V}_1^2 - \mathbf{V}_2^2) + g(z_1 \overset{0}{\cancel{-} z_2}) = 0$$

or

$$q + (h_1 - h_2) + \tfrac{1}{2}(\mathbf{V}_1^2 - \mathbf{V}_2^2) = 0$$

Furthermore, as the fluid velocity in going through these devices is usually very large, there is not enough time for effective heat transfer between the fluid and the passage walls. Thus, nozzle and diffuser processes in many applications can be assumed to be adiabatic ($q = 0$). Under these conditions, the energy equation reduces to

$$(h_1 - h_2) + \tfrac{1}{2}(\mathbf{V}_1^2 - \mathbf{V}_2^2) = 0$$

Notice also that for a converging or a converging–diverging nozzle, the inlet velocity V_1 is usually small compared to the exit velocity V_2; thus, the inlet KE is sometimes neglected. Similarly, for a diverging or a converging–diverging diffuser, the exit velocity V_2 is usually small compared to the inlet velocity V_1; thus, the exit KE is sometimes neglected.

The continuity equation, Eq. (5-8), for a nozzle or diffuser is written as

$$\frac{A_1 V_1}{v_1} = \frac{A_2 V_2}{v_2}$$

This equation can be used simultaneously with the energy equation to solve for two unknowns.

Example 5-2

Air enters an adiabatic nozzle at a pressure of 150 psia and a temperature of 200°F with a velocity of 100 ft/s. It leaves the nozzle at a pressure of 15 psia and a temperature of 80°F. For a flow rate of 1.2 lbm/s, determine the inlet and exit areas required. Assume that $c_p = 0.24$ Btu/lbm·°F for air.

Solution   The given conditions are $p_1 = 150$ psia, $T_1 = 200°F = 660°R$, $V_1 = 100$ ft/s, $p_2 = 15$ psia, $T_2 = 80°F = 540°R$, $\dot{m} = 1.2$ lbm/s, and $q = 0$.

Air can be considered as an ideal gas at the specified pressure and temperature conditions. Because $w_{cv} = 0$ and the change in potential energy can be neglected, the steady-flow, steady-state energy equation, Eq. (5-10), reduces to

$$(h_1 - h_2) + \tfrac{1}{2}(V_1^2 - V_2^2) = 0$$

Thus,

$$V_2 = [2(h_1 - h_2) + V_1^2]^{1/2}$$

$$= [2c_p(T_1 - T_2) + V_1^2]^{1/2}$$

$$= \left[2\left(32.2\,\frac{\text{ft·lbm}}{\text{lbf·s}^2}\right)\left(778\,\frac{\text{ft·lbf}}{\text{Btu}}\right)\left(0.24\,\frac{\text{Btu}}{\text{lbm·°F}}\right)(200°F - 80°F) + (100^2\ \text{ft}^2/\text{s}^2)\right]^{1/2}$$

$$= 1205\ \text{ft/s}$$

The specific volumes of air at the inlet and exit are

$$v_1 = \frac{RT_1}{p_1} = \left(\frac{1545}{29}\,\frac{\text{ft·lbf}}{\text{lbm·°R}}\right)\frac{660°R}{150 \times 144\ \text{lbf/ft}^2}$$

$$= 1.628\ \text{ft}^3/\text{lbm}$$

$$v_2 = \frac{RT_2}{p_2} = \left(\frac{1545}{29}\right)\left(\frac{540}{15 \times 144}\right) = 13.32\ \text{ft}^3/\text{lbm}$$

From the continuity equation,

$$\dot{m} = \frac{A_1 V_1}{v_1} = \frac{A_2 V_2}{v_2}$$

we obtain

$$A_1 = \frac{\dot{m}v_1}{V_1} = \frac{(1.2\ \text{lbm/s})(1.628\ \text{ft}^3/\text{lbm})}{100\ \text{ft/s}} = 0.0195\ \text{ft}^2$$

$$A_2 = \frac{\dot{m}v_2}{V_2} = \frac{(1.2)(13.32)}{1205} = 0.0133\ \text{ft}^2$$

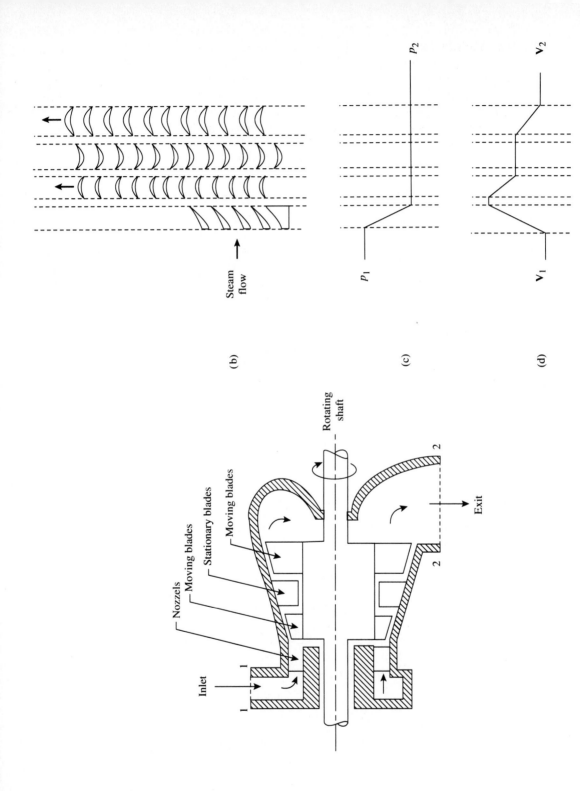

Figure 5-6 A two-velocity-stage impulse steam turbine: (a) schematic of the turbine, (b) schematic of the blades, (c) pressure variation, and (d) velocity variation.

5-5 STEAM AND GAS TURBINES

In a *steam* or *gas turbine,* the high-energy working fluid expands and does work against a series of curved blades attached on a rotating shaft. Figure 5-6 shows schematically a two-velocity-stage impulse steam turbine. An impulse turbine is characterized by the use of nozzles to provide high-velocity streams that, in turn, do work on the moving blades. The turbine shown in Fig. 5-6 has two rows of moving blades, each absorbing a portion of the kinetic energy of the high-velocity jets, thereby producing shaft work. The intermediate guide passages are not nozzles; they are there simply to ensure the direction change of the fluid streams. The variations in pressure and absolute velocity of the fluid are also shown in the figure. Notice carefully that the velocity increase in the nozzle is due to the drop in pressure, whereas the velocity decrease in the moving blades is responsible for the work done on the rotating shaft while keeping the pressure unchanged.

The preceding paragraph is a brief introduction on how a turbine works. As far as the first-law analysis of this device is concerned, we are interested only in the energy and mass quantities that cross the control surface. By denoting the inlet and exit sections of a turbine by the subscripts 1 and 2, respectively, as shown in Fig. 5-6, the steady-flow, steady-state energy equation, Eq. (5-10), can be used directly. The potential-energy change is normally negligible because of small changes in elevation. Furthermore, in a well-insulated turbine, the heat loss through the turbine housing is normally small compared to the shaft work. Under these conditions, Eq. (5-10) reduces to

$$w_{cv} = \cancel{q}^{\,0} + (h_1 - h_2) + \tfrac{1}{2}(V_1^2 - V_2^2) + g(z_1 \cancel{-} z_2)^0$$

or

$$w_{cv} = w_{out} = (h_1 - h_2) + \tfrac{1}{2}(V_1^2 - V_2^2)$$

Notice also that for many turbine applications, the change of velocity of the fluid at the inlet and the exit is small. When the kinetic-energy change is neglected, the energy equation further reduces to

$$w_{cv} = w_{out} = h_1 - h_2$$

Example 5-3

Steam at 5 MPa and 400°C enters a turbine with a velocity of 50 m/s. It leaves the turbine at 50 kPa and 90% quality with a velocity of 200 m/s. The mass-flow rate of steam is 1000 kg/min. The elevation decreases by 1 m from inlet to exit. The rate of heat loss to the surroundings amounts to 10 MJ/min. Determine the power output of the turbine.

Solution The process is shown on a p–v diagram in Fig. 5-7. The given data are $p_1 = 5$ MPa, $T_1 = 400$°C, $V_1 = 50$ m/s, $p_2 = 50$ kPa, $x_2 = 90\%$, $V_2 = 200$ m/s, $z_1 - z_2 = 1$ m, $\dot{m} = 1000$ kg/min, and $\dot{Q} = -10$ MJ/min.

From the steam tables (Tables A-19M and A-20M), at p_1 and T_1, we get $h_1 = 3195.7$ kJ/kg, and at p_2 and x_2, we have

$$h_2 = h_f + x_2 h_{fg} = 340.49 + 0.90(2305.4) = 2415.4 \text{ kJ/kg}$$

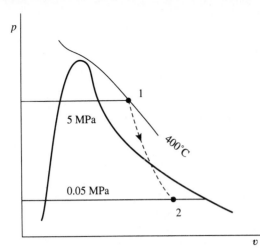

v **Figure 5-7** $p-v$ diagram for Example 5-3.

The work done by the steam is calculated from the steady-flow, steady-state energy equation, Eq. (5-10), on a unit-mass basis

$$w_{cv} = (h_1 - h_2) + \tfrac{1}{2}(\mathbf{V}_1^2 - \mathbf{V}_2^2) + g(z_1 - z_2) + q$$

$$= 3195.7 \text{ kJ/kg} - 2415.4 \text{ kJ/kg}$$

$$+ \frac{(50 \text{ m/s})^2 - (200 \text{ m/s})^2}{2}\left(\frac{1 \text{ N}}{1 \text{ kg·m/s}^2}\right)\left(\frac{1 \text{ kJ}}{1000 \text{ N·m}}\right)$$

$$+ (9.807 \text{ m/s}^2)(1 \text{ m})\left(\frac{1 \text{ N}}{1 \text{ kg·m/s}^2}\right)\left(\frac{1 \text{ kJ}}{1000 \text{ N·m}}\right) + \frac{(-10,000 \text{ kJ/min})}{(1000 \text{ kg/min})}$$

$$= 780.3 - 18.75 + 0.0098 - 10.0$$

$$= 751.6 \text{ kJ/kg}$$

The power output of the turbine is then

$$\dot{W}_{cv} = \dot{m}w_{cv} = \left(\frac{1000}{60} \text{ kg/s}\right)(751.6 \text{ kJ/kg}) = 12,530 \text{ kW}$$

When ΔKE, ΔPE, and heat loss are all neglected, the work done would be

$$w_{cv} = h_1 - h_2 = 780.3 \text{ kJ/kg}$$

and the power output would be

$$\dot{W}_{cv} = \dot{m}w_{cv} = \frac{1000}{60}(780.3) = 13,010 \text{ kW}$$

The error introduced by this calculation amounts to

$$\frac{13,010 - 12,530}{12,530} \times 100 = 3.8\%$$

5-6 COMPRESSORS AND PUMPS

Compressors and pumps are machines that are used to increase the pressure of fluid passing through them. The machine is called a *compressor* when the fluid is a gas or vapor. It is called a *pump* when the fluid is a liquid. Figure 5-8(a) shows a schematic dia-

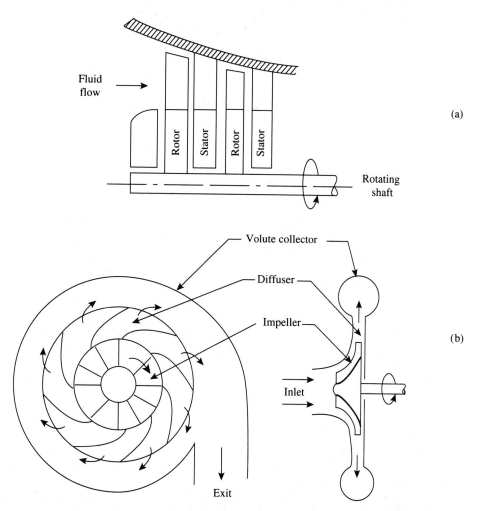

Figure 5-8 Schematics of rotating compressors: (a) axial-flow and (b) centrifugal.

gram of a section of an axial-flow rotating compressor. The function of this device is in a manner opposite to that of a turbine with expenditure of work on the working fluid. Each stage of the compressor consists of a rotor (one set of moving blades) and a stator (one set of fixed blades). The rotor does work on the fluid, thus raising the kinetic energy and pressure of the fluid, whereas the stator serves as a diffuser to reduce the fluid velocity and raise the fluid pressure, and, at the same time, redirect the fluid into the next rotor. Figure 5-8(b) shows a schematic diagram of a centrifugal compressor. Gas enters the rotating impeller axially and is compressed while developing a velocity in the radial direction because of centrifugal force. As the gas flows radially through the stationary diffuser, its velocity decreases, thus causing further increases in pressure. The contour of the impeller vanes can be radial, backward-bent, or forward-bent. Figure 5-8(b) shows radial vanes.

The work required to operate a compressor or pump can be calculated by the steady-flow, steady-state energy equation, (Eq. 5-10), on a unit-mass basis:

$$w_{in} = -w_{cv} = (h_2 - h_1) + \tfrac{1}{2}(\mathbf{V}_2^2 - \mathbf{V}_1^2) + g(z_2 - z_1) + (-q)$$

where the $-q$ term represents heat rejection from the working fluid. In Sec. 10-7, we will demonstrate that it is advantageous to cool the gas as it is compressed. Cooling a gas in a rotating compressor, however, is difficult, thus making the compression process for the compressor more or less adiabatic. In many applications, the ΔKE and ΔPE terms in the energy equation are negligible. In the pumping of a liquid against a large elevation difference, however, the ΔPE term can be significant.

Example 5-4

Nitrogen is compressed in a steady-flow process from 100 kPa, 300 K, to 500 kPa, 500 K. The volume-flow rate at the initial conditions is 50 m³/min. There is a heat loss of 5 kJ/kg to the environment. The changes in kinetic and potential energies are negligible. Determine the power requirement.

Solution   Given: $p_1 = 100$ kPa, $T_1 = 300$ K, $\dot{V}_1 = 50$ m³/min, $p_2 = 500$ kPa, $T_2 = 500$ K, $q = -5$ kJ/kg, ΔKE $= 0$, and ΔPE $= 0$.
   At the given pressure and temperature, nitrogen can be considered an ideal gas. The specific volume at the initial conditions is

$$v_1 = \frac{RT_1}{p_1} = \left(\frac{8.314 \text{ kJ/kgmole·K}}{28 \text{ kg/kgmole}}\right)\left(\frac{300 \text{ K}}{100 \text{ kN/m}^2}\right) = 0.891 \text{ m}^3/\text{kg}$$

The mass-flow rate of nitrogen is then

$$\dot{m} = \frac{\dot{V}_1}{v_1} = \frac{50 \text{ m}^3/\text{min}}{0.891 \text{ m}^3/\text{kg}} = 56.1 \text{ kg/min}$$

From the ideal gas property table for nitrogen (Table A-10), we obtain: at $T_1 = 300$ K,

$$h_1 = 8724.1 \text{ kJ/kgmole} = \frac{8724.1}{28} = 311.6 \text{ kJ/kg}$$

at $T_2 = 500$ K,

$$h_2 = 14581.0 \text{ kJ/kgmole} = \frac{14581.0}{28} = 520.8 \text{ kJ/kg}$$

The steady-flow, steady-state energy equation, Eq. (5-10), gives

$$w_{in} = -w_{cv} = (h_2 - h_1) + \tfrac{1}{2}(V_2^2 \cancel{-} V_1^2)^0 + g(z_2 \cancel{-} z_1)^0 - q$$

or

$$w_{in} = (h_2 - h_1) - q = 520.8 - 311.6 + 5 = 214.2 \text{ kJ/kg}$$

The power requirement is then

$$\dot{W}_{in} = \dot{m}w_{in} = (56.1/60 \text{ kg/s})(214.2 \text{ kJ/kg}) = 200 \text{ kW}$$

5-7 HEAT EXCHANGERS AND MIXING CHAMBERS

A *heat exchanger* is a device designed for the transfer of heat between fluids. Boilers, condensers, hot-water heaters, and car radiators are familiar examples of heat exchangers. Figure 5-9 shows some basic types of heat exchangers. Parts (a) and (b) of this figure are simple double-tube heat exchangers with one fluid flowing through the inner tube and a second fluid flowing through the annular space between the tubes. Heat is

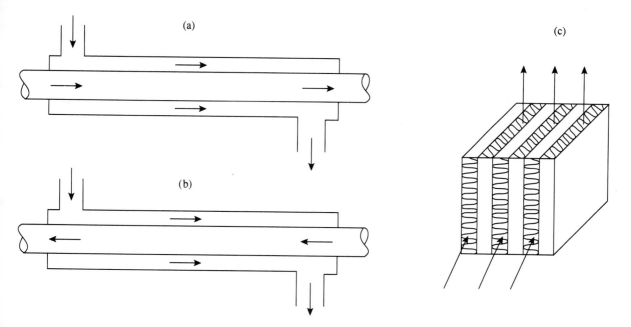

Figure 5-9 Basic types of heat exchangers: (a) tube-within-a-tube parallel flow, (b) tube-within-a-tube counterflow, and (c) crossflow with both flows unmixed.

transferred between the two fluids through the wall separating them. If both fluids flow in the same direction, the exchanger is a parallel-flow type; if the fluids flow in opposite directions, it is a counterflow type. When the two fluids flow at right angles to each other, the heat exchanger is of the crossflow type. In crossflow heat exchangers, each of the fluids can be either unmixed or mixed as it flows through the exchanger. Figure 5-9(c) shows a crossflow heat exchanger with both fluids unmixed. In order to increase the capacity and the heat-transfer area, the concept of a tube-within-a-tube design is usually applied to form many different shell-and-tube heat-exchange varieties, with one large shell containing many small tubes. Two such heat exchangers are shown in Fig. 5-10.

Heat exchangers typically involve no shaft work ($W_{cv} = 0$) and negligible change in potential energy. In most cases, the change in kinetic energy can also be neglected. Even though the transfer of heat is the main concern of a heat exchanger, but whether or not to include the heat transfer rate in an energy analysis will depend on the selection of a control volume. For example, when the control volume as shown in Fig. 5-11(a) is selected, the heat transfer between the two fluids occurring inside the control volume does not concern us when we write an energy equation for the control volume. Thus, if the heat exchanger is well insulated on the outside, the $\dot{Q}$ term in the energy equation is zero. If, however, the control volume as shown in Fig. 5-11(b) is selected, then heat will cross the control surface as it flows from one fluid to the other and $\dot{Q}$ is nonzero. Thus, according to Eq. (5-7), the steady-flow, steady-state energy equation for the control volume shown in Fig. 5-11(a) is

$$\cancel{\dot{Q}}^{0} - \cancel{\dot{W}_{cv}}^{0} + \dot{m}_a[(h_1 - h_2) + \tfrac{1}{2}(V_1^2 \cancel{-} \cancel{V_2^2}^{0}) + g(z_1 \cancel{-} \cancel{z_2}^{0})]$$

$$+ \dot{m}_b[(h_3 - h_4) + \tfrac{1}{2}(V_3^2 \cancel{-} \cancel{V_4^2}^{0}) + g(z_3 \cancel{-} \cancel{z_4}^{0})] = 0$$

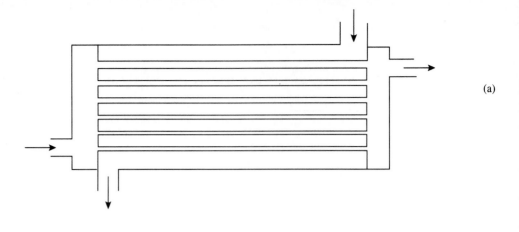

(a)

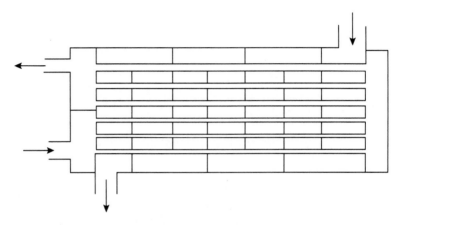

(b)

Figure 5-10 Shell-and-tube heat exchangers: (a) one tube pass, one shell pass; and (b) two tube passes, one shell pass with baffles.

or

$$\dot{m}_a(h_1 - h_2) = \dot{m}_b(h_4 - h_3)$$

However, the energy equation for the control volume shown in Fig. 5-11(b) is

$$\dot{Q} - \cancel{\dot{W}_{cv}}^{0} + \dot{m}_a[(h_1 - h_2) + \tfrac{1}{2}(V_1^2 \cancel{-} V_2^2)^{0} + g(z_1 \cancel{-} z_2)^{0}] = 0$$

or

$$\dot{Q} = \dot{m}_a(h_2 - h_1)$$

Combining the preceding two equations, we must have

$$\dot{Q} = \dot{m}_a(h_2 - h_1) = \dot{m}_b(h_3 - h_4)$$

which is the heat-transfer rate from fluid b to fluid a.

In engineering applications, there is often a need to mix two fluid streams to form a mixture stream in a steady-flow mixing process. By referring to the schematic dia-

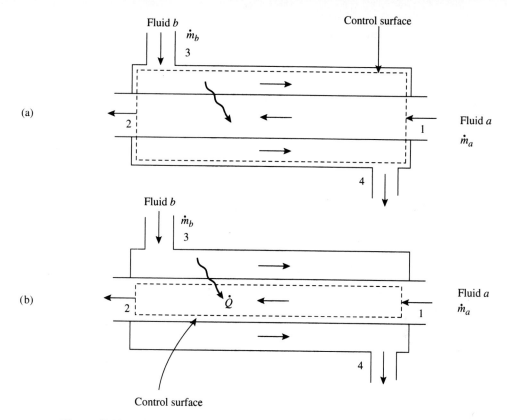

Figure 5-11 Two different control volumes for an insulated heat exchanger.

gram of a mixing chamber shown in Fig. 5-12, there is no work interaction across the control surface, and no heat interaction if the chamber is well insulated. From Eq. (5-7), the steady-flow, steady-state energy equation is written as

$$\cancel{\dot{Q}}^{0} - \cancel{\dot{W}}_{cv}^{0} + \dot{m}_1\left(h_1 + \frac{V_1^2}{2}\right) + \dot{m}_2\left(h_2 + \frac{V_2^2}{2}\right) - \dot{m}_3\left(h_3 + \frac{V_3^2}{2}\right) = 0$$

where the changes in potential energy are neglected. This equation can be solved simultaneously with the mass-balance equation, which is

$$\dot{m}_1 + \dot{m}_2 = \dot{m}_3$$

Control surface

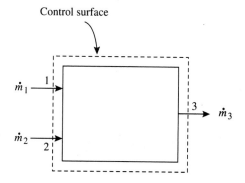

Figure 5-12 A mixing chamber.

Example 5-5

A heat exchanger as depicted diagrammatically in Fig. 5-13 is used to transfer heat between helium (He) and Refrigerant-12 (R-12). The heat exchanger is insulated. Helium gas flows in at section 1 through a 0.5-ft² pipe with a pressure of 100 psia, a temperature of 950°F, and a velocity of 500 ft/s. Helium leaves at section 2 through a 1.0-ft² pipe at 50 psia and 250°F. Refrigerant-12 flows in at section 3 with a pressure of 20 psia and a temperature of 80°F, and leaves at section 4 at 10 psia and 200°F. The flow velocities of Refrigerant-12 are low. Determine (a) the exit velocity of helium and (b) the mass-flow rates of helium and Refrigerant-12.

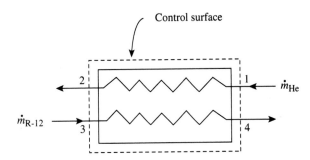

Figure 5-13 Schematic diagram for Example 5-5.

Solution   Given data: $p_1 = 100$ psia, $T_1 = 950°F = 1410°R$, $A_1 = 0.5$ ft², $V_1 = 500$ ft/s, $p_2 = 50$ psia, $T_2 = 250°F = 710°R$, $A_2 = 1.0$ ft², $p_3 = 20$ psia, $T_3 = 80°F = 540°R$, $p_4 = 10$ psia, $T_4 = 200°F = 660°R$, $V_3 \approx 0$, $V_4 \approx 0$, and $\dot{Q} = 0$.

At the given conditions, helium can be considered an ideal gas. From the continuity equation for the helium gas,

$$\dot{m}_{He} = \frac{A_1 V_1}{v_1} = \frac{A_2 V_2}{v_2} \qquad \text{or} \qquad \frac{A_1 V_1 p_1}{RT_1} = \frac{A_2 V_2 p_2}{RT_2}$$

so that

$$V_2 = V_1 \left(\frac{A_1}{A_2}\right)\left(\frac{p_1}{p_2}\right)\left(\frac{T_2}{T_1}\right)$$

$$= (500 \text{ ft/s})\left(\frac{0.5}{1.0}\right)\left(\frac{100}{50}\right)\left(\frac{710}{1410}\right) = 252 \text{ ft/s}$$

The mass-flow rate of helium is

$$\dot{m}_{He} = \frac{A_1 V_1 p_1}{RT_1} = \frac{(0.5 \text{ ft}^2)(500 \text{ ft/s})(100 \times 144 \text{ lbf/ft}^2)}{(1545/4 \text{ ft·lbf/lbm·°R})(1410°R)}$$

$$= 6.61 \text{ lbm/s}$$

From the Refrigerant-12 table (Table A-23E), we obtain the following:

$$h_3 = 89.168 \text{ Btu/lbm at } p_3 = 20 \text{ psia and } T_3 = 80°F$$

$$h_4 = 107.957 \text{ Btu/lbm at } p_4 = 10 \text{ psia and } T_4 = 200°F$$

For helium, $c_p = 5.0$ Btu/lbmole·°F $= 5.0/4 = 1.25$ Btu/lbm·°F and $h_1 - h_2 = c_p(T_1 - T_2)$.

By neglecting changes in potential energy, with $\dot{Q} = 0$ and $\dot{W}_{cv} = 0$, the steady-flow, steady-state energy equation for the control volume gives

$$\cancel{\dot{Q}}^{0} - \cancel{\dot{W}_{cv}}^{0} + \dot{m}_{He}[(h_1 - h_2) + \tfrac{1}{2}(\mathbf{V}_1^2 - \mathbf{V}_2^2) + g(z_1 \cancel{-z_2})^{0}]$$

$$+ \dot{m}_{R-12}[(h_3 - h_4) + \tfrac{1}{2}(\cancel{\mathbf{V}_3^2} \cancel{-\mathbf{V}_4^2}) + g(z_3 \cancel{-z_4})^{0}] = 0$$

or

$$\dot{m}_{R-12}(h_4 - h_3) = \dot{m}_{He}[(h_1 - h_2) + \tfrac{1}{2}(\mathbf{V}_1^2 - \mathbf{V}_2^2)]$$

so that

$$\dot{m}_{R-12} = \frac{\dot{m}_{He}[c_{p,He}(T_1 - T_2) + \tfrac{1}{2}(\mathbf{V}_1^2 - \mathbf{V}_2^2)]}{h_4 - h_3}$$

$$= \frac{6.61 \text{ lbm/s}}{(107.96 - 98.17) \text{ Btu/lbm}}$$

$$\times \left[(1.25 \text{ Btu/lbm} \cdot °F)(950°F - 250°F) + \frac{(500 \text{ ft/s})^2 - (252 \text{ ft/s})^2}{2(32.2 \text{ ft} \cdot \text{lbm/lbf} \cdot \text{s}^2)(778 \text{ ft} \cdot \text{lbf/Btu})}\right]$$

$$= 593.3 \text{ lbm/s}$$

Example 5-6

Water at 0.5 MPa and 15°C flows through a large pipe at a rate of 40 kg/min into an insulated tank at section 1 (see Fig. 5-12). Steam at 0.5 MPa and 200°C flows into the same tank at section 2 through a 2.5-cm² pipe at a rate of 6 kg/min. The steam then condenses and mixes with the water to form heated water, which leaves the tank at section 3 through a large pipe at 0.5 MPa. Determine the temperature of the heated water at the tank exit point.

Solution The given data are $\dot{Q} = 0$; for the incoming water, $p_1 = 0.5$ MPa, $T_1 = 15°C$, $\dot{m}_1 = 40$ kg/min, and $\mathbf{V}_1 \approx 0$; for the incoming steam, $p_2 = 0.5$ MPa, $T_2 = 200°C$, $A_2 = 2.5 \times 10^{-4}$ m², and $\dot{m}_2 = 6$ kg/min; and for the heated water, $p_3 = 0.5$ MPa and $\mathbf{V}_3 \approx 0$.

Water at 0.5 MPa and 15°C is a subcooled (or compressed) liquid. But 0.5 MPa is a low pressure, so that from the saturated steam table (Table A-18M), we obtain

$$h_1 \approx h_f \text{ at } 15°C = 62.99 \text{ kJ/kg}$$

Steam at 0.5 MPa and 200°C is a superheated vapor; thus, from the superheated steam table (Table A-20M), we obtain

$$h_2 = 2855.4 \text{ kJ/kg} \qquad \text{and} \qquad v_2 = 0.4249 \text{ m}^3/\text{kg}$$

The flow velocity of steam at section 2 is calculated as

$$\mathbf{V}_2 = \frac{\dot{m}_2 v_2}{A_2} = \frac{(6 \text{ kg/min})(0.4249 \text{ m}^3/\text{kg})}{(2.5 \times 10^{-4} \text{ m}^2)(60 \text{ s/min})} = 170 \text{ m/s}$$

With $\dot{Q} = 0$ and $\dot{W}_{cv} = 0$, by neglecting changes in potential energy, the steady-flow, steady-state energy equation, Eq. (5-7), for the control volume is

$$\dot{m}_1(h_1) + \dot{m}_2\left(h_2 + \frac{\mathbf{V}_2^2}{2}\right) = (\dot{m}_1 + \dot{m}_2)(h_3)$$

or

$$h_3 = \frac{1}{\dot{m}_1 + \dot{m}_2}\left[\dot{m}_1 h_1 + \dot{m}_2\left(h_2 + \frac{\mathbf{V}_2^2}{2}\right)\right]$$

$$= \frac{1}{(40 + 6)\ \text{kg/min}}\Bigg\{(40\ \text{kg/min})(62.99\ \text{kJ/kg})$$

$$+ \left(6\frac{\text{kg}}{\text{min}}\right)\left[\left(2855.4\frac{\text{kJ}}{\text{kg}}\right) + \tfrac{1}{2}(170\ \text{m/s})^2\left(\frac{1\ \text{N}}{1\ \text{kg·m/s}^2}\right)\left(\frac{1\ \text{kJ}}{1000\ \text{N·m}}\right)\right]\Bigg\}$$

$$= 429.1\ \text{kJ/kg}$$

At $p_3 = 0.5$ MPa and $h_3 = 429.1$ kJ/kg, water is in a subcooled (or compressed) liquid state. But p_3 is a low pressure, so that $h_3 \approx h_f$ at T_3. From the saturated steam table (Table A-18M), by interpolation, we obtain

$$T_3 = 102.4°C$$

5-8 THROTTLING PROCESS AND PIPE FLOW

A *throttling process* occurs when a fluid flows steadily through a restriction such as a partially opened valve, a porous plug, or a long capillary tube, with a resulting drop in pressure. Figure 5-14 depicts a throttle valve as the restriction, through which the fluid flows from the high-pressure region 1 to the low-pressure region 2. When the flow is sufficiently low, the fluid has well-defined pressure and temperature on both sides of the restriction. By the nature of the process, no shaft work is involved. In addition, in most applications, the process is carried out adiabatically. By neglecting the change in potential energy, if any, the steady-flow, steady-state energy equation, Eq. (5-10), becomes

$$\cancel{q}^{0} - \cancel{w}_{\text{cv}}^{0} + (h_1 - h_2) + \tfrac{1}{2}(\mathbf{V}_1^2 - \mathbf{V}_2^2) + g(z_1 \cancel{-} z_2)^{0} = 0$$

or

$$h_1 + \frac{\mathbf{V}_1^2}{2} = h_2 + \frac{\mathbf{V}_2^2}{2}$$

The kinetic-energy terms can be omitted if they are kept equal or approximately equal by making area 2 larger than area 1, or else by simply assuming that the volume rate of flow is small and the difference in kinetic energy is negligible. Accordingly, the steady-

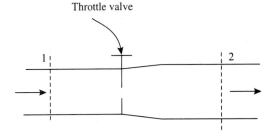

Figure 5-14   Throttling process.

flow equation becomes simply

$$h_1 = h_2 \qquad (5\text{-}11)$$

In other words, in a throttling process, the initial and final enthalpies are equal. Note that because the fluid passes through nonequilibrium states on its way from the initial equilibrium state to the final equilibrium state, the process is nonquasiequilibrium. The process is not at constant enthalpy, but only satisfies the condition that initial enthalpy equals final enthalpy.

We now turn our attention to the second topic of this section: pipe flow. For the transportation of fluids in engineering applications, the pipes or ducts may have different sizes at different sections, and the transportation may involve large elevation changes. Accordingly, it is essential to include the changes in kinetic and potential energies in the flow equation. In addition, the pipe system may not be insulated, so that heat transfer between the system and the surroundings may be considerable. If there is a pump or a turbine located between the inlet and exit sections of the pipe, the shaft work input or output should be taken care of. Denoting the pipe inlet and exit sections by the subscripts 1 and 2, respectively, the steady-flow, steady-state energy equation, Eq. (5-10), on a unit mass-flow basis is

$$q - w_{cv} = (h_2 - h_1) + \tfrac{1}{2}(\mathbf{V}_2^2 - \mathbf{V}_1^2) + g(z_2 - z_1)$$

For incompressible fluids (e.g., liquids), v = constant, and from Eq. (4-11), we have $du = c\,dT$, or for a constant c value,

$$u_2 - u_1 = c(T_2 - T_1)$$

Thus,

$$h_2 - h_1 = (u_2 - u_1) + v(p_2 - p_1) = c(T_2 - T_1) + v(p_2 - p_1)$$

Substituting this equation into Eq. (5-10) for the special case of adiabatic flow with no shaft work involved ($q = 0$, $w_{cv} = 0$), we have

$$c(T_2 - T_1) + v(p_2 - p_1) + \tfrac{1}{2}(\mathbf{V}_2^2 - \mathbf{V}_1^2) + g(z_2 - z_1) = 0$$

When energy dissipation due to friction and turbulence is absent in the steady adiabatic flow of an incompressible fluid, the fluid temperature will be constant. Accordingly, the preceding equation reduces to

$$v(p_2 - p_1) + \tfrac{1}{2}(\mathbf{V}_2^2 - \mathbf{V}_1^2) + g(z_2 - z_1) = 0 \qquad (5\text{-}12)$$

This is the well-known *Bernoulli equation* in fluid mechanics. It is imperative to notice that the Bernoulli equation is derived in fluid mechanics from the momentum principle based on Newton's second law for frictionless incompressible fluid flow with no shaft work involved. It is not a special case of the steady-flow energy equation as the deduction used here might suggest. We present the Bernoulli equation here because of its usefulness, not because of its relation with the steady-flow energy equation.

Example 5-7

Because pressure p and temperature T are dependent variables of a pure substance in a two-phase region, the measurement of p and T is inadequate for the determination of the quality of a wet steam. A throttling calorimeter can be used to determine the quality of a liquid–vapor mixture by throttling it to the superheated region and measuring its pressure and temperature. A sample of wet steam at 1 MPa is throttled in a calorimeter to 50 kPa and 100°C. Determine the quality of the original sample of steam.

Solution Denoting the initial and final states of the steam by the subscripts 1 and 2, respectively, for the throttling process, we have

$$h_1 = h_2$$

From the superheated steam table (Table A-20M), at $p_2 = 50$ kPa $= 0.05$ MPa and $T_2 = 100°C$, we obtain $h_2 = 2682.5$ kJ/kg. Thus,

$$h_1 = h_2 = 2682.5 = (h_f + x_1 h_{fg}) \text{ at } p_1 = 1 \text{ MPa}$$

$$= 762.8 + x_1(2015.3)$$

Therefore, the quality of the original sample of steam is

$$x_1 = \frac{2682.5 - 762.8}{2015.3} = 0.953 = 95.3\%$$

Figure 5-15 shows the throttling process on a p–v diagram.

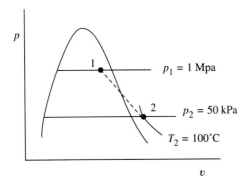

Figure 5-15 p–v diagram for Example 5-7.

Example 5-8

Brine having a density of 75 lbm/ft³ is pumped up a 10-ft hill at a flow rate of 4.5 ft³/s. The pump inlet pipe is 12 in. in diameter, and the exit pipe at the top of the hill is 8 in. in diameter. The pressure at the pump inlet is 12 psia and the pressure at the uphill pipe outlet is 35 psia. Neglecting heat loss, fluid temperature change, and friction in the piping, determine the pump input power.

Solution Figure 5-16 shows a schematic of the pump–piping system, where the pump inlet section and the pipe outlet section at the top of the hill are denoted by the subscripts 1 and 2, respectively. The given data are $\rho = 75$ lbm/ft³ $=$ constant, $z_2 - z_1 = 10$ ft, $\dot{V} = 4.5$ ft³/s, $D_1 = 12$ in., $D_2 = 8$ in., $p_1 = 12$ psia, and $p_2 = 35$ psia. The pipe areas are

$$A_1 = \frac{\pi}{4}\left(\frac{12}{12}\right)^2 = 0.785 \text{ ft}^2$$

$$A_2 = \frac{\pi}{4}\left(\frac{8}{12}\right)^2 = 0.349 \text{ ft}^2$$

First Law Analysis for Control Volumes   Chap. 5

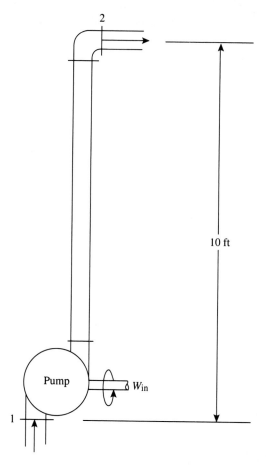

Figure 5-16   Schematic diagram for Example 5-8.

By the continuity equation for incompressible fluid,

$$\dot{V} = A_1 \mathbf{V}_1 = A_2 \mathbf{V}_2$$

the flow velocities are calculated as

$$\mathbf{V}_1 = \frac{\dot{V}}{A_1} = \frac{4.5 \text{ ft}^3/\text{s}}{0.785 \text{ ft}^2} = 5.73 \text{ ft/s}$$

$$\mathbf{V}_2 = \frac{\dot{V}}{A_2} = \frac{4.5}{0.349} = 12.89 \text{ ft/s}$$

By neglecting heat loss ($q = 0$), fluid temperature change ($T_1 = T_2$), and friction in the piping, for the control volume between sections 1 and 2 of the pump-pipe system, the steady-flow, steady-state energy equation is

$$\cancel{q}^{\,0} - w_{cv} = v(p_2 - p_1) + \tfrac{1}{2}(\mathbf{V}_2^2 - \mathbf{V}_1^2) + g(z_2 - z_1)$$

or

$$w_{in} = -w_{cv} = \frac{1}{\rho}(p_2 - p_1) + \tfrac{1}{2}(\mathbf{V}_2^2 - \mathbf{V}_1^2) + g(z_2 - z_1)$$

Sec. 5-8   Throttling Process and Pipe Flow

143

where w_{in} is the pump input work. Substituting numerical values, we obtain

$$w_{in} = \frac{1}{75 \text{ lbm/ft}^3}[(35 - 12)144 \text{ lbf/ft}^2]$$

$$+ \frac{1}{2(32.2 \text{ ft·lbm/lbf·s}^2)}[(12.89 \text{ ft/s})^2 - (5.73 \text{ ft/s})^2]$$

$$+ \frac{32.2 \text{ ft/s}^2}{32.2 \text{ ft·lbm/lbf·s}^2}(10 \text{ ft})$$

$$= 44.16 + 2.07 + 10 = 56.23 \text{ ft·lbf/lbm}$$

The pump power requirement is then

$$\dot{W}_{in} = \frac{(4.5 \text{ ft}^3/\text{s})(75 \text{ lbm/ft}^3)(56.23 \text{ ft·lbf/lbm})}{550 \text{ ft·lbf/s·hp}} = 34.5 \text{ hp}$$

5-9 UNIFORM-FLOW, UNIFORM-STATE TRANSIENT ANALYSIS

In Sec. 5-2, we presented the following general first-law equation, Eq. (5-5), for a control volume on a rate basis:

$$\dot{Q} - \dot{W}_{cv} + \sum_{\text{inlet}} \dot{m}\left(h + \frac{\mathbf{V}^2}{2} + gz\right) - \sum_{\text{exit}} \dot{m}\left(h + \frac{\mathbf{V}^2}{2} + gz\right) = \frac{dE_{cv}}{dt}$$

For an unsteady-flow process, the properties of the mass at each inlet and exit may be changing with time. As a result, the energy transport with mass flow at the crossing of the control surface in an unsteady-flow process can only be determined by integrating the energy content associated with differential mass flows. Thus, writing the preceding energy equation in differential form, we have

$$đQ - đW_{cv} + \sum_{\text{inlet}}\left(h + \frac{\mathbf{V}^2}{2} + gz\right)đm - \sum_{\text{exit}}\left(h + \frac{\mathbf{V}^2}{2} + gz\right)đm = dE_{cv}$$

Integrating this equation over the time interval of interest gives

$$Q - W_{cv} + \sum_{\text{inlet}}\int\left(h + \frac{\mathbf{V}^2}{2} + gz\right)đm - \sum_{\text{exit}}\int\left(h + \frac{\mathbf{V}^2}{2} + gz\right)đm = \Delta E_{cv} \qquad (5\text{-}13)$$

Notice that in the preceding equations, we use $đm$ instead of dm, because the amount of mass crossing the control surface is not a property, even though the mass within the control volume is a property.

In the transient-flow energy equation, the integration of the terms $(h + \mathbf{V}^2/2 + gz)\, đm$ for each inlet and exit stream requires the knowledge of the property in relation with mass flow. This will be, in general, a rather complicated undertaking. Some transient-flow processes, however, can be simplified with reasonable accuracy by adopting a model, called the *uniform-flow* and *uniform-state* process, which includes the following assumptions:

1. The state of the fluid within the control volume is uniform throughout even though this uniform state may change with time.
2. Although the mass-flow rate may change with time, the state of the fluid crossing the control surface is invariable.

Notice that the uniform-state assumption is different from that of steady state. In steady state, the thermodynamic state can vary throughout the control volume, but the state of the fluid at a particular point in the control volume does not change with time.

According to the uniform-flow, uniform-state model, for each inlet and exit stream, we can simply write

$$\int \left(h + \frac{\mathbf{V}^2}{2} + gz \right) dm = m \left(h + \frac{\mathbf{V}^2}{2} + gz \right)$$

and Eq. (5-13) becomes

$$Q - W_{cv} + \sum_{\text{inlet}} m \left(h + \frac{\mathbf{V}^2}{2} + gz \right) - \sum_{\text{exit}} m \left(h + \frac{\mathbf{V}^2}{2} + gz \right)$$

$$= (m_f u_f - m_i u_i) + \tfrac{1}{2}(m_f \mathbf{V}_f^2 - m_i \mathbf{V}_i^2) + g(m_f z_f - m_i z_i) \tag{5-14}$$

where the subscripts i and f denote, respectively, the initial and final uniform states of the entire control volume. We will illustrate the use of this equation in the next two sections.

5-10 CHARGING AND DISCHARGING RIGID TANKS

Under some restricted conditions, the charging and discharging of rigid tanks can be considered to occur under a uniform-flow, uniform-state model. First, consider the charging of an insulated rigid tank from a large gas supply line, as depicted schematically in Fig. 5-17. As the supply line is large and is continuously supplied with gas at constant pressure and temperature, the properties of the gas at the inlet to the tank do not change during the charging. Assume that the fluid flows into the tank slowly so that the fluid properties inside the tank change uniformly throughout the tank. There are no heat and work interactions between the control volume and its surroundings and no exit flow from the control volume. By neglecting charges in kinetic and potential energies,

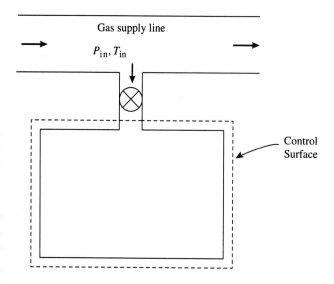

Figure 5-17 Charging a rigid tank.

Eq. (5-14) becomes

$$\cancel{\phi}^{\,0} - \cancel{W}_{cv}^{\,0} + m_{in}\left(h_{in} + \cancel{\frac{\mathbf{V}_{in}^2}{2}}^{\,0} + \cancel{gz_{in}}^{\,0}\right) - \cancel{m_{exit}}^{\,0}\left(h_{exit} + \frac{\mathbf{V}_{exit}^2}{2} + gz_{exit}\right)$$

$$= (m_f u_f - m_i u_i) + \tfrac{1}{2}(m_f \mathbf{V}_f^2 \cancel{- m_i \mathbf{V}_i^2})^{\,0} + g(m_f z_f \cancel{- m_i z_i})^{\,0}$$

or

$$m_{in} h_{in} = m_f u_f - m_i u_i$$

Because

$$m_{in} = m_f - m_i$$

we have, therefore,

$$(m_f - m_i)h_{in} = m_f u_f - m_i u_i \qquad (5\text{-}15)$$

In the special case when the tank is initially evacuated, $m_i = 0$, the preceding equation is further simplified to read

$$h_{in} = u_f \qquad (5\text{-}16)$$

That is, the final internal energy of the fluid in the tank is equal to the enthalpy of the fluid entering the tank. Equation (5-16) is applicable to any fluid, such as steam or real gas. To gain some physical significance of this equation, let us apply it to the case of an ideal gas with constant specific heats. Because $h = c_p T$ and $u = c_v T$ for an ideal gas based on $h = 0$ and $u = 0$ at the absolute zero temperature, Eq. (5-16) can be written as

$$c_p T_{in} = c_v T_f$$

or

$$T_f = \frac{c_p}{c_v} T_{in} = k T_{in}$$

where k is the specific-heat ratio. For many gases, the value of k ranges from 1.2 to 1.4, making the absolute temperature of the gas in the tank greater than that of the inlet gas by that ratio. Because for an ideal gas $u = f(T)$ only, the temperature of the gas before the inlet to the tank is a measure of its internal energy u_{in}, whereas the temperature of the gas after the inlet to the tank is a measure of u_f, which, in this case, is equal to $h_{in} = u_{in} + (pv)_{in}$. Thus, the original flow work $(pv)_{in}$ becomes part of the gas internal energy in the tank, thereby indicating a greater absolute temperature T_f.

Now consider the discharging of a fluid from a storage tank through a throttling valve into the environment, as depicted schematically in Fig. 5-18. Assume that the discharging rate is slow so that the properties of the fluid within the tank are uniform throughout at all times. There is no work interaction involved in the process and no inflow to the tank. All changes in kinetic and potential energies are neglected. As the pressure within the tank drops, the discharging mass-flow rate $\dot{m}_{exit}$ decreases and the enthalpy h_{exit} of the escaping fluid also decreases. To satisfy the uniform-flow, uniform-state requirement, one can use an average h_{exit} or add heat to the fluid to maintain a

First Law Analysis for Control Volumes   Chap. 5

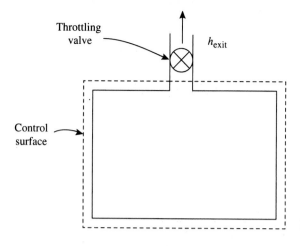

Throttling valve

h_{exit}

Control surface

Figure 5-18   Discharging process.

constant h_{exit}. Under these conditions, Eq. (5-14) can be simplified to read

$$\cancel{Q}^{\,0} - \cancel{W_{\text{cv}}}^{\,0} + \cancel{m_{\text{in}}}^{\,0}\left(h_{\text{in}} + \frac{V_{\text{in}}^2}{2} + gz_{\text{in}}\right) - m_{\text{exit}}\left(h_{\text{exit}} + \cancel{\frac{V_{\text{exit}}^2}{2}}^{\,0} + \cancel{gz_{\text{exit}}}^{\,0}\right)$$

$$= (m_f u_f - m_i u_i) + \tfrac{1}{2}(m_f V_f^2 \cancel{-} m_i V_i^2)^{\,0} + g(m_f z_f \cancel{-} m_i z_i)^{\,0}$$

or

$$Q - (m_i - m_f)h_{\text{exit}} = m_f u_f - m_i u_i \tag{5-17}$$

where $m_{\text{exit}} = m_i - m_f$ has been used.

Example 5-9

Consider the inflation of an automobile tire. Air at 115 psia and 100°F flows through the air hose into the tire, which initially contains 0.22 lbm of air at 100°F. If the tire recevies an additional 0.11 lbm of air from the hose, determine the temperature of the air in the tire at the instant the inflation ceases. Assume that the tire walls are adiabatic and inflexible, and kinetic and potential energies associated with the process are negligible. Assume that for air, $h = 0.24T$ and $u = 0.171T$, where T is the temperature in °R, and h and u are the specific enthalpy and internal energy, respectively, in Btu/lbm.

Solution   Figure 5-19 shows a schematic for tire inflation. The conditions in the air-hose line are

$$p_{\text{in}} = 115 \text{ psia} \qquad \text{and} \qquad T_{\text{in}} = 100°F = 100 + 460 = 560°R$$

For the initial mass of air in the tire,

$$m_i = 0.22 \text{ lbm} \qquad \text{and} \qquad T_i = 100°F = 560°R$$

For the final mass of air in the tire, $m_f = 0.22 + 0.11 = 0.33$ lbm. With $Q = 0$, $W_{\text{cv}} = 0$, $m_{\text{exit}} = 0$, and negligible changes in KE and PE, Eq. (5-15) is applicable. Thus,

$$(m_f - m_i)h_{\text{in}} = m_f u_f - m_i u_i$$

or

$$(m_f - m_i)(0.24T_{\text{in}}) = m_f(0.171T_f) - m_i(0.171T_i)$$

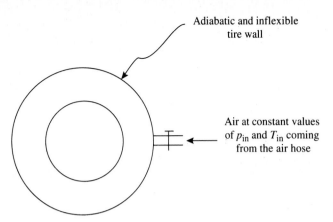

Figure 5-19 Inflation of an automobile tire.

so that

$$T_f = \frac{(m_f - m_i)(0.24T_{in}) + m_i(0.171T_i)}{m_f(0.171)}$$

Substituting numerical values yields

$$T_f = \frac{(0.33 - 0.22)(0.24 \times 560) + 0.22(0.171 \times 560)}{0.33(0.171)}$$

$$= 635°R = 175°F$$

Example 5-10

A 1-m³ pressure vessel initially contains a wet steam mixture of 50% liquid and 50% vapor by volume at 300°C. Liquid is withdrawn slowly from the bottom of the vessel, and heat is added to the vessel in order to maintain constant temperature. When half of the mass in the tank has been withdrawn, determine the amount of heat addition.

Solution Figure 5-20 shows a schematic for the liquid-withdrawal process. In this example, the subscripts i and f denote, respectively, the initial and final states of the steam in the vessel; and the subscripts F and G denote, respectively, the saturated liquid and saturated vapor states.

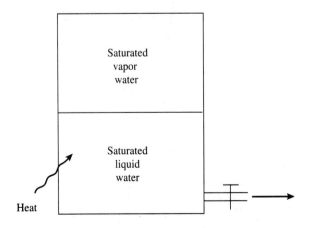

Figure 5-20 Schematic diagram for Example 5-10.

First Law Analysis for Control Volumes   Chap. 5

From the steam table (Table A-18M), at $T_i = 300°C$, we have $v_F = 0.0014036$ m³/kg and $v_G = 0.02167$ m³/kg. The mass of liquid and vapor initially in the vessel are, respectively,

$$m_{\text{liq}\,i} = \frac{0.5 \text{ m}^3}{0.0014036 \text{ m}^3/\text{kg}} = 356.2 \text{ kg}$$

$$m_{\text{vap}\,i} = \frac{0.5}{0.02167} = 23.1 \text{ kg}$$

The total mass in the vessel at the initial state is

$$m_i = m_{\text{liq}\,i} + m_{\text{vap}\,i} = 356.2 + 23.1 = 379.3 \text{ kg}$$

The total mass in the vessel at the final state is

$$m_f = \tfrac{1}{2} m_i = 189.7 \text{ kg}$$

For the wet steam at the initial state, the quality is calculated as

$$x_i = \frac{m_{\text{vap}\,i}}{m_{\text{vap}\,i} + m_{\text{liq}\,i}} = \frac{23.1}{23.1 + 356.2} = 0.0609 = 6.09\%$$

Thus,

$$u_i = u_F + x_i u_{FG} \qquad (\text{at } T_i = 300°C)$$
$$= 1332.0 + 0.0609(1231.0) = 1407.0 \text{ kJ/kg}$$

For the wet steam at the final state, the specific volume is

$$v_f = \frac{1 \text{ m}^3}{189.7 \text{ kg}} = 0.005272 \text{ m}^3/\text{kg}$$

Also

$$v_f = v_F + x_f v_{FG} \qquad (\text{at } T_f = 300°C)$$
$$= 0.0014036 + x_f(0.02167 - 0.0014036)$$

so that

$$x_f = 0.1909 = 19.09\%$$

Thus,

$$u_f = u_F + x_f u_{FG} \qquad (\text{at } T_f = 300°C)$$
$$= 1332.0 + 0.1909(1231.0) = 1567.0 \text{ kJ/kg}$$

For the saturated liquid withdrawn, $h_{\text{exit}} = h_F$ at 300°C = 1344.0 kJ/kg. Accordingly, Eq. (5-17) is applicable. It yields

$$Q = m_f u_f - m_i u_i + (m_i - m_f) h_{\text{exit}}$$
$$= (189.7 \text{ kg})(1567.0 \text{ kJ/kg}) - (379.3)(1407.0) + (379.3 - 189.7)(1344.4)$$
$$= 18,410 \text{ kJ}$$

This is the amount of heat addition to the vessel during the liquid-withdrawal process.

5-11 TRANSIENT PROCESS INVOLVING BOUNDARY WORK

Consider an insulated piston–cylinder assembly with gas entering through a throttling valve from a large supply line, as depicted schematically in Fig. 5-21. As the piston moves, the upper control surface moves with it and boundary work is involved in the process. Assume that the piston motion is slow and frictionless and a constant pressure p is maintained inside the cylinder by the combined effort of the regulating effect of the valve and the restraining force of the weight on top of the piston. The boundary work done by the gas is then expressed as

$$W_{cv} = W_{boundary} = p(V_f - V_i)$$

where V_i and V_f are the initial and final total volume, respectively, of the gas within the cylinder.

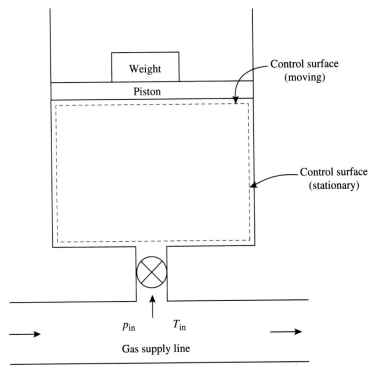

Figure 5-21 Transient process involving boundary work.

As the supply line is large and is continuously supplied with gas at constant pressure and temperature, the properties of the gas at the inlet to the cylinder do not change during the inflow, thus satisfying the condition of uniform flow. Because the piston motion is assumed to occur slowly, the entering gas from the supply line can quickly attain equilibrium with the gas already in the cylinder, thus satisfying the condition of uniform state. There is no heat transfer and no exit mass flow. Changes in kinetic and potential energies can be neglected. Applying the energy equation for uniform-flow, uniform-state, Eq. (5-14), to this unsteady-flow problem, we have

$$\cancel{Q} - W_{cv} + m_{in}\left(h_{in} + \frac{\cancel{V_{in}^2}}{2} + \cancel{gz_{in}}\right) - \cancel{m_{exit}}\left(h_{exit} + \frac{V_{exit}^2}{2} + gz_{exit}\right)$$

$$= (m_f u_f - m_i u_i) + \tfrac{1}{2}(m_f V_f^2 \cancel{- m_i V_i^2}) + g(m_f z_f \cancel{- m_i z_i})$$

or

$$-p(V_f - V_i) + (m_f - m_i)h_{in} = m_f u_f - m_i u_i \qquad (5\text{-}18)$$

where the subscripts i and f denote, respectively, the initial and final uniform states of the entire control volume. The term $m_f - m_i = m_{in}$ represents the mass inflow through the valve, and h_{in} is the constant specific enthalpy of the incoming gas.

Example 5-11

An insulated weighted piston-and-cylinder assembly (see Fig. 5-21) is attached through a valve to a large air-supply line that is at a constant pressure of 600 kPa and a constant temperature of 100°C. The cylinder initially contains 0.1 m³ of air at 150 kPa and 30°C. When the valve is opened, air enters the cylinder and pushes the piston upward smoothy in such a manner that the air pressure in the cylinder remains constant at 150 kPa. The filling process is terminated when the final volume of air in the cylinder is twice its initial volume. Assuming air to be an ideal gas with $c_p = 1.007$ kJ/kg·K and $c_v = 0.72$ kJ/kg·K, determine the final temperature of the air in the cylinder and the mass of air added through the valve.

Solution The given data are $Q = 0$, $p_{in} = 600$ kPa = constant, $T_{in} = 100°C = 373.15$ K = constant, $V_i = 0.1$ m³, $p_i = p_f = p = 150$ kPa = constant, $T_i = 30°C = 303.15$ K, $V_f = 0.2$ m³, $h = c_p T = 1.007T$, and $u = c_v T = 0.72T$ (based on $h = 0$ and $u = 0$ at 0 K). Assume ΔKE and ΔPE negligible. The gas constant of air is $R = c_p - c_v = 1.007 - 0.72 = 0.287$ kJ/kg·K.

The mass of air initially in the cylinder is

$$m_i = \frac{p_i V_i}{RT_i} = \frac{(150 \text{ kN/m}^2)(0.1 \text{ m}^3)}{(0.287 \text{ kJ/kg·K})(303.15 \text{ K})} = 0.1724 \text{ kg}$$

The mass of air finally in the cylinder is

$$m_f = \frac{p_f V_f}{RT_f} = \frac{150 \times 0.2}{0.287T_f} = \frac{104.53}{T_f}$$

Equation (5-18) is applicable, which is

$$-p(V_f - V_i) + (m_f - m_i)h_{in} = m_f u_f - m_i u_i$$

where p is in kPa, V in m³, m in kg, and h and u in kJ/kg. For this example, the preceding equation becomes

$$-p(V_f - V_i) + \left(\frac{104.53}{T_f} - 0.1724\right)(1.007T_{in}) = \frac{104.53}{T_f}(0.72T_f) - 0.1724(0.72T_i)$$

Substituting the given values of p, V_f, V_i, T_{in}, and T_i results in

$$T_f = 334.5 \text{ K} = 61.3°C$$

Thus,

$$m_f = \frac{104.53}{T_f} = \frac{104.53}{334.5} = 0.3125 \text{ kg}$$

and the mass of air added through the valve is

$$m_f - m_i = 0.3125 - 0.1724 = 0.1401 \text{ kg}$$

5-12 SUMMARY

A control volume (cv) is a region of space specified in order to focus study. The boundary of a control volume is called a control surface. The volume-flow rate of fluid crossing a control surface is given by

$$\dot{V} = A\mathbf{V}$$

The mass-flow rate of fluid crossing a control surface is given by

$$\dot{m} = \frac{\dot{V}}{v} = \frac{A\mathbf{V}}{v}$$

The general equation of conservation of mass (commonly known as the continuity equation) for a control volume is

$$\sum_{\text{inlet}} \dot{m} - \sum_{\text{exit}} \dot{m} = \frac{dm_{\text{cv}}}{dt}$$

The general equation of conservation of energy for a control volume is

$$\dot{Q} - \dot{W}_{\text{cv}} + \sum_{\text{inlet}} \dot{m}\left(h + \frac{\mathbf{V}^2}{2} + gz\right) - \sum_{\text{exit}} \dot{m}\left(h + \frac{\mathbf{V}^2}{2} + gz\right) = \frac{dE_{\text{cv}}}{dt}$$

Steady-flow, steady-state process

The flow is steady when no mass is accumulating within the control volume, and the state is steady when the properties at each point within the control volume are unchanging in time. For a steady-flow, steady-state system, the conservation of mass and conservation of energy, respectively, are

$$\sum_{\text{inlet}} \dot{m} - \sum_{\text{exit}} \dot{m} = 0$$

$$\dot{Q} - \dot{W}_{\text{cv}} + \sum_{\text{inlet}} \dot{m}(h + \mathbf{V}^2/2 + gz) - \sum_{\text{exit}} \dot{m}(h + \mathbf{V}^2/2 + gz) = 0$$

For a steady-flow, steady-state system with a single inlet at section 1 and a single exit at section 2, we have

$$\dot{m}_1 = \dot{m}_2 = \dot{m}$$

$$\dot{Q} - \dot{W}_{\text{cv}} + \dot{m}(h_1 + \mathbf{V}_1^2/2 + gz_1) - \dot{m}(h_2 + \mathbf{V}_2^2/2 + gz_2) = 0$$

or

$$q - w_{\text{cv}} + (h_1 + \mathbf{V}_1^2/2 + gz_1) - (h_2 + \mathbf{V}_2^2/2 + gz_2) = 0$$

Except for the short-term transient startup or shutdown, the long-term operation of many mechanical devices, such as nozzles, diffusers, turbines, compressors, pumps, and heat exchangers, can be analyzed by the steady-flow, steady-state equations.

Uniform-flow, uniform-state transient process

Assumptions: (1) The state of the fluid within the control volume is uniform throughout even though this uniform state can change with time. (2) Although the mass-flow rate can change with time, the state of the fluid crossing the control surface is invariable.

The energy equation for the uniform-flow, uniform-state model is

$$Q - W_{cv} + \sum_{\text{inlet}} m(h + V^2/2 + gz) - \sum_{\text{exit}} m(h + V^2/2 + gz)$$

$$= (m_f u_f - m_i u_i) + \tfrac{1}{2}(m_f V_f^2 - m_i V_i^2) + g(m_f z_f - m_i z_i)$$

where the subscripts i and f denote, respectively, the initial and final uniform states of the entire control volume.

Under some restricted conditions, the charging and discharging of tanks can be considered to occur under a uniform-flow, uniform-state model.

PROBLEMS

5-1. Helium at a flow rate of 0.1 kg/s is heated in a variable-area duct from an inlet condition of 300 kPa, 20°C, with a velocity of 150 m/s, to an exit condition of 285 kPa, 140°C, with a velocity of 130 m/s. Determine (a) the rate of heat addition, and (b) the inlet and exit flow areas.

5-2. Air at 50 kPa and 7°C enters a steady-flow flying device (jet engine) with a velocity of 300 m/s and a flow rate of 40 kg/s. See Fig. P5-2. In the device, the air is heated at the rate of 40,000 kJ/s. The heated air leaves the device at 50 kPa and 427°C. Determine the exit velocity (m/s) of the air. Use air tables for the solution of this problem.

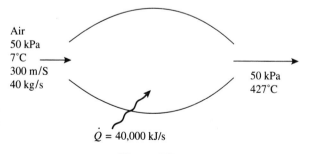

Air
50 kPa
7°C
300 m/S
40 kg/s

50 kPa
427°C

$\dot{Q} = 40,000$ kJ/s

Figure P5-2

5-3. Steam flows through a long insulated pipe. At one point, where the diameter is 4 in., the pressure and temperature are 300 psia and 600°F, respectively.

Further downstream, where the diameter is 3 in., the pressure and temperature are 250 psia and 550°F, respectively. Determine (a) the mass-flow rate of the stream, in lbm/s, and (b) the downstream velocity, in ft/s.

5-4. Steam at 1 MPa and 300°C enters a long horizontal pipe at a velocity of 6 m/s. See Fig. P5-4. At this inlet section, the pipe diameter is 10 cm. At a downstream location, where the pipe diameter is 8 cm, the steam conditions are 800 kPa and 250°C. Determine the mass-flow rate of steam and the rate of heat transfer.

5-5. Two liters per second of water enter a horizontal tube of 20-mm constant diameter at a pressure of 10 MPa and a temperature of 30°C, and leaves the tube as a saturated vapor at 9 MPa. Find the heat transfer rate to the water.

5-6. Air enters a long horizontal pipe at 20 psia and 60°F, with a flow rate of 30 lbm/min. It leaves at 16 psia and 180°F. If the enthaply increases by 28.9 Btu/lbm, and if the pipe has a cross-sectional area of 1.5 in²., at what rate is heat being supplied to the air?

5-7. Water enters the insulated pipe-pumping system from the reservoir, as shown in Fig. P5-7 at 14.0 psia, 50°F, and 8 ft/s. The pump discharges the water at 60 psia and 18 ft/s. The end nozzle finally discharges the water to the atmosphere at 14.7

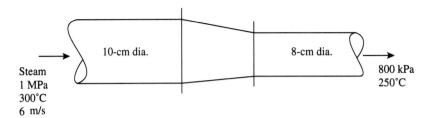

Figure P5-4

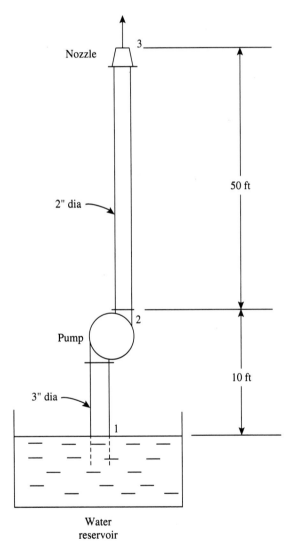

Figure P5-7

psia. Assuming that the flow is isothermal and the water is incompressible, determine (a) the power requirement of the pump, in hp, (b) the nozzle exit diameter, in in., and (c) the nozzle exit velocity, in ft/s.

5-8. Saturated steam at 200 kPa enters the pipe of a heating system at the ground floor of a tall skyscraper in New York City. See Fig. P5-8. The steam pressure in the pipe drops to 100 kPa at the top floor, which is 200 m above the ground. Assuming the steam pipe to be well insulated and the steam flow to be at a constant velocity, what will be the state of the steam at the top floor?

5-9. Water at 1.5 MPa, 150°C, is throttled adiabatically through a valve to 0.2 MPa. See Fig. P5-9. The inlet velocity is 5 m/s. The inlet and exit cross-sectional areas are equal. Determine the state and the velocity of the water at the exit.

5-10. An ideal gas enters a heater at 101 kPa and 300 K and leaves at 505 kPa and 600 K, while receiving heat of the amount of 275 kJ/kg during the process. The exit area is ten times larger than that of the inlet. Determine the inlet and exit velocities in m/s. Use $c_p = 1.021$ kJ/kg·K for the gas.

5-11. Refrigerant-12 flows through a long, horizontal, 5.25-cm-diameter pipe. The fluid enters at 15°C with a quality of 20% and a velocity of 3 m/s. The fluid leaves the pipe at 0.4 MPa and 20°C. Determine (a) the mass-flow rate, kg/s, (b) the exit velocity, m/s, and (c) the heat transfer rate, kJ/s.

5-12. A calorimeter is a device for measuring the state of a liquid–vapor mixture. Steam from a pipe where the pressure is 110 psia flows steadily through an electric calorimeter and comes out at 100 psia, 430°F. See Fig. P5-12. The electric power input to the calorimeter is 1 kW and the amount of steam that flows through the calorimeter in 5 min is 4.1 lbm. Find the quality of the steam taken from the pipe.

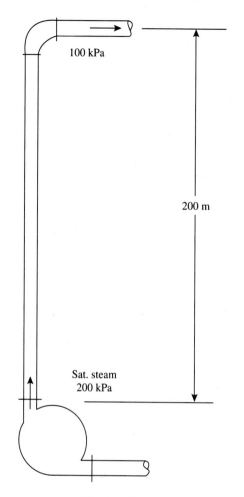

100 kPa

200 m

Sat. steam
200 kPa

Figure P5-8

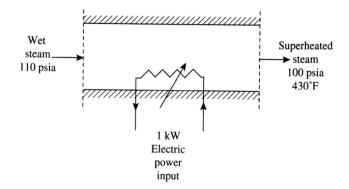

Wet
steam
110 psia

Superheated
steam
100 psia
430°F

1 kW
Electric
power
input

Figure P5-12

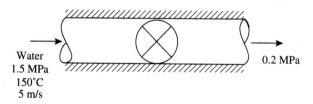

Water
1.5 MPa
150°C
5 m/s

0.2 MPa

Figure P5-9

5-13. Steam at 3 MPa, 350°C, with a flow rate of 0.5 kg/s enters an adiabatic nozzle slowly. It leaves the nozzle at 1.5 MPa with a velocity of 550 m/s. Determine the quality or temperature of the steam leaving the nozzle and the exit area of the nozzle.

5-14. Air expands through a nozzle from 25 psia, 200°F, to 15 psia, 80°F. If the inlet velocity is 100 ft/s and the heat loss is 2.0 Btu/lbm, determine (a) the exit velocity, in ft/s, and (b) the ratio of the inlet area to the exit area. For air, $c_p = 0.24$ Btu/lbm·°F and $c_v = 0.171$ Btu/lbm·°F.

5-15. Water enters a well-insulated nozzle at 320 kPa, 20°C, and 4 m/s, and leaves at 150 kPa. The inlet and exit areas of the nozzle are 16 cm² and 4.0 cm², respectively. Considering water to be incompressible with $v = 1.002$ cm³/g and $c = 4.19$ kJ/kg·K, determine the mass-flow rate (kg/s), the exit velocity (m/s), and the change in temperature (°C).

5-16. Steam enters an adiabatic nozzle at 300 psia, 500°F, with a velocity of 200 ft/s. At the outlet, the enthalpy of the steam is 1200 Btu/lbm. What is the exit velocity from the nozzle? If the flow rate is 2 lbm/s, what is the inlet area of the nozzle?

5-17. Steam at 5 MPa, 500°C, flows into a nozzle steadily with a velocity of 80 m/s. It leaves at 2 MPa and 400°C. The steam looses heat at a rate of 8 kJ/s as it passes through the nozzle. The inlet area of the nozzle is 38 cm². Determine the mass-flow rate and exit velocity of the steam, and the exit area of the nozzle.

5-18. Steam is accelerated from a negligible velocity in a well-insulated nozzle to an exit velocity of 1600 ft/s. The pressure and temperature at the inlet are 400 psia and 600°F, respectively, and the pressure at the exit is 265 psia. The mass-flow rate of steam is 3000 lbm/h. Determine the temperature of the steam leaving the nozzle and the exit area of the nozzle.

5-19. A well-insulated nozzle operates at steady state with steam entering slowly at a pressure of 0.4

MPa and a flow rate of 0.33 kg/s. Steam exits the nozzle with a pressure of 0.2 MPa and a temperature of 170°C through an area of 6.5 cm². Determine the inlet temperature of the steam.

5-20. Water enters the nozzle on a garden hose at 95 psia and 60°F with a mass-flow rate of 0.8 lbm/s. See Fig. P5-20. The nozzle has an inlet inside diameter of 1.0 in. Assuming water to be incompressible and flow to be isothermal, determine the outlet velocity from the nozzle. When the nozzle is pointed straight up, to what height will the stream of water rise? Neglect any frictional effects and air resistance in this calculation.

5-21. Water at 70°F flows out steadily in the form of a free jet through a small hole at the bottom of a large open tank. See Fig. P5-21. The tank is constantly filled to a depth of 12 ft above the hole. Assuming that no heat transfer or internal energy change occurs for the water, determine the jet velocity, in ft/s.

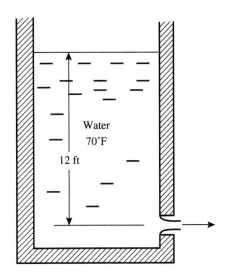

Figure P5-21

5-22. An ideal gas with $c_p = 1.008$ kJ/kg·°C enters an adiabatic horizontal diffuser at 80 kPa and 60°C with a velocity of 200 m/s. It leaves the diffuser at 100 kPa. The exit area of the diffuser is 20% greater than the inlet area. Determine the exit temperature and velocity of the gas.

5-23. Air enters a diffuser at a pressure of 0.1 MPa, a temperature of 60°C, with a velocity of 200 m/s. It exits at a pressure of 0.14 MPa. The outlet area is 20% greater than that of the inlet. The fluid gains 40 kJ/kg in heat transfer as it passes through. Determine the outlet temperature and velocity. Use the air table to solve this problem.

5-24. Saturated vapor Refrigerant-12 at 110 psia enters a diffuser steadily with a velocity of 380 ft/s. It leaves at 125 psia and 120°F. The refrigerant gains heat at the rate of 10 Btu/s as it passes through the

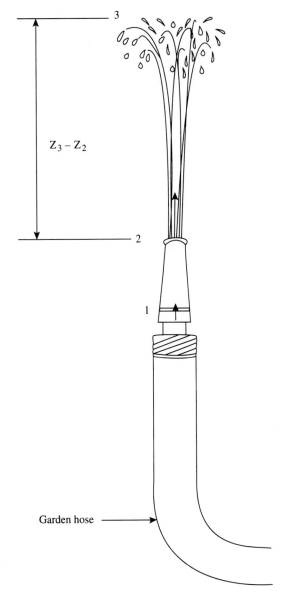

Figure P5-20

diffuser. The exit area is 30% greater than the inlet area. Determine the exit velocity and the mass-flow rate of the refrigerant.

5-25. Steam at 1.0 MPa and 200°C enters a diffuser with a velocity of 250 m/s. It leaves the diffuser as saturated vapor at 300 kPa with a velocity of 75 m/s. The inlet diameter of the diffuser is 30 cm. Determine the mass-flow rate of the steam (kg/s), the ratio of the inlet and exit areas of the diffuser, and the rate of heat loss from the steam (kW).

5-26. Steam enters a turbine at 4 MPa and 440°C and leaves at 20 kPa and 90% quality. The mass-flow rate of the steam is 2×10^4 kg/h and the heat loss is 20 kJ/kg. The inlet steam pipe has a 0.12-m diameter and the exhaust section is a 0.6 m by 0.7 m rectangle. Determine the kinetic-energy change (in kJ/kg) and the power output (in kW).

5-27. A hydraulic turbine works between an elevation difference of 8 ft with a flow rate of 6 ft³/s. At the inlet, the water pressure is 45.3 psia and the inlet pipe has a cross-sectional area of 0.50 ft². At the outlet, the water pressure is 18.7 psia and the flow velocity is 2.5 ft/s. Assuming that the flow is adiabatic with negligible internal energy change of the fluid, determine the power output.

5-28. Steam enters a turbine at 6 MPa and 500°C with a velocity of 110 m/s, and leaves as a saturated vapor at 30 kPa. The turbine inlet pipe has a diameter of 0.60 m and the outlet diameter is 4.5 m. Determine the mass-flow rate (in kg/h) and the exit velocity (in m/s).

5-29. A steady stream of a fluid enters a turbine with a velocity of 10 ft/s through a pipe of 3-in. diameter. At entry, the specific enthalpy is 1227.5 Btu/lbm and the specific volume is 4.934 ft³/lbm. At exit from the turbine, the specific enthalpy is 1000 Btu/lbm. The changes in kinetic and potential energies are small. Determine the power (in hp) generated by the turbine if the process is adiabatic.

5-30. A turbine is supplied with a steady flow of 5 kg/min of steam at 5.0 MPa and 400°C. Exhaust steam is saturated at 140 kPa. Heat transfer from the turbine to the surrounding is 200 kJ/min. Calculate the power output of the turbine.

5-31. An apartment building on the average uses 2500 gallons of domestic water per hour. The city water has a pressure of 90 psig. It is suggested that a hydraulic turbine be installed to utilize the water pressure to run an electric motor and charge a battery. See Fig. P5-31. Assuming an exit pressure of 10 psig for the turbine, determine the power that could be produced. Neglect contributions from kinetic and potential energies.

5-32. Air is supplied to a turbine at the rate of 5 lbm/min. It enters the turbine at 60 psia, 90°F, and leaves at 14.7 psia, 10°F. Its enthalpy decreases by 19.3 Btu/lbm as it flows through the turbine. If the cross-sectional area of the inlet and discharge passages are 0.40 and 0.60 in.², respectively, and if the turbine operates adiabatically, how much horsepower does it deliver?

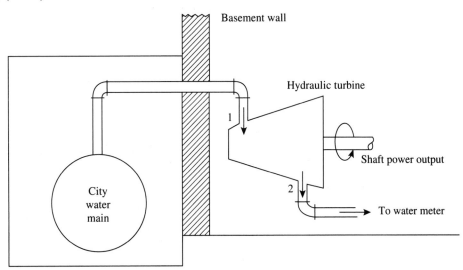

Figure P5-31

5-33. Argon gas enters a turbine at 350 kPa and 150°C, and leaves at 140 kPa. A heat loss of 4.5 kJ/kg occurs, and the measured work output is 55 kJ/kg. If the kinetic-energy and potential-energy changes are negligible, calculate the outlet temperature in °C. Use $c_v = 12.5$ kJ/kgmole·K and $c_p = 20.8$ kJ/kgmole·K for argon.

5-34. An air turbine operates steadily and adiabatically with $p \cdot v^k = $ constant and delivers 100 horsepower. Inlet conditions are 100 psia and 230°F, and the exit pressure is 20 psia. If the exit area is 0.1 ft², what is the exit velocity? Assume that the inlet kinetic energy can be ignored. Use $c_p = 0.24$ Btu/lbm·°R and $k = 1.4$ for air.

5-35. Air at 100 kPa and 20°C enters a steady-flow compressor and is compressed to a pressure of 1 MPa. Rate of heat loss from the compressor equals 10% of the power input to the compressor. The air enters at 50 m/s through an area of 90 cm² and leaves at 120 m/s through an area of 5 cm². Determine the exit air temperature and the power input to the compressor. Use $c_p = 1.005$ kJ/kg·K for air.

5-36. An adiabatic compressor receives air from the surrounding atmosphere at 20°C. It discharges the air at a pressure of 3 atm, a temperature of 180°C, and a velocity of 120 m/s. The compressor power input is 3 MW. Determine the mass-flow rate of the air by using (a) $c_p = 1.0035$ kJ/kg·°C for the air and (b) air table data.

5-37. In a test of a water-jacketed air compressor, it is found that the shaft work required to drive the compressor is 60,000 ft·lbf/lbm of air delivered, that the enthalpy of the air leaving is 30 Btu/lbm greater than that entering, and that the increase in enthalpy of the circulating water is 40.5 Btu/lbm of air. From these data, compute the amount of heat passing to the atmosphere from the compressor per lbm of air.

5-38. Water at 0.5 MPa and 15°C flows through a large pipe at a rate of 60 kg/min into an insulated tank. Steam at 0.5 MPa and 200°C flows into the same tank through another pipe with negligible velocity. See Fig. P5-38. The steam then condenses and mixes with the water to form heated water that leaves the tank through a 25-cm² pipeline at a pressure of 0.5 MPa and 80°C. Calculate the exit velocity of the heated water.

5-39. Water at 80°F flows through a large pipe at a rate of 10,000 lbm/h into an insulated tank. Steam having an enthalpy of 1195 Btu/lbm and a specific volume of 9 ft³/lbm flows into the same tank through a 1.5-in.² pipe at a rate of 2000 lbm/h. The steam then condenses and mixes with the water to form heated water that leaves the tank through a large pipe. The enthalpy of water is approximately equal to $(T - 32)$ Btu/lbm, where T denotes the temperature of water in °F. Determine the temperature of the heated water leaving the tank.

5-40. Fifty pounds per minute of water flowing through a line at 2000 psia, 100°F, is throttled across a valve to 60 psia, and then enters a mixing chamber. Steam at 100 lbm/min, 60 psia, and 900°F enters the mixing chamber through another line. The steam and water are mixed and then leave the mixing chamber at 60 psia. Assuming the entire process to be adiabatic, determine the quality of the exiting fluid.

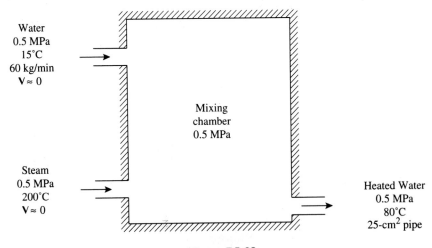

Water
0.5 MPa
15°C
60 kg/min
$V \approx 0$

Mixing
chamber
0.5 MPa

Steam
0.5 MPa
200°C
$V \approx 0$

Heated Water
0.5 MPa
80°C
25-cm² pipe

Figure P5-38

5-41. Steam at 700 kPa and 280°C with a flow rate of 2 kg/min is mixed with 4 kg/min of steam at 700 kPa and 50% quality in an adiabatic steady-flow device. Calculate the temperature (if superheated) or quality (if a mixture) of the outlet stream.

5-42. A stream of saturated steam at 50 psia and a stream of water at 50 psia and 60°F are mixed in an adiabatic chamber to form a stream of hot water at 50 psia and 180°F with a mass-flow rate of 500 lbm/min. Neglect all changes in kinetic and potential energies. Determine the mass-flow rate of the incoming steam in lbm/min.

5-43. An air stream at 15 psia and 60°F flows at a velocity of 200 ft/s through a 2.5-ft² duct. A second air stream at 15 psia and 160°F flows at a velocity of 300 ft/s through a 1.5-ft² duct. The two streams mix, and the resultant stream flows at a velocity of 100 ft/s. If the air in the first stream has an enthalpy of 14.5 Btu/lbm and the air in the second has an enthalpy of 38.6 Btu/lbm, what is the enthalpy of each pound of air in the final stream? The ducts are perfectly insulated.

5-44. A vortex tube is a straight length of separation chamber into which pressurized air is admitted tangentially at the outer radius, and in the swirling motion, the air at the central core becomes cold relative to that at the periphery. The cold and hot air streams are let to different exit ports. See Fig. P5-44. Calculate the cold-stream temperature T_c based on the following adiabatic vortex-tube test data:

For the inlet stream, $p_{in} = 0.70$ MPa and $T_{in} = 20°C$.

For the hot stream, $\dot{m}_h = 0.14$ kg/min and $T_h = 80°C$.

For the cold stream, $\dot{m}_c = 0.32$ kg/min.

5-45. Water at 100°F and 50 psia flows through a large pipe at a rate of 100 lbm/min into an insulated tank. Steam at 15 lbm/min, 300°F, and 50 psia flows into the same tank through another pipe that has a cross-sectional area of 0.7 in². The steam then condenses and mixes with the water to form a stream of heated water that leaves the tank through a large pipe at a pressure of 50 psia. Calculate the temperature of the heated water leaving the tank in steady flow.

5-46. A bathroom shower mixes hot water at 60°C and cold water at 10°C to become warm water at 43°C for a bath. See Fig. P5-46. The pressure of both hot and cold water is 120 kPa. Determine the ratio of mass-flow rates of hot and cold water.

5-47. It is desired to have a steady flow of 110°F warm water from an ordinary shower by mixing the hot water supply at 140°F and the cold water supply at 50°F. The mixing process takes place adiabatically at 20 psia. Determine the ratio of mass-flow rates of the hot to cold water.

5-48. In a counterflow heat exchanger, Refrigerant-12 enters slowly at 0.2 MPa, 20°C, and exits at 0.1 MPa, 80°C; meanwhile helium enters at 0.7 MPa, 525°C, through a 0.15-m² pipe with a velocity of 150 m/s, and exits at 0.35 MPa, 125°C, through a 0.3-m² pipe. See Fig. P5-48. Determine (a) the exit velocity of the helium and (b) the mass-flow rate of Refrigerant-12. Consider helium to be an ideal gas with $c_p = 5.1926$ kJ/kg·K and $R = 2.077$ kJ/kg·K.

5-49. A refrigerator compressor admits ammonia at −32°C, 95% quality, and discharges it at 1000 kPa, 100°C. See Fig. P5-49. The compression is adiabatic and power delivered to the ammonia is 22.0 kW. Determine the mass-flow rate of ammonia.

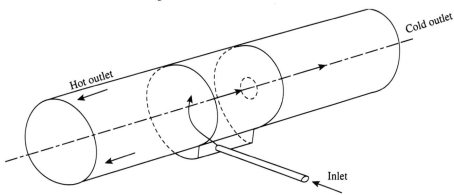

Figure P5-44

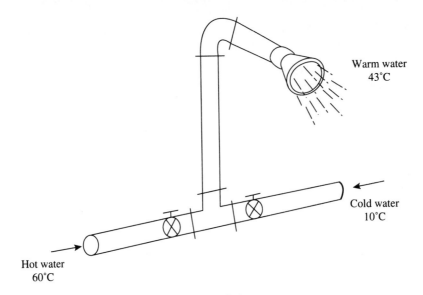

Figure P5-46

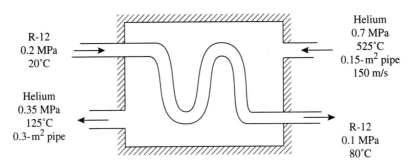

Figure P5-48

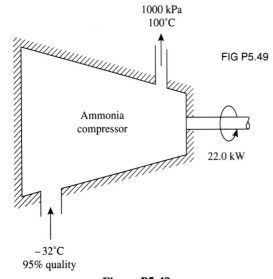

Figure P5-49

FIG P5.49

5-50. Ammonia vapor enters a refrigerator condenser at 1400 kPa, 40°C, and leaves at the same pressure and 30°C. Determine the rate of heat rejection from the ammonia for a mass-flow rate of 1.0 kg/s.

5-51. Ammonia is throttled from 1200 kPa, 30°C, to 190 kPa. Determine the quality of the ammonia leaving the throttling valve.

5-52. Ammonia enters the evaporator of a refrigeration system at 158.8 kPa with a quality of 20%. It leaves the evaporator at 150 kPa, −20°C. Determine the rate of heat addition to the ammonia for a mass-flow rate of 1.0 kg/s.

5-53. A nuclear-reactor power plant uses helium gas as the primary coolant in the reactor. See Fig. P5-53. A helium–water counterflow heat exchanger is used to transfer heat from helium to water, which, in turn, in a closed loop, transfers heat to an air stream in a water–air counterflow heat exchanger.

First Law Analysis for Control Volumes   Chap. 5

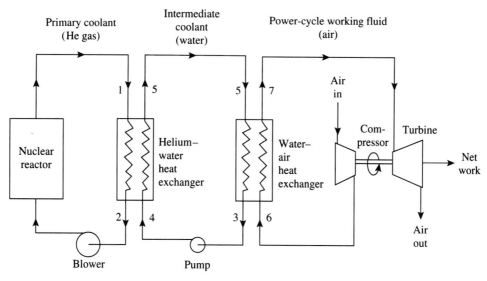

Figure P5-53 Schematic of a nuclear-reactor power plant.

The conditions for the two heat exchangers are as follows:

Helium–water heat exchanger:

Helium enters at 230°C, leaves at 40°C
Helium flow rate: 7×10^6 kg/h
Water enters at 35°C
Water flow rate: 12×10^6 kg/h

Water–air heat exchanger:

Water flow rate: 12×10^6 kg/h
Water leaves at 35°C
Air enters at 25°C, leaves at 75°C

Determine (a) the water temperature at exit from the helium–water heat exchanger, and (b) the required air flow rate. Use the following values of c_p: for helium, 5.2 kJ/kg·°C; for air, 1.0 kJ kg·°C; for water, 4.2 kJ/kg·°C.

5-54. A 400-m³ internal volume house has an electric heating system that consists of a 30-kW electric resistance heater placed in an insulated duct and a 250-W fan to circulate air through the duct. See Fig. P5-54. As air flows through the duct, its temperature is increased by 5°C. The house has a heat loss of 450 kJ/min. The initial temperature of the house is 14°C and the local atmospheric pressure is 95 kPa. (a) How long will it take for the air inside the house to reach 24°C? (b) What is the mass-flow rate of air in the duct?

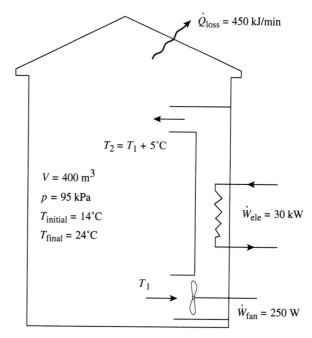

Figure P5-54

5-55. A solar collector system is designed to supply energy to provide hot water for a home. See Fig. P5-55. The solar collector has an area of 3.0 m² per panel. The rate of solar energy received by the collectors is 1.7 MJ/h·m², of which 40% is lost to the surroundings. The useful portion of the incoming solar energy can heat liquid water from 55°C to 70°C. By neglecting kinetic- and potential-energy

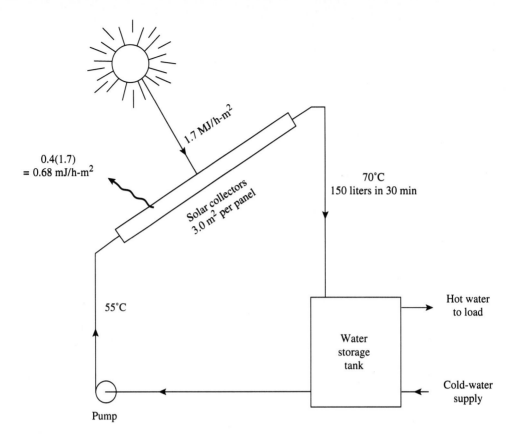

Figure P5-55

effects in steady-flow operation, how many collector panels would be required to provide 150 liters of hot water at 70°C in 30 min?

5-56. Air flows in a large pipeline at a pressure of 10 atm and a temperature of 30°C. Connected to this pipeline is a vessel that originally contains 1000 cm³ of air at 5 atm and 30°C. See Fig. P5-56. When the valve on this vessel is opened, air flows from the pipeline into the vessel until the pressure in the vessel becomes 10 atm. If the process occurs adiabatically, what is the final temperature of the air in the vessel? For air, $pv = 2.83T$, $h = 0.24T$, and $u = 0.171T$, where p is in atm, v in cm³/g, T in K, and h and u in cal/g.

5-57. Solve Problem 5-56 if the vessel is initially evacuated.

5-58. A 0.5-m³ insulated tank contains air initially at 0.2 MPa and 25°C. The tank is connected through a valve to a large compressed air line, in which air at 0.7 MPa and 120°C is flowing. If the valve is opened and air is allowed to flow into the tank until the pressure reaches 0.5 MPa, how much mass, in kg, has entered, and what is the final temperature in the tank, in °C?

5-59. A 0.2-m³ insulated tank is to be filled from a compressed air line in which air at 700 kPa, 95°C, is flowing. Initially, the tank contains air at 103 kPa, 30°C. Determine the final temperature and final mass of air in the tank when the tank pressure reaches 700 kPa.

5-60. A 0.2-m³ small insulated tank initially contains 0.2 kg of nitrogen at 27°C. The small tank is connected through a valve to a very large reservoir containing nitrogen at 1.2 MPa. See Fig. P5-60. The valve is opened and nitrogen is allowed to flow from the reservoir into the small tank until the pressure in the small tank becomes 0.3 MPa with a temperature of 180°C. What is the temperature of the large reservoir?

5-61. A storage tank is filled with 2 kg of air at 250 kPa and 42°C. It is discharged until the pressure reaches 150 kPa. During the discharging process,

First Law Analysis for Control Volumes   Chap. 5

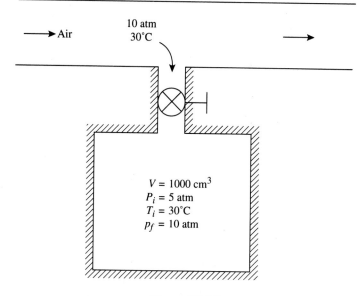

Figure P5-56

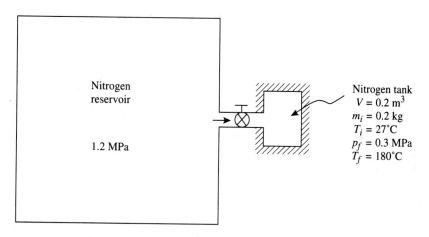

Figure P5-60

heat is added to the air within the tank to keep it at constant temperature. Determine the amount of heat added (in kJ) during the process. Use $c_p = 1.005$ kJ/kg·K and $c_v = 0.718$ kJ/kg·K for air.

5-62. A 2-m³ insulated rigid tank is filled with air initially at 500 kPa and 52°C. See Fig. P5-62. Air is discharged through a valve opening until the pressure in the tank reaches 200 kPa. The air temperature in the tank during the discharging process is maintained constant by an electric resistance heater placed in the tank. Determine the electric

work done (in kJ) during the process. Use the air table to solve this problem.

5-63. A 10-ft³ insulated rigid tank contains air initially at 14.0 psia and 100°F. An electric heating element within the tank adds energy at a constant rate of 0.1 Btu/s. Air is discharged from the tank to hold the pressure constant. Assuming conditions to be uniform throughout the tank at any instant, determine the time required for the air to reach a temperature of 300°F.

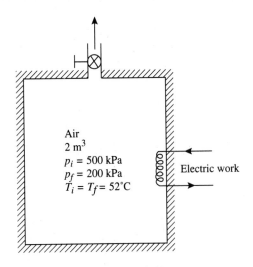

Figure P5-62

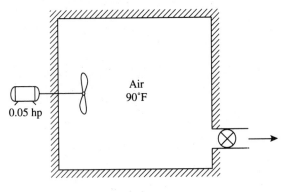

Figure P5-65

Calculate the mass-flow rate of air out of the tank. For air, $h = 0.24T$ and $u = 0.171T$, where h and u are in Btu/lbm and T in °R.

5-64. A 12-ft³ rigid tank is initially evacuated. See Fig. P5-64. Atmospheric air seeps into the tank through a porous plug slowly such that there is ample time for heat transfer to keep the temperature of the air inside the tank equal to the atmospheric temperature of 60°F. Finally, the pressure inside the tank is equal to the atmospheric pressure of 14.0 psia. Determine the amount of heat removed from the air in the tank.

5-65. Air is discharged through a small valve from an insulated rigid tank while a paddle wheel in the tank is running to keep the tank temperature constant at 90°F. See Fig. P5-65. The power input from the paddle wheel is 0.05 hp. Assume that the pressure and temperature are uniform throughout the tank.

5-66. An insulated tank initially contains 100 ft³ of water vapor and 500 ft³ of liquid water at a pressure of 200 psia. See Fig. P5-66. For how long can the steam be withdrawn through a pressure-reducing valve at a pressure of 50 psia and at a rate of 2000 lbm/h? State and justify any assumption you make.

5-67. A 0.5-m³ tank is half filled with liquid water and the other half is filled with water vapor in equilibrium at 3 MPa. See Fig. P5-67. Heat is added until half of the liquid mass is evaporated while an automatic valve lets saturated vapor escape at such a rate that the pressure remains constant. Determine the amount of heat added in kJ.

5-68. An insulated piston-and-cylinder assembly is attached through a valve to a large air-supply line

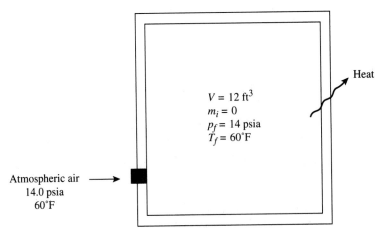

Figure P5-64

First Law Analysis for Control Volumes   Chap. 5

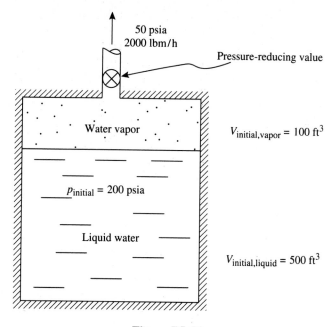

50 psia
2000 lbm/h

Pressure-reducing value

Water vapor

$V_{initial,vapor} = 100 \text{ ft}^3$

$p_{initial} = 200 \text{ psia}$

Liquid water

$V_{initial,liquid} = 500 \text{ ft}^3$

Figure P5-66

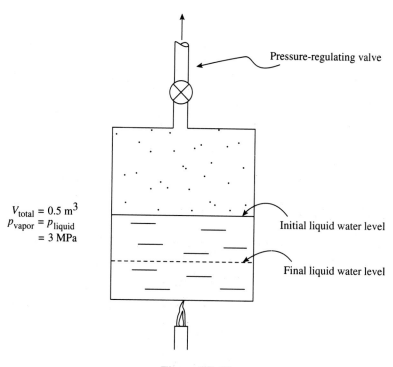

Pressure-regulating valve

$V_{total} = 0.5 \text{ m}^3$
$p_{vapor} = p_{liquid}$
$= 3 \text{ MPa}$

Initial liquid water level

Final liquid water level

Figure P5-67

that is at a constant pressure of 700 kPa and a constant temperature of 90°C. See Fig. P5-68. The cylinder initially contains 0.1 m³ of air at 100 kPa and 30°C. When the valve is opened, air enters the cylinder and pushes the piston upward smoothly in such a manner that the air pressure in the cylinder remains constant at 100 kPa. The filling process is terminated when the final volume of air in the cylinder is twice its initial volume. Assuming air to be an ideal gas with $c_p = 1.008$ kJ/kg·K and $c_v = 0.72$ kJ/kg·K, determine the final temperature of the air in the cylinder and the mass of air added through the valve.

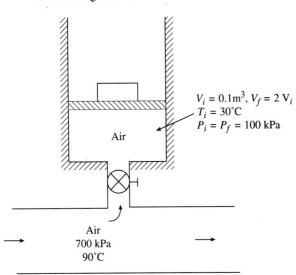

$V_i = 0.1 \text{m}^3, V_f = 2 V_i$
$T_i = 30°C$
$P_i = P_f = 100$ kPa

Air

Air
700 kPa
90°C

Figure P5-68

5-69. A weighted piston–cylinder device is attached through a valve to a large steam line that is at a constant pressure of 2.0 MPa and a constant temperature of 500°C. See Fig. P5-69. The cylinder initially contains 0.10 kg of saturated water vapor at 1.0 MPa. When the valve is opened, steam enters the cylinder and pushes the weighted piston upward smoothly until the contents within the cylinder reach 300°C, while a heat loss of 90 kJ occurs through the cylinder walls. Determine the amount of mass entering the cylinder from the line.

5-70. Write a computer program to solve the following problem. Water at 80°F flows through a large pipe into an insulated tank. Steam having an enthalpy of 1195 Btu/lbm and a specific volume of 9 ft³/lbm flows into the same tank through a 1.5-in.² pipe. The heated water leaves the tank through a large pipe. The enthalpy of water is approximately equal to $(T - 32)$ Btu/lbm, where T denotes the temperature of water in °F. If the mass-flow rate of the incoming water is 10,000 lbm/h and that of the incoming steam is 2000 lbm/h, determine the temperature of the heated water leaving the tank.

5-71. Solve Problem 5-70 for the following cases:

	Incoming water flow rate (lbm/h)	Incoming steam flow rate (lbm/h)
(a)	10,000	1,000
(b)	10,000	3,000
(c)	10,000	5,000
(d)	8,000	2,000
(e)	12,000	2,000
(f)	16,000	2,000

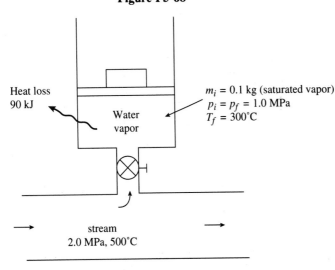

Heat loss
90 kJ

Water
vapor

$m_i = 0.1$ kg (saturated vapor)
$P_i = P_f = 1.0$ MPa
$T_f = 300°C$

stream
2.0 MPa, 500°C

Figure P5-69

First Law Analysis for Control Volumes   Chap. 5

The Second Law of Thermodynamics

Although the first law of thermodynamics allows unrestricted convertibility from one form of energy to another, as long as the overall quantity is conserved, experimental evidence tells that in certain types of energy conversion, restriction must be placed to the direction and extent of transformation. For example, it is a fact that although work can be completely converted into heat, heat cannot be completely converted into work, no matter how ideal the process may be. It is the second law of thermodynamics that imposes restrictions to the direction and extent of energy transformation processes. In this chapter, we study the second law of thermodynamics and some consequences of it.

6-1 THE SECOND LAW OF THERMODYNAMICS

Before starting the main topic of this section, we need to describe the functions of two cyclic devices: a heat engine and a heat pump, or refrigerator. This is shown schematically in Fig. 6-1. A *heat engine* receives heat from a high-temperature body, converts part of this heat to work, and rejects the remaining waste heat to a low-temperature body. A *heat pump,* or *refrigerator,* receives an amount of heat from a low-temperature body and rejects a larger amount of heat to a high-temperature body. The pumping of heat is accomplished with the help of a work input that is equal to the difference between the two heat quantities.

It is well known that physical processes in nature proceed toward equilibrium spontaneously. Liquids flow from a region of high elevation to one of low elevation; gases expand from a region of high pressure to one of low pressure; heat flows from a region of high temperature to one of low temperature; and material diffuses from a region of high concentration to one of low concentration. A spontaneous process can proceed only in a particular direction. Energy from an external source is required to reverse such processes. The second law of thermodynamics epitomizes our experiences with respect to the unidirectional nature of thermodynamic processes.

The second law of thermodynamics can be stated in many different forms, one of which is known as the *Clausius statement of the second law* and is as follows: It is impossible for any device to operate in a cycle in such a manner that the sole effect is the

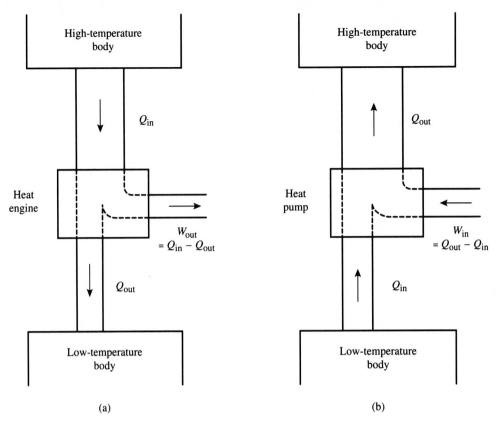

Figure 6-1 Schematic diagrams for (a) a heat engine and (b) a heat pump or refrigerator.

transfer of heat from one body to another body at a higher temperature. In the case of a heat pump, or refrigerator, heat does flow from a region of low temperature to one of high temperature, but only when work is added to the machine from an outside source.

The second law is one of the fundamental laws of nature. It cannot be derived from any other law, but when one statement of it is accepted as a postulate, all other statements of it can then be proved. With reference to another statement of the second law, let us investigate whether heat can be converted completely into work. Consider two energy reservoirs with a heat pump and a hypothetical heat engine, as shown in Fig. 6-2. A *thermal-energy reservoir* (or simply *energy reservoir*) is a body with a relatively large thermal capacity (mass × specific heat) that can supply or absorb finite amounts of heat without undergoing any temperature change. Assume that the hypothetical heat engine could convert all the heat it received into work. This work could then be used to drive the heat pump. Thus, the system comprised of both devices could cause the sole effect of transferring heat from the energy reservoir at a lower temperature to the one at a higher temperature while operating cyclically. This is a violation of the Clausius statement of the second law. We are led to the belief that the hypothetical engine we have postulated cannot exist. Thus, we conclude that no process is possible

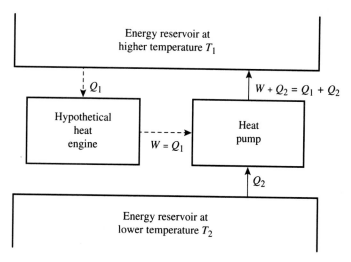

Figure 6-2 Impossibility of converting heat completely into work in a cycle.

whose sole result is the absorption of heat from a reservoir and the conversion of this heat into work. This is the *Kelvin–Planck statement of the second law*.

The Kelvin–Planck statement as stated before excludes the possibility of producing net work by a cyclic device that exchanges heat only with a single energy reservoir at a fixed temperature. This means that a heat engine needs two or more energy reservoirs at different levels of temperature.

Example 6-1

Demonstrate the equivalence of the Clausius and Kelvin–Planck statements of the second law of thermodynamics.

Proof The proof of the equivalence of the Clausius and Kelvin–Planck statements can be demonstrated by showing that the violation of either statement can result in a violation of the other one. The first part of the demonstration—violation of Kelvin–Planck statement leading to violation of Clausius statement—has already been done in the preceding paragraphs in reference to Fig. 6-2.

We now demonstrate the second part—violation of Clausius statement leading to violation of Kelvin–Planck statement. Referring to Fig. 6-3, if the transfer of a quantity of heat Q_2 occurring from the lower-temperature energy reservoir at T_2 to the higher-temperature energy reservoir at T_1 were possible, a legitimate heat engine can be constructed to take in a larger quantity of heat Q_1 from the higher-temperature energy reservoir and reject the smaller quantity of heat Q_2 back to the lower-temperature energy reservoir, thus performing a net work output of the amount $W = Q_1 - Q_2$. Accordingly, the assumed direct heat transfer and the legitimate heat engine, together with the higher-temperature energy reservoir, constitute a device that operates continuously from a single energy reservoir and converts heat completely into work. This is a violation of the Kelvin–Planck statement.

It is seen that a violation of the Kelvin–Planck statement leads to a violation of the Clausius statement, and a violation of the Clausius statement leads to a violation of the Kelvin–Planck statement. The two statements are, therefore, completely equivalent.

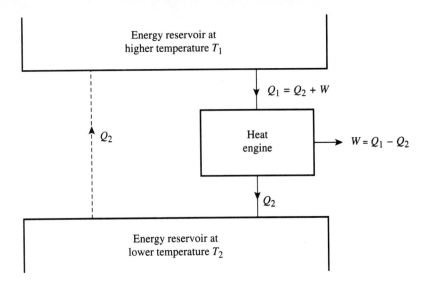

Energy reservoir at
higher temperature T_1

$Q_1 = Q_2 + W$

Q_2

Heat
engine

$W = Q_1 - Q_2$

Q_2

Energy reservoir at
lower temperature T_2

Figure 6-3 Impossibility of transferring heat from one body to another at a higher temperature.

6-2 REVERSIBLE PROCESSES

If at the end of a thermodynamic process, the initial states of the system and the surroundings can be restored without leaving any net change anywhere, the system is said to have undergone an internally and externally or *totally reversible* process. In many engineering analyses, we are interested in the reversibility of the system only, but not necessarily that of the surroundings. During an *internally reversible* process, a system proceeds through a series of equilibrium states, and, if the process direction is reversed at any stage, the system can be returned to its initial equilibrium state without leaving any permanent changes in the system, irrespective of any changes that may occur in the surroundings. Calculations of a system's behavior are the same for an internally reversible process, whether the process is externally reversible or not. In this book, although we often use the complete term "internally reversible" to emphasize that condition, the term "reversible" without a modifier implies internally reversible. In cases where total reversibility is required, it will be explicitly stated.

During a reversible process, the unbalanced potential (mechanical, thermal, chemical, and electrical) within the system or between it and its surroundings that promotes the occurrence of the process must be infinitesimal; and there must be no dissipative effects, such as solid or fluid friction, electrical resistance, and inelasticity, within the system for internal reversibility and in the system and its surroundings for total reversibility. Any finite unbalanced potential or dissipative effect will render the process irreversible. Among the most common irreversible processes are the following:

- Heat transfer across a finite temperature difference
- Unrestrained expansion of a fluid
- Mixing of fluids of different compositions and states
- Solid and fluid friction
- Spontaneous chemical reaction
- Inelastic deformation

- Electric resistance
- Hysteresis effects in magnetization and polarization

For a reversible process, all heat and work interactions that occurred across the boundaries during the original forward process are reversed in direction but equal in magnitude during the reverse process. Because we are most concerned with work and heat interactions between a fluid system and its surroundings, it is important to remind ourselves that for a process to be reversible, any work interaction must be promoted only by an infinitesimal pressure difference, and any heat interaction must be promoted only by an infinitesimal temperature difference. Thus, if heat is transferred to a system at varying temperatures, the temperature of the surroundings would have to vary during the heat exchange. Only at a constant-temperature heat transfer, the temperature of the surroundings can be that of a single energy reservoir.

So far until this section, we have been using the term quasiequilibrium process to mean an ideal process that passes through a series of equilibrium states. We know now a reversible process must be in quasiequilibrium so that the process can be made to traverse in the reverse order the series of equilibrium states passed through during the original process, with no change in magnitude of any energy transfer, but only a change in direction.

Reversible processes are idealized processes that can never be realized in the laboratory but can be approximated as closely as one wishes. For example, a gas confined in a cylinder with a well-lubricated piston can be made to undergo an approximately reversible process by pushing or pulling the piston in slow motion or by dividing the process into very small steps. This is true because in the limit in any stage of the process, it could be turned into the opposite direction by an infinitesimal change of the external conditions.

As an example of a reversible heat-transfer process, let us consider a vapor–liquid mixture of water at 14.7 psia and 212°F contained in a cylinder–piston device, as shown in Fig. 6-4, to be the system and a heat reservoir to be a part of the surround-

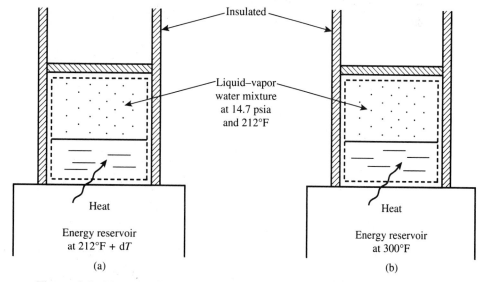

Figure 6-4 Heat transfer to a two-phase mixture: (a) internally and externally (or totally) reversible; and (b) internally reversible, externally irreversible.

ings. In both parts of the figure, the water mixture passes through a series of equilibrium states in a constant-pressure and constant-temperature phase transition, and the processes are therefore internally reversible. In Fig. 6-4(a), because the heat-transfer process is caused by an infinitesimal temperature difference dT, this process is therefore externally reversible. That is, Fig. 6-4(a) shows a totally reversible heat-transfer process. In Fig. 6-4(b), however, the heat-transfer process is caused by a finite temperature difference $\Delta T = (300 - 212)°F$, and this process is externally irreversible.

6-3 THE CARNOT CYCLE—A TOTALLY REVERSIBLE CYCLE

A cycle composed entirely of reversible processes (or internally reversible processes) is called a *reversible cycle* (or internally reversible cycle). If all the processes are totally reversible, the cycle is called a *totally reversible cycle*. Of particular importance in thermodynamics is an idealized totally reversible cycle, known as the *Carnot cycle*, which employs two energy reservoirs at different temperatures and operates on the following totally reversible processes (Fig. 6-5):

Process 1–2: A totally reversible isothermal expansion at T_H, absorbing a quantity of heat Q_H from the high-temperature energy reservoir.

Process 2–3: A totally reversible adiabatic expansion, lowering temperature from T_H to T_L.

Process 3–4: A totally reversible isothermal compression at T_L rejecting a quantity of heat Q_L to the low-temperature energy reservoir.

Process 4–1: A totally reversible adiabatic compression, raising temperature from T_L to T_H.

A heat engine designed to operate on a Carnot cycle is called a *Carnot engine*. The Carnot engine depicted in Fig. 6-5 is for a closed system, and the p–V diagram of Fig. 6-5 is for a gaseous working substance. Figure 6-6 depicts a steady-flow Carnot engine with a liquid–vapor mixture as the working substance. Notice that in the two-phase region, a constant-temperature process is also a constant-pressure process.

For the engine cycle, according to the first law, the net work output is

$$\text{net } W_{\text{out}} = W_{\text{out}} - W_{\text{in}} = Q_H - Q_L \tag{6-1}$$

The *thermal efficiency* η_{th} of a heat-engine cycle is defined as the ratio of the net work output to the heat input, or

$$\eta_{\text{th}} = \frac{\text{net } W_{\text{out}}}{Q_H} = \frac{Q_H - Q_L}{Q_H} = 1 - \frac{Q_L}{Q_H} \tag{6-2}$$

According to the Kelvin–Planck statement of the second law, we must have $Q_L > 0$. Thus, from Eq. (6-2), we conclude that $\eta_{\text{th}} < 1$ even for the totally reversible cycle.

Because the transfer of heat across an infinitesimal temperature difference is not feasible and totally reversible processes never occur in nature, a Carnot heat engine is hardly a practical device. We will study practical heat engines in the latter part of the book. We will see shortly, however, that the Carnot engine is the most efficient heat engine in theory that could operate between two specific temperatures.

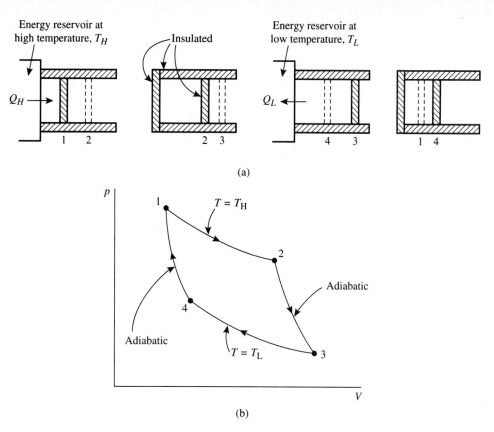

Figure 6-5 Carnot cycle: (a) schematic of a piston–cylinder device, and (b) p–V diagram.

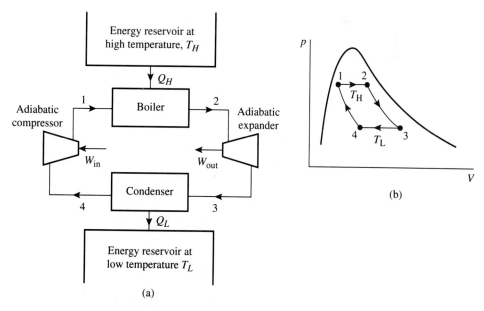

Figure 6-6 A Carnot cycle using a wet vapor in steady flow: (a) schematic, and (b) p–V diagram.

6-4 THE REVERSED CARNOT CYCLE

Because a Carnot cycle is a totally reversible cycle, it can truly be carried out in the reversed direction, thus acting as a refrigerator, or heat pump. A steady-flow version of a reversed Carnot cycle is shown in Fig. 6-7. The individual totally reversible processes are performed in the following manner:

Process 4–3: Isothermal expansion at T_L, absorbing a quantity of heat Q_L from the low-temperature energy reservoir.

Process 3–2: Adiabatic compression, raising the temperature from T_L to T_H.

Process 2–1: Isothermal compression at T_H, rejecting a quantity of heat Q_H to the high-temperature energy reservoir.

Process 1–4: Adiabatic expansion, lowering the temperature from T_H to T_L.

The net work input required to operate the cycle is given by

$$\text{net } W_{\text{in}} = W_{\text{in}} - W_{\text{out}} = Q_H - Q_L \tag{6-3}$$

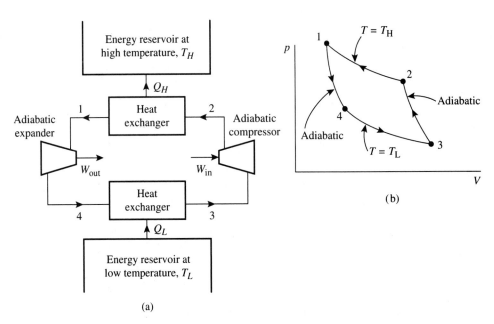

Figure 6-7   Reversed Carnot cycle: (a) schematic, and (b) p–V diagram.

Thus, a quantity of heat Q_L is transferred from the low-temperature energy reservoir to the high-temperature energy reservoir with the help of the net work input $(Q_H - Q_L)$. When the purpose of the device is to maintain a substance or space at a given level of low temperature by removing heat from it, the device is called a *refrigerator*. On the other hand, when the purpose of the device is to maintain a substance or space at a given level of high temperature by supplying heat to it, the device is called a *heat pump*.

The performance of a refrigerator or heat pump is expressed in terms of the *coefficient of performance*, which is abbreviated COP. For a refrigerator, the COP is defined as

$$(COP)_{refrigerator} = \frac{\text{refrigerating effect}}{\text{net work input}} = \frac{Q_L}{Q_H - Q_L} \qquad (6\text{-}4)$$

For a heat pump, the COP is defined as

$$(COP)_{heat \ pump} = \frac{\text{heating effect}}{\text{net work input}} = \frac{Q_H}{Q_H - Q_L} \qquad (6\text{-}5)$$

From Eqs. (6-4) and (6-5), it follows that

$$1 + \frac{Q_L}{Q_H - Q_L} = \frac{Q_H}{Q_H - Q_L}$$

or

$$1 + (COP)_{refrigerator} = (COP)_{heat \ pump} \qquad (6\text{-}6)$$

In addition, as $Q_L \geq 0$ and $Q_H \geq Q_H - Q_L$ for a reversed Carnot cycle, from Eq. (6-5), we see that $(COP)_{heat \ pump} \geq 1$.

It should be noted that a refrigerator or heat pump patented on the reversed Carnot cycle is not a practical device, but this idealized device has the merit of giving the highest performance in theory, with which a real machine can be compared to determine its degree of perfection.

6-5 THE CARNOT PRINCIPLE

The second law imposes the condition that $\eta_{th} < 1$ for any heat engine. Moreover, we will see presently that the efficiency of the Carnot engine is an upper limit that cannot be exceeded by any engine operating between the same two temperatures. This is stated in the following corollary of the second law, known as the *Carnot principle:* No engine can be more efficient than a totally reversible engine operating between the same temperature limits, and all totally reversible engines operating between the same temperature limits have the same efficiency. The following is a proof of this principle.

Consider any heat engine A and a totally reversible heat engine R operating simultaneously between the same temperature limits, as depicted in Fig. 6-8(a). Assume at first that $\eta_A > \eta_R$. But $\eta_A = W_A/Q_H$ and $\eta_R = W_R/Q_H$, the amounts of heat supplied to both engines are the same. Therefore, we have $W_A > W_R$. Now let the totally reversible engine R be reversed to operate as a heat pump, as depicted in Fig. 6-8(b). Because the work W_R required to drive the heat pump R is less in magnitude than the work output W_A of engine A, engine A can drive heat pump R and still have a net work output of $W_A - W_R$. The energy reservoir at temperature T_H could be eliminated by having heat pump R discharge heat directly into engine A. Thus, the composite system comprising the heat engine and the heat pump constitutes a device that could perform the sole result of absorbing heat from a single energy reservoir (the one at T_L) and converting the heat into work. This is a violation of the Kelvin–Planck statement of the second law. Consequently, our original assumption that $\eta_A > \eta_R$ is false and the first point of the Carnot principle is proved.

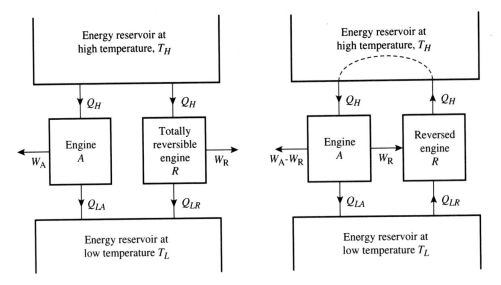

Figure 6-8 Proof of the Carnot principle.

To prove the second point of the principle, let both A and R be totally reversible engines. Assume that these two totally reversible engines have different efficiencies. By reversing the less efficient engine to operate as a heat pump and following the same procedure used earlier, we again have a situation that violates the Kelvin–Planck statement of the second law. We conclude that our assumption that two totally reversible engines operating between the same temperature limits can have different efficiencies is false. Thus, the second point of the Carnot principle is proved.

6-6 THERMODYNAMIC TEMPERATURE SCALE

Because, according to the Carnot principle, the efficiencies of all Carnot engines operating between the same temperatures T_H and T_L are the same regardless of the kind of working substance, these efficiencies must be a function of the two temperatures alone. Therefore, from Eq. (6-2), we can write

$$\frac{Q_H}{Q_L} = \phi(T_H, T_L) \tag{a}$$

where ϕ denotes some function. To determine the nature of the function ϕ, consider three Carnot engines, A, B, and C, operating between the pairs of temperatures (T_H, T_L), (T_H, T_m), and (T_m, T_L), respectively, as shown in Fig. 6-9. For these three engines, Eq. (a) becomes

$$\frac{Q_H}{Q_L} = \phi(T_H, T_L) \qquad \text{for engine } A$$

$$\frac{Q_H}{Q_m} = \phi(T_H, T_m) \qquad \text{for engine } B$$

$$\frac{Q_m}{Q_L} = \phi(T_m, T_L) \qquad \text{for engine } C$$

The Second Law of Thermodynamics   Chap. 6

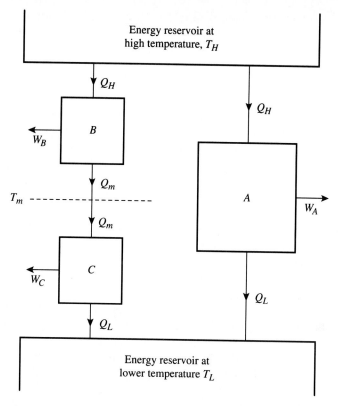

Figure 6-9 Establishment of thermodynamic temperature scale by means of Carnot engines.

Because

$$\frac{Q_H}{Q_L} = \frac{Q_H}{Q_m}\frac{Q_m}{Q_L}$$

we have

$$\phi(T_H, T_L) = \phi(T_H, T_m)\,\phi(T_m, T_L)$$

Because T_H, T_m, and T_L are independent, the previous equation can be satisfied only when the function is of the form

$$\phi(T_H, T_L) = \frac{f(T_H)}{f(T_L)} \tag{b}$$

where f denotes another function, so that

$$\frac{f(T_H)}{f(T_L)} = \frac{f(T_H)}{f(T_m)}\frac{f(T_m)}{f(T_L)}$$

Combining Eqs.(a) and (b), we obtain

$$\frac{Q_H}{Q_L} = \frac{f(T_H)}{f(T_L)} \tag{c}$$

Sec. 6-6   Thermodynamic Temperature Scale   **177**

The mathematical form of $f(T)$ is completely arbitrary. An infinite number of choices could be made, but obviously the simplest choice is to set $f(T) = T$, whereupon Eq. (c) becomes

$$\frac{Q_H}{Q_L} = \frac{T_H}{T_L} \tag{6-7}$$

Equation (6-7) was proposed by Kelvin to define a temperature scale called the *thermodynamic*, or *absolute temperature*, scale. This scale is based on the amounts of heat transferred to and from a Carnot engine and is independent of the nature of the thermometric substance. Assigning a numerical value of 273.16 to the triple point of water results in the Kelvin scale, whereupon we can write

$$T = 273.16 \frac{Q}{Q_{TP}} \tag{6-8}$$

where Q and Q_{TP} denote the amounts of heat exchange with an energy reservoir at any Kelvin temperature T and an energy reservoir at the Kelvin temperature of the triple point of water, respectively. Equation (6-8) introduces the concept of absolute zero, that is, $T = 0$ when $Q = 0$, a condition that cannot be physically accomplished, although it can be approached as a limiting case.

The thermodynamic temperature scale we just defined is equivalent to the ideal gas temperature scale defined in Sec. 1-7. The equivalence of these two temperature scales will be illustrated in an example in Sec. 7-5.

With the absolute temperature scale as defined by Eq. (6-7), the thermal efficiency of a Carnot engine, Eq. (6-2), can be expressed as follows:

$$\eta_{\text{th, Carnot}} = 1 - \frac{Q_L}{Q_H} = 1 - \frac{T_L}{T_H} \tag{6-9}$$

Similarly, the coefficients of performance of a Carnot refrigerator and heat pump can be expressed as

$$(\text{COP})_{\text{Carnot refrigerator}} = \frac{Q_L}{Q_H - Q_L} = \frac{T_L}{T_H - T_L} \tag{6-10}$$

$$(\text{COP})_{\text{Carnot heat pump}} = \frac{Q_H}{Q_H - Q_L} = \frac{T_H}{T_H - T_L} \tag{6-11}$$

Example 6-2

A heat engine receives 2000 kJ of heat from a reservoir at 550 K and rejects 1300 kJ of heat to the surrounding air at 300 K. Determine the net work output and the thermal efficiency. If this engine is replaced by a Carnot engine to produce the same work output, what would be the thermal efficiency, the heat addition, and the heat rejection?

Solution

(a) For the actual heat engine, the net work output and thermal efficiency are

$$\text{net } W_{\text{out}} = Q_H - Q_L = 2000 - 1300 = 700 \text{ kJ}$$

$$\eta_{\text{th}} = \frac{\text{net } W_{\text{out}}}{Q_H} = \frac{700}{2000} = 0.35 = 35\%$$

(b) For the Carnot engine, the thermal efficiency is

$$\eta_{\text{th, Carnot}} = 1 - \frac{T_L}{T_H} = 1 - \frac{300 \text{ K}}{550 \text{ K}} = 0.455 = 45.5\%$$

As

$$\eta_{th,\,Carnot} = \frac{net\ W_{out}}{Q_{H,\,Carnot}}$$

the heat addition to the Carnot engine is then

$$Q_{H,\,Carnot} = \frac{net\ W_{out}}{\eta_{th,\,Carnot}} = \frac{700\ kJ}{0.455} = 1538\ kJ$$

The heat rejection from the Carnot engine is

$$Q_{L,\,Carnot} = Q_{H,\,Carnot} - net\ W_{out} = 1538 - 700 = 838\ kJ$$

The heat rejection from the Carnot engine can also be determined by the equation

$$Q_{L,\,Carnot} = Q_{H,\,Carnot}\left(\frac{T_L}{T_H}\right) = (1538\ kJ)\left(\frac{300\ K}{550\ K}\right) = 839\ kJ$$

The difference in the values of $Q_{L,\,Carnot}$ obtained by the preceding two methods is due to roundoff error in the calculations.

Example 6-3

An ocean thermal-energy conversion plant near the equator is designed to operate between the ocean surface temperature of 80°F and the deep-water temperature of 40°F. Determine the maximum thermal efficiency of the plant.

Solution A schematic of the ocean thermal energy plant is shown in Fig. 6-10. The maximum thermal efficiency a heat engine operating between two specified temperature limits can have is the Carnot efficiency. In this ocean thermal-energy plant, the maximum

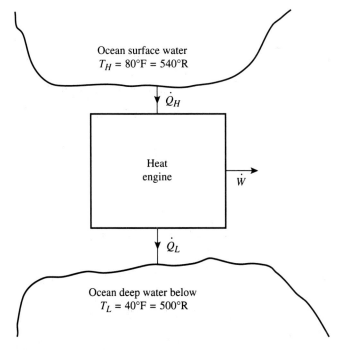

Figure 6-10   Schematic of an ocean thermal-energy plant.

thermal efficiency is

$$\eta_{th,\,max} = \eta_{th,\,Carnot} = 1 - \frac{T_L}{T_H}$$

$$= 1 - \frac{500}{540} = 0.0741 = 7.41\%$$

Notice that as the actual thermal efficiency would be less than this value, the economic viability of such a system would require careful study.

Example 6-4

A heat pump is used to maintain a house at 20°C during the winter. On a typical day when the average outdoor temperature is 2°C, the house loses heat at a steady rate of 70,000 kJ/h. The minimum power consumption is 10 kW if the heat pump follows the reversed Carnot cycle. Determine (a) the coefficient of performance of the heat pump, (b) how long the heat pump ran on the typical day to supply the required heating, (c) the total heating cost of a day, assuming a price of 9.0 ¢/kWh for electricity, and (d) the heating cost for the same day if resistance heating is used instead of a heat pump.

Solution A schematic of the heat-pump installation is shown in Fig. 6-11.

(a) For the Carnot heat pump, the coefficient of performance is given by Eq. (6-11):

$$(COP)_{Carnot\ heat\ pump} = \frac{T_H}{T_H - T_L} = \frac{293}{293 - 275} = 16.3$$

(b) The amount of heat loss in a day is

$$Q_H = \dot{Q}_H(1\ day) = (70,000\ kJ/h)(24\ h) = 1,680,000\ kJ$$

Because the COP of the heat pump is defined as

$$(COP)_{heat\ pump} = \frac{Q_H}{Q_H - Q_L} = \frac{Q_H}{net\ W_{in}}$$

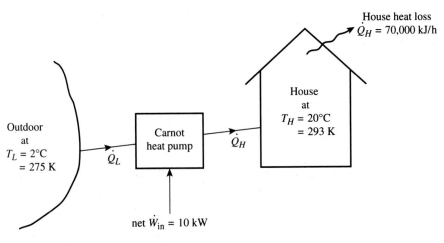

Figure 6-11 Heating a house with a heat pump.

the required net work input to the heat pump is calculated as

$$\text{net } W_{in} = \frac{Q_H}{(COP)_{\text{heat pump}}} = \frac{1,680,000 \text{ kJ}}{16.3} = 103,067 \text{ kJ}$$

Thus, the length of time (Δt) the heat pump ran for the day is

$$\Delta t = \frac{\text{net } W_{in}}{\text{net } \dot{W}_{in}} = \frac{103,067 \text{ kJ}}{10 \text{ kJ/s}} = 10,307 \text{ s}$$

$$= \frac{10,307 \text{ s}}{3600 \text{ s/h}} = 2.86 \text{ h}$$

(c) The total heating cost of a day is

$$\text{Heat-pump cost} = (\text{net } \dot{W}_{in})(\Delta t)(\text{price})$$

$$= (10 \text{ kW})(2.86 \text{ h})(0.09 \text{ \$/kW·h})$$

$$= \$2.57$$

(d) If resistance heating were used to supply the entire heating load of 1,680,000 kJ, the cost would be

$$\text{Resistance heating cost} = (1,680,000 \text{ kJ})\left(\frac{1}{3600 \text{ kJ/kW·h}}\right)(0.09 \text{ \$/kW·h})$$

$$= \$42.0$$

6-7 THE CLAUSIUS INEQUALITY

An important corollary of the second law known as the *Clausius inequality* states that the cyclic integral of the quantity $đQ/T$ for a closed system must be less than zero or equal to zero in the limit when the cycle is internally reversible. To establish the Clausius inequality, consider a closed system (see Fig. 6-12) undergoing a cycle that can be either internally reversible or internally irreversible. During a portion of the cycle, the

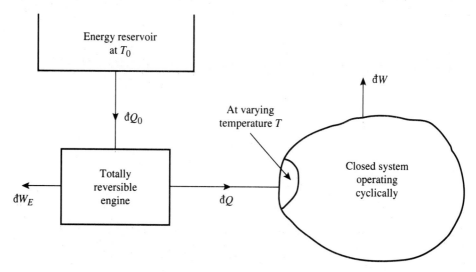

Figure 6-12 Schematic used for the development of the Clausius inequality.

system receives heat $đQ$ at a location where the temperature is T, which is allowed to vary. At the same time, work $đW$ is done by the system. In order to eliminate any source of irreversibility outside the system, a totally reversible engine is placed between the heat reservoir at T_0 and the portion of the closed system at T. This totally reversible engine receives heat $đQ_0$ from the reservoir, delivers heat $đQ$ to the closed system, and performs work $đW_E$ while executing a whole number of cycles for each cycle completed by the closed system. Assume that all works are done reversibly. During the closed-system cycle, the direction of $đQ$ on some other parts of the boundary or at other times in the cycle can be opposite to that shown in the figure and $đW$ can be negative for other parts of the cycle. They are immaterial to the proof we are undertaking.

When we consider the totally reversible engine and the closed system together as a composite system, because it interacts only with a single energy reservoir, the Kelvin–Planck statement of the second law can be satisfied only if

$$\oint đW_E + \oint đW \leq 0 \tag{a}$$

But for the totally reversible engine, we have

$$đW_E = đQ_0 - đQ$$

and

$$\frac{đQ_0}{T_0} = \frac{đQ}{T}$$

so that

$$đW_E = \frac{T_0}{T} đQ - đQ \tag{b}$$

Substituting Eq. (b) into Eq. (a) results in

$$\oint \frac{T_0}{T} đQ - \oint đQ + \oint đW \leq 0 \tag{c}$$

However, application of the first law to the closed system reveals that

$$\oint đW = \oint đQ \tag{d}$$

Combining Eqs. (c) and (d) gives

$$\oint T_0 \frac{đQ}{T} \leq 0$$

Because T_0 is a fixed positive value on the absolute temperature scale, we conclude finally that

$$\oint \frac{đQ}{T} \leq 0 \tag{6-12}$$

In the preceding development of Eq. (6-12) based on the schematic arrangement of Fig. 6-12, the only place irreversibility can occur is within the closed system itself.

This means that we are concerned only with the question of internal reversibility and internal irreversibility of the closed system.

Equation (6-12) is the mathematical expression of the Clausius inequality, in which the equality holds for internally reversible cycles and the inequality holds for internally irreversible cycles.

6-8 ENTROPY: A THERMODYNAMIC PROPERTY

According to the Clausius inequality relation, Eq. (6-12), for an internally reversible closed-system cycle, we have

$$\oint \left(\frac{dQ}{T} \right)_{\text{int rev}} = 0 \tag{6-13}$$

Notice that even though we used the arrangement depicted by Fig. 6-12 with no external irreversibility to develop the preceding equation, the evaluation of dQ and T is solely for the fixed mass in the closed system, and any event occurring in the surroundings is irrelevant to this evaluation. Therefore, Eq. (6-13) is valid as long as the cycle is internally reversible.

Equation (6-13) states that the integral of dQ/T for a fixed mass of material when carried out over an internally reversible cycle is equal to zero. It follows that the differential $(dQ/T)_{\text{int rev}}$ is an exact differential in mathematical terms and the integral $\int (dQ/T)_{\text{int rev}}$ is the change of a thermodynamic property (see Sec. 1-3).

The fact that the integral $\int (dQ/T)_{\text{int rev}}$ is a thermodynamic property change can be demonstrated further by referring to three arbitrary internally reversible processes, A, B and C, connecting two equilibrium states, 1 and 2, as shown in Fig. 6-13. From Eq. (6-13) for the internally reversible cycle 1–A–2–B–1 formed by the internally reversible processes A and B, we must have

$$\oint \frac{dQ}{T} = \int_{1,A}^{2} \frac{dQ}{T} + \int_{2,B}^{1} \frac{dQ}{T} = 0$$

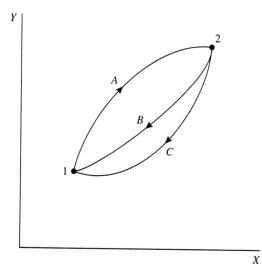

Figure 6-13   Internally reversible paths.

Similarly, for the internally reversible cycle 1–A–2–C–1 formed by the internally reversible processes A and C, we must have

$$\oint \frac{đQ}{T} = \int_{1,A}^{2} \frac{đQ}{T} + \int_{2,C}^{1} \frac{đQ}{T} = 0$$

Subtracting the preceding two expressions gives

$$\int_{2,B}^{1} \frac{đQ}{T} - \int_{2,C}^{1} \frac{đQ}{T} = 0$$

or

$$\int_{2,B}^{1} \frac{đQ}{T} = \int_{2,C}^{1} \frac{đQ}{T}$$

Because B and C are two arbitrary internally reversible processes connecting two equilibrium states, 1 and 2, it follows that the value of $\int (đQ/T)$ is the same for all internally reversible paths between the two states. Accordingly, we conclude that the value of $\int (đQ/T)_{\text{int rev}}$ depends only on the end states of any process, thus representing the change of a thermodynamic property.

The new thermodynamic property introduced here is called *entropy S*, as defined by the expression:

$$dS = \left(\frac{đQ}{T}\right)_{\text{int rev}} \tag{6-14}$$

Integrating along an internally reversible path between two equilibrium states 1 and 2 gives

$$\Delta S = S_2 - S_1 = \int_1^2 \left(\frac{đQ}{T}\right)_{\text{int rev}} \tag{6-15}$$

Because entropy is a thermodynamic property and, therefore, a point function, it does not matter what particular internally reversible path is followed in the integration as long as it is internally reversible. After the value of ΔS has been evaluated along some internally reversible path, this value will be the same even if the end states 1 and 2 are connected by an irreversible process.

Equations (6-14) and (6-15) hold for any closed system or any fixed quantity of matter flowing through an open system. On the basis of per unit mass or per mole, they are written as

$$ds = \left(\frac{đq}{T}\right)_{\text{int rev}} \tag{6-14a}$$

$$\Delta s = s_2 - s_1 = \int_1^2 \left(\frac{đq}{T}\right)_{\text{int rev}} \tag{6-15a}$$

where s is called specific entropy. The common units for s are kJ/kg·K (or kJ/kgmole·K) in the SI system and Btu/lbm·°R (or Btu/lbmole·°R) in the English system.

According to Eq. (6-14), the small quantity of heat transferred to a closed system in an infinitesimal internally reversible process is

$$đQ_{\text{int rev}} = T\, dS \tag{6-16}$$

The Second Law of Thermodynamics   Chap. 6

The heat transfer to the closed system in an internally reversible process between states 1 and 2 is then

$$Q_{\text{int rev}} = \int_1^2 T \, dS \tag{6-17}$$

This last integral is represented by the area under path 1–2 of an internally reversible process plotted on a temperature versus entropy diagram, as shown in Fig. 6-14. Because heat is a path function, meaning that the amount of heat transfer depends on the path of the process, the integral of Eq. (6-17) will be different for different internally reversible processes.

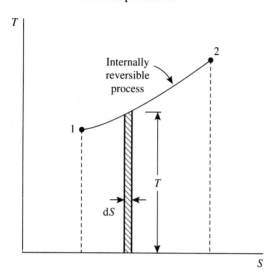

Figure 6-14 Area under the curve on T–S plane represents heat.

For an internally reversible isothermal (or constant-temperature) process, Eq. (6-17) yields

$$Q_{\text{int rev isothermal}} = T(\Delta S) = T(S_2 - S_1) \tag{6-18}$$

For an internally reversible adiabatic process, $\text{d}Q_{\text{int rev}} = 0$, then, from Eq. (6-14), $dS = 0$. This means that during an internally reversible adiabatic process, there is no change in entropy. Such a process is called an *isentropic* process.

It is convenient for many applications to use T and S as the two coordinates to describe thermodynamic processes. Of particular interest is the representation of the Carnot cycle on the T–S plane. Figure 6-15 depicts two cases of such a plane corresponding to the p–V diagrams of Figs. 6-5 and 6-6. For a Carnot cycle:

Heat added $= Q_H = T_H \Delta S$
Heat rejected $= Q_L = T_L \Delta S$
Net work output $= Q_H - Q_L = (T_H - T_L) \Delta S$

$$\text{Thermal efficiency } \eta_{\text{th, Carnot}} = \frac{Q_H - Q_L}{Q_H} = \frac{(T_H - T_L) \Delta S}{T_H \Delta S}$$

$$= \frac{T_H - T_L}{T_H} = 1 - \frac{T_L}{T_H} \tag{6-19}$$

This thermal efficiency expression is the same as Eq. (6-9).

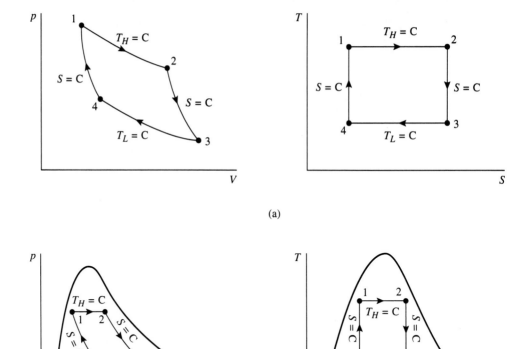

Figure 6-15 *p–V* and *T–S* diagrams for two Carnot cycles: (a) an ideal gas (such as air) as the working substance, and (b) a pure substance (such as steam) as the working substance.

It is to be noted that Eq. (6-15a) gives only the difference in specific entropy between two states. With the arbitrary assignment of zero to the specific entropy of a particular state, the values of specific entropy of other states of a pure substance can be evaluated. For example, in steam tables, the specific entropy of saturated liquid at the triple-point temperature (0.01°C or 32.02°F) is assigned zero. Other reference states are chosen for other substances.

Example 6-5

One kilogram of oxygen initially at 200 kPa and 350 K undergoes an irreversible process to a final state at 100 kPa and 300 K. Considering oxygen as an ideal gas with $c_p = 0.923$ kJ/kg·K and $c_v = 0.663$ kJ/kg·K, calculate the entropy change.

Solution Because entropy is a thermodynamic property, the change of entropy is the same no matter what process, reversible or irreversible, has taken place between the two end states of the fixed quantity of gas, either contained in a closed system or flowing through an open system. We will use three internally reversible paths (Fig. 6-16) in the evaluation of $\int_1^2 dq/T$ to obtain Δs.

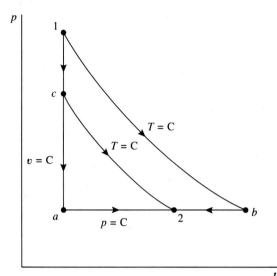

v **Figure 6-16** p–v diagram for Example 6-5.

The gas constant of oxygen is calculated as

$$R = c_p - c_v = 0.923 - 0.663 = 0.260 \text{ kJ/kg·K}$$

or

$$R = \frac{\mathscr{R}}{M} = \frac{8.314 \text{ kJ/kgmole·K}}{32 \text{ kg/kgmole}} = 0.260 \text{ kJ/kg·K}$$

The specific volumes of the end states are

$$v_1 = \frac{RT_1}{p_1} = \frac{(0.260 \text{ kJ/kg·K})(350 \text{ K})}{200 \text{ kN/m}^2} = 0.455 \text{ m}^3/\text{kg}$$

$$v_2 = \frac{RT_2}{p_2} = \frac{0.260 \times 300}{100} = 0.780 \text{ m}^3/\text{kg}$$

(a) Internally reversible path 1–a–2:

$$\Delta s = s_2 - s_1 = \int_1^2 \left(\frac{đq}{T}\right)_{\text{int rev}} = \int_1^a \frac{đq}{T} + \int_a^2 \frac{đq}{T}$$

$$= \int_1^a \frac{c_v\, dT}{T} + \int_a^2 \frac{c_p\, dT}{T}$$

$$= c_v \ln \frac{T_a}{T_1} + c_p \ln \frac{T_2}{T_a}$$

But

$$\frac{p_a v_a}{T_a} = \frac{p_1 v_1}{T_1} \qquad \text{or} \qquad \frac{T_a}{T_1} = \frac{p_a}{p_1} = \frac{p_2}{p_1}$$

and

$$\frac{p_a v_a}{T_a} = \frac{p_2 v_2}{T_2} \qquad \text{or} \qquad \frac{T_2}{T_a} = \frac{v_2}{v_a} = \frac{v_2}{v_1}$$

Therefore,

$$\Delta s = c_v \ln \frac{p_2}{p_1} + c_p \ln \frac{v_2}{v_1}$$

$$= (0.663 \text{ kJ/kg·K}) \ln \frac{100 \text{ kPa}}{200 \text{ kPa}}$$

$$+ (0.923 \text{ kJ/kg·K}) \ln \frac{0.780 \text{ m}^3/\text{kg}}{0.455 \text{ m}^3/\text{kg}}$$

$$= 0.0379 \text{ kJ/kg·K}$$

(b) Internally reversible path 1–b–2:

$$\Delta s = \int_1^2 \left(\frac{đq}{T}\right)_{\text{int rev}} = \int_1^b \frac{đq}{T} + \int_b^2 \frac{đq}{T}$$

$$= \frac{1}{T_1} \int_1^b đq + \int_b^2 \frac{c_p \, dT}{T} = \frac{1}{T_1} \int_1^b đq + c_p \ln \frac{T_2}{T_b}$$

$$= \frac{1}{T_1} \int_1^b đq + c_p \ln \frac{T_2}{T_1}$$

For the isothermal portion of the process (1–b), for convenience, we will use a closed-system equation, that is,

$$đq = du + đw = c_v \, dT + đw = đw = p \, dv$$

Thus,

$$\int_1^b đq = \int_1^b p \, dv = RT_1 \int_1^b \frac{dv}{v} = RT_1 \ln \frac{v_b}{v_1}$$

$$= RT_1 \ln \frac{p_1}{p_b} = RT_1 \ln \frac{p_1}{p_2}$$

Therefore,

$$\Delta s = R \ln \frac{p_1}{p_2} + c_p \ln \frac{T_2}{T_1}$$

$$= (0.260 \text{ kJ/kg·K}) \ln \frac{200 \text{ kPa}}{100 \text{ kPa}}$$

$$+ (0.923 \text{ kJ/kg·K}) \ln \frac{300 \text{ K}}{350 \text{ K}}$$

$$= 0.0379 \text{ kJ/kg·K}$$

(c) Internally reversible path 1–c–2:

$$\Delta s = \int_1^2 \left(\frac{đq}{T}\right)_{\text{int rev}} = \int_1^c \frac{đq}{T} + \int_c^2 \frac{đq}{T}$$

$$= \int_1^c \frac{c_v \, dT}{T} + \frac{1}{T_2} \int_c^2 đq$$

$$= c_v \ln \frac{T_c}{T_1} + \frac{1}{T_2} \int_c^2 đq = c_v \ln \frac{T_2}{T_1} + \frac{1}{T_2} \int_c^2 đq$$

But

$$\int_c^2 đq = \int_c^2 p \, dv = RT_2 \int_c^2 \frac{dv}{v}$$

$$= RT_2 \ln \frac{v_2}{v_c} = RT_2 \ln \frac{v_2}{v_1}$$

Therefore,

$$\Delta s = c_v \ln \frac{T_2}{T_1} + R \ln \frac{v_2}{v_1}$$

$$= 0.663 \ln \frac{300}{350} + 0.260 \ln \frac{0.780}{0.455}$$

$$= 0.0379 \text{ kJ/kg·K}$$

Notice that the value of Δs along all the three paths chosen for the calculation is the same. This Δs value is the specific entropy change of oxygen between the given end states, irrespective of the connecting process.

6-9 THE PRINCIPLE OF INCREASE IN ENTROPY

It cannot be overemphasized that the entropy of a system of fixed mass is a state function; it depends only on the state that the system is in, and not on how that state is reached. If a system goes from state 1 to state 2, its entropy changes from S_1 to S_2. However, only when the system travels along an internally reversible path between the two end states, the entropy change $S_2 - S_1$ equals $\int_1^2 (đQ/T)_{\text{int rev}}$. If the path is internally irreversible, $\int_1^2 (đQ/T)_{\text{int irr}}$ has a different value, although the change in entropy is the same as before. The relation that does exist between the change in entropy and the integral $\int (đQ/T)$ along any arbitrary path can be obtained in the following manner.

Let a system change from state 1 to state 2 by an internally reversible process A and return to state 1 by either an internally reversible process B or an internally irreversible process C, as depicted in Fig. 6-17. The cycle made up of the internally re-

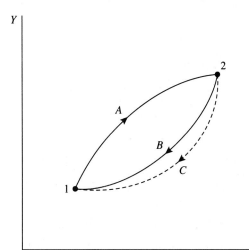

Figure 6-17   Reversible and irreversible cycles.

versible processes A and B is an internally reversible cycle. Following Eq. (3-13), we write

$$\int_{1,A}^{2} \frac{dQ}{T} + \int_{2,B}^{1} \frac{dQ}{T} = 0$$

The cycle made up of the internally reversible process A and the internally irreversible process C is an internally irreversible cycle. From Eq. (6-12),

$$\int_{1,A}^{2} \frac{dQ}{T} + \int_{2,C}^{1} \frac{dQ}{T} < 0$$

Subtracting the first equation from the second gives

$$\int_{2,C}^{1} \frac{dQ}{T} - \int_{2,B}^{1} \frac{dQ}{T} < 0$$

Transposing, we have

$$\int_{2,B}^{1} \frac{dQ}{T} > \int_{2,C}^{1} \frac{dQ}{T}$$

Because process B is internally reversible, dQ/T can be replaced by dS in the first integral. Thus,

$$\int_{2,B}^{1} \frac{dQ}{T} = \int_{2,B}^{1} dS$$

But because entropy is a property,

$$\int_{2,B}^{1} dS = \int_{2,C}^{1} dS$$

Therefore,

$$\int_{2,C}^{1} dS > \int_{2,C}^{1} \frac{dQ}{T}$$

For the general case, we can then write

$$dS \geq \frac{dQ}{T} \tag{6-20}$$

or

$$\Delta S = S_2 - S_1 \geq \int_{1}^{2} \frac{dQ}{T} \tag{6-20a}$$

where the equality holds for an internally reversible process and the inequality for an internally irreversible process. This is one of the most important equations of thermodynamics. It expresses the influence of irreversibility on the entropy of a closed system.

For an isolated system, $dQ = 0$; then, according to Eq. (6-20)

$$dS_{\text{isolated system}} \geq 0 \tag{6-21}$$

Since any system and its surroundings together constitute an isolated system, the preceding equation can be written as

$$dS_{system} + dS_{surr} \geq 0 \qquad (6\text{-}21a)$$

where dS_{surr} is the entropy change of every part of the surroundings that is affected by changes in the system of interest. This is the *principle of increase of entropy*.

In accordance with the first law, an isolated system can only assume those states for which the total internal energy remains constant. Now in accordance with the second law as expressed by Eq. (6-21), for the states of equal energy, only those with greater or equal values of entropy can be attained by the system. Because all natural processes occur spontaneously and irreversibly, an important conclusion one can draw from the increase of entropy principle is that an isolated system always proceeds continuously from states of low entropy to states of higher entropy until it attains a maximum entropy at the final equilibrium state.

According to Eq. (6-20), for an internally irreversible process, $dS > đQ/T$, that is, the entropy change of a fixed mass of material is bigger than the entropy transfer accompanying heat transfer. The additional entropy change is due to irreversibility effects that produce entropy. With the introduction of an *entropy production* term, $đS_{prod}$, Eq. (6-20) can be rewritten as

$$dS = \frac{đQ}{T} + đS_{prod} \qquad (6\text{-}22)$$

or

$$\Delta S = \int \frac{đQ}{T} + S_{prod} \qquad (6\text{-}22a)$$

where $S_{prod} = 0$ for internally reversible processes
$S_{prod} > 0$ for internally irreversible processes

Notice that S_{prod} is not a property of the system and its differential is written as $đS_{prod}$. The value of S_{prod} depends on how the process is executed, and thus it is a path function just as heat and work.

Example 6-6

A 100-W incandescent light bulb in a room has an isothermal surface temperature of 110°C in steady conditions. Determine the rate of entropy production for the light bulb.

Solution   Consider the light bulb (the glass bulb and its content together) as a simple incompressible closed system (see Fig. 6-18). The bulb in steady conditions does not change its thermodynamic state, so its properties must remain constant. For the bulb, we then have

$$U = \text{constant} \qquad \text{and} \qquad S = \text{constant}$$

The first law of thermodynamics is written as

$$Q - W = \Delta U^{0} = 0$$

or

$$\dot{Q} = \dot{W} = -100 \text{ W}$$

noting that electric work is done on the system and heat is rejected from the system.

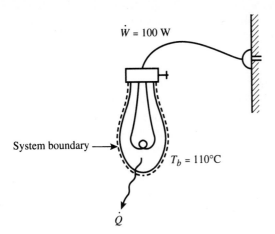

$\dot{W} = 100$ W

System boundary

$T_b = 110°C$

$\dot{Q}$

Figure 6-18   Schematic for Example 6-6.

The second law of thermodynamics is written as

$$\Delta S^{0}_{\text{system}} = \int \frac{dQ}{T} + S_{\text{prod}} = 0$$

or

$$S_{\text{prod}} = -\int \frac{dQ}{T} = -\frac{Q}{T_b}$$

so that

$$\dot{S}_{\text{prod}} = -\frac{\dot{Q}}{T_b} = -\frac{-100 \text{ W}}{(110 + 273) \text{ K}} = 0.261 \text{ W/K}$$

The cause of the entropy production is the dissipative effects of electrical resistance of the lighting element that render the process irreversible.

6-10 STATISTICAL DEFINITION OF ENTROPY

The thermodynamic function entropy introduced in Sec. 6-8 is a macroscopic property of matter and was defined through a macroscopic approach. This macroscopic property is clearly interrelated to a microscopic parameter, the thermodynamic probability. Such a treatment, however, is a topic of microscopic, or statistical, thermodynamics. We include a brief discussion of this topic here only to provide a physical interpretation and thus to gain some insight to the property entropy.

Let us consider a simplified case of an ideal-gas system in a well-insulated rigid container. The container is divided into two equal parts by a partition that has a small hole so that gas molecules can move between the two halves. The gas molecules are all in random motion and the residence of the molecules in either half of the container is equally probable. The mathematical probability or expectation of a molecule being located in one half, say, the left-hand half of the container is $\frac{1}{2}$. The mathematical probability of the event in which two molecules are located in the same half would be $(\frac{1}{2})(\frac{1}{2}) = (\frac{1}{2})^2$. Similarly, for three molecules, it would be $(\frac{1}{2})^3$, and for four molecules, it would be $(\frac{1}{2})^4$. If there are N molecules in the container, the mathematical probability of all the molecules being located in one-half of the container is $(\frac{1}{2})^N$. Because N is always a very large number and $(\frac{1}{2})^N$ is always negligibly small, it follows that the probability of having a spontaneous process that results in all the molecules being located in one of two halves of the container is indeed very low.

The preceding example indicates that the occurrence of spontaneous compression of gases is highly improbable. This means that a process of spontaneous expansion or free expansion of gases is irreversible. In addition, it is a common knowledge that other spontaneous processes such as the transfer of heat from a high-temperature body to one at a lower temperature and the mixing or diffusion of fluids are all irreversible. By determining the mathematical probability of these irreversible processes occurring in the reverse direction, one can be sure that they are all highly improbable. These confirm a general statement of the second law: spontaneous processes are irreversible. The second law also dictates that all irreversibility effects tend to promote the increase of entropy. Thus, it seems possible that there is some interrelationship between probability and entropy.

In order to examine the problem at hand, we need to introduce three additional terms: macroscopic state, microscopic state, and thermodynamic probability. The macroscopic state of a system or the *macrostate* is defined by the average characteristics of a large aggregation of molecules in terms of macroscopic properties such as pressure, temperature, density, and internal energy, without any detailed inquiry into the activities of individual molecules. For example, the pressure exerted by a gas depends only on how many molecules have specified velocities, not at all on which molecules have those velocities. On the other hand, the specification of the microscopic state of a system or *microstate* calls for a complete statement of the position and velocity (both magnitude and direction) of each of its molecules. With position and momentum exchanges among the gas molecules, there may be a large number of microstates that realize exactly the same macrostate. *Thermodynamic probability* is defined as the number of microstates corresponding to a given macrostate. Notice that a thermodynamic probability is never less than one, whereas a mathematical probability is never greater than one.

When an isolated system undergoes a change in the macrostate through a spontaneous process, the new macrostate must have a greater number of corresponding microstates as compared to the preceding macrostate. In other words, an isolated system tends spontaneously to shift from less probable states to more probable states, resulting in an increase in thermodynamic probability of the system. At the same time, an isolated system tends spontaneously to shift to states of higher entropy. The simultaneous increase of entropy and thermodynamic probability of an isolated system toward their maximum values at the equilibrium state leads to the conclusion that entropy is proportional to thermodynamic probability. We must realize, however, that entropy is an extensive thermodynamic property, meaning that if two independent systems are combined, their entropies are additive. On the contrary, if two independent systems are combined, the thermodynamic probability is multiplicative. We, therefore, take the entropy S as proportional to the logarithm of the thermodynamic probability Ω and set

$$S = k \ln \Omega \qquad (6\text{-}23)$$

where k is a proportional constant and is identified in statistical thermodynamics to be the Boltzmann constant. The preceding equation is the definition of entropy from the microscopic viewpoint. It is consistent with the macroscopic definition of entropy from the second law of thermodynamics.

From Eq. (6-23), it is clear that the greater the degree of randomness or disorder in a system, the larger the thermodynamic probability and therefore the larger the entropy. In general, for the same substance at the same pressure, Ω and S increase in the sequence: crystal to liquid to gas. Zero entropy is obtained when there is only one microstate possible. This condition exists at the absolute zero temperature. When nuclear spins are ignored, we can assign $S = 0$ at $T = 0$ K for all perfect crystals of a single-element pure isotope.

6-11 ENTROPY OF A PURE SUBSTANCE

For pure substances such as water and Refrigerant-12 in the liquid–vapor saturation, superheated-vapor, and compressed-liquid regions, the entropy data are tabulated in exactly the same manner as the properties v, u, and h. These tables are given in Appendix 3 (Tables A-18 to A-27). Specifically, in the saturation region, the entropy s is related to the quality x by the expression

$$s = s_f + x s_{fg} = s_f + x(s_g - s_f) \tag{6-24}$$

In the superheated-vapor and compressed-liquid regions, the entropy is tabulated along with other properties as a function of temperature and pressure. In the absence of compressed-liquid data in the low-pressure range, the entropy of a compressed liquid can be approximated by using the entropy of the saturated liquid at the temperature of the compressed liquid.

Example 6-7

Water at 3000 lbm/h, 100 psia, and 70°F, and steam at 1000 lbm/h, 100 psia, and 400°F are mixed in an adiabatic chamber. Assuming a steady-flow, steady-state process and no pressure drop in the mixing chamber, determine the rate of total entropy increase for the process.

Solution Denoting the incoming water, the incoming steam, and the outgoing steam by the subscripts 1, 2, and 3, respectively (Fig. 6-19), the given data are

$$\dot{m}_1 = 3000 \text{ lbm/h}, \qquad p_1 = 100 \text{ psia}, \qquad T_1 = 70°F,$$
$$\dot{m}_2 = 1000 \text{ lbm/h}, \qquad p_2 = 100 \text{ psia}, \qquad T_2 = 400°F$$

Since no pressure drop in the chamber is assumed, $p_3 = 100$ psia. The mass-flow rate of the outgoing steam is

$$\dot{m}_3 = \dot{m}_1 + \dot{m}_2 = 3000 + 1000 = 4000 \text{ lbm/h}$$

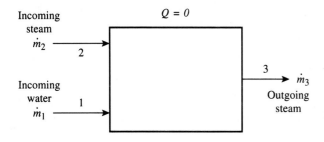

Figure 6-19 Schematic flow diagram for Example 6-7.

The Second Law of Thermodynamics   Chap. 6

The incoming water at 100 psia and 70°F is a compressed liquid. Because no compressed liquid data are available in this low-pressure range, use of the saturated steam table (Table A-18E) gives

$$h_1 = h_f \text{ (at 70°F)} = 38.09 \text{ Btu/lbm}$$

$$s_1 = s_f \text{ (at 70°F)} = 0.07463 \text{ Btu/lbm·°R}$$

The incoming steam at 100 psia and 400°F is a superheated vapor. The superheated steam table (Table A-20E) gives

$$h_2 = 1227.5 \text{ Btu/lbm} \qquad \text{and} \qquad s_2 = 1.6517 \text{ Btu/lbm·°R}$$

With $\dot{Q} = 0$, $\dot{W}_{cv} = 0$, and changes in kinetic and potential energies neglected, an energy balance for the steady-flow mixing process gives

$$\cancel{\dot{Q}}^0 - \cancel{\dot{W}_{cv}}^0 + \dot{m}_1\left(h_1 + \cancel{\frac{V_1^2}{2}}^0 + \cancel{gz_1}^0\right) + \dot{m}_2\left(h_2 + \cancel{\frac{V_2^2}{2}}^0 + \cancel{gz_2}^0\right) = \dot{m}_3\left(h_3 + \cancel{\frac{V_3^2}{2}}^0 + \cancel{gz_3}^0\right)$$

or

$$\dot{m}_1 h_1 + \dot{m}_2 h_2 = \dot{m}_3 h_3$$

so that

$$h_3 = \frac{\dot{m}_1 h_1 + \dot{m}_2 h_2}{\dot{m}_3} = \frac{3000(38.09) + 1000(1227.5)}{4000}$$

$$= 335.4 \text{ Btu/lbm}$$

This value of h is between $h_f = 298.61$ Btu/lbm and $h_g = 1187.8$ Btu/lbm at 100 psia (see Table A-19E), so that state 3 must be in the liquid–vapor two-phase region. Accordingly, we must have

$$T_3 = T_{sat} \text{ (at 100 psia)} = 327.86°F$$

and

$$h_3 = h_f + x_3(h_g - h_f) \qquad \text{(at 100 psia)}$$

or

$$335.4 = 298.61 + x_3(1187.8 - 298.61)$$

so that

$$x_3 = 0.0414 = 4.14\%$$

Thus, we have

$$s_3 = s_f + x_3(s_g - s_f) \qquad \text{(at 100 psia)}$$

$$= 0.47439 + 0.0414(1.6034 - 0.47439)$$

$$= 0.52113 \text{ Btu/lbm·°R}$$

The rate of total entropy increase is then

$$\frac{dS_{total}}{dt} = \cancel{\frac{dS_{system}}{dt}}^0 + \frac{dS_{surr}}{dt} = \frac{dS_{surr}}{dt} \qquad \text{(for a steady-flow, steady-state process)}$$

Therefore,

$$\frac{dS_{\text{total}}}{dt} = \dot{m}_3 s_3 - \dot{m}_1 s_1 - \dot{m}_2 s_2$$

$$= 4000(0.52113) - 3000(0.07463) - 1000(1.6517)$$

$$= 208.9 \text{ Btu/h·°R}$$

or

$$\frac{dS_{\text{total}}}{dt} = \dot{m}_1(s_3 - s_1) + \dot{m}_2(s_3 - s_2)$$

$$= 3000(0.52113 - 0.07463) + 1000(0.52113 - 1.6517)$$

$$= 208.9 \text{ Btu/h·°R}$$

Notice that this entropy increase is produced because of the irreversible mixing process.

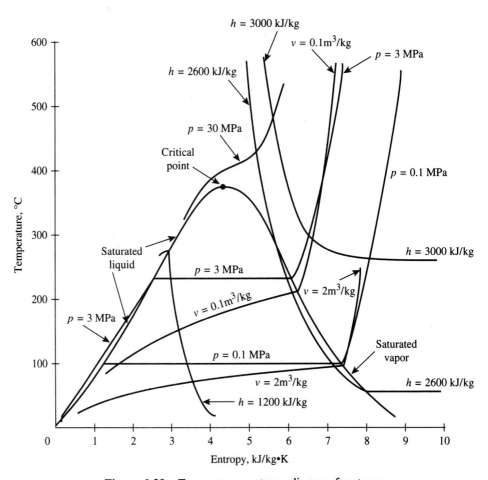

Figure 6-20 Temperature–entropy diagram for steam.

6-12 TEMPERATURE–ENTROPY DIAGRAM FOR PURE SUBSTANCES

From the macroscopic definition of entropy, for the calculation of heat transfer to a unit mass of a system during an internally reversible process, we have the following equation:

$$q_{\text{int rev}} = \int T \, ds \tag{6-25}$$

On a temperature–entropy diagram, this heat transfer is represented by the area under the path of the internally reversible process. This is one of the reasons why a T–s diagram is useful. Of course, presenting thermodynamic data on the T–s plot is highly desirable. In addition, drawings of thermodynamic processes on the T–s diagram are extremely informative in visualizing the processes.

In Appendix 4, we present the T–s diagrams for several substances. To illustrate the general feature of such a plot, a T–s diagram for steam in simplified form is shown in Fig. 6-20. The reader is encouraged to become familiar with the various characteristic lines shown there.

In Fig. 6-20, the lines in the liquid–vapor and superheated-vapor regions of steam are drawn approximately to scale. In the compressed liquid–water region, however, the scale has been magnified to separate the constant-pressure line from the saturated-liquid line. Notice that the constant-pressure line crosses the saturated-liquid line at the point of maximum density, which is about 4°C (39°F) for water at 1 atmospheric pressure. Notice also that in Fig. 6-20, the specific entropy of saturated liquid at the triple-point temperature (0.01°C or 32.02°F) is arbitrarily assigned the value of zero.

Example 6-8

A reversed Carnot cycle heat pump utilizes Refrigerant-12 (Freon-12) as the working fluid. At the beginning of the isentropic compression process, the fluid is a saturated vapor at 0°C. At the end of the isothermal heat-rejection process, the fluid is a saturated liquid at 40°C. Calculate the net work input per cycle, kJ/kg, and the coefficient of performance.

Solution The reversed Carnot cycle is shown in Fig. 6-21 on both T–s and p–v coordinates. The T–s diagram is more useful.

From the Freon-12 table (Table A-22M), at $T_1 = 0°C$, we get $s_1 = s_g = 0.6960$ kJ/kg·K $= s_2$, and at $T_3 = 40°C$, we get $s_3 = s_f = 0.2716$ kJ/kg·K $= s_4$. Thus,

$$\Delta s = s_1 - s_4 = s_2 - s_3 = 0.6960 - 0.2716 = 0.4244 \text{ kJ/kg·K}$$

The heat rejected from the fluid is represented by the area 2–3–a–b on the T–s diagram and has the value of

$$q_H = T_H \, \Delta s = (40 + 273.15)0.4244 = 132.9 \text{ kJ/kg}$$

The heat added to the fluid is represented by the area 4–1–b–a on the T–s diagram and has the value of

$$q_L = T_L \, \Delta s = (0 + 273.15)0.4244 = 115.9 \text{ kJ/kg}$$

The net work input to the fluid is represented by the area 1–2–3–4 on the T–s diagram and has the value of

$$\text{net } w_{\text{in}} = q_H - q_L = 132.9 - 115.9 = 17.0 \text{ kJ/kg}$$

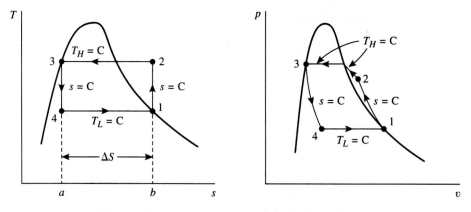

Figure 6-21 T–s and p–v diagrams for Example 6-8.

The coefficient of performance (COP) of the heat pump is calculated by

$$(\text{COP})_{\text{heat pump}} = \frac{q_H}{\text{net } w_{\text{in}}} = \frac{132.9}{17.0} = 7.8$$

or

$$(\text{COP})_{\text{heat pump}} = \frac{T_H}{T_H - T_L} = \frac{40 + 273.15}{(40 + 273.15) - (0 + 27.15)} = 7.8$$

Example 6-9

One pound of steam undergoes a thermodynamic cycle composed of the following internally reversible processes in a closed system:

Process 1–2: The steam, initially at 14.7 psia and having a quality of 40%, is heated at constant volume until its pressure rises to 60 psia.

Process 2–3: The steam is expanded isothermally to 14.7 psia.

Process 3–1: The steam is cooled at constant pressure back to its initial state.

Determine the work and heat interactions for each of the three processes, and the cycle thermal efficiency.

Solution   The T–s and p–v diagrams for the cycle are shown in Fig. 6-22. The given data are $p_1 = 14.7$ psia, $x_1 = 40\%$, $p_2 = 60$ psia, $v_1 = v_2$, $T_2 = T_3$, and $p_1 = p_3$. From the saturated steam table (Table A-19E) at 14.7 psia, we obtain

$$v_1 = v_f + x_1(v_g - v_f)$$

$$= 0.016715 + 0.40(26.80 - 0.016715) = 10.73 \text{ ft}^3/\text{lbm}$$

$$u_1 = u_f + x_1 u_{fg}$$

$$= 180.10 + 0.40(897.5) = 539.1 \text{ Btu/lbm}$$

$$h_1 = h_f + x_1 h_{fg}$$

$$= 180.15 + 0.40(970.4) = 568.3 \text{ Btu/lbm}$$

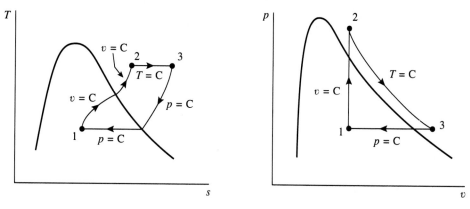

Figure 6-22 T–s and p–v diagrams for Example 6-9.

At $p_2 = 60$ psia and $v_2 = v_1 = 10.73$ ft³/lbm, by interpolation from the superheated steam table (Table A-20E), we obtain

$$T_2 = 630°F = 630 + 460 = 1090°R$$

$$u_2 = 1227.7 \text{ Btu/lbm}$$

$$s_2 = 1.8302 \text{ Btu/lbm·°R}$$

At $T_3 = T_2 = 630°F$ and $p_3 = p_1 = 14.7$ psia, the superheated steam table (Table A-20E) gives

$$v_3 = 44.08 \text{ ft³/lbm}, \qquad u_3 = 1229.8 \text{ Btu/lbm},$$

$$h_3 = 1349.7 \text{ Btu/lbm}, \qquad s_3 = 1.9872 \text{ Btu/lbm·°R}$$

The work and heat interactions are calculated as follows:

$$w_{12} = \int_1^2 p \, dv = 0$$

$$q_{12} = \Delta u_{12} + w_{12} = \Delta u_{12} = u_2 - u_1 = 1227.7 - 539.1 = 688.6 \text{ Btu/lbm}$$

$$q_{23} = T_2(s_3 - s_2) = (1090°R)(1.9872 - 1.8302 \text{ Btu/lbm·°R})$$

$$= 171.1 \text{ Btu/lbm}$$

$$w_{23} = q_{23} - \Delta u_{23} = q_{23} - (u_3 - u_2)$$

$$= 171.1 - (1229.8 - 1227.7) = 169.0 \text{ Btu/lbm}$$

$$w_{31} = \int_3^1 p \, dv = p_1(v_1 - v_3)$$

$$= (14.7 \times 144 \text{ lbf/ft}^2)(10.73 - 44.08 \text{ ft³/lbm})/(778 \text{ ft·lbf/Btu})$$

$$= -90.7 \text{ Btu/lbm}$$

$$q_{31} = \Delta u_{31} + w_{31} = (u_1 - u_3) + w_{31}$$

$$= (539.1 - 1229.8) + (-90.7) = -781.4 \text{ Btu/lbm}$$

or

$$q_{31} = (u_1 - u_3) + (p_1 v_1 - p_3 v_3) = h_1 - h_3$$
$$= 568.3 - 1349.7 = -781.4 \text{ Btu/lbm}$$

$$\text{net } w_{out} = w_{12} + w_{23} + w_{31}$$
$$= 0 + 169.0 - 90.7 = 78.3 \text{ Btu/lbm}$$

$$\text{net } q_{in} = q_{12} + q_{23} + q_{31}$$
$$= 688.6 + 171.1 - 781.4 = 78.3 \text{ Btu/lbm}$$

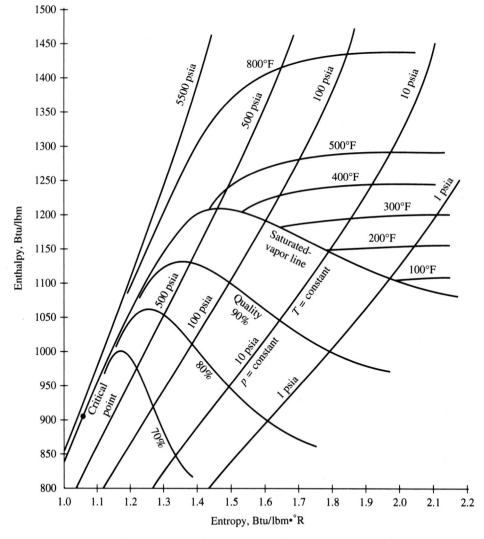

Figure 6-23 Enthalpy–entropy diagram for steam.

The cycle thermal efficiency is

$$\eta_{th} = \frac{\text{net } w_{out}}{q_{in}} = \frac{\text{net } w_{out}}{q_{12} + q_{23}}$$

$$= \frac{78.3}{688.6 + 171.1} = 0.0911 = 9.11\%$$

6-13 ENTHALPY–ENTROPY, OR MOLLIER, DIAGRAM

Thermodynamic property data of pure substances are often shown on an enthalpy–entropy (h–s) diagram, commonly known as *Mollier diagram*. This kind of graph is particularly popular in storing steam property data. The steam tables by Keenan, Keyes, Hill, and Moore, and by ASME* include very large h–s diagrams for steam that are convenient in solving steam problems. Reduced-scale Mollier diagrams for steam in both SI and English units are included in Appendix 4. The essential elements of an h-s diagram for steam is shown approximately to scale in Fig. 6-23. Be sure to notice that in the liquid–vapor two-phase region, a line of constant pressure is also a line of constant temperature. In the single-phase vapor region, however, constant-pressure and constant-temperature lines are different. This diagram is not useful in the single-phase liquid region.

Example 6-10

One kilogram of steam at 500 kPa and 300°C (state 1) enters an adiabatic turbine and expands to 100 kPa (state 2). It is then reheated at constant pressure of 100 kPa to 240°C (state 3). It is finally expanded in another adiabatic turbine to a pressure of 2 kPa (state 4). All processes are internally reversible, and kinetic- and potential-energy changes are negligible. Using data taken from the SI-units Mollier diagram, (Fig. A-11M), determine the work done by the turbines and the heat added in the reheater. Draw the processes on a Mollier diagram.

Solution   Schematic and Mollier diagrams for the processes are shown in Fig. 6-24. The given conditions are $p_1 = 500$ kPa, $T_1 = 300°C$, $p_2 = p_3 = 100$ kPa, $T_3 = 240°C$, and $P_4 = 2$ kPa.

The enthalpy and entropy data taken from Fig. A-11M (the original enlarged version) for the state points of this problem are indicated in Fig. 6-24. h_1 and s_1 are taken at the intersection of the lines for $p_1 = 500$ kPa and $T_1 = 300°C$. Because the turbine process 1–2 is adiabatic and internally reversible, it is an isentropic process with $s_1 = s_2$. State 2 is located at the intersection of the line for $p_2 = 100$ kPa and the vertical line from state 1, thus giving the values of h_2 and T_2. h_3 and s_3 are taken at the intersection of the lines for $p_3 = 100$ kPa and $T_3 = 240°C$. Because the turbine process 3–4 is adiabatic and internally reversible, it is an isentropic process with $s_3 = s_4$. State 4 is located at the intersection of the line for $p_4 = 2$ kPa and the vertical line from state 3, thus giving the value of h_4. It is noted that state 2 is in the superheated region and state 4 is a mixture of liquid and vapor. The value of x_4 is indicated in Fig. 6-24, but this quality value is not needed in the calculations of this problem.

* (1) J. H. Keenan, F. G. Keyes, P. G. Hill, and J. G. Moore, "Steam Tables", John Wiley, New York, 1969 and 1978. (2) ASME "Steam Tables", American Society of Mechanical Engineers, New York, 1967.

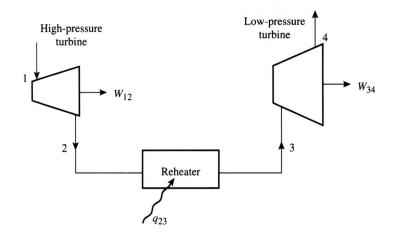

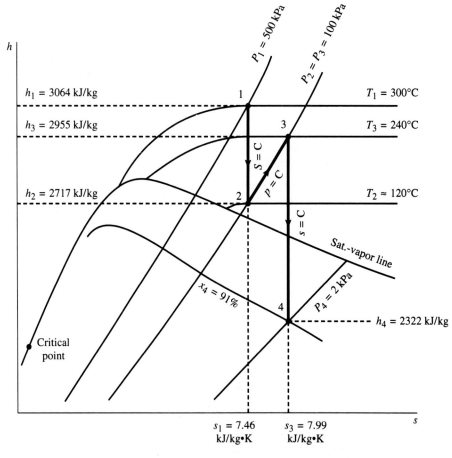

Figure 6-24 Schematic and Mollier diagrams for Example 6-10.

For the high-pressure turbine, because it is adiabatic ($q = 0$) with negligible changes in KE and PE, the steady-flow energy equation on a unit-mass basis gives

$$\cancel{q_{12}}^{\,0} - w_{12} + (h_1 + \cancel{\tfrac{1}{2}V_1^2}^{\,0} + \cancel{gz_1}^{\,0}) - (h_2 + \cancel{\tfrac{1}{2}V_2^2}^{\,0} + \cancel{gz_2}^{\,0}) = 0$$

or

$$w_{12} = h_1 - h_2 = 3064 - 2717 = 347 \text{ kJ/kg}$$

Similarly, for the low-pressure turbine, because it is adiabatic ($q = 0$) with negligible changes in KE and PE, the steady-flow energy equation on a unit-mass basis gives

$$\cancel{q_{34}}^{\,0} - w_{34} + (h_3 + \cancel{\tfrac{1}{2}V_3^2}^{\,0} - \cancel{gz_3}^{\,0}) - (h_4 + \cancel{\tfrac{1}{2}V_4^2}^{\,0} + \cancel{gz_4}^{\,0}) = 0$$

or

$$w_{34} = h_3 - h_4 = 2955 - 2322 = 633 \text{ kJ/kg}$$

For the reheater, as there is no work involved and no changes in KE and PE, the steady-flow energy equation on a unit-mass basis gives

$$q_{23} - \cancel{w_{23}}^{\,0} + (h_2 + \cancel{\tfrac{1}{2}V_2^2}^{\,0} + \cancel{gz_2}^{\,0}) - (h_3 + \cancel{\tfrac{1}{2}V_3^2}^{\,0} + \cancel{gz_3}^{\,0}) = 0$$

or

$$q_{23} = h_3 - h_2 = 2955 - 2717 = 238 \text{ kJ/kg}$$

6-14 SUMMARY

The Clausius statement of the second law: It is impossible for any device to operate in a cycle in such a manner that the sole effect is the transfer of heat from one body to another body at a higher temperature.

The Kelvin–Planck statement of the second law: No process is possible whose sole result is the absorption of heat from a reservoir and the conversion of this heat into work.

If at the end of a thermodynamic process, the initial states of the system and the surroundings can be restored without any net change anywhere, the system is said to have undergone an internally and externally, or totally reversible, process. During an internally reversible process, a system proceeds through a series of equilibrium states, and, if the process direction is reversed at any stage, the system can be returned to its initial equilibrium state without leaving any permanent changes in the system, irrespective of any changes that may occur in the surroundings.

The Carnot cycle is a totally reversible cycle that is composed of four totally reversible processes, two isothermal and two adiabatic.

The Carnot principle states that no engine can be more efficient than a totally reversible engine operating between the same temperature limits, and all totally reversible engines operating between the same temperature limits have the same efficiency.

The Carnot principle forms the basis for establishing a thermodynamic temperature scale by relating the temperatures of the energy reservoirs and the quantities of heat transfer for a totally reversible engine:

$$\frac{T_H}{T_L} = \frac{Q_H}{Q_L}$$

The thermal efficiency of a Carnot heat engine can be expressed as

$$\eta_{th,\,Carnot} = 1 - \frac{Q_L}{Q_H} = 1 - \frac{T_L}{T_H}$$

The performance of a refrigerator or heat pump is expressed in terms of the coefficient of performance (COP) as defined by

$$(COP)_{refrigerator} = \frac{\text{refrigerating effect}}{\text{net work input}} = \frac{Q_L}{Q_H - Q_L}$$

$$(COP)_{heat\,pump} = \frac{\text{heating effect}}{\text{net work input}} = \frac{Q_H}{Q_H - Q_L}$$

For a Carnot refrigerator and heat pump, we have

$$(COP)_{Carnot\,refrigerator} = \frac{Q_L}{Q_H - Q_L} = \frac{T_L}{T_H - T_L}$$

$$(COP)_{Carnot\,heat\,pump} = \frac{Q_H}{Q_H - Q_L} = \frac{T_H}{T_H - T_L}$$

The Clausius inequality is expressed mathematically as

$$\oint \frac{đQ}{T} \leq 0$$

where the equality holds for internally reversible cycles and the inequality holds for internally irreversible cycles.

Based on the equality part of the Clausius inequality expression, a new thermodynamic property called entropy is defined as

$$dS = \left(\frac{đQ}{T}\right)_{internally\,reversible}$$

This is the macroscopic (or classical) definition of entropy. Microscopically (or statistically) entropy is defined by

$$S = k \ln \Omega$$

where k is the Boltzmann constant, and Ω is the thermodynamic probability. Entropy can be interpreted as a quantitative measure of microscopic disorder for a system.

The entropy change during a process is obtained by integrating $(đQ/T)$ along an internally reversible path between the two end equilibrium states of the process:

$$\Delta S = S_2 - S_1 = \int_1^2 \left(\frac{đQ}{T}\right)_{int\,rev}$$

It does not matter what particular internally reversible path is followed in the integration as long as it is internally reversible. After the value of ΔS has been evaluated along some internally reversible path, this value will be the same even if the end states are connected by an irreversible process.

For an internally reversible adiabatic process, $đQ_{\text{int rev}} = 0$, it follows that $dS = 0$. This means that during an internally reversible adiabatic process, there is no change in entropy. Such a process is called an isentropic process.

The heat transfer of any internally reversible process is given by

$$Q_{\text{int rev}} = \int T\, dS$$

Thus, the heat transfer of any internally reversible process is represented by the area under the process curve on a T–S diagram.

The principle of increase of entropy states that

$$dS \geq \frac{đQ}{T}$$

For an isolated system, $đQ = 0$ and

$$dS_{\text{isolated system}} \geq 0$$

or

$$dS_{\text{system}} + dS_{\text{surr}} \geq 0$$

where the equality holds for an internally reversible process and the inequality for an internally irreversible process.

The entropy balance for a fixed mass of substance can be written as

$$dS = \frac{đQ}{T} + đS_{\text{prod}}$$

or

$$\Delta S = \int \frac{đQ}{T} + S_{\text{prod}}$$

where S_{prod} represents the entropy production
$S_{\text{prod}} = 0$ for internally reversible processes
$S_{\text{prod}} > 0$ for internally irreversible processes

PROBLEMS

6-1. An ideal gas with $c_p = 0.24$ Btu/lbm·°R and $c_v = 0.171$ Btu/lbm·°R initially at 30 psia, 340°F, undergoes an irreversible process in a closed system to a final state at 15 psia, 240°F. Determine the entropy change by following three different internally reversible paths.

6-2. An ideal gas with $c_p = 1.063$ kJ/kg·K and $c_v = 0.766$ kJ/kg·K expands irreversibly in a closed system from 2 atm, 300°C, to 1 atm, 250°C. Calculate the entropy change by following three different internally reversible paths.

6-3. A mass of hydrogen in a closed system originally having a volume of 1 ft³ and a pressure of 180 psia expands in an internally reversible process at constant temperature of 400°F. The final pressure is 15 psia. Determine (a) the work done, (b) the heat transfer, and (c) the change of entropy.

6-4. (a) For a certain internally reversible process of a system, the rate of heat transfer per unit temperature rise is 0.5 Btu/°R. Determine the increase in entropy of the system if its temperature rises from 500°R to 600°R. (b) In a second process between the same end states, the temperature rise is accomplished by stirring accompanied by a heat addition half as great as in (a). What is the increase in entropy in this case?

6-5. A system undergoes a certain internally reversible cycle in the following manner: At first, 100 Btu of heat is received at a temperature of 500 K; then an adiabatic expansion occurs to 400 K, at which temperature 50 Btu of heat is received; then a further adiabatic expansion to 300 K, after which 100 Btu of heat is rejected at constant temperature; and, finally, returns to the initial state by some unknown processes. (a) Determine the change in entropy that occurs as the system is restored to its initial state through the final unknown processes. (b) If during the final unknown processes, heat is transferred only at 400 K, how much heat is transferred, and in which direction?

6-6. An empirical equation for the specific heat of oxygen at constant low pressure can be expressed as

$$c_p = 4.186(6.0954 + 3.2533 \times 10^{-3}T - 1.0171 \times 10^{-6}T^2)$$

where c_p is in kJ/kgmole·K, and T in K. Calculate the heat added and the increase in entropy for 1 kilogram-mole of the gas during an internally reversible heating process from 300 to 400 K at a constant low pressure.

6-7. An empirical equation for the specific heat of air at constant low pressure is

$$c_p = 0.2398 - 4.42 \times 10^{-6}T + 9.24 \times 10^{-9}T^2$$

where c_p is in Btu/lbm·°R, and T in °R. Determine the heat added to 2 lbm of air during an internally reversible, constant low-pressure process from an initial temperature of 500°R to a final temperature of 600°R. What is the increase in entropy of the air during the heating process? Show the area representing the heat addition on an appropriate diagram.

6-8. Ten pounds of water at a temperature of 80°F is converted under a constant pressure of 14.7 psia to superheated steam at 480°F. Calculate the change in entropy. Given data:

$$c_{p(\text{liquid})} = 1.02 \text{ Btu/lbm·°R}$$
$$c_{p(\text{sup. vapor})} = 1.1 - 33.2T^{-1/2} + 417T^{-1},$$
where c_p is in Btu/lbm·°R, and T in °R.
$$h_{fg} \text{ (at 212°F)} = 970.3 \text{ Btu/lbm}$$

Use of the steam table is not permitted in solving this problem.

6-9. A heat engine completes its cycle in three steps: process 1–2, adiabatic compression from $T_1 = 100°F$ to $T_2 = 300°F$; process 2–3, isothermal expansion at 300°F; process 3–1, $dT/ds = $ constant. Assuming each step in the cycle to be internally reversible, calculate the thermal efficiency of the engine.

6-10. A Carnot engine R and an irreversible engine I operate between the same energy-source reservoir at T_H and the same energy-sink reservoir at T_L. See Fig. P6-10. Both engines produce the same quantity of work. Answer the following questions for engine R as compared to engine I:
(a) Is the quantity of heat supplied greater, the same, or less?
(b) Is the quantity of heat rejected greater, the same, or less?
(c) Is the thermal efficiency greater, the same, or less?
(d) Is the cyclic integral of $T \, dS$ greater, the same, or less?
(e) Is the cyclic integral of dS greater, the same, or less?

6-11. A Carnot engine operates between 1227°C and 227°C. For a heat input of 150,000 kJ/h, determine: (a) the amount of heat rejected in kW, (b) the cycle thermal efficiency, (c) the power output in kW, (d) the rate of entropy change for the heat-addition process, and (e) the rate of entropy change for the heat-rejection process.

6-12. A Carnot heat engine has an efficiency of 40% and rejects heat to a sink at 25°C. Find (a) the net power output in kW, and (b) the temperature of the source in °C, if the heat supplied is 4000 kJ/h.

6-13. A Carnot engine producing 10 Btu of work for one cycle has a thermal efficiency of 50%. The working fluid is 1 lbm of air, and the pressure and volume at the beginning of the isothermal expansion process are 100 psia and 4 ft³, respectively. Find (a) the heat transfer and work for each of the four individual processes, (b) the temperature at the end state of each process, and (c) the volume at the end of the isothermal expansion process.

6-14. An array of solar collectors is used to collect solar energy for the operation of a heat engine. See Fig. P6-14. Experiments indicate that 190 Btu/h·ft² of energy can be collected when the collector plate is operating at 180°F. Using the atmosphere at 70°F as the low-temperature energy reservoir, estimate the minimum collector area required to produce 1 kW of engine power.

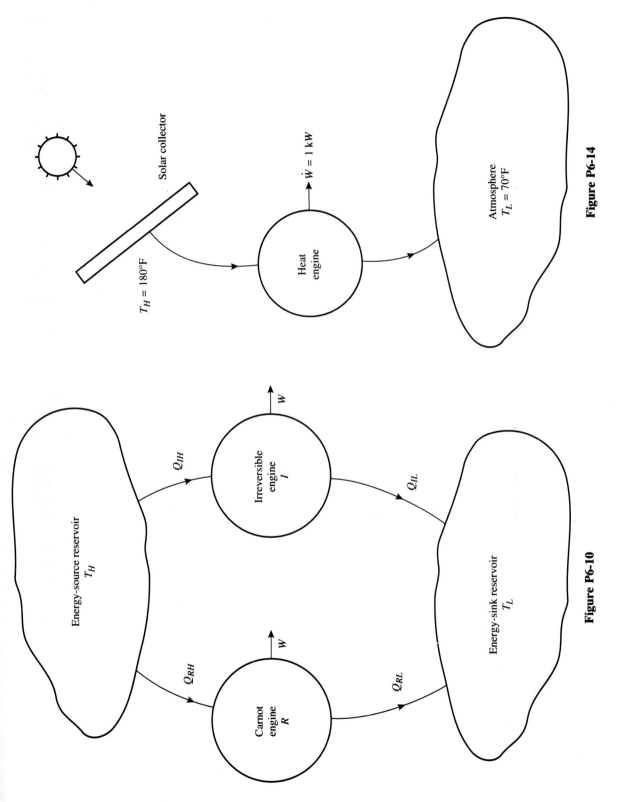

Solar collector

$T_H = 180°F$

Heat engine

$\dot{W} = 1\,kW$

Atmosphere
$T_L = 70°F$

Figure P6-14

Energy-source reservoir
T_H

Q_{IH}

Irreversible engine
I

W

Q_{IL}

Q_{RH}

Carnot engine
R

W

Q_{RL}

Energy-sink reservoir
T_L

Figure P6-10

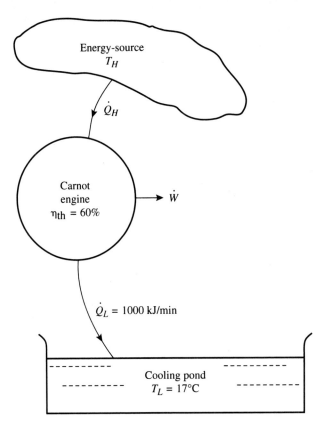

Energy-source
T_H

$\dot{Q}_H$

Carnot
engine
$\eta_{th} = 60\%$

$\dot{W}$

$\dot{Q}_L = 1000$ kJ/min

Cooling pond
$T_L = 17°C$

Figure P6-15

6-15. The efficiency of a Carnot heat engine is 60%. A cooling pound at 17°C receives 1000 kJ/min of heat from the engine. See Fig. P6-15. Determine the power output of the engine in kW, and the temperature of the high-temperature reservoir in °C.

6-16. If 50,000 Btu/h of heat were supplied to a totally reversible heat engine from a source at 500°F, and the engine delivered 10 hp, what would the temperature of the receiver to which it rejected heat have to be?

6-17. A Carnot refrigerator uses Refrigerant-12 as the working fluid. Heat is rejected at 40°C, and during this process, the refrigerant changes from saturated vapor to saturated liquid. Heat transfer to the refrigerant is at −20°C. (a) Show this cycle on a T–s diagram. (b) Determine the quality at the beginning of the heat-addition process. (c) Determine the coefficient of performance for the cycle.

6-18. A Carnot refrigerator is suggested to be used to produce ice at 32°F. The heat-rejection tempera-

ture is 78°F, and the enthalpy change of water during freezing is 144 Btu/lbm. How many pounds of ice can be formed per hour per horsepower of power input?

6-19. A reversed Carnot cycle uses 1 lbm of steam as the working fluid. At the beginning of the isothermal heat-addition process, the steam is at 0.5 psia, 35% quality; at the end of the isothermal heat-addition process, the steam is dry and saturated. Heat is rejected at 250°F. Determine (a) the change in internal energy of the steam during the isothermal heat-addition process, and (b) the work input per cycle.

6-20. A Carnot engine operating between the temperature limits of 230°C and 60°C uses steam as the working substance. At the beginning of the isothermal heat-addition process, the steam has a specific volume of 0.002 m³/kg. After isothermal expansion, the steam is a saturated vapor. Determine (a) p, v, T values for the four state points of

The Second Law of Thermodynamics   Chap. 6

the cycle, (b) the heat addition and heat rejection, kJ/kg, and (c) the thermal efficiency of the cycle.

6-21. A reversed Carnot cycle is used to pump heat from a cold room at 0°F and discharges heat at 95°F. What horsepower is required to remove 2000 Btu/min from the cold room? If the machine operates at 200 cycles per minute, what is the entropy change of the medium during the absorption?

6-22. A Carnot heat pump is used to supply heat to a building at 140°F by removing heat from well water at 40°F. See Fig. P6-22. The building requires 600,000 Btu/h for heating. Determine the kilowatt input required by the machine and the heat removed per hour from the well water. If electric resistance heaters are used, what kW input is needed?

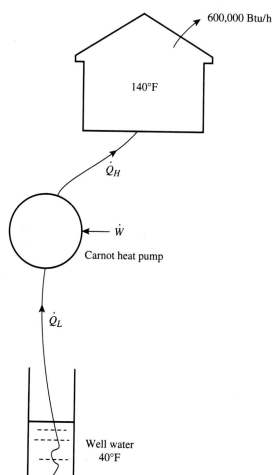

Figure P6-22

6-23. A solar pond at 80°C supplies heat to a thermoelectric generator (see Figs. 1-9 and P6-23) at a rate of 10^6 kJ/h. The waste heat is rejected to the environment at 30°C. Determine the maximum power this thermoelectric generator can produce.

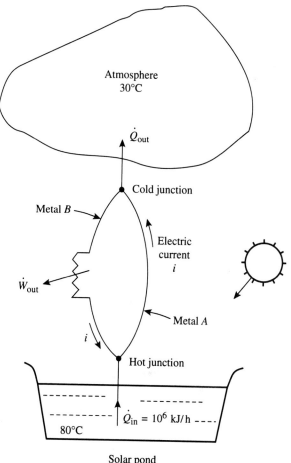

Figure P6-23

6-24. A refrigerator unit is used to maintain a cooled space at −40°C in a 20°C environment. The cooling capacity of the unit is 3 kW·h with a coefficient of performance of 3.5. Determine (a) the energy required to operate this refrigeration unit and (b) the minimum energy requirement of any refrigeration unit under the given temperature conditions.

6-25. The ocean surface water at 30°C and the deep seawater at 3°C are used as the two energy reservoirs to operate a power cycle. Calculate the maximum thermal efficiency of the cycle.

6-26. A refrigeration machine maintains the cooled space at 2°C when the ambient temperature is 25°C. Calculate the maximum coefficient of performance of the machine.

6-27. A thermoelectric refrigerator (Fig. 1-9 working as a power-consuming device) removes 200 W of heat from a space at −5°C and rejects heat to the environment at 20°C. See Fig. P6-27. Determine the minimum power input required.

6-28. An engineer claims that he has developed an automobile engine that produces 60 hp for a steady fuel flow of 10 lbm/h, with the working fluid receiving heat from combustion gases at 2500°F and rejecting heat to ambient air at 40°F. How do you evaluate this claim? Assume that for each pound of fuel

burned, 18,000 Btu of heat are transferred into the working fluid.

6-29. An inventor claims to have developed an engine that takes in 100,000 Btu at a temperature of 700°R, rejects 40,000 Btu at a temperature of 385°R, and delivers 15 kW·h of mechanical work. How do you evaluate his claim?

6-30. Saturated water vapor at 200°C is contained in a cylinder fitted with a piston. The initial volume of the steam is 0.01 m³. The steam expands in an internally reversible isothermal process until the final pressure is 0.2 MPa. Determine the work done and heat transfer during the process.

6-31. One-tenth kilogram of steam at 0.3 MPa and 200°C is compressed isothermally to 1.5 MPa in a piston–cyclinder device. See Fig. P6-31. During this process, 20 kJ of work is done on the steam,

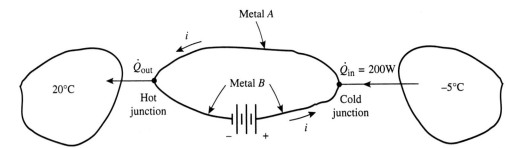

Figure P6-27

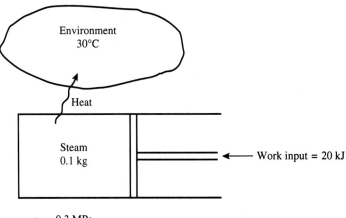

$p_i = 0.3$ MPa

$p_f = 1.5$ MPa

$T_i = T_f = 200°C$

Figure P6-31

and heat is lost to the environment at 30°C. Determine (a) the heat transfer, in kJ, (b) the entropy change of the steam, in kJ/K, and (c) the total change of entropy for the overall process, in kJ/K. Is the process reversible, irreversible, or impossible?

6-32. A piston–cylinder device initially contains Refrigerant-12 at 6.0 bars and 80°C. See Fig. P6-32. It is compressed quasistatically at constant pressure, with a boundary-work input of 13.63 kJ/kg. Determine (a) the final specific volume, in cm³/g, (b) the final specific entropy, in kJ/kg·K, and (c) the heat transfer, in kJ/kg. If the surrounding temperature is 20°C, determine the total entropy change for the overall process, in kJ/kg·K. Draw the process on a T–s diagram. Is the process reversible, irreversible, or impossible?

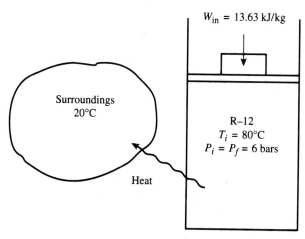

Figure P6-32

6-33. One pound of saturated steam at a pressure of 100 psia is expanded isentropically to a temperature of 200°F. What are the final quality and enthalpy?

6-34. Steam at a pressure of 1 MPa and a temperature of 250°C flows into an expander at a rate of 1350 kg/h. It leaves the expander at 25 kPa. The process is internally reversible and adiabatic. Find the power delivered.

6-35. In the evaporator of a refrigeration system, a liquid–vapor mixture of Refrigerant-12 at 200 kPa absorbs 70 kJ of heat and changes to the saturated vapor state. The source of heat is a cooled space at −5°C. Determine (a) the change of entropy of the refrigerant, (b) the change of entropy of the cooled space, and (c) the total change of entropy for this process.

6-36. An insulated piston–cylinder device contains 0.01 m³ of saturated Refrigerant-12 vapor at a pressure of 0.8 MPa. The refrigerant undergoes a quasiequilibrium process to 0.4 MPa. Determine the final temperature of the refrigerant and the work done by the refrigerant.

6-37. A rigid tank contains 2 kilograms of Refrigerant-12 at 0.15 MPa and 20°C. See Fig. P6-37. The refrigerant is cooled and stirred until its pressure drops to 0.10a MPa. Determine the entropy change of the refrigerant.

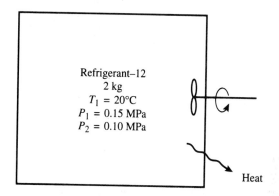

Figure P6-37

6-38. Steam in a cylinder–piston device initially at 3.0 MPa and 300°C expands isentropically to 1.0 MPa. Determine the work done if the cylinder volume at the start of expansion is 3.2 liters.

6-39. An inventor claims that he has developed a steam turbine that delivers 3800 hp, with the steam entering the turbine at 100 psia, 500°F, and leaving the turbine at 2 psia. The required rate of steam flow is 30,000 lbm/h. (a) How do you evaluate his claim? (b) Suppose he changed his claim and said the required steam flow was 34,000 lbm/h.

6-40. A salesman claims that he has a steam turbine available that delivers 3000 kW, with steam entering at 1 MPa and 250°C and leaving at 25 kPa. The required rate of steam flow is said to be 13,500 kg/h. How do you evaluate his claim?

6-41. Steam enters a turbine at 6 MPa and 600°C with a velocity of 80 m/s. It leaves the turbine at 10 kPa with a velocity of 140 m/s. See Fig. P6-41. The entropy of the steam increases by 0.80 kJ/kg·K as it passes through the turbine. The mass-flow rate of steam is 1000 kg/min and the power output of the turbine is 18,000 kW. Determine the quantity of heat transfer, in kJ/kg, and the direction of heat transfer.

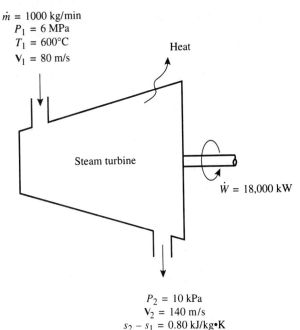

$\dot{m} = 1000$ kg/min
$P_1 = 6$ MPa
$T_1 = 600°C$
$V_1 = 80$ m/s

Heat

Steam turbine

$\dot{W} = 18,000$ kW

$P_2 = 10$ kPa
$V_2 = 140$ m/s
$s_2 - s_1 = 0.80$ kJ/kg·K

Figure P6-41

6-42. Steam at a pressure of 10 MPa and a temperature of 400°C enters a turbine. At the exit of the turbine, the pressure is 0.1 MPa and the entropy is 0.6 kJ/kg·K greater than at the inlet. The process is adiabatic and changes in KE and PE may be neglected. Find the work done by the steam in kJ/kg. What is the mass-flow rate of steam in kg/s that is required to produce a power output of 1000 kW?

6-43. Steam at a pressure of 5 MPa and a temperature of 500°C expands adiabatically in a turbine to an exit pressure of 10 kPa. The entropy at the exit is 0.5 kJ/kg·K greater than at the inlet. The changes in KE and PE can be neglected. (a) Find the work done by the steam in kJ/kg. (b) What is the mass-flow rate of steam in kg/min that is required to produce a power output of 100 kW?

6-44. Steam enters an adiabatic turbine at 800 psia and 1000°F. At the exit of the turbine, the pressure is 2 psia and the entropy is 0.2 Btu/lbm·°R greater than that at the inlet. Neglect changes in KE and PE. (a) Determine the work done by the steam in Btu/lbm. (b) What is the mass-flow rate of steam in lbm/min that is required to produce a power output of 20,000 hp?

6-45. Steam is throttled from 0.7 MPa, 75% quality, to 0.1 MPa, with negligible change in kinetic energy. Calculate the increase in entropy.

6-46. In a refrigeration plant, ammonia enters the expansion valve at 200 psia, 80°F, and leaves at 40 psia. The changes in kinetic energy are negligible. What is the increase in entropy per lbm of ammonia? Show this process on a *T–s* diagram.

6-47. Refrigerant-12 flows through a capillary tube in a refrigerator, entering at 150 psia, 80°F, and leaving at 30 psia. The process is adiabatic. What is the increase in entropy per pound of the fluid flowing through the capillary tube? Show the inlet and final states on a *T–s* diagram.

6-48. Steam at 60 psia and 320°F expands adiabatically to 14.7 psia and 94% quality. Draw the process on a *T–s* diagram and calculate the entropy change during the process. Is the process reversible, irreversible, or impossible?

6-49. A refrigeration compressor requires an input of 4.5 hp when compressing 90 lbm/h of ammonia in steady flow from saturated vapor at 35 psia to superheated vapor at 170 psia. See Fig. P6-49. If the heat transferred from the compressor to cooling water and the surroundings is 5000 Btu/h, find the final temperature of the ammonia.

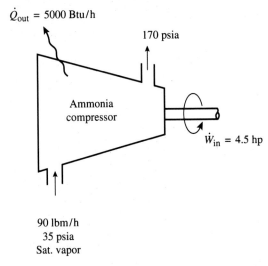

$\dot{Q}_{out} = 5000$ Btu/h

170 psia

Ammonia compressor

$\dot{W}_{in} = 4.5$ hp

90 lbm/h
35 psia
Sat. vapor

Figure P6-49

6-50. A tank containing 10 pounds of water at 32°F is brought into contact with an energy reservoir at 212°F. (a) When the water reaches 212°F, what is the change in entropy of the water, of the energy reservoir, and of the universe? (b) If the water is heated from 32°F to 212°F by first bringing it in contact with a reservoir at 122°F and then with a reservoir at 212°F, what is the change in entropy

The Second Law of Thermodynamics   Chap. 6

of the universe? (c) Explain how the water might be heated from 32°F to 212°F with no change in the entropy of the universe.

6-51. A 10-ft³ vessel contains steam at 200 psia and 500°F. A valve at the top is opened and steam is expelled until the pressure drops to 60 psia and the temperature to 350°F. Determine the total change in entropy within the vessel for the process.

6-52. A long insulated vertical tube of 0.125-m diameter contains 0.015 m³ of steam at 5 MPa and 400°C at the bottom of the tube held by a metal plug of 40-kg mass. See Fig. P6-52. The metal plug is held in position initially by a pin. When the pin is released, the steam expands slowly and forces the metal plug to accelerate upward. When the metal plug leaves the top of the tube, the steam in the tube has a pressure of 600 kPa. Determine the exit velocity of the metal plug.

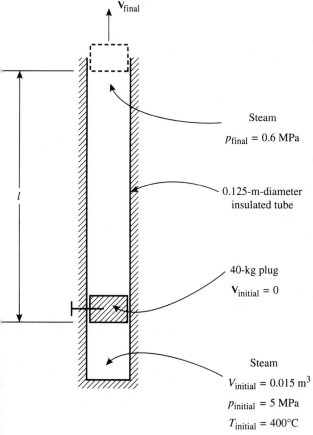

V_{final}

Steam

$p_{final} = 0.6$ MPa

0.125-m-diameter insulated tube

l

40-kg plug

$V_{initial} = 0$

Steam

$V_{initial} = 0.015$ m³

$p_{initial} = 5$ MPa

$T_{initial} = 400°C$

Figure P6-52

6-53. A transmission gear box operating at steady conditions without any temperature change receives 100 kW of power from an engine and delivers 97 kW to the output shaft. See Fig. P6-53. The surface temperature of the gear box is 50°C and the surrounding temperature is 20°C. Determine the rate of entropy production for the gear box.

6-54. A 100-W incandescent light bulb and a 20-W fluorescent tube will produce the same amount of light. When it operates at steady conditions, the surface temperature of the 100-W incandescent light bulb is 60°C, whereas the surface temperature of the fluorescent tube is 30°C. Determine the rate of entropy production of each light source and comment on which is more efficient.

6-55. An insulated piston–cylinder device contains water initially as a saturated liquid at 100°C. See Fig. P6-55. A paddle wheel in the cylinder is turned on until the water becomes a saturated vapor at 100°C. During the process, the piston moves freely in the cylinder. Determine the net work done on the water in kJ/kg and the entropy production in kJ/kg·K.

6-56. In a closed system, 1 kilogram of saturated steam at 700 kPa is heated in a constant-volume process until its pressure is 1.2 MPa. It is then expanded isothermally to 700 kPa and finally cooled to the saturation temperature at constant pressure. All processes are internally reversible. Calculate work and heat for each of the three processes and the thermal efficiency.

6-57. In a closed system, 1 pound of saturated steam at 100 psia is heated in a constant-volume process until its pressure is 175 psia. It is then expanded isothermally to 100 psia, and finally cooled to the saturation temperature at constant pressure. All processes are internally reversible. Calculate the work and heat for each of the three processes and the thermal efficiency. Draw the p–v and T–s diagrams for the cycle.

6-58. One pound of steam undergoes a thermodynamic cycle composed of the following internally reversible processes in a closed system:

Process 1–2: The steam initially at 40 psia and having a quality of 50% is heated at constant volume until its pressure rises to 210 psia.

Process 2–3: It is then expanded isothermally to 40 psia.

Process 3–1: It is finally cooled at constant pressure back to its initial state.

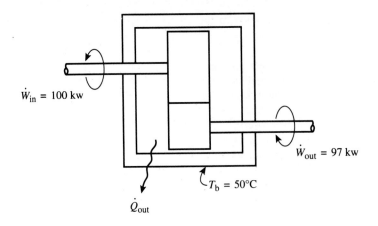

$\dot{W}_{in} = 100$ kw

$\dot{W}_{out} = 97$ kw

$T_b = 50°C$

$\dot{Q}_{out}$

$T_0 = 20°C$

Figure P6-53

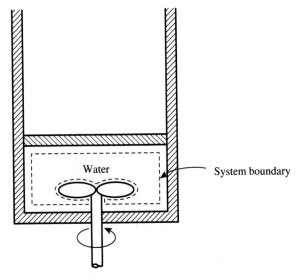

Water

System boundary

Figure P6-55

Draw the $p-v$ and $T-s$ diagrams for the cycle, and determine the work done and the heat transferred for each of the three processes. What is the thermal efficiency of the cycle?

6-59. One kilogram of steam in a closed system undergoes a thermodynamic cycle composed of the following internally reversible processes:

Process 1–2: The steam initially at 1.0 MPa and having a quality of 40% is heated at constant volume until its pressure rises to 3.5 MPa.

Process 2–3: It is then expanded isothermally to 1.0 MPa.

Process 3–1: It is finally cooled at constant pressure back to its initial state.

Draw the $T-s$ and $p-v$ diagrams for the cycle and calculate the work done, the heat transferred, and the change of entropy for each of the three processes. What is the thermal efficiency of the cycle?

6-60. One pound of Refrigerant-12 in a piston–cylinder machine undergoes a thermodynamic cycle composed of the following internally reversible processes:

Process 1–2: Saturated Refrigerant-12 vapor at 40°F is expanded isothermally to 5 psia.

Process 2–3: It is then compressed adiabatically back to its initial volume.

Process 3–1: It is finally cooled at constant volume back to the initial saturated vapor state.

Draw the $p-v$, $T-s$, and $p-h$ diagrams for the cycle, and calculate the work done, the heat transferred, and the entropy change for each of the three processes. When this machine is used as a heat pump, determine the coefficient of performance (COP) as defined by

$$COP = \text{heat rejected/net work input}$$

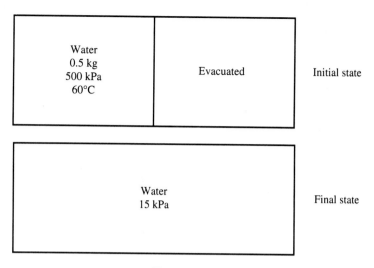

Figure P6-61

6-61. A rigid tank is divided in two equal halves by a partition. See Fig. P6-61. One side of the tank contains 0.5 kg of water at 500 kPa and 60°C, while the other side is evacuated. The partition is removed, and the water expands to fill the entire tank. Determine the entropy change of the water if the final pressure in the tank is 15 kPa.

6-62. A 2-liter insulated rigid can contains Refrigerant-12 initially at 700 kPa and 20°C. See Fig. P6-62. The refrigerant leaks out slowly through a crack.

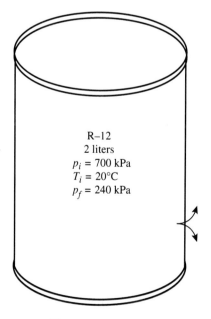

R–12
2 liters
$p_i = 700$ kPa
$T_i = 20°C$
$p_f = 240$ kPa

Figure P6-62

Assuming that the refrigerant remaining in the can undergoes an internally reversible, adiabatic process, determine the final mass in the can when the pressure drops to 240 kPa.

6-63. Write a computer program to solve Problem 6-8.

6-64. Write a computer program to solve the following problem. An empirical equation for the specific heat of oxygen at constant low pressure can be expressed as

$$c_p = 4.186(6.0954 + 3.2533 \times 10^{-3}T - 1.0171 \times 10^{-6}T^2)$$

where c_p is in kJ/kgmole·K and T in K. Calculate the heat addition and the increase in entropy per kilogram-mole of the gas during an internally reversible heating process at a constant low pressure (a) from 300 to 400 K, (b) from 300 to 500 K, (c) from 300 to 600 K, (d) from 400 to 600 K, (e) from 400 to 800 K, and (f) from 400 to 1000 K.

7

Entropy Change During Processes

In the last chapter, we introduced the property entropy and made some use of this property. In this chapter, we expand the topic and develop procedures pertaining to the determination of entropy changes during processes for various thermodynamic systems.

7-1 TWO BASIC RELATIONS OF PROPERTIES

In the previous chapters, we studied the first and second laws separately. We now combine them to produce two very important relations for simple compressible systems of fixed mass. The first law for an infinitesimal process is

$$đQ = dU + đW$$

If the process is internally reversible, then according to the second law,

$$đQ = T \, dS$$

Furthermore, for an internally reversible process of a simple compressible closed system, we have

$$đW = p \, dV$$

Therefore,

$$T \, dS = dU + p \, dV \tag{7-1}$$

This is the very basic relation that combines the first and second laws as applied to a simple compressible system of fixed mass.

The combined first and second laws as expressed by Eq. (7-1) can be used in connection with the definition of enthalpy to form another basic relation. Thus,

$$H = U + pV$$

It follows that

$$dH = dU + p \, dV + V \, dp$$

Substituting Eq. (7-1) into this relation gives

$$T \, dS = dH - V \, dp \tag{7-2}$$

Although we used the condition of an internally reversible process of a closed system to derive Eq. (7-1), we see that once it is written, it is a relation solely among the properties of a fixed quantity of material. If this material passes from one equilibrium state to another, the property changes are the same whether the process taking place is reversible or irreversible. Therefore, we conclude that Eqs. (7-1) and (7-2) are valid for a fixed mass either contained in a closed system or passing through an open system irrespective of the kind of process. If the process is not internally reversible, however, the term $T \, dS$ in the equations is not the heat added $đQ$ and the term $p \, dV$ is not the work done $đW$ even for a closed system. Nevertheless, it is imperative to notice that one must follow an internally reversible path between the initial and final equilibrium states to perform the integration of Eq. (7-1) or (7-2).

On a unit mass or unit mole basis, the preceding two basic relations are written as

$$T \, ds = du + p \, dv \tag{7-1a}$$

$$T \, ds = dh - v \, dp \tag{7-2a}$$

7-2 SECOND LAW FOR AN OPEN SYSTEM

For an open system, in addition to heat and work interactions, there are mass transfers across its boundary. To establish the expression of dS for such a system, let us consider the open system or control volume as shown in Fig. 7-1, where for convenience, only one inlet and one exit are shown. All work interactions of the open system are assumed to be reversible and thus have no bearing on our present analysis. Because, in general, the temperature of an open system is not uniform, the temperature T shown in the figure is the temperature of the particular part of the system to which the heat $đQ$ is transferred. The entropy change associated with the heat transfer is given in the form $đQ/T$. If the inlet and exit masses at sections 1 and 2 have specific entropies s_1 and s_2, respectively, the entropies transported in and out the open system are given by $s_1 \, đm_1$ and $s_2 \, đm_2$, respectively. In addition, the entropy of the open system can be further increased due to irreversibility effects within the system, such as mixing and other fac-

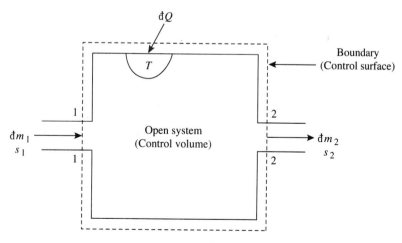

Figure 7-1 Entropy change of an open system.

tors, as represented by a term dS_{prod}, called *entropy production*. Summing up all the terms, we obtain the general expression of dS for an open system with multiple inlets and exits and multiple heat supplies over the control surface:

$$dS_{open\,sys} = \sum \left(\frac{dQ}{T}\right) + \sum_{inlet} (s\,dm) - \sum_{exit} (s\,dm) + dS_{prod}$$

or

$$dS_{prod} = dS_{open\,sys} - \sum \left(\frac{dQ}{T}\right) - \sum_{inlet} (s\,dm) + \sum_{exit} (s\,dm) \geq 0 \qquad (7\text{-}3)$$

where the equality applies to internally reversible processes and the inequality to internally irreversible processes. The entropy production is a positive quantity for an irreversible process; it is zero for an internally reversible process.

When an open system has only a single heat supply, a single inlet at section 1 and a single exit at section 2 as shown in Fig. 7-1, application of Eq. (7-3) to an internally reversible process yields

$$dS_{open\,sys} = \left(\frac{dQ}{T}\right)_{int\,rev} + s_1\,dm_1 - s_2\,dm_2 \qquad (7\text{-}4)$$

For an internally reversible, single-stream, steady-state, steady-flow process, $dS_{open\,sys} = 0$ and $dm_1 = dm_2 = dm$, so that Eq. (7-4) becomes

$$\left(\frac{dQ}{T}\right)_{int\,rev} - (s_2 - s_1)\,dm = 0$$

For steady flow, $\int dQ/T)_{int\,rev}$ for the system is the same as $\int (dQ/T)_{int\,rev}$ for the mass flowing through the system. Thus, this integration can be performed between the entering and leaving states of the fluid. Consequently, on a unit mass-flow basis, we have

$$s_2 - s_1 = \int_1^2 \left(\frac{dq}{T}\right)_{int\,rev} \qquad (7\text{-}5)$$

or

$$ds = \left(\frac{dq}{T}\right)_{int\,rev} \qquad (7\text{-}6)$$

which confirms that the definition of ds holds for any fixed mass, including a mass flowing through an open system.

From Eq. (7-5) or (7-6), it is clear that for a single-stream, steady-state, steady-flow, internally reversible, adiabatic process, $s_2 - s_1 = 0$, that is, the entering and leaving specific entropies are equal. For a nonadiabatic internally reversible process, the heat transfer is

$$q_{int\,rev} = \int_1^2 T\,ds \qquad (7\text{-}7)$$

For an internally reversible isothermal process, the previous equation can be integrated and the heat transfer is

$$q_{int\,rev,\,isothermal} = T(s_2 - s_1) \qquad (7\text{-}8)$$

Notice that these equations are identical to those corresponding equations for closed systems.

Example 7-1

An insulated rigid tank of 1 ft³ volume (see Fig. 7-2) is connected through a valve to a large steam-supply line that is maintained at a constant pressure of 100 psia and a constant temperature of 450°F. The tank is initially evacuated. The valve is opened until the tank is filled with steam at a pressure of 100 psia. Kinetic and potential energy changes are negligible. Determine the final temperature of the steam in the tank (in °F) and the amount of entropy production within the tank (in Btu/°R).

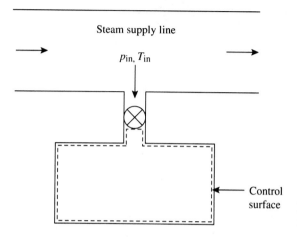

Steam supply line

p_{in}, T_{in}

Control surface

Figure 7-2 Entropy production in charging an insulated rigid tank.

Solution This is an internally irreversible unsteady-flow problem. The given data are $p_{in} = 100$ psia, $T_{in} = 450°F$, $V_i = V_f = 1$ ft³, $m_i = 0$, $p_f = 100$ psia, $Q = 0$, and $W_{cv} = 0$. The subscript "in" denotes the inlet condition to the tank, and the subscripts i and f denote the initial and final conditions, respectively, in the tank.

Because p_{in} and T_{in} remain constant during the filling process, the values of enthalpy (h_{in}) and entropy (s_{in}) of the steam entering the control volume are constants. From the superheated steam table (Table A-20E), at $p_{in} = 100$ psia and $T_{in} = 450°F$, we have

$$h_{in} = 1253.6 \text{ Btu/lbm} \qquad \text{and} \qquad s_{in} = 1.6812 \text{ Btu/lbm·°R}$$

Assume that the steam enters the tank slowly so that the steam properties inside the tank change uniformly throughout the tank, thus satisfying the requirement of a uniform-flow, uniform-state model. Accordingly, from Eq. (5-14), with $m_{in} = m_f - m_i$, we have

$$\cancel{Q}^0 - \cancel{W_{cv}}^0 + (m_f - m_i)(h_{in} + \cancel{\tfrac{1}{2}V_{in}^2}^0 + \cancel{gz_{in}}^0) - \cancel{m_{exit}}^0(h_{exit} + \tfrac{1}{2}V_{exit}^2 + gz_{exit})$$

$$= (m_f u_f - \cancel{m_i u_i}^0) + \tfrac{1}{2}(m_f V_f^2 \cancel{/} m_i V_i^2)^0 + g(m_f z_f \cancel{/} m_i z_i)^0$$

or

$$m_f h_{in} = m_f u_f$$

that is,

$$u_f = h_{in} = 1253.6 \text{ Btu/lbm}$$

Therefore, the final state of the steam in the tank is fixed by $p_f = 100$ psia and $u_f = 1253.6$ Btu/lbm. From the superheated steam table (Table A-20E), at these values of p_f and u_f, by interpolation, we have

$$T_f = \text{final temperature of steam in tank} = 702°F$$

$$s_f = 1.8042 \text{ Btu/lbm·°R}$$

and

$$v_f = 6.847 \text{ ft}^3/\text{lbm}$$

As there is no heat transfer and no exit flow, Eq. (7-3) for the irreversible process reduces to

$$dS_{prod} = dS_{open\ system} - \overset{0}{\frac{dQ}{T}} - (s\ dm)_{in} + \overset{0}{(s\ dm)_{exit}} > 0$$

or

$$S_{prod} = \Delta S_{open\ system} - s_{in} m_{in} > 0$$

where

$$m_{in} = m_f - \overset{0}{m_i} = m_f$$

$$\Delta S_{open\ system} = m_f s_f - \overset{0}{m_i s_i} = m_f s_f$$

and

$$m_f = \frac{V_f}{v_f}$$

Finally, we have

$$S_{prod} = m_f(s_f - s_{in}) = \frac{V_f}{v_f}(s_f - s_{in})$$

$$= \left(\frac{1 \text{ ft}^3}{6.847 \text{ ft}^3/\text{lbm}}\right)(1.8042 - 1.6812) \text{ Btu/lbm·°R}$$

$$= 0.018 \text{ Btu/°R}$$

Notice that the entropy production for the process can be attributed to friction and pressure unbalance associated with the flow of the fluid. The process is clearly irreversible because the tank cannot become evacuated again by simply opening the valve again.

Example 7-2

An insulated counterflow heat exchanger (see Fig. 7-3) operates under steady-flow, steady-state conditions. Liquid water at 1 MPa and 15°C enters the heat exchanger and leaves at 25°C. Refrigerant-12 enters the heat exchanger at 1.4 MPa, 80°C, with a mass-flow rate of 5 kg/min and leaves as a saturated liquid at 50°C. All kinetic and potential energy changes

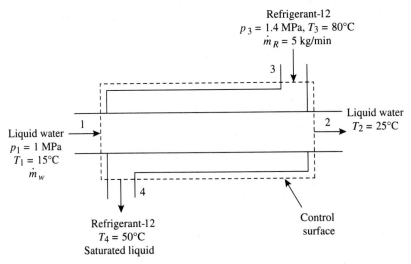

Figure 7-3 Entropy production in irreversible heat transfer.

are negligible. Determine the mass-flow rate of the water stream (in kg/min) and the rate of entropy production due to the irreversible heat transfer within the heat exchanger (in kW/K).

Solution   Referring to Fig. 7-3, the given data are $p_1 = 1$ MPa, $T_1 = 15°C$, $T_2 = 25°C$, $p_3 = 1.4$ MPa, $T_3 = 80°C$, $\dot{m}_R = 5$ kg/min, $T_4 = 50°C$, $\dot{Q}$ (for the control volume) $= 0$, and $\dot{W}_{cv} = 0$. Both water and Refrigerant-12 are in steady flow, with negligible changes in KE and PE.

The incoming water at 1 MPa and 15°C is a compressed liquid. As 1 MPa is a relatively low pressure, the saturated water table (Table A-18M) gives

$$h_1 \approx h_f \text{ at } 15°C = 62.99 \text{ kJ/kg}$$

and

$$s_1 \approx s_f \text{ at } 15°C = 0.2245 \text{ kJ/kg·K}$$

The outgoing water must be at a pressure equal to or a little less than 1 MPa. At $T_2 = 25°C$, this liquid is in a compressed liquid state. Table A-18M then gives $h_2 \approx h_f$ at 25°C $= 104.89$ kJ/kg and $s_2 \approx s_f$ at 25°C $= 0.3674$ kJ/kg·K.

The incoming Refrigerant-12 is a superheated vapor. At $p_3 = 1.4$ MPa and $T_3 = 80°C$, Table A-23M gives $h_3 = 227.89$ kJ/kg and $s_3 = 0.7355$ kJ/kg·K. At $T_4 = 50°C$, the saturated Refrigerant-12 table (Table A-22M) gives $h_4 = h_f = 84.87$ kJ/kg and $s_4 = s_f = 0.3034$ kJ/kg·K.

Because the heat exchanger is insulated, there is no heat interaction with the surroundings, that is, $\dot{Q} = 0$. The heat exchange between the two fluid streams occurring inside the control volume does not concern us when we write an energy equation for the control volume. An energy balance for the control volume in steady-flow, steady-state conditions gives

$$\cancel{\dot{Q}}^0 - \cancel{\dot{W}_{cv}}^0 + \dot{m}_w[(h_1 - h_2) + \tfrac{1}{2}(V_1^2 \cancel{\neq} V_2^2)^0 + g(z_1 \cancel{\neq} z_2)^0]$$

$$+ \dot{m}_R[(h_3 - h_4) + \tfrac{1}{2}(V_3^2 \cancel{\neq} V_4^2)^0 + g(z_3 \cancel{\neq} z_4)^0] = 0$$

or

$$\dot{m}_w(h_1 - h_2) + \dot{m}_R(h_3 - h_4) = 0$$

so that

$$\dot{m}_w = \frac{\dot{m}_R(h_3 - h_4)}{h_2 - h_1} = \frac{(5 \text{ kg/min})(227.89 - 84.87) \text{ kJ/kg}}{(104.89 - 62.99) \text{ kJ/kg}}$$

$$= 17.07 \text{ kg/min}$$

Because there is no heat exchange with the surroundings, $\int dQ/T = 0$; and for a steady-flow, steady-state process, $\Delta S_{\text{open system}} = 0$. From Eq. (7-3), the rate of entropy production for the irreversible process of the heat exchanger is given by

$$\dot{S}_{\text{prod}} = \cancel{\Delta \dot{S}_{\text{open system}}}^0 - \int \cancel{\frac{dQ}{T}}^0 - \dot{m}_w(s_1 - s_2) - \dot{m}_R(s_3 - s_4)$$

$$= -\dot{m}_w(s_1 - s_2) - \dot{m}_R(s_3 - s_4)$$

$$= -(17.07 \text{ kg/min})(0.2245 - 0.3674) \text{ kJ/kg·K}$$

$$- (5 \text{ kg/min})(0.7355 - 0.3034) \text{ kJ/kg·K}$$

$$= 0.2788 \text{ kJ/min·K}$$

$$= (0.2788 \text{ kJ/min·K})/(60 \text{ s/min})$$

$$= 0.00465 \text{ kW/K} = 4.65 \times 10^{-3} \text{ kW/K}$$

Notice that this entropy production is caused by the irreversible heat transfer due to finite temperature differences between the two fluid streams within the heat exchanger.

7-3 INTERNALLY REVERSIBLE STEADY-STATE, STEADY-FLOW PROCESS

For a simple compressible steady-state, steady-flow system with a single inlet and a single exit, the energy equation on the basis of a unit mass (Eq. 5-10) can be written in differential form as

$$đw_{cv} = đq - dh - dKE - dPE \tag{7-9}$$

If the process is internally reversible, from Eq. (7-2a),

$$đq = T \, ds = dh - v \, dp$$

Combining these two equations yields

$$đw_{cv} = (dh - v \, dp) - dh - dKE - dPE$$

or

$$đw_{cv} = -v \, dp - dKE - dPE \tag{7-10}$$

This is a very important equation for a single-stream, simple-compressible, open system on a unit mass basis that undergoes a steady-state, steady-flow, internally reversible process. Although this equation is expressed explicitly for mechanical work, it is valid for processes involving heat transfer. It is imperative to notice that this equation can be developed solely from principles of mechanics based on Newton's second law. It is a momentum equation, independent of the laws of thermodynamics. Thus, the momentum equation (7-10) and the energy equation (7-9) can be used simultaneously in analyzing internally reversible, steady-state, steady-flow processes.

If the changes in kinetic and potential energies are negligible, Eq. (7-10) reduces to

$$đw_{cv} = -v \, dp \tag{7-11}$$

The total work done by the fluid as it undergoes an internally reversible process from state 1 to state 2 with negligible changes in kinetic and potential energies is

$$w_{cv} = -\int_1^2 v \, dp \tag{7-12}$$

This work is represented by the area $1-2-b-a-1$ on the left of the process curve $1-2$ on a $p-v$ diagram, as shown in Fig. 7-4. For comparison, the total work done in a closed system, $\int_1^2 p \, dv$, for the same process $1-2$ is represented by the area under the curve $1-2$ (area $1-2-d-c-1$) on the $p-v$ diagram shown in Fig. 7-4.

The evaluation of the integral of Eq. (7-12) requires a knowledge of the functional relationship between p and v during an internally reversible process. Let us perform the integration for a polytropic process,

$$pv^n = c$$

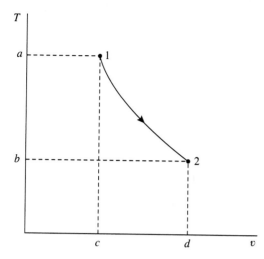

Figure 7-4 $\int v\,dp$ and $\int p\,dv$ on a $p\text{–}v$ diagram.

where n is the polytropic constant having any value from $-\infty$ to $+\infty$, and c represents a constant. Integrating Eq. (7-12) with this $p\text{–}v$ relation yields

$$w_{cv} = -\int_1^2 v\,dp = -c^{1/n}\int_1^2 \frac{dp}{p^{1/n}}$$

$$= -\frac{n}{n-1}c^{1/n}(p_2^{(n-1)/n} - p_1^{(n-1)/n})$$

Substituting $p_1 v_1^n = p_2 v_2^n = c$, we have

$$w_{cv} = \frac{n}{1-n}(p_2 v_2 - p_1 v_1) \tag{7-13}$$

Recall that for a closed system, we have Eq. (2-4),

$$w = \int_1^2 p\,dv = \frac{1}{1-n}(p_2 v_2 - p_1 v_1)$$

Notice that Eq. (7-13) is valid for any value of n except $n = 1$. For the case when $n = 1$, we have

$$pv = c = p_1 v_1 = p_2 v_2$$

so that

$$w_{cv} = -\int_1^2 v\,dp = -c\int_1^2 \frac{dp}{p} = -p_1 v_1 \ln\frac{p_2}{p_1} \tag{7-14}$$

In general, the integral of $v\,dp$ for Eq. (7-10) represents shaft work done plus changes in kinetic and potential energies. In some mechanical devices, there is no shaft work involved, but there is a change in kinetic or potential energy. For example, in a frictionless horizontal nozzle, $đw_{cv} = 0$ and $dPE = 0$, so that $dKE = -v\,dp$ and the area $1\text{–}2\text{–}b\text{–}a\text{–}1$ shown in Fig. 7-4 then represents the change in kinetic energy.

In Sec. 5-8, we introduced the Bernoulli equation, which is

$$v(p_2 - p_1) + \tfrac{1}{2}(\mathbf{V}_2^2 - \mathbf{V}_1^2) + g(z_2 - z_1) = 0$$

It should be recognized now that this equation is simply the integral form of Eq. (7-10) with no shaft work involved for an internally reversible steady-state, steady-flow process of an incompressible fluid.

Example 7-3

Carbon monoxide expands in an internally reversible steady-state, steady-flow process through a nozzle from 500 kPa, 200°C, to 100 kPa at a rate of 0.1 kg/s. The expansion follows the polytropic relation $pv^{1.2}$ = constant. Determine the exit velocity and the rate of heat transfer, the inlet velocity to the nozzle being negligibly small. Use c_p = 1.041 kJ/kg·K for CO.

Solution Given p_1 = 500 kPa, T_1 = 200°C = 473 K, p_2 = 100 kPa, $\dot{m}$ = 0.1 kg/s, $pv^{1.2} = p_1v_1^{1.2} = p_2v_2^{1.2} = c$, and c_p = 1.041 kJ/kg·K. For nozzle, w_{cv} = 0. The gas constant for CO is

$$R = \frac{\mathcal{R}}{M} = \frac{8.314 \text{ kJ/kgmole·K}}{28 \text{ kg/kgmole}} = 0.297 \text{ kJ/kg·K}$$

From the polytropic relation, Eq. (3-20), the final temperature T_2 is

$$T_2 = T_1\left(\frac{p_2}{p_1}\right)^{(n-1)/n} = 473\left(\frac{100}{500}\right)^{(1.2-1)/1.2}$$

$$= 362 \text{ K} = 89°C$$

Because the process is internally reversible steady state and steady flow, Eq. (7-10) is applicable. With the inlet kinetic energy and the change in potential energy neglected, we have

$$\cancel{\dot{w}_{cv}}^{0} = -\int_1^2 v \, dp - [(KE)_2 - \cancel{(KE)_1}^{0}] - [(PE)_2 \cancel{-} \cancel{(PE)_1}^{0}]$$

or

$$-\int_1^2 v \, dp - (KE)_2 = 0$$

so that

$$(KE)_2 = -\int_1^2 v \, dp = -c^{1/n}\int_1^2 \frac{dp}{p^{1/n}}$$

$$= -\frac{n}{n-1}c^{1/n}\left(p_2^{(n-1)/n} - p_1^{(n-1)/n}\right)$$

$$= \frac{n}{1-n}(p_2v_2 - p_1v_1) = \frac{n}{1-n}R(T_2 - T_1)$$

$$= \frac{1.2}{1-1.2} \times 0.297(362 - 473) = 198 \text{ kJ/kg}$$

Additionally,

$$(KE)_2 = \tfrac{1}{2}V_2^2$$

so that

$$V_2 = [2(KE)_2]^{1/2} = [2(1000 \text{ J/kJ})(198 \text{ kJ/kg})]^{1/2} = 629 \text{ m/s}$$

noting that

$$1 \text{ J} = 1 \text{ N·m} = 1 \text{ (kg·m/s}^2)\text{·m} = 1 \text{ kg·(m/s)}^2$$

Now with $w_{cv} = 0$, $(KE)_1 = 0$, and $\Delta PE = 0$, the steady-state, steady-flow energy equation gives

$$q - \cancel{w_{cv}}^{0} - (h_2 - h_1) - [(KE)_2 - \cancel{(KE)_1}^{0}] - [(PE)_2 - \cancel{(PE)_1}^{0}] = 0$$

or

$$q = h_2 - h_1 + (KE)_2$$

so that the heat addition is

$$q = c_p(T_2 - T_1) + (KE)_2$$

$$= (1.041 \text{ kJ/kg·K})(362 \text{ K} - 473 \text{ K}) + 198 \text{ kJ/kg}$$

$$= 82.4 \text{ kJ/kg}$$

The rate of heat addition is calculated as

$$\dot{q} = \dot{m}q = (0.1 \text{ kg/s})(82.4 \text{ kJ/kg})$$

$$= 8.24 \text{ kW}$$

7-4 ENTROPY CHANGE OF IDEAL GASES

In Chapter 4, we learned that for an ideal gas, the properties u, h, c_p, and c_v are all functions of temperature only, and Eqs. (4-10) and (4-11) declare that for any process of an ideal gas, we have

$$du = c_v \, dT \qquad \text{and} \qquad dh = c_p \, dT$$

These two relations can be used in connection with Eqs. (7-1) and (7-2) to obtain two "ds" equations for the calculation of entropy changes of ideal gases. Thus, from Eq. (7-1a),

$$ds = \frac{1}{T} \, du + \frac{p}{T} \, dv$$

we have

$$ds = c_v \frac{dT}{T} + R \frac{dv}{v} \tag{7-15}$$

and from Eq. (7-2a),

$$ds = \frac{1}{T} \, dh - \frac{v}{T} \, dp$$

we have

$$ds = c_p \frac{dT}{T} - R \frac{dp}{p} \tag{7-16}$$

where the ideal gas relation, $pv = RT$, has been incorporated.

In addition, when the ideal gas relation is differentiated,

$$p \, dv + v \, dp = R \, dT$$

Dividing through by $pv = RT$, we have

$$\frac{dv}{v} + \frac{dp}{p} = \frac{dT}{T}$$

Substituting this relation together with the relation $c_p - c_v = R$ into Eq. (7-15) yields

$$ds = c_v\left(\frac{dv}{v} + \frac{dp}{p}\right) + (c_p - c_v)\frac{dv}{v}$$

or

$$ds = c_v \frac{dp}{p} + c_p \frac{dv}{v} \qquad (7\text{-}17)$$

The three equations (7-15 to 7-17) express the entropy change of ideal gases in terms of T and v, T and p, and p and v, respectively.

Although the specific heats of ideal gases generally vary with temperature, they are sometimes considered constant in simplified calculations involving small temperature ranges. For such cases, integrating Eqs. (7-15) to (7-17) between two end states 1 and 2 leads to

$$\Delta s = s_2 - s_1 = c_v \ln \frac{T_2}{T_1} + R \ln \frac{v_2}{v_1} \qquad (7\text{-}18)$$

$$\Delta s = s_2 - s_1 = c_p \ln \frac{T_2}{T_1} - R \ln \frac{p_2}{p_1} \qquad (7\text{-}19)$$

$$\Delta s = s_2 - s_1 = c_v \ln \frac{p_2}{p_1} + c_p \ln \frac{v_2}{v_1} \qquad (7\text{-}20)$$

These equations can be used to evaluate Δs between two equilibrium states of an ideal gas without any knowledge of the process connecting them. In many engineering calculations when the temperature range is small, arithmetically averaged specific heats can be used in these equations with little error. The variation of c_p with temperature for several gases are shown in Fig. 2-13. This figure can be used as a reference guide to determine whether or not an average c_p value can be used for a particular gas in a particular range of temperatures.

It is to be noted that Eqs. (7-15) to (7-20) are applicable on a unit-mass or unit-mole basis. When used on a unit-mole basis, the universal gas constant $\mathscr{R}$ replaces the gas constant R in these equations.

7-5 INTERNALLY REVERSIBLE ADIABATIC PROCESS FOR AN IDEAL GAS WITH CONSTANT SPECIFIC HEATS

An important relation describing an internally reversible adiabatic (isentropic) process for an ideal gas with constant specific heats can be derived directly by assigning $ds = 0$ in Eq. (7-17). Thus,

$$c_v \frac{dp}{p} + c_p \frac{dv}{v} = 0$$

or

$$\frac{dp}{p} + k\frac{dv}{v} = 0$$

If $k = c_p/c_v = $ constant, integrating the preceding equation yields

$$\ln p + k \ln v = \text{constant}$$

or

$$\ln (pv^k) = \text{constant}$$

so that

$$pv^k = \text{constant} \qquad (7\text{-}21)$$

Combining this equation with the ideal gas equation gives two more useful relations. Thus,

$$\left(\frac{RT}{v}\right)v^k = \text{constant}$$

or

$$Tv^{k-1} = \text{constant} \qquad (7\text{-}22)$$

and

$$p^{1/k}\left(\frac{RT}{p}\right) = \text{constant}$$

or

$$\frac{T}{p^{(k-1)/k}} = \text{constant} \qquad (7\text{-}23)$$

When Eqs. (7-21) to (7-23) are written for a process between states 1 and 2, we have

$$\left(\frac{T_2}{T_1}\right)_s = \left(\frac{p_2}{p_1}\right)_s^{(k-1)/k} = \left(\frac{v_1}{v_2}\right)_s^{k-1} \qquad (7\text{-}24)$$

where the subscript s emphasizes the isentropic condition. Be sure to notice that Eqs. (7-21) to (7-24) hold only for internally reversible adiabatic (isentropic) processes of ideal gases with constant specific heats.

Recall that in Sec. 4-4, from the definition of the quasiequilibrium (now called internally reversible) polytropic process, $pv^n = $ constant, we identified some familiar thermodynamic processes for ideal gases as special cases with various values of n. We now see that when the polytropic constant n equals the constant specific-heat ratio k, the process is isentropic ($s = $ constant). It is appropriate for us to illustrate the polytropic processes with various values of n on a T–s diagram for ideal gases as a counterpart of Fig. 4-4, which is a p–v diagram. This is shown in Fig. 7-5.

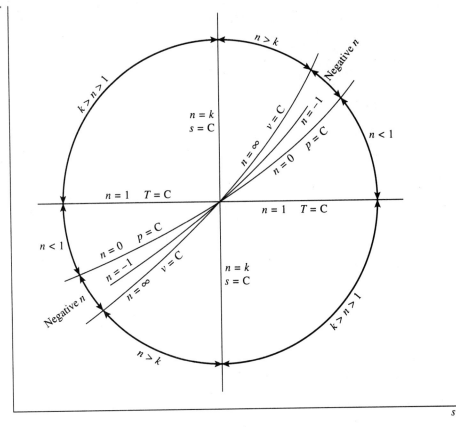

Figure 7-5 Polytropic processes on T–s diagram for an ideal gas.

Example 7-4

An ideal gas with a constant value of c_p = 29.6 kJ/kgmole·K is made to undergo a cycle consisting of the following internally reversible processes in a closed system:

Process 1–2: The gas expands adiabatically from 5000 kPa and 550 K to 1000 kPa.

Process 2–3: The gas is heated at constant volume until 550 K.

Process 3–1: The gas is compressed isothermally back to its initial condition.

Calculate the work, heat, and change of entropy per kgmole of the gas for each of the three processes.

Solution The processes of this example problem are shown on the p–v and T–s diagrams of Fig. 7-6. The given data are p_1 = 5000 kPa, p_2 = 1000 kPa, and $T_1 = T_3$ = 550 K.

Because for an ideal gas, $c_p - c_v = \mathcal{R}$, we have

$$c_v = c_p - \mathcal{R} = 29.6 - 8.31 = 21.3 \text{ kJ/kgmole·K}$$

and

$$k = \frac{c_p}{c_v} = \frac{29.6}{21.3} = 1.39$$

Entropy Change During Processes   Chap. 7

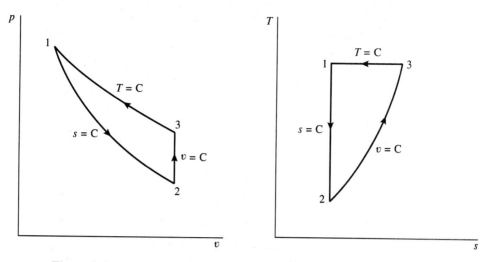

Figure 7-6 p–v and T–s diagrams showing the processes of Example 7-4.

From Eq. (7-24), we obtain

$$T_2 = T_1\left(\frac{p_2}{p_1}\right)^{(k-1)/k} = 550\left(\frac{1000}{5000}\right)^{(1.39-1)/1.39}$$

$$= 350 \text{ K}$$

From the ideal gas equation, we obtain

$$v_1 = \frac{\mathscr{R}T_1}{p_1} = \frac{(8.31 \text{ kJ/kgmole·K})(550 \text{ K})}{5000 \text{ kN/m}^2}$$

$$= 0.914 \text{ m}^3/\text{kgmole}$$

$$v_2 = \frac{\mathscr{R}T_2}{p_2} = \frac{8.31 \times 350}{1000} = 2.91 \text{ m}^3/\text{kgmole}$$

Now, because $v_3 = v_2$, we then have

$$p_3 = \frac{\mathscr{R}T_3}{v_3} = \frac{8.31 \times 550}{2.91} = 1570 \text{ kPa}$$

The value of p_3 can also be obtained by

$$p_3 = p_2\left(\frac{T_3}{T_2}\right) = 1000 \times \frac{550}{350} = 1570 \text{ kPa}$$

Because the process from 1 to 2 is internally reversible and adiabatic, it follows that

$$q_{12} = 0 \qquad \text{and} \qquad \Delta s_{12} = s_2 - s_1 = 0$$

By the first law, we then have

$$w_{12} = \cancel{q_{12}}^{\,0} - \Delta u_{12} = -\Delta u_{12} = u_1 - u_2 = c_v(T_1 - T_2)$$

$$= 21.3(550 - 350) = 4260 \text{ kJ/kgmole}$$

As process 1–2 follows the relation pv^k = constant, the preceding value of w_{12} can also be obtained by the formula

$$w_{12} = \int_1^2 p\, dv = \frac{p_2 v_2 - p_1 v_1}{1 - k} = \frac{\Re(T_2 - T_1)}{1 - k}$$

$$= \frac{8.31(350 - 550)}{1 - 1.39} = 4260 \text{ kJ/kgmole}$$

Because the process from 2 to 3 is at constant volume, it follows that

$$w_{23} = \int_2^3 p\, dv = 0$$

$$q_{23} = \Delta u_{23} + \cancel{w_{23}}^{0} = \Delta u_{23} = u_3 - u_2 = c_v(T_3 - T_2)$$

$$= 21.3(550 - 350) = 4260 \text{ kJ/kgmole}$$

The entropy change Δs_{23} is given by Eq. (7-18),

$$\Delta s_{23} = s_3 - s_2 = c_v \ln \frac{T_3}{T_2} + \Re \ln \cancel{\frac{v_3}{v_2}}^{0} = c_v \ln \frac{T_3}{T_2} = 21.3 \ln \frac{550}{350}$$

$$= 9.62 \text{ kJ/kgmole·K}$$

For the isothermal process 3 to 1, Eq. (7-19) gives

$$\Delta s_{31} = s_1 - s_3 = c_p \ln \cancel{\frac{T_1}{T_3}}^{0} - \Re \ln \frac{p_1}{p_3} = -\Re \ln \frac{p_1}{p_3}$$

$$= -8.31 \ln \frac{5000}{1570} = -9.62 \text{ kJ/kgmole·K}$$

which is equal to $-\Delta s_{23}$. The heat rejection is given by

$$q_{31} = T_1(\Delta s_{31}) = 550(-9.62) = -5290 \text{ kJ/kgmole}$$

Because $\Delta u_{31} = 0$ for the isothermal process, the first law leads to

$$w_{31} = q_{31} = -5290 \text{ kJ/kgmole}$$

As process 3–1 follows the relation $pv = \Re T$ = constant, the value of w_{31} can also be obtained by the formula

$$w_{31} = \int_3^1 p\, dv = p_1 v_1 \ln \frac{v_1}{v_3} = \Re T_1 \ln \frac{p_3}{p_1}$$

$$= 8.31 \times 550 \ln \frac{1570}{5000}$$

$$= -5290 \text{ kJ/kgmole}$$

The minus sign for q_{31} and w_{31} indicates that heat is removed from the gas and work is done on the gas.

Both the net work and the net heat of the cycle have the same value of

$$4260 - 5290 = -1030 \text{ kJ/kgmole}$$

Example 7-5

An ideal gas whose equation of state is $pV = mRT'$ and whose specific heats are constants is used as the working substance in a closed system to operate as a Carnot engine (see Fig.

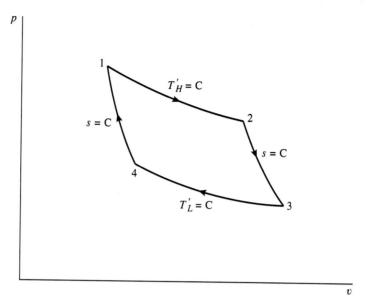

Figure 7-7 Equivalence of the ideal gas and the thermodynamic temperature scales.

7-7). Show that the ideal gas temperature T' as defined in Sec. 1-7 is equal to the thermodynamic temperature T as defined in Sec. 6-6.

Solution The first law for a process is

$$Q = \Delta U + W$$

For the isothermal process 1–2, because T'_H = constant,

$$\Delta U_{12} = mc_v(\Delta T_{12}) = 0$$

we have

$$Q_H = Q_{12} = W_{12} = \int_1^2 p \, dV$$

$$= mRT'_H \int_{V_1}^{V_2} \frac{dV}{V} = mRT'_H \ln \frac{V_2}{V_1}$$

Similarly,

$$Q_L = Q_{43} - mRT'_L \ln \frac{V_3}{V_4}$$

But for a reversible-adiabatic process, $T'V^{k-1}$ = constant, where $k = c_p/c_v$ = constant, we have for the processes 2–3 and 4–1,

$$\frac{T'_H}{T'_L} = \left(\frac{V_3}{V_2}\right)^{k-1} = \left(\frac{V_4}{V_1}\right)^{k-1}$$

from which

$$\frac{V_2}{V_1} = \frac{V_3}{V_4}$$

Therefore,

$$\frac{Q_H}{Q_L} = \frac{mRT'_H \ln \dfrac{V_2}{V_1}}{mRT'_L \ln \dfrac{V_3}{V_4}} = \frac{T'_H}{T'_L}$$

Comparison of the previous equation with Eq. (6-7), $Q_H/Q_L = T_H/T_L$, reveals that

$$\frac{T_H}{T_L} = \frac{T'_H}{T'_L}$$

In addition, because the same numerical value was assigned to the fixed point (triple point of water), it follows that

$$T = T'$$

or the thermodynamic temperature is numerically equal to the absolute temperature as measured by an ideal gas thermometer.

7-6 IDEAL GASES WITH VARIABLE SPECIFIC HEATS

Because the specific heats of an ideal gas are functions of temperature, these variations in temperature should be incorporated in the evaluation of entropy changes when more accurate results are required. For example, integration of Eq. (7-16) gives

$$\Delta s = s_2 - s_1 = \int_1^2 c_p \frac{dT}{T} - R \ln \frac{p_2}{p_1} \tag{7-25}$$

An analytical expression of c_p as a function of T, such as that given in Table A-7 in Appendix 3, can be introduced in the preceding equation. The result from this procedure will be more precise than that obtained with the assumption of constant specific heats.

An alternative means of taking care of the temperature variations of specific heats is to tabulate the integration of Eq. (7-25) in the manner similar to the tabulations of h and u in ideal-gas property tables. Because entropy is not a function of temperature alone even for ideal gases, it cannot be tabulated directly with temperature as the single entry. It is tabulated indirectly in such tables through the introduction of a new variable ϕ as defined by

$$\phi = \int_{T_0}^T c_p \frac{dT}{T}$$

where T_0 is some selected reference temperature. ϕ is a function of temperature T only and is tabulated in the ideal-gas property tables. Then from Eq. (7-25),

$$\Delta s = s_2 - s_1 = \int_{T_1}^{T_2} c_p \frac{dT}{T} - R \ln \frac{p_2}{p_1}$$

$$= \int_{T_0}^{T_2} c_p \frac{dT}{T} - \int_{T_0}^{T_1} c_p \frac{dT}{T} - R \ln \frac{p_2}{p_1}$$

or

$$\Delta s = s_2 - s_1 = \phi_2 - \phi_1 - R \ln \frac{p_2}{p_1} \tag{7-26}$$

In the ideal-gas property tables for air and other gases given in Appendix 3 (Tables A-9 to A-16), the reference temperature T_0 is assigned the value of 0 K (0°R). The function ϕ and entropy s are expressed in the same units, which are kJ/kg·K (or kJ/kgmole·K) in the SI system and Btu/lbm·°R (or Btu/1bmole·°R) in the English system.

As Eq. (7-26) is valid for any process of an ideal gas with variable specific heats, this equation can be used to establish a temperature–pressure and a temperature–volume relationship for the special case of an internally reversible adiabatic process for which $\Delta s = 0$. Thus, from Eq. (7-26) for an internally reversible adiabatic process, we have

$$\Delta s = s_2 - s_1 = 0 = \phi_2 - \phi_1 - R \ln \frac{p_2}{p_1}$$

Rearranging,

$$\left(\frac{p_2}{p_1}\right)_s = \exp\left(\frac{\phi_2 - \phi_1}{R}\right) = \frac{\exp(\phi_2/R)}{\exp(\phi_1/R)}$$

where the subscript s indicates the isentropic condition. This equation suggests the introduction of a new variable p_r, called the relative pressure as defined by the expression

$$p_r = \exp\left(\frac{\phi}{R}\right)$$

which is a function of temperature only and is tabulated in the gas tables. Thus,

$$\left(\frac{p_2}{p_1}\right)_s = \frac{p_{r2}}{p_{r1}} \tag{7-27}$$

for an internally reversible adiabatic process.

Equation (7-27) can be converted into a volume ratio by substituting $p = RT/v$. Thus,

$$\left(\frac{T_2/v_2}{T_1/v_1}\right)_s = \frac{p_{r2}}{p_{r1}} \quad \text{or} \quad \left(\frac{v_2}{v_1}\right)_s = \frac{T_2/p_{r2}}{T_1/p_{r1}}$$

This equation suggests the introduction of a new variable v_r, called the relative volume as defined by the expression

$$v_r = \frac{RT}{p_r}$$

which is a function of temperature only and is tabulated in the gas tables. Thus,

$$\left(\frac{v_2}{v_1}\right)_s = \frac{v_{r2}}{v_{r1}} \tag{7-28}$$

for an internally reversible adiabatic process.

Example 7-6

One kilogram of air is made to undergo the following internally reversible processes in a closed system:

Process 1–2: The gas is compressed adiabatically from $p_1 = 100$ kPa and $T_1 = 300$ K to $p_2 = 2p_1$.

Process 2–3: It is then heated at constant pressure $p_2 = p_3$ until $T_3 = 560$ K.

Calculate the specific volumes of the three state points, and calculate the work and heat interactions and the changes in entropy for each of the two processes, using (a) the air table and (b) $c_p = 1.013$ kJ/kg·K and $c_v = 0.726$ kJ/kg·K.

Solution The p–v and T–s diagrams for this example problem is shown in Fig. 7-8. The gas constant of the air is calculated by

$$R = \frac{\mathscr{R}}{M} = \frac{8.314 \text{ kJ/kgmole·K}}{29 \text{ kg/kgmole}} = 0.287 \text{ kJ/kg·K}$$

The specific volumes of state points 1 and 3 are calculated as

$$v_1 = \frac{RT_1}{p_1} = \frac{(0.287 \text{ kJ/kg·K})(300 \text{ K})}{100 \text{ kN/m}^2} = 0.861 \text{ m}^3/\text{kg}$$

$$v_3 = \frac{RT_3}{p_3} = \frac{0.287 \times 560}{200} = 0.804 \text{ m}^3/\text{kg}$$

(a) *Air-table solution.* From the air table (Table A-9M), at $T_1 = 300$ K, we obtain:

$$h_1 = 300.43 \text{ kJ/kg}, \qquad p_{r1} = 1.3801, \qquad u_1 = 214.32 \text{ kJ/kg},$$

$$v_{r1} = 62.393, \qquad \phi_1 = 5.7016 \text{ kJ/kg·K}$$

and at $T_3 = 560$ K, we obtain:

$$h_3 = 565.42 \text{ kJ/kg}, \qquad p_{r3} = 12.608, \qquad u_3 = 404.68 \text{ kJ/kg},$$

$$v_{r3} = 12.749, \qquad \phi_3 = 6.3366 \text{ kJ/kg·K}$$

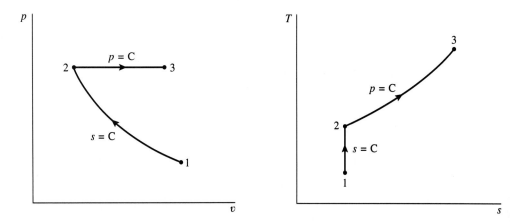

Figure 7-8 p–v and T–s diagrams for Example 7-6.

Because process 1–2 is internally reversible and adiabatic, Eq. (7-27) yields

$$p_{r2} = p_{r1}\left(\frac{p_2}{p_1}\right) = 1.3801(2) = 2.7602$$

At this p_r value, the air table (Table A-9M) gives (by interpolation)

$$h_2 = 366.4 \text{ kJ/kg}, \qquad u_2 = 261.5 \text{ kJ/kg}, \qquad T_2 = 365.6 \text{ K},$$

$$v_{r2} = 38.01, \qquad \phi_2 = 5.901 \text{ kJ/kg·K}$$

The specific volume at state point 2 is calculated by

$$v_2 = v_1\left(\frac{v_{r2}}{v_{r1}}\right) = 0.861\left(\frac{38.01}{62.39}\right) = 0.525 \text{ m}^3/\text{kg}$$

or

$$v_2 = \frac{RT_2}{p_2} = \frac{0.287(365.6)}{200} = 0.525 \text{ m}^3/\text{kg}$$

The heat and work interactions are

$$q_{12} = 0 \qquad \text{(because process 1–2 is adiabatic)}$$

$$w_{12} = \cancel{q_{12}}^{\,0} - \Delta u_{12} = -\Delta u_{12} = u_1 - u_2$$

$$= 214.3 - 261.5 = -47.2 \text{ kJ/kg}$$

where the minus sign indicates that work is done on the air.

$$w_{23} = \int_2^3 p \, dv = p_2(v_3 - v_2) = p_3 v_3 - p_2 v_2 = R(T_3 - T_2)$$

$$= (0.287 \text{ kJ/kg·K})(560 \text{ K} - 365.6 \text{ K})$$

$$= 55.8 \text{ kJ/kg}$$

indicating work is done by the air.

$$q_{23} = w_{23} + \Delta u_{23} = w_{23} + (u_3 - u_2)$$

$$= 55.8 + (404.7 - 261.5) = 199.0 \text{ kJ/kg}$$

or

$$q_{23} = (p_3 v_3 - p_2 v_2) + (u_3 - u_2) = h_3 - h_2$$

$$= 565.4 - 366.4 = 199.0 \text{ kJ/kg}$$

indicating heat is added to the air.

The changes in entropy of the two processes are

$$\Delta s_{12} = 0 \qquad \text{(because process 1–2 is internally reversible and adiabatic)}$$

and by Eq. (7-26),

$$\Delta s_{23} = s_3 - s_2 = \phi_3 - \phi_2 - R \cancel{\ln\frac{p_3}{p_2}}^{\,0} = \phi_3 - \phi_2$$

$$= 6.337 - 5.901 = 0.436 \text{ kJ/kg·K}$$

(b) *Constant-specific-heat solution.* The ratio of specific heats is

$$k = \frac{c_p}{c_v} = \frac{1.013}{0.726} = 1.395$$

Equation (7-24) gives

$$T_2 = T_1 \left(\frac{p_2}{p_1} \right)^{(k-1)/k} = 300(2)^{(1.395-1)/1.395} = 365.1 \text{ K}$$

Equation (7-21) gives

$$v_2 = v_1 \left(\frac{p_1}{p_2} \right)^{1/k} = 0.861(\tfrac{1}{2})^{1/1.395} = 0.524 \text{ m}^3/\text{kg}$$

or

$$v_2 = \frac{RT_2}{p_2} = \frac{0.287 \times 365.1}{200} = 0.524 \text{ m}^3/\text{kg}$$

The heat and work interactions are

$$q_{12} = 0$$
$$w_{12} = \overset{0}{\cancel{q_{12}}} - \Delta u_{12} = -\Delta u_{12} = u_1 - u_2 = c_v(T_1 - T_2)$$
$$= (0.726 \text{ kJ/kg·K})(300 \text{ K} - 365.1 \text{ K}) = -47.3 \text{ kJ/kg}$$

$$w_{23} = \int_2^3 p \, dv = p_3 v_3 - p_2 v_2 = R(T_3 - T_2)$$
$$= 0.287(560 - 365.1) = 55.9 \text{ kJ/kg}$$

$$q_{23} = w_{23} + \Delta u_{23} = w_{23} + (u_3 - u_2) = w_{23} + c_v(T_3 - T_2)$$
$$= 55.9 + 0.726(560 - 365.1)$$
$$= 197.4 \text{ kJ/kg}$$

or

$$q_{23} = h_3 - h_2 = c_p(T_3 - T_2)$$
$$= 1.013(560 - 365.1)$$
$$= 197.4 \text{ kJ/kg}$$

The changes in entropy are

$$\Delta s_{12} = 0$$

and by Eq. (7-19),

$$\Delta s_{23} = s_3 - s_2 = c_p \ln \frac{T_3}{T_2} - R \, \overset{0}{\cancel{\ln \frac{p_3}{p_2}}} = c_p \ln \frac{T_3}{T_2}$$

$$= 1.013 \ln \frac{560}{365.1} = 0.433 \text{ kJ/kg·K}$$

7-7 ENTROPY CHANGE OF INCOMPRESSIBLE SUBSTANCES

For an incompressible substance, Eq. (3-11) declares that

$$du = c \, dT$$

where $c = c_v = c_p$ is the specific heat of the incompressible substance. Because $dv = 0$ by definition for an incompressible substance, Eq. (7-1a) gives

$$T \, ds = du + p \, dv = du$$

or

$$ds = \frac{du}{T}$$

Thus, we conclude that the entropy change of an incompressible substance is given by

$$ds = c\frac{dT}{T} \tag{7-29}$$

When the specific heat c is assumed to be a constant or when an appropriate average value of the specific heat is used, the preceding equation can be integrated to obtain

$$s_2 - s_1 = c \ln \frac{T_2}{T_1} \tag{7-30}$$

Example 7-7

A block of copper having a mass of 200 lbm and a temperature of 200°F is immersed into a tank containing 100 lbm of water at 60°F. Neglecting the thermal capacity of the tank, determine the changes of entropy of the copper, water, and the combined system. For copper, $c = 0.0915$ Btu/lbm·°F; for water, $c = 1$ Btu/lbm·°F.

Solution Let T_f represent the final temperature of the combined system. An energy balance for the process of immersion gives

$$m_{copper}c_{copper}(T_{copper} - T_f) = m_{water}c_{water}(T_f - T_{water})$$

or

$$(200 \text{ lbm})(0.0915 \text{ Btu/lbm·°F})(200°F - T_f°F)$$

$$= (100 \text{ lbm})(1 \text{ Btu/lbm·°F})(T_f°F - 60°F)$$

so that

$$T_f = \frac{200 \times 0.0915 \times 200 + 100 \times 1 \times 60}{200 \times 0.0915 + 100 \times 1} = 82°F$$

$$= 82 + 460 = 542°R$$

Application of Eq. (7-30) to the copper block and water, respectively, yields

$$\Delta S_{copper} = m_{copper}c_{copper} \ln \frac{T_f}{T_{copper}}$$

$$= (200 \text{ lbm})(0.0915 \text{ Btu/lbm·°R}) \ln \frac{542°R}{(200 + 460)°R}$$

$$= -3.60 \text{ Btu/°R}$$

$$\Delta S_{water} = m_{water}c_{water} \ln \frac{T_f}{T_{water}} = (100)(1) \ln \frac{542}{60 + 460}$$

$$= 4.14 \text{ Btu/°R}$$

The entropy change of the combined system is then

$$\Delta S_{comb\,sys} = \Delta S_{copper} + \Delta S_{water}$$

$$= -3.60 + 4.14$$

$$= 0.54 \text{ Btu/°R}$$

Notice that during the process, the entropy of the copper decreases and that of the water increases. The change in total entropy is positive, in accordance with the increase-in-entropy principle for the irreversible process.

7-8 EFFICIENCIES OF STEADY-FLOW DEVICES

As we know, the reversibility assumption in thermodynamic processes is basically only an idealization. During the flow of real fluids, frictional losses and other dissipative effects always result in the production of entropy and render the process irreversible. The concept of efficiency of a device is often used to account for the degree of perfection of the real process. For many steady-flow devices, the actual flow processes are approximately adiabatic, thus making the internally reversible and adiabatic, or isentropic, process an appropriate standard of performance comparison.

For gas and steam turbines, the *adiabatic turbine efficiency* η_T is defined as the ratio of the actual work w_a to the isentropic work w_s, or

$$\eta_T = \frac{w_a}{w_s} \tag{7-31}$$

where w_s is defined as the work that could be performed by the turbine if it is operated isentropically between the same initial state and the same final pressure of the actual turbine. Figure 7-9 shows the T–s diagrams of typical actual and ideal expansion processes for gas and steam turbines. If changes in potential energy are neglected, with $q = 0$, the steady-flow energy equation written for the actual and ideal turbines, respectively, on a unit mass-flow basis are

$$w_a = (h_1 - h_{2'}) + \frac{\mathbf{V}_1^2 - \mathbf{V}_{2'}^2}{2}$$

$$w_s = (h_1 - h_2) + \frac{\mathbf{V}_1^2 - \mathbf{V}_2^2}{2}$$

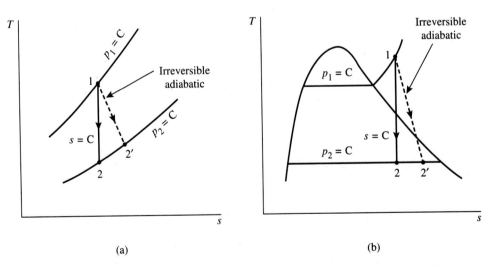

(a)   (b)

Figure 7-9 Reversible and irreversible adiabatic expansion for (a) a gas turbine, and (b) a steam turbine.

When changes in kinetic energy are negligible, substitution of the preceding equations into Eq. (7-31) yields

$$\eta_T = \frac{h_1 - h_{2'}}{h_1 - h_2} \tag{7-32}$$

As demonstrated in Fig. 7-9, irreversibility effects always result in an increase in entropy and a decrease in work output. The typical value of η_T for actual turbines ranges from 80 to 90%.

For gas and vapor compressors, the *adiabatic compressor efficiency* η_c is defined as the ratio of the isentropic work input w_s to the actual work input w_a, or

$$\eta_c = \frac{w_s}{w_a} \tag{7-33}$$

where w_s is defined as the work input to the compressor if it is operated isentropically between the same initial state and the same final pressure of the actual compressor. By referring to Fig. 7-10(a), with ΔKE and ΔPE neglected, the adiabatic compressor efficiency can be written as

$$\eta_c = \frac{h_2 - h_1}{h_{2'} - h_1} \tag{7-34}$$

Typical values of η_c for actual compressors range from 75 to 85%.

For liquid pumps, the *adiabatic pump efficiency* η_P can be written similar to Eq. (7-34) with reference to Fig. 7-10(b).

It is significant to notice that although turbines are usually insulated to prevent heat loss to the surroundings that would reduce the work output, it is advantageous, however, to cool the gas during compression, and isothermal compression requires minimum work input for the same pressure ratio (see Sec. 10-7). When heat is deliberately transferred out in a gas-compression process, the desirable idealized process may

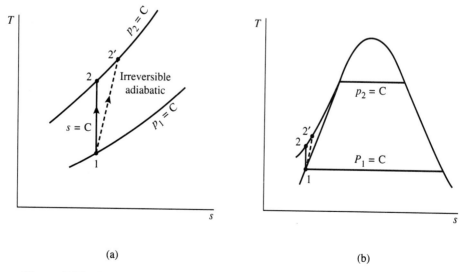

(a)   (b)

Figure 7-10 Reversible and irreversible adiabatic compression for (a) a gas or vapor compressor, and (b) a liquid pump.

Sec. 7-8   Efficiencies of Steady-Flow Devices

be an internally reversible isothermal process rather than an internally reversible adiabatic process. In this case, the *isothermal compressor efficiency* $\eta_{c,\,\text{isothermal}}$ as defined in what follows can be used as a measure of achievement:

$$\eta_{c,\,\text{isothermal}} = \frac{w_t}{w_a} \qquad (7\text{-}35)$$

where w_t is the work required for internally reversible isothermal compression from the actual initial state to the actual final pressure. T–s and p–v diagrams showing an actual cooled-compression process and its corresponding internally reversible isothermal compression process are given in Fig. 7-11.

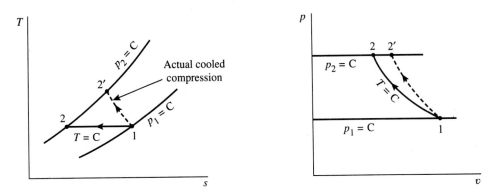

Figure 7-11 Isothermal compression as a standard of comparison.

A *nozzle* is a flow channel that increases the velocity of a fluid at the expense of a decrease in the enthalpy accompanied by a pressure drop in the direction of flow. As the channel is usually short and the flow is usually quick, a nozzle can be considered as an adiabatic device. The *adiabatic nozzle efficiency* η_N is defined as the ratio of the actual kinetic energy at the nozzle exit, $\mathbf{V}_{2'}^2/2$ (see Fig. 7-12), to the kinetic energy that could be achieved at the exit of an isentropic nozzle, $\mathbf{V}_2^2/2$, for the same inlet condition and the same exit pressure, or

$$\eta_N = \frac{\mathbf{V}_{2'}^2/2}{\mathbf{V}_2^2/2} \qquad (7\text{-}36)$$

By the use of the steady-flow energy equation with $q = 0$, $w_{cv} = 0$, and ΔPE neglected, this equation can be written as

$$\eta_N = \frac{(h_1 - h_{2'}) + \mathbf{V}_1^2/2}{(h_1 - h_2) + \mathbf{V}_1^2/2} \qquad (7\text{-}37)$$

If the inlet velocity is small compared with the exit velocity and the enthalpy change, the adiabatic nozzle efficiency can be expressed as

$$\eta_N = \frac{h_1 - h_{2'}}{h_1 - h_2} \qquad (7\text{-}38)$$

Values of η_N range from 90 to 95%.

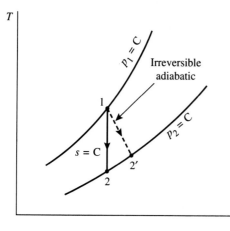

Figure 7-12 Reversible and irreversible adiabatic nozzle processes.

A *diffuser* is a flow channel that increases the pressure of a fluid stream at the expense of a decrease in velocity. There is no mechanical work involved and the flow process can be considered as adiabatic. Assuming that the velocity at the exit of the diffuser is negligible, with $\Delta \text{PE} = 0$, $q = 0$, and $w_{cv} = 0$, an energy balance for the actual diffuser process yields (see Fig. 7-13),

$$\Delta h_a = h_{2'} - h_1 = \mathbf{V}_1^2/2$$

Notice that for an ideal gas, the T–s and h–s diagrams have the same appearance so that Fig. 7-13 can be either a T–s diagram or an h–s diagram. By referring to Fig. 7-13(a), the *adiabatic diffuser efficiency* η_D is defined as the ratio of the isentropic enthalpy change Δh_s to the actual enthalpy change Δh_a, or

$$\eta_D = \frac{\Delta h_s}{\Delta h_a} = \frac{h_2 - h_1}{h_{2'} - h_1} = \frac{h_2 - h_1}{\mathbf{V}_1^2/2} \tag{7-39}$$

where Δh_s is the enthalpy change if the diffuser is operated isentropically between the same initial state and the same final pressure of the actual diffuser.

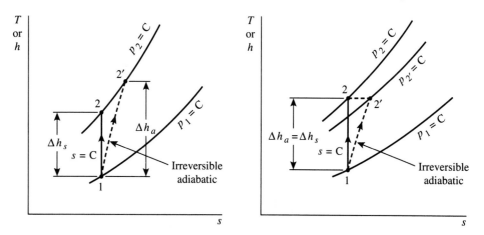

Figure 7-13 Reversible and irreversible diffuser processes for an ideal gas for (a) the same pressure increase, and (b) the same enthalpy change.

The performance of a diffuser can be alternatively measured in terms of the *pressure coefficient* K_p as defined by the ratio of the actual pressure rise Δp_a to the isentropic pressure rise Δp_s for the same enthalpy change occurring in both the actual and ideal diffusers, or, referring to Fig. 7-13(b),

$$K_p = \frac{\Delta p_a}{\Delta p_s} = \frac{p_{2'} - p_1}{p_2 - p_1} \tag{7-40}$$

For an ideal gas with constant specific heats, because

$$\Delta h_a = \Delta h_s \qquad \text{or} \qquad h_{2'} - h_1 = h_2 - h_1$$

it follows that

$$c_p(T_{2'} - T_1) = c_p(T_2 - T_1) \qquad \text{or} \qquad T_{2'} = T_2$$

Accordingly, for an ideal gas, the pressure rise Δp_s in Eq. (7-40) is the pressure rise if the diffuser is operated isentropically between the same initial state and the same final temperature of the actual diffuser.

Example 7-8

Steam at 5 MPa and 600°C enters an adiabatic turbine and expands to 50 kPa at a mass-flow rate of 10,000 kg/h. The turbine has an adiabatic efficiency of 85%. Determine the outlet steam temperature, the turbine power output, and the specific entropy change for the turbine process, neglecting ΔKE and ΔPE.

Solution By referring to Fig. 7-9(b), the given data are $\dot{Q} = 0$, $p_1 = 5$ MPa, $T_1 = 600°C$, $p_2 = p_{2'} = 50$ kPa, $\dot{m} = 10,000$ kg/h, and $\eta_T = 85\%$.

At $p_1 = 5$ MPa and $T_1 = 600°C$, the superheated steam table (Table A-20M) gives

$$h_1 = 3666.5 \text{ kJ/kg} \qquad \text{and} \qquad s_1 = 7.2589 \text{ kJ/kg·K}$$

For the isentropic process 1–2,

$$s_2 = s_1 = s_f + x_2(s_g - s_f) \qquad \text{(at 50 kPa)}$$

Taking data from the saturated steam table (Table A-19M), we have

$$7.2589 = 1.0910 + x_2(7.5939 - 1.0910)$$

so that

$$x_2 = 0.948$$

Therefore,

$$h_2 = h_f + x_2(h_g - h_f) \qquad \text{(at 50 kPa)}$$

$$= 340.49 + 0.948(2645.9 - 340.49) = 2526.0 \text{ kJ/kg}$$

By the definition of adiabatic turbine efficiency, we have

$$h_{2'} = h_1 - \eta_T(h_1 - h_2)$$

$$= 3666.5 - 0.85(3666.5 - 2526.0)$$

$$= 2697.1 \text{ kJ/kg}$$

Because this $h_{2'}$ value is greater than h_g (at 50 kPa) = 2645.9 kJ/kg, state 2' is in the superheated region. From the superheated steam table (Table A-20M) at $p_{2'} = 50$ kPa and

$h_{2'} = 2697.1$ kJ/kg, by interpolation, we obtain

$$T_{2'} = 107.4°C \qquad \text{and} \qquad s_{2'} = 7.733 \text{ kJ/kg·K}$$

The specific entropy change of the turbine process is

$$s_{2'} - s_1 = 7.733 - 7.259 = 0.474 \text{ kJ/kg·K}$$

With ΔKE and ΔPE neglected and $q = 0$, the steady-flow energy equation yields

$$w_a = h_1 - h_{2'} = 3666.5 - 2697.1 = 969.4 \text{ kJ/kg}$$

or

$$w_a = \eta_T w_s = \eta_T(h_1 - h_2)$$
$$= 0.85(3666.5 - 2526.0) = 969.4 \text{ kJ/kg}$$

The power output is then

$$\dot{W}_a = (969.4 \text{ kJ/kg}) \frac{10{,}000 \text{ kg/h}}{3600 \text{ s/h}} = 2693 \text{ kW}$$

Example 7-9

Air at 100 kPa and 350 K is compressed adiabatically to 400 kPa in a steady-flow process. The compressor has an adiabatic efficiency of 82%. Determine the work requirement and the entropy change per kilogram of air using (a) the air table and (b) $c_p = 1.020$ kJ/kg·K. ΔKE and ΔPE can be neglected.

Solution   Referring to Fig. 7-10(a), the given data are $p_1 = 100$ kPa, $T_1 = 350$ K, $p_2 = p_{2'} = 400$ kPa, $\eta_c = 82\%$, and $Q = 0$. For air, the gas constant is

$$R = \frac{\mathcal{R}}{M} = \frac{8.314}{29} = 0.287 \text{ kJ/kg·K}$$

(a) *Air-table solution.* At $T_1 = 350$ K, the air table (Table A-9M) gives

$$h_1 = 350.73 \text{ kJ/kg}, \qquad p_{r1} = 2.3689 \qquad \text{and} \qquad \phi_1 = 5.8567 \text{ kJ/kg·K}$$

For the isentropic process 1–2, Eq. (7-27) gives

$$p_{r2} = p_{r1}\left(\frac{p_2}{p_1}\right) = 2.3689\left(\frac{400}{100}\right) = 9.476$$

At this value of p_{r2}, the air table gives by interpolation,

$$T_2 = 517.4 \text{ K} \qquad \text{and} \qquad h_2 = 521.3 \text{ kJ/kg}$$

By the definition of the adiabatic compressor efficiency, we have

$$h_{2'} = h_1 + \frac{h_2 - h_1}{\eta_c}$$
$$= 350.73 + \frac{521.3 - 350.73}{0.82} = 558.7 \text{ kJ/kg}$$

At this value of $h_{2'}$ the air table gives by interpolation,

$$T_{2'} = 553.5 \text{ K} \qquad \text{and} \qquad \phi_{2'} = 6.3245 \text{ kJ/kg·K}$$

From Eq. (7-26), the entropy change of the air is calculated as

$$s_{2'} - s_1 = \phi_{2'} - \phi_1 - R \ln \frac{p_{2'}}{p_1}$$

$$= 6.3245 - 5.8567 - 0.287 \ln \frac{400}{100}$$

$$= 0.0699 \text{ kJ/kg·K}$$

The work requirement is

$$w_a = h_{2'} - h_1 = 558.7 - 350.7 = 208.0 \text{ kJ/kg}$$

or

$$w_a = \frac{w_s}{\eta_c} = \frac{h_2 - h_1}{\eta_c} = \frac{521.3 - 350.7}{0.82} = 208.0 \text{ kJ/kg}$$

(b) *Constant-specific-heat solution.* $c_p = 1.020$ kJ/kg·K is given, then

$$c_v = c_p - R = 1.020 - 0.287 = 0.733 \text{ kJ/kg·K}$$

$$k = \frac{c_p}{c_v} = \frac{1.020}{0.733} = 1.39$$

For the isentropic process 1–2, Eq. (7-24) gives

$$T_2 = T_1 \left(\frac{p_2}{p_1}\right)^{(k-1)/k} = 350 \left(\frac{400}{100}\right)^{(1.39-1)/1.39} = 516.4 \text{ K}$$

By the definition of η_c, we have

$$\eta_c = \frac{h_2 - h_1}{h_{2'} - h_1} = \frac{c_p(T_2 - T_1)}{c_p(T_{2'} - T_1)}$$

so that

$$T_{2'} = T_1 + \frac{T_2 - T_1}{\eta_c} = 350 + \frac{516.4 - 350}{0.82} = 552.9 \text{ K}$$

From Eq. (7-19), the entropy change of the air is calculated as

$$s_{2'} - s_1 = c_p \ln \frac{T_{2'}}{T_1} - R \ln \frac{p_{2'}}{p_1}$$

$$= 1.020 \ln \frac{552.9}{350} - 0.287 \ln \frac{400}{100} = 0.0685 \text{ kJ/kg·K}$$

The work requirement is

$$w_a = h_{2'} - h_1 = c_p(T_{2'} - T_1) = 1.020(552.9 - 350)$$

$$= 207.0 \text{ kJ/kg}$$

or

$$w_a = \frac{w_s}{\eta_c} = \frac{h_2 - h_1}{\eta_c} = \frac{c_p(T_2 - T_1)}{\eta_c}$$

$$= \frac{1.020(516.4 - 350)}{0.82}$$

$$= 207.0 \text{ kJ/kg}$$

Example 7-10

Argon enters an adiabatic nozzle at a low velocity with a pressure of 50 psia and a temperature of 1800°F. At the nozzle exit, the gas pressure is 18 psia and temperature is 1100°F. Determine the exit velocity of argon and the adiabatic nozzle efficiency, using $k = 1.667$ for argon.

Solution By referring to Fig. 7-12, the given data are $p_1 = 50$ psia, $T_1 = 1800°F = 1800 + 460 = 2260°R$, $V_1 \approx 0$, $p_2 = p_{2'} = 18$ psia, $T_{2'} = 1100°F = 1560°R$, $Q = 0$, and $k = 1.667$. The gas constant of argon is

$$R = \frac{\mathcal{R}}{M} = \frac{1.986 \text{ Btu/lbmole·°R}}{39.95 \text{ lbm/lbmole}} = 0.0497 \text{ Btu/lbm·°R}$$

Because

$$k = 1.667 = c_p/c_v \qquad \text{and} \qquad R = 0.0497 \text{ Btu/lbm·°R} = c_p - c_v$$

it follows that

$$1.667 = \frac{c_p}{c_p - 0.0497}$$

or

$$c_p = 0.124 \text{ Btu/lbm·°R}$$

There is no mechanical work involved. With $V_1 \approx 0$, $q = 0$, and ΔPE neglected, the steady-flow energy equation on a unit mass-flow basis for the actual nozzle process is

$$\cancel{q}^{0} - \cancel{w_{cv}}^{0} + (h_1 + \cancel{\tfrac{1}{2}V_1^2}^{0} + \cancel{gz_1}^{0}) - (h_{2'} + \tfrac{1}{2}V_{2'}^2 + \cancel{gz_{2'}}^{0}) = 0$$

or

$$\tfrac{1}{2}V_{2'}^2 = h_1 - h_{2'}$$

Therefore, the velocity of argon at the nozzle exit is

$$V_{2'} = [2(h_1 - h_{2'})]^{1/2} = [2c_p(T_1 - T_{2'})]^{1/2}$$
$$= [2(32.2 \text{ ft·lbm/lbf·s}^2)(778 \text{ ft·lbf/Btu})(0.124 \text{ Btu/lbm·°R})$$
$$\times (2260°R - 1560°R)]^{1/2}$$
$$= 2085 \text{ ft/s}$$

For the isentropic process 1–2, Eq. (7-24) gives

$$T_2 = T_1\left(\frac{p_2}{p_1}\right)^{(k-1)/k} = 2260\left(\frac{18}{50}\right)^{(1.667-1)/1.667}$$
$$= 1502°R = 1042°F$$

The adiabatic nozzle efficiency is

$$\eta_N = \frac{h_1 - h_{2'}}{h_1 - h_2} = \frac{c_p(T_1 - T_{2'})}{c_p(T_1 - T_2)} = \frac{2260 - 1560}{2260 - 1502}$$
$$= 0.92 = 92\%$$

7-9 AVAILABLE AND UNAVAILABLE ENERGY

The first law of thermodynamics allows the transformation completely from one form of energy into another. The second law of thermodynamics, however, forbids the complete transformation from heat into work. The portion of the heat supplied by a source that could be converted to work by a totally reversible heat engine operating between the source and the surroundings is called the *available energy* or available part of energy. The portion of heat not converted to work by the totally reversible heat engine is called the *unavailable energy* or unavailable part of energy.

Consider an energy reservoir at temperature T_H from which a quantity of heat Q is extracted to operate a Carnot heat engine, as depicted by the T–S diagram of Fig. 7-14. The total heat Q supplied by the energy reservoir is represented by the area 1–2–b–a in the figure. The lowest sink temperature is that of the atmosphere at T_0. The unavailable energy is given by

$$Q_{unav} = T_0 \, \Delta S = \text{area } 3\text{–}4\text{–}a\text{–}b$$

The available energy is then

$$Q_{av} = Q - T_0 \, \Delta S = \text{area } 1\text{–}2\text{–}3\text{–}4$$

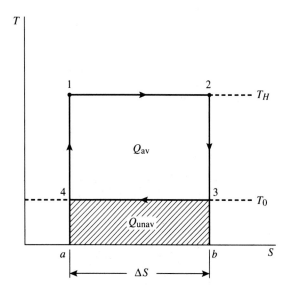

Figure 7-14 Available and unavailable energy for a constant-temperature heat supply.

If the reversible heat supply is at varying temperatures, as depicted by process 1–2 shown in the T–S diagram of Fig. 7-15, one can break up the heat-addition process into a series of infinitesimal processes each supplying a differential amount of heat $đQ$ at a temperature T to an elemental Carnot engine operating between the temperature T and the sink temperature T_0. The unavailable and available energies, respectively, for the differential heat supply $đQ$ are given by

$$đQ_{unav} = T_0 \, dS$$

$$đQ_{av} = đQ - T_0 \, dS$$

Entropy Change During Processes   Chap. 7

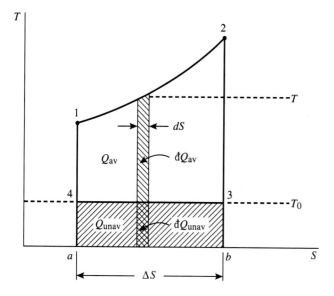

Figure 7-15 Available and unavailable energy for a varying-temperature heat supply.

Integrating the preceding equations for the entire process 1–2 results in the unavailable and available energies, respectively, for the total heat supply Q as follows:

$$Q_{unav} = T_0 \, \Delta S = \text{area } 3\text{–}4\text{–}a\text{–}b \qquad (7\text{-}41)$$

$$Q_{av} = Q - T_0 \, \Delta S = \text{area } 1\text{–}2\text{–}3\text{–}4 \qquad (7\text{-}42)$$

When heat is transferred from a higher-temperature body to a lower-temperature body, its available energy decreases. This is demonstrated in the $T\text{–}S$ diagram of Fig. 7-16, where a quantity of heat Q from an energy reservoir at temperature T_H as represented by area $1\text{–}2\text{–}b\text{–}a$ is transferred to another energy reservoir at a lower-temperature T_L so that

$$Q = \text{area } 1\text{–}2\text{–}b\text{–}a = \text{area } 5\text{–}6\text{–}d\text{–}c$$

or

$$T_H \, \Delta S_{12} = T_L \, \Delta S_{56}$$

Because

$$T_H > T_L$$

it follows that

$$\Delta S_{12} < \Delta S_{56}$$

so that

$$T_0 \, \Delta S_{12} < T_0 \, \Delta S_{56}$$

or

$$Q_{unav, \, 12} < Q_{unav, \, 56}$$

Therefore, we conclude that

$$Q_{av, \, 12} > Q_{av, \, 56}$$

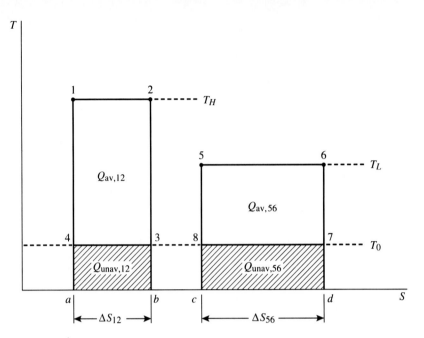

Figure 7-16 Heat transfer across a finite temperature difference.

The conclusion we arrived at here is an important one that implies that the usefulness of heat energy for producing work depends not only on the amount of heat supplied from a given source, but also on the temperature of the source.

Example 7-11

Two hundred Btu of heat are transferred from a heat reservoir at 550°F to 6 lbm of air initially at 15 psia and 150°F trapped in a closed rigid tank. The temperature of the surroundings is 40°F. (a) Determine the changes of entropy of the reservoir and the air. (b) How much of the energy removed from the reservoir is available energy? How much is unavailable? (c) How much of the energy added to the air in the tank is available energy? How much is unavailable? For air, $c_v = 0.173$ Btu/lbm·°F.

Solution The T–S diagrams for the reservoir (process 1–2) and for the air (process 3–4) are shown together in Fig. 7-17. The given data are $Q = 200$ Btu, $T_H = 550°F = 1010°R$, $m = 6$ lbm, $T_3 = 150°F = 610°R$, $p_3 = 15$ psia, $V_3 = V_4$, and $T_0 = 40°F = 500°R$.

For the energy reservoir: $Q = T_H \Delta S_{12}$, so that

$$\Delta S_{12} = \frac{Q}{T_H} = \frac{-200 \text{ Btu}}{1010°R} = -0.198 \text{ Btu/°R}$$

which means that the entropy of the energy reservoir decreases by 0.198 Btu/°R. The unavailable and available energies, respectively, of the heat removed from the energy reservoir are

$$Q_{\text{unav}} = T_0 \Delta S_{21} = 500(0.198) = 99.0 \text{ Btu}$$

$$Q_{\text{av}} = Q - Q_{\text{unav}} = 200 - 99 = 101 \text{ Btu}$$

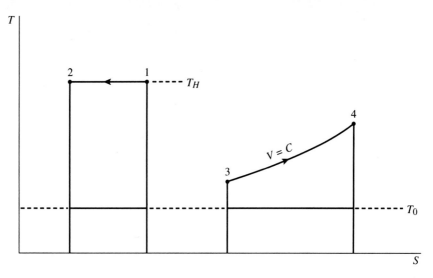

Figure 7-17 *T–S* diagram for Example 7-11.

For the air in the tank: Considering air as an ideal gas and the constant-volume heating process to be internally reversible, we have

$$\text{đ}Q = dU + p\cancel{dV}^{0} = dU$$

or

$$Q = \Delta U_{34} = mc_v(T_4 - T_3)$$

so that

$$T_4 = T_3 + \frac{Q}{mc_v} = 610 + \frac{200}{6(0.173)}$$

$$= 803°R = 343°F$$

From Eq. (7-18) for the constant-volume process,

$$\Delta S_{34} = m\left(c_v \ln \frac{T_4}{T_3} + R \cancel{\ln \frac{v_4}{v_3}}^{0}\right) = mc_v \ln \frac{T_4}{T_3}$$

$$= 6(0.173) \ln \frac{803}{610} = 0.285 \text{ Btu/°R}$$

which means that the entropy of the air increases by 0.285 Btu/°R. The unavailable and available energies, respectively, of the heat added to the air are

$$Q_{\text{unav}} = T_0 \Delta S_{34} = 500(0.285) = 142.5 \text{ Btu}$$

$$Q_{\text{av}} = Q - Q_{\text{unav}} = 200 - 142.5 = 57.5 \text{ Btu}$$

Notice that although the total amount of energy before leaving the reservoir and after entering the air is the same, the available part of the energy before leaving the reservoir is greater than the available part of the energy after entering the air. This means that more work can be obtained by the same amount of thermal energy when this energy is associated with the higher-temperature reservoir than with the lower-temperature air.

7-10 SUMMARY

There are two basic relations that combine the first and second laws as applied to a simple compressible system of fixed mass. They are

$$T \, dS = dU + p \, dV$$

$$T \, dS = dH - V \, dp$$

The general expression of dS for an open system or control volume can be written as

$$dS_{\text{open sys}} = \sum \left(\frac{dQ}{T}\right) + \sum_{\text{inlet}} (s \, dm) - \sum_{\text{exit}} (s \, dm) + dS_{\text{prod}}$$

In terms of entropy production, we write

$$dS_{\text{prod}} = dS_{\text{open sys}} - \sum \left(\frac{dQ}{T}\right) - \sum_{\text{inlet}} (s \, dm) + \sum_{\text{exit}} (s \, dm) \geq 0$$

where the equality applies to internally reversible processes and the inequality to internally irreversible processes. Notice that for a steady-flow, steady-state process, we have $dS_{\text{open sys}} = 0$.

For a single-stream, simple-compressible open system undergoing a steady-flow, steady-state internally reversible process, we have the following momentum equation:

$$dw_{\text{cv}} = -v \, dp - dKE - dPE$$

With negligible changes in kinetic and potential energies, after integration between states 1 and 2, this equation becomes

$$w_{\text{cv}} = -\int_{1}^{2} v \, dp$$

This work is represented by the area on the left of the process curve on a p–v diagram.

There are three important ds equations for ideal gases. They are

$$ds = c_v \frac{dT}{T} + R \frac{dv}{v}$$

$$ds = c_p \frac{dT}{T} - R \frac{dp}{p}$$

$$ds = c_v \frac{dp}{p} + c_p \frac{dv}{v}$$

When the specific heats of an ideal gas are considered constant, the preceding equations can be written in the following forms between states 1 and 2:

$$\Delta s = s_2 - s_1 = c_v \ln \frac{T_2}{T_1} + R \ln \frac{v_2}{v_1}$$

$$\Delta s = s_2 - s_1 = c_p \ln \frac{T_2}{T_1} - R \ln \frac{p_2}{p_1}$$

$$\Delta s = s_2 - s_1 = c_v \ln \frac{p_2}{p_1} + c_p \ln \frac{v_2}{v_1}$$

These equations can be used to evaluate Δs between two equilibrium states of an ideal gas without any knowledge of the process connecting them.

The $p–v$ relation for an internally reversible adiabatic process (or isentropic process) of an ideal gas with constant specific heats is

$$pv^k = \text{constant}$$

where $k = c_p/c_v$. Written between states 1 and 2, we have

$$\left(\frac{T_2}{T_1}\right)_{s=\text{const}} = \left(\frac{p_2}{p_1}\right)^{(k-1)/k}_{s=\text{const}} = \left(\frac{v_1}{v_2}\right)^{k-1}_{s=\text{const}}$$

For ideal gases with variable specific heats, the entropy change is written as

$$\Delta s = s_2 - s_1 = \phi_2 - \phi_1 - R \ln \frac{p_2}{p_1}$$

In the special case of an isentropic process, we have

$$\left(\frac{p_2}{p_1}\right)_{s=\text{const}} = \frac{p_{r2}}{p_{r1}}$$

and

$$\left(\frac{v_2}{v_1}\right)_{s=\text{const}} = \frac{v_{r2}}{v_{r1}}$$

The variables ϕ, p_r, and v_r (all functions of temperature only) are listed in the ideal-gas property tables for many gases in Appendix 3.

The entropy change of an incompressible substance is written as

$$\Delta s = s_2 - s_1 = c \ln \frac{T_2}{T_1}$$

The adiabatic efficiencies of steady-flow devices are defined by the following:
For turbines,

$$\eta_T = \frac{\text{actual turbine work}}{\text{isentropic turbine work}}$$

$$= \frac{h_1 - h_{2'}}{h_1 - h_2} \quad \text{when } \Delta KE = \Delta PE = 0$$

For compressors (or pumps),

$$\eta_c = \frac{\text{isentropic compressor work}}{\text{actual compressor work}}$$

$$= \frac{h_2 - h_1}{h_{2'} - h_1} \quad \text{when } \Delta KE = \Delta PE = 0$$

For nozzles,

$$\eta_N = \frac{\text{actual KE at nozzle exit}}{\text{isentropic KE at nozzle exit}}$$

$$= \frac{V_{2'}^2/2}{V_2^2/2} = \frac{h_1 - h_{2'}}{h_1 - h_2} \quad \text{when } V_1 \ll V_{2'}$$

Sec. 7-10   Summary

For diffusers,

$$\eta_D = \frac{\text{isentropic } \Delta h}{\text{actual } \Delta h} = \frac{h_2 - h_1}{h_{2'} - h_1}$$

In the preceding equations for adiabatic efficiencies, h_2 and $h_{2'}$ are the enthalpy values at the exit state for isentropic and actual processes, respectively, between the same initial state and the same final pressure.

The pressure coefficient K_p of a diffuser is defined as

$$K_p = \frac{\text{actual } \Delta p}{\text{isentropic } \Delta p} = \frac{p_{2'} - p_1}{p_2 - p_1}$$

where the actual Δp and isentropic Δp are pressure rises for the same enthalpy change occurring in both the actual and isentropic processes. For an ideal gas, the isentropic Δp is the pressure rise if the diffuser is operated isentropically between the same initial state and the same final temperature of the actual diffuser.

The portion of the heat supplied by a source that could be converted to work by a totally reversible heat engine operating between the source and the surroundings is called the available energy or available part of energy. The portion of heat not converted to work by the totally reversible heat engine is called the unavailable energy or unavailable part of energy. The unavailable energy is given by

$$Q_{\text{unav}} = T_0 \, \Delta S$$

The available energy is given by

$$Q_{\text{av}} = Q - T_0 \, \Delta S$$

where Q is the total heat supply, T_0 is the lowest temperature in the surroundings, and ΔS is the entropy change of the heat-supply process.

PROBLEMS

7-1. A movie projector uses a 500-W light bulb. See Fig. P7-1. The bulb is cooled by a fan that blows room air at 20°C at a rate of 1.0 kg/min. Assuming the equilibrium surface temperature of the bulb to be 350°C, determine (a) the outlet temperature of the cooling air, and (b) the entropy production rate due to irreversibility effects.

7-2. One-half kilogram per second of saturated water vapor at 100°C and 0.2 kg/s of saturated liquid water at 100°C are mixed in an uninsulated chamber under constant pressure. See Fig. P7-2. During the mixing process, there are 75 kJ/s of heat rejected from the chamber, and 1.0 hp input to the

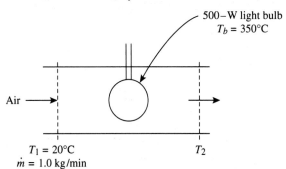

Figure P7-1

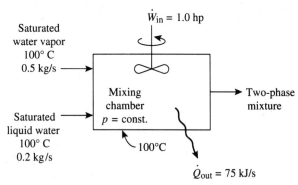

Figure P7-2

Entropy Change During Processes   Chap. 7

chamber. Determine the condition of the exit flow and the entropy production rate for the mixing process. Assume that the system boundary is isothermal at 100°C.

7-3. An open heater mixes steam and liquid water to form heated water. See Fig. P7-3. The incoming steam is at 100 psia and 98% quality; the incoming water is at 100 psia and 80°F; and the outgoing heated water has a flow rate of 25,000 lbm/h and is at 95 psia and 290°F. Assuming a steady-flow, steady-state adiabatic mixing process, determine the hourly entropy production for the mixing process. Changes in KE and PE can be neglected.

7-4. Steam enters a nozzle at 1000 psia and 1200°F and discharges at 14.7 psia and 400°F. The process is adiabatic. For steady flow and steady state, calculate the rate of entropy production for the process. The mass rate of flow is 1 lbm/s.

7-5. Steam enters an adiabatic turbine at 1.5 MPa and 700°C and exits the turbine at 0.2 MPa and 400°C. The process is steady flow and steady state. The

mass-flow rate of steam is 6.3 kg/s. Determine the output power and the entropy production rate of the turbine.

7-6. Air at 101.3 kPa and 20°C enters a compressor at a rate of 5 kg/min and exits at 1.0 MPa. The compression process is adiabatic and follows the relation $pv^{1.47}$ = constant. Neglecting ΔKE and ΔPE, determine (a) the exit temperature, (b) the power requirement, and (c) the entropy production rate for the process.

7-7. Helium is cooled in a piping system from 200°C to 132°C at a flow rate of 2.5 kg/s. Due to friction effects in the pipe, the pressure drops from 200 kPa to 180 kPa. Determine the cooling required and the rate of entropy change of the helium.

7-8. An insulated rigid tank of 2-m³ volume is connected through a valve to a large argon-supply line that is maintained at a constant pressure of 500 kPa and a constant temperature of 27°C. See Fig. P7-8. The tank is initially evacuated. The valve is opened until the tank is filled with argon at

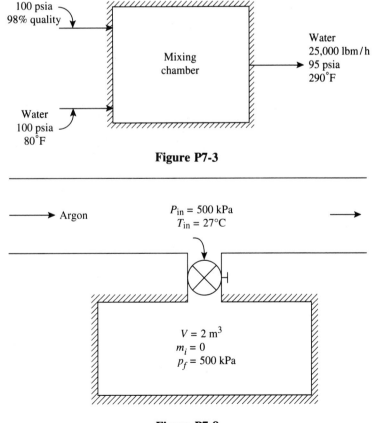

Figure P7-3

Figure P7-8

the supply-line pressure. Determine the final temperature of the argon in the tank and the amount of entropy production within the tank.

7-9. Solve Prob. 7-8 if the tank initially contains argon at 100 kPa and 27°C. See Fig. P7-9.

7-10. An 0.6-m³ insulated tank (tank A) is connected with an 0.3-m³ uninsulated tank (tank B) by a pipeline with a valve. See Fig. P7-10. Tank A is initially filled with steam at 1.4 MPa and 300°C. Tank B is initially filled with steam at 0.2 MPa and 200°C. When the valve is opened, steam flows slowly from tank A to tank B until the temperature of tank A is 250°C, at which time the valve is closed. During this time, heat is transferred from

tank B to the surroundings at 23°C such that the temperature in tank B remains at 200°C. Assume that the steam remaining in tank A has undergone an internally reversible adiabatic process. Determine (a) the final pressure in each tank, (b) the final mass in tank B, (c) the total heat transfer during the process, (d) the entropy change of the system, and (e) the entropy change of the surroundings.

7-11. A 0.25-m³ uninsulated cylinder containing 0.3 kg of nitrogen initially at 530 K has an adiabatic, frictionless piston of negligible volume and weight initially at its lowest position. See Fig. P7-11. The bottom of the cylinder is attached through a

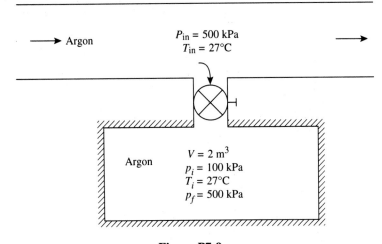

Argon

$P_{in} = 500$ kPa
$T_{in} = 27°C$

Argon

$V = 2\ \text{m}^3$
$p_i = 100$ kPa
$T_i = 27°C$
$p_f = 500$ kPa

Figure P7-9

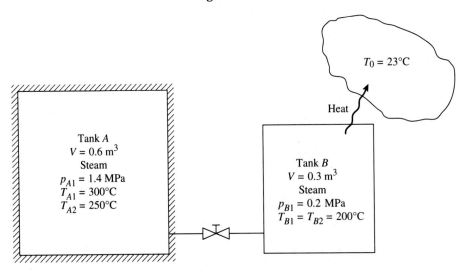

$T_0 = 23°C$

Heat

Tank A
$V = 0.6\ \text{m}^3$
Steam
$p_{A1} = 1.4$ MPa
$T_{A1} = 300°C$
$T_{A2} = 250°C$

Tank B
$V = 0.3\ \text{m}^3$
Steam
$p_{B1} = 0.2$ MPa
$T_{B1} = T_{B2} = 200°C$

Figure P7-10

Entropy Change During Processes   Chap. 7

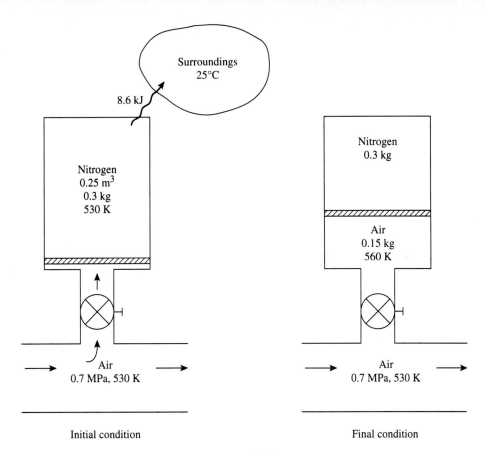

Surroundings
25°C

8.6 kJ

Nitrogen
0.25 m³
0.3 kg
530 K

Air
0.7 MPa, 530 K

Initial condition

Nitrogen
0.3 kg

Air
0.15 kg
560 K

Air
0.7 MPa, 530 K

Final condition

Figure P7-11

valve to a large air-supply line that is at a constant pressure of 0.7 MPa and a constant temperature of 530 K. When the valve is opened, air enters the cylinder and pushes the piston upward until 0.15 kg of air has entered with a final temperature of 560 K. There is 8.6 kJ of heat transfer from the nitrogen to the surroundings at 25°C. Determine the entropy production of the process. Use ideal-gas property tables for nitrogen and air in your solution.

7-12. An uninsulated rigid tank of 0.1 m³ volume containing Refrigerant-12 initially at 1 MPa and 100% quality is connected through a valve to a large Refrigerant-12 supply line that is maintained at a constant pressure of 1.4 MPa and a constant temperature of 30°C. See Fig. P7-12. The valve is opened until the tank is filled with saturated liquid Refrigerant-12 at 1.2 MPa. Determine (a) the amount of the refrigerant that entered the tank, (b) the amount of heat transfer from the surroundings at 50°C, and (c) the entropy production during the process.

7-13. Air at 96.0 kPa, 20°C, and 0.876 m³/kg with a low velocity enters a wind tunnel and flows steadily through a nozzle into the text section. See Fig. P7-13. The expansion of the air in the nozzle follows the relation $pv^{1.4} = $ constant. What pressure must the air be expanded in the nozzle to achieve a velocity of 220 m/s in the test section?

7-14. Water is pumped in a steady internally reversible process at a volume flow rate of 0.5 ft³/s from an open tank into a vessel that is maintained at a pressure of 50 psig. See Fig. P7-14. The height at entry to the vessel is 20 ft above the water surface of the tank, and the velocity of the water at entry is 10 ft/s. The density of water is 62.4 lbm/ft³. Calculate the horsepower required to drive the pump.

7-15. Saturated liquid Refrigerant-12 at 0°C enters a 0.675-cm diameter pipe with a velocity of 20 m/s, and leaves the pipe at 20°C with a specific volume of 0.016 m³/kg. See Fig. P7-15. Heat is transferred to the flowing refrigerant from the surroundings at 50°C. Determine (a) the state of the refrigerant at exit from the pipe, (b) the velocity at

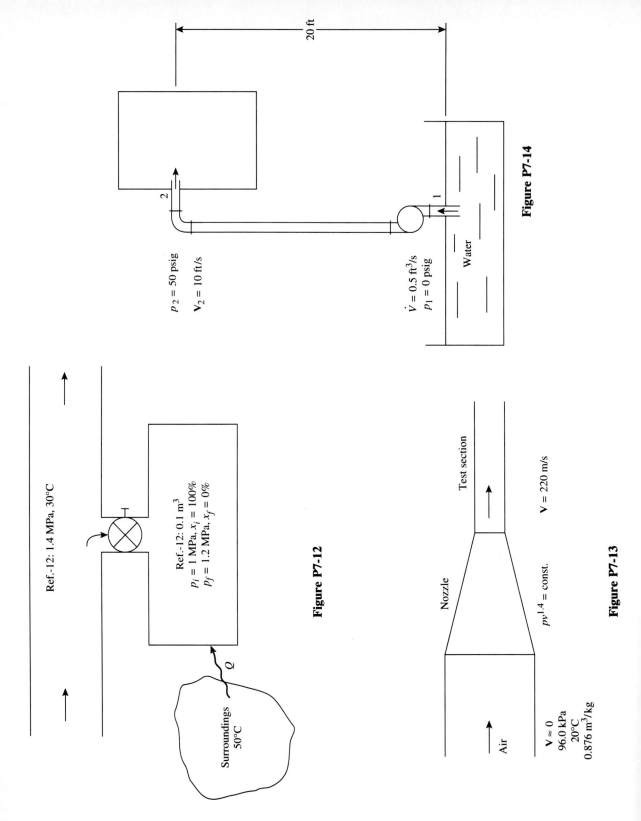

Figure P7-14

$p_2 = 50$ psig

$\mathbf{v}_2 = 10$ ft/s

20 ft

2

1

$\dot{V} = 0.5$ ft^3/s

$p_1 = 0$ psig

Water

Ref.-12: 1.4 MPa, 30°C

Ref.-12: 0.1 m^3

$p_i = 1$ MPa, $x_i = 100\%$

$p_f = 1.2$ MPa, $x_f = 0\%$

Figure P7-12

Q

Surroundings 50°C

Nozzle

Test section

$pv^{1.4} = $ const.

$\mathbf{V} = 220$ m/s

Air

$\mathbf{V} \approx 0$

96.0 kPa

20°C

0.876 m^3/kg

Figure P7-13

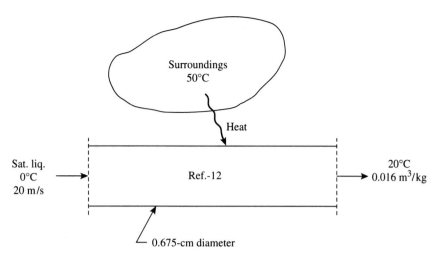

Figure P7-15

the exit, in m/s, (c) the heat transfer rate, in kJ/s, (d) the entropy change of the refrigerant, in kJ/kg·K, and (e) the total entropy change of the refrigerant and the surroundings together, in kJ/kg·K. Is the process reversible, irreversible, or impossible?

7-16. Oil with a constant specific volume of 0.001177 m³/kg is pumped from a pressure of 0.70 bar to a pressure of 1.20 bars, and the outlet lies 2.0 m above the inlet. See Fig. P7-16. The fluid temper-

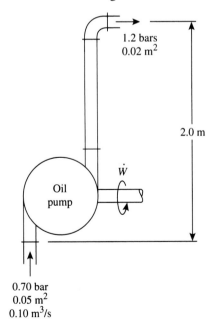

Figure P7-16

ature is 15°C. It flows at a rate of 0.10 m³/s through an inlet cross-sectional area of 0.050 m². The outlet area is 0.020 m². The local gravity is 9.8 m/s². Assume adiabatic and frictionless flow. Determine the required input power to the pump, in kilowatts.

7-17. Five pounds of steam occupying 10.466 ft³ at 250 psia expand to a pressure of 10 psia according to the relation $pV^{1.25}$ = constant, where p is pressure, and V is total volume. Calculate (a) ΔV, (b) quality at the final condition, (c) ΔH, (d) ΔS, (e) ΔU, (f) W and Q if the expansion is an internally reversible closed-system process, and (g) W and Q if the expansion is an internally reversible steady-flow process with negligible changes in KE and PE.

7-18. Air contained in a piston–cylinder arrangement is originally at 500 kPa and 30°C and has a volume of 0.10 liter. The air expands irreversibly to occupy a volume of 0.19 liter. The final equilibrium temperature is 30°C. Determine the change of entropy of the air.

7-19. Carbon dioxide at 50 MPa and 207°C expands through a horizontal uninsulated nozzle to 1.5 MPa in an isothermal steady-flow, steady-state process. See Fig. P7-19. The surface temperature of the nozzle is 307°C. The rate of entropy production has a value of 10% of the rate of heat transfer. Assuming carbon dioxide to be an ideal gas with constant specific heats, determine (a) the rate of heat transfer, and (b) the change in kinetic energy of the CO_2, per kg of CO_2 flowing through the nozzle.

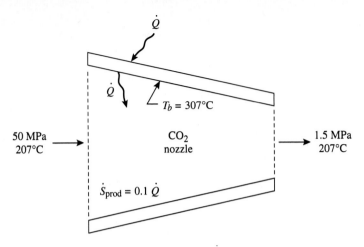

$\dot{Q}$

$T_b = 307°C$

50 MPa
207°C

CO_2
nozzle

1.5 MPa
207°C

$\dot{S}_{prod} = 0.1\,\dot{Q}$

Figure P7-19

7-20. A concentrating solar collector directs sunlight onto a long straight pipe, giving the pipe 0.15 kW of power input per meter of the pipe. See Fig. P7-20. Water flows through the pipe at the mass-flow rate of 2.3 kg/h, entering as a saturated liquid at 150°C and leaving as a vapor at 140 kPa and 150°C. In steady-flow, steady-state conditions, (a) how long is the pipe, and (b) what is the rate of entropy production? Show whether this process violates the second law of thermodynamics.

7-21. A 15-cm³ cylinder contains nitrogen at 70 atm and 20°C. This cylinder is placed in a 10,000 cm³ evacuated insulated chamber. See Fig. P7-21. A crack develops on the small cylinder and the nitrogen ex-

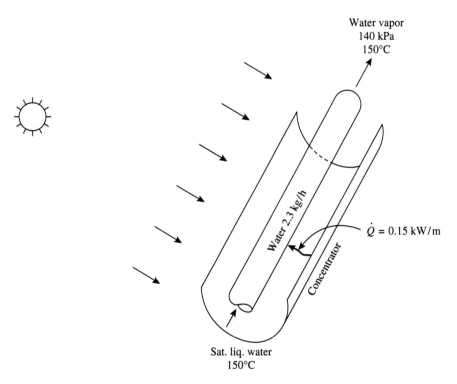

Water vapor
140 kPa
150°C

Water 2.3 kg/h

$\dot{Q} = 0.15$ kW/m

Concentrator

Sat. liq. water
150°C

Figure P7-20

Entropy Change During Processes   Chap. 7

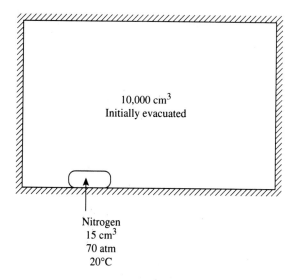

10,000 cm³
Initially evacuated

Nitrogen
15 cm³
70 atm
20°C

Figure P7-21

pands to fill the entire chamber. Determine (a) the entropy change for the nitrogen and (b) the work required to compress the nitrogen reversibly back to the initial state. Assume that nitrogen behaves as an ideal gas.

7-22. Air at 14.7 psia and 70°F fills a 300-ft³ insulated room. A 1-kW electric heater is placed in the room and is allowed to operate for 5 min. See Fig. P7-22. Determine the final pressure and temperature of the room air. What is the change in entropy of the air? For air, $c_p = 0.24$ Btu/lbm·°R and $c_v = 0.1715$ Btu/lbm·°R.

7-23. A closed system consists of 1 lbm of an ideal gas with $c_p = 0.24$ Btu/lbm·°R and $c_v = 0.171$ Btu/lbm·°R initially at 20 psia and 600°R. The volume

doubles as the system undergoes each of the following internally reversible processes: (a) constant pressure, (b) constant temperature, (c) constant internal energy, (d) constant enthalpy, (e) constant entropy, (f) $pv^2 = $ constant, and (g) $p/v = $ constant. Determine for each process, (1) work done, in ft·lbf, (2) heat transferred, in Btu, (3) change in internal energy, in Btu, (4) change in enthalpy, in Btu, and (5) change in entropy, in Btu/°R. Record the results in tabular form.

7-24. An ideal gas with a molecular mass of 39 has a specific heat at constant pressure c_p that depends upon the temperature T in accordance with the equation

$$c_p = 0.120 + 10/T$$

where c_p is in Btu/lbm·°R, and T in °R. If a pound of this gas initially at 20 psia and 90°F changes to a state where its pressure and temperature are 60 psia and 220°F, by how much does its entropy change?

7-25. The specific heat of air at constant pressure is given by

$$c_p = 0.342 - 1.25T^{-1/2} - 82.42T^{-1} + 31,200T^{-2}$$

where T is the temperature in °R, and c_p is in Btu/lbm·°R. How many Btu are required to heat 10 lbm of air reversibly at constant pressure in a closed system from 140°F to 220°F? What is the entropy increase in this process? Check your answers by means of the air table.

7-26. A rigid tank contains 1.35 kg of air ($c_v = 0.716$ kJ/kg·K). If adding 90 kJ of heat to the air causes its entropy to increase by 0.188 kJ/K, what are the initial and final temperatures of the air?

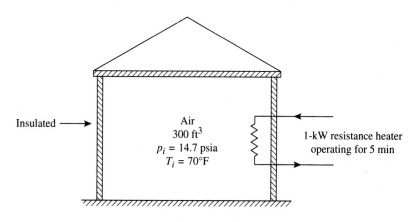

Insulated

Air
300 ft³
$p_i = 14.7$ psia
$T_i = 70°F$

1-kW resistance heater
operating for 5 min

Figure P7-22

7-27. An ideal gas with a molecular mass of 29.5 and $c_p = 0.23$ Btu/lbm·°R is compressed in an engine cylinder from 36 to 5.5 in.³ Its initial pressure is 14 psia and initial temperature is 80°F. Determine the work required if compression is internally reversible and adiabatic.

7-28. Oxygen expands isentropically in a closed system from 45 psia, 60°F, to 30 psia. Determine the final temperature, the change of enthalpy, the heat transferred, and the work done per pound of the gas.

7-29. It is suggested to expand air from 30 psia, 180°F, to 15 psia, 40°F, by means of an adiabatic process. (a) Is the expansion possible? (b) For the same initial state and the same final pressure, what is the lowest temperature that could be achieved by an adiabatic expansion? (c) Is there a maximum final temperature, and, if so, what is it?

7-30. Helium enters a diffuser at 4.91 psia and 40°F with a velocity of 4200 ft/s. For an isentropic process to 14.7 psia, what are the final velocity and the final temperature?

7-31. In a gas turbine, 1000 kg/min of air are expanded in an internally reversible adiabatic process from 1100 K, 1.5 MPa, to 100 kPa. Neglecting changes in kinetic and potential energies, determine the power developed. Consider air as an ideal gas with $c_p = 1.005$ kJ/kg·K and $k = 1.4$.

7-32. Oxygen (considered as an ideal gas with $k = 1.4$) expands in an internally reversible adiabatic process in a steady-flow system from 45 psia, 60°F, to 30 psia. Determine the final temperature, the change of specific internal energy, the heat transferred per pound of gas, and the shaft work done per pound of gas. Changes in elevation and in kinetic energy are negligible.

7-33. One kg of an ideal gas with $c_p = 841.8$ J/kg·K and $R = 188.9$ J/kg·K expands from 750 kPa, 30°C, to 120 kPa in an internally reversible polytropic process with $n = 1.3$. Determine the change in entropy. Draw the p–v and T–s diagrams for the process.

7-34. Air enters a diffuser at 100 kPa and 5°C with a velocity of 900 m/s. For an internally reversible and adiabatic flow to 200 kPa, what are the final velocity and the final temperature? Use (a) $k = 1.4$ and $c_p = 1.005$ kJ/kg·K, and (b) the air-table values.

7-35. Air at 30°C and 101.3 kPa is compressed in an internally reversible adiabatic process to 1.013 MPa. Determine its initial specific volume and final temperature. For a steady-state, steady-flow process,

determine the work per kilogram in kJ and the power required for 1 kg/s of flow. Use (a) air-table data and (b) $c_p = 1.005$ kJ/kg·K and $k = 1.4$.

7-36. Air initially at 100 psia and 1000°F expands reversibly and adiabatically to 15 psia. Determine the change of enthalpy and specific volume. (a) Use $k = 1.4$ and $c_p = 0.24$ Btu/lbm·°R. (b) Use the air tables.

7-37. Air enters a diffuser at a pressure of 8 psia, a temperature of 20°F, and a velocity of 800 ft/s. It leaves the diffuser at a velocity of 400 ft/s. For isentropic flow, determine (a) the temperature at the outlet, in °F, (b) the outlet pressure, in psia, and (c) the ratio of the outlet area to the inlet area. Use air tables to solve this problem.

7-38. Air at 14.0 psia, 80°F, is drawn into a compressor and is compressed isentropically to 42 psia to fill a 20-ft³ insulated tank, which initially contains air at 14.0 psia, 80°F. See Fig. P7-38. Determine the total work required.

7-39. A small high-speed turbine operating adiabatically by compressed air produces 0.1 kW. See Fig. P7-39. The turbine has an adiabatic efficiency of 90%. The air enters the turbine at 4 bars, 80°C, and exits at 1 bar. Assuming the change in KE to be small, determine the required mass-flow rate.

7-40. Air at 75 psia and 100°F expands in an adiabatic engine to 15 psia. (a) If the expansion is internally reversible, what horsepower is developed when the air flow is 100 lbm/min? (b) If the air engine has an efficiency of 65%, what air flow per minute is required to develop the same horsepower as in (a), and what is the temperature of the exhaust air?

7-41. Steam at a pressure of 2 MPa, a temperature of 450°C, with a velocity of 80 m/s enters an adiabatic turbine and expands to 70 kPa at a mass-flow rate of 10,000 kilograms per hour. The adiabatic efficiency of the turbine is 80%. Determine (a) the turbine inlet area, (b) the steam exit temperature, and (c) the power output.

7-42. Saturated ammonia vapor at −10°C is compressed in a steady-flow adiabatic compressor to 400 kPa and 20°C. Determine the adiabatic compressor efficiency and the entropy change, per kilogram of the fluid.

7-43. An ideal gas with a molecular mass of 28.7 and an average $c_p = 0.270$ Btu/lbm·°R expands in an adiabatic turbine from 80 psia, 1200°F, to 15 psia. The turbine has an adiabatic efficiency of 84%. Determine the cubic feet per minute entering and leaving if the turbine produces 10,000 hp.

Entropy Change During Processes   Chap. 7

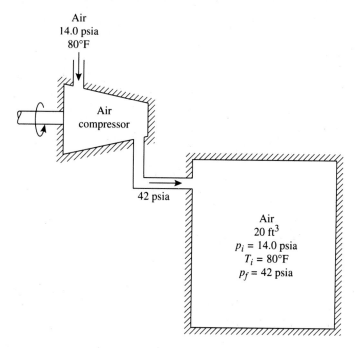

Figure P7-38

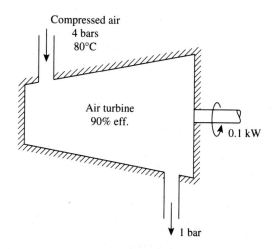

Figure P7-39

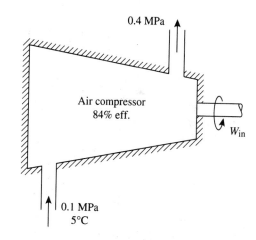

Figure P7-44

7-44. Air enters an adiabatic compressor at 0.1 MPa, 5°C, and leaves at 0.4 MPa. See Fig. P7-44. The compressor has an efficiency of 84%. Determine the work of compression per kg of air. Neglect changes in KE and PE. Determine also the exit air temperature and the change in entropy.

7-45. Carbon monoxide at 10 psia and 100°F is compressed adiabatically to 115 psia in an 80% effi-

cient compressor. Determine the ft³ per min entering and leaving if the compressor required 496 hp.

7-46. Determine the adiabatic compression efficiency and the entropy change for a steady-flow adiabatic machine in which air is compressed from 0.1 MPa and 27°C to 0.5 MPa and 250°C. See Fig. P7-46. Consider air as an ideal gas with variable specific heats.

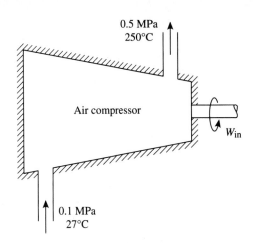

0.5 MPa
250°C

Air compressor

W_{in}

0.1 MPa
27°C

Figure P7-46

7-47. Air at 14.7 psia and 100°F is compressed adiabatically to 50 psia in a compressor with 80% efficiency. Calculate the work required per pound of air. (a) Use c_p = 0.24 Btu/lbm·°R and k = 1.4. (b) Use the air tables.

7-48. Air at 10 atm and 1040°F is expanded adiabatically to 1 atm in a turbine with 80% adiabatic efficiency. For a constant mass-flow rate of 10 lbm/s, determine the horsepower output of the turbine. (a) Use the air tables. (b) Use constant values of c_p and k at 100°F, which are c_p = 0.240 Btu/lbm·°R and k = 1.40. (c) Use constant values of c_p and k at 1000°F, which are c_p = 0.263 Btu/lbm·°R and k = 1.353.

7-49. Air initially at 100 psia, 500°F, with a total volume of 1 ft³ undergoes the following internally reversible processes in a closed system: In process 1–2, it expands at constant pressure to 3 ft³. In process 2–3, it expands according to $pv^{1.50}$ = constant to 500°F. In process 3–1, it follows a constant-temperature process to complete the cycle. Draw the cycle in p–v and T–s diagrams and determine (a) the heat received and heat rejected in the cycle, (b) the cycle thermal efficiency, and (c) if the gas had been helium instead of air, describe qualitatively any differences in the solution of this problem.

7-50. Argon (considered as an ideal gas with c_p = 0.125 Btu/lbm·°R and k = 1.67) undergoes the following three-process cycle in a closed system: Process 1–2: An internally reversible adiabatic expansion from 100 psia, 540°F, 0.5 ft³ to 2 ft³. Process 2–3: An internally reversible isothermal compression back to 100 psia. Process 3–1: An irreversible

constant-pressure expansion with zero heat addition. Draw the cycle in the p–v and T–s planes and determine the work done, in ft·lbf, for each of the three processes, and the total work of the cycle.

7-51. Nitrogen gas with c_p = 29.1 kJ/kgmole·K is made to undergo the following internally reversible processes in a closed system: Process 1–2: The gas initially at 2 atm and 350 K is heated at constant volume to a temperature of 500 K. Process 2–3: It is then expanded adiabatically back to its initial temperature of 350 K. Process 3–1: It is finally compressed isothermally back to its initial pressure of 2 atm. Draw the p–v and T–s diagrams for the cycle. For 1 kgmole of nitrogen, calculate the work done, the heat transferred, the internal energy change, and the entropy change for each of the three processes.

7-52. A pound of air undergoes an internally reversible expansion at a constant temperature of 120°F. If at the end of the process, the air were compressed isentropically back to its initial pressure, its temperature would rise to 200°F. How much heat is added to the air during the isothermal process?

7-53. Helium gas contained in a closed system initially at 100 kPa and 20°C is compressed in an internally reversible isothermal process to 600 kPa, after which the gas is expanded back to the initial pressure of 100 kPa in an internally reversible adiabatic process. Show the process on a T–s diagram and calculate the final temperature and the net work per kilogram of helium.

7-54. One pound of an ideal gas with c_p = 0.2397 Btu/lbm·°R and c_v = 0.1711 Btu/lbm·°R occupying 1 ft³ at 200 psia changes to 100 psia at constant volume and then expands in an internally reversible and adiabatic process to 4 ft³. Draw the processes on T–s and p–v diagrams and determine the final temperature and the total change in entropy.

7-55. Air in a closed system expands isentropically from 3 MPa and 200°C to two times its initial volume, and then cools at constant volume until the pressure drops to 0.8 MPa. Calculate the work done and the heat transferred per kg of air.

7-56. One-tenth pound of air is compressed isentropically in a closed system from 12 psia, 140°F, to 60 psia, and is then expanded at constant pressure to the original volume. Draw the processes on p–v and T–s planes. Calculate the heat transfer and the work for the total path using (a) constant specific heats and (b) air-table values.

7-57. One gram of air initially at a pressure of 0.2 MPa and a temperature of 25°C is heated at constant volume until its pressure reaches 0.4 MPa. The air is then cooled at constant pressure until its entropy is the same as it was when the air was at 0.2 MPa and 25°C. Determine the final temperature of the air. All processes are assumed to be internally reversible.

7-58. A cycle involving 1 lbm of an ideal gas with $k = 1.4$ and $R = 53.3$ ft·lbf/lbm·°R is made up of the following internally reversible processes in a closed system: In process 1–2, the gas is compressed isentropically from $p_1 = 14.7$ psia, $v_1 = 14$ ft^3/lbm, to $p_2 = 2p_1$. In process 2–3, the gas is heated at constant pressure p_2 until $v_3 = v_1$. In process 3–1, the gas is cooled at constant volume. Calculate work and heat for each of the three processes and the thermal efficiency of the cycle. Draw the p–v and T–s diagrams.

7-59. A variable specific-heat air cycle in a closed system is composed of the following four processes: Process 1–2: Isentropic compression from $p_1 = 110$ kPa, $T_1 = 310$ K, $v_1 = 0.810$ m^3/kg, to $p_2 = 320$ kPa. Process 2–3: Constant-pressure heating to $T_3 = 1060$ K. Process 3–4: Constant-volume cooling to $p_4 = 110$ kPa. Process 4–1: Constant pressure back to state 1. Draw the p–v and T–s diagrams of the cycle and determine (a) p, T, and v of the four state points, (b) heat and work quanti-

ties for each of the four processes, and (c) the cycle thermal efficiency.

7-60. The properties of saturated liquid mercury at $p_0 = 0.01$ psia are $T_0 = 233$°F, $h_0 = 6.668$ Btu/lbm, $s_0 = 0.01137$ Btu/lbm·°R, and $v_0 = 0.00121$ ft^3/lbm. Estimate the enthalpy and entropy of liquid mercury at 1 atm and 100°F. The specific heat of liquid mercury at 1 atm and between 50°F and 600°F is 0.0325 Btu/lbm·°R.

7-61. Water enters the nozzle on a garden hose at 95 psia and 60°F with a mass-flow rate of 0.2 lbm/s and exits at 14.7 psia. The nozzle has an inlet diameter of 1.0 inch and an exit diameter of 0.25 inch. See Fig. P7-61. Assuming water to be incompressible and flow to be adiabatic, determine the exit temperature and the rate of entropy change.

7-62. A 30-kg iron block and a 20-kg copper block, both initially at 80°C, are dropped at the same time into a large lake that is at 20°C. See Fig. P7-62. Determine the total entropy change when thermal equilibrium is established between the blocks and the lake water.

7-63. A 20-kg copper block initially at 80°C is immersed in an insulated tank that contains 150 kg of water initially at 25°C. See Fig. P7-63. Determine the final equilibrium temperature and the total entropy change for this process.

7-64. A 4.0-g lead bullet at 80°C traveling at 900 m/s impacts a very rigid wall adiabatically. See Fig. P7-64. The specific heat of lead at the mean temperature of the bullet is 167 J/kg·K. Neglecting any energy imparted to the wall, determine the temperature of the bullet immediately after the impact, and the entropy change of the bullet.

7-65. One kilogram of steam contained in a closed system is heated in an internally reversible process at a constant pressure of 0.70 MPa from 180°C to 250°C. The source of heat is a reservoir at 300°C. (a) What is the available energy of the heat before it leaves the energy reservoir? (b) What is the

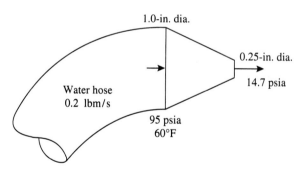

1.0-in. dia.

0.25-in. dia.

14.7 psia

Water hose
0.2 lbm/s

95 psia
60°F

Figure P7-61

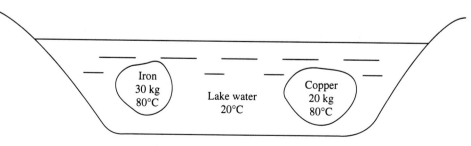

Iron
30 kg
80°C

Lake water
20°C

Copper
20 kg
80°C

Figure P7-62

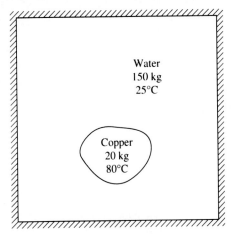

Water
150 kg
25°C

Copper
20 kg
80°C

Figure P7-63

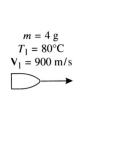

$m = 4$ g
$T_1 = 80°C$
$\mathbf{V}_1 = 900$ m/s

$\mathbf{V}_2 = 0$

Figure P7-64

available energy of the heat after being received by the steam? The temperature of the surroundings is 27°C.

7-66. After having done 400 Btu of work per pound of gas in an internal combustion engine, the combustion products leave the engine at 1500°F and 1 atm. The constant-pressure specific heat of the gases is 0.26 Btu/lbm·°F. The atmosphere is at 80°F. Determine (a) the loss of available energy in throwing away the exhaust gases, and (b) the ratio of the lost available energy to the engine work.

7-67. Write a computer program to solve Prob. 7-25, using the c_p equation given in the problem.

7-68. Write a computer program to determine the entropy production rate inside an insulated counterflow heat exchanger when the mass-flow rate, temperature, and specific heat of all incoming and outgoing fluid streams in steady flow, steady state are known. See Fig. P7-68. One fluid is an incompressible liquid and the other fluid is a constant specific-heat ideal gas. Neglect changes in kinetic and potential energies. The professor may assign different input variables with proper units (either SI or English units) to different students in the class for test runs of their programs.

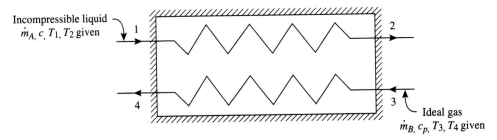

Incompressible liquid
$\dot{m}_A, c, T_1, T_2$ given

1

2

4

3

Ideal gas
$\dot{m}_B, c_p, T_3, T_4$ given

Figure P7-68

Entropy Change During Processes   Chap. 7

8

Availability and Irreversibility

In the last chapter, we defined available energy to be the portion of a quantity of heat that could be converted to work. That definition is particularly convenient in determining the usefulness of energy transfer as heat. In this chapter, we introduce the more general concept of availability of energy regarding the maximum work available from a substance in any specified state. Together with the introduction of the concept and procedure for evaluating maximum work and irreversibility of processes, this chapter emphasizes the optimization of energy utilization based on the second law of thermodynamics. There is another common name for availability: *exergy*. If you wish, this chapter can be renamed *exergy analysis*.

8-1 MAXIMUM WORK

Consider an open system as depicted in Fig. 8-1 that exchanges heat only with the atmosphere at T_0 and p_0. Let T be the local temperature of that part of the system where heat enters and leaves as shown in the figure. For maximum work output from the system (or minimum work input to the system), all interactions undertaken by the system must be totally reversible. This requires that the system itself must undergo an internally reversible process and the heat interchange with the atmosphere must be accomplished with a totally reversible engine (a Carnot engine) placed between the part of the system at T and the atmosphere at T_0, such as shown schematically in the figure. Notice that with the local temperature T varying over the open-system boundary, many Carnot engines would be required.

For the Carnot engine shown in Fig. 8-1, we have

$$\frac{đQ}{T} = \frac{đQ_0}{T_0}$$

and

$$đW_{\text{engine}} = đQ_0 - đQ = \left(\frac{T_0}{T} - 1\right) đQ$$

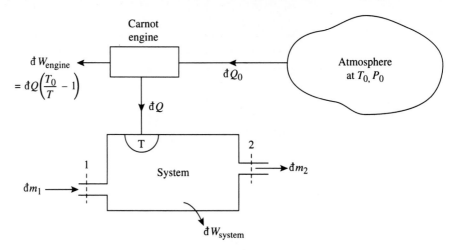

Figure 8-1 Reversible heat transfer between an open system and the atmosphere.

As the system process is internally reversible, the work performed by the system is a maximum. Together with the work done by the Carnot engine, the total maximum work is given by

$$\text{đ}W_{max} = \text{đ}W_{system} + \text{đ}W_{engine}$$

$$= \text{đ}W_{system} - \text{đ}Q + T_0 \frac{\text{đ}Q}{T}$$

(a)

Now, for a differential process of the system, the first law of thermodynamics gives

$$\text{đ}Q - \text{đ}W_{system} + (e_1 + p_1v_1)\,\text{đ}m_1 - (e_2 + p_2v_2)\,\text{đ}m_2 = dE_{system}$$

or

$$\text{đ}W_{system} - \text{đ}Q = -dE_{system} + (e_1 + p_1v_1)\,\text{đ}m_1 - (e_2 + p_2v_2)\,\text{đ}m_2$$

(b)

where e_1 and e_2 are the specific total energies of fluid crossing the boundary at the inlet section 1 and the exit section 2, respectively, and E_{system} represents the total energy of the system. In the meantime, for the system undergoing an internally reversible process, the second law of thermodynamics gives

$$dS_{system} = \left(\frac{\text{đ}Q}{T}\right)_{system} + s_1\,\text{đ}m_1 - s_2\,\text{đ}m_2$$

or

$$T_0\left(\frac{\text{đ}Q}{T}\right)_{system} = T_0\,dS_{system} - T_0s_1\,\text{đ}m_1 + T_0s_2\,\text{đ}m_2$$

(c)

where s_1 and s_2 are the specific entropies of fluid crossing the boundary at sections 1 and 2, respectively, and S_{system} represents the entropy of the system. Substituting Eqs. (b) and (c) into Eq. (a) results in

$$\text{đ}W_{max} = -dE_{system} + (e_1 + p_1v_1)\,\text{đ}m_1 - (e_2 + p_2v_2)\,\text{đ}m_2$$
$$+ T_0\,dS_{system} - T_0s_1\,\text{đ}m_1 + T_0s_2\,\text{đ}m_2$$

or

$$\mathrm{d}W_{\max} = -\mathrm{d}(E - T_0 S)_{\text{system}} + (e_1 + p_1 v_1 - T_0 s_1)\, \mathrm{d}m_1$$
$$- (e_2 + p_2 v_2 - T_0 s_2)\, \mathrm{d}m_2 \tag{8-1}$$

This equation has been developed for a single inlet stream and a single exit stream. For multiple streams, it can be written as follows:

$$\mathrm{d}W_{\max} = -\mathrm{d}(E - T_0 S)_{\text{system}} + \sum_{\text{inlet}} [(e + pv - T_0 s)\, \mathrm{d}m]$$
$$- \sum_{\text{exit}} [(e + pv - T_0 s)\, \mathrm{d}m] \tag{8-2}$$

Equation (8-1) or (8-2) is the final form of maximum work expression for an open system that exchanges heat only with the atmosphere.

Now consider the case in which an open system exchanges heat with an energy reservoir at a temperature T_R in addition to the atmosphere. For a reversible heat transfer, a Carnot engine must be interposed between the reservoir and the part of the system where heat transfer occurs. As the heat-transfer surface of the system also interchanges heat with the atmosphere, a Carnot engine must be interposed between them. These two Carnot engines act in series, as shown schematically in Fig. 8-2. Because the two Carnot engines working between the two extreme temperatures T_R and T_0 with an intermediate temperature level at T between them is equivalent to a single Carnot engine working between T_R and T_0, the total work done by the combined Carnot engine is given by

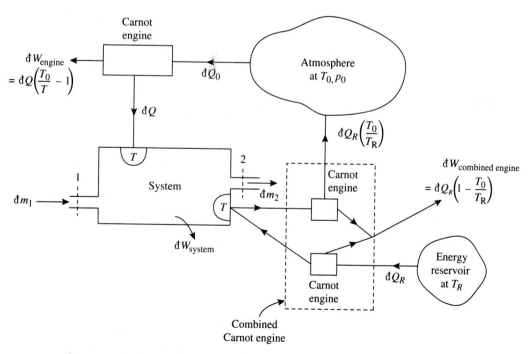

Figure 8-2 Reversible heat transfer between an open system and an energy reservoir plus the atmosphere.

$$\text{d}W_{\text{combined engine}} = \text{d}Q_R - \text{d}Q_R\left(\frac{T_0}{T_R}\right) = \text{d}Q_R\left(1 - \frac{T_0}{T_R}\right) \tag{d}$$

in which $\text{d}Q_R$ is the heat supply from the energy reservoir, and $\text{d}Q_R(T_0/T_R)$ is the heat rejection to the atmosphere by the combined Carnot engine. Adding Eq. (d) to Eq. (8-1) results in

$$\text{d}W_{\text{max}} = -\text{d}(E - T_0 S)_{\text{system}} + (e_1 + p_1 v_1 - T_0 s_1)\,\text{d}m_1$$
$$- (e_2 + p_2 v_2 - T_0 s_2)\,\text{d}m_2 + \text{d}Q_R\left(1 - \frac{T_0}{T_R}\right) \tag{8-3}$$

This is the general expression for maximum work for an open system that exchanges heat with an energy reservoir at T_R in addition to the atmosphere at T_0.

Under steady-state and steady-flow conditions, $dE_{\text{system}} = 0$, $dS_{\text{system}} = 0$, and $\text{d}m_1 = \text{d}m_2 = \text{d}m$, and Eq. (8-3) reduces to

$$\text{d}W_{\text{max}} = [(e_1 + p_1 v_1 - T_0 s_1) - (e_2 + p_2 v_2 - T_0 s_2)]\,\text{d}m + \text{d}Q_R\left(1 - \frac{T_0}{T_R}\right) \tag{8-4}$$

For a unit mass of fluid flowing through a steady-state, steady-flow system, the maximum work equation can be written as

$$w_{\text{max}} = (e_1 + p_1 v_1 - T_0 s_1) - (e_2 + p_2 v_2 - T_0 s_2) + q_R\left(1 - \frac{T_0}{T_R}\right) \tag{8-5}$$

or

$$w_{\text{max}} = \left(h_1 + \frac{\mathbf{V}_1^2}{2} + gz_1 - T_0 s_1\right) - \left(h_2 + \frac{\mathbf{V}_2^2}{2} + gz_2 - T_0 s_2\right) + q_R\left(1 - \frac{T_0}{T_R}\right) \tag{8-6}$$

where

$$e + pv = u + \frac{\mathbf{V}^2}{2} + gz + pv = h + \frac{\mathbf{V}^2}{2} + gz$$

has been substituted.

Equation (8-3) can be modified for the case of a closed system, for which $\text{d}m_1 = \text{d}m_2 = 0$, and the total energy E can be replaced by the internal energy U if there are no kinetic and potential energy changes involved for the system. The result is

$$\text{d}W_{\text{max}} = -\text{d}(U - T_0 S)_{\text{system}} + \text{d}Q_R\left(1 - \frac{T_0}{T_R}\right) \tag{8-7}$$

For a change from initial state i to final state f, we have for the closed system,

$$W_{\text{max}, i-f} = (U_i - T_0 S_i) - (U_f - T_0 S_f) + Q_R\left(1 - \frac{T_0}{T_R}\right) \tag{8-8}$$

where Q_R is the heat transfer from the energy reservoir during the process. Notice that when the last term involving Q_R (or q_R) in Eqs. (8-3) to (8-8) is not included, these equations are for systems that exchange heat with the atmosphere only.

Example 8-1

A solar pond is an artificial lake of a few meters deep for the storage of solar energy. While the surface water cools due to heat transfer to the atmosphere, the water at the bottom can keep the energy it stores, thereby maintaining a higher temperature. A solar pond of 3 m deep produces hot water corresponding to the saturation condition at the local pressure. Determine the maximum work (in kJ/kg) of the bottom hot water in changing to the atmospheric condition of 101 kPa and 25°C for (a) a steady-flow system and (b) a closed system.

Solution   Assuming a specific volume of $v = 1.0 \times 10^{-3}$ m³/kg for water, the local pressure p at the pond bottom is

$$p = p_0 + gz/v$$

where p_0 = atmospheric pressure at water surface = 101 kPa
$\quad g$ = gravitational acceleration = 9.81 m/s²
$\quad z$ = depth of water = 3 m

Thus,

$$p = 101 \text{ kPa} + \frac{(9.81 \text{ m/s}^2)(3 \text{ m})}{1.0 \times 10^{-3} \text{ m}^3/\text{kg}} \times \frac{1}{1000 \text{ Pa/kPa}} = 130 \text{ kPa}$$

At this pressure, from the saturated steam table (Table A-19M), we have

$$T = T_{\text{sat}} = 107.13°C, \qquad u = u_f = 449.02 \text{ kJ/kg},$$
$$h = h_f = 449.15 \text{ kJ/kg}, \qquad s = s_f = 1.3867 \text{ kJ/kg·K}$$

At $p_0 = 101$ kPa and $T_0 = 25°C$, water is a compressed liquid. Because 101 kPa is a low pressure, we use the saturated liquid values (Table A-18M):

$$u_0 \approx u_f \text{ at } 25°C = 104.88 \text{ kJ/kg}$$
$$h_0 \approx h_f \text{ at } 25°C = 104.89 \text{ kJ/kg}$$
$$s_0 \approx s_f \text{ at } 25°C = 0.3674 \text{ kJ/kg·K}$$

(a) For a steady-flow system, the maximum work is calculated from Eq. (8-6) with ΔKE and ΔPE neglected and heat exchange with only the atmosphere. Thus,

$$w_{\text{max}} = (h + \tfrac{1}{2}\cancel{V^2} + \cancel{gz} - T_0 s) - (h_0 + \tfrac{1}{2}\cancel{V_0^2} + \cancel{gz_0} - T_0 s_0)$$
$$= (h - h_0) - T_0(s - s_0)$$
$$= [(449.15 - 104.89) \text{ kJ/kg}] - [(25 + 273) \text{ K}][(1.3867 - 0.3674) \text{ kJ/kg·K}]$$
$$= 40.5 \text{ kJ/kg}$$

(b) For a closed system, the maximum work is calculated from Eq. (8-8) with heat transfer only to the atmosphere. Thus,

$$w_{\text{max}} = (u - u_0) - T_0(s - s_0)$$
$$= (449.02 - 104.88) \text{ kJ/kg}$$
$$\quad - [(25 + 273) \text{ K}][(1.3867 - 0.3674) \text{ kJ/kg·K}]$$
$$= 40.4 \text{ kJ/kg}$$

The difference between the values of w_{max} obtained in parts (a) and (b) is very small. This is because $h = u + pv$ and the product pv is small for the given conditions of the problem. Notice that w_{max} is the maximum work that could be obtained. In reality, the work output will be smaller due to irreversible effects involved in an actual process.

8-2 IRREVERSIBILITY

Due to dissipative effects (such as solid or fluid friction) or unbalanced potentials (such as heat transfer across a finite temperature difference), actual processes are always irreversible and actual work output from a system is always less than the maximum work as given by the equations in Sec. 8-1. The difference between the maximum work W_{max} and the actual work W_{actual} is called *irreversibility I*. Thus,

$$I = W_{max} - W_{actual} \qquad (8\text{-}9)$$

or

$$đI = đW_{max} - đW_{actual} \qquad (8\text{-}10)$$

Consider an open system that undergoes an actual process with irreversible heat transfers from the atmosphere at T_0 and an energy reservoir at T_R, as depicted in Fig. 8-3. The first law of thermodynamics for this system gives

$$đQ_0 + đQ_R - đW_{actual} + (e_1 + p_1v_1)\, đm_1 - (e_2 + p_2v_2)\, đm_2 = dE_{system}$$

or

$$đW_{actual} = -dE_{system} + (e_1 + p_1v_1)\, đm_1 - (e_2 + p_2v_2)\, đm_2 + đQ_0 + đQ_R$$

On the other hand, if the process is totally reversible, Eq. (8-3) could be applied, or

$$đW_{max} = -dE_{system} + T_0\, dS_{system} + (e_1 + p_1v_1 - T_0s_1)\, đm_1$$

$$- (e_2 + p_2v_2 - T_0s_2)\, đm_2 + đQ_R - T_0\left(\frac{đQ_R}{T_R}\right)$$

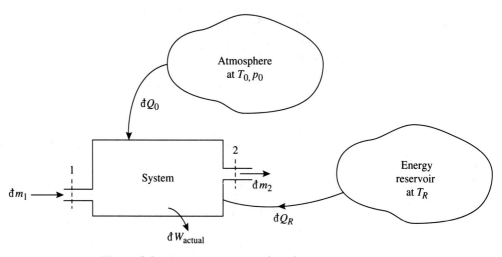

Figure 8-3 An open system undergoing an actual process.

Substituting the preceding two equations in Eq. (8-10) yields

$$\dbar I = T_0 \left(dS_{\text{system}} - s_1 \, \dbar m_1 + s_2 \, \dbar m_2 - \frac{\dbar Q_R}{T_R} - \frac{\dbar Q_0}{T_0} \right)$$

Now, referring to Fig. 8-3, the change of entropy of the surroundings is given by

$$dS_{\text{surr}} = -s_1 \, \dbar m_1 + s_2 \, \dbar m_2 - \frac{\dbar Q_R}{T_R} - \frac{\dbar Q_0}{T_0}$$

Accordingly, the preceding $\dbar I$ equation can be written as

$$\dbar I = T_0(dS_{\text{system}} + dS_{\text{surr}}) \qquad (8\text{-}11)$$

As T_0 is a constant, integrating the preceding equation yields

$$I = T_0(\Delta S_{\text{system}} + \Delta S_{\text{surr}}) \qquad (8\text{-}12)$$

The quantity within the parentheses in this equation represents the change of entropy of the universe, which includes the system and all parts of the surroundings having interaction with the system. This term is always greater than zero and approaches zero as the overall process becomes totally reversible.

Equation (8-11) or (8-12) is the general expression for irreversibility. It can be readily applied for the case of a closed system or an open system under steady-state, steady-flow conditions. Thus, for a closed system of mass m undergoing a process from state 1 to state 2 while receiving a quantity of heat Q_R from an energy reservoir at T_R, the entropy change of the system is given by

$$\Delta S_{\text{system}} = m(s_2 - s_1)$$

and the entropy change of the surroundings is given by

$$\Delta S_{\text{surr}} = \frac{-Q_R}{T_R}$$

The irreversibility of the process as given by Eq. (8-12) is then

$$I = T_0(\Delta S_{\text{system}} + \Delta S_{\text{surr}})$$

$$= T_0 \left[m(s_2 - s_1) + \frac{-Q_R}{T_R} \right]$$

If the preceding closed system is insulated, the irreversibility of the process would be

$$I = T_0(\Delta S_{\text{system}} + \overset{0}{\cancel{\Delta S_{\text{surr}}}}) = T_0[m(s_2 - s_1)]$$

For an open system undergoing a steady-flow, steady-state process from inlet condition 1 to exit condition 2, while receiving a quantity of heat q_R per unit mass flow from an energy reservoir at T_R, we have

$$\Delta s_{\text{system}} = 0 \quad \text{(because there is no change in any thermodynamic property of a steady-flow, steady-state system)}$$

and

$$\Delta s_{\text{surr}} = (s_2 - s_1) + \frac{-q_R}{T_R}$$

The specific irreversibility of the process as given by Eq. (8-11) is then

$$i = T_0(\cancel{\Delta s_{\text{system}}}^{\,0} + \Delta s_{\text{surr}})$$

$$= T_0\left[(s_2 - s_1) + \frac{-q_R}{T_R}\right]$$

If the preceding steady-flow, steady-state system is insulated, the specific irreversibility of the process would be

$$i = T_0(s_2 - s_1)$$

Example 8-2

A rigid tank having a volume of 50 ft³ contains air initially at 20 psia and 60°F. The pressure in the tank is increased to 35 psia by (a) stirring with a paddle wheel while keeping the tank insulated, and (b) heating from a reservoir at 500°F. The atmosphere is at 77°F and 14.7 psia. Determine, for each case, the actual work, the maximum work, and the irreversibility. For air, $c_v = 0.172$ Btu/lbm·°F and $R = 53.3$ ft·lbf/lbm·°R.

Solution Schematic diagrams for the two cases of this example are shown in Fig. 8-4. The given conditions are $V_1 = V_2 = 50$ ft³, $p_1 = 20$ psia, $T_1 = 60°F = 520°R$, $p_2 = 35$ psia, $T_0 = 77°F = 537°R$, and $p_0 = 14.7$ psia.

At the low pressures involved in this problem, air can be considered as an ideal gas. The mass of air in the tank is calculated by the ideal-gas equation,

$$m = \frac{p_1 V_1}{RT_1} = \frac{(20 \times 144 \text{ lbf/ft}^2)(50 \text{ ft}^3)}{(53.3 \text{ ft·lbf/lbm·°R})(520°R)} = 5.196 \text{ lbm}$$

For the constant-volume process from the ideal-gas equation, we have

$$T_2 = T_1\left(\frac{p_2}{p_1}\right) = 520\left(\frac{35}{20}\right) = 910°R = 450°F$$

(a) Stirring by a paddle wheel with $Q = 0$. The actual work W_{act} is given by the first law for the closed system,

$$\cancel{Q}^{\,0} = \Delta U + W_{\text{act}}$$

or

$$W_{\text{act}} = -\Delta U = m(u_1 - u_2) = mc_v(T_1 - T_2)$$

$$= (5.196 \text{ lbm})(0.172 \text{ Btu/lbm·°R})(520°R - 910°R) = -348.5 \text{ Btu}$$

where the negative sign indicates that work is done on the system by the paddle wheel.

Had the system's state change taken place under a totally reversible condition with heat exchange only between the system and the atmosphere, the maximum work output (for this case, the minimum work input) would be, by Eq. (8-8) (without the Q_R term),

$$W_{\text{max}} = (U_1 - U_2) - T_0(S_1 - S_2)$$

$$= m[(u_1 - u_2) - T_0(s_1 - s_2)]$$

where

$$u_1 - u_2 = c_v(T_1 - T_2)$$

$$s_1 - s_2 = c_v \ln\frac{T_1}{T_2} + R \ln\cancel{\frac{v_1}{v_2}}^{\,0} = c_v \ln\frac{T_1}{T_2}$$

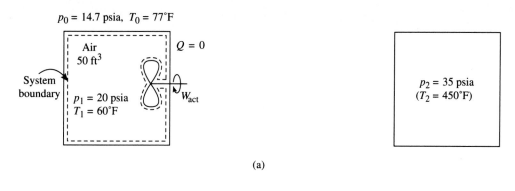

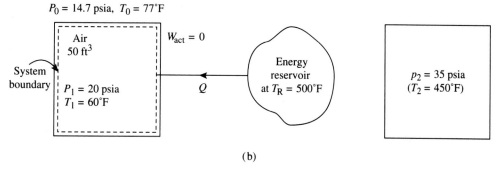

Figure 8-4 Schematic diagrams for Example 8-2: (a) paddle-wheel work, and (b) heat transfer.

Thus,

$$W_{\max} = mc_v\left[(T_1 - T_2) - T_0 \ln\frac{T_1}{T_2}\right]$$

$$= (5.196 \text{ lbm})(0.172 \text{ Btu/lbm·°R})$$

$$\times\left[(520°R - 910°R) - (537°R)\ln\frac{520°R}{910°R}\right]$$

$$= -80.0 \text{ Btu}$$

The irreversibility of the process is given by Eq. (8-9),

$$I = W_{\max} - W_{\text{act}} = (-80.0) - (-348.5) = 268.5 \text{ Btu}$$

The irreversibility can also be calculated by Eq. (8-12),

$$I = T_0(\Delta S_{\text{system}} + \Delta S_{\text{surr}})$$

Because for the surroundings there is no change in entropy during the adiabatic process of the closed system, it follows that

$$I = T_0(\Delta S_{\text{system}}) = T_0 m(s_2 - s_1) = T_0 mc_v \ln\frac{T_2}{T_1}$$

$$= (537°R)(5.196 \text{ lbm})(0.172 \text{ Btu/lbm·°R})\ln\frac{910°R}{520°R} = 268.5 \text{ Btu}$$

(b) Heating from a reservoir at $T_R = 500°F = 960°R$. Because the actual work $W_{act} = 0$, the heat transfer Q is given by the first law as

$$Q = \Delta U + \cancel{W_{act}}^{0} = \Delta U = m(u_2 - u_1) = mc_v(T_2 - T_1)$$

$$= (5.196 \text{ lbm})(0.172 \text{ Btu/lbm·°R})(910°R - 520°R) = 348.5 \text{ Btu}$$

With Q added to the system, the maximum work output would be, by Eq. (8-8),

$$W_{max} = (U_1 - U_2) - T_0(S_1 - S_2) + Q\left(1 - \frac{T_0}{T_R}\right)$$

$$= m[(u_1 - u_2) - T_0(s_1 - s_2)] + Q\left(1 - \frac{T_0}{T_R}\right)$$

$$= mc_v\left[(T_1 - T_2) - T_0 \ln \frac{T_1}{T_2}\right] + Q\left(1 - \frac{T_0}{T_R}\right)$$

$$= (5.196)(0.172)\left[(520 - 910) - 537 \ln \frac{520}{910}\right] + 348.5\left(1 - \frac{537}{960}\right)$$

$$= 73.6 \text{ Btu}$$

As the actual work $W_{act} = 0$, the irreversibility given by Eq. (8-9) is

$$I = W_{max} - \cancel{W_{act}}^{0} = W_{max} = 73.6 \text{ Btu}$$

The irreversibility can also be calculated by Eq. (8-12),

$$I = T_0(\Delta S_{system} + \Delta S_{surr})$$

$$= T_0\left[m(s_2 - s_1) + \frac{-Q}{T_R}\right] = T_0\left(mc_v \ln \frac{T_2}{T_1} - \frac{Q}{T_R}\right)$$

$$= 537\left(5.196 \times 0.172 \ln \frac{910}{520} - \frac{348.5}{960}\right)$$

$$= 73.6 \text{ Btu}$$

Notice that the irreversibility of 73.6 Btu in the heat-addition process is much smaller than the irreversibility of 268.5 Btu in the paddle-wheel process. The clear choice is the heat-addition process when the second-law analysis is considered. Conversion of work to heat as in the paddle-wheel process is not a wise use of energy resources. But be aware that the irreversibility in the heat-addition process would increase if the reservoir temperature were increased causing a larger temperature difference between the reservoir and the air in the tank.

Example 8-3

A piston–cylinder device contains 0.5 kg of water initially at 160°C and 1 MPa. 1200 kJ of heat is transferred to the water from an energy reservoir at 600°C while the system undergoes an internally reversible, isothermal expansion process. The atmosphere is at 298 K and 0.1 MPa. Determine (a) the actual work, (b) the maximum work, and (c) the irreversibility of the process, in kJ.

Solution   A schematic diagram for this example is shown in Fig. 8-5. The given data are $m = 0.5$ kg, $T_1 = T_2 = 160°C = 433$ K, $p_1 = 1$ MPa, $Q = 1200$ kJ, $T_R = 600°C = 873$ K, $T_0 = 298$ K, and $p_0 = 0.1$ MPa.

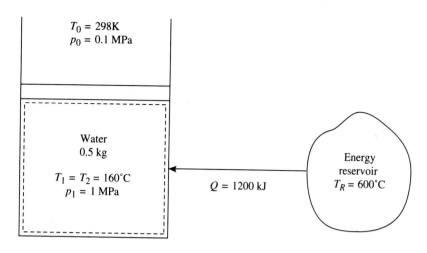

Figure 8-5 Schematic diagram for Example 8-3.

At $T_1 = 160°C$ and $p_1 = 1$ MPa, water is in a compressed liquid state. Using the saturated steam table (Table A-18M) at $T_1 = 160°C$, we obtain

$$u_1 \approx u_f = 674.87 \text{ kJ/kg} \qquad \text{and} \qquad s_1 \approx s_f = 1.9427 \text{ kJ/Kg·K}$$

As

$$ds = \left(\frac{dq}{T}\right)_{\text{int rev}}$$

we have for the internally reversible, isothermal process 1–2,

$$s_2 - s_1 = \frac{1}{T_1}(q)$$

or

$$s_2 = s_1 + \frac{Q}{mT_1} = 1.9427 \text{ kJ/kg·K} + \frac{1200 \text{ kJ}}{(0.5 \text{ kg})(433 \text{ K})}$$

$$= 7.485 \text{ kJ/kg·K}$$

which is greater than s_g at 160°C, so that state 2 is superheated. At $T_2 = 160°C$ and $s_2 = 7.485$ kJ/kg·K from the superheated steam table (Table A-20M), we obtain by interpolation:

$$p_2 = 144.4 \text{ kPa} \qquad \text{and} \qquad u_2 = 2595.5 \text{ kJ/kg}$$

(a) The actual work done is calculated from the first law:

$$W_{\text{act}} = Q - \Delta U = Q - m(u_2 - u_1)$$

$$= (1200 \text{ kJ}) - (0.5 \text{ kg})(2595.5 - 674.9 \text{ kJ/kg}) = 239.7 \text{ kJ}$$

(b) The maximum work is calculated from Eq. (8-8):

$$W_{\text{max}} = (U_1 - U_2) - T_0(S_1 - S_2) + Q\left(1 - \frac{T_0}{T_R}\right)$$

$$= m[(u_1 - u_2) - T_0(s_1 - s_2)] + Q\left(1 - \frac{T_0}{T_R}\right)$$

$$= (0.5 \text{ kg})[(674.9 - 2595.5 \text{ kJ/kg})$$

$$- (298 \text{ K})(1.9427 - 7.485 \text{ kJ/kg·K})$$

$$+ (1200 \text{ kJ})\left(1 - \frac{298 \text{ K}}{873 \text{ K}}\right)$$

$$= 655.9 \text{ kJ}$$

(c) The irreversibility I is calculated from Eq. (8-9),

$$I = W_{\text{max}} - W_{\text{act}} = 655.9 - 239.7 = 416.2 \text{ kJ}$$

The irreversibility can also be calculated from Eq. (8-12),

$$I = T_0(\Delta S_{\text{system}} + \Delta S_{\text{surr}})$$

where

$$\Delta S_{\text{system}} = m(s_2 - s_1) \qquad \text{and} \qquad \Delta S_{\text{surr}} = \frac{-Q}{T_R}$$

Thus,

$$I = T_0\left[m(s_2 - s_1) + \left(\frac{-Q}{T_R}\right)\right]$$

$$= 298\left[0.5(7.485 - 1.9427) + \left(\frac{-1200}{873}\right)\right]$$

$$= 416.2 \text{ kJ}$$

Notice that although the water undergoes an internally reversible process, the heat transfer from the energy reservoir to the water is through a finite temperature difference. The overall process is, therefore, not totally reversible, thus promoting irreversibility.

Example 8-4

Air enters an adiabatic compressor at 14.2 psia and 60°F, with a velocity of 300 ft/s. The air leaves the compressor at 65 psia, 400°F, and a velocity of 200 ft/s. Based on data from the air tables, determine the actual work, the maximum work, and the irreversibility per pound of air flow. The atmosphere is at 14.2 psia and 60°F.

Solution   A schematic for the adiabatic compressor of the problem is shown in Fig. 8-6. The given data are $p_1 = 14.2$ psia, $T_1 = 60°F = 520°R$, $V_1 = 300$ ft/s, $p_2 = 65$ psia, $T_2 = 400°F = 860°R$, $V_2 = 200$ ft/s, $p_0 = 14.2$ psia, $T_0 = 60°F = 520°R$, and $q = 0$.

From the air table (Table A-9E) at $T_1 = 520°R$, we get $h_1 = 124.36$ Btu/lbm and $\phi_1 = 0.59160$ Btu/lbm·°R; and at $T_2 = 860°R$, we get $h_2 = 206.57$ Btu/lbm and $\phi_2 = 0.71314$ Btu/lbm·°R. The change of entropy of the air is calculated as

$$s_2 - s_1 = \phi_2 - \phi_1 - R \ln\frac{p_2}{p_1}$$

$$= (0.71314 - 0.59160 \text{ Btu/lbm·°R})$$

$$- \frac{53.3 \text{ ft·lbf/lbm·°R}}{778 \text{ ft·lbf/Btu}} \ln\frac{65 \text{ psia}}{14.2 \text{ psia}}$$

$$= 0.01733 \text{ Btu/lbm·°R}$$

With $q = 0$ and ΔPE neglected, application of the steady-flow energy equation yields for the actual work w_{act},

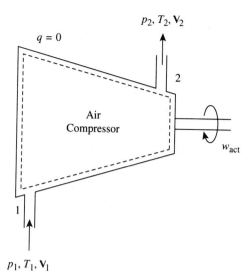

$p_2, T_2, \mathbf{V}_2$

$q = 0$

2

Air
Compressor

w_act

1

$p_1, T_1, \mathbf{V}_1$

Figure 8-6 Schematic diagram for Example 8-4.

$$w_\text{act} = \cancel{q}^{\,0} + (h_1 - h_2) + \tfrac{1}{2}(\mathbf{V}_1^2 - \mathbf{V}_2^2) + g(z_1 \cancel{- z_2})^{\,0}$$

$$= (h_1 - h_2) + \tfrac{1}{2}(\mathbf{V}_1^2 - \mathbf{V}_2^2)$$

$$= (124.36 - 206.57) \text{ Btu/lbm} + \frac{(300^2 - 200^2) \text{ ft}^2/\text{s}^2}{2(32.2 \text{ ft·lbm/lbf·s}^2)(778 \text{ ft·lbf/Btu})}$$

$$= -81.2 \text{ Btu/lbm}$$

which means that the actual work input is 81.2 Btu/lbm.

The maximum work output (minimum work input for this case) is calculated from Eq. (8-6) (without the q_R term):

$$w_\text{max} = (h_1 - h_2) + \frac{\mathbf{V}_1^2 - \mathbf{V}_2^2}{2} - T_0(s_1 - s_2)$$

$$= (124.36 - 206.57) + \frac{(300)^2 - (200)^2}{2 \times 32.2 \times 778} - 520(-0.01733)$$

$$= -72.2 \text{ Btu/lbm}$$

which means that the minimum work input is 72.2 Btu/lbm.

The specific irreversibility i is calculated from Eq. (8-9),

$$i = w_\text{max} - w_\text{act} = (-72.2) - (-81.2) = 9.0 \text{ Btu/lbm}$$

The specific irreversibility can also be calculated from Eq. (8-12):

$$i = T_0(\Delta s_\text{system} + \Delta s_\text{surr})$$

where, for steady flow, $\Delta s_\text{system} = 0$. The change of entropy of the surroundings is given by

$$\Delta s_\text{surr} = s_2 - s_1 = 0.01733 \text{ Btu/lbm·°R}$$

Therefore,

$$i = T_0(\Delta s_\text{surr}) = 520(0.01733) = 9.0 \text{ Btu/lbm}$$

Thus, for the compression process, the work input destroyed by irreversibilities within the compressor amounts to $9.0/81.2 = 11.1\%$.

Example 8-5

The thermal efficiency of a steam power plant can be improved when the feedwater is preheated by extracted steam from the turbine before it enters the steam generator. As illustrated in Fig. 8-7, extracted steam at section 1 is used to preheat the incoming feedwater at section 2 to obtain a higher-temperature feedwater at section 3. The conditions of the fluids are indicated in the figure. Determine the irreversibility per kilogram of feedwater at section 3. The atmosphere is at 298 K and 101 kPa.

Solution From the steam tables, we obtain the following: At $p_1 = 150$ kPa, Table A-19M gives

$$h_1 = h_g = 2693.6 \text{ kJ/kg}, \qquad s_1 = s_g = 7.2233 \text{ kJ/kg·K},$$

$$h_a = h_f = 467.11 \text{ kJ/kg}, \qquad v_a = v_f = 0.0010528 \text{ m}^3\text{/kg}$$

For the compressed liquid at section 2, at $p_2 = 2.5$ MPa, and $T_2 = 40°C$, Table A-21M gives

$$h_2 = 169.77 \text{ kJ/kg} \qquad \text{and} \qquad s_2 = 0.5715 \text{ kJ/kg·K}$$

For the compressed liquid at section 3, Table A-18M gives

$$h_3 \approx h_f \text{ (at } 90°C) = 376.92 \text{ kJ/kg}$$

$$s_3 \approx s_f \text{ (at } 90°C) = 1.1925 \text{ kJ/kg·K}$$

By assuming adiabatic operation and neglecting kinetic and potential energies, the steady-flow energy equation gives the following pump power requirement:

$$\dot{W}_{\text{pump}} = \dot{m}_1(h_b - h_a)$$

In addition, if the pump process can be assumed to be reversible and adiabatic, Eq. (7-2a) gives

$$T \, ds = 0 = dh - v \, dp$$

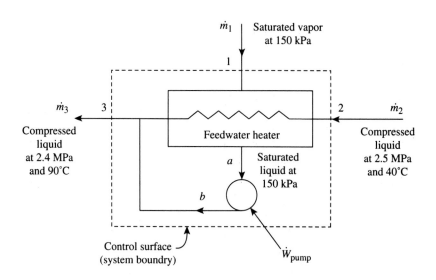

Figure 8-7 Schematic diagram for Example 8-5.

Availability and Irreversibility   Chap. 8

or

$$h_b - h_a = \int_a^b v \, dp = v_a(p_b - p_a)$$

in which the liquid specific volume $v_a = v_b$ has been assumed. Therefore,

$$\dot{W}_{pump} = \dot{m}_1 v_a(p_b - p_a) = \dot{m}_1 v_a(p_3 - p_1)$$

$$= (\dot{m}_1 \text{ kg/s})(0.0010528 \text{ m}^3\text{/kg})(2400 - 150 \text{ kPa})$$

$$= 2.369\dot{m}_1 \text{ kW}$$

For the control volume around the heater and the pump, with $\dot{Q} = 0$ and KE and PE neglected, an energy balance yields

$$\dot{m}_1 h_1 + \dot{m}_2 h_2 + \dot{W}_{pump} = \dot{m}_3 h_3$$

But

$$\dot{m}_1 + \dot{m}_2 = \dot{m}_3$$

so that

$$\dot{m}_1 h_1 + (\dot{m}_3 - \dot{m}_1)h_2 + 2.369\dot{m}_1 = \dot{m}_3 h_3$$

Dividing through by $\dot{m}_3$,

$$\left(\frac{\dot{m}_1}{\dot{m}_3}\right)h_1 + \left(1 - \frac{\dot{m}_1}{\dot{m}_3}\right)h_2 + 2.369\left(\frac{\dot{m}_1}{\dot{m}_3}\right) = h_3$$

Solving for $\dot{m}_1/\dot{m}_3$, we have

$$\frac{\dot{m}_1}{\dot{m}_3} = \frac{h_3 - h_2}{h_1 - h_2 + 2.369} = \frac{376.92 - 169.77}{2693.6 - 169.77 + 2.369} = 0.0820$$

$$\frac{\dot{m}_2}{\dot{m}_3} = \frac{\dot{m}_3 - \dot{m}_1}{\dot{m}_3} = 1 - \frac{\dot{m}_1}{\dot{m}_3} = 1 - 0.0820 = 0.9180$$

Considering the heater and the pump together as the system, the irreversibility per unit mass flow at section 3, $\dot{I}/\dot{m}_3$, is calculated by Eq. (8-12):

$$\frac{\dot{I}}{\dot{m}_3} = \frac{T_0}{\dot{m}_3}(\Delta\dot{S}_{system} + \Delta\dot{S}_{surr})$$

where

$$\Delta\dot{S}_{system} = 0 \quad \text{(for steady state, steady flow)}$$

$$\Delta\dot{S}_{surr} = \dot{m}_3 s_3 - \dot{m}_1 s_1 - \dot{m}_2 s_2$$

Thus,

$$\frac{\dot{I}}{\dot{m}_3} = T_0\left(\frac{\Delta\dot{S}_{surr}}{\dot{m}_3}\right) = T_0\left(s_3 - \frac{\dot{m}_1}{\dot{m}_3}s_1 - \frac{\dot{m}_2}{\dot{m}_3}s_2\right)$$

$$= 298(1.1925 - 0.0820 \times 7.2233 - 0.9180 \times 0.5715)$$

$$= 22.5 \text{ kJ per kg at section 3}$$

The irreversibility for the process can be attributed to the heat transfer across the finite temperature differences inside the heater and the mixing of the two liquid streams outside the heater, as well as friction caused by fluid flow.

8-3 AVAILABILITY OF A CLOSED SYSTEM

For a system that exchanges heat with the atmosphere only, if its temperature equals the atmosphere temperature and its pressure equals the atmosphere pressure, then there would be no unbalanced temperature and pressure that could be used to realize any work from the system–atmosphere combination. Accordingly, it is clear that a maximum possible amount of work can be done if the system goes reversibly from its present state to a state in which its temperature and pressure equal those of the atmosphere. When the system temperature T equals the atmosphere temperature T_0 and the system pressure p equals the atmosphere pressure p_0, the system is said to be in the *dead state*. If a system is in the dead state, it would be at rest and possess no kinetic energy, but it does possess potential energy.

When a system goes from one specified state to another, some of the work done by the system may be done on the atmosphere and thus serves no useful purpose. The work done by a system exclusive of that done on the atmosphere as the system boundary expands or contracts is called the *useful work*. The useful work W_{useful} can be expressed symbolically as

$$W_{useful} = W - p_0(V_f - V_i) \tag{8-13}$$

where W is the work done by the system, and V_i and V_f are, respectively, the initial and final volumes of the system.

The *availability* of a closed system in a given state is defined as the maximum useful work that can be obtained from the system–atmosphere combination as the system proceeds reversibly from that state to its dead state while exchanging heat only with the atmosphere. The availability of a closed system will be denoted by the symbol Φ for the total mass and φ for a unit mass, and $\Phi = m\varphi$. Let the subscripts 1 and 0 denote, respectively, a given state and the dead state of a system. The closed system availability Φ_1 at state 1 can be expresses as

$$\Phi_1 = W_{max\,useful,\,1-0} = W_{max,\,1-0} - p_0(V_0 - V_1)$$

Substituting Eq. (8-8) (without the Q_R term) into the preceding equation yields

$$\begin{aligned}
\Phi_1 &= (U_1 - T_0 S_1) - (U_0 - T_0 S_0) - p_0(V_0 - V_1) \\
&= (U_1 + p_0 V_1 - T_0 S_1) - (U_0 + p_0 V_0 - T_0 S_0)
\end{aligned} \tag{8-14}$$

For a unit mass, the specific availability φ_1 at state 1 is written as

$$\varphi_1 = (u_1 + p_0 v_1 - T_0 s_1) - (u_0 + p_0 v_0 - T_0 s_0) \tag{8-15}$$

It is imperative to notice that although we speak of the availability of a system in a given state, its value depends on T_0 and p_0 of the atmosphere as well as on the state of the system. The value of φ can be positive or negative depending on the state of the system in relation to the dead state.

When a closed system that exchanges heat only with the atmosphere undergoes a reversible process from state 1 to state 2, the maximum useful work obtainable is the decrease in availability. Thus, for a unit mass, we have

$$\begin{aligned}
W_{max\,useful,\,1-2} &= -\Delta\varphi_{12} = \varphi_1 - \varphi_2 \\
&= [(u_1 + p_0 v_1 - T_0 s_1) - (u_0 + p_0 v_0 - T_0 s_0)] \\
&\quad - [(u_2 + p_0 v_2 - T_0 s_2) - (u_0 + p_0 v_0 - T_0 s_0)]
\end{aligned} \tag{8-16}$$

$$= (u_1 + p_0 v_1 - T_0 s_1) - (u_2 + p_0 v_2 - T_0 s_2)$$

If the system also receives heat q_R from an energy reservoir at T_R, then

$$w_{\text{max useful, 1-2}} = (\varphi_1 - \varphi_2) + q_R\left(1 - \frac{T_0}{T_R}\right) \tag{8-17}$$

Example 8-6

A rigid tank of 10-ft³ capacity contains nitrogen (N_2) at 500 psia and 77°F. The nitrogen escapes through a small valve opening until the pressure in the tank drops to 200 psia when the valve is again closed. Assume that no heat transfer occurs during the short period of time when the gas is escaping and the gas remaining in the tank undergoes an internally reversible adiabatic expansion process. The atmosphere is at 14.7 psia and 77°F. Using data from the ideal-gas property table for N_2 in Appendix 3, determine (a) the availability of the N_2 in the tank before the valve leakage, (b) the availability of the N_2 in the tank immediately after the expansion process when the valve is again closed, and (c) the availability of the N_2 in the tank a long time later when its temperature has returned to the initial value of 77°F due to heat exchange with the atmosphere.

Solution   Denote the three states of the N_2 (i.e., before the leakage, immediately after the leakage, and a long time later) by the subscripts 1, 2, and 3, respectively. Given $V_1 = V_2 = V_3 = 10$ ft³, $p_1 = 500$ psia, $T_1 = 77°F = 537°R$, $p_2 = 200$ psia, $Q_{12} = 0$, $T_3 = 77°F = 537°R$, $p_0 = 14.7$ psia, and $T_0 = 77°F = 537°R$. Thus, at the dead state,

$$v_0 = \frac{RT_0}{p_0} = \frac{(1545/28 \text{ ft·lbf/lbm·°R})(537°R)}{14.7 \times 144 \text{ lbf/ft}^2} = 14.0 \text{ ft}^3/\text{lbm}$$

(a) For the gas initially in the tank:

$$m_1 = \frac{p_1 V_1}{RT_1} = \frac{(500 \times 144 \text{ lbf/ft}^2)(10 \text{ ft}^3)}{(1545/28 \text{ ft·lbf/lbm·°R})(537°R)} = 24.30 \text{ lbm}$$

$$v_1 = \frac{V_1}{m_1} = \frac{10}{24.30} = 0.4115 \text{ ft}^3/\text{lbm}$$

From the ideal-gas property table for N_2 (Table A-16), at $T_1 = T_3 = T_0 = 537°R$, we get

$$u_1 = u_3 = u_0 = 2663.4 \text{ Btu/lbmole} = 95.12 \text{ Btu/lbm}$$

$$\phi_1 = \phi_3 = \phi_0 = 45.735 \text{ Btu/lbmole·°R} = 1.6334 \text{ Btu/lbm·°R}$$

Thus,

$$s_1 - s_0 = \phi_1 - \phi_0 - R \ln \frac{p_1}{p_0} = -R \ln \frac{p_1}{p_0}$$

$$= -\left(\frac{1.986}{28} \text{ Btu/lbm·°R}\right) \ln \frac{500 \text{ psia}}{14.7 \text{ psia}}$$

$$= -0.2501 \text{ Btu/lbm·°R}$$

The availability of the gas is calculated by Eq. (8-15):

$$\Phi_1 = W_{\text{max useful, 1-0}} = m_1 \varphi_1 = m_1[(u_1 - u_0) + p_0(v_1 - v_0) - T_0(s_1 - s_0)]$$

$$= (24.30 \text{ lbm})\left[0 + \frac{14.7 \times 144 \text{ lbf/ft}^2}{778 \text{ ft·lbf/Btu}}(0.4115 - 14.0 \text{ ft}^3/\text{lbm})\right.$$

$$\left. - (537°R)(-0.2501 \text{ Btu/lbm·°R})\right]$$

$$= (24.30 \text{ lbm})(97.33 \text{ Btu/lbm}) = 2365 \text{ Btu}$$

(b) For the gas in the tank immediately after the internally reversible adiabatic expansion: For the internally reversible adiabatic process, we have

$$s_2 - s_1 = 0 = \phi_2 - \phi_1 - R \ln \frac{p_2}{p_1}$$

so that

$$\phi_2 = \phi_1 + R \ln \frac{p_2}{p_1}$$

$$= 45.735 \text{ Btu/lbmole·°R} + (1.986 \text{ Btu/lbmole·°R}) \ln \frac{200 \text{ psia}}{500 \text{ psia}}$$

$$= 43.92 \text{ Btu/lbmole·°R}$$

At this ϕ_2 value, the N_2 ideal-gas table (Table A-16) gives

$$T_2 = 413.6°R \qquad \text{and} \qquad u_2 = 2051 \text{ Btu/lbmole} = 73.25 \text{ Btu/lbm}$$

Therefore,

$$m_2 = \frac{p_2 V_2}{R T_2} = \frac{(200 \times 144)(10)}{(1545/28)(413.6)} = 12.62 \text{ lbm}$$

$$v_2 = \frac{V_2}{m_2} = \frac{10}{12.62} = 0.7924 \text{ ft}^3/\text{lbm}$$

$$s_2 - s_0 = \phi_2 - \phi_0 - R \ln \frac{p_2}{p_0}$$

$$= 43.92 \text{ Btu/lbmole·°R} - 45.735 \text{ Btu/lbmole·°R}$$

$$- (1.986 \text{ Btu/lbmole·°R}) \ln \frac{200 \text{ psia}}{14.7 \text{ psia}}$$

$$= -6.999 \text{ Btu/lbmole·°R} = -0.2500 \text{ Btu/lbm·°R}$$

$$\Phi_2 = W_{\text{max useful, 2-0}} = m_2 \varphi_2 = m_2[(u_2 - u_0) + p_0(v_2 - v_0) - T_0(s_2 - s_0)]$$

$$= 12.62 \left[(73.25 - 95.12) + \frac{14.7 \times 144}{778}(0.7924 - 14.0) - 537(-0.2500) \right]$$

$$= 12.62(76.44) = 965 \text{ Btu}$$

(c) For the gas in the tank a long time after the tightening of the valve: For this case, we have

$$m_3 = m_2 = 12.62 \text{ lbm}$$

$$v_3 = v_2 = 0.7924 \text{ ft}^3/\text{lbm}$$

$$T_3 = T_0 = 77°F = 537°R$$

In addition, for the constant-volume process 2–3, the ideal-gas equation gives

$$p_3 = p_2 \left(\frac{T_3}{T_2} \right) = 200 \left(\frac{537}{413.6} \right) = 259.7 \text{ psia}$$

so that

$$s_3 - s_0 = \phi_3 - \phi_0 - R \ln \frac{p_3}{p_0} = - R \ln \frac{p_3}{p_0}$$

$$= -\left(\frac{1.986}{28} \text{ Btu/lbm·°R}\right) \ln \frac{259.7 \text{ psia}}{14.7 \text{ psia}}$$

$$= -0.2037 \text{ Btu/lbm·°R}$$

$$\Phi_3 = W_{\text{max useful, 3-0}} = m_3 \varphi_3$$

$$= m_3[(u_3 - u_0) + p_0(v_3 - v_0) - T_0(s_3 - s_0)]$$

$$= 12.62\left[0 + \frac{14.7 \times 144}{778}(0.7924 - 14.0) - 537(-0.2037)\right]$$

$$= 12.62(73.45) = 927 \text{ Btu}$$

Because, by definition, the availability of a closed system in a given state is the maximum useful work that can be obtained from the system–atmosphere combination as the system proceeds reversibly from that state to the dead state while exchanging heat only with the atmosphere, the three availability values calculated for this problem represent the following maximum useful work:

Φ_1 (before the leakage):	From state 1 (500 psia, 77°F) to the dead state (14.7 psia, 77°F)
Φ_2 (immediately after the leakage):	From state 2 (200 psia, 46.4°F) to the dead state (14.7 psia, 77°F)
Φ_3 (a long time later):	From state 3 (259.7 psia, 77°F) to the dead state (14.7 psia, 77°F)

As a result of the leakage, the system lost $\Phi_1 - \Phi_3 = 2365 - 927 = 1438$ Btu of work potential.

8-4 AVAILABILITY OF A STEADY-FLOW SYSTEM

For open systems of fixed boundary, the volume of the system does not change and there is no work done against the atmosphere. Accordingly, for such systems, the maximum useful work is equal to the maximum work. The availability of a steady-flow fluid stream is called the *stream availability* and is defined as the maximum work that can be obtained as the fluid changes reversibly from the given state to the dead state while exchanging heat with only the atmosphere.

The stream availability is denoted by the symbol Ψ for the total mass and ψ for a unit mass, and $\Psi = m\psi$. From Eq. (8-5) (without the q_R term), the specific stream availability at state 1 is given by the expression

$$\psi_1 = w_{\text{max, 1-0}} = (e_1 + p_1 v_1 - T_0 s_1) - (e_0 + p_0 v_0 - T_0 s_0) \qquad (8\text{-}18)$$

Or, from Eq. (8-6) (without the q_R term), we write

$$\psi_1 = w_{\text{max, 1-0}} = \left(h_1 + \frac{V_1^2}{2} + gz_1 - T_0 s_1\right) - (h_0 + gz_0 - T_0 s_0) \qquad (8\text{-}19)$$

Notice that although we speak of the stream availability of a fluid in a certain state, its value depends on T_0 and p_0 of the local atmosphere as well as on the state of the fluid. The value of ψ can be greater than or less than zero for any state of the fluid other than the dead state.

When a steady-flow fluid stream that exchanges heat only with the atmosphere proceeds reversibly from state 1 to state 2, the maximum work obtainable is the decrease in stream availability. Thus, for a unit mass,

$$w_{max, 1-2} = -\Delta\psi_{12} = \psi_1 - \psi_2$$

$$= \left[\left(h_1 + \frac{V_1^2}{2} + gz_1 - T_0 s_1\right) - (h_0 + gz_0 - T_0 s_0)\right]$$

$$- \left[\left(h_2 + \frac{V_2^2}{2} + gz_2 - T_0 s_2\right) - (h_0 + gz_0 - T_0 s_0)\right] \tag{8-20}$$

$$= \left(h_1 + \frac{V_1^2}{2} + gz_1 - T_0 s_1\right) - \left(h_2 + \frac{V_2^2}{2} + gz_2 - T_0 s_2\right)$$

If the fluid stream also receives heat q_R from an energy reservoir at T_R, then

$$w_{max, 1-2} = (\psi_1 - \psi_2) + q_R\left(1 - \frac{T_0}{T_R}\right) \tag{8-21}$$

If there is more than one flow into and out of an open system in a steady-state, steady-flow process, we can write on a rate basis in the following form:

$$\dot{W}_{max} = \sum_{inlet} \dot{m}\psi - \sum_{exit} \dot{m}\psi + \dot{Q}_R\left(1 - \frac{T_0}{T_R}\right) \tag{8-22}$$

where the last term is for heat interaction with an energy reservoir other than the atmosphere. Omit this term if there is no such energy reservoir involved.

Example 8-7

One way to increase the temperature of feedwater leading to the steam generator in a steam power plant is to use an open tank (see Fig. 8-8) to mix steam (section 1) extracted from the steam turbine with the cooler feedwater (section 2) to obtain a hotter feedwater (section 3). The conditions of the fluids are indicated in the figure. Determine the rate of stream availability change and irreversibility of the process. The atmosphere is at $T_0 = 20°C$ and $p_0 = 0.1$ MPa.

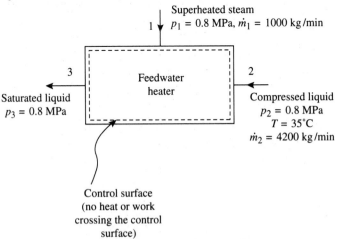

Figure 8-8 Schematic diagram for Example 8-7.

Availability and Irreversibility   Chap. 8

Solution From the saturated steam tables (Tables A-18M and 1-19M), the following data are obtained:

$$h_2 \approx h_f \text{ at } 35°C = 146.68 \text{ kJ/kg}, \qquad s_2 \approx s_f \text{ at } 35°C = 0.5053 \text{ KJ/kg·K},$$

$$T_3 = T_{sat} \text{ at } 0.8 \text{ MPa} = 170.43°C, \qquad h_3 = h_f \text{ at } 0.8 \text{ MPa} = 721.11 \text{ kJ/kg},$$

$$s_3 = s_f \text{ at } 0.8 \text{ MPa} = 2.0462 \text{ kJ/kg·K}$$

At the dead state, $T_0 = 20°C = 293°K$, $p_0 = 0.1$ MPa, and

$$h_0 \approx h_f \text{ at } 20°C = 83.96 \text{ kJ/kg} \qquad \text{and} \qquad s_0 \approx s_f \text{ at } 20°C = 0.2966 \text{ kJ/kg·K}$$

A mass balance for the heater gives

$$\dot{m}_3 = \dot{m}_1 + \dot{m}_2 = 1000 + 4200 = 5200 \text{ kg/min}$$

By neglecting kinetic and potential energies, an energy balance for the heater gives

$$\dot{m}_1 h_1 + \dot{m}_2 h_2 = \dot{m}_3 h_3$$

so that

$$h_1 = \frac{1}{\dot{m}_1}(\dot{m}_3 h_3 - \dot{m}_2 h_2)$$

$$= \frac{1}{1000}(5200 \times 721.11 - 4200 \times 146.68)$$

$$= 3133.7 \text{ kJ/kg}$$

From the superheated steam table (Table A-20M), at $p_1 = 0.8$ MPa and $h_1 = 3133.7$ kJ/kg, we get

$$T_1 = 336.7°C \qquad \text{and} \qquad s_1 = 7.3634 \text{ kJ/kg·K}$$

By neglecting kinetic and potential energies, the stream availability is calculated by Eq. (8-19):

$$\psi_1 = w_{max,1-0} = (h_1 - h_0) - T_0(s_1 - s_0)$$

$$= (3133.7 - 83.96) - 293(7.3634 - 0.2966) = 979.2 \text{ kJ/kg}$$

$$\psi_2 = w_{max,2-0} = (h_2 - h_0) - T_0(s_2 - s_0)$$

$$= (146.68 - 83.96) - 293(0.5053 - 0.2966) = 1.571 \text{ kJ/kg}$$

$$\psi_3 = w_{max,3-0} = (h_3 - h_0) - T_0(s_3 - s_0)$$

$$= (721.11 - 83.96) - 293(2.0462 - 0.2966) = 124.5 \text{ kJ/kg}$$

The rate of stream availability change is then given by

$$\Delta\dot{\Psi} = \dot{m}_3\psi_3 - \dot{m}_1\psi_1 - \dot{m}_2\psi_2$$

$$= 5200(124.5) - 1000(979.2) - 4200(1.571)$$

$$= -388,000 \text{ kJ/min}$$

The rate of stream availability change can also be calculated by considering $\dot{m}_1$ to change from state 1 to state 3 and $\dot{m}_2$ to change from state 2 to state 3. Thus,

$$\Delta\dot{\Psi} = \dot{m}_1(\psi_3 - \psi_1) + \dot{m}_2(\psi_3 - \psi_2)$$

$$= 1000(124.5 - 979.2) + 4200(124.5 - 1.571)$$

$$= -338,000 \text{ kJ/min}$$

Sec. 8-4   Availability of a Steady-Flow System

The irreversibility rate is given by Eq. (8-12):

$$\dot{I} = T_0(\Delta\dot{S}_{system} + \Delta\dot{S}_{surr})$$

where $\Delta\dot{S}_{system} = 0$ for a steady-flow, steady-state process

$$\Delta\dot{S}_{surr} = \dot{m}_3 s_3 - \dot{m}_1 s_1 - \dot{m}_2 s_2$$
$$= 5200(2.0462) - 1000(7.3634) - 4200(0.5053)$$
$$= 1154.6 \text{ kJ/min·K}$$

Thus,

$$\dot{I} = 293(1154.6) = 338,000 \text{ kJ/min}$$

The irreversibility rate can also be calculated by Eq. (8-9):

$$\dot{I} = \dot{W}_{max} - \dot{W}_{actual}$$

where

$$\dot{W}_{actual} = 0$$

$$\dot{W}_{max} = \sum_{inlet} \dot{m}\psi - \sum_{exit} \dot{m}\psi \qquad \text{(Eq. (8-22) without the } \dot{Q}_R \text{ term)}$$

$$= \dot{m}_1\psi_1 + \dot{m}_2\psi_2 - \dot{m}_3\psi_3$$

$$= -\Delta\dot{\Psi} = 338,000 \text{ kJ/min}$$

Thus,

$$\dot{I} = 338,000 - 0 = 338,000 \text{ kJ/min}$$

Notice that the irreversibility is due to mixing the two incoming streams. The process is irreversible as the effect cannot be reversed by simply reversing the direction of the process. Because no work is done during the process, the irreversibility is equal in magnitude to the maximum work that could have been produced if the process were reversible.

8-5 EFFECTIVENESS OF PROCESSES AND CYCLES

For the efficient use of energy, we are concerned not only in the quantity of the energy, but also the quality of it. The equipment efficiency of a device and the thermal efficiency of a power cycle as frequently employed in engineering analysis are *first-law efficiencies* that are based solely on the quantities of energy according to the first law of thermodynamics. We now introduce a new measure of performance based on the second law of thermodynamics, involving the "quality" of energy in terms of availability. The definition of the *second-law effectiveness* (or simply called *effectiveness*) can best be illustrated in separate cases as follows.

(I) Gas and steam turbines

As indicated in the *T–s* diagrams of Fig. 8-9, process 1–2 represents the actual irreversible adiabatic expansion and process 1–2' represents the ideal internally reversible adiabatic expansion from the same initial state to the same final pressure as the actual case. By neglecting kinetic and potential energies, the actual and ideal work output of the turbine per unit mass flow are given, respectively, by

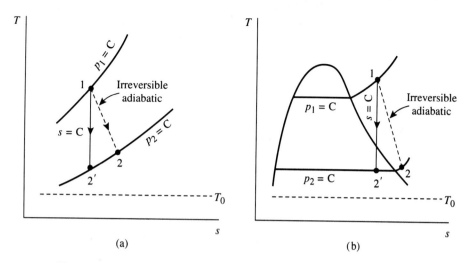

Figure 8-9 Expansion processes: (a) gas turbine, and (b) steam turbine.

$$w_{\text{actual, 12}} = h_1 - h_2$$

$$w_{\text{ideal, 12}'} = h_1 - h_{2'}$$

The first-law adiabatic efficiency of the turbine is defined as

$$\eta_T = \frac{w_{\text{actual, 12}}}{w_{\text{ideal, 12}'}} = \frac{h_1 - h_2}{h_1 - h_{2'}} \tag{8-23}$$

On the other hand, the second-law effectiveness of the turbine is defined as

$$\epsilon_T = \frac{w_{\text{actual, 12}}}{w_{\text{max, 12}}} = \frac{h_1 - h_2}{\psi_1 - \psi_2} \tag{8-24}$$

where

$$w_{\text{max, 12}} = -\Delta\psi_{12} = \psi_1 - \psi_2$$

is the maximum work obtainable from the steady-flow fluid stream in terms of the decrease in stream availability as expressed by Eq. (8-20).

(II) Gas and vapor compressors

As indicated in the T–s diagram of Fig. 8-10, process 1–2 represents the actual irreversible adiabatic compression and process 1–2' represents the ideal internally reversible adiabatic compression from the same initial state to the same final pressure as the actual case. By neglecting kinetic and potential energies, the actual and ideal work input of the compressor per unit mass flow are given, respectively, by

$$-w_{\text{actual,12}} = h_2 - h_1$$

$$-w_{\text{ideal,12}'} = h_{2'} - h_1$$

The first-law adiabatic efficiency of the compressor is defined as

$$\eta_c = \frac{-w_{\text{ideal,12}'}}{-w_{\text{actual,12}}} = \frac{h_{2'} - h_1}{h_2 - h_1} \tag{8-25}$$

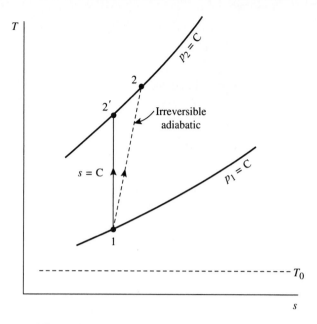

Figure 8-10 Gas or vapor compression processes.

On the other hand, the second-law effectiveness of the compressor is defined as

$$\epsilon_c = \frac{-w_{\text{max},12}}{-w_{\text{actual},12}} = \frac{\psi_2 - \psi_1}{h_2 - h_1} \qquad (8\text{-}26)$$

where

$$-w_{\text{max},12} = \Delta\psi_{12} = \psi_2 - \psi_1$$

is the minimum work requirement attributable to the increase in stream availability.

(III) Heat exchanger

Consider the heat exchanger shown in the schematic diagram of Fig. 8-11(a) in which heat is transferred from the hotter fluid a to the colder fluid b. The T–s diagram of Fig. 8-11(b) is for the particular case in which fluid a is an ideal gas (such as hot products of combustion) in a constant-pressure cooling process, and fluid b is a pure substance (such as steam) in a constant-pressure heating, vaporization, and superheating process. Assuming no heat transfer across the control surface and neglecting changes in kinetic and potential energies, the first law of thermodynamics gives the energy balance equation,

$$\dot{m}_a(h_1 - h_2) = \dot{m}_b(h_4 - h_3)$$

As the processes occurring within the two fluid streams can be considered internally reversible, we must have on the T–s diagram

$$\text{area } 1\text{--}2\text{--}a\text{--}b\text{--}1 = \text{area } 3\text{--}4\text{--}d\text{--}c\text{--}3$$

Even though the first law dictates that the energy is conserved between the two fluid streams, the availability change of the hotter fluid stream is greater than that of

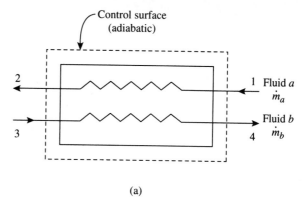

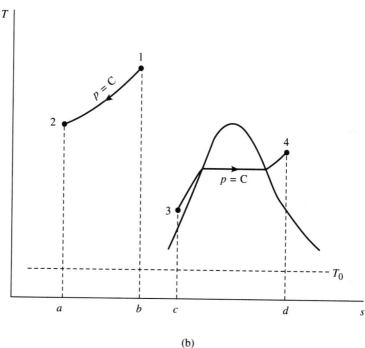

(b)

Figure 8-11 Heat exchanger: (a) schematic diagram, and (b) T–s diagram.

the colder fluid stream due to the transfer of heat across finite temperature differences. Expressed mathematically,

$$\dot{m}_a(\psi_1 - \psi_2) > \dot{m}_b(\psi_4 - \psi_3)$$

The second-law effectiveness of the heat exchanger is defined as

$$\epsilon_{\text{heat exchanger}} = \frac{\dot{m}_b(\psi_4 - \psi_3)}{\dot{m}_a(\psi_1 - \psi_2)} \tag{8-27}$$

(IV) Thermodynamic cycle

Consider a steady-flow steam power cycle as depicted in the T–s diagram of Fig. 8-12, all processes being internally reversible. Heat is transferred from an energy reservoir at

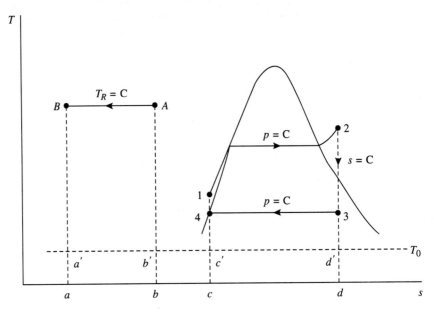

Figure 8-12 T–s diagram for a steam cycle.

T_R to the steam generator, thus heating the steam under a constant pressure from state 1 to state 2. A turbine then expands the steam from state 2 to state 3 in an isentropic process. The steam exhausted from the turbine is subsequently condensed to saturated liquid at state 4, giving up waste heat to the cooling water in the condenser. A pump is used to raise the water pressure isentropically back to that of the steam generator.

By neglecting changes in kinetic and potential energies, the steady-flow energy balances, per unit mass flow, written for the steam generator, the condenser, the turbine, and the pump, respectively, are

$$q_{in} = q_{\text{steam generator}} = h_2 - h_1$$

$$q_{out} = q_{\text{condenser}} = h_3 - h_4$$

$$w_{out} = w_{\text{turbine}} = h_2 - h_3$$

$$w_{in} = w_{\text{pump}} = h_1 - h_4$$

Thus,

$$\text{net work out} = w_{out} - w_{in} = (h_2 - h_3) - (h_1 - h_4)$$

$$= \text{net heat in} = q_{in} - q_{out} = (h_2 - h_1) - (h_3 - h_4)$$

The preceding equation is true according to the first law of thermodynamics. The cycle thermal efficiency is defined as

$$\eta_{th} = \frac{\text{net work out}}{\text{heat in}} = \frac{w_{out} - w_{in}}{q_{in}} = \frac{q_{in} - q_{out}}{q_{in}} = 1 - \frac{q_{out}}{q_{in}}$$

$$= 1 - \frac{h_3 - h_4}{h_2 - h_1}$$

(8-28)

The second-law effectiveness of the power cycle can be defined as

$$\epsilon'_{\text{power cycle}} = \frac{\text{net work out}}{\text{stream availability increase due to heat addition}}$$

$$= \frac{w_{\text{out}} - w_{\text{in}}}{\Delta\psi_{\text{heat in}}} = \frac{q_{\text{in}} - q_{\text{out}}}{\Delta\psi_{\text{heat in}}} \quad (8\text{-}29)$$

$$= \frac{(h_2 - h_1) - (h_3 - h_4)}{\psi_2 - \psi_1}$$

where

$$\Delta\psi_{\text{heat in}} = \psi_2 - \psi_1 = (h_2 - h_1) - T_0(s_2 - s_1)$$

is the increase in stream availability of steam in the steam generator due to heating.

If the available energy in the terminology of Sec. 7-9 based on the energy reservoir at T_R is used as the standard of performance comparison, the second-law effectiveness of the overall system including the energy reservoir and all the cycle components can be defined as

$$\epsilon_{\text{power cycle}} = \frac{\text{net work out}}{q_{\text{av, reservoir}}}$$

$$= \frac{(h_2 - h_1) - (h_3 - h_4)}{q_R\left(1 - \dfrac{T_0}{T_R}\right)} \quad (8\text{-}30)$$

where

$$q_{\text{av, reservoir}} = q_R\left(1 - \frac{T_0}{T_R}\right)$$

is the available energy of the heat q_R (per unit mass of steam flow of the cycle) transferred from the energy reservoir at T_R with a magnitude

$$|q_R| = q_{\text{in}} = q_{12}$$

as represented by the areas in Fig. 8-12 such that

$$\text{area } A\text{--}B\text{--}a\text{--}b\text{--}A = \text{area } 1\text{--}2\text{--}d\text{--}c\text{--}1$$

Notice that the difference between ϵ and ϵ' as defined before is due to the irreversible heat transfer from the reservoir to the steam.

It is noted that

$$\eta_{\text{th, actual}} = \frac{\text{actual net work out}}{q_R}$$

and

$$\eta_{\text{th, Carnot}} = 1 - \frac{T_0}{T_R}$$

where q_R is the heat addition to the cycle from the high-temperature reservoir at T_R, and

T_0 is the temperature of the low-temperature reservoir. Equation (8-30) can then be rewritten as

$$\epsilon_{\text{power cycle}} = \frac{\eta_{\text{th, actual}}}{\eta_{\text{th, Carnot}}} \qquad (8\text{-}31)$$

or

$$\epsilon_{\text{power cycle}} = \frac{w_{\text{net, actual}}}{w_{\text{net, Carnot}}} \qquad (8\text{-}32)$$

where

$$w_{\text{net, actual}} = q_R \eta_{\text{th, actual}}$$

and

$$w_{\text{net, Carnot}} = q_R \eta_{\text{th, Carnot}}$$

Note that in Eq. (8-31), the second-law effectiveness is defined as the ratio of the thermal efficiency of the actual cycle to the thermal efficiency of the Carnot cycle operating between the same temperature limits. The sink temperature can be defined to have a value other than the atmospheric temperature T_0. Thus, we can write

$$\eta_{\text{th, Carnot}} = 1 - \frac{T_L}{T_H}$$

where T_H and T_L are the temperatures of the high- and low-energy reservoirs, respectively.

Analogous to Eq. (8-31), the second-law effectiveness of a refrigerator and a heat pump can be defined as

$$\epsilon_{\text{refrigerator}} = \frac{(\text{COP})_{\text{actual ref.}}}{(\text{COP})_{\text{Carnot ref.}}} = \frac{W_{\text{in, Carnot ref.}}}{W_{\text{in, actual ref.}}} \qquad (8\text{-}33)$$

$$\epsilon_{\text{heat pump}} = \frac{(\text{COP})_{\text{actual heat pump}}}{(\text{COP})_{\text{Carnot heat pump}}} = \frac{W_{\text{in, Carnot heat pump}}}{W_{\text{in, actual heat pump}}} \qquad (8\text{-}34)$$

where

$$(\text{COP})_{\text{Carnot refrigerator}} = \frac{T_L}{T_H - T_L}$$

$$(\text{COP})_{\text{Carnot heat pump}} = \frac{T_H}{T_H - T_L}$$

in which T_H and T_L are the temperatures of the high- and low-energy reservoirs, respectively.

It should be mentioned that there are numerous different definitions of second-law effectiveness that appear in the literature. What we defined here is by no means exclusive.

Example 8-8

Steam enters an adiabatic turbine at 3 MPa and 400°C and expands to 0.1 MPa and 120°C. The atmosphere is at 0.1 MPa and 27°C. Neglecting changes in kinetic and potential energies, determine, per kilogram of steam, the stream availability at the inlet and exit conditions, the maximum work, and the irreversibility. What are the first-law efficiency and the second-law effectiveness of the turbine?

Solution Referring to Fig. 8-9(b), p_1 = 3 MPa, T_1 = 400°C, p_2 = 0.1 MPa, T_2 = 120°C, p_0 = 0.1 MPa, and T_0 = 27°C = 300 K. From the superheated steam table (Table A-20M), at p_1 = 3 MPa and T_1 = 400°C, we get h_1 = 3230.9 kJ/kg, s_1 = 6.9212 kJ/kg·K; and at p_2 = 0.1 MPa and T_2 = 120°C, we get h_2 = 2716.6 kJ/kg and s_2 = 7.4668 kJ/kg·K. At the dead state of p_0 = 0.1 MPa and T_0 = 27°C, the water is a compressed liquid and its properties are essentially equal to those of saturated liquid at T_0. Thus, from the saturated steam table (Table A-18M), we have

$$h_0 \approx h_f \text{ at } 27°C = 113.25 \text{ kJ/kg} \qquad \text{and} \qquad s_1 \approx s_f \text{ at } 27°C = 0.3954 \text{ kJ/kg·K}$$

For the internally reversible adiabatic process 1–2′,

$$s_{2'} = s_1 = s_f + x_{2'} s_{fg} \text{ at } 0.1 \text{ MPa}$$

taking data from Table A-19M, we have

$$6.9212 = 1.3026 + x_{2'}(7.3594 - 1.3026)$$

so that

$$x_{2'} = 0.928$$

Accordingly,

$$h_{2'} = (h_f + x_{2'} h_{fg}) \text{ at } 0.1 \text{ MPa}$$
$$= 417.46 + 0.928(2258.0)$$
$$= 2513 \text{ kJ/kg}$$

The stream availability of the steam at the inlet and exit conditions are

$$\psi_1 = h_1 - h_0 - T_0(s_1 - s_0)$$
$$= 3230.9 - 113.25 - 300(6.9212 - 0.3954)$$
$$= 1160 \text{ kJ/kg}$$
$$\psi_2 = h_2 - h_0 - T_0(s_2 - s_0)$$
$$= 2716.6 - 113.25 - 300(7.4668 - 0.3954) = 482 \text{ kJ/kg}$$

The actual work and maximum work of the turbine are, respectively,

$$w_{\text{actual, 12}} = h_1 - h_2 = 3230.9 - 2716.6 = 514 \text{ kJ/kg}$$
$$w_{\text{max, 12}} = \psi_1 - \psi_2 = 1160 - 482 = 678 \text{ kJ/kg}$$

The irreversibility of the turbine is

$$i = w_{\text{max, 12}} - w_{\text{actual, 12}} = 678 - 514 = 164 \text{ kJ/kg}$$

or

$$i = T_0(\Delta s_{\text{system}} + \Delta s_{\text{surr}}) = T_0(\Delta s_{\text{surr}}) \text{ for steady flow}$$
$$= T_0(s_2 - s_1) = 300(7.4668 - 6.9212) = 164 \text{ kJ/kg}$$

The first-law adiabatic turbine efficiency is

$$\eta_T = \frac{h_1 - h_2}{h_1 - h_{2'}} = \frac{3230.9 - 2716.6}{3230.9 - 2513} = 0.716$$

The second-law effectiveness of the turbine is

$$\epsilon_T = \frac{W_{actual,\,12}}{W_{max,\,12}} = \frac{h_1 - h_2}{\psi_1 - \psi_2} = \frac{514}{678} = 0.758$$

Notice that the first-law adiabatic turbine efficiency compares the actual turbine work with the maximum work that could be produced for the given initial state and the given exit pressure. On the other hand, the second-law effectiveness compares the actual turbine work with the maximum work that could be produced for the given initial state and exit state.

Example 8-9

In the heat-exchanger section of a steam generator as depicted in Fig. 8-11, heat is transferred from the products of combustion (fluid *a*) to the water (fluid *b*). The products of combustion can be considered as air with variable specific heats. It is cooled at a constant pressure of 101 kPa from 1350 K (1077°C) to 800 K (527°C). The water enters the heat exchanger at 1 MPa and 140°C and leaves at 1 MPa and 250°C. The atmosphere is at 27°C and 101 kPa. Neglecting kinetic and potential energies, determine, per kilogram of water flow, the irreversibility and the second-law effectiveness of the heat exchanger.

Solution   Given $p_1 = p_2 = 101$ kPa, $T_1 = 1350$ K, $T_2 = 800$ K, $p_3 = p_4 = 1$ MPa, $T_3 = 140°C$, $T_4 = 250°C$, $T_0 = 27°C = 300$ K, $p_0 = 101$ kPa, and $m_b = 1$. From the air table (Table A-9M), at $T_1 = 1350$ K, we get $h_1 = 1455.35$ kJ/kg and $\phi_1 = 7.3177$ kJ/kg·K; and at $T_2 = 800$ K, we get $h_2 = 822.15$ kJ/kg and $\phi_2 = 6.7175$ kJ/kg·K. Thus,

$$s_1 - s_2 = \phi_1 - \phi_2 - R \ln (p_1/p_2)$$

$$= 7.3177 - 6.7175 - 0 = 0.6002 \text{ kJ/kg·K}$$

From the saturated steam table (Table A-18M), we get for approximation

$$h_3 \approx h_f \text{ at } 140°C = 589.13 \text{ kJ/kg}$$

$$s_3 \approx s_f \text{ at } 140°C = 1.7391 \text{ kJ/kg·K}$$

From the superheated steam table (Table A-20M), at $p_4 = 1$ MPa and $T_4 = 250°C$, we get $h_4 = 2942.6$ kJ/kg and $s_4 = 6.9247$ kJ/kg·K.

An energy balance for the adiabatic control volume of Fig. 8-11(a) gives

$$\frac{m_a}{m_b} = \frac{h_4 - h_3}{h_1 - h_2} = \frac{2942.6 - 589.13}{1455.35 - 822.15} = 3.717 \text{ kg air/kg water}$$

The increase in stream availability for the water is given by

$$m_b(\psi_4 - \psi_3) = m_b[(h_4 - h_3) - T_0(s_4 - s_3)]$$

$$= 1[(2942.6 - 589.13) - 300(6.9247 - 1.7391)]$$

$$= 797.8 \text{ kJ/kg water}$$

The decrease in stream availability for the air (per kilogram of water) is given by

$$\frac{m_a}{m_b}(\psi_1 - \psi_2) = \frac{m_a}{m_b}[(h_1 - h_2) - T_0(s_1 - s_2)]$$

$$= 3.717[(1455.35 - 822.15) - 300(0.6002)]$$

$$= 1684 \text{ kJ/kg water}$$

By Eq. (8-22), the maximum work per kilogram of water flow is

$$w_{max} = m_b(\psi_3 - \psi_4) + \frac{m_a}{m_b}(\psi_1 - \psi_2)$$

$$= -797.8 + 1684 = 886 \text{ kJ/kg water}$$

Because the actual work is zero, the irreversibility is given by Eq. (8-9):

$$i = w_{max} - w_{actual} = w_{max} = 886 \text{ kJ/kg water}$$

The irreversibility can also be calculated by Eq. (8-12):

$$i = T_0(\Delta s_{system} + \Delta s_{surr}) = T_0(\Delta s_{surr}) \text{ for steady flow}$$

$$= T_0\left[m_b(s_4 - s_3) + \frac{m_a}{m_b}(s_2 - s_1)\right]$$

$$= 300[1(6.9247 - 1.7391) + 3.717(-0.6002)]$$

$$= 886 \text{ kJ/kg water}$$

From Eq. (8-27), the second-law effectiveness of the heat exchanger is

$$\epsilon_{\text{heat exchanger}} = \frac{m_b(\psi_4 - \psi_3)}{m_a(\psi_1 - \psi_2)} = \frac{797.8}{1684} = 0.474 = 47.4\%$$

Notice that because there is no actual work produced, the maximum work is equal to the irreversibility. If the process described in the example could be achieved by using reversible heat engines and heat pumps, 886 kJ/kg water of work could be produced while the combustion products and water experienced the same state changes as in the problem description.

Example 8-10

A simple steam power cycle as depicted in the T–s diagram of Fig. 8-12 has the following conditions: $p_1 = p_2 = 1000$ psia, $T_2 = 900°F$, and $p_3 = p_4 = 2$ psia. The energy reservoir is at $T_R = 1000°F$, and the atmosphere is at $T_0 = 40°F$ and $p_0 = 14.7$ psia. Determine the thermal efficiency and the second-law effectiveness for the power cycle.

Solution   From Table A-20E for superheated steam, at $p_2 = 1000$ psia and $T_2 = 900°F$, we get $h_2 = 1448.1$ Btu/lbm and $s_2 = 1.6120$ Btu/lbm·°R. For the internally reversible adiabatic process 2–3,

$$s_3 = s_2 = s_f + x_3 s_{fg} \text{ at 2 psia}$$

Taking data from Table A-19E for saturated steam, we have

$$1.6120 = 0.17499 + x_3(1.7448)$$

so that

$$x_3 = 0.8236$$

Accordingly,

$$h_3 = h_f + x_3 h_{fg} \text{ at 2 psia}$$

$$= 94.02 + 0.8236(1022.1) = 935.8 \text{ Btu/lbm}$$

Also from Table A-19E,

$$h_4 = h_f \text{ at 2 psia} = 94.02 \text{ Btu/lbm}$$

and

$$s_4 = s_f \text{ at 2 psia} = 0.17499 \text{ Btu/lbm·°R}$$

From Table A-21E for compressed liquid, at $p_1 = 1000$ psia and $s_1 = s_4 = 0.17499$ Btu/lbm·°R, we get by interpolation

$$T_1 = 127°F \qquad \text{and} \qquad h_1 = 97.5 \text{ Btu/lbm}$$

The heat added to the steam in the steam generator and the heat rejected from the steam in the condenser are, respectively,

$$q_{in} = q_{12} = h_2 - h_1 = 1448.1 - 97.5 = 1350.6 \text{ Btu/lbm}$$

$$q_{out} = -q_{34} = h_3 - h_4 = 935.8 - 94.0 = 841.8 \text{ Btu/lbm}$$

The net work out is

$$\text{net } w_{out} = q_{in} - q_{out} = 1350.6 - 841.8 = 508.8 \text{ Btu/lbm}$$

The thermal efficiency is

$$\eta_{th} = \frac{\text{net } w_{out}}{q_{in}} = \frac{508.8}{1350.6} = 0.377$$

The increase of stream availability of the steam in the heating process is

$$\psi_2 - \psi_1 = (h_2 - h_1) - T_0(s_2 - s_1)$$
$$= (1448.1 - 97.5) - 500(1.6120 - 0.17499) = 632.1 \text{ Btu/lbm}$$

The second-law effectiveness of the cycle is given by Eq. (8-29):

$$\epsilon'_{\text{power cycle}} = \frac{\text{net } w_{out}}{\psi_2 - \psi_1} = \frac{508.8}{632.1} = 0.805$$

With q_R and q_{in} numerically equal, Eq. (8-30) gives the second-law effectiveness of the overall system as

$$\epsilon_{\text{power cycle}} = \frac{\text{net } w_{out}}{q_R\left(1 - \dfrac{T_0}{T_R}\right)} = \frac{508.8}{1350.6\left(1 - \dfrac{500}{1460}\right)}$$
$$= 0.573$$

The preceding expression can be written as

$$\epsilon_{\text{power cycle}} = \frac{\eta_{th}}{1 - \dfrac{T_0}{T_R}} = \frac{0.377}{1 - \dfrac{500}{1460}} = 0.573$$

It is instructive to indicate the preceding values of thermal efficiency and second-law effectiveness graphically in reference to Fig. 8-12 as follows:

$$\eta_{th} = 0.377 = (\text{area } 1-2-3-4-1)/(\text{area } 1-2-d-c-1)$$

$$\epsilon'_{\text{power cycle}} = 0.805 = (\text{area } 1-2-3-4-1)/(\text{area } 1-2-d'-c'-1)$$

$$\epsilon_{\text{power cycle}} = 0.573 = (\text{area } 1-2-3-4-1)/(\text{area } A-B-a'-b'-A)$$

where area $1-2-d'-c'-1 <$ area $A-B-a'-b'-A$ due to irreversible heat transfer from the reservoir to the steam.

8-6 SUMMARY

The general expression for maximum work for an open system that exchanges heat with an energy reservoir at T_R in addition to the atmosphere at T_0 is

$$\text{d}W_{\max} = -\text{d}(E - T_0 S)_{\text{system}} + \sum_{\text{inlet}} [(e + pv - T_0 s)\,\text{d}m]$$

$$- \sum_{\text{exit}} [(e + pv - T_0 s)\text{d}m] + \text{d}Q_R\left(1 - \frac{T_0}{T_R}\right)$$

For a single inlet stream at section 1 and a single exit stream at section 2, this equation becomes

$$\text{d}W_{\max} = -\text{d}(E - T_0 S)_{\text{system}} + (e_1 + p_1 v_1 - T_0 s_1)\,\text{d}m_1$$

$$-(e_2 + p_2 v_2 - T_0 s_2)\,\text{d}m_2 + \text{d}Q_R\left(1 - \frac{T_0}{T_R}\right)$$

Under steady-state and steady-flow conditions, $dE_{\text{system}} = 0$, $dS_{\text{system}} = 0$, and $\text{d}m_1 = \text{d}m_2 = \text{d}m$, and the maximum work equation reduces to

$$\text{d}W_{\max} = [(e_1 + p_1 v_1 - T_0 s_1) - (e_2 + p_2 v_2 - T_0 s_2)]\,\text{d}m + \text{d}Q_R\left(1 - \frac{T_0}{T_R}\right)$$

For a unit mass of fluid flowing through a steady-state, steady-flow system, the maximum work equation can be written as

$$w_{\max} = (h_1 + \tfrac{1}{2}\mathbf{V}_1^2 + gz_1 - T_0 s_1)$$

$$= -(h_2 + \tfrac{1}{2}\mathbf{V}_2^2 + gz_2 - T_0 s_2) + q_R\left(1 - \frac{T_0}{T_R}\right)$$

For a closed system, $\text{d}m_{\text{inlet}} = \text{d}m_{\text{exit}} = 0$, and the total energy E can be replaced by the internal energy U if there are no kinetic and potential energy changes for the system. The maximum-work equation in differential form is then

$$\text{d}W_{\max} = -\text{d}(U - T_0 S)_{\text{system}} + \text{d}Q_R\left(1 - \frac{T_0}{T_R}\right)$$

For a change from initial state i to final state f for a closed system, we have

$$W_{\max, i-f} = (U_i - T_0 S_i) - (U_f - T_0 S_f) + Q_R\left(1 - \frac{T_0}{T_R}\right)$$

Notice that when the last term involving Q_R (or q_R) in the preceding equations is not included, these equations are for systems that exchange heat with the atmosphere only.

The difference between the maximum work $W_{\max}$ and the actual work W_{actual} is called irreversibility I. Thus,

$$I = W_{\max} - W_{\text{actual}}$$

or

$$\text{d}I = \text{d}W_{\max} - \text{d}W_{\text{actual}}$$

The value of irreversibility for a process can be obtained by the expression

$$I = T_0(\Delta S_{\text{system}} + \Delta S_{\text{surr}})$$

or

$$dI = T_0(dS_{\text{system}} + dS_{\text{surr}})$$

For a closed system of mass m undergoing a process from state 1 to state 2 while receiving a quantity of heat Q_R from an energy reservoir at T_R, the irreversibility of the process is given by

$$I = T_0(\Delta S_{\text{system}} + \Delta S_{\text{surr}}) = T_0\left[m(s_2 - s_1) + \frac{-Q_R}{T_R}\right]$$

If the preceding closed system is insulated, $Q_R = 0$, and

$$I = T_0[m(s_2 - s_1)]$$

For an open system undergoing a steady-flow, steady-state process from inlet condition 1 to exit condition 2, while receiving a quantity of heat q_R per unit mass flow from an energy reservoir at T_R, the specific irreversibility of the process is given by

$$i = T_0(\overset{0}{\cancel{\Delta s_{\text{system}}}} + \Delta s_{\text{surr}}) = T_0\left[(s_2 - s_1) + \frac{-q_R}{T_R}\right]$$

If the preceding steady-flow, steady-state system is insulated, $q_R = 0$, and

$$i = T_0(s_2 - s_1)$$

When the system temperature T equals the atmosphere temperature T_0 and the system pressure p equals the atmosphere pressure p_0, such a system is said to be in the dead state.

The work done by a system exclusive of that done on the atmosphere as the system boundary expands or contracts is called the useful work. Symbolically,

$$W_{\text{useful}} = W_{\text{system}} - p_0(V_f - V_i)$$

where V_i and V_f are, respectively, the initial and final volumes of the system.

The availability Φ of a closed system in a given state is defined as the maximum useful work that can be obtained from the system–atmosphere combination as the system proceeds reversibly from that state to its dead state while exchanging heat only with the atmosphere. Let the subscripts 1 and 0 denote, respectively, a given state and the dead state of a system. The closed-system availability Φ_1 at state 1 is expressed as

$$\Phi_1 = W_{\text{max useful}, 1-0} = W_{\text{max}, 1-0} - p_0(V_0 - V_1)$$
$$= (U_1 + p_0 V_1 - T_0 S_1) - (U_0 + p_0 V_0 - T_0 S_0)$$

For a unit mass, the specific availability φ_1 at state 1 is

$$\varphi_1 = (u_1 + p_0 v_1 - T_0 s_1) - (u_0 + p_0 v_0 - T_0 s_0)$$

When a closed system that exchanges heat only with the atmosphere undergoes a reversible process from state 1 to state 2, the maximum useful work obtainable is the decrease in availability. Thus, for a unit mass, we have

$$w_{\text{max useful, }1-2} = \varphi_1 - \varphi_2$$

$$= (u_1 + p_0 v_1 - T_0 s_1) - (u_2 + p_0 v_2 - T_0 s_2)$$

If the system also receives heat q_R from an energy reservoir at T_R, then

$$w_{\text{max useful, }1-2} = (\varphi_1 - \varphi_2) + q_R\left(1 - \frac{T_0}{T_R}\right)$$

The stream availability Ψ of a steady-flow fluid stream is defined as the maximum work that can be obtained as the fluid changes reversibly from the given state to the dead state while exchanging heat with only the atmosphere. The specific stream availability at state 1 is given by the expression

$$\psi_1 = w_{\text{max, }1-0}$$

$$= (h_1 + \tfrac{1}{2}V_1^2 + gz_1 - T_0 s_1) - (h_0 + gz_0 - T_0 s_0)$$

When a steady-flow fluid stream that exchanges heat only with the atmosphere proceeds reversibly from state 1 to state 2, the maximum work obtainable is the decrease in stream availability. Thus, for a unit mass,

$$w_{\text{max, }1-2} = \psi_1 - \psi_2$$

$$= (h_1 + \tfrac{1}{2}V_1^2 + gz_1 - T_0 s_1) - (h_2 + \tfrac{1}{2}V_2^2 + gz_2 - T_0 s_2)$$

If the fluid stream also receives heat q_R from an energy reservoir at T_R, then

$$w_{\text{max, }1-2} = (\psi_1 - \psi_2) + q_R\left(1 - \frac{T_0}{T_R}\right)$$

If there is more than one flow into and out of an open system in a steady-flow, steady-state process, we write on a rate basis that

$$\dot{W}_{\text{max}} = \sum_{\text{inlet}} \dot{m}\psi - \sum_{\text{exit}} \dot{m}\psi + \dot{Q}_R\left(1 - \frac{T_0}{T_R}\right)$$

where the last term is for the heat interaction with an energy reservoir other than the atmosphere.

The second-law effectiveness of steady-flow devices is defined as follows:
For gas and steam turbines,

$$\epsilon_T = \frac{w_{\text{actual, }1-2}}{w_{\text{max, }1-2}} = \frac{h_1 - h_2}{\psi_1 - \psi_2}$$

For gas and vapor compressors,

$$\epsilon_c = \frac{-w_{\text{max, }1-2}}{-w_{\text{actual, }1-2}} = \frac{\psi_2 - \psi_1}{h_2 - h_1}$$

For heat exchangers,

$$\epsilon_{\text{heat exchanger}} = \frac{\text{availability change of the colder fluid stream}}{\text{availability change of the hotter fluid stream}}$$

The second-law effectiveness of a power cycle is defined as

$$\epsilon'_{power\ cycle} = \frac{net\ work\ out}{(stream\ availability\ increase\ due\ to\ heat\ addition)} = \frac{W_{out} - W_{in}}{\Delta\psi_{heat\ in}}$$

or

$$\epsilon_{power\ cycle} = \frac{net\ work\ out}{q_{av,\ reservoir}} = \frac{W_{out} - W_{in}}{q_R(1 - T_0/T_R)}$$

The second-law effectiveness of a power cycle can be redefined as

$$\epsilon_{power\ cycle} = \frac{\eta_{th}}{\eta_{th,\ Carnot}}$$

where

$$\eta_{th,\ Carnot} = 1 - \frac{T_L}{T_H}$$

Similarly, the second-law effectiveness of a refrigerator and a heat pump can be defined as

$$\epsilon_{refrigerator} = \frac{(COP)_{refrigerator}}{(COP)_{Carnot\ refrigerator}}$$

$$\epsilon_{heat\ pump} = \frac{(COP)_{heat\ pump}}{(COP)_{Carnot\ heat\ pump}}$$

where

$$(COP)_{Carnot\ regrigerator} = \frac{T_L}{T_H - T_L}$$

$$(COP)_{Carnot\ heat\ pump} = \frac{T_H}{T_H - T_L}$$

in which T_H and T_L are the temperatures of the high- and low-energy reservoirs, respectively.

PROBLEMS

8-1. A 1.2-m³ insulated rigid tank contains 2.13 kg of carbon dioxide initially at 100 kPa. See Fig. P8-1. Paddle-wheel work is done on the gas until the pressure increases to 120 kPa. Determine (a) the actual paddle-wheel work and (b) the minimum work between the two end states. The atmosphere is at 25°C and 100 kPa.

8-2. A storage battery is capable of delivering 1 kW·h of energy. In order to have the same work capability, what would be the volume needed, in ft³, of air stored in a tank at 80°F and 750 psia? The atmosphere is at 80°F and 14.7 psia.

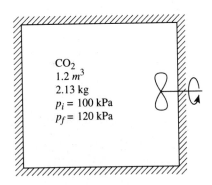

CO_2
1.2 m³
2.13 kg
$p_i = 100$ kPa
$p_f = 120$ kPa

Figure P8-1

Availability and Irreversibility   Chap. 8

8-3. Air in a closed system expands from 30 psia, 140°F, to 15 psia, 100°F, while receiving 10 Btu/lbm of heat from a reservoir at 250°F. The atmosphere is at 14.0 psia, 80°F. Determine the maximum work in Btu/lbm. How much of this work would be done on the atmosphere?

8-4. A 12.0-ft³ insulated rigid tank contains saturated steam at 200°F. A paddle wheel in the tank is turned on until the steam is at 200 psia and 1000°F. (a) Determine the work input in Btu/lbm. (b) After the process has occurred, how much of the work input by the paddle wheel could possibly be reconverted to work again?

8-5. As a young engineer, you are asked to compare the performance of two adiabatic nozzle processes. Both nozzles have an adiabatic efficiency of 95%. Both nozzles take in nitrogen at 2 MPa with the same mass-flow rate and negligible inlet velocity. The inlet temperatures of the two nozzles are different: they are 300°C for nozzle A and 400°C for nozzle B. From a thermodynamics viewpoint, which of these nozzle processes is superior? Assume $T_0 = 25$°C.

8-6. Refrigerant-12 enters an insulated compressor at 200 kPa, 0°C, with a mass-flow rate of 90 kg/h and exits at 1.2 MPa in steady-flow operation. The compressor has an isentropic efficiency of 83%. Determine (a) the temperature of the refrigerant at exit from the compressor, (b) the power input to the compressor, in kW, and (c) the irreversibility rate, in kW. Neglect kinetic and potential energies. The atmosphere is at 101 kPa and 20°C.

8-7. Steam contained in a cylinder–piston device initially at 1.0 MPa, 230°C, with a total volume of 1.20 m³, is compressed to 2.8 MPa, 90% quality in an internally reversible isothermal process. See Fig. P8-7. The heat removed from the steam goes directly to the atmosphere at 100 kPa, 15°C, thus making the process externally irreversible. Determine, per kilogram of steam, (a) the required work, and (b) the irreversibility of the process.

8-8. One kilogram of steam is contained in a frictionless cylinder–piston assembly initially at 1.6 MPa and 500°C. See Fig. P8-8. The steam is compressed isothermally until the pressure reaches 2.5 MPa, while heat is transferred to the atmosphere, which is at 25°C and 100 kPa. Determine the heat transfer, the work, and the irreversibility for the process.

8-9. Air enters a nozzle steadily at 300 kPa and 87°C with a velocity of 50 m/s and exits at 95 kPa with a velocity of 300 m/s. See Fig. P8-9. There is heat loss to the surrounding atmosphere equal to the amount of 4 kJ/kg of air. The atmospheric temperature is 17°C. Determine the exit temperature of the air and the irreversibility for the process. Consider air as an ideal gas with variable specific heats.

8-10. Steam at 200 psia, 600°F, is supplied to a turbine and exhausted at 14.7 psia. The flow is adiabatic, the power output of the turbine is 10,000 kW, and the turbine efficiency is 65%. The atmosphere is at

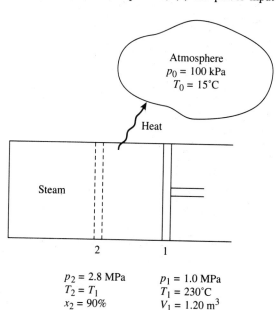

Atmosphere
$p_0 = 100$ kPa
$T_0 = 15$°C

Heat

Steam

2   1

$p_2 = 2.8$ MPa $p_1 = 1.0$ MPa
$T_2 = T_1$ $T_1 = 230$°C
$x_2 = 90\%$ $V_1 = 1.20$ m³

Figure P8-7

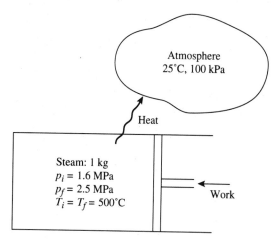

Atmosphere
25°C, 100 kPa

Heat

Steam: 1 kg
$p_i = 1.6$ MPa
$p_f = 2.5$ MPa
$T_i = T_f = 500$°C

Work

Figure P8-8

Chap. 8   Problems

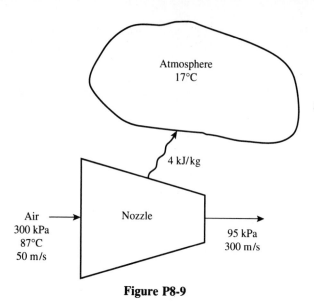

Atmosphere
17°C

4 kJ/kg

Air
300 kPa
87°C
50 m/s

Nozzle

95 kPa
300 m/s

Figure P8-9

14.7 psia and 80°F. Determine (a) the steam rate in lbm/kW·h, and (b) the amount of irreversibility during the turbine expansion in Btu/lbm.

8-11. A desuperheater is an insulated chamber in which superheated steam is desuperheated by cold water spray. See Fig. P8-11. Superheated steam at 700 kPa, 270°C, entering a desuperheater with a mass-flow rate of 9000 kg/h, is sprayed by water at 90°C to become saturated steam at 700 kPa. The atmosphere is at 15°C. Determine, per kilogram of steam entering the desuperheater, (a) the amount of water needed, and (b) the irreversibility of the process.

8-12. Argon enters an insulated turbine at 2 MPa and 1000°C with a mass-flow rate of 0.5 kg/s and exits at 350 kPa in steady-flow operation. The turbine

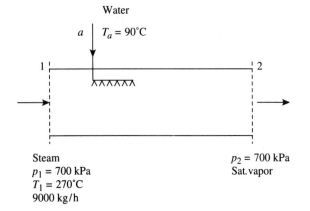

Water

a $T_a = 90°C$

1

2

Steam
$p_1 = 700$ kPa
$T_1 = 270°C$
9000 kg/h

$p_2 = 700$ kPa
Sat. vapor

Figure P8-11

develops 120 kW. Determine (a) the temperature of argon at the exit from the turbine, (b) the irreversibility rate, in kW, and (c) the turbine second-law effectiveness. Neglect kinetic and potential energies. The atmosphere is at 101 kPa. and 20°C.

8-13. Two liters of saturated liquid water are contained in a piston–cylinder device at 150 kPa. See Fig. P8-13. An electric resistance heater inside the cylinder supplies 2200 kJ of energy to the steam, making it undergo a constant-pressure process. Assuming the surrounding atmosphere to be at 100 kpa and 25°C, determine the maximum useful work and the irreversibility for this process.

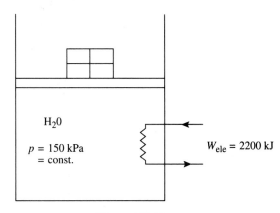

H_2O

$p = 150$ kPa
= const.

$W_{ele} = 2200$ kJ

Figure P8-13

8-14. Steam enters a two-stage adiabatic turbine at 7 MPa and 500°C with a flow rate of 15 kg/s and exits from the first stage at 1 MPa. See Fig. P8-14. Ten percent of the exiting steam is extracted for industrial use and the remainder enters the second stage for further expansion to 50 kPa. The turbine has an adiabatic efficiency of 88%. Determine the irreversibility for the process. The atmosphere is at 25°C.

8-15. Saturated liquid Refrigerant-12 at 100 psia is throttled to 30 psia. Determine the irreversibility of the process, in Btu/lbm, if (a) the process is adiabatic, and (b) the fluid receives 4.0 Btu/lbm of heat from the atmosphere, which is at 14.7 psia and 70°F.

8-16. Saturated-liquid Refrigerant-12 at 0.6 MPa enters an expansion valve and leaves at 0.2 MPa. Determine the irreversibility of the process in kJ/kg, if (a) the process is adiabatic, and (b) the fluid receives 4.0 kJ/kg of heat from the atmosphere, which is at 27°C and 0.1 MPa.

8-17. Atmospheric air is compressed adiabatically from 15 psia, 40°F, to 45 psia at a rate of 2.78 lbm/s.

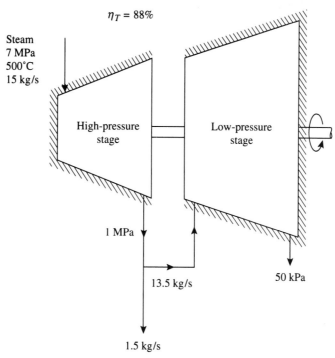

$\eta_T = 88\%$

Steam
7 MPa
500°C
15 kg/s

High-pressure
stage

Low-pressure
stage

1 MPa

13.5 kg/s

50 kPa

1.5 kg/s

Figure P8-14

Kinetic and potential energy changes are negligible. Work input is 60 Btu/lbm. Calculate the irreversibility of the process in Btu/lbm. Explain the physical significance of the answer.

8-18. Air enters a compressor at 14.0 psia, 60°F, with a velocity of 400 ft/s and leaves the compressor at 60 psia, 400°F, with a velocity of 200 ft/s. The process is adiabatic. Calculate the maximum work and irreversibility per pound of air for the process. The atmosphere is at 60°F. Use (a) $c_p = 0.24$ Btu/lbm·°F, and (b) the air-table values.

8-19. An insulated tank containing 0.2 m³ of steam at 400 kPa and 80% quality and an uninsulated tank containing 3 kg of steam at 200 kPa and 250°C are connected through a valve. See Fig. P8-19. The valve is opened and steam flows from the insulated tank to the uninsulated tank until the pressure of the insulated tank drops to 300 kPa. During this process, 600 kJ of heat is transferred from the uninsulated tank to the surroundings at 0°C. Assuming the steam remaining inside the insulated tank to have undergone an internally reversible adiabatic process, determine the final temperature in each tank and the irreversibility for the process.

8-20. A steady-flow heat exchanger operates with the conditions as follows: Water enters the heat exchanger at 10 MPa, 40°C, with a mass-flow rate of 1 kg/s and leaves at 10 MPa and 180°C. Steam enters the heat exchanger at 1.5 MPa and 95% quality and leaves at 1.5 MPa as saturated liquid. Determine the time rates of maximum work and irreversibility. ΔKE and ΔPE can be neglected. The atmosphere is at 20°C.

8-21. A 0.85-m³ rigid tank initially contains air at 35 kPa and −20°C. See Fig. P8-21. Atmospheric air at 100 kPa and 20°C gradually seeps into the tank until the tank contents come to equilibrium with the atmosphere. Determine the maximum work in kJ.

8-22. An industrial plant has steam available at 2.5 MPa and 70°C. This plant needs 3 kg/s of steam at 1.8 MPa and 400°C for an industrial process. The atmosphere is at 100 kPa and 25°C. Devise a system that could satisfy the needs for the industrial process and at the same time produce maximum extra power if it operates ideally. Draw your proposed system and show the processes on a T–s diagram.

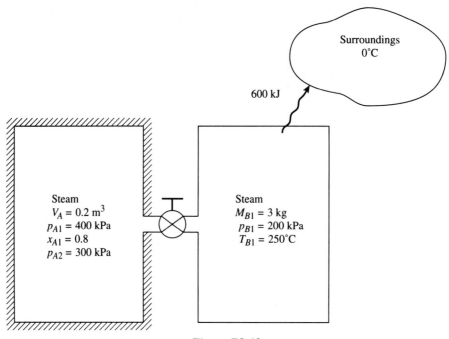

Figure P8-19

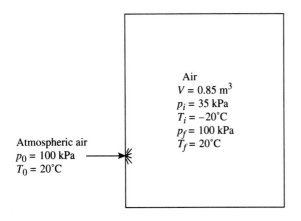

Air
$V = 0.85$ m^3
$p_i = 35$ kPa
$T_i = -20°C$
$p_f = 100$ kPa
$T_f = 20°C$

Atmospheric air
$p_0 = 100$ kPa
$T_0 = 20°C$

Figure P8-21

8-23. A hydrocarbon oil with an average specific heat of 2.30 kJ/kg·K is to be cooled in an adiabatic counterflow heat exchanger in steady flow from 440 K to 320 K by cooling water coming in at 20°C. See Fig. P8-23. The mass-flow rate of oil is 750 kg/h and that of the water is 3000 kg/h. The atmosphere is at 17°C. Determine (a) the outlet temperature of the water flow, in °C, (b) the rate of change in stream availability for the oil, for the water, and for the overall process, in kJ/h, and (c) the rate of irreversibility for the process, in kJ/h.

8-24. Refrigerant-12 enters the evaporator of a window air conditioner at 120 kPa and 30% quality with a mass-flow rate of 2 kg/min and leaves as saturated vapor. See Fig. P8-24. Air enters the evaporator section at 100 kPa, 27°C, with a volume-flow rate of 6 m^3/min. Determine the exit temperature of the air and the irreversibility for the process, assuming (a) the outer surfaces of the air conditioner are insulated, and (b) heat is transferred to the evaporator from the atmosphere at 32°C at a rate of 30 kJ/min.

8-25. An unknown mass of iron block at 85°C is dropped into an insulated tank containing 0.1 m^3 of water at 20°C. See Fig. P8-25. At the same time, a stirring device driven by a 200-W motor is activated. After 20 min, thermal equilibrium is established with a final temperature of 24°C. Determine the mass of the iron block and the irreversibility for the process.

8-26. A 2.5-kg copper block initially at 90°C and a 2.5-kg lead block initially at 70°C are dropped into an uninsulated tank that contains 0.028 m^3 of water initially at 20°C and 100 kPa. See Fig. P8-26. In order to increase the cooling rate, an impeller in the tank is activated to stir the water. During the cooling process, the impeller work input amounts 330 kJ, and the heat loss to the atmosphere

Availability and Irreversibility   Chap. 8

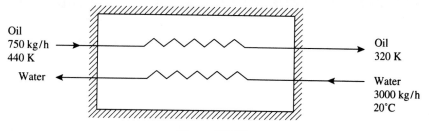

Figure P8-23

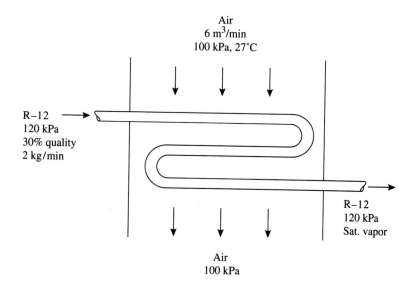

Figure P8-24

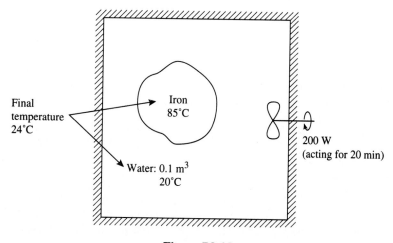

Figure P8-25

Chap. 8   Problems

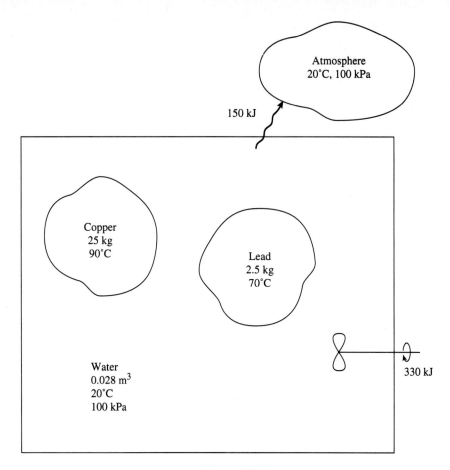

Figure P8-26

amounts 150 kJ. Determine the final equilibrium temperature of the tank contents, the change in entropy of the blocks and the water, and the irreversibility of the process. The atmosphere is at 20°C and 100 kPa.

8-27. A 4.2-ft³ uninsulated cylinder containing 0.7 lbm of nitrogen at 950°R has an adiabatic frictionless piston of negligible volume and weight initially at its lowest position. See Fig. P8-27. The bottom of the cylinder is attached through a valve to a large air-supply line that is at a constant pressure of 100 psia and a constant temperature of 950°R. When the valve is opened, air enters the cylinder and pushes the piston upward until 0.35 lbm of air has entered with a final air temperature of 1000°R. There is 8.2 Btu of heat transfer from the nitrogen to the surroundings at 75°F. Determine the irreversibility of the process. Use ideal-gas property tables for nitrogen and air in your solution. (*Hint:*

Consider nitrogen and air in the cylinder as the system and use the equation $I = T_0(\Delta S_{system} + \Delta S_{surr})$, where $\Delta S_{system} \neq 0$ for unsteady flow.)

8-28. 0.05 kg of steam is contained in a piston–cylinder device initially at 1 MPa and 300°C. The steam is expanded to a final state of 200 kPa, 150°C, while losing 2 kJ of heat to the atmosphere, which is at 25°C and 100 kPa. Determine (a) the maximum work, (b) the irreversibility, and (c) the availability of the steam at the initial and final states.

8-29. An adiabatic compressor having an efficiency of 90% compresses Refrigerant-12 from 200 kPa, 5°C, to 1 MPa under steady-flow conditions. The atmosphere is at 25°C. Determine (a) the exit temperature of the refrigerant, (b) the work input to the compressor, in kJ/kg, (c) the irreversibility of the process, in kJ/kg, and (d) the change in stream availability of the refrigerant for the process, in kJ/kg.

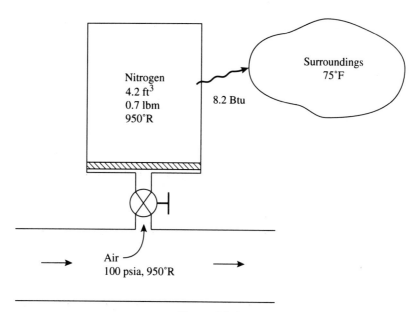

Figure P8-27

8-30. A 2-m³ rigid tank initially contains 0.05 m³ of liquid water and 1.95 m³ of water vapor at 125 kPa (state 1). See Fig. P8-30. Heat is transferred to the tank from a reservoir at 300°C until the pressure in the tank reaches 4 MPa (state 2). Heat is then transferred from the tank to the environment at 25°C, 100 kPa, until the water in the tank returns to its initial state (state 1 again). Determine (a) the amount of heat transferred (in kJ) from the reservoir at 300°C, and (b) the irreversibility (in kJ) of the cycle.

8-31. Three kilograms per second of Refrigerant-12 at 600 kPa and 120°C enter an uninsulated nozzle with negligible velocity. The refrigerant leaves the nozzle at 150 kPa and 40°C with a velocity of 300 m/s. The surrounding atmosphere is at 27°C and 100 kPa. (a) Determine the magnitude and direction of the heat transfer (in kW). (b) Determine the irreversibility rate based on the given data and verify the validity of the data.

8-32. A rigid vessel of 10-ft³ capacity contains air originally at 500 psia and 77°F. The air escapes through a small valve until the pressure in the vessel drops to 200 psia when the valve is closed. Assume that the gas remaining in the vessel undergoes an internally reversible adiabatic expansion process. The atmosphere is at 14.7 psia and 77°F. By the use of air tables, calculate (a) the initial availability (in

Btu) of the air in the vessel before the valve leakage, and (b) the temperature and availability (in Btu) of the air in the vessel immediately after the expansion process when the valve is closed.

8-33. Air is stored in a tank at 27°C and 2 MPa. Determine the volume of the tank needed to store enough air having an availability of 1 kW·h. The atmosphere is at 27°C and 0.1 MPa.

8-34. One-half pound of air, initially at 60 psia, 140°F, is expanded in a closed system until its volume is doubled and its temperature equals that of the surrounding atmosphere. Calculate (a) the maximum work, (b) the change in availability, and (c) the irreversibility. The atmosphere is at 14.7 psia and 40°F.

8-35. Air in a closed rigid tank initially at 120°F and 20 psia receives energy and changes to a pressure of 75 psia. If the environmental conditions are 15 psia and 60°F, determine the change of availability per pound of the air.

8-36. An insulated rigid tank has a partition in it. See Fig. P8-36. One side of the tank contains 50 kg of water at 95 kPa and 0°C, and the other side contains 30 kg of water at 95 kPa and 80°C. When the partition is removed, what is the temperature of the entire tank? Determine the availability change and the irreversibility for the overall process. The atmosphere is at 20°C and 0.1 MPa.

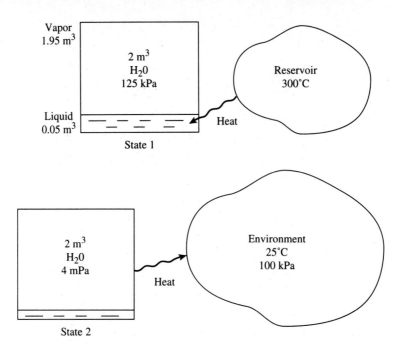

State 1

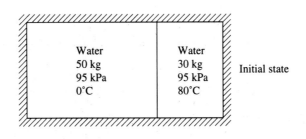

State 2

Figure P8-30

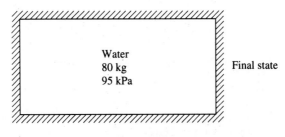

Figure P8-36

Availability and Irreversibility   Chap. 8

8-37. A 0.30-m³ tank contains air at 600 kPa and 600 K. The atmosphere is at 95 kPa and 300 K. (a) Determine the availability of the air in the tank, in kJ. (b) If the air expands freely until its volume is doubled, determine the availability change of the air, in kJ.

8-38. A geothermal well can produce steam at 1 MPa and 220°C. See Fig. P8-38. Determine the stream availability and maximum work per kilogram steam produced by this energy source. The atmosphere is at 101.3 kPa and 25°C.

8-39. A geothermal hot-water well can supply hot water at 200°C and 2 MPa. Determine the stream availability of this energy resource with respect to atmospheric conditions of 25°C and 0.1 MPa.

8-40. A geothermal well can supply steam at 400°F and 150 psia. Determine the stream availability and maximum work for this energy source with respect to atmospheric conditions of 77°F and 14.7 psia.

8-41. Steam at a pressure of 400 psia and a temperature of 600°F is throttled through a well-insulated valve to 300 psia. Calculate the stream availability per pound of steam before and after this process, and the maximun work and irreversibility per pound of steam for this process. Show the initial and final states of the steam on a T–s diagram. $T_0 = 77°F$ and $p_0 = 14.7$ psia.

8-42. Steam enters a turbine at 3 MPa, 400°C, with a velocity of 160 m/s and leaves the turbine as a saturated vapor at 100°C with a velocity of 100 m/s. See Fig. P8-42. The steady-state work output is 550 kJ/kg of steam. There is heat transfer between

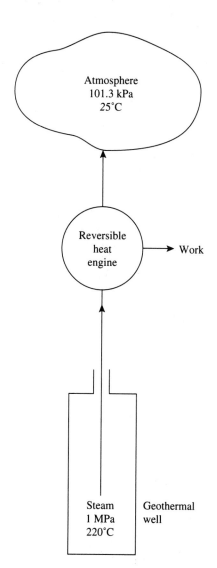

Figure P8-38

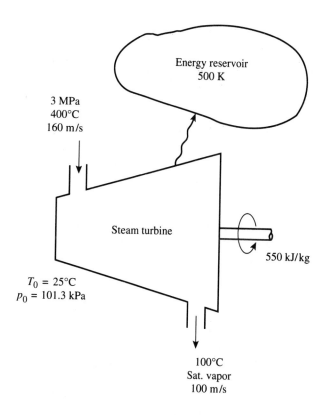

Figure P8-42

the turbine and a reservoir at 500 K. The atmosphere is at 25°C and 101.3 kPa. Determine the change of stream availability and the irreversibility, in kJ/kg.

8-43. In an adiabatic steady-flow device, 3 kg/s of water at 100 kPa and 20°C are mixed with 5 kg/s of water at 100 kPa and 80°C. The outlet water stream is also at 100 kPa. The atmosphere is at 20°C and 100kPa. Determine the rate of stream availability change and the irreversibility of the process.

8-44. The hot products of combustion from a furnace are used to heat the air to be used for combustion. See Fig. P8-44. The products with a mass-flow rate of 100,000 lbm/h are cooled from 600°F to 400°F at constant pressure of 14.7 psia. The mass-flow rate of air is 90,000 lbm/h, and the air is heated from an initial tremperature of 100°F at constant pressure of 14.7 psia. Use $c_p = 0.26$ Btu/lbm·°R for the products and $c_p = 0.24$ Btu/lbm·°R for the air. Determine the initial and final stream availability of the products and the irreversibility for the process, in Btu/h. The atmosphere is at 77°F and 14.7 psia.

8-45. In a large steam power plant, an economizer is often used to preheat the water before entering the boiler by the hot combustion gases before being sent to the chimney. See Fig. P8-45. 68 kg/s of hot gases at 400°C with $c_p = 1.05$ kJ/kg·K is used to heat 50 kg/s of water from 120°C to 200°C. Determine (a) the change in stream availability of the

gas, the water, and the overall system, and (b) the irreversibility for the overall system. The atmosphere is at 20°C.

8-46. Air at 45 psia and 400°F enters a turbine and expands to 15 psia and 300°F. Heat transfer to the atmosphere at 14.7 psia and 80°F amounts to 3 Btu/lbm of air passing through the turbine. Calculate the entering stream availability, the leaving stream availability, and the maximum work, in Btu/lbm. What is the physical meaning of the maximum work? Does it refer to an adiabatic turbine or to one with a heat loss of 3 Btu/lbm?

8-47. Ammonia vapor at 100 kPa and −10°C enters an adiabatic compressor and leaves at 250 kPa. See Fig. P8-47. The compressor has an isentropic efficiency of 70%. Determine, in kJ/kg, (a) the actual work, (b) the maximum work, (c) the flow availability of the gas entering the compressor, (d) the flow availability of the gas leaving the compressor, and (e) the irreversibility of the process. The atmosphere is at 25°C and 100 kPa.

8-48. A gaseous system under a pressure of 1 atm and having a constant heat capacity at constant pressure (defined as mass times c_p) of 3 Btu/°F is used in a steady-flow process to supply heat for the evaporation of water from a saturated liquid state to a saturated vapor state at a pressure of 1 atm. If the initial temperature of the gaseous system is 2000°F, determine the values of the change in stream availability for the gas and for the water, and the

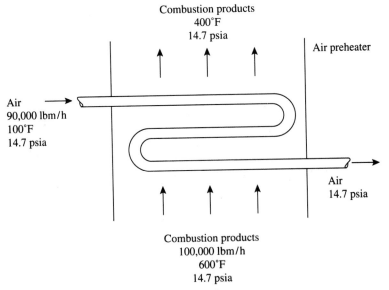

Combustion products
400°F
14.7 psia

Air preheater

Air
90,000 lbm/h
100°F
14.7 psia

Air
14.7 psia

Combustion products
100,000 lbm/h
600°F
14.7 psia

Figure P8-44

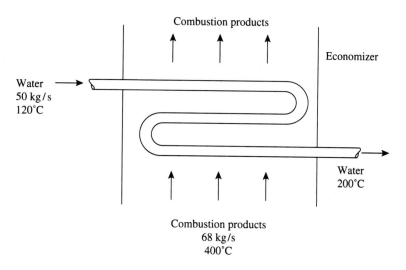

Combustion products

Economizer

Water
50 kg/s
120°C

Water
200°C

Combustion products
68 kg/s
400°C

Figure P8-45

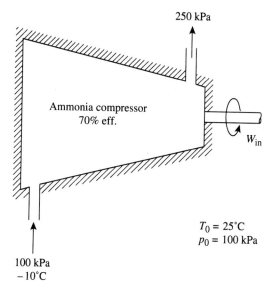

250 kPa

Ammonia compressor
70% eff.

W_{in}

$T_0 = 25°C$
$p_0 = 100$ kPa

100 kPa
−10°C

Figure P8-47

irreversibility of the process as a whole, per pound of water evaporated. The atmosphere is at 50°F.

8-49. A solar pond 10 ft deep produces hot water corresponding to the saturation temperature at the pond bottom. Assume that the specific volume of the pond water can be taken as 1.335×10^{-2} ft³/lbm. Determine the stream availability and maximum work (in Btu/lbm) of the pond-bottom hot water with reference to the atmospheric conditions of 14.7 psia and 77°F.

8-50. An ocean thermal energy conversion system operates as a heat engine using surface water at 28°C as the high-temperature source and the deep water at 14°C as the low-temperature sink. See Fig. P8-50. (a) What is the stream availability of the surface water, in kJ/kg, with respect to the dead state at 10°C? (b) What is the maximum work of the conversion system, in kJ/kg? Consider water as incompressible with $c = 4.213$ kJ/kg·K.

8-51. An ocean thermal energy conversion system operates as a heat engine using surface water at 80°F as the high-temperature source and the deep water at 56°F as the low-temperature sink. See Fig. P8-50. (a) What is the stream availability of the source water, in Btu/lbm, with respect to the dead state at 50°F? (b) What is the maximum work of the conversion system, in Btu/lbm? Consider water as incompressible with $c = 1.0$ Btu/lbm·°R.

8-52. A house is maintained at 22°C by electric resistance heaters. See Fig. P8-52. The house loses heat at a rate of 60,000 kJ/h when the outside temperature is 15°C. What is the irreversibility rate as compared to a totally reversible (Carnot) heat pump working between the two temperature limits?

8-53. A 300-liter water storage tank is placed in a sunny location inside a hosue to absorb and store solar energy during the day, and to provide heating by releasing its stored energy at night. See Fig. P8-53. The water in the tank can be heated to 45°C during the day, and the house is maintained at 22°C. (a) Determine the amount of heating (in kJ)

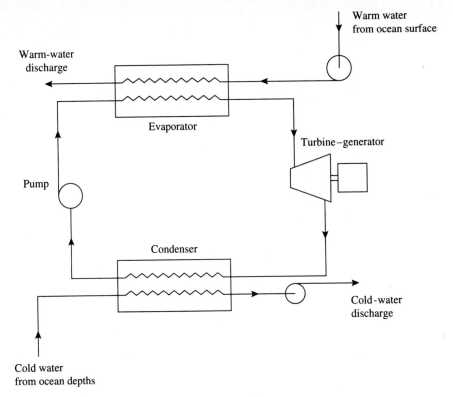

Warm water
from ocean surface

Warm-water
discharge

Evaporator

Turbine–generator

Pump

Condenser

Cold-water
discharge

Cold water
from ocean depths

Figure P8-50

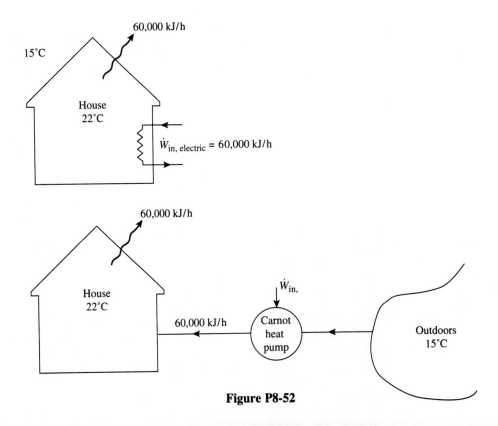

60,000 kJ/h

15°C

House
22°C

$\dot{W}_{in,\ electric} = 60{,}000$ kJ/h

60,000 kJ/h

House
22°C

$\dot{W}_{in,}$

60,000 kJ/h

Carnot
heat
pump

Outdoors
15°C

Figure P8-52

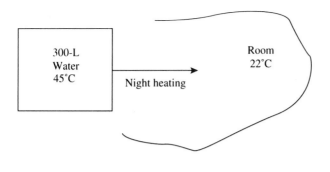

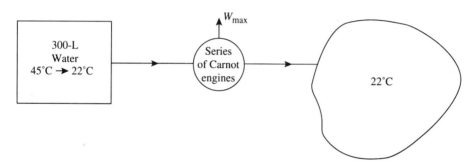

Figure P8-53

the water tank can provide at night. (b) In bringing the tank temperature down to the room temperature, if a series of imaginary Carnot engines are introduced between the two temperatures, what would be the maximum work (in kJ) these Carnot engines can yield? (c) Determine the irreversibility (in kJ) of the simple cooling process as compared to the imaginary Carnot engines.

8-54. A heat engine receives 700 kJ/s of heat from a high-temperature source at 1500 K, and rejects the waste heat to the atmosphere at 300 K. See Fig. P8-54. The power output of the heat engine is 350 kW. Determine the maximum power, the rate of irreversibility, and the second-law effectiveness of this heat engine.

8-55. A heat engine cycle has a thermal efficiency of 35%. The heat source is the hot gases from a furnace at 1100°C, and the heat sink is the water in a river at 20°C. Determine the second-law effectiveness of the power cycle.

8-56. Heat is removed from a refrigerated space at a rate of 180 kJ/min by a refrigerator that has a second-

law effectiveness of 45%. The refrigerated space is maintained at 3°C and the surrounding atmosphere is at 27°C. Determine the power input to the refrigerator.

8-57. Heat is removed from a freezer box at a rate of 75 kJ/min in order to maintain it at a constant temperature of −6°C. See Fig. P8-57. The power input to the refrigeration machine is 0.50 kW and the surrounding air is at 24°C. Determine (a) the COP of the refrigeration machine, (b) the COP and power input if the machine follows the reversed Carnot cycle, (c) the second-law effectiveness of the refrigeration machine, and (d) the irreversibility of the machine as compared to the reversed Carnot cycle.

8-58. Helium is expanded in an adiabatic turbine from 1.0 MPa, 300°C, to 100 kPa at a mass-flow rate of 1.8 kg/s. The turbine has an adiabatic efficiency of 92%. Determine the maximum work, the irreversibility, and the second-law effectiveness. $T_0 = 293$ K. Use $c_p = 20.8$ kJ/kgmole·K and $c_v = 12.5$ kJ/kgmole·K for helium.

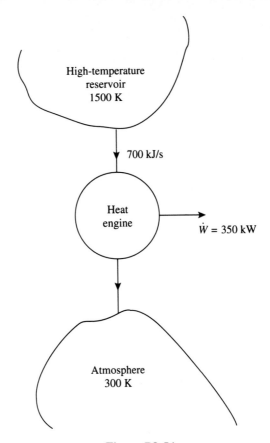

Figure P8-54

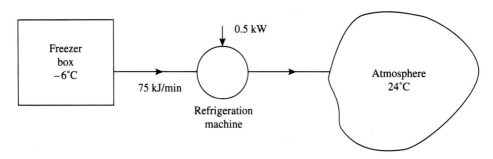

Figure P8-57

8-59. An insulated rigid tank contains 2 kg of a liquid–vapor water mixture at 100 kPa with a quality of 0.25. See Fig. P8-59. An electric resistance heater in the tank is turned on until all the liquid in the tank is vaporized. The atmosphere is at 25°C and 100 kPa. Determine the irreversibility and the second-law effectiveness for the process.

8-60. An insulated counterflow heat exchanger has steam entering as saturated vapor at 150 kPa and leaving as saturated liquid at the same pressure. See Fig. P8-60. Oxygen enters the heat exchanger at 27°C and 110 kPa and leaves at 100 kPa. The terminal temperature difference of the heat exchanger is 8°C. Oxygen enters the heat exchanger at the rate

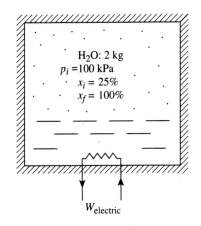

$T_0 = 25°C$
$p_0 = 100$ kPa

Figure P8-59

of 15 m³/min. Neglecting kinetic and potential energies, determine the irreversibility in kJ/min and the second-law effectiveness of the heat exchanger. The atmosphere is at 27°C, 101 kPa.

8-61. A simple ideal steam power cycle (see Fig. 8-12) has the following conditions: $p_1 = p_2 = 6$ MPa, $T_2 = 400°C$, $p_3 = p_4 = 8$ kPa, $h_1 = 179.9$ kJ/kg, and $s_1 = 0.5926$ kJ/kg·K. The atmosphere is at 25°C and 101 kPa. Determine the cycle thermal efficiency and the second-law effectiveness.

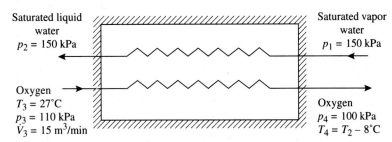

Figure P8-60

Ideal Gas and Gas–Vapor Mixtures

So far, our discussions have been limited mainly to single-component systems. There are, however, many engineering applications that involve multicomponent systems of various substances. In this chapter, we extend the treatment to multicomponent systems of nonreactive ideal gases. Because an unlimited number of compositions are possible for ideal-gas mixtures, it is vital to develop methods of evaluating the properties of any mixture in terms of the properties of its constituents. This chapter considers the behavior of ideal-gas mixtures in general, and includes air–water vapor mixtures as a special case. It should be mentioned that although in the last part of this chapter, we are particularly interested in air–water vapor mixtures, the basic principles involved are applicable to any other gas–vapor mixtures.

9-1 MASS ANALYSIS OF GAS MIXTURES

When the analysis of a gas mixture is made on the basis of mass or weight, it is called a *mass* or *gravimetric analysis*. The total mass m of a mixture is the sum of the masses of its constituents. That is,

$$m = m_1 + m_2 + m_3 + \cdots = \sum_i m_i \tag{9-1}$$

where m_i refers to the mass of the ith component. The *mass fraction y_i* of component i is defined as

$$y_i = \frac{m_i}{m} \tag{9-2}$$

It follows that

$$\sum_i y_i = 1$$

When the analysis of a gas mixture is made on the basis of the number of moles of each component present, it is called a *molar analysis*. The total number of moles n of a mixture is the sum of the number of moles of its constituents. That is,

$$n = n_1 + n_2 + n_3 + \cdots = \sum_i n_i \qquad (9\text{-}3)$$

where n_i refers to the number of moles of the ith component. The *mole fraction* x_i of component i is defined as

$$x_i = \frac{n_i}{n} \qquad (9\text{-}4)$$

It follows that

$$\sum_i x_i = 1$$

The mass m, the number of moles n, and the molar mass or molecular weight M of a substance are related by

$$m = nM \qquad (9\text{-}5)$$

Applying this equation to each component of a mixture and to the mixture as a whole, we have, from Eq. (9-1),

$$nM = n_1 M_1 + n_2 M_2 + n_3 M_3 + \cdots = \sum_i n_i M_i$$

or

$$M = x_1 M_1 + x_2 M_2 + x_3 M_3 + \cdots = \sum_i x_i M_i \qquad (9\text{-}6)$$

where M is the average, or apparent, molar mass or molecular weight of the mixture. Accordingly, the apparent gas constant R of the mixture is given by the equation

$$R = \frac{\mathcal{R}}{M}$$

where $\mathcal{R}$ is the universal gas constant.

Example 9-1

Consider air as a mixture of approximately 79% nitrogen and 21% oxygen on a molar basis. Determine (a) the apparent molecular weight of air and (b) the gas constant of air.

Solution   The apparent molecular weight of air is

$$M = x_{N_2} M_{N_2} + x_{O_2} M_{O_2}$$

$$= 0.79(28) + 0.21(32)$$

$$= 28.84 \text{ kg/kgmole} = 28.84 \text{ lbm/lbmole}$$

It should be noted that when traces of other gases are included in the composition of air, the commonly quoted value of the apparent molecular weight of air is 28.97 kg/kgmole = 28.97 lbm/lbmole.

When the commonly quoted value of M is used, the apparent gas constant of air is

$$R = \frac{\mathcal{R}}{M} = \frac{8.314 \text{ kg/kgmole·K}}{28.97 \text{ kg/kgmole}} = 0.287 \text{ kJ/kg·K}$$

$$= \frac{1545 \text{ ft·lbf/lbmole·°R}}{28.97 \text{ lbm/lbmole}} = 53.3 \text{ ft·lbf/lbm·°R}$$

9-2 p–V–T PROPERTIES OF IDEAL-GAS MIXTURES

It is an empirical fact that a mixture of ideal gases behaves also as an ideal gas. Let us consider a homogeneous mixture of inert ideal gases with r components; each component occupies the entire volume V of the container at a common temperature T (Fig. 9-1). Applying the ideal gas equation to the mixture as a whole and to any component i, we have

$$pV = n\mathcal{R}T \tag{9-7}$$

$$p_i V = n_i \mathcal{R}T \tag{9-8}$$

where p is the total pressure of the mixture, and p_i is the partial pressure of component i. The *partial pressure* p_i of component i is defined as the pressure that this component would exert if it occupied the volume alone at the mixture temperature.

Since $n = \Sigma_{i=1}^{r} n_i$, it follows that

$$p = \sum_{i=1}^{r} p_i \tag{9-9}$$

This is *Dalton's law of additive pressures*. It states that the total pressure of a mixture of ideal gases is equal to the sum of the partial pressures of the components, if each existed alone at the temperature and volume of the mixture. Taking the ratio between Eqs. (9-8) and (9-7), we obtain

$$p_i = \frac{n_i}{n}p = x_i p \tag{9-10}$$

where x_i is the mole fraction of component i.

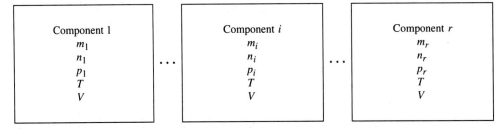

Figure 9-1 Illustration of Dalton's law of additive pressures.

An alternate approach in the analysis of ideal-gas mixtures is to consider the case in which each component exists at the temperature and pressure of the mixture (Fig. 9-2). The *partial volume* V_i of component i is defined as the volume that would be occupied by this component at the same temperature T and pressure p as that of the mixture. Applying the ideal gas equation then gives

$$pV_i = n_i \mathscr{R} T \tag{9-11}$$

Summing up all V_i's and comparing the result with Eq. (9-7) leads to

$$V = \sum_{i=1}^{r} V_i \tag{9-12}$$

This is *Amagat's law of additive volumes*. It states that the volume of a mixture of ideal gases is equal to the sum of the partial volumes of the components when the partial volumes are determined at the pressure and temperature of the mixture. Taking the ratio between Eqs. (9-11) and (9-7), we obtain

$$V_i = \frac{n_i}{n} V = x_i V \tag{9-13}$$

Accordingly,

$$x_i = \frac{n_i}{n} = \frac{V_i}{V} \tag{9-14}$$

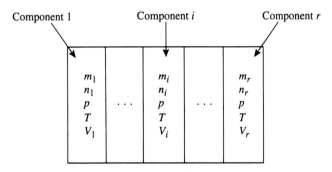

Gaseous mixture

$m = m_1 + \cdots + m_i + \cdots + m_r$
$n = n_1 + \cdots + n_i + \cdots + n_r$
p
T
$V = V_1 + \cdots + V_i + \cdots + V_r$

Component 1   Component i Component r

m_1		m_i		m_r
n_1		n_i		n_r
p	$\cdots$	p	$\cdots$	p
T		T		T
V_1		V_i		V_r

Figure 9-2   Illustration of Amagat's law of additive volumes.

where V_i/V is the volume fraction of component i. Thus, for an ideal-gas mixture, the volume fraction and mole fraction are equal, and the volumetric analysis and molar analysis are the same.

Referring to Fig. 9-2, the application of the ideal-gas equation to any component i and to the mixture as a whole in terms of mass, we have

$$pV_i = m_i RT = m_i \frac{\mathcal{R}}{M_i} T$$

$$pV = mRT = m \frac{\mathcal{R}}{M} T$$

Taking the ratio of the preceding equations yields

$$\frac{V_i}{V} = \left(\frac{m_i}{m}\right)\left(\frac{M}{M_i}\right) \tag{9-15}$$

or

$$x_i = y_i \frac{M}{M_i} \tag{9-15a}$$

This equation is useful in conversion from volume fraction (or mole fraction) to mass fraction, or vice versa. The molecular weight of the mixture needed in the equation can be calculated by either of the following two equations:

$$M = \sum_i x_i M_i$$

$$M = \left(\sum_i \frac{y_i}{M_i}\right)^{-1} \tag{9-16}$$

Example 9-2

A 0.825-m³ tank contains 0.95 kg of oxygen, 0.82 kg of nitrogen, 1.32 kg of carbon dioxide, and 0.09 kg of carbon monoxide at a temperature of 30°C. Assuming the mixture to behave as an ideal gas, determine (a) the volumetric analysis, (b) the apparent molecular weight of the mixture, (c) the total pressure of the mixture, and (d) the partial pressures of the constituents.

Solution   The total mass of the mixture is

$$m = m_{O_2} + m_{N_2} + m_{CO_2} + m_{CO}$$

$$= 0.95 + 0.82 + 1.32 + 0.09 = 3.18 \text{ kg}$$

The mass fractions of the constituents are

$$y_{O_2} = \frac{m_{O_2}}{m} = \frac{0.95}{3.18} = 0.299$$

$$y_{N_2} = \frac{m_{N_2}}{m} = \frac{0.82}{3.18} = 0.258$$

$$y_{CO_2} = \frac{m_{CO_2}}{m} = \frac{1.32}{3.18} = 0.415$$

$$y_{CO} = \frac{m_{CO}}{m} = \frac{0.09}{3.18} = 0.028$$

The apparent molecular weight of the mixture is

$$M = \left(\frac{y_{O_2}}{M_{O_2}} + \frac{y_{N_2}}{M_{N_2}} + \frac{y_{CO_2}}{M_{CO_2}} + \frac{y_{CO}}{M_{CO}}\right)^{-1}$$

$$= \left(\frac{0.299}{32} + \frac{0.258}{28} + \frac{0.415}{44} + \frac{0.028}{28}\right)^{-1} = 34.5 \text{ kg/kgmole}$$

The volume fractions, or mole fractions, of the constituents are

$$x_{O_2} = \left(\frac{V_{O_2}}{V}\right) = y_{O_2}\left(\frac{M}{M_{O_2}}\right) = 0.299\left(\frac{34.5}{32}\right) = 0.322$$

$$x_{N_2} = \left(\frac{V_{N_2}}{V}\right) = y_{N_2}\left(\frac{M}{M_{N_2}}\right) = 0.258\left(\frac{34.5}{28}\right) = 0.318$$

$$x_{CO_2} = \left(\frac{V_{CO_2}}{V}\right) = y_{CO_2}\left(\frac{M}{M_{CO_2}}\right) = 0.415\left(\frac{34.5}{44}\right) = 0.325$$

$$x_{CO} = \left(\frac{V_{CO}}{V}\right) = y_{CO}\left(\frac{M}{M_{CO}}\right) = 0.028\left(\frac{34.5}{28}\right) = 0.035$$

The total pressure of the mixture is

$$p = \frac{m(\mathcal{R}/M)T}{V} = \frac{(3.18 \text{ kg})(8.314/34.5 \text{ kJ/kg·K})(303 \text{ K})}{0.825 \text{ m}^3}$$

$$= 281 \text{ kPa}$$

The partial pressures of the constituents are

$$p_{O_2} = x_{O_2}p = 0.322(281) = 90.5 \text{ kPa}$$

$$p_{N_2} = x_{N_2}p = 0.318(281) = 89.4 \text{ kPa}$$

$$p_{CO_2} = x_{CO_2}p = 0.325(281) = 91.3 \text{ kPa}$$

$$p_{CO} = x_{CO}p = 0.035(281) = 9.8 \text{ kPa}$$

9-3 ENERGY PROPERTIES OF IDEAL-GAS MIXTURES

The energy-related properties, such as internal energy, enthalpy, entropy, and specific heat of ideal-gas mixtures, can be calculated from the properties of the individual gases by means of the *Gibbs–Dalton law*. It states that in a mixture of ideal gases, each component of the mixture acts as if it alone were occupying the volume of the mixture at the temperature of the mixture. Thus, on a molar basis, the internal energy U, enthalpy H, and entropy S of an ideal-gas mixture are given by

$$U = nu = \sum_i n_i u_i \qquad \text{or} \qquad u = \sum_i x_i u_i \qquad (9\text{-}17)$$

$$H = nh = \sum_i n_i h_i \qquad \text{or} \qquad h = \sum_i x_i h_i \qquad (9\text{-}18)$$

$$S = ns = \sum_i n_i s_i \qquad \text{or} \qquad s = \sum_i x_i s_i \qquad (9\text{-}19)$$

On the other hand, on a mass basis, these expressions become

$$U = mu = \sum_i m_i u_i \qquad \text{or} \qquad u = \sum_i y_i u_i \qquad (9\text{-}20)$$

$$H = mh = \sum_i m_i h_i \qquad \text{or} \qquad h = \sum_i y_i h_i \qquad (9\text{-}21)$$

$$S = ms = \sum_i m_i s_i \qquad \text{or} \qquad s = \sum_i y_i s_i \qquad (9\text{-}22)$$

Because the internal energy and enthalpy for an ideal gas are functions only of temperature, the u_i's and h_i's in the foregoing equations are to be evaluated at the temperature of the mixture. The entropy, however, is a function of two independent properties even for an ideal gas. Therefore, the s_i's in Eqs. (9-19) and (9-22) are to be evaluated either at T and V of the mixture, or at T of the mixture and the partial pressure p_i of the component.

The specific heats of a mixture are defined as

$$c_v = \left(\frac{\partial u}{\partial T}\right)_v \qquad \text{and} \qquad c_p = \left(\frac{\partial h}{\partial T}\right)_p$$

Applying Eqs. (9-17) and (9-18) for a constant composition of the mixture, we have

$$c_v = \sum_i x_i \left(\frac{\partial u_i}{\partial T}\right)_v = \sum_i x_i c_{vi} \qquad \text{or} \qquad nc_v = \sum_i n_i c_{vi} \qquad (9\text{-}23)$$

$$c_p = \sum_i x_i \left(\frac{\partial h_i}{\partial T}\right)_p = \sum_i x_i c_{pi} \qquad \text{or} \qquad nc_p = \sum_i n_i c_{pi} \qquad (9\text{-}24)$$

where c_{vi} and c_{pi} are the molar specific heats of component i. Similarly, on a mass basis, from Eqs. (9-20) and (9-21), we have

$$c_v = \sum_i y_i \left(\frac{\partial u_i}{\partial T}\right)_v = \sum_i y_i c_{vi} \qquad \text{or} \qquad mc_v = \sum_i m_i c_{vi} \qquad (9\text{-}25)$$

$$c_p = \sum_i y_i \left(\frac{\partial h_i}{\partial T}\right)_p = \sum_i y_i c_{pi} \qquad \text{or} \qquad mc_p = \sum_i m_i c_{pi} \qquad (9\text{-}26)$$

where c_{vi} and c_{pi} are the specific heats for unit mass of component i.

The gas constant of a mixture is given by

$$R = c_p - c_v = \sum_i y_i (c_{pi} - c_{vi}) = \sum_i y_i R_i$$

or

$$mR = \sum_i m_i R_i \qquad (9\text{-}27)$$

where R_i is the gas constant of component i.

In the energy analysis of most thermodynamic systems, changes in internal energy, enthalpy, and entropy are important. Thus, between states 1 and 2 on a mass basis,

$$\Delta U = U_2 - U_1 = \sum_i m_i (u_{i2} - u_{i1}) \qquad (9\text{-}28)$$

$$\Delta H = H_2 - H_1 = \sum_i m_i(h_{i2} - h_{i1}) \tag{9-29}$$

$$\Delta S = S_2 - S_1 = \sum_i m_i(s_{i2} - s_{i1}) \tag{9-30}$$

Similar equations can be written on a molar basis.

Because the internal energy and enthalpy of ideal gases are functions of temperature alone, the data for u_i and h_i needed in Eqs. (9-28) and (9-29) can be obtained from ideal-gas property tables for the individual gases at the two temperatures, T_1 and T_2, specified. If the specific heats of the gases in the temperature range are constants, or their average values are used, then the changes in internal energy and enthalpy of the mixture can be calculated from the equations

$$\Delta U = U_2 - U_1 = \sum_i m_i c_{vi}(T_2 - T_1) = m c_v(T_2 - T_1) \tag{9-31}$$

$$\Delta H = H_2 - H_1 = \sum_i m_i c_{pi}(T_2 - T_1) = m c_p(T_2 - T_1) \tag{9-32}$$

where the symbols m_i, c_{vi}, and c_{pi} are for the ith component, and m, c_v, and c_p are for the mixture as a whole. These equations are written on a mass basis. Similar equations can be written on a molar basis.

As mentioned previously, the entropy is a function of two independent properties even for an ideal gas. The entropy change of an ideal gas is usually expressed as a function of temperature and pressure. For a component i in an ideal-gas mixture, the entropy change can be calculated in terms of the temperature T of the mixture and the partial pressure p_i of the component by either of the following two equations:

$$\Delta s_i = s_{i2} - s_{i1} = \phi_{i2} - \phi_{i1} - R_i \ln \frac{p_{i2}}{p_{i1}} \tag{9-33}$$

$$\Delta s_i = s_{i2} - s_{i1} = c_{pi} \ln \frac{T_2}{T_1} - R_i \ln \frac{p_{i2}}{p_{i1}} \tag{9-34}$$

Equation (9-33) is for variable specific heats for which ideal-gas property tables are used, whereas Eq. (9-34) is for constant specific heats, both on a mass basis. When the equations are written on a molar basis, the universal gas constant $\mathcal{R}$ should be used instead of the gas constant R_i of component i.

Example 9-3

An ideal-gas mixture consisting of 0.20 kg of nitrogen, 0.20 kg of carbon dioxide, and 0.10 kg of carbon monoxide is compressed in a closed system through an internally reversible and adiabatic process from 100 kPa and 300 K to 500 kPa. Determine the final temperature, the work requirement, and the entropy changes of the component gases by the use of (a) ideal-gas property tables and (b) constant specific heats.

Solution   The numbers of moles of the component gases are

$$n_{N_2} = \frac{0.20 \text{ kg}}{28 \text{ kg/kgmole}} = 0.00714 \text{ kgmole}$$

$$n_{CO_2} = \frac{0.20}{44} = 0.00455 \text{ kgmole}$$

$$n_{CO} = \frac{0.10}{28} = 0.00357 \text{ kgmole}$$

(a) *Variable specific-heats solution.* Because the process is internally reversible and adiabatic, we must have $\Delta S = 0$ for the mixture. Using Eq. (9-33) for the component Δs_i, we have

$$\Delta S = 0 = n_{N_2}\left(\phi_{N_2,2} - \phi_{N_2,1} - \Re \ln \frac{p_2}{p_1}\right)$$

$$+ n_{CO_2}\left(\phi_{CO_2,2} - \phi_{CO_2,1} - \Re \ln \frac{p_2}{p_1}\right)$$

$$+ n_{CO}\left(\phi_{CO,2} - \phi_{CO,1} - \Re \ln \frac{p_2}{p_1}\right)$$

where for each gas, the ratio p_{i2}/p_{i1} has been replaced by p_2/p_1, because the mole fractions of the components remain constant during the process. Substituting numerical values, including the ϕ_{i1}'s obtained at $T_1 = 300$ K from the Keenan–Chao–Kaye Gas Tables [20] or the ideal-gas property tables in Appendix 3 (Tables A-10, A-13, and A-15), we have

$$\Delta S = 0.00714\left(\phi_{N_2,2} - 191.646 - 8.314 \ln \frac{500}{100}\right)$$

$$+ 0.00455\left(\phi_{CO_2,2} - 213.908 - 8.314 \ln \frac{500}{100}\right)$$

$$+ 0.00357\left(\phi_{CO,2} - 197.697 - 8.314 \ln \frac{500}{100}\right) = 0$$

or

$$0.00714\phi_{N_2,2} + 0.00455\phi_{CO_2,2} + 0.00357\phi_{CO,2} = 3.252 \text{ kJ/K}$$

The solution of this equation is found by assuming a value for T_2 and obtaining the ϕ_{i2}'s from the ideal-gas property tables in an iteration process. Thus:

At 450 K,

$$\Delta S = 0.00714(203.490) + 0.00455(230.168) + 0.00357(209.568) = 3.248 \text{ kJ/K}$$

At 460 K,

$$\Delta S = 0.00714(204.136) + 0.00455(231.119) + 0.00357(210.218) = 3.260 \text{ kJ/K}$$

By interpolation, we obtain $T_2 = 453$ K.

For the adiabatic compression process in a closed system,

$$Q = \Delta U + W = 0$$

Thus, the work requirement is

$$W = -\Delta U = U_1 - U_2$$

$$= n_{N_2}(u_{N_2,1} - u_{N_2,2}) + n_{CO_2}(u_{CO_2,1} - u_{CO_2,2}) + n_{CO}(u_{CO,1} - u_{CO,2})$$

Obtaining the u's at $T_1 = 300$ K and $T_2 = 453$ K from the ideal-gas property tables and substituting, we have

$$W = 0.00714(6229.7 - 9428.8) + 0.00455(6939.4 - 11842.1)$$
$$+ 0.00357(6230.7 - 9441.0)$$
$$= -56.6 \text{ kJ}$$

where the minus sign indicates work input.

The entropy changes of the component gases in the mixture are calculated from Eq. (9-33):

$$\Delta S_i = n_i \left(\phi_{i2} - \phi_{i1} - \mathcal{R} \ln \frac{p_2}{p_1} \right)$$

Obtaining the ϕ_{i2}'s at $T_2 = 453$ K and substituting all known numerical values, we have

$$\Delta S_{N_2} = n_{N_2} \left(\phi_{N_2,2} - \phi_{N_2,1} - \mathcal{R} \ln \frac{p_2}{p_1} \right)$$

$$= 0.00714 \left(203.684 - 191.646 - 8.314 \ln \frac{500}{100} \right)$$

$$= -0.00959 \text{ kJ/K}$$

$$\Delta S_{CO_2} = n_{CO_2} \left(\phi_{CO_2,2} - \phi_{CO_2,1} - \mathcal{R} \ln \frac{p_2}{p_1} \right)$$

$$= 0.00455 \left(230.453 - 213.908 - 8.314 \ln \frac{500}{100} \right)$$

$$= 0.01440 \text{ kJ/K}$$

$$\Delta S_{CO} = n_{CO} \left(\phi_{CO,2} - \phi_{CO,1} - \mathcal{R} \ln \frac{p_2}{p_1} \right)$$

$$= 0.00357 \left(209.763 - 197.697 - 8.314 \ln \frac{500}{100} \right)$$

$$= -0.00469 \text{ kJ/K}$$

Accordingly,

$$\Delta S = \Delta S_{N_2} + \Delta S_{CO_2} + \Delta S_{CO}$$

$$= -0.00959 + 0.01440 - 0.00469$$

$$= 0.00012 \text{ kJ/K}$$

which should be equal to zero for the reversible adiabatic process. The slight error is due to roundoff.

(b) *Constant specific-heats solution.* In general, the constant specific heats should be evaluated at the average temperature of the process, For this example, the final temperature T_2 is unknown. We can simply use the specific-heat data at the initial temperature T_1 for an approximate solution, or we can guess the final temperature T_2 and use the average specific-heat data in an iteration process. For this example in the final trial, we assume that the final temperature is about 450 K and use an average temperature of 375 K to obtain the specific-heat values. Accordingly, from the table on specific heats in Appendix 3 (Table A-6M), we have the following:

For nitrogen, $c_p = 1.043$ kJ/kg·K and $c_v = 0.746$ kJ/kg·K

For carbon dioxide, $c_p = 0.917$ kJ/kg·K and $c_v = 0.728$ kJ/kg·K

For carbon monoxide, $c_p = 1.045$ kJ/kg·K and $c_v = 0.749$ kJ/kg·K

The specific heats for the mixture are then

$$c_p = \frac{0.20(1.043) + 0.20(0.917) + 0.10(1.045)}{0.20 + 0.20 + 0.10} = 0.993 \text{ kJ/kg·K}$$

$$c_v = \frac{0.20(0.746) + 0.20(0.728) + 0.10(0.749)}{0.20 + 0.20 + 0.10} = 0.739 \text{ kJ/kg·K}$$

The specific-heat ratio for the mixture is

$$k = \frac{c_p}{c_v} = \frac{0.993}{0.739} = 1.34$$

For the reversible adiabatic process of the mixture, we have

$$T_2 = T_1 \left(\frac{p_2}{p_1}\right)^{(k-1)/k} = 300 \left(\frac{500}{100}\right)^{(1.34-1)/1.34} = 451 \text{ K}$$

which is approximately the same as assumed. Therefore, we conclude that the final temperature is $T_2 = 451$ K.

The compression work requirement is

$$W = U_1 - U_2 = \sum_i m_i(u_{i1} - u_{i2}) = \sum_i m_i c_{vi}(T_1 - T_2)$$

$$= (0.20 \times 0.746 + 0.20 \times 0.728 + 0.10 \times 0.749)(300 - 451) = -55.8 \text{ kJ}$$

The entropy changes of the component gases are calculated from Eq. (9-34):

$$\Delta S_i = m_i \left(c_{pi} \ln \frac{T_2}{T_1} - R_i \ln \frac{p_2}{p_1}\right)$$

Thus,

$$\Delta S_{N_2} = m_{N_2} \left(c_{p,N_2} \ln \frac{T_2}{T_1} - \frac{\mathcal{R}}{M_{N_2}} \ln \frac{p_2}{p_1}\right)$$

$$= 0.20 \left[1.043 \left(\frac{451}{300}\right) - \frac{8.314}{28} \ln \frac{500}{100}\right] = -0.01053 \text{ kJ/K}$$

$$\Delta S_{CO_2} = m_{CO_2} \left(c_{p,CO_2} \ln \frac{T_2}{T_1} - \frac{\mathcal{R}}{M_{CO_2}} \ln \frac{p_2}{p_1}\right)$$

$$= 0.20 \left(0.917 \ln \frac{451}{300} - \frac{8.314}{44} \ln \frac{500}{100}\right) = 0.01395 \text{ kJ/K}$$

$$\Delta S_{CO} = m_{CO} \left(c_{p,CO} \ln \frac{T_2}{T_1} - \frac{\mathcal{R}}{M_{CO}} \ln \frac{p_2}{p_1}\right)$$

$$= 0.10 \left(1.045 \ln \frac{451}{300} - \frac{8.314}{28} \ln \frac{500}{100}\right) = -0.00519 \text{ kJ/K}$$

Now,

$$\Delta S = \Delta S_{N_2} + \Delta S_{CO_2} + \Delta S_{CO}$$

$$= -0.01053 + 0.01395 - 0.00519 = -0.00177 \text{ kJ/K}$$

which should be equal to zero for the reversible adiabatic process. The error is mostly due to the inconsistency in the specific-heat data.

9-4 MIXING PROCESSES

So far in this chapter, we have studied mixtures that have already been formed. We now turn our attention to the mixing of different gases to form a mixture. Consider a number of ideal gases initially at different pressures and temperatures in different compartments of a well-insulated rigid tank, as shown in Fig. 9-3(a). The thin partitions separating the compartments are broken open and the gases form a mixture with the properties shown in Fig. 9-3(b). For mass conservation, we must have

$$m = \sum_i m_i \qquad \text{and} \qquad n = \sum_i n_i$$

As the tank is rigid,

$$V = \sum_i V_i$$

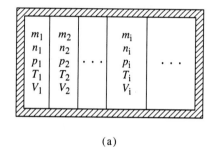
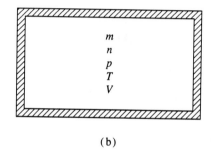

(a)   (b)

Figure 9-3 Adiabatic mixing of ideal gases in a closed system: (a) initial, before mixing, and (b) final, after mixing.

Furthermore, because the tank is rigid and insulated, neither work nor heat transfer is involved. From the first law, it is apparent that $\Delta U = 0$ for the total system. In terms of the individual components, this takes the form

$$\sum_i \Delta U_i = \sum_i m_i (\Delta u_i) = \sum_i m_i (u_{i,\,\text{final}} - u_{i,\,\text{initial}}) = 0 \qquad (9\text{-}35)$$

on a mass basis. On the other hand, on a molar basis, this equation becomes

$$\sum_i \Delta U_i = \sum_i n_i (\Delta u_i) = \sum_i n_i (u_{i,\,\text{final}} - u_{i,\,\text{initial}}) = 0 \qquad (9\text{-}36)$$

When ideal-gas property tables are used, with m_i and $u_{i,\,\text{initial}}$ on a mass basis, or n_i and $u_{i,\,\text{initial}}$ on a molar basis known for the various components, the final temperature of the mixture can be found from either Eq. (9-35) or Eq. (9-36) through an iterative process.

If c_v for each component is constant in the temperature range between the component's initial temperature and the mixture's final temperature, Eq. (9-35) becomes

$$\sum_i m_i c_{vi} (T - T_i) = 0$$

so that the mixture final temperature T is given as

$$T = \frac{\sum\limits_i (m_i c_{vi} T_i)}{\sum\limits_i (m_i c_{vi})} \qquad (9\text{-}37)$$

where T_i is the initial temperature of component i before mixing. Similarly, on a molar basis, this equation becomes

$$T = \frac{\sum\limits_i (n_i c_{vi} T_i)}{\sum\limits_i (n_i c_{vi})} \qquad (9\text{-}38)$$

With the mixture temperature known, the pressure p of the mixture can be calculated from the ideal gas equation:

$$p = \frac{mRT}{V} = \frac{n\mathcal{R}T}{V}$$

where R is the gas constant of the mixture, and $\mathcal{R}$ is the universal gas constant.

If the initial temperatures of the components are the same before mixing, then from Eq. (9-37) or Eq. (9-38), it is clear that

$$T = T_i$$

meaning that the mixture temperature equals the initial temperature of the components. In addition, if the initial pressure of the components are also the same before mixing, then

$$p = \frac{mRT}{V} = \frac{T}{V}\sum_i m_i R_i = \frac{T}{V}\sum_i \left(\frac{p_i V_i}{T_i}\right) = \frac{T}{V}\frac{p_i}{T_i}\sum_i V_i = \frac{T}{V}\frac{p_i}{T_i}(V) = p_i$$

meaning that the mixture pressure equals the initial pressure of the components.

The process of mixing of gases to form a mixture is highly irreversible. For the adiabatic mixing process we are considering (Fig. 9-3), the entropy change must be positive. Thus,

$$\Delta S = \sum_i \Delta S_i > 0$$

The entropy change of each component can be calculated by either of the following two equations:

$$\Delta S_i = n_i\left(\phi_{i,\,final} - \phi_{i,\,initial} - \mathcal{R} \ln \frac{p_{i,\,final}}{p_i}\right) \qquad (9\text{-}39)$$

$$\Delta S_i = n_i\left(c_{pi} \ln \frac{T}{T_i} - \mathcal{R} \ln \frac{p_{i,\,final}}{p_i}\right) \qquad (9\text{-}40)$$

depending on whether variable specific heats or constant specific heats are used. In these equations, p_i is the pressure of component i before mixing, $p_{i,\,final}$ is the partial pressure of component i in the mixture, and $\phi_{i,\,initial}$ and $\phi_{i,\,final}$ are evaluated at the component's initial temperature T_i and the mixture final temperature T, respectively. The preceding two equations are written on a molar basis. Similar equations on a mass basis can be written.

Let us now examine a special case in which the gases in the system shown in Fig. 9-3 (a) are initially at the same temperature T_i and pressure p_i so that the final mixture temperature T and pressure p must equal the initial temperature and pressure of the gases, or $T = T_i$ and $p = p_i$. For each component, from Eq. (9-39) or Eq. (9-40), we have

$$\Delta S_i = n_i \, \Delta S_i = n_i \left(-\mathcal{R} \ln \frac{p_{i,\text{final}}}{p} \right)$$

But from the Dalton's law for component i in the final mixture,

$$p_{i,\text{final}} = x_i p$$

where x_i is the mole fraction of component i in the mixture. From which

$$\Delta S_i = -\mathcal{R} n_i \ln \frac{x_i p}{p} = -\mathcal{R} n_i \ln x_i$$

For the total system, the entropy change for the mixing process is then

$$\Delta S = -\mathcal{R} \sum_i n_i \ln x_i \tag{9-41}$$

Accordingly, the entropy of the final mixture is given by

$$S_{\text{final mixture}} = \sum_i n_i s_i - \mathcal{R} \sum_i n_i \ln x_i \tag{9-42}$$

in which the term $\sum_i n_i s_i$ represents the total entropy of the gases before mixing.

Because mixing of gases is an irreversible process, the total entropy change as given by Eq. (9-41) is positive. From this equation, it is worth noticing that the increase in entropy due to mixing depends only on the number of moles of the component gases, and is independent of the composition of the gases. Thus, when equal moles of nitrogen and oxygen are mixed, the increase in entropy due to mixing would be exactly the same as when equal moles of carbon dioxide and carbon monoxide are mixed. However, when two portions of the same gas are allowed to mix, there is no increase in entropy. This can be easily seen from Eq. (9-41), because for this case, we have $x = 1$ and $\Delta S = 0$.

Example 9-4

A stream of nitrogen at 15 psia and 120°F flows at a rate of 300 ft³/min. A stream of oxygen at 20 psia and 200°F flows at a rate of 50 lbm/min. The two streams mix adiabatically to form a stream of mixture at 17 psia. Determine for the mixture stream the temperature, the mass-flow rate, the volume-flow rate, and the mole fraction of each gas. Determine also the rate of change of entropy for nitrogen and oxygen and the rate of irreversibility for the mixing process. The constant-pressure specific heats for nitrogen and oxygen will be assumed to be constant for the process, and are 0.248 Btu/lbm·°R and 0.222 Btu/lbm·°R, respectively. The atmosphere is at $p_{\text{atm}} = 14.7$ psia and $T_{\text{atm}} = 77°F = 537°R$.

Solution Let us denote nitrogen by the subscript N and oxygen by the subscript O; and symbols without such a subscript denote the mixture. The mixing process is depicted in Fig. 9-4, in which the given data are $p_N = 15$ psia, $T_N = 120°F = 580°R$, $\dot{V}_N = 300$ ft³/min, $p_O = 20$ psia, $T_O = 200°F = 660°R$, $\dot{m}_O = 50$ lbm/min, $p = 17$ psia, $c_{pN} = 0.248$ Btu/lbm·°R, and $c_{p,O} = 0.222$ Btu/lbm·°R.

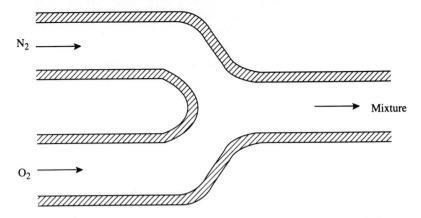

Figure 9-4 Adiabatic mixing of ideal gases in steady flow for Example 9-4.

The mass-flow rate of nitrogen is

$$\dot{m}_N = \frac{p_N \dot{V}_N}{R_N T_N} = \frac{15 \times 144 \times 300}{(1545/28)580} = 20.2 \text{ lbm/min}$$

The mass-flow rate of the mixture is then

$$\dot{m} = \dot{m}_N + \dot{m}_O = 20.2 + 50 = 70.2 \text{ lbm/min}$$

The mass-flow rates of nitrogen and oxygen in moles per unit time are

$$\dot{n}_N = \frac{\dot{m}_N}{M_N} = \frac{20.2}{28} = 0.721 \text{ lbmole/min}$$

$$\dot{n}_O = \frac{\dot{m}_O}{M_O} = \frac{50}{32} = 1.56 \text{ lbmole/min}$$

The mole fractions of the component gases in the mixture are

$$x_N = \frac{\dot{n}_N}{\dot{n}_N + \dot{n}_O} = \frac{0.721}{0.721 + 1.56} = 0.316$$

$$x_O = \frac{\dot{n}_O}{\dot{n}_N + \dot{n}_O} = \frac{1.56}{0.721 + 1.56} = 0.684$$

With $\dot{Q} = 0$, $\dot{W}_{cv} = 0$, and ΔKE and ΔPE neglected, the application of the steady-flow energy equation to the mixing process leads to

$$(\dot{m}_N h_N + \dot{m}_O h_O)_{\text{before mixing}} = (\dot{m}_N h_N + \dot{m}_O h_O)_{\text{after mixing}}$$

or

$$\dot{m}_N(h_{N,\text{ after}} - h_{N,\text{ before}}) + \dot{m}_O(h_{O,\text{ after}} - h_{O,\text{ before}}) = 0$$

Because c_p values are constant, this equation can be written as

$$\dot{m}_N c_{p,N}(T - T_N) + \dot{m}_O c_{p,O}(T - T_O) = 0$$

Therefore, the mixture temperature is given by

$$T = \frac{\dot{m}_N c_{p,N} T_N + \dot{m}_O c_{p,O} T_O}{\dot{m}_N c_{p,N} + \dot{m}_O c_{p,O}}$$

$$= \frac{20.2(0.248)(580) + 50(0.222)(660)}{20.2(0.248) + 50(0.222)} = 635°R = 175°F$$

The volume-flow rate of mixture is

$$\dot{V} = \frac{\dot{n}\mathscr{R}T}{p} = \frac{(\dot{n}_N + \dot{n}_O)\mathscr{R}T}{p} = \frac{(0.721 + 1.56)(1545)(635)}{17 \times 144}$$

$$= 914 \text{ ft}^3/\text{min}$$

The partial pressures of nitrogen and oxygen in the mixture are

$$p_{N,\text{partial}} = x_N p = 0.316(17) = 5.372 \text{ psia}$$

$$p_{O,\text{partial}} = x_O p = 0.684(17) = 11.628 \text{ psia}$$

The rates of entropy change of nitrogen and oxygen are calculated by the equations:

$$\Delta\dot{S}_N = \dot{m}_N\left(c_{p,N} \ln\frac{T}{T_N} - R_N \ln\frac{p_{N,\text{partial}}}{p_N}\right)$$

$$= 20.2\left(0.248 \ln\frac{635}{580} - \frac{1545}{778 \times 28} \ln\frac{5.372}{15}\right)$$

$$= 1.92 \text{ Btu/°R·min}$$

$$\Delta\dot{S}_O = \dot{m}_O\left(c_{p,O} \ln\frac{T}{T_O} - R_O \ln\frac{p_{O,\text{partial}}}{p_O}\right)$$

$$= 50\left(0.222 \ln\frac{635}{660} - \frac{1545}{778 \times 32} \ln\frac{11.628}{20}\right)$$

$$= 1.25 \text{ Btu/°R·min}$$

These values can also be obtained by the following equations:

$$\Delta\dot{S}_N = \dot{n}_N\left(c_{p,N} \ln\frac{T}{T_N} - \mathscr{R} \ln\frac{p_{N,\text{partial}}}{p_N}\right)$$

$$= 0.721\left(0.248 \times 28 \ln\frac{635}{580} - \frac{1545}{778} \ln\frac{5.372}{15}\right) = 1.92 \text{ Btu/°R·min}$$

$$\Delta\dot{S}_O = \dot{n}_O\left(c_{p,O} \ln\frac{T}{T_O} - \mathscr{R} \ln\frac{p_{O,\text{partial}}}{p_O}\right)$$

$$= 1.56\left(0.222 \times 32 \ln\frac{635}{660} - \frac{1545}{778} \ln\frac{11.628}{20}\right) = 1.25 \text{ Btu/°R·min}$$

The irreversibility rate is given by Eq. (8-12),

$$\dot{i} = T_{\text{atm}}(\Delta\dot{S}_{\text{system}} + \Delta\dot{S}_{\text{surr}})$$

where

$$\Delta\dot{S}_{\text{system}} = 0 \text{ for the steady-flow, steady-state process}$$

$$\Delta\dot{S}_{\text{surr}} = \Delta\dot{S}_N + \Delta\dot{S}_O = 1.92 + 1.25 = 3.17 \text{ Btu/°R·min}$$

Thus,

$$\dot{i} = 537(3.17) = 1700 \text{ Btu/min}$$

The irreversibility rate can be also calculated by Eq. (8-9),

$$\dot{I} = \dot{W}_{max} - \dot{W}_{actual}$$

where

$$\dot{W}_{actual} = 0$$

$$\dot{W}_{max} = \sum_{inlet} \dot{m}\psi - \sum_{exit} \dot{m}\psi \qquad \text{(Eq. (8-22) without the } \dot{Q}_R \text{ term)}$$

$$= \dot{m}_N[(h_{N, before} - h_{N, after}) - T_{atm}(s_{N, before} - s_{N, after})]$$

$$+ \dot{m}_O[(h_{O, before} - h_{O, after}) - T_{atm}(s_{O, before} - s_{O, after})]$$

But

$$\dot{m}_N(h_{N, before} - h_{N, after}) + \dot{m}_O(h_{O, before} - h_{O, after}) = 0$$

so that

$$\dot{W}_{max} = T_{atm}(\Delta \dot{S}_N + \Delta \dot{S}_O)$$

$$= 537(1.92 + 1.25)$$

$$= 537(3.17) = 1700 \text{ Btu/min}$$

Therefore, the irreversibility rate is then

$$\dot{I} = \dot{W}_{max} - \dot{W}_{actual} = 1700 - 0 = 1700 \text{ Btu/min}$$

Because no actual work is produced, the irreversibility rate is equal to the maximum power. The irreversibility is due to mixing of two streams of different gases. The effect cannot be reversed by simply reversing the direction of the process. The result of this example shows that mixing processes are highly irreversible.

9-5 PROPERTIES OF AIR–WATER VAPOR MIXTURES

In general, the water vapor in atmospheric air is a superheated vapor at the temperature of the mixture, as depicted by state v in Fig. 9-5. The mixture temperature in most applications involving air–water-vapor mixtures are low with a corresponding low saturation pressure for the water vapor. At a constant mixture temperature, any attempt to increase the water-vapor partial pressure to the saturation value (p_g in Fig. 9-5) will promote partial condensation of the vapor, thereby tending to reduce the mole fraction of water vapor in the mixture. Because the partial pressure of water vapor is low, the vapor behaves as an ideal gas. Accordingly, atmospheric air can be considered as an ideal-gas mixture of dry air (mostly nitrogen and oxygen) and water vapor, satisfying Dalton and Gibbs–Dalton laws.

If the partial pressure of water vapor in atmospheric air equals the saturation pressure of water at the mixture temperature, the water vapor is a saturated vapor and the air is said to be saturated. When unsaturated air is cooled at constant pressure, the temperature at which the vapor begins to condense is called the *dew point* of the mixture. The dew point corresponding to a water-vapor partial pressure p_v is depicted by state d in Fig. 9-5.

The *dry-bulb* temperature of air is the temperature indicated by an ordinary thermometer placed in the air. The *wet-bulb* temperature is the temperature indicated by a thermometer with its temperature-sensitive element covered by a water-saturated wick

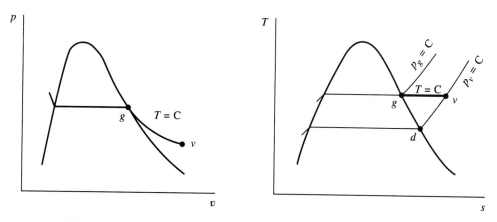

Figure 9-5 $p-v$ and $T-s$ diagrams for water vapor in atmospheric air.

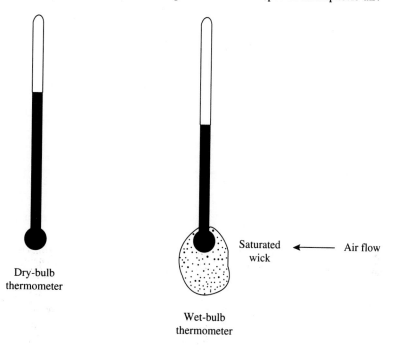

Dry-bulb
thermometer

Saturated
wick

Air flow

Wet-bulb
thermometer

Figure 9-6 Measurement of dry-bulb and wet-bulb temperatures.

and placed in the moist air stream (see Fig. 9-6). If the air is not saturated, evaporation of water from the wick will occur, resulting in a drop of temperature of the wick and the thermometer itself. Eventually, a steady condition is reached in which there is no more change in the temperature of the thermometer. This final temperature is the wet-bulb temperature of the moist air.

Relative humidity ϕ of moist air is defined as the ratio of the mass of water vapor m_v in a given volume to the mass of water vapor m_{sat} required to produce a saturated mixture occupying the same volume at the same temperature, or

$$\phi = \frac{m_v}{m_{\text{sat}}} \tag{9-43}$$

Because the vapor is considered an ideal gas, the definition of relative humidity reduces to the ratio of the partial pressure p_v of the vapor as it exits in the mixture to the saturation pressure p_g of the vapor at the same temperature. Thus,

$$\phi = \frac{m_v}{m_{\text{sat}}} = \frac{p_v V/R_v T}{p_g V/R_v T}$$

or

$$\phi = \frac{p_v}{p_g} \qquad (9\text{-}44)$$

Specific humidity or the humidity ratio ω of an air–vapor mixture is defined as the ratio of the mass of water vapor m_v in a given volume of mixture to the mass of dry air m_a in the same volume, or

$$\omega = \frac{m_v}{m_a} \qquad (9\text{-}45)$$

Considering water vapor as an ideal gas, we can write

$$\omega = \frac{m_v}{m_a} = \frac{p_v V/R_v T}{p_a V/R_a T} = \frac{p_v}{p_a}\frac{R_a}{R_v}$$

$$= \frac{p_v}{p_a}\frac{\mathcal{R}/M_a}{\mathcal{R}/M_v} = \frac{p_v}{p_a}\frac{M_v}{M_a} = \left(\frac{18.02}{28.97}\right)\left(\frac{p_v}{p_a}\right)$$

or

$$\omega = 0.622\frac{p_v}{p_a} = 0.622\frac{p_v}{p - p_v} \qquad (9\text{-}46)$$

where p_v is the partial pressure of water vapor, p_a the partial pressure of dry air, and p the pressure of the mixture.

Assuming that the Gibbs–Dalton law is valid, the enthalpy H of an air–vapor mixture is the sum of the enthalpies of the dry air and the water vapor, or

$$H = H_{\text{dry air}} + H_{\text{water vapor}} = m_a h_a + m_v h_v$$

Because in most processes involving air–water-vapor mixtures (such as heating and humidification or cooling and dehumidification) the amount of dry air does not change, it is convenient to express the specific enthalpy h of an air–vapor mixture on the basis of a unit mass of dry air. Dividing the preceding equation by m_a, we have

$$h = h_a + \omega h_v \qquad (9\text{-}47)$$

where h = specific enthalpy of mixture, per unit mass of dry air
h_a = specific enthalpy of dry air, per unit mass of dry air
h_v = specific enthalpy of water vapor, per unit mass of water vapor

In the temperature range of air-conditioning processes, the specific enthalpy of dry air can be approximated by the expression,

$$h_a = c_{pa} T$$

where the specific heat of dry air c_{pa} can be taken to be 1.005 kJ/kg·°C or 0.240 Btu/lbm·°F, and the dry-bulb temperature T is in °C or °F. Due to the low partial pressure of water vapor in atmospheric air, the specific enthalpy h_v is approximately equal to h_g for a saturated vapor at the given dry-bulb temperature. Accordingly, Eq. (9-47) can be expressed as

$$h = h_a + \omega h_g = c_{pa} T + \omega h_g \qquad (9\text{-}47a)$$

The entropy of an air–water-vapor mixture is the sum of the entropies of the dry air and the water vapor. On the basis of a unit mass of dry air, the specific entropy of an air–water-vapor mixture can be written as

$$s = s_a + \omega s_v$$

where the specific entropy of dry air s_a and the specific entropy of water vapor s_v are at the mixture temperature and the partial pressures of dry air and water vapor, respectively. For water vapor in the mixture, according to Eq. (7-19), we can write

$$s_v(T, p_v) - s_v(T, p_g) = -R_v \ln (p_v/p_g)$$

As

$$s_v(T, p_g) = s_g(T)$$

and

$$p_v/p_g = \phi = \text{relative humidity}$$

we then have

$$s_v(T, p_v) = s_g(T) - R_v \ln \phi \qquad (9\text{-}48)$$

Example 9-5

Atmospheric air at a barometric pressure of 98 kPa and a dry-bulb temperature of 25°C has a water-vapor partial pressure of 2.5 kPa. Determine (a) the relative humidity, (b) the specific humidity, (c) the dew-point temperature, (d) the enthalpy of the air, and (e) the specific volume of the air.

Solution   From the steam table (Table A-18M), at $T = 25°C$, we get $p_g = 3.169$ kPa and $h_g = 2547.2$ kJ/kg. The relative humidity is

$$\phi = \frac{p_v}{p_g} = \frac{2.5}{3.169} = 0.789 = 78.9\%$$

and the specific humidity is

$$\omega = 0.622 \frac{p_v}{p - p_v} = 0.622 \left(\frac{2.5}{98 - 2.5} \right) = 0.0163 \frac{\text{kg water vapor}}{\text{kg dry air}}$$

The dew-point temperature is the saturation temperature at $p_v = 2.5$ kPa. The steam table (Table A-19M) gives

$$T_{\text{dew point}} = 21.1°C$$

The enthalpy of the atmospheric air is given by

$$h = c_{pa} T + \omega h_g = 1.005(25) + 0.0163(2547.2)$$

$$= 66.6 \text{ kJ/kg dry air}$$

As the volume occupied by the mixture is the same as the volume occupied by the dry-air component, the specific volume of the atmospheric air per unit mass of dry air is given by

$$v = v_a = \frac{R_a T}{p_a} = \frac{(8.314/29)(25 + 273)}{98 - 2.5}$$

$$= 0.895 \text{ m}^3/\text{kg dry air}$$

9-6 ADIABATIC SATURATION

Consider a stream of unsaturated atmospheric air at a dry-bulb temperature T_1 entering an insulated long flow chamber, as depicted in Fig. 9-7(a). As the unsaturated air comes in contact with the body of water lying in the bottom of the chamber, some of the water evaporates, thus increasing the humidity of the air. The energy for evaporation comes from both the air–vapor stream and the water in the chamber, resulting in the lowering of the temperature of the air stream. If the air leaving the chamber is saturated at temperature T_2 and makeup water is supplied at T_2, this final equilibrium temperature is known as the *adiabatic saturation temperature*. The states of the water vapor in the mixture during the adiabatic saturation process are shown on the T–s diagram of Fig. 9-7(b). During the process, the water-vapor partial pressure increases and the temperature decreases; as a result, the adiabatic saturation temperature is lower than the dry-bulb temperature and higher than the dew-point temperature.

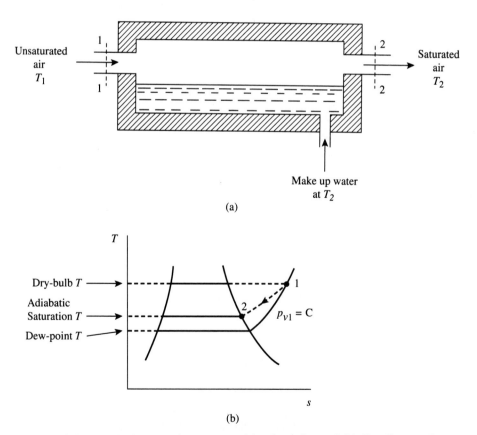

(a)

(b)

Figure 9-7 Adiabatic saturation process: (a) schematic, and (b) T–s diagram for water.

Neglecting changes in kinetic and potential energy, a steady-flow energy balance for the adiabatic saturation process per unit mass-flow of dry air yields

$$h_1 + (\omega_2 - \omega_1)h_{f2} = h_2 \qquad (9\text{-}49a)$$

where h_1 and h_2, respectively, are the enthalpies of the incoming and outgoing mixtures per unit mass of dry air. In terms of the enthalpies of the dry-air and water-vapor components, this equation is rewritten as

$$(h_{a1} + \omega_1 h_{g1}) + (\omega_2 - \omega_1)h_{f2} = h_{a2} + \omega_2 h_{g2} \qquad (9\text{-}49b)$$

where h_a is the enthalpy of dry air, and h_g and h_f are enthalpies of water in the saturated vapor and liquid states, respectively. Solving for ω_1, we have

$$\omega_1 = \frac{h_{a2} - h_{a1} + \omega_2(h_{g2} - h_{f2})}{h_{g1} - h_{f2}}$$

or

$$\omega_1 = \frac{c_{pa}(T_2 - T_1) + \omega_2 h_{fg2}}{h_{g1} - h_{f2}} \qquad (9\text{-}50)$$

Because the air leaving the chamber is saturated, $\phi_2 = 100\%$ and $p_{v2} = p_{g2} = p_{sat}$ at T_2, it follows that

$$\omega_2 = 0.622\left(\frac{p_{g2}}{p - p_{g2}}\right) \qquad (9\text{-}51)$$

Therefore, ω_2 can be evaluated when the mixture pressure p is measured. With T_1, T_2, and ω_2 known, Eq. (9-50) can be used to calculate the specific humidity (and hence the relative humidity) of unsaturated atmospheric air.

Although the technique of adiabatic saturation leads to the desired results, its execution is not practical. It is found, however, that for air–water-vapor mixtures at atmospheric pressure and temperature, the wet-bulb temperature of the mixture is very nearly equal to the adiabatic saturation temperature. They differ slightly because the wet-bulb temperature is influenced by heat- and mass-transfer rates, whereas the adiabatic saturation temperature simply involves equilibrium between the entering air–vapor mixture and water at the adiabatic saturation temperature. Equation (9-50) can be used to calculate the specific humidity of moist air from the measurements of wet- and dry-bulb temperatures. In these calculations, the wet-bulb temperature is taken equal to the adiabatic saturation temperature T_2, and the dry-bulb temperature equal to the inlet temperature T_1.

Example 9-6

Atmospheric air at 14.7 psia has a dry-bulb temperature of 80°F and a wet-bulb temperature of 70°F. Determine (a) specific humidity, (b) the relative humidity, (c) the dew-point temperature, (d) the enthalpy of the mixture, and (e) the specific volume of the mixture.

Solution From the steam table (Table A-18E), the following values are obtained:

At $T_1 = 80°F$, $p_{g1} = p_{sat} = 0.5073$ psia, and $h_{g1} = 1096.4$ Btu/lbm
At $T_2 = 70°F$, $p_{g2} = p_{sat} = 0.3632$ psia, $h_{f2} = 38.09$ Btu/lbm, and h_{g2}
$\qquad = 1092.0$ Btu/lbm

The specific humidity of saturated air at the wet-bulb temperature is

$$\omega_2 = 0.622\left(\frac{p_{g2}}{p - p_{g2}}\right) = 0.622\left(\frac{0.363}{14.7 - 0.363}\right)$$

$$= 0.0157 \text{ lbm water vapor/lbm dry air}$$

The specific humidity of the unsaturated atmospheric air is then

$$\omega_1 = \frac{c_{pa}(T_2 - T_1) + \omega_2 h_{fg2}}{h_{g1} - h_{f2}}$$

$$= \frac{0.24(70 - 80) + 0.0157(1092.0 - 38.1)}{1096.4 - 38.1}$$

$$= 0.0134 \text{ lbm water vapor/lbm dry air}$$

Now,

$$\omega_1 = 0.622\left(\frac{p_{v1}}{p - p_{v1}}\right)$$

or

$$0.0134 = 0.622\left(\frac{p_{v1}}{14.7 - p_{v1}}\right)$$

so that

$$p_{v1} = 0.310 \text{ psia}$$

The dew-point temperature is then obtained from the steam table (Table A-19E):

$$T_{\text{dew point}} = T_{\text{sat}}(\text{at } p_{v1}) = 65.4°F$$

The relative humidity of the atmospheric air is calculated as

$$\phi_1 = \frac{p_{v1}}{p_{g1}} = \frac{0.310}{0.507} = 0.611 = 61.1\%$$

The enthalpy of the mixture is given by

$$h_1 = c_{pa}T_1 + \omega_1 h_{g1}$$

$$= 0.240(80) + 0.0134(1096.4) = 33.9 \text{ Btu/lbm dry air}$$

The specific volume of the mixture is given by

$$v_1 = v_{a1} = \frac{R_a T_1}{p_{a1}} = \frac{R_a T_1}{p - p_{v1}}$$

$$= \frac{(1545/29)(80 + 460)}{(14.7 - 0.310)144}$$

$$= 13.9 \text{ ft}^3/\text{lbm dry air}$$

Notice that when the dry-bulb and wet-bulb temperatures of a moist air are known (as in this example), Eq. (9-50) can be used to obtain the specific humidity.

9-7 PSYCHROMETRIC CHART

For computations involving atmospheric air in air-conditioning problems, a psychrometric chart is often used for property determination and process visulization. A *psychrometric chart* is a graphical correlation of the thermodynamic properties of moist air. For a given mixture pressure (usually one standard atmosphere), the properties represented are the specific humidity, relative humidity, dry-bulb temperature, wet-bulb temperature (for adiabatic saturation temperature), specific volume, and enthalpy. Such a chart is shown schematically in Fig. 9-8 with specific humidity as the ordinate and dry-bulb temperature as the abscissa.

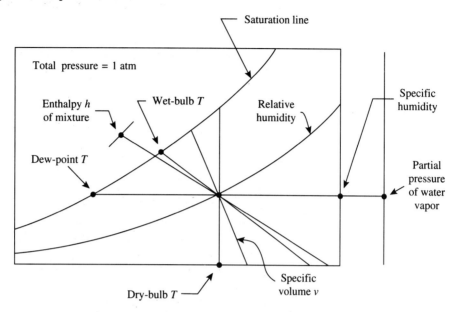

Figure 9-8 Components of a psychrometric chart.

As shown by Eq. (9-46), at a given mixture pressure, the partial pressure of water vapor is a function of the specific humidity. Therefore, the partial pressure of water vapor is also frequently plotted on the ordinate. Accordingly, a horizontal line of constant specific humidity is also a line of constant partial pressure of water vapor and a line of constant dew point.

The saturation line is the line of 100% relative humidity. Other relative-humidity lines follow the same general shape to the right of the saturation line according to the definition given by Eq. (9-44). The wet-bulb temperature shown in the psychrometric chart is the adiabatic saturation temperature, with relation to the dry-bulb temperature and the specific humidity according to Eq. (9-50). The lines of constant specific volume are given as volume of mixture (or, identically, of dry air or of water vapor) per unit mass of dry air.

For an adiabatic saturation process, if the small enthalpy change due to the addition of water to the air stream is neglected, Eq. (9-49a) would yield $h_1 = h_2$. Because h_2 is a function of the adiabatic saturation temperature only and the adiabatic saturation temperature is constant during an adiabatic saturation process, we see that lines of constant enthalpy are essentially parallel to the lines of adiabatic saturation (wet-bulb) temperature. It should be noted that the enthalpy values given in the chart are for the mixture but on the basis of unit mass of dry air.

The choice of coordinates for a psychrometric chart is arbitrary. Most common coordinates are specific humidity versus dry-bulb temperature. The psychrometric charts in Appendix 4 are those from the American Society of Heating, Refrigerating and Air-Conditioning Engineers (ASHRAE). The ASHRAE psychrometric charts use oblique-angle coordinates of enthalpy and specific humidity. In these charts, the dry-bulb temperature lines are straight lines, not precisely parallel to each other, and inclined slightly from the vertical position.

Example 9-7

Repeat Example 9-6 using a psychrometric chart.

Solution   At the intersection of the lines for dry-bulb temperature of 80°F and wet-bulb temperature of 70°F, we have from the English-unit psychrometric chart (Fig. A-16E):

$\phi = 61.0\%$

$\omega = 0.0135$ lbm water vapor/lbm dry air

dew-point temperature $= 65.5°F$

$v = 13.9$ ft³/lbm dry air

$h = 34.0$ Btu/lbm dry air

9-8   HEATING AND HUMIDIFICATION

In a simple heating process, as depicted in Fig. 9-9, there is no change in the moisture content of the air, thus constituting a constant-specific-humidity process. By neglecting changes in kinetic and potential energies, an energy balance for the process on the basis of unit mass flow of dry air yields

$$h_1 + q_{in} = h_2 \tag{9-52a}$$

where h_1 and h_2, respectively, are the enthalpies of the incoming and outgoing moist air per unit mass of dry air, and q_{in} is the heat addition per unit mass of dry air. In terms of the enthalpies of the dry–air and water-vapor components, this equation becomes

$$(h_{a1} + \omega h_{g1}) + q_{in} = h_{a2} + \omega h_{g2} \tag{9-52b}$$

where h_a and h_g, respectively, are the enthalpies of dry air and water vapor, and ω is the constant specific humidity of the moist air.

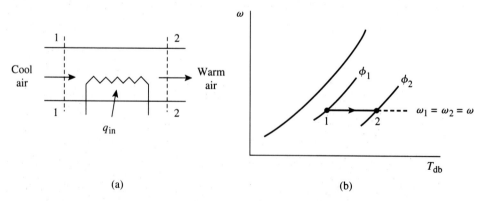

Figure 9-9   Simple heating: (a) schematic diagram, and (b) process diagram on a psychrometric chart.

It can be seen from Fig. 9-9 for a simple heating process that the relative humidity of the air after heating is smaller than that before heating. Even though the moisture content is the same throughout the heating process, the outgoing hotter air has a higher ability to hold more water vapor. The addition of moisture to the dry hot air is necessary in order to achieve the conditions for human comfort. The injection of water or steam to the hot-air stream, as shown in Fig. 9-10, is one method of humidification of the air. In Fig. 9-10, the heating process 1–2 is well-defined, but the humidifying process 2–3 varies with the conditions of the injecting water or steam. If cool water is sprayed to the air, state 3 will be at a lower temperature than state 2, possibly following an evaporative cooling process (see Sec. 9-10). On the other hand, the injection of steam to the air stream will result in humidification with additional heating, as shown by process 2–3'. The energy equation for process 2–3 (or 2–3') can be written as

$$h_2 + (\omega_3 - \omega_2)h_w = h_3 \tag{9-53a}$$

or

$$(h_{a2} + \omega_2 h_{g2}) + (\omega_3 - \omega_2)h_w = (h_{a3} + \omega_3 h_{g3}) \tag{9-53b}$$

wherein h_w is the enthalpy of the injecting water or steam. The energy equation for the combined process 1–2–3 (or 1–2–3') can be written as

$$h_1 + q_{in} + (\omega_3 - \omega_1)h_w = h_3 \tag{9-54a}$$

or

$$(h_{a1} + \omega_1 h_{g1}) + q_{in} + (\omega_3 - \omega_1)h_w = h_{a3} + \omega_3 h_{g3} \tag{9-54b}$$

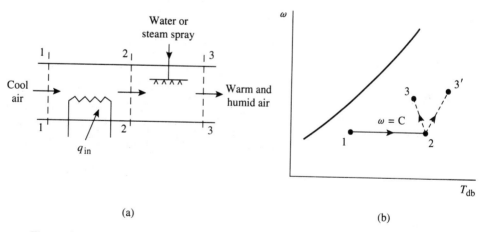

Figure 9-10 Heating and humidification: (a) schematic diagram, and (b) process diagram on a psychrometric chart.

Example 9-8

Outdoor air at a total pressure of 101 kPa, a dry-bulb temperature of 4°C, and a relative humidity of 90% is first heated and then humidified so that it can be delivered to an indoor space at a dry-bulb temperature of 23°C and a relative humidity of 45%. Saturated steam at 120°C is used in the humidifier section. Determine the heating and steam requirements per unit mass of dry air. Determine also the temperature and relative humidity of the air at the exit from the heating section.

Solution Referring to Fig. 9-10, from the steam table (Table A-18M), we have the following data:

$$\text{At } T_1 = 4°C, p_{g1} = p_{sat} = 0.8131 \text{ kPa, and } h_{g1} = 2508.7 \text{ kJ/kg}$$
$$\text{At } T_{3'} = 23°C, p_{g3'} = p_{sat} = 2.810 \text{ kPa, and } h_{g3'} = 2543.5 \text{ kJ/kg}$$
$$\text{At } T_{steam} = 120°C, h_{steam} = h_g = 2706.3 \text{ kJ/kg}$$

For the inlet air,

$$p_{v1} = \phi_1 p_{g1} = 0.90(0.8131) = 0.732 \text{ kPa}$$

$$\omega_1 = 0.622\frac{p_{v1}}{p - p_{v1}} = 0.622\left(\frac{0.732}{101 - 0.732}\right)$$

$$= 0.00454 \text{ kg water vapor/kg dry air}$$

$$h_1 = c_{pa}T_1 + \omega_1 h_{g1} = 1.005(4) + 0.00454(2508.7)$$

$$= 15.4 \text{ kJ/kg dry air}$$

For the exit air after humidification,

$$p_{v3'} = \phi_{3'} p_{g3'} = 0.45(2.810) = 1.26 \text{ kPa}$$

$$\omega_{3'} = 0.622\left(\frac{p_{v3'}}{p - p_{v3'}}\right) = 0.622\left(\frac{1.26}{101 - 1.26}\right)$$

$$= 0.00786 \text{ kg water vapor/kg dry air}$$

$$h_{3'} = c_{pa}T_{3'} + \omega_{3'} h_{g3'} = 1.005(23) + 0.00786(2543.5)$$

$$= 43.1 \text{ kJ/kg dry air}$$

The steam requirement is calculated as

$$\omega_{3'} - \omega_1 = 0.00786 - 0.00454$$

$$= 0.00332 \text{ kg steam/kg dry air}$$

The heating requirement is calculated from an energy balance for the combined heating and humidification system:

$$q_{in} = h_{3'} - h_1 - (\omega_{3'} - \omega_1)h_{steam}$$

$$= 43.1 - 15.4 - 0.00332(2706.3)$$

$$= 18.7 \text{ kJ/kg dry air}$$

For the heating process only, an energy balance gives

$$h_2 = h_1 + q_{in} = 15.4 + 18.7 = 34.1 \text{ kJ/kg dry air}$$

For the condition at point 2, we also have

$$h_2 = c_{pa}T_2 + \omega_2 h_{g2} = 1.005T_2 + 0.00454h_{g2}$$

From the preceding two equations, with the value of h_g at T_2 from the steam table (Table A-18M), by trial and error, we obtain $T_2 = 22.4°C$.
    Because

$$p_{v2} = p_{v1} = 0.732 \text{ kPa}$$

and

$$p_{g2} = p_{sat}(\text{at } T_2 = 22.4°C) = 2.711 \text{ kPa}$$

we have

$$\phi_2 = \frac{p_{v2}}{p_{g2}} = \frac{0.732}{2.711} = 0.270 = 27.0\%$$

9-9 COOLING AND DEHUMIDIFICATION

In a simple cooling process, such as depicted in Fig. 9-11, the specific humidity of the air remains unchanged as the temperature decreases. Even though the air is cooler, the relative humidity at the exit may become too high and cause discomfort.

In order to reduce the moisture content, the air is cooled below its dew point so that enough of the water vapor in the air condenses. The cooling-plus-dehumidifying process $1-d-2$ is as shown in Fig. 9-12. The first part of the process $(1-d)$ is a simple cooling. When the dew-point temperature (state d) is reached, vapor begins to con-

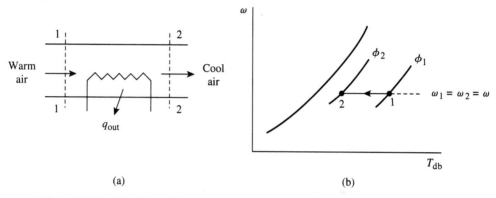

(a)   (b)

Figure 9-11 Simple cooling: (a) schematic diagram, and (b) process diagram on a psychrometric chart.

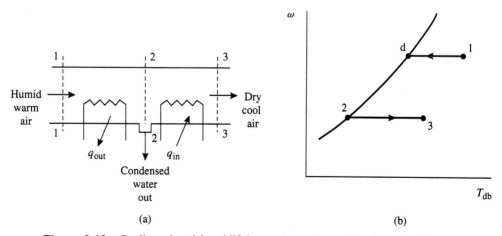

(a)   (b)

Figure 9-12 Cooling plus dehumidifying and reheating: (a) schematic diagram, and (b) process diagram on a psychrometric chart.

dense. Upon further cooling, condensation of moisture continues along the saturation line until the lowest temperature T_2 of the cooling process is reached. The condensate is assumed to be cooled to this final lowest temperature before leaving. The quantity of water removed per unit mass of dry air is given by the difference in the specific humidity between the inlet and exit states of the cooling section, or, referring to Fig. 9-12,

$$\text{Water removed} = \omega_1 - \omega_2 \qquad (9\text{-}55)$$

The heat removed from the air per unit mass flow of dry air is given by

$$q_{\text{out}} = h_1 - h_2 - (\omega_1 - \omega_2)h_{f2} \qquad (9\text{-}56a)$$

or

$$q_{\text{out}} = (h_{a1} + \omega_1 h_{g1}) - (h_{a2} + \omega_2 h_{g2}) - (\omega_1 - \omega_2)h_{f2} \qquad (9\text{-}56b)$$

The cooling coil for this heat-removal process is usually the evaporator of a refrigeration machine.

The temperature of the air leaving the cooling-plus-dehumidifying section is usually too low for direct use in human-comfort applications. The air can be reheated to a more desirable temperature as illustrated by process 2–3 in Fig. 9-12, or it can be mixed with a warm-air stream to obtain a suitable conditioned air.

Example 9-9

Ten thousand ft³/min of outdoor air at 14.7 psia, 90°F, and 60% relative humidity is to be conditioned to a final temperature of 70°F and a final relative humidity of 45%. The air is first cooled below its dew point and water condenses from the air stream until the desired specific humidity is reached. It is then reheated to the final desired conditions. Determine the lowest temperature of the cooling process, the rate of water removal, the rate of heat rejection in the cooling section, and the rate of heat addition in the reheating section.

Solution By referring to Fig. 9-12, $\dot{V}_1 = 10{,}000$ ft³/min, $p = 14.7$ psia, $T_1 = 90°F$, $\phi_1 = 60\%$, $T_3 = 70°F$, and $\phi_3 = 45\%$. From the steam table (Table A-18E), we have the following:

$$\text{At } T_1 = 90°F, \; p_{g1} = p_{\text{sat}} = 0.6988 \text{ psia and } h_{g1} = 1100.7 \text{ Btu/lbm}$$
$$\text{At } T_3 = 70°F, \; p_{g3} = p_{\text{sat}} = 0.3632 \text{ psia and } h_{g3} = 1092.0 \text{ Btu/lbm}$$

For the incoming air,

$$p_{v1} = \phi_1 p_{g1} = 0.60(0.6988) = 0.419 \text{ psia}$$

$$\omega_1 = 0.622 \frac{p_{v1}}{p - p_{v1}} = 0.622\left(\frac{0.419}{14.7 - 0.419}\right)$$

$$= 0.0182 \text{ lbm water vapor/lbm dry air}$$

$$h_1 = c_{pa} T_1 + \omega_1 h_{g1} = 0.24(90) + 0.0182(1100.7)$$

$$= 41.6 \text{ Btu/lbm dry air}$$

$$v_1 = v_{a1} = \frac{R_a T_1}{p_{a1}} = \frac{53.3(90 + 460)}{(14.7 - 0.419)144} = 14.3 \text{ ft}^3/\text{lbm dry air}$$

For the air in the final condition,

$$p_{v3} = \phi_3 p_{g3} = 0.45(0.3632) = 0.163 \text{ psia}$$

$$\omega_3 = 0.622 \left(\frac{p_{v3}}{p - p_{v3}} \right) = 0.622 \left(\frac{0.163}{14.7 - 0.163} \right)$$

$$= 0.00697 \text{ lbm water vapor/lbm dry air}$$

$$h_3 = c_{pa} T_3 + \omega_3 h_{g3} = 0.24(70) + 0.00697(1092.0)$$

$$= 24.4 \text{ Btu/lbm dry air}$$

The lowest temperature of the process is the dew point for the air in the final conditions, that is,

$$T_2 = T_{sat} \text{ (at } p_{v2} = p_{v3} = 0.163 \text{ psia)} = 47.6°F$$

At $T_2 = 47.6°F$, the steam table (Table A-18E) gives

$$h_{f2} = 15.7 \text{ Btu/lbm} \qquad \text{and} \qquad h_{g2} = 1082.2 \text{ Btu/lbm.}$$

Thus,

$$h_2 = c_{pa} T_2 + \omega_2 h_{g2} \qquad \text{(noting that } \omega_2 = \omega_3)$$

$$= 0.24(47.6) + 0.00697(1082.2) = 19.0 \text{ Btu/lbm dry air}$$

The mass-flow rate of dry air is

$$\dot{m}_a = \frac{\dot{V}_1}{v_1} = \frac{10,000}{14.3} = 699.3 \text{ lbm dry air/min}$$

$$= 42,000 \text{ lbm dry air/h}$$

The rate of water removal is given by

$$\dot{m}_a (\omega_1 - \omega_2) = 42,000(0.0182 - 0.00697)$$

$$= 472 \text{ lbm water/h}$$

The rate of heat rejection in the cooling section is given by

$$\dot{Q}_{out} = \dot{m}_a [h_1 - h_2 - (\omega_1 - \omega_2) h_{f2}]$$

$$= 42,000[41.6 - 19.0 - (0.0182 - 0.00697)15.7]$$

$$= 942,000 \text{ Btu/h}$$

The rate of heat addition in the reheating section is given by

$$\dot{Q}_{in} = \dot{m}_a (h_3 - h_2) = 42,000(24.4 - 19.0)$$

$$= 227,000 \text{ Btu/h}$$

9-10 EVAPORATIVE COOLING

For regions of the world where the climate is hot and dry, it is possible to take advantage of the low humidity to achieve cooling. This is accomplished by spraying water into the dry hot air stream, thus causing part of the water to evaporate. The energy for the evaporation of water comes from the air stream, so that the air becomes cooler and more humid. This is the so-called *evaporative cooling* process. Figure 9-13 shows an elementary evaporative cooling system together with a psychrometric chart for the process. An evaporative cooling process is essentially an adiabatic saturation process carried short of completion. It therefore follows a constant wet-bulb temperature line

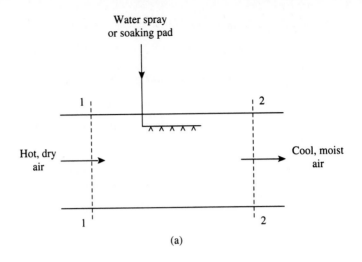

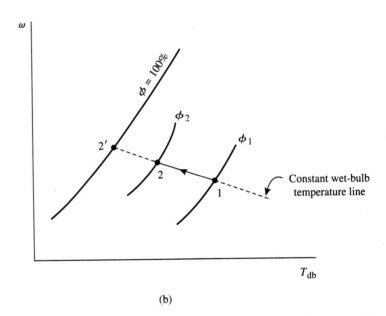

Figure 9-13 Evaporative cooling: (a) schematic diagram, and (b) process diagram on a psychrometric chart.

on a psychrometric chart. On the basis of unit mass flow of dry air, the energy balance for the process is

$$h_1 + (\omega_2 - \omega_1)h_w = h_2 \tag{9-57a}$$

or

$$(h_{a1} + \omega_1 h_{g1}) + (\omega_2 - \omega_1)h_w = h_{a2} + \omega_2 h_{g2} \tag{9-57b}$$

where h_w is the enthalpy of the spaying water. Because both h_w and $(\omega_2 - \omega_1)$ are small quantities, their product is quite small. It follows that evaporative cooling can be considered as a constant-enthalpy process.

It should be mentioned that in evaporative cooling, the cool humid air can be utilized to cool drier air through a heat exchanger; in this manner, the discomfort that would otherwise result from the increased humidity inherent in the process can be avoided.

Example 9-10

Air at 101 kPa total pressure, 32°C dry-bulb temperature, and 10% relative humidity passes through an evaporative cooler until the final dry-bulb temperature is 20°C. Water is added at 20°C. Determine how much water is added to the air per kilogram of dry air and the final relative humidity. What is the minimum temperature that could be achieved by this process? Use (a) data taken from a psychrometric chart, and (b) constant c_p for dry air and steam-table values for water vapor.

Solution Referring to Fig. 9-13, $p = 101$ kPa, $T_1 = T_{db1} = 32°C$, $\phi_1 = 10\%$, $T_2 = T_{db2} = 20°C$, $T_w = 20°C$.

(a) Use the metric-unit psychrometric chart: At the intersection of the lines for $T_1 = T_{db1} = 32°C$ and $\phi_1 = 10\%$, the chart gives $\omega_1 = 0.003$ kg water vapor/kg dry air, $T_{wb1} = 14.4°C$, and $h_1 = 40.0$ kJ/kg dry air.

At the intersection of the lines for $T_{wb2} = T_{wb1} = 14.4°C$ and $T_2 = T_{db2} = 20°C$, the chart gives $\phi_2 = 54\%$, $\omega_2 = 0.008$ kg water vapor/kg dry air, and $h_2 = 40.0$ kJ/kg dry air. Thus, the final relative humidity is $\phi_2 = 54\%$. The water added to the air is given by

$$\omega_2 - \omega_1 = 0.008 - 0.003 = 0.005 \text{ kg water/kg dry air}$$

The minimum temperature that could be achieved is

$$T_2' = T_{wb1} = T_{wb2} = 14.4°C$$

(b) Calculation by the use of the steam table (Table A-18M): From the steam table at $T_1 = 32°C$, we get $p_{g1} = p_{sat} = 4.759$ kPa, $h_{g1} = 2559.9$ kJ/kg; at $T_2 = 20°C$, we get $p_{g2} = p_{sat} = 2.339$ kPa, $h_{g2} = 2538.1$ kJ/kg; at $T_w = 20°C$, we get $h_w = h_f = 83.96$ kJ/kg.

For the initial state,

$$p_{v1} = \phi_1 p_{g1} = 0.10(4.759) = 0.476 \text{ kPa}$$

$$\omega_1 = 0.622 \frac{p_{v1}}{p - p_{v1}} = 0.622\left(\frac{0.476}{101 - 0.476}\right)$$

$$= 0.00295 \text{ kg water vapor/kg dry air}$$

$$h_1 = c_{pa}T_1 + \omega_1 h_{g1} = 1.005(32) + 0.00295(2559.9)$$

$$= 39.7 \text{ kJ/kg dry air}$$

The energy balance for the evaporative cooling process 1–2 is given by

$$(c_{pa}T_1 + \omega_1 h_{g1}) + (\omega_2 - \omega_1)h_w = c_{pa}T_2 + \omega_2 h_{g2}$$

so that

$$\omega_2 = \frac{c_{pa}(T_1 - T_2) + \omega_1(h_{g1} - h_w)}{h_{g2} - h_w}$$

$$= \frac{1.005(32 - 20) + 0.00295(2559.9 - 83.96)}{2538.1 - 83.96}$$

$$= 0.00789 \text{ kg water vapor/kg dry air}$$

The amount of water added is

$$\omega_2 - \omega_1 = 0.00789 - 0.00295$$

$$= 0.00494 \text{ kg water/kg dry air}$$

The specific humidity of the air at the exit from the cooler is given by

$$\omega_2 = 0.622\left(\frac{p_{v2}}{p - p_{v2}}\right)$$

so that

$$p_{v2} = \frac{p\omega_2}{0.622 + \omega_2} = \frac{101(0.00789)}{0.622 + 0.00789} = 1.265 \text{ kPa}$$

thus,

$$\phi_2 = \frac{p_{v2}}{p_{g2}} = \frac{1.265}{2.339} = 0.541 = 54.1\%$$

In order to find the minimum temperature $T_{2'}$ (Fig. 9-13), we apply the adiabatic saturation equation, Eq. (9-50):

$$\omega_1 = \frac{c_{pa}(T_{2'} - T_1) + \omega_{2'}\,h_{fg2'}}{h_{g1} - h_{f2'}}$$

or

$$0.00295 = \frac{1.005(T_{2'} - 32) + \omega_{2'}h_{fg2'}}{2559.9 - h_{f2'}}$$

In addition, for the final saturation state,

$$\omega_{2'} = 0.622\left(\frac{p_{g2'}}{p - p_{g2'}}\right)$$

or

$$\omega_{2'} = 0.622\left(\frac{p_{\text{sat}} \text{ at } T_{2'}}{101 - p_{\text{sat}} \text{ at } T_{2'}}\right)$$

Solving the preceding two equations by a trial-and-error solution, we have the following: Assume $T_{2'} = 14°C$:

$$\omega_{2'} = 0.622\left(\frac{1.5983}{101 - 1.5983}\right)$$

$$= 0.01000 \text{ kg water vapor/kg dry air}$$

so that

$$\frac{1.005(T_{2'} - 32) + \omega_{2'}h_{fg2'}}{2559.9 - h_{f2'}} = \frac{1.005(14 - 32) + 0.01000(2468.3)}{2559.9 - 58.80} = 0.002636$$

Assume $T_{2'} = 15°C$:

$$\omega_{2'} = 0.622\left(\frac{1.7051}{101 - 1.7051}\right)$$

$$= 0.01068 \text{ kg water vapor/kg dry air}$$

so that

$$\frac{1.005(15 - 32) + 0.01068(2465.9)}{2559.9 - 62.99} = 0.003705$$

Assuming linear interpolation is valid, we conclude that the value of the minimum temperature $T_{2'}$ is about 14.3°C.

9-11 ADIABATIC MIXING

In air-conditioning systems, it is often necessary to mix two moist air streams to form a mixture stream of a desired quality. One common example is the mixing of outdoor fresh-air to indoor recirculated-air in central heating or cooling applications. Figure 9-14(a) shows schematically the adiabatic mixing process in a simple duct system. Part (b) of this figure shows a typical process diagram on a psychrometric chart. Through the use of the principles of conservation of mass and energy, we have the following

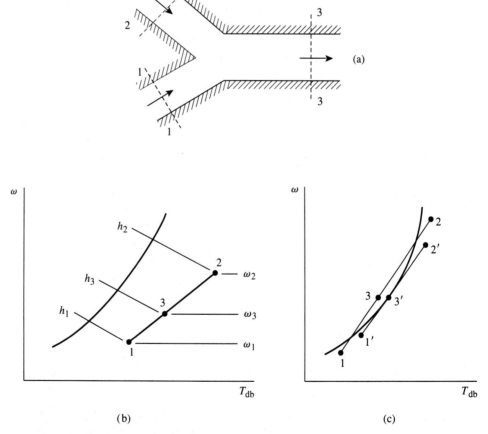

Figure 9-14 Adiabatic mixing: (a) schematic diagram, (b) process diagram on a psychrometric chart, and (c) special mixing cases.

three basic relations on a rate basis:

The mass balance of dry air:

$$\dot{m}_{a1} + \dot{m}_{a2} = \dot{m}_{a3} \tag{9-58}$$

The mass balance of water vapor:

$$\dot{m}_{a1}\omega_1 + \dot{m}_{a2}\omega_2 = \dot{m}_{a3}\omega_3 \tag{9-59}$$

The energy balance for adiabatic mixing:

$$\dot{m}_{a1}h_1 + \dot{m}_{a2}h_2 = \dot{m}_{a3}h_3 \tag{9-60}$$

Combining Eqs. (9-58) and (9-59), we have

$$\frac{\dot{m}_{a1}}{\dot{m}_{a2}} = \frac{\omega_2 - \omega_3}{\omega_3 - \omega_1}$$

Combining Eqs. (9-58) and (9-60), we have

$$\frac{\dot{m}_{a1}}{\dot{m}_{a2}} = \frac{h_2 - h_3}{h_3 - h_1}$$

Therefore, we have

$$\frac{\dot{m}_{a1}}{\dot{m}_{a2}} = \frac{\omega_2 - \omega_3}{\omega_3 - \omega_1} = \frac{h_2 - h_3}{h_3 - h_1} \tag{9-61}$$

This equation implies that on the psychrometric chart, the resulting state 3 must lie on a straight line connecting the initial states 1 and 2 and divide this straight line into segments inversely proportional to the relative masses of the original two streams.

Figure 9-14(c) shows two special cases of the mixing results. If states 1 and 2 lie close to the saturation line, then state 3 may lie to the left of the saturation line, meaning that condensation occurs as a result of mixing and the condensed water may remain suspended as foglike droplets in the exit flow stream. The other special case shown in Fig. 9-14(c) is the mixing of two streams at states $1'$ and $2'$. With a certain mass ratio $\dot{m}_{a1}/\dot{m}_{a2}$, the final mixture may be saturated at state $3'$.

Example 9-11

In an air-conditioning system, the air from the refrigeration cooling coils at 10°C and 100% relative humidity is to be mixed adiabatically with 20% outside fresh air at 36°C and 30% relative humidity. Determine the final temperature and relative humidity of the mixture stream by the use of (a) a psychrometirc chart and (b) the steam table values. The total pressure is 101 kPa.

Solution   The process diagram on a psychrometric chart is similar to that shown in Fig. 9-14(b), except that point 1 is on the saturation line. The given data are $T_1 = T_{db1} = 10°C$, $\phi_1 = 100\%$, $T_2 = T_{db2} = 36°C$, $\phi_2 = 30\%$, $p = 101$ kPa, and $\dot{m}_{a1}/\dot{m}_{a2} = 1/0.2 = 5$.

(a) *By the psychrometric chart:* Locate states 1 and 2 on the psychrometric chart from the given values of dry-bulb temperature and relative humidity and draw a straight line between these points. Locate the final state 3 so that $\overline{13}/\overline{23} = 1/5$. At point 3, we get the required answers: $T_3 = T_{db3} = 14.4°C$ and $\phi_3 = 80\%$.

(b) *By calculation from steam-table data:* From the steam table (Table A-18M), we have the following values: at $T_1 = 10°C$, $p_{g1} = p_{sat} = 1.2276$ kPa and $h_{g1} = 2519.8$ kJ/kg; at $T_2 = 36°C$, $p_{g2} = p_{sat} = 5.947$ kPa and $h_{g2} = 2567.1$ kJ/kg.

For state 1,

$$p_{v1} = \phi_1 p_{g1} = 1.00(1.2276) = 1.228 \text{ kPa}$$

$$\omega_1 = 0.622\left(\frac{p_{v1}}{p - p_{v1}}\right) = 0.622\left(\frac{1.228}{101 - 1.228}\right)$$

$$= 0.007656 \text{ kg water vapor/kg dry air}$$

$$h_1 = c_{pa}T_1 + \omega_1 h_{g1} = 1.005(10) + 0.007656(2519.8)$$

$$= 29.34 \text{ kJ/kg dry air}$$

For state 2,

$$p_{v2} = \phi_2 p_{g2} = 0.30(5.947) = 1.784 \text{ kPa}$$

$$\omega_2 = 0.622\left(\frac{p_{v2}}{p - p_{v2}}\right) = 0.622\left(\frac{1.784}{101 - 1.784}\right)$$

$$= 0.01118 \text{ kg water vapor/kg dry air}$$

$$h_2 = c_{pa}T_2 + \omega_2 h_{g2} = 1.005(36) + 0.01118(2567.1)$$

$$= 64.88 \text{ kJ/kg dry air}$$

From Eq. (9-61),

$$\frac{\dot{m}_{a1}}{\dot{m}_{a2}} = \frac{\omega_2 - \omega_3}{\omega_3 - \omega_1} = \frac{h_2 - h_3}{h_3 - h_1}$$

thus,

$$5 = \frac{0.01118 - \omega_3}{\omega_3 - 0.007656} = \frac{64.88 - h_3}{h_3 - 29.34}$$

so that

$$\omega_3 = 0.00824 \text{ kg water vapor/kg dry air}$$

$$h_3 = 35.3 \text{ kJ/kg dry air}$$

To find the temperature T_3, we use the expression for h_3:

$$h_3 = c_{pa}T_3 + \omega_3 h_{g3} \qquad \text{or} \qquad 35.3 = 1.005T_3 + 0.00824h_{g3}$$

where h_{g3} is a function of T_3. A trial-and-error solution gives the value of T_3 to be approximately equal to 14.4°C.

For the final state 3, we have

$$\omega_3 = 0.622\left(\frac{p_{v3}}{p - p_{v3}}\right)$$

or

$$p_{v3} = \frac{\omega_3 p}{0.622 + \omega_3} = \frac{0.00824(101)}{0.622 + 0.00824}$$

$$= 1.321 \text{ kPa}$$

At $T_3 = 14.4$°C, the steam table gives

$$p_{g3} = p_{sat} = 1.641 \text{ kPa}$$

Therefore,

$$\phi_3 = \frac{p_{v3}}{p_{g3}} = \frac{1.321}{1.641} = 0.805 = 80.5\%$$

9-12 WET COOLING TOWER

The condenser of a steam power plant or a refrigeration system requires the dissipation of waste energy. Typically, water is the cooling agent that carries the waste energy out from the plant. To avoid thermal pollution of river water caused by dumping waste energy into the river, the cooling-water effluent from a plant can be cooled in a cooling tower for reuse. There are two types of cooling tower. A *dry cooling tower* is simply a closed-type heat exchanger in which the water is cooled by the air due to temperature difference between the two. A *wet cooling tower* involves evaporative cooling and is therefore more efficient.

Figure 9-15 shows a schematic of a wet cooling tower. The warm water is introduced at the top of the tower and is sprayed down over baffles to provide a large area for evaporation and heat transfer to a rising mass of atmospheric air coming in from the bottom. The evaporation process requires energy, and this energy transfer results in a cooling of the remaining water stream. The cooled water stream is then returned to the plant to pick up additional waste energy. The air is assumed to leave the cooling tower in a saturated state. Makeup water must be added to the cycle to replace the water lost by evaporation.

Considering a control volume (as indicated in Fig. 9-15), which includes the entire cooling tower in steady-flow operation, a mass balance for the dry air gives

$$\dot{m}_{a1} = \dot{m}_{a2} = \dot{m}_a$$

and a mass balance for the water gives

$$\dot{m}_{w3} - \dot{m}_{w4} = \dot{m}_a(\omega_2 - \omega_1) \tag{9-62}$$

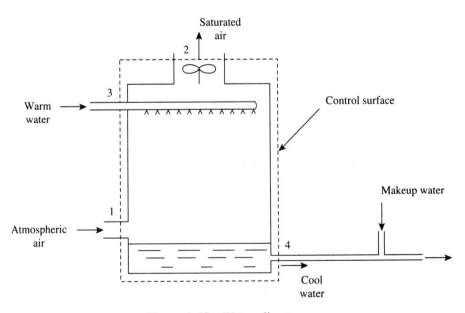

Figure 9-15   Wet cooling tower.

where $\dot{m}_{w3}$ and $\dot{m}_{w4}$ are the mass-flow rates of water at section 3 and 4, respectively, and $\dot{m}_a$ is the mass-flow rate of dry air through the process, noting that $\dot{m}_{w3} - \dot{m}_{w4}$ is the mass-flow rate of the required makeup water. By assuming the control volume to be adiabatic and neglecting the fan work and changes in kinetic and potential energies, an energy balance for the control volume on a rate basis yields

$$\dot{m}_a h_1 + \dot{m}_{w3} h_{f3} = \dot{m}_a h_2 + \dot{m}_{w4} h_{f4} \qquad (9\text{-}63a)$$

or

$$\dot{m}_a(h_{a1} + \omega_1 h_{g1}) + \dot{m}_{w3} h_{f3} = \dot{m}_a(h_{a2} + \omega_2 h_{g2}) + \dot{m}_{w4} h_{f4} \qquad (9\text{-}63b)$$

where h_{f3} and h_{f4} are to be taken as saturated liquid enthalpies at the given temperatures.

Example 9-12

A cooling tower is used to cool water from an initial temperature of 110°F to a final temperature of 88°F. Atmospheric air enters the cooling tower at 80°F and 40% relative humidity and leaves at 105°F saturated. The volume-flow rate of the entering water is 250 ft³/min. Determine the air required per minute and the mass of water evaporated per minute. Determine also the irreversibility rate for the cooling tower process, in Btu/min. The atmosphere is at $T_0 = 80°F$ and $p_0 = 14.7$ psia.

Solution Referring to Fig. 9-15, the given data are $T_1 = T_{db1} = 80°F$, $\phi_1 = 40\%$, $T_2 = T_{db2} = 105°F$, $\phi_2 = 100\%$, $T_3 = 110°F$, $T_4 = 88°F$, $\dot{V}_3 = 250$ ft³/min, and $p = 14.7$ psia.

From the steam table (Table A-18E), we have the following:
At $T_1 = 80°F$, we get

$$p_{g1} = p_{sat} = 0.5073 \text{ psia}$$

$$h_{g1} = 1096.4 \text{ Btu/lbm} \qquad \text{and} \qquad s_{g1} = 2.0356 \text{ Btu/lbm·°R}$$

At $T_2 = 105°F$, we get

$$p_{g2} = p_{sat} = 1.1029 \text{ psia}$$

$$h_{g2} = 1107.2 \text{ Btu/lbm} \qquad \text{and} \qquad s_{g2} = 1.9697 \text{ Btu/lbm·°R}$$

At $T_3 = 110°F$, we get

$$v_{f3} = 0.016166 \text{ ft}^3/\text{lbm}$$

$$h_{f3} = 78.02 \text{ Btu/lbm} \qquad \text{and} \qquad s_{f3} = 0.14730 \text{ Btu/lbm·°R}$$

At $T_4 = 88°F$, we get

$$h_{f4} = 56.07 \text{ Btu/lbm} \qquad \text{and} \qquad s_{f4} = 0.10801 \text{ Btu/lbm·°R}$$

For the inlet air,

$$p_{v1} = \phi_1 p_{g1} = 0.40(0.5073) = 0.203 \text{ psia}$$

$$\omega_1 = 0.622\left(\frac{p_{v1}}{p - p_{v1}}\right) = 0.622\left(\frac{0.203}{14.7 - 0.203}\right)$$

$$= 0.00871 \text{ lbm water vapor/lbm dry air}$$

$$h_1 = c_{pa} T_1 + \omega_1 h_{g1} = 0.24(80) + 0.00871(1096.4)$$

$$= 28.75 \text{ Btu/lbm dry air}$$

For the exit air,

$$p_{v2} = \phi_2 p_{g2} = 1.00(1.1029) = 1.103 \text{ psia}$$

$$\omega_2 = 0.622\left(\frac{p_{v2}}{p - p_{v2}}\right) = 0.622\left(\frac{1.103}{14.7 - 1.103}\right)$$

$$= 0.05046 \text{ lbm water vapor/lbm dry air}$$

$$h_2 = c_{pa} T_2 + \omega_2 h_{g2} = 0.24(105) + 0.05046(1107.2)$$

$$= 81.07 \text{ Btu/lbm dry air}$$

The mass-flow rate of water inlet is

$$\dot{m}_{w3} = \frac{\dot{V}_3}{v_{f3}} = \frac{250}{0.016166} = 15.460 \text{ lbm water/min}$$

The mass-flow rate of water outlet is

$$\dot{m}_{w4} = \dot{m}_{w3} - \dot{m}_a(\omega_2 - \omega_1)$$

An energy balance for the cooling tower yields

$$\dot{m}_a h_1 + \dot{m}_{w3} h_{f3} = \dot{m}_a h_2 + [\dot{m}_{w3} - \dot{m}_a(\omega_2 - \omega_1)]h_{f4}$$

so that the mass-flow rate of dry air is

$$\dot{m}_a = \frac{\dot{m}_{w3}(h_{f3} - h_{f4})}{h_2 - h_1 - (\omega_2 - \omega_1)h_{f4}}$$

$$= \frac{15,460(78.02 - 56.07)}{81.07 - 28.75 - (0.05046 - 0.00871)56.07}$$

$$= 6790 \text{ lbm dry air/min}$$

The rate of water evaporation is given by

$$\dot{m}_{w3} - \dot{m}_{w4} = \dot{m}_a(\omega_2 - \omega_1)$$

$$= 6790(0.05046 - 0.00871)$$

$$= 283 \text{ lbm water/min}$$

The mass-flow rate of water outlet is then

$$\dot{m}_{w4} = 15,460 - 283 = 15,180 \text{ lbm water/min}$$

The irreversibility rate of the steady-state, steady-flow process of the cooling tower system as indicated by the control volume shown in Fig. 9-15 can be calculated from Eq. (8-12):

$$\dot{I} = T_0(\Delta\dot{S}_{\text{system}} + \Delta\dot{S}_{\text{surr}})$$

where for steady flow, $\Delta\dot{S}_{\text{system}} = 0$. The rate of entropy change of the surroundings is given by

$$\Delta\dot{S}_{\text{surr}} = \dot{m}_a[s_a(T_2, p_{a2}) + \omega_2 s_v(T_2, p_{v2})]$$
$$- \dot{m}_a[s_a(T_1, p_{a1}) + \omega_1 s_v(T_1, p_{v1})]$$
$$+ \dot{m}_{w4} s_{f4} - \dot{m}_{w3} s_{f3} \tag{9-64}$$

$$= \dot{m}_a[s_a(T_2, p_{a2}) - s_a(T_1, p_{a1}) + \omega_2 s_v(T_2, p_{v2}) - \omega_1 s_v(T_1, p_{v1})]$$
$$+ \dot{m}_{w4} s_{f4} - \dot{m}_{w3} s_{f3}$$

in which, for the dry air, we have from Eq. (9-34):

$$s_a(T_2, p_{a2}) - s_a(T_1, p_{a1}) = c_{pa} \ln \frac{T_2}{T_1} - \frac{\mathcal{R}}{M_a} \ln \frac{p_{a2}}{p_{a1}}$$

$$= c_{pa} \ln \frac{T_2}{T_1} - \frac{\mathcal{R}}{M_a} \ln \frac{p - p_{v2}}{p - p_{v1}}$$

$$= (0.24 \text{ Btu/lbm·°R}) \ln \frac{105 + 460}{80 + 460}$$

$$- \frac{1.986 \text{ Btu/lbmole·°R}}{29 \text{ lbm/lbmole}} \ln \frac{14.7 - 1.103}{14.7 - 0.203}$$

$$= 0.01525 \text{ Btu/lbm·°R}$$

$$s_v(T_2, p_{v2}) = s_g(T_2, p_{g2}) = s_{g2} = 1.9697 \text{ Btu/lbm·°R}$$

$$s_v(T_1, p_{v1}) = s_v(\text{at } 80°\text{F}, 0.203 \text{ psia}) = 2.1385 \text{ Btu/lbm·°R}$$

(This value was taken from the original Keenan–Keyes–Hill–Moore, *Steam Tables,* 1969 ed. Table A-20E in Appendix 3 does not include low-pressure data.)

The value of $s_v(T_1, p_{v1})$ can also be evaluated by the use of Eq. (9-48). Thus,

$$s_v(T_1, p_{v1}) = s_g(T_1) - R_v \ln \phi_1$$

$$= 2.0356 - \frac{1.9858}{18} \ln (0.40)$$

$$= 2.1367 \text{ Btu/lbm·°R}$$

Substituting all known values in Eq. (9-64) gives

$$\Delta \dot{S}_{\text{surr}} = (6790 \text{ lbm da/min})[(0.01525 \text{ Btu/lbm da·°R})$$

$$+ (0.05046 \text{ lbm wv/lbm da})(1.9697 \text{ Btu/lbm wv·°R})$$

$$- (0.00871 \text{ lbm wv/lbm da})(2.1385 \text{ Btu/lbm wv·°R})]$$

$$+ (15{,}180 \text{ lbm w/min})(0.10801 \text{ Btu/lbm w·°R})$$

$$- (15{,}460 \text{ lbm w/min})(0.14730 \text{ Btu/lbm w·°R})$$

$$= 14.27 \text{ Btu/min·°R}$$

The irreversibility rate of the cooling tower process is then

$$\dot{i} = (80 + 460)(14.27) = 7710 \text{ Btu/min}$$

Notice that because no work is done during the process, the irreversibility rate is equal in magnitude to the maximum power that would have been produced if the process were reversible. Notice also that the makeup water mass-flow rate is $283/15{,}460 = 0.0183$, or 1.83%, of the inlet water mass-flow rate. This makeup water must be supplied to the power plant to compensate for the loss of water to the surrounding atmosphere in the cooling-tower process.

9-13 SUMMARY

The total mass m of a mixture is the sum of the masses m_i of its constitutes:

$$m = \sum m_i$$

The mass fraction y_i of component i is defined as

$$y_i = m_i/m$$

The total number of moles n of a mixture is the sum of the number of moles n_i of its constitutes:

$$n = \sum n_i$$

The mole fraction x_i of component i is defined as

$$x_i = n_i/n$$

The average, or apparent, molar mass or molecular weight M of a mixture is given by

$$M = \sum x_i M_i \qquad \text{or} \qquad M = \left(\sum \frac{y_i}{M_i} \right)^{-1}$$

where M_i is the molar mass of component i. The apparent gas constant R of a mixture is given by

$$R = \mathcal{R}/M$$

A mixture of ideal gases behaves also as an ideal gas. Dalton's law states that the total pressure p of an ideal-gas mixture is equal to the sum of the partial pressure p_i of the components:

$$p = \sum p_i$$

where the partial pressure p_i of component i is defined as the pressure that this component would exert if it occupied the volume alone at the mixture temperature, and is given by

$$p_i = x_i p$$

Amagat's law states that the volume V of an ideal-gas mixture is equal to the sum of the partial volumes V_i of the components:

$$V = \sum V_i$$

where the partial volume V_i of component i is defined as the volume that would be occupied by this component at mixture temperature T and mixture pressure p, and is given by

$$V_i = x_i V$$

For an ideal-gas mixture, the volume fraction (V_i/V), mole fraction $(x_i = n_i/n)$, and mass fraction $(y_i = m_i/m)$ of a component are related by the equation:

$$\frac{V_i}{V} = \frac{n_i}{n} = \left(\frac{m_i}{m}\right)\frac{M}{M_i} \qquad \text{or} \qquad x_i = y_i\frac{M}{M_i}$$

Gibbs–Dalton law states that in a mixture of ideal gases, each component acts as if it alone were occupying the volume of the mixture at the temperature of the mixture. This is the basis for the calculation of energy-related properties of ideal-gas mixtures. Thus,

$$U = nu = \sum n_i u_i \qquad \text{or} \qquad U = mu = \sum m_i u_i$$

$$H = nh = \sum n_i h_i \qquad \text{or} \qquad H = mh = \sum m_i h_i$$

$$S = ns = \sum n_i s_i \qquad \text{or} \qquad S = ms = \sum m_i s_i$$

$$nc_p = \sum n_i c_{pi} \qquad \text{or} \qquad mc_p = \sum m_i c_{pi}$$

$$nc_v = \sum n_i c_{vi} \qquad \text{or} \qquad mc_v = \sum m_i c_{vi}$$

Atmospheric air is a mixture of dry air (considered as N_2 and O_2) and water vapor. In general, the water vapor in atmospheric air is a superheated vapor. When the water vapor is a saturated vapor, the air is said to be saturated. The partial pressure of water vapor in atmospheric air is very low, so that the water vapor can be treated as an ideal gas.

Relative humidity ϕ of an air–vapor mixture is defined as the ratio of the partial pressure p_v of the vapor as it exits in the mixture to the saturation pressure p_g of the vapor at the mixture temperature, or

$$\phi = p_v/p_g$$

Specific humidity or humidity ratio ω of an air–vapor mixture is defined as the ratio of the mass of water vapor m_v to the mass of dry air m_a, or

$$\omega = \frac{m_v}{m_a} = 0.622\left(\frac{p_v}{p - p_v}\right)$$

where p is the total pressure of the mixture.

The specific enthalpy h of an air–vapor mixture per unit mass of dry air can be expressed as

$$h = c_{pa}T + \omega h_g$$

where c_{pa} is the constant-pressure specific heat of dry air, and h_g is the saturated-vapor enthalpy at the mixture temperature T.

The specific entropy s of an air–vapor mixture per unit mass of dry air can be written as

$$s = s_a + \omega s_v$$

where the specific entropy of dry air s_a and the specific entropy of water vapor s_v are at the mixture temperature T and the partial pressures of dry air p_a and water vapor p_v, respectively. For the water vapor in the mixture, the specific entropy can be written as

$$s_v(T, p_v) = s_g(T) - R_v \ln \phi$$

where

$$\phi = \text{relative humidity} = p_v/p_g$$

The dry-bulb temperature of air is the temperature indicated by an ordinary thermometer placed in the air. The wet-bulb temperature is the temperature indicated by a thermometer with its temperature-sensitive element covered by a water-saturated wick and placed in the moist air stream. For air-water vapor mixtures at atmospheric pressure and temperature, the wet-bulb temperature of the mixture is very nearly equal to the adiabatic saturation temperature (see Sec. 9-6 for this temperature). When the dry-bulb temperature T_{db} and wet-bulb temperature T_{wb} are known, the specific humidity ω of an atmospheric air can be calculated by the equation:

$$\omega = \frac{c_{pa}(T_{wb} - T_{db}) + \omega_{wb}h_{fg,\,wb}}{h_{g,\,db} - h_{f,\,wb}}$$

where $h_{f,\,wb}$ and $h_{fg,\,wb}$ are the steam-table values based on wet-bulb temperature, $h_{g,\,db}$ is the steam-table value based on dry-bulb temperature, and

$$\omega_{wb} = 0.622\left(\frac{p_{g,\,wb}}{p - p_{g,\,wb}}\right)$$

where $p_{g,\,wb}$ is the saturation pressure of water vapor at the wet-bulb temperature.

When unsaturated air is cooled at constant pressure, the temperature at which the vapor begins to condense is called the dew-point temperature. Dew-point temperature T_{dp} of a gas–vapor mixture can be evaluated by the expression:

$$T_{dp} = T_{sat} \text{ at partial pressure of the vapor}$$

A psychrometric chart is a graphical correlation of the thermodynamic properties of atmospheric air.

PROBLEMS

9-1. A gaseous mixture consists of 1.3 kg of CO, 0.7 kg of CO_2, and 3 kg of N_2. Determine the mole fraction of the component gases. What volume is occupied by the mixture if $p = 101$ kPa and $T = 27°C$?

9-2. Air consists of 78.03% nitrogen, 20.99% oxygen, and 0.98% argon by volume. Determine (a) percent by mass of components, (b) molecular mass of the mixture, and (c) gas constant of the mixture.

9-3. A 0.05-kg ideal-gas mixture consisting of 64% O_2 and 36% H_2 by mass is at a pressure of 0.3 MPa and occupies a volume of 0.10 m³. Determine (a)

the apparent molar mass of the mixture, (b) the gas constant of the mixture, (c) the partial pressure of the component gases, and (d) the temperature.

9-4. A mixture of 56 lbm of N_2, 8 lbm of H_2, and 132 lbm of CO_2 is at 280 psia and 120°F. Find the volume of the mixture and the partial pressure of the component gases.

9-5. A gaseous mixture consisting of 2 lbm of oxygen and 5 lbm of nitrogen initially at 70°F and 60 psia total pressure is cooled at constant pressure to 40°F. Determine the changes in internal energy, enthalpy, and entropy. For oxygen, $c_p = 0.219$

Btu/lbm·°F and $c_v = 0.156$ Btu/lbm·°F. For nitrogen, $c_p = 0.248$ Btu/lbm·°F and $c_v = 0.177$ Btu/lbm·°F.

9-6. An ideal-gas mixture of N_2, CO_2, and H_2O in a molar ratio of 4 : 1 : 1 is compressed isentropically in a steady-flow process to six times its original pressure. The original temperature is 300 K. Using ideal-gas property tables, determine the final temperature and the work requirement per kgmole of the mixture.

9-7. An ideal-gas mixture of N_2, CO_2, and H_2O having a molar ratio of 4 : 1 : 1 expands isentropically in a turbine from 1000 K to 1/6 of its initial pressure. Determine the final temperature of the mixture and the work done by the mixture, in kJ/kgmole of mixture. Use the ideal-gas property tables for this problem.

9-8. A mixture of 1 kg of methane (CH_4) and 20 kg of air at 27°C and 101 kPa is compressed to one-tenth its original volume in an internally reversible adiabatic process. Determine the final temperature and pressure.

9-9. An ideal-gas mixture consisting of 1 lbm air and 0.94 lbm water vapor is compressed in a closed system from 50 psia and 250°F to 100 psia in an internally reversible and adiabatic process. Determine the final temperature and the change in entropy for each component. Use $c_p = 0.243$ Btu/lbm·°R for air and $c_p = 0.446$ Btu/lbm·°R for water vapor.

9-10. An ideal-gas mixture consisting of 4.55 kg of argon, 0.45 kg of hydrogen, and 7.70 kg of dry air is contained in a closed system initially at 88 kPa and 270 K. See Fig. P9-10. Heat is added to the mixture, making it to undergo a constant-pressure process until a final temperature of 360 K. Determine the partial pressure of the component gases in the mixture (in kPa), and the heat and work interactions (in kJ).

9-11. An ideal-gas mixture consisting of 2 kg of H_2 and 14 kg of N_2 is contained in a closed system initially at 160 K and 2 MPa. Heat is added to the mixture, making it to undergo a constant-pressure process until a final temperature of 200 K. Determine the heat added, the total entropy change (called entropy production), and the irreversibility associated with the process. The atmosphere is at 25°C. Use ideal-gas property tables for this problem.

9-12. A rigid tank contains 2 kilograms of nitrogen at 700 kPa and 177°C, and another rigid tank contains 1 kilogram of oxygen at 300 kPa and 27°C.

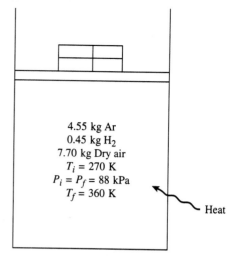

4.55 kg Ar
0.45 kg H_2
7.70 kg Dry air
$T_i = 270$ K
$P_i = P_f = 88$ kPa
$T_f = 360$ K

Heat

Figure P9-10

See Fig. P9-12. The two tanks are connected by a pipe with a valve. When the valve is opened the two gases mix. The final temperature of the mixture is 102°C. Determine the heat transfer to the surroundings, and the final pressure of the mixture.

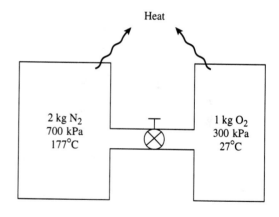

Heat

2 kg N_2
700 kPa
177°C

1 kg O_2
300 kPa
27°C

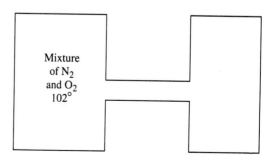

Mixture
of N_2
and O_2
102°

Figure P9-12

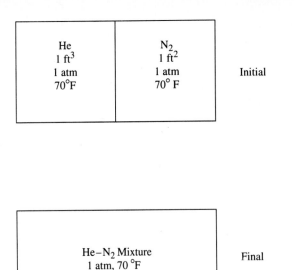

He	N₂	
1 ft³	1 ft²	
1 atm	1 atm	Initial
70°F	70° F	

He–N₂ Mixture	
1 atm, 70 °F	Final

Figure P9-13

9-13. A 2-ft³ tank has a partition at the center. One side of the tank contains helium at 1 atm and 70°F, and the other side contains nitrogen at 1 atm and 70°F. When the partition is taken off, a mixture at 1 atm and 70°F is formed. See Fig. P9-13. For the mixture, determine (a) the mole fractions of the components, (b) the mass fractions of the components, (c) the partial volumes of the components, (d) the partial pressures of the components, (e) the c_p, c_v, and k values of the mixture, (f) the molecular mass of the mixture, and (g) the gas constant of the mixture. The c_p, c_v, R, and molecular mass of helium and nitrogen are as follows:

For helium, $c_p = 1.25$ Btu/lbm·°R, $c_v = 0.75$ Btu/lbm·°R, $R = 386.2$ ft·lbf/lbm·°R and $M = 4$ lbm/lbmole.

For nitrogen, $c_p = 0.248$ Btu/lbm·°R, $c_v = 0.177$ Btu/lbm·°R, $R = 55.16$ ft·lbf/lbm·°R, and $M = 28$ lbm/lbmole.

9-14. Oxygen and carbon monoxide are contained in the two equal halves of an insulated 1-m³ tank divided by a thin wall. See Fig. P9-14. The two gases are initially at 350 kPa and 65°C. The partition is re-

moved and the gases mix. What is the change of entropy?

9-15. A tank containing 2 lbm of methane at 20 psia, 50°F, and another tank containing 4 1bm of oxygen at 100 psia, 20°F, are connected through a valve. See Fig. P9-15. The valve is opened and the gases mix adiabatically. Atmospheric temperature is 50°F. Determine (a) the mixture pressure, (b) the mixture temperature, (c) the volumetric analysis of the mixture, (d) the total change in entropy, and (e) the irreversibility of the process.

9-16. A tank containing 15 ft³ of helium at 100 psia and 50°F and another tank containing 10 ft³ of nitrgen at 50 psia and 150°F are connected through a valve. The valve is opened and the gases mix adiabatically. Determine the resulting pressure, temperature, and the change in entropy.

9-17. A tank contains 3.77 moles of nitrogen at 20 psia and 100°F, and another tank contains 1 mole of oxygen at 20 psia and 100°F. A valve connecting the tanks is opened and the gases mix adiabatically. The temperature of the surroundings is 40°F. Determine the irreversibility of the process.

9-18. Hydrogen and nitrogen in separate streams enter an adiabatic mixing chamber in a 3-to-1 molar ratio. See Fig. P9-18. At the inlet, the hydrogen is at 0.2 MPa and 70°C, and the nitrogen is at 0.2 MPa and 270°C. The mixture leaves at 0.19 MPa. Determine the temperature of the mixture and the net entropy change per kg of mixture.

9-19. A stream of oxygen at 300 kPa and 25°C flows at a rate of 8 kg/min. A stream of nitrogen at 300 kPa and 80°C flows at a rate of 14 kg/min. See Fig. P9-19. The two streams mix adiabatically to form a stream of mixture at 300 kPa. Determine the changes of entropy for oxygen and nitrogen per minute and the rate of irreversibility for the mixing process. The atmosphere is at 101 kPa and 298 K.

9-20. A stream of nitrogen at 15 psia and 120°F flows at a rate of 300 ft³/min. A stream of oxygen at 20 psia and 200°F flows at a rate of 50 1bm/min. The two streams are mixed in a nonadiabatic mixing chamber. The mixture leaves the chamber at 17 psia and 130°F. Determine (a) the mass-flow rate of the nitrogen, (b) the mole fraction of each gas in the mixture, (c) the volume-flow rate of the mixture, (d) the rates of entropy change for nitrogen and oxygen in the mixing process, and (e) the rate of heat transfer during the process.

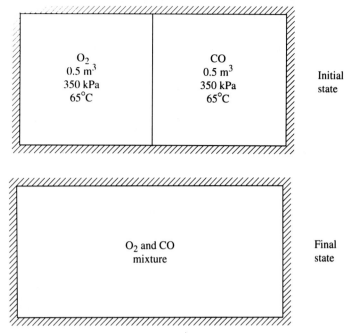

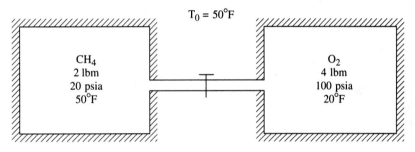

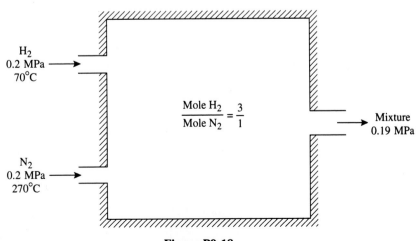

Figure P9-14

Figure P9-15

Figure P9-18

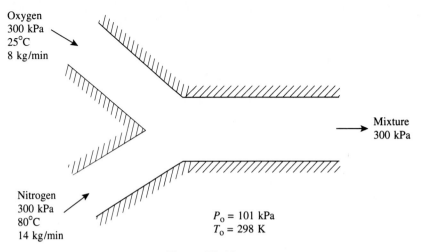

Oxygen
300 kPa
25°C
8 kg/min

Mixture
300 kPa

Nitrogen
300 kPa
80°C
14 kg/min

$P_0 = 101$ kPa
$T_0 = 298$ K

Figure P9-19

9-21. A well-insulated tank 2 ft³ in volume containing air at 14.7 psia and 70°F is connected through a valve to a pipeline in which helium at 50 psia and 120°F is flowing. See Fig. P9-21. The valve is opened and helium is allowed to flow into the tank until the total pressure in the tank becomes 30 psia. Assuming that no air flows from the tank and that the process is adiabatic, determine the mass of helium that flows into the tank and the final temperature of the mixture in the tank.

9-22. A tank of 2 ft³ volume contains air initially at 14.7 psia and 60°F. The tank is connected through a valve to a pipeline in which argon at 90 psia and 90°F is flowing steadily. The valve is opened and argon is allowed to flow into the tank until the total pressure in the tank is 45 psia. Assuming that no air leaves the tank, determine the heat transfer that is required to maintain the contents of the tank at 60°F.

9-23. A mixture of 60 mole percent nitrogen and 40 mole percent carbon dioxide enters an insulated compressor at 0.15 MPa, 30°C, with a mass-flow rate of 2 kg/s. It is compressed to 0.45 MPa, 142°C in steady-flow operation. See Fig. P9-23.

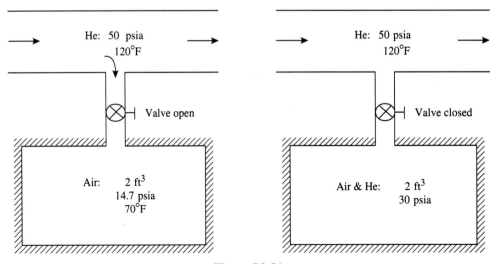

He: 50 psia
120°F

Valve open

Air: 2 ft³
14.7 psia
70°F

He: 50 psia
120°F

Valve closed

Air & He: 2 ft³
30 psia

Figure P9-21

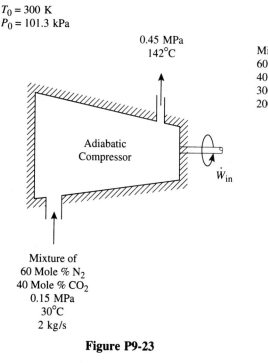

$T_0 = 300$ K
$P_0 = 101.3$ kPa

0.45 MPa
142°C

Adiabatic
Compressor

$\dot{W}_{in}$

Mixture of
60 Mole % N_2
40 Mole % CO_2
0.15 MPa
30°C
2 kg/s

Figure P9-23

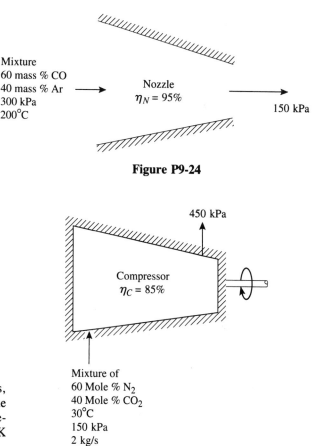

Mixture
60 mass % CO
40 mass % Ar
300 kPa
200°C

Nozzle
$\eta_N = 95\%$

150 kPa

Figure P9-24

450 kPa

Compressor
$\eta_C = 85\%$

Mixture of
60 Mole % N_2
40 Mole % CO_2
30°C
150 kPa
2 kg/s

Figure P9-25

Neglecting kinetic and potential energy changes, determine (a) the power required, in kW, (b) the isentropic compressor efficiency, and (c) the irreversibility rate, in kW. The atmosphere is at 300 K and 101.3 kPa.

9-24. A gaseous mixture of 60% carbon monoxide and 40% argon on a mass basis is expanded in a nozzle from 300 kPa, 200°C, to 150 kPa. See Fig. P9-24. The nozzle is adiabatic and has an efficiency of 95%. The inlet velocity is negligible. Determine (a) the exit temperature, and (b) the irreversibility per unit mass of the mixture. Assume $T_0 = 25°C$.

9-25. A gaseous mixture consisting of 60 mole percent N_2 and 40 mole percent CO_2 at 30°C, 150 kPa, enters a steady-flow adiabatic compressor at a mass-flow rate of 2 kg/s. See Fig. P9-25. The mixture is compressed to 450 kPa. The compressor has an adiabatic efficiency of 85%. Neglecting changes in kinetic and potential energy, determine (a) the exit temperature of the mixture, and (b) the irreversibility of the process. Assume $T_0 = 25°C$.

9-26. The specific humidity of atmospheric air is 0.0428 1bm water vapor per 1bm dry air when the barometer reading is 14.25 psia. For a dry-bulb temperature of 114°F, determine (a) the partial pressure of the water vapor, (b) the partial pressure of the dry air, (c) the relative humidity, and (d) the dew

point. Show the thermodynamic properties of the water vapor on the $T–s$ plane.

9-27. A room humidifier is used to increase the relative humidity of air from 40% to 60% while maintaining a dry-bulb temperature of 21°C. The volume-flow rate of air entering the humidifier is 1 m³/s. Calculate the mass-flow rate of water in kilograms per day. Room pressure is 101.3 kPa.

9-28. Atmospheric air at 75°F, 14.0 psia, and 75% relative humidity enters an air compressor and exits at 70 psia, 280°F. See Fig. P9-28. At the compressor discharge, find (a) the specific humidity and (b) the relative humidity.

9-29. The air discharged from the compressor in Fig. P9-28 is passed through an aftercooler in which heat is transferred to cooling water. Some of the moisture in the air condenses and the liquid is drained off separately from the air. The air then leaves the cooler as saturated air at 68 psia and

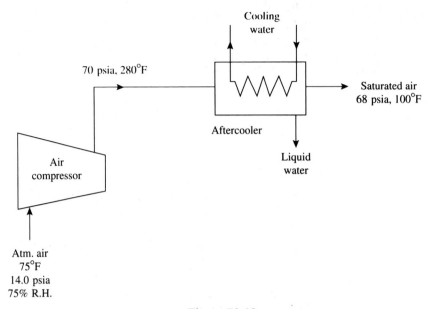

Figure P9-28

100°F. How much water is removed from each pound of dry air flowing through the cooler?

9-30. Air at a temperature of 60°F and a relative humidity of 60% with a flow rate of 1000 ft³/min enters a warm-air furnace and humidifier unit. The air leaves the unit at a temperature of 120°F and a relative humidity of 12%. If the air flows through the unit at a pressure of 14.5 psia, and if water is supplied to the humidifier at a temperature of 65°F, (a) how many pounds of water must be supplied per hour, and (b) at what rate must heat be supplied?

9-31. Atmospheric air at 69°F dry-bulb temperature and 60% relative humidity with a volume-flow rate of 3000 ft³/min is humidified by passing through a water spray chamber. It exits at 69°F dry-bulb temperature and 80% relative humidity. See Fig. P9-31. The water in the spray chamber is continuously recirculated through the sprays and back to a reservoir, heat is supplied to keep the reservoir temperature constant, and makeup water is supplied at 60°F to compensate for evaporation. Assuming that external heat losses and the work of the spray pump are negligible, calculate (a) the rate of makeup water supply and (b) the rate of heat supply. The process occurs at constant total pressure of 14.7 psia.

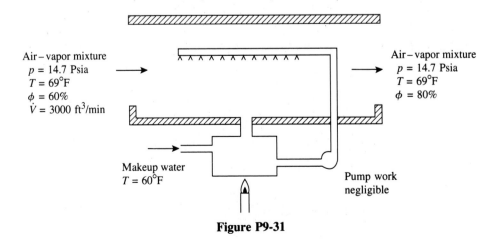

Figure P9-31

9-32. Air at $-5°C$ dry-bulb temperature and 90% relative humidity is brought to $22°C$ and 35% relative humidity by being preheated to $2°C$ and then humidified by being washed with spray water held at a temperature corresponding to the dew point for the final condition. See Fig. P9-32. Reheating is used to give the final desired temperature and relative humidity. Taking data from a psychrometric chart, determine the heat and moisture added per kilogram of dry air. The barometric pressure is 101 kPa. Makeup water enters the water heater at $4°C$.

9-33. Air at 101 kPa, $5°C$, and 20% relative humidity is delivered to a heating chamber at a rate of 0.8 m^3/s. In the heating chamber the air is also sprayed with water at $25°C$. The air discharges from the chamber at $25°C$ and 50% relative humidity. See Fig. P9-33. Calculate the amount of heat that must be supplied and the mass of water spray needed by (a) using data taken from a psychrometric chart, and (b) calculating without the use of any data from a psychrometric chart.

9-34. An insulated humidification device takes in atmospheric air at $95°F$ with a specific humidity of

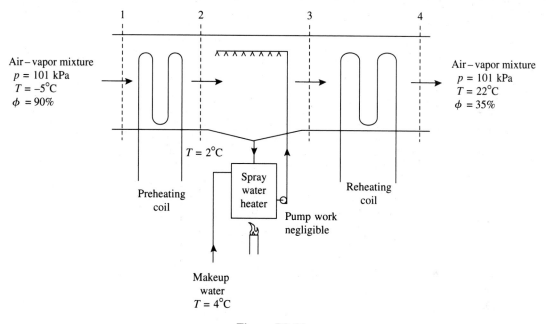

Figure P9-32

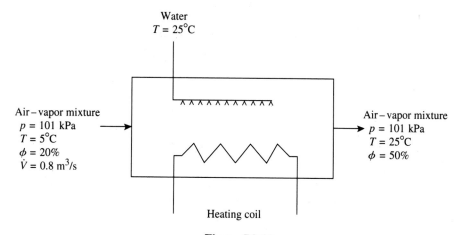

Figure P9-33

0.0141 lbm water vapor/lbm dry air. Saturated water vapor at 230°F and a flow rate of 1.40 lbm/min is sprayed into the air, and the air comes out from the device at 95.7°F with a specific humidity of 0.0176 lbm water vapor/lbm dry air. The mass-flow rate of dry air passing through the device in steady flow is 400 lbm/min. The atmosphere is at 14.7 psia and 95°F. Determine the irreversibility rate for the process, in Btu/min.

9-35. Atmospheric air at 36°C and 10% relative humidity enters an evaporative cooler. What is the minimum temperature that could be achieved by the process? If the final specific humidity is 0.009 kilograms of water vapor per kilogram of dry air, how much water is added in the cooler, and what is the final temperature?

9-36. A heating-and-humidifying system consists of a heating section and an evaporative cooler. Atmospheric air at 10°C and 70% relative humidity enters the heating section at a volume-flow rate of 70 m³/min, and it leaves the evaporative cooler at 20°C and 60% relative humidity. See Fig. P9-36. Determine (a) the temperature and relative humidity of the air when it leaves the heating section, (b) the rate of heat transfer in the heating section, and (c) the rate of water added to the air in the evaporative cooler.

9-37. In a heating and ventilating system, fresh air is taken in at 40°F, 60% relative humidity. See Fig. P9-37. The air is split into two streams. One stream is heated and humidified to saturation, while the other stream is simply heated. The two streams are then mixed to obtain the desired final state of 70°F and 65% relative humidity. Taking data from a psychrometric chart, determine (a) the

temperature to which the first stream is heated, if the second stream is heated to 85°F, and (b) the fractions into which the total stream is split. The barometric pressure is 14.7 psia.

9-38. An air–water-vapor mixture at 20°C, 100 kPa, 40% relative humidity is compressed to 60°C, 400 kPa, and is then cooled at constant pressure. At what temperature will water begin to condense?

9-39. An air washer comprises a chamber into which a water spray is introduced and from which liquid water is removed at the bottom. See Fig. P9-39. As air–water vapor passes through the chamber, dust particles and odors are thus removed. Five thousand ft³/min of an air–vapor mixture at 65°F dry-bulb and 70% relative humidity enters the air washer, and leaves as a saturated mixture at 60°F. Water is supplied to the chamber at 70°F with a flow rate of 30 lbm/min. Assume that the process occurs adiabatically at a constant total pressure of 14.7 psia. Calculate the temperature at which water leaves the chamber.

9-40. Atmospheric air at 100°F and 40% relative humidity is cooled to 80°F by spraying water at 60°F into it. Mixture pressure remains constant at 14.7 psia. Assuming that the mixing occurs in an insulated pipe, calculate the mass of water added per pound of dry air.

9-41. A 30-m³ insulated vessel contains 1 m³ liquid water and a mixture of air and saturated water vapor. See Fig. P9-41. The initial state of the contents inside the vessel is at 35°C and 0.1 MPa. The vessel is connected to a boiling-water nuclear reactor. During a short period of emergency blowdown of the reactor, 55 kg of water vapor and liquid mixture enters the vessel at an average state of 0.7

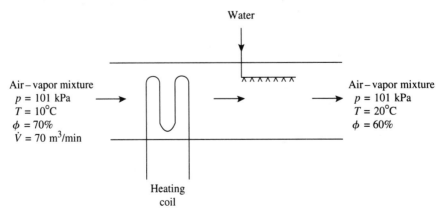

Water

Air–vapor mixture
$p = 101$ kPa
$T = 10°C$
$\phi = 70\%$
$\dot{V} = 70$ m³/min

Heating coil

Air–vapor mixture
$p = 101$ kPa
$T = 20°C$
$\phi = 60\%$

Figure P9-36

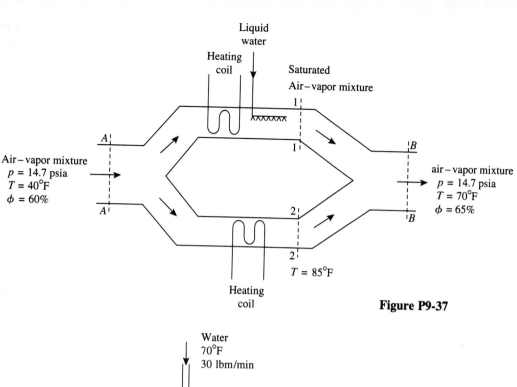

Figure P9-37

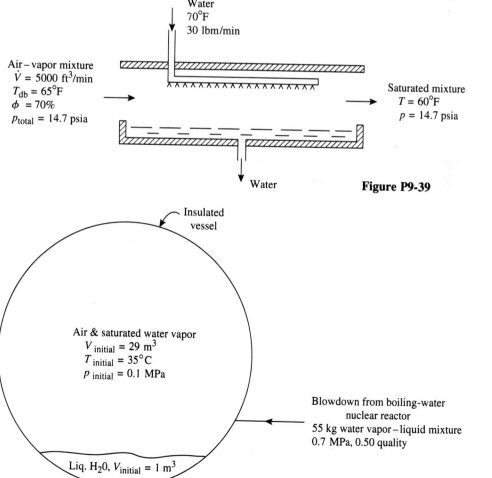

Figure P9-39

Insulated
vessel

Air & saturated water vapor
$V_{initial}$ = 29 m³
$T_{initial}$ = 35°C
$p_{initial}$ = 0.1 MPa

Blowdown from boiling-water
nuclear reactor
55 kg water vapor–liquid mixture
0.7 MPa, 0.50 quality

Liq. H₂0, $V_{initial}$ = 1 m³

Figure P9-41

MPa and 0.50 quality. Determine the temperature and pressure in the vessel at the end of this period.

9-42. Air at 75°F dry-bulb temperature, 70% relative humidity, with a volume-flow rate of 4000 ft³/min is cooled and dehumidified to 68°F dry-bulb, 60% relative humidity, by cooling and dehumidifying a portion of the total air flow to 45°F dry-bulb and then reheating this portion of air to a proper temperature before mixing with the remainder of the air to obtain the final desired condition. See Fig. P9-42. Taking data from a psychrometric chart, determine (a) the fraction of the entire flow that passes through the cooler, (b) the Btu/min of heat rejected in the cooling process, and (c) the Btu/min of heat added in the reheating process. The barometric pressure is 14.7 psia.

9-43. Atmospheric air at T_{1db}, ϕ_1 relative humidity, is to be cooled and dehumidified to T_{3db}, ϕ_3 relative humidity, by cooling and reheating. For (i) $T_{1db} = 75°F$, $\phi_1 = 70\%$, $T_{3db} = 68°F$, and $\phi_3 = 60\%$,

and (ii) $T_{1db} = 80°F$, $\phi_1 = 60\%$, $T_{3db} = 75°F$, and $\phi_3 = 40\%$, determine (a) the temperature to which the air must be cooled before reheating, (b) the heat removed in the cooling process, Btu/lbm dry air, and (c) the heat required in the reheating process, Btu/lbm dry air. $P_{atm} = 14.7$ psia.

9-44. Atmospheric air at 101 kPa, 35°C, and 90% relative humidity is to be cooled and dehumidified to 20°C and 30% relative humidity by cooling and reheating. The volume-flow rate of the air-vapor mixture at the outlet is 0.01 m³/s. See Fig. P9-44. Find (a) the temperature to which the air must be cooled before reheating, (b) the time rate of heat rejection in the cooling process, in kW, (c) the time rate of heat addition in the reheating process, in kW, and (d) the mass of water condensed, in kg/s. Use a psychrometric chart to obtain the required property data.

9-45. Rework Prob. 9-44 without the use of data from any psychrometric chart.

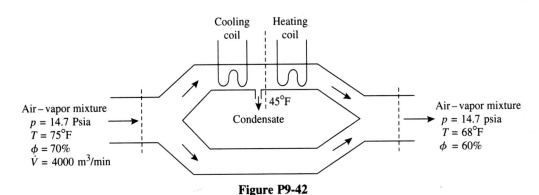

Cooling coil   Heating coil

45°F

Condensate

Air–vapor mixture
$p = 14.7$ Psia
$T = 75°F$
$\phi = 70\%$
$\dot{V} = 4000$ m³/min

Air–vapor mixture
$p = 14.7$ psia
$T = 68°F$
$\phi = 60\%$

Figure P9-42

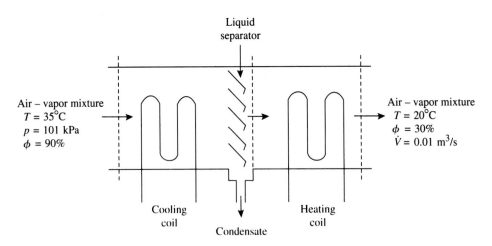

Liquid separator

Air – vapor mixture
$T = 35°C$
$p = 101$ kPa
$\phi = 90\%$

Air – vapor mixture
$T = 20°C$
$\phi = 30\%$
$\dot{V} = 0.01$ m³/s

Cooling coil

Condensate

Heating coil

Figure P9-44

Ideal Gas and Gas–Vapor Mixtures   Chap. 9

9-46. Air at 75°F dry-bulb temperature and 70% relative humidity with a flow rate of 4000 ft³/min is to be cooled and dehumidified to 68°F dry-bulb temperature and 60% relative humidity by cooling and reheating. Find (a) the temperature to which the air must be cooled before reheating, (b) the heat removed, in Btu/min and in tons of refrigeration required, and (c) the heat required in reheating, in Btu/min. Barometric pressure is 15 psia.

9-47. Two streams of atmospheric air at 0.1 MPa are mixed in an adiabatic, steady-flow process. Stream 1 has a dry-bulb temperature of 29°C and a wet-bulb temperature of 21°C. Stream 2 has a dry-bulb temperature of 14°C and a relative humidity of 80%. The volume-flow rates of stream 2 to stream 1 is 2.4. Find, for the resultant mixture, (a) the dry-bulb temperature, (b) the specific humidity, and (c) the relative humidity. Solve this problem without the use of data from any psychrometric chart.

9-48. A mixture of methane (CH₄), dry air, and water vapor is to be produced that has mass proportions of 1 part methane to 15 parts dry air. See Fig. P9-48. The mixing is accomplished in a steady-flow process with both methane and air–vapor mixture entering at 27°C. The inlet air–vapor mixture has a relative humidity of 70%. The gases are heated during the mixing process to produce an outlet temperature of 93°C. The process occurs at a constant pressure of 101.3 kPa. How much heating must be supplied for an inlet volume flow of methane of 28 m³/min, and what is the relative humidity of the outlet mixture? For dry air, $c_p = 1.004$ kJ/kg·°C; for methane, $c_p = 2.254$ kJ/kg·°C; and for water vapor, $c_p = 1.872$ kJ/kg·°C.

9-49. A heating-and-humidifying system operates at a total pressure of 1 atm with saturated water vapor at 100°C as the moisture source. See Fig. P9-49. Air enters the heating section at 10°C and 70% relative humidity at a rate of 70 m³/min. It leaves the humidifying section at 20°C and 60% relative humidity. Determine (a) the temperature and relative humidity of air as it leaves the heating section, (b) the heat transfer rate in the heating section, and (c) the rate of moisture added in the humidifying section.

9-50. Atmospheric air at 95°F, 14.7 psia, with a specific humidity of 0.0141 enters a steady-flow humidifier at a dry-air mass-flow rate of 400 lbm/min. See Fig. P9-50. It is sprayed with saturated water vapor at 230°F at a rate of 1.40 lbm/min. The resulting air-vapor mixture exits from the humidifier at 95.7°F, 14.7 psia, with a specific humidity of 0.0176. There are 20 Btu/min of heat loss to the atmosphere during the process. The kinetic and potential energy effects can be ignored. Determine the irreversibility rate of the process. The atmosphere is at 95° and 14.7 psia.

9-51. Atmospheric air at 35°C, 100 kPa, and 10% relative humidity enters a steady-flow evaporative cooler. See Fig. P9-51. It is sprayed with liquid water at 20°C. The resulting moist air exits the cooler at 25°C, 100 kPa. The process is adiabatic and kinetic and potential energy effects can be ignored. Determine (a) the rate of water spray, in kg/kg dry air, (b) the relative humidity of the moist air at the exit, and (c) the irreversibility rate, in kJ/kg of dry air, using $T_0 = 20°C$.

9-52. A stream of moist air at 36°C, 100 kPa, and 40% relative humidity, flows at a rate of 2.5 kg/min.

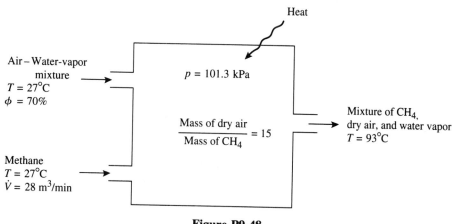

Figure P9-48

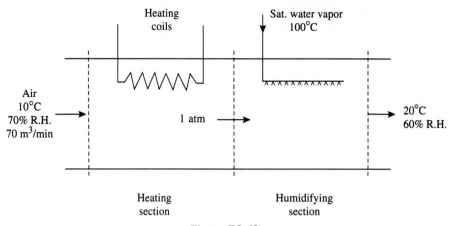

Heating
coils

Sat. water vapor
100°C

Air
10°C
70% R.H.
70 m³/min

1 atm

20°C
60% R.H.

Heating
section

Humidifying
section

Figure P9-49

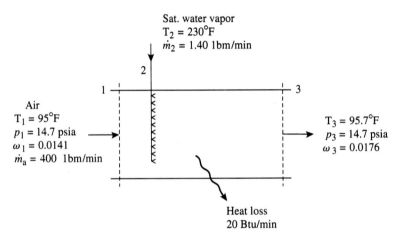

Sat. water vapor
$T_2 = 230°F$
$\dot{m}_2 = 1.40$ 1bm/min

2

1

3

Air
$T_1 = 95°F$
$p_1 = 14.7$ psia
$\omega_1 = 0.0141$
$\dot{m}_a = 400$ 1bm/min

$T_3 = 95.7°F$
$p_3 = 14.7$ psia
$\omega_3 = 0.0176$

Heat loss
20 Btu/min

Figure P9-50

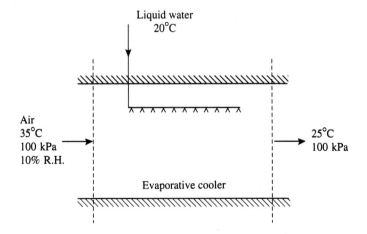

Liquid water
20°C

Air
35°C
100 kPa
10% R.H.

25°C
100 kPa

Evaporative cooler

Figure P9-51

Ideal Gas and Gas–Vapor Mixtures   Chap. 9

Another stream of moist air at 5°C, 100 kPa, and 100% relative humidity, flows at a rate of 5 kg/min. The two streams mix in an adiabatic steady-flow chamber and exit as a single stream at 100 kPa. See Fig. P9-52. Neglect kinetic and potential energy effects. Determine (a) the temperature and relative humidity of the exiting stream, and (b) the irreversibility rate of the mixing process, in kW, using $T_0 = 20°C$.

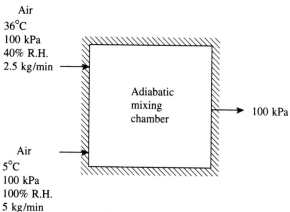

Air
36°C
100 kPa
40% R.H.
2.5 kg/min

Adiabatic mixing chamber

100 kPa

Air
5°C
100 kPa
100% R.H.
5 kg/min

Figure P9-52

9-53. Water at a temperature of 38°C enters a wet-cooling tower at a mass-flow rate of 800 kg/min and leaves at 24°C. See Fig P9-53. Atmospheric air at a dry-bulb temperature of 20°C and a relative humidity of 40% enters the tower at a volume-flow rate of 500 m³/min. This process requires 15 kg/min of 22°C makeup water. Determine (a) the dry-bulb temperature and relative humidity of the air leaving the tower, and (b) the irreversibility of the process. The atmosphere is at 101.3 kPa and 20°C.

9-54. A wet cooling tower (see Fig. 9-15) receives 250,000 ft³/min of atmospheric air at 14.5 psia, 84°F, and 45% relative humidity and discharges the air saturated at 98°F. If the tower receives 3500 gal/min of water at 104°F, what will be the exit temperature of the cooled water?

9-55. Atmospheric air enters a cooling tower at 14.7 psia pressure, 72°F dry-bulb temperature, and 60°F wet-bulb temperature; it leaves the tower at 14.3 psia pressure, 90°F dry-bulb temperature, and 80% relative humidity. See Fig. P9-55. 20,000 lbm/h of water at 105°F enters the tower, and cool water at 65°F leaves the tower. Assume the process to be adiabatic. Calculate the mass-flow rate of dry air, in lbm/min, and the fraction of the incoming water that evaporates.

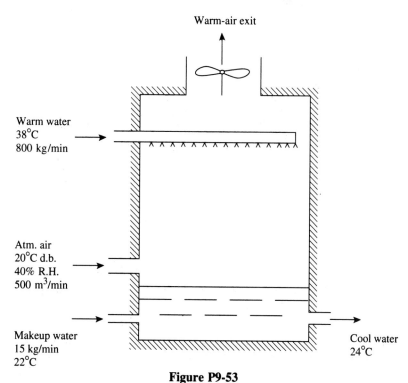

Warm-air exit

Warm water
38°C
800 kg/min

Atm. air
20°C d.b.
40% R.H.
500 m³/min

Makeup water
15 kg/min
22°C

Cool water
24°C

Figure P9-53

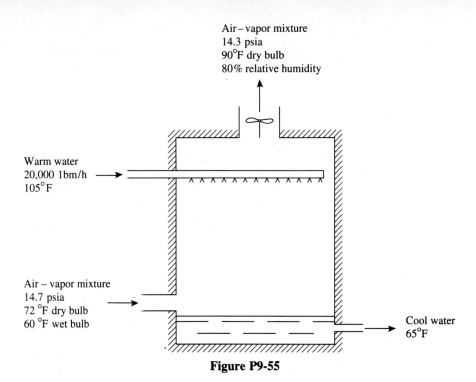

Air – vapor mixture
14.3 psia
90°F dry bulb
80% relative humidity

Warm water
20,000 1bm/h
105°F

Air – vapor mixture
14.7 psia
72 °F dry bulb
60 °F wet bulb

Cool water
65°F

Figure P9-55

9-56. Water at a temperature of 28°C enters a cooling tower at the rate of 36 m³/min. See Fig. P9-56. Part of it evaporates within the tower and the remainder leaves at 20°C. If the air entering the tower has a dry-bulb temperature of 21°C and a wet-bulb temperature of 15°C, and if the air leav- ing is at 28°C and is saturated, how many cubic meters of air enter the tower per minute? What is the rate of water evaporation, in kg/min? Determine also the irreversibility rate for the cooling-tower process, in kJ/min. Assume the cooling-tower process to be adiabatic. The atmosphere is at 21°C and 101.3 kPa.

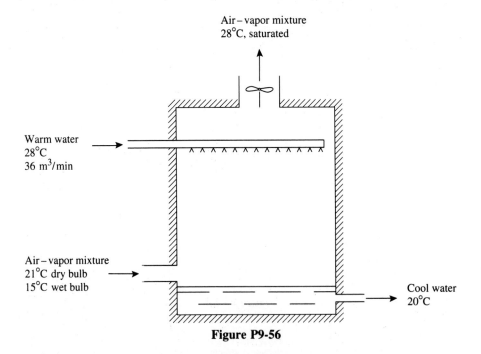

Air – vapor mixture
28°C, saturated

Warm water
28°C
36 m³/min

Air – vapor mixture
21°C dry bulb
15°C wet bulb

Cool water
20°C

Figure P9-56

10

Gas Power Cycles

From an engineering standpoint, one of the most important uses of thermodynamics is the study of cyclic devices for power production. Power cycles can be classified according to whether the working substance changes phase. When the working substance remains in the gaseous phase throughout the cyclic process, it is termed a gas cycle. On the other hand, when the working substance changes alternately between vapor and liquid phases, it is termed a vapor cycle. We study gas cycles in this chapter, leaving the treatment of vapor cycles to the next chapter.

10-1 AIR-STANDARD ANALYSIS

In Chapter 6, we introduced a totally reversible cycle, namely, the Carnot cycle. This cycle is composed of four totally reversible processes: two isothermal processes and two adiabatic processes. Schematic, $p-V$, and $T-S$ diagrams of Carnot cycles are shown in Fig. 6-15. The Carnot cycle is the most efficient cycle working between a heat source at temperature T_H and a sink at temperature T_L, with an efficiency of (see Eq. (6-19))

$$\eta_{\text{th, Carnot}} = 1 - \frac{T_L}{T_H}$$

Reversible isothermal heat transfer, however, is very difficult to achieve in reality, and, as a matter of fact, totally reversible processes can never occur in real life. Thus, to build an engine to approximate the Carnot cycle is not practical. We now study some more practical heat engines, aiming to achieve an efficiency as high as physically possible.

In this chapter, we study actual gas cycles. In actual gas power devices, such as reciprocating internal-combustion engines and gas-turbine power plants, fuel is burned with atmospheric air to form combustion products that then undergo the remaining processes and eventually exhaust to the atmosphere. The processes occurring in such actual devices are not cyclic and naturally complex. In order to examine the influence of major parameters on performance of the device, considerable simplifications must be made in an elementary analysis. A convenient procedure is the use of an *air-standard analysis* in which the following idealizations are made:

1. The working fluid is a constant mass of air in the cyclic operation of the device and air behaves as an ideal gas.
2. The combustion process is replaced by heat addition from a high-temperature reservoir.
3. The exhaust process is replaced by heat rejection to a low-temperature reservoir.

When air is treated as an ideal gas with constant specific heats, such an analysis is called the *cold-air-standard analysis*. In this analysis, the constant specific heats of air are often taken at their room-temperature values. Because specific heats of air are functions of temperature, the great temperature variations in air cycles sometimes make the assumption of constant specific heats questionable. A more realistic analysis is the use of variable specific heats according to the property data given in the Air Table in Appendix 3 (Table A-9).

10-2 OTTO CYCLE

The *Otto Cycle* is the idealized prototype of the spark-ignition reciprocating internal-combustion engine. For each cycle in a four-stroke engine, the piston undergoes four strokes within the cylinder while the crankshaft completes two revolutions, as depicted in the schematic diagram of Fig. 10-1. The sequence of processes of an actual engine is as follows:

Intake stroke: The intake valve is open and the piston moves down from the head-end dead center (HDC) to the crank-end dead center (CDC), taking in a fuel–air mixture.

Compression stroke: The intake valve is closed and the piston moves upward, compressing the fuel-air mixture. As the piston approaches the HDC, the fuel is ignited by the spark, causing a rapid increase in pressure and temperature of the combustion products.

Expansion (or power) stroke: The high-pressure gases expand, performing useful power on the crankshaft.

Exhaust stroke: The exhaust valve is open and the piston moves upward, expelling the combustion gases.

A typical p–v diagram for an actual spark-ignition engine is shown in Fig. 10-2, with the series of events corresponding to that of Fig. 10-1 so indicated. As the intake stroke and the main portion of the exhaust stroke are essentially at atmospheric pressure, their contributions to the generation of power cancel each other. The net work output is the difference between the work done by the gas in the expansion stroke and the work done on the gas in the compression stroke.

In an air-standard cycle, the same mass of air undergoes the internally reversible cyclic processes, with no intake or exhaust taking place. The combustion process is replaced by constant-volume heat addition, the initial portion of the exhaust process is replaced by a constant-volume heat rejection, and the compression and expansion processes are internally reversible and adiabatic, that is, isentropic. This theoretical cycle is shown in the p–v and T–s diagrams in Fig. 10-3.

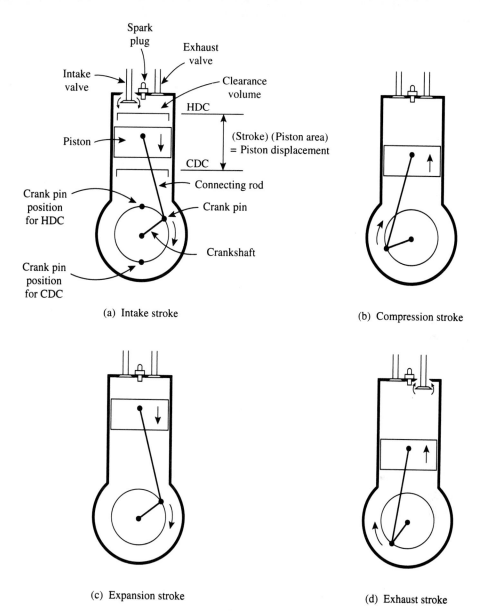

Figure 10-1 Schematic diagram of a four-stroke spark-ignition engine.

Because the same mass of air undergoes the cyclic processes in an air-standard Otto cycle, the air acts as a closed system. For a unit mass of air, the first law for the adiabatic compression and expansion processes gives

$$đq = du + đw = 0 \qquad \text{or} \qquad đw = -du$$

Thus,

$$w_{in} = -w_{12} = \Delta u_{12} = u_2 - u_1$$

$$w_{out} = w_{34} = -\Delta u_{34} = u_3 - u_4$$

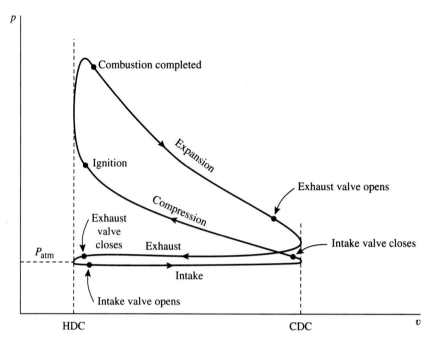

Figure 10-2 Typical p–v diagram for an actual spark-ignition engine.

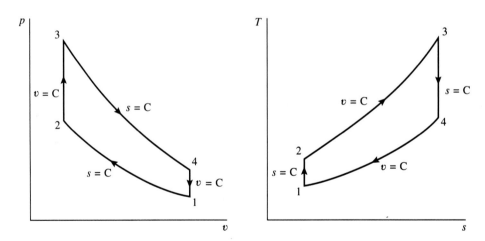

Figure 10-3 p–v and T–s diagrams for an air-standard Otto cycle.

and

$$\text{net } w_{\text{out}} = w_{\text{out}} - w_{\text{in}} = (u_3 - u_4) - (u_2 - u_1)$$

For the constant-volume heat-addition and heat-rejection processes, because đ$w = p \, dv = 0$, we have

$$đq = du$$

Gas Power Cycles   Chap. 10

Thus,

$$q_{in} = q_{23} = \Delta u_{23} = u_3 - u_2$$

$$q_{out} = -q_{41} = -\Delta u_{41} = u_4 - u_1$$

and

$$\text{net } q_{in} = q_{in} - q_{out} = (u_3 - u_2) - (u_4 - u_1)$$

The thermal efficiency is given by

$$\eta_{th} = \frac{\text{net } w_{out}}{q_{in}} = \frac{\text{net } q_{in}}{q_{in}} = \frac{q_{in} - q_{out}}{q_{in}} = 1 - \frac{q_{out}}{q_{in}}$$

or

$$\eta_{th} = 1 - \frac{u_4 - u_1}{u_3 - u_2} \tag{10-1}$$

For the cold-air-standard cycle, the preceding energy quantities become

$$w_{in} = -w_{12} = c_v(T_2 - T_1)$$

$$w_{out} = w_{34} = c_v(T_3 - T_4)$$

$$q_{in} = q_{23} = c_v(T_3 - T_2)$$

$$q_{out} = -q_{41} = c_v(T_4 - T_1)$$

and the thermal efficiency becomes

$$\eta_{th} = 1 - \frac{T_4 - T_1}{T_3 - T_2} \tag{10-2}$$

As

$$\Delta s_{23} = \Delta s_{14} \qquad \text{or} \qquad \int_2^3 \frac{\dbar q}{T} = \int_1^4 \frac{\dbar q}{T}$$

thus,

$$c_v \int_2^3 \frac{dT}{T} = c_v \int_1^4 \frac{dT}{T} \qquad \text{or} \qquad \ln \frac{T_3}{T_2} = \ln \frac{T_4}{T_1}$$

so that

$$\frac{T_3}{T_2} = \frac{T_4}{T_1} \qquad \text{or} \qquad \frac{T_3 - T_2}{T_2} = \frac{T_4 - T_1}{T_1}$$

therefore,

$$\eta_{th} = 1 - \frac{T_1}{T_2}$$

For the reversible-adiabatic process 1–2, we have

$$\frac{T_1}{T_2} = \left(\frac{v_2}{v_1}\right)^{k-1} = \frac{1}{r^{k-1}}$$

Therefore, finally, we have

$$\eta_{th} = 1 - \frac{1}{r^{k-1}} \qquad (10\text{-}3)$$

where r is the *compression ratio* as defined by

$$r = \frac{\text{volume of fluid at CDC}}{\text{volume of fluid at HDC}} = \frac{\text{clearance volume} + \text{piston displacement}}{\text{clearance volume}}$$

(referring to Fig. 10-1a)

or

$$r = v_1/v_2 \qquad \text{(referring to Fig. 10-3)}$$

Equation (10-3) indicates that for the cold-air-standard Otto cycle, the thermal efficiency is a function of the compression ratio and the specific-heat ratio, as depicted in Fig. 10-4. For a given specific-heat ratio, an increase in the compression ratio results in an increase in thermal efficiency. There is, however, an upper limit of the compression ratio as set by the ignition temperature of the fuel. As the compression ratio is increased, the fuel–air mixture during the compression process can reach a temperature that is sufficiently high to ignite itself prematurely, giving rise to engine knock or detonation. This is a phenomenon characterized by the presence of strong pressure waves due to the extremely rapid burning of fuel.

One way to provide a comparative measure of the performance of a reciprocating engine is the concept of the mean effective pressure. *Mean effective pressure* (mep) is defined as the pressure that, if exerted on the piston for one expansion stroke, would produce the same net work output as the cycle, or

$$\text{mep} = \frac{\text{net work output of the cycle}}{\text{piston displacement}} \qquad (10\text{-}4)$$

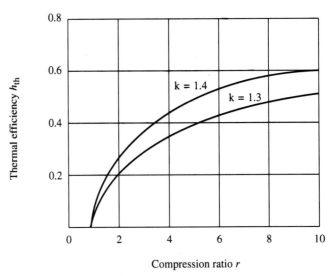

Figure 10-4 Thermal efficiency of a cold-air-standard Otto cycle in relation to the compression ratio and the specific-heat ratio.

The piston displacement is the volume swept by the piston for one stroke. As an illustration for the ideal Otto cycle as depicted in Fig. 10-5, the mep can be expressed as

$$\text{mep} = \frac{\text{area } 1\text{--}2\text{--}3\text{--}4}{v_1 - v_2}$$

For two engines of equal size, the one with the larger value of mep would produce a greater net work output.

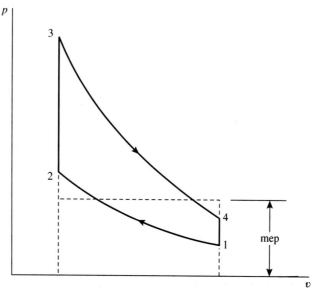

Figure 10-5 Representation of mean effective pressure on a p–v diagram.

Many small spark-ignition engines, such as those used in lawn mowers and motorcycles, operate on a *two-stroke cycle*, in which there is one power stroke per crankshaft revolution. Figure 10-6 shows a schematic diagram of a two-stroke Otto engine. The operation of this engine is as follows. As the piston moves upward, the fuel–air mixture in the cylinder is compressed. Toward the end of the compression stroke when intake port A is uncovered by the piston, fresh fuel–air mixture is admitted to the crankcase. At the end of the compression stroke, the fuel in the cylinder is ignited by the spark, causing a rapid combustion of the fuel and initiating the expansion or power stroke. Toward the end of the power stroke when the exhaust port E is uncovered by the piston, most of the products of combustion rush out of the cylinder by virtue of their high pressure. During the downward motion of the piston, the fuel–air mixture in the crankcase is compressed, and at the end of the power stroke, this mixture is admitted to the cylinder when intake port B is uncovered by the piston. The incoming charge in effect scavenges the exhaust gases from the cylinder.

Although a two-stroke engine has twice as many power strokes as an equivalent four-stroke engine, the power developed by the former is not doubled as compared to the latter. This is mainly because some fresh fuel–air mixture may escape from the cylinder through the exhaust port prior to combustion and some products of combustion may still remain in the cylinder after closing of the exhaust port.

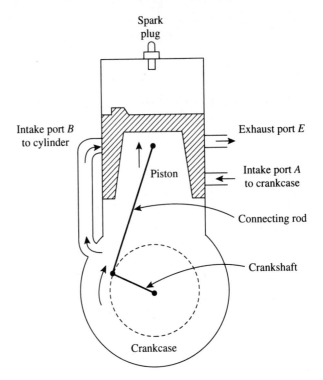

Figure 10-6 Schematic diagram of a two-stroke Otto engine.

Example 10-1

An air-standard four-stroke Otto cycle operates with a compression ratio of 8. At the beginning of the compression process, the air is at 100 kPa and 17°C. The heat added to the air is 1600 kJ/kg during each cycle. Based on the values taken from the air table, determine the pressure and temperature at the end of each process of the cycle, the thermal efficiency, and the mean effective pressure. Determine also the power output at 2200 rpm for an engine displacement of 2000 cm³.

Solution Referring to Fig. 10-3, the given conditions are $p_1 = 100$ kPa, $T_1 = 17°C = 290$ K, $r = v_1/v_2 = 8$, and $q_{in} = 1600$ kJ/kg. From the air table (Table A-9M), at $T_1 = 290$ K, we have $u_1 = 207.2$ kJ/kg, $p_{r1} = 1.226$, and $v_{r1} = 67.91$. By the ideal gas equation,

$$v_1 = \frac{RT_1}{p_1} = \frac{8.314 \text{ kJ/kgmole·K}}{29 \text{ kg/kgmole}} \frac{290 \text{ K}}{100 \text{ kPa}} = 0.831 \text{ m}^3/\text{kg}$$

Accordingly,

$$v_2 = \frac{v_1}{r} = \frac{0.831}{8} = 0.104 \text{ m}^3/\text{kg}$$

For the reversible adiabatic process 1–2, we have

$$v_{r2} = v_{r1}\left(\frac{v_2}{v_1}\right) = 67.91/8 = 8.489$$

At this v_{r2} value, the air table gives $T_2 = 652$ K, $u_2 = 475.2$ kJ/kg, and $p_{r2} = 22.04$.

Thus,

$$p_2 = p_1\left(\frac{p_{r2}}{p_{r1}}\right) = 100\left(\frac{22.04}{1.226}\right) = 1798 \text{ kPa}$$

Notice that p_2 can also be calculated by the ideal-gas relation:

$$p_2 = p_1\left(\frac{T_2}{T_1}\right)\left(\frac{v_1}{v_2}\right)$$

Because

$$q_{in} = u_3 - u_2$$

we have

$$u_3 = u_2 + q_{in} = 475.2 + 1600 = 2075 \text{ kJ/kg}$$

At this u_3 value, the air table gives $T_3 = 2409$ K, $v_{r3} = 0.1489$, and $p_{r3} = 4643$. For the constant-volume process 2–3, the ideal-gas relation gives

$$p_3 = p_2\left(\frac{T_3}{T_2}\right) = 1798\left(\frac{2409}{652}\right) = 6643 \text{ kPa}$$

For the reversible-adiabatic process 3–4, we have

$$v_{r4} = v_{r3}\left(\frac{v_4}{v_3}\right) = v_{r3}\left(\frac{v_1}{v_2}\right) = 0.1489(8) = 1.191$$

At this v_{r4} value, the air table gives $T_4 = 1280$ K, $u_4 = 1004$ kJ/kg, and $p_{r4} = 308.5$. Thus,

$$p_4 = p_3\left(\frac{p_{r4}}{p_{r3}}\right) = 6643\left(\frac{308.5}{4643}\right) = 441 \text{ kPa}$$

Notice that p_4 can also be calculated by the ideal-gas relation:

$$p_4 = p_3\left(\frac{T_4}{T_3}\right)\left(\frac{v_3}{v_4}\right)$$

For the constant-volume process 4–1,

$$q_{out} = -q_{41} = u_4 - u_1 = 1004 - 207 = 797 \text{ kJ/kg}$$

Therefore,

$$\eta_{th} = 1 - \frac{q_{out}}{q_{in}} = 1 - \frac{797}{1600} = 0.502 = 50.2\%$$

The net work output of the cycle is

$$\text{net } w_{out} = \text{net } q_{in} = q_{in} - q_{out} = 1600 - 797 = 803 \text{ kJ/kg}$$

Therefore,

$$\text{mep} = \frac{\text{net } w_{out}}{v_1 - v_2} = \frac{803 \text{ kJ/kg}}{(0.831 - 0.104) \text{ m}^3/\text{kg}} = 1105 \text{ kPa}$$

The power output of the engine is calculated as

$$\text{Power} = (\text{mep})(\text{piston displacement})\left(\frac{\text{rpm}}{2 \times 60}\right)$$

$$= (1105 \text{ kN/m}^2)(2000 \times 10^{-6} \text{ m}^3)\left(\frac{2200}{2 \times 60} \text{ s}^{-1}\right) = 40.5 \text{ kW}$$

Example 10-2

Rework Example 10-1 using the cold-air standard with $k = 1.4$ and $c_v = 0.718$ kJ/kg·K.

Solution By the ideal-gas equation,

$$v_1 = \frac{RT_1}{p_1} = \left(\frac{8.314}{29}\right)\left(\frac{290}{100}\right) = 0.831 \text{ m}^3/\text{kg}$$

Thus,

$$v_2 = \frac{v_1}{r} = \frac{0.831}{8} = 0.104 \text{ m}^3/\text{kg}$$

For the reversible adiabatic process 1–2, we have

$$T_2 = T_1\left(\frac{v_1}{v_2}\right)^{k-1} = 290(8)^{1.4-1} = 666 \text{ K}$$

$$p_2 = p_1\left(\frac{v_1}{v_2}\right)^{k} = 100(8)^{1.4} = 1838 \text{ kPa}$$

Notice that p_2 can also be calculated by the ideal-gas relation

$$p_2 = p_1\left(\frac{T_2}{T_1}\right)\left(\frac{v_1}{v_2}\right)$$

Because

$$q_{\text{in}} = u_3 - u_2 = c_v(T_3 - T_2)$$

we have

$$T_3 = T_2 + \frac{q_{\text{in}}}{c_v} = 666 + \frac{1600}{0.718} = 2894 \text{ K}$$

For the constant-volume process 2–3, the ideal-gas relation gives

$$p_3 = p_2\left(\frac{T_3}{T_2}\right) = 1838\left(\frac{2894}{666}\right) = 7987 \text{ kPa}$$

For the reversible adiabatic process 3–4, we have

$$T_4 = T_3\left(\frac{v_3}{v_4}\right)^{k-1} = T_3\left(\frac{v_2}{v_1}\right)^{k-1} = 2894\left(\frac{1}{8}\right)^{1.4-1} = 1260 \text{ K}$$

$$p_4 = p_3\left(\frac{v_3}{v_4}\right)^{k} = p_3\left(\frac{v_2}{v_1}\right)^{k} = 7987\left(\frac{1}{8}\right)^{1.4} = 435 \text{ kPa}$$

Notice that p_4 can also be calculated by the relation

$$p_4 = p_3\left(\frac{T_4}{T_3}\right)\left(\frac{v_3}{v_4}\right)$$

For the constant-volume process 4–1,

$$q_{\text{out}} = -q_{41} = u_4 - u_1 = c_v(T_4 - T_1)$$

$$= 0.718(1260 - 290) = 696 \text{ kJ/kg}$$

Therefore,

$$\eta_{th} = 1 - \frac{q_{out}}{q_{in}} = 1 - \frac{696}{1600} = 0.565 = 56.5\%$$

The thermal efficiency can also be calculated by the equation

$$\eta_{th} = 1 - \frac{1}{r^{k-1}} = 1 - \frac{1}{8^{1.4-1}} = 0.565 = 56.5\%$$

The net work output of the cycle is

$$\text{net } w_{out} = \text{net } q_{in} = q_{in} - q_{out} = 1600 - 696 = 904 \text{ kJ/kg}$$

Therefore,

$$\text{mep} = \frac{\text{net } w_{out}}{v_1 - v_2} = \frac{904}{0.831 - 0.104} = 1243 \text{ kPa}$$

The power output of the engine is calculated as

$$\text{Power} = (\text{mep})(\text{piston displacement})\left(\frac{\text{rpm}}{2 \times 60}\right)$$

$$= (1243 \text{ kPa})(2000 \times 10^{-6} \text{ m}^3)\left(\frac{2200}{2 \times 60} \text{ s}^{-1}\right) = 45.6 \text{ kW}$$

It is worth noting that there are considerable differences in the results obtained in Examples 10-1 and 10-2. The results in Example 10-1 should be more accurate because the variation in specific heats has been taken care of in that solution.

10-3 DIESEL CYCLE

The *Diesel cycle* is the idealized prototype of the compression-ignition reciprocating internal-combustion engine. In the Diesel engine, air alone is admitted into the cylinder in the intake stroke and is then compressed to a high pressure, thereby reaching a high temperature. As air alone is compressed in a Diesel engine, it is operated at a much higher compression ratio than an Otto engine. At the end of the compression process, fuel is sprayed into the hot air and is ignited spontaneously. The rate of burning can be controlled by the rate of fuel injection, with the aim of achieving constant-pressure combustion.

The Diesel engine can operate on either a four-stroke or a two-stroke cycle. In a four-stroke Diesel cycle, the basic components are the same as those of the Otto engine shown in Fig. 10-1, except that the spark plug is replaced by a fuel injector.

In an air-standard Diesel cycle, the idealization is essentially the same as in the air-standard Otto cycle, except that the combustion process is simulated by heat addition at constant pressure instead of at constant volume. The p–v and T–s diagrams for an air-standard Diesel cycle are illustrated in Fig. 10-7.

Because the air acts as a closed system in the air-standard Diesel cycle, the first-law equation for each of the four internally reversible processes of the cycle is

$$đq = du + đw \qquad \text{with} \qquad đw = p \, dv$$

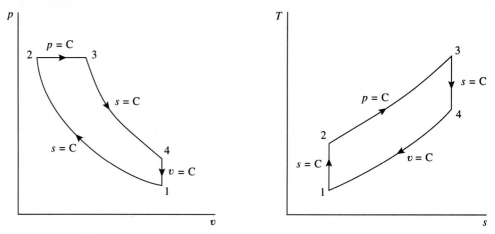

Figure 10-7 p–v and T–s diagrams for an air-standard Diesel cycle.

Thus,

$$w_{in} = -w_{12} = \Delta u_{12} = u_2 - u_1$$

$$w_{out} = w_{23} + w_{34} = \int_2^3 p \, dv - \Delta u_{34} = (p_3v_3 - p_2v_2) + (u_3 - u_4)$$

$$q_{in} = q_{23} = \Delta u_{23} + w_{23} = (u_3 - u_2) + (p_3v_3 - p_2v_2) = h_3 - h_2$$

$$q_{out} = -q_{41} = -\Delta u_{41} = u_4 - u_1$$

The net work output is

$$\text{net } w_{out} = w_{out} - w_{in} = (p_3v_3 - p_2v_2) + (u_3 - u_4) - (u_2 - u_1)$$
$$= (h_3 - h_2) - (u_4 - u_1)$$

which is also equal to

$$\text{net } q_{in} = q_{in} - q_{out}$$

The thermal efficiency is given by

$$\eta_{th} = \frac{\text{net } w_{out}}{q_{in}} = \frac{\text{net } q_{in}}{q_{in}} = \frac{q_{in} - q_{out}}{q_{in}} = 1 - \frac{q_{out}}{q_{in}}$$

or

$$\eta_{th} = 1 - \frac{u_4 - u_1}{h_3 - h_2} \tag{10-5}$$

For the cold-air standard, the preceding energy quantities become

$$w_{in} = -w_{12} = c_v(T_2 - T_1)$$

$$w_{out} = w_{23} + w_{34} = (p_3v_3 - p_2v_2) + c_v(T_3 - T_4)$$

$$q_{in} = q_{23} = c_p(T_3 - T_2)$$

$$q_{out} = -q_{41} = c_v(T_4 - T_1)$$

and the thermal efficiency becomes

$$\eta_{th} = 1 - \frac{c_v(T_4 - T_1)}{c_p(T_3 - T_2)} = 1 - \frac{T_4 - T_1}{k(T_3 - T_2)} \tag{10-6}$$

The thermal efficiency of a cold-air-standard Diesel cycle can be expressed in terms of the following two factors:

$$r = \text{compression ratio} = \frac{v_1}{v_2}$$

$$r_c = \text{cutoff ratio} = \frac{v_3}{v_2}$$

which is the ratio of the cylinder volumes after and before the heat-addition (combustion) process.

Thus,

$$T_1 = T_2\left(\frac{v_2}{v_1}\right)^{k-1} = \frac{T_2}{r^{k-1}}$$

$$T_3 = T_2\left(\frac{v_3}{v_2}\right) = T_2 r_c$$

$$T_4 = T_3\left(\frac{v_3}{v_4}\right)^{k-1} = T_3\left(\frac{v_3/v_2}{v_1/v_2}\right)^{k-1} = T_3\left(\frac{r_c}{r}\right)^{k-1}$$

or

$$T_4 = T_2\left(\frac{r_c^k}{r^{k-1}}\right)$$

Therefore,

$$\eta_{th} = 1 - \frac{1}{r^{k-1}}\left[\frac{r_c^k - 1}{k(r_c - 1)}\right] \tag{10-7}$$

Equation (10-7) indicates that for the cold-air-standard Diesel cycle, with a given specific-heat ratio, the thermal efficiency increases as the compression ratio increases and the cutoff ratio decreases, as depicted in Fig. 10-8. Equation (10-7) also indicates that, for a given compression ratio, the efficiency of a Diesel cycle is lower than that of an Otto cycle, because the term in square brackets is always greater than unity. The curve with $r_c = 1$ in Fig. 10-8 corresponds to the Otto cycle. Because a Diesel engine can operate at a much higher compression ratio than an Otto engine, the former can have a better overall fuel economy than the latter.

Example 10-3

An air-standard four-stroke Diesel cycle operates with a compression ratio of 18. At the beginning of the compression process, the air is at 100 kPa and 20°C. The maximum temperature of the cycle is 1900 K. Determine the pressure, temperature, and specific volume at the end of each process of the cycle, the cutoff ratio, the thermal efficiency, and the mean effective pressure. Determine also the power output at 2000 rpm for an engine displacement of 5000 cm³. What is the Carnot efficiency for a heat engine with the same maximum and minimum temperatures as the Diesel cycle? Use data from the air table.

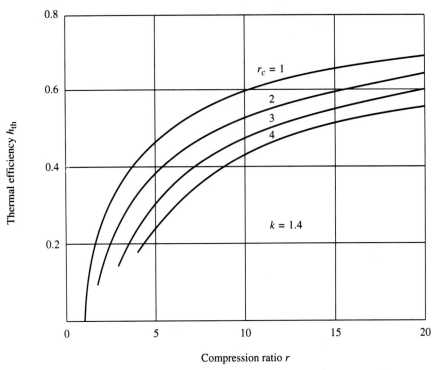

Figure 10-8 Thermal efficiency of cold-air-standard Diesel cycle.

Solution Referring to Fig. 10-7, the given conditions are $r = v_1/v_2 = 18$, $p_1 = 100$ kPa, $T_1 = 20\ °C = 293$ K, and $T_3 = 1900$ K. From the air table (Table A-9M), at $T_1 = 293$ K, we have $u_1 = 209.3$ kJ/kg, $p_{r1} = 1.271$, $v_{r1} = 66.19$; and at $T_3 = 1900$ K, we have $h_3 = 2126.7$ kJ/kg, $p_{r3} = 1643.0$, $v_{r3} = 0.3319$. By the ideal-gas equation,

$$v_1 = \frac{RT_1}{p_1} = \frac{8.314\ \text{kJ/kgmole·K}}{29\ \text{kg/kgmole}} \frac{293\ \text{K}}{100\ \text{kPa}} = 0.8400\ \text{m}^3/\text{kg}$$

which is also the value of v_4. Accordingly,

$$v_2 = \frac{v_1}{r} = \frac{0.840}{18} = 0.0467\ \text{m}^3/\text{kg}$$

For the reversible adiabatic process 1–2, we have

$$v_{r2} = v_{r1}\frac{v_2}{v_1} = \frac{66.19}{18} = 3.677$$

At this v_{r2} value, the air table gives $T_2 = 880$ K, $h_2 = 910.9$ kJ/kg, and $p_{r2} = 68.71$. Thus,

$$p_2 = p_1\left(\frac{p_{r2}}{p_{r1}}\right) = 100\left(\frac{68.71}{1.271}\right) = 5406\ \text{kPa}$$

or

$$p_2 = p_1\left(\frac{T_2}{T_1}\right)\left(\frac{v_1}{v_2}\right) = 100\left(\frac{880}{293}\right)(18) = 5406\ \text{kPa}$$

which is also the value of p_3. For the constant-pressure process 2–3, the ideal-gas relation gives

$$v_3 = v_2\left(\frac{T_3}{T_2}\right) = 0.0467\left(\frac{1900}{880}\right) = 0.1008 \text{ m}^3/\text{kg}$$

For the reversible adiabatic process 3–4, we have

$$v_{r4} = v_{r3}\left(\frac{v_4}{v_3}\right) = 0.3319\left(\frac{0.8400}{0.1008}\right) = 2.766$$

At this v_{r4} value, the air table gives $T_4 = 970$ K, $u_4 = 733.8$ kJ/kg, and $p_{r4} = 100.7$. Thus,

$$p_4 = p_3\left(\frac{p_{r4}}{p_{r3}}\right) = 5406\left(\frac{100.7}{1643.0}\right) = 331 \text{ kPa}$$

or

$$p_4 = p_3\left(\frac{T_4}{T_3}\right)\left(\frac{v_3}{v_4}\right) = 5406\left(\frac{970}{1900}\right)\left(\frac{0.1008}{0.8400}\right) = 331 \text{ kPa}$$

The cutoff ratio is calculated as

$$r_c = \frac{v_3}{v_2} = \frac{0.1008}{0.0467} = 2.16$$

or

$$r_c = \frac{T_3}{T_2} = \frac{1900}{880} = 2.16$$

For the constant-pressure process 2–3,

$$q_{in} = q_{23} = h_3 - h_2 = 2126.7 - 910.9 = 1215.8 \text{ kJ/kg}$$

For the constant-volume process 4–1,

$$q_{out} = -q_{41} = u_4 - u_1 = 733.8 - 209.3 = 524.5 \text{ kJ/kg}$$

Therefore,

$$\eta_{th} = 1 - \frac{q_{out}}{q_{in}} = 1 - \frac{524.5}{1215.8} = 0.569 = 56.9\%$$

The net work output of the cycle is

$$\text{net } w_{out} = \text{net } q_{in} = q_{in} - q_{out} = 1215.8 - 524.5 = 691.3 \text{ kJ/kg}$$

Therefore,

$$\text{mep} = \frac{\text{net } w_{out}}{v_1 - v_2} = \frac{691.3}{0.8400 - 0.0467} = 871 \text{ kPa}$$

The power output of the engine is calculated as

$$\text{Power} = (\text{mep})(\text{piston displacement})\left(\frac{\text{rpm}}{2 \times 60}\right)$$

$$= (871 \text{ kPa})(5000 \times 10^{-6} \text{ m}^3)\left(\frac{2000}{2 \times 60} \text{ s}^{-1}\right) = 72.6 \text{ kW}$$

A Carnot cycle operating between the given temperature limits would have an efficiency of

$$\eta_{\text{Carnot}} = 1 - \frac{T_L}{T_H} = 1 - \frac{T_1}{T_3} = 1 - \frac{293}{1900} = 0.846 = 84.6\%$$

Notice that this solution uses the air tables to account for the variation of the specific heats with temperature. Equation (10-7) is based on the assumption of constant specific heats and not valid for a variable specific-heats solution. This is why Eq. (10-7) has not been used in the present solution to determine the thermal efficiency of the Diesel cycle.

10-4 DUAL CYCLE

In modern compression-ignition engines, injection of fuel starts before the end of the compression stroke and continues during the early part of the return stroke, thus forming a rounded top in the actual p–v plot. These operating conditions are closely approximated by the air-standard *dual cycle* shown in Fig. 10-9 on the p–v and T–s planes.

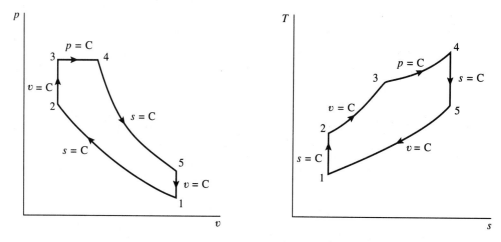

Figure 10-9 p–v and T–s diagrams for an air-standard dual cycle.

For a unit mass of air in the air-standard dual cycle, the total heat addition during processes 2–3 and 3–4 is given by

$$q_{\text{in}} = q_{23} + q_{34} = \Delta u_{23} + \Delta h_{34} = (u_3 - u_2) + (h_4 - h_3)$$

and the heat rejection during process 5–1 is given by

$$q_{\text{out}} = -q_{51} = -\Delta u_{51} = u_5 - u_1$$

The cycle thermal efficiency is then

$$\eta_{\text{th}} = 1 - \frac{q_{\text{out}}}{q_{\text{in}}} = 1 - \frac{u_5 - u_1}{(u_3 - u_2) + (h_4 - h_3)} \tag{10-8}$$

For the cold-air standard, the heat-addition and heat-rejection terms become

$$q_{\text{in}} = c_v(T_3 - T_2) + c_p(T_4 - T_3)$$

$$q_{\text{out}} = c_v(T_5 - T_1)$$

Gas Power Cycles   Chap. 10

and the thermal efficiency becomes

$$\eta_{th} = 1 - \frac{c_v(T_5 - T_1)}{c_v(T_3 - T_2) + c_p(T_4 - T_3)}$$

$$= 1 - \frac{T_5 - T_1}{(T_3 - T_2) + k(T_4 - T_3)} \tag{10-9}$$

The thermal efficiency of a cold-air-standard dual cycle can be expressed in terms of the following three factors:

$$r = \text{compression ratio} = \frac{v_1}{v_2}$$

$$r_c = \text{cutoff ratio} = \frac{v_4}{v_3}$$

and

$$r_p = \text{constant-volume pressure ratio} = \frac{p_3}{p_2}$$

Thus,

$$T_1 = T_2\left(\frac{v_2}{v_1}\right)^{k-1} = \frac{T_2}{r^{k-1}}$$

$$T_3 = T_2\left(\frac{p_3}{p_2}\right) = T_2 r_p$$

$$T_4 = T_3\left(\frac{v_4}{v_3}\right) = T_3 r_c = T_2 r_p r_c$$

$$T_5 = T_4\left(\frac{v_4}{v_5}\right)^{k-1} = T_4\left(\frac{v_4/v_3}{v_1/v_2}\right)^{k-1} = T_4\left(\frac{r_c}{r}\right)^{k-1}$$

$$= T_2 r_p\left(\frac{r_c^k}{r^{k-1}}\right)$$

Therefore,

$$\eta_{th} = 1 - \frac{1}{r^{k-1}}\left[\frac{r_p r_c^k - 1}{(r_p - 1) + kr_p(r_c - 1)}\right] \tag{10-10}$$

It can be stated that with the same compression ratio and the same heat input, the thermal efficiency of the Otto cycle is the highest and the Diesel cycle the lowest, with the dual cycle falling in between.

Example 10-4

An air-standard dual cycle operates with a compression ratio of 15. At the beginning of the compression process, the air is at 14.6 psia, 70°F, and has a total volume of 240 in³. The total heat addition per cycle is 7.0 Btu, of which 30% is added at constant volume. Determine the pressure, temperature, and specific volume at the end of each process of the cycle, the thermal efficiency, and the mean effective pressure. Use (a) air-table data and (b) constant specific heats, $c_p = 0.240$ Btu/lbm·°F and $c_v = 0.171$ Btu/lbm·°F.

Sec. 10-4   Dual Cycle

Solution Referring to Fig. 10-9, the given conditions are $r = v_1/v_2 = 15$, $p_1 = 14.6$ psia, $T_1 = 70°F = 530°R$, $V_1 = 240$ in^3. $= 0.139$ ft^3, $Q_{23} = 0.30 \times 7.0 = 2.1$ Btu, and $Q_{34} = 0.70 \times 7.0 = 4.9$ Btu. Thus,

$$v_1 = v_5 = \frac{RT_1}{p_1}$$

$$= \frac{1545 \text{ ft lbf/lbmole·°R}}{29 \text{ lbm/lbmole}} \frac{530°R}{14.6 \times 144 \text{ psfa}} = 13.43 \text{ ft}^3/\text{lbm}$$

$$m = \frac{V_1}{v_1} = \frac{0.139}{13.43} = 0.0104 \text{ lbm}$$

$$v_2 = v_3 = \frac{v_1}{r} = \frac{13.43}{15} = 0.895 \text{ ft}^3/\text{lbm}$$

(a) *Air-table solution*. At $T_1 = 530°R$, we get $u_1 = 90.4$ Btu/lbm, $p_{r1} = 1.293$, and $v_{r1} = 151.9$. For the internally reversible adiabatic process 1–2,

$$v_{r2} = v_{r1}\left(\frac{v_2}{v_1}\right) = 151.9\left(\frac{1}{15}\right) = 10.13$$

At this v_{r2} value, the air table gives $T_2 = 1493°R$, $u_2 = 265.0$ Btu/lbm, and $p_{r2} = 54.60$. Accordingly,

$$p_2 = p_1\left(\frac{p_{r2}}{p_{r1}}\right) = 14.6\left(\frac{54.60}{1.293}\right) = 617 \text{ psia}$$

or

$$p_2 = p_1\left(\frac{T_2}{T_1}\right)\left(\frac{v_1}{v_2}\right) = 14.6\left(\frac{1493}{530}\right)15 = 617 \text{ psia}$$

Because

$$q_{23} = \frac{Q_{23}}{m} = \frac{2.1}{0.0104} = 201.9 \text{ Btu/lbm} = u_3 - u_2$$

Therefore,

$$u_3 = u_2 + 201.9 = 265.0 + 201.9 = 466.9 \text{ Btu/lbm}$$

At this u_3 value, the air table gives $T_3 = 2466°R$, $h_3 = 636.0$ Btu/lbm, $p_{r3} = 409.0$, and $v_{r3} = 2.234$. Accordingly,

$$p_3 = p_4 = p_2\left(\frac{T_3}{T_2}\right) = 617\left(\frac{2466}{1493}\right) = 1019 \text{ psia}$$

The constant-volume pressure ratio is then

$$r_p = \frac{p_3}{p_2} = \frac{1019}{617} = 1.65$$

Because

$$q_{34} = \frac{Q_{34}}{m} = \frac{4.9}{0.0104} = 471.2 \text{ Btu/lbm} = h_4 - h_3$$

Therefore,

$$h_4 = h_3 + 471.2 = 636.0 + 471.2 = 1107 \text{ Btu/lbm}$$

At this h_4 value, the air table gives $T_4 = 4064°R$, $v_{r4} = 0.4317$, and $p_{r4} = 3488$. For the constant-pressure process 3–4, we have

$$v_4 = v_3\left(\frac{T_4}{T_3}\right) = 0.895\left(\frac{4064}{2466}\right) = 1.47 \text{ ft}^3/\text{lbm}$$

The cutoff ratio is then

$$r_c = \frac{v_4}{v_3} = \frac{1.47}{0.895} = 1.64$$

For the internally reversible adiabatic process 4–5,

$$v_{r5} = v_{r4}\left(\frac{v_5}{v_4}\right) = 0.4317\left(\frac{13.43}{1.47}\right) = 3.944$$

At this v_{r5} value, the air table gives $T_5 = 2055°R$, $u_5 = 379.0$ Btu/lbm, and $p_{r5} = 193.0$. Accordingly,

$$p_5 = p_4\left(\frac{p_{r5}}{p_{r4}}\right) = 1019\left(\frac{193.0}{3488}\right) = 56.4 \text{ psia}$$

or

$$p_5 = p_4\left(\frac{T_5}{T_4}\right)\left(\frac{v_4}{v_5}\right) = 1019\left(\frac{2055}{4064}\right)\left(\frac{1.47}{13.43}\right) = 56.4 \text{ psia}$$

For the constant-volume process 5–1,

$$Q_{\text{out}} = -Q_{51} = m(u_5 - u_1) = 0.0104(379.0 - 90.4) = 3.001 \text{ Btu}$$

Therefore,

$$\eta_{\text{th}} = 1 - \frac{Q_{\text{out}}}{Q_{\text{in}}} = 1 - \frac{3.001}{7.0} = 0.571 = 57.1\%$$

The net work output of the cycle is

$$\text{net } W_{\text{out}} = \text{net } Q_{\text{in}} = Q_{\text{in}} - Q_{\text{out}} = 7.0 - 3.001 = 3.999 \text{ Btu}$$

Therefore,

$$\text{mep} = \frac{\text{net } W_{\text{out}}}{m(v_1 - v_2)} = \frac{3.999 \times 778}{0.0104(13.43 - 0.895)} = 23,900 \text{ psfa}$$

$$= 166 \text{ psia}$$

(b) *Constant-specific-heats solution.*

$$k = \frac{c_p}{c_v} = \frac{0.240}{0.171} = 1.4$$

$$T_2 = T_1\left(\frac{v_1}{v_2}\right)^{k-1} = 530(15)^{1.4-1} = 1566°R$$

$$p_2 = p_1\left(\frac{v_1}{v_2}\right)^{k} = 14.6(15)^{1.4} = 647 \text{ psia}$$

or

$$p_2 = p_1\left(\frac{T_2}{T_1}\right)\left(\frac{v_1}{v_2}\right) = 14.6\left(\frac{1566}{530}\right)(15) = 647 \text{ psia}$$

Because

$$q_{23} = \frac{2.1}{0.0104} = 201.9 \text{ Btu/lbm} = c_v(T_3 - T_2) = 0.171(T_3 - 1566)$$

So that

$$T_3 = 2747°\text{R}$$

$$p_3 = p_4 = p_2\left(\frac{T_3}{T_2}\right) = 647\left(\frac{2747}{1566}\right) = 1135 \text{ psia}$$

$$r_p = \frac{p_3}{p_2} = \frac{1135}{647} = 1.75$$

Because

$$q_{34} = \frac{4.9}{0.0104} = 471.2 \text{ Btu/lbm} = c_p(T_4 - T_3) = 0.240(T_4 - 2747)$$

So that

$$T_4 = 4710°\text{R}$$

$$v_4 = v_3\left(\frac{T_4}{T_3}\right) = 0.895\left(\frac{4710}{2747}\right) = 1.53 \text{ ft}^3/\text{lbm}$$

$$r_c = \frac{v_4}{v_3} = \frac{1.53}{0.895} = 1.71$$

or

$$r_c = \frac{T_4}{T_3} = \frac{4710}{2747} = 1.71$$

$$T_5 = T_4\left(\frac{v_4}{v_5}\right)^{k-1} = 4710\left(\frac{1.53}{13.43}\right)^{1.4-1} = 1975°\text{R}$$

$$p_5 = p_4\left(\frac{v_4}{v_5}\right)^{k} = 1135\left(\frac{1.53}{13.43}\right)^{1.4} = 54.2 \text{ psia}$$

or

$$p_5 = p_4\left(\frac{T_5}{T_4}\right)\left(\frac{v_4}{v_5}\right) = 1135\left(\frac{1975}{4710}\right)\left(\frac{1.53}{13.43}\right) = 54.2 \text{ psia}$$

$$Q_{\text{out}} = mc_v(T_5 - T_1)$$

$$= 0.0104 \times 0.171(1975 - 530) = 2.570 \text{ Btu}$$

$$\eta_{\text{th}} = 1 - \frac{Q_{\text{out}}}{Q_{\text{in}}} = 1 - \frac{2.570}{7.0} = 0.633 = 63.3\%$$

or

$$\eta_{\text{th}} = 1 - \frac{1}{r^{k-1}}\left[\frac{r_p r_c^k - 1}{(r_p - 1) + kr_p(r_c - 1)}\right]$$

$$= 1 - \frac{1}{(15)^{1.4-1}}\left[\frac{1.75(1.71)^{1.4} - 1}{(1.75 - 1) + 1.4(1.75)(1.71 - 1)}\right]$$

$$= 0.632 = 63.2\%$$

$$\text{net } W_\text{out} = \text{net } Q_\text{in} = Q_\text{in} - Q_\text{out} = 7.0 - 2.570 = 4.43 \text{ Btu}$$

$$\text{mep} = \frac{\text{net } W_\text{out}}{m(v_1 - v_2)} = \frac{4.43 \times 778}{0.0104(13.43 - 0.895)}$$

$$= 26{,}440 \text{ psfa} = 184 \text{ psia}$$

Of the two solutions presented for this example, the one using variable specific heats gives more reliable results. In that solution, the mean effective pressure has a value of 166 psia. That is, a constant pressure of 166 psia during the expansion stroke would produce the same net work output as the entire dual cycle.

10-5 GAS-TURBINE CYCLE

A gas-turbine engine is a rotary type of internal-combustion engine consisting of a compressor, a combustion chamber, and a turbine, as shown schematically in Fig. 10-10(a). The engine components are assumed to operate under steady-state, steady-flow conditions. Atmospheric air is drawn into the adiabatic compressor to be raised to a condition of higher pressure and temperature. The air then enters the combustion chamber where fuel is injected and burned at essentially constant pressure. The resulting high-pressure, high-temperature products of combustion are finally allowed to expand through the turbine and discharge to the atmosphere. Some of the work produced by the turbine is used to drive the compressor, with the balance as the net work output. As fresh air is continually drawn into the engine and hot products of combustion are continually discharged from the engine, the actual gas-turbine engine is operated on an open cycle.

An air-standard gas-turbine engine is analyzed on a closed-cycle basis, shown schematically in Fig. 10-10(b), where a heat exchanger is used to bring the condition of the turbine exhaust to that of the compressor inlet through a constant-pressure process. Furthermore, the combustion chamber is replaced by a heat exchanger to which heat is added under constant pressure from an external source. In the ideal case, the compression and expansion processes are assumed to be internally reversible and adiabatic. Such a simple air-standard gas-turbine cycle is commonly called the *Brayton cycle,* for

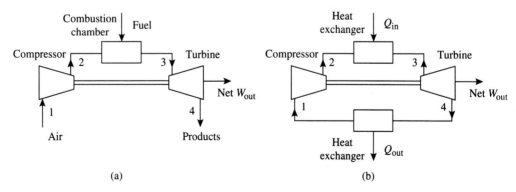

(a)   (b)

Figure 10-10 Schematic diagrams of gas-turbine engines: (a) actual engine (open cycle), and (b) air-standard engine (closed cycle).

which the p–v and T–s diagrams are depicted in Fig. 10-11. By referring to this figure, the energy quantities of each process of the cycle can be obtained through the use of the steady-flow energy equation. Thus, on a unit-mass-flow basis for the adiabatic turbine,

$$w_{out} = w_{turb} = \left(h_3 + \frac{V_3^2}{2}\right) - \left(h_4 + \frac{V_4^2}{2}\right) \tag{10-11}$$

Similarly, for the adiabatic compressor,

$$w_{in} = w_{comp} = \left(h_2 + \frac{V_2^2}{2}\right) - \left(h_1 + \frac{V_1^2}{2}\right) \tag{10-12}$$

For the two heat exchangers, heat addition and rejection are

$$q_{in} = \left(h_3 + \frac{V_3^2}{2}\right) - \left(h_2 + \frac{V_2^2}{2}\right) \tag{10-13}$$

$$q_{out} = \left(h_4 + \frac{V_4^2}{2}\right) - \left(h_1 + \frac{V_1^2}{2}\right) \tag{10-14}$$

In the preceding equations, the changes in potential energies are neglected. If, in addition, the changes in kinetic energies are also neglected, we have

$$w_{out} = w_{turb} = h_3 - h_4 \tag{10-11a}$$

$$w_{in} = w_{comp} = h_2 - h_1 \tag{10-12a}$$

$$q_{in} = h_3 - h_2 \tag{10-13a}$$

$$q_{out} = h_4 - h_1 \tag{10-14a}$$

The cycle thermal efficiency is

$$\eta_{th} = \frac{w_{out} - w_{in}}{q_{in}} = \frac{q_{in} - q_{out}}{q_{in}}$$

$$= 1 - \frac{q_{out}}{q_{in}} = 1 - \frac{h_4 - h_1}{h_3 - h_2} \tag{10-15}$$

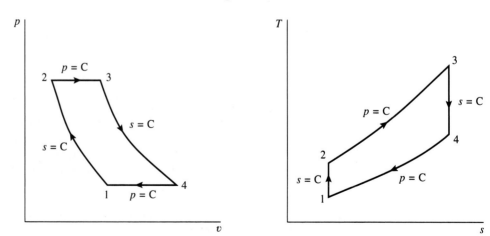

Figure 10-11 p–v and T–s diagrams for an air-standard Brayton cycle.

Gas Power Cycles   Chap. 10

In a cold-air-standard Brayton cycle, the energy quantities are expressed in terms of constant specific heats. Accordingly,

$$w_{out} = w_{turb} = c_p(T_3 - T_4) \tag{10-16}$$

$$w_{in} = w_{comp} = c_p(T_2 - T_1) \tag{10-17}$$

$$q_{in} = c_p(T_3 - T_2) \tag{10-18}$$

$$q_{out} = c_p(T_4 - T_1) \tag{10-19}$$

The cycle thermal efficiency is

$$\eta_{th} = 1 - \frac{T_4 - T_1}{T_3 - T_2} \tag{10-20}$$

As

$$\Delta s_{23} = \Delta s_{14} \qquad \text{or} \qquad \int_2^3 \frac{đq}{T} = \int_1^4 \frac{đq}{T}$$

thus,

$$c_p \int_2^3 \frac{dT}{T} = c_p \int_1^4 \frac{dT}{T} \qquad \text{or} \qquad \ln \frac{T_3}{T_2} = \ln \frac{T_4}{T_1}$$

so that

$$\frac{T_3}{T_2} = \frac{T_4}{T_1} \qquad \text{or} \qquad \frac{T_3 - T_2}{T_2} = \frac{T_4 - T_1}{T_1}$$

therefore,

$$\eta_{th} = 1 - \frac{T_1}{T_2}$$

For the internally reversible adiabatic process 1–2,

$$\frac{T_1}{T_2} = \left(\frac{p_1}{p_2}\right)^{(k-1)/k} = \frac{1}{r_p^{(k-1)/k}}$$

Finally, we have

$$\eta_{th} = 1 - \frac{1}{r_p^{(k-1)/k}} \tag{10-21}$$

where r_p is the compressor pressure ratio as defined by

$$r_p = \frac{p_2}{p_1}$$

Consequently, it is apparent that the thermal efficiency of an air-standard Brayton cycle is primarily a function of the pressure ratio.

Because the Brayton cycle operates under steady-state, steady-flow conditions, the inlet to the turbine is continuous at a high temperature, the value of which, however, must be limited due to metallurgical reasons. As shown in Fig. 10-12, considering the three cycles, 1–2′–3′–4′, 1–2–3–4, and 1–2″–3″–4″, with the same maximum per-

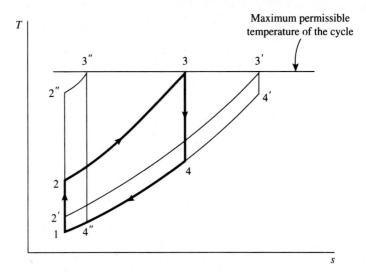

Figure 10-12 Effect of the pressure ratio on the thermal efficiency and the net work output of the air-standard Brayton cycle.

missible temperature, although the cycle thermal efficiency always increases as the pressure ratio increases, the net work output at first increases and then decreases. With T_1 and T_3 fixed, there must be an optimum value of T_2 that maximizes the net work output. This value of T_2 is found by differentiating the equation for the net work with respect to T_2 and setting the resulting expression equal to zero. Thus, from Eqs. (10-18) and (10-19), the net work output w_{net} of the cycle is given by

$$w_{net} = q_{in} - q_{out} = c_p(T_3 - T_2) - c_p(T_4 - T_1)$$

$$= c_p T_1 \left(\frac{T_3}{T_1} - \frac{T_2}{T_1} - \frac{T_4}{T_1} + 1 \right)$$

But for the internally reversible and adiabatic processes 1–2 and 3–4, we have

$$\frac{T_2}{T_1} = \left(\frac{p_2}{p_1} \right)^{(k-1)/k} = \left(\frac{p_3}{p_4} \right)^{(k-1)/k} = \frac{T_3}{T_4}$$

so that

$$\frac{T_4}{T_1} = \frac{T_3}{T_2}$$

The net work expression then becomes

$$w_{net} = c_p T_1 \left(\frac{T_3}{T_1} - \frac{T_2}{T_1} - \frac{T_3}{T_2} + 1 \right)$$

Because T_1, T_3, and c_p are fixed values, the condition

$$\frac{đw_{net}}{dT_2} = 0$$

yields

$$-\frac{1}{T_1} - T_3(-T_2^{-2}) = 0$$

or

$$T_2 = (T_1 T_3)^{1/2}$$

so that

$$\frac{T_2}{T_1} = \left(\frac{T_3}{T_1}\right)^{1/2}$$

On the other hand, for the internally reversible adiabatic process 1–2, we have

$$\frac{T_2}{T_1} = \left(\frac{p_2}{p_1}\right)^{(k-1)/k}$$

Therefore, the optimum isentropic pressure ratio for the cold-air-standard Brayton cycle must be

$$\frac{p_2}{p_1} = \left(\frac{T_3}{T_1}\right)^{k/(2k-2)} \tag{10-22}$$

When irreversible effects in the turbine and compressor are taken care of in an air-standard Brayton cycle, its T–s diagram is modified as depicted in Fig. 10-13, where the actual turbine and compressor processes are assumed to be irreversible and adiabatic. The exit states $4'$ and $2'$ of the actual turbine and compressor, respectively, are related to the adiabatic turbine efficiency η_T and adiabatic compressor efficiency η_C as follows:

$$\eta_T = \frac{h_3 - h_{4'}}{h_3 - h_4} \quad \text{and} \quad \eta_C = \frac{h_2 - h_1}{h_{2'} - h_1}$$

It is imperative to notice that in a gas-turbine cycle, the compressor consumes a

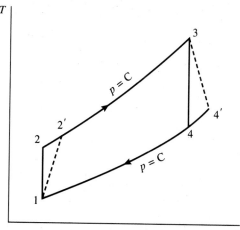

Figure 10-13 Air-standard Brayton cycle with irreversible compression and expansion.

very large amount of the turbine work. The ratio between the compressor work and the turbine work, known as the *back work ratio*, of a Brayton cycle typically ranges from 40 to 80%. Severe irreversible effects in the compressor and turbine may cause the back work ratio to become even larger.

Example 10-5

A gas-turbine power plant working on the air-standard Brayton cycle with a compressor pressure ratio of 10 is to produce 10 MW net power output. The air enters the compressor at 100 kPa and 30°C. The maximum temperature of the cycle is 1000°C. The turbine and compressor processes are adiabatic with a turbine efficiency of 90% and a compressor efficiency of 85%. Based on the air-table data, determine the temperature at the end of each process of the cycle, the turbine power output, the compressor power input, and the thermal efficiency of the cycle.

Solution Referring to Fig. 10-13, the given conditions are $p_1 = 100$ kPa, $T_1 = 30°C = 303$ K, $p_2/p_1 = 10$, $T_3 = 1000°C = 1273$ K, $\eta_T = 0.90$, and $\eta_C = 0.85$. From the air table, at $T_1 = 303$ K, we get $h_1 = 303.4$ kJ/kg, $p_{r1} = 1.429$, and $\phi_1 = 5.712$ kJ/kg·K; and at $T_3 = 1273$ K, we get $h_3 = 1363.8$ kJ/kg, $p_{r3} = 301.66$, and $\phi_3 = 7.248$ kJ/kg·K. For the internally reversible adiabatic process 1–2, we have

$$p_{r2} = p_{r1}\left(\frac{p_2}{p_1}\right) = 1.429(10) = 14.29$$

At this p_{r2} value, the air table gives $T_2 = 580$ K, $h_2 = 585.9$ kJ/kg, and $\phi_2 = 6.373$ kJ/kg·K. Notice that the value of ϕ_2 can also be calculated as follows:

$$s_2 - s_1 = 0 = \phi_2 - \phi_1 - R \ln\frac{p_2}{p_1} = \phi_2 - 5.712 - \frac{8.314}{29} \ln 10$$

whereupon

$$\phi_2 = 6.372 \text{ kJ/kg·K}$$

From this ϕ_2 value, the air table can be used to obtain the value of T_2 and h_2. For the internally reversible adiabatic process 3–4, we have

$$p_{r4} = p_{r3}\left(\frac{p_4}{p_3}\right) = p_{r3}\left(\frac{p_1}{p_2}\right) = \frac{301.66}{10} = 30.166$$

At this p_{r4} value, the air table gives $T_4 = 710$ K, $h_4 = 723.7$ kJ/kg, and $\phi_4 = 6.587$ kJ/kg·K. Or, alternatively,

$$s_4 - s_3 = 0 = \phi_4 - \phi_3 - R \ln\frac{p_4}{p_3}$$

$$= \phi_4 - 7.248 - \frac{8.314}{29} \ln \frac{1}{10}$$

whereupon

$$\phi_4 = 6.588 \text{ kJ/kg·K}$$

From this ϕ_4 value, the air table can be used to obtain the value of T_4 and h_4.
For the irreversible adiabatic process 3–4', we have

$$h_{4'} = h_3 - \eta_T(h_3 - h_4)$$

$$= 1363.8 - 0.90(1363.8 - 723.7)$$

$$= 787.7 \text{ kJ/kg}$$

At this value of $h_{4'}$, the air table gives $T_{4'} = 769$ K. For the irreversible-adiabatic process 1–2', we have

$$h_{2'} = h_1 + \frac{h_2 - h_1}{\eta_c} = 303.4 + \frac{585.9 - 303.4}{0.85} = 635.8 \text{ kJ/kg}$$

At this value of $h_{2'}$, the air table gives $T_{2'} = 627$ K.

The turbine work output and the compressor work input are

$$w_{\text{out}} = w_{\text{turb}} = h_3 - h_{4'} = 1363.8 - 787.7 = 576.1 \text{ kJ/kg}$$

$$w_{\text{in}} = w_{\text{comp}} = h_{2'} - h_1 = 635.8 - 303.4 = 332.4 \text{ kJ/kg}$$

Consequently,

$$\text{net } w_{\text{out}} = w_{\text{out}} - w_{\text{in}} = 576.1 - 332.4 = 243.7 \text{ kJ/kg}$$

For a net output power of 10 MW, the mass-flow rate of air is

$$\dot{m}_{\text{air}} = \frac{10^4 \text{ kW}}{243.7 \text{ kJ/kg}} = 41.0 \text{ kg/s}$$

The turbine power output and the compressor power input are, respectively,

$$\dot{W}_{\text{turb}} = (41.0 \text{ kg/s})(576.1 \text{ kJ/kg}) = 23{,}600 \text{ kW} = 23.6 \text{ MW}$$

$$\dot{W}_{\text{comp}} = (41.0 \text{ kg/s})(332.4 \text{ kJ/kg}) = 13{,}600 \text{ kW} = 13.6 \text{ MW}$$

The heat input of the cycle is

$$q_{\text{in}} = q_{2'3} = h_3 - h_{2'} = 1363.8 - 635.8 = 728.0 \text{ kJ/kg}$$

The thermal efficiency of the cycle is then

$$\eta_{\text{th}} = \frac{\text{net } w_{\text{out}}}{q_{\text{in}}} = \frac{243.7}{728.0} = 0.335 = 33.5\%$$

The back work ratio of the cycle is given by $w_{\text{comp}}/w_{\text{turb}} = 332.4/576.1 = 0.577$. This means that 57.7% of the turbine work is used to operate the compressor, resulting in a relatively low thermal efficiency of the gas-turbine cycle.

10-6 REGENERATIVE GAS-TURBINE CYCLE

It is clear from the study of the simple Brayton cycle that the exhaust gases from the turbine are usually at a high temperature. Discharging these high-temperature gases to the atmosphere is a waste of energy. In the meantime, the high-pressure air from the compressor is at a low temperature, requiring heat addition. It is highly desirable to take advantage of these circumstances by the use of a regenerator in which the hot turbine exhaust gases are used to preheat the high-pressure air from the compressor before it enters the combustion chamber, thereby reducing the fuel consumption. Figure 10-14 shows the schematic diagram of a regenerative gas-turbine engine, together with the T–s diagram of the air-standard cycle.

Assuming that ΔKE and ΔPE are negligible and that no heat is lost to the surroundings, the first law as applied to the regenerator gives

$$h_5 - h_6 = h_3 - h_2 \tag{10-23}$$

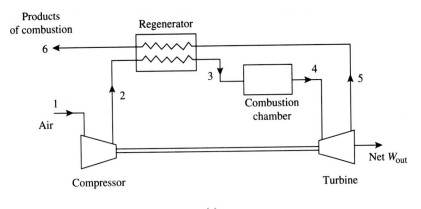

Products of combustion

Regenerator

6

1

Air

2

3

Combustion chamber

4

5

Compressor

Turbine

Net W_{out}

(a)

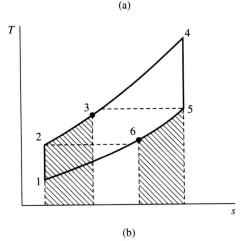

(b)

Figure 10-14 A regenerative gas-turbine cycle: (a) schematic diagram, and (b) T–s diagram.

In the air-standard cycle, it is assumed that the two fluid streams undergo internally reversible constant-pressure processes through the regenerator. Accordingly, the two crosshatched areas on the T–s diagram of Fig. 10-14 are equal in magnitude. The *regenerator effectiveness* is defined as

$$\eta_{\text{reg}} = \frac{h_3 - h_2}{h_5 - h_2} \tag{10-24}$$

For the cold-air-standard cycle, Eqs. (10-23) and (10-24) become

$$T_5 - T_6 = T_3 - T_2 \tag{10-23a}$$

$$\eta_{\text{reg}} = \frac{T_3 - T_2}{T_5 - T_2} \tag{10-24a}$$

If the regenerator were ideal (meaning $\eta_{\text{reg}} = 100\%$), T_3 would equal to T_5, and state 3 in Fig. 10-14(b) would lie horizontally across from state 5.

In the regenerative cycle of Fig. 10-14, neglecting ΔKE and ΔPE, the heat addition and heat rejection per unit-mass flow are

$$q_{in} = h_4 - h_3$$

$$q_{out} = h_6 - h_1$$

Consequently, the thermal efficiency is

$$\eta_{th} = 1 - \frac{q_{out}}{q_{in}} = 1 - \frac{h_6 - h_1}{h_4 - h_3} \qquad (10\text{-}25)$$

For the cold-air-standard cycle, the thermal efficiency becomes

$$\eta_{th} = 1 - \frac{T_6 - T_1}{T_4 - T_3} \qquad (10\text{-}25a)$$

Example 10-6

A regenerative air-standard gas-turbine cycle uses air with $c_p = 0.25$ Btu/lbm·°F and $k = 1.37$ as the working fluid. The air enters the compressor at 14.5 psia and 80°F, leaves the compressor at 85.0 psia and 520°F, enters the combustion chamber at 84.9 psia and 900°F, enters the turbine at 83.0 psia and 1550°F, and leaves the turbine at 14.6 psia and 960°F. Determine the regenerator effectiveness, the turbine adiabatic efficiency, the compressor adiabatic efficiency, and the cycle thermal efficiency.

Solution Referring to Fig. 10-15, the given conditions are $p_1 = 14.5$ psia, $T_1 = 80°F = 540°R$, $p_{2'} = 85.0$ psia, $T_{2'} = 520°F = 980°R$, $p_3 = 84.9$ psia, $T_3 = 900°F = 1360°R$, $p_4 = 83.0$ psia, $T_4 = 1550°F = 2010°R$, $p_{5'} = 14.6$ psia, $T_{5'} = 960°F = 1420°R$. In addition, $c_p = 0.25$ Btu/lbm·°F and $k = 1.37$.
 The regenerator effectiveness is

$$\eta_{reg} = \frac{h_3 - h_{2'}}{h_{5'} - h_{2'}} = \frac{T_3 - T_{2'}}{T_{5'} - T_{2'}} = \frac{900 - 520}{960 - 520} = 0.864 = 86.4\%$$

For the internally reversible adiabatic process 4–5,

$$T_5 = T_4 \left(\frac{p_5}{p_4}\right)^{(k-1)/k} = 2010\left(\frac{14.6}{83.0}\right)^{(1.37-1)/1.37} = 1257°R = 797°F$$

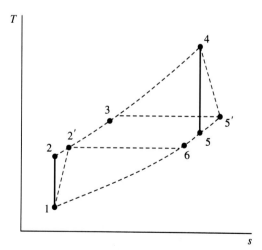

Figure 10-15 *T–s* diagram for an irreversible regenerative gas-turbine cycle.

Accordingly, the adiabatic turbine efficiency is

$$\eta_T = \frac{h_4 - h_{5'}}{h_4 - h_5} = \frac{T_4 - T_{5'}}{T_4 - T_5} = \frac{1550 - 960}{1550 - 797} = 0.784 = 78.4\%$$

For the internally reversible adiabatic process 1–2,

$$T_2 = T_1\left(\frac{p_2}{p_1}\right)^{(k-1)/k} = 540\left(\frac{85.0}{14.5}\right)^{(1.37-1)/1.37} = 871°R = 411°F$$

Accordingly, the adiabatic compressor efficiency is

$$\eta_C = \frac{h_2 - h_1}{h_{2'} - h_1} = \frac{T_2 - T_1}{T_{2'} - T_1} = \frac{411 - 80}{520 - 80} = 0.752 = 75.2\%$$

An energy balance for the regenerator gives

$$h_{5'} - h_6 = h_3 - h_{2'} \qquad \text{or} \qquad T_{5'} - T_6 = T_3 - T_{2'}$$

so that

$$T_6 = T_{5'} + T_{2'} - T_3 = 960 + 520 - 900 = 580°F = 1040°R$$

The heat added to the cycle is

$$q_{in} = h_4 - h_3 = c_p(T_4 - T_3) = 0.25(1550 - 900) = 162.5 \text{ Btu/lbm}$$

The heat rejected from the cycle is

$$q_{out} = h_6 - h_1 = c_p(T_6 - T_1) = 0.25(580 - 80) = 125.0 \text{ Btu/lbm}$$

The work done by the turbine is

$$w_{out} = w_{turb} = h_4 - h_{5'} = c_p(T_4 - T_{5'}) = 0.25(1550 - 960) = 147.5 \text{ Btu/lbm}$$

The work done on the compressor is

$$w_{in} = w_{comp} = h_{2'} - h_1 = c_p(T_{2'} - T_1) = 0.25(520 - 80) = 110 \text{ Btu/lbm}$$

Thus,

$$\text{net } w_{out} = w_{out} - w_{in} = 147.5 - 110 = 37.5 \text{ Btu/lbm}$$

$$= \text{net } q_{in} = q_{in} - q_{out} = 162.5 - 125.0 = 37.5 \text{ Btu/lbm}$$

Consequently, the cycle thermal efficiency is

$$\eta_{th} = \frac{\text{net } w_{out}}{q_{in}} = \frac{37.5}{162.5} = 0.231 = 23.1\%$$

or

$$\eta_{th} = 1 - \frac{q_{out}}{q_{in}} = 1 - \frac{125.0}{162.5} = 0.231 = 23.1\%$$

In order to illustrate the benefit of the addition of a regenerator, let us consider briefly the present example again by omitting the regenerator, thus forming the cycle 1–2′–4–5′–1, as shown in Fig. 10-15. The values for work of the compressor and turbine are unchanged. That is,

$$w_{in} = w_{comp} = h_{2'} - h_1 = 110 \text{ Btu/lbm}$$

$$w_{out} = w_{turb} = h_4 - h_{5'} = 147.5 \text{ Btu/lbm}$$

and

$$\text{net } w_{out} = w_{out} - w_{in} = 37.5 \text{ Btu/lbm}$$

The values of heat addition and heat rejection are changed as follows:

$$q_{in} = h_4 - h_{2'} = c_p(T_4 - T_{2'})$$

$$= 0.25(1550 - 520) = 257.5 \text{ Btu/lbm}$$

$$q_{out} = h_{5'} - h_1 = c_p(T_{5'} - T_1)$$

$$= 0.25(960 - 80) = 220.0 \text{ Btu/lbm}$$

and

$$\text{net } q_{in} = q_{in} - q_{out} = 257.5 - 220.0 = 37.5 \text{ Btu/lbm}$$

$$= \text{net } w_{out} \text{ (satisfying the first law of thermodynamics)}$$

The thermal efficiency of the simple cycle is then

$$\eta_{th} = \frac{\text{net } w_{out}}{q_{in}} = \frac{37.5}{257.5} = 0.146 = 14.6\%$$

as compared with the thermal efficiency of 23.1% for the regenerative cycle. The thermal efficiencies of both cycles are very low, because the back work ratio based on the given data is very high:

$$\text{Back work ratio} = \frac{w_{comp}}{w_{turb}} = \frac{110}{147.5} = 0.746 = 74.6\%$$

meaning that 74.6% of the turbine work is used to operate the compressor and only 25.4% of the turbine work becomes the net work output of the cycle.

10-7 MULTISTAGE COMPRESSION AND EXPANSION

In Chapter 7, an expression of shaft work for an internally reversible steady-flow process was developed. Written in differential form for a unit mass, it is

$$đw_{in} = v \, dp + dKE + dPE$$

If the changes in kinetic and potential energies are negligible, the equation reduces to

$$đw_{in} = v \, dp$$

Integrating this equation between the inlet state 1 and the exit state 2 of an internally reversible steady-flow process gives

$$w_{in} = \int_1^2 v \, dp \tag{10-26}$$

Geometrically, Eq. (10-26) states that the area on the left side of the process curve on a p–v diagram represents the work per unit mass during an internally reversible steady-flow process.

Consider the steady-flow compression of an ideal gas from a given state 1 to a higher pressure p_2 by three different internally reversible processes, as depicted in Fig. 10-16. Application of Eq. (10-26) to these processes shows that the isothermal compression requires the minimum work and the adiabatic compression the maximum

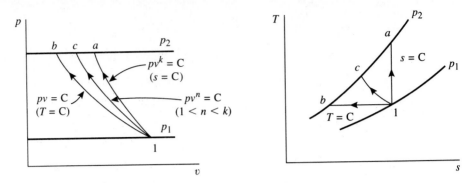

Figure 10-16 Compression of a gas along three internally reversible paths.

work, with the polytropic compression falling in between. This result suggests that it is advantageous to cool the gas as it is compressed. Cooling a gas during compression in a flow process, however, is difficult. In practice, in order to achieve the benefits of cooling, the gas is compressed in stages, and between stages it is cooled in heat exchangers called intercoolers.

Figure 10-17 shows a two-stage compression process with intercooling. In the first stage, the gas is compressed in a polytropic process 1–a. It is then cooled in a constant-pressure process a–b back to its initial temperature (a case of so-called perfect cooling with $T_b = T_1$). The gas is finally compressed in the second stage through a polytropic process b–2. Compared to single-stage polytropic compression 1–a–c, the savings in compression work by the intercooled two-stage compressor is represented by the crosshatched area on the p–v diagram. Clearly, there is a particular intercooler pressure that would maximize the savings in compression work. To find the optimum interstage pressure p_i, let us write the total work of compression for the two stages, ne-

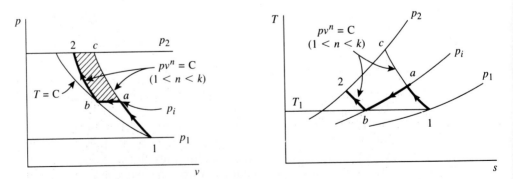

Figure 10-17 Two-stage compression with intercooling to the initial temperature.

glecting kinetic-energy and potential-energy changes:

$$w_{in} = \int_{p_1}^{p_i} v\, dp + \int_{p_i}^{p_2} v\, dp$$

$$= \frac{n}{n-1}(p_i v_a - p_1 v_1) + \frac{n}{n-1}(p_2 v_2 - p_i v_b)$$

$$= \frac{n}{n-1} p_1 v_1\left(\frac{p_i}{p_1}\frac{v_a}{v_1} - 1\right) + \frac{n}{n-1} p_i v_b\left(\frac{p_2}{p_i}\frac{v_2}{v_b} - 1\right)$$

$$= \frac{n}{n-1} RT_1\left[\left(\frac{p_i}{p_1}\right)^{(n-1)/n} - 1\right] + \frac{n}{n-1} RT_1\left[\left(\frac{p_2}{p_i}\right)^{(n-1)/n} - 1\right]$$

$$= \frac{n}{n-1} RT_1\left[\left(\frac{p_i}{p_1}\right)^{(n-1)/n} + \left(\frac{p_2}{p_i}\right)^{(n-1)/n} - 2\right]$$

Differentiating the preceding expression with respect to p_i and then setting it to zero (i.e., $dw_{in}/dp_i = 0$), we find that the condition for minimum work is given by

$$p_i = (p_1 p_2)^{1/2} \qquad \text{or} \qquad \frac{p_i}{p_1} = \frac{p_2}{p_i} = \left(\frac{p_2}{p_i}\right)^{1/2}$$

Therefore,

$$w_{in,\,1st.\,stage} = w_{in,\,2nd.\,stage} = \frac{n}{n-1} RT_1\left[\left(\frac{p_2}{p_1}\right)^{(n-1)/2n} - 1\right]$$

and

$$w_{in,\,2\,stages} = \frac{2n}{n-1} RT_1\left[\left(\frac{p_2}{p_1}\right)^{(n-1)/2n} - 1\right] \qquad (10\text{-}27)$$

In general, for N stages,

$$w_{in,\,N\,stages} = \frac{Nn}{n-1} RT_1\left[\left(\frac{p_2}{p_1}\right)^{(n-1)/Nn} - 1\right] \qquad (10\text{-}28)$$

The preceding equation implies that if intercooling brings the gas to its initial temperature after each stage, the minimum work is required when the pressure ratio is the same for each stage and the total work requirement is divided equally among the stages.

In the case of expansion, the multistage concept can be employed to obtain a larger work output. Figure 10-18 shows the p–v and T–s diagrams of a two-stage expander with reheating to the initial temperature. Because expansion machines are usually insulated to prevent heat loss to the surroundings that would reduce the work output, the expansion process in each stage in Fig. 10-18 is assumed to be adiabatic.

Isentropic expansion occurs in the first turbine stage from 1 to a. The gas is withdrawn from the turbine and reheated at constant pressure from a to b back to its initial temperature ($T_b = T_1$). It then enters the second turbine stage for isentropic expansion from b to 2. Compared to the single-stage isentropic expansion 1–a–c, the increase in expansion work by the reheated two-stage expander is represented by the crosshatched area on the p–v diagram. Analogous to the procedure used in the two-stage intercooled compression process, the optimum interstage pressure p_i for maximum work output in the case of a two-stage reheated expansion process is $p_i = (p_1 p_2)^{1/2}$. This means that the work output of each stage will be equal for the ideal situation.

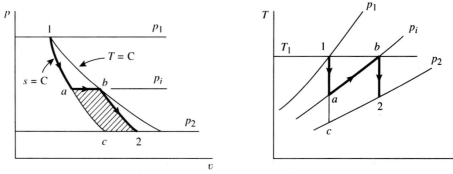

Figure 10-18 Two-stage expansion with reheating to the initial temperature.

10-8 GAS-TURBINE CYCLE WITH INTERCOOLING AND REHEATING

As shown in Sec. 10-7, multistage intercooling reduces the work requirement of a gas compressor operating between a given initial state and a specified final pressure. Figure 10-19 shows an installation schematic and a T–s diagram for a gas-turbine cycle with an intercooled two-stage compressor. Because the thermal efficiency of the added cycle 3–4–a–2 due to intercooling is less than that of the basic Brayton cycle 1–a–6–7 without intercooling, the effect of intercooling is always the lowering of the thermal efficiency of the combined cycle. To increase the thermal efficiency and at the same time decrease the compressor work requirement, a regenerator must be incorporated in the intercooled multistage-compression gas-turbine cycle, such as shown in Fig. 10-19, in its entire installation. The addition of a regenerator to an intercooled gas-turbine cycle increases the cycle thermal efficiency because the heat required for process 4–5 comes from the hot turbine exhaust in going through the regenerator (process 7–8).

Recall from Sec. 10-7 that multistage expansion with reheating in between enhances work output. By reasoning similar to that used in connection with intercooling, one can show that the effect of reheating is always the lowering of the cycle thermal efficiency. In order to increase the thermal efficiency, a regenerator must be added to the reheated multistage-expansion gas-turbine cycle, such as shown in Fig. 10-20.

It is possible to compine in one cycle regeneration, multistage compression with intercooling, and multistage expansion with reheating. Figure 10-21 shows such an installation and the corresponding T–s diagram for the case of irreversible two-stage compression with intercooling to its initial temperature and irreversible two-stage expansion with reheating to its initial temperature.

Assuming steady flow and negligible kinetic and potential energy changes, application of the first law to each of the components of the air-standard cycle shown in Fig. 10-21 yields, per unit mass of fluid flow, the following.

For the regenerator,

$$h_5 - h_{4'} = h_{9'} - h_{10} \quad \text{and} \quad \eta_{\text{reg}} = \frac{h_5 - h_{4'}}{h_{9'} - h_{4'}}$$

For the compressors,

$$w_{\text{in}} = (h_{2'} - h_1) + (h_{4'} - h_3)$$

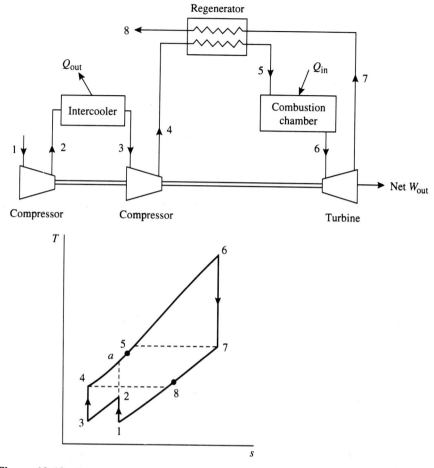

Figure 10-19 Intercooled two-stage compression gas-turbine cycle with regeneration.

For the turbines,

$$w_{out} = (h_6 - h_{7'}) + (h_8 - h_{9'})$$

For the combustion chamber and reheater,

$$q_{in} = (h_6 - h_5) + (h_8 - h_{7'})$$

For the intercooler and the atmosphere (for process 10–1),

$$q_{out} = (h_{2'} - h_3) + (h_{10} - h_1)$$

The thermal efficiency of the cycle is

$$\eta_{th} = \frac{w_{out} - w_{in}}{q_{in}} = \frac{q_{in} - q_{out}}{q_{in}}$$

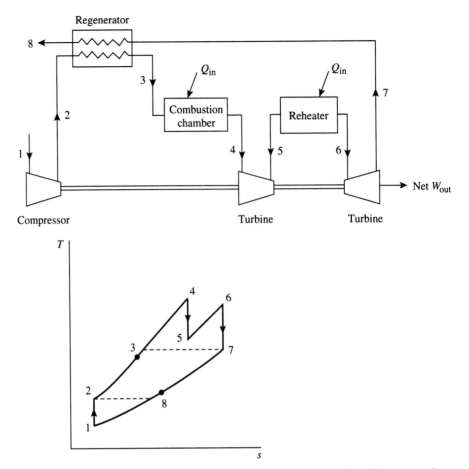

Figure 10-20 Reheated two-stage expansion gas-turbine cycle with regeneration.

It is imperative to notice that if the number of intercooled compression stages, the number of reheated expansion stages, and the regenerator effectiveness are increased, the gas-turbine cycle approaches the Ericsson cycle (see Sec. 10-11) and its thermal efficiency approaches that of the Carnot cycle. Figure 10-22 shows how a regenerative gas-turbine cycle with many stages of compression and expansion approaches the Ericsson cycle.

Example 10-7

A gas-turbine unit is equipped with a regenerator. It has two stages of compression and two stages of expansion with a pressure ratio across each stage of 2.5. The air enters the first stage of the compressor at 100 kPa and 22°C, and its temperature is reduced by intercooling back to 22°C before entering the second stage. The inlet temperature is 827°C to each stage of expansion. The compressor and turbine adiabatic efficiencies are 78 and 84%, respectively, and the regenerator has an effectiveness of 70%. Based on variable specific-heats air-standard analysis, determine the temperature of the air leaving the regenerator and entering the atmosphere, the compressor work and turbine work, in kJ/kg, and the thermal efficiency.

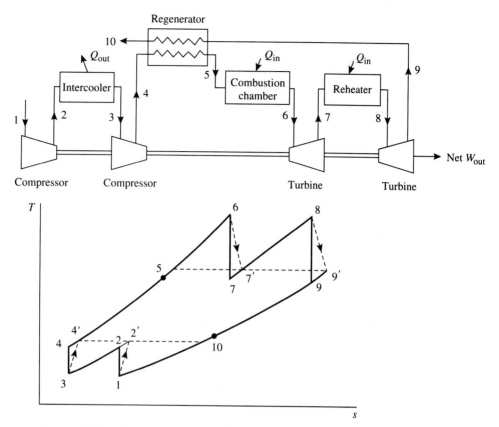

Figure 10-21 Two-stage compression and two-stage expansion gas-turbine cycle with regeneration.

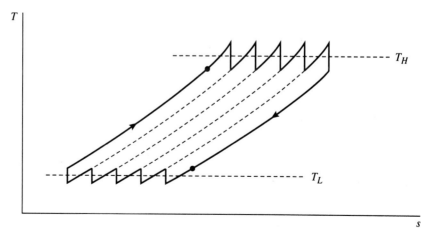

Figure 10-22 Gas-turbine cycle having many stages of compression and expansion with regeneration.

Solution Referring to Fig. 10-21, the given data are $p_2/p_1 = p_4/p_3 = p_6/p_7 = p_8/p_9 = 2.5$, $p_1 = 100$ kPa, $T_1 = T_3 = 22°C = 295$ K, $T_6 = T_8 = 827°C = 1100$ K, $\eta_C = 78\%$, $\eta_T = 84\%$, and $\eta_{\text{reg}} = 70\%$. From the air table (Table A-9M), at $T_1 = T_3 = 295$ K, we have $h_1 = h_3 = 295.4$ kJ/kg and $p_{r1} = p_{r3} = 1.301$. At $T_6 = T_8 = 1100$ K, we have $h_6 = h_8 = 1161.1$ kJ/kg and $p_{r6} = p_{r8} = 166.2$.

For the internally reversible adiabatic processes 1–2 and 3–4, we have

$$p_{r2} = p_{r4} = p_{r1}\left(\frac{p_2}{p_1}\right) = 1.301 \, (2.5) = 3.253$$

At this p_r value, the air table gives $h_2 = h_4 = 384.1$ kJ/kg.

For the internally reversible adiabatic processes 6–7 and 8–9, we have

$$p_{r7} = p_{r9} = p_{r6}\left(\frac{p_7}{p_6}\right) = 166.2/2.5 = 66.48$$

At this p_r value, the air table gives $h_7 = h_9 = 902.6$ kJ/kg. Accordingly,

$$h_{2'} = h_{4'} = h_1 + \frac{h_2 - h_1}{\eta_C} = 295.4 + \frac{384.1 - 295.4}{0.78} = 409.1 \text{ kJ/kg}$$

$$h_{7'} = h_{9'} = h_6 - \eta_T(h_6 - h_7) = 1161.1 - 0.84(1161.1 - 902.6) = 944.0 \text{ kJ/kg}$$

For the regenerator,

$$\eta_{\text{reg}} = \frac{h_5 - h_{4'}}{h_{9'} - h_{4'}}$$

or

$$h_5 = h_{4'} + \eta_{\text{reg}}(h_{9'} - h_{4'}) = 409.1 + 0.70(944.0 - 409.1) = 783.5 \text{ kJ/kg}$$

Also,

$$h_{9'} - h_{10} = h_5 - h_{4'}$$

or

$$h_{10} = h_{9'} - h_5 + h_{4'} = 944.0 - 783.5 + 409.1 = 569.6 \text{ kJ/kg}$$

At this h_{10} value, the air table gives $T_{10} = 564$ K $= 291°C$.

The compressor work requirement is

$$w_{\text{in}} = (h_{2'} - h_1) + (h_{4'} - h_3) = 2(409.1 - 295.4) = 227.4 \text{ kJ/kg}$$

The turbine work output is

$$w_{\text{out}} = (h_6 - h_{7'}) + (h_8 - h_{9'}) = 2(1161.1 - 944.0) = 434.2 \text{ kJ/kg}$$

The heat addition to the cycle is

$$q_{\text{in}} = (h_6 - h_5) + (h_8 - h_{7'}) = (1161.1 - 783.5) + (1161.1 - 944.0)$$
$$= 594.7 \text{ kJ/kg}$$

The cycle thermal efficiency is then

$$\eta_{\text{th}} = \frac{w_{\text{out}} - w_{\text{in}}}{q_{\text{in}}} = \frac{434.2 - 227.4}{594.7} = 0.348$$

10-9 TURBOJET PROPULSION

Because space and weight factors are of overwhelming importance in aircraft engine design, the gas-turbine power cycle is most suitable for propelling high-speed flight. A gas-turbine engine can provide large amounts of power while occupying a small space and having a low weight.

The most widely used gas-turbine device for aircraft propulsion is the *turbojet* engine. Figure 1-6 (in Chapter 1) shows a cutaway view of a typical turbojet engine and Fig. 10-23 shows a schematic of it. The main elements of the engine are the inlet diffuser, the compressor, the combustor, the turbine, and the exit nozzle. The inlet diffuser is utilized to slow down the entering air and create a small pressure rise (called the ram effect). The pressure of the air is further increased in the compressor. After leaving the compressor, the high-pressure air flow enters the combustor, where fuel is added and the resultant mixture burned. The hot products of combustion from the combustor are expanded in the turbine to such a pressure far enough to generate sufficient power to drive the compressor and auxiliary equipment, so that no net shaft power is available. The remaining energy in the gases exhausted from the turbine is then converted in the exit nozzle to high-velocity kinetic energy. The change in momentum of the gases entering and leaving the gas-turbine unit causes a force to be exerted on the aircraft.

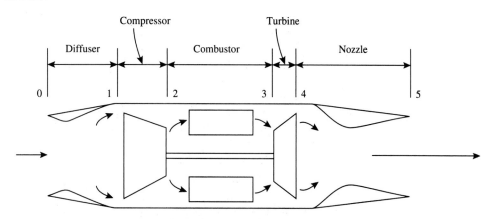

Figure 10-23   Turbojet engine schematic.

The general thermodynamic characteristics of an ideal turbojet cycle, with internally reversible adiabatic compressor, turbine, diffuser, and nozzle processes and no pressure drops in the combustor or between components, is shown on a T–s diagram in Fig. 10-24. Based on air-standard analysis and under steady-flow conditions with negligible changes in potential energy, the energy-balance equations for the engine components can be written as follows:

For the diffuser with no work and no heat involved,

$$h_0 + \tfrac{1}{2}\mathbf{V}_0^2 = h_1 + \tfrac{1}{2}\mathbf{V}_1^2$$

For the adiabatic compressor and turbine, by neglecting ΔKE,

$$w_{\text{comp}} = w_{\text{turb}} \qquad \text{or} \qquad h_2 - h_1 = h_3 - h_4$$

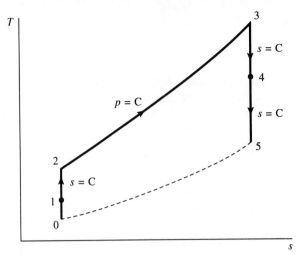

Figure 10-24 *T–s* diagram for an ideal turbojet cycle.

For the combustor with ΔKE neglected and no work,

$$q_{in} = h_3 - h_2$$

For the nozzle with no work and no heat involved,

$$h_4 + \tfrac{1}{2}\mathbf{V}_4^2 = h_5 + \tfrac{1}{2}\mathbf{V}_5^2$$

According to Newton's second law, the *thrust* or driving force produced by the jet engine is equal to the change of momentum of the fluid passing through the engine. For the case in which the nozzle expands the gases completely to the local ambient pressure, so that there are no unbalanced pressure forces acting on the flow, the thrust produced by the engine is given by

$$F = \dot{m}(\mathbf{V}_5 - \mathbf{V}_0) \qquad (10\text{-}29)$$

where $\dot{m}$ is the mass-flow rate of air. In the actual case when the mass-flow rate of fuel, $\dot{m}_f$, is included, the thrust equation is written as

$$F = \dot{m}(\mathbf{V}_5 - \mathbf{V}_0) + \dot{m}_f\mathbf{V}_5 \qquad (10\text{-}30)$$

The thrust force per unit mass-flow rate is called the *specific impulse*.

When the jet moves forward in still air at velocity $\mathbf{V}_0$, the power developed from the thrust is given by

$$\dot{W} = F\mathbf{V}_0 = \dot{m}(\mathbf{V}_5 - \mathbf{V}_0)\mathbf{V}_0 \qquad (10\text{-}31)$$

and the work done per unit mass-flow is given by

$$w = (\mathbf{V}_5 - \mathbf{V}_0)\mathbf{V}_0 \qquad (10\text{-}32)$$

The *propulsive efficiency* η_{prop} of a turbojet engine is defined as the ratio of the work done in flight and the sum of this work quantity and the kinetic energy lost in the exhaust jet, or

$$\eta_{prop} = \frac{(\mathbf{V}_5 - \mathbf{V}_0)\mathbf{V}_0}{(\mathbf{V}_5 - \mathbf{V}_0)\mathbf{V}_0 + \dfrac{(\mathbf{V}_5 - \mathbf{V}_0)^2}{2}} = \frac{2}{1 + \dfrac{\mathbf{V}_5}{\mathbf{V}_0}} \qquad (10\text{-}33)$$

This equation reveals that when the plane is standing still ($V_0 = 0$), the propulsive efficiency is zero and the thrust is a maximum for the given value of mass flow. However, when the plane speed equals the exhaust velocity ($V_0 = V_5$), the propulsive efficiency is 100% and the thrust is zero.

When irreversibility effects of the engine components are considered, the T–s diagram of an actual turbojet cycle 0–$1'$–$2'$–3–$4'$–$5'$ is as shown in Fig. 10-25. As explained in Sec. 7-8, in reference to Fig. 7-13(b), for the inlet diffuser we have

$$T_1 = T_{1'}, \qquad h_1 = h_{1'}, \qquad \text{and} \qquad p_1 > p_{1'}$$

The diffuser pressure coefficient K_p is defined as

$$K_p = \frac{p_{1'} - p_0}{p_1 - p_0}$$

With the condition that

$$w_{\text{turb } 3-4'} = w_{\text{comp } 1'-2'}$$

we must have

$$w_{\text{turb } 3-4} > w_{\text{comp } 1'-2}$$

Figure 10-25 also shows the pressure drop in the combustor.

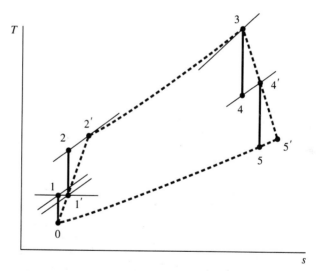

Figure 10-25 T–s diagram for an actual turbojet cycle.

It should be noted that, in general, turbojet engines operate at higher pressure ratios than stationary gas-turbine engines. In turbojets, pressure ratios of from 10 to 25 are common.

In order to provide greater nozzle–exhaust velocities so as to enhance the engine thrust, an afterburner can be added to a turbojet engine, as shown in Fig. 10-26. Although afterburning is the most effective means of thrust augmentation, its use will cause a great increase in fuel consumption. As a result, afterburning is usually used only for short periods of high thrust as in takeoffs and in military operations.

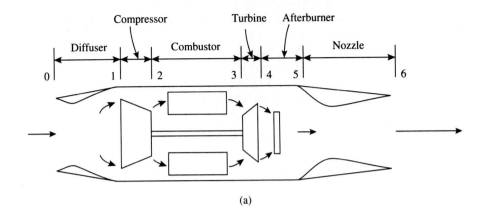

(a)

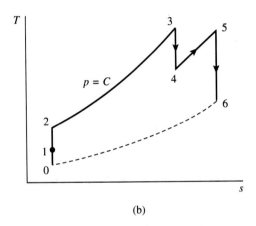

(b)

Figure 10-26 Turbojet engine with afterburner: (a) schematic diagram, and (b) idealized T–s diagram.

Example 10-8

A turbojet engine flies at a steady speed of 900 km/h at an altitude where the atmospheric pressure is 50 kPa and the atmospheric temperature is 250 K. The maximum cycle pressure is 300 kPa and the maximum cycle temperature is 1390 K. Assume internally reversible adiabatic flow in the diffuser, compressor, turbine, and nozzle, and neglect pressure drop in the combustor. Find the pressures and temperatures at various locations in the jet engine, and determine the air flow required to produce a thrust of 68 kN. Use air-standard analysis based on (a) air table data and (b) constant specific heats with $c_p = 1.003$ kJ/kg·K and $k = 1.40$.

Solution Referring to Fig. 10-24, the given data are $p_0 = p_5 = 50$ kPa, $T_0 = 250$ K, $p_2 = p_3 = 300$ kPa, $T_3 = 1390$ K, $F = 68,000$ N, and $\mathbf{V}_0 = 900$ km/h $= 900 \times 1000/3600 = 250$ m/s.

(a) *Air-table solution.* From the air table (Table A-9M), at $T_0 = 250$ K, we have $h_0 = 250.25$ kJ/kg and $p_{r0} = 0.7296$; and at $T_3 = 1390$ K, we have $h_3 = 1503.19$ kJ/kg and $p_{r3} = 434.52$.

By neglecting the exit kinetic energy, the steady-flow equation for the diffuser gives

$$h_1 = h_0 + \tfrac{1}{2}V_0^2 = 250.25 \text{ kJ/kg} + \tfrac{1}{2}(250 \text{ m/s})^2 \frac{1}{1000 \text{ J/kJ}}$$

$$= 281.50 \text{ kJ/kg}$$

At this h_1 value, the air table gives $T_1 = 281.1$ K and $p_{r1} = 1.0998$. For the internally reversible adiabatic process 0–1, we have

$$p_1 = p_0\left(\frac{p_{r1}}{p_{r0}}\right) = 50\left(\frac{1.0998}{0.7296}\right) = 75.4 \text{ kPa}$$

For the internally reversible adiabatic process 1–2, we have

$$p_{r2} = p_{r1}\left(\frac{p_2}{p_1}\right) = 1.0998\left(\frac{300}{75.4}\right) = 4.376$$

At this p_{r2} value, the air table gives $T_2 = 416.6$ K and $h_2 = 418.09$ kJ/kg. For the jet engine,

$$w_{comp} = w_{turb} \qquad \text{or} \qquad h_2 - h_1 = h_3 - h_4$$

therefore,

$$h_4 = h_3 - h_2 + h_1 = 1503.19 - 418.09 + 281.50 = 1366.60 \text{ kJ/kg}$$

At this h_4 value, the air table gives $T_4 = 1275.4$ K and $p_{r4} = 303.98$. For the internally reversible adiabatic process 3–4,

$$p_4 = p_3\left(\frac{p_{r4}}{p_{r3}}\right) = 300\left(\frac{303.98}{434.52}\right) = 209.9 \text{ kPa}$$

For the internally reversible adiabatic process 4–5,

$$p_{r5} = p_{r4}\left(\frac{p_5}{p_4}\right) = 303.98\left(\frac{50}{209.9}\right) = 72.41$$

At this p_{r5} value, the air table gives

$$T_5 = 892.1 \text{ K} \qquad \text{and} \qquad h_5 = 924.27 \text{ kJ/kg}$$

By neglecting the inlet velocity, the energy equation for the nozzle yields

$$V_5 = [2(h_4 - h_5)]^{1/2}$$

$$= [2(1000 \text{ J/kJ})(1366.60 - 924.27 \text{ kJ/kg})]^{1/2} = 940.6 \text{ m/s}$$

Accordingly, the mass-flow rate of air is given by

$$\dot{m} = \frac{F}{V_5 - V_0} = \frac{68000}{940.6 - 250} = 98.5 \text{ kg/s}$$

(b) Constant $c_p = 1.003$ kJ/kg·K and $k = 1.40$.

$$T_1 = T_0 + \frac{V_0^2}{2c_p} = (250 \text{ K}) + \frac{(250 \text{ m/s})^2}{2(1000 \text{ J/kJ})(1.003 \text{ kJ/kg·K})} = 281.2 \text{ K}$$

$$p_1 = p_0\left(\frac{T_1}{T_0}\right)^{k/(k-1)} = 50\left(\frac{281.2}{250}\right)^{1.4/0.4} = 75.5 \text{ kPa}$$

$$T_2 = T_1\left(\frac{p_2}{p_1}\right)^{(k-1)/k} = 281.2\left(\frac{300}{75.5}\right)^{0.4/1.4} = 417.1 \text{ K}$$

$$T_4 = T_3 - T_2 + T_1 = 1390 - 417.1 + 281.2 = 1254.1 \text{ K}$$

$$p_4 = p_3\left(\frac{T_4}{T_3}\right)^{k/(k-1)} = 300\left(\frac{1254.1}{1390}\right)^{1.4/0.4} = 209.3 \text{ kPa}'$$

$$T_5 = T_4\left(\frac{p_5}{p_4}\right)^{(k-1)/k} = 1254.1\left(\frac{50}{209.3}\right)^{0.4/1.4} = 833.1 \text{ K}$$

$$V_5 = [2c_p(T_4 - T_5)]^{1/2}$$

$$= [2(1000 \text{ J/kJ})(1.003 \text{ kJ/kg·K})(1254.1 - 833.1 \text{ K})]^{1/2}$$

$$= 919.0 \text{ m/s}$$

$$\dot{m} = \frac{F}{V_5 - V_0} = \frac{68,000}{919.0 - 250} = 101.6 \text{ kg/s}$$

Example 10-9

A turbojet aircraft flies at a steady speed of 800 ft/s in still air at an altitude where the atmospheric pressure and temperature are 8.0 psia and 0°F, respectively. The compressor pressure ratio is 11 and the maximum temperature in the cycle is 1920°F. The diffuser pressure coefficient is 90%, the adiabatic compressor efficiency is 84%, the adiabatic turbine efficiency is 88%, and the adiabatic nozzle efficiency is 95%. There is no pressure drop in the combustor and the nozzle exits at the local atmospheric pressure. Using air-standard analysis based on air-table data, determine the pressures and temperatures at the various locations in the jet engine and the exit-jet velocity, in ft/s.

Solution Referring to Fig. 10-25, the given data are $V_0 = 800$ ft/s, $p_0 = p_{5'} = 8.0$ psia, $T_0 = 0°F = 460°R$, $p_{2'}/p_{1'} = p_2/p_{1'} = 11$, $p_3 = p_{2'}$, $T_3 = 1920°F = 2380°R$, $K_p = 90\%$, $\eta_C = 84\%$, $\eta_T = 88\%$, and $\eta_N = 95\%$.

At $T_0 = 460°R$, the air table (Table A-9E) gives $h_0 = 109.98$ Btu/lbm and $p_{r0} = 0.7878$. At $T_3 = 2380°R$, the air table gives $h_3 = 611.45$ Btu/lbm, and $p_{r3} = 352.9$.

For the diffuser, assuming that the outlet velocity $V_{1'}$ (also V_1) is negligible compared to the inlet velocity V_0, the energy balances for the reversible adiabatic and irreversible adiabatic diffuser processes 0–1 and 0–1', respectively, are

$$h_1 + \tfrac{1}{2}\cancel{V}_1^{2\,0} = h_0 + \tfrac{1}{2}V_0^2$$

$$h_{1'} + \tfrac{1}{2}\cancel{V}_{1'}^{2\,0} = h_0 + \tfrac{1}{2}V_0^2$$

Combining these equations, we have

$$h_1 = h_{1'} = h_0 + \tfrac{1}{2}V_0^2$$

$$= 109.98 \text{ Btu/lbm} + \frac{(800 \text{ ft/s})^2}{2(32.2 \text{ ft·lbm/lbf·s}^2)(778 \text{ ft·lbf/Btu})}$$

$$= 122.75 \text{ Btu/lbm}$$

At this value of h, the air table gives

$$T_1 = T_{1'} = 513°R \qquad \text{and} \qquad p_{r1} = p_{r1'} = 1.1534$$

Therefore,

$$p_1 = p_0\left(\frac{p_{r1}}{p_{r0}}\right) = 8.0\left(\frac{1.1534}{0.7878}\right) = 11.7 \text{ psia}$$

The diffuser pressure coefficient K_p is defined as

$$K_p = \frac{p_{1'} - p_0}{p_1 - p_0}$$

so that

$$p_{1'} = p_0 + K_p(p_1 - p_0) = 8.0 + 0.90(11.7 - 8.0)$$

$$= 11.3 \text{ psia}$$

For the compressor,

$$p_{r2} = p_{r1'}\left(\frac{p_2}{p_{1'}}\right) = 1.1534 \times 11 = 12.69$$

At this p_{r2} value, the air table gives

$$T_2 = 1010°R \qquad \text{and} \qquad h_2 = 243.58 \text{ Btu/lbm}$$

Accordingly,

$$h_{2'} = h_{1'} + \frac{h_2 - h_{1'}}{\eta_C} = 122.75 + \frac{243.58 - 122.75}{0.84}$$

$$= 266.60 \text{ Btu/lbm}$$

At this $h_{2'}$ value, the air table gives $T_{2'} = 1102°R$. The pressure at the compressor outlet is

$$p_2 = p_{2'} = 11 \times p_{1'} = 11 \times 11.3 = 124.3 \text{ psia}$$

which is also the turbine inlet pressure p_3.

For the turbine, because

$$w_{\text{turb } 3\,4'} = w_{\text{comp } 1'\,2'}$$

or

$$h_3 - h_{4'} = h_{2'} - h_{1'}$$

we have

$$h_{4'} = h_3 - h_{2'} + h_{1'} = 611.45 - 266.60 + 122.75$$

$$= 467.60 \text{ Btu/lbm}$$

At this $h_{4'}$ value, the air table gives

$$T_{4'} = 1865°R \qquad \text{and} \qquad p_{r4'} = 130.71$$

We now use the value of η_T to evaluate the condition at the ideal turbine exit. Thus,

$$h_4 = h_3 - \frac{h_3 - h_{4'}}{\eta_T} = 611.45 - \frac{611.45 - 467.60}{0.88}$$

$$= 447.98 \text{ Btu/lbm}$$

At this h_4 value, the air table gives

$$T_4 = 1793°R \qquad \text{and} \qquad p_{r4} = 111.73$$

Accordingly, the pressure at the turbine outlet is

$$p_{4'} = p_4 = p_3\left(\frac{p_{r4}}{p_{r3}}\right) = 124.3\left(\frac{111.73}{352.9}\right) = 39.4 \text{ psia}$$

For the nozzle,

$$p_{r5} = p_{r4'}\left(\frac{p_5}{p_{4'}}\right) = 130.71\left(\frac{8.0}{39.4}\right) = 26.54$$

At this p_{r5} value, the air table gives

$$T_5 = 1234°R \qquad \text{and} \qquad h_5 = 300.09 \text{ Btu/lbm}$$

Accordingly,

$$h_{5'} = h_{4'} - \eta_N(h_{4'} - h_5)$$

$$= 467.60 - 0.95(467.60 - 300.09) = 308.47 \text{ Btu/lbm}$$

At this $h_{5'}$ value, the air table gives $T_{5'} = 1267°R$.

The pressures and temperatures at various locations in the engine cycle are

State point	Pressure, psia	Temperature,°R
0	8.0	460
1'	11.3	513
2'	124.3	1102
3	124.3	2380
4'	39.4	1865
5'	8.0	1267

By neglecting the inlet kinetic energy to the nozzle, an energy balance for the nozzle yields

$$h_{4'} + \tfrac{1}{2}\cancel{\mathbf{V}^2_{4'}}^{\,0} = h_{5'} + \tfrac{1}{2}\mathbf{V}^2_{5'}$$

so that

$$\mathbf{V}_{5'} = [2(h_{4'} - h_{5'})]^{1/2}$$

$$= [2(32.2 \text{ ft·lbm/lbf·s}^2)(778 \text{ ft·lbf/Btu})(467.60 - 308.47 \text{ Btu/lbm})]^{1/2}$$

$$= 2824 \text{ ft/s}$$

10-10 TURBOPROP, TURBOFAN, RAMJET, AND ROCKET ENGINES

In gas-turbine aircraft applications, there are a number of commonly used variations of design, one of which is the *turboprop* engine. As illustrated schematically in Fig. 10-27(a), the turbine drives not only the compressor, but also the propeller through a reduction gear. In general, the thrust of a turboprop engine comes mainly from the propeller, aided by only a small amount of jet effect. The proportion of contribution to the total thrust by the propeller and the jet, however, can be controlled by the designer. Figure 10-27(b) shows the T–s diagram for an ideal turboprop cycle, for which the following energy equation can be written:

$$\dot{W}_{turb} = \dot{W}_{comp} + \dot{W}_{prop}$$

or, for an air-standard analysis,

$$\dot{m}(h_3 - h_4) = \dot{m}(h_2 - h_1) + \dot{W}_{prop} \qquad (10\text{-}34)$$

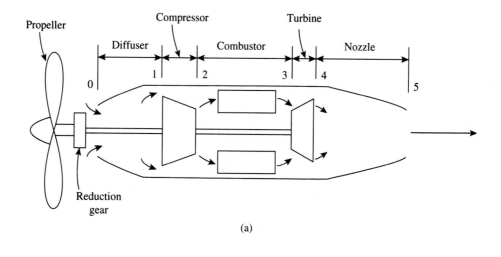

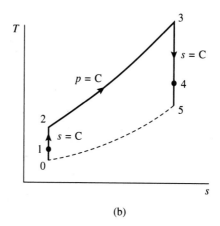

(b)

Figure 10-27 Turboprop engine: (a) schematic diagram, and (b) idealized T–s diagram.

It is significant to notice that propeller systems are capable of providing strong thrust at low flight speeds because these systems accelerate relatively large amounts of air through only modest velocities. On the other hand, turbojets are efficient at high flight speeds because these systems provide thrust by accelerating smaller amounts of air to much higher velocities. In intermediate speed ranges, the combination of turbojet and propeller is economical, particularly for long flights. Figure 10-28 shows schematically three hybrid arrangements, known as the turbofan engine. Figure 10-29 shows the cutaway view of a commercial turbofan. A *turbofan* is a jet engine with additional thrust from an axial-flow bypass or ducted fan. The air stream passing through the fan contributes to the thrust due to both the propeller action and the jet effect.

At very high flight speeds, the ram effect can be substantial and capable of being the sole compression effort for a jet engine. A machine operating on this principle is called a *ramjet*, in which there is no compressor or turbine. As illustrated in Fig.

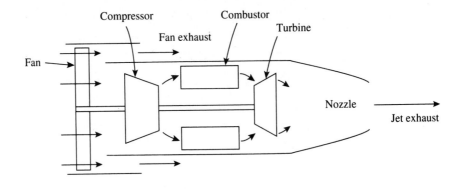

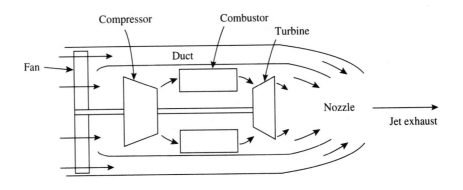

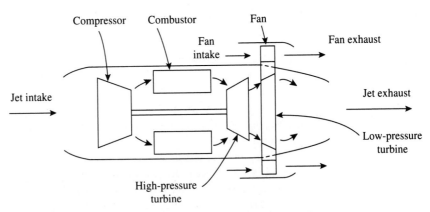

Figure 10-28 Schematic diagrams of turbofan arrangements.

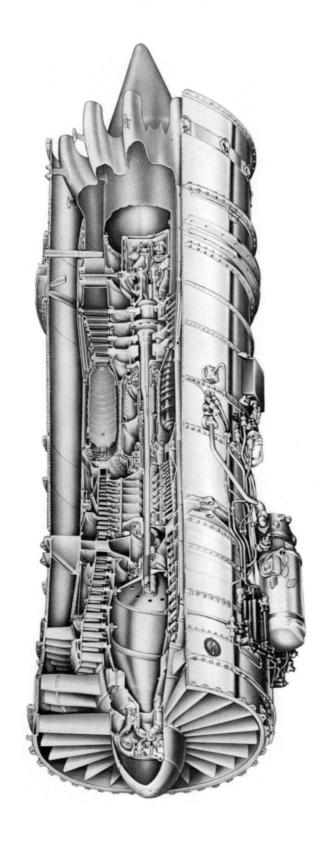

Figure 10-29 Cutaway view of a commercial turbofan. (Pratt & Whitney JT8D-200 Series turbofan, used for aircrafts such as Boeing 727 and 737 and McDonnell Douglas DC-9. Courtesy of United Technologies—Pratt & Whitney Co.)

10-30, atmospheric air is compressed in the diffuser and subsequently admitted into the combustion chamber, where liquid fuel is injected and burned continuously. The high-temperature products of combustion are then expanded through the propelling nozzle. The thrust of the engine is given by

$$F = \dot{m}(\mathbf{V}_3 - \mathbf{V}_0) \tag{10-35}$$

where $\dot{m}$ is the air mass-flow rate in the air-standard analysis. The ramjet cannot be operated until it has been accelerated by an external power supply. It can be carried aloft by a turbojet aircraft and ignited once the craft reaches operating speeds for high-altitude, long-distance cruising.

A *rocket* is a jet-propulsion device that carries both fuel and an oxidizer required to ensure fuel combustion. The fuel and oxidizer (called propellants) are supplied to the combustion chamber at high pressure. The high-temperature products of combustion are then expanded through a nozzle to high velocities, thereby producing thrust.

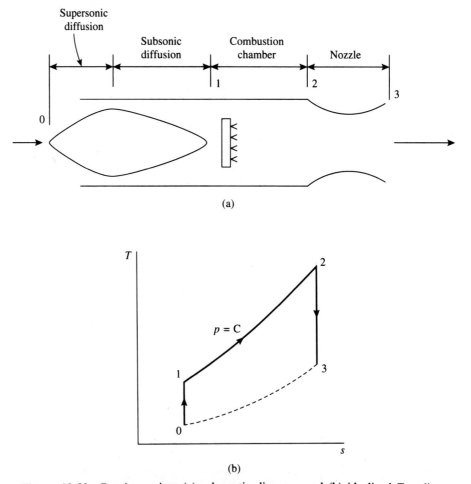

Figure 10-30 Ramjet engine: (a) schematic diagram, and (b) idealized T–s diagram.

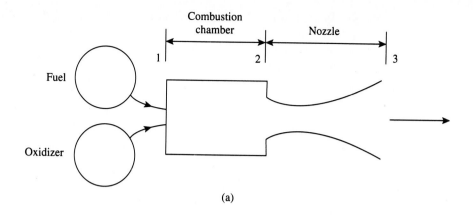

(a)

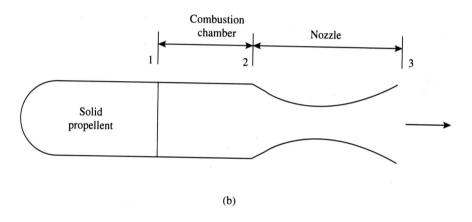

(b)

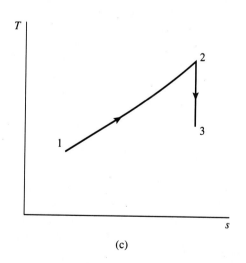

(c)

Figure 10-31 Rockets: (a) liquid-propellant, (b) solid-propellant, and (c) T–s diagram.

Rocket engines can be operated with either liquid propellants or solid propellants. Figure 10-31(a) shows diagrammatically the elements of a liquid-propellant rocket engine, where the fuel and oxidizer propellants are stored in separate tanks. Figure 10-31(b) shows diagrammatically a solid-propellant rocket. Solid propellants are mixtures that contain the fuel and the oxidizer required for combustion. The solid propellant is stored within the combustion chamber. Figurte 10-31(c) shows an idealized $T–s$ diagram for a rocket engine. The thrust produced by the rocket engine is given by

$$F = \dot{m}\mathbf{V}_3 \tag{10-36}$$

where $\dot{m}$ is the total mass-flow rate leaving the exhaust nozzle. It is significant to note that because a rocket does not depend on atmospheric oxygen for its operation and its thrust is independent of its environment, it can power a flight in the vacuum of space.

10-11 STIRLING AND ERICSSON CYCLES

In Chapter 6, we learned that the Carnot cycle is a totally reversible heat-engine cycle with the highest possible thermal efficiency when operating between two fixed temperatures. This cycle, however, is highly theoretical and not practical. We now study two other totally reversible cycles of somewhat more practical importance. They are external-combustion engines capable of achieving the Carnot efficiency by means of regeneration.

First, consider the *Stirling cycle,* as depicted in Fig. 10-32 on the idealized $p–v$ and $T–s$ diagram, together with a possible schematic arrangement of operation. The working fluid is air and the machine is sometimes called a hot-air engine. The engine consists of a cylinder with a piston at each end and a regenerator in the middle. The regenerator can be a porous metallic or ceramic matrix that has a large internal surface area for transverse heat transfer and a large heat capacity for energy storage. The design of the regenerator should be such that the longitudinal heat conduction is minimum so as to maintain a temperature gradient in the direction of the cylinder axis. The four different piston-position diagrams in Fig. 10-32(c) identified as 1, 2, 3, and 4 correspond to the four state points similarly marked on the $p–v$ and $T–s$ diagrams in the figure. In process 1–2, the air is expanded isothermally, receiving heat from the high-temperature energy reservoir at T_H. In process 2–3, the air is passed from left to right through the regenerator to which it transfers sufficient heat to lower its temperature to T_L. In process 3–4, the air is compressed isothermally, rejecting heat to the low-temperature reservoir at T_L. In process 4–1, the air is passed from right to left through the regenerator from which it receives sufficient heat to bring it to temperature T_H. In this cycle, the heat rejected by the air to the regenerator in process 2–3 is equal in magnitude to the heat received by the air from the regenerator in process 4–1, as shown by the cross-hatched areas under process curves 2–3 and 4–1 on the $T–s$ diagram. Thus, the only exchanges of heat between the air in the engine and the surroundings are those involving the heat source at T_H and the heat sink at T_L. As a result, the thermal efficiency of a Stirling cycle will equal that of a Carnot cycle operating between the same two fixed temperatures T_H and T_L, or

$$\eta_{\text{th}} = 1 - \frac{T_L}{T_H}$$

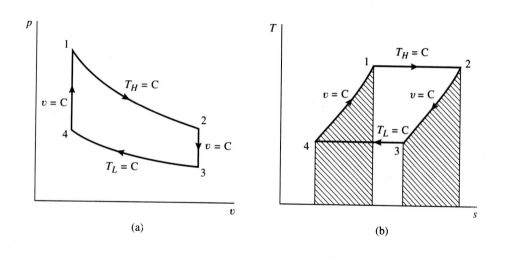

(a)

(b)

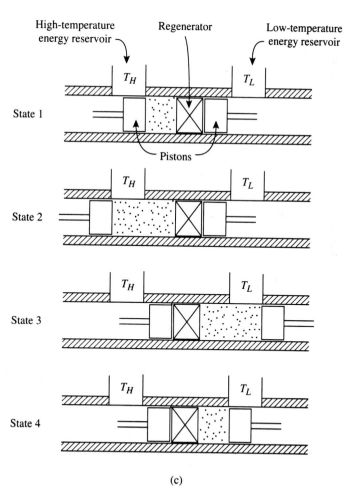

High-temperature
energy reservoir

Regenerator

Low-temperature
energy reservoir

T_H

T_L

State 1

Pistons

T_H

T_L

State 2

T_H

T_L

State 3

T_H

T_L

State 4

(c)

Figure 10-32 Stirling cycle: (a) p–v diagram, (b) T–s diagram, and (c) schematic diagram.

The *Ericsson cycle* is similar to the Stirling cycle except that the heat exchanges in the regenerator take place at constant pressures, as depicted by the schematic and idealized $p–v$ and $T–s$ diagrams of Fig. 10-33. Again, because the only heat exchanges with the surroundings are those involving the heat source at T_H and the heat sink at T_L, the thermal efficiency of an Ericsson cycle equals that of a Carnot cycle operating between the same two fixed temperatures. The application of the Ericsson cycle is most appropriate to rotating machinery. As mentioned in Sec. 10-8, this cycle can be approached

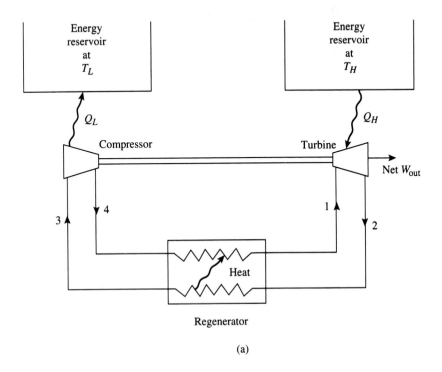

(a)

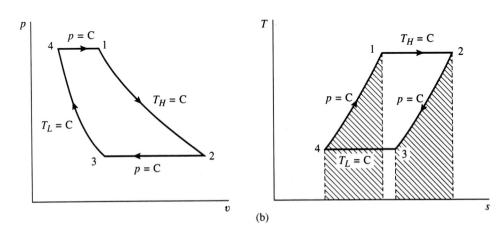

(b)

Figure 10-33 Ericsson cycle: (a) schematic diagram, and (b) $p–v$ and $T–s$ diagrams.

by an ideal regenerative gas-turbine cycle with many numbers of reheated expansion stages and intercooled compression stages.

10-12 AVAILABILITY ANALYSIS FOR GAS POWER CYCLES

The determination of availability change and irreversibility for various processes in a thermodynamic cycle will reveal the degree of perfection of the processes, thus dictating where efforts should be made in improving the system performance.

The ideal Carnot, Stirling, and Ericsson cycles are totally reversible cycles and do not involve any irreversibilities. The ideal Otto, Diesel, and Brayton cycles, however, are only internally reversible cycles and may involve irreversibilities external to the system. Actual gas cycles of any kind may involve internal as well as external irreversibilities.

Equations for availability and irreversibility have been developed in Chapter 8. Here is a summary of those equations.

The irreversibility for a process of a closed system in changing from initial state i to final state f while receiving a quantity of heat Q_R from an energy reservoir at T_R is given by

$$I = T_0(\Delta S_{\text{system}}^{0} + \Delta S_{\text{surr}})$$

$$= T_0\left[m(s_f - s_i) + \frac{-Q_R}{T_R} \right]$$

For an open system undergoing a steady-flow, steady-state process from inlet condition 1 to exit condition 2, while receiving a quantity of heat q_R per unit mass flow from an energy reservoir at T_R, the specific irreversibility is given by

$$i = T_0(\Delta s_{\text{system}}^{0} + \Delta s_{\text{surr}})$$

$$= T_0\left[(s_2 - s_1) + \frac{-q_R}{T_R} \right]$$

The irreversibility of a cycle is the sum of the irreversibilities of the processes that form the cycle. That is,

$$i_{\text{cycle}} = \sum i_{\text{process}}$$

The irreversibility of a cycle can also be determined by considering the entire cycle as a single process. Because the initial and final states of a cycle are identical, $\Delta s_{\text{system}} = 0$, we then have

$$i_{\text{cycle}} = T_0(\Delta s_{\text{system}}^{0} + \Delta s_{\text{surr}})$$

$$= T_0\left[\sum \left(\frac{-q_R}{T_R} \right) \right]$$

where q_R is positive when heat is added to the system.

The specific availability φ of a closed system in a given state is given by

$$\varphi = (u + p_0 v - T_0 s) - (u_0 + p_0 v_0 - T_0 s_0)$$

The specific stream availability ψ of a steady-flow fluid stream in a given state is given by

$$\psi = (h + \tfrac{1}{2}\mathbf{V}^2 + gz - T_0 s) - (h_0 + g s_0 - T_0 s_0)$$

where the subscript 0 denotes the dead state.

The second-law effectiveness of a power cycle can be defined as

$$\epsilon'_{\text{power cycle}} = \frac{\text{net work out}}{\text{availability increase due to heat addition}}$$

The procedure of calculation and the implication of the calculated results for the case of gas power cycles can be illustrated in the following examples.

Example 10-10

An air-standard intercooled regenerative gas-turbine cycle has the temperature data as indicated in the schematic diagram of Fig. 10-34. The initial pressure is 101.3 kPa and the pressure ratio of each compressor is 2.5. The turbine and compressors are adiabatic, and there is no pressure drop in the intercooler, the regenerator, and the combustion chamber. The atmosphere is at 27°C and 101.3 kPa. Neglecting kinetic and potential energies, (a) determine the cycle thermal efficiency and (b) make an availability accounting per kilogram of air-flow.

Solution Given are $T_1 = 27°C = 300$ K, $p_1 = 101.3$ kPa, $T_2 = 147°C = 420$ K, $T_3 = 60°C = 333$ K, $T_4 = 190°C = 463$ K, $T_5 = 400°C = 673$ K, $T_6 = 870°C = 1143$K, $T_7 = 510°C = 783$ K, $p_2/p_1 = p_4/p_3 = 2.5$ $T_0 = 27°C = 300$ K, and $p_0 = 101.3$ kPa. Thus,

$$p_2 = 2.5 p_1 = 2.5(101.3) = 253.3 \text{ kPa}$$

$$p_4 = 2.5 p_3 = 2.5 p_2 = 2.5(253.3) = 633.3 \text{ kPa}$$

$$p_6/p_7 = p_4/p_1 = 633.3/101.3 = 6.25$$

From the air table (Table A-9M) the following data are obtained:

at $T_1 = T_0 = 300$ K, $h_1 = h_0 = 300.43$ kJ/kg, $\phi_1 = \phi_0 = 5.7016$ kJ/kg·K
at $T_2 = 420$ K, $h_2 = 421.54$ kJ/kg, $\phi_2 = 6.0411$ kJ/kg·K
at $T_3 = 333$ K, $h_3 = 333.61$ kJ/kg, $\phi_3 = 5.8065$ kJ/kg·K
at $T_4 = 463$ K, $h_4 = 465.35$ kJ/kg, $\phi_4 = 6.1404$ kJ/kg·K
at $T_5 = 673$ K, $h_5 = 684.59$ kJ/kg, $\phi_5 = 6.5303$ kJ/kg·K
at $T_6 = 1143$ K, $h_6 = 1211.07$ kJ/kg, $\phi_6 = 7.1213$ kJ/kg·K
at $T_7 = 783$ K, $h_7 = 803.52$ kJ/kg, $\phi_7 = 6.6940$ kJ/kg·K

An energy balance for the regenerator gives

$$h_7 - h_8 = h_5 - h_4$$

or

$$h_8 = h_7 - h_5 + h_4 = 803.52 - 684.59 + 465.35 = 584.28 \text{ kJ/kg}$$

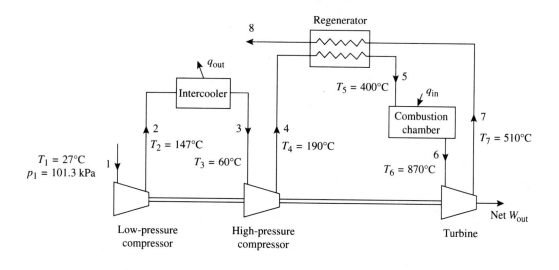

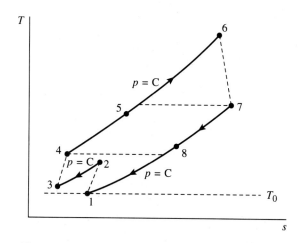

Figure 10-34 Schematic and T–s diagrams for Example 10-10.

At this h_8 value, Table A-9M gives

$$T_8 = 578 \text{ K} = 305°\text{C} \quad \text{and} \quad \phi_8 = 6.3697 \text{ kJ/Kg·K}$$

For air, the gas constant is

$$R = 8.314/29 = 0.287 \text{ kJ/kg·K}$$

The entropy changes for the processes are calculated as

$$s_2 - s_1 = \phi_2 - \phi_1 - R \ln (p_2/p_1)$$

$$= 6.0411 - 5.7016 - 0.287 \ln 2.5 = 0.07652 \text{ kJ/kg·K}$$

$$s_4 - s_3 = \phi_4 - \phi_3 - R \ln (p_4/p_3)$$

$$= 6.1404 - 5.8065 - 0.287 \ln 2.5 = 0.07092 \text{ kJ/kg·K}$$

$$s_2 - s_3 = \phi_2 - \phi_3 - R \ln (p_2/p_3)$$
$$= 6.0411 - 5.8065 - 0 = 0.2346 \text{ kJ/kg·K}$$

$$s_5 - s_4 = \phi_5 - \phi_4 - R \ln (p_5/p_4)$$
$$= 6.5303 - 6.1404 - 0 = 0.3899 \text{ kJ/kg·K}$$

$$s_6 - s_5 = \phi_6 - \phi_5 - R \ln (p_6/p_5)$$
$$= 7.1213 - 6.5303 - 0 = 0.5910 \text{ kJ/kg·K}$$

$$s_7 - s_6 = \phi_7 - \phi_6 - R \ln (p_7/p_6)$$
$$= 6.6940 - 7.1213 - 0.287 \ln (1/6.25) = 0.09865 \text{ kJ/kg·K}$$

$$s_7 - s_8 = \phi_7 - \phi_8 - R \ln (p_7/p_8)$$
$$= 6.6940 - 6.3697 - 0 = 0.3243 \text{ kJ/kg·K}$$

$$s_8 - s_1 = \phi_8 - \phi_1 - R \ln (p_8/p_1)$$
$$= 6.3697 - 5.7016 - 0 = 0.6681 \text{ kJ/kg·K}$$

(a) The net work output of the cycle is

$$\text{net } w_{\text{out}} = (h_6 - h_7) - (h_2 - h_1) - (h_4 - h_3)$$
$$= (1211.07 - 803.52) - (421.54 - 300.43) - (465.35 - 333.61)$$
$$= 154.7 \text{ kJ/kg}$$

or

$$\text{net } w_{\text{out}} = \text{net } q_{\text{in}} = (h_6 - h_5) - (h_8 - h_1) - (h_2 - h_3)$$
$$= (1211.07 - 684.59) - (584.28 - 300.43) - (421.54 - 333.61)$$
$$= 154.7 \text{ kJ/kg}$$

The heat added to the cycle is

$$q_{\text{in}} = h_6 - h_5 = 1211.07 - 684.59 = 526.5 \text{ kJ/kg}$$

The cycle thermal efficiency is

$$\eta_{\text{th}} = \frac{\text{net } w_{\text{out}}}{q_{\text{in}}} = \frac{154.7}{526.5} = 0.294 = 29.4\%$$

(b) Because changes in kinetic and potential energies are neglected, the changes in stream availability for a process is given by Eq. (8-20):

$$\Delta\psi = \Delta h - T_0 \, \Delta s$$

Thus,

$$\Delta\psi_{56} = h_6 - h_5 - T_0(s_6 - s_5)$$
$$= 1211.07 - 684.59 - 300(0.5910) = 349.2 \text{ kJ/kg}$$

$$\Delta\psi_{18} = h_8 - h_1 - T_0(s_8 - s_1)$$
$$= 584.28 - 300.43 - 300(0.6681) = 83.4 \text{ kJ/kg}$$

$$\Delta\psi_{32} = h_2 - h_3 - T_0(s_2 - s_3)$$
$$= 421.54 - 333.61 - 300(0.2346) = 17.6 \text{ kJ/kg}$$

The irreversibility of the turbine, the compressors, and the regenerator are calculated by Eq. (8-12):

$$i = T_0(\Delta s_{\text{system}} + \Delta s_{\text{surr}})$$

$$= T_0(\Delta s_{\text{surr}}) \text{ for steady-flow processes}$$

Thus,

$$i_{\text{turbine}} = i_{67} = T_0(s_7 - s_6) = 300(0.09865)=29.6 \text{ kJ/kg}$$

$$i_{\text{LP comp}} = i_{12} = T_0(s_2 - s_1)$$

$$= 300(0.07652) = 22.9 \text{ kJ/kg}$$

$$i_{\text{HP comp}} = i_{34} = T_0(s_4 - s_3)$$

$$= 300(0.07092) = 21.3 \text{ kJ/kg}$$

$$i_{\text{reg}} = T_0[(s_5 - s_4) + (s_8 - s_7)]$$

$$= 300[0.3899 + (-0.3243)]$$

$$= 19.7 \text{ kJ/kg}$$

The increase in stream availability $\Delta\psi_{56} = 349.2$ kJ/kg in the combustion chamber is disposed of as the sum:

net $w_{\text{out}} + \Delta\psi_{18} + \Delta\psi_{32} + i_{67} + i_{12} + i_{34} + i_{\text{reg}}$

$$= 154.7 + 83.4 + 17.6 + 29.6 + 22.9 + 21.3 + 19.7$$

$$= 349.2 \text{ kJ/kg}$$

The availability accounting is shown graphically in Fig. 10-35. It is seen that a large portion of the availability increase from the combustion process is carried away and wasted in the exhaust gases, or is simply destroyed by irreversibilities in the various machine parts. Possible improvements in the cycle performance should include utilization of the energy contained in the exhaust gases and better design of the machine elements to reduce irreversibility.

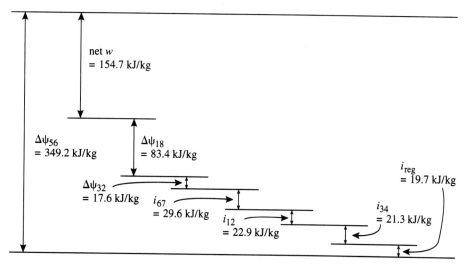

Figure 10-35   Availability accounting for Example 10-10.

Example 10-11

For the air-standard Otto cycle described in Example 10-1, determine (a) the irreversibility associated with each of the four processes, (b) the irreversibility of the cycle, and (c) the availability of the exhaust gases at the end of the power stroke. Assume that the high-temperature source is at 2600 K and the low-temperature sink is at 290 K. The atmosphere is at 100 kPa and 290 K.

Solution From Example 10-1, with reference to Fig. 10-3, we have the following data: $p_1 = 100$ kPa, $T_1 = 290$ K, $v_1 = 0.831$ m³/kg, $u_1 = 207.2$ kJ/kg, $p_2 = 1798$ kPa, $T_2 = 652$ K, $v_2 = 0.104$ m³/kg, $u_2 = 475.2$ kJ/kg, $p_3 = 6643$ kPa, $T_3 = 2409$ K, $u_3 = 2075$ kJ/kg, $p_4 = 441$ kPa, $T_4 = 1280$ K, $u_4 = 1004$ kJ/kg, $q_{in,23} = u_3 - u_2 = 1600$ kJ/kg, $q_{out,41} = u_4 - u_1 = 797$ kJ/kg, and net $w_{out} = 803$ kJ/kg. Additional data for this problem: $T_H = 2600$ K, $T_L = 290$ K, $p_0 = 100$ kPa, and $T_0 = 290$ K. From the air table (Table A-9M), we obtain the following:

$$\phi_1 = 5.6676 \text{ kJ/kg·K at } T_1 = 290 \text{ K}$$

$$\phi_2 = 6.4966 \text{ kJ/kg·K at } T_2 = 652 \text{ K}$$

$$\phi_3 = 8.0327 \text{ kJ/kg·K at } T_3 = 2409 \text{ K}$$

$$\phi_4 = 7.2544 \text{ kJ/kg·K at } T_4 = 1280 \text{ K}$$

Thus,

$$
\begin{aligned}
s_3 - s_2 &= \phi_3 - \phi_2 - R \ln (p_3/p_2) \\
&= 8.0327 - 6.4966 - (8.314/29) \ln (6643/1798) \\
&= 1.1614 \text{ kJ/kg·K} \\
s_1 - s_4 &= \phi_1 - \phi_4 - R \ln (p_1/p_4) \\
&= 5.6676 - 7.2544 - (8.314/29) \ln (100/441) \\
&= -1.1614 \text{ kJ/kg·K}
\end{aligned}
$$

Notice that $s_3 - s_2 = s_4 - s_1$.

(a) Processes 1–2 and 3–4 are internally reversible and adiabatic (isentropic with $s_1 = s_2$, $s_3 = s_4$, $q_{12} = 0$, and $q_{34} = 0$). They do not involve any internal or external irreversibilities, that is, $i_{12} = 0$ and $i_{34} = 0$.

Processes 2–3 and 4–1 are constant-volume heat-addition and heat-rejection processes, respectively. They are internally reversible. But the heat transfer between the working fluid and the thermal reservoirs involves finite temperature differences, thus making both processes externally irreversible. The irreversibility associated with each of these processes is calculated by Eq. (8-12):

$$i = T_0(\Delta s_{system} + \Delta s_{surr})$$

Thus,

$$= T_0 \left[(s_{final} - s_{initial})_{system} + \frac{q_{reservoir}}{T_{reservoir}} \right]$$

$$
\begin{aligned}
i_{23} &= T_0 (s_3 - s_2 - q_{23}/T_H) \\
&= 290 (1.1614 - 1600/2600) = 158.3 \text{ kJ/kg} \\
i_{41} &= T_0 (s_1 - s_4 + q_{41}/T_L) \\
&= 290 (-1.1614 + 797/290) = 460.2 \text{ kJ/kg}
\end{aligned}
$$

(b) The irreversibility of the cycle is calculated by

$$i_{cycle} = i_{12} + i_{23} + i_{34} + i_{41}$$

$$= 0 + 158.3 + 0 + 460.2 = 618.5 \text{ kJ/kg}$$

The irreversibility of the cycle can also be calculated by considering the entire cycle as a single process. Equation (8-12) then gives

$$i_{cycle} = T_0(\Delta s_{surr}) = T_0 \left(-\frac{q_{23}}{T_H} + \frac{q_{41}}{T_L} \right)$$

$$= 290 \left(-\frac{1600}{2600} + \frac{797}{290} \right) = 618.5 \text{ kJ/kg}$$

(c) The availability of the gases at the end of the power stroke (state 4) is calculated by Eq. (8-15):

$$\varphi_4 = (u_4 - u_0) + p_0(v_4 - v_0) - T_0(s_4 - s_0)$$

For this problem, $p_1 = p_0$ and $T_1 = T_0$, we have then $u_0 = u_1$, $v_0 = v_1$, and $s_0 = s_1$. Therefore,

$$\varphi_4 = (u_4 - u_1) + p_0(v_4 - v_1) - T_0(s_4 - s_1)$$

$$= (1004 - 207.2) + 0 - 290(1.1614) = 460.0 \text{ kJ/kg}$$

Because state 1 in this problem is a dead state, the availability φ_4 is equivalent to the irreversibility of process 4–1. It is imperative to notice that the availability of the exhaust gases (460 kJ/kg) is significant when compared to the net work output of the cycle (803 kJ/kg). The work potential of 460 kJ/kg in this cycle is simply lost.

Example 10-12

For the air-standard dual cycle described in Example 10-4, determine the irreversibility of the cycle, in Btu/lbm, assuming that the high-temperature source is at 5000°R and the low-temperature sink is at 530°R. Give an account for the available energy supplied from the high-temperature reservoir. The atmosphere is at 14.6 psia and 530°R. Use $c_p = 0.240$ Btu/lbm·°F and $c_v = 0.171$ Btu/lbm·°F for air.

Solution Referring to Fig. 10-9, from Example 10-4, we have the following data: $q_{in, 23} = 201.9$ Btu/lbm, $q_{in, 34} = 471.2$ Btu/lbm, $q_{out, 51} = 0.171(1975 - 530) = 247.1$ Btu/lbm, and

$$\text{net } w_{out} = q_{in, 23} + q_{in, 34} - q_{out, 51} = 201.9 + 471.2 - 247.1$$

$$= 426.0 \text{ Btu/lbm}$$

Considering the entire cycle as a single process, the irreversibility of the cycle is given by Eq. (8-12):

$$i_{cycle} = T_0(\Delta s_{system}^{\,0} + \Delta s_{surr})$$

$$= T_0 \left(-\frac{q_{23}}{T_H} - \frac{q_{34}}{T_H} + \frac{q_{51}}{T_L} \right)$$

$$= 530 \left(-\frac{201.9}{5000} - \frac{471.2}{5000} + \frac{247.1}{530} \right)$$

$$= 175.8 \text{ Btu/lbm}$$

The available energy of the heat supplied from the high-temperature reservoir is

$$(q_{23} + q_{34})\left(1 - \frac{T_0}{T_H}\right) = (201.9 + 471.2)\left(1 - \frac{530}{5000}\right)$$

$$= 601.8 \text{ Btu/lbm}$$

This available energy is disposed of as the sum:

$$\text{net } w_{\text{out}} + i_{\text{cycle}} = 426.0 + 175.8 = 601.8 \text{ Btu/lbm}$$

10-13 SUMMARY

In an air-standard analysis, the working substance is assumed to be air, which behaves as an ideal gas, and the combustion and exhaust processes are replaced by heat-addition and heat-rejection processes, respectively. When air is treated as an ideal gas with constant specific heats at room temperature, such an analysis is called the cold-air-standard analysis.

The Otto cycle is the idealized prototype of the speark-ignition reciprocating internal-combustion engine, consisting of four internally reversible processes: isentropic compression, constant-volume heat addition, isentropic expansion, and constant-volume heat rejection. For a cold-air-standard analysis, the thermal efficiency of an ideal Otto cycle is

$$\eta_{\text{th}} = 1 - \frac{1}{r^{k-1}}$$

where r is the compression ratio, defined as the ratio of maximum volume to minimum volume, and k is the specific-heat ratio c_p/c_v. Notice that this equation is good only when the specific heats are constants, and is not valid in variable-specific-heat (air-table) air-standard analyses.

The Diesel cycle is the idealized prototype of the compression-ignition reciprocating internal-combustion engine, consisting of four internally reversible processes: isentropic compression, constant-pressure heat addition, isentropic expansion, and constant-volume heat rejection. For a cold-air-standard analysis, the thermal efficiency of an ideal Diesel cycle is

$$\eta_{\text{th}} = 1 - \frac{1}{r^{k-1}}\left[\frac{r_c^k - 1}{k(r_c - 1)}\right]$$

where r_c is the cutoff ratio, defined as the ratio of the volumes after and before the combustion process. Notice that this equation is good only when the specific heats are constants.

The mean effective pressure (mep) of reciprocating engines is defined as

$$\text{mep} = \frac{\text{net work output of the cycle}}{\text{piston displacement}}$$

The Brayton cycle is a simple air-standard gas-turbine cycle, consisting of four internally reversible processes: isentropic compression in a compressor, constant-pressure heat addition, isentropic expansion in a turbine, and constant-pressure heat rejection. For a cold-air-standard analysis, the thermal efficiency of an ideal Brayton cycle is

$$\eta_{th} = 1 - \frac{1}{r_p^{(k-1)/k}}$$

where r_p is the compressor pressure ratio as defined by

$$r_p = p_{max}/p_{min}$$

and k is the specific-heat ratio, which must be a constant for the thermal efficiency equation given to be valid.

Gas-turbine cycles have many variations, including regeneration, multistage compression and expansion, intercooling and reheating, and turbojet propulsion.

Stirling and Ericsson cycles are two totally reversible cycles. They differ from the Carnot cycle in that the two isentropic processes are replaced by two constant-volume regeneration processes in the Stirling cycle and two constant-pressure regeneration processes in the Ericsson cycle. The thermal efficiencies of Carnot, Stirling, and Ericsson cycles are all the same, that is,

$$\eta_{th, \text{ Carnot, Stirling, or Ericsson}} = 1 - \frac{T_L}{T_H}$$

where T_H and T_L are the heat source and heat sink temperatures, respectively.

The irreversibility of a cycle can be determined by either of the following equations:

$$i_{cycle} = \sum i_{process}$$

$$i_{cycle} = T_0 \left[\sum \left(\frac{-q_R}{T_R} \right) \right]$$

The second-law effectiveness of a power cycle can be defined as

$$\epsilon'_{\text{power cycle}} = \frac{\text{net work out}}{\text{availability increase due to heat addition}}$$

PROBLEMS

10-1. At the beginning of the compression process in a cold-air-standard Otto cycle, the air is at 100 kPa and 300 K. The compression ratio is 5 and the heat added to the air is 750 kJ/kg. Considering air as an ideal gas with $c_v = 0.718$ kJ/kg·K and $k = 1.4$, determine the pressure, temperature, and specific volume for the four stages of the cycle. Determine also the heat transfer and work done per kilogram of air for each process and the cycle thermal efficiency.

10-2. An Otto cycle uses nitrogen as the working fluid. It operates with a compression ratio of 7 and the compression begins at 100°F and 14 psia. The maximum temperature of the cycle is 2000°F. Determine (a) the pressure and temperature of each state point, (b) the heat supplied per pound of the gas, (c) the work done per pound of the gas, (d) the thermal efficiency, and (e) the change in entropy of the heat-addition and heat-rejection processes. Use the ideal-gas nitrogen table for this problem.

10-3. An air-standard Otto cycle operates with a compression ratio of 7, and compression begins at 100°F, 14 psia. The maximum temperature of the cycle is 3000°F. Based on the values taken from the air table, determine (a) the heat supplied per pound of air, (b) the work done per pound of air, (c) the thermal efficiency, (d) the temperature at the end of the isentropic expansion, and (e) the maximum pressure of the cycle.

10-4. An ideal Otto cycle with air as the working fluid has a maximum temperature of 3000°C. At the beginning of the compression process, the air is at 100 kPa and 20°C. Determine the net work output per kilogram of air and the cycle thermal efficiency for a compression ratio of 5. Use $c_v = 0.718$ kJ/kg·K and $k = 1.4$.

10-5. At the beginning of the compression in an air-standard Otto cycle, the air is at 14 psia and 80°F. The compression ratio is 9. The heat added to the air is 800 Btu/lbm per cycle. Determine the maximum temperature and pressure, the thermal efficiency, and the mean effective pressure.

10-6. An ideal Otto cycle with air as the working fluid operates under the following conditions. At the beginning of the compression process, the air is at 14.5 psia, 80°F. The heat added to the air is 370 Btu/lbm during each cycle. The maximum temperature is 3400°R. Using $c_v = 0.172$ Btu/lbm·°R and $k = 1.4$ for air, determine (a) the pressure, temperature, and specific volume at the four state points of the cycle, (b) the compression ratio, (c) the thermal efficiency, and (d) the mean effective pressure.

10-7. A Diesel cycle uses nitrogen as the working fluid. At the beginning of the compression process, the gas is at 100 kPa and 20°C, and the maximum temperature of the cycle is 2000 K. Determine the net work output per kilogram of the gas and the cycle thermal efficiency for a compression ratio of 15.

10-8. A Diesel cycle uses nitrogen as the working fluid. The compression ratio is 18, and the heat input is 600 kJ/kg. The conditions at the beginning of the compression are 300 K and 100 kPa. Calculate the temperature, pressure, and specific volume at all points in the cycle, the cutoff ratio, and the thermal efficiency. Use the ideal-gas nitrogen table for this problem.

10-9. The pressure and temperature before compression in an air-standard Diesel cycle are 14.2 psia and 75°F, and the temperature before and after the heat-addition process are 800°F and 3000°F, respectively. Determine the thermal efficiency.

10-10. An air-standard Diesel cycle operates with a compression ratio of 15. The air before compression is at 14.7 psia and 60°F. The heat addition to the air per cycle is 800 Btu/lbm. Determine (a) the pressure and temperature at each point of the cycle, (b) the cutoff ratio, and (c) the thermal efficiency.

10-11. In an ideal Diesel cycle with air as the working fluid, the compression ratio is 18 and the heat input is 800 kJ/kg. The conditions at the start of compression are 25°C and 100 kPa. Calculate the temperature, pressure, and specific volume at all points in the cycle, the cutoff ratio, and the thermal efficiency.

10-12. Argon is used as the working substance in an ideal Diesel cycle. It operates with a compression ratio of 5, and the compression begins at 15 psia and 140°F. The maximum temperature of the cycle is 2540°F. Determine (a) the heat supplied per pound of gas, (b) the work done per pound of gas, (c) the cycle efficiency, (d) the temperature at the end of the isentropic expansion, (e) the cutoff ratio, and (f) the maximum pressure of the cycle.

10-13. An air-standard Diesel cycle operates with a compression ratio of 13 and compression begins at 15 psia and 140°F. The maximum temperature of the cycle is 2540°F. Using data from the air table, determine (a) the heat supplied per pound of air, (b) the work done per pound of air, (c) the thermal efficiency, (d) the temperature at the end of the isentropic expansion, (e) the cutoff ratio, and (f) the maximum pressure of the cycle.

10-14. At the beginning of the compression process in an ideal air-standard Diesel cycle, the air is at 100 kPa and 300 K. The compression ratio is 16 and the maximum temperature of the cycle is 2000 K. Considering air as an ideal gas with $c_p = 1.005$ KJ/kg·K and $c_v = 0.718$ kJ/kg·K, determine the pressure, temperature, and specific volume for the four stages of the cycle. Determine also the heat transfer and work done per kilogram of air for each process and the cycle thermal efficiency.

10-15. An air-standard four-stroke dual cycle operates with a compression ratio of 7. The cylinder diameter of the engine is 10 inches, and the stroke is 12 inches. The air at the start of compression is at 14.7 psia, 70°F. The pressure at the end of the constant-volume heating process is 800 psia. If heat is added at constant pressure during 3% of the stroke, determine (a) the net work of the cycle, (b) the thermal efficiency, (c) the amount of heat added that is available energy, and (d) the amount of heat rejected that is available energy. The atmosphere is at 14.7 psia and 70°F.

10-16. Nitrogen is used as the working fluid for an ideal dual cycle having a compression ratio of 14. At the beginning of the compression process, the gas is at 100 kPa and 300 K. At the end of the heat-addition process, the gas is at 2200 K. The heat added to the gas is 1550 kJ/kg. Taking data from the ideal-gas nitrogen table, determine (a) the fraction of heat addition at constant volume, and (b) the thermal efficiency of the cycle.

10-17. Air at 1.0 bar and 17°C enters the compressor of a simple ideal Brayton cycle that has a pressure ratio of 6. The turbine is limited to a temperature of 1000 K and the mass-flow rate is 3.5 kg/s. Determine (a) the compressor and turbine work, in kJ/kg, (b) the thermal efficiency, (c) the net power output, in kW, and (d) the volume-flow rate at the compressor inlet, in m^3/min. Use air-table data.

10-18. An air-standard Brayton cycle operates between a minimum temperature of 40°F and a maximum temperature of 1540°F. (a) Find the pressure ratio at which the cycle efficiency equals the Carnot efficiency. (b) Find the pressure ratio at which the work per pound of air is maximum. (c) Find the cycle efficiency for the condition as stated in (b). (d) Find the Carnot cycle efficiency working between the temperature limits of the Brayton cycle.

10-19. In a Brayton cycle using air as the working fluid, the compressor inlet pressure and temperature are 100 kPa and 300 K, the compressor pressure ratio is 12, and the compressor and turbine isentropic efficiencies are 85% and 90%, respectively. Determine the cycle thermal efficiency and the net work output per kilogram for a maximum cycle temperature of 1000 K.

10-20. A gas-turbine plant is assumed to operate on the simple Brayton closed cycle with 1 lbm of air ($k = 1.4$) as the working fluid. The air enters the compressor at 540°R and 14.7 psia. The pressure ratio of compression is 6 and the temperature of the air entering the turbine is 2000°R. The adiabatic-compressor efficiency is 0.82 and the adiabatic-turbine efficiency is 0.85. Find (a) the work of the compressor, (b) the work of the turbine, (c) the heat supplied, (d) the thermal efficiency, (e) the temperature of the air leaving the compressor, and (f) the temperature of the air leaving the turbine.

10-21. Air enters a simple Brayton cycle at 15 psia and 80°F. The pressure after compression is 85 psia and the temperature after combustion is 1250°F. The turbine is adiabatic and 82% efficient. The compressor is adiabatic and 84% efficient. Determine (a) the thermal efficiency, (b) the rate of air flow (in lbm/min) to generate 3500 kW net, (c) the cubic feet per minute entering the compressor, (d) the gross kilowatt output of the turbine, and (e) the kilowatts absorbed by the compressor.

10-22. Nitrogen is used as the working fluid in a simple Brayton cycle. The pressure and temperature be-

fore compression are 15 psia and 80°F, respectively, the compressor pressure ratio is 6.25, and the maximum temperature of the cycle is 1440°F. The turbine and compressor adiabatic efficiencies are each 80%. Find the turbine exhaust temperature, the compressor work (Btu/lbm), the turbine work (Btu/lbm), the heat added (Btu/lbm), and the thermal efficiency. (a) Use constant specific heats for nitrogen. (b) Use the ideal-gas nitrogen table.

10-23. A two-stage compressor receives air at 14.2 psia, 80°F, and delivers air at 150 psia with an intercooler pressure of 46.2 psia. See Fig. P10-23. Water jacketing is used and it is assumed that compression follows the equation $pv^{1.32}$ = constant in each state. The temperature leaving the intercooler is 100°F. The capacity of the compressor is 200 ft^3/min measured at the compressor inlet. What is the compressor horsepower required?

10-24. Air enters a steady-flow, two-stage compressor at 14.5 psia, 60°F, with a volumetric flow rate of 100 ft^3/min, and exits at 150 psia. Compression is done under the following conditions: It is first compressed isentropically to 40 psia in the first stage; it is then cooled at constant pressure to 90°F in the intercooler; and it is finally compressed isentropically to 150 psia in the second stage. Use $k = 1.4$ for air. Determine the horsepower required.

10-25. An air-standard Brayton cycle has a regenerator with 75% effectiveness. The turbine and compressor processes are isentropic. The inlet temperature to the compressor is 60°F, the inlet temperature to the turbine is 1200°F, and the pressure ratio is 5. Calculate the turbine work, compressor work, heat added, heat rejected per pound of air, and the cycle efficiency. Use the air tables.

10-26. An air-standard Brayton regenerative cycle has a regenerator effectiveness of 80%. See Fig. P10-26. The turbine has an adiabatic efficiency of 85%, and the compressor has an adiabatic efficiency of 80%. The inlet temperature to the compressor is 300 K and the inlet temperature to the turbine is 1300 K. The pressure ratio is 8. Determine the exit temperature of the turbine and the thermal efficiency. Use the air tables to solve this problem.

10-27. The following enthalpy data (in kJ/kg) were taken on a test of an air standard Brayton regenerative cycle with a pressure ratio of 5.41. See Fig. P10-27. Determine (a) the thermal efficiency

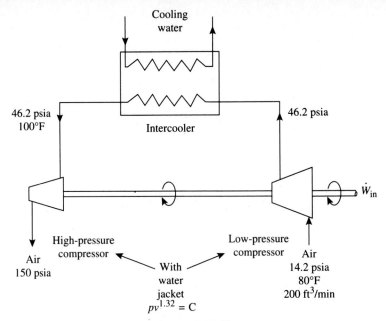

Figure P10-23

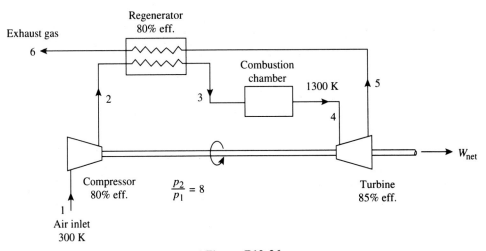

Figure P10-26

of the cycle, (b) the effectiveness of the regenerator, (c) the adiabatic efficiency of the compressor, and (d) the adiabatic efficiency of the turbine.

	Entering	Leaving
Compressor	$h_1 = 290.2$	$h_2 = 505.0$
Regenerator	$h_2 = 505.0$	$h_3 = 629.4$
Combustor	$h_3 = 629.4$	$h_4 = 1046.0$
Turbine	$h_4 = 1046.0$	$h_5 = 713.7$
Regenerator	$h_5 = 713.7$	$h_6 = 590.1$

10-28. A Brayton-cycle power plant takes in air at 15 psia and 80°F. See Fig. P10-28. Compression is in two stages with intercooling, and each stage has a pressure ratio of 2.38 and an efficiency of 84%, with intercooling to 80°F. The maximum temperature of the cycle is 1250°F. Turbine efficiency is 82%. The turbine and compressor stages are adiabatic. A regenerator heats the compressed air to 620°F before sending it to the combustor. Calculate the turbine work, compressor work, heat added, and heat rejected, per pound of air. What is the thermal efficiency of the cycle?

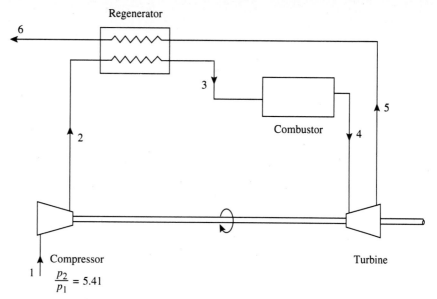

Figure P10-27

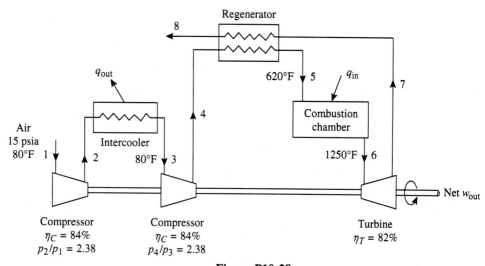

Figure P10-28

Use (a) constant specific heats for air, and (b) the ideal-gas air table.

10-29. A closed-cycle gas turbine using helium has an overall pressure ratio of 9. See Fig. P10-29. Compression is adiabatic in two stages, each with a pressure ratio of 3 and with intercooling between stages. Expansion is adiabatic in two stages, each with a pressure ratio of 3 and with reheating between stages. A regenerator with an effectiveness of 70% is included in the cycle. Gas enters the first compressor at 80°F and 100 psia, the second compressor at 80°F, and each turbine

at 1350°F. Compressors and turbines are all 86% efficient. For a net output of 50,000 kW, calculate (a) the thermal efficiency, (b) the cubic feet per minute entering the first compressor, and (c) the cubic feet per minute leaving the second turbine. Neglect all pressure losses. For helium, use $c_p = 1.25$ Btu/lbm·°R, $k = 1.66$, and $R = 386$ ft·lbf/lbm·°R.

10-30. A turbojet engine is flying at an altitude where the atmospheric pressure is 95 kPa and the atmospheric temperature is 7°C. See Fig. P10-30. The speed of the aircraft is 150 m/s. Assume isen-

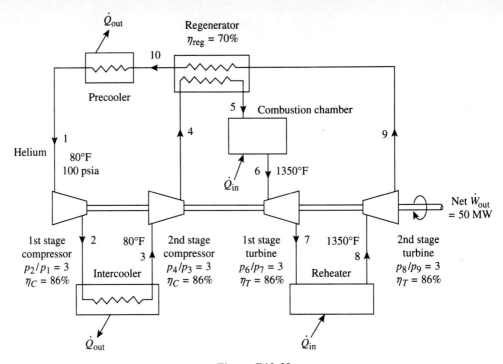

Figure P10-29

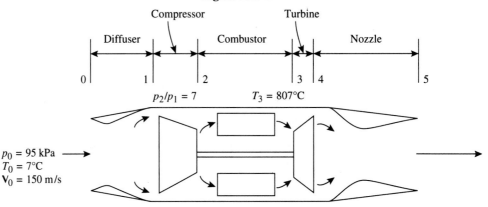

Figure P10-30

tropic flow in the diffuser, compressor, turbine, and nozzle, and neglect the pressure drop in the combustor. The turbine inlet temperature is 807°C. The compressor pressure ratio is 7. Using data from air tables, determine the pressures and temperatures at various locations in the jet engine, the engine exhaust velocity, and the thrust per unit-mass flow.

10-31. A turbojet engine is flying at an altitude where the atmospheric pressure and temperature are 50 kPa and 250 K, respectively. The speed of the aircraft is 280 m/s. The air is decelerated in the diffuser to zero velocity before entering the compressor.

The pressure ratio across the compressor is 11, and the maximum cycle temperature is 1400 K. Devise an variable-specific-heat, air-standard cycle (using the air tables) that approximates this cycle, and determine (a) the compressor work, in kJ/kg, (b) the pressure at the turbine outlet, in kPa, and (c) the velocity (relative to the aircraft) of the air leaving the exhaust nozzle, in m/s, assuming that it has been expanded to the atmospheric pressure.

10-32. A turbojet aircraft is flying at an altitude where the atmospheric pressure and temperature are 60 kPa and −13°C, respectively. The speed of the aircraft is 300 m/s, and the incoming air is decelerated to zero velocity by a diffuser before entering the compressor. The pressure ratio across the compressor is 14. The air is heated to 1500 K in the combustion chamber. It is then expanded in the turbine to such a pressure that the turbine work is just equal to the compressor work. Compression and expansion are all ideal. Determine the velocity of the air leaving the exhaust nozzle at 60 kPa, and the thrust per unit-mass flow. For a schematic diagram of a turbojet engine, see Fig. 10-23.

10-33. A turbojet engine on ground takes air at 14.7 psia and 40°F. The pressure of air at the exit from the compressor is 80 psia and the maximum temperature is 1140°F. The air expands in the turbine to such a pressure that the turbine work is just equal to the compressor work. Compression and expansion are all ideal (internally reversible and adiabatic). There is no pressure drop in the burner. Determine the velocity of the air leaving the nozzle. Use $c_p = 0.24$ Btu/lbm·°R and $k = 1.4$ for air. See Fig. 10-23 for a schematic of a turbojet.

10-34. A turbojet engine takes in air at 14.0 psia and 60°F while in a stationary position on the ground. See Fig. P10-34. After passing through the combustion chamber, the air enters the turbine at 60 psia and 1600°F. Compression and expansion processes are internally reversible and adiabatic. There is no pressure drop in the combustion chamber. Using data from the air tables, determine the velocity of the air leaving the exhaust nozzle at 14.0 psia.

10-35. A turbojet engine on the ground takes in air at 40°F and 14.7 psia. The pressure of the air at the exit from the compressor is 80 psia and the maximum temperature of the cycle is 1600°F. On leaving the turbine, the air expands in a nozzle to 14.7 psia. Compression and expansion processes are internally reversible and adiabatic. There is no pressure drop in the burner. Determine the velocity of the air leaving the nozzle and the specific impulse of the engine. For a schematic diagram of a turbojet engine, see Fig. 10-23.

10-36. While in a stationary position on the ground, a turbojet engine takes in air at 100 kPa and 22°C. The pressure ratio in the compressor is 5 and the upper temperature limit in the cycle is 987°C. The compressor and turbine adiabatic efficiencies are 80% and 85%, respectively, and the expansion in the exhaust nozzle can be assumed isentropic. There is no pressure drop in the combustor and the nozzle exhaust pressure is 100 kPa. Neglecting the inlet velocity, determine the thrust and the heat input per kilogram per second of air flow. (a) Use the air-table values. (b) Use $k = 1.4$ and $c_p = 1.004$ kJ/kg·K for air. For a schematic of a turbojet, see Fig. 10-23.

10-37. An air-standard turboprop engine that provides 1000 lbf of thrust in addition to 1000 hp to a propeller takes in air at 8 psia, 0°F, with a velocity of 600 ft/s. See Fig. P10-37. The velocity is also 600 ft/s at the compressor outlet, turbine inlet, and turbine exhaust. The maximum temperature is 1600°F, and the flow rate is not to exceed 20 lbm/s. Assuming that all processes are internally reversible, determine the required compressor pressure ratio.

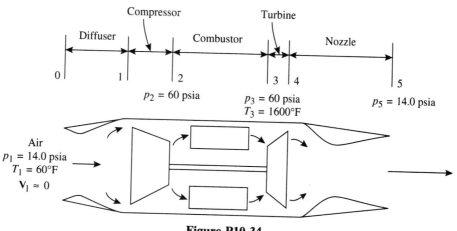

Figure P10-34

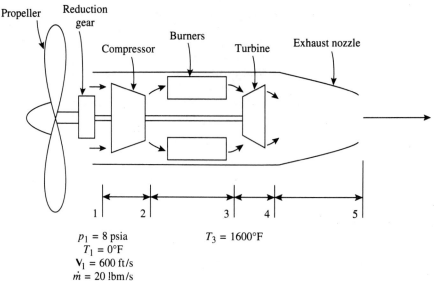

$p_1 = 8$ psia $T_3 = 1600°F$
$T_1 = 0°F$
$V_1 = 600$ ft/s
$\dot{m} = 20$ lbm/s

Figure P10-37

10-38. In a Stirling cycle with an ideal regenerator, the volume varies between 0.03 and 0.06 m³, the maximum pressure is 2 atm, and the temperature varies between 500 and 250°C. The working fluid is air. Calculate the net work, in Joules/cycle, and the cycle thermal efficiency. See Fig. 10-31 for a schematic of a Stirling cycle.

10-39. In a Stirling cycle, the volume varies between 1 and 2 ft³, the temperature varies between 1000 and 500°F, and the maximum pressure is 30 psia. The working fluid is air. (a) Find the thermal efficiency and the work per cycle for the simple cycle without a regenerator. (b) Find the thermal efficiency and the work per cycle for the cycle with an ideal regenerator, and compare with the Carnot cycle having the same isothermal heat supply process and the same temperature range. For a schematic of the cycle, see Fig. 10-31.

10-40. An air-standard Ericsson cycle operates under steady-flow conditions. See Fig. P10-40. Air is at 120 kPa and 300 K at the beginning of the isothermal compression process, during which 150 kJ/kg of heat is rejected. The air is at 1200 K during the isothermal heat-addition process. Determine the maximum pressure in the cycle, the net work output per unit mass of air, and the thermal efficiency of the cycle.

10-41. An Otto cycle utilizes carbon dioxide as the working fluid. At the beginning of the compression process, the fluid is at 100 KPa and 300 K. At the beginning of the expansion process, the fluid is at 3 MPa and 1400 K. The high-temperature and low-temperature reservoirs are at 1500 K and 300 K, respectively. The atmosphere is at 300 K. Determine (a) p, v, and T of each state point, (b) the heat addition and rejection, kJ/kg, (c) the net work output, kJ/kg, (d) the thermal efficiency, and (e) the irreversibility for the four processes of the cycle, kJ/kg.

10-42. An Otto cycle uses nitrogen as the working fluid and has a compression ratio of 8. At the end of the expansion process, the gas is at 0.6 MPa and 500 K. 150 KJ/kg of heat is transferred to the atmosphere during the heat-rejection process. The high-temperature and low-temperature reservoirs are at 1300 K and 300 K, respectively. Determine, in kJ/kg, the heat transfer during the heat-addition process, the net work output, the irreversibility for each of the four processes of the cycle, and the irreversibility of the cycle.

10-43. A Diesel cycle with nitrogen as the working fluid has a compression ratio of 18 and a cutoff ratio of 2.5. At the beginning of the compression process, the gas is at 100 kPa and 37°C. Determine, per kg of the gas, (a) the net work of the cycle, the heat added and heat rejected, and the thermal efficiency of the cycle, and (b) the irreversibility of each process of the cycle, and the irreversibility of the cycle, assuming that the high-temperature source is at 2500 K and the low-temperature sink is at 270 K. The atmosphere is at 270 K. Use ideal-gas nitrogen table to solve this problem.

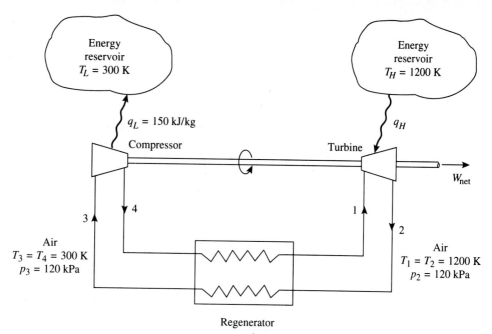

Energy reservoir $T_L = 300$ K

Energy reservoir $T_H = 1200$ K

$q_L = 150$ kJ/kg

q_H

Compressor

Turbine

W_{net}

4

1

3

2

Air
$T_3 = T_4 = 300$ K
$p_3 = 120$ kPa

Air
$T_1 = T_2 = 1200$ K
$p_2 = 120$ kPa

Regenerator

Figure P10-40

10-44. For the ideal dual cycle using nitrogen as the working fluid, described in Prob. 10-16, determine (a) the availability change of the various processes of the cycle, in kJ/kg, and (b) the irreversibility of the processes and the cycle, in kJ/kg, assuming that the high-temperature source is at 2500 K and the low-temperature sink is at 300 K. The atmosphere is at 100 kPa and 300 K. Use the ideal-gas nitrogen table to solve this problem.

10-45. An ideal Brayton cycle using nitrogen as the working fluid has a minimum pressure and temperature of 120 kPa and 27°C, respectively, and a maximum pressure and temperature of 1.2 MPa and 1400 K, respectively. Determine (a) the net work output per kg of the gas and the thermal efficiency, and (b) the irreversibility of the cycle if the high-temperature and low-temperature reservoirs are at 1500 K and 250 K, respectively.

10-46. An air-standard ideal Brayton cycle has a minimum pressure and temperature of 14.0 psia and 520°R, respectively, and a maximum pressure and temperature of 140 psia and 2700°R, respectively. The power output of the cycle is 20 MW. Determine (a) the power of the turbine, (b) the cycle thermal efficiency, and (c) the irreversibility of each process in the cycle if the high-temperature reservoir is at 2700°R and the low-temperature reservoir (the atmosphere) is at 520°R.

10-47. A Brayton cycle with a regenerator has a pressure

ratio of 14.5 and a regenerator effectiveness of 45%. At the compressor inlet, the air is at 300 K and 130 kPa, and at the turbine inlet, the air is at 1600 K. Determine, per kg of air, (a) the net work output and the thermal efficiency, and (b) the irreversibility of each process and the irreversibility of the cycle, if the high-temperature and low-temperature reservoirs are at 1800 K and 250 K, respectively.

10-48. A Brayton cycle with a regenerator uses nitrogen as the working fluid and has a pressure ratio of 15. See Fig. P10-48. The regenerator has an effectiveness of 60%. At the compressor inlet, the gas is at 300 K and 130 kPa, and at the turbine inlet, the gas is at 1600 K. The high-temperature and low-temperature reservoirs are at 1800 K and 300 K, respectively. The atmosphere is at 300 K. Determine (a) the heat addition and rejection, in kJ/kg, (b) the net work, in kJ/kg, (c) the cycle thermal efficiency, (d) the irreversibility of each process of the cycle, in kJ/kg, and (e) make an availability accounting per kgmole of nitrogen flow. Use the ideal-gas property table for nitrogen to solve this problem.

10-49. Make an availability accounting per pound of air flow for the intercooled two-stage compression, gas-turbine cycle as given in Prob. 10-28. The atmosphere is at 15 psia and 80°F. Use the ideal-gas air table to solve this problem.

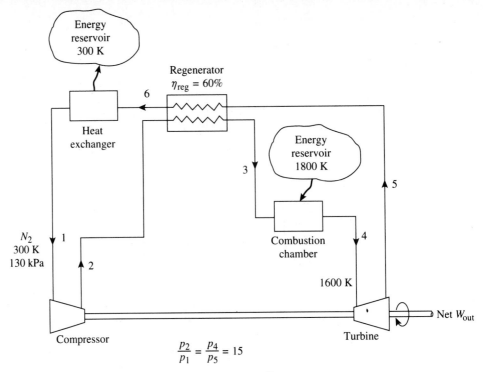

Regenerator
$\eta_{reg} = 60\%$

6

Heat
exchanger

Energy
reservoir
1800 K

3

5

N_2
300 K
130 kPa

1

2

Combustion
chamber

4

1600 K

Net W_{out}

Compressor

Turbine

$$\frac{p_2}{p_1} = \frac{p_4}{p_5} = 15$$

Figure P10-48

10-50. An air-standard Otto cycle has a maximum temperature of 3000°C. At the beginning of the compression process, the gas is at 100 kPa and 20°C. Write a computer program to determine the net work output per kilogram of air and the cycle thermal efficiency for a compression ratio of (a) 5, (b) 6, (c) 7, (d) 8, (e) 9, (f) 10, (g) 11, and (h) 12. Use a constant specific heat of $c_v = 0.718$ kJ/kg·K with $k = 1.4$ for air.

10-51. If computerized air-property data are available to you, rework Prob. 10-50 by the use of those data to take care of the variations of specific heats.

10-52. In an air-standard Brayton cycle, the compressor inlet pressure and temperature are 100 kPa and 300 K, respectively, the compressor pressure ratio is 12, and the compressor and turbine isentropic efficiencies are 85% and 90%, respectively. Write a computer program to determine the cycle thermal efficiency and the net work output per kilogram of air for a maximum cycle temperature of (a) 1000 K, (b) 1200 K, (c) 1400 K, (d) 1600 K, and (e) 1800 K. Assume $c_p = 1.005$ kJ/kg·K and $k = 1.40$ for air.

10-53. If computerized air property data are available to you, rework Prob. 10-52 by the use of those data to take care of the variations of specific heats.

10-54. A Diesel cycle uses nitrogen as the working fluid. At the beginning of the compression process, the gas is at 100 kPa and 20°C, and the maximum temperature of the cycle is 2000 K. Write a computer program to determine the net work output per kilogram of the gas and the cycle thermal efficiency for a compression ratio of (a) 10, (b) 11, (c) 12, (d) 13, (e) 14, (f) 15, (g) 16, (h) 17, (i) 18, (j) 19, and (k) 20. Use constant specific heats $c_p = 1.039$ kJ/kg·K and $c_v = 0.743$ kJ/kg·K for nitrogen.

10-55. In an air-standard Brayton cycle, the compressor inlet pressure and temperature are 100 kPa and 300 K, respectively. Write a computer program to determine the effects of the compressor pressure ratio, the maximum cycle temperature, and the compressor and turbine isentropic efficiency on the cycle thermal efficiency and the net work output per kilogram air. Consider all combinations of the following variables: (a) Compressor pressure ratio: 6, 10, and 14. (b) Maximum cycle temperature: 1200 K, 1600 K, and 2000 K. (c) Compressor and turbine isentropic efficiency: each at 70%, 80%, and 90%. Assume a constant specific heat of $c_p = 1.005$ kJ/kg·K and $k = 1.40$.

Vapor Power Cycles

This chapter treats vapor power cycles in which the working fluids change alternately between vapor and liquid phases. Vapor power cycles employing water as the working fluid are used extensively to generate electricity. These cycles are called steam cycles. The energy source of a steam cycle can be either from the burning of a fuel in a conventional furnace or from the fission process in a nuclear reactor.

11-1 RANKINE CYCLE

Steam power generation is based on the *Rankine cycle* or its modification. The Rankine cycle for a simple steam power plant is as shown schematically in Fig. 11-1, together with the idealized $p-v$ and $T-s$ property diagrams. Starting at state 1, saturated steam is admitted into the turbine, where it executes an internally reversible adiabatic expansion process to state 2, thereby delivering work. The wet vapor at state 2 is next liquefied at constant pressure in the condenser to the saturated liquid state 3 by transferring heat to the circulating cooling water. The saturated liquid is then pumped isentropically to the compressed liquid state 4. In the boiler, this compressed liquid is heated at constant pressure to saturated liquid and then vaporized back to the saturated vapor state 1.

Assuming steady flow and negligible kinetic and potential energy changes, application of the first law to each of the four components of a Rankine cycle yields, per unit mass of fluid flow, the following equations for work and heat quantities:

Process 4–1,

$$q_{in} = q_{boiler} = h_1 - h_4 \tag{11-1}$$

Process 2–3,

$$q_{out} = q_{cond} = h_2 - h_3 \tag{11-2}$$

Process 1–2,

$$w_{out} = w_{turb} = h_1 - h_2 \tag{11-3}$$

Process 3–4,

$$w_{in} = w_{pump} = h_4 - h_3$$

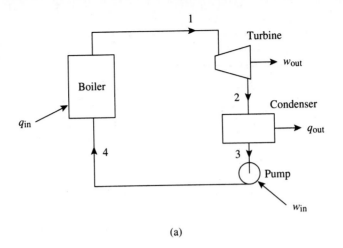

(a)

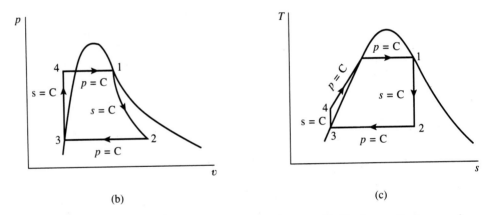

(b)   (c)

Figure 11-1 Rankine cycle: (a) schematic diagram, (b) ideal p–v diagram, and (c) ideal T–s diagram.

Because state 4 is a compressed liquid, its enthalpy value may not be readily available in standard property tables. To resolve this difficulty, we turn to the basic property relation, Eq. (7-2a):

$$T\, ds = dh - v\, dp$$

As $ds = 0$ for the internally reversible adiabatic pump process, we have

$$h_4 - h_3 = \int_3^4 dh = \int_3^4 v\, dp$$

Inasmuch as the specific volume of a liquid undergoes very little change for moderate changes in pressure, we can write

$$w_{\text{in}} = w_{\text{pump}} = h_4 - h_3 = v_3(p_4 - p_3) \tag{11-4}$$

The thermal efficiency of the Rankine cycle is

$$\eta_{th} = \frac{net\ w_{out}}{q_{in}} = \frac{w_{out} - w_{in}}{q_{in}} = \frac{(h_1 - h_2) - (h_4 - h_3)}{h_1 - h_4}$$

$$= \frac{net\ q_{in}}{q_{in}} = \frac{q_{in} - q_{out}}{q_{in}} = \frac{(h_1 - h_4) - (h_2 - h_3)}{h_1 - h_4} \tag{11-5}$$

Carefully note that the value of net w_{out} = net q_{in} is proportional to the area 1–2–3–4 in both the p–v and T–s diagrams.

Based on knowledge of the Carnot cycle, the larger the temperature difference between the heat source and the heat sink, the higher will be the cycle thermal efficiency. Accordingly, the thermal efficiency of a simple Rankine cycle can be increased, either by decreasing the temperature of the heat-rejection process or by increasing the average temperature of the heat-addition process. Modern steam power plants usually utilize cooling towers (see Sec. 9-12) to make available cooling water at the lowest possible temperatures for condenser heat rejection. The best way of increasing the average temperature of the heat-addition process, without unduly increasing the steam-generating pressure, is the use of a superheater to superheat the steam after boiling. Modern steam-generators usually include both boiler and superheater. Figure 11-2 shows the ideal T–s diagram of a Rankine cycle using superheated steam at the turbine inlet.

When the pressure in the steam generator is increased beyond the critical pressure of the working fluid, no phase change of the fluid will occur during heat addition, such as depicted by process 4–1 in the supercritical Rankine cycle of Fig. 11-3. A *supercritical Rankine cycle* is one with the heat-addition process occurring at a constant pressure that is higher than the critical pressure. Because the average temperature of the heat-addition process in a supercritical cycle is higher than that in a subcritical one, the thermal efficiency of a supercritical cycle will certainly be enhanced.

When irreversible effects in the turbine and pump are taken care of in a Rankine steam cycle, its T–s diagram is modified as depicted in Fig. 11-4, in which the actual turbine and pump processes are assumed to be irreversible and adiabatic. The exit states 2' and 4' of the actual turbine and pump, respectively, are related to the adia-

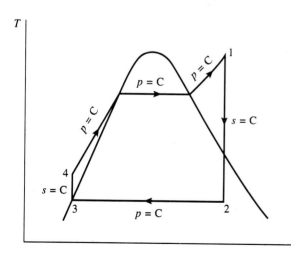

Figure 11-2 Ideal Rankine cycle with superheated steam.

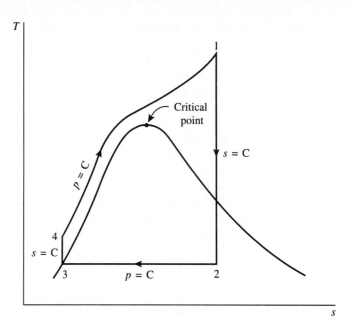

Figure 11-3 Supercritical Rankine cycle.

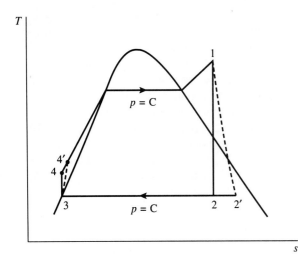

Figure 11-4 Rankine cycle with irreversible turbine and pump processes.

batic turbine efficiency η_T and adiabatic pump efficiency η_p as follows:

$$\eta_T = \frac{h_1 - h_{2'}}{h_1 - h_2} \quad \text{and} \quad \eta_p = \frac{h_4 - h_3}{h_{4'} - h_3}$$

Example 11-1

In a simple Rankine cycle, steam enters the turbine at 600 psia and 1000°F and leaves at 1.0 psia. The adiabatic turbine and pump efficiencies are 0.90 and 0.70, respectively. Determine (a) the quality of steam at the turbine exit, (b) the turbine work, pump work, heat added, and heat rejected, per pound of steam, (c) the cycle thermal efficiency, and (d) the steam rate in lbm/kW·h.

Solution The T-s property diagram is as shown in Fig. 11-4. From the superheated steam table (Table A-20E), at $p_1 = 600$ psia and $T_1 = 1000°F$, we get $h_1 = 1517.8$ Btu/lbm and $s_1 = 1.7155$ Btu/lbm·°R. From the saturated steam table (Table A-19E), at $p_3 = 1.0$ psia, we get $h_3 = h_f = 69.74$ Btu/lbm and $v_3 = v_f = 0.016136$ ft³/lbm.

For the internally reversible adiabatic process 1–2, we have

$$s_1 = s_2 = s_f + x_2 s_{fg} \text{ at } 1.0 \text{ psia}$$

or

$$1.7155 = 0.13266 + x_2(1.8453)$$

so that

$$x_2 = 0.858$$

Therefore,

$$h_2 = h_f + x_2 h_{fg} \text{ at } 1.0 \text{ psia} = 69.7 + 0.858(1036.0)$$

$$= 958.6 \text{ Btu/lbm}$$

For the irreversible adiabatic process 1-2′, we have

$$h_{2'} = h_1 - \eta_T(h_1 - h_2) = 1517.8 - 0.90(1517.8 - 958.6)$$

$$= 1014.5 \text{ Btu/lbm}$$

Also

$$h_{2'} = h_f + x_{2'} h_{fg} \text{ at } 1.0 \text{ psia}$$

or

$$1014.5 = 69.7 + x_{2'}(1036.0)$$

so that

$$x_{2'} = 0.912$$

which is the quality of steam at the turbine exit.

For the internally reversible adiabatic process 3–4, we have

$$h_4 - h_3 = v_3(p_4 - p_3)$$

or

$$h_4 = h_3 + v_3(p_4 - p_3) = 69.7 + 0.016136(600 - 1)\left(\frac{144}{778}\right) = 71.5 \text{ Btu/lbm}$$

For the irreversible adiabatic process 3-4′, we have

$$h_{4'} = h_3 + \frac{h_4 - h_3}{\eta_P} = 69.7 + \frac{71.5 - 69.7}{0.70} = 72.3 \text{ Btu/lbm}$$

The work and heat quantities are calculated as

$$w_{out} = w_{turb} = h_1 - h_{2'} = 1517.8 - 1014.5 = 503.3 \text{ Btu/lbm}$$

$$w_{in} = w_{pump} = h_{4'} - h_3 = 72.3 - 69.7 = 2.6 \text{ Btu/lbm}$$

$$q_{in} = q_{boiler} = h_1 - h_{4'} = 1517.8 - 72.3 = 1445.5 \text{ Btu/lbm}$$

$$q_{out} = q_{cond} = h_{2'} - h_3 = 1014.5 - 69.7 = 944.8 \text{ Btu/lbm}$$

Consequently,

$$\text{net } w_{out} = w_{out} - w_{in} = 503.3 - 2.6 = 500.7 \text{ Btu/lbm}$$

$$= \text{net } q_{in} = q_{in} - q_{out} = 1445.5 - 944.8 = 500.7 \text{ Btu/lbm}$$

The thermal efficiency of the cycle is then

$$\eta_{th} = \frac{\text{net } w_{out}}{q_{in}} = \frac{500.7}{1445.5} = 0.346 = 34.6\%$$

The steam rate is calculated by the equation

$$\text{steam rate} = \frac{3413 \text{ Btu/kW·h}}{\text{net } w_{out} \text{ Btu/lbm}} = \frac{3413}{500.7} = 6.82 \text{ lbm/kW·h}$$

Notice that the back work ratio for this steam cycle is

$$w_{pump}/w_{turb} = 2.6/503.3 = 0.0052$$

or only 0.52% of the work produced by the turbine is required to operate the pump. The back work ratio of a Rankine cycle is low as compared with a Brayton cycle. In a Rankine cycle, a liquid is compressed, whereas in a Brayton cycle, a gas is compressed. Compressing a liquid through a given pressure difference requires a smaller work input than compressing a gas through the same pressure difference.

Notice also that for this example, the ratio

$$q_{cond}/q_{boiler} = 944.8/1445.5 = 0.654$$

or 65.4% of the energy transferred to the water in the boiler is eventually rejected to the cooling water in the condenser. Although considerable amount of energy is carried away by the cooling water, this waste energy is of very low grade, because it is at a temperature close to the atmosphere.

11-2 REHEAT CYCLE

The thermal efficiency of a simple Rankine cycle can be increased by increasing the boiler pressure and the superheater temperature. For a given maximum temperature in the steam generater, however, the sole increase of generater pressure will result in a decrease in the quality of the steam leaving the turbine. No more than 10% moisture content can be tolerated in the last stages of a turbine, because liquid droplets can cause serious erosion of the turbine blades. The moisture content of the steam flowing through a turbine can be kept within an acceptable limit by reheating. As shown in Fig. 11-5, after the steam has partly expanded in the high-pressure turbine, it is withdrawn from the turbine and returned to the steam-generating unit to be resuperheated at constant pressure. It is then further expanded through the low-pressure turbine to the condenser pressure. It is significant to notice that the chief advantage of using reheat is not so much in the gain of thermal efficiency as in maintaining a low moisture content at the exhaust. As a matter of fact, the efficiency of the cycle may or may not be improved because the average temperature at which heat is added is not necessarily increased by reheating.

Assuming steady flow and negligible kinetic and potential energy changes, application of the first law to the components of the cycle shown in Fig. 11-5, per unit mass of steam, we have

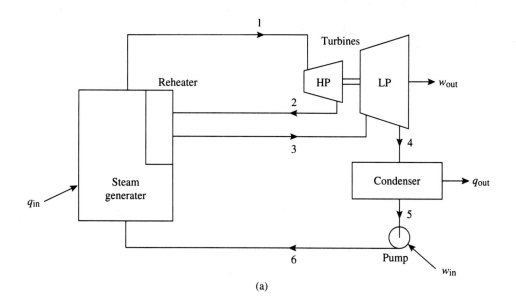

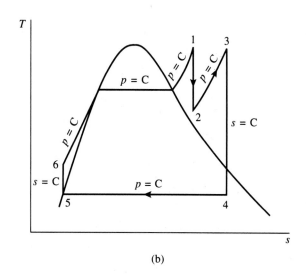

Figure 11-5 Reheat vapor cycle: (a) schematic diagram, and (b) ideal T–s diagram.

$$w_{out} = w_{turb} = (h_1 - h_2) + (h_3 - h_4) \tag{11-6}$$

$$w_{in} = w_{pump} = h_6 - h_5 \tag{11-7}$$

$$q_{in} = q_{gen} + q_{reh} = (h_1 - h_6) + (h_3 - h_2) \tag{11-8}$$

$$q_{out} = q_{cond} = h_4 - h_5 \tag{11-9}$$

The cycle thermal efficiency is then

$$\eta_{th} = \frac{w_{out} - w_{in}}{q_{in}} = \frac{(h_1 - h_2) + (h_3 - h_4) - (h_6 - h_5)}{(h_1 - h_6) + (h_3 - h_2)} \tag{11-10}$$

Example 11-2

A 100-MW reheat steam power cycle has the first-stage supply condition of 7.0 MPa and 450°C. Reheat occurs at 0.5 MPa to 450°C. The condenser pressure is 3 kPa. The adiabatic efficiency of the turbines is 85%, and that of the pump is 72%. Determine the mass-flow rate of steam and the cycle thermal efficiency.

Solution   The T–s diagram of this reheat steam cycle is shown in Fig. 11-6. The given data are net $\dot{W}_{\text{out}} = 100$ MW, $p_1 = p_6 = 7.0$ MPa, $T_1 = 450°C$, $p_2 = p_3 = 0.5$ MPa, $T_3 = 450°C$, $p_4 = p_5 = 3$ kPa, $\eta_T = 85\%$, and $\eta_p = 72\%$.

From the steam tables (Tables A-19M and 20M), at p_1 and T_1, we obtain $h_1 = 3287.1$ kJ/kg, $s_1 = 6.6327$ kJ/kg·K; and at p_3 and T_3, we obtain $h_3 = 3377.2$ kJ/kg, $s_3 = 7.9445$ kJ/kg·K. Also,

$$h_5 = h_f \text{ at } 3 \text{ kPa} = 101.1 \text{ kJ/kg}$$

$$v_5 = v_f \text{ at } 3 \text{ kPa} = 1.0027 \times 10^{-3} \text{ m}^3/\text{kg}$$

For the internally reversible adiabatic process 1–2, we have

$$s_1 = s_2 = s_f + x_2 s_{fg} \text{ at } 0.5 \text{ MPa}$$

or

$$6.6327 = 1.8607 + x_2(4.9606)$$

so that

$$x_2 = 0.962$$

Therefore,

$$h_2 = h_f + x_2 h_{fg} \text{ at } 0.5 \text{ MPa} = 640.2 + 0.962(2108.5)$$

$$= 2668.6 \text{ kJ/kg}$$

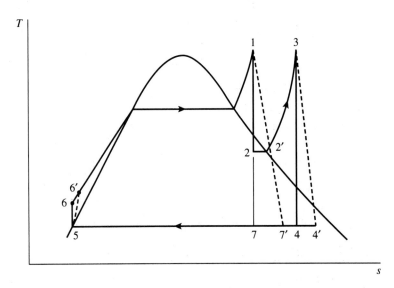

Figure 11-6 T–s diagram of a reheat steam cycle with irreversible turbine and pump processes.

For the irreversible adiabatic process 1–2', we have

$$h_{2'} = h_1 - \eta_T(h_1 - h_2)$$

$$= 3287.1 - 0.85(3287.1 - 2668.6) = 2761.4 \text{ kJ/kg}$$

This $h_{2'}$ is bigger than h_g at 0.5 MPa, we conclude that state 2' is superheated. For the internally reversible adiabatic process 3–4, we have

$$s_3 = s_4 = s_f + x_4 s_{fg} \text{ at 3 kPa}$$

or

$$7.9445 = 0.3545 + x_4(8.2231)$$

so that

$$x_4 = 0.923$$

Therefore,

$$h_4 = h_f + x_4 h_{fg} \text{ at 3 kPa}$$

$$= 101.1 + 0.923(2444.5) = 2357.4 \text{ kJ/kg}$$

For the irreversible adiabatic process 3–4', we have

$$h_{4'} = h_3 - \eta_T(h_3 - h_4)$$

$$= 3377.2 - 0.85(3377.2 - 2357.4) = 2510.4 \text{ kJ/kg}$$

Also

$$h_{4'} = h_f + x_{4'} h_{fg} \text{ at 3 kPa}$$

or

$$2510.4 = 101.1 + x_{4'}(2444.5)$$

so that

$$x_{4'} = 0.986$$

For the internally reversible adiabatic process 5–6, we have

$$h_6 = h_5 + v_5(p_6 - p_5)$$

$$= (101.1 \text{ kJ/kg}) + (0.0010027 \text{ m}^3/\text{kg})(7000 - 3 \text{ kN/m}^2)$$

$$= 108.1 \text{ kJ/kg}$$

For the irreversible adiabatic process 5–6', we have

$$h_{6'} = h_5 + \frac{h_6 - h_5}{\eta_p} = 101.1 + \frac{108.1 - 101.1}{0.72} = 110.8 \text{ kJ/kg}$$

The work and heat quantities are calculated as

$$w_{\text{out}} = w_{\text{turb}} = (h_1 - h_{2'}) + (h_3 - h_{4'})$$

$$= (3287.1 - 2761.4) + (3377.2 - 2510.4)$$

$$= 1392.5 \text{ kJ/kg}$$

$$w_{\text{in}} = w_{\text{pump}} = h_{6'} - h_5 = 110.8 - 101.1 = 9.7 \text{ kJ/kg}$$

$$q_{\text{in}} = q_{\text{gen}} + q_{\text{reh}} = (h_1 - h_{6'}) + (h_3 - h_{2'})$$

$$= (3287.1 - 110.8) + (3377.2 - 2761.4)$$

$$= 3792.1 \text{ kJ/kg}$$

$$q_{out} = q_{cond} = h_{4'} - h_5 = 2510.4 - 101.1 = 2409.3 \text{ kJ/kg}$$

Consequently,

$$\text{net } w_{out} = w_{out} - w_{in} = 1392.5 - 9.7 = 1382.8 \text{ kJ/kg}$$

$$= \text{net } q_{in} = q_{in} - q_{out} = 3792.1 - 2409.3 = 1382.8 \text{ kJ/kg}$$

The cycle thermal efficiency is then

$$\eta_{th} = \frac{\text{net } w_{out}}{q_{in}} = \frac{1382.8}{3792.1} = 0.365 = 36.5\%$$

The mass-flow rate of steam is calculated by the relation,

$$\text{net } \dot{W}_{out} = \dot{m}(\text{net } w_{out})$$

or

$$\dot{m} = \frac{\text{net } \dot{W}_{out}}{\text{net } w_{out}} = \frac{100,000 \text{ kW}}{1382.8 \text{ kJ/kg}} = 72.3 \text{ kg/s}$$

If no reheating is involved in this steam cycle, the simple Rankine cycle would follow the processes 1–7–5–6–1 in the ideal cycle, and the processes 1–7'–5–6'–1 in the actual cycle. For the internally reversible adiabatic process 1–7, we have

$$s_1 = s_7 = s_f + x_7 s_{fg} \text{ at 3 kPa}$$

$$6.6327 = 0.3545 + x_7(8.2231)$$

so that

$$x_7 = 0.763$$

Therefore,

$$h_7 = h_f + x_7 h_{fg} \text{ at 3 kPa}$$

$$= 101.1 + 0.763(2444.5) = 1966.3 \text{ kJ/kg}$$

For the irreversible adiabatic process 1–7', we have

$$h_{7'} = h_1 - \eta_T(h_1 - h_7)$$

$$= 3287.1 - 0.85(3287.1 - 1966.3)$$

$$= 2164.4 \text{ kJ/kg}$$

Also

$$h_{7'} = h_f + x_{7'} h_{fg} \text{ at 3 kPa}$$

$$2164.4 = 101.1 + x_{7'}(2444.5)$$

so that

$$x_{7'} = 0.844$$

The previous calculations reveal that the moisture content at both state 7 (ideal case) and state 7' (actual case) is too high. These unacceptable conditions at the turbine exit are remitted by reheating with the results of $x_4 = 0.923$ (ideal case) and $x_{4'} = 0.986$ (actual case).

The thermal efficiency of the actual simple cycle $1-7'-5-6'-1$ is given by

$$\eta_{th} = \frac{(h_1 - h_{7'}) - (h_{6'} - h_5)}{h_1 - h_{6'}}$$

$$= \frac{(3287.1 - 2164.4) - (110.8 - 101.1)}{3287.1 - 110.8}$$

$$= 0.350 = 35.0\%$$

The thermal efficiency of the actual reheating cycle $1-2'-3-4'-5-6'-1$ is 36.5%. Therefore, for this particular case, the thermal efficiency of the reheating cycle is a little higher than that of the simple cycle. It should be noted, however, that improvement in thermal efficiency is not guaranteed by reheating.

11-3 REGENERATIVE CYCLE

In the simple Rankine cycle, the initial portion of the heat addition to the compressed liquid leaving the pump is at a very low temperature. The second law tells us that this low-temperature heat-addition process greatly impairs the thermal efficiency of the cycle. The thermal efficiency could be improved if the compressed liquid were preheated before it entered the steam generator by the use of internal heat transfer within the cycle. This is precisely what a *regenerative cycle* tries to accomplish.

Figure 11-7 shows the schematic and ideal T–s diagrams of a regenerative cycle with one *open feedwater heater*. Consider a unit mass (1 kg) of steam upon its entrance to the turbine (state 1). After expansion to state 2, a fraction of the steam, m kg, is bled or extracted from the turbine and fed into the open feedwater heater. The steam remaining in the turbine, $(1 - m)$ kg, expands to the condenser pressure (state 3) and then condenses to saturated liquid at state 4. The condensate pump is used to increase the pressure of the liquid from the condenser to the same pressure as that of the extracted steam ($p_5 = p_2$). The compressed liquid at state 5 and the extracted steam at state 2 are mixed directly in the open heater to form saturated liquid at state 6. In the ideal situation, the amount of steam extracted (m kg) is calculated to be just sufficient to cause the liquid leaving the feedwater heater to be in the saturated state. The feed pump finally raises the pressure of the feedwater to that of the steam generator ($p_7 = p_1$), and boiler heating commences at state 7.

For the cycle of Fig. 11-7, the amount of steam extracted at state 2 per unit mass at the turbine inlet can be calculated from a steady-flow energy balance for the open heater with ΔKE and ΔPE neglected. Thus,

$$mh_2 + (1 - m)h_5 = 1h_6 \tag{11-11}$$

or

$$m = \frac{h_6 - h_5}{h_2 - h_5}$$

Assuming steady flow and negligible kinetic and potential energy changes, applying the first law to the components of the cycle shown in Fig. 11-7, per unit mass at the

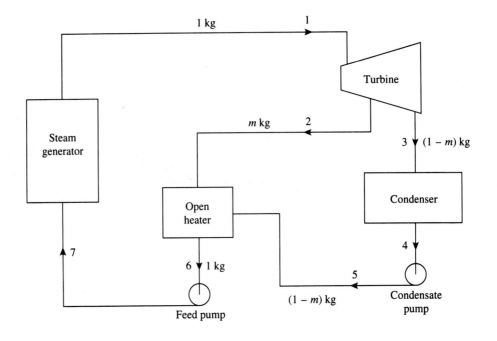

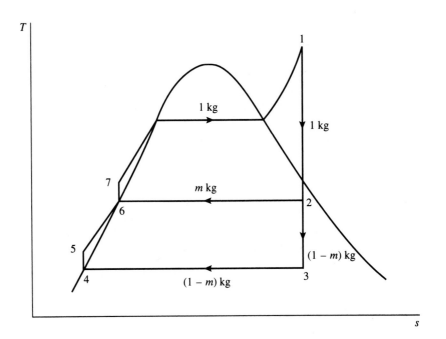

Figure 11-7 A regenerative cycle with one open feedwater heater.

turbine inlet, we have

$$w_{out} = w_{turb} = 1(h_1 - h_2) + (1 - m)(h_2 - h_3) \qquad (11\text{-}12)$$

$$w_{in} = w_{pump} = (1 - m)(h_5 - h_4) + 1(h_7 - h_6) \qquad (11\text{-}13)$$

$$q_{in} = q_{gen} = 1(h_1 - h_7) \qquad (11\text{-}14)$$

$$q_{out} = q_{cond} = (1 - m)(h_3 - h_4) \qquad (11\text{-}15)$$

The thermal efficiency is given by

$$\eta_{th} = \frac{w_{out} - w_{in}}{q_{in}} = \frac{q_{in} - q_{out}}{q_{in}} \qquad (11\text{-}16)$$

Figure 11-8 shows the schematic and ideal T–s diagrams of a regenerative cycle with one *closed feedwater heater*. In the closed heater, the extracted steam and the feedwater do not mix, but exchange heat through the metal-tube walls. The extracted steam condenses and leaves the heater as a saturated liquid (state 7), and the feedwater heats up and leaves the heater as a compressed liquid (state 6). The temperature difference $(T_7 - T_6)$, called the *terminal temperature difference*, is usually some 4 or 5 degrees Celsius. The heater drain after being pumped to the boiler pressure (state 8) is mixed with the feedwater at state 6 to form the feedwater at state 9, leading to the steam generator.

For the cycle of Fig. 11-8, the amount of steam extracted at state 2 per unit mass at the turbine inlet can be calculated from an energy balance for the closed heater. Thus,

$$m(h_2 - h_7) = (1 - m)(h_6 - h_5) \qquad (11\text{-}17)$$

The condition at state 9 can be found from an energy balance for the mixing of $(1 - m)$ kg at state 6 and m kg at state 8,

$$(1 - m)h_6 + mk_8 = 1h_9 \qquad (11\text{-}18)$$

The work and heat quantities per unit mass at the turbine inlet are

$$w_{out} = w_{turb} = 1(h_1 - h_2) + (1 - m)(h_2 - h_3) \qquad (11\text{-}19)$$

$$w_{in} = w_{pump} = (1 - m)(h_5 - h_4) + m(h_8 - h_7) \qquad (11\text{-}20)$$

$$q_{in} = q_{gen} = 1(h_1 - h_9) \qquad (11\text{-}21)$$

$$q_{out} = q_{cond} = (1 - m)(h_3 - h_4) \qquad (11\text{-}22)$$

There are two methods of handling the drain from a closed feedwater heater: either pump ahead to the next higher-pressure line (as in Fig. 11-8) or dump back to the next lower-pressure region by the use of a trap. A trap is a device with a float-operated valve that permits the passage of liquid but does not permit the passage of vapor. Figure 11-9 shows the schematic and ideal T–s diagrams of a regenerative cycle with a closed feedwater heater in dump-back arrangement as part of the scheme. The liquid condensed in the closed heater is throttled through the trap into the lower-pressure open heater. Application of the first law to the trap results in $h_9 = h_{10}$.

For the cycle of Fig. 11-9, the amounts of steam extracted at states 2 and 3 per unit mass at the turbine inlet can be calculated from the energy equations for the closed and open heaters. Thus, for the closed heater,

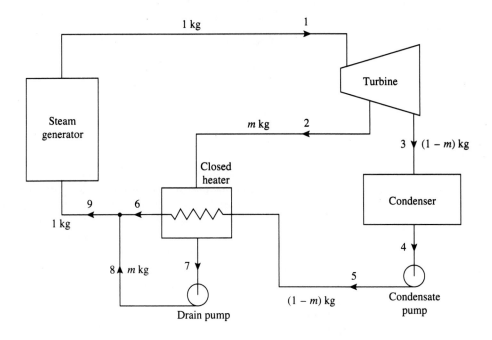

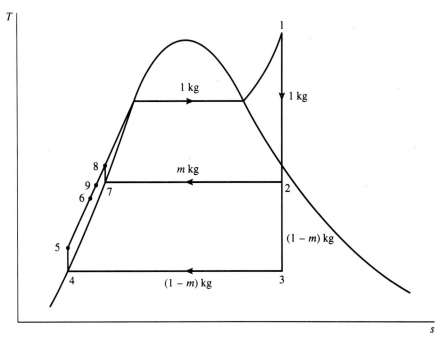

Figure 11-8 A regenerative cycle with one closed feedwater heater in pump-ahead arrangement.

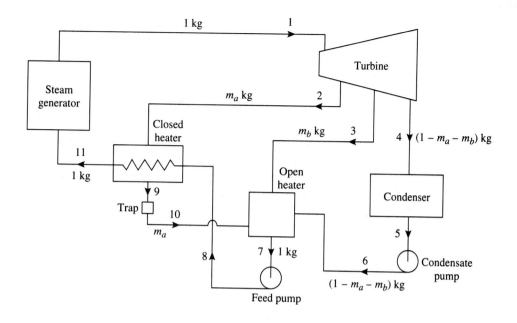

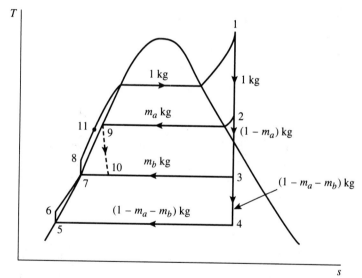

Figure 11-9 A regenerative cycle with one open feedwater heater and one closed feedwater heater in dump-back arrangement.

$$m_a(h_2 - h_9) = 1(h_{11} - h_8) \qquad (11\text{-}23)$$

and for the open heater,

$$m_a h_{10} + m_b h_3 + (1 - m_a - m_b)h_6 = 1h_7 \qquad (11\text{-}24)$$

The work and heat quantities per unit mass at the turbine inlet are

$$w_{out} = w_{turb} = 1(h_1 - h_2) + (1 - m_a)(h_2 - h_3) + (1 - m_a - m_b)(h_3 - h_4)$$

$$= 1h_1 - m_a h_2 - m_b h_3 - (1 - m_a - m_b)h_4 \qquad (11\text{-}25)$$

$$w_{in} = w_{pump} = (1 - m_a - m_b)(h_6 - h_5) + 1(h_8 - h_7) \qquad (11\text{-}26)$$

$$q_{in} = q_{gen} = 1(h_1 - h_{11}) \qquad (11\text{-}27)$$

$$q_{out} = q_{cond} = (1 - m_a - m_b)(h_4 - h_5) \qquad (11\text{-}28)$$

It is imperative to notice that the chief advantage of open heaters is their effective heat transfer between the extracted steam and the feedwater because of the intimate mixing of the fluids. The drawback of open heaters is that each heater must be provided with a separate pump in its operation. On the contrary, closed heaters are less effective in transferring heat from the extracted steam to the feedwater, but they do not require a separate pump for each heater. Modern high-pressure steam power plants usually employ six to eight feedwater heaters; most of them are the closed type. At least one open heater, however, must be incorporated in the chain of closed heaters so as to facilitate the ejection of air and other gases from the feedwater to avoid corrosion of the boiler and piping.

Example 11-3

The steam generator of a 300-MW regenerative cycle produces steam at 8 MPa and 500°C. A closed feedwater heater receives steam from the turbine at 1 MPa and an open feedwater heater operates at 0.125 MPa. The condenser operates at 5 kPa. The liquid condensate from the closed heater is dumped back into the open heater through a trap. The steam turbine is adiabatic with efficiency of 80% throughout. Assume that the pumps operate isentropically. The terminal temperature difference of the closed heater is 5°C. Determine (a) the mass-flow rates of steam extracted for the heaters, (b) the total mass-flow rate of steam at the turbine throttle (turbine inlet), (c) the thermal efficiency, and (d) the quality of steam at the turbine exit.

Solution The T–s diagram of the cycle is shown in Fig. 11-10. The given data are net $\dot{W}_{out} = 300$ MW, $p_1 = 8$ MPa, $T_1 = 500$°C, $p_2 = 1$ MPa, $p_3 = 0.125$ MPa, $p_4 = 5$ kPa, $\eta_T = 80\%$, and $T_9 - T_{11} = 5$°C.

From the steam tables (Tables A-19M and 20M), at p_1 and T_1, we have $h_1 = 3398.3$ kJ/kg and $s_1 = 6.7240$ kJ/kg·K. At $s_1 = s_2 = 6.7240$ kJ/kg·K and $p_2 = 1$ MPa, we have $h_2 = 2842.3$ kJ/kg. Also

$$s_1 = s_3 = s_f + x_3 s_{fg} \text{ at } 0.125 \text{ MPa} = 1.3740 + x_3(5.9104)$$

so that

$$x_3 = 0.905$$

Therefore,

$$h_3 = h_f + x_3 h_{fg} \text{ at } 0.125 \text{ MPa}$$
$$= 444.3 + 0.905(2241.0) = 2472.4 \text{ kJ/kg}$$

and

$$s_1 = s_4 = s_f + x_4 s_{fg} \text{ at } 5 \text{ kPa} = 0.4764 + x_4(7.9187)$$

so that

$$x_4 = 0.789$$

Therefore,

$$h_4 = h_f + x_4 h_{fg} \text{ at } 5 \text{ kPa}$$
$$= 137.8 + 0.789(2423.7) = 2050.1 \text{ kJ/kg}$$

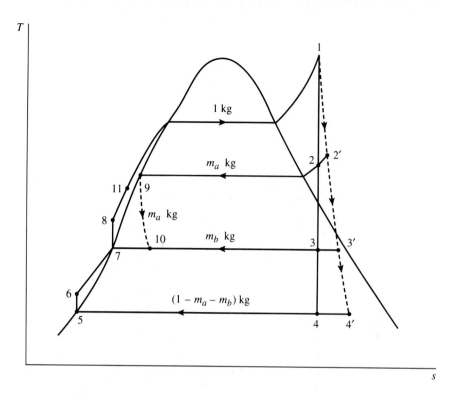

Figure 11-10 *T–s* diagram for the cycle of Example 11-3.

By the use of the adiabatic turbine efficiency, we calculate the enthalpy values of the actual states:

$$h_{2'} = h_1 - \eta_T(h_1 - h_2) = 3398.3 - 0.80(3398.3 - 2842.3) = 2953.5 \text{ kJ/kg}$$

$$h_{3'} = h_1 - \eta_T(h_1 - h_3) = 3398.3 - 0.80(3398.3 - 2472.4) = 2657.6 \text{ kJ/kg}$$

$$h_{4'} = h_1 - \eta_T(h_1 - h_4) = 3398.3 - 0.80(3398.3 - 2050.1) = 2319.7 \text{ kJ/kg}$$

The quality of steam at the turbine exit is calculated by the relation

$$h_{4'} = 2319.7 \text{ kJ/kg} = h_f + x_{4'}h_{fg} \text{ at 5 kPa} = 137.8 + x_{4'}(2423.7)$$

It follows that

$$x_{4'} = 0.900 = 90.0\%$$

Again from the steam table (Table A-19M),

$$h_5 = h_f \text{ at 5 kPa} = 137.8 \text{ kJ/kg}$$

$$v_5 = v_f \text{ at 5 kPa} = 0.001005 \text{ m}^3/\text{kg}$$

$$h_7 = h_f \text{ at 0.125 MPa} = 444.3 \text{ kJ/kg}$$

$$v_7 = v_f \text{ at 0.125 MPa} = 0.001048 \text{ m}^3/\text{kg}$$

$$h_9 = h_{10} = h_f \text{ at 1 MPa} = 762.8 \text{ kJ/kg}$$

$$T_9 = T_{sat} \text{ at 1 MPa} = 179.9°C$$

Accordingly, we have

$$h_6 = h_5 + v_5(p_6 - p_5)$$

$$= 137.8 \text{ kJ/kg} + (0.001005 \text{ m}^3\text{/kg})(125 - 5 \text{ kPa})$$

$$= 137.9 \text{ kJ/kg}$$

$$h_8 = h_7 + v_7(p_8 - p_7)$$

$$= 444.3 + 0.001048(8000 - 125) = 452.6 \text{ kJ/kg}$$

From the compressed-liquid steam table (Table A-21M), at $p_{11} = 8$ MPa and $T_{11} = T_9 - 5 = 179.9 - 5 = 174.9°C$, we obtain $h_{11} = 744.7$ kJ/kg. Carefully note that if we assume $h_{11} = h_f$ at T_{11}, the value of h_{11} obtained from the saturated steam table is 740.7 kJ/kg.

The amounts of steam extracted, m_a and m_b, per unit mass at the turbine throttle are calculated from the energy equations for the closed and open heaters. Thus,

$$m_a = \frac{h_{11} - h_8}{h_{2'} - h_9} = \frac{744.7 - 452.6}{2953.5 - 762.8} = 0.133 \text{ kg/kg throttle}$$

$$m_b = \frac{h_7 - h_6 - m_a(h_{10} - h_6)}{h_{3'} - h_6}$$

$$= \frac{444.3 - 137.9 - 0.133(762.8 - 137.9)}{2657.6 - 137.9}$$

$$= 0.0886 \text{ kg/kg throttle}$$

The turbine work output and pump work input, per unit mass at the turbine throttle, are

$$w_{\text{out}} = w_{\text{turb}} = 1h_1 - m_a h_{2'} - m_b h_{3'} - (1 - m_a - m_b)h_{4'}$$

$$= 3398.3 - 0.133(2953.5) - 0.0886(2657.6)$$

$$- (1 - 0.133 - 0.0886)(2319.7)$$

$$= 964.4 \text{ kJ/kg throttle}$$

$$w_{\text{in}} = w_{\text{pump}} = (1 - m_a - m_b)(h_6 - h_5) + 1(h_8 - h_7)$$

$$= (1 - 0.133 - 0.0886)(137.9 - 137.8) + 1(452.6 - 444.3)$$

$$= 8.4 \text{ kJ/kg throttle}$$

The mass-flow rate at the turbine throttle is calculated as

$$\dot{m}_{\text{throttle}} = \frac{\text{net } \dot{W}_{\text{out}}}{\text{net } w_{\text{out}}} = \frac{\text{net } \dot{W}_{\text{out}}}{w_{\text{out}} - w_{\text{in}}}$$

$$= \frac{300,000 \text{ kJ/s}}{(964.4 - 8.4) \text{ kJ/kg throttle}}$$

$$= 313.8 \text{ kg throttle/s}$$

Accordingly, the mass-flow rates of extracted steam are

$$\dot{m}_a = \dot{m}_{\text{throttle}} m_a$$

$$= (313.8 \text{ kg throttle/s})(0.133 \text{ kg/kg throttle})$$

$$= 41.7 \text{ kg/s}$$

$$\dot{m}_b = \dot{m}_{\text{throttle}} m_b = 313.8(0.0886) = 27.8 \text{ kg/s}$$

The heat added in the steam generator per unit mass at the turbine throttle is

$$q_{\text{in}} = q_{\text{gen}} = 1(h_1 - h_{11}) = 1(3398.3 - 744.7)$$
$$= 2653.6 \text{ kJ/kg throttle}$$

The thermal efficiency is then

$$\eta_{\text{th}} = \frac{w_{\text{out}} - w_{\text{in}}}{q_{\text{in}}} = \frac{964.4 - 8.4}{2653.6} = 0.360 = 36.0\%$$

11-4 REHEAT-REGENERATIVE CYCLE

In order to secure a low-moisture content at the turbine exit, modern high-pressure steam power plants invariably include reheating in regenerative cycle. Figure 11-11 depicts a reheat-regenerative cycle with one reheater and five regenerative feedwater heaters. There are four closed heaters in the cycle: three are in the dump-back arrangement and one in the pump-ahead arrangement. One open heater is used as a deaerator to facilitate the ejection of dissolved and entrained gases such as oxygen and carbon dioxide from the feedwater system.

It should be noted that the exact location of the state points shown in the T–s diagram of Fig. 11-11 will not be known unless the property data are given. The amounts of steam extracted from the turbine can be obtained by solving the following energy equations for the feedwater heaters in the order written:

For heater a,

$$m_a(h_2 - h_{20}) = 1(h_{19} - h_{18}) \tag{11-29}$$

For heater b,

$$m_b h_4 + m_a h_{21} + 1h_{17} = 1h_{18} + (m_a + m_b)h_{22} \tag{11-30}$$

For heater c,

$$m_c h_5 + (m_a + m_b)h_{23} + 1h_{16} = 1h_{17} + (m_a + m_b + m_c)h_{24} \tag{11-31}$$

For heater d,

$$m_d h_6 + (m_a + m_b + m_c)h_{25} + (1 - m_a - m_b - m_c - m_d)h_{14} = 1h_{15} \tag{11-32}$$

For heater e,

$$m_e(h_7 - h_{12}) = (1 - m_a - m_b - m_c - m_d - m_e)(h_{11} - h_{10}) \tag{11-33}$$

The work and heat quantities are given by the expressions:

$$w_{\text{out}} = w_{\text{turb}} = 1(h_1 - h_2) + (1 - m_a)(h_3 - h_4) + (1 - m_a - m_b)(h_4 - h_5)$$
$$+ (1 - m_a - m_b - m_c)(h_5 - h_6)$$
$$+ (1 - m_a - m_b - m_c - m_d)(h_6 - h_7)$$
$$+ (1 - m_a - m_b - m_c - m_d - m_e)(h_7 - h_8)$$
$$= 1(h_1 - h_2) + (1 - m_a)h_3 - m_b h_4 - m_c h_5 - m_d h_6 - m_e h_7$$
$$- (1 - m_a - m_b - m_c - m_d - m_e)h_8 \tag{11-34}$$

Sec. 11-4   Reheat-Regenerative Cycle

463

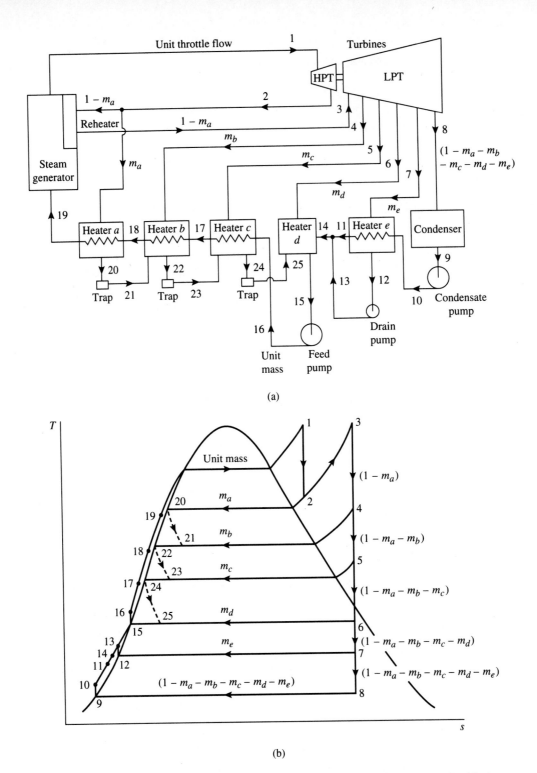

Figure 11-11 A reheat-regenerative cycle: (a) schematic diagram, and (b) ideal T–s diagram.

$$w_{in} = w_{pump} = (1 - m_a - m_b - m_c - m_d - m_e)(h_{10} - h_9)$$
$$+ m_e(h_{13} - h_{12}) + 1(h_{16} - h_{15})$$

(11-35)

$$q_{in} = q_{gen} + q_{reh} = 1(h_1 - h_{19}) + (1 - m_a)(h_3 - h_2)$$

(11-36)

$$q_{out} = q_{cond} = (1 - m_a - m_b - m_c - m_d - m_e)(h_8 - h_9)$$

(11-37)

Example 11-4

In a reheating-regenerative supercritical steam power cycle (see Fig. 11-12), the feedwater enters the steam generator at 30 MPa while the steam enters the turbine at 25 MPa and 600°C. The steam is expanded through the high-pressure turbine to 3 MPa. Part of the exhaust steam from this turbine is supplied to a closed feedwater heater, and the remainder is reheated to 540°C. The reheated steam is then expanded in the low-pressure turbine to an exhaust pressure of 5 kPa. Both turbines have an adiabatic efficiency of 86%. The heater drain is pumped ahead to join the main feedwater line. The terminal temperature difference of the closed heater is 4°C. The pump processes are assumed to be isentropic. Determine (a) the enthalpies of the state points around the cycle, (b) the quality of steam at the exit of the low-pressure turbine, (c) the fraction of the total mass-flow rate that is extracted to the heater, and (d) the thermal efficiency of the cycle.

Solution By referring to Fig. 11-12, the given data are $p_{10} = 30$ MPa, $p_1 = 25$ MPa, $T_1 = 600°C$, $p_2 = p_3 = 3$ MPa, $T_3 = 540°C$, $p_4 = 5$ kPa, and $\eta_T = 86\%$. In the T–s diagram of Fig. 11-12, the pressure drop in the steam line is assumed to occur at the turbine throttle valve. From the steam tables (Tables A-19M and A-20M), at p_1 and T_1, we get $h_1 = h_{1'} = 3491.4$ kJ/kg, $s_1 = 6.3602$ kJ/kg·K; and at p_3 and T_3, we get $h_3 = 3546.6$ kJ/kg and $s_3 = 7.3474$ kJ/kg·K. Also

$$h_5 = h_f \text{ at } 5 \text{ kPa} = 137.8 \text{ kJ/kg}$$

$$v_5 = v_f \text{ at } 5 \text{ kPa} = 0.001005 \text{ m}^3/\text{kg}$$

$$h_8 = h_f \text{ at } 3 \text{ MPa} = 1008.4 \text{ kJ/kg}$$

$$v_8 = v_f \text{ at } 3 \text{ MPa} = 0.001217 \text{ m}^3/\text{kg}$$

$$T_8 = T_{sat} \text{ at } 3 \text{ MPa} = 233.9°C$$

For the internally reversible adiabatic process 1–2, $s_1 = s_2 = 6.3602$ kJ/kg·K. Also, as given in the problem, $p_2 = 3$ MPa. Thus, from the superheated steam table (Table A-20M), we have $h_2 = 2894.5$ kJ/kg. Accordingly,

$$h_{2'} = h_1 - \eta_T(h_1 - h_2)$$
$$= 3491.4 - 0.86(3491.4 - 2894.5) = 2978.1 \text{ kJ/kg}$$

For the internally reversible adiabatic process 3–4, we have

$$s_3 = s_4 = s_f + x_4 s_{fg} \text{ at } 5 \text{ kPa}$$

or

$$7.3474 = 0.4764 + x_4(7.9187)$$

so that

$$x_4 = 0.868$$

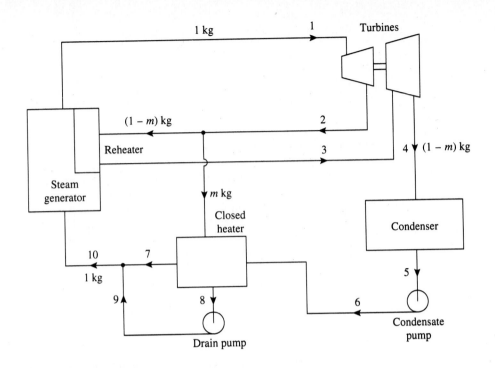

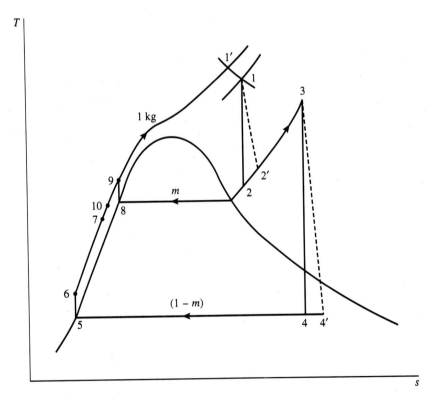

Figure 11-12 Schematic and *T–s* diagrams for the cycle of Example 11-4.

Therefore,

$$h_4 = h_f + x_4 h_{fg} \text{ at 5 kPa}$$

$$= 137.8 + 0.868(2423.7) = 2241.6 \text{ kJ/kg}$$

Accordingly,

$$h_{4'} = h_3 - \eta_T(h_3 - h_4)$$

$$= 3546.6 - 0.86(3546.6 - 2241.6) = 2424.3 \text{ kJ/kg}$$

Furthermore,

$$h_{4'} = h_f + x_{4'} h_{fg} \text{ at 5 kPa}$$

or

$$2424.3 = 137.8 + x_{4'}(2423.7)$$

so that

$$x_{4'} = 0.943 = 94.3\%$$

which is the quality of steam at the low-pressure turbine exit.

For the internally reversible adiabatic process 5–6,

$$h_6 = h_5 + v_5(p_6 - p_5)$$

$$= (137.8 \text{ kJ/kg}) + (0.001005 \text{ m}^3/\text{kg})(30,000 - 5 \text{ kPa})$$

$$= 167.9 \text{ kJ/kg}$$

For the internally reversible adiabatic process 8–9,

$$h_9 = h_8 + v_8(p_9 - p_8)$$

$$= 1008.4 + 0.001217(30,000 - 3000) = 1041.3 \text{ kJ/kg}$$

Because $T_7 = T_8 - 4°C = 233.9 - 4 = 229.9°C$ and $p_7 = 30$ MPa, from the compressed-liquid steam table (Table A-21M), by interpolation, we have $h_7 = 997.4$ kJ/kg.

An energy balance for the closed heater gives

$$m(h_{2'} - h_8) = (1 - m)(h_7 - h_6)$$

or

$$m = \frac{h_7 - h_6}{(h_{2'} - h_8) + (h_7 - h_6)}$$

$$= \frac{997.4 - 167.9}{(2978.1 - 1008.4) + (997.4 - 167.9)}$$

$$= 0.296 \text{ kg/kg throttle}$$

An energy balance for the mixing of the heater drain and the main feedwater yields

$$h_{10} = mh_9 + (1 - m)h_7$$

$$= 0.296(1041.3) + (1 - 0.296)997.4 = 1010.4 \text{ kJ/kg}$$

The work and heat quantities are calculated as

$$w_{out} = w_{turb} = 1(h_1 - h_{2'}) + (1 - m)(h_3 - h_{4'})$$

$$= 1(3491.4 - 2978.1) + (1 - 0.296)(3546.6 - 2424.3)$$

$$= 513.3 + 790.1 = 1303.4 \text{ kJ/kg throttle}$$

$$w_{in} = w_{pump} = (1 - m)(h_6 - h_5) + m(h_9 - h_8)$$

$$= (1 - 0.296)(167.9 - 137.8) + 0.296(1041.3 - 1008.4)$$

$$= 30.9 \text{ kJ/kg throttle}$$

$$q_{in} = q_{gen} + q_{reh} = 1(h_{1'} - h_{10}) + (1 - m)(h_3 - h_{2'})$$

$$= 1(3491.4 - 1010.4) + (1 - 0.296)(3546.6 - 2978.1)$$

$$= 2881.2 \text{ kJ/kg throttle}$$

The cycle thermal efficiency is

$$\eta_{th} = \frac{w_{out} - w_{in}}{q_{in}} = \frac{1303.4 - 30.9}{2881.2} = 0.442 = 44.2\%$$

11-5 BINARY VAPOR CYCLE

The critical temperature and pressure of water are, respectively, 374.14°C (705.44°F) and 22.09 MPa (3203.6 psia). This critical-temperature value is well below the permissible structural limit of steam-generator materials. The use of a pressure higher than the water critical pressure in a steam cycle, however, demands high-grade expansive materials for the steam generator, piping, and related equipment. On the other hand, there are other working fluids, such as mercury and potassium, that can operate at high temperatures, but at the same time maintain low pressures. A *binary vapor cycle* uses two working fluids, one for the high-temperature range and the other for the low-temperature range. Mercury–water and potassium–water are examples of suitable combinations.

Figure 11-13 depicts the schematic and ideal *T–s* diagrams of a potassium–water binary vapor cycle. The data indicated on the *T–s* diagram are only a possible suggestion. The mass ratio of potassium and water can be obtained by writing an energy balance for the heat exchanger that serves as the potassium condenser and steam generator. Thus,

$$m(h_b - h_c) = 1(h_1 - h_4) \tag{11-38}$$

The work and heat quantities are

$$w_{out} = w_{turb} = m(h_a - h_b) + 1(h_1 - h_2) \tag{11-39}$$

$$w_{in} = w_{pump} = m(h_d - h_c) + 1(h_4 - h_3) \tag{11-40}$$

$$q_{in} = q_{potassium\,boiler} = m(h_a - h_d) \tag{11-41}$$

$$q_{out} = q_{steam\,condenser} = 1(h_2 - h_3) \tag{11-42}$$

11-6 LOW-TEMPERATURE RANKINE CYCLE

In a low-temperature Rankine cycle, the thermal efficiencies obtainable when water is used as the working fluid are too low to be practical. Some organic fluids, such as Refrigerant-113 (CCl_2FCClF_2) and Refrigerant-114 ($CClF_2CClF_2$), are found to possess

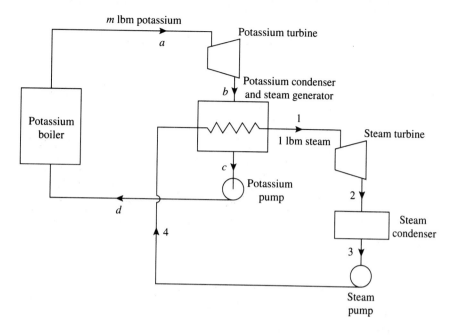

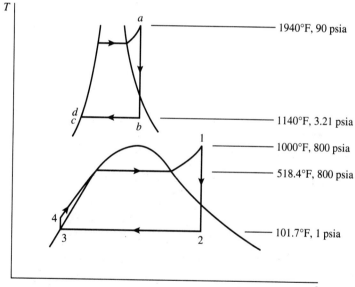

m lbm potassium
a
Potassium turbine
b
Potassium condenser
and steam generator
Potassium
boiler
1
1 lbm steam
Steam turbine
c
Potassium
pump
d
2
4
Steam
condenser
3
Steam
pump
T
a
1940°F, 90 psia
d
c
b
1140°F, 3.21 psia
1
1000°F, 800 psia
518.4°F, 800 psia
4
3
2
101.7°F, 1 psia
s

Figure 11-13 Potassium–water binary vapor cycle.

the thermodynamic properties and stability characteristics suitable for such applications.

The saturated vapor line of water on *T–s* coordinates has a negative slope, such as depicted in Fig. 11-2, which is duplicated in Fig. 11-14 for comparison. A *p–h* diagram of the same cycle is also shown in this figure. Contrary to the case of water, Refrigerants-113 and -114 have positively sloped saturated vapor lines on *T–s* coordinates, such as depicted in Fig. 11-15(b). Although dry vapor throughout the expansion

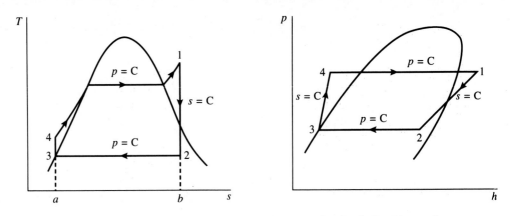

Figure 11-14 Idealized T–s and p–h diagrams of a simple Rankine cycle.

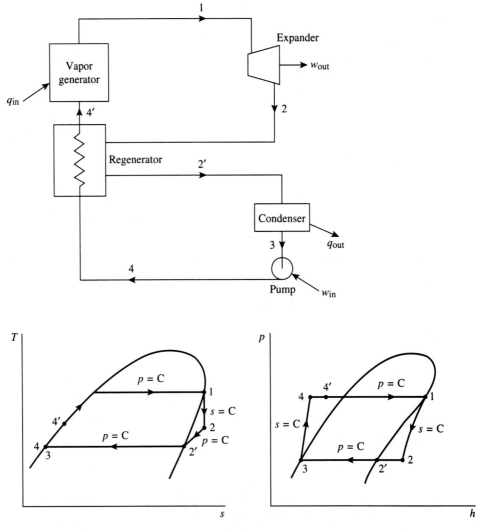

Figure 11-15 Rankine cycle with regeneration for a working fluid having a positively sloped saturated-vapor line on a T–s plot.

process (process 1–2) can be guaranteed in using these fluids in power cycles, the temperature at the end of expansion (T_2) is higher than the temperature of condensation ($T_{2'} = T_3$). Thus, an amount of heat rejection is necessary to desuperheat the exhaust vapor before condensing if a simple Rankine cycle is used. However, the energy to desuperheat the exhaust vapor can be used to our advantage by means of regeneration. This is done as shown in Fig. 11-15(a) by adding a heat exchanger, called a regenerator, to make the condensate exchange energy with the superheated exhaust vapor.

For the cycle of Fig. 11-15, assuming steady-flow operations and neglecting changes in kinetic and potential energies, the equations for the energy balances of the component parts are

$$h_2 - h_{2'} = h_{4'} - h_4 \tag{11-43}$$

$$w_{out} = w_{exp} = h_1 - h_2 \tag{11-44}$$

$$w_{in} = w_{pump} = h_4 - h_3 \tag{11-45}$$

$$q_{in} = q_{gen} = h_1 - h_{4'} \tag{11-46}$$

$$q_{out} = q_{cond} = h_{2'} - h_3 \tag{11-47}$$

The low-temperature Rankine cycle is most appropriate in solar-energy applications when flat-plate solar collectors are used to supply energy for the vapor generator. It is also applicable in ocean thermal-energy conversion in which the small temperature difference between the surface water and the water deep below the ocean surface is used to operate the power cycle.

Example 11-5

A low-temperature Rankine cycle uses Refrigerant-113 as the working fluid. At the inlet to the expander, the fluid is saturated at 200°F. It expands adiabatically in the expander to a pressure of 10 psia and a temperature of 140°F; and is then cooled in the regenerator at constant pressure to the saturation state before being condensed in the condenser. Pressure drops in the generator, condenser, regenerator, and pipelines are negligible. Determine the temperature at all state points of the cycle and calculate, per unit mass of working fluid, the expander work, pump work, heat added in the generator, heat rejected in the condenser, and the thermal efficiency of the cycle.

Solution The T–s and p–h diagrams are as depicted in Fig. 11-15(b), except that the expander is adiabatic but not reversible, resulting in increased entropy in the expansion process 1–2.

Referring to Fig. 11-15, the given data are $T_1 = 200°F$, $T_2 = 140°F$, and $p_2 = 10$ psia $= p_{2'} = p_3$. From the saturated Ref.-113 table (Table A-26), we obtain

$$p_1 = p_{sat} \text{ at } 200°F = 54.0 \text{ psia} = p_4 = p_{4'}$$

$$h_1 = h_g \text{ at } 200°F = 108.3 \text{ Btu/lbm}$$

$$s_1 = s_g \text{ at } 200°F = 0.183 \text{ Btu/lbm·°R}$$

$$T_{2'} = T_3 = T_{sat} \text{ at } 10 \text{ psia} = 97.2°F$$

$$h_{2'} = h_g \text{ at } 10 \text{ psia} = 92.7 \text{ Btu/lbm}$$

$$h_3 = h_f \text{ at } 10 \text{ psia} = 29.2 \text{ Btu/lbm}$$

and

$$v_3 = v_f \text{ at 10 psia} = 0.0104 \text{ cu ft/lbm}$$

From Table A-27 for superheated Ref.-113, at $p_2 = 10$ psia and $T_2 = 140°F$, we obtain

$$h_2 = 99.9 \text{ Btu/lbm} \qquad \text{and} \qquad s_2 = 0.186 \text{ Btu/lbm·°R}$$

As compressed-liquid data for Ref.-113 is lacking, let us assume the pump process to be internally reversible and adiabatic, so that we may write

$$h_4 = h_3 + v_3(p_4 - p_3) = 29.2 + (0.0104)(54 - 10)(144)/778$$

$$= 29.2 + 0.085 = 29.3 \text{ Btu/lbm}$$

Assuming $h_4 = h_f$ at T_4, we obtain from Table A-26, $T_4 \approx 97.7°F$.

Then an energy balance for the regenerator yields

$$h_{4'} = h_4 + h_2 - h_{2'} = 29.3 + 99.9 - 92.7 = 36.5 \text{ Btu/lbm}$$

Assuming $h_{4'} = h_f$ at $T_{4'}$, we obtain from Table A-26, $T_{4'} \approx 129°F$.

The temperature at the various state points are

State point	1	2	2'	3	4	4'
Temperature, °F	200	140	97.2	97.2	97.7	129

With the enthalpy value at every point known, we now calculate the work and heat quantities and the thermal efficiency.

$$w_{out} = w_{exp} = h_1 - h_2 = 108.3 - 99.9 = 8.4 \text{ Btu/lbm}$$

$$w_{in} = w_{pump} = h_4 - h_3 = 29.3 - 29.2 = 0.1 \text{ Btu/lbm}$$

$$q_{in} = q_{gen} = h_1 - h_{4'} = 108.3 - 36.5 = 71.8 \text{ Btu/lbm}$$

$$q_{out} = q_{cond} = h_{2'} - h_3 = 92.7 - 29.2 = 63.5 \text{ Btu/lbm}$$

$$\eta_{th} = \frac{w_{out} - w_{in}}{q_{in}} = \frac{q_{in} - q_{out}}{q_{in}}$$

$$= \frac{8.4 - 0.1}{71.8} = \frac{71.8 - 63.5}{71.8}$$

$$= 11.6\%$$

11-7 COMBINED GAS AND STEAM POWER CYCLE

The temperature of the exhaust from a nonregenerative gas turbine is typically very high. Instead of using a regenerator, the hot exhaust gases can be employed in a heat-recovery boiler to generate steam for a Rankine cycle, thus forming a gas–steam-turbine *combined power cycle*. Such a combined cycle will result in a higher thermal efficiency than either of the subcycles, because it takes advantage of both the high-temperature heat addition for a gas-turbine cycle and the low-temperature heat rejection for a vapor cycle. Figure 11-16 shows the schematic and ideal T–s diagrams for the gas and steam of a combined cycle. Because the gas-turbine exhaust contains a significant

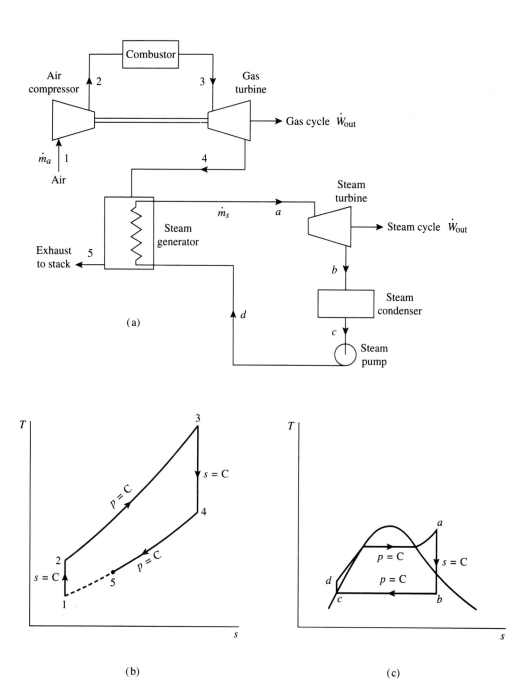

Figure 11-16 A gas-steam-turbine combined cycle: (a) schematic diagram, (b) ideal T–s diagram for the gas cycle, and (c) ideal T–s diagram for the steam cycle.

amount of free oxygen, additional fuel may be burned in the steam generator as needed.

If there is no additional energy input to the steam generator, an energy balance between the gas and the steam yields

$$\dot{m}_a(h_4 - h_5) = \dot{m}_s(h_a - h_d) \tag{11-48}$$

where $\dot{m}_a$ and $\dot{m}_s$ are the mass-flow rates of air and steam, respectively. The total net power output of the combined cycle is

$$\text{net } \dot{W}_{\text{out}} = \dot{m}_a[(h_3 - h_4) - (h_2 - h_1)] + \dot{m}_s[(h_a - h_b) - (h_d - h_c)] \tag{11-49}$$

The rate of heat addition to the combined cycle is

$$\dot{Q}_{\text{in}} = \dot{m}_a(h_3 - h_2) \tag{11-50}$$

The rate of heat rejection from the combined cycle is

$$\dot{Q}_{\text{out}} = \dot{m}_a(h_5 - h_1) + \dot{m}_s(h_b - h_c) \tag{11-51}$$

From the first law, we must have

$$\text{net } \dot{W}_{\text{out}} = \dot{Q}_{\text{in}} - \dot{Q}_{\text{out}}$$

The thermal efficiency of the combined cycle is

$$\begin{aligned}
\eta_{\text{th, comb}} &= \frac{\text{net } \dot{W}_{\text{out}}}{\dot{Q}_{\text{in}}} = \frac{\dot{Q}_{\text{in}} - \dot{Q}_{\text{out}}}{\dot{Q}_{\text{in}}} = 1 - \frac{\dot{Q}_{\text{out}}}{\dot{Q}_{\text{in}}} \\
&= 1 - \frac{\dot{m}_a(h_5 - h_1) + \dot{m}_s(h_b - h_c)}{\dot{m}_a(h_3 - h_2)}
\end{aligned} \tag{11-52}$$

For the gas cycle alone, the thermal efficiency is

$$\eta_{\text{th, gas}} = \frac{(h_3 - h_4) - (h_2 - h_1)}{h_3 - h_2} \tag{11-53}$$

For the steam cycle alone, the thermal efficiency is

$$\eta_{\text{th, steam}} = \frac{(h_a - h_b) - (h_d - h_c)}{h_a - h_d} \tag{11-54}$$

Example 11-6

In a gas–steam-turbine combined cycle, air enters the compressor at 100 kPa, 22°C, and leaves at 1100 kPa. The cold-air-standard gas-turbine cycle with a maximum temperature of 1200°C is to produce 8 MW of net power output. On the Rankine-cycle subsystem, steam enters the turbine at 3 MPa, 400°C, and leaves at 7.5 kPa. There is no additional heat input to the steam generator and the stack gas temperature is 200°C. Assuming all processes to be ideal, determine (a) the air and steam mass-flow rates, (b) the thermal efficiency of the combined cycle, and (c) the thermal efficiencies of the gas and steam sub-cycles. For air, $c_p = 1.005$ kJ/kg·K and $k = 1.40$.

Solution Referring to Fig. 11-16, the given data are $p_1 = 100$ kPa, $T_1 = 22°C = 295$ K, $p_2 = 1100$ kPa, $T_3 = 1200°C = 1473$ K, $T_5 = 200°C = 473$ K, net $\dot{W}_{\text{out, gas}} = 8$ MW, $p_a = 3$ MPa, $T_a = 400°C$, and $p_b = 7.5$ kPa.

For the internally reversible adiabatic ideal-gas processes 1–2 and 3–4,

$$T_2 = T_1\left(\frac{p_2}{p_1}\right)^{(k-1)/k} = 295\left(\frac{1100}{100}\right)^{(1.4-1)/1.4} = 585 \text{ K} = 312°C$$

$$T_4 = T_3\left(\frac{p_4}{p_3}\right)^{(k-1)/k} = 1473\left(\frac{100}{1100}\right)^{(1.4-1)/1.4} = 742 \text{ K} = 469°C$$

The net power output of the gas cycle is given by

$$\text{net } \dot{W}_{\text{out, gas}} = \dot{m}_a[(h_3 - h_4) - (h_2 - h_1)]$$

$$= \dot{m}_a c_p[(T_3 - T_4) - (T_2 - T_1)]$$

or

$$8000 = \dot{m}_a(1.005)[(1473 - 742) - (585 - 295)]$$

Therefore, $\dot{m}_a = 18.05$ kg/s.

From the steam tables (Tables A-19M and A-20M), at $p_a = 3$ MPa and $T_a = 400°C$, we get $h_a = 3230.9$ kJ/kg and $s_a = 6.9212$ kJ/kg·K. Also

$$h_c = h_f \text{ at } 7.5 \text{ kPa} = 168.8 \text{ kJ/kg}$$

$$v_c = v_f \text{ at } 7.5 \text{ kPa} = 0.001008 \text{ m}^3/\text{kg}$$

For the internally reversible adiabatic process a–b, we have

$$s_a = s_b = s_f + x_b s_{fg} \text{ at } 7.5 \text{ kPa}$$

or

$$6.9212 = 0.5764 + x_b(7.6750)$$

so that

$$x_b = 0.827$$

Therefore,

$$h_b = h_f + x_b h_{fg} \text{ at } 7.5 \text{ kPa}$$

$$= 168.8 + 0.827(2406.0) = 2158.6 \text{ kJ/kg}$$

For the internally reversible adiabatic process c–d, we have

$$h_d = h_c + v_c(p_d - p_c)$$

$$= 168.8 + 0.001008(3000 - 7.5) = 171.8 \text{ kJ/kg}$$

An energy balance for the steam generator gives

$$\dot{m}_s = \dot{m}_a\left(\frac{h_4 - h_5}{h_a - h_d}\right) = \dot{m}_a\left[\frac{c_p(T_4 - T_5)}{h_a - h_d}\right]$$

$$= 18.05\left[\frac{1.005(742 - 473)}{3230.9 - 171.8}\right] = 1.60 \text{ kg/s}$$

The thermal efficiency of the combined cycle is

$$\eta_{\text{th, comb}} = 1 - \frac{\dot{m}_a(h_5 - h_1) + \dot{m}_s(h_b - h_c)}{\dot{m}_a(h_3 - h_2)}$$

$$= 1 - \frac{\dot{m}_a c_p(T_5 - T_1) + \dot{m}_s(h_b - h_c)}{\dot{m}_a c_p(T_3 - T_2)}$$

$$= 1 - \frac{18.05(1.005)(473 - 295) + 1.60(2158.6 - 168.8)}{18.05(1.005)(1473 - 585)}$$

$$= 0.602 = 60.2\%$$

The thermal efficiency of the gas cycle is

$$\eta_{th,\,gas} = \frac{(h_3 - h_4) - (h_2 - h_1)}{h_3 - h_2}$$

$$= \frac{(T_3 - T_4) - (T_2 - T_1)}{T_3 - T_2} = \frac{(1473 - 742) - (585 - 295)}{1473 - 585}$$

$$= 0.497 = 49.7\%$$

The thermal efficiency of the steam cycle is

$$\eta_{th,\,steam} = \frac{(h_a - h_b) - (h_d - h_c)}{h_a - h_d}$$

$$= \frac{(3230.9 - 2158.6) - (171.8 - 168.8)}{3230.9 - 171.8}$$

$$= 0.350 = 35.0\%$$

Notice that the thermal efficiency of this combined cycle (60.2%) is considerably higher than the thermal efficiency of the gas cycle (49.7%) or the steam cycle (35.0%) operating alone.

11-8 COGENERATION

Cogeneration is the generation of steam for simultaneous use in electricity production and thermal energy supply for space or industrial-process heating. It is a highly energy-conservative scheme suitable for large college campuses and industrial complexes where a steam power station on site also serves as the source of steam for space hot-water heating or industrial requirement. The performance of a cogeneration system can be expressed in term of its effectiveness η_{cog} as defined by

$$\eta_{cog} = \frac{\text{electric energy delivered} + \text{thermal energy delivered}}{\text{heat added from combustion}} \qquad (11\text{-}55)$$

We now proceed to illustrate several possible cogeneration arrangements. Figure 11-17 shows a cogeneration system using a back-pressure turbine that operates on a high exhaust pressure to satisfy the temperature demand of the heating load. The condenser of the simple Rankine cycle is replaced by the heating load, making the turbine and heating loads act in series operation. The bypass line through the pressure-reducing valve provides flexibility in serving varying power and heating requirements.

Figure 11-18 shows a cogeneration system with the turbine and heating loads in parallel operation. The Rankine-cycle condenser operates on a usual low vacuum pressure so as to achieve a high thermal efficiency for the cycle. However, the condensing pressure of the heating steam will depend on the temperature demand of the load. The mass-flow rates for turbine ($\dot{m}_T$) and for heating ($\dot{m}_H$) can be easily adjusted, making this system more flexible in meeting the varying load demands.

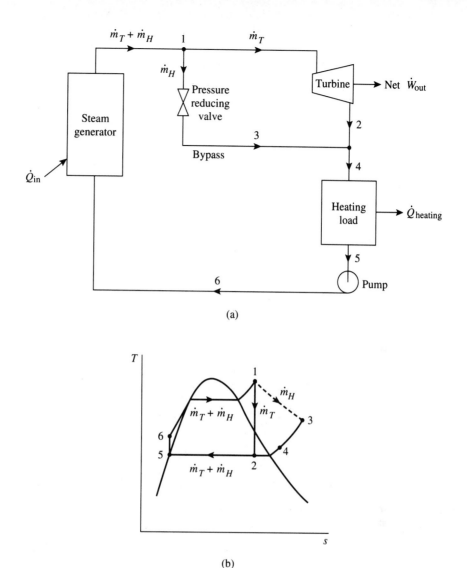

(a)

(b)

Figure 11-17 Cogeneration with a back-pressure turbine: turbine and heating loads in series operation: (a) schematic diagram, and (b) ideal T–s diagram.

Figure 11-19 shows a cogeneration system with an extraction turbine. Steam is extracted from an intermediate pressure in the turbine to provide energy for the heating load. The steam that is not extracted continues to expand to the usual vacuum pressure of the condenser.

Example 11-7

A cogeneration plant is designed to deliver 8 MW of electric power and 50 GJ/h of thermal energy for heating. Steam is generated at 4.5 MPa and 450°C. The heating steam is supplied at 100 kPa and the condenser pressure is 7.5 kPa. For each of the cogeneration systems as depicted in Figs. 11-18 and 11-19 (assuming ideal processes), determine (a)

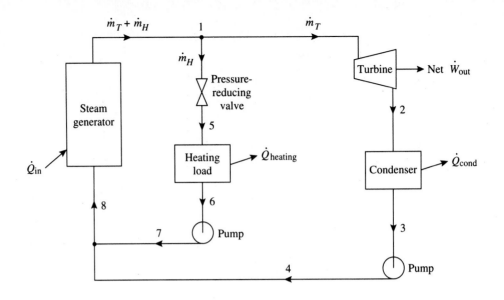

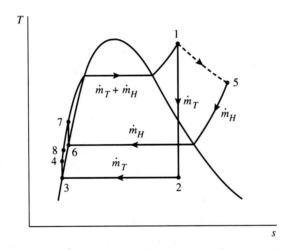

Figure 11-18 Cogeneration system with turbine and heating loads in parallel operation: (a) schematic diagram, and (b) ideal T–s diagram.

the amounts of steam required per hour for power and for heating, (b) the heat addition per hour to the steam generator, (c) the heat rejection per hour from the condenser, and (d) the effectiveness of the cogeneration system.

Solution

$$\text{Electric load} = 8 \text{ MW} = 8000 \text{ kW}$$

$$\text{Heating load} = 50 \times 10^6 \text{ kJ/h} = 13890 \text{ kW}$$

(A) For the cogeneration system shown in Fig. 11-18, the given data are $p_1 =$ 4.5 MPa, $T_1 = 450°C$, $p_5 = 100$ kPa, and $p_2 = 7.5$ kPa. From the steam tables (Tables A-19M and A-20M), at p_1 and T_1, we get $h_1 = h_5 = 3323.3$ kJ/kg, and $s_1 = 6.8746$ kJ/kg·K. Also

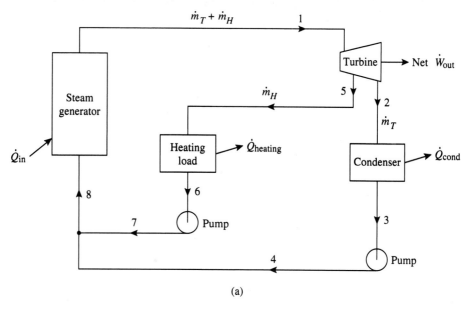

(a)

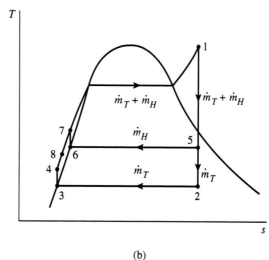

(b)

Figure 11-19 Cogeneration with an extraction turbine: (a) schematic diagram, and (b) ideal T–s diagram.

$$h_3 = h_f \text{ at } 7.5 \text{ kPa} = 168.8 \text{ kJ/kg}$$

$$v_3 = v_f \text{ at } 7.5 \text{ kPa} = 0.001008 \text{ m}^3/\text{kg}$$

$$h_6 = h_f \text{ at } 100 \text{ kPa} = 417.5 \text{ kJ/kg}$$

$$v_6 = v_f \text{ at } 100 \text{ kPa} = 0.001043 \text{ m}^3/\text{kg}$$

Because

$$s_1 = s_2 = 6.8746 = s_f + x_2 s_{fg} \text{ at } 7.5 \text{ kPa} = 0.5764 + x_2(7.6750)$$

so that

$$x_2 = 0.821$$

Therefore,

$$h_2 = h_f + x_2 h_{fg} \text{ at } 7.5 \text{ kPa}$$
$$= 168.8 + 0.821(2406.0) = 2144.1 \text{ kJ/kg}$$

Furthermore,

$$h_4 = h_3 + v_3(p_4 - p_3)$$
$$= 168.8 + 0.001008(4500 - 7.5) = 173.3 \text{ kJ/kg}$$
$$h_7 = h_6 + v_6(p_7 - p_6)$$
$$= 417.5 + 0.001043(4500 - 100) = 422.1 \text{ kJ/kg}$$

An energy balance for the heating load gives the value of the mass-flow rate for heating:

$$\dot{m}_H = \frac{\dot{Q}_{\text{heating}}}{h_5 - h_6} = \frac{13890 \text{ kJ/s}}{(3323.3 - 417.5) \text{ kJ/kg}}$$
$$= 4.78 \text{ kg/s} = 17,210 \text{ kg/h}$$

Because

$$\text{net } \dot{W}_{\text{out}} = \dot{m}_T(h_1 - h_2) - \dot{m}_T(h_4 - h_3) - \dot{m}_H(h_7 - h_6)$$

the value of the mass-flow rate for power is

$$\dot{m}_T = \frac{\text{net } \dot{W}_{\text{out}} + \dot{m}_H(h_7 - h_6)}{(h_1 - h_2) - (h_4 - h_3)}$$
$$= \frac{8000 + 4.78(422.1 - 417.5)}{(3323.3 - 2144.1) - (173.3 - 168.8)}$$
$$= 6.83 \text{ kg/s} = 24,590 \text{ kg/h}$$

An energy balance for the mixing point where $\dot{m}_T$ and $\dot{m}_H$ meet yields

$$h_8 = \frac{\dot{m}_T h_4 + \dot{m}_H h_7}{\dot{m}_T + \dot{m}_H}$$
$$= \frac{6.83(173.3) + 4.78(422.1)}{6.83 + 4.78} = 275.7 \text{ kJ/kg}$$

The heat addition to the steam generator is

$$\dot{Q}_{\text{in}} = \dot{Q}_{\text{st gen}} = (\dot{m}_T + \dot{m}_H)(h_1 - h_8)$$
$$= (6.83 + 4.78)(3323.3 - 275.7)$$
$$= 35,380 \text{ kJ/s} = 1.274 \times 10^8 \text{ kJ/h}$$

The heat rejection from the condenser is

$$\dot{Q}_{\text{cond}} = \dot{m}_T(h_2 - h_3) = 6.83(2144.1 - 168.8)$$
$$= 13,490 \text{ kJ/s} = 4.86 \times 10^7 \text{ kJ/h}$$

The cogeneration system effectiveness is

$$\eta_{cog} = \frac{\text{electric energy delivered + thermal energy delivered}}{\text{heat added to steam generator}}$$

$$= \frac{8000 + 13,890}{35,380} = 0.619 = 61.9\%$$

(B) For the cogeneration system shown in Fig. 11-19, the known values are $p_1 = 4.5$ MPa, $T_1 = 450°C$, $p_5 = 100$ kPa, $p_2 = 7.5$ kPa, $h_1 = 3323.3$ kJ/kg, $s_1 = 6.8746$ kJ/kg·K, $h_2 = 2144.1$ kJ/kg, $h_3 = 168.8$ kJ/kg, $h_4 = 173.3$ kJ/kg, $h_6 = 417.5$ kJ/kg, and $h_7 = 422.1$ kJ/kg.

Because

$$s_1 = s_5 = 6.8746 = s_f + x_5 s_{fg} \text{ at } 100 \text{ kPa}$$

$$= 1.3026 + x_5(6.0568)$$

so that

$$x_5 = 0.920$$

Therefore,

$$h_5 = h_f + x_5 h_{fg} \text{ at } 100 \text{ kPa}$$

$$= 417.46 + 0.920(2258.0) = 2494.8 \text{ kJ/kg}$$

An energy balance for the heating load gives

$$\dot{m}_H = \frac{\dot{Q}_{heating}}{h_5 - h_6} = \frac{13,890}{2494.8 - 417.5}$$

$$= 6.69 \text{ kg/s} = 24,080 \text{ kg/h}$$

Because

$$\text{net } \dot{W}_{out} = (\dot{m}_T + \dot{m}_H)(h_1 - h_5) + \dot{m}_T(h_5 - h_2) - \dot{m}_T(h_4 - h_3) - \dot{m}_H(h_7 - h_6)$$

so that

$$\dot{m}_T = \frac{\text{net } \dot{W}_{out} - \dot{m}_H[(h_1 - h_5) - (h_7 - h_6)]}{(h_1 - h_5) + (h_5 - h_2) - (h_4 - h_3)}$$

$$= \frac{8000 - 6.69[(3323.3 - 2494.8) - (422.1 - 417.5)]}{(3323.3 - 2144.1) - (173.3 - 168.8)}$$

$$= 2.12 \text{ kg/s} = 7630 \text{ kg/h}$$

An energy balance for the mixing point where $\dot{m}_T$ and $\dot{m}_H$ meet yields

$$h_8 = \frac{\dot{m}_T h_4 + \dot{m}_H h_7}{\dot{m}_T + \dot{m}_H}$$

$$= \frac{2.12(173.3) + 6.69(422.1)}{2.12 + 6.69} = 362.2 \text{ kJ/kg}$$

The heat addition to the steam generator is

$$\dot{Q}_{in} = \dot{Q}_{st\,gen} = (\dot{m}_T + \dot{m}_H)(h_1 - h_8)$$

$$= (2.12 + 6.69)(3323.3 - 362.2)$$

$$= 26,090 \text{ kJ/s} = 9.39 \times 10^7 \text{ kJ/h}$$

The heat rejection from the condenser is

$$\dot{Q}_{cond} = \dot{m}_T(h_2 - h_3) = 2.12(2144.1 - 168.8)$$

$$= 4190 \text{ kJ/s} = 1.51 \times 10^7 \text{ kJ/h}$$

The cogeneration system effectiveness is

$$\eta_{cog} = \frac{8000 + 13{,}890}{26{,}090} = 0.839 = 83.9\%$$

Therefore, this cogeneration system puts 83.9% of the energy supplied to it into useful purposes.

11-9 AVAILABILITY ANALYSIS FOR VAPOR POWER CYCLES

A vapor cycle based on an ideal Carnot cycle is totally reversible, and thus it does not involve any irreversibilities. The ideal Rankine cycles, including simple, reheat, and regenerative variations, however, are only internally reversible, and thus may involve irreversibilities external to the system. Actual vapor cycles of any kind may involve internal as well as external irreversibilities.

Equations for availability and irreversibility were developed in Chapter 8. Because steam-cycle processes are usually working on steady-flow, steady-state conditions, it follows that $\Delta s_{system} = 0$ and the specific irreversibility for each process is given by

$$i = T_0(\Delta s_{system}^{\;0} = \Delta s_{surr})$$

$$= T_0\left[(s_2 - s_1) + \frac{-q_R}{T_R} \right]$$

for one inlet at section 1 and one exit at section 2. q_R is the heat added to the fluid stream from an energy reservoir at T_R.

The irreversibility of a cycle is the sum of the irreversibilities of the processes that form the cycle. That is,

$$i_{cycle} = \sum i_{process}$$

The irreversibility of a cycle can also be determined by considering the entire cycle as a single process. Because the initial and final states of a cycle are identical, $\Delta s_{system} = 0$, we then have

$$i_{cycle} = T_0(\Delta s_{system}^{\;0} + \Delta s_{surr})$$

$$= T_0\left[\sum \left(\frac{-q_R}{T_R} \right) \right]$$

where q_R is positive when heat is added to the system.

The specific stream availability ψ of a steady-flow fluid stream in a given state is given by

$$\psi = (h + \tfrac{1}{2}\mathbf{V}^2 + gz - T_0 s) - (h_0 + gz_0 - T_0 s_0)$$

where the subscript 0 denotes the dead state.

The second-law effectiveness of a steam power cycle can be defined as

$$\epsilon'_{\text{power cycle}} = \frac{\text{net } w_{\text{out}}}{\Delta\psi_{\text{steam generator and reheater}}}$$

The following examples are given to illustrate the procedure of calculation for second-law analyses of vapor power cycles, whereby inefficient processes of certain components can be singled out for performance improvement.

Example 11-8

A two-feedwater-heater regenerative steam power plant shown in the schematic diagram of Fig. 11-20 has the pressure p, temperature T, and quality x at various points as indicated. The turbine and pumps operate adiabatically. The atmosphere is at 60°F and 14.7 psia. Neglecting kinetic and potential energies, (a) determine the cycle thermal efficiency, and (b) make an availability accounting per pound of steam at the turbine throttle.

Solution From the superheated steam table (Table A-20E), at $p_1 = 1150$ psia and $T_1 = 920$°F, we get

$$h_1 = 1454.3 \text{ Btu/lbm} \quad \text{and} \quad s_1 = 1.6020 \text{ Btu/lbm·°R}$$

and at $p_2 = 145$ psia and $T_2 = 490$°F, we get

$$h_2 = 1269.3 \text{ Btu/lbm} \quad \text{and} \quad s_2 = 1.6584 \text{ Btu/lbm·°R}$$

From the saturated steam table (Table A-19E), at $p_3 = p_7 = 18$ psia, we get

$$T_3 = T_7 = T_{\text{sat}} = 222.40\text{°F}$$

$$h_7 = h_f = 190.64 \text{ Btu/lbm}$$

$$s_7 = s_f = 0.32762 \text{ Btu/lbm·°R}$$

$$h_3 = h_f + x_3 h_{fg} = 190.64 + 0.99(963.7) = 1144.7 \text{ Btu/lbm}$$

$$s_3 = s_f + x_3 s_{fg} = 0.32762 + 0.99(1.4128) = 1.7263 \text{ Btu/lbm·°R}$$

Also from Table A-19E, at $p_4 = p_5 = 0.7$ psia, we get

$$T_4 = T_5 = T_{\text{sat}} = 90.05\text{°F}$$

$$h_5 = h_f = 58.12 \text{ Btu/lbm}$$

$$s_5 = s_f = 0.11174 \text{ Btu/lbm·°R}$$

$$h_4 = h_f + x_4 h_{fg} = 58.12 + 0.90(1042.6) = 996.5 \text{ Btu/lbm}$$

$$s_4 = s_f + x_4 s_{fg} = 0.11174 + 0.90(1.8964) = 1.8185 \text{ Btu/lbm·°R}$$

Also from Table A-19E, at $p_9 = 145$ psia, we get

$$h_9 = h_f = 327.93 \text{ Btu/lbm} \quad \text{and} \quad s_9 = s_f = 0.51079 \text{ Btu/lbm·°R}$$

For the throttling process 9–10, we have

$$h_{10} = h_9 = 327.93 \text{ Btu/lbm} \quad \text{and} \quad h_{10} = h_f + x_{10} h_{fg} \text{ at 18 psia}$$

Thus, Table A-19E gives

$$327.93 = 190.64 + x_{10}(963.7)$$

so that

$$x_{10} = 0.1425$$

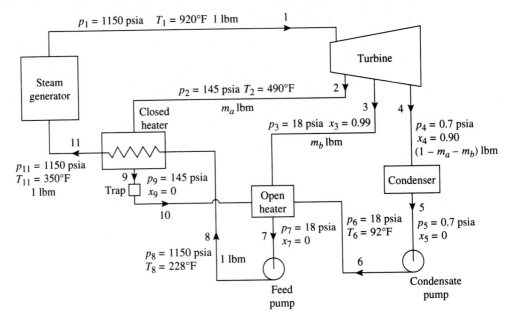

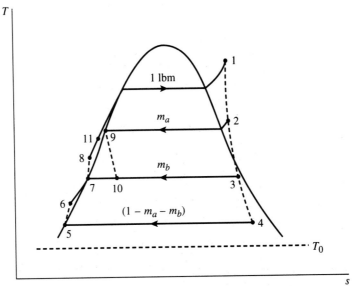

Figure 11-20   Schematic and T–s diagrams for Example 11-8.

Therefore,

$$s_{10} = s_f + x_{10}s_{fg} \text{ at 18 psia}$$

$$= 0.32762 + 0.1425(1.4128)$$

$$= 0.52894 \text{ Btu/lbm·°R}$$

As state 6 is a compressed liquid at a low pressure, from the saturated-steam table (Table A-18E), at $T_6 = 92°F$, we have

$$h_6 \approx h_f = 60.06 \text{ Btu/lbm} \qquad \text{and} \qquad s_6 \approx s_f = 0.11527 \text{ Btu/lbm·°R}$$

On the other hand, states 8 and 11 are compressed liquids at a high pressure, so the compressed liquid table (Table A-21E) must be used. From this table, by interpolation, at $p_8 = 1150$ psia and $T_8 = 228°F$, we get

$$h_8 = 198.8 \text{ Btu/lbm} \qquad \text{and} \qquad s_8 = 0.33375 \text{ Btu/lbm·°R}$$

and at $p_{11} = 1150$ psia and $T_{11} = 350°F$, we get

$$h_{11} = 323.4 \text{ Btu/lbm} \qquad \text{and} \qquad s_{11} = 0.50108 \text{ Btu/lbm·°R}$$

An energy balance for the closed heater yields

$$m_a = \frac{h_{11} - h_8}{h_2 - h_9} = \frac{323.4 - 198.8}{1269.3 - 327.93} = 0.1324 \text{ lbm/lbm throttle}$$

An energy balance for the open heater yields

$$m_b h_3 + m_a h_{10} + (1 - m_a - m_b)h_6 = 1h_7$$

or

$$m_b = \frac{h_7 - h_6 - m_a(h_{10} - h_6)}{h_3 - h_6}$$

$$= \frac{190.64 - 60.06 - 0.1324(327.93 - 60.06)}{1144.7 - 60.06}$$

$$= 0.0877 \text{ lbm/lbm throttle}$$

(a) The heat added and heat rejected for the cycle are

$$q_{in} = 1(h_1 - h_{11}) = 1(1454.3 - 323.4) = 1130.9 \text{ Btu/lbm throttle}$$

$$q_{out} = (1 - m_a - m_b)(h_4 - h_5) = (1 - 0.1324 - 0.0877)(996.5 - 58.12)$$

$$= 731.8 \text{ Btu/lbm throttle}$$

The net work output of the cycle is

$$\text{net } w_{out} = \text{net } q_{in} = q_{in} - q_{out}$$

$$= 1130.9 - 731.8 = 399.1 \text{ Btu/lbm throttle}$$

The cycle thermal efficiency is then

$$\eta_{th} = \frac{\text{net } w_{out}}{q_{in}} = \frac{399.1}{1130.9} = 0.353 = 35.3\%$$

(b) Because changes in kinetic and potential energies are neglected, the change in stream availability for a process is given by Eq. (8-20):

$$\Delta\psi = \Delta h - T_0 \, \Delta s$$

Thus, with $T_0 = 60°F = 520°R$,

$$\Delta\psi_{st\,gen} = \Delta\psi_{11-1} = 1[(h_1 - h_{11}) - T_0(s_1 - s_{11})]$$

$$= 1[(1454.3 - 323.4) - 520(1.6020 - 0.50108)]$$

$$= 558.4 \text{ Btu/lbm throttle}$$

$$\Delta\psi_{cond} = \Delta\psi_{5-4} = (1 - m_a - m_b)[(h_4 - h_5) - T_0(s_4 - s_5)]$$

$$= (1 - 0.1324 - 0.0877)[(996.5 - 58.12) - 520(1.8185 - 0.11174)]$$

$$= 39.7 \text{ Btu/lbm throttle}$$

The irreversibilities of the turbine, the pumps, and the feedwater heaters are calculated by Eq. (8-12),

$$i = T_0(\Delta s_{\text{system}} + \Delta s_{\text{surr}})$$

$$= T_0(\Delta s_{\text{surr}}) \text{ for steady-flow processes}$$

Thus,

$$i_{\text{turbine}} = T_0[m_a s_2 + m_b s_3(1 - m_a - m_b)s_4 - 1 s_1]$$

$$= 520[0.1324(1.6584) + 0.0877(1.7263)$$

$$+ (1 - 0.1324 - 0.0877)(1.8185) - 1(1.6020)]$$

$$= 97.3 \text{ Btu/lbm throttle}$$

$$i_{\text{condensate pump}} = T_0[(1 - m_a - m_b)(s_6 - s_5)]$$

$$= 520[(1 - 0.1324 - 0.0877)(0.11527 - 0.11174)]$$

$$= 1.4 \text{ Btu/lbm throttle}$$

$$i_{\text{feed pump}} = T_0[1(s_8 - s_7)]$$

$$= 520[1(0.33375 - 0.32762)]$$

$$= 3.2 \text{ Btu/lbm throttle}$$

$$i_{\text{open heater}} = T_0[1 s_7 - m_a s_{10} - m_b s_3 - (1 - m_a - m_b)s_6]$$

$$= 520[1(0.32762) - 0.1324(0.52894) - 0.0877(1.7263)$$

$$- (1 - 0.1324 - 0.0877)(0.11527)]$$

$$= 8.5 \text{ Btu/lbm throttle}$$

$$i_{\text{closed heater and trap}} = T_0[1(s_{11} - s_8) + m_a(s_{10} - s_2)]$$

$$= 520[1(0.50108 - 0.33375)$$

$$+ 0.1324(0.52894 - 1.6584)]$$

$$= 9.2 \text{ Btu/lbm throttle}$$

The increase in stream availability

$$\Delta \psi_{\text{st gen}} = 558.4 \text{ Btu/lbm throttle}$$

is disposed of as the sum:

$$\text{net } w_{\text{out}} + \Delta \psi_{\text{cond}} + i_{\text{turb}} + i_{\text{condensate pump}}$$

$$+ i_{\text{feed pump}} + i_{\text{open heater}} + i_{\text{closed heater and trap}}$$

$$= 399.1 + 39.7 + 97.3 + 1.4 + 3.2 + 8.5 + 9.2$$

$$= 558.4 \text{ Btu/lbm throttle}$$

The availability accounting is shown graphically in Fig. 11-21. It is seen that for this steam power cycle, the biggest irreversibility occurs in the turbine, which requires special attention. The condenser performance requires improvement, too.

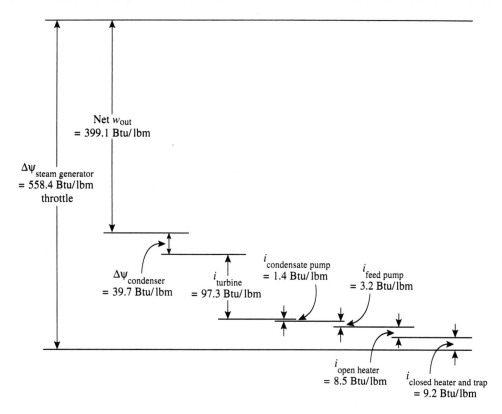

Figure 11-21 Availability accounting for Example 11-8 (per pound of throttle steam).

Example 11-9

For the reheat steam cycle described in Example 11-2, determine (a) the changes in stream availability for the steam generator and reheater and the condenser, (b) the irreversibility for the turbines and the pump, and (c) the second-law effectiveness of the cycle. The atmosphere is at $T_0 = 290$ K.

Solution The $T-s$ diagram of this reheat steam cycle is shown in Fig. 11-22, which is the same as Fig. 11-6. From Example 11-2, we have the following data: $p_1 = 7.0$ MPa, $T_1 = 450°C$, $h_1 = 3287.1$ kJ/kg, $s_1 = 6.6327$ kJ/kg·K, $p_{2'} = 0.5$ MPa, $h_{2'} = 2761.4$ kJ/kg, $p_3 = 0.5$ MPa, $T_3 = 450°C$, $h_3 = 3377.2$ kJ/kg, $s_3 = 7.9445$ kJ/kg·K, $p_{4'} = 3$ kPa, $h_{4'} = 2510.4$ kJ/kg, $x_{4'} = 0.986$, $p_5 = 3$ kPa, $h_5 = 101.1$ kJ/kg, $p_{6'} = 7.0$ MPa, $h_{6'} = 110.8$ kJ/kg, net $\dot{W}_{out} = 100$ MW, net $w_{out} = 1392.5 - 9.7 = 1382.8$ kJ/kg, $\dot{m} = 72.3$ kg/s, and $\eta_{th} = 36.5\%$.

At $p_{2'} = 0.5$ MPa and $h_{2'} = 2761.4$ kJ/kg, the steam table gives $s_{2'} = 6.8508$ kJ/kg·K. At $p_{4'} = 3$ kPa and $x_{4'} = 0.986$, we have

$$s_{4'} = s_f + x_{4'} s_{fg} = 0.3545 + 0.986(8.2231)$$

$$= 8.4625 \text{ kJ/kg·K}$$

At $p_5 = 3$ kPa, $s_5 = s_f = 0.3545$ kJ/kg·K. At $p_{6'} = 7.0$ MPa and $h_{6'} = 110.8$ kJ/kg, from the compressed liquid water table, by interpolation, we have $s_{6'} = 0.2629$ kJ/kg·K.

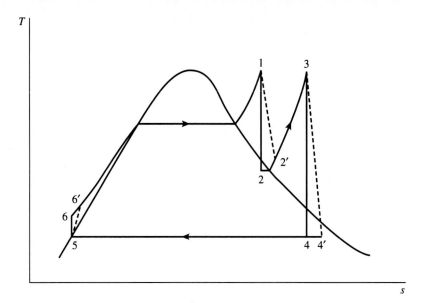

Figure 11-22 T–s diagram for Example 11-9.

(a) Neglecting changes in KE and PE, the change in stream availability for a process is given by

$$\Delta\psi = \Delta h - T_0 \Delta s$$

Thus,

$$\Delta\psi_{\text{st gen and reheater}} = [(h_1 - h_{6'}) + (h_3 - h_{2'})] - T_0[(s_1 - s_{6'}) + (s_3 - s_{2'})]$$

$$= [(3287.1 - 110.8) + (3377.2 - 2761.4)]$$

$$\quad - 290[(6.6327 - 0.3629) + (7.9445 - 6.8508)]$$

$$= 1656.7 \text{ kJ/kg}$$

$$= (1656.7 \text{ kJ/kg})(72.3 \text{ kg/s}) = 119,800 \text{ kJ/s}$$

$$\Delta\psi_{\text{cond}} = (h_{4'} - h_5) - T_0(s_{4'} - s_5)$$

$$= (2510.4 - 101.1) - 290(8.4625 - 0.3545)$$

$$= 58.0 \text{ kJ/kg} = 4190 \text{ kJ/s}$$

(b) The irreversibilities of the turbines and the pump are given by Eq. (8-12):

$$i = T_0(\Delta s_{\text{system}} + \Delta s_{\text{surr}})$$

$$= T_0(\Delta s_{\text{surr}}) \text{ for steady-flow, steady-state processes}$$

Thus,

$$i_{\text{turbines}} = T_0[(s_{2'} - s_1) + (s_{4'} - s_3)]$$

$$= 290[(6.8508 - 6.6327) + (8.4625 - 7.9445)]$$

$$= 213.5 \text{ kJ/kg} = 15,440 \text{ kJ/s}$$

$$i_{\text{pump}} = T_0(s_{6'} - s_5) = 290(0.3629 - 0.3545)$$

$$= 2.4 \text{ kJ/kg} = 170 \text{ kJ/s}$$

The increase in stream availability $\Delta\psi_{\text{st gen and reheater}} = 1656.7 \text{ kJ/kg} = 119,800 \text{ kJ/s}$ is disposed of as the sum:

$$\text{net } w_{\text{out}} + \Delta\psi_{\text{cond}} + i_{\text{turbines}} + i_{\text{pump}}$$

$$= 1382.8 + 58.0 + 213.5 + 2.4 = 1656.7 \text{ kJ/kg}$$

$$= 100,000 + 4190 + 15,440 + 170 = 119,800 \text{ kJ/s}$$

(c) The second-law effectiveness of the cycle is given by Eq.(8-29):

$$\epsilon'_{\text{power cycle}} = \text{net } w_{\text{out}}/\Delta\psi_{\text{st gen and reheater}}$$

$$= 1382.8/1656.7 = 0.835 = 83.5\%$$

$$= 100,000/119,800 = 0.835 = 83.5\%$$

Note that the first-law efficiency of the cycle as calculated in Example 11-2 is $\eta_{\text{th}} = 0.365 = 36.5\%$.

Example 11-10

For the reheat steam cycle described in Example 11-9 (also Example 11-2), determine the irreversibilities associated with each of the processes, in kJ/kg, assuming a source temperature 1400 K and a sink temperature of 290 K. Determine also the irreversibility of the cycle, in kJ/kg. Give an account for the available energy supplied from the high-temperature reservoir.

Solution Given $T_H = 1400$ K and $T_L = T_0 = 290$ K. Referring to Fig. 11-22, from Example 11-9, we have the following data: net w_{out} 1382.8 kJ/kg, $h_1 = 3287.1$ kJ/kg, $s_1 = 6.6327$ kJ/kg·K, $h_{2'} = 2761.4$ kJ/kg, $s_{2'} = 6.8508$ kJ/kg·K, $h_3 = 3377.2$ kJ/kg, $s_3 = 7.9445$ kJ/kg·K, $h_{4'} = 2510.4$ kJ/kg, $s_{4'} = 8.4625$ kJ/kg·K, $h_5 = 101.1$ kJ/kg, $s_5 = 0.3545$ kJ/kg·K, $h_{6'} = 110.8$ kJ/kg, and $s_{6'} = 0.3629$ kJ/kg·K.

The values of heat transfer in the steam generator, the reheater, and the condenser are calculated as

$$q_{\text{in}, 6'1} = h_1 - h_{6'} = 3287.1 - 110.8 = 3176.3 \text{ kJ/kg}$$

$$q_{\text{in}, 2'3} = h_3 - h_{2'} = 3377.2 - 2761.4 = 615.8 \text{ kJ/kg}$$

$$q_{\text{out}, 4'5} = h_{4'} - h_5 = 2510.4 - 101.1 = 2409.3 \text{ kJ/kg}$$

The irreversibilities associated with the processes are given by Eq. (8-12):

$$i = T_0(\Delta s_{\text{system}} + \Delta s_{\text{surr}})$$

$$= T_0(\Delta s_{\text{surr}}) \text{ for steady flow and steady state}$$

Thus,

$$i_{\text{st gen}, 6'1} = T_0(s_1 - s_{6'} - q_{6'1}/T_H)$$

$$= 290(6.6327 - 0.3629 - 3176.3/1400) = 1160.3 \text{ kJ/kg}$$

$$i_{\text{reheater}, 2'3} = T_0(s_3 - s_{2'} - q_{2'3}/T_H)$$

$$= 290(7.9445 - 6.8508 - 615.8/1400) = 189.6 \text{ kJ/kg}$$

$$i_{\text{cond}, 4'5} = T_0(s_5 - s_{4'} + q_{4'5}/T_L)$$

$$= 290(0.3545 - 8.4625 + 2409.3/290) = 58.0 \text{ kJ/kg}$$

$$i_{\text{turbines}} = 213.5 \text{ kJ/kg (as calculated in Example 11-9)}$$

$$i_{\text{pump}} = 2.4 \text{ kJ/kg (as calculated in Example 11-9)}$$

The irreversibility of the cycle is calculated by

$$i_{\text{cycle}} = i_{\text{st gen}, 6'1} + i_{\text{reheater}, 2'3} + i_{\text{cond}, 4'5} + i_{\text{turbines}} + i_{\text{pump}}$$

$$= 1160.3 + 189.6 + 58.0 + 213.5 + 2.4$$

$$= 1623.8 \text{ kJ/kg}$$

The irreversibility of the cycle can also be calculated by considering the entire cycle as a single steady-flow, steady-state process. Equation (8-12) then gives

$$i_{\text{cycle}} = T_0(\Delta s_{\text{surr}})$$

$$= T_0\left(-\frac{q_{6'1}}{T_H} - \frac{q_{2'3+}}{T_H} + \frac{q_{4'5}}{T_L}\right)$$

$$= 290\left(-\frac{3176.3}{1400} - \frac{615.8}{1400} + \frac{2409.3}{290}\right)$$

$$= 1623.8 \text{ kJ/kg}$$

The available part of the heat supply from the high-temperature reservoir is

$$(q_{\text{in}, 6'1} + q_{\text{in}, 2'3})\left(1 - \frac{T_0}{T_H}\right) = (3176.3 + 615.8)\left(1 - \frac{290}{1400}\right) = 3006.6 \text{ kJ/kg}$$

This available energy is disposed of as the sum:

$$\text{net } w_{\text{out}} + i_{\text{cycle}} = 1382.8 + 1623.8 = 3006.6 \text{ kJ/kg}$$

11-10 SUMMARY

An ideal simple Rankine cycle consists of four internally reversible processes: constant-pressure heat addition in a steam generator, isentropic expansion in a turbine, constant-pressure heat rejection in a condenser, and isentropic compression in a pump. Modern steam generators usually include both a boiler and a superheater.

Regeneration is used to improve the efficiency of a steam power cycle. In a regeneration process, some steam is bled off the turbine after it has partially expanded to preheat the feedwater going to the steam generator. In open feedwater heaters, the bled steam is mixed with the feedwater in the heater. In closed heaters, heat is transferred from the bled steam to the feedwater without mixing.

Reheating is used to reduce the moisture content in the low-pressure stages of the turbine. In a reheat cycle, steam is withdrawn from the turbine after it has partly expanded and is returned to the steam-generating unit to be resuperheated before being sent back for further expansion in the low-pressure stages of the turbine.

A binary vapor cycle uses two working fluids: one for the high-temperature range and the other for the low-temperature range. In a combined gas-and-steam power cycle, hot exhaust gases from a nonregenerative gas turbine are used in a heat-recovery boiler to generate steam for a Rankine cycle. Cogeneration is the generation of steam for simultaneous use in electricity production and thermal-energy supply for space or industrial-process heating.

A low-temperature Rankine cycle uses some organic fluid (such as Refrigerants-113 and -114) as the working fluid. It is suitable for power production when the temperature of the energy source is low.

The irreversible effects in turbines and pumps are accounted for by the adiabatic turbine efficiency η_T and adiabatic pump efficiency η_P. They are defined as

$$\eta_T = \frac{\Delta h_{\text{actual}}}{\Delta h_{\text{isentropic}}} \quad \text{and} \quad \eta_P = \frac{\Delta h_{\text{isentropic}}}{\Delta h_{\text{actual}}}$$

The first-law efficiency of a steam power cycle is defined as

$$\eta_{\text{th}} = \frac{\text{net work output}}{\text{heat input}}$$

The second-law effectiveness of a steam power cycle can be defined as

$$\epsilon'_{\text{power cycle}} = \frac{\text{net work output}}{\left(\begin{array}{c} \text{change in stream availability in steam} \\ \text{generator and reheater} \end{array}\right)}$$

The specific stream availability ψ of a steady-flow fluid stream in a given state is given by

$$\psi = (h + \tfrac{1}{2}\mathbf{V}^2 + gz - T_0 s) - (h_0 + gz_0 - T_0 s_0)$$

The specific irreversibility of a cycle can be calculated by either of the following equations:

$$i_{\text{cycle}} = \sum i_{\text{process}}$$

$$i_{\text{cycle}} = T_0 \left[\sum \left(\frac{-q_R}{T_R} \right) \right]$$

PROBLEMS

11-1. In an ideal simple Rankine cycle, steam enters the turbine at 6 MPa and 500°C and leaves at 10 kPa. Determine the quality of steam at the exit from the turbine and the cycle thermal efficiency.

11-2. In an ideal Rankine cycle, steam enters the turbine at 800 psia and 1000°F. Determine the net work output, in Btu/lbm, and the thermal efficiency of the cycle with a condenser pressure of 2 psia.

11-3. An ideal Rankine steam power cycle that produces 125 MW of turbine power has the following conditions: turbine inlet, 1000 psia and 1000°F; condenser pressure, 1.0 psia. Determine: (a) the heat input in Btu/lbm, (b) the mass-flow rate of steam in lbm/h, (c) the thermal efficiency of the cycle, and (d) the mass-flow rate of cooling water in the condenser in lbm/h if the cooling water experiences a 12°F temperature rise.

11-4. At latitudes close to the equator, the temperature difference between surface seawater and deep seawater can be big enough to run a heat engine. Figure P11-4 shows a schematic diagram of an ocean thermal power plant based on the Rankine cycle using Refrigerant-12 as the working fluid. The conditions at various points of the plant are: surface seawater, $T_A = 27°C$, $T_B = 24°C$; deep seawater, $T_C = 7°C$, $T_D = 10°C$; and Refrigerant-12, $T_1 = 22°C$ saturated vapor, $T_3 = 12°C$ saturated liquid. Turbine and pump processes are ideal. The power output of the turbine is 10 MW. Determine (a) the mass-flow rate of Refrigerant-

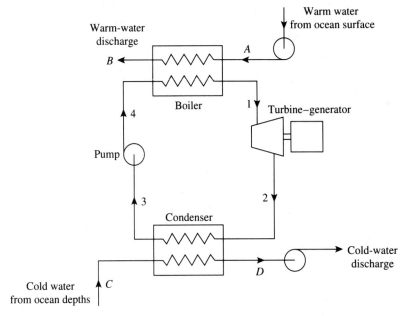

Figure P11-4

12, kg/min, (b) the boiler heat-transfer rate, kJ min, (c) the condenser heat-transfer rate, kJ/min, (d) the thermal efficiency of the Rankine cycle, (e) the mass-flow rate of the surface seawater, kg/min, and (f) the mass-flow rate of the deep seawater, kg/min. Use pure-water properties for the seawater.

11-5. Referring to the ocean thermal power plant shown in the schematic diagram of Fig. P11-4, the conditions at various points of the plant are surface seawater, $T_A = 80°F$, $T_B = 75°F$; deep seawater, $T_C = 40°F$, $T_D = 45°F$; and Refrigerant-12, $T_1 = 70°F$ saturated vapor, $T_3 = 50°F$ saturated liquid. Turbine and pump processes are ideal. The power output of the turbine is 10,000 hp. Determine (a) the mass-flow rate of Refrigerant-12, lbm/min, (b) the heat added and heat rejected for the Rankine cycle, Btu/min, (c) the thermal efficiency, and (d) the mass-flow rates of the surface seawater and deep seawater, lbm/min. Use pure-water properties for seawater.

11-6. Steam enters the turbine of a simple Rankine cycle at 6 MPa and 500°C and leaves at 10 kPa. The pump process is ideal. The turbine process has an isentropic efficiency of 85%. Determine the quality of steam at the exit from the turbine and the cycle thermal efficiency.

11-7. In a simple Rankine cycle, steam enters the turbine at 1000 psia and 800°F and leaves at 1.0 psia. The adiabatic turbine and pump efficiencies are 0.90 and 0.70, respectively. Determine (a) the turbine work, pump work, heat added, and heat rejected, per pound of steam, (b) the cycle thermal efficiency, and (c) the steam rate in lbm/kW·h.

11-8. A steam power plant operates on a simple Rankine cycle. Steam enters the turbine at 2000 psia and 1000°F and leaves at 1.0 psia. The adiabatic turbine and pump efficiencies are 0.90 and 0.70, respectively. Determine (a) the turbine work, pump work, heat added, and heat rejected, per pound of steam, (b) the cycle thermal efficiency, and (c) the steam rate in lbm/kW·h.

11-9. Steam enters the turbine of a steam power plant that operates on a simple Rankine cycle at 5 MPa, 300 K, and leaves at 4 kPa. The turbine has an adiabatic efficiency of 90%, and the pump has an adiabatic efficiency of 80%. Determine the quality of steam at the turbine exit and the cycle thermal efficiency.

11-10. A Rankine steam cycle has a condenser pressure of 1 psia and a turbine inlet state of 2500 psia and 1000°F. A single open feedwater heater operates at 140°F. Determine (a) the fraction of the steam bled to the feedwater heater, (b) the turbine work, (c) the pump work, and (d) the cycle thermal efficiency.

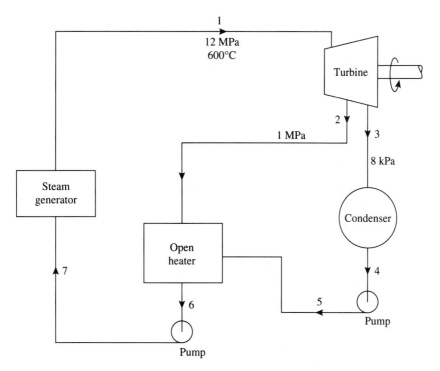

1
12 MPa
600°C

Turbine

2

1 MPa

3

8 kPa

Steam
generator

Condenser

Open
heater

7

6

5

Pump

Pump

Figure P11-11

11-11. Steam enters the turbine of an ideal power cycle at 12 MPa and 600°C and expands to 1 MPa, where a portion is bled to an open feedwater heater. The remaining of the steam expands to the condenser at 8 kPa. See Fig. P11-11. Determine the thermal efficiency.

11-12. Steam at 5.0 MPa and 600°C enters the turbine of an ideal regenerative power plant. See Fig. P11-12. Steam is extracted from the turbine at 0.30 MPa and delivered to an ideal closed feedwater heater. The condensate from the heater is pumped ahead. The condenser pressure is 2 kPa. Determine, per kg steam at the turbine inlet, the heat added, heat rejected, turbine work, pump work, and thermal efficiency.

11-13. Steam at 5.0 MPa and 600°C enters the turbine of an ideal regenerative power plant. See Fig. P11-13. Steam is extracted from the turbine at 0.30 MPa and delivered to an ideal closed feedwater heater. The condensate from the heater is trapped back to the condenser. The condenser pressure is 2 kPa. Determine, per kg steam at the turbine inlet, the heat added, heat rejected, turbine work, pump work, and thermal efficiency.

11-14. A steam power plant has a high-pressure turbine and a low-pressure turbine, a reheater, and a closed-type feedwater heater. Steam at 450 psia and 700°F enters the high-pressure turbine and leaves at 80 psia. Part of the steam leaving this turbine flows into the reheater, is reheated to 600°F, and then flows into the low-pressure turbine. The rest of the steam leaving the high-pressure turbine flows into the feedwater heater and condenses. The condensate from the feedwater heater flows through a trap into the condenser. The pressure in the condenser is 1 psia. Assume that the turbines and the pump are ideal, and there is no terminal temperature difference in the feedwater heater. Calculate the thermal efficiency. Draw the $T-s$ diagram for the cycle.

11-15. A steam power plant has a high-pressure turbine, a low-pressure turbine, a reheater, and a closed feedwater heater. See Fig. P11-15. Steam at 15 MPa and 500°C enters the high-pressure turbine and leaves at 1.1 MPa. Part of the steam leaving this turbine flows into the reheater, is reheated to 500°C, and then flows into the low-pressure turbine. The rest of the steam leaving the high-pressure turbine flows into the feedwater heater and condenses. The condensate from the feedwater heater flows through a trap into the condenser. The pressure in the condenser is

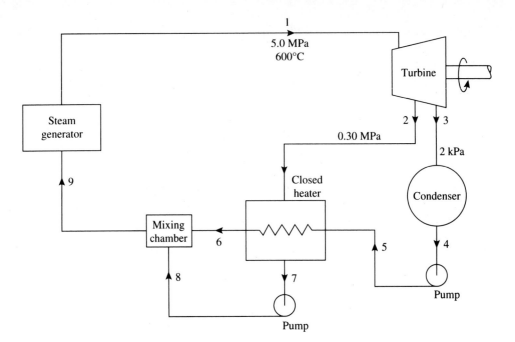

Figure P11-12

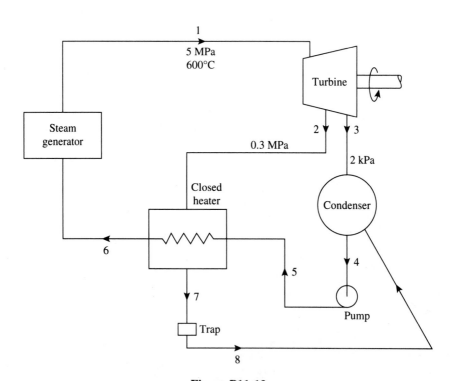

Figure P11-13

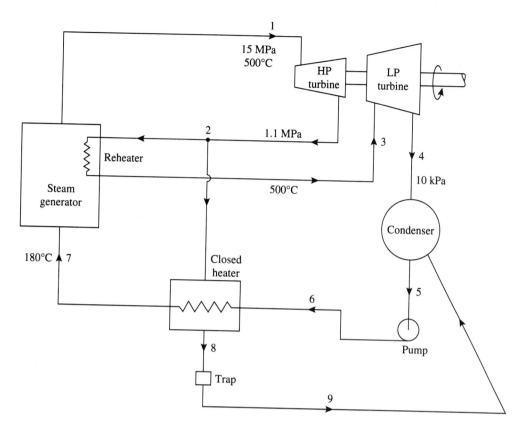

Figure P11-15

10 kPa. The feedwater entering the steam generator is at 180°C. The turbines and the pump are ideal. Neglect pressure drops and heat losses in the piping. Calculate, per kg of throttling steam, the turbine work, pump work, heat added, heat rejected, and thermal efficiency of the cycle.

11-16. An ideal reheat-regenerative steam cycle is to produce 150,000 kW at the turbine coupling if the throttle conditions are 2400 psia and 1100°F. Reheat is at 300 psia to 1050°F. See Fig. P11-16. A single closed feedwater heater with a 5°F terminal-temperature difference is supplied with steam from the high-pressure turbine exhaust, the heater drain being pumped into the feed line. The condenser pressure is 1 psia. Determine the cycle thermal efficiency and the flow rate, in lbm/h.

11-17. A reheat-regenerative steam cycle operates with a turbine inlet condition of 500 psia and 900°F. See Fig. P11-17. Reheat occurs at 100 psia, back to 900°F. At 50 psia, steam is bled to an open feedwater heater. Condenser is at 1 psia. Determine

the cycle thermal efficiency if the turbine and pump processes are isentropic.

11-18. In a certain steam power plant, the original installation operated between 400 psia, 750°F, and 1 psia. See Fig. P11-18(a). In order to increase the power output, a new high-pressure boiler and turbine were installed. See Fig. P11-18(b). The new turbine receives steam at 2000 psia, 950°F, and exhausts at 420 psia. The exhaust steam is reheated and supplied to the old turbine at 400 psia, 750°F. (a) With a turbine efficiency of 80%, how much power (kW) is delivered, per lbm/h of steam, by the new turbine? (b) If the old turbine has 75% efficiency, how much power does it deliver per lbm/h of steam? (c) What is the total turbine output, new and old, if the new turbine exhausts enough steam for the old turbine to produce its rated output of 70,000 kW? (d) Under conditions in part (c), what fraction of the total heat supplied to the steam is transferred in the reheater? (e) What are the thermal efficiencies of the old and new cycles?

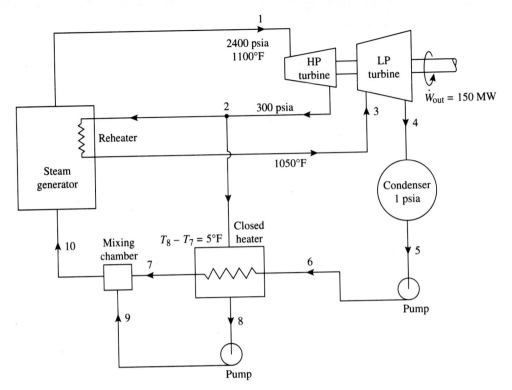

Figure P11-16

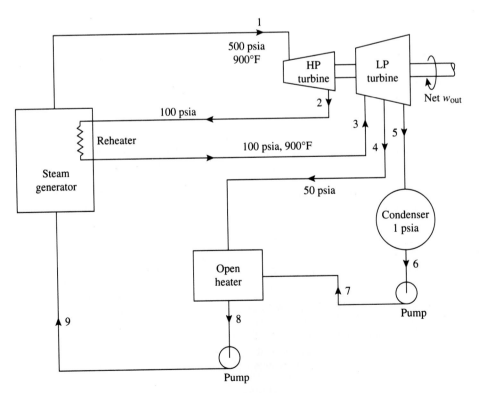

Figure P11-17

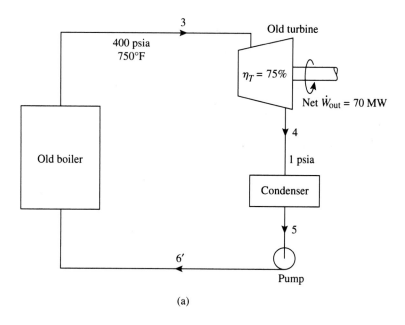

(a)

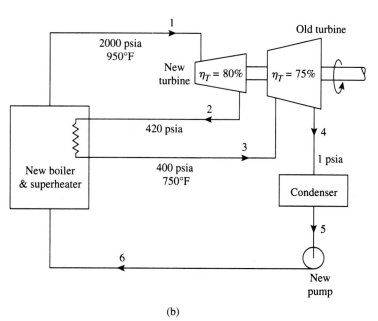

(b)

Figure P11-18 (a) Old installation, and (b) new installation.

11-19. An ideal regenerative steam cycle operates with throttle condition of p_1 and T_1; one closed feedwater heater at p_2; one open feedwater heater at p_3; and a condenser at p_4, where $p_1 > p_2 > p_3 > p_4$. Steam at states 1 and 2 is superheated, and steam at states 3 and 4 is in the two-phase region. The drain from the closed heater is trapped to the open heater, and the terminal temperature difference of the closed heater is ΔT. Draw the flow diagram and the appropriate T–s diagram. Describe the procedures for finding the enthalpy values needed in the solution of this problem, and write equations for the calculation of (a) the amount of steam extracted for the closed heater, (b) the amount of steam extracted for the open heater, (c) the turbine work and the pump work, (d) the heat added and the heat rejected, and (e) the cycle thermal efficiency.

11-20. An ideal reheat-regenerative steam cycle has an extraction for feedwater heating first, then later a single reheating, and, finally, two extraction points for feedwater heating. Draw the flow diagram and draw a T–s diagram for the cycle, with all feedwater heaters being of the open type. Write equations for calculating (a) the amount of steam extracted at each point of the turbine, (b)

the turbine work, (c) the pump work, (d) the heat supplied, (e) the heat rejected, and (f) the cycle thermal efficiency.

11-21. The flow diagram of a regenerative steam cycle is shown in Fig. P11-21. The turbine process is adiabatic with a given efficiency. The pump processes are ideal. Terminal temperature differences for the closed heaters are known. Draw an appropriate T–s diagram for the cycle and write the equations for the calculation of (a) m_a, m_b, m_c, and m_d, (b) the turbine work and the pump work, and (c) the heat added and the heat rejected, assuming that all enthalpy values are known.

11-22. Steam enters the turbine of a two-heater ideal regenerative steam cycle at 400 psia, 760°F, and exhausts at 2 psia. The high-pressure heater is a closed heater using steam extracted at 120 psia; both the feedwater and the drip leave this heater at the saturation temperature for 120 psia. The low-pressure heater is an open heater operating at 20 psia. The drip from the high-pressure heater is dumped through a trap to the low-pressure heater. Determine the turbine work, pump work, heat added, heat rejected, and the cycle thermal efficiency. Determine also the steam rate, in lbm/hp·h, for this cycle.

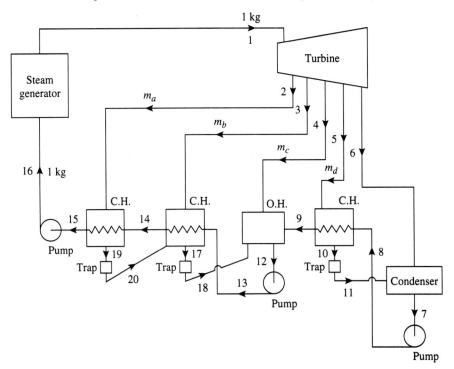

Figure P11-21

11-23. Steam enters the turbine of a two-feedwater-heater regenerative cycle at 10 MPa and 450°C to generate 2 MW of electric power. The high-pressure heater is an open heater using steam extracted at 1.8 MPa. The low-pressure heater is a closed heater using steam extracted at 0.18 MPa, with the drip being dumped through a trap to the condenser. The condenser operates at 5 kPa. The turbine operates adiabatically with an efficiency of 85%, and the pumps are assumed to operate isentropically. The closed heater has a 3°C terminal temperature difference. Determine the flow rates, the heat input to the steam generator, the heat output from the condenser, and the thermal efficiency.

11-24. In a two-feedwater-heater steam plant steam enters the turbine at 1000 psia and 800°F, and the exhaust pressure is 1 psia. See Fig. P11-24. The high-pressure heater is a closed heater using steam extracted at 260 psia. The low-pressure heater is an open heater operating at 30 psia. The drip from the high-pressure heater is passed through a trap to the low-pressure heater. The terminal-temperature difference of the closed heater is 5°F. The turbine efficiency for each section of the turbine is 75%. The throttling flow is 50,000 lbm/h. Neglect pump work and friction

loss in the piping. Calculate the turbine output in kW and the turbine heat rate, in Btu/kW·h.

11-25. In a two-feedwater steam plant, steam enters the turbine at 1000 psia and 800°F, and the exhaust pressure is 1 psia. See Fig. P11-25. The high-pressure heater is a closed heater using steam extracted at 260 psia. The low-pressure heater is an open heater operating at 30 psia. The drain from the high-pressure heater is passed through a trap to the low-pressure heater. The terminal temperature difference of the closed heater is 5°F. The turbine efficiency for each section of the turbine is 75%. The throttling flow is 50,000 lbm/h. There are 1000 lbm/h boiler blowdown and steam losses (assumed to be live steam at the outlet of the steam generator). These losses are made up in an evaporator that takes steam from the same extraction opening as does the closed heater. The makeup water is at 60°F and 30 psia. Both the condensed liquid and the evaporated vapor from the evaporator are collected in the open heater. Neglect pump work and friction loss in the piping. Calculate the turbine output in kW and the turbine heat rate in Btu/kW·h.

11-26. A low-temperature Rankine cycle uses Refrigerant-114 as the working fluid. See Fig. P11-26. Saturated vapor at 1.0 MPa enters the expander

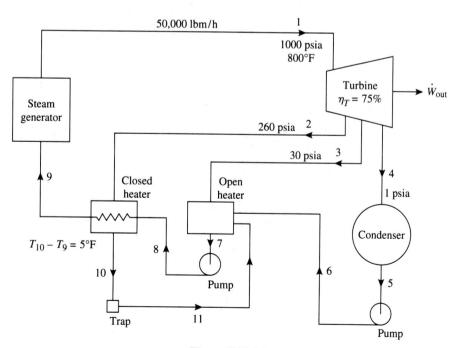

Figure P11-24

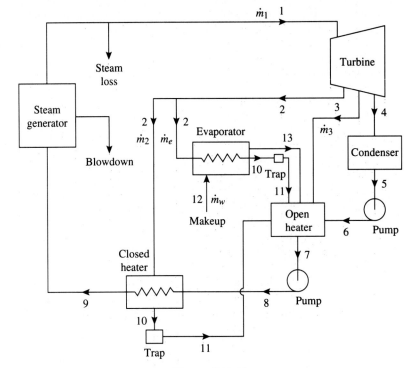

Figure P11-25

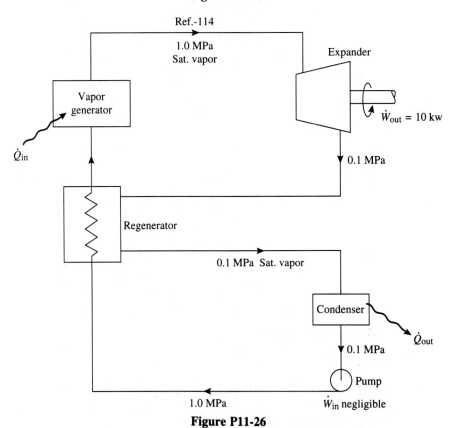

Figure P11-26

and expands to 0.1 MPa. It is then cooled in the regenerator at constant pressure to the saturated vapor state before being condensed in the condenser. The expander power output is 10 kW and the pump work can be neglected. Determine the mass-flow rate of the fluid and the cycle thermal efficiency if (a) the expander process is internally reversible and adiabatic, and (b) the expander process is adiabatic with an efficiency of 85%.

11-27. A low-temperature Rankine cycle uses Refrigerant-113 as the working fluid. See Fig. P11-27. Saturated vapor at 0.4 MPa enters the expander and expands isentropically to 0.05 MPa. It is then cooled in the regenerator at constant pressure to the saturated vapor state before being condensed in the condenser. Calculate, per unit mass of the fluid, the expander work, the heat added in the boiler, the heat rejected in the condenser, and the cycle thermal efficiency.

11-28. One efficient means of concentrating and collecting solar energy for electricity generation is the use of a central receiver located in a high tower to intersect solar rays reflected by tracking mirrors on ground. As depicted in Fig. P11-28, hot liquid sodium is directed down the tower to serve as the energy source for a reheating-regenerative 100-MW steam power plant. The pressure and temperature data at various points of the plant are given in the figure. The specific heat of liquid sodium is assumed to have a constant value of 1.26 kJ/kg·K. Calculate the various sodium and steam mass-flow rates and the steam-cycle thermal efficiency. The turbine and pump processes are internally reversible and adiabatic.

11-29. Referring to the solar central receiver power plant shown in the schematic diagram of Fig. P11-28, for a certain installation, the conditions at various points of the plant are as follows: Liquid sodium is heated in the central receiver from 640°F to 1150°F. The low-temperature sodium storage tank is maintained at 570°F. The high-temperature sodium storage tank is maintained at 1070°F. The specific heat of liquid sodium is assumed to have a constant value of 0.302

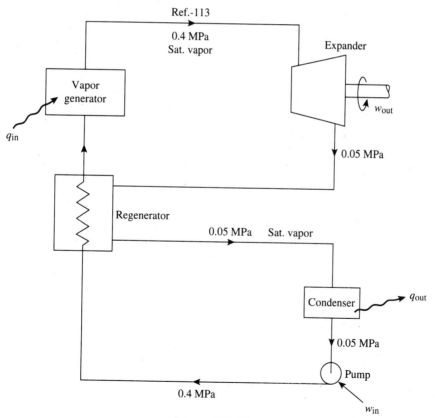

Figure P11-27

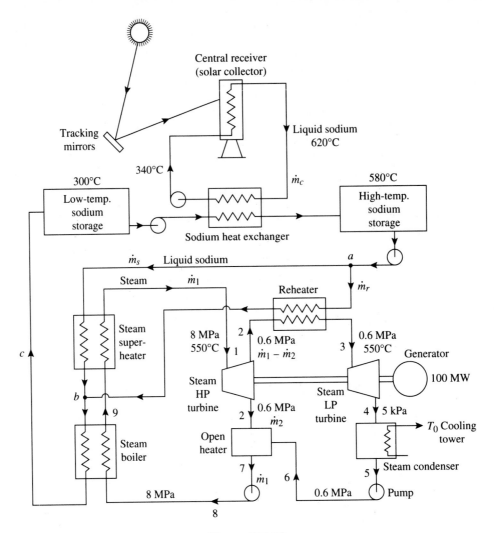

Figure P11-28

Btu/lbm · °F. Steam enters the high-pressure turbine at 1160 psia and 1020°F. After expansion to 87 psia, some of the steam goes to an open feedwater heater and the balance is reheated to 1020°F, after which it expands to 0.7 psia. The adiabatic efficiency of the turbines is 86% and the adiabatic efficiency of the pumps is 78%. For a turbine power output of 50 MW, determine the various sodium and steam mass-flow rates and the steam-cycle thermal efficiency.

11-30. Figure 11-13 depicts the schematic and *T–s* diagrams of a potassium–water binary vapor cycle. Taking the conditions shown in the *T–s* diagram, determine (a) the pounds of potassium circulated per pound of steam, and (b) the thermal efficiency of the binary cycle. For potassium, at 1940°F = 2400°R and 90 psia, $h = 1238.5$ Btu/lbm, $s = 1.0559$ Btu/lbm·°R; and at 1140°F = 1600°R, $p_{sat} = 3.210$ psia, $v_f = 0.0229$ ft³/lbm, $h_f = 310.4$ Btu/lbm, $h_g = 1176.6$ Btu/lbm, $s_f = 0.6400$ Btu/lbm·°R, and $s_g = 1.1813$ Btu/lbm·°R.

11-31. An air-standard Brayton cycle and a steam Rankine cycle are to be combined in a power plant in such a way that heat transferred from the air leaving the turbine is used to evaporate the water. The Brayton cycle has a minimum pressure of 15 psia and a maximum pressure of 70 psia, and a minimum temperature of 60°F and a maximum temperature of 1540°F. The Rankine cycle has a maximum pressure of 400 psia, a maximum temperature of 600°F, and a minimum pressure of 1 psia. After transferring heat to the steam in the air–steam heat exchanger, the air is exhausted at 600°F and 15 psia. Draw a flow diagram of this power plant and determine the thermal efficiency of the combined cycle. Use the air table to evaluate the properties of air.

11-32. In a gas–steam-turbine combined cycle, air enters the compressor at 300 K and 101 kPa and leaves at a pressure 14 times that of the inlet. See Fig. P11-32. The air is heated to 1400 K in the combustion chamber and then expanded in the turbine. The exhaust gases from the turbine are used to heat the steam in a heat exchanger, from which air exits at 460 K and steam exits at 400°C and 8 MPa. An open feedwater heater working at 0.6 MPa is included in the steam cycle. The con-

denser pressure of the steam cycle is 20 kPa. The adiabatic efficiency of the gas compressor is 82% and that of the gas turbine is 86%. The adiabatic efficiency of the steam pumps is 100% and that of the steam turbine is 86%. The net power output of the combined cycle is 600 MW. Taking data from the air and steam tables, find (a) the mass-flow-rate ratio of air to steam, (b) the steam-cycle net work output in kJ/kg steam, (c) the gas-cycle net work output in kJ/kg air, (d) the rate of heat input in the combustion chamber in kJ/s, and (e) the thermal efficiency of the combined cycle.

11-33. An ideal Rankine cycle using Refrigerant-12 and an ideal air-standard Diesel cycle are the two subcycles for a combined cycle in which the exhaust gas from the Diesel cycle supplies the energy required for the Refrigerant-12 boiler.

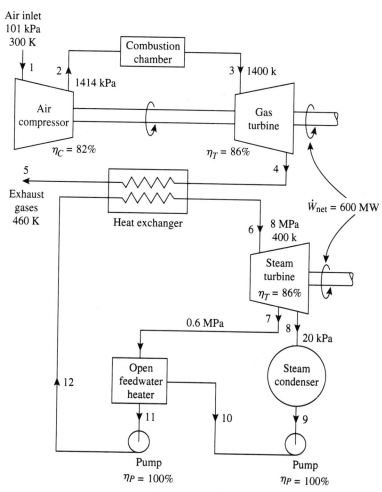

Figure P11-32

Diesel inlet conditions are 14.7 psia and 70°F, the compression ratio is 18, and the maximum temperature is 4500°R. Saturated vapor Refrigerant-12 leaves the Rankine cycle boiler at 230°F, and the condenser temperature is 90°F. The power output of the Diesel engine is 1 MW. Determine (a) the flow rate required in the Diesel cycle, and (b) the power output of the Rankine cycle, assuming that the Diesel exhaust is cooled to 400°F in the Refrigerant-12 boiler.

11-34. An ideal Rankine cycle using Refrigerant-12 and an ideal air-standard Diesel cycle are the two subcycles for a combined cycle in which the exhaust gas from the Diesel cycle supplies the energy required for the Refrigerant-12 boiler. Diesel inlet conditions are 95 kPa and 20°C, the compression ratio is 18, and the maximum temperature is 2500°C. Saturated vapor Refrigerant-12 leaves the Rankine cycle boiler at 60°C, and the condenser temperature is 20°C. The power

output of the Diesel engine is 1 MW. Determine (a) the flow rate required in the Diesel cycle, and (b) the power output of the Rankine cycle, assuming that the Diesel exhaust is cooled to 200°C in the Refrigerant-12 boiler.

11-35. A cogeneration steam cycle is designed to serve a large food processing plant while generating electricity. See Fig. P11-35. The steam boiler generates steam at 8 MPa and 500°C at a rate of 5 kg/s. The food processing requires 2 kg/s of saturated or slightly superheated steam at 0.4 MPa. Condenser pressure is 15 kPa. The saturated liquid leaving the food processing plant and the pumped condensate from the condenser are mixed and then pumped back to the boiler. Assuming that the turbine and the pumps all have an adiabatic efficiency of 84%, determine (a) the rate of heat supply to the boiler, (b) the power output of the turbine, and (c) the cogeneration system effectiveness.

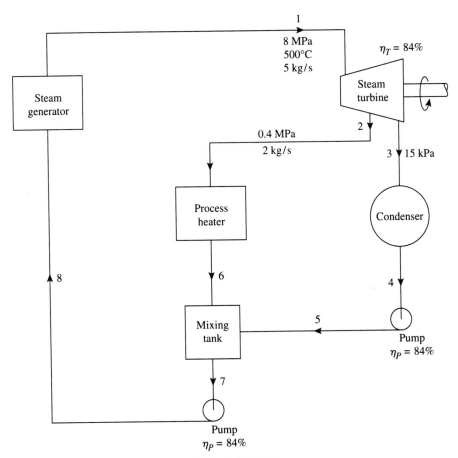

Figure P11-35

11-36. A cogeneration plant is designed to deliver 50 MW of turbine output power and 5×10^6 Btu/min of thermal energy for heating. Steam is supplied to the first stage of the turbine at 1800 psia and 800°F and leaves at 100 psia. A portion of the steam is then sent to the heating load and the remaining portion is reheated to 700°F before expansion in the second stage of the turbine to the condenser pressure of 1 psia. The water leaves the heating load as a saturated liquid at 60 psia and then mixes with the water at 60 psia coming out from the condensate pump. The mixed water is finally pumped to the steam generator. Both stages of the turbine have an isentropic efficiency of 85%, and both pumps have an isentropic efficiency of 60%. Determine (a) the mass-flow rates, in lbm/min, through the heating load and through the condenser, (b) the heat input to the steam generator, in Btu/min, and (c) the effectiveness of the cogeneration system.

11-37. A cogeneration plant is designed to deliver 50 MW of turbine output power and 5000 MJ/min of thermal energy for heating. See Fig. P11-37. Steam is supplied to the first stage of the turbine at 12 MPa, 440°C, and leaves at 0.7 MPa. A portion of the steam is then sent to the heating load and the remaining portion is reheated to 400°C before expansion in the second stage of the turbine to the condenser pressure of 8 kPa. The water leaves the heating load as a saturated liquid at 0.5 MPa and then mixes with the water at 0.5 MPa coming out from the condensate pump. The mixed water is finally pumped back to the steam generator. Both stages of the turbine have an isentropic efficiency of 85%, and both pumps have an isentropic efficiency of 60%. Determine (a) the mass-flow rates, in kg/min, through the heating load and through the condenser, (b) the heat input to the steam generator, in kJ/min, and (c) the effectiveness of the cogeneration system.

11-38. Saturated steam at 700 kPa is generated in the boiler of an industrial plant for manufacturing processes. See Fig. P11-38. The plant has a heating load of 20×10^6 kJ/h that is met by drawing steam from the main line through a pressure-reducing valve at an outlet pressure of 140 kPa. Condensate leaves the heating system at 140 kPa and 82°C. It is proposed to modify the present set-up to form a cogeneration system by the use of a 60% efficient turbine in place of the pressure-

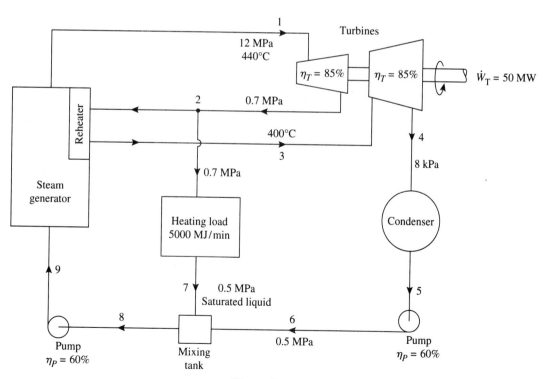

Figure P11-37

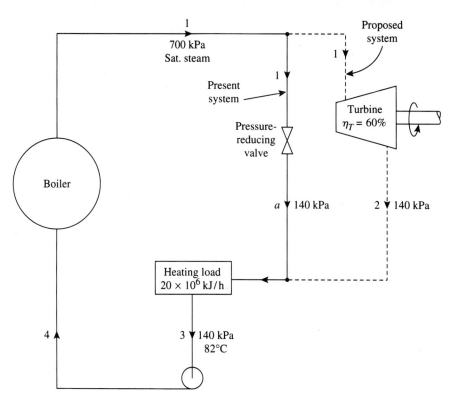

Figure P11-38

reducing valve. Determine the amount of power that can be obtained by the proposed system and the change of steam flow rate, in kg/h, resulting from the adoption of the proposed system.

11-39. Saturated steam at 100 psia is generated in the boiler of an industrial plant for manufacturing processes. The plant has a heating load of 20×10^6 Btu/h that is met by drawing steam from the main line through a pressure-reducing valve set at an outlet pressure of 20 psia. Condensate leaves the heating system at 20 psia and 180°F. It is proposed to modify the present setup to form a co-generation system by the use of a 60% efficient turbine in place of the pressure-reducing valve. Determine the amount of power that can be obtained by the proposed system and the change of steam flow rate, in lbm/h, resulting from the adoption of the proposed system.

11-40. In a cogeneration plant, steam enters an extraction turbine at 200 psia and 500°F; extraction occurs at 20 psia, $h = 1140$ Btu/lbm; and steam entering the condenser is at 1 psia, $h = 1005$ Btu/lbm. See Fig. P11-40. Part of the extracted steam goes to an open feedwater heater that oper-

ates at 20 psia, and the rest goes to a heating system that provides 30 million Btu/h and from which condensate at 152°F is pumped into the open heater. Saturated liquid leaves the open heater. The turbine power output is 3500 kW. Determine (a) the flow rate of steam into the turbine, (b) the efficiency of the turbine between the throttle and the extraction point, and (c) the irreversibility of the turbine expansion upstream of the extraction point. The atmosphere is at 70°F.

11-41. A nuclear reactor is cooled by liquid sodium in the primary circuit, which in turn is cooled in a heat exchanger by an intermediate sodium circuit. See Fig. P11-41. The sodium in the intermediate circuit is finally cooled in another heat exchanger by a steam circuit, which supplies energy to run a steam Rankine cycle. Liquid sodium leaves the nuclear reactor at 20 psia, 700°F, and returns at 600°F, with a flow rate of 100,000 lbm/h. Saturated steam at 1000 psia enters the steam turbine and expands adiabatically to 1 psia. The isentropic efficiency of the turbine is 60%. Neglecting the work required by sodium pumps and assuming the specific heat of sodium to be constant at

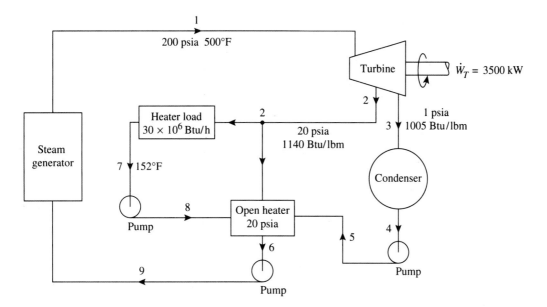

Figure P11-40

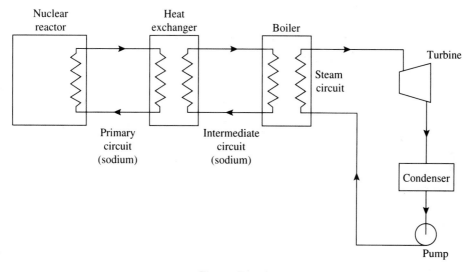

Figure P11-41

0.30 Btu/lbm·°F, determine (a) the power output in kW of the plant, and (b) the irreversibility in Btu/h of the process of transferring energy from the primary sodium circuit through the intermediate circuit to the water. The atmosphere is at 70°F.

11-42. In an ideal simple Rankine cycle, steam enters the turbine at 3.5 MPa and 700°C and leaves at 20 kPa. The mass-flow rate of steam is 2.0 kg/s. Determine (a) the power generated and the ther-

mal efficiency, and (b) the irreversibility rate for each process and for the entire cycle, if the high-temperature and low-temperature reservoirs are at 1000 K and 300 K, respectively. The atmosphere is at 300 K.

11-43. In a 200-MW reheat Rankine power cycle, steam enters the high-pressure turbine at 10 MPa and 500°C. Reheat occurs at 1 MPa to 500°C. The condenser pressure is 10 kPa. The adiabatic efficiency of the turbine is 80% and that of the pump is 85%. Determine (a) the mass-flow rate of steam, in kg/s, (b) the changes in stream availability for the steam generator and the condenser, in kJ/kg, and (c) the irreversibility for the turbines and the pump, in kJ/kg. Make an availability accounting for the cycle. The atmosphere is at 290 K.

11-44. For the reheat Rankine cycle described in Prob. 11-43, determine the irreversibility, in kJ/kg, associated with each of the processes, assuming a source temperature of 1400 K and a sink temperature of 290 K. Determine also the irreversibility of the cycle, in kJ/kg. Give an account for the available energy supplied from the high-temperature reservoir. The atmosphere is at 290 K.

11-45. For the reheat Rankine cycle described in Prob. 11-43, determine the first-law efficiency and the second-law effectiveness. Describe the difference between the two in reference to the T–s diagram for the cycle.

11-46. A reheat steam power cycle has the first stage supply conditions of 4.0 MPa and 700°C, and the second stage supply condition of 700°C. The condenser pressure is 10 kPa. Both turbine stages and the pump operate ideally, with a steam quality of 95% at the exit from the low-pressure stage turbine. Determine (a) the constant pressure in the reheater, (b) the work output for each turbine stage, kJ/kg, (c) the heat transfer in the steam generator and the condenser, kJ/kg, and (d) the irreversibility for each process of the cycle, and the irreversibility of the entire cycle, kJ/kg, if the high-temperature and low-temperature reservoirs are at 900°C and 20°C, respectively. The atmosphere is at 20°C.

11-47. A regenerative steam power cycle with one open heater has the turbine inlet conditions of 6.0 MPa and 900°C, the open-heater pressure of 1 MPa, and the condenser pressure of 2 kPa. The net power output is 2.5 MW. The turbine and the pumps operate ideally. Determine (a) the mass-flow rate at the inlet to the turbine, (b) the mass-flow rate extracted for the feedwater heater, (c) the turbine power output, (d) the cycle thermal efficiency, and (e) the irreversibility of each process and of the entire cycle, if the high-temperature reservoir is at 1000°C and the low-temperature reservoir (the atmosphere) is at 15°C.

11-48. For the open-heater regenerative steam cycle described in Prob. 11-47, determine (a) the stream availability rate change for the steam generator and for the condenser, and (b) the second-law cy-

cle effectiveness. Make an availability accounting for the cycle.

11-49. For the regenerative cycle described in Prob. 11-11, determine the irreversibility of the cycle, assuming a source temperature of 1400 K and a sink temperature of 290 K.

11-50. For the one closed feedwater heater regenerative steam power plant described in Prob. 11-13, determine the irreversibility for each of the component processes and for the entire power cycle, assuming a source temperature of 1400 K and a sink (the atmosphere) temperature of 280 K.

11-51. In a Rankine cycle, steam enters the turbine at 5 MPa and leaves the turbine at 4 kPa. The turbine has an adiabatic efficiency of 90%, and the pump has an adiabatic efficiency of 80%. Write a computer program to determine the quality of steam at the turbine exit, the net work output, and the cycle thermal efficiency if the turbine inlet temperature is (a) 300°C, (b) 400°C, (c) 500°C, (d) 600°C, and (e) 700°C. Take the required data from a steam table and use them as input to your computer program. The required steam data are v_f, h_f, h_g, s_f, and s_g at 4 kPa; h and s at 5 MPa and the various turbine inlet temperatures.

11-52. If computerized steam property data are available to you, rework Prob. 11-51 by the use of those data.

11-53. In an ideal Rankine cycle, steam enters the turbine at 800 psia and 1000°F. Write a computer program to determine the net work and the thermal efficiency of the cycle with the following condenser pressures: (a) 2 psia, (b) 1.5 psia, (c) 1 psia, and (d) 0.5 psia. Take the required data from a steam table and use them as input to your computer program. The required steam data are: h and s at 800 psia and 1000°F; and v_f, h_f, h_g, s_f, and s_g at the various condenser pressures.

11-54. If computerized steam property data are available to you, rework Prob. 11-53 by the use of those data.

11-55. In an ideal simple Rankine cycle, steam enters the turbine at 500°C and leaves at 10 kPa. Write a computer program to determine the quality of steam at the exit from the turbine and the cycle thermal efficiency if the steam generator pressure is (a) 1 MPa, (b) 2 MPa, (c) 3 MPa, (d) 4 MPa, (e) 5 MPa, (f) 6 MPa, (g) 7 MPa, (h) 8 MPa, (i) 9 MPa, and (j) 10 MPa. Take the required data from a steam table and use them as input to your computer program. The required steam data are v_f, h_f, h_g, s_f, and s_g at 10 kPa; and h and s at 500°C and the various boiler pressures.

11-56. In a simple Rankine cycle, steam enters the turbine at 6 MPa and 500°C and leaves at 10 kPa. The pump process is ideal. Write a computer program to determine the quality of steam at the exit from the turbine and the cycle thermal efficiency,

if the turbine process has an isentropic efficiency of (a) 70%, (b) 75%, (c) 80%, (d) 85%, (e) 90%, (f) 95%, and (g) 100%. Take the required data from a steam table and use them as input to your computer program. The required steam data are v_f, h_f, h_g, s_f, and s_g at 10 kPa; and h and s at 6 MPa and 500°C.

Refrigeration and Cryogenics

Refrigeration is the production of a temperature lower than that of the natural environment. The working fluid within a refrigeration cycle is called the *refrigerant*. The refrigerant can remain in the gaseous phase throughout the cyclic process (as in gas refrigeration), or can change alternately between vapor and liquid phases (as in vapor-compression refrigeration). By supplying work to the refrigerant in a refrigeration cycle, it is possible to absorb heat from a low-temperature source and to deliver heat to a high-temperature sink. Thus, a refrigeration machine is essentially the reverse of a heat engine. When the purpose is the absorption of heat from a low-temperature source, the machine is called a *refrigerator*. When the purpose is the delivery of heat to a high-temperature sink, it is called a *heat pump*.

The study of the phenomena of very low temperatures, say, below $-150°C$ or $-240°F$, is referred to as *cryogenics*. This chapter includes topics on gas liquefaction. Methods of achieving temperatures lower than 1 K will be found in the next chapter.

12-1 VAPOR-COMPRESSION REFRIGERATION

There are many ways in which a cooling effect can be achieved. Compression cycles using vapor refrigerants are by far the most common refrigeration method. The basic operations involved in a *vapor-compression refrigeration* cycle are illustrated in Fig. 12-1 by a flow diagram along with the idealized *T–s* and *p–h* diagrams. At state 1, the low-pressure saturated vapor refrigerant enters the compressor and, in the ideal case, is reversibly and adiabatically compressed to state 2. This is followed by a constant-pressure heat-rejection process in the condenser, with the refrigerant leaving the condenser as saturated liquid at state 3. The liquid refrigerant then expands through an expansion valve and is partially vaporized to the two-phase state 4. A constant-pressure heat-addition process in the evaporator returns the refrigerant to its starting state (point 1), thus completing the cycle. The process occurring in the evaporator is responsible for the removal of heat from the substance or space to be cooled.

For the purpose of analysis, the refrigeration cycle is assumed to be carried out in a steady-flow operation. By applying the first law for steady flow to the processes of the cycle, the equations for the various work and heat quantities can be written for a unit mass rate of flow, assuming that the changes in potential and kinetic energies are negli-

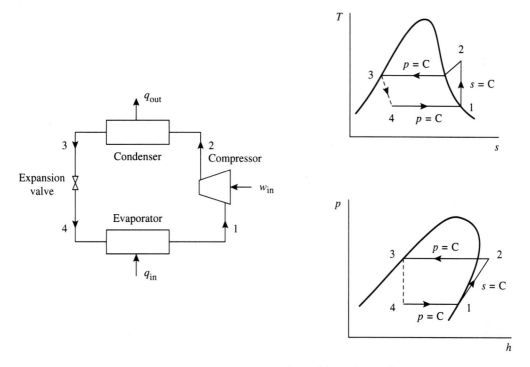

Figure 12-1 Vapor-compression refrigeration cycle.

gible. By referring to Fig. 12-1, because the compression process is adiabatic, the work required to operate the compressor is given by

$$w_{in} = w_{comp} = h_2 - h_1 \tag{12-1}$$

The heat rejected from the refrigerant in the condenser is given by

$$q_{out} = q_{cond} = h_2 - h_3 \tag{12-2}$$

The heat absorbed by the refrigerant in the evaporator is given by

$$q_{in} = q_{evap} = h_1 - h_4 \tag{12-3}$$

For the throttling process through the expansion valve, we have

$$h_3 = h_4 \tag{12-4}$$

The *coefficient of performance* (COP) of a refrigeration cycle is defined as the amount of cooling produced per unit work supplied, or

$$COP = \frac{q_{evap}}{w_{comp}} \tag{12-5}$$

For the cycle shown in Fig. 12-1 with an adiabatic compressor, this equation becomes

$$COP = \frac{h_1 - h_4}{h_2 - h_1} \tag{12-6}$$

The COP of a mechanical vapor-compression refrigeration machine is characteristically about 2 and can be as high as 5.

The amount of heat absorbed by the refrigerant in the evaporator is usually referred to as the *refrigeration effect*. A widely used unit of cooling capacity is the *ton of refrigeration,* which is defined as 200 Btu/min or 211 kJ/min of heat-removal rate from the space to be cooled or heat-absorption rate by the refrigerant in its travel through the evaporator. The mass-flow rate of refrigerant through the cycle can be determined when the refrigeration effect and the number of tons of refrigeration are known. Thus, by referring to Fig. 12-1,

$$\dot{m} = \left(\frac{200}{h_1 - h_4}\right)(\text{number of tons of refrigeration}) \tag{12-7}$$

or

$$\dot{m} = \left(\frac{211}{h_1 - h_4}\right)(\text{number of tons of refrigeration}) \tag{12-7a}$$

where $\dot{m}$ is the mass-flow rate of refrigerant, lbm/min in Eq. (12-7) or kg/min in Eq. (12-7a), and $h_1 - h_4$ is the refrigeration effect, Btu/lbm in Eq. (12-7) or kJ/kg in Eq. (12-7a).

With the mass-flow rate of refrigerant known, the compressor power requirement and the condenser coolant requirement can then be determined. Thus, by referring again to Fig. 12-1, the compressor power requirement is given by

$$\dot{W}_{in} = \dot{W}_{comp} = \dot{m}(h_2 - h_1) \tag{12-8}$$

For the condenser, if water is used as the coolant, we have

$$\dot{Q}_{out} = \dot{Q}_{cond} = \dot{m}(h_2 - h_3) = \dot{m}_{water}(1.0)\,\Delta T_{water} \tag{12-9}$$

where $\dot{m}_{water}$ is the mass-flow rate of cooling water through the condenser, ΔT_{water} is the temperature increase of cooling water in the condenser, and the specific heat of water is taken as 1.0 Btu/lbm·°F = 1.0 kcal/kg·°C.

In an actual vapor compression cycle, due to irreversibility and heat transfer to the compressor coolant, the compression process is neither reversible nor adiabatic. This process might approach either line 1–2a or line 1–2b shown in Fig. 12-2, depending on whether frictional effects or heat-transfer effects dominate. In this figure, state 1 is shown slightly superheated and state 3 slightly subcooled. The superheating at

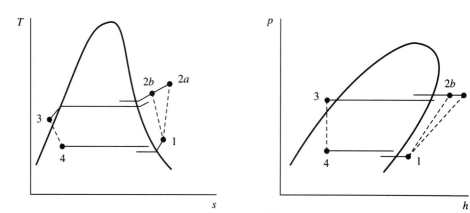

Figure 12-2 Practical vapor-compression cycle.

state 1 is advisable to ensure complete vaporization at the entrance to the compressor. The subcooling at state 3 is advisable as it would result in an increased refrigeration effect. The pressure differences between points 2 and 3 and between points 4 and 1 shown in Fig. 12-2 are simply to indicate that there usually are pressure drops in the condenser and evaporator coils, in compressor suction and exhaust valves, and in pipelines.

Equations (12-1) through (12-9) are applicable to the practical cycles shown in Fig. 12-2 with the following modifications:

1. In cycle $1-2a-3-4$, all the equations are valid with point $2a$ in place of point 2 if process $1-2a$ is irreversible but adiabatic.
2. In cycle $1-2b-3-4$, because process $1-2b$ is not adiabatic, a quantity of heat, q_{comp} transferred out to the coolant in the compression process should be included in the first-law equation. Thus,

$$w_{comp} = h_{2b} - h_1 + q_{comp} \qquad (12\text{-}10)$$

As mentioned previously, subcooling of the liquid before throttling and superheating of the vapor before compression are highly advisable from thermodynamic and practical points of view. A heat exchanger, called a liquid subcooler, is often added to the system to achieve these purposes. This addition is shown schematically in Fig. 12-3 along with the idealized T–s and p–h diagrams. It should be noted that although the refrigeration effect is increased due to subcooling, it is decreased due to superheating by the amount of the sensible heating between points 1 and $1'$.

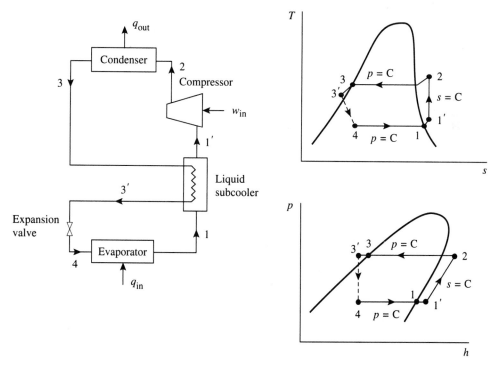

Figure 12-3 Vapor-compression refrigeration cycle with liquid subcooling by suction vapor.

Several factors should be considered in the selection of a refrigerant for a compression cycle. Some of the more important factors are high latent heat of vaporization; positive evaporating pressure; moderately low condensing pressure; low specific volumes for reciprocating compressors and high specific volumes for centrifugal compressors; nontoxicity and nonflammability; inertness and stability; and low cost. The most frequently employed refrigerants are the various halocarbons, such as Refrigerant-11 (CCl_3F), Refrigerant-12 (CCl_2F_2), and Refrigerant-22 ($CHClF_2$).

It is worthwhile drawing attention here that the expansion valve used in a vapor compression refrigeration machine can be replaced by an expansion engine. With this replacement, the T–s diagram of the simple cycle shown in Fig. 12-1 would be modified to that as depicted in Fig. 12-4, with the throttling process 3–4 being replaced by the internally reversible adiabatic expansion process 3–4' in the ideal situation. The work output of the expansion engine could be used to help drive the compressor. Moreover, the refrigeration effect per unit mass of refrigerant is increased by the amount $(h_4 - h_{4'})$. As a result, the coefficient of performance of the refrigeration machine would be higher when an expansion engine is used, in comparison with the use of a throttling valve. Unfortunately, the work output of an expansion engine is usually very small because the working fluid in the engine is mainly a liquid with a small specific volume. Furthermore, an expansion engine is more expensive and requires more maintenance care than a throttling device. Therefore, in practice, throttling devices are always used in vapor-compression refrigeration machines.

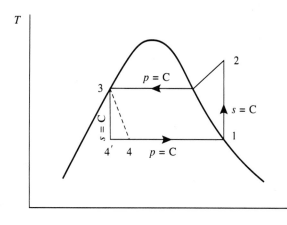

Figure 12-4 Vapor-compression refrigeration cycle with an internally reversible adiabatic expansion process.

Example 12-1

A 5-ton vapor-compression refrigeration machine uses Refrigerant-12 (Freon-12) as the working fluid and operates with an evaporator temperature of $-15°C$ and a condenser pressure of 0.80 MPa. There is no superheating at the compressor inlet and no subcooling at the condenser outlet. Determine (a) the temperature of the fluid leaving the isentropic compressor, (b) the coefficient of performance, (c) the mass-flow rate of the refrigerant, (d) the power input to the compressor, and (e) the percent increase in refrigeration capacity if an isentropic expansion engine were used instead of a free-expansion device.

Solution By referring to Fig. 12-1, the given data are $T_1 = T_4 = -15°C$ and $p_2 = p_3 = 0.80$ MPa. From the tables for Refrigerant-12 in Appendix 3 (Tables A-22M and A-

23M), we obtain

$$h_1 = h_g \text{ at } -15°C = 180.85 \text{ kJ/kg}$$

$$s_1 = s_g \text{ at } -15°C = 0.7046 \text{ kJ/kg·K}$$

$$h_3 = h_4 = h_f \text{ at } 0.80 \text{ MPa} = 67.19 \text{ kJ/kg}$$

$$s_3 = s_f \text{ at } 0.80 \text{ MPa} = 0.2483 \text{ kJ/kg·K}$$

Furthermore, at $p_2 = 0.80$ MPa and $s_2 = s_1 = 0.7046$ kJ/kg·K, from superheated Ref.-12 table, by interpolation, we obtain $h_2 = 206.88$ kJ/kg and $T_2 = 41.3°C$, which is the temperature of the fluid leaving the isentropic compressor.

The coefficient of performance is

$$\text{COP} = \frac{h_1 - h_4}{h_2 - h_1} = \frac{180.85 - 67.19}{206.88 - 180.85} = 4.37$$

The mass-flow rate of the refrigerant is

$$\dot{m} = \frac{211}{h_1 - h_4} \text{ (number of tons of refrigeration)}$$

$$= \frac{211}{180.85 - 67.19}(5) = 9.28 \text{ kg/min}$$

The power input to the compressor is

$$\dot{W}_{comp} = \dot{m}(h_2 - h_1)$$

$$= \left(\frac{9.28}{60}\text{kg/s}\right)(206.88 - 180.85 \text{ kJ/kg}) = 4.03 \text{ kW}$$

Referring to Fig. 12-4, when an isentropic expander is used, we have

$$s_{4'} = s_3 = s_f + x_{4'}s_{fg} \text{ at } -15°C$$

or

$$0.2483 = 0.0906 + x_{4'}(0.6141)$$

so that

$$x_{4'} = 0.257$$

Therefore,

$$h_{4'} = h_f + x_{4'}h_{fg} \text{ at } -15°C$$

$$= 22.312 + 0.257(158.534) = 63.06 \text{ kJ/kg}$$

The percent increase in refrigeration capacity due to the use of the isentropic expander is equal to

$$\frac{h_4 - h_{4'}}{h_1 - h_4} = \frac{67.19 - 63.06}{180.85 - 67.19} = 0.036 = 3.6\%$$

The COP of the refrigeration cycle with expander is given by

$$(\text{COP})_{\text{with expander}} = \frac{h_1 - h_{4'}}{h_2 - h_1} = \frac{180.85 - 63.06}{206.88 - 180.85} = 4.53$$

Thus, due to the use of an isentropic expander, the COP of the refrigerator is increased from 4.37 to 4.53, an increase of $(4.53 - 4.37)/4.37 = 0.037$, or 3.7%.

12-2 HEAT PUMP

A *heat pump* is a refrigeration machine used in reverse. Instead of pumping heat away from a space in order to maintain it at a low temperature, as in space cooling, a heat pump is used to pump heat into a space in order to raise its temperature to a high level. A heat-pump system can be utilized for space heating as well as for space cooling. The dual operating modes of a vapor-compression heat-pump system are illustrated schematically in Fig. 12-5.

Figure 12-5(a) depicts the system in the heating mode, in which heat from an energy source is added to the refrigerant in the outdoor coil and usable heat is delivered from the indoor coil to the space to be heated. Figure 12-5(b) depicts the system in the cooling mode, in which heat is transferred from the space to be cooled to the refrigerant in the indoor coil and waste heat is dissipated from the outdoor coil. Both modes of operation follow the basic principle introduced in Sec. 12-1. The coefficient of performance (COP) of the heating cycle is defined by

$$(COP)_{heating} = \frac{q_{cond}}{w_{comp}} \tag{12-11}$$

and the COP of the cooling cycle is defined by

$$(COP)_{cooling} = \frac{q_{evap}}{w_{comp}} \tag{12-12}$$

Note that this is the same as Eq. (12-5).

For an adiabatic steady-flow compressor, the required work input is given by

$$w_{in} = w_{comp} = h_2 - h_1$$

The heat rejected from the refrigerant in the condenser is given by

$$q_{out} = q_{cond} = h_2 - h_3$$

The heat absorbed by the refrigerant in the evaporator is given by

$$q_{in} = q_{evap} = h_1 - h_4$$

Consequently, Eqs. (12-11) and (12-12) can be expressed as

$$(COP)_{heating} = \frac{h_2 - h_3}{h_2 - h_1} \tag{12-13}$$

and

$$(COP)_{cooling} = \frac{h_1 - h_4}{h_2 - h_1} = \frac{h_1 - h_3}{h_2 - h_1} \tag{12-14}$$

Notice that $h_3 = h_4$ for the throttling process. From the preceding two equations, we have

$$(COP)_{heating} = (COP)_{cooling} + 1 \tag{12-15}$$

This relation is satisfied approximately in actual machine operations.

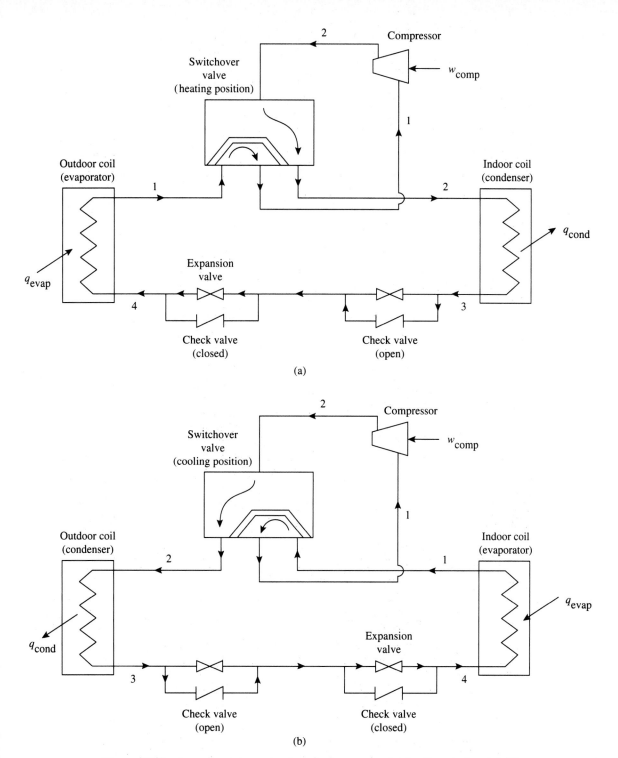

Figure 12-5 A vapor-compression heat-pump system: (a) heating mode, and (b) cooling mode.

Example 12-2

A Refrigerant-12 compression-cycle air-to-air heat pump is used to heat a building. An energy addition of 6000 kJ/min is needed to maintain the interior air supply at 34°C when the outdoor temperature is $-10°C$. The heat-pump evaporator operates at a temperature 10°C below the outdoor temperature and its condenser operates at 1 MPa. The fluid leaving the evaporator and entering the compressor is a saturated vapor and it is a subcooled liquid at 40°C as it leaves the condenser. The compressor has an adiabatic efficiency of 74%. Determine (a) the pressure in the evaporator, (b) the temperature of the fluid at the compressor outlet, (c) the temperature difference in the condenser between the refrigerant before subcooling and the heated air supply, (d) the coefficient of performance, (e) the mass-flow rate of the refrigerant, (f) the power input to the compressor, and (g) the percent savings in input power as compared to direct electric-resistance heating.

Solution A flow schematic and the T–s and p–h diagrams are shown in Fig. 12-6. The given data are

$$T_{\text{indoor air supply}} = 34°C, \qquad\qquad T_{\text{outdoor}} = -10°C$$

$$T_1 = T_4 = -10 - 10 = -20°C \qquad p_2 = p_3 = 1 \text{ MPa}$$

$$T_3 = 40°C \qquad\qquad\qquad \eta_{\text{comp}} = 74\%$$

$$\text{Building energy requirement} = \frac{6000 \text{ kJ/min}}{60 \text{ s/min}} = 100 \text{ kJ/s}$$

From the saturated Refrigerant-12 table (Table A-22M), we obtain $p_1 = p_4 = p_{\text{sat}}$ at $-20°C = 0.1509$ MPa, which is the pressure in the evaporator. In addition, we have

$$h_1 = h_g \text{ at } -20°C = 178.61 \text{ kJ/kg}$$

$$s_1 = s_g \text{ at } -20°C = 0.7082 \text{ kJ/kg·K}$$

$$h_3 = h_4 \approx h_f \text{ at } 40°C = 74.53 \text{ kJ/kg}$$

$$T_{3'} = T_{\text{sat}} \text{ at } 1 \text{ MPa} = 41.6°C$$

Thus, the temperature difference between the refrigerant before subcooling and the heated air supply is given by

$$T_{3'} - T_{\text{indoor air supply}} = 41.6 - 34 = 7.6°C$$

Notice that the refrigerant has been subcooled $41.6 - 40 = 1.6°C$ before it leaves the condenser.

For the internally reversible adiabatic process 1–2', $s_{2'} = s_1 = 0.7082$ kJ/kg·K, also as given in the problem, $p_{2'} = 1$ MPa. At these conditions, from Table A-23M for superheated Refrigerant-12, we obtain $h_{2'} = 212.16$ kJ/kg. For the irreversible adiabatic process 1–2, we then have

$$h_2 = h_1 + \frac{h_{2'} - h_1}{\eta_{\text{comp}}}$$

$$= 178.61 + \frac{212.16 - 178.61}{0.74} = 223.95 \text{ kJ/kg}$$

At $p_2 = 1$ MPa and $h_2 = 223.95$ kJ/kg, from Table A-23M for superheated Refrigerant-12, we obtain $T_2 = 68.2°C$. This is the temperature of the fluid at the outlet of the compressor.

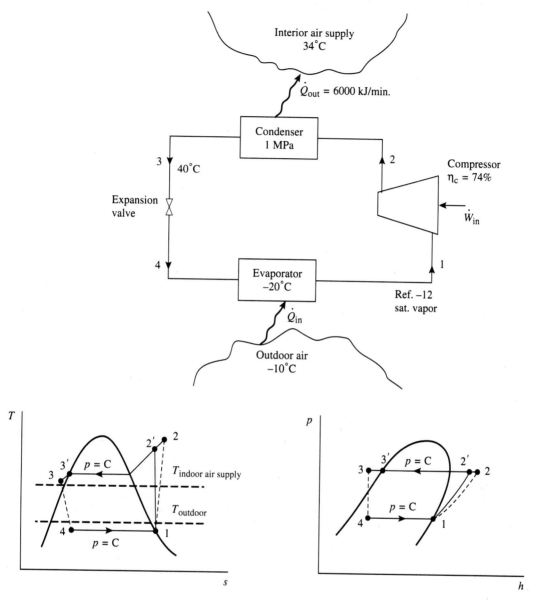

Figure 12-6 The heat-pump cycle of Example 12-2.

The mass-flow rate of Refrigerant-12 is given by

$$\dot{m} = \frac{\left(\begin{array}{c} \text{Energy delivered from the refrigerant} \\ \text{in the condenser to the building} \end{array} \right)}{(h_2 - h_3)}$$

$$= \frac{100}{223.95 - 74.53} = 0.669 \text{ kg/s}$$

The power input to the compressor is

$$\dot{W}_{comp} = \dot{m}(h_2 - h_1) = 0.669(223.95 - 178.61) = 30.3 \text{ kW}$$

The coefficient of performance of the heat pump is

$$COP = \frac{\text{rate of energy delivery to building}}{\text{compressor power input}} = \frac{100}{30.3} = 3.30$$

or

$$COP = \frac{h_2 - h_3}{h_2 - h_1} = \frac{223.95 - 74.53}{223.95 - 178.61} = 3.30$$

The percent savings in input power as compared to direct electric-resistance heating is

$$\frac{100 - 30.3}{100} = 0.697 = 69.7\%$$

12-3 MULTISTAGE VAPOR-COMPRESSION REFRIGERATION

Recall from Sec. 10-7 that multistage compression with intercooling reduces the work requirement in bringing a gas from a given initial state to a specific final pressure. In applying this method to the vapor compressor of a refrigeration system, instead of water or air at the environmental temperature, the low-temperature circulating refrigerant itself can be relied on to serve as the coolant. Figure 12-7 depicts a two-stage vapor-compression refrigeration cycle with intercooling by flashed refrigerant vapor. Saturated refrigerant vapor at state 1 is compressed in the low-pressure compressor to an intermediate pressure (state 2), and is cooled by the lower-temperature flashed vapor at state 7 coming from the flash chamber. The resultant vapor at state 3 from the mixing of the two streams is compressed in the high-pressure compressor to the condenser pressure at state 4 and is then condensed to saturated liquid at state 5. The liquid refrigerant is subsequently expanded through an expansion valve and partially vaporized to the two-phase state 6 with saturated vapor at state 7 and saturated liquid at state 8 separated in the flash chamber. The saturated vapor at state 7 is led to the intercooler to serve as the coolant and the saturated liquid at state 8 is again partially vaporized to state 9 in passing through another expansion valve. The evaporating process 9–1 is responsible for the cooling effect of the cycle.

For the refrigeration system shown in Fig. 12-7, let

$\dot{m}_H$ = mass-flow rate through the high-pressure compressor

$= \dot{m}_3 = \dot{m}_4 = \dot{m}_5 = \dot{m}_6$

$\dot{m}_L$ = mass-flow rate through the low-pressure compressor

$= \dot{m}_8 = \dot{m}_9 = \dot{m}_1 = \dot{m}_2$

An energy balance for the mixing intercooler yields

$$(\dot{m}_H - \dot{m}_L)h_7 + \dot{m}_L h_2 = \dot{m}_H h_3 \tag{12-16}$$

Refrigeration and Cryogenics   Chap. 12

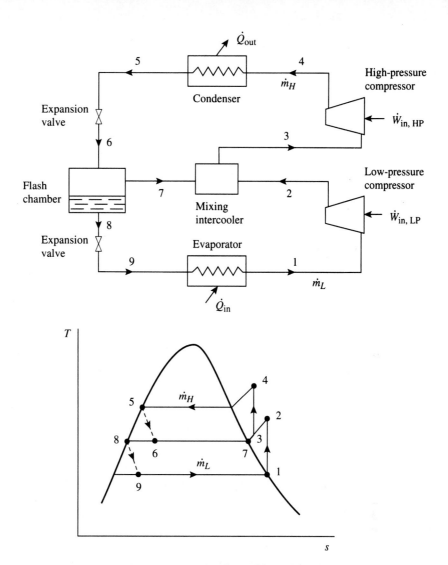

Figure 12-7 Two-stage vapor-compression refrigeration cycle with flash-chamber intercooling: (a) schematic diagram, and (b) ideal T–s diagram.

or

$$\dot{m}_H(h_3 - h_7) = \dot{m}_L(h_2 - h_7) \tag{12-16a}$$

The refrigeration effect is given by

$$\dot{Q}_{in} = \dot{Q}_{evap} = \dot{m}_L(h_1 - h_9) \tag{12-17}$$

The power input to produce this cooling effect is

$$\dot{W}_{in} = \dot{W}_{comp} = \dot{W}_{in, LP} + \dot{W}_{in, HP}$$
$$= \dot{m}_L(h_2 - h_1) + \dot{m}_H(h_4 - h_3) \tag{12-18}$$

The coefficient of performance is given by

$$\text{COP} = \frac{\dot{Q}_{\text{in}}}{\dot{W}_{\text{in}}} = \frac{\dot{Q}_{\text{evap}}}{\dot{W}_{\text{comp}}}$$

$$= \frac{\dot{m}_L(h_1 - h_9)}{\dot{m}_L(h_2 - h_1) + \dot{m}_H(h_4 - h_3)} \tag{12-19}$$

$$= \frac{h_1 - h_9}{(h_2 - h_1) + (\dot{m}_H/\dot{m}_L)(h_4 - h_3)}$$

Another variation of the two-stage vapor-compression refrigeration cycle is shown in Fig. 12-8, in which a direct-contact heat exchanger serves as both the evaporator of the high-pressure subcycle and the condenser of the low-pressure subcycle. From this heat exchanger, the flashed saturated vapor at state 5 goes to the high-pressure compressor intake and the saturated liquid at state 3 goes to the low-pressure throttling valve. For this cycle, let

$$\dot{m}_H = \text{mass-flow rate through the high-pressure compressor}$$

$$= \dot{m}_5 = \dot{m}_6 = \dot{m}_7 = \dot{m}_8$$

$$\dot{m}_L = \text{mass-flow rate through the low-pressure compressor}$$

$$= \dot{m}_1 = \dot{m}_2 = \dot{m}_3 = \dot{m}_4$$

An energy balance for the direct-contact heat exchanger yields

$$\dot{m}_H h_8 + \dot{m}_L h_2 = \dot{m}_H h_5 + \dot{m}_L h_3 \tag{12-20}$$

or

$$\dot{m}_H(h_5 - h_8) = \dot{m}_L(h_2 - h_3) \tag{12-20a}$$

The two-stage refrigeration system shown in Fig. 12-8 is actually a cascade cycle in which the same refrigerant is used in both the high-pressure and low-pressure subcycles. A *cascade cycle* is an arrangement of simple vapor-compression cycles in series, such that the condenser of a lower-temperature cycle provides the heat input to the evaporator of a higher-temperature cycle. If two different refrigerants are used in the two-stage unit shown in Fig. 12-8, the evaporator-condenser heat exchanger should be of the closed type with the two fluid flows separated by a metal wall. In such a case, it is necessary to provide a temperature difference across the solid barrier between the fluids. A cascade system can have three or four units in series, thereby reaching a very low evaporator temperature at the bottom unit.

Example 12-3

A two-stage vapor-compression refrigeration cycle with flash-chamber intercooling as depicted in Fig. 12-7 has pressures of 30, 60, and 150 psia in the evaporator, flash chamber, and condenser, respectively. The compressor processes are assumed to be isentropic. Refrigerant-12 is used to produce 5 tons of refrigeration. Determine (a) the mass-flow rate of refrigerant through the low-pressure compressor, (b) the mass-flow rate of refrigerant through the high-pressure compressor, (c) the power input to the low-pressure compressor, (d) the power input to the high-pressure compressor, and (e) the coefficient of performance of the cycle.

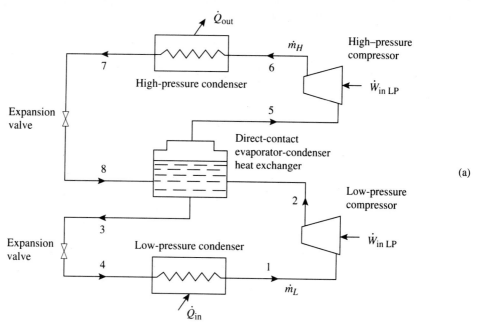

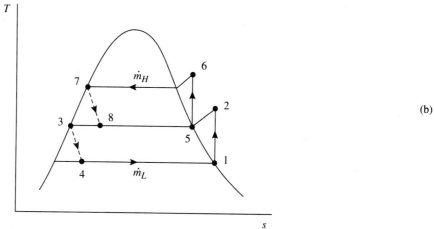

Figure 12-8 Two-stage cascade vapor-compression refrigeration cycle with direct-contact evaporator–condenser heat exchanger: (a) schematic diagram, and (b) ideal T–s diagram.

Solution By referring to Fig. 12-7, $p_1 = p_9 = 30$ psia, $p_2 = p_3 = p_6 = p_7 = p_8 = 60$ psia, $p_4 = p_5 = 150$ psia, $\dot{m}_L = \dot{m}_1 = \dot{m}_2 = \dot{m}_8 = \dot{m}_9$, and $\dot{m}_H = \dot{m}_3 = \dot{m}_4 = \dot{m}_5 = \dot{m}_6$.

From the saturated Refrigerant-12 table (Table A-22E), at $p_1 = 30$ psia, by interpolation, we have

$$h_1 = 78.44 \text{ Btu/lbm} \qquad \text{and} \qquad s_1 = 0.1679 \text{ Btu/lbm·°R}$$

at $p_7 = p_8 = 60$ psia, we have

$$h_7 = 82.29 \text{ Btu/lbm} \qquad \text{and} \qquad h_8 = 19.19 \text{ Btu/lbm} = h_9$$

and at $p_5 = 150$ psia, we have

$$h_5 = 33.39 \text{ Btu/lbm} = h_6$$

From the superheated Refrigerant-12 table (Table A-23E) at $p_2 = 60$ psia and $s_2 = s_1 = 0.1679$ Btu/lbm·°R, by interpolation, we have

$$h_2 = 83.60 \text{ Btu/lbm}$$

(a) For 5 tons of refrigeration, the mass-flow rate through the low-pressure compressor is given by

$$\dot{m}_L = \frac{(5 \text{ tons})(200 \text{ Btu/min·ton})}{(h_1 - h_9) \text{ Btu/lbm}} = \frac{5(200)}{78.44 - 19.19} = 16.88 \text{ lbm/min}$$

(b) State 6 is in the two-phase region with an enthalpy of

$$h_6 = h_8 + x_6(h_7 - h_8) \text{ at } p_6 = 60 \text{ psia}$$

so that the quality at state 6 is given by

$$x_6 = \frac{h_6 - h_8}{h_7 - h_8} = \frac{33.39 - 19.19}{82.29 - 19.19} = 0.2250$$

The mass-flow rate at state 7 is

$$\dot{m}_7 = \dot{m}_H - \dot{m}_L$$

Then

$$x_6 = \frac{\dot{m}_7}{\dot{m}_6} = \frac{\dot{m}_H - \dot{m}_L}{\dot{m}_H}$$

so that the mass-flow rate through the high-pressure compressor is given by

$$\dot{m}_H = \frac{\dot{m}_L}{1 - x_6} = \frac{16.88}{1 - 0.2250} = 21.78 \text{ lbm/min}$$

(c) By neglecting kinetic and potential energies, the power input to the low-pressure compressor is calculated as

$$\dot{W}_{\text{in, LP}} = \dot{m}_L(h_2 - h_1)$$

$$= (16.88 \text{ lbm/min})(83.60 - 78.44 \text{ Btu/lbm})\frac{60 \text{ min/h}}{2545 \text{ Btu/hp·h}}$$

$$= 2.05 \text{ hp}$$

(d) An energy balance for the mixing intercooler yields

$$(\dot{m}_H - \dot{m}_L)h_7 + \dot{m}_L h_2 = \dot{m}_H h_3$$

so that

$$h_3 = \frac{1}{\dot{m}_H}[(\dot{m}_H - \dot{m}_L)h_7 + \dot{m}_L h_2]$$

$$= \frac{1}{21.78}[(21.78 - 16.88)82.29 + 16.88 \times 83.60]$$

$$= 83.31 \text{ Btu/lbm}$$

From the superheated Refrigerant-12 table (Table A-23E), at $p_3 = 60$ psia and $h_3 = 83.31$

Btu/lbm, by interpolation, we obtain

$$s_3 = 0.1673 \text{ Btu/lbm·°R}$$

and at $p_4 = 150$ psia and $s_4 = s_3 = 0.1673$ Btu/lbm·°R, we obtain

$$h_4 = 90.40 \text{ Btu/lbm}$$

By neglecting kinetic and potential energies, the power input to the high-pressure compressor is calculated as

$$\dot{W}_{\text{in, HP}} = \dot{m}_H(h_4 - h_3)$$

$$= (21.78 \text{ lbm/min})(90.40 - 83.31 \text{ Btu/lbm})\frac{60 \text{ min/h}}{2545 \text{ Btu/hp·h}}$$

$$= 3.64 \text{ hp}$$

(e) From Eq. (12-19), the coefficient of performance is calculated as

$$\text{COP} = \frac{\dot{Q}_{\text{evap}}}{\dot{W}_{\text{comp}}} = \frac{\dot{m}_L(h_1 - h_9)}{\dot{m}_L(h_2 - h_1) + \dot{m}_H(h_4 - h_3)}$$

$$= \frac{16.88(78.44 - 19.19)}{16.88(83.60 - 78.44) + 21.78(90.40 - 83.31)} = 4.14$$

The COP can also be calculated by

$$\text{COP} = \frac{\dot{Q}_{\text{evap}}}{\dot{W}_{\text{comp}}} = \frac{(5 \text{ tons})(200 \text{ Btu/min·ton})(60 \text{ min/h})}{(2.05 + 3.64 \text{ hp})(2545 \text{ Btu/hp·h})} = 4.14$$

12-4 GAS-CYCLE REFRIGERATION

Because the temperature of a gas decreases as it expands adiabatically under a restraining force, a reversed Brayton cycle can be modeled for the production of refrigeration. This method of cooling is referred to as *gas-cycle refrigeration*. Although less efficient than a vapor-compression refrigeration system, gas refrigeration units offer the advantage of weight and space savings, particularly suitable for aircraft cabin cooling.

Figure 12-9 shows the schematic and idealized p–v and T–s diagrams of a reversed-Brayton-cycle refrigeration system. Outside air is first compressed (process 1–2) and then cooled in the energy-dissipation heat exchanger (process 2–3). The air is next expanded through an expander (process 3–4), thus lowering its temperature. The work output of the expander can be used to supply part of the work input required for the compressor. The heat addition (process 4–1) occurring in the refrigeration-space heat exchanger is responsible for the removal of heat from the space to be cooled.

By referring to Fig. 12-9, with ΔKE and ΔPE neglected, application of the steady-flow energy equation per unit mass flow to each of the four pieces of component gives

$$w_{\text{in}} = w_{\text{comp}} = h_2 - h_1 \tag{12-21}$$

$$w_{\text{out}} = w_{\text{exp}} = h_3 - h_4 \tag{12-22}$$

$$q_{\text{out}} = q_{\text{dissipation}} = h_2 - h_3 \tag{12-23}$$

$$q_{\text{in}} = q_{\text{ref space}} = h_1 - h_4 \tag{12-24}$$

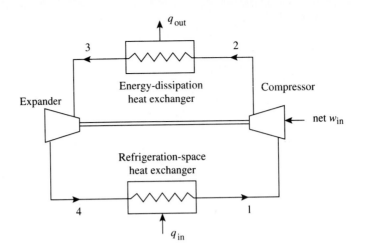

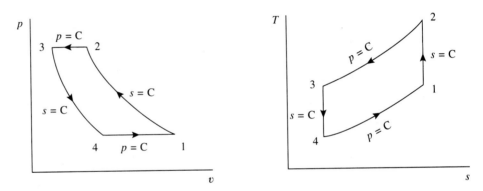

Figure 12-9 Reversed Brayton cycle: (a) schematic diagram, and (b) ideal p–v and T–s diagrams.

The coefficient of performance as a refrigerator is

$$\text{COP} = \frac{q_{\text{in}}}{w_{\text{in}} - w_{\text{out}}} = \frac{h_1 - h_4}{(h_2 - h_1) - (h_3 - h_4)} \qquad (12\text{-}25)$$

If the temperature achieved in a simple reversed Brayton cycle is not low enough for a particular application, a regenerator can be inserted into the cycle to reduce the turbine inlet temperature, thereby providing a means of lowering the turbine discharge temperature. Figure 12-10 depicts the schematic diagram and idealized T–s diagram of a reversed Brayton cycle with regeneration. After being cooled to state 3 by a coolant, the gas is further cooled in the regenerator to state 4, before expanding through the turbine to reach the lowest temperature of the cycle at state 5.

By referring to Fig. 12-10, an energy balance for the regenerator gives

$$h_3 - h_4 = h_1 - h_6 \qquad (12\text{-}26)$$

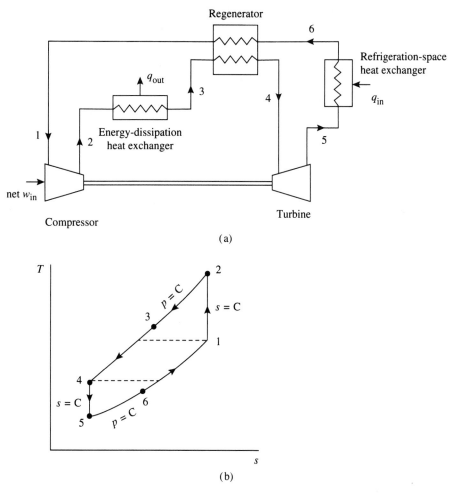

Figure 12-10 Reversed Brayton cycle with regeneration: (a) schematic diagram, and (b) ideal T–s diagram.

The work and heat quantities, per unit mass flow, are given by

$$w_{in} = w_{comp} = h_2 - h_1 \tag{12-27}$$

$$w_{out} = w_{turb} = h_4 - h_5 \tag{12-28}$$

$$q_{out} = q_{dissipation} = h_2 - h_3 \tag{12-29}$$

$$q_{in} = q_{ref\ space} = h_6 - h_5 \tag{12-30}$$

The coefficient of performance of the refrigeration cycle is then

$$\text{COP} = \frac{q_{in}}{w_{in} - w_{out}} = \frac{h_6 - h_5}{(h_2 - h_1) - (h_4 - h_5)} \tag{12-31}$$

Figure 12-11 shows a version of a jet aircraft cooling system. Air bled from the main compressor of the gas-turbine engine is cooled in the first heat exchanger before

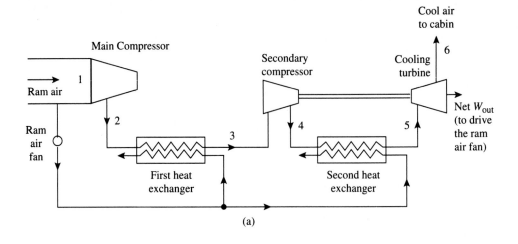

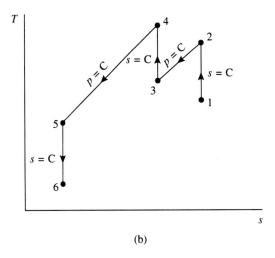

Figure 12-11   A jet aircraft cooling system: (a) schematic diagram, and (b) ideal *T–s* diagram.

entering the secondary compressor for a further increase in pressure. It is then further cooled in the second heat exchanger, and finally expanded through the cooling turbine, reaching a low temperature before entering the cabin. Here ram air is the coolant for the heat exchangers. The ram pressure of the moving airplane forces air through the heat exchangers. When the airplane is on the ground and there is no ram pressure available, the ram-air fan will be relied on to move the coolant.

Example 12-4

A reversed Brayton cycle with regeneration operates with air entering the compressor at 14.5 psia, 80°F, and leaving at 60 psia. The air is precooled in the energy-dissipation heat exchanger to 140°F and it is 80°F at the entrance to the turbine. For an ideal cycle, determine the minimum temperature of the cycle and the coefficient of performance using (a) $c_p = 0.24$ Btu/lbm·°F and $k = 1.4$ for air, and (b) the air table data.

Solution By referring to Fig. 12-10, the given data are $p_1 = 14.5$ psia, $T_1 = 80°F = 540°R$, $p_2 = 60$ psia, $T_3 = 140°F = 600°R$, and $T_4 = 80°F = 540°R$.

(a) $c_p = 0.24$ Btu/lbm·°F and $k = 1.4$. For the internally reversible adiabatic process 1–2,

$$T_2 = T_1 \left(\frac{p_2}{p_1}\right)^{(k-1)/k} = 540 \left(\frac{60}{14.5}\right)^{0.4/1.4} = 810°R = 350°F$$

For the internally reversible adiabatic process 4–5,

$$T_5 = T_4 \left(\frac{p_5}{p_4}\right)^{(k-1)/k} = 540 \left(\frac{14.5}{60}\right)^{0.4/1.4} = 360°R = -100°F$$

which is the lowest temperature of the cycle.

An energy balance for the regenerator gives

$$h_3 - h_4 = h_1 - h_6 \qquad \text{or} \qquad T_3 - T_4 = T_1 - T_6$$

so that

$$T_6 = T_1 - T_3 + T_4$$

$$= 540 - 600 + 540 = 480°R = 20°F$$

The coefficient of performance of the refrigeration cycle is then

$$\text{COP} = \frac{h_6 - h_5}{(h_2 - h_1) - (h_4 - h_5)} = \frac{T_6 - T_5}{(T_2 - T_1) - (T_4 - T_5)}$$

$$= \frac{480 - 360}{(810 - 540) - (540 - 360)} = 1.33$$

(b) *Air-table solution.* From the air table (Table A-9E), at $T_1 = T_4 = 540°R$, we get

$$h_1 = h_4 = 129.16 \text{ Btu/lbm} \qquad \text{and} \qquad p_{r1} = p_{r4} = 1.3801$$

and at $T_3 = 600°R$, we get $h_3 = 143.57$ Btu/lbm.

For the internally reversible adiabatic process 1–2,

$$p_{r2} = p_{r1} \left(\frac{p_2}{p_1}\right) = 1.3801 \left(\frac{60}{14.5}\right) = 5.711$$

At this p_{r2} value, the air table gives

$$h_2 = 194.0 \text{ Btu/lbm} \qquad \text{and} \qquad T_2 = 808°R = 348°F$$

For the internally reversible adiabatic process 4–5,

$$p_{r5} = p_{r4} \left(\frac{p_5}{p_4}\right) = 1.3801 \left(\frac{14.5}{60}\right) = 0.3335$$

At this p_{r5} value, the air table gives

$$h_5 = 85.95 \text{ Btu/lbm} \qquad \text{and} \qquad T_5 = 360°R = -100°F$$

which is the lowest temperature of the cycle.

An energy balance for the regenerator gives

$$h_3 - h_4 = h_1 - h_6$$

or

$$h_6 = h_1 - h_3 + h_4 = 129.16 - 143.57 + 129.16 = 114.75 \text{ Btu/lbm}$$

At this h_6 value, the air table gives $T_6 = 480°R = 20°F$.

The coefficient of performance of the refrigeration cycle is

$$\text{COP} = \frac{h_6 - h_5}{(h_2 - h_1) - (h_4 - h_5)}$$

$$= \frac{114.75 - 85.95}{(194.0 - 129.16) - (129.16 - 85.95)}$$

$$= 1.33$$

12-5 ABSORPTION REFRIGERATION

Because the heat-addition and heat-rejection processes occur at different temperatures in refrigeration cycles, a vapor-compression cycle requires the use of a compressor to create the pressure difference between the evaporator and the condenser. A relatively larger power input per unit mass flow is needed to compress a gas than to pump a liquid between the same pressure difference. An absorption refrigeration cycle is a scheme whereby the vapor compressor is replaced by a liquid pump for creating the required pressure difference.

In addition to a liquid pump, an *absorption refrigeration cycle* employs a generator, an absorber, along with a pressure-reducing valve, as shown schematically in the right half of Fig. 12-12, to form the replacement of a vapor compressor. The solution heat exchanger shown in the figure between the generator and the absorber is an energy

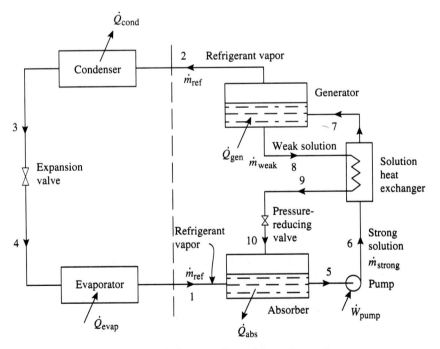

Figure 12-12 Absorption refrigeration cycle.

saver but is not essential to the successful operation of the cycle. This portion of the cycle involves the use of a binary vapor–liquid mixture, as, for example, a mixture of ammonia and water with ammonia as the refrigerant and water as the absorbent, or a mixture of water and lithium bromide with water as the refrigerant and lithium bromide as the absorbent. The left half of Fig. 12-12 is identical with that of a regular vapor-compression cycle through which the refrigerant alone travels. At the end of its lonely journey through this regular part of the cycle, the refrigerant at state 1 enters the absorber and is absorbed by the weak solution at state 10 coming back from the generator via the solution heat exchanger and the pressure-reducing valve. By "weak solution," we mean a liquid solution of low refrigerant concentration. The dissolving process occurring in the absorber is exothermic, thus, heat must be removed from the absorber in order to keep it at a constant low temperature. The strong solution (liquid solution of high refrigerant concentration) emerging from the absorber at state 5 is pumped through the solution heat exchanger to extract some energy from the weak solution and then flows into the generator at state 7. In the generator, heat is added to the strong solution to boil off some of the refrigerant. The remaining weak solution at state 8 flows back to the absorber. The pure (or nearly pure) refrigerant vapor leaves the generator at state 2 to travel through the condenser, expansion valve, and evaporator, and finally returns to the absorber at state 1 to complete the cycle.

In order to understand the underlying principles involved in the operation of an absorption refrigeration cycle, it is essential for the reader to study the behavior of binary vapor–liquid mixtures. We defer the analysis of absorption refrigeration until Sec. 16-6.

12-6 STEAM-JET REFRIGERATION

In a *steam-jet refrigeration* system, water is the refrigerant and the evaporation of water provides the refrigeration. As depicted in Fig. 12-13, water to be cooled at state A is piped into the flash chamber (evaporator), where it flashes into a vapor–liquid mixture at state B with liquid at saturated liquid state C and vapor of high quality at state D. High-pressure motive steam at state 1 expands through a converging-diverging nozzle and rushes out at state 2 with a supersonic speed. This high-speed steam-jet entrains the slow-moving vapor from the flash chamber and together they move on to the diffuser to be compressed to the condenser pressure in process 4–5. The entrained vapor carries away the latent heat of vaporization, thereby cooling the water in the flash chamber. The chilled water is piped out and pumped to a higher pressure to serve the cooling load. The vapor flashed from the evaporator is replenished by makeup water through a float-valve control. The condenser must be equipped with an air ejector to remove air from the system.

It is imperative to recognize that the transfer of kinetic energy from the motive steam to the flashed vapor is inherently inefficient. Some of the kinetic energy delivered from the expansion of the motive steam in the nozzle is dissipated to the form of thermal energy that is no longer available for the purpose of compression. To account for this deficiency, a factor called the entrainment efficiency is introduced as

$$\eta_{ent} = \frac{h_1 - h_3}{h_1 - h_2} \tag{12-32}$$

wherein the enthalpy change $(h_1 - h_3)$ is equivalent to the kinetic energy of the motive

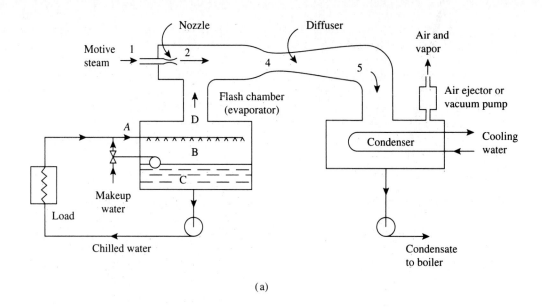

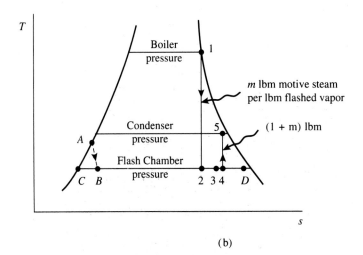

Figure 12-13 Steam-jet refrigeration: (a) schematic diagram, and (b) ideal T-s diagram (partial).

steam available for the entrainment of the flashed vapor. In Eq. (12-32), the enthalpy change $(h_1 - h_2)$ is based on ideal expansion through the nozzle. When the irreversibility effect of the nozzle is taken care of, the actual change in enthalpy should be used in place of the ideal one.

Because the required work of compression comes from the available energy delivered from the expansion of the motive steam in the nozzle after the deficiency of entrainment and the irreversibility effect of nozzle process having been taken care of, thus, we have

$$m(h_1 - h_3) = (1 + m)(h_5 - h_4) \tag{12-33}$$

Refrigeration and Cryogenics   Chap. 12

where m is the mass flow of motive steam per unit mass of entrained vapor. The refrigeration effect per unit mass of vapor flashed from the evaporator is

$$\text{Refrigeration effect} = h_D - h_B \tag{12-34}$$

The efficiency of a steam-jet refrigeration system can be expressed in terms of steam consumption, or the mass-flow rate of motive steam per ton refrigeration. Thus,

Steam consumption, lbm motive steam/min·ton

$$= \frac{(200 \text{ Btu/min·ton})(m \text{ lbm motive steam/lbm flashed vapor})}{(h_D - h_B) \text{ Btu/lbm flashed vapor}} \tag{12-35a}$$

Steam consumption, kg motive steam/min·ton

$$= \frac{(211 \text{ hJ/min·ton})(m \text{ kg motive steam/kg flashed vapor})}{(h_D - h_B) \text{ kJ/kg flashed vapor}} \tag{12-35b}$$

Example 12-5

A steam-jet refrigeration system is supplied with motive steam at 80 psia saturated. The condenser is operated at 1 psia and the flash-chamber water is at 40°F. Makeup water, including water returned from the load, is supplied to the flash chamber at 65°F. The entrainment efficiency is 65%. The nozzle has an adiabatic efficiency of 90%. The diffuser process is assumed to be ideal. The quality of the motive steam and entrained vapor at the beginning of compression is 91%. Determine (a) the mass of motive steam required per pound of entrained vapor, (b) the quality of vapor entrained from the flash chamber, (c) the refrigeration effect per pound of entrained vapor, and (d) the mass of motive steam required per hour for each ton of refrigeration.

Solution By referring to Fig. 12-13, the given data are $p_1 = 80$ psia, $p_5 = 1$ psia, $T_2 = T_4 = 40°F$, $T_A = 65°F$, $\eta_{ent} = 0.65$, $\eta_{nozzle} = 0.90$, and $x_4 = 91\%$. From the steam tables (Tables A-18E and A-19E), we have

$$h_1 = h_g \text{ at 80 psia} = 1183.6 \text{ Btu/lbm}$$

$$s_1 = s_g \text{ at 80 psia} = 1.6214 \text{ Btu/lbm·°R}$$

$$p_2 = p_{sat} \text{ at 40°F} = 0.12166 \text{ psia}$$

For the internally reversible adiabatic process 1–2,

$$s_2 = s_1 = s_f + x_2 s_{fg} \text{ at 40°F}$$

or

$$1.6214 = 0.01617 + x_2(2.1430)$$

so that

$$x_2 = 0.749$$

Therefore,

$$h_2 = h_f + x_2 h_{fg} \text{ at 40°F}$$

$$= 8.02 + 0.749(1070.9) = 810.1 \text{ Btu/lbm}$$

By the definition of adiabatic nozzle efficiency,

$$\eta_{nozzle} = \frac{h_1 - h_{2'}}{h_1 - h_2}$$

Sec. 12-6 Steam-Jet Refrigeration

or

$$h_1 - h_{2'} = \eta_{\text{nozzle}}(h_1 - h_2)$$

where state $2'$ is the actual condition at the nozzle exit (not shown in Fig. 12-13). In addition, by the definition of entrainment efficiency,

$$\eta_{\text{ent}} = \frac{h_1 - h_3}{h_1 - h_{2'}} \qquad \text{or} \qquad h_1 - h_3 = \eta_{\text{ent}}(h_1 - h_{2'})$$

Thus,

$$h_1 - h_3 = \eta_{\text{ent}}\,\eta_{\text{nozzle}}(h_1 - h_2)$$

so that

$$h_3 = h_1 - \eta_{\text{ent}}\,\eta_{\text{nozzle}}(h_1 - h_2)$$

$$= 1183.6 - 0.65 \times 0.90(1183.6 - 810.1)$$

$$= 965.1 \text{ Btu/lbm}$$

From the steam tables, we have

$$h_4 = h_f + x_4 h_{fg} \text{ at } 40°F = 8.02 + 0.91(1070.9) = 982.5 \text{ Btu/lbm}$$

$$s_4 = s_f + x_4 s_{fg} \text{ at } 40°F = 0.01617 + 0.91(2.1430) = 1.9663 \text{ Btu/lbm·°R}$$

For the internally reversible adiabatic process 4–5,

$$s_5 = s_4 = s_f + x_5 s_{fg} \text{ at } 1 \text{ psia}$$

or

$$1.9663 = 0.13266 + x_5(1.8453)$$

so that

$$x_5 = 0.994$$

Therefore,

$$h_5 = h_f + x_5 h_{fg} \text{ at } 1 \text{ psia}$$

$$= 69.74 + 0.994(1036.0) = 1099.5 \text{ Btu/lbm}$$

From Eq. (12-33),

$$m(h_1 - h_3) = (1 + m)(h_5 - h_4)$$

or

$$m = \frac{h_5 - h_4}{(h_1 - h_3) - (h_5 - h_4)}$$

$$= \frac{1099.5 - 982.5}{(1183.6 - 965.1) - (1099.5 - 982.5)}$$

$$= 1.15 \text{ lbm motive steam per lbm entrained vapor}$$

An energy balance for the mixing of the motive steam with the entrained vapor leads to

$$mh_3 + 1h_D = (1 + m)h_4$$

or

$$h_D = (1 + m)h_4 - mh_3$$

$$= (1 + 1.15)982.5 - 1.15 \times 965.1 = 1002.5 \text{ Btu/lbm}$$

Furthermore,

$$h_D = h_f + x_D h_{fg} \text{ at } 40°F$$

or

$$1002.5 = 8.02 + x_D(1070.9)$$

so that

$$x_D = 0.929$$

which is the quality of vapor entrained from the flash chamber. From the steam tables,

$$h_A = h_B = h_f \text{ at } 65°F = 33.1 \text{ Btu/lbm}$$

The refrigeration effect is then calculated as

$$\text{Refrigeration effect} = h_D - h_B = 1002.5 - 33.1$$

$$= 969.4 \text{ Btu per lbm of entrained vapor}$$

The steam consumption is calculated as

$$\text{Steam consumption} = \frac{(200 \times 60)m}{h_D - h_B} = \frac{12000(1.15)}{1002.5 - 33.1}$$

$$= 14.2 \text{ lbm motive steam/hour·ton refrigeration}$$

12-7 GAS LIQUEFACTION

There are a few common methods of producing cryogenic temperatures. These are (a) evaporation under reduced pressure, (b) Joule–Thomson free expansion (or throttling process), (c) adiabatic expansion against a restraining force, and (d) adiabatic demagnetization of paramagnetic materials.

When a gas at a temperature lower than its critical temperature and higher than its triple-point temperature is compressed isothermally, the gas will liquefy when the pressure reaches the vapor pressure at the given temperature. Once it is in liquid form, it can be allowed to evaporate in an insulated container under reduced pressure, the vapor formed being constantly pumped away, carrying with it the latent heat of vaporization. When the pumping of the vapor is rapid enough, the triple-point temperature of the substance can be achieved and the substance solidified. Because the vapor pressure of a solid falls off rapidly as the temperature goes below the triple point, this method fairly soon becomes impractical.

If different substances having consecutively lower triple-point temperatures are used in a series of compression-evaporation processes, very low temperatures can be achieved. In such a series of processes, the gas with the higher triple point is liquefied first; this liquid is then allowed to evaporate under reduced pressure, thus cooling itself. This cooled liquid is then used as the cooling fluid during isothermal compression of a second gas, thus liquefying the second gas.

The thermodynamic process of Joule–Thomson free expansion process will be analyzed in Sec. 13-5, where a term called the maximum inversion temperature is introduced. When the Joule–Thomson free expansion process is used to produce a cooling effect, the initial temperature of the gas must be below its maximum inversion temperature. The maximum inversion temperature of most gases is above the normal room temperature. This temperature for hydrogen and helium, however, is below the normal room temperature; thus, precooling is necessary before these gases can be throttled to produce a cooling effect.

In contrast to a Joule–Thomson expansion, an adiabatic expansion against a restraining force always produces a cooling, regardless of the initial temperature. From a thermodynamic point of view, an internally reversible adiabatic expansion is superior to a Joule–Thomson expansion. Between any two given pressures, the former always gives a greater temperature drop than does the latter. However, the temperature drop due to an internally reversible adiabatic expansion decreases as the initial temperature decreases for a given pressure drop. However, in the case of a Joule–Thomson expansion, the reverse is true. Furthermore, at low temperatures, there are lubrication difficulties for a moving expansion mechanism, such as an engine. On the contrary, in a Joule–Thomson expansion device, there are no moving parts and so lubrication is no problem. Because of these reasons, a combination of these two methods, namely, the use of an adiabatic expansion against a restraining force in the higher-temperature range followed by a Joule–Thomson expansion in the lower-temperature range, is in common use to take advantage of both methods.

In order to illustrate the use of Joule–Thomson and internally reversible adiabatic expansion in cryogenic applications, we proceed now to discuss a few basic systems in gas liquefaction. The simplest liquefaction system is the Linde–Hampson system, shown schematically in Fig. 12-14(a). The idealized cycle of this system in the temperature–entropy plane for the liquefaction of air is shown in Fig. 12-14(b). Here the gas is assumed to be compressed from state 1 to state 2 reversibly and isothermally. In actual operations, the compression process could be carried out in multistages with intercooling and aftercooling. The gas is then cooled in a counterflow heat exchanger, where it gives heat to the returning low-pressure stream of gas. Upon throttling through an expansion valve, the gas cools itself and then returns to the heat exchanger to cool the incoming gas so that the gas arriving at the expansion valve gets cooler and cooler. After many repetitions, eventually a low-temperature state such as indicated by point 3 is reached, from which a Joule–Thomson expansion to state 4 results in partial liquefaction of the gas. At state 4, the gas is in the two-phase region, consisting of saturated liquid at state 5 and saturated vapor at state 6. The liquid is withdrawn from the liquid receiver and the vapor leaves the liquid receiver and returns through the heat exchanger toward the compressor. The returned vapor at state 7 is mixed with an amount of makeup air from the outside equal to the amount of liquid withdrawn, and this mixture at state 1 is admitted to the compressor.

Referring to Fig. 12-14, an energy balance on the control volume, which includes the heat exchanger, expansion valve, and liquid receiver, yields

$$\dot{m}h_2 = \dot{m}_f h_5 + (\dot{m} - \dot{m}_f)h_7$$

or

$$\dot{m}_f(h_7 - h_5) = \dot{m}(h_7 - h_2)$$

so that

$$\frac{\dot{m}_f}{\dot{m}} = \frac{h_7 - h_2}{h_7 - h_5} \tag{12-36}$$

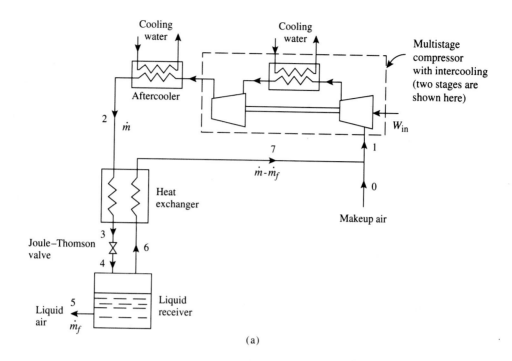

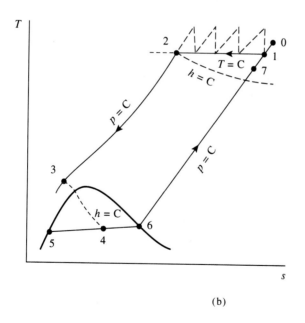

(b)

Figure 12-14 Linde–Hampson system for air liquefaction: (a) schematic diagram, and (b) ideal T–s diagram.

where

$$\dot{m}_f = \text{liquid discharge rate from the liquid receiver}$$

$$\dot{m} = \text{mass-flow rate of gas through the compressor}$$

$$\dot{m}_f/\dot{m} = \text{liquid yield per unit mass of gas compressed}$$

According to Eq. (12-36), there is yield only when h_7 is greater than h_2. But T_7 should be less than T_2 for good heat transfer in the heat exchanger.

Again, referring to Fig. 12-14, for the internally reversible isothermal compression process 1–2, the heat rejected from the gas is given by

$$\frac{\dot{Q}_{\text{out}}}{\dot{m}} = T_1(s_1 - s_2) \tag{12-37}$$

and the work done on the gas per unit mass compressed is

$$\frac{\dot{W}_{\text{in}}}{\dot{m}} = \frac{\dot{Q}_{\text{out}}}{\dot{m}} + (h_2 - h_1) \tag{12-38}$$

As mentioned previously, there are advantages in the use of an adiabatic expansion against a restraining force in a higher temperature range followed by a Joule–Thomson expansion in a lower temperature range. This idea is utilized in the Claude system shown in Fig. 12-15. Here the gas is first compressed and then cooled in passing through the first heat exchanger. The gas is then divided into two streams. A major part of the gas stream is diverted to an expander, where the gas gets cooler by expanding against a restraining force. The remainder of the gas stream is sent to the second heat exchanger on its way to the third heat exchanger and the throttling valve to be partially liquefied. The liquid portion is withdrawn from the liquid receiver, and the vapor portion leaves the liquid receiver and joins the cooled gas from the expander to serve as the cooling agency on its return trip to the compressor.

Because the maximum inversion temperatures of neon, hydrogen, and helium are below normal room temperature, these gases must be precooled before a Joule–Thomson expansion can be used to produce a cooling effect. The precoolant must have a triple-point temperature below that of the maximum inversion temperature of the gas to be liquefied. From a practical standpoint, nitrogen would be a good choice as the precoolant for the liquefaction of hydrogen or neon. Figure 12-16(a) shows a schematic diagram of a precooled Linde–Hampson system for hydrogen or neon, using liquid nitrogen as the precoolant. Figure 12-16(b) shows the idealized cycle in the T–s plane for the main gas only.

Because an adiabatic expansion against a restraining force always produces a cooling regardless of the initial temperature, a Claude system, as shown in Fig. 12-15, without modification, can be used to liquefy hydrogen or neon. However, if a liquid-nitrogen precooling is used with a Claude system the performance of the system can certainly be improved.

For the liquefaction of helium, a compact system known as the Collins helium liquefier, as shown in Fig. 12-17, can be used. It is essentially an extension of the Claude system with two or more expansion engines.

Temperature–entropy diagrams of several gases (air, nitrogen, oxygen, carbon dioxide, carbon monoxide, hydrogen, helium, argon, and methane) are given in Appendix 4 (Figs. A-1 to A-9). These figures are on a reduced scale, good for the purpose

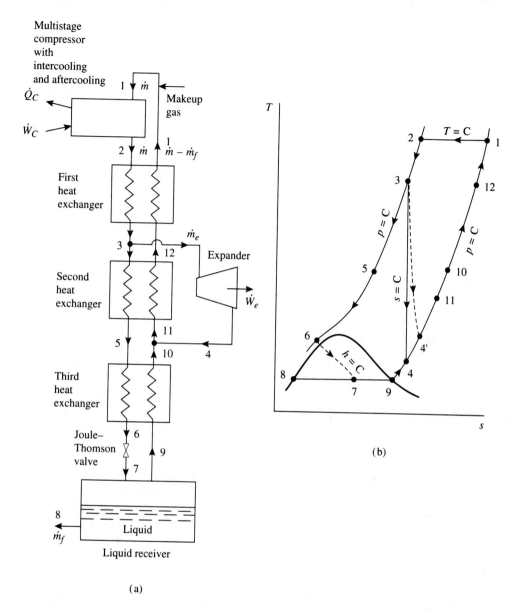

Multistage compressor with intercooling and aftercooling

$\dot{Q}_C$

$\dot{W}_C$

1 $\dot{m}$

Makeup gas

2 $\dot{m}$ 1 $\dot{m} - \dot{m}_f$

First heat exchanger

3   12

$\dot{m}_e$

Expander

$\dot{W}_e$

Second heat exchanger

5   10   11   4

Third heat exchanger

Joule–Thomson valve   6   9   7

8

$\dot{m}_f$

Liquid

Liquid receiver

(a)

(b)

Figure 12-15 Claude system: (a) schematic diagram, and (b) ideal T–s diagram.

of preliminary design of cryogenic systems. For professional use in more accurate calculations, one should employ the original large graphs, such as those given in the references for the reduced figures.

Example 12-6

In a Claude liquefaction system as depicted in Fig. 12-15, nitrogen gas at 25°C and 1 atm is compressed reversibly and isothermally to 50 atm. The gas is then cooled to 0°C in passing through a heat exchanger. At the outlet of this heat exchanger, 60% of the total flow is diverted and is expanded adiabatically to 1 atm through an expander that has an adiabatic efficiency of 75%. The expander work is utilized as part of the compressor work

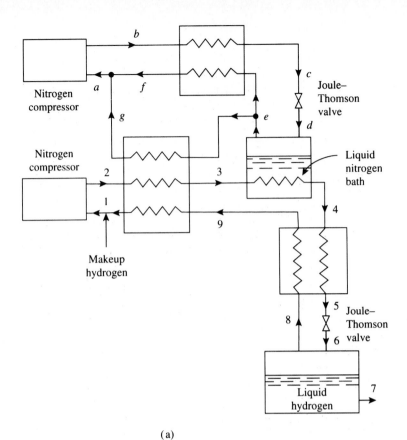

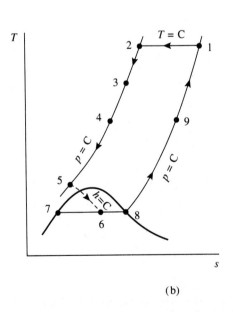

(a)

(b)

Figure 12-16 Precooled Linde–Hampson system for hydrogen (or neon) liquefaction, using liquid nitrogen as the precoolant: (a) schematic diagram, and (b) ideal T–s diagram for the hydrogen (or neon) cycle.

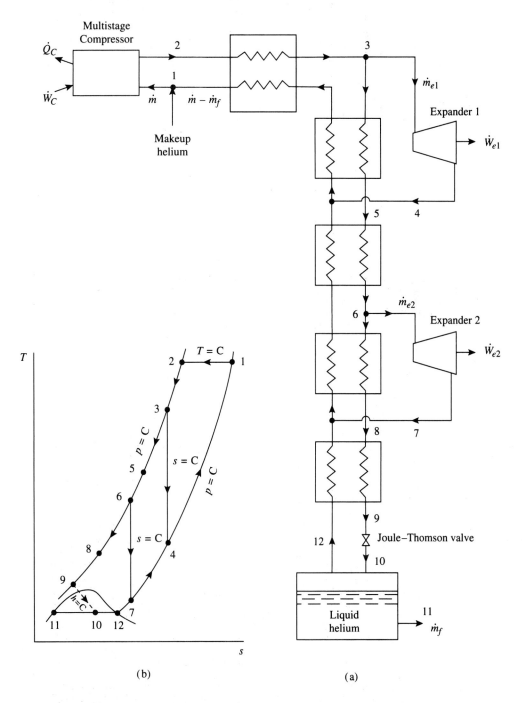

Figure 12-17 Collins helium-liquefaction system: (a) schematic diagram, and (b) ideal T–s diagram.

requirement. All the heat exchangers are working under ideal conditions. Determine the liquid yield per unit mass of gas compressed and the net work required to liquefy a unit mass of gas.

Solution In Fig. 12-15, $\dot{m}$ and $\dot{m}_e$ denote the mass-flow rates through the compressor and the expander, respectively, and $\dot{m}_f$ denotes the liquid discharge rate from the liquid receiver.

From the T–s diagram for nitrogen in Appendix 4 (Fig. A-2), we obtain the following values:

At $p_1 = 1$ atm and $T_1 = 298$ K,

$$h_1 = 460 \text{ kJ/kg} \qquad \text{and} \qquad s_1 = 4.414 \text{ kJ/kg·K}$$

At $p_2 = 50$ atm and $T_2 = 298$ K,

$$h_2 = 450 \text{ kJ/kg} \qquad \text{and} \qquad s_2 = 3.222 \text{ kJ/kg·K}$$

At $p_3 = 50$ atm and $T_3 = 273$ K,

$$h_3 = 422 \text{ kJ/kg} \qquad \text{and} \qquad s_3 = 3.123 \text{ kJ/kg·K}$$

At $p_4 = 1$ atm and $s_4 = s_3 = 3.123$ kJ/kg·K

$$h_4 = 293 \text{ kJ/kg}$$

At $p_8 = 1$ atm and saturated liquid,

$$h_8 = 29.4 \text{ kJ/kg}$$

The adiabatic efficiency of the expander is defined as

$$\eta_{\text{exp}} = \frac{h_3 - h_{4'}}{h_3 - h_4}$$

From which

$$h_{4'} = h_3 - \eta_{\text{exp}}(h_3 - h_4) = 422 - 0.75(422 - 239)$$
$$= 285 \text{ kJ/kg}$$

The work done by the gas during expansion is

$$\frac{\dot{W}_e}{\dot{m}} = \frac{\dot{W}_{34'}}{\dot{m}} = \frac{\dot{m}_e}{\dot{m}}(h_3 - h_{4'}) = 0.60(422 - 285)$$
$$= 82.2 \text{ kJ/kg compressed}$$

Because the compression process is reversible and isothermal, the heat rejected from the gas during compression is

$$\frac{\dot{Q}_c}{\dot{m}} = \frac{-\dot{Q}_{12}}{\dot{m}} = T_1(s_1 - s_2) = 298(4.414 - 3.222)$$
$$= 355 \text{ kJ/kg compressed}$$

From the first law for steady flow, the work required to compress the gas is

$$\frac{\dot{W}_c}{\dot{m}} = \frac{-\dot{W}_{12}}{\dot{m}} = \frac{\dot{Q}_c}{\dot{m}} + h_2 - h_1 = 355 + 450 - 460$$
$$= 345 \text{ kJ/kg compressed}$$

The net work required is then

$$\frac{\dot{W}_{\text{net}}}{\dot{m}} = \frac{\dot{W}_c}{\dot{m}} - \frac{\dot{W}_e}{\dot{m}} = 345 - 82.2$$

$$= 263 \text{ kJ/kg compressed}$$

Applying the first law for steady flow to the system as a whole, we have

$$\dot{W}_c - \dot{W}_e - \dot{Q}_c + \dot{m}_f h_1 - \dot{m}_f h_8 = 0$$

assuming that the condition of the makeup gas is the same as that of state 1. Dividing through by $\dot{m}$ and rearranging, we have

$$\frac{\dot{m}_f}{\dot{m}} = \frac{-\dot{W}_c/\dot{m} + \dot{W}_e/\dot{m} + \dot{Q}_c/\dot{m}}{h_1 - h_8}$$

$$= \frac{-345 + 82.2 + 355}{460 - 29.4} = 0.214 \text{ kg/kg compressed}$$

This is the liquid yield per unit mass of gas compressed.

The net work required to liquefy a unit mass of gas is then

$$\frac{\dot{W}_{\text{net}}}{\dot{m}_f} = \frac{\dot{W}_{\text{net}}/\dot{m}}{\dot{m}_f/\dot{m}} = \frac{263}{0.214}$$

$$= 1230 \text{ kJ/kg liquefied}$$

Example 12-7

For the ideal two-expander Collins helium-liquefaction system as depicted in Fig. 12-17, show that the liquid yield per unit mass of gas compressed can be expressed by the equation

$$\frac{\dot{m}_f}{\dot{m}} = \frac{1}{h_1 - h_{11}} \left[(h_1 - h_2) + \frac{\dot{m}_{e1}}{\dot{m}}(h_3 - h_4) + \frac{\dot{m}_{e2}}{\dot{m}}(h_6 - h_7) \right] \qquad (12\text{-}39)$$

wherein $\dot{m}$, $\dot{m}_{e1}$, and $\dot{m}_{e2}$ denote the mass-flow rates through the compressor, expander 1, and expander 2, respectively; and $\dot{m}_f$ denotes the liquid discharge rate from the liquid receiver.

For such a system, let $p_1 = 1$ atm, $p_2 = 15$ atm, $T_1 = 300$ K, $T_3 = 60$ K, $T_6 = 15$ K, $\dot{m}_{e1}/\dot{m} = 0.25$, and $\dot{m}_{e2}/\dot{m} = 0.50$. Calculate the liquid helium yielded per unit mass of gaseous helium compressed.

Solution Applying the first law for steady flow to the system as a whole except the helium compressor, we have

$$\dot{W}_{e1} + \dot{W}_{e2} + \dot{m}_f h_{11} + (\dot{m} - \dot{m}_f)h_1 - \dot{m}h_2 = 0$$

But

$$\dot{W}_{e1} = \dot{m}_{e1}(h_3 - h_4)$$

and

$$\dot{W}_{e2} = \dot{m}_{e2}(h_6 - h_7)$$

Therefore,

$$\dot{m}_{e1}(h_3 - h_4) + \dot{m}_{e2}(h_6 - h_7) + \dot{m}_f(h_{11} - h_1) + \dot{m}(h_1 - h_2) = 0$$

Dividing through by $\dot{m}$ and rearranging, we obtain Eq. (12-39).

Now from the T–s diagram for helium in Appendix 4 (Fig. A-7), we obtain the following values:

> At $p_1 = 1$ atm and $T_1 = 300$ K,
>
> $$h_1 = 1574 \text{ kJ/kg}$$
>
> At $p_2 = 15$ atm and $T_2 = 300$ K,
>
> $$h_2 = 1579 \text{ kJ/kg}$$
>
> At $p_3 = 15$ atm and $T_3 = 60$ K,
>
> $$h_3 = 328.5 \text{ kJ/kg} \qquad \text{and} \qquad s_3 = 17.41 \text{ kJ/kg·K}$$
>
> At $p_6 = 15$ atm and $T_6 = 15$ K,
>
> $$h_6 = 80.5 \text{ kJ/kg} \qquad \text{and} \qquad s_6 = 9.62 \text{ kJ/kg·K}$$
>
> At $p_4 = 1$ atm and $s_4 = s_3 = 17.41$ kJ/kg·K,
>
> $$h_4 = 119.6 \text{ kJ/kg}$$
>
> At $p_7 = 1$ atm and $s_7 = s_6 = 9.62$ kJ/kg·K,
>
> $$h_7 = 36.2 \text{ kJ/kg}$$
>
> At $p_{11} = 1$ atm and saturated liquid,
>
> $$h_{11} = 9.76 \text{ kJ/kg}$$

Substituting numerical values in Eq. (12-39) leads to

$$\frac{\dot{m}_f}{\dot{m}} = \frac{1}{1574 - 9.76}[(1574 - 1579) + 0.25(328.5 - 119.6) + 0.50(80.5 - 36.2)]$$

$$= 0.044 \text{ kg/kg compressed}$$

This is the liquid yield per unit mass of gas compressed.

12-8 AVAILABILITY ANALYSIS FOR REFRIGERATION CYCLES

The following examples are given to illustrate the procedure of calculation for an availability accounting of refrigeration cycles, whereby inefficient processes of certain components can be singled out for performance improvement.

Example 12-8

A vapor-compression heat pump uses Refrigerant-12 as the working fluid and operates under the following conditions:

> Evaporator exit: saturated vapor at 280 kPa
> Compressor inlet: 280 kPa, 0°C
> Compressor exit and condenser inlet: 1.40 MPa, 80°C
> Condenser exit and expansion valve inlet: 1.35 MPa, 50°C
> Expansion valve exit: 300 kPa
> Refrigerant mass-flow rate: 280 kg/h
> Compressor power inlet: 4.89 kW

The atmosphere is at 1°C and 101.3 kPa. Determine (a) the changes in specific stream availability of each component of the cycle, in kJ/kg, and (b) the irreversibility rate in the compressor, the expansion valve, and the suction line, in kJ/s. Make an availability accounting for the cycle.

Solution Schematic and T–s diagrams for the heat-pump system are shown in Fig. 12-18. By referring to this figure, the given data are $p_1 = p_2 = 0.28$ MPa, $T_2 = 0°C$, $p_3 = 1.40$ MPa, $T_3 = 80°C$, $p_4 = 1.35$ MPa, $T_4 = 50°C$, $p_5 = 0.30$ MPa, $\dot{m} = 280$ kg/h, $\dot{W}_{comp} = 4.89$ kW, $T_0 = 1°C = 274$ K, and $p_0 = 101.3$ kPa.

From the Refrigerant-12 tables in Appendix 3 (Tables A-22M and A-23M), we have the following values:

At $p_1 = 0.28$ MPa, we get $T_1 = T_{sat} = -3.0°C$, $h_1 = h_g = 186.1$ kJ/kg, and $s_1 = s_g = 0.6976$ kJ/kg·K.

At $p_2 = 0.28$ MPa and $T_2 = 0°C$, we get $h_2 = 188.0$ kJ/kg and $s_2 = 0.7046$ kJ/kg·K.

At $p_3 = 1.40$ MPa, we get $T_{sat} = 56°C$.

At $p_3 = 1.40$ MPa and $T_3 = 80°C$, we get $h_3 = 227.9$ kJ/kg and $s_3 = 0.7355$ kJ/kg·K.

At $T_4 = 50°C$ and $p_4 = 1.35$ MPa (compressed liquid), we get $h_4 \approx h_f$ at $50°C = 84.87$ kJ/kg and $s_4 \approx s_f$ at $50°C = 0.3034$ kJ/kg·K.

At $p_5 = 0.30$ MPa, by interpolation, we get $T_5 = -0.90°C$, $h_f = 35.19$ kJ/kg, $h_g = 187.0$ kJ/kg, $s_f = 0.1388$ kJ/kg·K, and $s_g = 0.6965$ kJ/kg·K. Then

$$h_5 = h_4 = h_f + x_5(h_g - h_f) \qquad \text{at} \qquad p_5 = 0.30 \text{ MPa}$$

or

$$84.87 = 35.19 + x_5(187.0 - 35.19)$$

so that

$$x_5 = 0.3273$$

Therefore,

$$s_5 = s_f + x_5(s_g - s_f) \text{ at } p_5 = 0.30 \text{ MPa}$$
$$= 0.1388 + 0.3273(0.6965 - 0.1388)$$
$$= 0.3213 \text{ kJ/kg·K}$$

(a) By neglecting changes in kinetic and potential energies, the changes in specific stream availability for a process as given by Eq. (8-20) reduces to

$$\Delta\psi = \Delta h - T_0\,\Delta s$$

Thus, for the suction line 1–2,

$$\psi_2 - \psi_1 = (h_2 - h_1) - T_0(s_2 - s_1)$$
$$= (188.0 - 186.1) - 274(0.7046 - 0.6976)$$
$$= -0.018 \text{ kJ/kg}$$

where the decrease in stream availability is due to the heat transfer to the suction line from the atmosphere, which is at a higher temperature.

For the compressor, there is a rate of heat transfer of the amount as given by

$$\dot{Q}_{out, comp} = \dot{W}_{in, comp} + \dot{m}(h_2 - h_3)$$
$$= 4.89 \text{ kJ/s} + \frac{280 \text{ kg/h}}{3600 \text{ s/h}}(188.0 - 227.9 \text{ kJ/kg})$$
$$= 1.787 \text{ kJ/s}$$

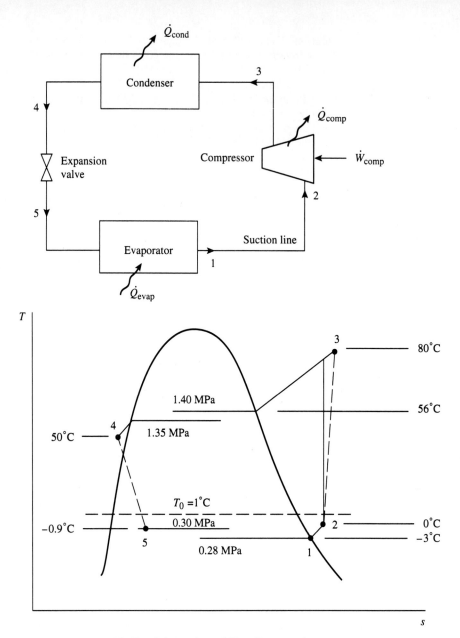

Figure 12-18 Schematic and T–s diagrams for Example 12-8.

The availability change for the compressor process 2–3 is

$$\psi_3 - \psi_2 = (h_3 - h_2) - T_0(s_3 - s_2)$$
$$= (227.9 - 188.0) - 274(0.7355 - 0.7046)$$
$$= 31.433 \text{ kJ/kg}$$

where the net increase in stream availability is due to work input, coupled with heat output and internal irreversibility effect.

For the condenser process 3–4,

$$\psi_4 - \psi_3 = (h_4 - h_3) - T_0(s_4 - s_3)$$
$$= (84.87 - 227.9) - 274(0.3034 - 0.7355)$$
$$= -24.635 \text{ kJ/kg}$$

where the decrease in stream availability is due to heat transfer out, together with irreversibility associated with frictional pressure drop.

For the expansion-valve process 4–5,

$$\psi_5 - \psi_4 = (h_5 - h_4) - T_0(s_5 - s_4)$$
$$= (84.87 - 84.87) - 274(0.3213 - 0.3034)$$
$$= -4.905 \text{ kJ/kg}$$

where the decrease in stream availability is due to internal irreversibility.

For the evaporator process 5–1,

$$\psi_1 - \psi_5 = (h_1 - h_5) - T_0(s_1 - s_5)$$
$$= (186.1 - 84.87) - 274(0.6976 - 0.3213)$$
$$= -1.876 \text{ kJ/kg}$$

where the decrease in stream availability is due to heat transfer to the evaporator from the higher-temperature atmosphere, together with irreversibility associated with frictional pressure drop.

(b) According to Eqs. (8-9) and (8-22), the irreversibility rate of a steady-flow process having heat exchange with only the atmosphere is given by

$$\dot{I} = \dot{W}_{\max} - \dot{W}_{\text{act}} = \dot{m}(\psi_{\text{inlet}} - \psi_{\text{exit}}) - \dot{W}_{\text{act}}$$

Thus, for the compressor, the irreversibility rate is

$$\dot{I}_{\text{comp}} = \dot{m}(\psi_2 - \psi_3) - \dot{W}_{\text{act}}$$
$$= \left(\frac{280}{3600} \text{ kg/s}\right)(-31.433 \text{ kJ/kg}) - (-4.89 \text{ kJ/s})$$
$$= 2.445 \text{ kJ/s}$$

For the expansion valve, the irreversibility rate (with $\dot{W}_{\text{act}} = 0$) is

$$\dot{I}_{\text{exp valve}} = \dot{m}(\psi_4 - \psi_5)$$
$$= \frac{280}{3600}(4.905) = 0.382 \text{ kJ/s}$$

For the suction line, the irreversibility rate (with $\dot{W}_{\text{act}} = 0$) is

$$\dot{I}_{\text{suction line}} = \dot{m}(\psi_1 - \psi_2)$$
$$= \frac{280}{3600}(0.018) = 0.001 \text{ kJ/s}$$

The rate of availability loss from the condenser is

$$\Delta\dot{\Psi}_{\text{cond}} = \Delta\dot{\Psi}_{43} = \dot{m}(\psi_3 - \psi_4)$$
$$= \frac{280}{3600}(24.635) = 1.916 \text{ kJ/s}$$

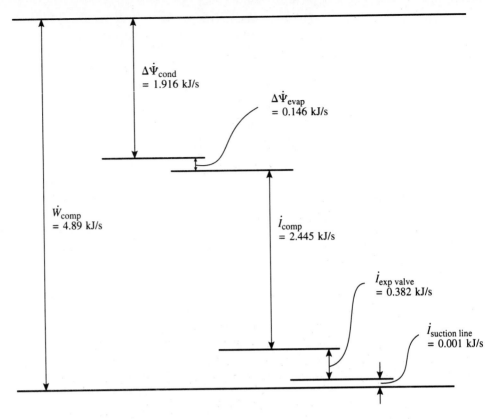

Figure 12-19   Availability accounting for Example 12-8.

The rate of availability loss from the evaporator is

$$\Delta\dot{\Psi}_{evap} = \Delta\dot{\Psi}_{15} = \dot{m}(\psi_5 - \psi_1)$$

$$= \frac{280}{3600}(1.876) = 0.146 \text{ kJ/s}$$

The power input to the compressor, $\dot{W}_{comp} = 4.89$ kW, is disposed of as follows:

$$\Delta\dot{\Psi}_{cond} + \Delta\dot{\Psi}_{evap} + \dot{I}_{comp} + \dot{I}_{exp\,valve} + \dot{I}_{suction\,line}$$

$$= 1.916 + 0.146 + 2.445 + 0.382 + 0.001$$

$$= 4.89 \text{ kJ/s}$$

The availability accounting is shown graphically in Fig. 12-19. It can be seen that a very large availability destruction occurs in the compressor due to irreversibility. It represents $2.445/4.89 = 50\%$ of the availability input to the compressor as work that is destroyed in the compressor. The percentage of the availability input to the compressor that is destroyed in the expansion valve is $0.382/4.89 = 7.8\%$.

Example 12-9

A Refrigerant-12 vapor-compression refrigeration machine is used to make ice. The refrigeration capacity is 125 kW. Saturated vapor at $-10°C$ enters the compressor and leaves at 1.1 MPa after an internally reversible and adiabatic process. Determine (a) the mass-flow rate of the refrigerant, (b) the power requirement, and (c) the irreversibility rate for each of the processes and for the cycle, if the high-temperature reservoir (the atmosphere) is at 30°C and the low-temperature reservoir is at −5°C.

Solution By referring to Fig. 12-20, the given data are $T_1 = T_4 = -10°C$, $p_2 = p_3 = 1.1$ MPa, $\dot{Q}_L = 125$ kW, $T_H = T_0 = 30°C = 303$ K, and $T_L = -5°C = 268$ K.

From the saturated Refrigerant-12 table in Appendix 3 (Table A-22M), at $T_1 = -10°C$, $p_1 = p_{sat} = 0.2191$ MPa, $h_1 = h_g = 183.06$ kJ/kg, and $s_1 = s_g = 0.7014$ kJ/kg·K. At $p_3 = 1.1$ MPa, $T_3 = T_{sat} = 45.6°C$, $h_3 = h_f = 80.26$ kJ/kg, and $s_3 = s_f = 0.2893$ kJ/kg·K. For the throttling process 3–4,

$$h_4 = h_3 = h_f + x_4 h_{fg} \qquad \text{at} \qquad T_4 = -10°C$$

or

$$80.26 = 26.85 + x_4(183.06 - 26.85)$$

so that

$$x_4 = 0.3419$$

Therefore,

$$s_4 = s_f + x_4 s_{fg} \text{ at } T_4 = -10°C$$

$$= 0.1079 + 0.3419(0.7014 - 0.1079)$$

$$= 0.3108 \text{ kJ/kg·K}$$

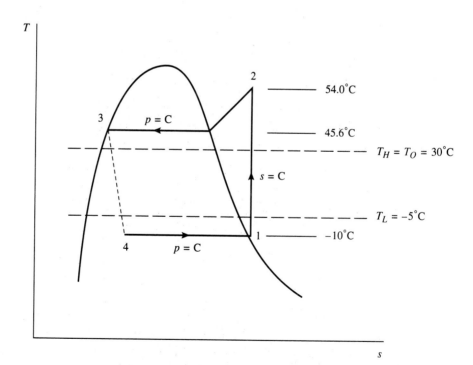

Figure 12-20 *T–s* diagram for Example 12-9.

From the superheated Refrigerant-12 table in Appendix 3 (Table A-23M), at $p_2 = 1.1$ MPa and $s_2 = s_1 = 0.7014$ kJ/kg·K, by interpolation, we have $T_2 = 54.0°C$ and $h_2 = 211.60$ kJ/kg.

(a) For the evaporator,

$$\dot{Q}_L = \dot{m}(h_1 - h_4)$$

Therefore,

$$\dot{m} = \frac{\dot{Q}_L}{h_1 - h_4} = \frac{125 \text{ kJ/s}}{(183.06 - 80.26) \text{ kJ/kg}} = 1.216 \text{ kg/s}$$

(b) The compressor power input is

$$\dot{W}_{in} = \dot{m}(h_2 - h_1) = (1.216 \text{ kg/s})(211.60 - 183.06 \text{ kJ/kg})$$

$$= 34.7 \text{ kW}$$

(c) For the isentropic compression process 1–2, the irreversibility rate is zero, or $\dot{I}_{12} = 0$.

For the condenser process in steady flow, the irreversibility rate is given by

$$\dot{I}_{23} = T_0(\Delta\dot{S}^0_{system} + \Delta\dot{S}_{surr})$$

$$= T_0(\Delta\dot{S}_{surr})$$

$$= T_0[\dot{m}(s_3 - s_2) + \dot{Q}_H/T_H]$$

where

$$\dot{Q}_H = \dot{m}(h_2 - h_3) = 1.216(211.60 - 80.26) = 159.7 \text{ kW}$$

Therefore,

$$\dot{I}_{23} = (303 \text{ K})[(1.216 \text{ kg/s})(0.2893 - 0.7014 \text{ kJ/kg·K})$$

$$+ (159.7 \text{ kJ/s})/(303 \text{ K})]$$

$$= 7.86 \text{ kW}$$

$$\dot{I}_{34} = T_0[\dot{m}(s_4 - s_3)] = 303[1.216(0.3108 - 0.2893)]$$

$$= 7.92 \text{ kW}$$

$$\dot{I}_{41} = T_0[\dot{m}(s_1 - s_4) - \dot{Q}_L/T_L]$$

$$= 303[1.216(0.7014 - 0.3108) - 125/268]$$

$$= 2.59 \text{ kW}$$

The irreversibility rate for the cycle is given by

$$\dot{I}_{cycle} = \dot{I}_{12} + \dot{I}_{23} + \dot{I}_{34} + \dot{I}_{41}$$

$$= 0 + 7.86 + 7.92 + 2.59$$

$$= 18.37 \text{ kW}$$

Example 12-10

A reversed Brayton refrigeration cycle operates between two thermal reservoirs at 310 K and 270 K, respectively, with air entering the compressor at 100 kPa and 270 K, and leaving at 300 kPa. The turbine inlet temperature is 310 K. The compressor has an adiabatic efficiency of 80%, and the turbine has an adiabatic efficiency of 85%. The lowest naturally occurring temperature, that is, the atmospheric temperature, is 310 K. Taking data from an ideal-gas air table, determine (a) the coefficient of performance and (b) the irreversibilities in the compressor and the turbine, per unit mass of air flow.

Solution By referring to Fig. 12-21, the given data are $p_1 = 100$ kPa, $T_1 = 270$ K, $p_2 = 300$ kPa, $T_3 = 310$ K, $\eta_C = 80\%$, $\eta_T = 85\%$, $T_H = 310$ K, $T_L = 270$ K, and $T_0 = 310$ K.

From the air table in Appendix 3 (Table A-9M), we have the following:

At $T_1 = 270$ K, we get $h_1 = 270.31$ kJ/kg, $p_{r1} = 0.9547$, and $\phi_1 = 5.5958$ kJ/kg·K

At $T_3 = 310$ K, we get $h_3 = 310.48$ kJ/kg, $p_{r3} = 1.5480$, and $\phi_3 = 5.7346$ kJ/kg·K

For the internally reversible adiabatic process 1-2', we have

$$p_{r2'} = p_{r1}(p_2/p_1) = 0.9547(300/100) = 2.8641$$

At this $p_{r2'}$ value, the air table gives $h_{2'} = 370.33$ kJ/kg.

For the internally reversible adiabatic process 3-4', we have

$$p_{r4'} = p_{r3}(p_4/p_3) = 1.5480(100/300) = 0.5160$$

At this $p_{r4'}$ value, the air table gives $h_{4'} = 226.59$ kJ/kg.

Using values of η_C and η_T, we have

$$h_2 = h_1 + \frac{h_{2'} - h_1}{\eta_C} = 270.31 + \frac{370.33 - 270.31}{0.80} = 395.34 \text{ kJ/kg}$$

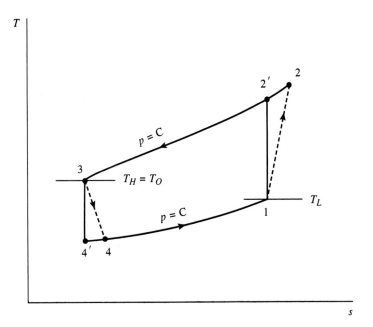

Figure 12-21 *T–s* diagram for Example 12-10.

$$h_4 = h_3 - \eta_T(h_3 - h_{4'}) = 310.48 - 0.85(310.48 - 226.59)$$
$$= 239.17 \text{ kJ/kg}$$

(a) The heat added during process 4–1 is

$$q_{in} = h_1 - h_4 = 270.31 - 239.17 = 31.14 \text{ kJ/kg}$$

The net work input for the cycle is

$$\text{net } w_{in} = (h_2 - h_1) - (h_3 - h_4)$$
$$= (395.34 - 270.31) - (310.48 - 239.17) = 53.72 \text{ kJ/kg}$$

The coefficient of performance of the refrigeration cycle is

$$\text{COP} = \frac{q_{in}}{\text{net } w_{in}} = \frac{31.14}{53.72} = 0.580 = 58.0\%$$

(b) At $h_2 = 395.34$ kJ/kg, the air table gives $\phi_2 = 5.9767$ kJ/kg·K. At $h_4 = 239.17$ kJ/kg, the air table gives $\phi_4 = 5.4733$ kJ/kg·K. Thus,

$$s_2 - s_1 = \phi_2 - \phi_1 - R \ln (p_2/p_1)$$
$$= 5.9767 - 5.5958 - (8.314/29) \ln (300/100)$$
$$= 0.06594 \text{ kJ/kg·K}$$

$$s_4 - s_3 = \phi_4 - \phi_3 - R \ln (p_4/p_3)$$
$$= 5.4733 - 5.7346 - (8.314/29) \ln (100/300)$$
$$= 0.05366 \text{ kJ/kg·K}$$

The irreversibility for the compressor is

$$i_{comp} = T_0(\Delta\cancelto{0}{s}_{system} + \Delta s_{surr}) = T_0(\Delta s_{surr}) \text{ for steady state}$$
$$= T_0(s_2 - s_1) = 310(0.06594) = 20.44 \text{ kJ/kg}$$

The irreversibility for the turbine is

$$i_{turb} = T_0(\Delta s_{surr}) = T_0(s_4 - s_3)$$
$$= 310(0.05366) = 16.63 \text{ kJ/kg}$$

12-9 SUMMARY

Refrigeration is the production of a temperature lower than that of its natural environment. When the purpose is the absorption of heat from a low-temperature source, the machine is called a refrigerator. However, when the purpose is the delivery of heat to a high-temperature sink, it is called a heat pump. The study of the phenomena of very low temperatures, say, below $-150°C$ or $-240°F$, is referred to as cryogenics.

Compression cycles using vapor refrigerants are by far the most common refrigeration method. In an ideal simple vapor-compression refrigeration cycle, saturated (or slightly superheated) vapor refrigerant is compressed isentropically to a higher pressure in the compressor, followed by a constant-pressure heat-rejection process in the condenser to the saturated liquid state (or a slightly subcooled liquid state). It is then throttled to the evaporator pressure and vaporizes as it absorbs heat from the refrigerated space.

The vapor-compression refrigeration cycle can have many variations, including multistage compression and multistage throttling with two or more condensers.

The amount of heat absorbed by the refrigerant in the evaporator is referred to as the refrigeration effect. A widely used unit of cooling capacity is the ton of refrigeration as defined by

$$1 \text{ ton of refrigeration} = 200 \text{ Btu/min} = 211 \text{ kJ/min}$$

The coefficient of performance (COP) of a refrigerator is defined as

$$(\text{COP})_{\text{refrigerator}} = \frac{q_{\text{evap}}}{w_{\text{comp}}}$$

The coefficient of performance of a heat pump is defined as

$$(\text{COP})_{\text{heat pump}} = \frac{q_{\text{cond}}}{w_{\text{comp}}}$$

Reversed Brayton cycles can be modeled for the production of refrigeration in the so-called gas-cycle refrigeration. Gas refrigeration units offer advantages of weight and space savings, particularly suitable for aircraft cabin cooling.

An absorption refrigeration cycle employs a generator, an absorber, and a liquid pump to replace a vapor compressor, thus saving electrical power consumption. This refrigeration method is economically attractive when inexpensive heat-energy sources at relatively low temperatures, such as solar energy and waste heat, are used to operate the machine.

A steam-jet refrigeration system relies on the evaporation of water in a flash chamber to provide the refrigeration.

Joule–Thomson free expansion and adiabatic expansion against a restraining force are the common methods of producing cryogenic temperatures and liquefying gases. Linde–Hampson, Claude, and Collins gas-liquefaction systems are studied in this chapter.

Examples of availability analysis for vapor-compression and gas refrigeration cycles are given in the chapter.

PROBLEMS

12-1. An ideal vapor-compression refrigeration cycle uses Refrigerant-12 as the working fluid. Saturated vapor enters the compressor and saturated liquid leaves the condenser. The evaporator pressure is maintained at 120 kPa, and the condenser pressure is maintained at 500 kPa. Determine the coefficient of performance of the cycle.

12-2. In a Refrigerant-12 vapor-compression refrigeration cycle, the temperature of the condensing vapor is 110°F and the temperature during evaporation is 10°F. Saturated vapor enters the compressor and saturated liquid leaves the condenser. Determine the coefficient of performance and the cooling capacity (in Btu/lbm) of the cycle.

12-3. The pressures in the evaporator and condenser of a 5-ton vapor-compression refrigeration plant operating on Refrigerant-12 are 0.20 MPA and 0.70 MPa, respectively. For the ideal cycle, the fluid enters the compressor as a saturated vapor, and no subcooling occurs in the condenser. Draw the T–s and p–h diagrams and determine (a) the temperature of the fluid leaving the isentropic compressor, in °C, (b) the coefficient of performance, (c) the compressor displacement, in liters/min, and (d) the power input, in kW.

12-4. A Refrigerant-12 vapor-compression refrigeration system with a compressor adiabatic efficiency of 0.75 operates with saturated vapor leaving the evaporator at 30°F and saturated liquid leaving the condenser at 150°F. Determine the coefficient of performance for the system.

12-5. A vapor-compression refrigeration cycle of 1-ton capacity operates with Refrigerant-12 at an evaporator temperature of −20°C and a condenser pressure of 0.8 MPa. Saturated vapor enters the compressor and saturated liquid leaves the condenser. See Fig. P12-5. At the end of the compression, the vapor is at a temperature of 60°C. The compressor is water-cooled with a net heat rejection of 5 kJ/kg. Determine (a) the coefficient of performance, and (b) the condenser cooling water required if the water temperature rise is 6°C. For water, $c = 4.18$ kJ/kg·°C.

12-6. A Refrigerant-12 vapor-compression refrigeration unit has the following conditions: evaporator temperature 38°F, condenser temperature 105°F, compressor inlet temperature 40°F, isentropic compression in a compressor with a 3% pressure drop in the inlet and discharge valves. Determine (a) the coefficient of performance, (b) the flow rate of refrigerant for a 0.75-ton capacity, and (c)

the horsepower of the driving motor, assuming 95% electrical and mechanical efficiency.

12-7. A 10-ton vapor-compression refrigeration plant uses ammonia as the refrigerant and operates between a cold region at 10°F and a hot region at 80°F with saturated vapor entering the compressor and saturated liquid entering the expansion valve. Find (a) the coefficient of performance, (b) the refrigerant flow rate, lbm/min, (c) the volume flow rate entering the compressor, ft³/min, and (d) the maximum and minimum pressures of the cycle.

12-8. In a vapor-compression refrigeration cycle using ammonia as the refrigerant, saturated vapor at 20 psia enters the compressor, and leaves at 180 psia and 280°F. Subcooled liquid at 75°F and 180 psia leaves the condenser. The refrigeration effect is 20 tons and the cooling water passing through the compressor-cylinder water jacket picks up heat at a rate of 250 Btu/min. (a) Calculate the power input, in hp. (b) Is the compression process reversible?

12-9. A vapor-compression refrigeration machine,

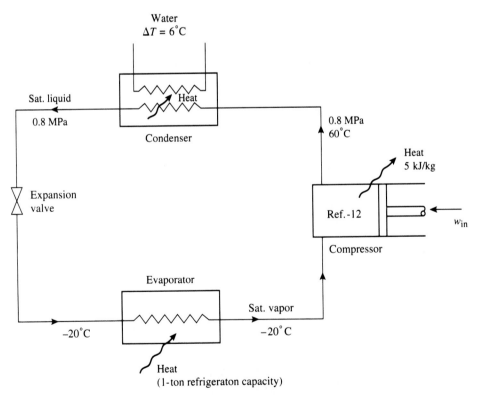

Water
$\Delta T = 6°C$

Sat. liquid
0.8 MPa
Heat
Condenser

0.8 MPa
60°C

Heat
5 kJ/kg

Ref.-12
w_{in}

Expansion
valve

Compressor

Evaporator

Sat. vapor

−20°C
−20°C

Heat
(1-ton refrigeraton capacity)

Figure P12-5

which includes a liquid subcooler, uses Refrigerant-12 as the working substance. See Fig. P12-9. The liquid subcooler cools the saturated liquid coming from the condenser from 40°C to 35°C while receiving saturated vapor coming from the evaporator at −20°C. Pressure drops in the components and connecting pipelines are negligible. Compression of the refrigerant is internally reversible and adiabatic and the compressor is capable of handling 1.2 m³/min of refrigerant measured at its inlet conditions. Determine (a) the mass-flow rate of the refrigerant, (b) the refrigeration capacity in tons, (c) the COP of the machine, (d) the compressor power requirement, and (e) the rate of energy removal from the condenser.

12-10. A Refrigerant-12 vapor-compression system includes a liquid-to-suction heat exchanger. The heat exchanger cools saturated liquid coming from the condenser from 90°F to 70°F with vapor coming from the evaporator at 10°F. The compression is isentropic and the compressor is capable of pumping 50 ft³/min measured at the compressor suction. Calculate the refrigeration capacity in tons and the coefficient of performance of the system.

12-11. In a Refrigerant-12 vapor-compression cycle, the maximum demand for winter heating is expected to be 1000 cfm of 40°F outside air heated to 86°F. The evaporator temperature is 20°F and the condenser pressure is 140 psia. Saturated vapor enters the compressor and no subcooling occurs in the condenser. For the ideal cycle, find (a) the mass flow of refrigerant, lbm/min, (b) the horsepower input, (c) the COP for heating, (d) the cost of heating at 10 cents/kW·h, (e) the heating

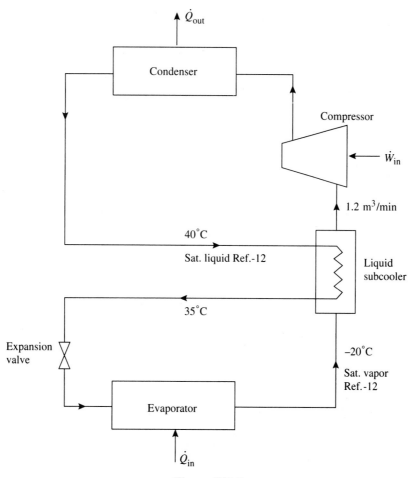

Figure P12-9

cost when an electrical heating element is used, and (f) the heating cost when fuel oil at 75 cents/gal is used. The heating value of the oil is 135,000 Btu/gal and the efficiency of the oil heater is 80%.

12-12. A Refrigerant-12 ideal vapor-compression heat pump operates between an evaporator temperature of 30°F and a condenser pressure of 140 psia. Saturated vapor enters the compressor and no sub-cooling occurs in the condenser. The winter heating demand for the heat pump is 1200 Btu/min. Determine (a) the temperature at the exit of the isentropic compressor, in °F, (b) the heating coefficient of performance, (c) the effective displacement of the compressor, in ft³/min, (d) the power input to the compressor, in kW, and (e) the power input if electric-resistance heating is used, in kW.

12-13. A Refrigerant-12 ideal vapor-compression heat pump operates between an evaporator temperature of 0°C and a condenser pressure of 0.8 MPa. Saturated vapor enters the compressor and no sub-cooling occurs in the condenser. The winter heating demand for the heat pump is 1000 kJ/min. Determine (a) the temperature at the exit of the isentropic compressor, °C, (b) the heating coefficient of performance, (c) the effective displacement of the compressor, liters/min, (d) the power input to the compressor, kW, and (e) the power input required if electric resistance heating is used, kW.

12-14. A heat pump using Refrigerant-12 as its working substance is to supply 100,000 kJ/h of heat to a building. See Fig. P12-14. The evaporator oper-

ates at 0.3 MPa and the condenser operates at 1 MPa. The fluid leaving the evaporator is at the condition of 0.3 MPa and 10°C. Saturated liquid leaves the condenser. The compression process is internally reversible and adiabatic. Determine (a) the COP of the heat pump, (b) the mass-flow rate of the refrigerant, in kg/min, and (c) the kW input to the compressor.

12-15. A Refrigerant-12 heat pump operates between outdoor air at 30°F and air in a domestic heating system at 105°F with a temperature difference in the evaporator and in the condenser of 15°F. See Fig. P12-15. The compressor adiabatic efficiency is 80%, the compression begins with saturated vapor, and the expansion begins with saturated liquid. The combined efficiency of the electric motor and the belt drive for the compressor is 75%. If the required heat supply to the warm air is 150,000 Btu/h, what will be the electrical load in kW?

12-16. An ammonia-compression refrigeration system operates with saturated vapor entering the compressor and has a rated capacity of 6 tons of refrigeration. The temperature of the ammonia leaving the condenser is 90°F and the temperature of the evaporator is 20°F. The condenser pressure is 180 psia. The compression process is isentropic. (a) What is the mass of ammonia to be circulated per minute? (b) What is the coefficient of performance of the refrigeration system? (c) If the system operates under the same conditions as a heat pump, what is its coefficient of performance? (d) Is there any relation between the cooling-cycle and the heating-cycle coefficients of performance for the system? Show it.

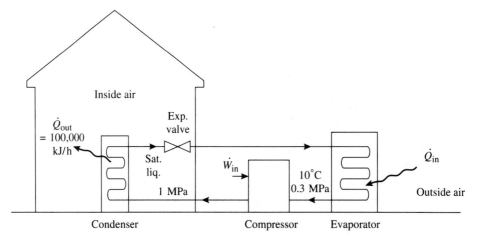

Figure P12-14

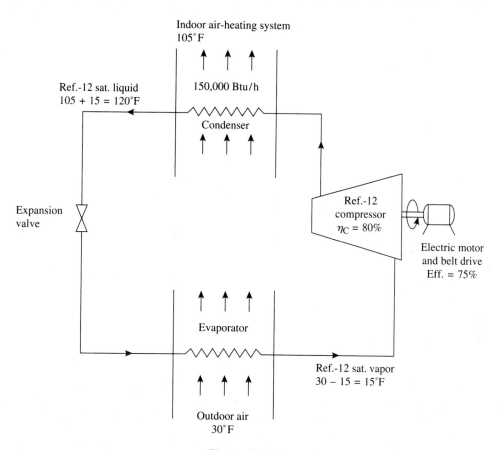

Indoor air-heating system
105°F

150,000 Btu/h

Condenser

Ref.-12 sat. liquid
105 + 15 = 120°F

Expansion
valve

Ref.-12
compressor
$\eta_C = 80\%$

Electric motor
and belt drive
Eff. = 75%

Evaporator

Ref.-12 sat. vapor
30 − 15 = 15°F

Outdoor air
30°F

Figure P12-15

12-17. A Refrigerant-12 vapor-compression system has refrigerant-to-air heat exchangers. See Fig. P12-17. Assume that there is a 12°F difference between the temperature of the refrigerant and that of the air leaving each exchanger, and that the compressor is adiabatic with 68% efficiency. (a) For summer operation with an air temperature leaving the evaporator of 55°F and an air temperature leaving the condenser of 105°F, find the coefficient of performance. (b) If electrical energy costs 10 cents/kW·h, find the energy cost per hour per ton of refrigeration under the operating conditions of part (a). (c) For winter operation with an air temperature leaving the condenser of 120°F and an air temperature leaving the evaporator of 25°F, find the coefficient of performance as a heat pump. (d) Find the hourly cost of electrical energy for a heating load of 100,000 Btu/h under the conditions of part (c). Assume that saturated vapor leaves the evaporator and saturated liquid leaves the condenser.

12-18. A refrigeration unit uses nitrogen as the working fluid and operates as a reversed Brayton cycle with a pressure ratio of 5. The inlet conditions to the compressor are 200 kPa and 30°C. The minimum temperature of the cycle is 10°C. Determine the coefficient of performance and net work input. Assume $c_p = 1.04$ kJ/kg·K and $k = 1.40$ for nitrogen.

12-19. An air-refrigeration system using an ideal reversed Brayton cycle of pressure ratio 5 and minimum pressure 1 atm is required to produce 10 tons of refrigeration. The temperature of the air leaving the energy dissipative space is 80°F, and the air leaving the refrigeration space is at 10°F. Find (a) the mass of air circulated per minute and (b) the coefficient of performance.

12-20. In an air-refrigeration cycle, air at 0.1 MPa and −13°C enters an 80% efficient adiabatic compressor and leaves at 0.5 MPa. See Fig. P12-20. The air is then cooled by ambient air in a heat ex-

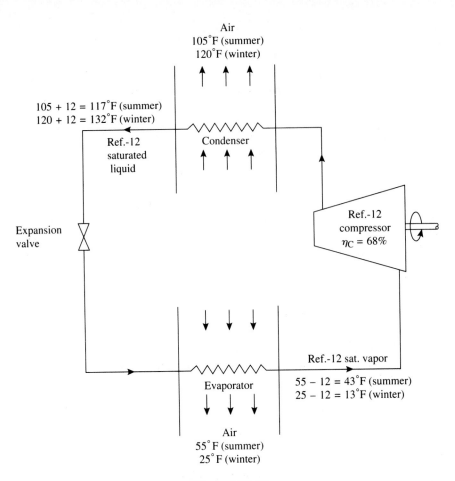

Air
105°F (summer)
120°F (winter)

↑ ↑ ↑

105 + 12 = 117°F (summer)
120 + 12 = 132°F (winter)

Ref.-12
saturated
liquid

Condenser

↑ ↑ ↑

Ref.-12
compressor
$\eta_C = 68\%$

Expansion
valve

Ref.-12 sat. vapor

↓ ↓ ↓

Evaporator

55 − 12 = 43°F (summer)
25 − 12 = 13°F (winter)

↓ ↓ ↓

Air
55° F (summer)
25° F (winter)

Figure P12-17

changer to 0.49 MPa, 47°C, before entering an 80% efficient adiabatic turbine. The air exits from the turbine at 0.11 MPa. On the basis of air-table data, determine (a) the cooling capacity in kJ/kg and (b) the net work input in kJ/kg.

12-21. An open air cycle of refrigeration is considered for passenger-car cooling. See Fig. P12-21. The working fluid is dry air with $c_p = 0.24$ Btu/lbm·°F, and the required rate of circulation is 2310 lbm/h. Air enters the compressor at 14.7 psia and 555°R, air enters the air turbine at 570°R and leaves the turbine at 470°R and 14.7 psia. The adiabatic efficiency of the compressor is 0.8 and that of the turbine is 0.65. Ignore the pressure drop in the air-to-air heat exchanger and heat transfer in the compressor and turbine. Find (a) the pressure of the heat-rejection process, (b) the required engine horsepower, and (c) the tons

of refrigeration available for cooling if the air leaving the turbine is warmed to 535°R in the car.

12-22. An air-refrigeration system working between a refrigeration space at 15 psia and an energy-dissipation space at 55 psia has a mass-flow rate of 100 lbm/min in steady flow. Air enters the expander at 75°F and enters the compressor at 30°F. The compression process follows the relation $pv^{1.35} = $ constant, and the expansion process is assumed to be isentropic with $k = 1.41$. Neglecting heat losses and pressure drops in pipelines, determine (a) the heat absorbed in the refrigeration space per minute, (b) the heat released in the energy-dissipation space per minute, (c) the horsepower developed by the air expander, (d) the horsepower required to operate the compressor, (e) the horsepower required of the auxiliary

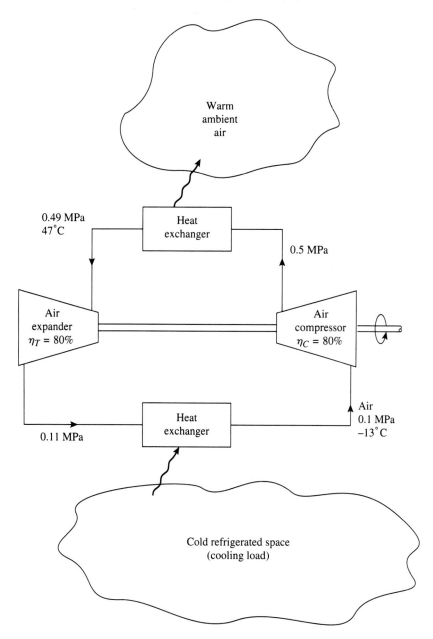

Figure P12-20

motor, and (f) the COP. Draw the cycle in the $p-v$ and $T-s$ planes.

12-23. Air at 10 psia, 40°F, is to be supplied to the cockpit of an aircraft at a rate of 0.2 lbm/s by bleeding air at 50 psia, 400°F, from the aircraft jet-engine compressor, cooling it in a heat exchanger, and expanding it isentropically through a turbine before exhausting it into the cockpit. Neglecting

friction and KE and PE changes, determine (a) the heat transfer rate in the heat exchanger for cooling, in Btu/s, and (b) the power output of the turbine, in hp.

12-24. In a jet-aircraft air-conditioning unit using a reversed Brayton cycle, 11 lbm/min of air at 51 psia and 527°F are bled from the main air compressor serving the jet engine. See Fig. P12-24. The air

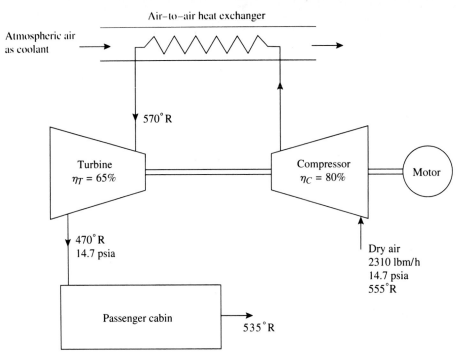

Figure P12-21

then passes through a heat exchanger, leaving at 49 psia and 167°F, at which point it is expanded through a small turbine to 11 psia and 15.8°F. Ultimately, the air leaves the plane at 90°F. Assume that both the compressor and the turbine are adiabatic machines. Find (a) the tons of refrigeration with respect to the 90°F condition. (b) If the main air compressor receives air at the state of 15 psia and 209°F, and if the work done by the small air turbine is used to drive a centrifugal fan for passing coolant air through the heat exchanger, find the horsepower input to the air-conditioning unit. (c) What is the coefficient of performance of the air-conditioning unit? (d) If no

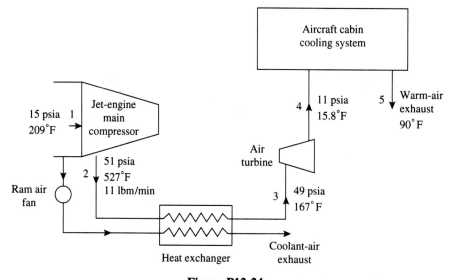

Figure P12-24

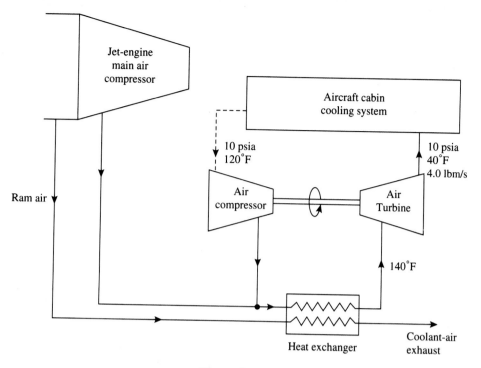

Figure P12-25

fan is needed for the heat-exchanger coolant, what is the horsepower input and what is the COP for this case?

12-25. In a jet-aircraft air-conditioning system using reversed Brayton cycle, the turbine is to supply cooled air to the cabin cooling system at 10 psia, 40°F, at a rate of 4.0 lbm/s. See Fig. P12-25. The turbine drives a compressor that takes in air at 10 psia, 120°F. The air discharged from this compressor is mixed with air at the same pressure and temperature bled from the main jet-engine com-

pressor. It is then cooled to 140°F before entering the cooling-system turbine. Assume that compression and expansion are isentropic, and neglect changes in kinetic energy. Determine the pressure and the flow rate of the air to be bled from the main jet-engine compressor.

12-26. A reversed Brayton cycle with regeneration operates with helium entering the compressor at 100 kPa and −10°C, and leaving at 300 kPa. See Fig. P12-26. Helium is then cooled to 20°C by water, and subsequently cooled further in the re-

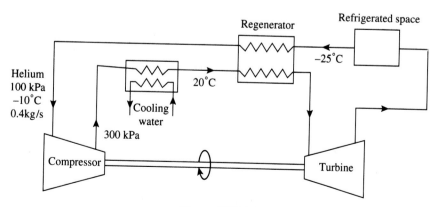

Figure P12-26

generator. It leaves the refrigerated space and enters the regenerator at $-25°C$. Assuming the compression and expansion processes to be internally reversible and adiabatic, determine (a) the temperature of the helium at the entrance to the turbine, (b) the coefficient of performance of the cycle, and (c) the net power input required for a mass-flow rate of 0.4 kg/s.

12-27. A reversed Brayton cycle with regeneration, see Fig. 12-10(a), operates with air entering the compressor at 100 kPa, 15°C, and leaving at 1.4 MPa. At entrance to the expander the air is at 1.4 MPa, $-50°C$. The compression and expansion are internally reversible and adiabatic. The regenerator process is ideal. Determine the coefficient of performance for cooling. For air, $c_p = 1.005$ kJ/kg·K and $k = 1.4$.

12-28. A reversed Brayton cycle with regeneration, see Fig. 12-10(a), operates with air entering the compressor at 100 kPA, 20°C, and leaving at 1.5 MPa. At the entrance to the expander, the air is at 1.5 MPa, $-50°C$. The compressor and expander processes are adiabatic with 80% efficiency. The regenerator has an effectiveness of 90%. Determine the coefficient of performance for cooling. Use $c_p = 1.005$ kJ/kg·K and $k = 1.4$ for air.

12-29. It is suggested that the setup shown in Fig. P12-29 be used for air-conditioning processes of de-

humidification and reheating. Air at 60°F and 100% relative humidity is dehumidified in passing over the evaporator coils of a vapor-compression refrigeration unit. The moisture condensed is drained away and the air then passes over the compressor jacket and the condenser coils of the refrigeration unit. The heat rejected by the refrigeration unit thus goes into reheating the air, which emerges at 85°F and 30% relative humidity. The flow rate is 10 lbm of dry air per minute. Barometric pressure is 14.0 psia. Determine the power input to the refrigeration unit.

12-30. A steam-jet refrigeration system is supplied with motive steam at 90 psia saturated. See Fig. 12-13(a) for a schematic of the system. The condenser is operated at 1 psia and the flash-chamber water is at 45°F. Makeup water, including water returned from the load, is supplied to the flash chamber at 65°F. The entrainment efficiency is 70%. The quality of the motive steam and entrained vapor at the beginning of compression is 90%. Assuming that the nozzle and diffuser processes are ideal, determine (a) the mass of motive steam required per pound of entrained vapor, (b) the quality of vapor entrained from the flash chamber, (c) the refrigeration effect per pound of entrained vapor, and (d) the mass of motive steam required per hour for each ton of refrigeration.

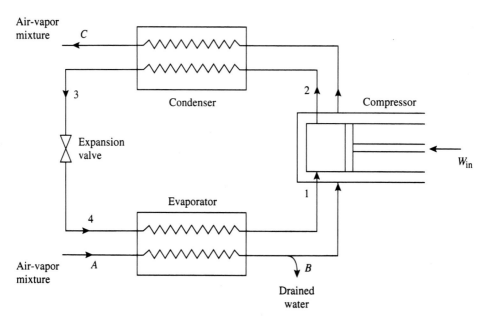

Figure P12-29

12-31. In a steam-jet refrigeration system, 50,000 lbm of water per hour at 45°F is to be delivered, the return water is at 70°F, and the makeup water is at 55°F. See Fig. P12-31. Assume that the ejector entrains only vapor from the flash chamber. Determine the makeup water flow rate.

12-32. In a two-stage vapor-compression refrigeration cycle with a flash-chamber intercooler as shown in Fig. 12-7, the condenser, flash chamber, and evaporator pressures are 100 psia, 35.7 psia, and 12 psia, respectively. Saturated liquid enters each expansion valve. The refrigerant used is Refrigerant-12 and the refrigeration effect is 20 tons. Calculate the power input, assuming that both compressors are ideal. Show the processes on T–s and p–h planes.

12-33. A Refrigerant-12 vapor-compression system consists of one condenser, one compressor, and two evaporators with individual expansion valves, as shown schematically in Fig. P12-33. The pressures are 120 psia in the condenser, 45 psia in the high-temperature evaporator, and 30 psia in the low-temperature evaporator. There is an evaporator pressure-regulating valve at the outlet of the high-temperature evaporator. Saturated liquid leaves the condenser and saturated vapor leaves each evaporator. The refrigeration load on the high-temperature evaporator is 10 tons and that

on the low-temperature evaporator is 5 tons. Determine (a) the mass-flow rate of refrigerant through each evaporator, (b) the specific enthalpy of the vapor entering the compressor, and (c) the horsepower required to drive the compressor in an internally reversible adiabatic process. Show the processes on the p–h plane.

12-34. Air initially at 1 atm and 300 K is liquefied in an ideal simple Linde–Hampson system. See Fig. 12-14 for a schematic diagram of the system. A liquid yield of 0.09 kg liquid per kg gas compressed is desired. Determine the pressure to which the air must be compressed, and the net work required per kg of air compressed. For the given initial condition, is there any upper limit in the liquid yield per unit mass of gas compressed? Explain.

12-35. An ideal Linde–Hampson air-liquefaction system compresses air isothermally from 1 atm, 300 K, to 200 atm. Calculate the liquid yield per unit mass of air compressed and the kW·h work per unit mass of air liquefied. What is the air temperature at the entrance to the expansion valve?

12-36. In order to improve the performance of a Linde–Hampson air liquefaction system, a Refrigerant-12 vapor-compression refrigeration cycle is superimposed on it, as shown in Fig. P12-36. The

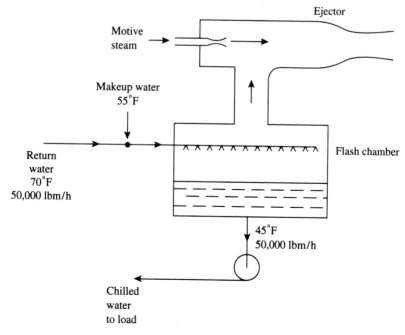

Figure P12-31

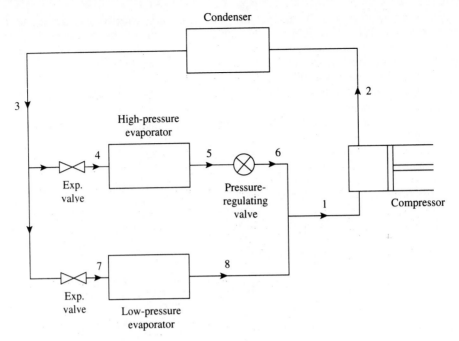

Figure P12-33

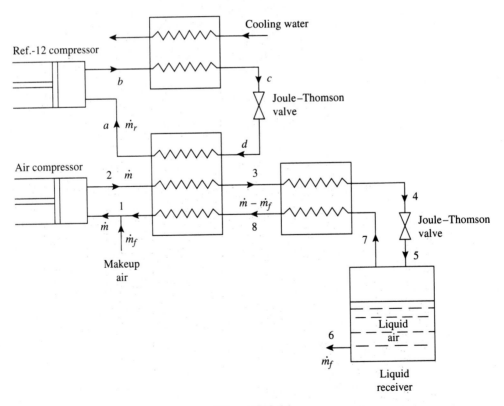

Figure P12-36

Refrigeration and Cryogenics   Chap. 12

air portion of the system operates between 1 atm and 200 atm, and the compression process is internally reversible and isothermal at 300 K. The Refrigerant-12 portion of the system operates between 2 atm and 5 atm; saturated liquid enters the expansion valve; vapor at 2 atm and 300 K enters the compressor; and the compression process is internally reversible and adiabatic. The mass-flow rate of Refrigerant-12 is one-tenth of the mass-flow rate of air. All heat exchangers are perfectly effective. Show the state points of air on a T–s diagram and those of Refrigerant-12 on T–s and an h–s diagrams. Calculate the liquid air yielded per unit mass of air compressed and the kW·h work per unit mass of air liquefied. What are the temperatures at points 3 and 4? Compare the results obtained in this problem with those obtained in Problem 12-35.

12-37. Figure P12-37 shows the flow diagram of a Linde–Hampson liquefaction system. Helium gas at $p_1 = 1$ atm and $T_1 = 15$ K is compressed isentropically in a steady-flow process to a pressure $p_2 = 30$ atm. It is then precooled by an external coolant back to its initial temperature $T_3 = T_1 = 15$ K at constant pressure $p_3 = p_2 = 30$ atm. The gas then passes through a counterflow heat exchanger, reaching the throttle valve at point 4. It is throttled in a Joule–Thomson expansion to atmospheric pressure and becomes partly liquefied (state 5). Liquefied gas is discharged through the valve at point 6, and the cool vapor at state 7 is returned to the compressor via the counterflow heat exchanger. The makeup gas is at state 1. Determine the mass of gas liquefied per unit mass of gas compressed. Also determine the compressor work required to liquefy 1 kg of helium. What is the quality of the mixture at state 5? Show the processes on a T–s diagram.

12-38. Oxygen gas initially at 14.7 psia and 60°F is liquefied in a Claude system (see Fig. 12-15). It is compressed reversibly and isothermally to 900 psia and then cooled to 10°F in passing through a heat exchanger. At the outlet of this heat exchanger, 60% of the total flow is diverted and expanded adiabatically to 14.7 psia through an expander, which has an adiabatic efficiency of 75%. The expander work is utilized as a part of the compressor work requirement. All the heat exchangers are working under ideal conditions. Calculate the liquid yield per unit mass of gas compressed and the net work required to liquefy a unit mass of gas.

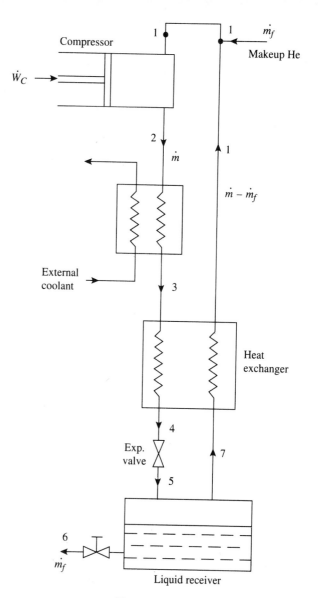

Figure P12-37

12-39. In order to reduce the total work requirement per unit mass of liquid yield, a dual-pressure Linde system, as shown schematically in Fig. P12-39, can be used. Methane initially at 1 atm and 300 K is compressed reversibly and isothermally first to an intermediate pressure of 20 atm; and after joining a return stream from the intermediate-pressure liquid receiver, it is then compressed to a high pressure of 160 atm. The high-pressure stream is then cooled in passing through the three-channel heat exchanger on its way to be

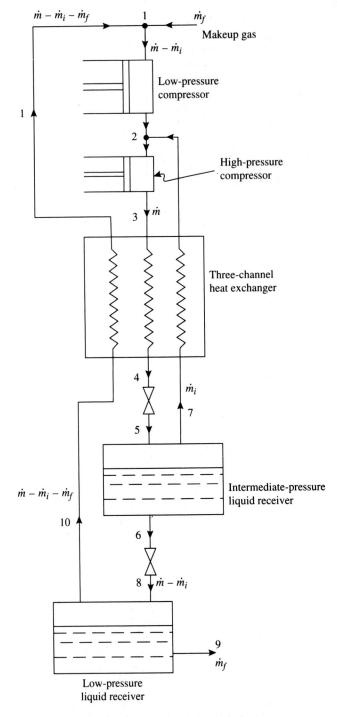

Figure P12-39 Dual-pressure Linde system.

throttled to the intermediate-pressure liquid receiver. From this receiver, the saturated vapor is returned to the high-pressure compressor through the three-channel heat exchanger, and the saturated liquid is further throttled to the low-pressure liquid receiver. The mass-flow rate returning from the intermediate-pressure liquid receiver to the high-pressure compressor is 0.8 of the total mass-flow rate handled by this compressor. All processes are ideal. Show the processes on a T–s diagram. Calculate the liquid yield per unit mass compressed in the high-pressure compressor, and the work requirement per unit mass liquefied. If the gas is compressed directly from 1 atm to 160 atm in a single compressor, what is the work requirement per unit mass liquefied?

12-40. The dual-pressure Linde system shown schematically in Fig. P12-39 is used for the liquefaction of nitrogen under the following conditions: $p_1 = 1$ atm, $T_1 = 70°F$, $p_2 = 20$ atm, $T_2 = 70°F$, $p_3 = 200$ atm, $T_3 = 70°F$, and $\dot{m}_i/\dot{m} = 0.8$. Calculate the liquid yield per unit mass compressed in the

high-pressure compressor ($\dot{m}_f/\dot{m}$ in lbm/lbm), and the work requirement per unit mass liquefied (in Btu/lbm).

12-41. A vapor-compression refrigeration system is used to produce solid carbon dioxide (dry ice). See Fig. P12-41. Precooled carbon dioxide vapor at 4 atm is compressed isentropically to 50 atm. The vapor then flows into the condenser where heat is removed to a suitable coolant to condense the vapor to a saturated liquid at constant pressure. It finally expands through the expansion valve into the evaporator (called the snow chamber in this case), which is maintained at 4 atm. Solid carbon dioxide is removed from the snow chamber at a rate of 100 kg/h, and vapor carbon dioxide is returned to the compressor. Makeup vapor is at 0°C and 4 atm. Determine the power requirement. Show the processes on a T–s diagram.

12-42. A Refrigerant-12 vapor-compression heat pump is used for home heating to raise 0.50 m³/s of room air at 5°C, 101.3 kPa, to 30°C. See Fig. P12-42. The saturated-liquid refrigerant at exit from the

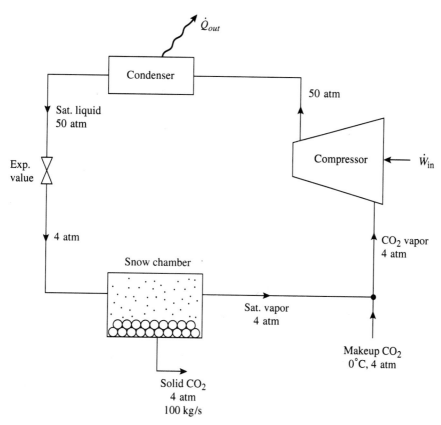

Figure P12-41

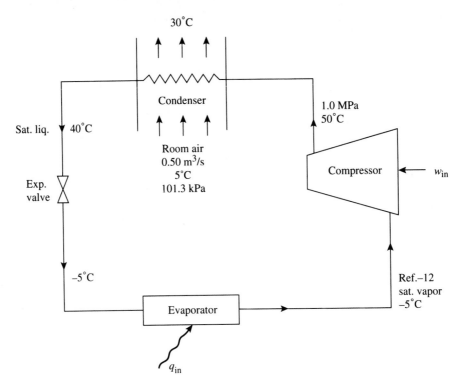

Figure P12-42

condenser is at a temperature 10°C higher than the maximum room temperature. The evaporator operates at −5°C. The compression process is adiabatic with saturated-vapor refrigerant entering the compressor. The refrigerant vapor exits from the compressor at 1.0 MPa and 50°C. Determine (a) the refrigerant mass-flow rate, in kg/s, (b) the power required, in kW, (c) the coefficient of performance of the heat pump, and (d) the second-law effectiveness of the heat-pump cycle.

12-43. In a compression refrigeration cycle using Refrigerant-12 as the working fluid, the mass-flow rate is 6 kg/min. See Fig. P12-43. The refrigerant enters the compressor at −10°C, 0.15 MPa, and leaves at 0.7 MPa. The compressor is adiabatic with an isentropic efficiency of 67%. The refrigerant leaves the condenser as a saturated liquid at 0.7 MPa. Determine (a) the coefficient of performance, (b) the refrigerating capacity, in tons, (c) the irreversibility rates of the compressor and the expansion valve, in kW, and (d) the availability changes of the refrigerant in the evaporator and condenser, in kJ/kg. The atmosphere is at 21°C and 0.1 MPa.

12-44. Helium gas initially at 1 bar and 290 K is liquefied in a two-expander Collins liquefaction system as depicted in the flow diagram of Fig. 12-17. The compression process is internally reversible and isothermal with the high-pressure side at 15 bars. The temperatures at the intakes of the two expanders are 55 K and 14 K, respectively. The respective mass-flow rates through the two expanders are 0.25 and 0.5 times the mass-flow rate through the compressor. Both expanders are adiabatic with an efficiency of 80%. The expander work outputs are utilized as parts of the compressor work requirement. All the heat exchangers are perfectly effective. Calculate the liquid yield per unit mass of gas compressed and the net work requirement per unit mass liquefied.

12-45. In a compression refrigeration cycle using Refrigerant-12 as the working fluid, the rate of flow of refrigerant is 300 lbm/h. The refrigerant enters the compressor at 25 psia, 20°F, and leaves at 175 psia, 170°F. There is heat rejection to the atmosphere of 3.3 Btu/lbm during compression. The refrigerant enters the expansion valve at 165 psia, 100°F, leaves the expansion valve at 27 psia, and leaves the evaporator at 27 psia, 10°F. Determine

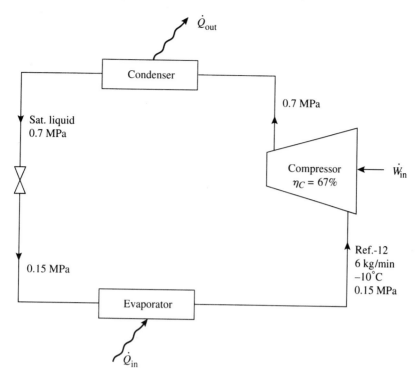

$\dot{Q}_{out}$

Condenser

Sat. liquid
0.7 MPa

0.7 MPa

Compressor
$\eta_C = 67\%$

$\dot{W}_{in}$

0.15 MPa

Ref.-12
6 kg/min
$-10°C$
0.15 MPa

Evaporator

$\dot{Q}_{in}$

Figure P12-43

(a) the refrigeration capacity in tons, (b) the coefficient of performance for the cycle, and (c) the irreversibility during the compression process. The atmosphere is at 77°F.

12-46. For the Refrigerant-12 vapor-compression refrigeration cycle described in Example 12-9, determine the rate of availability change for each component process. Make an availability accounting for the cycle.

12-47. A Refrigerant-12 vapor-compression refrigeration cycle for an air-conditioning application has a mass-flow rate of 5.0 kg/min, with a condenser pressure of 1.2 MPa and an evaporator pressure of 0.32 MPa. Saturated vapor enters the compressor and saturated liquid leaves the condenser. The compression process is isentropic. Determine (a) the cooling capacity and the power requirement, and (b) the irreversibility rate for each of the processes and for the cycle, if the high-temperature reservoir (the atmosphere) is at 35°C and the low-temperature reservoir is at 20°C.

12-48. For the Refrigerant-12 vapor-compression refrigeration cycle described in Prob. 12-47, determine the rate of availability change for each component

process, and make an availability accounting for the cycle.

12-49. A Refrigerant-12 vapor-compression heat pump is used to provide 18 kW of heating to a residence. The refrigerant enters the compressor at 300 kPa and 0°C to undergo an internally reversible and adiabatic process to 900 kPa. It is then condensed to saturated liquid in the condenser. Determine (a) the mass-flow rate of refrigerant, (b) the rate of heat addition from the atmosphere, (c) the power requirement, (d) the coefficient of performance of the cycle, and (e) the irreversibility rate for each process of the cycle if the high-temperature reservoir is at 30°C and the low-temperature reservoir (the atmosphere) is at 5°C.

12-50. For the Refrigerant-12 vapor-compression heat pump described in Prob. 12-49, determine the rate of stream availability change in the condenser and the evaporator, and the irreversibility rate in the compressor and the expansion valve. Make an availability accounting for the cycle.

12-51. An ideal vapor-compression refrigeration cycle uses Refrigerant-12 as the working fluid. Saturated vapor enters the compressor and saturated liquid leaves the condenser. The evaporator pressure is maintained at 120 kPa. Write a computer program to determine the coefficient of performance of the cycle if the condenser pressure is maintained at (a) 0.4 MPa, (b) 0.5 MPa, (c) 0.6 MPa, (d) 0.7 MPa, (e) 0.8 MPa, (f) 0.9 MPa, (g) 1.0 MPa, (h) 1.2 MPa, and (i) 1.4 MPa. Take the required data from the Refrigerant-12 tables and use them as input in your computer program. The required data are T_{sat}, h_g, and s_g at 120 kPa; T_{sat} and h_f at the various condenser pressures; T and h at $s = s_g$ (at 120 kPa) and the various condenser pressures.

12-52. If computerized Refrigerant-12-property data are available to you, rework Prob. 12-51 by the use of those data.

12-53. A reversed Brayton cycle with regeneration operates with air entering the compressor at 100 kPa, 20°C, and entering the expander at −50°C. The compressor and expander processes are adiabatic with 80% efficiency. The regenerator has an effectiveness of 90%. Write a computer program to determine the coefficient of performance for refrigeration for a maximum cycle pressure of (a) 0.5 MPa, (b) 1.0 MPa, (c) 1.5 MPa, and (d) 2.0 MPa. Use the constant specific heat $c_p = 1.005$ kJ/kg·K and $k = 1.4$ for air.

12-54. If computerized air-property data are available to you, rework Prob. 12-53 by the use of those data to take care of the variations of specific heats.

Thermodynamic Relations for Simple Compressible and Simple Paramagnetic Systems

In previous chapters, we defined and used several thermodynamic properties. Some of them (such as pressure, temperature, and volume) are directly measurable, but others (such as internal energy, enthalpy, and entropy) cannot be measured and must be calculated from data on other properties and quantities that can be measured. We now develop some useful general relations between thermodynamic properties that facilitate such calculations. We restrict our attention to simple systems that require only two independent properties to determine their thermodynamic states. The major emphasis of this chapter is the study of property relations for simple compressible systems. Simple paramagnetic systems are included here in order to study thermodynamic processes of magnetic cooling.

13-1 FUNDAMENTALS OF PARTIAL DERIVATIVES

Most general thermodynamic relations for simple systems involve the rate of change of one property with respect to another while a third property is held constant. Thus, the methods of partial differential calculus play an important part in the development of thermodynamic relations and a brief review of the rules of partial differentiation is in order.

Let z denote a dependent property of a simple system as defined by two independent properties x and y. The functional dependence can be stated symbolically as

$$z = z(x, y)$$

According to the calculus, the *total differential* of the dependent variable z is given by the equation

$$dz = \left(\frac{\partial z}{\partial x}\right)_y dx + \left(\frac{\partial z}{\partial y}\right)_x dy \qquad (13\text{-}1)$$

The partial derivatives in this equation are written with subscripts to indicate the independent variable that is held constant in forming the derivative. In thermodynamics, such a notation is necessary in order to show clearly the independent variables of which the dependent variable is considered to be a function.

Equation (13-1) can be written as

$$dz = M \, dx + N \, dy$$

where

$$M = \left(\frac{\partial z}{\partial x}\right)_y \qquad \text{and} \qquad N = \left(\frac{\partial z}{\partial y}\right)_x$$

then

$$\left(\frac{\partial M}{\partial y}\right)_x = \frac{\partial^2 z}{\partial y \, \partial x} \qquad \text{and} \qquad \left(\frac{\partial N}{\partial x}\right)_y = \frac{\partial^2 z}{\partial x \, \partial y}$$

The order of differentiation is immaterial for thermodynamic properties because they are continuous point functions. Therefore, we must have

$$\left(\frac{\partial M}{\partial y}\right)_x = \left(\frac{\partial N}{\partial x}\right)_y \tag{13-2}$$

This is the necessary and sufficient condition that $dz = M \, dx + N \, dy$ be an *exact differential* and that z be a point function.

Because of the exactness of dz in Eq. (13-1), the line integral $\int dz$ depends only on the two end states and is independent of the path or curve connecting the end states. For a complete cycle or closed curve where the two end states have identical coordinates, it is clear that

$$\oint dz = \oint \left[\left(\frac{\partial z}{\partial x}\right)_y dx + \left(\frac{\partial z}{\partial y}\right)_x dy\right] = 0$$

Usually, independent variables of a partial derivative are so chosen as to be particularly convenient for a given case. In thermodynamic analyses, there is often a need to interchange the variables of a given partial derivative or to transform it to new sets of variables. We now review some mathematical relations that are useful in the transformation of state variables.

Let us consider four state variables: x, y, z, and α; any two of them might be selected as the independent variables for a given simple system. When x is considered to be a function of y and α, then

$$dx = \left(\frac{\partial x}{\partial y}\right)_\alpha dy + \left(\frac{\partial x}{\partial \alpha}\right)_y d\alpha \tag{13-3}$$

When y is considered to be a function of z and α, then

$$dy = \left(\frac{\partial y}{\partial z}\right)_\alpha dz + \left(\frac{\partial y}{\partial \alpha}\right)_z d\alpha \tag{13-3a}$$

Substitution of Eq. (13-3a) into Eq. (13-3) gives

$$dx = \left[\left(\frac{\partial x}{\partial y}\right)_\alpha\left(\frac{\partial y}{\partial z}\right)_\alpha\right] dz + \left[\left(\frac{\partial x}{\partial \alpha}\right)_y + \left(\frac{\partial x}{\partial y}\right)_\alpha\left(\frac{\partial y}{\partial \alpha}\right)_z\right] d\alpha \tag{13-3b}$$

On the other hand, when x is considered to be a function of z and α, we have

$$dx = \left(\frac{\partial x}{\partial z}\right)_\alpha dz + \left(\frac{\partial x}{\partial \alpha}\right)_z d\alpha \tag{13-3c}$$

Because α and z are independent, the coefficients of $d\alpha$ and dz in Eqs. (13-3b) and (13-3c) must be respectively equal, then

$$\left(\frac{\partial x}{\partial \alpha}\right)_z = \left(\frac{\partial x}{\partial \alpha}\right)_y + \left(\frac{\partial x}{\partial y}\right)_\alpha \left(\frac{\partial y}{\partial \alpha}\right)_z$$

$$\left(\frac{\partial x}{\partial z}\right)_\alpha = \left(\frac{\partial x}{\partial y}\right)_\alpha \left(\frac{\partial y}{\partial z}\right)_\alpha$$

(13-4)

The last equation can be put in the following more symmetrical form:

$$\left(\frac{\partial x}{\partial y}\right)_\alpha \left(\frac{\partial y}{\partial z}\right)_\alpha \left(\frac{\partial z}{\partial x}\right)_\alpha = 1$$

(13-5)

This equation is sometimes called the *chain relation*, each partial derivative in the equation being a link of the chain. This relation can be extended or shortened by adding or subtracting links. Thus, if there are five state variables, x, y, z, γ, and α, involved in the transformation, the chain relation becomes

$$\left(\frac{\partial x}{\partial y}\right)_\alpha \left(\frac{\partial y}{\partial z}\right)_\alpha \left(\frac{\partial z}{\partial \gamma}\right)_\alpha \left(\frac{\partial \gamma}{\partial x}\right)_\alpha = 1$$

Likewise, for three variables, x, y, and α, the chain relation reduces to

$$\left(\frac{\partial x}{\partial y}\right)_\alpha \left(\frac{\partial y}{\partial x}\right)_\alpha = 1$$

or

$$\left(\frac{\partial x}{\partial y}\right)_\alpha = \frac{1}{\left(\frac{\partial y}{\partial x}\right)_\alpha}$$

(13-6)

which is a simple reciprocal relation.

When there are only three state variables involved in a transformation, there is a very useful partial differential relation that can be obtained as follows: The total differentials of $x = x(y, z)$ and $y = y(x, z)$ are, respectively,

$$dx = \left(\frac{\partial x}{\partial y}\right)_z dy + \left(\frac{\partial x}{\partial z}\right)_y dz$$

and

$$dy = \left(\frac{\partial y}{\partial x}\right)_z dx + \left(\frac{\partial y}{\partial z}\right)_x dz$$

Combining the last two equations gives

$$\left[1 - \left(\frac{\partial x}{\partial y}\right)_z \left(\frac{\partial y}{\partial x}\right)_z\right] dx = \left[\left(\frac{\partial x}{\partial y}\right)_z \left(\frac{\partial y}{\partial z}\right)_x + \left(\frac{\partial x}{\partial z}\right)_y\right] dz$$

Because x and z are independent, the last equation must be true for all values of dx and dz. Letting $dz = 0$ and $dx \neq 0$ results in the simple reciprocal relation, Eq. (13-6). However, letting $dx = 0$ and $dz \neq 0$ results in

$$\left(\frac{\partial x}{\partial y}\right)_z \left(\frac{\partial y}{\partial z}\right)_x + \left(\frac{\partial x}{\partial z}\right)_y = 0$$

which can be arranged in the following symmetrical form

$$\left(\frac{\partial x}{\partial y}\right)_z \left(\frac{\partial y}{\partial z}\right)_x \left(\frac{\partial z}{\partial x}\right)_y = -1 \tag{13-7}$$

This equation is sometimes called the *cyclic relation* because the variables are permuted cyclically. Notice that unlike the chain relation, this equation cannot be extended or shortened.

13-2 SIMPLE COMPRESSIBLE SYSTEMS

In Sec. 7-1, we derived the following two important basic relations of properties for a simple compressible system of fixed mass:

$$dU = T \, dS - p \, dV \tag{13-8}$$

$$dH = T \, dS + V \, dp \tag{13-9}$$

For a unit mass or a unit mole, they are

$$du = T \, ds - p \, dv \tag{13-8a}$$

$$dh = T \, ds + v \, dp \tag{13-9a}$$

Two additional relations can be developed by defining two thermodynamic properties of great utility. These are the *Helmholtz function A* and the *Gibbs function G*, as defined by the equations

$$A = U - TS \tag{13-10}$$

$$G = H - TS \tag{13-11}$$

For a unit mass or a unit mole, lowercase letters are used. Thus,

$$a = u - Ts \tag{13-10a}$$

$$g = h - Ts \tag{13-11a}$$

Differentiation of Eqs. (13-10) and (13-11) gives

$$dA = dU - T \, dS - S \, dT$$

$$dG = dH - T \, dS - S \, dT$$

Substitution of Eqs. (13-8) and (13-9) into the last two equations yields

$$dA = -S \, dT - p \, dV \tag{13-12}$$

$$dG = -S \, dT + V \, dp \tag{13-13}$$

For a unit mass or a unit mole, they are

$$da = -s \, dT - p \, dv \tag{13-12a}$$

$$dg = -s \, dT + v \, dp \tag{13-13a}$$

Equations (13-8), (13-9), (13-12), and (13-13) are the four basic relations of properties. They are applicable for any process, reversible or irreversible, between equilibrium states of a simple compressible system of fixed mass.

A number of useful partial-derivative relations can be readily obtained from the four basic relations. Thus, the total differential of U in terms of S and V is

$$dU = \left(\frac{\partial U}{\partial S}\right)_V dS + \left(\frac{\partial U}{\partial V}\right)_S dV$$

Comparing this equation with Eq. (13-8) gives

$$\left(\frac{\partial U}{\partial S}\right)_V = T \quad \text{and} \quad \left(\frac{\partial U}{\partial V}\right)_S = -p \qquad (13\text{-}14a)$$

Similar relations are obtained from the other three basic relations. They are as follows:

$$\left(\frac{\partial H}{\partial S}\right)_p = T \quad \text{and} \quad \left(\frac{\partial H}{\partial p}\right)_S = V \qquad (13\text{-}14b)$$

$$\left(\frac{\partial A}{\partial T}\right)_V = -S \quad \text{and} \quad \left(\frac{\partial A}{\partial V}\right)_T = -p \qquad (13\text{-}14c)$$

$$\left(\frac{\partial G}{\partial T}\right)_p = -S \quad \text{and} \quad \left(\frac{\partial G}{\partial p}\right)_T = V \qquad (13\text{-}14d)$$

Because U, H, A, and G are all thermodynamic properties, they are point functions. Therefore, dU, dH, dA, and dG are exact differentials. Applying the condition of exactness to the four basic relations, one obtains

$$\left(\frac{\partial T}{\partial V}\right)_S = -\left(\frac{\partial p}{\partial S}\right)_V \qquad (13\text{-}15a)$$

$$\left(\frac{\partial T}{\partial p}\right)_S = \left(\frac{\partial V}{\partial S}\right)_p \qquad (13\text{-}15b)$$

$$\left(\frac{\partial S}{\partial V}\right)_T = \left(\frac{\partial p}{\partial T}\right)_V \qquad (13\text{-}15c)$$

$$\left(\frac{\partial S}{\partial p}\right)_T = -\left(\frac{\partial V}{\partial T}\right)_p \qquad (13\text{-}15d)$$

These equations are known as the four *Maxwell relations*. They are of great usefulness in the transformation of state variables, and particularly in the determination of changes in entropy, which are not experimentally measurable, in terms of the measurable properties P, V, and T.

If we regard the volume of a system as a function of temperature and pressure, we have

$$dV = \left(\frac{\partial V}{\partial T}\right)_p dT + \left(\frac{\partial V}{\partial p}\right)_T dp$$

The differential coefficients in the preceding equation have important physical significance when defined as follows:

Coefficient of thermal expansion

$$\alpha = \frac{1}{V}\left(\frac{\partial V}{\partial T}\right)_p \qquad (13\text{-}16)$$

Isothermal compressibility

$$\kappa = -\frac{1}{V}\left(\frac{\partial V}{\partial p}\right)_T \tag{13-17}$$

We accordingly write

$$dV = \alpha V\,dT - \kappa V\,dp$$

or

$$d\ln V = \alpha\,dT - \kappa\,dp \tag{13-18}$$

Application of the condition of exactness to the foregoing equation results in

$$\left(\frac{\partial \alpha}{\partial p}\right)_T = -\left(\frac{\partial \kappa}{\partial T}\right)_p \tag{13-19}$$

which indicates that the coefficients α and κ are not independent of each other. From the cyclic relation, we have

$$\left(\frac{\partial p}{\partial T}\right)_V\left(\frac{\partial T}{\partial V}\right)_p\left(\frac{\partial V}{\partial p}\right)_T = -1$$

Thus,

$$\frac{\alpha}{\kappa} = -\frac{(\partial V/\partial T)_p}{(\partial V/\partial p)_T} = \left(\frac{\partial p}{\partial T}\right)_V \tag{13-20}$$

Similar in form to Eq. (13-17) is another coefficient defined as follows

Adiabatic compressibility

$$\kappa_S = -\frac{1}{V}\left(\frac{\partial V}{\partial p}\right)_S \tag{13-21}$$

Notice that all equations given in this chapter for simple compressible systems involving the partial derivatives $(\partial V/\partial T)_p$, $(\partial V/\partial p)_T$, and/or $(\partial V/\partial p)_S$ can be written in terms of α, κ, and/or κ_S.

Example 13-1

Use the first Maxwell relation, Eq. (13-15a), to demonstrate the correctness of the data given by Keenan, Keyes, Hill, and Moore [21] for superheated steam at 600 psia and 800°F.

Solution At 600 psia and 800°F, the steam tables give $v = 1.1900$ cu ft/lbm and $s = 1.6343$ Btu/lbm·°R. From the steam tables, select values of pressure and entropy corresponding to a constant specific volume of 1.1900 ft³/lbm. Plot these data on a p–s diagram and draw the constant-volume curve. Then measure the slope of this curve at 600 psia. The value obtained from this measurement is

$$\left(\frac{\partial p}{\partial s}\right)_v = 1670\,\frac{\text{lbf/in.}^2}{\text{Btu/lbm·°R}}$$

$$= 309\,\frac{°R}{\text{ft}^3/\text{lbm}}$$

Next, from the steam tables, select values of temperature and specific volume corresponding to a constant entropy of 1.6343 Btu/lbm·°R. Plot these data on a T–v diagram and draw the constant-entropy curve. Then measure the slope of this curve at 800°F. The value obtained from this measurement is

$$\left(\frac{\partial T}{\partial v}\right)_s = -309 \frac{°R}{ft^3/lbm}$$

These two slopes are equal in magnitude but opposite in sign, as predicted by Eq. (13-15a).

Example 13-2

The p–v–T relation for a gas is given as

$$p = \frac{RT}{v - b} - \frac{a}{v^2}$$

where a and b are two specific constants for the gas, and R is the gas constant. Derive the expressions for the coefficient of thermal expansion α and the isothermal compressibility κ for the gas.

Solution Differentiation of the given p–v–T relation yields

$$\left(\frac{\partial p}{\partial T}\right)_v = \frac{R}{v - b}$$

and

$$\left(\frac{\partial p}{\partial v}\right)_T = \frac{2a}{v^3} - \frac{RT}{(v - b)^2}$$

From the cyclic relation, we have

$$\left(\frac{\partial v}{\partial T}\right)_p = -\frac{(\partial p/\partial T)_v}{(\partial p/\partial v)_T} = -\frac{R/(v - b)}{(2a/v^3) - RT/(v - b)^2}$$

$$= \frac{Rv^3(v - b)}{RTv^3 - 2a(v - b)^2}$$

Therefore,

$$\alpha = \frac{1}{v}\left(\frac{\partial v}{\partial T}\right)_p = \frac{Rv^2(v - b)}{RTv^3 - 2a(v - b)^2}$$

$$\kappa = -\frac{1}{v}\left(\frac{\partial v}{\partial p}\right)_T = -\frac{1}{v}\frac{1}{(2a/v^3) - RT/(v - b)^2}$$

$$= \frac{v^2(v - b)^2}{RTv^3 - 2a(v - b)^2}$$

Also

$$\frac{\alpha}{\kappa} = -\left(\frac{\partial v}{\partial T}\right)_p\left(\frac{\partial p}{\partial v}\right)_T = \left(\frac{\partial p}{\partial T}\right)_v = \frac{R}{v - b}$$

13-3 RELATIONS FOR SPECIFIC HEATS

The specific heat at constant volume c_v and the specific heat at constant pressure c_p of a homogeneous simple compressible system have been defined in Sec. 2-8 as

$$c_v = \left(\frac{\partial u}{\partial T}\right)_v \qquad (13\text{-}22)$$

$$c_p = \left(\frac{\partial h}{\partial T}\right)_p \qquad (13\text{-}23)$$

These specific heats can be expressed in terms of the entropy through the use of Eqs. (13-14a) and (13-14b). Thus, because

$$\left(\frac{\partial u}{\partial s}\right)_v = \left(\frac{\partial u}{\partial T}\right)_v \left(\frac{\partial T}{\partial s}\right)_v = T$$

and

$$\left(\frac{\partial h}{\partial s}\right)_p = \left(\frac{\partial h}{\partial T}\right)_p \left(\frac{\partial T}{\partial s}\right)_p = T$$

whereupon we obtain

$$c_v = \left(\frac{\partial u}{\partial T}\right)_v = T\left(\frac{\partial s}{\partial T}\right)_v \qquad (13\text{-}24)$$

and

$$c_p = \left(\frac{\partial h}{\partial T}\right)_p = T\left(\frac{\partial s}{\partial T}\right)_p \qquad (13\text{-}25)$$

Because specific heats are related to energy measurements, they cannot be calculated from data on p–v–T alone. However, the isothermal changes in c_v with respect to volume and in c_p with respect to pressure can be calculated when adequate p–v–T data are available. We are now to derive the required relations that facilitate such calculations. Differentiating Eqs. (13-24) and (13-25) with respect to v and p, respectively, while holding T constant in both cases, we have

$$\left(\frac{\partial c_v}{\partial v}\right)_T = T\frac{\partial^2 s}{\partial v\,\partial T}$$

and

$$\left(\frac{\partial c_p}{\partial p}\right)_T = T\frac{\partial^2 s}{\partial p\,\partial T}$$

Differentiating the Maxwell relations,

$$\left(\frac{\partial p}{\partial T}\right)_v = \left(\frac{\partial s}{\partial v}\right)_T \qquad \text{and} \qquad \left(\frac{\partial v}{\partial T}\right)_p = -\left(\frac{\partial s}{\partial p}\right)_T$$

yields

$$\left(\frac{\partial^2 p}{\partial T^2}\right)_v = \left(\frac{\partial^2 s}{\partial T\,\partial v}\right) \qquad \text{and} \qquad \left(\frac{\partial^2 v}{\partial T^2}\right)_p = -\frac{\partial^2 s}{\partial T\,\partial p}$$

Therefore, we finally have the desired relations:

$$\left(\frac{\partial c_v}{\partial v}\right)_T = T\left(\frac{\partial^2 p}{\partial T^2}\right)_v \qquad (13\text{-}26)$$

and

$$\left(\frac{\partial c_p}{\partial p}\right)_T = -T\left(\frac{\partial^2 v}{\partial T^2}\right)_p \tag{13-27}$$

There are two more important equations for specific heats in relation to p–v–T data: one of them is the ratio of c_p and c_v and the other is their difference. The relation for the ratio of specific heats, denoted by the symbol k, is obtained by dividing Eq. (13-25) by Eq. (13-24). Thus,

$$k = \frac{c_p}{c_v} = \frac{\left(\dfrac{\partial s}{\partial T}\right)_p}{\left(\dfrac{\partial s}{\partial T}\right)_v}$$

But by the cyclic relation of Eq. (13-7), we write

$$\left(\frac{\partial s}{\partial T}\right)_p = -\left(\frac{\partial p}{\partial T}\right)_s\left(\frac{\partial s}{\partial p}\right)_T$$

and

$$\left(\frac{\partial s}{\partial T}\right)_v = -\left(\frac{\partial v}{\partial T}\right)_s\left(\frac{\partial s}{\partial v}\right)_T$$

Consequently,

$$k = \frac{c_p}{c_v} = \frac{\left(\dfrac{\partial p}{\partial v}\right)_s}{\left(\dfrac{\partial p}{\partial v}\right)_T} \tag{13-28}$$

The preceding equation can also be written in terms of the isothermal compressibility κ and the adiabatic compressibility κ_s as follows:

$$k = \frac{c_p}{c_v} = \frac{\kappa}{k_S} \tag{13-29}$$

The difference between c_p and c_v is obtained by subtracting Eq. (13-24) from Eq. (13-25). Thus,

$$c_p - c_v = T\left(\frac{\partial s}{\partial T}\right)_p - T\left(\frac{\partial s}{\partial T}\right)_v$$

But by Eq. (13-14), we have

$$\left(\frac{\partial s}{\partial T}\right)_p = \left(\frac{\partial s}{\partial T}\right)_v + \left(\frac{\partial s}{\partial v}\right)_T\left(\frac{\partial v}{\partial T}\right)_p$$

Therefore,

$$c_p - c_v = T\left(\frac{\partial s}{\partial v}\right)_T\left(\frac{\partial v}{\partial T}\right)_p$$

Upon using the third Maxwell relation,

$$\left(\frac{\partial s}{\partial v}\right)_T = \left(\frac{\partial p}{\partial T}\right)_v$$

we obtain

$$c_p - c_v = T\left(\frac{\partial p}{\partial T}\right)_v\left(\frac{\partial v}{\partial T}\right)_p \tag{13-30}$$

Moreover, from the cyclic relation, we have

$$\left(\frac{\partial p}{\partial T}\right)_v = -\left(\frac{\partial v}{\partial T}\right)_p\left(\frac{\partial p}{\partial v}\right)_T$$

Finally,

$$c_p - c_v = -T\left(\frac{\partial v}{\partial T}\right)_p^2\left(\frac{\partial p}{\partial v}\right)_T = -\frac{T\left(\frac{\partial v}{\partial T}\right)_p^2}{\left(\frac{\partial v}{\partial p}\right)_T} \tag{13-31}$$

Equation (13-30) or (13-31) can be used for the conversion of c_p into c_v, or vice versa, when adequate p–v–T data are available. In the case of solids and liquids, it is very difficult to determine c_v experimentally; it must be converted from measured values of c_p by the use of these equations.

Some observations concerning the relative magnitudes of c_p and c_v can be made from Eq. (13-31). Although $(\partial v/\partial T)_p$ is usually positive, it can be zero or negative. For example, at atmospheric pressure, $(\partial v/\partial T)_p$ for water is zero at 4°C and is negative between 0°C and 4°C. However, the square of $(\partial v/\partial T)_p$ can never be negative. On the other hand, $(\partial p/\partial v)_T$ is always negative for all known substances. Consequently, $c_p - c_v$ can never be negative, or c_p can never be smaller than c_v. The two specific heats are equal in the particular case when $(\partial v/\partial T)_p = 0$. Furthermore, $c_p - c_v$ approaches zero as T approaches zero, or $c_p = c_v$ at the absolute zero temperature. For solids and liquids, both $(\partial v/\partial T)_p$ and $(\partial p/\partial v)_T$ are relatively small, and so at low temperatures, the difference between c_p and c_v is small, although this difference becomes appreciable at high temperatures. Also, because c_p can never be smaller than c_v, we observe from Eq. (13-28) that k can never be smaller than unity.

Example 13-13

For a gas obeying the p–v–T relation given in Example 13-2, derive the expression for $c_p - c_v$. Show that for this gas, c_v is a function of temperature only.

Solution Equation (13-31) for $c_p - c_v$ can be rewritten in terms of the coefficient of thermal expansion α and isothermal compressibility κ as follows:

$$c_p - c_v = \frac{Tv[(1/v)(\partial v/\partial T)_p]^2}{-(1/v)(\partial v/\partial p)_T} = \frac{Tv\alpha^2}{\kappa}$$

Substituting the expressions for α and α/κ as obtained in Example 13-2 leads to the desired expression

$$c_p - c_v = Tv\left[\frac{Rv^2(v-b)}{RTv^3 - 2a(v-b)^2}\right]\left(\frac{R}{v-b}\right)$$

$$= \frac{R}{1 - 2a(v-b)^2/RTv^3}$$

For the given p–v–T relation, we have

$$\left(\frac{\partial p}{\partial T}\right)_v = \frac{R}{v - b}$$

and

$$\left(\frac{\partial^2 p}{\partial T^2}\right)_v = 0$$

it follows that

$$\left(\frac{\partial c_v}{\partial v}\right)_T = T\left(\frac{\partial^2 p}{\partial T^2}\right)_v = 0$$

Hence, c_v is a function of temperature only for a gas obeying the given p–v–T relation. However, as can be seen from the following equation

$$\left(\frac{\partial c_p}{\partial p}\right)_T = -T\left(\frac{\partial^2 v}{\partial T^2}\right)_p$$

$$= R^2 T\left[\frac{2av^{-3} - 6abv^{-4}}{(p - av^{-2} + 2abv^{-3})^3}\right]$$

that for a gas obeying the given p–v–T relation, c_p is not a function of temperature only.

13-4 RELATIONS FOR ENTROPY, INTERNAL ENERGY, AND ENTHALPY

The entropy change for a simple compressible system can be expressed as a function of any pair of the variables p, v, and T, thus forming three general equations, the derivations of which are as follows. When s is considered as a function of T and v, we have

$$ds = \left(\frac{\partial s}{\partial T}\right)_v dT + \left(\frac{\partial s}{\partial v}\right)_T dv$$

But from Eq. (13-24),

$$\left(\frac{\partial s}{\partial T}\right)_v = \frac{c_v}{T}$$

and from Eq. (13-15c),

$$\left(\frac{\partial s}{\partial v}\right)_T = \left(\frac{\partial p}{\partial T}\right)_v$$

Consequently,

$$ds = \frac{c_v}{T} dT + \left(\frac{\partial p}{\partial T}\right)_v dv \tag{13-32}$$

This is the first ds equation. Now when s is considered as a function of T and p, we have

$$ds = \left(\frac{\partial s}{\partial T}\right)_p dT + \left(\frac{\partial s}{\partial p}\right)_T dp$$

But from Eq. (13-25),

$$\left(\frac{\partial s}{\partial T}\right)_p = \frac{c_p}{T}$$

and from Eq. (13-15d),

$$\left(\frac{\partial s}{\partial p}\right)_T = -\left(\frac{\partial v}{\partial T}\right)_p$$

Consequently,

$$ds = \frac{c_p}{T}\, dT - \left(\frac{\partial v}{\partial T}\right)_p dp \tag{13-33}$$

This is the second ds equation. Finally, when s is considered as a function of p and v, we have

$$ds = \left(\frac{\partial s}{\partial p}\right)_v dp + \left(\frac{\partial s}{\partial v}\right)_p dv$$

But by Eqs. (13-5), (13-24), and (13-25), we have

$$\left(\frac{\partial s}{\partial p}\right)_v = \left(\frac{\partial T}{\partial p}\right)_v\left(\frac{\partial s}{\partial T}\right)_v = \left(\frac{\partial T}{\partial p}\right)_v \frac{c_v}{T}$$

and

$$\left(\frac{\partial s}{\partial v}\right)_p = \left(\frac{\partial T}{\partial v}\right)_p\left(\frac{\partial s}{\partial T}\right)_p = \left(\frac{\partial T}{\partial v}\right)_p \frac{c_p}{T}$$

Consequently,

$$ds = \frac{c_v}{T}\left(\frac{\partial T}{\partial p}\right)_v dp + \frac{c_p}{T}\left(\frac{\partial T}{\partial v}\right)_p dv \tag{13-34}$$

This is the third ds equation. The three ds equations are useful in the evaluation of entropy changes. Because $đq = T\, ds$ in an internally reversible process, these equations also provide means to calculate heat transfer in an internally reversible process when specific heats and p–v–T data are available.

The three ds equations we just derived can be used to yield three equations for du as functions of each pair of the variables p, v, and T. Thus, from Eq. (13-8), the combined first and second laws for a simple compressible system is

$$du = T\, ds - p\, dv$$

Upon substituting the first ds equation into Eq. (13-8), we get the first du equation:

$$du = c_v\, dT + \left[T\left(\frac{\partial p}{\partial T}\right)_v - p\right] dv \tag{13-35}$$

Substituting the second ds equation into Eq. (13-8), and noting that

$$dv = \left(\frac{\partial v}{\partial T}\right)_p dT + \left(\frac{\partial v}{\partial p}\right)_T dp$$

we get the second du equation:

$$du = \left[c_p - p\left(\frac{\partial v}{\partial T}\right)_p\right] dT - \left[p\left(\frac{\partial v}{\partial p}\right)_T + T\left(\frac{\partial v}{\partial T}\right)_p\right] dp \tag{13-36}$$

Substituting the third ds equation into Eq. (13-8), we get the third du equation:

$$du = c_v \left(\frac{\partial T}{\partial p}\right)_v dp + \left[c_p \left(\frac{\partial T}{\partial v}\right)_p - p \right] dv \qquad (13\text{-}37)$$

It should be noted that out of the three du equations, only the one with T and v as independent variables, that is, Eq. (13-35), is most useful. This equation can be used to form the so-called *energy equation,* which is

$$\left(\frac{\partial u}{\partial v}\right)_T = T \left(\frac{\partial p}{\partial T}\right)_v - p \qquad (13\text{-}38)$$

The preceding equation can be used to demonstrate Joule's law, mentioned in Sec. 4-2. Thus, for an ideal gas, $pv = RT$, so that

$$\left(\frac{\partial p}{\partial T}\right)_v = \frac{R}{v} \qquad \text{and} \qquad \left(\frac{\partial u}{\partial v}\right)_T = T\left(\frac{R}{v}\right) - p = p - p = 0$$

which implies that the internal energy of an ideal gas is a function of temperature only.

As in the case of internal energy, the three ds equations can also be used to yield three equations for dh as functions of each pair of the variables p, v, and T. Thus, from Eq. (13-9), the basic equation for h is

$$dh = T\, ds + v\, dp$$

Upon substituting the first ds equation into Eq. (13-9) and noting that

$$dp = \left(\frac{\partial p}{\partial T}\right)_v dT + \left(\frac{\partial p}{\partial v}\right)_T dv$$

we get the first dh equation:

$$dh = \left[c_v + v\left(\frac{\partial p}{\partial T}\right)_v \right] dT + \left[T\left(\frac{\partial p}{\partial T}\right)_v + v\left(\frac{\partial p}{\partial v}\right)_T \right] dv \qquad (13\text{-}39)$$

Substituting the second ds equation into Eq. (13-9), we get the second dh equation:

$$dh = c_p\, dT + \left[v - T\left(\frac{\partial v}{\partial T}\right)_p \right] dp \qquad (13\text{-}40)$$

Substituting the third ds equation into Eq. (13-9), we get the third dh equation:

$$dh = \left[v + c_v\left(\frac{\partial T}{\partial p}\right)_v \right] dp + c_p\left(\frac{\partial T}{\partial v}\right)_p dv \qquad (13\text{-}41)$$

Among the three general equations for enthalpy, the one with T and p as independent variables, that is, Eq. (13-40), is most useful.

Example 13-4

For a gas that obeys the p–v–T relation given in Example 13-2, determine the relations between T and v and between p and v in an internally reversible adiabatic process, assuming $c_v = $ constant.

Solution For a gas that obeys the given p–v–T relation, we have

$$\left(\frac{\partial p}{\partial T}\right)_v = \frac{R}{v - b}$$

From Eq. (13-32),

$$ds = \frac{c_v}{T}\, dT + \frac{R}{v - b}\, dv$$

For an internally reversible adiabatic process, this equation becomes

$$\frac{c_v}{T}\, dT + \frac{R}{v - b}\, dv = 0$$

When c_v = constant, integrating gives

$$c_v \ln T + R \ln (v - b) = \text{constant}$$

Therefore,

$$T(v - b)^{R/c_s} = \text{constant}$$

Combining this equation with the given p–v–T relation leads to

$$\left[\left(p + \frac{a}{v^2}\right)\left(\frac{v - b}{R}\right)\right](v - b)^{R/c_v} = \text{constant}$$

Hence,

$$\left(p + \frac{a}{v_2}\right)(v - b)^{1 + R/c_v} = \text{constant}$$

13-5 JOULE–THOMSON COEFFICIENT

The throttling process introduced in Sec. 5-8 can be used in a Joule–Thomson expansion experiment to study the temperature variation of a gas in relation to changes in pressure. By referring to Fig. 5-14, a throttling process satisfies the relation

$$h_1 = h_2 \tag{13-42}$$

For an ideal gas, because enthalpy is a function of temperature alone, its temperature will not change upon throttling to a low pressure. For a real gas, however, its temperature may increase or decrease upon throttling, depending on the conditions of the initial and final states.

To investigate the temperature–pressure relationship of a real gas, a series of Joule–Thomson expansion experiments can be performed. In each experiment, the values of p_1 and T_1 for the higher-pressure side are kept the same, but different values of p_2, such as p_{2a}, p_{2b}, p_{2c}, etc., are maintained for the lower-pressure side and the corresponding temperature T_2's are measured. The data can then be plotted on a T–p diagram, giving the discrete points 1, 2a, 2b, 2c, etc., as shown in Fig 13-1(a). Because $h_1 = h_{2a} = h_{2b} = h_{2c} = \cdots$, a smooth curve drawn through these points is a curve of constant enthalpy. However, this curve does not represent the process executed by the gas in passing through the restriction. By performing other series of similar experiments, using different constant values of p_1 and T_1 in each series, a family of curves corresponding to different values of h can be obtained. Such a family of isenthalpic curves is depicted in Fig. 13-1(b).

The slope of an isenthalpic curve on a T–p diagram at any point is called the *Joule–Thomson coefficient* μ_J:

$$\mu_J = \left(\frac{\partial T}{\partial p}\right)_h \tag{13-43}$$

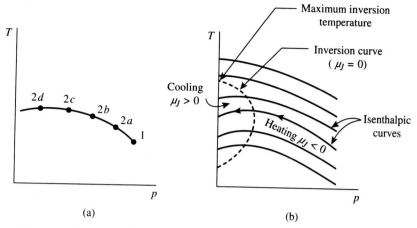

Figure 13-1 Isenthalpic and inversion curves for a Joule–Thomson experiment.

The value of μ_J can be positive, negative, or zero. The point at which $\mu_J = 0$ corresponds to the maximum of an isenthalpic curve and is called the *inversion point*. The locus of all the inversion points is called the *inversion curve*. In the region inside the inversion curve where $\mu_J > 0$, the temperature decreases as the pressure decreases upon throttling—a cooling effect. However, in the region outside the inversion curve where $\mu_J < 0$, the temperature increases as the pressure decreases upon throttling—a heating effect. The temperature at which the inversion curve intersects the temperature axis is called the *maximum inversion temperature*. When the initial temperature of a Joule–Thomson expansion is higher than the maximum inversion temperature, no cooling is possible. Table 13-1 gives the maximum inversion temperatures for a number of substances.

The expression for the Joule–Thomson coefficient μ_J in terms of p, v, T, and c_p is obtained by applying the second dh equation:

$$dh = c_p\, dT + \left[v - T\left(\frac{\partial v}{\partial T}\right)_p \right] dp$$

Rearranging,

$$dT = \frac{1}{c_p}\, dh + \frac{1}{c_p}\left[T\left(\frac{\partial v}{\partial T}\right)_p - v \right] dp$$

TABLE 13-1 MAXIMUM INVERSION TEMPERATURES

Gas	Maximum inversion temperature, K
Carbon dioxide	1500
Oxygen	761
Argon	723
Nitrogen	621
Air	603
Hydrogen	202
Helium	40

But the total differential of T in terms of h and p is

$$dT = \left(\frac{\partial T}{\partial h}\right)_p dh + \left(\frac{\partial T}{\partial p}\right)_h dp$$

Comparing the coefficients of dp in the foregoing two equations, we obtain

$$\mu_J = \left(\frac{\partial T}{\partial p}\right)_h = \frac{1}{c_p}\left[T\left(\frac{\partial v}{\partial T}\right)_p - v\right] \tag{13-44}$$

It is clear from this equation that the Joule–Thomson coefficient and the specific heats are equivalent sources of supplementary data, which when used in connection with p–v–T data facilitate the evaluation of such quantities as internal energy, enthalpy, and entropy.

Example 13-5

Derive the expression for the Joule–Thomson coefficient of a gas obeying the p–v–T relation given in Example 13-2, and determine the equation of the inversion curve.

Solution   Equation (13-44) for the Joule–Thomson coefficient can be rewritten in terms of the coefficient of thermal expansion α as follows:

$$\mu_J = \left(\frac{\partial T}{\partial p}\right)_h = \frac{v}{c_p}\left[T\frac{1}{v}\left(\frac{\partial v}{\partial T}\right)_p - 1\right] = \frac{v}{c_p}(\alpha T - 1)$$

For a gas obeying the given p–v–T relation, the coefficient of thermal expansion α, as obtained in Example 13-2, is

$$\alpha = \frac{Rv^2(v - b)}{RTv^3 - 2a(v - b)^2}$$

Therefore,

$$\mu_J = \frac{v}{c_p}\left[\frac{RTv^2(v - b)}{RTv^3 - 2a(v - b)^2} - 1\right]$$

$$= \frac{v}{c_p}\left[\frac{2a(v - b)^2 - RTbv^2}{RTv^3 - 2a(v - b)^2}\right]$$

The equation of the inversion curve is obtained by setting $\mu_J = 0$, which gives

$$T = \frac{2a}{Rb}\left(1 - \frac{b}{v}\right)^2$$

Substituting this equation into the given p–v–T relation gives

$$p = \frac{a}{bv}\left(2 - \frac{3b}{v}\right)$$

These two equations are the defining equations for the inversion curve in parametric form. The equation connecting the inversion temperature and the corresponding pressure can be obtained by eliminating v between the preceding two equations.

13-6 CLAPEYRON EQUATION

During a change of phase, such as vaporization, melting, and sublimation, of a pure substance, its temperature and pressure remain constant while its entropy and volume undergo changes. The temperature and pressure of a pure substance consisting of two phases in equilibrium are not independent variables. We now proceed to determine how the equilibrium pressure depends on the temperature. The desired relation can be obtained directly from the third Maxwell relation, Eq. (13-15c):

$$\left(\frac{\partial p}{\partial T}\right)_V = \left(\frac{\partial S}{\partial V}\right)_T$$

Because pressure is a function only of temperature and is independent of volume for coexisting phases, the partial derivative $(\partial p / \partial T)_V$ can be replaced by the total derivative dp/dT. Furthermore, at constant temperature, and incidentally also constant pressure, the entropy and volume are linear functions of each other. We can write accordingly

$$\left(\frac{\partial S}{\partial V}\right)_T = \frac{S^{(\beta)} - S^{(\alpha)}}{V^{(\beta)} - V^{(\alpha)}}$$

where the superscripts (α) and (β) denote the two coexisting phases. Consequently, Eq. (13-15c) becomes

$$\frac{dp}{dT} = \frac{S^{(\beta)} - S^{(\alpha)}}{V^{(\beta)} - V^{(\alpha)}}$$

or, in terms of quantities per unit mass or per mole,

$$\frac{dp}{dT} = \frac{s^{(\beta)} - s^{(\alpha)}}{v^{(\beta)} - v^{(\alpha)}}$$

But according to Eq. (13-14b),

$$\left(\frac{\partial h}{\partial s}\right)_p = T$$

we have

$$h^{(\beta)} - h^{(\alpha)} = T(s^{(\beta)} - s^{(\alpha)})$$

Therefore,

$$\frac{dp}{dT} = \frac{h^{(\beta)} - h^{(\alpha)}}{T[v^{(\beta)} - v^{(\alpha)}]} \qquad (13\text{-}45)$$

This is the famous *Clapeyron equation*.

Another method of arriving at the Clapeyron equation is by utilizing the fact that because from Eq. (13-13),

$$dg = -s\, dT + v\, dp$$

the Gibbs function remains constant during a reversible isothermal–isobaric phase transition. Thus, for a phase transition at T and p, we have

$$g^{(\alpha)} = g^{(\beta)}$$

and for a phase transition at $T + dT$ and $p + dp$, we have

$$g^{(\alpha)} + dg^{(\alpha)} = g^{(\beta)} + dg^{(\beta)}$$

from which

$$dg^{(\alpha)} = dg^{(\beta)}$$

Applying Eq. (13-13), we obtain

$$-s^{(\alpha)} \, dT + v^{(\alpha)} \, dp = -s^{(\beta)} \, dT + v^{(\beta)} \, dp$$

or

$$\frac{dp}{dT} = \frac{s^{(\beta)} - s^{(\alpha)}}{v^{(\beta)} - v^{(\alpha)}}$$

which leads to the Clapeyron equation.

The Clapeyron equation is one of the most useful equations in thermodynamics. It can be used in a variety of ways, including the prediction of the effect of pressure on transition temperatures and the computation of latent heats of phase transitions. It is applicable not only to equilibrium transitions between liquids and vapors, but also between solids and vapors, between solids and liquids, and between two different solid phases of a substance.

When the Clapeyron equation is applied to a liquid–vapor (or solid–vapor) equilibrium transition, we can generally neglect the volume of the liquid (or solid) compared with that of the vapor. Furthermore, if we approximate the volume of the vapor by the ideal gas law, the Clapeyron equation then simplifies to

$$\frac{dp}{dT} = \frac{h_{fg}}{T v_g} = \frac{p h_{fg}}{R T^2}$$

or

$$\frac{1}{p} \frac{dp}{dT} = \frac{d(\ln p)}{dT} = \frac{h_{fg}}{R T^2} \tag{13-46}$$

This equation is known as the *Clausius–Clapeyron equation*. It can be used to estimate the latent heat of vaporization h_{fg} at low pressures. When this equation is applied to a solid–vapor equilibrium transition (use h_{ig} in the equation), it can be used to calculate the latent heat of sublimation, because in this case, the vapor pressure is usually small.

For small temperature changes, the latent heat h_{fg} can be considered constant. Under this circumstance, Eq. (13-46) can be integrated to give

$$\ln p = -\frac{h_{fg}}{RT} + \text{constant} \tag{13-47}$$

which means that when $\ln p$ is plotted against $1/T$, one should obtain a straight line with a slope equal to $-h_{fg}/R$. In view of the approximations involved, this conclusion cannot be regarded as exact. However, for rough calculations, it does provide a means to obtain the vapor pressure of a substance at various temperatures from a small amount of data.

Example 13-6

The SI edition of steam tables by Keenan, Keyes, Hill, and Moore [21] gives the following data:

Saturation temperature, °C	99	100	101
Saturation pressure, kPa	97.78	101.35	105.02
v_f, m³/kg	0.0010427	0.0010435	0.0010443
v_g, m³/kg	1.7299	1.6729	1.6182

Determine the approximate enthalpy of vaporization of water at 100°C.

Thermodynamic Relations   Chap. 13

Solution   At 100°C, the approximate value of dp/dT is

$$\frac{105.02 - 97.78}{101 - 99} = 3.62 \text{ kPa/°C}$$

By the Clapeyron equation, the enthalpy of vaporization of water at 100°C is then

$$h_{fg} = T v_{fg} \frac{dp}{dT}$$

$$= (100 + 273.15)(1.6729 - 0.0010435)(3.62)$$

$$= 2258.3 \text{ kJ/kg}$$

as compared with the tabulated value of 2257.0 kJ/kg.

13-7 SIMPLE PARAMAGNETIC SYSTEMS*

Of the four methods of producing cryogenic temperatures mentioned in Sec. 12-7, adiabatic demagnetization of paramagnetic solids is the ultimate method that enables us to achieve very, very low temperatures approaching absolute zero. Before discussing this method, a study of the properties of the working material—paramagnetic solids—is in order.

Just as an electron current in a small loop produces a magnetic field, an electron revolving in its orbit around the nucleus and rotating about its own axis has an associated magnetic dipole with its motion. In the absence of an external magnetic field, all such dipoles cancel each other. In the presence of an external field, the frequencies and senses of orbiting and spinning of the electrons are changed in such a manner as to oppose the external field. This is the *diamagnetic* nature of all material substances. In some materials, however, there are permanent magnetic dipoles owing to unbalanced electron orbits or spins. These atoms behave like elementary dipoles which tend to align with an external field and strengthen it. When this effect in a material is greater than the diamagnetic tendency common to all atoms, this material is called *paramagnetic*. Paramagnetism is temperature-dependent. When temperature is lowered sufficiently, the atomic elementary dipoles are magnetically aligned within microscopic domains, which can be readily aligned by a relatively small external field to form a large induction. This is referred to as *ferromagnetic*. A ferromagnetic material becomes paramagnetic above a temperature known as the *Curie temperature*. The Curie temperatures of iron and nickel are far above room temperature, thus they are usually referred to as ferromagnetic; whereas the Curie temperatures of some metallic salts are below 1 K, they are usually referred to as paramagnetic. A paramagnetic material is not a magnet if there is no external magnetic field applied to it. In the presence of an external field, it becomes slightly magnetized, in contradistinction to a ferromagnetic material which would show very strong magnetic effects.

The magnetic effects in a ferromagnetic material are not reversible, because the reverse process of demagnetization forms a hysteresis loop with the forward process of magnetization. Thus, the state of a ferromagnetic system depends not only on its present condition, but also on the past history of the system. A ferromagnetic system

---

*Sections 13-7 and 13-8 can be omitted without loss of continuity. But be sure to study these two sections before studying Sec. 16-10.

is, therefore, not amenable to thermodynamic analyses. On the other hand, for a paramagnetic system, such as a paramagnetic salt, the process of magnetization is reversible and the state of the system can be described in terms of a few thermodynamic variables.

The expression* for reversible work in the magnetization of a paramagnetic system is given by

$$đW = -\mu_0 V \mathbf{H} \, d\mathbf{M} \tag{13-48}$$

where $\mathbf{H}$ is the *magnetic field* or *intensity* in A/m, $\mathbf{M}$ is the *magnetization* or *magnetic moment per unit volume* in A/m, V is the volume in m³, W is the work done in J, and μ_0 is the *permeability of free space* and has a value of $4\pi \times 10^{-7}$ N/A². The minus sign in the equation indicates that work input is required to increase the magnetization of a substance.

In thermodynamic analyses of paramagnetic solids, all changes in volume are small and can be ignored. Furthermore, the effects of pressure on the solids are also small and can also be ignored. The only work interaction is that due to the magnetization of the material. A system for which the only reversible work mode is the magnetization of the paramagnetic material is called a *simple paramagnetic system*.

For a simple paramagnetic system, the first law for a reversible process is written as

$$đQ = dU + đW$$

or

$$đQ = dU - \mu_0 V \mathbf{H} \, d\mathbf{M} \tag{13-49}$$

Combining this equation and the second law leads to

$$T \, ds = dU - \mu_0 V \mathbf{H} \, d\mathbf{M}$$

or

$$dU = T \, ds + \mu_0 V \mathbf{H} \, d\mathbf{M} \tag{13-50}$$

This is the basic equation that combines the first and second laws as applied to a simple paramagnetic system. Note that this equation is a relation solely among the properties of the system.

It is helpful to define two new properties, namely, the magnetic enthalpy H and magnetic Gibbs function G:

$$H = U - \mu_0 V \mathbf{H}\mathbf{M} \tag{13-51}$$

$$G = H - TS = U - TS - \mu_0 V \mathbf{H}\mathbf{M} \tag{13-52}$$

The Helmholtz function A takes the usual definition:

$$A = U - TS$$

From these definitions and Eq. (13-50), it follows that

$$dH = T \, dS - \mu_0 V \mathbf{M} \, d\mathbf{H} \tag{13-53}$$

$$dA = -S \, dT + \mu_0 V \mathbf{H} \, d\mathbf{M} \tag{13-54}$$

*For the derivation, see [14, Chapter 7].

and

$$dG = -S\,dT - \mu_0 V \mathbf{M}\,d\mathbf{H} \tag{13-55}$$

Application of the condition of exactness to the four basic relations, Eqs. (13-50), (13-53), (13-54), and (13-55), leads to the following four Maxwell relations for a simple paramagnetic system:

$$\left(\frac{\partial T}{\partial \mathbf{M}}\right)_S = \mu_0 V \left(\frac{\partial \mathbf{H}}{\partial S}\right)_{\mathbf{M}} \tag{13-56a}$$

$$\left(\frac{\partial T}{\partial \mathbf{H}}\right)_S = -\mu_0 V \left(\frac{\partial \mathbf{M}}{\partial S}\right)_{\mathbf{H}} \tag{13-56b}$$

$$\left(\frac{\partial S}{\partial \mathbf{M}}\right)_T = -\mu_0 V \left(\frac{\partial \mathbf{H}}{\partial T}\right)_{\mathbf{M}} \tag{13-56c}$$

and

$$\left(\frac{\partial S}{\partial \mathbf{H}}\right)_T = \mu_0 V \left(\frac{\partial \mathbf{M}}{\partial T}\right)_{\mathbf{H}} \tag{13-56d}$$

Analogous to C_v and C_p, the heat capacity at constant magnetic moment $C_\mathbf{M}$ and the heat capacity at constant magnetic field $C_\mathbf{H}$ for a simple paramagnetic system are defined as

$$C_\mathbf{M} = \left(\frac{\partial U}{\partial T}\right)_\mathbf{M} = T\left(\frac{\partial S}{\partial T}\right)_\mathbf{M} \tag{13-57}$$

$$C_\mathbf{H} = \left(\frac{\partial H}{\partial T}\right)_\mathbf{H} = T\left(\frac{\partial S}{\partial T}\right)_\mathbf{H} \tag{13-58}$$

The first and second dS equations for a simple paramagnetic system are

$$dS = \frac{C_\mathbf{M}}{T}\,dT - \mu_0 V\left(\frac{\partial \mathbf{H}}{\partial T}\right)_\mathbf{M}\,d\mathbf{M} \tag{13-59}$$

$$dS = \frac{C_\mathbf{H}}{T}\,dT + \mu_0 V\left(\frac{\partial \mathbf{M}}{\partial T}\right)_\mathbf{H}\,d\mathbf{H} \tag{13-60}$$

It is imperative to notice that all relations between thermodynamic properties for a simple paramagnetic system can be readily obtained from the corresponding equations for a simple compressible system by observing that $\mathbf{H}$ corresponds to p and $-\mu_0 V\,d\mathbf{M}$ corresponds to dV.

One simple form of the equation of state for a paramagnetic solid was established from the experimental fact that the magnetic moment of a paramagnetic solid is directly proportional to the external magnetic field $\mathbf{H}$ and inversely proportional to the temperature T. For small values of the ratio $\mathbf{H}/T$, this is expressed in the simple equation, known as *Curie's equation:*

$$V\mathbf{M} = C_c \frac{\mathbf{H}}{T} \tag{13-61}$$

where C_c is a proportional constant called the *Curie constant*. The Curie constants for a number of paramagnetic salts are given in Table 13-2.

TABLE 13-2 DATA FOR PARAMAGNETIC SALTS*

Paramagnetic salt	Gram-ionic mass of crystal, $\dfrac{\text{g}}{\text{g-ion}}$	C_c, $\dfrac{\text{cm}^3 \cdot \text{K}}{\text{g-ion}}$	A_c, $\dfrac{\text{J} \cdot \text{K}}{\text{g-ion}}$
$Cr_2(SO_4)_3 \cdot K_2SO_4 \cdot 24H_2O$ (chromium potassium alum)	499	1.84	0.150
$Fe_2(SO_4)_3 \cdot (NH_4)_2SO_4 \cdot 24H_2O$ (iron ammonium alum)	482	4.39	0.108
$Gd_2(SO_4)_3 \cdot 8H_2O$ (gadolinium sulfate)	373	7.80	2.91
$2Ce(NO_3)_3 \cdot 3Mg(NO_3)_2 \cdot 24H_2O$ (cerium magnesium nitrate)	765	0.317	5.07×10^{-5}

* Data from M. W. Zemansky, *Heat and Thermodynamics*, 5th ed., McGraw-Hill, New York, 1968.

Note (1): A mass of crystal containing exactly N_A (Avogadro's number) magnetic ions is known as 1 gram-ion.

Note (2): To obtain the value of C_c in SI units, the tabulated value is multiplied by $(4\pi \times 10^{-3})$, and the unit of C_c then becomes $m^3 \cdot K/(kg\text{-ion})$.

For paramagnetic salts that conform to Curie's equation, there are two useful specific heat–temperature relations of the form

$$C_M = \frac{A_c}{T^2} \tag{13-62}$$

and

$$C_H = \frac{A_c}{T^2} + \mu_0 C_c \frac{H^2}{T^2} \tag{13-63}$$

where A_c is a constant, whose values are listed in Table 13-2.

13-8 MAGNETIC COOLING

The method of *magnetic cooling* is based on the fact that when a thermally isolated paramagnetic material is demagnetized, its temperature drops. This can be verified analytically from the second dS equation for a simple paramagnetic system, Eq. (13-60):

$$dS = \frac{C_H}{T} dT + \mu_0 V \left(\frac{\partial M}{\partial T}\right)_H dH$$

Because at constant magnetic field an increase in temperature of a paramagnetic solid always causes a decrease in its magnetization, the derivative $(\partial M/\partial T)_H$ is always negative. It follows from Eq. (13-60) that for a reversible adiabatic demagnetization process, we have

$$\frac{dT}{T} = -\frac{\mu_0 V}{C_H}\left(\frac{\partial M}{\partial T}\right)_H dH \tag{13-64}$$

indicating that a negative dH would give a negative dT, thus a drop in temperature. This ideal process is depicted by the vertical line 2–3 in the T–S diagram of Fig. 13-2.

The paramagnetic material must be magnetized before the demagnetization process. An ideal isothermal magnetization process is depicted by the horizontal line 1–2

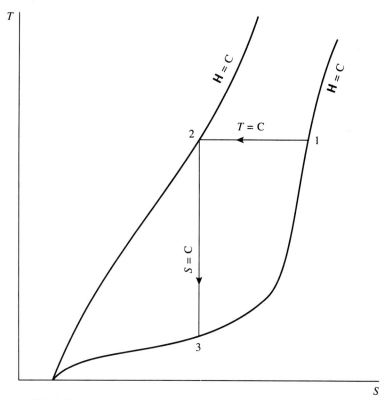

Figure 13-2 Temperature–entropy diagram of a paramagnetic salt showing reversible processes of isothermal magnetization and adiabatic demagnetization.

in Fig. 13-2. According to Eq. (13-60), for this process, we have

$$T \, dS = \mu_0 V T \left(\frac{\partial \mathbf{M}}{\partial T} \right)_{\mathbf{H}} d\mathbf{H} \tag{13-65}$$

Because $(\partial \mathbf{M}/\partial T)_{\mathbf{H}}$ is always negative, an increase in magnetic field results in an outflow of heat and a corresponding decrease in entropy. Because entropy is physically an index of disorder, it is apparent that an unmagnetized state in a paramagnetic material necessarily has a greater entropy or disorder than a magnetized state at the same temperature. The line of zero magnetic field must lie on the right of the line representing the constant magnetic field $\mathbf{H}$ on the T–S diagram. This confirms that heat must be rejected from the material while being isothermally magnetized.

In order to carry an experiment of magnetic cooling, a sample of paramagnetic salt crystals is cooled first to about 1 K. This is done by mounting the salt sample inside a chamber that is completely immersed in a bath of liquid helium that boils at reduced pressure, as shown diagrammatically in Fig. 13-3. Surrounding the liquid helium is liquid nitrogen, also boiling at reduced pressure. The intervening space between the two liquids is evacuated. Inside the sample chamber is some helium gas, called exchange gas, serving as a heat-exchanging agent. When the salt sample is cooled to a temperature as low as possible, a strong magnetic field is then applied either by a conventional electromagnet or a superconducting magnet. The heat evolved due to magne-

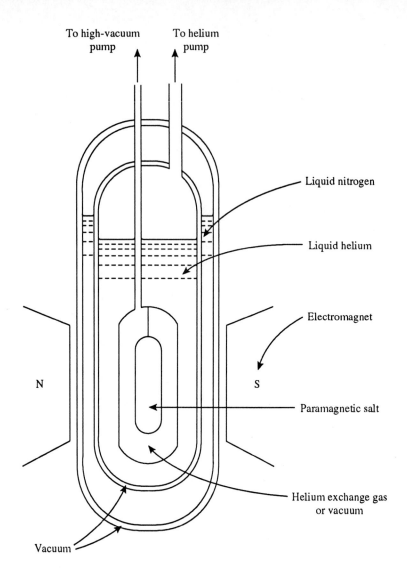

To high-vacuum
pump

To helium
pump

Liquid nitrogen

Liquid helium

Electromagnet

N

S

Paramagnetic salt

Helium exchange gas
or vacuum

Vacuum

Figure 13-3 Schematic diagram of a cryostat for ionic demagnetization.

tization is transported by the helium exchange gas to the surrounding liquid helium, causing some of the liquid to evaporate. The temperature of the salt sample is thus maintained essentially constant during magnetization. At the end of the magnetization process, the helium exchange gas in the sample chamber is pumped out, thereby breaking the thermal connection between the salt and its surroundings. The magnetic field is then reduced to zero and the temperature of the salt falls under adiabatic conditions. In such a cooling process, temperatures down to a few thousandths of a degree above absolute zero can be achieved without major difficulties.

Example 13-7

A gadolinium sulfate crystal at 1.5 K is magnetized reversibly and isothermally from zero field to 1×10^6 A/m. For two g-ions of the salt, what are the heat, work, and change in entropy for the process of magnetization? Assume that Curie's equation is obeyed. Compute the final temperature of the salt when it is then demagnetized reversibly and adiabatically.

Solution   From Eq. (13-60), we have

$$dS = \frac{C_\mathbf{H}}{T} dT + \mu_0 V \left(\frac{\partial \mathbf{M}}{\partial T}\right)_\mathbf{H} d\mathbf{H}$$

But according to the Curie's equation,

$$V\mathbf{M} = C_c \frac{\mathbf{H}}{T}$$

from which

$$V\left(\frac{\partial \mathbf{M}}{\partial T}\right)_\mathbf{H} = -C_c \frac{\mathbf{H}}{T^2}$$

Thus,

$$dS = \frac{C_\mathbf{H}}{T} dT - \mu_0 C_c \frac{\mathbf{H}}{T^2} d\mathbf{H} \qquad (13\text{-}66)$$

For the process of isothermal magnetization, the preceding equation reduces to

$$dS = -\mu_0 C_c \frac{\mathbf{H}}{T^2} d\mathbf{H}$$

which yields on integration (see Fig. 13-2),

$$\Delta S_{12} = S_2 - S_1 = -\mu_0 C_c (\mathbf{H}_2^2 - \mathbf{H}_1^2)/2T_1^2 \qquad (13\text{-}67)$$

Now in SI units, we have

$$\mu_0 = 4\pi \times 10^{-7} \text{ N/A}^2$$

$$C_c = \left(4\pi \times 7.80 \times 10^{-6} \frac{\text{m}^3 \cdot \text{K}}{\text{g-ion}}\right)(2 \text{ g-ions})$$

$$= 4\pi \times 15.6 \times 10^{-6} \text{ m}^3 \cdot \text{K}$$

$$\mathbf{H}_2 = 1 \times 10^6 \text{ A/m}$$

$$\mathbf{H}_1 = 0$$

and

$$T_1 = 1.5 \text{ K}$$

Upon substituting into Eq. (13-67), we obtain

$$\Delta S_{12} = S_2 - S_1$$

$$= -\frac{(4\pi \times 10^{-7})(4\pi \times 15.6 \times 10^{-6})}{2 \times 1.5^2}[(1 \times 10^6)^2 - 0]$$

$$= -54.7 \text{ J/K}$$

From which

$$Q_{12} = T_1(\Delta S_{12}) = 1.5(-54.7) = -82.1 \text{ J}$$

The minus sign indicates that heat is removed from the salt during the process of isothermal magnetization.

Now, because the internal energy is a function of temperature only for a paramagnetic salt obeying Curie's equation (see Problem 13-46), we have $\Delta U_{12} = 0$. Therefore, from the first law, we obtain

$$W_{12} = Q_{12} = -82.1 \text{ J}$$

Here the minus sign indicates that work is done on the salt during the process of isothermal magnetization.

For the process of reversible adiabatic demagnetization, Eq. (13-66) reduces to

$$C_\mathbf{H} \frac{dT}{T} - \mu_0 C_c \frac{\mathbf{H}}{T^2} d\mathbf{H} = 0 \qquad (13\text{-}68)$$

But from Eq. (13-63),

$$C_\mathbf{H} = \frac{1}{T^2}(A_c + \mu_0 C_c \mathbf{H}^2)$$

Thus, we have

$$(A_c + \mu_0 C_c \mathbf{H}^2)\frac{dT}{T} = \mu_0 C_c \mathbf{H} \, d\mathbf{H}$$

or

$$\frac{dT}{T} = \frac{\mu_0 C_c \mathbf{H} \, d\mathbf{H}}{A_c + \mu_0 C_c \mathbf{H}^2}$$

Integrating between states 2 and 3 yields

$$\ln \frac{T_3}{T_2} = \frac{1}{2} \ln \frac{A_c + \mu_0 C_c \mathbf{H}_3^2}{A_c + \mu_0 C_c \mathbf{H}_2^2}$$

Thus,

$$\frac{T_3}{T_2} = \left(\frac{A_c + \mu_0 C_c \mathbf{H}_3^2}{A_c + \mu_0 C_c \mathbf{H}_2^2} \right)^{1/2}$$

From Table 13-2, we have $A_c = 2.91$ J · K/g-ion for gadolinium sulfate. Therefore,

$$\frac{T_3}{T_2} = \left[\frac{2.91}{2.91 + (4\pi \times 10^{-7})(4\pi \times 7.80 \times 10^{-6})(1 \times 10^6)^2} \right]^{1/2}$$
$$= 0.152$$

Hence,

$$T_3 = 0.152 T_2 = 0.152 \times 1.5 = 0.23 \text{ K}$$

13-9 SUMMARY

For any process, reversible or irreversible, between equilibrium states of a simple compressible system of fixed mass, the four basic relations of properties in differential form are

$$dU = T \, dS - p \, dV$$
$$dH = T \, dS + V \, dp$$
$$dA = -S \, dT + p \, dV$$
$$dG = -S \, dT + V \, dp$$

The four Maxwell relations for simple compressible systems are

$$\left(\frac{\partial T}{\partial V}\right)_S = -\left(\frac{\partial p}{\partial S}\right)_V$$

$$\left(\frac{\partial T}{\partial p}\right)_S = \left(\frac{\partial V}{\partial S}\right)_p$$

$$\left(\frac{\partial S}{\partial V}\right)_T = \left(\frac{\partial p}{\partial T}\right)_V$$

$$\left(\frac{\partial S}{\partial p}\right)_T = -\left(\frac{\partial V}{\partial T}\right)_p$$

The coefficient of thermal expansion is defined as

$$\alpha = \frac{1}{V}\left(\frac{\partial V}{\partial T}\right)_p$$

Isothermal compressibility κ and adiabatic compressibility κ_S are defined as

$$\kappa = -\frac{1}{V}\left(\frac{\partial V}{\partial p}\right)_T$$

$$\kappa_S = -\frac{1}{V}\left(\frac{\partial V}{\partial p}\right)_S$$

Some important specific-heats relations for simple compressible systems are

$$c_v = \left(\frac{\partial u}{\partial T}\right)_v = T\left(\frac{\partial s}{\partial T}\right)_v$$

$$c_p = \left(\frac{\partial h}{\partial T}\right)_p = T\left(\frac{\partial s}{\partial T}\right)_p$$

$$c_p - c_v = -T\left(\frac{\partial v}{\partial T}\right)_p^2\left(\frac{\partial p}{\partial v}\right)_T$$

$$\left(\frac{\partial c_v}{\partial v}\right)_T = T\left(\frac{\partial^2 p}{\partial T^2}\right)_v$$

$$\left(\frac{\partial c_p}{\partial p}\right)_T = -T\left(\frac{\partial^2 v}{\partial T^2}\right)_p$$

$$k = \frac{c_p}{c_v} = \frac{\kappa}{\kappa_S}$$

The three ds equations are

$$ds = \frac{c_v}{T}\,dT + \left(\frac{\partial p}{\partial T}\right)_v\,dv$$

$$ds = \frac{c_p}{T}\,dT - \left(\frac{\partial v}{\partial T}\right)_p\,dp$$

$$ds = \frac{c_v}{T}\left(\frac{\partial T}{\partial p}\right)_v\,dp + \frac{c_p}{T}\left(\frac{\partial T}{\partial v}\right)_p\,dv$$

These ds equations are useful in the evaluation of entropy changes. Because $đq = T\,ds$ in an internally reversible process, these equations provide means to calculate heat transfers.

The three du equations are

$$du = c_v\,dT + \left[T\left(\frac{\partial p}{\partial T}\right)_v - p\right]dv$$

$$du = \left[c_p - p\left(\frac{\partial v}{\partial T}\right)_p\right]dT - \left[p\left(\frac{\partial v}{\partial p}\right)_T + T\left(\frac{\partial v}{\partial T}\right)_p\right]dp$$

$$du = c_v\left(\frac{\partial T}{\partial p}\right)_v dp + \left[c_p\left(\frac{\partial T}{\partial v}\right)_p - p\right]dv$$

Out of the preceding three du equations, the one with T and v as independent variables is most useful. The following energy equation is obtained from that equation:

$$\left(\frac{\partial u}{\partial v}\right)_T = T\left(\frac{\partial p}{\partial T}\right)_v - p$$

The three dh equations are

$$dh = \left[c_v + v\left(\frac{\partial p}{\partial T}\right)_v\right]dT + \left[T\left(\frac{\partial p}{\partial T}\right)_v + v\left(\frac{\partial p}{\partial v}\right)_T\right]dv$$

$$dh = c_p\,dT + \left[v - T\left(\frac{\partial v}{\partial T}\right)_p\right]dp$$

$$dh = \left[v + c_v\left(\frac{\partial T}{\partial p}\right)_v\right]dp + c_p\left(\frac{\partial T}{\partial v}\right)_p dv$$

Among the preceding three dh equations, the one with T and p as independent variables is most useful.

The Joule–Thomson coefficient is given by

$$\mu_J = \left(\frac{\partial T}{\partial p}\right)_h = \frac{1}{c_p}\left[T\left(\frac{\partial v}{\partial T}\right)_p - v\right]$$

The point at which $\mu_J = 0$ corresponds to the maximum of an isenthalpic curve on a T–p diagram and is called the inversion point. The locus of all the inversion points on a T–p diagram is called the inversion curve. From Fig. 13-1(b), the region inside the inversion curve where $\mu_J > 0$, the temperature decreases as the pressure decreases upon throttling—a cooling effect. In the region outside the inversion curve where $\mu_J < 0$, the temperature increases as the pressure decreases upon throttling—a heating effect. The temperature at which the inversion curve intersects the temperature axis in a T–p diagram is called the maximum inversion temperature.

The Clapeyron equation is

$$\frac{dp}{dT} = \frac{h^{(\beta)} - h^{(\alpha)}}{T[v^{(\beta)} - v^{(\alpha)}]}$$

where the superscripts (α) and (β) denote the two coexisting phases. This equation can be used in a variety of ways, including the prediction of the effect of pressure on transition temperatures and the computation of latent heats of phase transitions.

Thermodynamic Relations   Chap. 13

For a simple paramagnetic system, the four basic relations of properties are

$$dU = T\,dS + \mu_0 V\mathbf{H}\,d\mathbf{M}$$

$$dH = T\,dS - \mu_0 V\mathbf{M}\,d\mathbf{H}$$

$$dA = -S\,dT + \mu_0 V\mathbf{H}\,d\mathbf{M}$$

$$dG = -S\,dT - \mu_0 V\mathbf{M}\,d\mathbf{H}$$

where

$$H = U - \mu_0 V\mathbf{H}\mathbf{M}$$

$$A = U - TS$$

$$G = U - TS - \mu_0 V\mathbf{H}\mathbf{M}$$

The four Maxwell relations for a simple paramagnetic system are

$$\left(\frac{\partial T}{\partial \mathbf{M}}\right)_S = \mu_0 V\left(\frac{\partial \mathbf{H}}{\partial S}\right)_\mathbf{M}$$

$$\left(\frac{\partial T}{\partial \mathbf{H}}\right)_S = -\mu_0 V\left(\frac{\partial \mathbf{M}}{\partial S}\right)_\mathbf{H}$$

$$\left(\frac{\partial S}{\partial \mathbf{M}}\right)_T = -\mu_0 V\left(\frac{\partial \mathbf{H}}{\partial T}\right)_\mathbf{M}$$

$$\left(\frac{\partial S}{\partial \mathbf{H}}\right)_T = \mu_0 V\left(\frac{\partial \mathbf{M}}{\partial T}\right)_\mathbf{H}$$

For a simple paramagnetic system, the heat capacity at constant magnetic moment $C_\mathbf{M}$ and the heat capacity at constant magnetic field $C_\mathbf{H}$ are defined as

$$C_\mathbf{M} = \left(\frac{\partial U}{\partial T}\right)_\mathbf{M} = T\left(\frac{\partial S}{\partial T}\right)_\mathbf{M}$$

$$C_\mathbf{H} = \left(\frac{\partial H}{\partial T}\right)_\mathbf{H} = T\left(\frac{\partial S}{\partial T}\right)_\mathbf{H}$$

The first and second dS equations for a simple paramagnetic system are

$$dS = \frac{C_\mathbf{M}}{T}\,dT - \mu_0 V\left(\frac{\partial \mathbf{H}}{\partial T}\right)_\mathbf{M}\,d\mathbf{M}$$

$$dS = \frac{C_\mathbf{H}}{T}\,dT + \mu_0 V\left(\frac{\partial \mathbf{M}}{\partial T}\right)_\mathbf{H}\,d\mathbf{H}$$

Curie's equation is written as

$$V\mathbf{M} = C_c\frac{\mathbf{H}}{T}$$

where C_c is the Curie constant.

The method of magnetic cooling is based on the fact that when a thermally isolated paramagnetic material is demagnetized, its temperature drops. Temperatures down to a few thousandths of a degree above absolute zero can be achieved by magnetic cooling.

PROBLEMS

13-1. Determine whether the following differential expressions are exact or not. If they are exact, find the functions of which these expressions are the differentials.
(a) $y\,dx + x\,dy$
(b) $x\,dx + y\,dy$
(c) $-\dfrac{1}{x^2 y}\,dx + \dfrac{1}{xy^2}\,dy$
(d) $R(1 + B'p)\,dT + RTB'\,dp$ (where R and B' are constants)

13-2. The second $T\,ds$ equation is

$$T\,ds = c_p\,dT - Tv\alpha\,dp$$

where

$$\alpha = \frac{1}{v}\left(\frac{\partial v}{\partial T}\right)_p$$

Applying the condition of exactness to the $T\,ds$ equation, one finds

$$\left(\frac{\partial c_p}{\partial p}\right)_T = -\left[\frac{\partial(Tv\alpha)}{\partial T}\right]_p$$

For an ideal gas,

$$Tv\alpha = RT/p.$$

Therefore,

$$(\partial c_p/\partial p)_T = -R/p$$

But this result is false. Point out exactly the error that has been made and derive the correct expressions.

13-3. Using only the cyclic relation, Eq. (13-7), and the first Maxwell relation, Eq. (13-15a), derive the remaining three Maxwell relations, Eqs. (13-15b), (13-15c), and (13-15d), for a simple compressible system.

13-4. The p–v–T relation of a gas at low pressures is assumed to be

$$\frac{pv}{RT} = 1 + Bp$$

where B is a function of T only. Derive the expressions of $(\partial u/\partial p)_T$ and $(\partial c_p/\partial p)_T$ for the gas.

13-5. For a gas whose p–v–T relation can be expressed as

$$p(v - b) = RT$$

and its specific internal energy can be expressed as

$$u = c_v T + \text{constant}$$

where b and c_v are constants, show that the equation for an internally reversible adiabatic process is

$$p(v - b)^k = \text{constant}$$

where $k = c_p/c_v = a$ constant.

13-6. Find the coefficient of thermal expansion α and the isothermal compressibility κ based on each of the following p–v–T relations:
(a) Ideal gas equation,

$$p = RT/v$$

(b) Dieterici equation,

$$p = \frac{RT}{v - b}e^{-a/RTv}$$

(c) Saha–Bose equation,

$$p = -\frac{RT}{2b}e^{-a/RTv}\ln\left(\frac{v - 2b}{v}\right)$$

In these equations, R is the gas constant, and a and b are specific constants. Verify that for large values of T and v, all the expressions for α and κ go over into the corresponding expressions for an ideal gas.

13-7. Derive the expression for the difference of the isothermal and adiabatic compressibility $\kappa - \kappa_s$ in terms of T, v, c_p, and the coefficient of thermal expansion α.

13-8. (a) Derive the equation $dV/v = \alpha\,dT - \kappa\,dp$, where α is the coefficient of thermal expansion, and κ is the isothermal compressibility.
(b) A substance has the following expressions of α and κ:

$$\alpha = 3aT^3/V \qquad \text{and} \qquad \kappa = b/V$$

where a and b are constants. Find the equation of state.

13-9. For a simple compressible system show that

(a) $c_p - c_v = \left(\dfrac{\partial v}{\partial T}\right)_p\left[p + \left(\dfrac{\partial u}{\partial v}\right)_T\right]$

$\qquad = -\left(\dfrac{\partial p}{\partial T}\right)_v\left[p\left(\dfrac{\partial v}{\partial p}\right)_T + \left(\dfrac{\partial u}{\partial p}\right)_T\right]$

(b) $(c_p - c_v)\dfrac{\partial^2 T}{\partial p\,\partial v} + \left(\dfrac{\partial c_p}{\partial p}\right)_v\left(\dfrac{\partial T}{\partial v}\right)_p$

$\qquad - \left(\dfrac{\partial c_v}{\partial v}\right)_p\left(\dfrac{\partial T}{\partial p}\right)_v = 1$

13-10. Prove the following equations for a simple compressible system:

(a) $\left(\dfrac{\partial c_v}{\partial v}\right)_T = T\left[\dfrac{\partial}{\partial T}\left(\dfrac{\alpha}{\kappa}\right)\right]_v$

(b) $\left(\dfrac{\partial c_p}{\partial p}\right)_T = -T\left[\dfrac{\partial}{\partial T}(\alpha v)\right]_p$

(c) $c_p - c_v = \alpha^2 vT/\kappa$

where α and κ are the coefficient of thermal expansion and isothermal compressibility, respectively.

13-11. Derive the equations:

(a) $c_p = T\,(\partial v/\partial T)_p(\partial p/\partial T)_s$

(b) $\left(\dfrac{\partial p}{\partial T}\right)_s = \dfrac{c_p}{v\alpha T}$

(c) $\dfrac{(\partial p/\partial T)_s}{(\partial p/\partial T)_v} = \dfrac{k}{k-1}$

13-12. For a simple compressible system, show that

(a) $\left(\dfrac{\partial u}{\partial v}\right)_T = T^2\left[\dfrac{\partial(p/T)}{\partial T}\right]_v$

(b) $\left(\dfrac{\partial h}{\partial p}\right)_T = -T^2\left[\dfrac{\partial(v/T)}{\partial T}\right]_p$

13-13. Prove the following equations for a simple compressible system:

(a) $du = c_v\,dT + \left(\dfrac{\alpha T}{\kappa} - p\right)dv$

(b) $dh = c_p\,dT + (v - \alpha Tv)\,dp$

(c) $ds = \dfrac{\kappa c_v}{\alpha T}\,dp + \dfrac{c_p}{\alpha Tv}\,dv$

(d) $da = -(\alpha pv + s)\,dT + \kappa pv\,dp$

13-14. For a simple compressible system, show that

(a) $u = a - T\left(\dfrac{\partial a}{\partial T}\right)_v = -T^2\left[\dfrac{\partial(a/T)}{\partial T}\right]_v$

(b) $h = g - T\left(\dfrac{\partial g}{\partial T}\right)_p = -T^2\left[\dfrac{\partial(g/T)}{\partial T}\right]_p$

(c) $c_v = -T\,(\partial^2 a/\partial T^2)_v$

(d) $c_p = -T\,(\partial^2 g/\partial T^2)_p$

13-15. For an incompressible substance, show that the internal energy and entropy are functions of temperature only.

13-16. The relation between u, T, and v for a given simple compressible system is given by

$$u = A + BT + C/v$$

Find the corresponding p–v–T relation.

13-17. A p–V–T relation of helium gas is

$$pV = n\Re T + B(T)p$$

where $B(T)$ is a function of temperature only, n is the number of moles, and $\Re$ is the universal gas constant. Find (a) C_p as a function of p and T, and (b) H as a function of p and T. Given $c_p{}^*$ and H^* as the values of C_p and H at $p = 0$ and $T = T$.

13-18. The velocity of sound in a material is given by

$$V = (B_s/\rho)^{1/2}$$

where ρ is the density, and B_s is the adiabatic bulk modulus defined by the relation

$$B_s = -v(\partial p/\partial v)_s$$

Show that

$$V = (\partial p/\partial\rho)_s{}^{1/2}$$

13-19. The entropy of an ideal gas can be written as

$$s = c_v \ln p + c_p \ln v + \text{constant}$$

where c_v and c_p are constants whose ratio c_p/c_v is designated by k. Show that the adiabatic bulk modulus $B_s = -v(\partial p/\partial v)_s$ of an ideal gas is equal to kp.

13-20. Consider a gas for which $p(v - b) = RT$, where b and R are constants. Derive algebraic equations for internal energy, enthalpy, and entropy for the gas, assuming c_p and c_v to be constants. The properties of the gas at a reference state are p_0, T_0, u_0, h_0, and s_0.

13-21. For a gas that obeys the p–v–T relation

$$pv = \Re T + pA(T) + p^2 B(T)$$

where $A(T)$ and $B(T)$ are empirically determined functions of temperature. Show that

$$\left(\dfrac{\partial u}{\partial p}\right)_T = -T\dfrac{dA}{dT} - p\left(B + T\dfrac{dB}{dT}\right)$$

13-22. A hypothetical substance has the following isothermal compressibility κ, coefficient of thermal expansion α, and constant-volume specific heat c_v:

$$\kappa = a/v \qquad \alpha = 2bT/v \qquad \text{and} \qquad c_v = cTv$$

where a, b, and c are constants. Establish the p–v–T relation, the entropy and the internal energy functions, and the necessary relation between a, b, and c. The properties at a reference state are p_0, T_0, v_0, s_0, and u_0.

13-23. For an ideal gas, the constant-pressure specific heat c_p is a function of temperature T according to the equation

$$c_p = a + bT$$

where a and b are constants. (a) What is the expression for c_v of this gas? (b) Use these expressions for c_p and c_v to derive expressions for specific entropy s and specific enthalpy h of this gas at any pressure p and temperature T in terms of the specific entropy s_0 and specific enthalpy h_0 at some reference state with p_0 and T_0.

13-24. For an ideal gas, the ratio of specific heats k is observed to vary, over a certain range of temperatures, according to the equation

$$k = c_p/c_v = 1 + aT$$

where a is a constant. Find (a) the internal energy as a function of T, (b) the entropy as a function of T and v, and (c) the equation for an internally reversible adiabatic process. The properties of the gas at a reference state are p_0, T_0, v_0, u_0, and s_0.

13-25. The internal energy u of a gas is represented by the equation

$$u = aT - bp$$

where a and b are constants. The coefficient of thermal expansion α and isothermal compressibility κ are, respectively, $1/T$ and $1/p$. Find the specific heat at constant volume c_v in terms of a, b, p, and T.

13-26. The vapor pressure of many liquids can be represented by the equation

$$\ln p = B - A/T$$

where A and B are constants, and A/T is always greater than 5. (a) Assuming the saturated vapor to obey the ideal-gas equation, what is the equation of the vapor saturation curve on a p–v diagram? (b) Show that the slope of this curve is greater than that of an isotherm of an ideal gas, but less than that of an adiabatic-process line of an ideal gas whose value of $k = c_p/c_v$ is greater than 1.25.

13-27. Prove that on a Mollier diagram, (a) the slope of a curve representing an internally reversible isothermal process is $T - 1/\alpha$, and (b) the slope of a curve representing an internally reversible isometric process is $T + (k - 1)/\alpha$, where α is the coefficient of thermal expansion defined as $\alpha = (1/v)(\partial v/\partial T)_p$ and $k = c_p/c_v$.

13-28. (a) Show that all lines of constant pressure on a Mollier (h–s) chart must have the same slope at the same temperature, and these lines must be concave upward in the superheated region. (b) On what kind of chart would the slope of the lines of constant temperature be given by the pressure?

13-29. The steam tables published by the American Society of Mechanical Engineers (ASME) give the following data:

Saturation temperature (°F)	Saturation pressure (psia)	Specific-volume change, $v_{fg} = v_g - v_f$ (ft³/lbm)
201	11.766	32.979
200	11.526	33.622
199	11.290	34.280

Determine the approximate enthalpy of vaporization at 200°F, and compare with the tabulated value of 977.9 Btu/lbm.

13-30. The following data show the effect of pressure on the melting point of sodium:

p (atm)	T (°F)	$v_{if} \times 10^5$ (ft³/lbm)
5810	288.5	30.06
7740	310.6	27.46
9680	332.0	24.97

Determine the approximate enthalpy of fusion at 310.6°F, and compare with the experimental value of 5.16 Btu/lbm.

13-31. Determine the effect of pressure on the sublimation temperature of carbon dioxide at 1.328 atm. At this pressure, the saturation temperature is -75°C, the enthalpies of saturated solid and vapor are 37.391 and 173.16 kcal/kg, respectively, and the volumes of saturated solid and vapor are 0.643×10^{-3} and 269.68×10^{-3} m³/kg, respectively.

13-32. The liquid–vapor equilibrium curve for nitrogen over the range from the triple point to the normal boiling point can be expressed by the formula

$$\ln p = 2.303(A - BT - C/T)$$

where p is the vapor pressure in mm Hg, T is the temperature in K, and $A = 7.782$, $B = 0.006265$, and $C = 341.6$.
(a) Derive an expression for the enthalpy of vaporization h_{fg} in terms of the constants A, B, and C, the temperature T, and the change in specific volume between saturated liquid and saturated vapor v_{fg}.
(b) Calculate h_{fg} for nitrogen at 71.9 K. At

this temperature, v_{fg} is approximately 11,530 cm³/gmole.

13-33. The vapor pressure of water over a limited range of temperatures can be represented by the empirical equation

$$p = Ke^{(A+BT)/(C+DT)}$$

where K, A, B, C, and D are constants. If the specific volume of the liquid is negligible and the vapor can be considered as an ideal gas, find the form of the latent heat as a function of temperature that is implied by the equation.

13-34. Based on the triple point of water (0.01°C, 611.3 Pa) as the reference state, estimate the saturation pressure of ice in equilibrium with water vapor at $-20°C$ if $h_{ig} = 2834.8$ kJ/kg is a constant over this range.

13-35. The equation for the vapor-pressure curve of ammonia can be expressed as

$$\ln p = 2.303(A - B/T - DT + ET^2) - C \ln T$$

where $A = 25.57$, $B = 3.295 \times 10^3$, $C = 6.401$, $D = 4.148 \times 10^{-4}$, $E = 1.476 \times 10^{-6}$, p is the pressure in psia, and T is the temperature in °R.

(a) Derive an expression for h_{fg} in terms of the constants A, B, C, D, and E, the temperature T, and the change in specific volume between the saturated liquid and saturated vapor v_{fg}.

(b) Calculate the enthalpy of evaporation at 40°F for ammonia, given that $v_{fg} = 3.9457$ ft³/lbm.

13-36. For a single-component, two-phase system, show that the latent heat of transition L from phase (i) to phase (f) varies with temperature T according to the relation

$$\frac{dL}{dT} = c_p^{(f)} - c_p^{(i)} + \frac{L}{T} - L\left[\frac{v^{(f)}\alpha^{(f)} - v^{(i)}\alpha^{(i)}}{v^{(f)} - v^{(i)}}\right]$$

where c_p denotes the constant-pressure specific heat, v denotes the specific volume, and α denotes the coefficient of thermal expansion. Simplify the preceding relation if phase (i) is either a solid or a liquid and phase (f) is an ideal gas.

13-37. An equation of state for helium gas is given by

$$pv = \mathcal{R}T - \frac{ap}{T} + bp$$

where $a = 386.7$ cm³·K/gmole, $b = 15.29$ cm³/gmole, p is in atm, T is in K, and v is in cm³/gmole.

(a) Calculate the Joule–Thomson coefficient for helium at 150 K and 15 K.

(b) Find the inversion temperature in K.

(c) Estimate the temperature reached in an ideal throttling process from 24 atm and 15 K to 1 atm. For helium, $c_p = 2.06 \times 10^2$ cm³·atm/K·gmole.

13-38. The Joule–Thomson coefficient μ_J is a measure of the temperature change during a throttling process as defined by

$$\mu_J = (\partial T/\partial p)_h$$

A similar measure of the temperature change produced by an isentropic change of pressure is provided by the coefficient μ_s as defined by

$$\mu_s = (\partial T/\partial p)_s$$

Prove that

$$\mu_s - \mu_J = V/c_p$$

13-39. Estimate the Joule–Thomson coefficient for steam at 30 MPa and 550°C. Estimate the pressure drop required to cause a 50°C drop in temperature of steam when it is throttled from 30 MPa, 550°C, considering the Joule–Thomson coefficient to be a constant over the range.

13-40. Estimate the Joule–Thomson coefficient for Refrigerant-12 at 250 kPa and 60°C. When Refrigerant-12 is throttled from 250 kPa, 60°C, to 50 kPa, estimate the final temperature.

13-41. A certain gas obeys the following equation of state and specific heat c_p equation:

$$v = \frac{\mathcal{R}T}{p} + \alpha T^2$$

$$c_p = A + BT + Cp$$

where α, A, B, and C are constants. Derive the expressions for (a) the Joule–Thomson coefficient and (b) the specific heat c_v.

13-42. The p–v–T relationship of ammonia at low pressures can be represented by the equation

$$v = \frac{RT}{p} - 0.003 - \frac{0.34}{(0.01T)^3} - \frac{60}{(0.01T)^{11}}$$

where p is the pressure in atm, v is the specific volume in m³/kg, T is the temperature in K, and $R = 0.00483$ atm·m³/kg·K. The Joule–Thomson coefficient for ammonia at 400 K and 10 atm is 1.05 K/atm. Estimate the constant-pressure specific heat c_p of ammonia at 400 K and 10 atm.

13-43. The p–v–T relationship for a gas at moderate pressures can be written as

$$\frac{pv}{\mathcal{R}T} = 1 + B'p + C'p^2$$

where p is the pressure, v is the molar volume, T is the temperature, $\mathcal{R}$ is the universal gas constant, and B' and C' are functions of temperature only.

(a) Show that as the pressure approaches zero,

$$\mu_J c_p \longrightarrow \mathcal{R}T^2 \, (dB'/dT)$$

where μ_J is the Joule–Thomson coefficient, and c_p is the constant-pressure specific heat.

(b) Show that the equation of the inversion curve is

$$p = -\frac{dB'/dT}{dC'/dT}$$

13-44. Derive the following expression for Joule–Thomson coefficient:

$$\mu_J = \left(\frac{\partial T}{\partial p}\right)_h = \frac{v}{c_p}\left[\frac{v(\partial Z/\partial v)_T + T(\partial Z/\partial T)_v}{Z - v(\partial Z/\partial v)_T}\right]$$

where

$$Z = \frac{pv}{RT} = \text{compressibility factor}$$

13-45. The p–v–T relation for a substance can be given in the form

$$Z = \frac{pv}{\mathcal{R}T} = 1 + \frac{B}{v} + \frac{C}{v^2} + \cdots$$

where $\mathcal{R}$, B, and C are constants. Show that the locus of minimum points of Z–p isotherms for which $(\partial Z/\partial p)_T = 0$ corresponds to the inversion line of the Joule–Thomson coefficient for which $\mu_J = 0$.

13-46. For a paramagnetic solid that obeys Curie's law, show that its internal energy U and specific heat at constant magnetic moment C_M are functions of temperature only.

13-47. For a paramagnetic substance that obeys Curie's law, show that

$$S = \int C_M \frac{dT}{T} - \mu_0 \frac{I^2}{2C_c} + \text{constant}$$

where $I = V\mathbf{M}$.

13-48. For a paramagnetic substance that obeys Curie's law, show that

$$C_H - C_M = \mu_0 \frac{I^2}{C_c}$$

where $I = V\mathbf{M}$.

13-49. For a simple paramagnetic substance, starting with Eq. (13-50),

$$du = T \, ds + \mu_0 v \, \mathbf{H} \, d\mathbf{M}$$

show that the thermodynamic temperature is defined by

$$T = \left(\frac{\partial u}{\partial s}\right)_\mathbf{M}$$

13-50. For a simple paramagnetic system that obeys the Curie equation, show that for an internally reversible adiabatic process, $\mathbf{H}\mathbf{M}^{-\gamma} = \text{constant}$, where $\gamma = C_\mathbf{H}/C_\mathbf{M}$ is assumed to be a constant.

13-51. One kilogram of iron ammonium alum at 1.2 K is magnetized reversibily and isothermally from a zero field to 1.5×10^6 A/m. What are the heat, work, and change in entropy for the process of magnetization? Assume that Curie's law is valid. Calculate the final temperature of the salt when it is then demagnetized reversibly and adiabatically.

13-52. A crystal of 2×10^{-3} kg-ions of gadolinium sulfate is initially at 10 K and in a magnetic field of 1×10^6 A/m.

(a) When the field is reduced reversibly and isothermally to zero, calculate the heat transferred, the work done, and the entropy change.

(b) When the field is reduced reversibly and adiabatically to zero, calculate the final temperature reached. Assume that Curie's law is valid.

13-53. The $\mathbf{M}$–$\mathbf{H}$–T relationship for a paramagnetic solid can be expressed by the Curie–Weiss equation:

$$I = V\mathbf{M} = C_{cw}\frac{\mathbf{H}}{T - \theta}$$

where I is the total magnetic moment, and C_{cw} and θ are characteristic constants of the material. This equation is intended to apply for temperatures in excess of θ. For iron, $C_{cw} = 0.0155$ $m^3\cdot$K per kgmole and $\theta = 1093$ K. Iron at 1200 K is assumed to obey the Curie–Weiss equation. Determine **(a)** the entropy change when 1 kg of iron is placed in a magnetic field of 1×10^6 A/m at a constant temperature of 1200 K, and **(b)** the change in temperature when the iron is then adiabatically demagnetized, using $C_M = 22.2$ kJ/kgmole·K.

Real Gases and Real-Gas Mixtures

One of the chief uses of thermodynamics in engineering is in the analysis and prediction of the behavior of matter. As gases are the most often used working substances in thermodynamic systems, a knowledge of the p–v–T relationship and other physical behavior of gases and their mixtures is of prime importance. The ideal-gas equation of state, however, does not accurately portray the behavior of real gases. We are in need of a study of the behavior of real gases and real-gas mixtures. From a computational standpoint, it is convenient to have analytical expressions for the p–v–T behavior of real gases. We now explore a few real-gas equations of state and use these equations to evaluate other thermodynamic properties. Graphical correlations of thermodynamic properties are also treated.

14-1 VAN DER WAALS EQUATION OF STATE

In the study of a simple compressible system, the relation between the pressure p, the specific or molar volume v, and the temperature T is fundamental. In the case of an ideal gas, the equation of state is expressed by the simple relation, Eq. (4-1):

$$pv = \mathcal{R}T$$

This equation is essentially an experimental formulation; it can nevertheless be demonstrated theoretically on the basis of simple kinetic theory by making certain assumptions including these two major ones: the molecules are point masses and there are no intermolecular forces between molecules. These assumptions are reasonable for gases at vanishing pressures. However, as the pressure increases and the specific volume decreases, the volume occupied by the gas molecules themselves becomes increasingly important and intermolecular forces between molecules become increasingly significant. To take into account these effects, van der Waals in 1873 proposed an equation of state in accordance with the following argument.

Due to the finite size of the molecules, the space available for free motion of any molecule is decreased; consequently, the rate of molecular collisions and thereby the pressure are increased over the ideal-gas values. Let b denote the volume excluded from molecular motion for each mole of the gas having a molar volume v. The increased impact rate of the molecules in the restricted volume $v - b$ causes the kinetic pressure of the gas due to kinetic energy to increase to

$$\left(\frac{\mathcal{R}T}{v}\right)\left(\frac{v}{v-b}\right) = \frac{\mathcal{R}T}{v-b}$$

The cohesive force between molecules can be assumed to decrease rapidly with distance between them so that this force is appreciable only at small distances. A molecule far from the containing wall is attracted by its neighboring molecules from all directions with a balanced average net effect. However, for a molecule near the wall, while the force components paralled to the wall due to attractions from other molecules are balanced, there is an average net force normal to the wall, tending to pull it away from the wall, and the pressure on the wall is thereby reduced. The diminution of pressure should be proportional to the number of molecules being pulled and also proportional to the number of molecules that perform the pulling. Accordingly, the cohesive pressure on account of potential energy should be proportional to the square of the density of the gas and can be written as a/v^2, where a is a constant.

Subtracting the cohesive pressure from the kinetic pressure, we have for the net pressure

$$p = \frac{\mathcal{R}T}{v-b} - \frac{a}{v^2}$$

Rearranging,

$$\left(p + \frac{a}{v^2}\right)(v - b) = \mathcal{R}T \qquad (14\text{-}1)$$

This is the well-known *van der Waals equation of state,* in which p is the pressure, T the absolute temperature, v the molar volume, $\mathcal{R}$ the universal gas constant, and a and b are constants for the gas, in consistent units. It should be noted that the p–v–T relation given in Example 13-2 is the van der Waals equation.

The van der Waals equation is represented graphically as a family of isotherms on a p–v diagram in Fig. 14-1. At large volumes, the isotherms approximate rectangular hyperbolas in agreement with the ideal-gas equation. At small volumes, the gradients of the isotherms become very steep, indicating the behavior of incompressibility of a liquid. Because the equation is a cubic in v, it has three roots for any given values of p and T. At high temperatures, such as T_1, only one root is real for all pressures. At lower temperatures, such as T_2, all three roots are real over a certain pressure range. Thus, a maximum and a minimum are formed on the isotherm in the region where experimentally the phenomenon of condensation or vaporization takes place. For a fixed set of values of p and T corresponding to point c in Fig. 14-1, the three real roots become identical, and the isotherm labeled T_c has a point of inflection at c with a horizontal tangent. This point must be identified as the critical point and the isotherm labeled T_c is then the critical isotherm.

The van der Waals isotherms in the normally liquid–vapor two-phase region diverge from the stable paths of isobaric and isothermal phase changes. This divergence is, however, not entirely devoid in meaning. Thus in the isotherm labeled T_2, the portion AB represents a superheated liquid as it is at a temperature higher than the saturation temperature for the existing pressure. The portion FE represents a subcooled vapor as it is at a temperature lower than the saturation temperature for the existing pressure. It is possible to realize experimentally both these portions, but, as we will discuss in

Real Gases and Real-Gas Mixtures   Chap. 14

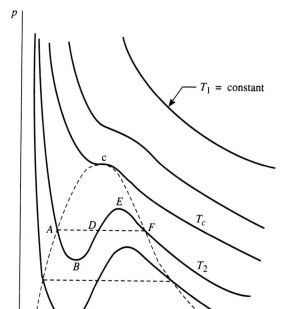

v **Figure 14-1**   van der Waals isotherms.

Sec. 16-3, they are states in so-called metastable equilibrium and can be easily disturbed by mechanical or other means to change suddenly to stable two-phase states. On the other hand, the portion BDE has a positive slope corresponding to negative compressibility, which is not physically possible.

The position of the isobaric–isothermal line AF in Fig. 14-1 that corresponds to states of stable equilibrium in which liquid and vapor coexist can be determined by the fact that the specific Gibbs functions of the two phases in equilibrium have the same value (see Eq. 13-13a). If A and F are saturated liquid and saturated vapor states, we must have

$$g_A = g_F$$

But for an isothermal process, Eq. (13-13a) reduces to

$$dg = v \, dp$$

Applying these relations to the isotherm $ABDEF$, we have

$$g_F - g_A = \int_A^F v \, dp = 0$$

which is readily seen to mean that the two areas enclosed between the van der Waals isotherm $ABDEF$ and the horizontal line AF are equal. The position of line AF is thus determined.

Because the critical isotherm has a point of inflection at the critical point, we have at this point

$$\left(\frac{\partial p}{\partial v}\right)_{T_c} = 0 \qquad \text{and} \qquad \left(\frac{\partial^2 p}{\partial v^2}\right)_{T_c} = 0$$

Sec. 14-1   van der Waals Equation of State

Differentiating the van der Waals equation and applying the results to the critical point lead to

$$\left(\frac{\partial p}{\partial v}\right)_{T_c} = -\frac{\Re T_c}{(v_c - b)^2} + \frac{2a}{v_c^3} = 0$$

and

$$\left(\frac{\partial^2 p}{\partial v^2}\right)_{T_c} = \frac{2\Re T_c}{(v_c - b)^3} - \frac{6a}{v_c^4} = 0$$

Solving the last two equations together with the van der Waals equation as applied to the critical point, we obtain the following expressions for the critical properties p_c, T_c, and v_c in terms of the van der Waals constants a and b:

$$p_c = \frac{a}{27b^2} \tag{14-2a}$$

$$T_c = \frac{8a}{27\Re b} \tag{14-2b}$$

$$v_c = 3b \tag{14-2c}$$

Conversely, the van der Waals constants a and b can be expressed in terms of the critical properties. Thus, the simultaneous solution of the foregoing expressions for p_c and T_c yields

$$a = \frac{27}{64}\frac{(\Re T_c)^2}{p_c} \tag{14-3a}$$

$$b = \frac{\Re T_c}{8p_c} \tag{14-3b}$$

It should be noted that the equation $v_c = 3b$ can be used to yield the value of b in terms of v_c. However, because the measurements of p_c and T_c are usually more reliable than that of v_c, the data on p_c and T_c are to be used to determine the values of a and b. Table A-30 in Appendix 3 gives the values of a and b for a number of gases as determined from Eqs. (14-3a) and (14-3b).

Despite its capability of describing the general behavior of real gases, the van der Waals equation holds only qualitatively. Any attempt to test the equation quantitatively is doomed to failure. One definite and clear discrepancy is seen in the value of the ratio $p_c v_c / \Re T_c$, which, according to van der Waals equation, is

$$\frac{p_c v_c}{\Re T_c} = \left(\frac{a}{27b^2}\right)(3b) \Big/ \Re\left(\frac{8a}{27\Re b}\right)$$

$$= \frac{3}{8} = 0.375$$

for all gases. But, as given in Table A-5 in Appendix 3, the actual values of $p_c v_c / \Re T_c$ are different for different gases, and for most gases, they are much smaller than the value predicted by the van der Waals equation. It is noted, however, that due to the simplified reasoning made by van der Waals, his equation is not expected to hold at high densities such as in the critical region. Testing of the equation for lower densities should give better correlation.

14-2 BEATTIE–BRIDGEMAN EQUATION OF STATE

Since the days of van der Waals, a large number of equations of state have been suggested to account for the behavior of real gases. Some are based on theoretical arguments; others are strictly empirical. In this section, we present the general reasoning of one of the best known equations, the Beattie–Bridgeman equation of state.*

The measured pressure of a gas can be written as the sum of the pressure due to kinetic energy and the pressure due to potential energy, or, symbolically,

$$p = p_k + p_p$$

where p_k is the kinetic pressure, and p_p is the cohesive pressure. For an ideal gas, p_k would be $\mathcal{R}T/v$. Due to the existence of intermolecular forces, some molecules that have just passed through a reference plane drawn anywhere in the midst of the gas will be reflected and pass through again, and thus introduce additional momentum transfer. But because the pressure acting across the reference plane can be defined as the net rate at which momentum normal to it is being transmitted across it per unit area in a chosen positive direction, the additional momentum transfer as a result of intermolecular forces would mean additional pressure. If the density of the gas is small so that the molecules act independently as reflectors, one can assume that

$$p_k = \frac{\mathcal{R}T}{v}\left(1 + \frac{B}{v}\right) = \frac{\mathcal{R}T}{v^2}(v + B)$$

where B is a constant. But if the density is such that the reflecting power per molecule is interfered with by those around it, B will not be constant but will be dependent on the density. The simplest form for this second-order correction is a linear function of the density. Accordingly,

$$B = B_0\left(1 - \frac{b}{v}\right)$$

then

$$p_k = \frac{\mathcal{R}T}{v^2}\left[v + B_0\left(1 - \frac{b}{v}\right)\right]$$

where B_0 and b are constants.

When slowly moving molecules collide, there is a tendency for them to move under the influence of each other; thus molecular association or aggregation is simulated. This has the same effect as a change in the average molecular weight of the gas and hence the gas constant may be considered to vary. Experimental data indicate that the variation in the number of independent aggregates due to the effect of the average time of encounter is directly proportional to the density and inversely proportional to the cu-

---

*J. A. Beattie and O. C. Bridgeman, *Proceedings of the American Academy of Arts and Sciences*, vol. 63, p. 229 (1928).

bic power of absolute temperature, so that the gas constant is modified to read

$$\mathscr{R}\left(1 - \frac{c}{vT^3}\right)$$

where $\mathscr{R}$ is the usual ideal-gas constant, and c is a constant. Whereupon

$$p_k = \frac{\mathscr{R}T\,(1 - c/vT^3)}{v^2}\left[v + B_0\left(1 - \frac{b}{v}\right)\right]$$

The cohesive pressure is $-A/v^2$ as in the van der Waals equation. However, A should not be a constant but varies with density. A simple linear variation is assumed for simplicity; thus,

$$A = A_0\left(1 - \frac{a}{v}\right)$$

where A_0 and a are constants. Consequently,

$$p_p = -\frac{A_0}{v^2}\left(1 - \frac{a}{v}\right)$$

The final form of the equation is then

$$p = \frac{\mathscr{R}T\,(1 - c/vT^3)}{v^2}\left[v + B_0\left(1 - \frac{b}{v}\right)\right] - \frac{A_0}{v^2}\left(1 - \frac{a}{v}\right) \qquad (14\text{-}4)$$

The preceding equation is the *Beattie–Bridgeman equation of state*. In addition to the gas constant $\mathscr{R}$, this equation has five specific constants, (A_0, a, B_0, b, and c) for a gas. These constants have been evaluated for a number of gases by curve fitting to the experimental data. Table A-31 in Appendix 3 gives the values of these constants for a number of gases.

The Beattie–Bridgeman equation was the best closed equation of state at the time it was proposed. In general, this equation is reasonably accurate when the volumes involved in the calculations are greater than twice the critical volume. In the critical region, it is inaccurate.

14-3 OTHER EQUATIONS OF STATE

Among the many equations of state proposed in the past, the *Benedict–Webb–Rubin equation*[*] is certainly one of the best. This equation is entirely empirical and was developed by fitting the experimental data of light hydrocarbons. It is written as follows:

$$p = \frac{\mathscr{R}T}{v} + \left(B_0\mathscr{R}T - A_0 - \frac{C_0}{T^2}\right)\frac{1}{v^2} + (b\mathscr{R}T - a)\frac{1}{v^3}$$
$$+ \frac{a\alpha}{v^6} + \frac{c\,(1 + \gamma/v^2)}{T^2}\frac{1}{v^3}e^{-\gamma/v^2} \qquad (14\text{-}5)$$

wherein the eight parameters B_0, A_0, C_0, b, a, c, α, and γ are numerical constants that are different for different gases. Their values for a number of gases are given in Table

---

[*] M. Benedict, G. B. Webb, and L. C. Rubin, *Journal of Chemical Physics*, vol. 8, p. 334 (1940).

A-32 in Appendix 3. These constants were obtained by curve fitting to the experimental p–v–T data. This equation is sufficiently accurate in describing the p–v–T relations of real gases up to densities of 1.8 times the critical density.

In general, the more specific constants used in an equation of state, the more accurate the equation can be. This explains why the Benedict–Webb–Rubin and other elaborate equations with many specific constants are so often used for calculations of real gas properties that require high precision.

However, this is not to infer that all simple equations are inaccurate. The types of gas and the ranges of pressure and temperature are factors to be considered in the selection of equations of state to use for any application. One of the simpler equations, namely, the *Redlich–Kwong equation** with only two specific constants has been found to have considerable accuracy over a wide range of p–v–T conditions. The Redlich–Kwong equation of state is as follows:

$$p = \frac{\mathscr{R}T}{v - b} - \frac{a}{T^{1/2}v(v + b)} \tag{14-6}$$

The constants a and b can be evaluated from critical data by the conditions

$$\left(\frac{\partial p}{\partial v}\right)_{T_c} = 0 \quad \text{and} \quad \left(\frac{\partial^2 p}{\partial v^2}\right)_{T_c} = 0$$

In terms of p_c and T_c, these constants are

$$a = \frac{0.42748\mathscr{R}^2 T_c^{2.5}}{p_c} \tag{14-7a}$$

$$b = \frac{0.08664\mathscr{R}T_c}{p_c} \tag{14.7b}$$

The numerical coefficients in Eqs. (14-7a) and (14-7b) are dimensionless, so they can be used with any consistent set of units. The Redlich–Kwong equation is said to furnish satisfactory results above the critical temperature for any pressure. The fair degree of accuracy combined with simplicity in form and in the evaluation of the constants make this equation attractive for industrial use.

Additional equations of state are presented along with the problems at the end of this chapter.

14-4 EQUATION OF STATE IN VIRIAL FORM

Kammerlingh Onnes in 1901 introduced the use of power series expansions to express the p–v–T relations of gases. The most usual form is an expansion of the product pv as a power series in density; thus,

$$pv = A\left(1 + \frac{B}{v} + \frac{C}{v^2} + \frac{D}{v^3} + \cdots\right)$$

We must have $A = \mathscr{R}T$ in the foregoing equation to satisfy the ideal gas law at zero density, so that we have

$$\frac{pv}{\mathscr{R}T} = 1 + \frac{B}{v} + \frac{C}{v^2} + \frac{D}{v^3} + \cdots \tag{14-8}$$

*O. Redlich and J. N. S. Kwong, *Chemical Reviews*, vol. 44, p. 233 (1949).

An equation of this type is known as the *virial equation of state,* and the coefficients B, C, D, etc., are called the second, third, fourth, etc., *virial coefficients.* These coefficients are functions of temperature, and also, of course, functions of the substance of interest.

At moderately low densities, only two-particle interactions are of significance in explaining the departure from ideality of the gas. This effect is expressed by the B/v term in the virial equation. As density is increased, three-particle interactions also become significant; this effect is expressed by the C/v^2 term in the equation. Similarly, higher virial terms are used when there are higher order molecular interactions. It should be noted that a true virial expansion applies to gases at low and medium densities only. As the series diverges at about the liquid density, it is unsuitable for high densities.

The virial equation can also be expressed in terms of a power series in pressure:

$$\frac{pv}{\mathscr{R}T} = 1 + B'p + C'p^2 + D'p^3 + \cdots \qquad (14\text{-}9)$$

where the coefficients B', C', D', etc., are functions of the temperature of the gas. The relationships between these coefficients and those of Eq. (14-8) are

$$B' = \frac{B}{\mathscr{R}T} \qquad C' = \frac{C - B^2}{(\mathscr{R}T)^2} \qquad D' = \frac{D - 3BC + 2B^3}{(\mathscr{R}T)^3}$$

Equation (14-9) converges less rapidly than does Eq. (14-8) and is useful mainly at low densities.

Because the virial equation is an infinite series, it has to be truncated before the determination of the virial coefficients is made. In general, the series is to be truncated at a point where the sum of higher-power terms is estimated to be within the experimental error over the full range studied. The coefficients of the truncated power series are then determined by a least-square or other type of curve fit of the experimental p–v–T data. In order to obtain an accurate second virial coefficient of a gas, the data used must extend to sufficiently low pressures. Note that the third and higher coefficients are extremely sensitive to the degree of the polynomial chosen, and care must be exercised in choosing the optimum density range from which to derive the higher virials.

When extremely accurate p–v–T data at very low densities are available, the second and third virials can be obtained by writing the virial equation in the form

$$\left(\frac{pv}{\mathscr{R}T} - 1\right) = \frac{B}{v} + \frac{C}{v^2} + \cdots$$

Thus, the second virial coefficient is given by

$$B = \lim_{p \to 0} \left(\frac{pv}{\mathscr{R}T} - 1\right)v$$

and the third virial coefficient is given by

$$C = \lim_{p \to 0} \left[\left(\frac{pv}{\mathscr{R}T} - 1\right)v - B\right]v$$

Any of the equations of state discussed in the preceding sections can be expanded into a virial form. The following is an example.

Example 14-1

Expand the van der Waals equation of state into a power series in density and in pressure.

Solution The van der Waals equation can be transformed to read

$$pv = \mathscr{R}T\left(1 - \frac{b}{v}\right)^{-1} - \frac{a}{v}$$

Use of the binomial theorem gives

$$pv = \mathscr{R}T\left(1 + \frac{b}{v} + \frac{b^2}{v^2} + \frac{b^3}{v^3} + \cdots\right) - \frac{a}{v}$$

or

$$\frac{pv}{\mathscr{R}T} = 1 + \frac{b - (a/RT)}{v} + \frac{b^2}{v^2} + \frac{b^3}{v^3} + \cdots \tag{14-10}$$

This is the van der Waals equation in virial form in terms of density. The virial coefficients are

$$B = b - \frac{a}{\mathscr{R}T}, \qquad C = b^2, \qquad D = b^3, \qquad \text{etc.}$$

When the equation is written in terms of a power series in pressure, we must have

$$B' = \frac{B}{\mathscr{R}T} = \frac{b}{\mathscr{R}T} - \frac{a}{\mathscr{R}^2 T^2}$$

$$C' = \frac{C - B^2}{\mathscr{R}^2 T^2} = \frac{2\,ab}{\mathscr{R}^3 T^3} - \frac{a^2}{\mathscr{R}^4 T^4}$$

$$D' = \frac{D - 3BC + 2B^3}{\mathscr{R}^3 T^3}$$

$$= -\frac{3ab^2}{\mathscr{R}^4 T^4} + \frac{6a^2 b}{\mathscr{R}^5 T^5} - \frac{2a^3}{\mathscr{R}^6 T^6}$$

Thus,

$$\frac{pv}{\mathscr{R}T} = 1 + \left(\frac{b}{\mathscr{R}T} - \frac{a}{\mathscr{R}^2 T^2}\right)p + \left(\frac{2ab}{\mathscr{R}^3 T^3} - \frac{a^2}{\mathscr{R}^4 T^4}\right)p^2$$

$$+ \left(-\frac{3ab^2}{\mathscr{R}^4 T^4} + \frac{6a^2 b}{\mathscr{R}^5 T^5} - \frac{2a^3}{\mathscr{R}^6 T^6}\right)p^3 + \cdots \tag{14-10a}$$

Because a and b are small corrections in the van der Waals equation, the terms involving the product of them (ab, a^2, etc.) in Eqs. (14-10) and (14-10a) can be neglected. Thus, we can write simply

$$\frac{pv}{\mathscr{R}T} = 1 + \left(b - \frac{a}{\mathscr{R}T}\right)\frac{1}{v}$$

and

$$\frac{pv}{\mathscr{R}T} = 1 + \left(\frac{b}{\mathscr{R}T} - \frac{a}{\mathscr{R}^2 T^2}\right)p$$

14-5 EVALUATION OF THERMODYNAMIC PROPERTIES FROM AN EQUATION OF STATE

An equation of state can be used not only for the calculation of pressure, volume, or temperature, but also the evaluation of other thermodynamic properties, such as internal energy, enthalpy, and entropy. The procedure calls for the integration of the differential relations for these properties with the aid of an equation of state and some supplementary data, such as specific heats or Joule–Thomson coefficients. For example, suppose that the change in enthalpy per unit mass of a gas from a reference state at p_0, T_0, to some other state at p, T, is to be calculated. We use the second dh equation, Eq. (13-40):

$$dh = c_p \, dT + \left[v - T\left(\frac{\partial v}{\partial T}\right)_p \right] dp$$

Because h is a property, dh is an exact differential; the line integral of dh is then a function of the end states only and independent of the path. Any process or combination of processes between the two end states can be chosen. The simple combinations are as shown in Fig. 14-2.

For the combination of processes as depicted by line 0–a–A in Fig. 14-2, Eq. (13-40) is integrated first at constant pressure p_0 from T_0 to T and then at constant temperature T from p_0 to p. The results are

$$h_a - h_0 = \left[\int_{T_0}^{T} c_p \, dT \right]_{p_0}$$

and

$$h - h_a = \left\{ \int_{p_0}^{p} \left[v - T\left(\frac{\partial v}{\partial T}\right)_p \right] dp \right\}_T$$

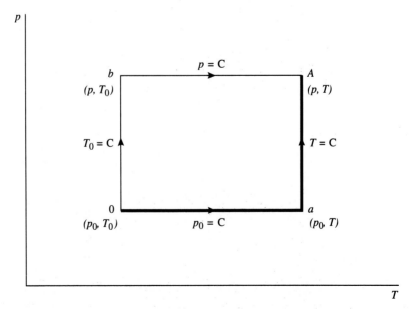

Figure 14-2 Two combinations of processes connecting states (p_0, T_0) and (p, T).

where h is the enthalpy at the final state (p, T), h_0 is the enthalpy at the initial state (p_0, T_0), and h_a is the enthalpy at the intermediate state (p_0, T). Combining these two equations yields

$$h - h_0 = \left[\int_{T_0}^{T} c_p \, dT \right]_{P0} + \left\{ \int_{P0}^{p} \left[v - T\left(\frac{\partial v}{\partial T}\right)_p \right] dp \right\}_T \tag{14-11}$$

For the combination of processes 0–b–A, Eq. (13-40) is integrated first at constant temperature T_0 from p_0 to p and then at constant pressure p from T_0 to T. This combination yields

$$h - h_0 = \left\{ \int_{P0}^{p} \left[v - T\left(\frac{\partial v}{\partial T}\right)_p \right] dp \right\}_{T_0} + \left[\int_{T_0}^{T} c_p \, dT \right]_p \tag{14-12}$$

Equation (14-11) requires data on c_p for all temperatures in the specified range at one pressure p_0, whereas Eq. (14-12) requires the c_p data at a higher pressure p. Because specific-heat measurements are relatively more convenient to make at low pressures, Eq. (14-11) is preferable.

If the equation of state is in the form $v = f(p, T)$, the second integrand in Eq. (14-11) can be integrated analytically, at least in principle, because the partial derivative $(\partial v/\partial T)_p$ can be obtained by differentiating the equation of state and the variable v can be replaced by substitution. Most equations of state, however, are in the form $p = f(T, v)$ as a consequence of kinetic-theory considerations. With this form of equation of state, the second integrand in Eq. (14-11) in the present form is usually difficult to handle. This integrand can be transformed as follows to facilitate easy integration. Because

$$d(pv) = p \, dv + v \, dp$$

we have

$$\left[\int_{P0}^{p} v \, dp \right]_T = (pv - p_0 v_a) - \left[\int_{v_a}^{v} p \, dv \right]_T$$

where v is the specific volume at p, T, and v_a is the specific volume at p_0, T. Also, because

$$\left(\frac{\partial v}{\partial T}\right)_p = -\left(\frac{\partial v}{\partial p}\right)_T \left(\frac{\partial p}{\partial T}\right)_v$$

we have

$$\left[\int_{P0}^{p} \left(\frac{\partial v}{\partial T}\right)_p dp \right]_T = -\left[\int_{v_a}^{v} \left(\frac{\partial p}{\partial T}\right)_v dv \right]_T$$

So that Eq. (14-11) becomes

$$h - h_0 = \left[\int_{T_0}^{T} c_p \, dT \right]_{P0} + (pv - p_0 v_a) - \left\{ \int_{v_a}^{v} \left[p - T\left(\frac{\partial p}{\partial T}\right)_v \right] dv \right\}_T \tag{14-13}$$

Let us now consider the integration of the second ds equation, Eq. (13-33):

$$ds = \frac{c_p}{T} \, dT - \left(\frac{\partial v}{\partial T}\right)_p dp$$

from the reference state (p_0, T_0) to the final state (p, T). Following the combination of processes $0-a-A$ in Fig. 14-2, we can write

$$s - s_0 = \left[\int_{T_0}^{T} \frac{c_p}{T} \, dT\right]_{p_0} - \left[\int_{p_0}^{p} \left(\frac{\partial v}{\partial T}\right)_p \, dp\right]_T$$

where s is the entropy at the final state (p, T), and s_0 is the entropy at the reference state (p_0, T_0). When the equation of state is in the form $p = f(T, v)$, the previous equation is best written as

$$s - s_0 = \left[\int_{T_0}^{T} \frac{c_p}{T} \, dT\right]_{p_0} + \left[\int_{v_a}^{v} \left(\frac{\partial p}{\partial T}\right)_v \, dv\right]_T \tag{14-14}$$

Example 14-2

By the use of the Benedict–Webb–Rubin equation of state, calculate values of v, h, s, c_v, and c_p for methane at $p = 200$ atm and $T = 400$ K. Given that $h_0 = 19260$ kJ/kgmole and $s_0 = 186.4$ kJ/kgmole·K at $p_0 = 1$ atm and $T_0 = 300$ K based on the value of zero for the perfect crystal at absolute zero temperature. The constant-pressure specific heat for methane at 1 atm can be expressed as

$$c_p = 4.186(4.750 + 12.00 \times 10^{-3}T + 3.030 \times 10^{-6}T^2 - 2.630 \times 10^{-9}T^3)$$

where c_p is in kJ/kgmole·K and T in K.

Solution The Benedict–Webb–Rubin equation is

$$p = \frac{\mathscr{R}T}{v} + \left(B_0\mathscr{R}T - A_0 - \frac{C_0}{T^2}\right)\frac{1}{v^2} + (b\mathscr{R}T - a)\frac{1}{v^3} + \frac{a\alpha}{v^6} + \frac{c(1 + \gamma/v^2)}{T^2}\frac{1}{v^3}e^{-\gamma/v^2}$$

where p is in atm, T in K, v in m³/kgmole, $\mathscr{R} = 0.08207$ atm·m³/kgmole·K, and for methane, the eight specific constants are

$$B_0 = 0.0426000, \ A_0 = 1.85500, \ C_0 = 0.0225700 \times 10^6$$

$$b = 0.00338004, \ a = 0.0494000, \ c = 0.00254500 \times 10^6$$

$$\alpha = 0.124359 \times 10^{-3}, \ \gamma = 0.60000 \times 10^{-2}$$

By differentiation of the Benedict–Webb–Rubin equation, we obtain

$$\left(\frac{\partial p}{\partial T}\right)_v = \frac{\mathscr{R}}{v} + \left(B_0\mathscr{R} + \frac{2C_0}{T^3}\right)\frac{1}{v^2} + \frac{b\mathscr{R}}{v^3} - \frac{2c}{T^3}\frac{1}{v^3}\left(1 + \frac{\gamma}{v^2}\right)e^{-\gamma/v^2} \tag{14-15a}$$

$$\left(\frac{\partial^2 p}{\partial T^2}\right)_v = -\frac{6C_0}{T^4}\frac{1}{v^2} + \frac{6c}{T^4}\frac{1}{v^3}\left(1 + \frac{\gamma}{v^2}\right)e^{-\gamma/v^2} \tag{14-15b}$$

$$\left(\frac{\partial p}{\partial v}\right)_T = -\frac{\mathscr{R}T}{v^2} - 2\left(B_0\mathscr{R}T - A_0 - \frac{C_0}{T^2}\right)\frac{1}{v^3}$$

$$- 3(b\mathscr{R}T - a)\frac{1}{v^4} - \frac{6a\alpha}{v^7} \tag{14-15c}$$

$$+ \frac{c}{T^2v^4}e^{-\gamma/v^2}\left(-3 - \frac{3\gamma}{v^2} + \frac{2\gamma^2}{v^4}\right)$$

The state points for this example are shown in Fig. 14-3. The volume of the gas at point A, where $p = 200$ atm and $T = 400°$K as determined from the Benedict–Webb–

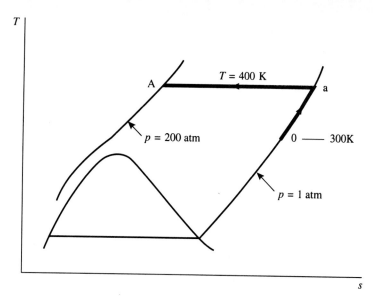

Figure 14-3 T–s diagram showing the state points for Example 14-2.

Rubin equation by a trial process, is

$$v = 0.161 \text{ m}^3/\text{kgmole}$$

Because $p_a = p_0 = 1$ atm, the volume of the gas at point a can be obtained with sufficient accuracy by the use of the ideal-gas equation. Thus,

$$v_a = \frac{RT}{p_0} = \frac{0.08207 \times 400}{1} = 32.8 \text{ m}^3/\text{kgmole}$$

To evaluate the value of enthalpy, we now calculate each term of Eq. (14-13).

$$\int_{T_0}^{T} c_p \, dT = 4.186 \int_{T_0}^{T} (4.75 + 12.0 \times 10^{-3}T + 3.03 \times 10^{-6}T^2 - 2.63 \times 10^{-9}T^3) \, dT$$

$$= 4.186[4.75(T - T_0) + 6.0 \times 10^{-3}(T^2 - T_0^2)$$

$$+ 1.01 \times 10^{-6}(T^3 - T_0^3) - 0.6575 \times 10^{-9}(T^4 - T_0^4)]$$

$$= 3855 \text{ kJ/kgmole}$$

after substituting numerical values of T and T_0.

$$(pv - p_0 v_a) = 200 \times 0.161 - 1 \times 32.8$$

$$= -0.6 \text{ atm·m}^3/\text{kgmole}$$

$$= -60.8 \text{ kJ/kgmole}$$

$$\left[\int_{v_a}^{v} p \, dv\right]_T = \left\{\int_{v_a}^{v} \left[\frac{\mathcal{R}T}{v} + \left(B_0 \mathcal{R}T - A_0 - \frac{C_0}{T^2}\right)\frac{1}{v^2} + (b\mathcal{R}T - a)\frac{1}{v^3} + \frac{a\alpha}{v^6}\right.\right.$$

$$\left.\left. + \frac{c}{T^2}\frac{1}{v^3}e^{-\gamma/v^2} + \frac{c}{T^2}\frac{\gamma}{V^5}e^{-\gamma/v^2}\right] dv\right\}_T$$

wherein

$$\int \frac{1}{v^3} e^{-\gamma/v^2}\, dv = \frac{1}{2\gamma} \int e^{-\gamma/v^2}\, d\left(-\frac{\gamma}{v^2}\right) = \frac{1}{2\gamma} e^{-\gamma/v^2}$$

and

$$\int \frac{\gamma}{v^5} e^{-\gamma/v^2}\, dv = \frac{1}{2} e^{-\gamma/v^2}\left(\frac{1}{v^2} + \frac{1}{\gamma}\right)$$

The last integration is done by the method of integration by parts. Hence we have

$$\left[\int_{v_a}^{v} p\, dv\right]_T = \mathcal{R}T \ln \frac{v}{v_a} - \left(B_0\mathcal{R}T - A_0 - \frac{C_0}{T^2}\right)\left(\frac{1}{v} - \frac{1}{v_a}\right)$$
$$- \frac{1}{2}(b\mathcal{R}T - a)\left(\frac{1}{v^2} - \frac{1}{v_a^2}\right) - \frac{a\alpha}{5}\left(\frac{1}{v^5} - \frac{1}{v_a^5}\right)$$
$$+ \frac{c}{2\gamma T^2}(e^{-\gamma/v^2} - e^{-r/v_a^2})$$
$$+ \frac{c}{2T^2}\left[e^{-\gamma/v^2}\left(\frac{1}{v^2} + \frac{1}{\gamma}\right) - e^{-\gamma/v_a^2}\left(\frac{1}{v_a^2} + \frac{1}{\gamma}\right)\right]$$

Also, we have

$$\left[\int_{v_a}^{v} T\left(\frac{\partial p}{\partial T}\right)_v dv\right]_T = \left\{\int_{v_a}^{v}\left[\frac{\mathcal{R}T}{v} + \left(B_0\mathcal{R}T + \frac{2C_0}{T^2}\right)\frac{1}{v^2} + \frac{b\mathcal{R}T}{v^3}\right.\right.$$
$$\left.\left. - \frac{2c}{T^2}\frac{1}{v^3}e^{-\gamma/v^2} - \frac{2c}{T^2}\frac{\gamma}{v^5}e^{-\gamma/v^2}\right] dv\right\}_T$$
$$= \mathcal{R}T \ln \frac{v}{v_a} - \left(B_0\mathcal{R}T + \frac{2C_0}{T^2}\right)\left(\frac{1}{v} - \frac{1}{v_a}\right)$$
$$- \frac{1}{2}b\mathcal{R}T\left(\frac{1}{v^2} - \frac{1}{v_a^2}\right) - \frac{c}{\gamma T^2}(e^{-\gamma/v^2} - e^{-\gamma/v_a^2})$$
$$- \frac{c}{T^2}\left[e^{-\gamma/v^2}\left(\frac{1}{v^2} + \frac{1}{\gamma}\right) - e^{-\gamma/v_a^2}\left(\frac{1}{v_a^2} + \frac{1}{\gamma}\right)\right]$$

Combining of the last two equations yields

$$\left\{\int_{v_a}^{v}\left[p - T\left(\frac{\partial p}{\partial T}\right)_v\right] dv\right\}_T = \left(A_0 + \frac{3C_0}{T^2}\right)\left(\frac{1}{v} - \frac{1}{v_a}\right) + \frac{a}{2}\left(\frac{1}{v^2} - \frac{1}{v_a^2}\right)$$
$$- \frac{a\alpha}{5}\left(\frac{1}{v^5} - \frac{1}{v_a^5}\right) + \frac{3}{2}\frac{c}{\gamma T^2}(e^{-\gamma/v^2} - e^{-\gamma/v_a^2})$$
$$+ \frac{3}{2}\frac{c}{T^2}\left[e^{-\gamma/v^2}\left(\frac{1}{v^2} + \frac{1}{\gamma}\right) - e^{-\gamma/v_a^2}\left(\frac{1}{v_a^2} + \frac{1}{\gamma}\right)\right]$$

where,

$$\left(\frac{1}{v} - \frac{1}{v_a}\right) = \frac{1}{0.161} - \frac{1}{32.8} = 6.18$$

$$\left(\frac{1}{v^2} - \frac{1}{v_a^2}\right) = \frac{1}{(0.161)^2} = \frac{1}{(32.8)^2} = 38.58$$

$$\left(\frac{1}{v^5} - \frac{1}{v_a^5}\right) = \frac{1}{(0.161)^5} - \frac{1}{(32.8)^5} = 9244$$

$$e^{-\gamma/v^2} - e^{-\gamma/v_a^2} = e^{-0.006/(0.161)^2} - e^{-0.006/(32.8)^2}$$

$$= 0.793 - 1 = -0.207$$

and

$$e^{-\gamma/v^2}\left(\frac{1}{v^2} + \frac{1}{\gamma}\right) - e^{-\gamma/v_a^2}\left(\frac{1}{v_a^2} + \frac{1}{\gamma}\right) = 0.793\left[\frac{1}{(0.161)^2} + \frac{1}{0.006}\right] - 1\left[\frac{1}{(32.8)^2} + \frac{1}{0.006}\right]$$

$$= -3.91$$

Thus, substituting the numerical values of the specific constants and temperature gives

$$\left\{\int_{v_a}^{v}\left[p - T\left(\frac{\partial p}{\partial T}\right)_v\right]dv\right\}_T = 14.10 \text{ atm·m}^3/\text{kgmole}$$

$$= 1429 \text{ kJ/kgmole}$$

Then, upon substituting in Eq. (14-13),

$$h - h_0 = 3855 - 60.8 - 1429 = 2365 \text{ kJ/kgmole}$$

Hence, the value of enthalpy at 200 atm and 400 K is

$$h = 19,260 + 2365 = 21,630 \text{ kJ/kgmole}$$

Now we proceed to evaluate the entropy by calculating each term of Eq. (14-14).

$$\int_{T_0}^{T}\frac{c_p}{T}dT = 4.186\int_{T_0}^{T}\left(\frac{4.75}{T} + 12.0 \times 10^{-3} + 3.03 \times 10^{-6}T\right.$$

$$\left. - 2.63 \times 10^{-9}T^2\right)dT$$

$$= 4.186\left[4.75\ln\frac{T}{T_0} + 12.0 \times 10^{-3}(T - T_0)\right.$$

$$+ 1.515 \times 10^{-6}(T^2 - T_0^2)$$

$$\left. - 0.8767 \times 10^{-9}(T^3 - T_0^3)\right]$$

$$= 11.1 \text{ kJ/kgmole·K}$$

after substituting numerical values of T and T_0.

$$\left[\int_{v_a}^{v}\left(\frac{\partial p}{\partial T}\right)_v dv\right]_T = \left\{\int_{v_a}^{v}\left[\frac{\mathcal{R}}{v} + \left(B_0\mathcal{R} + \frac{2C_0}{T^3}\right)\frac{1}{v^2} + \frac{b\mathcal{R}}{v^3}\right.\right.$$

$$\left.\left. - \frac{2c}{T^3}\frac{1}{v^3}e^{-\gamma/v^2} - \frac{2c}{T^3}\frac{\gamma}{v^5}e^{-\gamma/v^2}\right]dv\right\}_T$$

$$= \mathcal{R}\ln\frac{v}{v_a} - \left(B_0\mathcal{R} + \frac{2C_0}{T^3}\right)\left(\frac{1}{v} - \frac{1}{v_a}\right)$$

$$- \frac{1}{2}b\mathcal{R}\left(\frac{1}{v^2} - \frac{1}{v_a^2}\right) - \frac{c}{\gamma T^3}(e^{-\gamma/v^2} - e^{-\gamma/v_a^2})$$

$$- \frac{c}{T^3}\left[e^{-\gamma/v^2}\left(\frac{1}{v^2} + \frac{1}{\gamma}\right) - e^{-\gamma/v_a^2}\left(\frac{1}{v_a^2} + \frac{1}{\gamma}\right)\right]$$

$$= -0.466 \text{ atm·m}^3/\text{kgmole·K}$$

$$= -47.2 \text{ kJ/kgmole·K}$$

after substituting numerical values. Then upon substituting in Eq. (14-14),

$$s - s_0 = 11.1 - 47.2 = -36.1 \text{ kJ/kgmole·K}$$

Hence, the value of entropy at 200 atm and 400 K is

$$s = 186.4 - 36.1 = 150.3 \text{ kJ/kgmole·K}$$

The value of c_p at 1 atm and 400 K is

$$c_{pa} = 4.186[4.75 + 12.0 \times 10^{-3} \times 400$$
$$+ 3.03 \times 10^{-6} \times (400)^2 - 2.63 \times 10^{-9} \times (400)^3]$$
$$= 41.3 \text{ kJ/kgmole·K}$$

Because at 1 atm, the gas can be treated as an ideal gas, it follows that

$$c_{va} = c_{pa} - \mathscr{R} = 41.3 - 0.08207(101.325)$$
$$= 33.0 \text{ kJ/kgmole·K}$$

To calculate the value of c_v at 200 atm and 400 K, we use Eq. (13-26):

$$\left(\frac{\partial c_v}{\partial v}\right)_T = T\left(\frac{\partial^2 p}{\partial T^2}\right)_v$$

Integrating gives

$$c_v - c_{va} = \left[\int_{v_a}^{v} T\left(\frac{\partial^2 p}{\partial T^2}\right)_v dv\right]_T$$

$$= \left[\int_{v_a}^{v}\left(-\frac{6C_0}{T^3}\frac{1}{v^2} + \frac{6c}{T^3}\frac{1}{v^3}e^{-\gamma/v^2}\right.\right.$$
$$\left.\left. + \frac{6c}{T^3}\frac{\gamma}{v^5}e^{-\gamma/v^2}\right) dv\right]_T$$

$$= \frac{6}{T^3}\left\{C_0\left(\frac{1}{v} - \frac{1}{v_a}\right) + \frac{c}{2\gamma}(e^{-\gamma/v^2} - e^{-\gamma/v_a^2})\right.$$
$$\left. + \frac{c}{2}\left[e^{-\gamma/v^2}\left(\frac{1}{v^2} + \frac{1}{\gamma}\right) - e^{-\gamma/v_a^2}\left(\frac{1}{v_a^2} + \frac{1}{\gamma}\right)\right]\right\}$$

$$= 0.00849 \text{ atm·m}^3/\text{kgmole·K}$$

$$= 0.860 \text{ kJ/kgmole·K}$$

after substituting numerical values. Hence, the value of c_v at 200 atm and 400 K is

$$c_v = 33.0 + 0.860 = 33.9 \text{ kJ/kmole·K}$$

With the value of c_v at 200 atm and 400 K known, we can now calculate the corresponding value of c_p by the use of Eq. (13-30):

$$c_p - c_v = T\left(\frac{\partial p}{\partial T}\right)_v\left(\frac{\partial v}{\partial T}\right)_p$$

which, upon the use of the cyclic relation, can be transformed to read

$$c_p - c_v = -T\frac{(\partial p/\partial T)_v^2}{(\partial p/\partial v)_T}$$

At 200 atm, 400 K, the volume of the gas is 0.161 m³/kgmole. Substituting these values and the specific constants into Eqs. (14-15a) and (14-15c), we obtain

$$\left(\frac{\partial p}{\partial T}\right)_v = 0.719 \text{ atm/K}$$

and

$$\left(\frac{\partial P}{\partial v}\right)_T = -1330 \text{ atm·kgmole/m}^3$$

With these data,

$$c_p - c_v = (-400)\frac{(0.719)^2}{-1330}$$

$$= 0.155 \text{ atm·m}^3/\text{kgmole·K}$$

$$= 15.7 \text{ kJ/kgmole·K}$$

Hence, the value of c_p at 200 atm and 400°K is

$$c_p = c_v + 15.7 = 33.9 + 15.7$$

$$= 49.6 \text{ kJ/kgmole·K}$$

The results of the preceding calculations are summarized and compared with experimental data as follows:

p = 299 atm T = 400 K	$v,$ $\frac{m^3}{kgmole}$	$h,$ $\frac{kJ}{kgmole}$	$s,$ $\frac{kJ}{kgmole \cdot K}$	$c_v,$ $\frac{kJ}{kgmole \cdot K}$	$c_p,$ $\frac{kJ}{kgmole \cdot K}$
Calculated	0.161	21630	150.3	33.9	49.6
Experimental data*	0.1608	21510	149.9	33.2	49.2

*F. Din, *Thermodynamic Functions of Gases*, Vol. 3, Butterworths, London, 1961.

Example 14-3

Argon at 1 atm and 300 K enters a compressor at a rate of 100 kg/h. It is compressed reversibly and isothermally to 500 atm. Calculate the power needed to run the compressor and the amount of heat that must be removed per hour from the compressor. The gas is assumed to obey the Redlich–Kwong equation of state.

Solution Let us denote the initial and final states of the gas by the subscripts 1 and 2, respectively. Thus,

$$p_1 = 1 \text{ atm}, \qquad p_2 = 500 \text{ atm}, \qquad T_1 = T_2 = 300 \text{ K}$$

The Redlich–Kwong equation is

$$p = \frac{\mathcal{R}T}{v - b} - \frac{a}{T^{1/2}v(v + b)}$$

where

$$a = 0.42748 \frac{\mathscr{R}^2 T_c^{2.5}}{p_c}$$

$$b = 0.08664 \frac{\mathscr{R} T_c}{p_c}$$

For argon, the critical constants are

$$T_c = 151 \text{ K} \quad \text{and} \quad p_c = 48 \text{ atm}$$

Hence, for argon,

$$a = 0.42748 \frac{(0.08206)^2 (151)^{2.5}}{48}$$

$$= 16.8 \frac{(\text{atm})(\text{K}^{1/2})(\text{m}^6)}{(\text{kgmole})^2}$$

$$b = 0.08664 \frac{(0.08206)(151)}{48} = 0.0224 \text{ m}^3/\text{kgmole}$$

Substituting the numerical values of p_2, T_2, a, b, and $\mathscr{R}$ into the Redlich–Kwong equation leads to

$$v_2^3 - 49.24 v_2^2 + 335.6 v_2 - 43440 = 0$$

from which we obtain

$$v_2 = 0.0568 \text{ m}^3/\text{kgmole}$$

Because $p_1 = 1$ atm, the volume of the gas at the initial state can be obtained with sufficient accuracy by the use of the ideal-gas equation. Thus,

$$v_1 = \frac{\mathscr{R} T_1}{p_1} = \frac{0.08206 \times 300}{1} = 24.6 \text{ m}^3/\text{kgmole}$$

In order to calculate the work done and the heat transfer, first we need to calculate the changes in enthalpy and entropy between the initial and final states. From the second dh equation, Eq. (13-40), for the isothermal compression process, we have

$$\Delta h = h_2 - h_1 = \left\{ \int_{p_1}^{p_2} \left[v - T \left(\frac{\partial v}{\partial T} \right)_p \right] dp \right\}_T$$

Now because

$$d(pv) = p \, dv + v \, dp$$

we have

$$\left[\int_{p_1}^{p_2} v \, dp \right]_T = p_2 v_2 - p_1 v_1 - \left[\int_{v_1}^{v_2} p \, dv \right]_T$$

And because

$$\left(\frac{\partial v}{\partial T} \right)_p = - \left(\frac{\partial v}{\partial p} \right)_T \left(\frac{\partial p}{\partial T} \right)_v$$

we have

$$\left[\int_{p_1}^{p_2} \left(\frac{\partial v}{\partial T} \right)_p dp \right]_T = - \left[\int_{v_1}^{v_2} \left(\frac{\partial p}{\partial T} \right)_v dv \right]_T$$

Hence,

$$\Delta h = h_2 - h_1$$

$$= (p_2 v_2 - p_1 v_1) - \left\{ \int_{v_1}^{v_2} \left[p - T \left(\frac{\partial p}{\partial T} \right)_v \right] dv \right\}_T$$

But according to the Redlich–Kwong equation, we have

$$\left(\frac{\partial p}{\partial T} \right)_v = \frac{\mathcal{R}}{v - b} + \frac{a}{2T^{3/2} v (v + b)}$$

Thus,

$$\Delta h = h_2 - h_1$$

$$= (p_2 v_2 - p_1 v_1) - \left\{ \int_{v_1}^{v_2} \left[\frac{\mathcal{R}T}{v - b} - \frac{a}{T^{1/2} v (v + b)} \right. \right.$$

$$\left. \left. - \frac{\mathcal{R}T}{v - b} - \frac{a}{2T^{1/2} v (v + b)} \right] dv \right\}_T$$

$$= (p_2 v_2 - p_1 v_1) - \left\{ \int_{v_1}^{v_2} \left[\frac{-3a}{2T^{1/2} v (v + b)} \right] dv \right\}_T$$

$$= (p_2 v_2 - p_1 v_1) - \frac{1.5a}{T_1^{1/2}} \frac{1}{b} \ln \left[\left(\frac{v_2 + b}{v_2} \right) \middle/ \left(\frac{v_1 + b}{v_1} \right) \right]$$

$$= -17.7 \text{ atm·m}^3/\text{kgmole}$$

$$= -1790 \text{ kJ/kgmole}$$

after substituting numerical values.

We now use the second ds equation, Eq. (13-33), to calculate the entropy change. For the isothermal compression process, we have

$$\Delta s = s_2 - s_1$$

$$= -\left[\int_{p_1}^{p_2} \left(\frac{\partial v}{\partial T} \right)_p dp \right]_T = \left[\int_{v_1}^{v_2} \left(\frac{\partial p}{\partial T} \right)_v dv \right]_T$$

For the Redlich–Kwong equation, this becomes

$$\Delta s = s_2 - s_1$$

$$= \left\{ \int_{v_1}^{v_2} \left[\frac{\mathcal{R}}{v - b} + \frac{a}{2T^{3/2} v (v + b)} \right] dv \right\}_T$$

$$= \mathcal{R} \ln \left(\frac{v_2 - b}{v_1 - b} \right) - \frac{a}{2bT_1^{3/2}} \ln \left[\left(\frac{v_2 + b}{v_2} \right) \middle/ \left(\frac{v_1 + b}{v_1} \right) \right]$$

$$= -0.563 \text{ atm·m}^3/\text{kgmole·K} = -57.0 \text{ kJ/kgmole·K}$$

after substituting numerical values.

With the changes in entropy and enthalpy known, we are now ready to calculate the rate of heat transfer and the power requirement. The heat transfer for the reversible isothermal compression process is given by

$$\dot{Q} = \dot{m} T_1 \, \Delta s$$

$$= \frac{100 \text{ kg/h}}{39.9 \text{ kg/kgmole}} (300 \text{ K})(-57.0 \text{ kJ/kgmole·K})$$

$$= -42{,}900 \text{ kJ/h} = -11.9 \text{ kW}$$

where the minus sign indicates that heat is removed from the gas.

When the changes in kinetic and potential energies are neglected, the steady-flow energy balance gives the power requirement as

$$\dot{W} = \dot{Q} + \dot{m}(h_1 - h_2)$$

$$= -42{,}900 + \frac{100}{39.9} \times 1790$$

$$= -38{,}400 \text{ kJ/h} = -10.7 \text{ kW}$$

where the minus sign indicates that work is done on the gas.

Example 14-4

Nitrogen at a pressure of 250 atm and a temperature of 400 K expands reversibly and adiabatically in a turbine to an exhaust pressure of 5 atm. The flow rate is 1 kg/s. Calculate the power output if nitrogen obeys the Redlich–Kwong equation of state. The constant-pressure specific heat of nitrogen at 1 atm is given by

$$c_p = 4.186(6.903 - 0.3752 \times 10^{-3}T + 1.930 \times 10^{-6}T^2 - 6.861 \times 10^{-9}T^3)$$

where c_p is in kJ/kgmole·K, and T in K.

Solution   The state points of this example are shown in Fig. 14-4.
For nitrogen, the critical constants are

$$T_c = 126.2 \text{ K} \qquad \text{and} \qquad p_c = 33.5 \text{ atm}$$

Substituting these constants into Eqs. (14-7a) and (14-7b) leads to

$$a = 15.4 \text{ (atm)}(K^{1/2})(m^6)/(kgmole)^2$$

$$b = 0.0268 \text{ m}^3/\text{kgmole}$$

for the Redlich–Kwong equation for nitrogen.
From the Redlich–Kwong equation by a trial-and-error solution, at $p_1 = 250$ atm and $T_1 = 400$ K, we obtain

$$v_1 = 0.143 \text{ m}^3/\text{kgmole}$$

Because state 4 is at a low pressure, the ideal-gas equation gives

$$v_4 = \frac{\mathcal{R}T_4}{p_4} = \frac{0.08206 \times 400}{1} = 32.8 \text{ m}^3/\text{kgmole}$$

For the reversible and adiabatic process from state 1 to state 2, we must have $\Delta s_{12} = s_2 - s_1 = 0$. We now use the three-step path 1–4–3–2 as depicted in Fig. 14-4 for the evaluation of Δs_{12}. Accordingly, we write

$$\Delta s_{12} = \Delta s_{14} + \Delta s_{43} + \Delta s_{32}$$

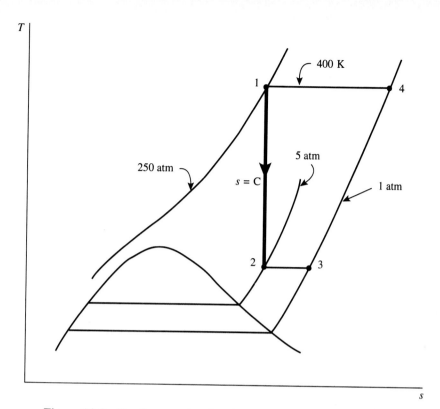

Figure 14-4 *T–s* diagram showing the state points for Example 14-4.

As in the preceding example, we have

$$\Delta s_{14} = s_4 - s_1$$

$$= \left\{ \left[\int_{v_1}^{v_4} \left[\frac{\mathcal{R}}{v - b} + \frac{a}{2T^{3/2}v(v + b)} \right] dv \right\}_T \right.$$

$$= \mathcal{R} \ln \left(\frac{v_4 - b}{v_1 - b} \right) - \frac{a}{2bT_1^{3/2}} \ln \left[\left(\frac{v_4 + b}{v_4} \right) \middle/ \left(\frac{v_1 + b}{v_1} \right) \right]$$

$$= 0.469 \text{ atm·m}^3/\text{kgmole·K} = 47.5 \text{ kJ/kgmole·K}$$

after substituting numerical values.

If the temperature at state 2 is known, the entropy changes Δs_{43} and Δs_{32} (in kJ/kgmole·K) can be calculated by the following two equations:

$$\Delta s_{43} = s_3 - s_4 = \int_{T_4 = T_1}^{T_3 = T_2} c_p \frac{dT}{T}$$

$$= 4.186 \int_{T_1 = 400 \text{ K}}^{T_2} \left(\frac{6.903}{T} - 0.3753 \times 10^{-3} + 1.93 \times 10^{-6}T \right.$$

$$\left. - 6.861 \times 10^{-9}T^2 \right) dT$$

$$= 4.186 \left\{ -6.903 \ln \frac{400}{T^2} + 0.3753 \times 10^{-3}(400 - T_2) \right.$$

$$\left. - 0.965 \times 10^{-6}[(400)^2 - T_2^2] + 2.287 \times 10^{-9}[(400)^3 - T_2^3] \right\}$$

and

$$\Delta s_{32} = s_2 - s_3$$

$$= 101.325 \left\{ \int_{v_3}^{v_2} \left[\frac{\mathcal{R}}{v - b} + \frac{a}{2T^{3/2}v(v + b)} \right] dv \right\}_T$$

$$= 101.325 \left\{ \mathcal{R} \ln \left(\frac{v_2 - b}{v_3 - b} \right) - \frac{a}{2bT^{1.5}} \ln \left[\left(\frac{v_2 + b}{v_2} \right) \middle/ \left(\frac{v_3 + b}{v_3} \right) \right] \right\}$$

$$= 101.325 \left\{ 0.08206 \ln \left(\frac{v_2 - 0.0268}{v_3 - 0.0268} \right) \right.$$

$$\left. - \frac{15.4}{2 \times 0.0268 T_2^{1.5}} \ln \left[\left(\frac{v_2 + 0.0268}{v_2} \right) \middle/ \left(\frac{v_3 + 0.0268}{v_3} \right) \right] \right\}$$

In order to find T_2 such that $\Delta s_{12} = 0$, a trial-and-error solution will be used. First, assume a value of T_2, calculate v_2 by trial and error from the Redlich–Kwong equation, calculate v_3 by the ideal-gas equation and then use the preceding two equations to calculate Δs_{43} and Δs_{32}. If $\Delta s_{12} = \Delta s_{14} + \Delta s_{43} + \Delta s_{32} = 0$, the assumed value of T_2 is the correct one. The value of T_2 as determined by the ideal-gas relations would be a good initial choice. After a few trials, the correct value of T_2 was determined to be

$$T_2 = 124 \text{ K}$$

The following is a final check of the correctness of this assumed value of T_2.
At $p_2 = 5$ atm and $T_2 = 124$ K, the Redlich–Kwong equation gives

$$v_2 = 1.92 \text{ m}^3/\text{kgmole}$$

At $p_3 = 1$ atm and $T_3 = T_2 = 124$ K, the ideal-gas equation gives

$$v_3 = 10.2 \text{ m}^3/\text{kgmole}$$

From which

$$\Delta s_{43} = -33.4 \text{ kJ/kgmole·K}$$

$$\Delta s_{32} = -14.2 \text{ kJ/kgmole·K}$$

Hence,

$$\Delta s_{12} = 47.5 - 33.4 - 14.2 \approx 0$$

Therefore, $T_2 = 124$ K is the correct value.

With the value of T_2 known, we are now ready to calculate $\Delta h_{12} = h_2 - h_1$. We again use the three-step path 1–4–3–2 to yield $\Delta h_{12} = \Delta h_{14} + \Delta h_{43} + \Delta h_{32}$. As in Example 14-3, we have

$$\Delta h_{14} = h_4 - h_1$$

$$= (p_4 v_4 - p_1 v_1) - \frac{1.5a}{T_1^{1/2}} \frac{1}{b} \ln \left[\left(\frac{v_4 + b}{v_4} \right) \middle/ \left(\frac{v_1 + b}{v_1} \right) \right]$$

$$= 4.40 \text{ atm·m}^3/\text{kgmole} = 446 \text{ kJ/kgmole}$$

Real Gases and Real-Gas Mixtures   Chap. 14

and

$$\Delta h_{32} = h_2 - h_3$$

$$= (p_2 v_2 - p_3 v_3) - \frac{1.5a}{T_2^{1/2}} \frac{1}{b} \ln \left[\left(\frac{v_2 + b}{v_2} \right) \bigg/ \left(\frac{v_3 + b}{v_3} \right) \right]$$

$$= -1.45 \text{ atm·m}^3/\text{kgmole} = -147 \text{ kJ/kgmole}$$

The value of Δh_{43} is given by

$$\Delta h_{43} = h_3 - h_4 = \int_{T_4=T_1}^{T_3=T_2} c_p \, dT$$

$$= 4.186 \int_{T_1=400\,\text{K}}^{T_2=124\,\text{K}} (6.903 - 0.3753 \times 10^{-3}T + 1.930 \times 10^{-6}T^2$$

$$- 6.861 \times 10^{-9}T^3) \, dT$$

$$= -7892 \text{ kJ/kgmole}$$

Therefore,

$$\Delta h_{12} = h_2 - h_1 = \Delta h_{14} + \Delta h_{43} + \Delta h_{32}$$

$$= 446 - 7892 - 147 = -7590 \text{ kJ/kgmole}$$

By neglecting the changes in kinetic and potential energies, an energy balance for the adiabatic turbine gives

$$\dot{W}_{12} = \dot{m}(h_1 - h_2) = 1 \times \frac{7590}{28} = 271 \text{ kW}$$

14-6 PRINCIPLE OF CORRESPONDING STATES

For an ideal gas, $pv = \mathcal{R}T$ under all conditions. For a real gas, a correction factor is introduced so that

$$pv = Z\mathcal{R}T \tag{14-16}$$

where Z is called the *compressibility factor*. It expresses the extent of deviation of the gas from an ideal gas. The value of Z is unity for an ideal gas under all conditions. For a real gas, Z is a function of its state.

Values of the compressibility factor of any gas can be determined experimentally. There is, however, some generalized information that shall allow us to introduce a very great simplification in the evaluation of approximate values of the compressibility factor. To do this, let us first define three new variables called the *reduced pressure* p_r, *reduced temperature* T_r, and *reduced volume* v_r as follows:

$$p_r = \frac{p}{p_c} \qquad T_r = \frac{T}{T_c} \qquad v_r = \frac{v}{v_c} \tag{14-17}$$

where p_c, T_c, and v_c are the critical properties.

At the same pressure and temperature, the specific or molar volumes of different gases are different. However, it has been found from experience that at the same reduced pressure and reduced temperature, the reduced volumes of different gases are approximately the same. This experimental fact was first used by van der Waals, who suggested that

$$v_r = f_1(p_r, T_r) \tag{14-18}$$

for all substances in the gaseous and liquid states. This is known as the van der Waals *principle of corresponding states*.

From the definitions of v_r and Z, we have

$$v_r = \frac{v}{v_c} = \frac{Z \mathcal{R} T p_c}{Z_c \mathcal{R} T_c p} = \frac{Z}{Z_c} \frac{T_r}{p_r} \tag{14-19}$$

where

$$Z_c = \frac{p_c v_c}{\mathcal{R} T_c}$$

and is called the critical compressibility factor. From Eqs. (14-18) and (14-19), it follows that

$$Z = f_2(p_r, T_r, Z_c) \tag{14-20}$$

Experimental values of Z_c for most substances fall within a narrow range of from 0.20 to 0.30. Therefore, as a first approximation, Z_c can be considered as a universal constant so that Eq. (14-20) can be simplified to the form

$$Z = f_3(p_r, T_r) \tag{14-21}$$

This last expression is often called the modified principle of corresponding states.

According to Eq. (14-21), if we plot Z in terms of p_r and T_r, we can use a single graph for all gases. A graph of this kind is called a *generalized compressibility chart*. Figure 14-5 shows a simplified chart. Examination of this figure reveals several important features of the volumetric behavior of real gases. It indicates clearly that all gases approach ideality ($z \to 1$) as the pressure approaches zero ($p \to 0$). The deviation from ideal-gas behavior is extreme in the vicinity of the critical point. A gas really behaves exactly as an ideal gas at low (but finite) pressures only at the Boyle temperature as defined by the expression

$$\lim_{p \to 0} \left(\frac{\partial Z}{\partial p} \right)_T = 0$$

The Boyle temperature is about $2.5 T_c$ for many fluids. At low pressures, the slope of isotherms on a Z–p plot increases as temperature increases, reaching a maximum value at about $5 T_c$. The isotherm of maximum slope as p approaches zero is sometimes called the foldback isotherm. As temperature increases beyond the foldback temperature, the slope of the isotherm decreases toward the $Z = 1$ line, indicating that all gases behave substantially as ideal gases at relatively high temperatures even under moderate pressures.

Figure A-19 in Appendix 4 shows a general chart for all fluids based on Eq. (14-21) for reduced pressures up to 40. Improved accuracy has been obtained by using Z_c as a third parameter according to Eq. (14-20). Appendix Figs. A-20 and A-21 present two different charts for $Z_c = 0.29$ and $Z_c = 0.27$.

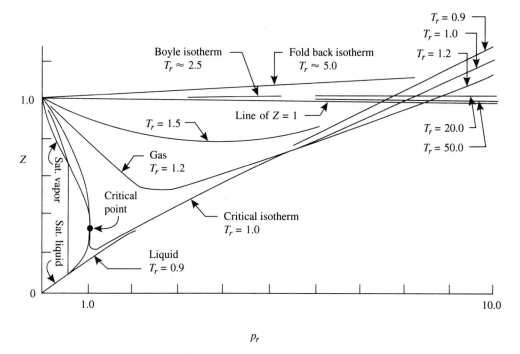

Figure 14-5 Generalized compressibility chart.

It is known that the light gases helium and hydrogen do not correlate very well on a generalized compressibility chart. Better correlation can be obtained by redefining the reduced pressure and temperature for these gases as follows:

$$p_r = \frac{p}{p_c + 8} \quad \text{and} \quad T_r = \frac{T}{T_c + 8}$$

where pressure is in atmospheres, and temperature is in K.

According to the principle of corresponding states, there is a single relationship between reduced coordinates that holds for all substances. We now demonstrate the validity of this statement through the use of the van der Waals equation of state. Upon substituting Eq. (14-17) into the van der Waals equation, we get

$$\left(p_r p_c + \frac{a}{v_r^2 v_c^2}\right)(v_r v_c - b) = \mathcal{R} T_r T_c$$

But

$$p_c = \frac{a}{27b^2} \qquad T_c = \frac{8a}{27\mathcal{R}b} \qquad v_c = 3b$$

so that

$$\left(p_r + \frac{3}{v_r^2}\right)(3v_r - 1) = 8T_r \tag{14-22}$$

This is the *van der Waals equation in reduced form*. It contains no specific constants for individual gases and is in a sense a universal equation for all gases. Of course, like the van der Waals equation itself, this reduced equation is only an approximate equation for real gases.

Example 14-5

By the use of a generalized compressibility chart, determine the molar volume of methane at 200 atm and 400 K.

Solution For methane, $p_c = 45.8$ atm, $T_c = 190.7$ K, and $Z_c = 0.290$. Hence,

$$p_r = \frac{p}{p_c} = \frac{200}{45.8} = 4.37$$

$$T_r = \frac{T}{T_c} = \frac{400}{190.7} = 2.10$$

From Appendix 4, Fig. A-20, we get

$$Z = 0.98$$

which yields

$$v = \frac{Z\mathcal{R}T}{p} = \frac{0.98 \times 0.08206 \times 400}{200}$$

$$= 0.161 \text{ m}^3/\text{kgmole}$$

Note that we obtained the same value of v in Example 14-2 by the use of the Benedict–Webb–Rubin equation of state.

Example 14-6

By the use of a generalized compressibility chart, determine the pressure of carbon dioxide at a temperature of 160°F and having a specific volume of 0.045 ft³/lbm.

Solution For carbon dioxide, $p_c = 1071$ psia, $T_c = 547.5$ R, and $Z_c = 0.275$. Hence,

$$T_r = \frac{T}{T_c} = \frac{619.7}{547.5} = 1.13$$

Upon substituting numerical values into the equation,

$$p = \frac{Z\mathcal{R}T}{v} \qquad \text{or} \qquad p_r p_c = \frac{Z\mathcal{R}T}{v}$$

we obtain

$$p_r(1071 \times 144) = \frac{Z \times 1545 \times 619.7}{0.045 \times 44.01}$$

or

$$p_r = 3.13Z$$

Because $Z_c = 0.275$, the compressibility chart shown in Fig. A-21 is used. The equation $p_r = 3.13Z$ represents a straight line on this chart. The intersection of this line with the reduced isotherm of $T_r = 1.13$ gives the solution as

$$Z = 0.53 \qquad \text{and} \qquad p_r = 1.66$$

from which

$$p = p_r p_c = 1.66 \times 1071 = 1780 \text{ psia}$$

Experimental data show that the preceding calculations are reasonably accurate. However, the ideal-gas equation gives the following highly inaccurate value:

$$p = \frac{\mathcal{R}T}{v} = \frac{1545 \times 619.7}{0.045 \times 44.01 \times 144} = 3360 \text{ psia}$$

14-7 GENERALIZED THERMODYNAMIC CORRECTION CHARTS

The principle of corresponding states in useful not only in the correlation of compressibility data, but also in the calculations of other thermodynamic properties, such as enthalpy, entropy, and specific heats. In this section, we present the Hougen and Watson's generalized charts for thermodynamic properties of gases and liquids [12]. These charts show the nonideality corrections to various thermodynamic properties. In the following discussions, the symbols marked with "*" denote properties of ideal gases for which $pv = \mathcal{R}T$, and the unmarked symbols denote the properties of real fluids.

To obtain the enthalpy correction, we use the second dh equation, Eq. (13-40):

$$dh = c_p \, dT + \left[v - T\left(\frac{\partial v}{\partial T}\right)_p \right] dp$$

Integrating this equation isothermally from $p = 0$ to some higher pressure p yields

$$(h^* - h)_T = \left\{ \int_0^p \left[T\left(\frac{\partial v}{\partial T}\right)_p - v \right] dp \right\}_T \tag{14-23}$$

where h^* is the enthalpy of the ideal gas, and h the enthalpy of the real fluid at the given elevated pressure p, both at the same temperature T. Note that for ideal gases, the enthalpy is a function of temperature only. From the equation of state

$$v = \frac{Z\mathcal{R}T}{p}$$

and its partial derivative

$$\left(\frac{\partial v}{\partial T}\right)_p = \frac{\mathcal{R}Z}{p} + \frac{\mathcal{R}T}{p}\left(\frac{\partial Z}{\partial T}\right)_p \tag{14-24}$$

Equation (14-23) can be rewritten as

$$(h^* - h)_T = \left\{ \int_0^p \left[\frac{\mathcal{R}T^2}{p}\left(\frac{\partial Z}{\partial T}\right)_p \right] dp \right\}_T$$

In terms of reduced coordinates, this equation becomes

$$(h^* - h)_T = \left\{ \int_0^{p_r} \left[\frac{\mathcal{R}T_r^2 T_c}{p_r}\left(\frac{\partial Z}{\partial T_r}\right)_{p_r} \right] dp_r \right\}_T$$

or

$$\left(\frac{h^* - h}{T_c}\right)_T = \left\{ \mathcal{R}T_r^2 \int_0^{p_r} \left(\frac{\partial Z}{\partial T_r}\right)_{p_r} d(\ln p_r) \right\}_T \tag{14-25}$$

Values of the foregoing integral can be obtained by graphical integration using data from a generalized compressibility chart. Equation (14-25) is the basis of the generalized enthalpy correction chart as shown in Fig. A-22 in Appendix 4.

A procedure similar to the previous is used to construct an entropy correction chart, utilizing the second ds equation, Eq. (13-33):

$$ds = \frac{c_p}{T}\,dT - \left(\frac{\partial v}{\partial T}\right)_p dp$$

Integrating this equation isothermally from $p = 0$ to some higher pressure p yields

$$(s_p - s_0^*)_T = -\left[\int_0^p \left(\frac{\partial v}{\partial T}\right)_p dp\right]_T \qquad (14\text{-}26)$$

where s_p is the entropy of the real fluid at p and T, and s_0^* is the entropy of the ideal gas at $p = 0$ and T. Because entropy is a function of both pressure and temperature even for ideal gases, the subscript 0 in the symbol s_0^* is necessary to indicate that p is equal to zero. Because the value of s_0^* is infinite, Eq. (14-26) is not directly useful. To get around this difficulty, Eq. (13-33) is integrated isothermally from $p = 0$ to the given higher pressure p, assuming that the ideal-gas equation is satisfied at all times. Thus,

$$(s_p^* - s_0^*)_T = -\left[\int_0^p \left(\frac{\partial v}{\partial T}\right)_p dp\right]_T = -\left(\mathcal{R}\int_0^p \frac{dp}{p}\right)_T \qquad (14\text{-}27)$$

where s_p^* is the entropy in the ideal-gas state at p and T. Subtracting Eq. (14-26) from Eq. (14-27) yields

$$(s_p^* - s_p)_T = -\left\{\int_0^p \left[\frac{\mathcal{R}}{p} - \left(\frac{\partial v}{\partial T}\right)_p\right] dp\right\}_T$$

Substituting Eq. (14-24) into the preceding equation leads to

$$(s_p^* - s_p)_T = -\left\{\mathcal{R}\int_0^p \left[\frac{1 - Z}{p} - \frac{T}{p}\left(\frac{\partial Z}{\partial T}\right)_p\right] dp\right\}_T$$

In terms of reduced coordinates, this equation becomes

$$(s_p^* - s_p)_T = -\left[\mathcal{R}\int_0^{p_r} (1 - Z)\frac{dp_r}{p_r}\right]_T + \left[\mathcal{R}T_r\int_0^{p_r} \left(\frac{\partial Z}{\partial T_r}\right)_{p_r}\frac{dp_r}{p_r}\right]_T$$

Comparison of the last term of this equation with Eq. (14-25) reveals that this term can be written in terms of $(h^* - h)_T$. We accordingly write

$$(s_p^* - s_p)_T = -\left[\mathcal{R}\int_0^{p_r} (1 - Z)\,d(\ln p_r)\right]_T + \left(\frac{h^* - h}{T_r T_c}\right)_T \qquad (14\text{-}28)$$

Values of the integral in the first term on the right of Eq. (14-28) can be obtained by graphical integration using data from a generalized compressibility chart, whereas a generalized enthalpy correction chart can be used directly for the last term. Equation (14-28) is the basis of the generalized entropy correction chart as shown in Fig. A-23 in Appendix 4.

A correction chart for the molar specific heat at constant pressure can be obtained by defining c_p and c_p^* for any substance when it is considered as a real fluid and as an ideal gas, respectively, as follows:

$$c_p = \left(\frac{\partial h}{\partial T}\right)_p \qquad \text{and} \qquad c_p^* = \left(\frac{\partial h^*}{\partial T}\right)_p$$

From which

$$(c_p - c_p^*)_T = \left[\frac{\partial}{\partial T}(h - h^*) \right]_p$$

In terms of reduced coordinates, this equation becomes

$$(c_p - c_p^*)_T = \left[\frac{\partial}{\partial T_r} \left(\frac{h - h^*}{T_c} \right) \right]_p \qquad (14\text{-}29)$$

where c_p and c_p^* are at the same temperature. Values of $(c_p - c_p^*)$ can be obtained by the use of the data from a generalized enthalpy correction chart. Figure A-24 in Appendix 4 shows a generalized c_p correction chart.

The generalized thermodynamic correction charts shown in Figs. A-22 to A-24 are for $Z_c = 0.27$. Corrections for Z_c other than 0.27 are available in the reference given in the beginning of this section. In the following examples, we use the charts without correction for Z_c.

Example 14-7

Using generalized thermodynamic correction charts, estimate the values of enthalpy and entropy for methane at 200 atm and 400 K. The values of enthalpy and entropy at 1 atm and 400 K are 23,070 kJ/kgmole and 197.4 kJ/kgmole·K, respectively.

Solution   Referring to Fig. 14-3, we have

$$p_A = 200 \text{ atm} \qquad p_a = 1 \text{ atm} \qquad T_A = T_a = 400 \text{ K}$$

$$p_{rA} = \frac{p_A}{p_c} = \frac{200}{45.8} = 4.37$$

$$p_{ra} = \frac{p_a}{p_c} = \frac{1}{45.8} = 0.0218$$

$$T_{rA} = T_{ra} = \frac{T_A}{T_c} = \frac{400}{190.7} = 2.10$$

From the generalized enthalpy correction chart (Fig. A-22), at $T_r = 2.10$ and $p_r = 4.37$,

$$\frac{h_A^* - h_A}{T_c} = 7.45 \text{ kJ/kgmole·K}$$

and at $T_r = 2.10$ and $p_r = 0.0218$,

$$\frac{h_a^* - h_a}{T_c} = 0$$

For an ideal gas, the enthalpy is a function of temperature only; thus, $h_A^* = h_a^*$. Then

$$\frac{h_A - h_a}{T_c} = \frac{h_a^* - h_a}{T_c} - \frac{h_A^* - h_A}{T_c} = -7.45 \text{ kJ/kgmole·K}$$

$$h_A - h_a = -7.45T_c = -7.45 \times 190.7 = -1420 \text{ kJ/kgmole}$$

Therefore,

$$h_A = h_a - 1420 = 23,070 - 1420 = 21,650 \text{ kJ/kgmole}$$

From the generalized entropy correction chart (Fig. A-23), we obtain, at $T_r = 2.10$

and $p_r = 4.37$,

$$s_A^* - s_A = 3.01 \text{ kJ/kgmole·K}$$

and at $T_r = 2.10$ and $p_r = 0.0218$,

$$s_a^* - s_a = 0$$

From Eq. (7-16),

$$ds = c_p \frac{dT}{T} - \mathcal{R} \frac{dp}{p} = -\mathcal{R} \frac{dp}{p}$$

for an isothermal process of an ideal gas. Thus,

$$s_A^* - s_a^* = -\mathcal{R} \ln \frac{p_A}{p_a} = -8.314 \ln 200$$

$$= -44.05 \text{ kJ/kgmole·K}$$

Then

$$s_A - s_a = (s_a^* - s_a) + (s_A^* - s_a^*) - (s_A^* - s_A)$$

$$= 0 - 44.05 - 3.01 = -47.1 \text{ kJ/kgmole·K}$$

Therefore,

$$s_A = s_a - 47.1 = 197.4 - 47.1 = 150.3 \text{ kJ/kgmole·K}$$

Because the given conditions are the same in Examples 14-2 and 14-7, the reader should compare the two results.

Example 14-8

A quantity of nitrogen gas contained in a piston–cylinder arrangement is initially at 200 atm and 200 K. Heat is added to the gas reversibly in such a way that the pressure of the gas remains constant until its temperature reaches 300 K. Determine the work and heat transfers per kgmole of the gas by the use of generalized charts.

Solution For nitrogen, $p_c = 33.5$ atm, $T_c = 126.2$ K, and $Z_c = 0.291$. Hence,

$$p_{r1} = p_{r2} = \frac{p_1}{p_c} = \frac{200}{33.5} = 5.97$$

$$T_{r1} = \frac{T_1}{T_c} = \frac{200}{126.2} = 1.58$$

$$T_{r2} = \frac{T_2}{T_c} = \frac{300}{126.2} = 2.38$$

From Fig. A-20, we get

$$Z_1 = 0.90 \text{ at } p_{r1} = 5.97 \text{ and } T_{r1} = 1.58$$

$$Z_2 = 1.05 \text{ at } p_{r2} = 5.97 \text{ and } T_{r2} = 2.38$$

which yield

$$v_1 = \frac{Z_1 \mathcal{R} T_1}{p_1} = \frac{0.90 \times 0.08206 \times 200}{200} = 0.0739 \text{ m}^3/\text{kgmole}$$

$$v_2 = \frac{Z_2 \mathcal{R} T_2}{p_2} = \frac{1.05 \times 0.08206 \times 300}{200} = 0.1292 \text{ m}^3/\text{kgmole}$$

Therefore, the work done by the gas is

$$w = \int_1^2 p \, dv = p_1(v_2 - v_1) = 200(0.1292 - 0.0739)$$

$$= 11.06 \text{ atm·m}^3/\text{kgmole} = 1120 \text{ kJ/kgmole}$$

From Fig. A-22, we get, at $p_{r1} = 5.97$ and $T_{r1} = 1.58$,

$$\frac{h_1^* - h_1}{T_c} = 15.3 \text{ kJ/kgmole·K}$$

and at $p_{r2} = 5.97$ and $T_{r2} = 2.38$,

$$\frac{h_2^* - h_2}{T_c} = 6.9 \text{ kJ/kgmole·K}$$

Hence,

$$h_1^* - h_1 = 15.3 T_c = 15.3 \times 126.2 = 1930 \text{ kJ/kgmole}$$

$$h_2^* - h_2 = 6.9 T_c = 6.9 \times 126.2 = 871 \text{ kJ/kgmole}$$

The ideal-gas state c_p for nitrogen in the temperature range of this problem is fairly constant. Thus, the ideal-gas state enthalpy change ($h_2^* - h_1^*$) is simply equal to this constant value of c_p times ($T_2 - T_1$). Nevertheless, for added accuracy, let us use Keenan and Kaye's gas tables [20]. From the table for nitrogen, we read

$$h_2^* - h_1^* = 8724.1 - 5812.8 = 2910 \text{ kJ/kgmole}$$

Therefore,

$$h_2 - h_1 = (h_1^* - h_1) + (h_2^* - h_1^*) - (h_2^* - h_2)$$

$$= 1930 + 2910 - 871 = 3970 \text{ kJ/kgmole}$$

Then, according to the first law, the heat added to the gas is

$$q = (u_2 - u_1) + w = (u_2 - u_1) + (p_2 v_2 - p_1 v_1)$$

$$= h_2 - h_1 = 3970 \text{ kJ/kgmole}$$

Note that if the gas is assumed to behave as an ideal gas, the results would be

$$v_1 = \frac{\mathscr{R} T_1}{p_1} = \frac{0.08206 \times 200}{200} = 0.0821 \text{ m}^3/\text{kgmole}$$

$$v_2 = \frac{\mathscr{R} T_2}{p_2} = \frac{0.08206 \times 300}{200} = 0.1231 \text{ m}^3/\text{kgmole}$$

$$w = p_1(v_2 - v_1) = 200(0.1231 - 0.0821)$$

$$= 8.20 \text{ atm · m}^3/\text{kgmole} = 831 \text{ kJ/kgmole}$$

$$q = h_2 - h_1 = h_2^* - h_1^* = 2910 \text{ kJ/kgmole}$$

which are, of course, all erroneous.

14-8 REAL-GAS MIXTURES

As noted in Chapter 9, Dalton's law of additive pressures and Amagat's law of additive volumes hold exactly only for mixtures of ideal gases. However, it was found experimentally that these laws do hold approximately for mixtures of real gases even in some ranges of pressure and temperature where the ideal-gas law itself is quite inaccurate. In this case, one is not to use the ideal-gas equation, but rather some suitable real-gas equation of state to find the p_i's or V_i's for the individual components to be used in Eq. (9-9) or (9-12). Here the p_i's and V_i's are usually renamed as the component pressures and component volumes, respectively. They are no longer defined by Eqs. (9-10) and (9-13).

Often there is a need to have an equation of state for a mixture of real gases. Because there is an infinite variety of compositions for mixtures, it is desirable to devise methods of developing an equation of state for a mixture from the equation of state for the pure components. This can be done empirically by the use of various combining rules to obtain the constants of an equation of state for a mixture from the constants of the pure components. The most common combining rules are as follows:

Linear combination:

$$k = \sum_i x_i k_i$$

Linear square-root combination:

$$k = \left(\sum_i x_i k_i^{1/2} \right)^2$$

Linear cube-root combination:

$$k = \left(\sum_i x_i k_i^{1/3} \right)^3$$

Lorentz combination:

$$k = \tfrac{1}{4} \sum_i x_i k_i + \tfrac{3}{4} \left(\sum_i x_i k_i^{1/3} \right) \left(\sum_i x_i k_i^{2/3} \right)$$

where k represents a constant in an equation of state for a mixture, k_i the corresponding constant in the equation of state for pure component i, and x_i the mole fraction of component i in the mixture.

When accurate p–v–T relations for the components of a mixture are not available, a generalized compressibility chart can be used. One way of using a compressibility chart in this case is to determine the compressibility factors for the pure components from the chart and then to devise a means of combining them to obtain the compressibility factor for the mixture. It is convenient to use a simple linear combination rule of the type

$$Z = \sum_i x_i Z_i \qquad (14\text{-}30)$$

where Z is the compressibility factor for the mixture, Z_i the compressibility factor for pure component i, and x_i the mole fraction. If the Z_i's are evaluated at p and T of the mixture, Eq. (14-30) reduces to Amagat's law of additive volumes. However, if the Z_i's are evaluated at the V and T of the mixture, this equation then reduces to Dalton's law of additive pressures.

A general compressibility chart also can be used to evaluate directly the compressibility factor for a mixture. However, as suggested by W. B. Kay* that instead of using the true critical point of the mixture, a fictitious point, called the *pseudocritical point*, should be used for calculating the reduced coordinates. Kay first proposed the use of a simple linear combination as follows:

$$p_{c'} = \sum_i x_i p_{ci} \tag{14-31a}$$

and

$$T_{c'} = \sum_i x_i T_{ci} \tag{14-31b}$$

where $p_{c'}$ and $T_{c'}$ are the pseudocritical pressure and temperature, respectively, of the mixture; p_{ci} and T_{ci} are the critical pressure and temperature, respectively, of component i; and x_i is the mole fraction of component i.

Example 14-9

A gaseous mixture containing 69.5 mole percent carbon dioxide and 30.5 mole percent ethylene (C_2H_4) has a molar volume of 0.111 m³/kgmole at a temperature of 100°C. Estimate the pressure of the mixture by the use of (1) the ideal-gas law, (2) the law of additive pressures and the van der Waals equation, (3) the law of additive volumes and the van der Waals equation, (4) the van der Waals equation for the mixture, using a linear square-root combination for constant a and linear combination for constant b, (5) the law of additive pressures and the generalized compressibility chart, (6) the law of additive volumes and the generalized compressibility chart, and (7) Kay's rule of pseudocritical point. The experimental value for the pressure of the mixture at the given temperature and molar volume is 175.8 atm.

Solution In this example, we denote carbon dioxide by the subscript 1 and ethylene by the subscript 2. The symbols without such a subscript are for the mixture as a whole. Thus, the given data are $x_1 = 0.695$, $x_2 = 0.305$, $v = 0.111$ m³/kgmole, and $T = 100°C = 373.15$ K. The critical constants are

$$T_{c1} = 304.2 \text{ K}, \qquad p_{c1} = 72.9 \text{ atm}, \qquad Z_{c1} = 0.275$$

$$T_{c2} = 283.06 \text{ K}, \qquad p_{c2} = 50.5 \text{ atm}, \qquad Z_{c2} = 0.270$$

The van der Waals constants are

$$a_1 = 3.606 \text{ atm·m}^6/(\text{kgmole})^2$$

$$b_1 = 0.04280 \text{ m}^3/\text{kgmole}$$

$$a_2 = 4.507 \text{ atm·m}^6/(\text{kgmole})^2$$

$$b_2 = 0.05749 \text{ m}^3/\text{kgmole}$$

The universal gas constant is

$$\mathcal{R} = 0.08206 \text{ atm·m}^3/\text{kgmole·K}$$

---

* W. B. Kay, *Industrial and Engineering Chemistry*, vol. 28, p. 1014 (1936).

(1) Application of the ideal-gas equation gives

$$p = \frac{\mathscr{R}T}{v} = \frac{0.08206 \times 373.15}{0.111} = 276 \text{ atm}$$

Let us define the percentage error in pressure by the equation

$$\text{percentage error} = \frac{(\text{calculated pressure}) - (\text{experimental pressure})}{\text{experimental pressure}} \times 100$$

Thus, the pressure calculated by the ideal-gas equation has an error of $100(276 - 175.8)/175.8 = 57.0\%$.

(2) In order to use the law of additive pressures together with the van der Waals equation, we first calculate the molar volumes of carbon dioxide and ethylene at $T = T_1 = T_2$ and $V = V_1 = V_2$, where V denotes total volume.

$$v_1 = \frac{0.111}{0.695} = 0.160 \text{ m}^3/\text{kgmole}$$

$$v_2 = \frac{0.111}{0.305} = 0.364 \text{ m}^3/\text{kgmole}$$

From the van der Waals equation, the component pressures of carbon dioxide and ethylene are then

$$p_1 = \frac{\mathscr{R}T}{v_1 - b_1} - \frac{a_1}{v_1^2} = \frac{0.08206 \times 373.15}{0.160 - 0.0428} - \frac{3.606}{0.160^2}$$

$$= 120.4 \text{ atm}$$

$$p_2 = \frac{\mathscr{R}T}{v_2 - b_2} - \frac{a_2}{v_2^2} = \frac{0.08206 \times 373.15}{0.364 - 0.05749} - \frac{4.507}{0.364^2}$$

$$= 65.9 \text{ atm}$$

Therefore, by the law of additive pressures, we have

$$p = p_1 + p_2 = 120.4 + 65.9 = 186 \text{ atm}$$

This calculated pressure has an error of $100(186 - 175.8)/175.8 = 5.8\%$.

(3) The law of additive volumes states that the total volume of a mixture is the sum of the component volumes, evaluated at the pressure and temperature of the mixture. We accordingly write

$$v = x_1 v_1' + x_2 v_2'$$

or

$$0.111 = 0.695 v_1' + 0.305 v_2' \tag{14-32}$$

wherein v_1' and v_2' are the molar volumes in m³/kgmole of carbon dioxide and ethylene, respectively, as evaluated at $T = T_1 = T_2$ and $p = p_1 = p_2$. From the van der Waals equation, we have

$$p = p_1 = \frac{\mathscr{R}T}{v_1' - b_1} - \frac{a_1}{(v_1')^2} = \frac{0.08206 \times 373.15}{v_1' - 0.0428} - \frac{3.606}{(v_1')^2} \tag{14-33}$$

$$p = p_2 = \frac{\mathscr{R}T}{v_2' - b_2} - \frac{a_2}{(v_2')^2} = \frac{0.08206 \times 373.15}{v_2' - 0.05749} - \frac{4.507}{(v_2')^2} \tag{14-34}$$

The value of p can best be obtained in an iteration process, which consists in assuming values of v_1', obtaining v_2' from Eq. (14-32), and then calculating p_1 and p_2 from Eqs. (14-33) and (14-34). For a correct solution, we must have $p_1 = p_2$, which would be the pressure of the mixture. After a few trials, the correct value of v_1' was found to be 0.105 m^3/kgmole. Thus, from Eq. (14-32), we have

$$v_2' = \frac{0.111 - 0.695v_1'}{0.305} = \frac{0.111 - 0.695 \times 0.105}{0.305}$$

$$= 0.125 \text{ m}^3\text{/kgmole}$$

and from Eqs. (14-33) and (14-34), we have

$$p_1 = \frac{0.08206 \times 373.15}{0.105 - 0.0428} - \frac{3.606}{(0.105)^2} = 165 \text{ atm}$$

and

$$p_2 = \frac{0.08206 \times 373.15}{0.125 - 0.05749} - \frac{4.507}{(0.125)^2} = 165 \text{ atm}$$

Hence,

$$p = p_1 = p_2 = 165 \text{ atm}$$

This calculated pressure has an error of $100(165 - 175.8)/175.8 = -6.1\%$.

(4) The van der Waals constants for the mixture are

$$a = (x_1 a_1^{1/2} + x_2 a_2^{1/2})^2$$

$$= [0.695 \times (3.606)^{1/2} + 0.305 \times (4.507)^{1/2}]^2$$

$$= 3.870 \text{ atm·m}^6\text{/(kgmole)}^2$$

$$b = x_1 b_1 + x_2 b_2$$

$$= 0.695 \times 0.0428 + 0.305 \times 0.05749$$

$$= 0.04728 \text{ m}^3\text{/kgmole}$$

Therefore, from the van der Waals equation as applied to the mixture as a whole, we get

$$p = \frac{\mathcal{R}T}{v - b} - \frac{a}{v^2} = \frac{0.08206 \times 373.15}{0.111 - 0.04728} - \frac{3.870}{(0.111)^2}$$

$$= 166 \text{ atm}$$

This calculated pressure has an error of $100(166 - 175.8)/175.8 = -5.6\%$

(5) When each gas exists at the temperature and total volume of the mixture, we have, as before,

$$v_1 = 0.160 \text{ m}^3\text{/kgmole}$$

$$v_2 = 0.364 \text{ m}^3\text{/kgmole}$$

To evaluate the component pressure of carbon dioxide by the use of a generalized compressibility chart, we first write

$$p_1 = p_{r1} p_{c1} = \frac{Z_1 \mathcal{R}T}{v_1}$$

or

$$p_{r1} = \frac{Z_1 \mathscr{R} T}{p_{c1} v_1} = \frac{0.08206 \times 373.15}{72.9 \times 0.160} Z_1 = 2.625 Z_1$$

Because $Z_{c1} = 0.275$, Fig. A-21 is to be used. The equation $p_{r1} = 2.625 Z_1$ represents a straight line on this chart. The intersection of this line with the reduced isotherm of

$$T_{r1} = \frac{T}{T_{c1}} = \frac{373.15}{304.2} = 1.23$$

gives the solution as

$$p_{r1} = 1.78 \qquad \text{and} \qquad Z_1 = 0.68$$

From which

$$p_1 = p_{r1} p_{c1} = 1.78 \times 72.9 = 130 \text{ atm}$$

Similarly for ethylene, solving the equations

$$p_{r2} = \frac{Z_2 \mathscr{R} T}{p_{c2} v_2} = \frac{0.08206 \times 373.15}{50.5 \times 0.364} Z_2 = 1.666 Z_2$$

and

$$T_{r2} = \frac{T}{T_{c2}} = \frac{373.15}{283.06} = 1.32$$

in connection with Fig. A-21 leads to

$$p_{r2} = 1.36 \qquad \text{and} \qquad Z_2 = 0.82$$

From which

$$p_2 = p_{r2} p_{c2} = 1.36 \times 50.5 = 69 \text{ atm}$$

Therefore, by the law of additive pressures, we have

$$p = p_1 + p_2 = 130 + 69 = 199 \text{ atm}$$

This calculated pressure has an error of $100(199 - 175.8)/175.8 = 13.2\%$.

(6) To conform to the law of additive volumes, we again have Eq. (14-32), which is

$$0.111 = 0.695 v_1' + 0.305 v_2'$$

But now this equation will be used in connection with the following two equations:

$$v_1' = \frac{Z_1 \mathscr{R} T}{p} = 0.08206 \times 373.15 Z_1 / p \tag{14-35}$$

$$v_2' = \frac{Z_2 \mathscr{R} T}{p} = 0.08206 \times 373.15 Z_2 / p \tag{14-36}$$

Because p is unknown, an iteration process must be employed. After a few trials, it was found that p should be 174 atm. Thus,

$$p_{r1} = \frac{p}{p_{c1}} = \frac{174}{72.9} = 2.39$$

$$p_{r2} = \frac{p}{p_{c2}} = \frac{174}{50.5} = 3.45$$

From Fig. A-21, we get

$$Z_1 = 0.61 \quad \text{at} \quad p_{r1} = 2.39 \quad \text{and} \quad T_{r1} = 1.23$$

$$Z_2 = 0.69 \quad \text{at} \quad p_{r2} = 3.45 \quad \text{and} \quad T_{r2} = 1.32$$

Substituting these values in Eqs. (14-35) and (14-36) leads to

$$v_1' = 0.08206 \times 373.15 \times 0.61/174$$

$$= 0.107 \ \text{m}^3/\text{kgmole}$$

$$v_2' = 0.02806 \times 373.15 \times 0.69/174$$

$$= 0.121 \ \text{m}^3/\text{kgmole}$$

Thus,

$$0.695 v_1' + 0.305 v_2' = 0.695 \times 0.107 + 0.305 \times 0.121$$

$$= 0.111 \ \text{m}^3/\text{kgmole}$$

which agrees with Eq. (14-32). Therefore, we conclude that

$$p = 174 \ \text{atm}$$

This calculated pressure has an error of $100(174 - 175.8)/175.8 = -1.0\%$.

(7) According to Kay's rule, the pseudocritical temperature and pressure of the mixture are

$$T_{c'} = x_1 T_{c1} + x_2 T_{c2}$$

$$= 0.695 \times 304.2 + 0.305 \times 283.06$$

$$= 297.8 \ \text{K}$$

$$p_{c'} = x_1 p_{c1} + x_2 p_{c2}$$

$$= 0.695 \times 72.9 + 0.305 \times 50.5$$

$$= 66.1 \ \text{atm}$$

Thus,

$$T_r = \frac{T}{T_{c'}} = \frac{373.15}{297.8} = 1.25$$

$$p_r = \frac{p}{p_{c'}} = \frac{ZRT}{p_{c'}v}$$

$$= \frac{0.02806 \times 373.15}{66.1 \times 0.111} Z = 4.173 Z$$

The intersection of the straight line $p_r = 4.173Z$ and the reduced isotherm $T_r = 1.25$ on the compressibility chart (Fig. A-21) gives the solution as

$$p_r = 2.59 \quad \text{and} \quad Z = 0.62$$

Therefore,

$$p = p_r p_{c'} = 2.59 \times 66.1 = 171 \ \text{atm}$$

This calculated pressure has an error of $100(171 - 175.8)/175.8 = -2.7\%$.

It should be mentioned that one certainly cannot draw any conclusion from a single illustration about the relative accuracy of the various methods employed in this example.

However, it is clear that the ideal-gas law often gives erroneous solutions when the pressure is not low.

14-9 SUMMARY

The equations of state of gases studied in this chapter are as follows:

(1) Van der Waals equation:

$$\left(p + \frac{a}{v^2}\right)(v - b) = \mathcal{R}T$$

where

$$a = \frac{27}{64} \frac{(\mathcal{R}T_c)^2}{p_c}$$

$$b = \frac{\mathcal{R}T_c}{8 p_c}$$

(2) Beattie–Bridgeman equation:

$$p = \frac{\mathcal{R}T\,(1 - c/vT^3)}{v^2}\left[v + B_0\left(1 - \frac{b}{v}\right)\right] - \frac{A_0}{v^2}\left(1 - \frac{a}{v}\right)$$

(3) Benedict–Webb–Rubin equation:

$$p = \frac{\mathcal{R}T}{v} + \left(B_0\mathcal{R}T - A_0 - \frac{C_0}{T^2}\right)\frac{1}{v^2} + (b\mathcal{R}T - a)\frac{1}{v^3}$$

$$+ \frac{a\alpha}{v^6} + \frac{c(1 + \gamma/v^2)}{T^2}\frac{1}{v^3}e^{-\gamma/v^2}$$

(4) Redlich–Kwong equation:

$$p = \frac{\mathcal{R}T}{v - b} - \frac{a}{T^{1/2}v(v + b)}$$

where

$$a = 0.42748\ \mathcal{R}^2 T_c^{2.5}/p_c$$

$$b = 0.08664\ \mathcal{R}T_c/p_c$$

(5) Equation of state in virial form:

$$\frac{pv}{\mathcal{R}T} = 1 + \frac{B}{v} + \frac{C}{v^2} + \frac{D}{v^3} + \cdots$$

or

$$\frac{pv}{\mathcal{R}T} = 1 + B'p + C'P^2 + D'p^3 + \cdots$$

where B, C, D, etc., are called the virial coefficients, and

$$B' = \frac{B}{\mathcal{R}T} \qquad C' = \frac{C - B^2}{(\mathcal{R}T)^2} \qquad D' = \frac{D - 3BC + 2B^3}{(\mathcal{R}T)^3}$$

One useful equation for the evaluation of the value of enthalpy h from an equation of state in the form $p = f(T, v)$ is

$$h - h_0 = \left[\int_{T_0}^{T} c_p \, dT \right]_{p_0} + \left\{ \int_{p_0}^{p} \left[v - T \left(\frac{\partial v}{\partial T} \right)_p \right] dp \right\}_T$$

where h and v are at the final state (p, T), h_0 is at the initial reference state (p_0, T_0), and c_p is a function of T at constant pressure p_0.

One useful equation for the evaluation of the value of entropy s from an equation of state in the form $p = f(T, v)$ is

$$s - s_0 = \left[\int_{T_0}^{T} \frac{c_p}{T} \, dT \right]_{p_0} + \left[\int_{v_a}^{v} \left(\frac{\partial p}{\partial T} \right)_v dv \right]_T$$

where s and v are at the final state (p, T), s_0 is at the initial reference state (p_0, T_0), v_a is at the state (p_0, T), and c_p is a function of T at constant pressure p_0.

The compressibility factor Z is defined as

$$Z = \frac{pv}{\mathcal{R}T}$$

This factor expresses the extent of deviation of a gas from an ideal gas for which Z is unity for all conditions.

The reduced pressure p_r, reduced temperature T_r, and reduced volume v_r are defined as

$$p_r = \frac{p}{p_c} \qquad T_r = \frac{T}{T_c} \qquad v_r = \frac{v}{v_c}$$

where p_c, T_c, and v_c are the critical properties. The critical compressibility factor Z_c is defined as

$$Z_c = \frac{p_c v_c}{\mathcal{R}T_c}$$

The principle of corresponding states can be written as

$$v_r = \text{function } (p_r, T_r)$$

or

$$Z = \text{function } (p_r, T_r) \tag{a}$$

When Z_c is included as an independent variable, we can write

$$Z = \text{function } (p_r, T_r, Z_c) \tag{b}$$

According to Eq. (a), if we plot Z in terms of p_r and T_r, we can use a single graph for all gases. A graph of this kind is called a generalised compressibility chart. Improved accuracy can be obtained by using Z_c as a third parameter according to Eq. (b).

Generalized thermodynamic correction charts are graphs showing the property differences in reduced coordinates between an ideal gas and a real gas as the pressure is increased from zero along an isotherm. We present in this book the charts for $(h^* - h)_T$, $(s_p^* - s_p)_T$, and $(c_p - c_p^*)_T$, where the symbols marked with "*" denote properties of ideal gases for which $pv = RT$, and the unmarked symbols denote the properties of real fluids.

Dalton's law and Amagat's law hold approximately for real-gas mixtures. In this case, however, one is not to use the ideal-gas equation, but rather some suitable real-gas equation of state to find the p_i's or V_i's for the individual component gases.

An equation of state for a mixture can be developed from the equation of state for the pure components by using proper combination rules to obtain the constants for the mixture equation from the constants for the pure-component equation.

A general compressibility chart can be used to evaluate directly the compressibility factor for a mixture by using a pseudocritical point to calculate the reduced coordinates. The pseudocritical pressure p_c' and pseudocritical temperature T_c' can be defined as

$$p_c' = \sum x_i p_{ci} \qquad \text{and} \qquad T_c' = \sum x_i T_{ci}$$

where p_{ci} and T_{ci} are the critical pressure and temperature, respectively, of component i, and x_i is the mole fraction of component i.

PROBLEMS

14-1. For each of the following two-constant equations of state, verify the expressions for the specific constants a and b in terms of the critical pressure p_c and the critical temperature T_c.

(a) Dieterici equation (proposed in 1899):

$$p = \frac{\Re T}{v - b} e^{-a/\Re T v}$$

for which

$$a = 4 \, \Re^2 T_c^2 / p_c e^2 \qquad \text{and} \qquad b = \Re T_c / p_c e^2$$

(b) Berthelot equation (proposed in 1899):

$$\left(p + \frac{a}{T v^2}\right)(v - b) = \Re T$$

for which

$$a = \frac{27}{64} \frac{\Re^2 T_c^3}{p_c} \qquad \text{and} \qquad b = \frac{\Re T_c}{8 p_c}$$

(c) Redlich–Kwong equation (proposed in 1949): See Eq. (14-6) for the equation of state, and Eqs. (14-7a) and (14-7b) for constants a and b.

14-2. The Clausius equation (proposed in 1880) is given as

$$\left[p + \frac{a}{T (v + c)^2}\right](v - b) = \Re T$$

Verify the following expressions for the specific constants a, b, and c in terms of the critical properties p_c, T_c, and v_c.

$$a = \frac{27}{64} \frac{\Re^2 T_c^3}{p_c}$$

$$b = v_c - \frac{\Re T_c}{4 p_c}$$

$$c = \frac{3}{8} \frac{\Re T_c}{p_c} - v_c$$

14-3. Refrigerant-12 gas has a specific volume of 0.02638 m³/kg at a pressure of 800 kPa. Estimate the temperature of the gas by the use of (a) the ideal-gas equation of state, (b) the van der Waals equation, (c) the Redlich–Kwong equation, and (d) the generalized compressibility chart. The value recorded in the Refrigerant-12 property table for the temperature of the gas at the given pressure and specific volume is 70°C.

14-4. Nitrogen has a specific volume of 0.005016 m³/kg at a temperature of 200 K. Estimate the pressure of the gas by the use of (a) the ideal-gas equation of state, (b) the Beattie–Bridgeman equation, and (c) the Benedict–Webb–Rubin equation. The value recorded in the nitrogen property table for the pressure of the gas at the given temperature and specific volume is 10.0 MPa.

14-5. Estimate the specific volume of steam at 14.0 MPa and 400°C by the use of (a) the ideal-gas equation of state, (b) the van der Waals equation, (c) the Redlich–Kwong equation, and (d) the compressibility factor. The value recorded in the steam table for the specific volume of the vapor at the given pressure and temperature is 0.01722 m³/kg.

14-6. A 10-ft³ tank contains nitrogen at 840 atm and 820°R. Estimate the mass of nitrogen in the tank by the use of (a) the ideal-gas equation of state, (b) the compressibility factor, (c) the van der Waals equation, and (d) the reduced van der Waals equation. Experimental data indicate that the mass of nitrogen in the tank is 250 lbm.

14-7. Two kilograms of ammonia vapor at a pressure of 5 atm and a temperature of 300 K are compressed in a closed system isentropically to 50 atm. See Fig. P14-7. Calculate the final temperature, the final volume, and the work required if (a) the ideal-gas equation is obeyed, and (b) the van der Waals equation is obeyed. Assume $c_v = 34.3$ kJ/kgmole·K for both cases.

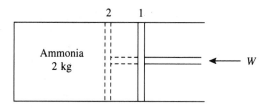

$P_1 = 5$ atm
$T_1 = 300$ K
$s_2 = s_1$
$P_2 = 50$ atm
(Use Van der Waals eq.)

Figure P14-7

14-8. Calculate the pressure of 3 lbmole of sulfur dioxide at 150°F when the total volume is 111 ft³ by the use of (a) the ideal-gas equation, and (b) the van der Waals equation.

14-9. Calculate the enthalpy of CO_2 at 50 atm and 100°C by the use of the van der Waals equation. The value of h for CO_2 at 100°C and 1 atm is 874.8 kJ/kg.

14-10. **(a)** Determine the values for the constants of the van der Waals equation corresponding to the critical pressure and temperature of steam. Plot the critical isotherms from this equation and from the steam tables on a z–p chart.
(b) Plot the isotherms from the van der Waals equation and from steam tables for 500°F. Locate the saturation states corresponding to the equation.

14-11. Ten pounds of argon at 70°F fill one-half of a 2-ft³ rigid, adiabatic tank. The other half is evacuated, and the two halves are separated by a membrane.

If the membrane should rupture, what final temperature and pressure would result (a) if the gas is ideal and (b) if the gas follows the van der Waals equation? Assume that $c_v = 3.0$ Btu/lbmole·°R for argon.

14-12. Ten kilograms of argon at 20°C fill one-half of a 0.1-m³ rigid, adiabatic tank. The other half is evacuated, and the two halves are separated by a membrane. See Fig. P14-12. If the membrane should rupture, what final temperature and pressure would result (a) if the gas is ideal, and (b) if the gas follows the van der Waals equation? Assume that $c_v = 12.5$ kJ/kgmole·K for argon.

14-13. Refrigerant-12 having a molecular mass of 121 expands irreversibly in a cylinder from 2 MPa, 80°C, to 0.2 MPa in an isothermal process. See Fig. P14-13. During this process, the heat transfer to the fluid is 45 kJ/kg. Determine the work per kg of Refrigerant-12 if it obeys the Redlich–Kwong equation of state.

14-14. Calculate the enthalpy of argon at 300 atm and 300 K by the use of the Redlich–Kwong equation of state. The value of h for argon at 1 atm and 300 K is 13,952 kJ/kgmole.

14-15. Calculate the entropy of argon at 300 atm and 300 K by the use of the Redlich–Kwong equation of state. The value of s for argon at 1 atm and 300 K is 154.56 kJ/kgmole·K.

14-16. Calculate the entropy of carbon monoxide at 400 atm and 100°C by the use of the Redlich–Kwong equation of state. The value of s for CO at 1 atm and 100°C is 855.6 kJ/kgmole·K.

14-17. One kilogram of carbon monoxide is compressed reversibly from 1 atm to 200 atm at a constant temperature of 150°C in a closed system. See Fig. P14-17. Determine the work done on the gas and the heat transferred from the gas, assuming that CO obeys the Clausius equation of state (see Prob. 14-2).

14-18. One kilogram of carbon monoxide is compressed reversibly from 1 atm to 100 atm at a constant temperature of 400 K in a closed system. See Fig. P14-18. Determine the work done on the gas and the heat transferred from the gas, assuming that CO obeys the Berthelot equation of state (see Prob. 14-1).

14-19. Carbon monoxide expands adiabatically in a steady-flow machine from 200 atm, 150°C, to 10 atm, 0°C. See Fig. P14-19. The constant-pressure specific heat of carbon monoxide at

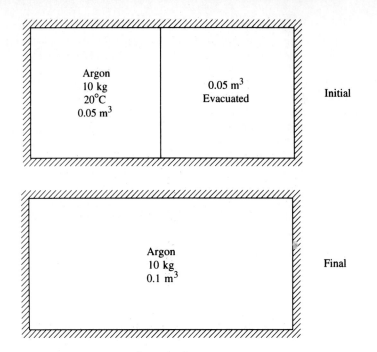

Figure P14-12

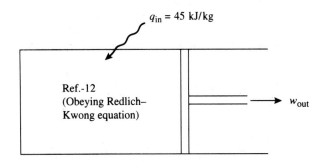

Irreversible expansion
$P_i = 2$ MPa $\quad T_i = 80°C$
$P_f = 0.2$ MPa $\quad T_f = T_i$

Figure P14-13

1 atm can be expressed as

$$c_p = 4.186(6.480 + 1.566 \times 10^{-3}T$$
$$- 0.2387 \times 10^{-6}T^2)$$

where c_p is in kJ/kgmole·K, and T in K. Using the Clausius equation of state (see Problem 14-2), determine the work done and the change in entropy per kg of the gas.

14-20. Argon is compressed reversibly and isothermally in a closed system from 1 atm and 300 K to 120 atm. Determine the work required and the heat transfer per kg of the gas, assuming that

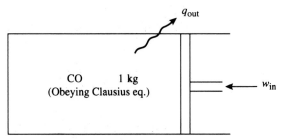

Reversible compression
$P_i = 1$ atm $\quad P_f = 200$ atm
$T_i = T_f = 150°$ C

Figure P14-17

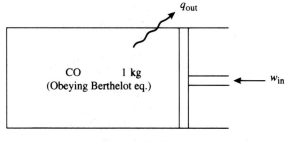

Reversible compression
$P_i = 1$ atm $\quad P_f = 100$ atm
$T_i = T_f = 400$ K

Figure P14-18

Real Gases and Real-Gas Mixtures Chap. 14

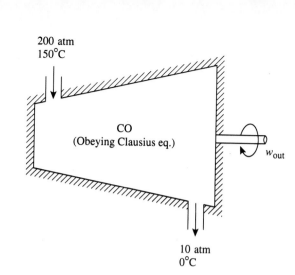

200 atm
150°C

CO
(Obeying Clausius eq.)

w_{out}

10 atm
0°C

Figure P14-19

argon obeys the Beattie equation, which is

$$v = \frac{\mathscr{R}T}{p} + \frac{\beta}{\mathscr{R}T} + \frac{\gamma}{\mathscr{R}^2 T^2}p + \frac{\delta}{\mathscr{R}^3 T^3}p^2$$

where

$$\beta = \mathscr{R}TB_0 - A_0 - \mathscr{R}c/T^2$$

$$\gamma = -\mathscr{R}TB_0 b + A_0 a - \mathscr{R}B_0 c/T^2$$

$$\delta = \mathscr{R}B_0 bc/T^2$$

in which constants A_0, B_0, a, b, and c are those of the Beattie–Bridgeman equation of state as given in Table A-31.

14-21. It is assumed that the following data on steam are available from experimental measurements. From these data, determine for 1 pound of steam, the entropy and enthalpy at a pressure of 550 psia and a temperature of 1500°F.

(1) The entropy and enthalpy for saturated liquid at 32.018°F is assumed to be zero.

(2) The average c_p for the saturated liquid between 32°F and 100°F is approximately 1 Btu/lbm·°F.

(3) The saturation pressure at 100°F is 0.9503 psia.

(4) The latent heat of vaporization at 100°F is 1037.0 Btu/lbm.

(5) The c_p values of superheated steam at low pressures are those shown by the curve designated by zero pressure in Fig. 4, p. 122, of Keenan, Keyes, Hill, and Moore, *Steam Tables* (John Wiley, New York, 1969).

(6) The values of specific volume of superheated steam in the required pressure and temperature ranges are those given in *Steam Tables*. (Use pressures, in psia, of 1, 2, 3, 4, 6, 8, 10, 20, 40, 80, 100, 150, 250, 300, 400, 500, 600, 800, and 1000, and each at temperatures, in °F, of 600, 700, 800, 900, 1000, 1100, 1200, 1300, 1400, 1500, and 1600.)

14-22. Determine the critical compressibility factor $Z_c = p_c v_c / \mathscr{R}T_c$ for a gas obeying (a) the Dieterici equation, (b) the Berthelot equation, and (c) the Redlich–Kwong equation. See Problem 14-1 for the equations.

14-23. Calculate the specific volume of oxygen in cm³/g at a pressure of 100 atm and a temperature of 170 K by (a) the ideal-gas equation, and (b) the compressibility chart.

14-24. Derive the virial form of the Redlich–Kwong equation of state.

14-25. One kilogram of argon is compressed reversibly from 1 atm to 40 atm at a constant temperature of 200 K in a closed system. Determine the work required and the heat transfer, assuming that argon obeys the following virial equation of state:

$$\frac{pv}{\mathscr{R}T} = 1 + \frac{B(T)}{v}$$

where

$$B(T) = \frac{9}{128} \frac{\mathscr{R}T_c}{p_c} \left(1 - 6\frac{T_c^2}{T^2}\right)$$

14-26. Verify that when Beattie–Bridgeman equation of state is written in virial form, the second, third, and fourth coefficients are, respectively,

$$B(T) = -\frac{A_0}{\mathscr{R}T} + B_0 - \frac{c}{T^3}$$

$$C(T) = \frac{aA_0}{\mathscr{R}T} - bB_0 - \frac{cB_0}{T^3}$$

$$D(T) = bcB_0/T^3$$

14-27. Show that the Redlich–Kwong equation in reduced form is

$$P_r = \frac{T_r}{Z_c(v_r - 0.08664/Z_c)}$$
$$- \frac{0.42748}{T_r^{0.5} v_r Z_c^2(v_r + 0.08664/Z_c)}$$

14-28. Write the Dieterici equation

$$p = \mathscr{R}T(v - b)^{-1} e^{-a/\mathscr{R}Tv}$$

Problems

in reduced form. Use this reduced equation to find the pressure dependence of the inversion point of the Joule–Thomson effect.

14-29. The van der Waals equation in reduced form is

$$\left(p_r + \frac{3}{v_r^2}\right)(3v_r - 1) = 8T_r$$

Derive the expression for the Joule–Thomson inversion line.

14-30. Using the van der Waals equation as a guide, estimate the temperature of inversion at zero pressure for steam and for hydrogen. Note that steam has a positive and hydrogen a negative Joule–Thomson coefficient at room temperature.

14-31. Derive an equation for calculating the effect of volume on the entropy of a gas when temperature is constant, using the generalized Beattie–Bridgeman equation of state, which is

$$P_r = \frac{T_r(1 - \epsilon')}{v_r^2}(v_r + B') - \frac{A'}{v_r^2}$$

where

$$\epsilon' = \frac{c'}{v_r T_r^3}, \qquad A' = A_0'\left(1 - \frac{a'}{v_r}\right),$$

$$B' = B_0'\left(1 - \frac{b'}{v_r}\right),$$

$$P_r = \frac{p}{p_c}, \qquad T_r = \frac{T}{T_c},$$

$$v_r = \frac{v}{v_{ci}}, \qquad v_{ci} = \frac{\mathcal{R}T_c}{p_c},$$

and A_0', B_0', a', b', and c' are universal constants (having the same values for all gases). The properties at a reference state on the isotherm T_r are p_0, v_0, s_0, p_{r0}, and v_{r0}.

14-32. Propane gas at 30 atm and 95°C enters a pipeline at a velocity of 20 m/s. See Fig. P14-32. The pipe is of constant cross section and is perfectly insulated. The gas leaves the pipe at a pressure of 8 atm. Use the generalized charts to find the temperature and velocity of the gas leaving the

pipe. Assume that the ideal-gas state, constant-pressure specific heat c_p^* has a constant value of 88.41 kJ/kgmole·K.

14-33. In a Joule–Thomson experiment, oxygen at 2000 psia and 60°F is expanded to 200 psia. By the use of a generalized chart, show by proper calculations that the exit temperature of the oxygen is 470°R.

14-34. To produce liquid oxygen, it is desired that the gas be first compressed and cooled to a pressure of 100 atm and a temperature of −90°C. The original oxygen gas is at a pressure of 1 atm and a temperature of 22°C. Calculate the volume (in m^3) of the compressed gas from 100 m^3 of the original gas by using (a) the ideal-gas equation, and (b) a compressibility chart.

14-35. Carbon monoxide expands adiabatically in a steady-flow machine from 200 atm, 150°C, to 10 atm, 0°C. The ideal-gas state, constant-pressure specific heat of CO can be expressed as

$$c_p^* = 4.186(6.480 + 1.566 \times 10^{-3}T$$
$$- 0.2387 \times 10^{-6}T^2)$$

where c_p^* is in kJ/kgmole·K, and T is in K. Use the generalized thermodynamic charts to determine the work done and the change in entropy per kg of the gas.

14-36. The equation of state of a geseous mixture and of each of its component gases are all assumed to be in the following virial form

$$pv = \mathcal{R}T + \frac{B}{v} + \frac{C}{v^2} + \frac{D}{v^3} + \cdots$$

where v denotes the molar volume. Derive the expressions for the virial coefficients in the equation of state of the mixture in terms of the virial coefficients in the equation of state of the component gases, assuming that Dalton's law holds for the mixture.

14-37. A gaseous mixture of 70 mole percent methane and 30 mole percent nitrogen at 100 atm and 300 K expands with a negligible initial velocity

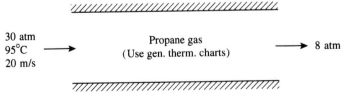

30 atm
95°C
20 m/s

Propane gas
(Use gen. therm. charts)

8 atm

Figure P14-32

Real Gases and Real-Gas Mixtures   Chap. 14

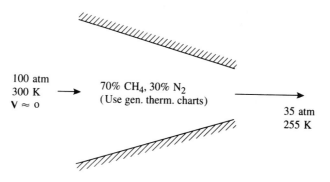

Figure P14-37

through an insulated nozzle to a final pressure of 35 atm. See Fig. P14-37. The final temperature is reported to be 255 K. Determine the possibility of this process and calculate the exit velocity by the use of (a) the ideal-gas relations and (b) Kay's pseudocritical rule and generalized thermodynamic charts. The ideal-gas state, constant-pressure specific heats at an average temperature of $\frac{1}{2}(300 + 255) = 277.5$ K for methane and nitrogen, respectively, are

$$c_p{}^*(CH_4) = 35.65 \text{ kJ/kgmole·K}$$

$$c_p{}^*(N_2) = 29.05 \text{ kJ/kgmole·K}$$

14-38. A gaseous mixture of 90 mole percent methane and 10 mole percent nitrogen at 70 atm and 340 K expands with a negligible initial velocity through an insulated nozzle to a final pressure of 7 atm. The final temperature is reported to be 280 K. Determine the possibility of this process and calculate the exit velocity by the use of the Kay's pseudocritical rule and the generalized thermodynamic charts. The ideal-gas state, constant-pressure specific heats at an average temperature of $\frac{1}{2}(340 + 280) = 310$ K for methane and nitrogen, respectively, are 36.5 kJ/kgmole·K and 29.2 kJ/kgmole·K.

14-39. A gaseous mixture contains 70 mole percent of methane and 30 mole percent of nitrogen. Calculate the specific volume of the mixture at 0°F and 1500 psia by using (a) the ideal-gas equation, and (b) the pseudocritical concept. The experimental value of the compressibility factor of this mixture at 0°F and 1500 psia is 0.82. What is the true value of the specific volume?

14-40. A rigid insulated tank of 0.2-m³ capacity contains a natural gas mixture of 80 mole percent methane and 20 mole percent ethane at a pressure of 3 MPa and a temperature of 40°C. See Fig. P14-

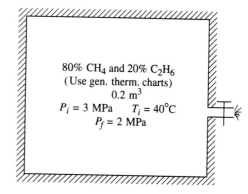

Figure P14-40

40. Due to leakage through a faulty valve, the pressure drops to 2 MPa before the valve is repaired. By the use of Kay's rule and generalized thermodynamic charts, calculate the mass of the mixture that leaks out from the tank, assuming that the mass that remains in the tank expands reversibly and adiabatically. The ideal-gas state, constant-pressure specific heat for methane and ethane are, respectively, 36.2 kJ/kgmole·K and 53.0 kJ/kgmole·K.

14-41. A mixture of 70 mole percent methane and 30 mole percent nitrogen is compressed irreversibly in an adiabatic steady-flow compressor from 35 atm, −100°C, to 100 atm, 10°C. See Fig. P14-41. The mass-flow rate is 1 kg/s. Calculate the rate of entropy increase and the power requirement, assuming the mixture to obey (a) the ideal-gas equation, and (b) the Redlich–Kwong equation, using the linear square-root combination for the constant *a* and the linear combination for the constant *b*. The constant-pressure specific heats of pure methane and pure nitrogen at 1 atm

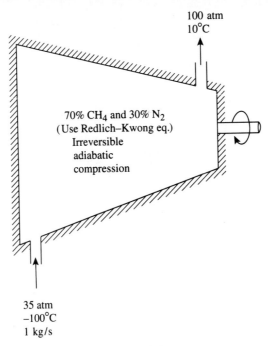

100 atm
10°C

70% CH$_4$ and 30% N$_2$
(Use Redlich–Kwong eq.)
Irreversible
adiabatic
compression

35 atm
−100°C
1 kg/s

Figure P14-41

can be assumed to be 34.4 kJ/kgmole·K and 29.0 kJ/kgmole·K, respectively.

14-42. Using the seven methods as stated in Example 14-9, estimate the pressure of a natural-gas mixture containing 70 mole percent of methane (CH$_4$), 20 mole percent of ethane (C$_2$H$_6$), and 10 mole percent of nitrogen, and having a temperature of 70°C and a molar volume of 0.2 liter/gmole.

14-43. One gmole of a mixture containing 50 mole percent of CO$_2$ and 50 mole percent of ethane (C$_2$H$_6$) is compressed in a cylinder in an internally reversible isothermal process. See Fig. P14-

43. The initial state is 0.7 MPa, 35°C, and the final pressure is 5.5 MPa. Calculate the heat transfer and work requirement for the process using (a) the ideal-gas behavior, (b) Kay's rule of the pseudocritical point and generalized thermodynamic charts, and (c) the van der Waals equation of state for the mixture, using the linear square-root combination for constant a and the linear combination for constant b.

14-44. A 0.142-m³ rigid insulated tank contains oxygen initially at 100 atm and 5°C. See Fig. P14-44. A paddle wheel inside the tank is turned on until the gas pressure becomes 300 atm. Taking data from

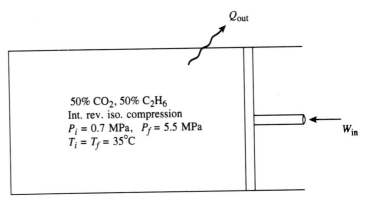

Q_{out}

50% CO$_2$, 50% C$_2$H$_6$
Int. rev. iso. compression
$P_i = 0.7$ MPa, $P_f = 5.5$ MPa
$T_i = T_f = 35$°C

W_{in}

Figure P14-43

Real Gases and Real-Gas Mixtures   Chap. 14

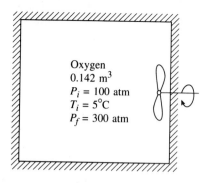

Figure P14-44

Oxygen
0.142 m³
$P_i = 100$ atm
$T_i = 5°C$
$P_f = 300$ atm

the generalized thermodynamic charts, determine (a) the final temperature of the gas, in°C, (b) the paddle-wheel work input, in kJ, and (c) the irreversibility for the process, in kJ, using $T_0 = 5°C$.

14-45. Carbon dioxide enters an adiabatic turbine at 5 MPa and 100°C, and exits at 1 MPa. The turbine has an isentropic efficiency of 75%. By the use of the generalized thermodynamic charts, determine (a) the exit temperature, and (b) the second-law effectiveness for this process, assuming $T_0 = 25°C$.

14-46. Two uninsulated rigid tanks are connected by a valve. See Fig. P14-46. One tank (tank A) initially contains 10 kg of liquid–vapor mixture of propane (C_3H_8) at 25°C and 80% quality. The other tank (tank B) is initially evacuated. The valve is now opened and the propane eventually occupies both tanks at a final temperature of 25°C and a final pressure of 90% of the initial tank A pressure. Taking data from the generalized thermodynamic charts, determine (a) the heat transfer, in kJ, and (b) the irreversibility for the process, in kJ, using $T_0 = 25°C$.

14-47. Carbon dioxide at a pressure of 800 psia and a temperature of 100°F expands isentropically in a nozzle to a pressure of 500 psia. By the use of the

$T_0 = 25°C$

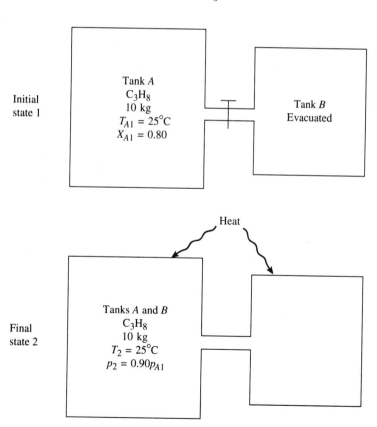

Initial
state 1

Tank A
C_3H_8
10 kg
$T_{A1} = 25°C$
$X_{A1} = 0.80$

Tank B
Evacuated

Heat

Final
state 2

Tanks A and B
C_3H_8
10 kg
$T_2 = 25°C$
$p_2 = 0.90p_{A1}$

Figure P14-46

van der Waals equation of state, compute the exit temperature and velocity, assuming the inlet velocity to be negligible. The constant-pressure specific heat of carbon dioxide at 14.7 psia can be expressed as

$$c_p = 6.85 + 0.00474T - 7.64 \times 10^{-7}T^2$$

where c_p is in Btu/lbmole·°R, and T is in °R. Use a computer to perform the calculations. Give your computer program.

14-48. Solve Prob. 14-47 by the use of the Benedict–Webb–Rubin equation of state.

14-49. Write a computer program to calculate the enthalpys h and entropys s of a gas at various pressures p and temperatures T, using an equation of state in the virial form:

$$\frac{pv}{\mathcal{R}T} = 1 + \frac{B(T)}{v} + \frac{C(T)}{v^2} + \frac{D(T)}{v^3}$$

and a constant-pressure specific heat equation

$$c_p = 4.186(\alpha + \beta T + \gamma T^2 + \delta T^3)$$

Use a reference point on the saturated-vapor line. Apply the Beattie–Bridgeman equation in virial form (see Prob. 14-26) for methane to perform a series of numerical calculations by your computer program. For methane at $p_0 = 1$ atm and $T_0 = T_{sat} = 111.36$ K, $h_0 = h_g = 12,795$ kJ/kgmole and $s_0 = s_g = 152.89$ kJ/kgmole·K. The coefficients in the c_p equation (with c_p in kJ/kgmole·K and T in K) are

$$\alpha = 4.750 \qquad \beta = 12.00 \times 10^{-3}$$
$$\gamma = 3.030 \times 10^{-6} \qquad \delta = -2.630 \times 10^{-9}$$

14-50. Write a computer program to calculate the values of v, h, s, c_p, and c_v for carbon monoxide at a pressure of 100 atm and a temperature of 400 K by the use of the Berthelot equation of state (see Prob. 14-1). Given that at $p_0 = 0.1$ atm and $T_0 = 273$ K, $h_0 = 16,049$ kJ/kgmole and $s_0 = 214.45$ kJ/kgmole·K based on the value of zero for the perfect crystal at absolute zero temperature. The ideal-gas state, constant-pressure specific heat for carbon monoxide can be expressed as

$$c_p = 4.186(6.726 + 0.4001 \times 10^{-3}T$$
$$+ 1.283 \times 10^{-6}T^2 - 0.5307 \times 10^{-9}T^3)$$

where c_p is in kJ/kgmole·K, and T is in K.

14-51. Carbon dioxide expands in an internally reversible isothermal steady-flow machine from a pressure of 50 atm and a temperature of 100°C to an exhaust pressure of 1 atm. The volume-flow rate at the inlet condition is 0.01 m³/s. Calculate the power output and the rate of heat transfer, assuming (a) the gas obeys the ideal-gas equation, and (b) the gas obeys the Dieterici equation (see Prob. 14-1). Use a computer to perform the calculations. Give your computer program.

14-52. An equation of state suggested by Martin and Hou in 1955 reads as follows:

$$p = \frac{\mathcal{R}T}{v - b} + \frac{A_2 + B_2T + C_2e^{-5.475T/T_c}}{(v - b)^2}$$
$$+ \frac{A_3 + B_3T + C_3e^{-5.475T/T_c}}{(v - b)^3}$$
$$+ \frac{A_4}{(v - b)^4} + \frac{B_5T}{(v - b)^5}$$

where p is in atm, T is in K, v is in cm³/gmole, and $R = 82.055$ cm³·atm/gmole·K. For nitrogen, the nine specific constants in units of atm, K, and cm³/gmole are

$b = 22.1466$	$A_2 = -1,592,238.2$
$B_2 = 3,221.616$	$C_2 = -22,350,930$
$A_3 = 89,845,367$	$B_3 = -134,024.05$
$C_3 = 1,518,821,653$	$A_4 = -2,467,297,480$
$B_5 = 244,915,183$	

Using the Martin–Hou equation of state, calculate the values of volume v, enthalpy h, entropy s, internal energy u, Gibbs function g, Helmholtz function a, constant-pressure specific-heat c_p, and constant-volume specific-heat c_v for 1 gram-mole of nitrogen at $T = 300$ K and $p = 300$ atm. Given that at $T_0 = 300$ K and $p_0 = 1$ atm, $h_0 = 15,590$ J/gmole, $s_0 = 191.4$ J/gmole·K, and $c_{p0} = 28.98$ J/gmole·K, where the values of h and s are based on the value of zero for the perfect crystal at 0 K. Use a computer to carry out the necessary calculations. Give your computer program.

Real Gases and Real-Gas Mixtures   Chap. 14

15

Chemical Reactions

All the preceding chapters were concerned with nonreactive systems. We now turn over attention to systems involving chemical reactions. Although the principles covered in this chapter apply to any chemical reaction, our interest is directed particularly toward an important class of chemical reaction, the combustion processes.

From a thermodynamic point of view, there are three major aspects of a chemical reaction: (1) a mass balance through the use of the principle of conservation of mass, thus establishing a chemical reaction equation; (2) an energy balance through the use of the first law of thermodynamics, thus determining the energy transfers and conversions; and (3) an equilibrium study through the use of the second law of thermodynamics, thus predicting the extent and direction of a given reaction. In addition, the third law of thermodynamics is used to compute entropies of different species involved in a reaction. In this chapter, we illustrate the mass-balance and energy-balance procedures. Topics on chemical equilibrium and second-law analysis for reacting systems will be taken up in the next chapter.

15-1 COMBUSTION REACTIONS

A chemical reaction is a process in which the interatomic bonds in the molecules of a collection of certain chemical constituents are broken, followed by the rearrangement of the atoms and electrons to form different constituents of new atomic combinations. Associated with every chemical reaction is a chemical equation. For example, the reaction between methane and oxygen to form carbon dioxide and water is expressed by the equation

$$CH_4 + 2O_2 \longrightarrow CO_2 + 2H_2O$$

where the constituents on the left are called the reactants and those on the right the products. The numerical coefficients that precede the chemical symbols give equal number of atoms of each chemical element on both sides of the equation and are called *stoichiometric coefficients*.

Many thermodynamic applications involve the use of heat generated by combustion reactions. *Combustion* is a rapid reaction between fuels and oxygen in which chemical energy is liberated. In a combustion process of a hydrocarbon fuel, if all the carbon present is converted into CO_2 and all the hydrogen is converted into H_2O, the process is described as *complete combustion*. If CO appears in the products, the combustion is incomplete.

In most combustion processes, the oxygen is supplied as air rather than pure oxygen. The composition of normal dry air on a molar basis is approximately 78% nitrogen, 21% oxygen, and 1% argon and carbon dioxide. For calculation purposes, we can treat the argon and carbon dioxide as nitrogen and the air is assumed to be 79% nitrogen and 21% oxygen. Thus, there are $79/21 = 3.76$ moles of nitrogen for each mole of oxygen in the air. In the normal temperature range of many combustion processes, the nitrogen is not involved in the chemical reaction. At high temperatures, such as those achieved in certain internal-combustion engines, some direct reactions of nitrogen and oxygen may take place to form oxides of nitrogen. In our analyses, unless otherwise stated, we will assume that the nitrogen is not involved in the chemical reaction. However, the temperature of nitrogen after the reaction process, in general, is not the same as before. This factor should be taken care of when the energy balance is made for the process.

The minimum amount of air that supplies sufficient oxygen for complete combustion of a fuel is called the *theoretical air* (or *stoichiometric air*). The amount of air actually supplied may be less than or in excess of the theoretical air. If the actual air supply is less than the theoretical air, the reaction will be incomplete and CO and even unburned fuel will appear in the products. If excess air is supplied, the excess oxygen in a complete combustion process will appear in the products unchanged. The amount of *excess air* usually expressed as a percentage of the theoretical air required for complete combustion of the fuel. Thus, 150% theoretical air is equivalent to 50% excess air. In practice, 20% excess air is usually regarded as a minimum in order to ensure that no CO and unburned fuel are contained in the products.

To illustrate the procedure of writing a correct chemical equation, let us consider the complete combustion of 1 mole of methane with a theoretical amount of dry air. We first write the equation with unknown coefficients of the various constituents in the following manner:

$$CH_4 + (aO_2 + 3.76aN_2) \longrightarrow xCO_2 + yH_2O + zN_2$$

The unknown quantities in the preceding equation can be determined by mass balances for the atomic species. Thus,

C balance:	$x = 1$
H_2 balance:	$y = 2$
O_2 balance:	$a = x + y/2 = 1 + 2/2 = 2$
N_2 balance:	$z = 3.76a = 3.76 \times 2 = 7.52$

Therefore, the stoichiometric or theoretical equation for the combustion of methane with dry air is

$$CH_4 + 2O_2 + 7.52N_2 \longrightarrow CO_2 + 2H_2O + 7.52N_2$$

If 20% excess air, or 120% theoretical air is used for the complete combustion of methane, we first write the following:

$$CH_4 + 1.2(aO_2 + 3.76aN_2) \longrightarrow xCo_2 + yH_2O + 0.2aO_2 + z'N_2$$

where $1(aO_2 + 3.76aN_2)$ is the theoretical air required, and $0.2(aO_2 + 3.76aN_2)$ is the excess air supplied. The coefficient $0.2a$ for O_2 in the products represents the 20% excess O_2. By writing mass balances for the atomic species, we obtain

$$x = 1 \qquad y = 2 \qquad a = 2 \qquad z' = 9.024$$

Therefore, the complete combustion of methane with 20% excess air is expressed by the chemical equation:

$$CH_4 + 2.4O_2 + 9.024N_2 \longrightarrow CO_2 + 2H_2O + 0.4O_2 + 9.024N_2$$

It should be mentioned that even when some excess air is supplied, due to uneven mixing and some other factors, the reaction may still be incomplete and CO_2, CO, and O_2 appear together in the products. Thus, an incomplete combustion of methane with 20% excess air might be as follows:

$$CH_4 + 1.2(2O_2 + 3.76 \times 2N_2)$$

$$\longrightarrow 0.95CO_2 + 0.05CO + 2H_2O + 0.425O_2 + 9.024N_2$$

The relationship between the air supply and the fuel can also be expressed in terms of the *air–fuel ratio* (AF). It is usually expressed on a mass basis and is defined as the ratio of the mass of air supplied to the mass of fuel. The air–fuel ratio can also be expressed on a mole basis as the ratio of the mole numbers of air to the mole numbers to fuel. When theoretical air is supplied to a combustion process, this ratio is referred to as the theoretical or stoichiometric air–fuel ratio.

As an example, for the complete combustion of methane with theoretical air as expressed by the chemical equation,

$$CH_4 + 2O_2 + 7.52N_2 \longrightarrow CO_2 + 2H_2O + 7.52N_2$$

the theoretical air–fuel ratio on a mole basis is

$$AF = (2 + 7.52)/1 = 9.52 \text{ kgmole air/kgmole fuel (or lbmole air/lbmole fuel)}$$

The air–fuel ratio on a mass basis is found by introducing the molecular weights of the air and the fuel. Thus,

$$AF = \frac{(9.52 \text{ kgmole air/kgmole fuel})(29 \text{ kg air/kgmole air})}{16 \text{ kg fuel/kgmole fuel}}$$

$$= 17.255 \text{ kg air/kg fuel (or lbm air/lbm fuel)}$$

As another example, for the complete combustion of methane with 20% excess air as expressed by the chemical equation,

$$CH_4 + 2.4O_2 + 9.024N_2 \longrightarrow CO_2 + 2H_2O + 0.4O_2 + 9.024N_2$$

the air–fuel ratio on a mole basis is

$$AF = (2.4 + 9.024)/1 = 11.424 \text{ kgmole air/kgmole fuel (or lbmole air/lbmole fuel)}$$

On a mass basis, it becomes

$$AF = \frac{(11.424 \text{ kgmole air/kgmole fuel})(29 \text{ kg air/kgmole air})}{16 \text{ kg fuel/kgmole fuel}}$$

$$= 20.706 \text{ kg air/kg fuel (or lbm air/lbm fuel)}$$

Sec. 15-1   Combustion Reactions

For this case, the percentage theoretical air is

$$\frac{20.706}{17.255} \times 100 = 120\%$$

and the percentage excess air is

$$\frac{20.706 - 17.255}{17.255} \times 100 = 20\% \text{ (as given)}$$

Fossil fuels can be classified as solid (such as coal), liquid (such as gasoline), and gaseous (such as natural gas). The combustible constituents of fuels are carbon, hydrogen, sulfur, and their compounds. The composition of solid and liquid fuels is usually given as an *ultimate analysis,* which specifies the percentages by mass of the various constituents (carbon, sulfur, hydrogen, nitrogen, oxygen, and ash if any). The composition of a gaseous fuel is usually given in terms of a *volumetric* (or *molar*) *analysis,* which specifies the percentages by volume (or mole) of the various gaseous components (methane, CH_4; ethane, C_2H_6; propane, C_3H_8; butane, C_4H_{10}; carbon monoxide, CO; hydrogen, H_2; nitrogen, N_2; oxygen, O_2; and others). Most liquid fossil fuels are mixtures of many different hydrocarbons. In combustion analysis of liquid fuels, however, it is sometimes convenient to express the composition in terms of a single hydrocarbon. For instance, gasoline is usually considered to be octane C_8H_{18}, and diesel fuel to be dodecane $C_{12}H_{26}$.

In the combustion of hydrocarbon fuels, one of the major products is water. When the products are cooled below the dew-point temperature, water vapor will condense. The dew-point temperature of the products is defined (see Sec. 9-5) as the saturation temperature of the water vapor corresponding to its partial pressure in the products. Considering the combustion product as an ideal-gas mixture, the partial pressure of water vapor, p_v, is given by Eq. (9-10),

$$p_v = x_v p$$

where p is the total pressure of the product mixture, and x_v is the mole fraction of water vapor in the mixture.

Example 15-1

A gaseous mixture of 70% CH_4, 20% CO, 5% O_2, and 5% N_2 on a mole basis is burned completely at 1 atm with 20% excess air. Write the chemical equation and determine (a) the volume of air supplied per unit volume of fuel, both being measured at the same pressure and temperature, (b) the mass of air supplied per kg of fuel, and (c) the volumetric analysis of the products of combustion.

Solution Let the number of moles of oxygen supply needed for the complete combustion of 1 mole of fuel be denoted by a. Thus, the theoretical air supplied is $(aO_2 + 3.76aN_2)$. With 20% excess air, the actual air supplied is then $1.2(aO_2 + 3.76aN_2)$ moles of air per mole of fuel. Therefore, the chemical equation for complete combustion per mole of fuel can be written as

$(0.70CH_4 + 0.20CO + 0.05O_2 + 0.05N_2) + 1.2(aO_2 + 3.76aN_2)$

$$\longrightarrow \quad xCO_2 + yH_2O + 0.2aO_2 + zN_2$$

where x, y, and z are the unknown numbers of moles of CO_2, H_2O, and N_2, respectively, in the products. The coefficient $0.2a$ for O_2 on the right-hand side of the equation represents the 20% excess O_2 that appears in the products unchanged.

The unknown quantities in the preceding equation can be determined by mass balances for the atomic species. Thus,

C balance :

$$x = 0.7 + 0.2 = 0.9 \text{ moles } CO_2/\text{mole fuel}$$

H_2 balance:

$$y = 2 \times 0.7 = 1.4 \text{ moles } H_2O/\text{mole fuel}$$

O_2 balance:

$$\tfrac{1}{2}x0.2 + 0.05 + 1.2a = x + \tfrac{1}{2}y + 0.2a$$

$$a = x + \tfrac{1}{2}y - 0.15$$

$$= 1.45 \text{ moles } O_2/\text{mole fuel (theoretical value)}$$

N_2 balance:

$$z = 0.05 + 1.2 \times 3.76a$$

$$= 6.59 \text{ moles } N_2/\text{mole fuel}$$

The chemical equation then becomes

$$(0.70CH_4 + 0.20CO + 0.05O_2 + 0.05N_2) + 1.2(1.45O_2 + 5.45N_2)$$
$$\longrightarrow \quad 0.9CO_2 + 1.4H_2O + 0.29O_2 + 6.59N_2$$

(a) According to the chemical equation, we have the ratio

$$\frac{\text{mole air}}{\text{mole fuel}} = 1.2(1.45 + 5.45) = 8.28$$

Because a mole of air and a mole of fuel at the same pressure and temperature occupy the same volume, the value 8.28 moles of air per mole of fuel also represents 8.28 m^3 of air per m^3 of fuel.

(b) The apparent or average molecular weight of air is 29. The apparent or average molecular weight of the fuel is

$$0.70 \times 16 + 0.20 \times 28 + 0.05 \times 32 + 0.05 \times 28 = 19.8 \text{ kg fuel/kgmole fuel}$$

Thus, the air–fuel mass ratio is

$$\left(8.28\frac{\text{kgmole air}}{\text{kgmole fuel}}\right)\frac{29 \text{ kg air/kgmole air}}{19.8 \text{ kg fuel/kgmole fuel}} = 12.1 \text{ kg air/kg fuel}$$

(c) According to Eq. (9-14), for an ideal-gas mixture the volume fraction of a constituent equals its mole fraction. There are a total of $0.9 + 1.4 + 0.29 + 6.59 = 9.18$ moles of products. The volumetric analysis of the products is then

CO_2: 0.9/9.18 = 9.80%
H_2O: 1.4/9.18 = 15.25%
O_2: 0.29/9.18 = 3.16%
N_2: 6.59/9.18 = 71.79%

Example 15-2

A power plant burns coal having an ultimate analysis in percentages as follows: C, 77.54; H, 4.28; S, 1.46; O, 7.72; N, 1.34; and ash, 7.66. For complete combustion with 20%

excess air, write the chemical reaction equation and determine the dew point of the products at 14.7 psia total pressure. If the product gases were cooled to 30°F below the dew-point temperature, determine the percent of H_2O formed that would be condensed.

Solution Because ash in the coal is composed of mineral and other inorganic matters deposited with organic materials during the compaction process, it does not take part in a combustion reaction process. We assume that all ashes go to the ashpit; thus, there is no flying ash in the combustion products. To write a chemical reaction equation, the ultimate analysis of the coal must be converted to a mole basis by using the appropriate atomic mass of the reactants. On the basis for the burning of 1 pound of coal, the results are as follows:

$$C: \frac{0.7754 \text{ lbm carbon/lbm fuel}}{12 \text{ lbm carbon/lbmole carbon}} = 0.06462 \text{ lbmole carbon/lbm fuel}$$

$$H: \frac{0.0428}{1} = 0.04280$$

$$S: \frac{0.0146}{32} = 0.00046$$

$$O: \frac{0.0772}{16} = 0.00483$$

$$N: \frac{0.0134}{14} = 0.00096$$

Thus, with 20% excess air, the chemical equation for complete combustion per pound of fuel can be written as

$$(0.06462C + 0.04280H + 0.00046S + 0.00483O + 0.00096N)$$
$$+ 1.2(aO_2 + 3.76aN_2) \longrightarrow xCO_2 + yH_2O + zSO_2 + 0.2aO_2 + \gamma N_2$$

The unknown quantities in the preceding equation can be determined by mass balances for the atomic species. Thus,

C balance: $x = 0.06462$
H_2 balance: $y = \frac{1}{2}(0.04280) = 0.02140$
S balance: $z = 0.00046$
O_2 balance: $a = x + \frac{1}{2}y + z - \frac{1}{2}(0.00483)$
$\qquad\qquad = 0.06462 + \frac{1}{2}(0.02140) + 0.00046 - \frac{1}{2}(0.00483)$
$\qquad\qquad = 0.07337$
N_2 balance: $\gamma = \frac{1}{2}(0.00096) + 1.2 \times 3.76a$
$\qquad\qquad = \frac{1}{2}(0.00096) + 1.2 \times 3.76 \times 0.07337$
$\qquad\qquad = 0.33153$

The chemical equation then becomes

$$(0.06462C + 0.04280H + 0.00046S + 0.00483O + 0.00096N)$$
$$+ (0.08804O_2 + 0.33105N_2)$$
$$\longrightarrow 0.06462CO_2 + 0.02140H_2O + 0.00046SO_2 + 0.01467O_2 + 0.33153N_2$$

The mole fraction of water vapor in the products is

$$x_v = \frac{0.02140}{0.02140 + (0.06462 + 0.00046 + 0.01467 + 0.33153)}$$

$$= \frac{0.02140}{0.02140 + 0.41128} = 0.0495$$

For a total pressure of $p = 14.7$ psia, the partial pressure of water vapor is

$$p_v = x_v p = 0.0495(14.7) = 0.728 \text{ psia}$$

The dew-point temperature of the products is obtained from the saturated steam table (Table A-19 E) as

$$T_{\text{dew-point}} = T_{\text{sat}}(\text{at } 0.728 \text{ psia}) = 91.3°F$$

At a temperature of $91.3 - 30 = 61.3°F$, the saturation pressure of water vapor as given by the steam table (Table A-18E) is

$$p_{\text{sat}} = 0.268 \text{ psia}$$

which is the partial pressure of saturated water vapor when the product mixture is cooled to 61.3°F. Accordingly, at 61.3°F, the mole fraction of water vapor in the mixture is

$$x_v' = \frac{0.268}{14.7} = 0.01823$$

In addition,

$$x_v' = \frac{n_v'}{n_v' + 0.41128}$$

where n_v' is the number of moles of water vapor in the mixture when the temperature is lowered to 61.3°F. Combination of the preceding two expressions for x_v' yields

$$0.01823 = \frac{n_v'}{n_v' + 0.41128}$$

so that

$$n_v' = 0.00764 \text{ lbmole}$$

Therefore,

$$\text{Number of moles of water condensed} = 0.02140 - 0.00764$$

$$= 0.01376$$

and

$$\text{Percent of water condensed} = \frac{0.01376}{0.02140} \times 100 = 64.3\%$$

15-2 ACTUAL COMBUSTION PROCESSES

In actual combustion processes, even though excess air is often supplied, combustion may still be incomplete mostly due to the uneven mixing of fuel and air, thus resulting in the presence of carbon monoxide and oxygen in the product gases. Furthermore, it is difficult to measure the air flow into a combustion chamber, making the prediction of an actual combustion process in detail formidable. A relatively easier way to obtain combustion information is to perform an analyses of the actual gaseous products, together with the usual analyses of the fuel and solid refuse (if any).

There are numerous modern experimental devices available for an accurate determination of the composition of combustion products. To obtain an approximate combustion analysis, however, one can rely on the well-established, simple, and inexpensive Orsat gas analyzer, shown in Fig. 15-1. It consists of a measuring burette and three reagent pipettes that are used to successively absorb carbon dioxide, oxygen, and carbon monoxide in the combustion products. A sample of the flue gas at room temperature is drawn into the measuring burette by lowering the leveling bottle. The gas is then forced into the first pipette, where carbon dioxide is absorbed. The volume of the remaining gas is measured and the gas is subsequently forced into the second pipette, where oxygen is absorbed. The procedure is repeated for the third pipette, where carbon monoxide is absorbed. Any gas remaining after the sample has passed through all three pipettes is assumed to be nitrogen.

The composition of gaseous combustion products can be reported on either a wet or a dry basis. A wet (or total) analysis includes water vapor in the composition, whereas a dry analysis does not. An Orsat analysis reports CO_2, O_2, CO, and N_2 on a dry basis. It should be mentioned that the sample used in an Orsat analysis does contain saturated water vapor at room temperature; but because the test is carried out at constant temperature, the amount of water vapor in the sample remains constant during the absorption process. Of course, the assumption that the remaining gas after CO_2, O_2, and CO have been removed is only N_2 apparently involves some error.

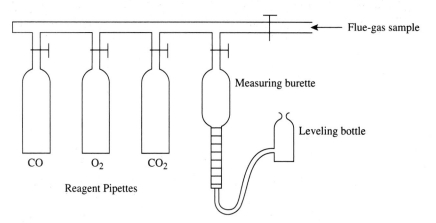

Figure 15-1   Schematic of an Orsat analyzer.

Example 15-3

A hydrocarbon fuel (C_xH_y) undergoes a combustion process such that the dry products have the following volumetric analysis: CO_2, 11.3%; CO, 1.1%; O_2, 3.5%; and N_2, 84.1%. Determine (a) the composition of the fuel and the chemical equation for the actual reaction, (b) the air–fuel mass ratio, and (c) the percent theoretical air.

Solution   The chemical equation for 100 moles of dry products can be written in the form:

$$C_xH_y + aO_2 + 3.76aN_2 \longrightarrow 11.3CO_2 + 1.1CO + 3.5O_2 + 84.1N_2 + bH_2O$$

The unknown quantities in the preceding equation can be determined by mass balances for the atomic species. Thus,

N_2 balance: $3.76a = 84.1$ or $a = 22.4$
O_2 balance: $a = 11.3 + \frac{1}{2}(1.1) + 3.5 + \frac{1}{2}b$
    or $b = 2(22.4 - 11.3 - \frac{1}{2} \times 1.1 - 3.5) = 14.1$
C balance: $x = 11.3 + 1.1 = 12.4$
H_2 balance: $y = 2b = 2(14.1) = 28.2$

Accordingly, the composition of the fuel could be written as $C_{12.4}H_{28.2}$. Notice that this is not the formula of a single chemical species, but the average formula for a mixture of many different hydrocarbons. The chemical equation can be written as

$$C_{12.4}H_{28.2} + 22.4O_2 + 84.1N_2$$

$$\longrightarrow 11.3CO_2 + 1.1CO + 3.5O_2 + 84.1N_2 + 14.1H_2O$$

The air–fuel mass ratio of the actual combustion process is then

$$\text{Actual air–fuel ratio} = \frac{22.4 \times 32 + 84.1 \times 28}{12.4 \times 12 + 28.2 \times 1}$$

$$= 17.4 \text{ kg air/kg fuel (or lbm air/lbm fuel)}$$

In order to determine the percent theoretical air used, we first need to write the theoretical complete combustion equation without any excess air. This equation can be written as

$$C_{12.4}H_{28.2} + \alpha O_2 + 3.76\alpha N_2 \longrightarrow \beta CO_2 + \gamma H_2O + 3.76\alpha N_2$$

Now by the use of mass balances to determine the unknown quantities, the results are

C balance: $\beta = 12.4$
H_2 balance: $\gamma = \frac{1}{2}(28.2) = 14.1$
O_2 balance: $\alpha = \beta + \frac{1}{2}\gamma = 12.4 + \frac{1}{2}(14.1) = 19.5$

The theoretical reaction equation is then

$$C_{12.4}H_{28.2} + 19.5O_2 + 73.3N_2 \longrightarrow 12.4CO_2 + 14.1H_2O + 73.3N_2$$

The theoretical air-fuel mass ratio is

$$\text{Theoretical air–fuel ratio} = \frac{19.5 \times 32 + 73.3 \times 28}{12.4 \times 12 + 28.2 \times 1}$$

$$= 15.1 \text{ kg air/kg fuel (or lbm air/lbm fuel)}$$

Therefore,

$$\text{Percent theoretical air} = \frac{17.4}{15.1} \times 100 = 115\%$$

or

$$\text{Percent excess air} = 15\%$$

The percent excess air can be also calculated by

$$\text{Percent excess air} = \frac{22.4 - 19.5}{19.5} \times 100 = 15\%$$

Example 15-4

A coal has the following ultimate analysis in percentages: C, 72; H, 5; O, 8; S, 3; N, 4; and ash, 8. The volumetric analysis of the products on a dry basis is as follows: CO_2, 13.8; O_2, 4.2; CO, 1.2; N_2, 80.8; SO_2 being included as CO_2. The solid refuse contains 20% unburned carbon. The atmospheric air supplied for combustion contains 0.030 lbm water vapor per lbm dry air, where dry air is assumed to be a mixture of N_2 and O_2 in a molar ratio of 79/21. The pressure and temperature of the products are 14.7 psia and 500°F, respectively. Write the chemical equation and determine the amount of water vapor in the products per pound of fuel.

Solution   The carbon burned is that portion of C in the coal that appears in the products as CO_2 and CO. Thus,

$$\frac{\text{lbm C burned}}{\text{lbm fuel}} = \frac{\text{lbm C in products}}{\text{lbm fuel}}$$

$$= \frac{\text{lbm C in fuel}}{\text{lbm fuel}} - \frac{\text{lbm C in refuse}}{\text{lbm fuel}}$$

$$= \left(0.72 \frac{\text{lbm C in fuel}}{\text{lbm fuel}}\right) - \left[\frac{0.20 \dfrac{\text{lbm C in refuse}}{\text{lbm refuse}}}{(1 - 0.20) \dfrac{\text{lbm ash}}{\text{lbm refuse}}} \left(0.08 \frac{\text{lbm ash}}{\text{lbm fuel}}\right)\right]$$

$$= 0.72 - 0.02 = 0.70$$

Because the SO_2 percentage is included in the CO_2 percentage in the given Orsat analysis, the mass of S in the fuel can be taken care of by adding 12/32 of the mass of S to the mass of C, thus treating S as additional C. Accordingly, we have

$$\frac{\text{lbm C burned (corrected)}}{\text{lbm fuel}} = 0.70 + \frac{12}{32} \times 0.03$$

$$= 0.70 + 0.011 = 0.71$$

Based on this amount of burned C, the combustion equation for the burning of 1 lbm of fuel with dry air can be written as

$$\left(\frac{0.71}{12}C + \frac{0.05}{1}H + \frac{0.08}{16}O + \frac{0.04}{14}N\right) + (aO_2 + 3.76aN_2)$$

$$\longrightarrow x(0.138CO_2 + 0.042O_2 + 0.012CO + 0.808N_2) + yH_2O$$

where x represents the number of moles of dry products per pound of fuel. We now use the mass balances to determine the unknown quantities in the preceding equation.

C balance: $\dfrac{0.71}{12} = x(0.138 + 0.012)$ or $x = 0.394$

H_2 balance: $y = \frac{1}{2} \times 0.05 = 0.025$

O_2 balance: $a = -\dfrac{0.08}{32} + x(0.138 + 0.042 + \frac{1}{2} \times 0.012) + \frac{1}{2}y$

$\qquad\qquad = -\dfrac{0.08}{32} + 0.394(0.138 + 0.042 + \frac{1}{2} \times 0.012) + \frac{1}{2} \times 0.025$

$\qquad\qquad = 0.0833$

N_2 balance for check:

$$\frac{0.04}{28} + 3.76a = \frac{0.04}{28} + 3.76 \times 0.0833 = 0.315$$

$$0.808x = 0.808 \times 0.394 = 0.318$$

which is as close a check as can be expected. The combustion equation for the burning of 1 lbm of fuel with dry air can then be written as

$$\left(\frac{0.71}{12}C + \frac{0.05}{1}H + \frac{0.08}{16}O + \frac{0.04}{14}N\right) + (0.0833O_2 + 0.313\ N_2)$$

$$\longrightarrow 0.394(0.138CO_2 + 0.042O_2 + 0.012CO + 0.808N_2) + 0.025H_2O$$

With 0.030 lbm water vapor per lbm dry air coming from the air supply, the amount of additional water vapor in the products is calculated as

$$\left(0.030\frac{\text{lbm }H_2O}{\text{lbm dry air}}\right)\left[\left(0.0833\frac{\text{lbmole }O_2}{\text{lbm fuel}}\right)\left(32\frac{\text{lbm }O_2}{\text{lbmole }O_2}\right)\right.$$

$$\left. + \left(0.313\frac{\text{lbmole }N_2}{\text{lbm fuel}}\right)\left(28\frac{\text{lbm }N_2}{\text{lbmole }N_2}\right)\right]$$

$$= \left(0.030\frac{\text{lbm }H_2O}{\text{lbm dry air}}\right)\left(11.43\frac{\text{lbm dry air}}{\text{lbm fuel}}\right)$$

$$= 0.343 \text{ lbm }H_2O/\text{lbm fuel}$$

$$= \frac{0.343}{18} = 0.0191 \text{ lbmole }H_2O/\text{lbm fuel}$$

The total number of moles of H_2O in the products is then

$$0.025 + 0.0191 = 0.0441 \text{ lbmole }H_2O/\text{lbm fuel}$$

Including the water vapor coming from the air supply, one can write the combustion equation as

$$\left(\frac{0.71}{12}C + \frac{0.05}{1}H + \frac{0.08}{16}O + \frac{0.04}{14}N\right) + (0.0833O_2 + 0.313N_2 + 0.0191H_2O)$$

$$\longrightarrow 0.394(0.138CO_2 + 0.042O_2 + 0.012CO + 0.808N_2) + 0.0441H_2O$$

The mole fraction of water vapor in the product mixture is then

$$x_v = \frac{0.0441}{0.0441 + 0.394} = 0.1007$$

The partial pressure of water vapor in the product mixture is

$$p_v = x_v p = 0.1007(14.7) = 1.48 \text{ psia}$$

From the saturated steam table (Table A-19E), we obtain

$$T_{\text{dew point}} = T_{\text{sat}} \text{ (at 1.48 psia)} = 115.2°F$$

Because the products temperature is given as 500°F, which is greater than the dew-point temperature, the water is in vapor form as part of the products. Therefore, the mass of water vapor in the products is calculated as

$$m_v = (0.0441 \text{ lbmole } H_2O/\text{lbm fuel})(18 \text{ lbm } H_2O/\text{lbmole } H_2O)$$
$$= 0.794 \text{ lbm } H_2O/\text{lbm fuel}$$

15-3 FIRST LAW ANALYSIS OF CHEMICAL REACTIONS

Chemical reactions are usually carried out either in a closed system at constant volume or constant pressure, or in an open system at steady-flow conditions. We now investigate the energy effects that accompany chemical changes in these processes on the basis of the first law of thermodynamics.

Consider first a constant-volume reaction in a closed, simple compressible system. Because for this process, $W = 0$, the first law becomes

$$Q = \Delta U + \overset{0}{\cancel{W}} \quad \text{or} \quad Q = \Delta U = U_2 - U_1$$

where subscript 1 denotes the initial state of the reactants, and subscript 2 denotes the final state of the products.

Because the chemical aggregations of the reactants and the products are different, the state of zero internal energy cannot be arbitrarily chosen for both. Instead, we must assign the zero-energy states so that for a constant-volume reaction, the change in internal energy between the reactants and the products equals the amount of heat transfer. The change in internal energy between the reactants and the products when both are at the same temperature is called the *internal energy of reaction* at that temperature. The temperature variations of internal energy in proper relative values for the reactants and the products are depicted in Fig. 15-2. For convenience, it is assumed in this plot that the internal energy of a chemical species is a function of temperature alone. This is, of course, exactly true only for ideal gases. In Fig. 15-2, the vertical distance at any temperature between the two curves is the internal energy of reaction at that temperature. In particular, the internal energy of reaction $\Delta U_R^\circ = U_y - U_x$ at a standard temperature T_0 is indicated in the figure.

From Fig. 15-2, for a process that begins with the reactants at state 1 and ends with the products at state 2, the transfer of heat is given by

$$Q = U_2 - U_1 = (U_2 - U_y) + (U_y - U_x) + (U_x - U_1)$$
$$= (U_2 - U_y) + \Delta U_R^\circ - (U_1 - U_x)$$

$$(15\text{-}1)$$

or

$$Q = \sum_{\text{products}} n(u_2 - u_y) + \Delta U_R^\circ - \sum_{\text{reactants}} n(u_1 - u_x) \qquad (15\text{-}2)$$

where the u's are molar internal energies, and n is the number of moles. ΔU_R° is taken at the standard condition of $T_0 = 25°C$ (77°F) and $p_0 = 1$ atm.

In the case of a steady-flow open system, the first law gives

$$Q = H_2 - H_1$$

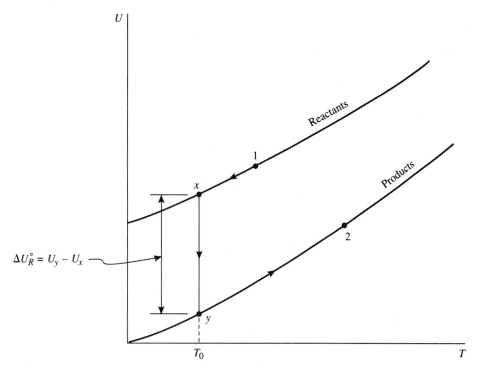

Figure 15-2 Internal energies of reactants and products as functions of temperature.

when changes in kinetic and potential energies are neglected and when there is no shaft work done. The preceding equation is also the expression of the first law for a constant-pressure reaction in a closed system, because for such a process

$$Q = U_2 - U_1 + p(V_2 - V_1) = H_2 - H_1$$

When the enthalpy is a function of temperature alone, such as in ideal gases, the *H–T* data in proper relative values for the reactants and the products can be shown as in Fig. 15-3. Referring to this figure, we can then write for the transfer of heat

$$Q = H_2 - H_1 = (H_2 - H_y) + (H_y - H_x) + (H_x - H_1)$$
$$= (H_2 - H_y) + \Delta H_R^\circ - (H_1 - H_x) \tag{15-3}$$

or

$$Q = \sum_{products} n(h_2 - h_y) + \Delta H_R^\circ - \sum_{reactants} n(h_1 - h_x) \tag{15-4}$$

where the *h*'s are molar enthalpies, and $\Delta H_R^\circ = H_y - H_x$ is the *enthalpy of reaction* at the standard condition of $T_0 = 25°C$ (77°F) and $p_0 = 1$ atm.

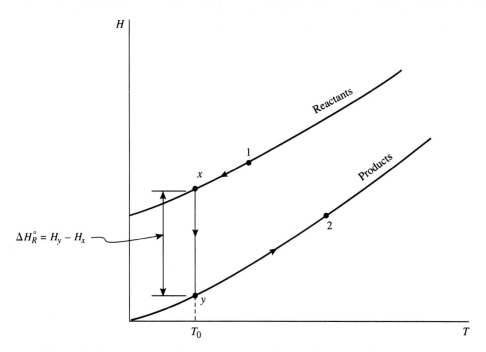

Figure 15-3 Enthalpies of reactants and products as functions of temperature.

15-4 ENTHALPY OF FORMATION

In order to establish an enthalpy scale for elements and compounds involved in chemical reactions, we now introduce the concept of an enthalpy of formation. The *enthalpy of formation* Δh_f° of a compound at the standard condition of 25°C (77°F) and 1 atm is defined as the enthalpy of reaction at that condition for the formation of the compound from its elements, which are originally in their most stable state at the standard condition. Using the formation of gaseous methane CH_4 as an example, the reaction equation is

$$C + 2H_2 \longrightarrow CH_4$$

The standard-state enthalpy of formation is the enthalpy of the products minus the enthalpy of the reactances, or

$$\Delta h_{f,\,CH_4}^\circ = h_{CH_4}^\circ - (h_C^\circ + 2h_{H_2}^\circ)$$

It is convenient to assign arbitrarily a value of zero to the enthalpy of all stable elements at 25°C and 1 atm. For example, the stable forms of elements such as hydrogen, nitrogen, and oxygen at room temperature and 1 atm are geseous H_2 (g), N_2 (g), and O_2 (g). However, the stable form of carbon under these same conditions is solid graphite C (s), but not diamond. Thus,

$$h_{H_2}^\circ = 0 \qquad \text{but} \qquad h_H^\circ \neq 0$$
$$h_{C\,(graphite)}^\circ = 0 \qquad \text{but} \qquad h_{C\,(diamond)}^\circ \neq 0$$

Accordingly, the enthalpy of formation of CH_4 equals the enthalpy of CH_4, both at the standard condition of 25°C and 1 atm, or

$$\Delta h^\circ_{f,CH_4} = h^\circ_{CH_4}$$

In general, the standard-state enthalpy of formation can be written symbolically as

$$\Delta h^\circ_f = h^\circ_{compound} - \sum_i (n_i h^\circ_i)_{elements} \tag{15-5}$$

where $h^\circ_{compound}$ is the molar enthalpy of the compound, and h°_i is the molar enthalpy of the ith element, all at the standard condition; n_i is the stoichiometric coefficient of the ith element in forming a single mole of the compound. With the enthalpy of all stable elements at the standard condition assigned the value of zero, Eq. (15-5) simplifies to

$$\Delta h^\circ_f = h^\circ_{compound} \tag{15-6}$$

which means that the molar enthalpy of any compound at the standard condition is merely its enthalpy of formation at that condition.

It should be noted here that properties at unit pressure (1 atm in this case) are symbolized by the superscript "°", whereas a temperature condition is symbolized with a subscript. Thus, the enthalpy of formation at the standard condition of 1 atm and 298 K (537°R) should be represented by the symbol $\Delta h^\circ_{f,298\,K}$ For convenience, however, we usually drop the subscript and use only the symbol Δh°_f.

Table A-28 in Appendix 3 gives values of Δh°_f for a number of substances. The values of Δh°_f listed in the table are for either the liquid or the gaseous phase. Values of Δh°_f between the liquid and vapor phases of a substance differ by the value of its latent heat of vaporization h_{fg} at the standard temperature of 25°C = 77°F. To obtain Δh°_f for a substance in the liquid phase, subtract h_{fg} from Δh°_f listed for the gaseous phase; and conversely to obtain Δh°_f for a substance in the vapor phase, add h_{fg} to Δh°_f listed for the liquid phase.

To evaluate the enthalpy of a pure substance at a specified state with temperature T and pressure p other than at the standard reference state with $T_0 = 298$ K (537°R) and $p_0 = 1$ atm, add to the value of standard-state Δh°_f the change in enthalpy between the reference state and the specified state. Thus, the enthalpy of a pure substance at T and p on a common basis in which the enthalpy of all stable elements at T_0 and p_0 is zero will be given by the expression

$$(\Delta h^\circ_f)_{T_0, p_0} + (h_{T,p} - h^\circ_{T_0, p_0})$$

For ideal gases, as enthalpy is a function of temperature only, this expression becomes

$$(\Delta h^\circ_f)_{T_0} + (h_T - h_{T_0})$$

where h_T is the enthalpy at the specified temperature T and h_{T_0} is the enthalpy at the standard reference temperature T_0. The ideal-gas enthalpy values as a function of temperature are given in the ideal-gas property tables in Appendix 3 (Tables A-10 to A-17). If tabular data are not available to evaluate the enthalpy difference $(h_T - h_{T_0})$, one can compute it from $\int_{T_0}^{T} c_p \, dT$, using an appropriate c_p equation.

15-5 ENTHALPY AND INTERNAL ENERGY OF REACTION

By definition, the enthalpy of reaction at the standard condition of 25°C (77°F) and 1 atm can be expressed symbolically as

$$\Delta H_R^\circ = \sum_{\text{products}} (nh^\circ) - \sum_{\text{reactants}} (nh^\circ)$$

where the n's are the stoichiometric coefficients of the individual constitutents based on a balanced chemical equation. Now because the molar enthalpy h° of any compound at the standard condition is merely its enthalpy of formation Δh_f° at that condition, we can substitute Δh_f° values for h° values in the preceding equation. Thus,

$$\Delta H_R^\circ = \sum_{\text{products}} (n\,\Delta h_f^\circ) - \sum_{\text{reactants}} (n\,\Delta h_f^\circ) \tag{15-7}$$

Thus, for any reaction, the value of ΔH_R° can be calculated from data on Δh_f° for each constituent involved in the reaction. This equation applies to reactions involving any mixture of reactants and to incomplete reactions as well.

The values of ΔU_R° are usually calculated from values of ΔH_R° using the definition of enthalpy. Thus, for a standard state reaction,

$$\Delta U_R^\circ = \Delta H_R^\circ - \Delta(pV)_R \tag{15-8}$$

where $\Delta(pV)_R = (pV)_{\text{products}} - (pV)_{\text{reactances}}$, all at the standard condition of $T_0 = 298$ K (537°R) and 1 atm. Because the molar volume of a liquid or solid constituent is negligibly small compared to that of a gaseous constituent, so that the volume change of the entire reaction process is practically due to that of the gaseous constituents only. If all the gaseous reactants and products can be considered as ideal gases, we have then

$$\Delta(pV)_R = \mathcal{R}T_0(n_{\text{gaseous products}} - n_{\text{gaseous reactants}})$$

and Eq. (15-8) reduces to

$$\Delta U_R^\circ = \Delta H_R^\circ - \mathcal{R}T_0(n_{\text{gaseous products}} - n_{\text{gaseous reactants}}) \tag{15-9}$$

Chemical reactions can be *exothermic,* wherein heat is released, or *endothermic,* wherein heat must be supplied to bring about a reaction. We are interested here in the exothermic reactions of combustion processes wherein the enthalpy of the products is less than that of the reactants, making ΔH_R° and ΔU_R° negative values. The enthalpy of reaction for a combustion process is frequently called the *enthalpy of combustion.* The enthalpy of combustion of a fuel is also frequently referred to as the *heating value* of the fuel. The heating value, however, is defined as the heat released from the combustion process and is always a positive value. Thus, the enthalpy of combustion and the heating value of a fuel have the same absolute numerical value but are opposite in sign. When the term heating value is used, two different values are often quoted. The *higher heating value* is the heat transfer based on liquid water in the products, and the *lower heating value* is the heat transfer based on vapor water in the products. They differ by the magnitude of the latent heat of vaporization of water at the given temperature.

Example 15-5

Calculate the enthalpy of reaction ΔH_R° and the internal energy of reaction ΔU_R° at 25°C for the following combustion processes:

1. Gaseous octane with gaseous water in the products.
2. Gaseous octane with liquid water in the products.
3. Liquid octane with gaseous water in the products.
4. Liquid octane with liquid water in the products.

Solution   From the table for enthalpy of formation (Table A-28M), we have

$$\Delta h^{\circ}_{f, C_8 H_{18}(g)} = -208,450 \text{ kJ/kgmole}$$

$$\Delta h^{\circ}_{f, C_8 H_{18}(l)} = -249,910 \text{ kJ/kgmole}$$

$$\Delta h^{\circ}_{f, H_2 O(g)} = -241,820 \text{ kJ/kgmole}$$

$$\Delta h^{\circ}_{f, H_2 O(l)} = -285,830 \text{ kJ/kgmole}$$

$$\Delta h^{\circ}_{f, CO_2(g)} = -393,520 \text{ kJ/kgmole}$$

$$\Delta h^{\circ}_{f, O_2(g)} = 0$$

where the symbols (g) and (l) after the chemical formula indicate the gas phase and liquid phase, respectively.

1. The combustion equation for gaseous octane with gaseous water in the products is

$$C_8 H_{18}(g) + 12.5 O_2 (g) \longrightarrow 8 CO_2(g) + 9 H_2 O (g)$$

$$\Delta H^{\circ}_R = 8 \Delta h^{\circ}_{f, CO_2(g)} + 9 \Delta h^{\circ}_{f, H_2 O(g)} - 1 \Delta h^{\circ}_{f, C_8 H_{18}(g)} - 12.5 \Delta h^{\circ}_{f, O_2 (g)}$$

$$= 8(-393,520) + 9(-241,820) - 1(-208,450) - 12.5(0)$$

$$= -5,116,090 \text{ kJ/kgmole } C_8 H_{18}$$

$$= (-5,116,090 \text{ kJ/kgmole})/(114 \text{ kg/kgmole})$$

$$= -44,878 \text{ kJ/kg } C_8 H_{18}$$

Thus, 44,878 kJ/kg is the lower heating value of gaseous octane.

$$\Delta U^{\circ}_R = \Delta H^{\circ}_R - \mathcal{R} T_0 (n_{\text{gaseous products}} - n_{\text{gaseous reactants}})$$

$$= (-5,116,090 \text{ kJ/kgmole}) - (8.314 \text{ kJ/kgmole·K})$$

$$\times (298 \text{ K})(17 - 13.5)$$

$$= -5,124,762 \text{ kJ/kgmole } C_8 H_{18}$$

$$= -5,124,762/114 = -44,954 \text{ kJ/kg } C_8 H_{18}$$

2. The combustion equation for gaseous octane with liquid water in the products is

$$C_8 H_{18} (g) + 12.5 O_2 (g) \longrightarrow 8 CO_2 (g) + 9 H_2 O (l)$$

$$\Delta H^{\circ}_R = 8 \Delta h^{\circ}_{f, CO_2(g)} + 9 \Delta h^{\circ}_{f, H_2 O(l)} - 1 \Delta h^{\circ}_{f, C_8 H_{18}(g)} - 12.5 \Delta h^{\circ}_{f, O_2(g)}$$

$$= 8(-393,520) + 9(-285,830) - 1(-208,450) - 12.5(0)$$

$$= -5,512,180 \text{ kJ/kg mole } C_8 H_{18}$$

$$= -5,512,180/114 = -48,352 \text{ kJ/kg } C_8 H_{18}$$

Thus, 48,352 kJ/kg is the higher heating value of gaseous octane.

$$\Delta U^{\circ}_R = \Delta H^{\circ}_R - \mathcal{R} T_0 (n_{\text{gaseous products}} - n_{\text{gaseous reactants}})$$

$$= -5,512,180 - (8.314)(298)(8 - 13.5)$$

$$= -5{,}498{,}553 \text{ kJ/kgmole } C_8H_{18}$$

$$= -5{,}498{,}553/114 = -48{,}233 \text{ kJ/kg } C_8H_{18}$$

3. The combustion equation for liquid octane with gaseous water in the products is

$$C_8H_{18}(l) + 12.5O_2 \text{ (g)} \longrightarrow 8CO_2 \text{ (g)} + 9H_2O \text{ (g)}$$

$$\Delta H_R^\circ = 8 \, \Delta h_{f,CO_2(g)}^\circ + 9 \, \Delta h_{f,H_2O(g)}^\circ - 1 \, \Delta h_{f,C_8H_{18}(l)}^\circ - 12.5 \, \Delta h_{f,O_2(g)}^\circ$$

$$= 8(-393{,}520) + 9(-241{,}820) - 1(-249{,}910) - 12.5(0)$$

$$= -5{,}074{,}630 \text{ kJ/kgmole } C_8H_{18}$$

$$= -5{,}074{,}630/114 = -44{,}514 \text{ kJ/kg } C_8H_{18}$$

Thus, 44,514 kJ/kg is the lower heating value of liquid octane.

$$\Delta U_R^\circ = \Delta H_R^\circ - \mathcal{R}T_0(n_{\text{gaseous products}} - n_{\text{gaseous reactants}})$$

$$= -5{,}074{,}630 - (8.314)(298)(17 - 12.5)$$

$$= -5{,}085{,}779 \text{ kJ/kgmole } C_8H_{18}$$

$$= -5{,}085{,}779/114 = -44{,}612 \text{ kJ/kg } C_8H_{18}$$

4. The combustion equation for liquid octane with liquid water in the products is

$$C_8H_{18} \text{ (l)} + 12.5O_2 \text{ (g)} \longrightarrow 8CO_2 \text{ (g)} + 9H_2O \text{ (l)}$$

$$\Delta H_R^\circ = 8 \, \Delta h_{f,CO_2(g)}^\circ + 9 \, \Delta h_{f,H_2O(l)}^\circ - 1 \, \Delta h_{f,C_8H_{18}(l)}^\circ - 12.5 \, \Delta h_{f,O_2(g)}^\circ$$

$$= 8(-393{,}520) + 9(-285{,}830) - 1(-249{,}910) - 12.5 (0)$$

$$= -5{,}470{,}720 \text{ kJ/kgmole } C_8H_{18}$$

$$= -5{,}470{,}720/114 = -47{,}989 \text{ kJ/kg } C_8H_{18}$$

Thus, 47,989 kJ/kg is the higher heating value of liquid octane.

$$\Delta U_R^\circ = \Delta H_R^\circ - \mathcal{R}T_0(n_{\text{gaseous products}} - n_{\text{gaseous reactants}})$$

$$= -5{,}470{,}720 - (8.314)(298)(8 - 12.5)$$

$$= -5{,}459{,}571 \text{ kJ/kgmole } C_8H_{18}$$

$$= -5{,}459{,}571/114 = -47{,}891 \text{ kJ/kg } C_8H_{18}$$

Notice that although we based the combustion equations with a stoichiometric amount of oxygen to evaluate ΔH_R°, the answers are independent of the amount of oxidant supplied to the reactions. If air is supplied, even in an excess amount, the oxygen and nitrogen enter and leave at the standard reference temperature with no change in enthalpy regardless of the quantity of each.

Example 15-6

A natural-gas mixture has the following volumetric analysis: methane (CH_4), 60.5%; ethane (C_2H_6), 14.7%; propane (C_3H_8), 13.3%; butane (C_4H_{10}), 4.2%; and nitrogen (N_2), 7.3%. Determine the enthalpy of reaction of the fuel mixture in Btu/lbmole fuel, Btu/lbm fuel, and Btu/ft³ fuel at 77°F and 1 atm with vapor water in the products.

Solution The balanced complete combustion equation with theoretical air on the basis of 1 mole of the fuel mixture is as follows:

$$(0.605CH_4 + 0.147C_2H_6 + 0.133C_3H_8 + 0.042C_4H_{10}$$
$$+ 0.073N_2) + (2.663O_2 + 10.013N_2)$$
$$\longrightarrow 1.466CO_2 + 2.393H_2O + 10.086N_2$$

where all the constituents are in the gaseous phase. The enthalpy of reaction of the fuel is given by the expression,

$$\Delta H_R^\circ = 1.466 \, \Delta h_{f,CO_2}^\circ + 2.393 \, \Delta h_{f,H_2O}^\circ + 10.086 \, \Delta h_{f,N_2}^\circ$$
$$- 0.605 \, \Delta h_{f,CH_4}^\circ - 0.147 \, \Delta h_{f,C_2H_6}^\circ - 0.133 \, \Delta h_{f,C_3H_8}^\circ$$
$$- 0.042 \, \Delta h_{f,C_4H_{10}}^\circ - 0.073 \, \Delta h_{f,N_2}^\circ - 2.663 \, \Delta h_{f,O_2}^\circ$$
$$- 10.013 \, \Delta h_{f,N_2}^\circ$$

Substituting numerical values from the Δh_f° English-units table (Table A-28E) yields

$$\Delta H_R^\circ = 1.466(-169,300) + 2.393(-104,040) + 10.086(0)$$
$$- 0.605(-32,210) - 0.147(-36,420) - 0.133(-44,680)$$
$$- 0.042(-54,270) - 0.073(0) - 2.663(0) - 10.013(0)$$
$$= -464,099 \text{ Btu/lbmole fuel}$$

The molecular weight of the fuel is calculated as

$$M_{fuel} = \sum x_i M_i$$
$$= 0.605(16) + 0.147(30) + 0.133(44) + 0.042(58) + 0.073(28)$$
$$= 24.422 \text{ lbm/lbmole}$$

Accordingly,

$$\Delta H_R^\circ = \frac{-464,099 \text{ Btu/lbmole}}{24.422 \text{ lbm/lbmole}} = -19,003 \text{ Btu/lbm fuel}$$

Because water formed is in the vapor phase and is present in the products, the lower heating value of the fuel is then 19,003 Btu/lbm.

At $p = 14.7$ psia and $T_0 = 77°F = 537°R$, the molar volume of the fuel is

$$v = \frac{\mathscr{R}T_0}{p} = \frac{(1545 \text{ ft·lbf/lbmole·°R})(537°R)}{(14.7 \times 144 \text{ lbf/ft}^2)}$$
$$= 392 \text{ ft}^3/\text{lbmole}$$

Therefore,

$$\Delta H_R^\circ = \frac{-464,099 \text{ Btu/lbmole}}{392 \text{ ft}^3/\text{lbmole}} = -1184 \text{ Btu/ft}^3 \text{ fuel}$$

15-6 ENERGY CALCULATIONS FOR COMBUSTION PROCESSES

With the enthalpy and internal energy of reaction calculated from the data of enthalpy of formation, or otherwise known through experimental measurements, one can than employ the first-law energy analysis established in Sec. 15-3 to calculate the heat liberation by combustion processes. For steady-flow, open-system or constant-pressure, closed-system combustion processes, by referring to Fig. 15-3, the first law is given by Eq. (15-4), which is

$$Q = \sum_{\text{products}} n(h_2 - h_y) + \Delta H_R^\circ - \sum_{\text{reactants}} n(h_1 - h_x)$$

where h_x and h_y are enthalpies of the reactants and products, respectively, at the standard condition of $T_0 = 298$ K ($537°$R) and 1 atm; and h_1 and h_2 are the enthalpies of the reactants and products, respectively, at the given temperature T and pressure p. Accordingly, one can rewrite the preceding equation in the form

$$Q = \sum_{\text{products}} n(h_{T,p} - h_{298\,\text{K, 1 atm}}) + \Delta H_R^\circ - \sum_{\text{reactants}} n(h_{T,p} - h_{298\,\text{K, 1 atm}}) \qquad (15\text{-}10)$$

Thermodynamic property tables (if available) for the constituents of the reactants and products should be used to find the values of $h_{T,p}$ and $h_{298\,\text{K, 1 atm}}$. For solid and liquid constitutents, specific-heat data are often used to calculate the enthalpy change; see Eq. (3-13). For real gases, if no table for the properties is available, generalized property charts as introduced in Sec. 14-7 are sometimes useful.

If the combustion constituents can be treated as ideal gases, their enthalpy values are functions of temperature only and independent of pressure. Accordingly, for ideal-gas reactions, Eq. (15-10) can be rewritten as

$$Q = \sum_{\text{products}} n(h_T - h_{298\,\text{K}}) + \Delta H_R^\circ - \sum_{\text{reactants}} n(h_T - h_{298\,\text{K}}) \qquad (15\text{-}11)$$

Substituting Eq. (15-7) into Eq. (15-11) yields

$$Q = \sum_{\text{products}} n(\Delta h_f^\circ + h_T - h_{298\,\text{K}}) - \sum_{\text{reactants}} n(\Delta h_f^\circ + h_T - h_{298\,\text{K}}) \qquad (15\text{-}12)$$

Ideal-gas property tables for N_2, O_2, H_2O, CO_2, H_2, and CO are included in Appendix 3 (Tables A-10 to A-16). These tables can be used to obtain the values of h_T and $h_{298\,\text{K}(537°\text{R})}$. If the equation for constant-pressure specific-heat c_p as a function of temperature T for a gas is available, the relation $dh = c_p\, dT$ can be integrated between $T_0 = 298$ K ($537°$R) and the given temperature T to obtain the value of $h_T - h_{298\,\text{K}}$. For simplified calculations, the use of an average c_p value between T_0 and T is convenient.

Analogous to Eqs. (15-10) and (15-11), expressions for a constant-volume combustion process can be written. Thus, from Eq. (15-2), we write in general

$$Q = \sum_{\text{products}} n(u_{T,p} - u_{298\,\text{K, 1 atm}}) + \Delta U_R^\circ - \sum_{\text{reactants}} n(u_{T,p} - u_{298\,\text{K, 1 atm}}) \qquad (15\text{-}13)$$

For ideal-gas reactions, this equation can be written as

$$Q = \sum_{\text{products}} n(u_T - u_{298\,\text{K}}) + \Delta U_R^\circ - \sum_{\text{reactants}} n(u_T - u_{298\,\text{K}}) \qquad (15\text{-}14)$$

where the u's are functions of temperature alone for ideal gases.

Example 15-7

The natural-gas mixture given in Example 15-6 is supplied to a steady-flow combustion chamber at $140°$F and burned with 30% excess air supplied at $240°$F. The gaseous products

Chemical Reactions   Chap. 15

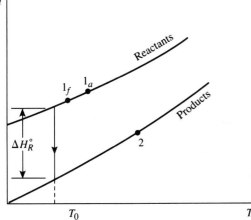

T_0 T **Figure 15-4** H–T diagram for Example 15-7.

leave the chamber at 940°F. Determine the amount of heat released per lbmole of fuel in a complete combustion process.

Solution The final temperature of 940°F is well above the dew point of the products, so the water must be in the gaseous state. Referring to the H–T diagram of Fig. 15-4, let the subscripts $1f$, $1a$, and 2 denote the initial state of fuel, the initial state of air, and the final state of the products, respectively. The temperatures are

$$T_{1f} = 140°F = 600°R$$

$$T_{1a} = 240°F = 700°R$$

$$T_2 = 940°F = 1400°R$$

The balanced complete combustion equation with 30% excess air on the basis of 1 mole of the fuel mixture is as follows:

$$(0.605CH_4 + 0.147C_2H_6 + 0.133C_3H_8 + 0.042C_4H_{10} + 0.073N_2)$$
$$+ (3.462O_2 + 13.017N_2) \longrightarrow 1.466CO_2 + 2.393H_2O + 0.799O_2 + 13.090N_2$$

where all the constituents are in the gaseous phase. Notice that except for the O_2 and N_2 terms in the air supply and in the products, the numerical coefficients in this chemical equation and in the chemical equation of Example 15-6 are all the same. Because $\Delta h^\circ_{f,O_2}$ and $\Delta h^\circ_{f,N_2}$ are zero, the different amounts of O_2 and N_2 that appear in the combustion process do not effect the value of ΔH°_R of the reaction. Accordingly we must have

$$\Delta H^\circ_R = -464{,}099 \text{ Btu/lbmole fuel}$$

which is the same as in Example 15-6.

From Eq. (15-11), the amount of heat released by the combustion process per mole of fuel is given by

$$Q = \sum_{\text{products}} n(h_2 - h_{537°R}) + \Delta H^\circ_R - \sum_{\text{reactants}} n(h_1 - h_{537°R})$$

$$= [1.466(h_2 - h_{537°R})_{CO_2} + 2.393(h_2 - h_{537°R})_{H_2O}$$
$$+ 0.799(h_2 - h_{537°R})_{O_2} + 13.090(h_2 - h_{537°R})_{N_2}] + \Delta H^\circ_R$$
$$- [0.605(h_{1f} - h_{537°R})_{CH_4} + 0.147(h_{1f} - h_{537°R})_{C_2H_6}$$

$$+ 0.133(h_{1f} - h_{537°R})_{C_3H_8} + 0.042(h_{1f} - h_{537°R})_{C_4H_{10}}$$
$$+ 0.073(h_{1f} - h_{537°R})_{N_2} + 3.462(h_{1a} - h_{537°R})_{O_2}$$
$$+ 13.017(h_{1a} - h_{537°R})_{N_2}]$$

Substituting numerical values obtained from the appropriate tables in Appendix 3 (Tables A-16 and A-17E) yields

$$Q = [1.466(13342.5 - 4029.1) + 2.393(11628.9 - 4260.5)$$
$$+ 0.799(10219.6 - 3735.4) + 13.090(9896.5 - 3729.8)] + (-464,099)$$
$$- [0.605(4853 - 4306) + 0.147(5929 - 5105)$$
$$+ 0.133(7499 - 6337) + 0.042(9841 - 8287)$$
$$+ 0.073(4168.2 - 3729.8) + 3.462(4890.3 - 3735.4)$$
$$+ 13.017(4865.2 - 3729.8)]$$
$$= -366,392 \text{ Btu/lbmole fuel}$$

Example 15-8

Gaseous benzene (C_6H_6) is mixed thoroughly with 200% theoretical air and ignited to burn completely in a closed rigid tank. The reactants are originally at 25°C and 1 atm. The final temperature of the products is 700 K. Determine the heat transfer per kgmole of benzene and the final pressure of the products, assume that all constituents of the reactants and products can be considered as ideal gases.

Solution   Referring to the U–T diagram of Fig. 15-5, let the subscripts 1 and 2 denote the initial state of the reactants and the final state of the products, respectively. The given data are $T_1 = 25°C = 298$ K, $p_1 = 1$ atm $= 101.3$ kPa, and $T_2 = 700$ K. The balanced complete combustion equation with 200% theoretical air on the basis of 1 kgmole of the fuel is as follows:

$$C_6H_6 + 15O_2 + 56.4N_2 \longrightarrow 6CO_2 + 3H_2O + 7.5O_2 + 56.4N_2$$

where all constituents are in the gaseous phase.

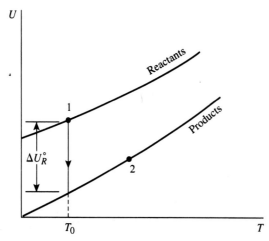

Figure 15-5 U–T diagram for Example 15-8.

Chemical Reactions   Chap. 15

From Eq. (15-7) with data obtained from Table A-28M, we calculate the value of ΔH_R° for the reaction. Thus,

$$\Delta H_R^\circ = \sum_{\text{products}} n\, \Delta h_f^\circ - \sum_{\text{reactants}} n\, \Delta h_f^\circ$$

$$= (6\, \Delta h_{f,CO_2}^\circ + 3\, \Delta h_{f,H_2O}^\circ + 7.5\, \Delta h_{f,O_2}^\circ + 56.4\, \Delta h_{f,N_2}^\circ)$$
$$- (1\, \Delta h_{f,C_6H_6}^\circ + 15\, \Delta h_{f,O_2}^\circ + 56.4\, \Delta h_{f,N_2}^\circ)$$

$$= 6(-393,520) + 3(-241,820) - 1(82,930)$$

$$= -3,169,510 \text{ kJ/kgmole } C_6H_6$$

The total number of moles in the reactants is

$$n_1 = 1 + 15 + 56.4 = 72.4$$

and the total number of moles in the products is

$$n_2 = 6 + 3 + 7.5 + 56.4 = 72.9$$

From Eq. (15-9), we calculate the value of ΔU_R°. Thus,

$$\Delta U_R^\circ = \Delta H_R^\circ - \mathcal{R}T_0(n_2 - n_1)$$

$$= -3,169,510 - (8.314)(298)(72.9 - 72.4)$$

$$= -3,170,749 \text{ kJ/kgmole } C_6H_6$$

The heat transfer in the constant-volume combustion process is given by Eq.(15-14):

$$Q = \sum_{\text{products}} n(u_2 - u_{298\,K}) + \Delta U_R^\circ - \sum_{\text{reactants}} n(u_1 - u_{298\,K})$$

Because $T_1 = T_0 = 298$ K, the terms for the reactants in the preceding equation drop out. Thus, we have

$$Q = [6(u_2 - u_{298\,K})_{CO_2} + 3(u_2 - u_{298\,K})_{H_2O} + 7.5(u_2 - u_{298\,K})_{O_2}$$
$$+ 56.4(u_2 - u_{298\,K})_{N_2}] + \Delta U_R^\circ$$

Taking data from the ideal-gas property tables (Tables A-10 to A-13), we have

$$Q = 6(21,300.8 - 6882.0) + 3(18,275.6 - 7421.3)$$
$$+ 7.5(15,361.9 - 6201.2) + 56.4(14,786.7 - 6188.1) + (-3,170,749)$$

$$= -2,498,007 \text{ kJ/kgmole } C_6H_6$$

For the ideal-gas mixtures of the reactants and the products, we have, respectively,

$$p_1 V = n_1 \mathcal{R} T_1 \qquad \text{and} \qquad p_2 V = n_2 \mathcal{R} T_2$$

It follows that the final pressure of the products is

$$p_2 = p_1 \left(\frac{n_2}{n_1}\right)\left(\frac{T_2}{T_1}\right) = 101.3\left(\frac{72.9}{72.4}\right)\left(\frac{700}{298}\right) = 239.6 \text{ kPa}$$

15-7 ADIABATIC FLAME TEMPERATURE

In a steady-flow combustion process, if no heat or work transfer is permitted and there is no kinetic- or potential-energy change, then all the energy released during the reaction will cause a significant increase in the temperature of the final products according to the condition

$$Q = H_{products} - H_{reactants} = 0 \quad \text{or} \quad H_{products} = H_{reactants}$$

The maximum temperature thus achieved is referred to as the *adiabatic flame tempera-ture* of the reacting mixture. This temperature will be highest when a stoichiometric mixture reacts to completion using no excess air. For the general case of a steady-flow, adiabatic combustion reaction, Eq. (15-10) becomes

$$\sum_{products} n(h_{T,p} - h_{298\,K,\,1\,atm}) + \Delta H_R^{\circ} - \sum_{reactants} n(h_{T,p} - h_{298\,K,\,1\,atm}) = 0 \qquad (15\text{-}15)$$

For the case of an ideal-gas reaction, Eq. (15-11) becomes

$$\sum_{products} n(h_T - h_{298\,K}) + \Delta H_R^{\circ} - \sum_{reactants} n(h_T - h_{298\,K}) = 0 \qquad (15\text{-}16)$$

Notice that Eqs. (15-15) and (15-16) are also valid for constant-pressure adiabatic com-bustion in a closed system.

For constant-volume adiabatic combustion, Eqs. (15-13) and (15-14), respec-tively, become

$$\sum_{products} n(u_{T,p} - u_{298\,K,\,1\,atm}) + \Delta U_R^{\circ} - \sum_{reactants} n(u_{T,p} - u_{298\,K,\,1\,atm}) = 0 \qquad (15\text{-}17)$$

$$\sum_{products} n(u_T - u_{298\,K}) + \Delta U_R^{\circ} - \sum_{reactants} n(u_T - u_{298\,K}) = 0 \qquad (15\text{-}18)$$

where Eq. (15-18) is for the case where ideal-gas behavior is assumed for the combus-tion constituents.

Example 15-9

A mixture of gaseous propane (C_3H_8) and air enters a steady-flow adiabatic combustion chamber at 77°F and reacts completely. Determine the air–fuel mass-ratio and the percent theoretical air required to give an adiabatic flame temperature of 1600°R.

Solution The combustion process is depicted on the H–T diagram in Fig. 15-6, where points 1 and 2 represent the initial and final states, respectively, of the process, with $Q = H_2 - H_1 = 0$ or $H_2 = H_1$. Because $T_2 = 1600$°R is such a high temperature, the water in the products must be in the gaseous phase. With all the constituents in the gaseous phase, the reaction equation can be written as

$$C_3H_8 + aO_2 + 3.76aN_2 \longrightarrow xCO_2 + yH_2O + zO_2 + 3.76aN_2$$

The mass balances for this chemical equation yield

C balance: $x = 3$
H_2 balance: $y = 4$
O_2 balance: $z = a - x - y/2 = a - 3 - 2 = a - 5$

Because the mass balances alone cannot evaluate all the unknown quantities, we must turn to the condition of energy balance for adiabatic combustion to determine the remaining unknown. By assuming an ideal-gas reaction, Eq. (15-16) gives

$$\sum_{products} n(h_2 - h_{537°R}) + \Delta H_R^{\circ} - \sum_{reactants} n(h_1 - h_{537°R}) = 0$$

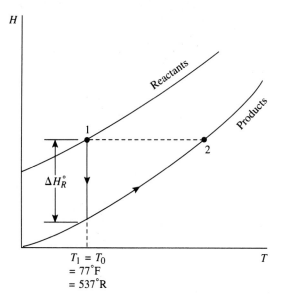

H

Reactants

Products

1

2

ΔH_R°

$T_1 = T_0$
$= 77°F$
$= 537°R$

T

Figure 15-6 *H–T* diagram for Example 15-9.

As $T_1 = T_0 = 537°R$, the preceding equation reduces to

$$\sum_{\text{products}} n(h_2 - h_{537°R}) + \Delta H_R^\circ = 0$$

or

$$[3(h_2 - h_{537°R})_{CO_2} + 4(h_2 - h_{537°R})_{H_2O} + (a - 5)(h_2 - h_{537°R})_{O_2}$$
$$+ 3.76a(h_2 - h_{537°R})_{N_2}] + \Delta H_R^\circ = 0$$

The value of ΔH_R° can be found based on the following theoretical reaction equation, using data for Δh_f° from Table A-28E:

$$C_3H_8 \text{ (g)} + 5O_2 \text{ (g)} \longrightarrow 3CO_2 \text{ (g)} + 4H_2O \text{ (g)}$$

Thus,

$$\Delta H_R^\circ = 3 \, \Delta h_{f, CO_2(g)}^\circ + 4 \, \Delta h_{f, H_2O(g)}^\circ - 1 \, \Delta h_{f, C_3H_8(g)}^\circ - 5 \, \Delta h_{f, O_2(g)}^\circ$$

$$= 3(-169,300) + (-104,040) - 1(-44,680) - 5(0)$$

$$= -879,380 \text{ Btu/lbmole } C_3H_8$$

Using the h values for CO_2, H_2O, O_2, and N_2 at $T_2 = 1600°R$ and $T_0 = 537°R$ from Table A-16, we finally have

$$3(15,825.2 - 4029.1) + 4(13,498.7 - 4260.5)$$
$$+ (a - 5)(11,841.3 - 3735.4) + 3.76a(11,408.6 - 3729.8) + (-879,380) = 0$$

so that

$$a = 22.92$$

The balanced complete-reaction equation for the actual process is then

$$C_3H_8 + 22.92O_2 + 86.18N_2 \longrightarrow 3CO_2 + 4H_2O + 17.92O_2 + 86.18N_2$$

The air–fuel mass ratio is given by

Air–fuel ratio = [(22.92 lbmole O_2/lbmole fuel) × (32 lbm O_2/lbmole O_2)
$$+ (86.18 \text{ lbmole } N_2\text{/lbmole fuel}) \times (28 \text{ lbm } N_2\text{/lbmole } N_2)]$$

$$\div (44 \text{ lbm fuel/lbmole fuel})$$

$$= 71.5 \text{ lbm air/lbm fuel}$$

The percent theoretical air is given by

$$\% \text{ theoretical air} = \frac{22.92}{5} \times 100 = 458.4\%$$

or

$$\% \text{ excess air} = \frac{22.92 - 5}{5} \times 100 = 358.4\%$$

Example 15-10

A 1.6-gram sample of liquid methyl alcohol (CH_3OH) is placed in a 0.03-m^3 constant-volume adiabatic bomb containing a stoichiometric amount of air initially at 25°C. The fuel is burned to completion. Determine the adiabatic flame temperature and the maximum pressure in the bomb.

Solution The combustion process is depicted on the U–T diagram of Fig. 15-7, where points 1 and 2 represent the initial and final states, respectively, of the process, satisfying the condition

$$Q = U_2 - U_1 = 0 \qquad \text{or} \qquad U_2 = U_1$$

The final temperature of the products will be high so that the water formed will be in the gaseous phase. The balanced complete combustion equation with theoretical air is given by

$$CH_3OH \text{ (l)} + 1.5O_2 \text{ (g)} + 5.64N_2 \text{ (g)} \longrightarrow 1CO_2 \text{ (g)} + 2H_2O \text{ (g)} + 5.64N_2 \text{ (g)}$$

From Eq. (15-7) and using data from Table A-28M, we have

$$\Delta H_R^\circ = 1 \, \Delta h_{f, CO_2(g)}^\circ + 2 \, \Delta h_{f, H_2O(g)}^\circ - 1 \, \Delta h_{f, CH_3OH(l)}^\circ$$

$$= 1(-393,520) + 2(-241,820) - 1(-238,810)$$

$$= -638,350 \text{ kJ/kgmole fuel}$$

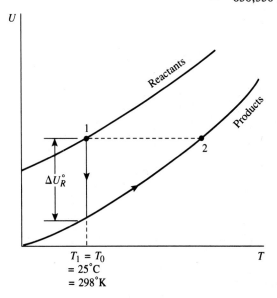

$T_1 = T_0$
$= 25°C$
$= 298°K$

Figure 15-7 *U–T* diagram for Example 15-10.

From Eq. (15-9), we have

$$\Delta U_R^{\circ} = \Delta H_R^{\circ} - \mathscr{R}T_0(n_{\text{gaseous products}} - n_{\text{gaseous reactants}})$$

$$= (-638{,}350) - (8.314)(298)(8.64 - 7.14)$$

$$= -642{,}066 \text{ kJ/kgmole fuel}$$

Because $T_1 = T_0$, Eq. (16-18) becomes

$$\sum_{\text{products}} n(u_2 - u_{298\,\text{K}}) + \Delta U_R^{\circ} = 0$$

or

$$1(u_2 - u_{298\,\text{K}})_{CO_2} + 2(u_2 - u_{298\,\text{K}})_{H_2O} + 5.64(u_2 - u_{298\,\text{K}})_{N_2} = -\Delta U_R^{\circ}$$

Substituting the known value of ΔU_R° and the u-values from Appendix 3 (Tables A-10 to A-13), we have

$$1(u_2 - 6882.0)_{CO_2} + 2(u_2 - 7421.3)_{H_2O} + 5.64(u_2 - 6188.1)_{N_2}$$

$$= 642{,}066 \text{ kJ/kgmole fuel}$$

As T_2 is unknown, calculations by iteration based on the ideal-gas values of u_2 at assumed T_2 for CO_2, H_2O, and N_2 from tables in Appendix 3 are made. Results of the final two trials are as follows:

At $T_2 = 2700$ K,

$$1(12{,}1161.2 - 6882.0) + 2(97{,}411.9 - 7421.3) + 5.64(67{,}841.4 - 6188.1)$$

$$= 641{,}985 \text{ kJ/kgmole fuel}$$

At $T_2 = 2710$ K,

$$1(121{,}695.9 - 6882.0) + 2(97{,}876.1 - 7421.3) + 5.64(68{,}126.0 - 6188.1)$$

$$= 645{,}053 \text{ kJ/kgmole fuel}$$

Accordingly, we conclude that $T_2 = 2700$ K (approximate value).

Because the mass of fuel is 1.6×10^{-3} kg and the molecular mass of fuel (CH_3OH) is 32 kg/kgmole, the number of kgmole of fuel is

$$\frac{1.6 \times 10^{-3} \text{ kg}}{32 \text{ kg/kgmole}} = 5 \times 10^{-5} \text{ kgmole fuel}$$

$$n_2 = \text{number of kgmole of products}$$

$$= (1 + 2 + 5.64)(5 \times 10^{-5})$$

$$= 4.32 \times 10^{-4} \text{ kgmole products}$$

The final maximum pressure of the bomb is then

$$p_2 = \frac{n_2 \mathscr{R} T_2}{V}$$

$$= \frac{(4.32 \times 10^{-4} \text{ kgmole})(8.314 \text{ kJ/kgmole·K})(2700 \text{ K})}{0.03 \text{ m}^3}$$

$$= 323 \text{ kPa}$$

Combustion is the rapid reaction between fuels and oxygen in which chemical energy is liberated. During a combustion process, the components that exist before the reaction are called reactants, and the components that exist after the reaction are called products. Chemical equations are balanced on the basis that the total mass of each element is conserved.

In a combustion process of a hydrocarbon fuel, if all the carbon present is converted into CO_2 and all the hydrogen is converted into H_2O, the process is described as complete combustion. The minimum amount of air that supplies sufficient oxygen for complete combustion of a fuel is called the theoretical or stoichiometric air. The air in excess of the stoichiometric amount is called the excess air. The ratio of the mass of air supplied to the mass of fuel during a combustion process is called the air–fuel ratio.

The dew-point temperature of the products is defined as the saturation temperature of the water vapor corresponding to its partial pressure in the products. Considering the products as a mixture of ideal gases, the partial pressure of water vapor p_v is given by

$$p_v = x_v p$$

where p is the total pressure of the product mixture, and x_v is the mole fraction of water vapor in the mixture. Then

$$T_{\text{dew point}} = T_{\text{sat}} \text{ (at } p_v)$$

The change of internal energy between the reactants and the products when both are at the same standard reference state of $T_0 = 25°C = 77°F$ and $p_0 = 1$ atm is the internal energy of reaction ΔU_R°. The change of enthalpy between the reactants and the products when both are at the same standard reference state of $T_0 = 25°C = 77°F$ and $p_0 = 1$ atm is the enthalpy of reaction ΔH_R°.

The enthalpy of formation Δh_f° of a compound at the standard condition of 25°C and 1 atm is defined as the enthalpy of reaction at that condition for the formation of the compound from its elements. When the enthalpies of all stable elements at 25°C and 1 atm are assigned the value of zero, we have the relation

$$\Delta h_f^\circ = h_{\text{compound}}^\circ$$

The enthalpy of reaction ΔH_R° can be expressed in terms of enthalpy of formation Δh_f° as follows:

$$\Delta H_R^\circ = \sum_{\text{products}} (nh^\circ) - \sum_{\text{reactants}} (nh^\circ)$$

$$= \sum_{\text{products}} (n \, \Delta h_f^\circ) - \sum_{\text{reactants}} (n \, \Delta h_f^\circ)$$

Neglecting the volume of liquid and solid constituents and considering gaseous constituents as ideal gases, we have

$$\Delta U_R^\circ = \Delta H_R^\circ - \mathcal{R} T_0(n_{\text{gaseous products}} - n_{\text{gaseous reactants}})$$

For a steady-flow, open-system or constant-pressure, closed system combustion processes, the first law for ideal-gas reactions can be written as

$$Q = \sum_{\text{products}} n(h_T - h_{298\,\text{K}}) + \Delta H_R^\circ - \sum_{\text{reactants}} n(h_T - h_{298\,\text{K}})$$

$$= \sum_{\text{products}} n(\Delta h_f^\circ + h_T - h_{298\,\text{K}}) - \sum_{\text{reactants}} n(\Delta h_f^\circ + h_T - h_{298\,\text{K}})$$

For constant-volume combustion processes, the first law for ideal-gas reactions can be written as

$$Q = \sum_{\text{products}} n(u_T - u_{298\,\text{K}}) + \Delta U_R^\circ - \sum_{\text{reactants}} n(u_T - u_{298\,\text{K}})$$

In the absence of any heat transfer ($Q = 0$), the temperature of the products will reach a maximum, which is called the adiabatic flame temperature of the reaction.

PROBLEMS

15-1. Using C_8H_{18} as the chemical equivalent of gasoline, determine the molar analysis of the total products of combustion if a gasoline engine is operated with 5% excess air. What is the dew point of the products? Total pressure of the products is 14.7 psia.

15-2. Ethane (C_2H_6) is burned completely with 50% excess air. Determine the dew-point temperature of the products if the mixture is at 0.1 MPa. Also determine the air–fuel ratio in kg air/kg fuel.

15-3. Liquid benzene (C_6H_6) is burned with 120% theoretical air at 15.0 psia. Calculate (a) the air–fuel mass ratio, (b) the mole fraction of N_2 in the total products, and (c) the dew point in °F of the products.

15-4. **(a)** Write the chemical equation for the combustion of propane (C_3H_8) with theoretical air. Show the relative masses and volumes, and calculate the air–fuel mass ratio.
(b) Repeat part (a) except that the combustion process now takes place in 15% excess air.
(c) Repeat part (a) except that the combustion process now takes place in 90% of theoretical air and CO is the only combustible in the products.

15-5. The complete combustion of ethane (C_2H_6) results in a mixture of gaseous products with a dew-point temperature of 50°C. The total pressure of the products is 101.3 kPa. What is the air–fuel mass ratio of the combustion process?

15-6. A gaseous fuel consists of a mixture of methane (CH_4) and carbon monoxide in equal parts by volume. Determine (a) the theoretical air, in lbm/lbm fuel, and (b) the mass fraction of the products of complete combustion of the fuel with 10% excess air.

15-7. A producer gas has the following volumetric analysis: CH_4, 3.0%; H_2, 14.0%; N_2, 50.9%; O_2, 0.6%; CO, 27.0%; and CO_2, 4.5%. The gas is burned with 20% excess air at 0.1 MPa. (a) Determine the dew point of the products. (b) How many kilograms of water will be condensed per kg of fuel if the products are cooled 10°C below the dew-point temperature?

15-8. A gaseous fuel has the following volumetric analysis: CH_4, 40%; CO, 25%; H_2, 10%; CO_2, 5%; O_2, 2%; and N_2, 18%. The fuel is burned completely in air with a molar air–fuel ratio of 7. Determine the volumetric analysis of the products, with the water formed being in the gaseous state.

15-9. Hexane gas (C_6H_{14}) is burned in dry air. A 100-mL sample of the products, on a dry basis, shows that the volumes of the component are 8.5 mL CO_2, 3.0 mL O_2, 5.2 mL CO, and 83.3 mL N_2. Determine the air–fuel ratio by volume for the combustion process and the dew-point temperature of the products, the total pressure of the products being 101.3 kPa. When the temperature of the products is lowered to 10°C below the dew point, determine the percent of water condensed.

15-10. A gaseous fuel mixture has the following volumetric analysis: CH_4, 20%; C_2H_6, 40%; N_2, 30%; and H_2O, 10%. The fuel mixture is supplied to the burner at 20 psia and 300°F and is burned completely with 115% theoretical air supplied at 15 psia and 100°F. Write the reaction equation and determine (a) the ft³ of dry air supplied per ft³ of fuel, and (b) the dew-point temperature of the products for a products pressure of 15 psia.

15-11. An unknown hydrocarbon fuel (C_xH_y) undergoes a combustion process such that the dry products have the following volumetric analysis: CO_2, 9.0%; CO, 1.1%; O_2, 8.8%; and N_2, 81.1%. Determine on a mass basis (a) the composition of the fuel, (b) the air–fuel ratio, and (c) the percent theoretical air.

15-12. The volumetric analysis of the dry products of combustion of a hydrocarbon fuel described by the general formula C_xH_y is CO_2, 13.6%; O_2, 0.4%; CO, 0.8%; CH_4, 0.4%; and N_2, 84.8%. There are 13.6 moles of CO_2 formed per mole of fuel. Determine (a) the values of x and y, and (b) the air–fuel mass ratio used.

15-13. A fuel-gas mixture has the following volumetric analysis: 75.5% of CH_4 and 24.5% of C_2H_6. For complete combustion with 20% excess air, write the chemical equation and calculate the volumetric analysis of the dry products. Calculate also the dew point of the products if the total pressure of the products is 1 atm.

15-14. A fuel oil has the following gravimetric analysis: carbon, 84%; hydrogen, 13%; sulfur, 2%; and nitrogen, 1%. The fuel is burned with 20% excess air, and the products of combustion contain 12.5% of CO_2 on dry basis. Write the reaction equation and determine the mass of air supply (in lbm/lbm oil), the mass of CO_2 in the products (in lbm/lbm dry products and in lbm/lbm total products), and the mass of H_2O in the products (in lbm/lbm total products).

15-15. A steam power plant burns coal having an ultimate analysis in percentages as follows: C, 71.04; H, 5.28; O, 6.72; S, 4.96; N, 1.34; and ash, 10.66. The combustion equipment operates efficiently with 20% excess air. For this case, determine (a) the pounds of air per pound of fuel, (b) the mass fraction of the flue gas, and (c) the dew point of the flue gas. Assume that combustion is complete and everything but the ash appears in the flue gas. Atmospheric pressure is 14.7 psia.

15-16. Ethyl alcohol (C_2H_5OH) is burned in dry air, and a volumetric analysis of the dry products gives 9.8% CO_2, 6.5% O_2, 0.6% CO, 0.6% HCHO, and 82.5% N_2. Determine per lbm of ethyl alcohol (a) the mass of air supplied, and (b) the mass of water formed.

15-17. A gaseous mixture containing 0.012 kg ethane (C_2H_6) and 0.022 kg propane (C_3H_8) is burned completely with 15% excess air. Determine the

air–fuel molar ratio and the volumetric analysis of the products; the H_2O formed is in the vapor state. Determine also the dew-point temperature for a total product pressure of 101 kPa.

15-18. A coal has the following ultimate analysis: C, 70%; H, 5%; O, 10%; N, 4%; S, 3%; and ash, 8%. The volumetric analysis of the products on a dry basis is as follows: CO_2, 13.6%; O_2, 4.5%; CO, 1.1%; and N_2, 80.8%; SO_2 being included as CO_2. Combustion air conditions are 29.2 in. Hg absolute pressure, 90°F dry-bulb temperature, with specific humidity of 0.030 lbm water vapor/lbm dry air. Products temperature is 510°F. Determine the amount of water in the gaseous products.

15-19. A gasoline with an ultimate analysis of 85% carbon and 15% hydrogen was burned in an automobile engine, and the products of combustion gave the following volumetric analysis on a dry basis: CO_2, 12.1%; O_2, 2.6%; and CO, 1.1%. The air supplied for combustion had a specific humidity of 0.03 lbm water vapor per lbm of dry air. Determine per pound of fuel (a) the amount of air supplied, and (b) the amount of water in the products of combustion.

15-20. Octane (C_8H_{18}) is burned with the theoretical amount of air in a complete combustion process. The products are cooled to 80°F at a pressure of 14.7 psia. How many pounds of water are condensed per pound of fuel burned (a) if the air used for combustion is dry, and (b) if the air used for combustion has a relative humidity of 90% and is at 80°F and 14.7 psia?

15-21. In an automobile engine fuel-combustion test, the results of a volumetric analysis were CO_2, 7.1%; CO, 0.8%; O_2, 9.9%; and N_2, 82.2%. The inlet air for combustion has a dry-bulb temperature of 90°F and a wet-bulb temperature of 76°F. The exhaust is at a total pressure of 14.7 psia. Determine the dew-point temperature of the exhaust products. Solve this problem without the use of data from any psychrometric chart.

15-22. A gaseous fuel has the following volumetric analysis: CH_4, 80%; N_2, 15%; and O_2, 5%. The fuel is burned with dry air that enters the steady-flow combustion chamber at 25°C and 100 kPa. See Fig. P15-22. The products of combustion have the following volumetric analysis on a dry basis: CO_2, 3.4%; CO, 0.1%; O_2, 14.9%; and N_2, 81.6%. Determine the air–fuel mass ratio, the percent theoretical air used, and the volume-flow

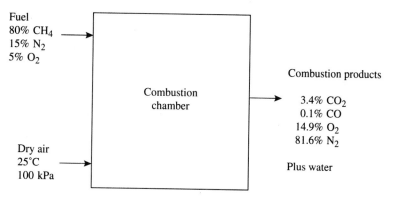

Figure P15-22

rate of air (in m³/min) for a fuel mass-flow rate of 1 kg/min.

15-23. A gaseous fuel with a volumetric analysis of 80% CH_4, 15% N_2, and 5% O_2 is burned to completion with 20% excess air, which enters the combustion chamber at 30°C, 100 kPa, and 60% relative humidity. See Fig. P15-23. Determine (a) the air–fuel mass ratio, and (b) the volume-flow rate of air required to burn fuel at a rate of 1 kg/min.

15-24. Gasoline (assumed to be octane C_8H_{18}) is burned to completion in an automobile engine with 120% theoretical air. (a) The exhaust temperature as it leaves the tail pipe on a cold day immediately after engine startup is 20°C. Will the water in the exhaust gases condense? (b) After the engine has warmed up, the temperature of the exhaust gases increases to 80°C. In this case, will the water in the exhaust gases condense?

15-25. A mixture of methane and oxygen is contained in a 1-m³ tank at a temperature of 35°C and a total

pressure of 200 kPa. The partial pressure of the methane is 50 kPa. The mixture is ignited and the combustion process is complete. Determine (a) the oxygen–methane mass ratio in kg O_2/kg CH_4 before combustion, and (b) the mole fraction of the combustion products.

15-26. A natural-gas mixture has the following volumetric analysis: 20% CH_4, 40% C_2H_6, and 40% C_3H_8. The fuel is burned with air, and the resulting products have the following volumetric analysis on a dry basis: 11.4% CO_2, 1.7% O_2, 1.2% CO, and 85.7% N_2. Determine the air–fuel molar ratio and mass ratio.

15-27. Gasoline (assumed to be octane C_8H_{18}) is burned to completion in an automobile engine with 20% excess air. The mass-flow rate of air through the combustor is 35 kg/h. Determine the fuel consumption rate in kilograms per hour.

15-28. An unknown hydrocarbon fuel undergoes a cataclysmic combustion process such that the dry products have the following volumetric analysis:

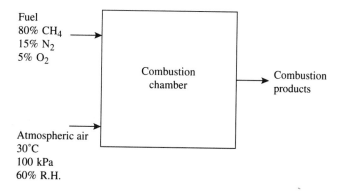

Figure P15-23

11.7% CO_2, 1.1% CO, 0% O_2, and 87.2% N_2. Determine (a) the composition of the fuel modeled by the formula C_nH_m, (b) the mass percentages of carbon and hydrogen in the fuel, (c) the percent deficit air used, and (d) the molar percentage of water vapor in the exhaust before it was dried.

15-29. A Diesel engine uses 14 kg of fuel per hour when the output power is 56 kW. The heating value of the fuel is 45,000 kJ/kg. Determine the thermal efficiency of the engine.

15-30. A coal has the following gravimetric analysis: carbon, 80.1%; hydrogen, 4.3%; oxygen, 4.2%; nitrogen, 1.7%; sulfur, 1.0%; and the rest is noncombustible ash. For complete combustion with the theoretical amount of air, write the chemical reaction equation, and determine the air–fuel mass ratio.

15-31. A gaseous fuel mixture has the following volumetric analysis: CH_4, 20%; C_2H_6, 10%; CO, 10%; CO_2, 10%; H_2, 10%; O_2, 10%; and N_2, 30%. For complete combustion with 30% excess air, write the chemical equation. Both the fuel and the air are supplied at 14.7 psia and 100°F. Determine (a) the volume of dry air per unit volume of fuel, (b) the volumetric analysis of the dry products, (c) the mass of the total products per unit mass of fuel, (d) the mass of water vapor in the products per unit mass of fuel, (e) the mass of dry air supplied per unit mass of fuel, (f) the dew point of the products if dry air is supplied for combustion, and (g) the dew point of the products if the air supplied for combustion contains 0.0295 lbm of water vapor per lbm of dry air.

15-32. Determine the value of the enthalpy of reaction ΔH_R° at 25°C for the following reaction:

$$CH_4 + C_2H_6 + 5O_2 \longrightarrow 0.1CH_4 + 0.1C_2H_6$$
$$+ 2.2CO_2 + 0.5CO + 4.5H_2O \text{ (l)} + 0.3O_2$$

when all the constituents except H_2O are in the gaseous state.

15-33. Determine the enthalpy of reaction ΔH_R° at 77°F of gaseous pentaborane (B_5H_9) with solid B_2O_3 and liquid H_2O in the products. The enthalpies of formation at 77°F are as follows:

B_5H_9 (g): 27,000 Btu/lbmole

B_2O_3 (s): −544,000 Btu/lbmole

Give the enthalpy of reaction in Btu/lbmole and Btu/lbm of the fuel.

15-34. Hydrogen is supplied to a steady-flow burner at 100°F and is completely burned with 30% excess air, which is also supplied at 100°F. The products leave at 900°F. Determine the amount of heat released per pound of hydrogen.

15-35. Propane gas (C_3H_8) is burned completely in steady flow with 50% excess air. The reactants enter the combustion chamber at 25°C and the products leave at 627°C. Determine the amount of heat released per kgmole of fuel.

15-36. Liquid octane is burned in an internal combustion engine with 20% excess air. Ninety percent of the carbon burns to CO_2 and the remainder burns to CO. The reactants enter the engine at 25°C, and the products leave at 627°C. The engine uses 2 grams of fuel per second. Determine the power output of the engine, assuming the heat transfer from the engine and the work done by the engine being numerically equal.

15-37. A cigarette lighter burns butane (C_4H_{10}) with 200% of theoretical air. Both fuel and air are at 25°C, and the products are at 127°C. Write the chemical equation and determine the air–fuel mass ratio and the heat released during combustion.

15-38. Propane gas (C_3H_8) is burned completely in steady flow with 110% theoretical air in an adiabatic furnace. The combustion air is preheated in an air preheater by heat transfer from the combustion products. See Fig. P15-38. The air enters the preheater at 25°C and the fuel enters the furnace at 25°C. Determine the temperature of the combustion products leaving the preheater.

15-39. Acetylene gas (C_2H_2) at 25°C and 0.16 MPa enters a reactor and burns completely with 50% excess air entering at 25°C and 0.16 MPa in a steady-flow operation. See Fig. P15-39. The combustion products are cooled and have the conditions of 25°C and 0.15 MPa at the exit from the reactor system. Neglecting kinetic and potential energies, determine the rate of heat transfer from the reaction system, in kJ/kgmole of fuel.

15-40. Liquid hydrogen peroxide (H_2O_2) is used as the source of oxygen for the combustion of carbon to CO_2 according to the reaction equation:

$$C + 2H_2O_2 \text{ (l)} \longrightarrow 2H_2O + CO_2$$

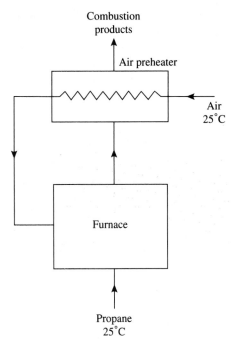

Combustion
products

Air preheater

Air
25°C

Furnace

Propane
25°C

Figure P15-38

The enthalpies of formation at 68°F are

$$H_2O_2 \text{ (l)}: -44,500 \text{ cal/gmole}$$
$$H_2O \text{ (l)}: -68,320 \text{ cal/gmole}$$
$$CO_2 \text{ (g)}: -94,050 \text{ cal/gmole}$$

Determine (a) the pounds of hydrogen peroxide needed per pound of carbon, and (b) the enthalpy of reaction at 68°F (in Btu/lbm·°C) for the reaction if liquid water is in the products.

15-41. A fuel oil with 84% carbon and 16% hydrogen by mass has a heating value of 19,000 Btu/lbm, based on gaseous products at 70°F. The oil is burned completely in an adiabatic, steady-flow combustion chamber. Air and oil are both supplied at 70°F, and the products temperature is 1450°F. Determine the air–fuel mass ratio.

15-42. Carbon monoxide is burned completely with 100% excess air in a steady-flow process at 1 atm. The carbon monoxide and the air are supplied at 77°F. Calculate the maximum adiabatic combustion temperature for the reaction.

15-43. A rigid, insulated vessel initially contains a mixture of gaseous octane (C_8H_{18}) and the theoretical amount of air at 25°C and 1 atm. The contents react completely. Determine the temperature and pressure of the combustion products.

15-44. Liquid gasoline (assumed C_8H_{18}) at 25°C is burned steadily with air at 25°C and 1 atm. Determine the highest possible temperature that can be obtained.

15-45. A liquid hydrocarbon fuel is burned with dry air in steady flow. The fuel and air are supplied at 25°C and the products are at 227°C. The fuel has a lower heating value of 43,000 kJ/kg. The volumetric analysis of the combustion products is as follows: 11.41% CO_2, 12.36% H_2O, 2.11% O_2, and 74.12% N_2. Determine the heat transfer per kilogram of fuel burned.

15-46. Liquid methyl alcohol (CH_3OH) is burned completely with no excess air in a steady-flow combustion. The fuel enters the combustor at the standard state of 25°C and 1 atm, and the air enters at 400 K and 1 atm. For adiabatic combus-

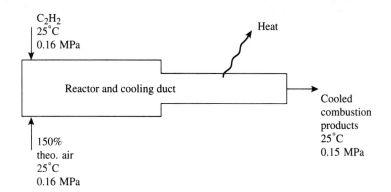

Figure P15-39

tion, determine the maximum temperature of the products.

15-47. A gaseous fuel has an equivalent chemical formula of $CH_{1.9}$ with a standard enthalpy of formation of $-23,000$ kJ/kgmole. The fuel is burned completely in a steady-flow combustor with 10% excess air. The fuel enters the combustor at the standard state of 25°C and 1 atm, and the air enters at 250 K and 1 atm. For an adiabatic combustion process, determine the maximum temperature of the products.

15-48. Liquid octane (C_8H_{18}) at 25°C is burned completely with 400% theoretical air supplied at 25°C in a low-pressure steady-flow process. No work or changes in kinetic and potential energies are involved. If the combustion process is adiabatic, determine the temperature of the products.

15-49. Diesel fuel (assumed to be $C_{12}H_{26}$) at 25°C is burned to completion in a steady-flow combustion chamber with 20% excess air entering at 25°C. See Fig. P15-49. The combustion products leave the chamber at 500 K. The heat transfer from the chamber is at the rate of 700 kJ/s. Determine the mass-flow rate of the fuel.

15-50. In an Otto engine, at the end of the compression stroke just prior to ignition (see Sec. 10-2), the air–gasoline mixture with an air–fuel mass ratio of 18 : 1 is at a pressure of 850 kPa and a temperature of 260°C. Neglecting dissociation and heat loss from the cylinder, determine the pressure and temperature after the constant-volume complete combustion.

15-51. A steam boiler test gave the following data when burning coal containing 80% carbon and 5% ash: coal consumption, 1.26 kg/s; coal heating value, 27,900 kJ/kg; Orsat analysis of products, 11%

CO_2, 8% O_2, 1% CO, and 80% N_2; refuse removed from the ash pit, 0.095 kg/s; stack temperature, 227°C; and ambient temperature, 25°C. Determine (a) the energy loss to the exiting combustion products, (b) the heat loss due to incomplete combustion, and (c) the energy loss due to carbon in the refuse.

15-52. Butane gas (C_4H_{10}) at 25°C and 5 atm enters the combustor of a simple Brayton engine. See Fig. P15-52. The fuel is burned with 30% excess air that was compressed in the compressor from 25°C and 1 atm before entering the combustor at 5 atm. The air compressor has an adiabatic efficiency of 70%. After the combustion process, the hot combustion products enter a 75%-efficiency turbine and leave at 1 atm. Determine the cycle thermal efficiency of the engine.

15-53. Gasoline (assumed to be pure octane C_8H_{18}) enters a jet engine at 25°C, 1 atm, with a low velocity, and burns completely with 300% of theoretical air entering at 25°C, 1 atm, and a low velocity. See Fig. P15-53. The combustion products exit from the engine at 990 K and 1 atm. Neglecting any heat transfer between the engine and its surroundings, determine the velocity of the exiting combustion products, in m/s.

15-54. In a simple Rankine steam power plant, methane (CH_4) at 25°C, 1 atm, enters the furnace and burns completely in steady flow with 110% of theoretical air entering at 25°C, 1 atm. See Fig. P15-54. The products of combustion exit the stack at 150°C, 1 atm. The thermal efficiency of the power plant is 33%. Neglecting kinetic and potential energies, determine the mass-flow rate of the fuel required in kg/h per MW of net power output of the plant.

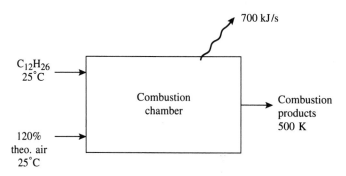

Figure P15-49

Chemical Reactions   Chap. 15

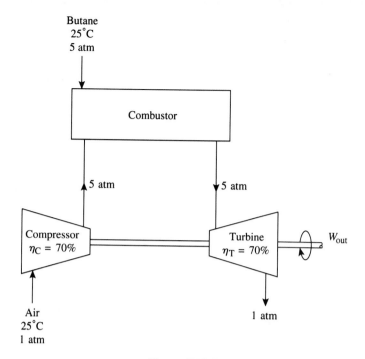

Butane
25°C
5 atm

Combustor

5 atm 5 atm

Compressor
$\eta_C = 70\%$

Turbine
$\eta_T = 70\%$

W_{out}

Air
25°C
1 atm

1 atm

Figure P15-52

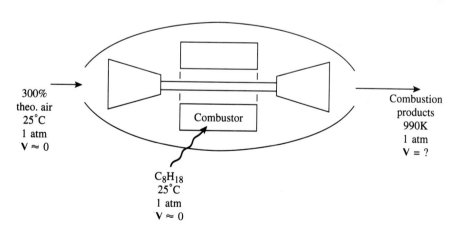

300%
theo. air
25°C
1 atm
$V \approx 0$

Combustor

Combustion
products
990K
1 atm
$V = ?$

C_8H_{18}
25°C
1 atm
$V \approx 0$

Figure P15-53

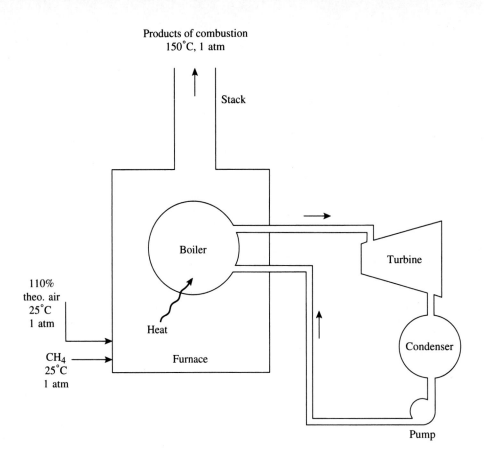

Products of combustion
150°C, 1 atm

Stack

Boiler

Turbine

110%
theo. air
25°C
1 atm

Heat

CH₄
25°C
1 atm

Furnace

Condenser

Pump

Figure P15-54

Chemical Reactions   Chap. 15

16

Phase and Chemical Equilibrium

In the last chapter, we treated mass and energy balances for chemical-reaction processes. We now aim our attention to the third aspect of chemical reactions—chemical equilibrium through the use of the second law of thermodynamics. We first examine the criteria of equilibrium and stability and investigate the condition of equilibrium between phases (using absorption–refrigeration two-phase mixtures as a special illustration). We then turn to the subject of equilibrium constant and equilibrium composition of reactive systems. The third law of thermodynamics is used to establish a common basis for absolute entropies. This chapter ends with an availability study of chemical reactions.

16-1 GIBBS EQUATIONS AND CHEMICAL POTENTIAL

According to the state principle, two independent properties are required to specify the thermodynamic state of a simple compressible system of fixed composition. However, for a system of variable composition, a specification of the composition is also required. Thus, if entropy and volume are selected as the independent variable, the energy of a system of fixed composition is

$$U = U(S, V)$$

In the case of a system of variable composition, we must express U as a function also of the amounts, for example, the numbers of moles $n_1, n_2, \ldots, n_r$ of the r different components. Thus,

$$U = U(S, V, n_1, n_2, \ldots, n_r)$$

The total differential of U is then

$$dU = \left(\frac{\partial U}{\partial S}\right)_{V,n} dS + \left(\frac{\partial U}{\partial V}\right)_{S,n} dV + \sum_{i=1}^{r} \left(\frac{\partial U}{\partial n_i}\right)_{S,V,n_j(j \neq i)} dn_i$$

in which the subscript n implies constancy of all the n's, that is, constant composition, and the subscript $n_j(j \neq i)$ implies constancy of all other n's except n_i. In light of Eq.

(13-14a), when all the n's are held constant during the differentiation, we have

$$\left(\frac{\partial U}{\partial S}\right)_{V,n} = T \quad \text{and} \quad \left(\frac{\partial U}{\partial V}\right)_{S,n} = -P \qquad (16\text{-}1)$$

Let us now introduce a new property, μ_i, to be called the *chemical potential* of component i, with the definition

$$\mu_i = \left(\frac{\partial U}{\partial n_i}\right)_{S,V,n_j(j\neq i)} \qquad (16\text{-}2)$$

Accordingly, the total differential of U becomes

$$dU = T\,dS - p\,dV + \sum_{i=1}^{r} \mu_i\,dn_i \qquad (16\text{-}3)$$

From this equation and the definitions of enthalpy H, Helmholtz function A, and Gibbs function G for simple compressible systems, we obtain

$$dH = T\,dS + V\,dp + \sum_{i=1}^{r} \mu_i\,dn_i \qquad (16\text{-}4)$$

$$dA = -S\,dT - p\,dV + \sum_{i=1}^{r} \mu_i\,dn_i \qquad (16\text{-}5)$$

$$dG = -S\,dT + V\,dp + \sum_{i=1}^{r} \mu_i\,dn_i \qquad (16\text{-}6)$$

Equations (16-3) to (16-6) are the fundamental relations for a homogeneous simple compressible system for changes between neighboring equilibrium states. They are known as the *Gibbs equations*. From these equations, it follows that μ_i can be related to U, H, A, and G by the expressions

$$\begin{aligned}
\mu_i &= \left(\frac{\partial U}{\partial n_i}\right)_{S,V,n_j(j\neq i)} \\[4pt]
&= \left(\frac{\partial H}{\partial n_i}\right)_{S,p,n_j(j\neq i)} \\[4pt]
&= \left(\frac{\partial A}{\partial n_i}\right)_{V,T,n_j(j\neq i)} \\[4pt]
&= \left(\frac{\partial G}{\partial n_i}\right)_{T,p,n_j(j\neq i)}
\end{aligned} \qquad (16\text{-}7)$$

The last of these expressions is sometimes used alone as the definition of the chemical potential μ_i.

Equations (16-3) to (16-6) can be manipulated to obtain many useful relations. Thus, since dG is an exact differential, it follows from Eq. (16-6) that

$$\left(\frac{\partial \mu_i}{\partial T}\right)_{p,n} = -\left(\frac{\partial S}{\partial n_i}\right)_{T,p,n_j(j\neq i)}$$

$$\left(\frac{\partial \mu_i}{\partial p}\right)_{T,n} = \left(\frac{\partial V}{\partial n_i}\right)_{T,p,n_j(j\neq i)}$$

$$\left(\frac{\partial \mu_i}{\partial n_k}\right)_{T, p, n_j(j \neq k)} = \left(\frac{\partial \mu_k}{\partial n_i}\right)_{T, p, n_j(j \neq i)}$$

Similarly, from Eq. (16-5), we obtain

$$\left(\frac{\partial \mu_i}{\partial T}\right)_{V, n} = -\left(\frac{\partial S}{\partial n_i}\right)_{T, V, n_j(j \neq i)}$$

$$\left(\frac{\partial \mu_i}{\partial V}\right)_{T, n} = -\left(\frac{\partial p}{\partial n_i}\right)_{T, V, n_j(j \neq i)}$$

$$\left(\frac{\partial \mu_i}{\partial n_k}\right)_{T, V, n_j(j \neq k)} = \left(\frac{\partial \mu_k}{\partial n_i}\right)_{T, V, n_j(j \neq i)}$$

Before returning to the study of more thermodynamic relations, we need now to state and prove a mathematical theorem known as Euler's theorem on homogeneous functions, which is of great use in thermodynamics. A function $f(z_1, z_2, \ldots, z_r)$ is said to be homogeneous of degree m in the z's if

$$f(\lambda z_1, \lambda z_2, \ldots, \lambda z_r) = \lambda^m f(z_1, z_2, \ldots, z_r)$$

where λ is a constant. *Euler's theorem* states that a homogeneous function of degree m obeys the equation

$$mf(z_1, z_2, \ldots, z_r) = \sum_i z_i \left(\frac{\partial f}{\partial z_i}\right)_{z_j(j \neq i)} \tag{16-8}$$

The proof of the theorem is straightforward upon defining the set of variables

$$y_1 = \lambda z_1 \quad \text{and} \quad y_2 = \lambda z_2, \ldots, \quad \text{and} \quad y_r = \lambda z_r$$

From which

$$f(y_1, y_2, \ldots, y_r) = \lambda^m f(z_1, z_2, \ldots, z_r)$$

Differentiating this equation with respect to λ at constant z's results in

$$\sum_i \left(\frac{\partial f}{\partial y_i}\right)_{y_j(j \neq i)} (z_i) = m\lambda^{m-1} f(z_1, z_2, \ldots, z_r)$$

This equation holds for all values of λ. If λ is set equal to 1, then $y_i = z_i$ and Eq. (16-8) follows.

Now we observe that U is homogeneous of first degree in S, V, and n_i. Applying Euler's theorem results in

$$U = \left(\frac{\partial U}{\partial S}\right)_{V, n} S + \left(\frac{\partial U}{\partial V}\right)_{S, n} V + \sum_{i=1}^{r} \left(\frac{\partial U}{\partial n_i}\right)_{S, V, n_j(j \neq i)} n_i$$

Inserting Eqs. (16-1) and (16-2) into this equation gives

$$U = TS - pV + \sum_{i=1}^{r} \mu_i n_i \tag{16-9}$$

Because by definition $G = U + pV - TS$, the preceding equation is equivalent to

$$G = \sum_{i=1}^{r} \mu_i n_i \tag{16-10}$$

Differentiating this equation gives

$$dG = \sum_{i=1}^{r} \mu_i \, dn_i + \sum_{i=1}^{r} n_i \, d\mu_i$$

Subtracting Eq. (16-6) from the preceding equation yields

$$S \, dT - V \, dp + \sum_{i=1}^{r} n_i \, d\mu_i = 0 \qquad (16\text{-}11)$$

For changes at constant T and p, this equation simplifies to

$$\sum_{i=1}^{r} n_i \, d\mu_i = 0 \qquad (16\text{-}12)$$

It is worth noticing that for a single-component system, Eq. (16-10) reduces to $G = \mu n$. Hence, $\mu = G/n = g$ for a pure component.

16-2 CRITERIA OF EQUILIBRIUM

We recall that one of the corollaries of the second law is the principle of increase of entropy, expressed symbolically by Eq. (6-22):

$$(dS)_{\text{isolated system}} \geq 0$$

in which the inequality holds for all natural processes where irreversible effects are inevitable. This equation provides a criterion for thermodynamic equilibrium. When an isolated system undergoes a natural process, only those states having higher entropies than the initial state are possible to attain. As a matter of fact, an isolated system always proceeds continuously from states of lower entropy to states of higher entropy, until it attains a state at which its entropy is a maximum. At this state, no further natural process is possible and the system is in a state of equilibrium.

Let us consider a closed simple-compressible system, containing various substances that can transport between phases and interact chemically. Such a system is not in chemical equilibrium. However, let us suppose that the system has already attained equality of temperature and pressure throughout; thus, it is in thermal and mechanical equilibrium. This system and its environment constitute an isolated system. Suppose that the environment is at a uniform temperature T and undergoes nothing but reversible processes. Let a quantity of heat $đQ$ be transferred from the environment into the system, causing a change in entropy of the environment:

$$dS_0 = \frac{-đQ}{T} \qquad (16\text{-}13)$$

Meanwhile, this heat flowing into the system causes the system to undergo some process, including phase or chemical changes, with a resulting increase in entropy dS. Applying Eq. (6-22) to the system–environment combination gives

$$dS + dS_0 \geq 0$$

When incorporated with Eq. (16-13), this equation becomes

$$dS - \frac{đQ}{T} \geq 0$$

or

$$\mathrm{d}Q - T\,\mathrm{d}S \leq 0$$

During the infinitesimal process, the system has a change in internal energy $\mathrm{d}U$ and performs an amount of work $p\,\mathrm{d}V$ on the environment, where p is the pressure exerted by the environment on the system. Hence, by the first law

$$\mathrm{d}Q = \mathrm{d}U + p\,\mathrm{d}V$$

Therefore, finally, we have

$$\mathrm{d}U + p\,\mathrm{d}V - T\,\mathrm{d}S \leq 0 \qquad (16\text{-}14)$$

where the equality holds if the system is already in complete equilibrium, and the inequality holds if it is not. Equation (16-14) is the general condition of equilibrium for closed simple-compressible systems.

Upon combining Eq. (16-14) and the defining equations of enthalpy, Helmholtz function, and Gibbs function, we have

$$\mathrm{d}H - V\,\mathrm{d}p - T\,\mathrm{d}S \leq 0 \qquad (16\text{-}15)$$

$$\mathrm{d}A + p\,\mathrm{d}V + S\,\mathrm{d}T \leq 0 \qquad (16\text{-}16)$$

$$\mathrm{d}G - V\,\mathrm{d}p + S\,\mathrm{d}T \leq 0 \qquad (16\text{-}17)$$

The condition of equilibrium as expressed in Eqs. (16-14) to (16-17) is too general to be of direct utility. More useful special forms, applicable to a system subject to additional restrictions, are given in what follows.

For a closed system at constant energy U and constant volume V, Eq. (16-14) reduces to

$$(\mathrm{d}S)_{U,V} \geq 0 \qquad (16\text{-}18)$$

meaning that for all processes at constant U and V, the entropy increases, or, in the limit, remains constant. Therefore, when a closed system is in equilibrium at constant U and V, its entropy must be a maximum. Note that Eq. (16-18) is equivalent to Eq. (6-22).

On the other hand, for a closed system at constant entropy S and constant volume V, Eq. (16-14) reduces to

$$(\mathrm{d}U)_{S,V} \leq 0 \qquad (16\text{-}19)$$

Thus, the energy of a closed system at constant S and V decreases during a natural process, becoming a minimum at the final equilibrium state for which the equality sign in Eq. (16-19) holds.

Similarly, from Eqs. (16-15) to (16-17), we have the following conditions:

$$(\mathrm{d}S)_{H,p} \geq 0 \qquad (16\text{-}20)$$

$$(\mathrm{d}H)_{S,p} \leq 0 \qquad (16\text{-}21)$$

$$(\mathrm{d}A)_{T,V} \leq 0 \qquad (16\text{-}22)$$

$$(\mathrm{d}G)_{T,p} \leq 0 \qquad (16\text{-}23)$$

It is imperative to notice that the inequality expressions in Eqs. (16-14) to (16-23) specify the conditions for natural processes to occur, thus carrying the system toward

Sec. 16-2   Criteria of Equilibrium

equilibrium. After the state of equilibrium has been reached, the equality sign in these equations then hold, and no more natural process is possible. Thus, for example, the expression $(dS)_{U,V} > 0$ is the condition for natual processes to occur at constant U and V for a system not in equilibrium. On the other hand, if we find $(\delta S)_{U,V} < 0$ for all possible variations away from a state, these variations are not really possible and no change in the system can really take place; this state must be an equilibrium state of the system. Therefore, the criterion for equilibrium under constant U and V can be rewritten in the following alternative formulation:

$$(\delta S)_{U,V} \leq 0 \qquad (16\text{-}18a)$$

Here we use the symbol δ to denote small but finite hypothetical or virtual variations away from equilibrium to distinguish from the symbol "d" for differential changes toward equilibrium.

Similarly, we rewrite the other criteria for equilibrium in the following alternative formulations in terms of virtual variations:

$$\delta U + p\,\delta V - T\,\delta S \geq 0 \qquad (16\text{-}14a)$$

$$(\delta U)_{S,V} \geq 0 \qquad (16\text{-}19a)$$

$$(\delta S)_{H,p} \leq 0 \qquad (16\text{-}20a)$$

$$(\delta H)_{S,p} \geq 0 \qquad (16\text{-}21a)$$

$$(\delta A)_{T,V} \geq 0 \qquad (16\text{-}22a)$$

$$(\delta G)_{T,p} \geq 0 \qquad (16\text{-}23a)$$

Because virtual variations are not necessarily infinitesimal, we need to retain the second-order terms in writing the differential relations for virtual variations based on the definitions of H, A, and G. For instance, from the definition that $A = U - TS$, we write

$$\delta A = \delta U - T\,\delta S - S\,\delta T - \delta S\,\delta T$$

or

$$\delta U - T\,\delta S = \delta A + S\,\delta T + \delta S\,\delta T$$

Substituting this relation into Eq. (16-14a) leads to

$$\delta A + p\,\delta V + S\,\delta T + \delta S\,\delta T \geq 0 \qquad (16\text{-}16a)$$

This is the alternative expression for Eq. (13-16). Notice that the second-order term introduced in this expression does not affect the condition expressed in Eq. (16-22a).

According to Eq. (16-18a), when we are to testify whether a system is in a state of equilibrium or not, the procedure is to choose a neighboring new state that has the same U and V as the state under test but differs in some other variables. Then the virtual variation $(\delta S)_{U,V}$ for the transition from the state under test to the proposed neighboring new state is calculated. This calculation is repeated for all possible neighboring states. If $(\delta S)_{U,V} < 0$ for all such neighboring states, then the state under test is truly an equilibrium state. In such calculations, one should note that unless the state under test and the neighboring state used in the test are both equilibrium states, their thermodynamic properties are not defined. Hence, one must imagine that the needed equilibrium conditions in the states can be established by some form of additional constraints, such as the insertion of semipermeable membranes or impervious partitions. The state under test is a true equilibrium state if $(\delta S)_{U,V} < 0$ for all possible variations after the additional contraints are removed.

16-3 STABILITY

The criteria as expressed in Eqs. (16-18a) to (16-23a) require that as a condition for equilibrium, the first-order variation of a given parameter under certain constraints be equal to zero. However, in order to ensure the *stability* of equilibrium, it is necessary that the condition of minimum or maximum be satisfied for second- (or higher-) order variations. Let us consider a homogeneous single-component, simple-compressible, closed system and start from the general condition as expressed by the inequality of Eq. (16-14a):

$$\delta U + p\, \delta V - T\, \delta S > 0$$

A virtual change of entropy and volume in the system would lead to a virtual change in its energy. To ensure stability, we need to include higher-order differentials in writing the virtual variation δU. Now a Taylor expansion of δU in powers of δS and δV gives

$$\delta U = \delta^1 U + \delta^2 U + \delta^3 U + \cdots$$

$$= (T\, \delta S - p\, \delta V)$$

$$+ \frac{1}{2!}\left[\left(\frac{\partial^2 U}{\partial S^2}\right)_V (\delta S)^2 + 2\frac{\partial^2 U}{\partial S\, \partial V}\delta S\, \delta V + \left(\frac{\partial^2 U}{\partial V^2}\right)_S (\delta V)^2\right]$$

$$+ \frac{1}{3!}\left[\left(\frac{\partial^3 U}{\partial S^3}\right)_V (\delta S)^3 + 3\frac{\partial^3 U}{\partial S^2\, \partial V}(\delta S)^2\, \delta V\right.$$

$$\left. + 3\frac{\partial^3 U}{\partial S\, \partial V^2}\delta S(\delta V)^2 + \left(\frac{\partial^3 U}{\partial V^3}\right)_S (\delta V)^3\right] + \cdots$$

where $\delta^1 U$, $\delta^2 U$, and $\delta^3 U$ represent the first-, second-, and third-order terms in the virtual variation δU, and

$$\left(\frac{\partial U}{\partial S}\right)_V = T \qquad \text{and} \qquad \left(\frac{\partial U}{\partial V}\right)_S = -p$$

Then from Eq. (16-14a), it follows that

$$\delta^2 U + \delta^3 U + \cdots > 0$$

When we choose the virtual changes δS and δV in a way such that the second-order terms outweigh all the higher terms in the Taylor expansion, we then have

$$\delta^2 U > 0 \tag{16-24}$$

or

$$\left(\frac{\partial^2 U}{\partial S^2}\right)_V (\delta S)^2 + 2\frac{\partial^2 U}{\partial S\, \partial V}\delta S\, \delta V + \left(\frac{\partial^2 U}{\partial V^2}\right)_S (\delta V)^2 > 0 \tag{16-24a}$$

The preceding expression is the criterion of stability. It requires that the second-order terms in a homogeneous quadratic differential form be positive for any conceivable virtual process with all possible pairs of δS and δV (except the trivial pair $\delta S = \delta V = 0$). This condition is referred to in mathematical terms that the quadratic form be positive definite, for which we must have

$$\left(\frac{\partial^2 U}{\partial S^2}\right)_v > 0 \qquad\qquad \left(\frac{\partial^2 U}{\partial V^2}\right)_s > 0$$

and

$$\left(\frac{\partial^2 U}{\partial S^2}\right)_v \left(\frac{\partial^2 U}{\partial V^2}\right)_s - \left(\frac{\partial^2 U}{\partial S\, \partial V}\right)^2 > 0 \qquad\qquad (16\text{-}25)$$

The condition for stable equilibrium as expressed by the first inequality in Eq. (16-25) can be rewritten as

$$\left(\frac{\partial^2 U}{\partial S^2}\right)_v = \left(\frac{\partial T}{\partial S}\right)_v > 0 \qquad \text{or} \qquad \left(\frac{\partial S}{\partial T}\right)_v = \frac{C_v}{T} > 0 \qquad (16\text{-}26)$$

The physical significance of this stability criterion is that when heat is added at constant volume to a stable system its temperature must increase. Because T is positive, it follows that the heat capacity at constant volume is positive, or $C_v > 0$.

Similarly, the condition as expressed by the second inequality in Eq. (16-25) can be rewritten as

$$\left(\frac{\partial^2 U}{\partial V^2}\right)_s = -\left(\frac{\partial p}{\partial V}\right)_s > 0 \qquad \text{or} \qquad \left(\frac{\partial p}{\partial V}\right)_s < 0 \qquad (16\text{-}27)$$

Thus, the adiabatic compressibility as defined by the expression $\kappa_S = -(1/V)(\partial V/\partial p)_s$ must be positive.

It is worth noting that there is no need to examine the third inequality in Eq. (16-25), because the physical result arising from it will be derived later in the consideration of other stability criteria. (See Example 16-2.)

Conditions for stable equilibrium imposed on other thermodynamic functions can be obtained by the use of alternative stability criteria. Thus, the condition that involves the variation of the Helmholtz function can be obtained by the inequality of Eq. (16-16a):

$$\delta A + p\, \delta V + S\, \delta T + \delta S\, \delta T > 0$$

Now a Taylor expansion of δA in powers of δV and δT gives

$$\delta A = (\text{first-order terms } \delta^1 A) + (\text{second-order terms } \delta^2 A) + \cdots$$

$$= (-p\, \delta V - S\, \delta T)$$

$$+ \frac{1}{2}\left[\left(\frac{\partial^2 A}{\partial V^2}\right)_T (\delta V)^2 + 2\frac{\partial^2 A}{\partial V\, \partial T}\delta V\, \delta T + \left(\frac{\partial^2 A}{\partial T^2}\right)_v (\delta T)^2\right] + \cdots$$

where

$$\left(\frac{\partial A}{\partial V}\right)_T = -p \qquad \text{and} \qquad \left(\frac{\partial A}{\partial T}\right)_v = -S$$

Then from Eq. (16-16a), it follows that

$$\delta S\ \delta T + \frac{1}{2}\left(\frac{\partial^2 A}{\partial V^2}\right)_T (\delta V)^2 + \frac{\partial^2 A}{\partial V\ \partial T}\delta V\ \delta T + \frac{1}{2}\left(\frac{\partial^2 A}{\partial T^2}\right)_V (\delta T)^2 > 0$$

in which the third- and higher-order terms have been neglected.

Because

$$S = -\left(\frac{\partial A}{\partial T}\right)_V, \qquad \left(\frac{\partial S}{\partial V}\right)_T = -\frac{\partial^2 A}{\partial V\ \partial T}$$

and

$$\left(\frac{\partial S}{\partial T}\right)_V = -\left(\frac{\partial^2 A}{\partial T^2}\right)_V$$

we have, therefore,

$$\delta S\ \delta T = \left(\frac{\partial S}{\partial V}\right)_T \delta V\ \delta T + \left(\frac{\partial S}{\partial T}\right)_V (\delta T)^2$$

$$= -\frac{\partial^2 A}{\partial V\ \partial T}\delta V\ \delta T - \left(\frac{\partial^2 A}{\partial T^2}\right)_V (\delta T)^2$$

Consequently, the criterion of stability is then

$$\left(\frac{\partial^2 A}{\partial V^2}\right)_T (\delta V)^2 - \left(\frac{\partial^2 A}{\partial T^2}\right)_V (\delta T)^2 > 0 \qquad (16\text{-}28)$$

for which we must have

$$\left(\frac{\partial^2 A}{\partial V^2}\right)_T > 0 \qquad \text{and} \qquad \left(\frac{\partial^2 A}{\partial T^2}\right)_V < 0 \qquad (16\text{-}29)$$

The condition for stable equilibrium as expressed by the second inequality in Eq. (16-29) leads to

$$-\left(\frac{\partial^2 A}{\partial T^2}\right)_V = \left(\frac{\partial S}{\partial T}\right)_V = \frac{C_v}{T} > 0$$

which is Eq. (16-26) as obtained previously. However, the first inequality in Eq. (16-29) leads to the following new condition:

$$\left(\frac{\partial^2 A}{\partial V^2}\right)_T = -\left(\frac{\partial p}{\partial V}\right)_T > 0 \qquad \text{or} \qquad \left(\frac{\partial p}{\partial V}\right)_T < 0 \qquad (16\text{-}30)$$

Thus, the isothermal compressibility as defined by the expression $\kappa = -(1/V)(\partial V/\partial p)_T$ must be positive.

The conditions as expressed by Eqs. (16-26) and (16-30) are the two basic conditions for stability for a homogeneous, single-component, simple-compressible, closed system. They are sometimes referred to as the conditions for thermal and mechanical stabilities, respectively. Other conditions, such as Eq. (16-27), are not independent, and can be derived from the two basic ones. We will illustrate this derivation in an example.

So far in this section, we are chiefly concerned with the conditions for stable equilibrium. There are, however, three other kinds of equilibrium that satisfy the equilibrium conditions, such as $(\delta^1 U)_{S,V} = 0$, for continuous changes of state. In a continuous change of state, all the intensive properties of a system are continuous functions of the independent variables. On the other hand, a discontinuous change of state involves a change of phase, where discontinuities appear in some of the intensive properties of the system. We now specify the conditions for all the four kinds of equilibrium for continuous changes of state, using internal energy as the parameter.

(a) Stable equilibrium:

$$(\delta^1 U)_{S,V} = 0$$

and

$$\delta^2 U > 0$$

for any conceivable virtual change with all possible pairs of δS and δV (except the trivial pair $\delta S = \delta V = 0$). These conditions are necessary but not sufficient for absolute stability.

(b) Neutral equilibrium:

$$(\delta^1 U)_{S,V} = 0$$

and

$$\delta^2 U = 0$$

(c) Unstable equilibrium:

$$(\delta^1 U)_{S,V} = 0$$

and

$$\delta^2 U < 0$$

(d) Metastable equilibrium:

$$(\delta^1 U)_{S,V} = 0$$

and

$$\delta^2 U > 0$$

These conditions are the same as that for stable equilibrium. The system is stable with respect to continuous changes, but not with respect to discontinuous changes. Superheated liquids and subcooled vapors (for example, the portions AB and EF in Fig. 14-1 in a van der Waals isotherm) are in metastable states.

It is instructive to note the similarity between equilibrium states in thermodynamics and in mechanics. A marble in four types of mechanical equilibrium is depicted in Fig. 16-1. In Fig. 16-1(a), the marble is in mechanical stable equilibrium, because it will always revert to that state after having been disturbed. In Fig. 16-1(b), the marble

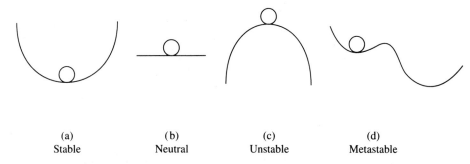

| (a) | (b) | (c) | (d) |
| Stable | Neutral | Unstable | Metastable |

Figure 16-1   Four types of mechanical equilibrium: (a) stable, (b) neutral, (c) unstable, and (d) metastable.

is in mechanical neutral equilibrium, because it will move to a new equilibrium position after disturbance. In Fig. 16-1(c), the marble is in mechanical unstable equilibrium, because it will leave its position under the influence of even a small disturbance. In Fig. 16-1(d), the marble is in mechanical metastable equilibrium, because it is stable with respect to small disturbances, but it will move to a new state of equilibrium when the disturbance exceeds a certain magnitude.

Example 16-1

From the two basic conditions expressed by Eqs. (16-26) and (16-30), show that for stability of a homogeneous, single-component, simple-compressible, closed system, we must have

$$\left(\frac{\partial S}{\partial T}\right)_p > 0 \qquad \text{and} \qquad \left(\frac{\partial p}{\partial V}\right)_S < 0$$

Solution   From Eq. (13-31), we have

$$C_p - C_v = -\frac{T\,(\partial V/\partial T)_p^2}{(\partial V/\partial p)_T}$$

Because $T \geq 0$, $(\partial V/\partial T)_p^2 \geq 0$, and, according to Eq. (16-30), $(\partial V/\partial p)_T < 0$, we must have

$$C_p - C_v \geq 0 \qquad \text{or} \qquad \frac{C_p}{C_v} \geq 1$$

Consequently, from Eqs. (13-24) and (13-25),

$$\frac{C_p}{C_v} = \frac{\left(\dfrac{\partial S}{\partial T}\right)_p}{\left(\dfrac{\partial S}{\partial T}\right)_V} \geq 1$$

But according to Eq. (16-26), $(\partial S/\partial T)_V = C_v/T > 0$. Therefore, we conclude that

$$\left(\frac{\partial S}{\partial T}\right)_p = \frac{C_p}{T} > 0 \qquad\qquad (16\text{-}31)$$

meaning that when heat is added at constant pressure to a stable system, its temperature must increase; in other words, the heat capacity at constant pressure is positive for positive T.

From the cyclic relation, we have

$$\left(\frac{\partial p}{\partial V}\right)_S = -\frac{\left(\frac{\partial S}{\partial V}\right)_p}{\left(\frac{\partial S}{\partial p}\right)_V} \tag{16-32a}$$

and

$$\left(\frac{\partial p}{\partial V}\right)_T = -\frac{\left(\frac{\partial T}{\partial V}\right)_p}{\left(\frac{\partial T}{\partial p}\right)_V} \tag{16-32b}$$

Dividing Eq. (16-32a) by Eq. (16-32b), we have

$$\frac{\left(\frac{\partial p}{\partial V}\right)_S}{\left(\frac{\partial p}{\partial V}\right)_T} = \frac{\left(\frac{\partial S}{\partial V}\right)_p \left(\frac{\partial T}{\partial p}\right)_V}{\left(\frac{\partial T}{\partial V}\right)_p \left(\frac{\partial S}{\partial p}\right)_V} = \frac{\left(\frac{\partial S}{\partial T}\right)_p}{\left(\frac{\partial S}{\partial T}\right)_V}$$

Substituting Eqs. (16-26), (16-30), and (16-31) into the preceding equation, we conclude that

$$\left(\frac{\partial p}{\partial V}\right)_S < 0$$

This is Eq. (16-27).

Example 16-2

Show that

$$\left(\frac{\partial^2 U}{\partial S^2}\right)_V \left(\frac{\partial^2 U}{\partial V^2}\right)_S - \left(\frac{\partial^2 U}{\partial S\, \partial V}\right)^2 = -\frac{T}{C_v}\left(\frac{\partial p}{\partial V}\right)_T$$

and then verify that the condition of stability as expressed by the third inequality in Eq. (16-25) is equivalent to that of Eqs. (16-26) and (16-30).

Solution   From Eq. (13-14a),

$$\left(\frac{\partial U}{\partial S}\right)_V = T \qquad \text{and} \qquad \left(\frac{\partial U}{\partial V}\right)_S = -p$$

we have

$$\left(\frac{\partial^2 U}{\partial S^2}\right)_V = \left(\frac{\partial T}{\partial S}\right)_V \qquad \text{and} \qquad \left(\frac{\partial^2 U}{\partial V^2}\right)_S = -\left(\frac{\partial p}{\partial V}\right)_S$$

$$\left(\frac{\partial^2 U}{\partial S\, \partial V}\right) = \left(\frac{\partial T}{\partial V}\right)_S = -\left(\frac{\partial T}{\partial S}\right)_V \left(\frac{\partial S}{\partial V}\right)_T$$

Thus,

$$\left(\frac{\partial^2 U}{\partial S^2}\right)_V \left(\frac{\partial^2 U}{\partial V^2}\right)_S - \left(\frac{\partial^2 U}{\partial S\, \partial V}\right)^2 = -\left(\frac{\partial T}{\partial S}\right)_V \left(\frac{\partial p}{\partial V}\right)_S - \left(\frac{\partial T}{\partial S}\right)_V^2 \left(\frac{\partial S}{\partial V}\right)_T^2$$

$$= -\left(\frac{\partial T}{\partial S}\right)_V \left[\left(\frac{\partial p}{\partial V}\right)_S + \left(\frac{\partial T}{\partial S}\right)_V \left(\frac{\partial S}{\partial V}\right)_T^2\right]$$

But by Eq. (13-4),

$$\left(\frac{\partial p}{\partial V}\right)_S = \left(\frac{\partial p}{\partial V}\right)_T + \left(\frac{\partial p}{\partial T}\right)_V\left(\frac{\partial T}{\partial V}\right)_S$$

By Eqs. (13-7) and (13-15c),

$$\left(\frac{\partial T}{\partial S}\right)_V = -\left(\frac{\partial T}{\partial V}\right)_S\left(\frac{\partial V}{\partial S}\right)_T = -\left(\frac{\partial T}{\partial V}\right)_S\left(\frac{\partial T}{\partial p}\right)_V$$

and

$$\left(\frac{\partial S}{\partial V}\right)_T = \left(\frac{\partial p}{\partial T}\right)_V$$

Thus,

$$\left(\frac{\partial^2 U}{\partial S^2}\right)_V\left(\frac{\partial^2 U}{\partial V^2}\right)_S - \left(\frac{\partial^2 U}{\partial S\,\partial V}\right)^2 = -\left(\frac{\partial T}{\partial S}\right)_V\left[\left(\frac{\partial p}{\partial V}\right)_T + \left(\frac{\partial p}{\partial T}\right)_V\left(\frac{\partial T}{\partial V}\right)_S - \left(\frac{\partial T}{\partial V}\right)_S\left(\frac{\partial T}{\partial p}\right)_V\left(\frac{\partial p}{\partial T}\right)_V^2\right]$$

$$= -\left(\frac{\partial T}{\partial S}\right)_V\left(\frac{\partial p}{\partial V}\right)_T$$

But from Eq. (13-24),

$$c_v = T\left(\frac{\partial S}{\partial T}\right)_V$$

Therefore,

$$\left(\frac{\partial^2 U}{\partial S^2}\right)_V\left(\frac{\partial^2 U}{\partial V^2}\right)_S - \left(\frac{\partial^2 U}{\partial S\,\partial V}\right)^2 = -\frac{T}{c_v}\left(\frac{\partial p}{\partial V}\right)_T$$

Now from the third inequality in Eq. (16-25),

$$\left(\frac{\partial^2 U}{\partial S^2}\right)_V\left(\frac{\partial^2 U}{\partial V^2}\right)_S - \left(\frac{\partial^2 U}{\partial S\,\partial V}\right)^2 > 0$$

or

$$-\frac{T}{C_v}\left(\frac{\partial p}{\partial V}\right)_T > 0 \tag{16-33}$$

From Eq. (16-26),

$$\frac{c_v}{T} > 0$$

And from Eq. (16-30),

$$-\left(\frac{\partial p}{\partial V}\right)_T > 0$$

Thus, Eq. (16-33) is the combination of Eqs. (16-26) and (16-30).

16-4 EQUATIONS OF PHASE EQUILIBRIUM

A *heterogeneous system* consists of two or more phases that are separated from each other by surfaces of phase transition. We now examine the conditions of equilibrium for such a system. In our present discussions, we neglect any surface effect as compared with the effects of bulk phases. Furthermore, let the system as a whole be a closed simple compressible system, allowing no chemical reactions. Assume that the system has already attained equality in temperature and pressure throughout; thus, it is in thermal and mechanical equilibrium. Our purpose here is to find the conditions that will ensure phase equilibrium, thereby achieving complete equilibrium.

Let the heterogeneous system under consideration have φ homogeneous phases and r components. In the absence of chemical reactions, the number of components of a system is identical with the number of molecular species present in the system. Let us assume, at first, that every component is present in every phase. Atlthough the system as a whole is closed, the phases are open because transportation of components between phases is allowed. For any one of the homogeneous simple-compressible open phases, the differential of the Gibbs function for changes between neighboring equilibrium states is expressed by Eq. (16-6):

$$dG^{(\alpha)} = -S^{(\alpha)}\,dT + V^{(\alpha)}\,dp + \sum_{i=1}^{r} \mu_i^{(\alpha)}\,dn_i^{(\alpha)}$$

Applying this equation to each phase of the system and summing up, we get the change in the Gibbs function of the system:

$$dG = \sum_{\alpha=1}^{\varphi}\left(-S^{(\alpha)}\,dT + V^{(\alpha)}\,dp + \sum_{i=1}^{r} \mu_i^{(\alpha)}\,dn_i^{(\alpha)}\right)$$

or

$$dG = -S\,dT + V\,dp + \sum_{\alpha=1}^{\varphi}\left(\sum_{i=1}^{r} \mu_i^{(\alpha)}\,dn_i^{(\alpha)}\right) \tag{16-34}$$

where the subscripts denote components and the superscripts denote phases; S and V are the entropy and volume, respectively, of the whole system. Because the system is assumed to be in thermal and mechanical equilibrium, it is obvious that temperature and pressure must be the same throughout the system.

At constant temperature and pressure, the condition for equilibrium, according to Eq. (16-23), is

$$(dG)_{T,p} = 0$$

Applying this condition to Eq. (16-34) results in

$$(dG)_{T,p} = \sum_{\alpha=1}^{\varphi}\left(\sum_{i=1}^{r} \mu_i^{(\alpha)}\,dn_i^{(\alpha)}\right) = 0 \tag{16-35}$$

Because the system as a whole is closed and no chemical reactions occur, the changes in the n_i's are due solely to the transfer of components between phases. Hence, the n_i's are subjected to the following equations of constraint:

$$\sum_{\alpha=1}^{\varphi} n_i^{(\alpha)} = \text{constant}, \qquad i = 1, 2, \ldots, r \tag{16-36}$$

Our task is to find the equations of phase equilibrium that satisfy the condition as expressed by Eq. (16-35), subject to the constraints as expressed by Eq. (16-36). To do this, it is convenient to use the *Lagrange's method of undertermined multipliers.* *

*See Appendix 1.

Thus, differentiating each equation of constraint and multiplying each of them by a different Lagrangian multiplier lead to the relations:

$$\sum_{\alpha=1}^{\varphi} \lambda_i \, dn_i^{(\alpha)} = 0, \qquad i=1, 2, \ldots, r$$

Adding this equation and Eq. (16-35) results in

$$\sum_{i=1}^{r} \sum_{\alpha=1}^{\varphi} \left(\mu_i^{(\alpha)} + \lambda_i \right) dn_i^{(\alpha)} = 0$$

Equating each coefficient to zero results in a set of equations:

$$\mu_i^{(\alpha)} = -\lambda_i, \qquad i = 1, 2, \ldots, r; \alpha = 1, 2, \ldots, \varphi$$

meaning that when a heterogeneous system is at equilibrium at constant temperature and pressure, the chemical potential of any given component must have a common value in all phases. Therefore, we conclude that

$$\mu_i^{(1)} = \mu_i^{(2)} = \cdots = \mu_i^{(\varphi)}, \qquad i = 1, 2, \ldots, r \qquad (16\text{-}37)$$

a total of $(\varphi - 1)r$ equations. These are the *equations of phase equilibrium*.

In view of the preceding discussion, it is evident that the chemical potential is the driving force for mass transfer between phases or between portions of a given phase, just as the temperature is the driving force for heat transfer and the pressure is the driving force for "work transfer." In other words, the chemical potential plays a role in chemical equilibrium similar to those played by the temperature in thermal equilibrium and the pressure in mechanical equilibrium.

In arriving at the series of Eq. (16-37), we have assumed that every component is present in every phase. However, if a component is not actually present in a phase but is present in an adjoining phase, the equilibrium condition would be seen from the virtual variation that $(\delta G)_{T,p} \geq 0$. For illustrative purposes, let us consider a component i in equilibrium between two phases. For this case, because $\delta n_i^{(1)} = -\delta n_i^{(2)}$, we have

$$(\mu_i^{(1)} - \mu_i^{(2)})\delta n_i^{(1)} \geq 0$$

If component i is present in phase 2 but not in phase 1, it is possible for $\delta n_i^{(1)}$ to be positive but not negative. Therefore, we must have for equilibrium

$$\mu_i^{(1)} \geq \mu_i^{(2)} \qquad (16\text{-}38)$$

Example 16-3

Liquid and vapor phases of a substance may coexist in equilibrium at different pressures. For example, if a liquid–vapor system is confined in a vessel that also contains some inert gas that is insoluble in the liquid, the pressure applied to the liquid can be appreciably greater than its partial vapor pressure. In order to study the effect of applied pressure on the vapor pressure of a liquid, consider the idealized setup shown in Fig. 16-2, in which the single-component liquid and vapor phases under different pressures are separated by a nondeformable, heat-conducting membrane permeable to vapor only.

(a) Show that

$$\left(\frac{\partial p^{(\beta)}}{\partial p^{(\alpha)}} \right)_T = \frac{v^{(\alpha)}}{v^{(\beta)}}$$

where the superscripts (α) and (β) denote the liquid and vapor phases, respectively.

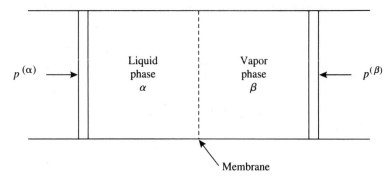

Figure 16-2 Effect of applied pressure on vapor pressure.

(b) Liquid water at 30 MPa and 200°C is in equilibrium with water vapor according to the setup shown in Fig. 16-2. Find the pressure of the water vapor.

Solution

(a) Because of the presence of the membrane, mechanical equilibrium is maintained in the system. To ensure phase equilibrium, according to Eq. (16-37), we must have

$$\mu^{(\alpha)} = \mu^{(\beta)}$$

From which

$$d\mu^{(\alpha)} = d\mu^{(\beta)}$$

But for a pure component; $\mu = g$, and

$$d\mu^{(\alpha)} = dg^{(\alpha)} = -s^{(\alpha)} \, dT + v^{(\alpha)} \, dp^{(\alpha)}$$

$$d\mu^{(\beta)} = dg^{(\beta)} = -s^{(\beta)} \, dT + v^{(\beta)} \, dp^{(\beta)}$$

Thus,

$$-s^{(\alpha)} \, dT + v^{(\alpha)} \, dp^{(\alpha)} = -s^{(\beta)} \, dT + v^{(\beta)} \, dp^{(\beta)}$$

At constant temperature, this equation becomes

$$v^{(\alpha)} \, dp^{(\alpha)} = v^{(\beta)} \, dp^{(\beta)}$$

Therefore,

$$\left(\frac{\partial p^{(\beta)}}{\partial p^{(\alpha)}}\right)_T = \frac{v^{(\alpha)}}{v^{(\beta)}} \tag{16-39}$$

This equation describes how the vapor pressure changes with applied pressure at constant temperature. Because at the same temperature, the specific volume of a liquid is much smaller than that of a vapor, we see from this equation that the effect of an applied pressure on the vapor pressure is very weak at ordinary conditions.

(b) Because the liquid and vapor are in thermal equilibrium, the vapor is also at 200°C. According to Eq. (16-39), we have

$$\left[\int_{p_{\text{sat}}}^{p^{(\alpha)}} v^{(\alpha)} \, dp^{(\alpha)}\right]_T = \left[\int_{p_{\text{sat}}}^{p^{(\beta)}} v^{(\beta)} \, dp^{(\beta)}\right]_T$$

where $p_{\text{sat}} = 1.5538$ MPa [21], which is the saturation pressure corresponding to 200°C. Integrating the foregoing equation can be performed by assuming the liquid to be incompressible and the vapor to behave as an ideal gas. Thus, $v^{(\alpha)} = v_f = 1.1565 \times 10^{-3}$ m³/kg, which is the specific volume of saturated liquid at 200°C; and $v^{(\beta)} = p_{\text{sat}} v_g / p^{(\beta)}$, where $v_g = 127.36 \times 10^{-3}$ m³/kg, which is the specific volume of saturated vapor at 200°C. We accordingly write

$$v_f(p^{(\alpha)} - p_{\text{sat}}) = p_{\text{sat}} v_g \ln \frac{p^{(\beta)}}{p_{\text{sat}}}$$

Hence,

$$p^{(\beta)} = p_{\text{sat}} \exp\left[\frac{v_f}{v_g}\left(\frac{p^{(\alpha)}}{p_{\text{sat}}} - 1\right)\right]$$

$$= 1.5538 \exp\left[\frac{1.1565 \times 10^{-3}}{127.36 \times 10^{-3}}\left(\frac{30}{1.5538} - 1\right)\right]$$

$$= 1.83 \text{ MPa}$$

Note that at 1.83 MPa, the saturation temperature is 207.6°C, which is greater than the existing temperature of 200°C. The vapor is therefore in a supersaturated metastable state.

Example 16-4

At extremely high pressures, graphite and diamond exist in equilibrium according to the requirement that their specific Gibbs functions are equal. Estimate the pressure required to make diamonds from graphite at a temperature of 25°C. At 25°C and 1 atm, the following data are available:

	Graphite	Diamond
Specific Gibbs function, kJ/kg	0	237
Specific volume, m³/kg	4.45×10^{-4}	2.85×10^{-4}
Isothermal compressibility κ, atm⁻¹	3.0×10^{-6}	0.16×10^{-6}

Solution   According to Eq. (13-17),

$$\kappa = -\frac{1}{v}\left(\frac{\partial v}{\partial p}\right)_T$$

For a constant-temperature process,

$$\frac{dv}{v} = -\kappa \, dp$$

If κ is considered constant,

$$\int_{v_0}^{v} \frac{dv}{v} = -\kappa \int_{p_0=1\text{ atm}}^{p} dp$$

$$\ln \frac{v}{v_0} = -\kappa(p - p_0) = -\kappa(p - 1)$$

Therefore,

$$v = v_0 e^\kappa e^{-\kappa p}$$

where in this example the subscript 0 is used for the property at 1 atm and 25°C. But from Eq. (13-13a),

$$dg = v \, dp - s \, dT$$

Then for a constant temperature process,

$$dg = v\, dp = v_0 e^{\kappa} e^{-\kappa p}\, dp$$

$$\int_{g_0}^{g} dg = g - g_0 = v_0 e^{\kappa} \int_{p_0=1\,\text{atm}}^{p} e^{-\kappa p}\, dp$$

$$= -\frac{v_0 e^{\kappa}}{\kappa}(e^{-\kappa p} - e^{-\kappa}) = -\frac{v_0}{\kappa}[e^{-\kappa(p-1)} - 1]$$

Applying this equation to graphite (indicated by the superscript G) and diamond (indicated by the superscript D) and setting $g^{(G)} = g^{(D)}$ for phase equilibrium, we have

$$g_0^{(G)} + \frac{v_0^{(G)}}{\kappa^{(G)}}[1 - e^{-\kappa^{(G)}(p-1)}] = g_0^{(D)} + \frac{v_0^{(D)}}{\kappa^{(D)}}[1 - e^{-\kappa^{(D)}(p-1)}]$$

Substituting numerical values gives

$$0 + \frac{4.45 \times 10^{-4}\ \text{m}^3/\text{kg}}{3.0 \times 10^{-6}\ \text{atm}^{-1}}[1 - e^{-(3.0\times10^{-6}\,\text{atm}^{-1})(p\,\text{atm}-1\,\text{atm})}]$$

$$= \frac{237\ \text{kJ/kg}}{101.325\ \text{kJ/atm}\cdot\text{m}^3} + \frac{2.85 \times 10^{-4}\ \text{m}^3/\text{kg}}{0.16 \times 10^{-6}\ \text{atm}^{-1}} \times [1 - e^{-(0.16\times10^{-6}\,\text{atm}^{-1})(p-1\,\text{atm})}]$$

From this expression, by trial and error, we obtain

$$p = 15{,}590\ \text{atm}$$

16-5 BINARY VAPOR–LIQUID SYSTEMS

It is clear that there are basic differences in phase behavior between a single-component system and a multicomponent system. First of all, in a single-component system, the compositions of different phases are always the same. In a multicomponent system, however, the compositions of different phases in equilibrium, in general, are different. Second, during a change of phase, say, in vaporization, in a single-component system, the temperature and the pressure always remain constant. However, for a mixture of two or more substances of different volatilities, the pressure does not remain constant but falls continuously during isothermal vaporization. These and other features in the phase behavior of multicomponent systems can be illustrated through a study of a binary vapor–liquid system with a miscible liquid.

The properties of a binary vapor–liquid system can be most readily seen by means of a three-dimensinal surface with pressure p, temperature T, and composition x as the coordinates. A plane of constant composition cutting through a p–T–x surface would give a p–T phase diagram, as depicted in Fig. 16-3. The vapor–liquid two-phase region in the figure is bounded by a *border curve A–B–C–D–E*. This curve is made up of two portions: the one on the left is the bubble-point line A–B–C, and the one on the right is the dew-point line E–D–C. A *bubble-point line* (saturated-liquid line) is the locus of states of initial vaporization, whereas a *dew-point line* (saturated-vapor line) is the locus of states of initial condensation. The point where the bubble-point and dew-point lines meet, or where the vapor and liquid phases become indistinguishable, is the *critical point* (C) of the mixture. It does not necessarily correspond to the maximum pressure and temperature at which the mixture can exist in two phases. Point B on the border curve represents the maximum pressure at which a system of the given composi-

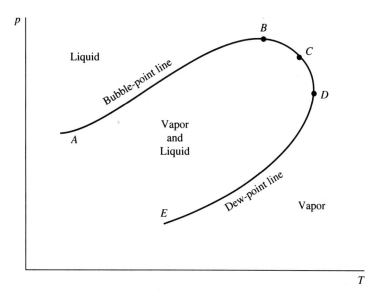

Figure 16-3 Pressure–temperature phase diagram at constant composition for a binary mixture.

tion can exist in two phases. Point D represents the maximum temperature at which the system can exist in two phases. Note that in the case of a single-component system, lines A–B–C and E–D–C coincide to form a single vapor-pressure line, and the points B,C, and D coincide to form a single critical point.

The border curve shown in Fig. 16-3 is for a single composition. To show the border curves of various compositions in a two-dimensional plane, we proceed to cut the p–T–x surface at a number of different constant compositions and project all the border curves onto a single p–T plane, as illustrated in Fig. 16-4. In this figure, each of the five loops represents the p–T border curve of a mixture of definite composition. These five mixtures are labeled a, b, . . . , e. The border curves converge at each end to single curves, labeled 1 and 2, representing the vapor-pressure curves of the two pure components. Any point, such as point F, where the vapor branch of one curve intersects the liquid branch of another, represents coexisting phases in equilibrium. The critical points of the five mixtures are labeled C_a, C_b, . . . , C_e in the figure, and the critical points of the two pure components are labeled C_1 and C_2. Line C_1–C_a– $\cdots$ C_e–C_2 represents the critical envelope, which is the locus of all critical points for all possible mixtures of the two substances.

Because many industrial processes are conducted under constant pressure, it is of great utility to have diagrams relating temperature and composition at constant pressure. Typical forms of T–x relations for a binary system at various pressures are shown in Fig. 16-5. These curves can be obtained by cutting a p–T–x surface by planes of constant pressure. The curves marked p_I, p_{II}, and p_{III} in Fig. 16-5 correspond to the indicted constant pressures in Fig. 16-4. The curve marked p_I is typical for a pressure that is below the critical pressures of both pure components; the curve marked p_{II} is for a pressure that is above the critical pressure of the less volatile component (component 2); and the curve marked p_{III} is for a pressure that is above the critical pressures of both pure components.

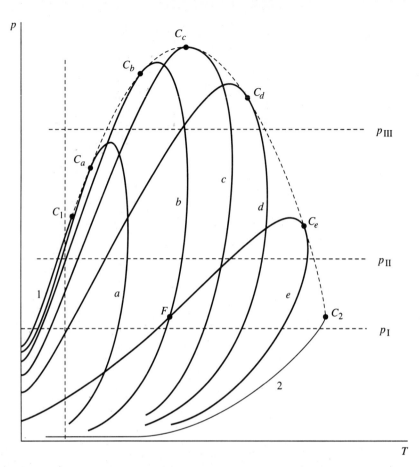

Figure 16-4 Pressure–temperature border curves for a binary system at various compositions. (After W. B. Kay, *Industrial Engineering Chemistry,* vol. 30, p. 459 [1938]; and *Journal of Chemical Engineering Data,* vol. 15, p. 46 [1970].)

To make the discussion easier, the p_I curve in Fig. 16-5 is duplicated in Fig. 16-6. The lens-shaped, vapor–liquid, two-phase area is bounded by the dew-point line on the top and the bubble-point line on the bottom. The area above the dew-point line represents states of superheated vapor mixtures, and the area below the bubble-point line represents states of subcooled liquid solutions. If heat is added at constant pressure to a liquid solution at state a, the temperature of the solution will increase while its composition remains unchanged until the bubble-point line at state b, is reached. Additional heat addition will cause vaporization of both components in such a proportion that the vapor appears in a composition corresponding to point b'. During vaporization, the composition of the vapor moves from b' toward d, while the composition of the liquid moves from b toward d''. At any state c in the two-phase region, the vapor has a composition corresponding to point c' and the liquid has a composition corresponding to point c''. If vaporization is carried to completion, the final vapor mixture at state d will have the same composition as the original liquid solution. Further heat addition beyond the dew-point line will, of course, superheat the vapor, say, to state e.

When processes are conducted under a constant-temperature condition, it is desirable to have a pressure-versus-composition diagram at constant temperature, as shown

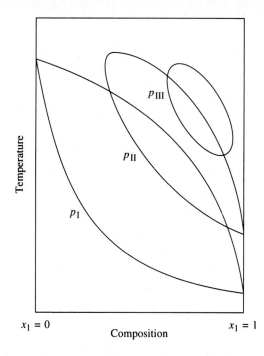

Figure 16-5 Temperature–composition diagram for a binary system at various pressures.

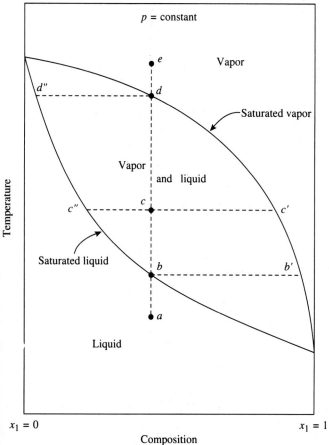

Figure 16-6 Temperature–composition diagram for a binary system at a constant pressure lower than the critical pressures of both pure components.

in Fig. 16-7. A p–x diagram can be obtained by cutting a p–T–x surface by a plane of constant temperature. The one shown in Fig. 16-7 is for a constant temperature that is below the critical temperatures of both pure components, corresponding to the constant-temperature line shown by the vertical dashed line in Fig. 16-4. In Fig. 16-7, the vapor–liquid two-phase area is bounded by the bubble-point line on the top and the dew-point line on the bottom. The area below the dew-point line represents states of superheated vapor mixtures, and the area above the bubble-point line represents states of compressed liquid solutions. Line a–b–c–d–e represents a constant-temperature compression process. During condensation between states b and d, the vapor composition moves from b toward d', while the liquid composition moves from b'' toward d.

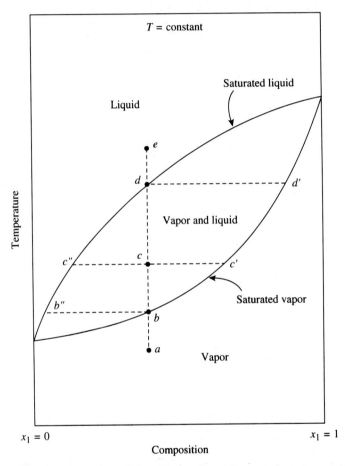

Figure 16-7 Pressure–composition diagram for a binary system at a constant temperature lower than the critical temperatures of both pure components.

Example 16-5

In a dilute liquid solution, the predominant component is called the solvent and the minor components are called solutes. If the vapor in equilibrium with an infinitely dilute solution is considered as an ideal gas, *Raoult's law* states that

$$p = p°x \qquad (16\text{-}40)$$

Phase and Chemical Equilibrium   Chap. 16

where p is the partial vapor pressure of the solvent, $p°$ is the vapor pressure of pure solvent at the temperature and pressure of the solution, and x is the mole fraction of the solvent. Because $p°$ is a strong function of temperature and a weak function of pressure, it can be considered as a function of temperature alone.

A vapor mixture of 20 mole percent ethane (C_2H_6) and 80 mole percent propylene (C_3H_6), initially at 14.7 psia and 40°F, is compressed isothermally to higher pressures. (a) What is the pressure when the first drop of liquid appears and what is the composition of that drop of liquid? (b) What is the pressure when the last portion of vapor condenses, and what is the composition of that portion of vapor? Assume that Raoult's law is obeyed by both components of the solution. Use Dalton's law for the vapor mixture. The vapor pressures of ethane and propylene at 40°F are 385 psia and 96.4 psia, respectively.

Solution   We denote ethane by subscript 1 and propylene by subscript 2. Superscripts (α) and (β) denote the liquid and vapor phases, respectively. Figure 16-7 shows schematically the pressure–composition diagram, except that in this example, because of the simplified assumption of Raoult's law, the bubble-point line would be a straight line.

(a) When the first drop of liquid appears at pressure p_b corresponding to point b, the vapor phase will be at point b with the given composition of $x_1^{(\beta)} = 0.20$ and $x_2^{(\beta)} = 0.80$. The liquid phase will be at point b'', the composition of which is to be determined. From Raoult's and Dalton's laws, the partial vapor pressures of the two components are

$$p_1 = x_1^{(\alpha)} p_1° = x_1^{(\beta)} p_b$$

$$p_2 = x_2^{(\alpha)} p_2° = x_2^{(\beta)} p_b = (1 - x_1^{(\alpha)}) p_2° = (1 - x_1^{(\beta)}) p_b$$

Dividing, we have

$$\frac{x_1^{(\alpha)}}{1 - x_1^{(\alpha)}} = \frac{p_2°}{p_1°} \frac{x_1^{(\beta)}}{1 - x_1^{(\beta)}} = \frac{96.4}{385} \frac{0.20}{0.80} = 0.0626$$

from which we obtain

$$x_1^{(\alpha)} = 0.059 \qquad \text{and} \qquad x_2^{(\alpha)} = 1 - x_1^{(\alpha)} = 0.941$$

Hence,

$$p_b = \frac{x_1^{(\alpha)}}{x_1^{(\beta)}} p_1° = \frac{0.059}{0.20} \times 385 = 113.6 \text{ psia}$$

or

$$p_b = p_1 + p_2 = x_1^{(\alpha)} p_1° + x_2^{(\alpha)} p_2°$$

$$= 0.059 \times 385 + 0.941 \times 96.4 = 113.4 \text{ psia}$$

(b) When the last portion of vapor condenses at pressure p_d corresponding to point d, the liquid phase will be at point d with the given composition of $x_1^{(\alpha)} = 0.20$ and $x_2^{(\alpha)} = 0.80$, while the vapor phase will be at point d', the composition of which is to be determined. Analogous to part (a), we have

$$\frac{x_1^{(\beta)}}{1 - x_1^{(\beta)}} = \frac{p_1°}{p_2°} \frac{x_1^{(\alpha)}}{1 - x_1^{(\alpha)}} = \frac{385}{96.4} \frac{0.20}{0.80} = 0.998$$

from which we obtain

$$x_1^{(\beta)} = 0.50 \qquad \text{and} \qquad x_2^{(\beta)} = 1 - x_1^{(\beta)} = 0.50$$

Hence,

$$p_d = \frac{x_1^{(\alpha)}}{x_1^{(\beta)}} p_1° = \frac{0.20}{0.50} \times 385 = 154 \text{ psia}$$

or

$$p_d = p_1 + p_2 = x_1^{(\alpha)} p_1^\circ + x_2^{(\alpha)} p_2^\circ$$

$$= 0.20 \times 385 + 0.80 \times 96.4 = 154 \text{ psia}$$

16-6 ABSORPTION REFRIGERATION ANALYSIS

As an illustration for multicomponent phase equilibrium, we devote this section to the study of absorption refrigeration analysis. In our discussion in this section, we denote component A as the refrigerant and component B as the absorbent. The *concentration* of component A is defined as

$$x_A = \frac{\text{mass of component } A}{\text{mass of mixture}} \tag{16-41}$$

Because the generator and the absorber of an absorption refrigeration system are kept at different levels of pressure, it is instructive to refer to Fig. 16-8 to describe the

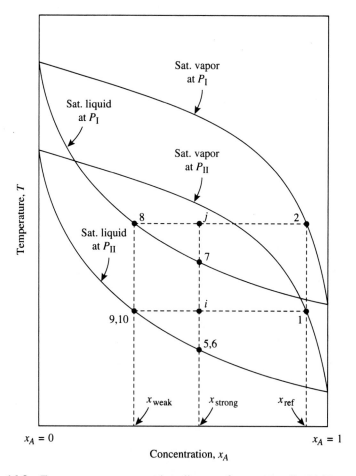

Figure 16-8 Temperature–concentration diagram for a vapor–liquid binary mixture at two pressures.

Phase and Chemical Equilibrium   Chap. 16

phase behavior of the mixture. In Fig. 16-8, two sets of vapor–liquid border curves are shown, one for the high pressure p_I and the other for the low pressure p_{II}. The state points in this figure correspond to those indicated in Fig. 12-12. The strong solution leaving the absorber with a refrigerant concentration of x_{strong}, as indicated in Fig. 16-8, might be a saturated liquid in state 5, which is at p_{II}. It is pumped to state 6 at p_I without change in concentration, the temperatures T_5 and T_6 being very nearly the same. The subcooled liquid at state 6 is then heated at constant pressure p_I in the solution heat exchanger to state 7, which might be a saturated liquid at p_I, as so indicated in the figure. As heat is added to the strong solution in the generator, the formation of the refrigerant-rich vapor reduces the concentration of refrigerant in the liquid phase, thus forming the weak solution. When the two-phase mixture reaches state j at p_I, the vapor phase reaches state 2, having a refrigerant concentration of x_{ref}, and the liquid phase reaches state 8, having a refrigerant concentration of x_{weak}. The refrigerant-rich vapor then flows to the condenser. In the meantime, the weak solution is subcooled in the solution heat exchanger at constant pressure p_I to state 9. It is then throttled to the lower pressure p_{II}, reaching a saturated or slightly subcooled state 10 with $T_{10} = T_9$. When the refrigerant-rich vapor coming back from the evaporator at state 1 meets the weak solution at state 10, the two-phase equilibrium mixture would be at state i, which when cooled in the absorber at constant pressure p_{II} results in the strong solution at state 5.

For the analysis and design of absorption refrigeration systems, the most frequently used property diagram is the enthalpy–concentration (h–x) diagram. The h–x diagrams for the ammonia–water (called aqua-ammonia) and the water–lithium bromide mixtures are given in Appendix 4. Figure 16-9 is a schematic h–x diagram for the two-phase region of a binary mixture at two pressures. The saturated vapor and saturated liquid lines for a given pressure are separated at $x_A = 0$ and $x_A = 1$ by a distance corresponding to the enthalpy of vaporization (h_{fg}) of the respective pure components. To facilitate the use of the graph, a series of fictitious saturated liquid lines, called equilibrium construction lines, are added. When the state of a saturated liquid is known, for example, at point a, a vertical line through this point is drawn to intersect the equilibrium construction line for the given pressure at point b; then the intersection (point c) of a horizontal line through point b and the corresponding saturated-vapor line represents the state of the saturated-vapor phase; and line a–c represents the vapor–liquid equilibrium temperature line, all at the given pressure.

By assuming steady-flow operations, the various machine component parts can be analyzed by writing the necessary energy- and mass-balance equations. Referring to Figs. 12-12 and 16-8, neglecting changes in potential and kinetic energies, an energy balance for the generator gives

$$\dot{Q}_{gen} + \dot{m}_{strong}h_7 = \dot{m}_{ref}h_2 + \dot{m}_{weak}h_8 \tag{16-42}$$

The overall mass balance for the generator is given by

$$\dot{m}_{strong} = \dot{m}_{ref} + \dot{m}_{weak} \tag{16-43}$$

and the mass balance for the refrigerant (component A) in the generator is given by

$$\dot{m}_{strong}x_{strong} = \dot{m}_{ref}x_{ref} + \dot{m}_{weak}x_{weak} \tag{16-44}$$

where the $\dot{m}$'s are the mass-flow rates, the h's are the enthalpies, and the x's are the concentrations of the refrigerant.

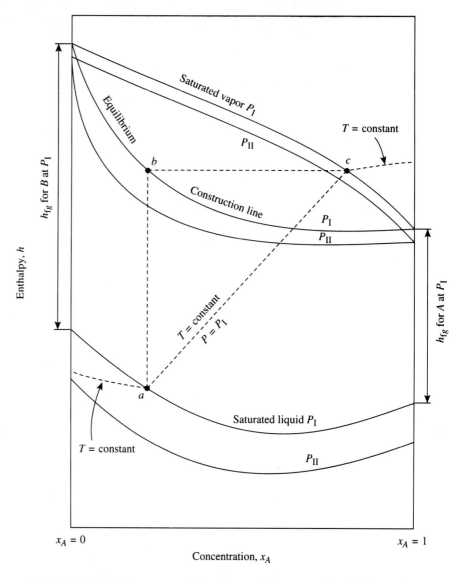

Figure 16-9 Enthalpy-concentration diagram for a vapor–liquid binary mixture.

For the absorber, the mass-balance equations are the same as Eqs. (16-43) and (16-44), and the energy-balance equation is given by

$$\dot{m}_{ref}h_1 + \dot{m}_{weak}h_{10} = \dot{m}_{strong}h_5 + \dot{Q}_{abs} \tag{16-45}$$

For the solution heat exchanger, an energy balance yields

$$\dot{m}_{strong}(h_7 - h_6) = \dot{m}_{weak}(h_8 - h_9) \tag{16-46}$$

and for the pressure-reducing valve, an energy balance yields

$$h_9 = h_{10}$$

Meanwhile, for the pump, the power requirement is given by

$$\dot{W}_{pump} = \dot{m}_{strong}(h_6 - h_5) \tag{16-47}$$

Note that the analyses for the condenser, the evaporator, and the expansion valve are the same as in the compression cycle given in Sec. 12-1.

With $\dot{Q}_{evap}$, $\dot{Q}_{gen}$, and $\dot{W}_{pump}$ known, the coefficient of performance of the cycle is given by

$$(COP)_{ref} = \frac{\dot{Q}_{evap}}{\dot{Q}_{gen} + \dot{W}_{pump}} \tag{16-48}$$

which, when the pump power is neglected, results in

$$(COP)_{ref} = \frac{\dot{Q}_{evap}}{\dot{Q}_{gen}} \tag{16-48a}$$

The COP of a lithium bromide–water absorption refrigeration machine is about 0.8 and more often operates in the range from 0.6 to 0.7. A water–ammonia absorption machine usually operates with lower COP than a corresponding lithium bromide–water absorption machine.

It must be recognized that absorption refrigeration systems are more sophisticated and less efficient than vapor-compression refrigeration systems. They are generally used only in applications where the energy source is at relatively low temperatures, such as the case of industrial waste heat. Absorption refrigeration is suitable for space-cooling purposes for which solar energy is the source of thermal energy input.*

16-7 GIBBS PHASE RULE

The thermodynamic state at equilibrium for a heterogeneous system of φ phases, each containing r components that do not react chemically, is a function of the temperature, the pressure, and the $(r - 1)$ values of mole fractions expressing the compositions of all the phases. Therefore, the total number of variables defining the intensive state of the system as a whole is $(r - 1)\varphi + 2$. However, these variables are not all independent. The equations of phase equilibrium as written in Eq. (16-37) impose a total of $(\varphi - 1)r$ conditions that must be satisfied. Therefore, the number of independent intensive variables required to fix the equilibrium state of a multicomponent, multiphase system is then

$$F = [(r - 1)\varphi + 2] - (\varphi - 1)r$$

or

$$F = r - \varphi + 2 \tag{16-49}$$

where F is called the variance or the degree of freedom of the system. Equation (16-49) is the *phase rule* first enunciated in 1875 by Gibbs.

In arriving at the phase rule, we have assumed that each of the φ phases of the system contains every one of the r components. Nevertheless, if any component is missing from any phase, there will be one less composition variable for that phase, and

*See, for example, [15, Chapter 9].

also there will be one less condition equation to be satisfied, so that the difference is the same and the phase rule remains unchanged.

A system for which there are no independent intensive properties is said to be invariant; one with one independent property is said to be univariant; one with two independent properties is said to be divariant; and so forth. We now apply the phase rule to determine the variance of a few systems.

For a single-component system, the phase rule reduces to $F = 3 - \varphi$. Thus, in a single-phase region, $F = 2$, the system is divariant. On the other hand, in a two-phase region, $F = 1$, the system is univariant. If either pressure or temperature is fixed, all the other intensive properties of each of the two phases in equilibrium are also fixed. At the triple point, $F = 0$, the system is invariant, that is, none of its intensive properties can be varied as long as the three phases exist together in equilibrium.

For a two-component system, such as the vapor–liquid system shown in Figs. 16-6 and 16-7, the phase rule reduces to $F = 4 - \varphi$. In a single-phase region, $F = 3$, the temperature, pressure, and composition may vary independently. In a two-phase region, $F = 2$, either the temperature and pressure, or the temperature (or pressure) and composition of one of the phases can be chosen as the independent variables.

So far in our discussions on vapor–liquid equilibrium of binary systems, we treat only miscible liquids. Two liquids are said to be miscible if when mixed in any concentration they form a single liquid phase. Many organic liquid pairs are miscible. On the other hand, if when mixed two liquids remain in two phases, they are said to be immiscible. Liquid mercury and water are practically immiscible. Intermediate between these two extremes are liquid pairs that are miscible for some concentrations and immiscible for the others; these liquids are said to be partially miscible. Most partially miscible mixtures are those with water or some other inorganic liquid as one of the pair. Some organic liquid pairs are also partially miscible over certain ranges. The degree of miscibility of two liquids may change greatly with temperature.

A phase diagram for two immiscible liquids is shown in Fig. 16-10, in which line A–B is the bubble-point line for the mixture of two liquid phases, and lines D–E and G–E are the dew-point lines in relation to pure liquids 2 and 1, respectively. In the liquid–liquid two-phase region, $F = 2$, the system is divariant. A liquid mixture at state a is actually a mixture of pure liquid 1 at state a' and pure liquid 2 at state a''. When heat is added at constant pressure to the system initially at state a, the system goes from state a toward state b, always consisting of two liquid phases, one of pure 1 and the other of pure 2. At state b, there are three phases coexisting (pure 1 liquid phase at state B, pure 2 liquid phase at state A, and a vapor phase at state E), and the system is univariant. Because the vapor formed at state E is richer in component 1 than the system as a whole is, more liquid 1 will be evaporated when more heat is added. As the last trace of liquid 1 is evaporated, the system is then comprised of two phases (pure 2 liquid phase at state A and a vapor phase at state E) and the system is again divariant. When the system changes from state b to state c, the pure 2 liquid phase changes from state A to state c'' and the vapor phase changes from state E to state c'. When the system reaches state d, the last trace of pure liquid 2 will be in state d''. Further heating superheats the vapor mixture toward state e.

A typical form of phase diagram for systems involving two partially miscible liquids is shown in Fig. 16-11, in which liquids I and II denote liquids that are rich in components 1 and 2, respectively. Lines D–A and D–E represent equilibrium between

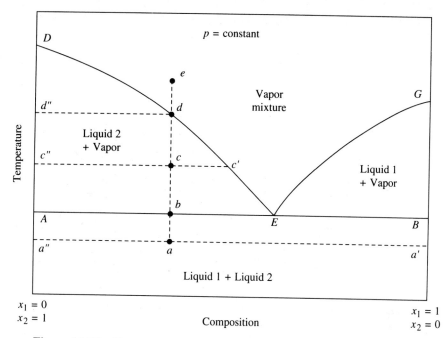

Figure 16-10 Temperature-composition diagram for two immiscible liquids.

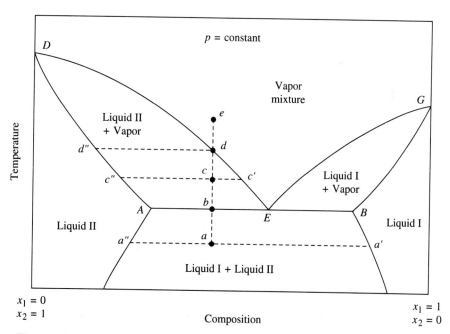

Figure 16-11 Temperature-composition diagram for two partially miscible liquids.

liquid II and vapor, with horizontal tie lines such as $c''-c'$ connecting the two phases that coexist in equilibrium. Similarly, lines $G-B$ and $G-E$ represent equilibrium between liquid I and vapor. The equilibrium between liquids I and II in the absence of a vapor phase is represented by lines $A-a''$ and $B-a'$, with a tie line $a''-a'$ as shown. Along line $A-B$ there are three phases coexisting: liquid II at state A, liquid I at state B, and a vapor mixture at state E. In the single-phase regions (liquid I, liquid II, and vapor mixture regions), the system is trivariant; in the two-phase regions (liquid I + liquid II, liquid I + vapor, and liquid II + vapor regions), the system is divariant; and in the three-phase region (along line $A-B$), the system is univariant. Line $a-b-c-d-e$ represents a constant-pressure heating process.

All the examples given in the preceding discussions are for vapor–liquid equilibrium. An example for solid–liquid equilibrium is shown in Fig. 16-12, which is a temperature–composition diagram for a binary system with miscible liquids and immiscible solids. The eutectic point E is the only state in which a liquid can coexist in equilibrium with both solid phases at a given pressure. If a liquid having the composition of state d is cooled at constant pressure, when state d is reached, component 2 will start to crystallize to form a solid phase at state d''. Component 1 will not crystallize until state b at the eutectic temperature is reached. It is clear that at a given pressure, only when the composition is at the eutectic point will the two components freeze uniformly. This is why a eutectic solution makes good alloy castings.

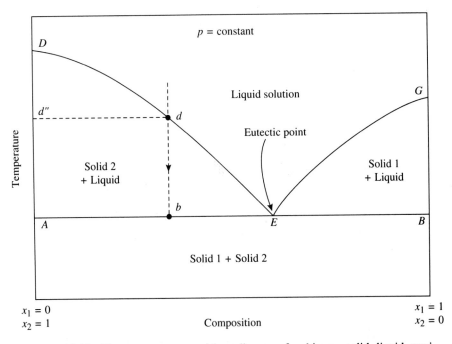

Figure 16-12 Temperature-composition diagram for binary solid–liquid equilibrium of miscible liquids and immiscible solids.

16-8 EQUATION OF REACTION EQUILIBRIUM

In a first-law analysis of a chemical reaction, it is assumed that the reaction under consideration simply takes place, and no question is asked about the possibility of the given reaction. The answers to the questions about the extent and direction of a reaction can only be furnished by the second law of thermodynamics. We now discuss the restrictions placed by the second law on the occurrence of chemical reactions.

We have established in Sec. 16-2 the criteria of equilibrium based on the second law of thermodynamics. The criteria as expressed by Eqs. (16-14) through (16-23) can now be applied to determine the possibility of chemical reactions. For instance, Eq. (16-22) states that the Helmholtz function must decrease and seek its minimum value when a chemical reaction occurs at constant temperature and volume. Once the minimum of the Helmholtz function is attained, a condition of chemical equilibrium is reached and the chemical reaction ceases—meaning really that the rates of reaction toward both directions are the same. In the case of a chemical reaction that occurs at constant temperature and pressure, Eq. (16-23) states that the Gibbs function must decrease and seek its minimum value, and once this minimum is attained, the chemical reaction ceases.

Let us consider the equilibrium condition at constant temperature and pressure of a reactive mixture capable of undergoing a chemical reaction according to the following balanced equation:

$$\nu_1 C_1 + \nu_2 C_2 \rightleftarrows \nu_3 C_3 + \nu_4 C_4 \tag{16-50}$$

where $C_1, \ldots, C_4$ denote the chemical constituents, and $\nu_1, \ldots, \nu_4$ denote the stoichiometric coefficients. From Eq. (16-6), the infinitesimal change in Gibbs function at constant T and p is

$$dG_{T,p} = \mu_1 \, dn_1 + \mu_2 \, dn_2 + \mu_3 \, dn_3 + \mu_4 \, dn_4$$

where $n_1, \ldots, n_4$ denote the numbers of moles of each constituent present, and $\mu_1, \ldots, \mu_4$ denote the respective chemical potentials. In a reacting mixture, the changes in the number of moles of the various constituents present are not independent of each other, but are proportional to the corresponding stoichiometric coefficients. Thus,

$$-\frac{dn_1}{\nu_1} = -\frac{dn_2}{\nu_2} = \frac{dn_3}{\nu_3} = \frac{dn_4}{\nu_4}$$

Accordingly, based on Eq. (16-23), we must have

$$dG_{T,p} = (\mu_1 \nu_1 + \mu_2 \nu_2 - \mu_3 \nu_3 - \mu_4 \nu_4) \frac{dn_1}{\nu_1} \leq 0$$

As ν_1 is a positive number, we conclude that

(a) if $\mu_1 \nu_1 + \mu_2 \nu_2 > \mu_3 \nu_3 + \mu_4 \nu_4$, then $dn_1 < 0$, and the reaction proceeds to the right; and

(b) if $\mu_1 \nu_1 + \mu_2 \nu_2 < \mu_3 \nu_3 + \mu_4 \nu_4$, then $dn_1 > 0$, and the reaction proceeds to the left.

Therefore, the condition of equilibrium at constant T and p is

$$\mu_1 \nu_1 + \mu_2 \nu_2 = \mu_3 \nu_3 + \mu_4 \nu_4 \qquad (16\text{-}51)$$

This is called the *equation of reaction equilibrium*. It provides a means for determining the equilibrium composition of a reactive mixture at a given temperature and pressure.

16-9 IDEAL-GAS REACTION AND EQUILIBRIUM CONSTANT

The reacting mixtures in most chemical reactions encountered in engineering applications can often be considered as ideal-gas mixtures. We use the equation of reaction equilibrium as derived in the last section to determine the equilibrium compositions of reacting ideal-gas mixtures. To do this, we need first to show how the chemical potential for an ideal gas is calculated.

Analogous to the property entropy, based on the Gibbs–Dalton law (see Sec. 9-3), the Gibbs function G of a mixture of ideal gases can be calculated as the sum of the Gibbs functions of the component gases evaluated either at the temperature and volume of the mixture or at the temperature of the mixture and the partial pressure of the component. Thus, we can write

$$G = \sum n_i g_i(T, p_i) \qquad (16\text{-}52)$$

where n_i is the number of moles of component i, and $g_i(T, p_i)$ is its molar Gibbs function to be evaluated at the temperature T of the mixture and the partial pressure p_i of the component. Taking the partial derivative of G with respect to n_i, and holding temperature T, pressure p, and the remaining n_i's constant, results in

$$\left(\frac{\partial G}{\partial n_i}\right)_{T, p, n_j(j \neq i)} = g_i(T, p_i)$$

But, according to Eq. (16-7),

$$\left(\frac{\partial G}{\partial n_i}\right)_{T, p, n_j(j \neq i)} = \mu_i$$

Therefore, we must have

$$\mu_i = g_i(T, p_i) \qquad (16\text{-}53)$$

Thus, the chemical potential of a component is equal to its molar Gibbs function evaluated at its partial pressure and the mixture temperature. However, by definition,

$$g_i(T, p_i) = h_i(T) - Ts_i(T, p_i)$$
$$g_i(T, p) = h_i(T) - Ts_i(T, p)$$

and from Eq. (7-16), we have

$$s_i(T, p_i) = s_i(T, p) - \Re \ln \frac{p_i}{p}$$

Therefore,

$$\mu_i = g_i(T, p_i) = h_i(T) - Ts_i(T, p) + \Re T \ln \frac{p_i}{p}$$

or

$$\mu_i = g_i(T, p) + \mathscr{R}T \ln x_i \qquad (16\text{-}54)$$

where μ_i is the chemical potential of component i with a mole fraction of x_i, and $g_i(T, p)$ is the molar Gibbs function of component i at T and p of the mixture.

Substituting Eq. (16-54) in Eq. (16-51) results in

$$[\nu_3 g_3(T, p) + \nu_4 g_4(T, p) - \nu_1 g_1(T, p) - \nu_2 g_2(T, p)]$$
$$+ \mathscr{R}T (\nu_3 \ln x_3 + \nu_4 \ln x_4 - \nu_1 \ln x_1 - \nu_2 \ln x_2) = 0$$

or

$$\Delta G_R(T, p) + \mathscr{R}T \ln \frac{x_3^{\nu_3} x_4^{\nu_4}}{x_1^{\nu_1} x_2^{\nu_2}} = 0 \qquad (16\text{-}55)$$

where the Gibbs function change for a complete reaction $\Delta G_R(T, p)$ is defined as

$$\Delta G_R(T, p) = \nu_3 g_3(T, p) + \nu_4 g_4(T, p) - \nu_1 g_1(T, p) - \nu_2 g_2(T, p) \qquad (16\text{-}56)$$

It is convenient to express ΔG_R in Eq. (16-55) in terms of a reference pressure p_0, which is commonly taken as 1 atm. Thus, applying Eq. (7-16) to the definition of the Gibbs function and noting that the enthalpy h is a function of temperature only for an ideal gas, we obtain

$$g(T, p) = g(T, p_0) + \mathscr{R}T \ln \left(\frac{p}{p_0} \right) \qquad (16\text{-}57)$$

Substituting this equation into Eq. (16-56) results in

$$\Delta G_R(T, p) = \Delta G_R(T, p_0) + \mathscr{R}T \ln \left(\frac{p}{p_0} \right)^{\nu_3 + \nu_4 - \nu_1 - \nu_2}$$

where

$$\Delta G_R(T, p_0) = \nu_3 g_3(T, p_0) + \nu_4 g_4(T, p_0) - \nu_1 g_1(T, p_0) - \nu_2 g_2(T, p_0) \qquad (16\text{-}58)$$

Accordingly, Eq. (16-55) can be written as

$$\Delta G_R(T, p_0) + \mathscr{R}T \ln \left[\left(\frac{x_3^{\nu_3} x_4^{\nu_4}}{x_1^{\nu_1} x_2^{\nu_2}} \right) \left(\frac{p}{p_0} \right)^{\nu_3 + \nu_4 - \nu_1 - \nu_2} \right] = 0$$

or

$$\left(\frac{x_3^{\nu_3} x_4^{\nu_4}}{x_1^{\nu_1} x_2^{\nu_2}} \right) \left(\frac{p}{p_0} \right)^{\nu_3 + \nu_4 - \nu_1 - \nu_2} = \exp \left[\frac{-\Delta G_R(T, p_0)}{\mathscr{R}T} \right] \qquad (16\text{-}59)$$

For a given chemical reaction and a given reference pressure p_0, the Gibbs function change $\Delta G_R(T, p_0)$ as defined by Eq. (16-58) is a function of temperature T only. As the right-hand side of Eq. (16-59) is a function of temperature T only, the left-hand side of this equation must depend only on temperature T and not on pressure p. For convenience, let us define this temperature function as

$$K_p = \frac{x_3^{\nu_3} x_4^{\nu_4}}{x_1^{\nu_1} x_2^{\nu_2}} \left(\frac{p}{p_0} \right)^{\nu_3 + \nu_4 - \nu_1 - \nu_2} \qquad (16\text{-}60)$$

and call it the *equilibrium constant*.

The equilibrium constant K_p can be expressed in terms of the partial pressures of components. According to Eq. (9-10), the partial pressure of component i in an ideal-gas mixture is given by

$$p_i = x_i p$$

Thus, in terms of partial pressures, Eq. (16-60) becomes

$$K_p = \frac{p_3^{\nu_3} p_4^{\nu_4}}{p_1^{\nu_1} p_2^{\nu_2}} (p_0)^{\nu_1 + \nu_2 - \nu_3 - \nu_4} \qquad (16\text{-}61a)$$

or

$$K_p = \frac{(p_3/p_0)^{\nu_3} (p_4/p_0)^{\nu_4}}{(p_1/p_0)^{\nu_1} (p_2/p_0)^{\nu_2}} \qquad (16\text{-}61b)$$

or

$$K_p = \frac{p_3^{\nu_3} p_4^{\nu_4}}{p_1^{\nu_1} p_2^{\nu_2}} \qquad (16\text{-}61c)$$

where, in Eq. (16-61c), $p_0 = 1$ atm has been incorporated.

With the equilibrium constant so defined, Eq. (16-59) can be rewritten to give the following relation between K_p and $\Delta G_R(T, p_0)$:

$$\Delta G_R(T, p_0) = -\mathscr{R} T \ln K_p \qquad (16\text{-}62)$$

It is worthwhile drawing attention again that for an ideal-gas mixture capable of undergoing the reaction as represented by Eq. (16-50), the equilibrium constant K_p is a function of temperature only and is a dimensionless property of the equilibrium mixture. Once the equilibrium values of the mole fractions or the partial pressures are determined, or if $\Delta G_R(T, p_0)$ is known, the equilibrium constant can be calculated at the desired temperature. Table A-29 in Appendix 3 gives the values of K_p for several reactions.

The equilibrium composition of a reacting ideal-gas mixture at a given temperature can be determined if the equilibrium constant is known. We will give a numerical example to illustrate the method employed in such a calculation.

The enthalpy of reaction $\Delta H_R(T, p_0)$ and the entropy change of reaction $\Delta S_R(T, p_0)$ for the complete reaction as expressed by Eq. (16-50) are defined as follows:

$$\Delta H_R(T, p_0) = \nu_3 h_3(T, p_0) + \nu_4 h_4(T, p_0) - \nu_1 h_1(T, p_0) - \nu_2 h_2(T, p_0) \qquad (16\text{-}63)$$

$$\Delta S_R(T, p_0) = \nu_3 s_3(T, p_0) + \nu_4 s_4(T, p_0) - \nu_1 s_1(T, p_0) - \nu_2 s_2(T, p_0) \qquad (16\text{-}64)$$

where $h(T, p_0)$ and $s(T, p_0)$ are the molar enthalpy and entropy, respectively, of individual components at temperature T and pressure p_0. The three quantities $\Delta H_R(T, p_0)$, $\Delta S_R(T, p_0)$, and $\Delta G_R(T, p_0)$ are related by the usual thermodynamic equation:

$$\Delta G_R(T, p_0) = \Delta H_R(T, p_0) - T \, \Delta S_R(T, p_0) \qquad (16\text{-}65)$$

Through the use of Eqs. (16-62) and (16-65), we can now derive a general expression for the variation of K_p with temperature. Substituting Eq. (16-65) in Eq. (16-62) leads to

$$-\mathscr{R} \ln K_p = \frac{1}{T} \Delta H_R(T, p_0) - \Delta S_R(T, p_0)$$

Differentiating with respect to temperature gives

$$-\mathcal{R}\frac{d[\ln K_p]}{dT} = -\frac{\Delta H_R(T, p_0)}{T^2} + \frac{1}{T}\frac{d[\Delta H_R(T, p_0)]}{dT} - \frac{d[\Delta S_R(T, p_0)]}{dT} \qquad (16\text{-}66)$$

Now applying the basic equation $T dS = dH - V dp$ at constant pressure separately to the products and the reactants, and taking the difference between the two resulting equations, we obtain

$$T \, d(S_{\text{products}} - S_{\text{reactants}}) = d(H_{\text{products}} - H_{\text{reactants}})$$

or

$$T \, d[\Delta S_R(T, p_0)] = d[\Delta H_R(T, p_0)]$$

Upon substituting this equation into Eq. (16-66), we have

$$\frac{d[\ln K_p]}{dT} = \frac{\Delta H_R(T, p_0)}{\mathcal{R}T^2} \qquad (16\text{-}67\text{a})$$

or

$$\frac{d[\ln K_p]}{d(1/T)} = -\frac{\Delta H_R(T, p_0)}{\mathcal{R}} \qquad (16\text{-}67\text{b})$$

Either of these equations is known as the *van't Hoff isobar equation*. Because $\Delta H_R(T, p_0)$ is approximately constant for a reaction over a wide temperature range, a plot of $\ln K_p$ versus $1/T$ according to Eq. (16-67b) is almost a straight line. Such a linear plot is useful in estimating the values of K_p at other temperatures when its values at two temperatures are known. Furthermore, if the values of K_p are found by measuring the equilibrium compositions at two different temperatures, a linear plot of Eq. (16-67b) can be used for evaluating the enthalpy of reaction at the standard temperature T_0.

Example 16-6

Gaseous propane (C_3H_8) is burned with 80% of theoretical air in a steady-flow process at 1 atm. Both the fuel and the air are supplied at 25°C. The products, which consist of CO_2, CO, H_2O, H_2, and N_2 in equilibrium, leave the combustion chamber at 1500 K. Determine the composition of the products and the amount of heat transfer in this process per kg of propane.

Solution   Let y and z be the numbers of moles of CO_2 and H_2O in the products, respectively. The chemical equation can then be written as

$$C_3H_8 + 0.8(5O_2 + 3.76 \times 5N_2)$$
$$\longrightarrow yCO_2 + (3 - y)CO + zH_2O + (4 - z)H_2 + 0.8 \times 3.76 \times 5N_2$$

where the coefficients for CO and H_2 are obtained by mass balances on C and H_2. A mass balance for O_2 then gives

$$0.8 \times 5 = y + \tfrac{1}{2}(3 - y) + \tfrac{1}{2}z$$

or

$$y + z = 5 \qquad (16\text{-}68)$$

Another relation between y and z can be obtained from the equilibrium condition for the reaction

$$CO_2 + H_2 \rightleftharpoons CO + H_2O \qquad (16\text{-}69)$$

According to Eq. (16-61c), the equilibrium constant for this reaction is

$$K_p = \frac{p_{CO}\, p_{H_2O}}{p_{CO_2}\, p_{H_2}}$$

where the p's are the partial pressures in the equilibrium mixture of

$$y\,CO_2 + (3 - y)CO + z\,H_2O + (4 - z)H_2 + 15.04\,N_2$$

For a total pressure of 1 atm, the values of partial pressures, in atm, are

$$p_{CO} = \frac{3 - y}{y + (3 - y) + z + (4 - z) + 15.04} = \frac{3 - y}{22.04}$$

$$p_{H_2O} = \frac{z}{22.04} \qquad p_{CO_2} = \frac{y}{22.04} \qquad p_{H_2} = \frac{4 - z}{22.04}$$

Thus,

$$K_p = \frac{(3 - y)z}{y(4 - z)}$$

From Table A-29 for the reaction as expressed by Eq. (16-69), we have at 1500 K,

$$\log_{10}[K_p] = 0.409 \qquad \text{or} \qquad K(T) = 2.56$$

Therefore,

$$\frac{(3 - y)z}{y(4 - z)} = 2.56 \qquad (16\text{-}70)$$

Solving Eqs. (16-68) and (16-70) simultaneously yields

$$y = 1.81 \qquad \text{and} \qquad z = 3.19$$

The chemical equation for the combustion process is then

$$C_3H_8 + 4O_2 + 15.04\,N_2$$
$$\longrightarrow 1.81\,CO_2 + 1.19\,CO + 3.19\,H_2O + 0.81\,H_2 + 15.04\,N_2 \qquad (16\text{-}71)$$

Thus, the equilibrium mole fractions of the products are

$$x_{CO_2} = \frac{1.81}{1.81 + 1.19 + 3.19 + 0.81 + 15.04}$$

$$= \frac{1.81}{22.04} = 0.0821 = 8.21\%$$

$$x_{CO} = \frac{1.19}{22.04} = 0.0540 = 5.40\%$$

$$x_{H_2O} = \frac{3.19}{22.04} = 0.1447 = 14.47\%$$

$$x_{H_2} = \frac{0.81}{22.04} = 0.0368 = 3.68\%$$

$$x_{N_2} = \frac{15.04}{22.04} = 0.6824 = 68.24\%$$

According to Eq. (15-7), the enthalpy of reaction ΔH_R^0 at 25°C and 1 atm for the reaction expressed by Eq. (16-71) with all the constituents in the gaseous phase is given by

$$\Delta H_R^0 = \sum_{products} (n\ \Delta h_f^\circ) - \sum_{reactants} (n\ \Delta h_f^\circ)$$

$$= 1.81\ \Delta h_{f,CO_2}^\circ + 1.19\ \Delta h_{f,CO}^\circ + 3.19\ \Delta h_{f,H_2O}^\circ - 1\Delta h_{f,C_3H_8}^\circ$$

in which the Δh_f° values of O_2, N_2, and H_2 have been assigned a value of zero. Upon substituting numerical values of Δh_f° from Table A-28M, we obtain

$$\Delta H_R^0 = 1.81(-393,520) + 1.19(-110,530) + 3.19(-241,820) - 1(-103,850)$$

$$= -1,511,358 \text{ kJ/kgmole } C_3H_8$$

Now applying Eq. (15-11) to the reaction represented by Eq. (16-17), we have

$$Q = \Delta H_R^0 + \sum_{products} n(h_T - h_{298\,K}) - \sum_{reactants} n(n_T - h_{298\,K})$$

$$= \Delta H_R^0 + 1.81(h_T - h_{298\,K})_{CO_2} + 1.19(h_T - h_{298\,K})_{CO}$$

$$+ 3.19(h_T - h_{298\,K})_{H_2O} + 0.81(h_T - h_{298\,K})_{H_2} + 15.04(h_T - h_{298\,K})_{N_2}$$

in which the terms for the reactants are zero, because the reactants enter the process at the reference temperature of 25°C. Substituting numerical values from the ideal-gas property tables (Tables A-10M to A-15M) for CO_2, CO, H_2O, H_2, and N_2 at the final products temperature of 1500 K for h_T and 298 K for $h_{298\,K}$, we have

$$Q = -1,511,358 + 1.81(71,071.5 - 9359.7)$$

$$+ 1.19(47,513.7 - 8666.7) + 3.19(58,054.5 - 9899.0)$$

$$+ 0.81(44,757.8 - 8463.1) + 15.04(47,073.2 - 8665.8)$$

$$= -592,770 \text{ kJ/kgmole } C_3H_8$$

$$= \frac{-592,770 \text{ kJ/kgmole}}{44 \text{ kg/kgmole}}$$

$$= -13,472 \text{ kJ/kg } C_3H_8$$

16-10 THE THIRD LAW OF THERMODYNAMICS AND ABSOLUTE ENTROPY

A basic law of thermodynamics was born from the attempt to calculate equilibrium constants of chemical reaction entirely from thermal data (i.e., enthalpies and heat capacities). What has come to be known as the third law of thermodynamics had its origin in the Nernst heat theorem, enunciated by Nernst in 1906. It states that there is no change in entropy if a chemical change takes place between pure crystalline solids at absolute zero temperature. This theorem was later modified and generalized by Planck, Simon, Lewis, Guggenheim, and others. One well-known version of the modified theorem, sometimes known as the *Nernst–Simon statement of the third law of thermodynamics* is as follows: The entropy change associated with any isothermal reversible process of a condensed system (i.e., a solid or a liquid) approaches zero as the temperature approaches absolute zero. Symbolically, we can write

$$\lim_{T \to 0} (\Delta S)_T = 0 \qquad (16\text{-}72)$$

In view of the empirical fact that in any cooling process, the lower the temperature achieved, the more difficult it is to go lower, another form of the third law has been advanced. It is stated as follows: It is impossible by means of any process, no matter how idealized, to reduce the temperature of a system to absolute zero in a finite number of steps. This is known as the *unattainability statement of the third law of thermodynamics*.

The Nernst–Simon and the unattainability statements of the third law are entirely equivalent to each other in their consequences. For the purpose of demonstrating this equivalence, it is convenient to use a paramagnetic system in typical magnetic cooling processes (see Sec. 13-8), such as shown in Fig. 16-13. According to the Nernst–

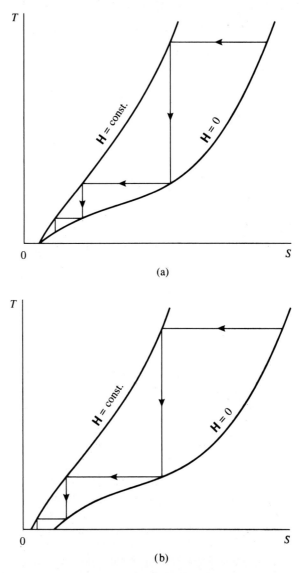

(a)

(b)

Figure 16-13 Temperature–entropy diagrams for magnetic cooling processes (a) in agreement with the third law, and (b) in violation of the third law.

Phase and Chemical Equilibrium   Chap. 16

Simon statement, the curves for constant magnetic fields must come together at the absolute zero temperature, so as to satisfy the condition that $(\Delta S)_T \rightarrow 0$ as $T \rightarrow 0$. Figure 16-13 (a) shows such correct curves. It is clear that the temperature drops in successive demagnetizations gradually decrease and an infinite number of demagnetization steps would be required to attain absolute zero of temperature, thus confirming the unattainability statement. On the other hand, if the Nernst–Simon statement is not obeyed and the curves for constant magnetic fields do not meet at the absolute zero, as depicted in Fig. 16-13 (b), then finite steps could suffice to lower the temperature down to the absolute zero. This is, of course, a violation of the unattainability statement.

We realize that both the Nernst–Simon and the unattainability statements call for the disappearance of the entropy differences between different states of a system at the absolute zero temperature. It is not necessary that the entropies themselves vanish. Planck in 1911 extended the Nernst theorem by making the additional postulate that the absolute value of the entropy of a pure solid or a pure liquid approaches zero as the temperature approaches absolute zero, or, symbolically,

$$\lim_{T \to 0} S = 0 \tag{16-73}$$

This is *Planck's postulate of the third law*, which permits the determination of entropies of every substance, including elements and compounds, on a common base, thus establishing a scale of "absolute" entropies for tabulational purposes.

The absolute entropies of a number of pure substances at 25°C (77°F) and 1 atm are listed in Table A-28 in Appendix 3. In addition, Tables A-10 to A-16 in Appendix 3 give the absolute entropy of several ideal gases (N_2, O_2, H_2, H_2O, CO_2, and CO) at 1 atm as a function of temperature. These ideal-gas tables are abridged from Keenan–Chao–Kaye Gas Tables, from which the tabulated function ϕ is used directly as absolute entropy $s^0(T, p_0)$ with the superscript "o" on the symbol representing the condition of a unit pressure ($p_0 = 1$ atm in this case). The absolute entropy $s(T, p)$ of an ideal gas at any temperature T (in K or °R) and any pressure p (in atm) can be calculated from the tabulated $s^0(T, p_0)$ (or simply designated as $s°$) value in these ideal-gas tables by the expression:

$$s(T, p) = s°(T, p_0) - \mathcal{R} \ln \frac{p}{p_0} \tag{16-74}$$

where $p_0 = 1$ atm.

For an ideal-gas mixture at T and p, its absolute entropy can be evaluated as the sum of the partial entropies of the component gases. Thus,

$$s_{\text{mixture}}(T, p) = \sum_i x_i s_i(T, p_i) \tag{16-75}$$

where

$$s_i(T, p_i) = s_i°(T, p_0) - \mathcal{R} \ln \frac{p_i}{p_0}$$
$$= s_i°(T, p_0) - \mathcal{R} \ln \frac{x_i p}{p_0} \tag{16-76}$$

in which

$$p_i = \text{partial pressure of component } i$$

$$x_i = \text{mole fraction of component } i$$

16-11 GIBBS FUNCTION OF FORMATION

Parallel to the definition of the enthalpy of formation, we now define the Gibbs function of formation to facilitate the second-law analysis of chemical reactions. The *Gibbs function of formation* Δg_f° of a compound at the standard condition of 25°C (77°F) and 1 atm is defined as the Gibbs function (change) of reaction at that condition for the formation of the compound from its elements that are originally in their most stable state at the standard condition. Using the formation of gaseous methane CH_4 as an example, the reaction equation is

$$C + 2H_2 \longrightarrow CH_4$$

The standard-state Gibbs function of formation is the Gibbs function of the products minus the Gibbs function of the reactances, or

$$\Delta g_{f,CH_4}^\circ = g_{CH_4}^\circ - (g_C^\circ + 2g_{H_2}^\circ)$$

Assigning arbitrarily a value of zero to the Gibbs function of all stable elements (C and H_2 in this example) at 25°C (77°F) and 1 atm results in

$$\Delta g_{f,CH_4}^\circ = g_{CH_4}^\circ$$

In general, the standard-state Gibbs function of formation can be written symbolically as

$$\Delta g_f^\circ = g_{compound}^\circ - \sum_i (n_i g_i^\circ)_{elements} \tag{16-77}$$

where $g_{compound}^\circ$ is the molar Gibbs function of the compound, and g_i° is the molar Gibbs function of the ith element, all at the standard condition; n_i is the stoichiometric coefficient of the ith element in forming a single mole of the compound. With the Gibbs function of all stable elements at the standard condition assigned the value of zero, Eq. (16-77) simplifies to

$$\Delta g_f^\circ = g_{compound}^\circ \tag{16-78}$$

which means that the molar Gibbs function of any compound at the standard condition is merely its Gibbs function of formation at that condition. Table A-28 in Appendix 3 gives values of Δg_f° for a number of substances.

Example 16-7

For the reaction

$$CH_4 \text{ (g)} + 2O_2 \text{ (g)} \longrightarrow CO_2 \text{ (g)} + 2H_2O \text{ (g)}$$

determine the change in the Gibbs function at 77°F and 1 atm by (a) using data on the Gibbs function of formation, and (b) using data on enthalpy of formation and absolute entropy.

Solution

$$T_0 = 77°F = 537°R \qquad \text{and} \qquad p_0 = 1 \text{ atm}$$

(a) *Using data on the Gibbs function of formation.* The change of Gibbs function for the reaction at the standard condition of T_0 and p_0 given by Eqs. (16-58) and (16-78) is

$$\Delta G_R(T_0, p_0) = \Delta G_R^\circ = g_{CO_2(g)}^\circ + 2g_{H_2O(g)}^\circ - g_{CH_4(g)}^\circ$$

$$= \Delta g_{f,CO_2(g)}^\circ + 2\,\Delta g_{f,H_2O(g)}^\circ - \Delta g_{f,CH_4(g)}^\circ$$

Taking data on the Gibbs function of formation from Table A-28E, we have

$$\Delta G_R^\circ = (-169{,}680) + 2(-98{,}350) - (-21{,}860)$$

$$= -344{,}500 \text{ Btu/lbmole } CH_4$$

(b) *Using data on enthalpy of formation and absolute entropy.* The change of Gibbs function for the reaction in terms of the enthalpy of reaction and absolute entropy at the standard condition of T_0 and p_0 is expressed by Eq. (16-65):

$$\Delta G_R(T_0, p_0) = \Delta H_R(T_0, p_0) - T_0 \Delta S_R(T_0, p_0)$$

or

$$\Delta G_R^\circ = \Delta H_R^\circ - T_0 \Delta S_R^\circ$$

where, from Eq. (15-7),

$$\Delta H_R^\circ = \Delta h_{f, CO_2(g)}^\circ + 2 \Delta h_{f, H_2O(g)}^\circ - \Delta h_{f, CH_4(g)}^\circ$$

and from Eq. (16-64),

$$\Delta S_R(T_0, p_0) = \Delta S_R^\circ = s_{CO_2(g)}^\circ + 2s_{H_2O(g)}^\circ - s_{CH_4(g)}^\circ - 2s_{O_2(g)}^\circ$$

Thus,

$$\Delta G_R^\circ = [\Delta h_{f, CO_2(g)}^\circ + 2 \Delta h_{f, H_2O(g)}^\circ - \Delta h_{f, CH_4(g)}^\circ] - T_0[s_{CO_2(g)}^\circ + 2s_{H_2O(g)}^\circ - s_{CH_4(g)}^\circ - 2s_{O_2(g)}^\circ]$$

Taking data on Δh_f° and s° from Table A-28E, we have

$$\Delta G_R^\circ = [(-169{,}300) + 2(-104{,}040) - (32{,}210)]$$

$$- 537[(51.07) + 2(45.11) - (44.49) - 2(49.00)]$$

$$= -344{,}500 \text{ Btu/lbmole } CH_4$$

16-12 AVAILABILITY ANALYSIS OF REACTING SYSTEMS

In Chapter 8, the concepts of maximum work, availability, and irreversibility were introduced. For a steady-state, steady-flow process, the maximum work is given by the change in the stream availability between the end states of the system, exchanging heat with only the atmosphere. Thus, from Eqs. (8-20) and (8-22), in the absence of changes in kinetic and potential energy, we have

$$W_{max} = \sum_{inlet} n\psi - \sum_{exist} n\psi$$

$$= \sum_{inlet} n(h - T_0 s) - \sum_{exit} n(h - T_0 s)$$

For a steady-state, steady-flow process involving a chemical reaction, this equation becomes

$$W_{max} = \sum_{reactants} n(h - T_0 s) - \sum_{products} n(h - T_0 s) \tag{16-79}$$

On a common basis in which the enthalpy of all stable elements at $T_0 = 298$ K ($537°R$) and $p_0 = 1$ atm is zero, the molar enthalpy of a constituent of the reactants or products of a chemical reaction, when considered as an ideal gas, is given by the expression

$$\Delta h_f^\circ + (h_T - h_{T_0})$$

where Δh_f° is the standard-state enthalpy of formation of a gas, h_T is its enthalpy at a given temperature T, and h_{T_0} is its enthalpy at the standard reference temperature T_0. Values of Δh_f°, h_T, and h_{T_0} are given in Tables A-28 and A-10 to A-17 in Appendix 3.

The molar entropy of a constituent of the reactants or products of a chemical reaction, when considered as an ideal gas at temperature T and partial pressure (xp), is given by the expression

$$s_T^\circ - \mathcal{R} \ln \left(\frac{xp}{p_0} \right)$$

or

$$s_T^\circ - \mathcal{R} \ln x$$

when $p = p_0 = 1$ atm. In the preceding expressions, s_T° is the absolute entropy of the gas at the given temperature T and pressure $p_0 = 1$ atm, p is the given pressure, p_0 is the reference pressure of 1 atm, and x is the mole fraction of the gas in the mixture of the reactants or products. Values of $s_T^\circ = \phi$ for a number of gases are given in Table A-10 to A-16 in Appendix 3.

Substituting the preceding expressions for h and s into Eq. (16-79) leads to

$$W_{\max} = \sum_{\text{reactants}} n[(\Delta h_f^\circ + h_T - h_{T_0}) - T_0(s_T^\circ - \mathcal{R} \ln x)]$$

$$- \sum_{\text{products}} n[(\Delta h_f^\circ + h_T - h_{T_0}) - T_0(s_T^\circ - \mathcal{R} \ln x)]$$

(16-80)

As defined by Eq. (8-9), the irreversibility of a process is given by

$$I = W_{\max} - W_{\text{actual}}$$

For a chemical reaction process, there is no actual work done, $W_{\text{actual}} = 0$; thus,

$$I = W_{\max} \qquad (16\text{-}81)$$

The irreversibility can also be calculated by Eq. (8-12),

$$I = T_0(\Delta S_{\text{system}} + \Delta S_{\text{surr}})$$

For a chemical-reaction process, ΔS_{system} is the change of entropy of reaction $\Delta S_{\text{reaction}}$. Because the total energy released by the reaction appears as a heat lost Q_{release}, we have

$$\Delta S_{\text{surr}} = \frac{-Q_{\text{release}}}{T_0}$$

Accordingly, for a chemical-reaction process,

$$I = T_0 \, \Delta S_{\text{reaction}} - Q_{\text{release}} \qquad (16\text{-}82)$$

For the special case where the chemical reaction occurs in such a manner that both the reactants and products are in temperature equilibrium with the atmosphere (i.e., $T_{\text{reactants}} = T_{\text{products}} = T_0 = 25°C$), and for each constituent

$$g = h - Ts$$

Eq. (16-79) then becomes

$$W_{max} = \sum_{reactants} ng - \sum_{products} ng \qquad (16-83)$$

Furthermore, if each component of the reactants and the products can be assumed to be at the standard condition of $p_0 = 1$ atm and $T_0 = 25°C$, we could replace the Gibbs function g in Eq. (16-83) by the Gibbs function of formation $\Delta g_f^°$ and write

$$W_{max} = \sum_{reactants} n \Delta g_f^° - \sum_{products} n \Delta g_f^° \qquad (16-84)$$

Example 16-8

Propane gas at 25°C and 1 atm is burned in steady flow with 40% excess air at 25°C and 1 atm. Assume the reaction is complete. The gaseous products (including water vapor) leave as a mixture at 25°C and 1 atm. The atmosphere is at 25°C and 1 atm. Determine (a) the heat released by the reaction, (b) the maximum work for the process, and (c) the irreversibility of the reaction, all in kJ/kgmole of fuel.

Solution The chemical equation for complete reaction with 40% excess air per mole of fuel can be written as

$$C_3H_8 + 1.4(aO_2 + 3.76aN_2) \longrightarrow xCO_2 + yH_2O + 0.4aO_2 + zN_2$$

where all the constituents, including water, are in the gaseous state. The unknown quantities in the preceding equation are determined by mass balances for the atomic species. Thus,

C balance: $x = 3$

H_2 balance: $y = 4$

O_2 balance: $a = x + \dfrac{y}{2} = 3 + \dfrac{4}{2} = 5$

N_2 balance: $z = 1.4 \times 3.76a = 1.4 \times 3.76 \times 5 = 26.32$

The chemical equation then becomes

$$C_3H_8 + (7O_2 + 26.32N_2) \longrightarrow 3CO_2 + 4H_2O + 2O_2 + 26.32N_2$$

(a) The heat release is given from Eq. (15-12):

$$Q = \sum_{products} n(\Delta h_f^° + h_T - h_{T_0}) - \sum_{reactants} n(\Delta h_f^° + h_T - h_{T_0})$$

With the temperatures of the reactants and products the same as the standard reference temperature of $T_0 = 298$ K, the preceding equation reduces to

$$Q = \sum_{products} n \Delta h_f^° - \sum_{reactants} n \Delta h_f^°$$

Taking $\Delta h_f^°$ data from Table A-28M, we obtain

$$Q = 3(\Delta h_{f,CO_2}^°) + 4(\Delta h_{f,H_2O}^°) - 1(\Delta h_{f,C_3H_8}^°)$$

$$= 3(-393,520) + 4(-241,820) - 1(-103,850)$$

$$= -2,043,990 \text{ kJ/kgmole fuel}$$

(b) With $T = T_0 = 298$ K for both the reactants and products, Eq. (16-80) for the maximum work reduces to

$$W_{max} = \sum_{reactants} n[\Delta h_f^\circ - T_0(s_T^\circ - \mathcal{R} \ln x)] - \sum_{products} n[\Delta h_f^\circ - T_0(s_T^\circ - \mathcal{R} \ln x)]$$

$$= \sum_{reactants} n\,\Delta h_f^\circ - \sum_{products} n\,\Delta h_f^\circ$$

$$- T_0\left[\sum_{reactants} n(s_T^\circ - \mathcal{R} \ln x) - \sum_{products} n(s_T^\circ - \mathcal{R} \ln x)\right]$$

in which

$$\sum_{reactants} n\,\Delta h_f^\circ - \sum_{products} n\,\Delta h_f^\circ = -Q = 2{,}043{,}990 \text{ kJ/kgmole fuel}$$

Taking s_T° data from Table A-28M, we have

$$\sum_{reactants} n(s_T^\circ - \mathcal{R} \ln x) - \sum_{products} n(s_T^\circ - \mathcal{R} \ln x)$$

$$= [1(s_T^\circ)_{C_3H_8} + 7(s_T^\circ - \mathcal{R} \ln x)_{O_2} + 26.32(s_T^\circ - \mathcal{R} \ln x)_{N_2}]$$

$$- [3(s_T^\circ - \mathcal{R} \ln x)_{CO_2} + 4(s_T^\circ - \mathcal{R} \ln x)_{H_2O}$$

$$+ 2(s_T^\circ - \mathcal{R} \ln x)_{O_2} + 26.32(s_T^\circ - \mathcal{R} \ln x)_{N_2}]$$

$$= \left[1(269.91) + 7\left(205.04 - 8.314 \ln \frac{7}{7 + 26.32}\right)\right.$$

$$\left. + 26.32\left(191.50 - 8.314 \ln \frac{26.32}{33.32}\right)\right]$$

$$- \left[3\left(213.67 - 8.314 \ln \frac{3}{3 + 4 + 2 + 26.32}\right)\right.$$

$$+ 4\left(188.72 - 8.314 \ln \frac{4}{35.32}\right)$$

$$+ 2\left(205.04 - 8.314 \ln \frac{2}{35.32}\right)$$

$$\left. + 26.32\left(191.50 - 8.314 \ln \frac{26.32}{35.32}\right)\right]$$

$$= -204.42 \text{ kJ/K·kgmole fuel}$$

Therefore, the maximum work that can be done through the combustion process is then

$$W_{max} = 2{,}043{,}990 - 298(-204.42)$$

$$= 2{,}104{,}910 \text{ kJ/kgmole fuel}$$

Notice that because in this problem both the products and the air in the reactants are at the same temperature and pressure as the atmosphere with zero stream availability, the value of $W_{max} = 2{,}104{,}910$ kJ/kgmole fuel may be considered as the stream availability of the fuel at 298 K and 1 atm for the case where water formed by combustion is in the vapor state.

(c) By Eq. (16-81), the irreversibility of the reaction is

$$I = W_{max} = 2{,}104{,}910 \text{ kJ/kgmole fuel}$$

The irreversibility of the reaction can also be calculated by Eq. (16-82):

$$I = T_0 \, \Delta S_{\text{reaction}} - Q_{\text{release}}$$

where

$$\Delta S_{\text{reaction}} = \sum_{\text{products}} n(s_T^\circ - \mathscr{R} \ln x) - \sum_{\text{reactants}} n(s_T^\circ - \mathscr{R} \ln x)$$

$$= 204.42 \text{ kJ/K·kgmole fuel}$$

Thus,

$$I = 298(204.42) - (-2,043,990)$$

$$= 2,104,910 \text{ kJ/kgmole fuel}$$

Notice that in this problem, if we assume that the constituents in the reactants and products were separated and each was at 1 atm, we could then calculate the maximum work by the use of Eq. (16-84). Thus,

$$W_{\text{max}} = \sum_{\text{reactants}} n \, \Delta g_f^\circ - \sum_{\text{products}} n \, \Delta g_f^\circ$$

$$= (1 \, \Delta g_{f,C_3H_8}^\circ) - (3 \, \Delta g_{f,CO_2}^\circ + 4 \, \Delta g_{f,H_2O}^\circ)$$

Taking Δg_f° data from Table A-28M, the result is

$$W_{\text{max}} = 1(-23,490) - 3(-394,380) - 4(-228,590)$$

$$= 2,074,010 \text{ kJ/kgmole fuel}$$

This value of the maximum work is essentially the same as when the pressure effect was included. In general, unless the desired accuracy of the problem demands it, the pressure effect can be neglected.

Example 16-9

Reconsider the same combustion process as in Example 16-8, but let it be carried out adiabatically. Determine (a) the maximum combustion temperature, (b) the change of entropy of the adiabatic process, (c) the maximum work for the adiabatic process, (d) the stream availability of the hot products, (e) the steam availability of the fuel, (f) the ratio of the availability of the hot products to that of the original reactants, and (g) the irreversibility of the reaction.

Solution The chemical equation is the same as in Example 16-8, which is

$$C_3H_8 + (7O_2 + 26.32N_2) \longrightarrow 3CO_2 + 4H_2O + 2O_2 + 26.32N_2$$

where all constituents are in the gaseous state.

Let the subscripts 1 and 2 denote the initial reactants and final products, respectively. For the adiabatic process, $Q = 0$, and Eq. (15-12) becomes

$$\sum_{\text{products}} n(\Delta h_f^\circ + h_{T_2} - h_{T_0}) - \sum_{\text{reactants}} n(\Delta h_f^\circ + h_{T_1} - h_{T_0}) = 0$$

With $T_1 = T_0 = 298$ K, this equation reduces to

$$\sum_{\text{products}} n(\Delta h_f^\circ + h_{T_2} - h_{T_0}) - \sum_{\text{reactants}} n \, \Delta h_f^\circ = 0$$

or

$$[3(\Delta h_f^\circ + h_{T_2} - h_{T_0})_{CO_2} + 4(\Delta h_f^\circ + h_{T_2} - h_{T_0})_{H_2O} + 2(h_{T_2} - h_{T_0})_{O_2}$$
$$+ 26.32(h_{T_2} - h_{T_0})_{N_2}] - 1(\Delta h_f^\circ)_{C_3H_8} = 0$$

Taking Δh_f° data from Table A-28M and considering the products to be an ideal-gas mixture with enthalpy values for its components given by Tables A-10 to A-13, we have

$$[3(-393,520 + h_{T_2} - 9359.7)_{CO_2} + 4(-241,820 + h_{T_2} - 9899.0)_{H_2O}$$
$$+ 2(h_{T_2} - 8676.9)_{O_2} + 26.32(h_{T_2} - 8665.8)_{N_2}] - 1(-103,850) = 0$$

or

$$(3h_{T_2,CO_2} + 4h_{T_2,H_2O} + 2h_{T_2,O_2} + 26.32h_{T_2,N_2}) = 2,357,100 \text{ kJ}$$

In this equation, as T_2 is unknown, calculations by iteration based on the ideal-gas enthalpy value at an assumed T_2 for CO_2, H_2O, O_2, and N_2 from the Keenan–Chao–Kaye Gas Tables* (or Tables A-10M to A-13M in Appendix 3 in abbreviated form). The results of the final two trials are as follows:

At $T_2 = 1910$ K,

$$[3(95,384.3) + 4(78,114.8) + 2(64,462.4) + 26.32(61,571.5)] = 2,348,100 \text{ kJ}$$

At $T_2 = 1920$ K,

$$[3(95,986.1) + 4(78,620.8) + 2(64,837.5) + 26.32(61,929.6)] = 2,362,100 \text{ kJ}$$

By interpolation, we conclude that the maximum combustion temperature is

$$T_2 = 1916 \text{ K}$$

(b) The change of entropy of the adiabatic combustion process is given by

$$\sum_{\text{products}} ns_{T_2} - \sum_{\text{reactants}} ns_{T_1} = \sum_{\text{products}} n(s_{T_2}^\circ - \mathcal{R} \ln x) - \sum_{\text{reactants}} n(s_{T_1}^\circ - \mathcal{R} \ln x)$$

$$= [3(s_{T_2}^\circ - \mathcal{R} \ln x)_{CO_2} + 4(s_{T_2}^\circ - \mathcal{R} \ln x)_{H_2O}$$
$$+ 2(s_{T_2}^\circ - \mathcal{R} \ln x)_{O_2} + 26.32(s_{T_2}^\circ - \mathcal{R} \ln x)_{N_2}]$$
$$- [1(s_{T_1}^\circ)_{C_3H_8} + 7(s_{T_1}^\circ - \mathcal{R} \ln x)_{O_2} + 26.32(s_{T_1}^\circ - \mathcal{R} \ln x)_{N_2}]$$

Taking data for s_T° at $T_2 = 1916$ K and $T_1 = 298$ K for CO_2, H_2O, O_2, and N_2 from Tables A-10 to A-13, and for $s_{T_1}^\circ$ for C_3H_8 from Table A-28M, the result is

$$\left[3\left(306.601 - 8.314 \ln \frac{3}{3 + 4 + 2 + 26.32} \right) + 4\left(262.467 - 8.314 \ln \frac{4}{35.32} \right) \right.$$

$$+ 2\left(266.973 - 8.314 \ln \frac{2}{35.32} \right) + 26.32\left(250.387 - 8.314 \ln \frac{26.32}{35.32} \right) \Big]$$

$$- \left[1(269.91) + 7\left(204.975 - 8.314 \ln \frac{7}{7 + 26.32} \right) \right.$$

$$+ 26.32\left(191.448 - 8.314 \ln \frac{26.32}{33.32} \right) \Big]$$

$$= 2453.8 \text{ kJ/K·kgmole fuel}$$

(c) The maximum work is given by Eq. (16.80) with $T_1 = T_0 = 298$ K and $T_2 = 1916$ K. Thus,

$$W_{\text{max}} = \sum_{\text{reactants}} n[(\Delta h_f^\circ) - T_0(s_{T_1}^\circ - \mathcal{R} \ln x)]$$

*J. H. Keenan, J. Chao, and J. Kaye, *Gas Tables (SI Units)*, John Wiley, New York, 1983.

$$- \sum_{\text{products}} n[(\Delta h_f^\circ + h_{T_2} - h_{T_0}) - T_0(s_{T_2}^\circ - \mathcal{R} \ln x)]$$

$$= \left[\sum_{\text{reactants}} n(\Delta h_f^\circ) - \sum_{\text{products}} n(\Delta h_f^\circ + h_{T_2} - h_{T_0}) \right]$$

$$- T_0 \left[\sum_{\text{reactants}} n(s_{T_1}^\circ - \mathcal{R} \ln x) - \sum_{\text{products}} n(s_{T_2}^\circ - \mathcal{R} \ln x) \right]$$

in which

$$\left[\sum_{\text{reactants}} n(\Delta h_f^\circ) - \sum_{\text{products}} n(\Delta h_f^\circ + h_{T_2} - h_{T_0}) \right] = 0$$

and

$$\left[\sum_{\text{reactants}} n(s_{T_1}^\circ - \mathcal{R} \ln x) - \sum_{\text{products}} n(s_{T_2}^\circ - \mathcal{R} \ln x) \right] = -2453.8 \text{ kJ/K}$$

Therefore,

$$W_{\max} = 0 - 298(-2453.8) = 731{,}232 \text{ kJ/kgmole fuel}$$

(d) In the absence of kinetic and potential energies, the stream availability of the hot products is given from Eq. (8-19):

$$\Psi_{\text{products}} = \sum_{\text{products}} n[(h_{T_2} - h_{T_0}) - T_0(s_{T_2} - s_{T_0})]$$

$$= \sum_{\text{products}} n(h_{T_2} - h_{T_0}) - T_0 \sum_{\text{products}} n(s_{T_2}^\circ - s_{T_0}^\circ)$$

$$= [3(h_{T_2} - h_{T_0})_{CO_2} + 4(h_{T_2} - h_{T_0})_{H_2O} + 2(h_{T_2} - h_{T_0})_{O_2} + 26.32(h_{T_2} - h_{T_0})_{N_2}]$$

$$- T_0[3(s_{T_2}^\circ - s_{T_0}^\circ)_{CO_2} + 4(s_{T_2}^\circ - s_{T_0}^\circ)_{H_2O} + 2(s_{T_2}^\circ - s_{T_0}^\circ)_{O_2}$$

$$+ 26.32(s_{T_2}^\circ - s_{T_0}^\circ)_{N_2}]$$

By taking data from Tables A-10 to A-13, the result is

$$\Psi_{\text{products}} = [3(95{,}745.4 - 9359.7) + 4(78{,}418.4 - 9899.0)$$

$$+ 2(64{,}687.5 - 8676.9) + 26.32(61{,}786.4 - 8665.8)]$$

$$- 298[3(306.601 - 213.657) + 4(262.467 - 188.697)$$

$$+ 2(266.973 - 204.975) + 26.32(250.387 - 191.448)]$$

$$= 1{,}373{,}130 \text{ kJ/kgmole fuel}$$

(e) For a steady-state, steady-flow, chemical-reaction process, we have

$$W_{\max} = \Psi_{\text{reactants}} - \Psi_{\text{products}}$$

where $\Psi_{\text{reactants}}$ and Ψ_{products} are the stream availability of the reactants and products, respectively. In this example, the air supplied for combustion is at the same pressure and temperature as the atmosphere and therefore has zero stream availability ($\Psi_{\text{air}} = 0$). Consequently, we must have

$$W_{\max} = \Psi_{\text{fuel}} - \Psi_{\text{products}}$$

where Ψ_{fuel} is the stream availability of the fuel and is given by

$$\Psi_{\text{fuel}} = W_{\max} + \Psi_{\text{products}} = 731{,}232 + 1{,}373{,}130$$

$$= 2{,}104{,}360 \text{ kJ/kgmole fuel}$$

Notice that in Example 16-8, we concluded that the stream availability of the same fuel is 2,104,910 kJ/kgmole. These two values are practically the same.

(f) The ratio of the stream availability of the hot products to that of the fuel is

$$\frac{\Psi_{products}}{\Psi_{fuel}} = \frac{1,373,130}{2,104,360} = 0.653$$

(g) The irreversibility of the reaction is given by Eq. (16-81),

$$I = W_{max} = 731,232 \text{ kJ/kgmole fuel}$$

The irreversibility can also be calculated by Eq. (16-82),

$$I = T_0 \Delta S_{reaction} - Q_{release}$$

$$= 298(2453.8) - 0 = 731,232 \text{ kJ/kgmole fuel}$$

16-13 SUMMARY

The Gibbs equations for simple compressible systems are

$$dU = T \, dS - p \, dV + \sum \mu_i \, dn_i$$

$$dH = T \, dS + V \, dp + \sum \mu_i \, dn_i$$

$$dA = -S \, dT - p \, dV + \sum \mu_i \, dn_i$$

$$dG = -S \, dT + V \, dp + \sum \mu_i \, dn_i$$

The chemical potential of component i can be written as

$$\mu_i = \left(\frac{\partial U}{\partial n_i}\right)_{S, V, n_j(j \neq i)}$$

$$= \left(\frac{\partial H}{\partial n_i}\right)_{S, p, n_j(j \neq i)}$$

$$= \left(\frac{\partial A}{\partial n_i}\right)_{V, T, n_j(j \neq i)}$$

$$= \left(\frac{\partial G}{\partial n_i}\right)_{T, p, n_j(j \neq i)}$$

The last of these expressions is sometimes used alone as the definition of the chemical potential μ_i.

Based on Euler's theorem, we have the following expressions:

$$G = \sum \mu_i n_i$$

$$S \, dT - V \, dp + \sum n_i \, d\mu_i = 0$$

For changes at constant T and p, we have

$$\sum n_i \, d\mu_i = 0$$

For a pure component,

$$\mu = G/n = g$$

The general condition of equilibrium for closed simple compressible systems can be written as

$$dU + p \, dV - T \, dS \leq 0$$

$$dH - V \, dp - T \, dS \leq 0$$

$$dA + p \, dV + S \, dT \leq 0$$

$$dG - V \, dp + S \, dT \leq 0$$

where the equality holds if the system is already in complete equilibrium, and the inequality holds if it is not.

The following are some special conditions of equilibrium for a closed simple compressible system:

$$(dS)_{U,V} \geq 0$$

$$(dU)_{S,V} \leq 0$$

$$(dG)_{T,p} \leq 0$$

In terms of virtual variations, these conditions can be written as

$$(\delta S)_{U,V} \leq 0$$

$$(\delta U)_{S,V} \geq 0$$

$$(\delta G)_{T,p} \geq 0$$

where the symbol δ is used to denote small but finite hypothetical or virtual variations away from equilibrium to distinguish from the symbol d for differential changes toward equilibrium.

The basic conditions for stability for a homogeneous, single-component, simple compressible closed system are

$$\left(\frac{\partial S}{\partial T}\right)_V = \frac{C_v}{T} > 0$$

and

$$\left(\frac{\partial p}{\partial V}\right)_T < 0$$

Two phases are said to be in phase equilibrium when there is no transformation from one phase to the other. For two phases of a pure substance, the condition for phase equilibrium is that they have the same specific Gibbs function. For a heterogeneous system consisting of φ homogeneous phases and r components, the condition of phase equilibrium is written as

$$\mu_i^{(1)} = \mu_i^{(2)} = \cdots = \mu_i^{(\varphi)}, \qquad i = 1, 2, \ldots, r$$

a total of $(\varphi - 1)r$ equations.

The Raoult's law states that for a dilute liquid solution,

$$p = p^\circ x$$

where p is the partial vapor pressure of the solvent, p° is the vapor pressure of pure solvent at the p and T of the solution, and x is the mole fraction of the solvent.

The number of independent intensive variables F required to fix the equilibrium state of a heterogeneous system of φ phases, each containing r components, is given by the Gibbs phase rule:

$$F = r - \varphi + 2$$

For the chemical reaction

$$\nu_1 C_1 + \nu_2 C_2 \rightleftharpoons \nu_3 C_3 + \nu_4 C_4$$

the condition of equilibrium at constant T and p is

$$\mu_1 \nu_1 + \mu_2 \nu_2 = \mu_3 \nu_3 + \mu_4 \nu_4$$

where C_1, C_2, C_3, and C_4 denote the chemical constituents, and ν_1, ν_2, ν_3, and ν_4 denote the stoichiometric coefficients. This equation of reaction equilibrium provides a means for determining the equilibrium composition of a reactive mixture at a given temperature and pressure.

Based on the equation of reaction equilibrium stated before, the equilibrium constant K_p for ideal-gas reactions can be expressed as

$$K_p = \frac{x_3^{\nu_3} x_4^{\nu_4}}{x_1^{\nu_1} x_2^{\nu_2}} \left(\frac{p}{p_0} \right)^{\nu_3 + \nu_4 - \nu_1 - \nu_2}$$

where the x's are mole fractions. When the reference pressure $p_0 = 1$ atm is introduced, the equilibrium constant can be written as

$$K_p = \frac{p_3^{\nu_3} p_4^{\nu_4}}{p_1^{\nu_1} p_2^{\nu_2}}$$

where the p's are the partial pressures.

The van't Hoff isobar equation expresses the variation of K_p with temperature:

$$\frac{d(\ln K_p)}{dT} = \frac{\Delta H_R(T, p_0)}{\mathscr{R} T^2}$$

where $\Delta H_R(T, p_0)$ is the enthalpy of reaction at temperature T and reference pressure p_0.

The various statements of the third law of thermodynamics are as follows.

1. Nernst–Simon statement: The entropy change associated with any isothermal reversible process of a condensed system (i.e., a solid or a liquid) approaches zero as the temperature approaches absolute zero, or, symbolically,

$$\lim_{T \to 0} (\Delta S)_T = 0$$

2. Unattainability statement: It is impossible by means of any process, no matter how idealized, to reduce the temperature of a system to absolute zero in a finite number of steps.

3. Planck's postulate: The absolute value of the entropy of a pure solid or a pure liquid approaches zero as the temperature approaches absolute zero, or, symbolically,

$$\lim_{T \to 0} S = 0$$

The Gibbs function of formation Δg_f^0 of a compound at the standard condition of 25°C and 1 atm is defined as the Gibbs function of reaction at that condition for the formation of the compound from its elements. When the Gibbs function of all stable elements at the standard condition are assigned the value of zero, we have the relation:

$$\Delta g_f^\circ = g^\circ_{\text{compound}}$$

For a steady-state, steady-flow process involving a chemical reaction, when changes in kinetic and potential energies are neglected, the maximum work can be written as

$$W_{\text{max}} = \sum_{\text{reactants}} n[(\Delta h_f^\circ + h_T - h_{T_0}) - T_0(s_T^\circ - \mathscr{R} \ln x)]$$

$$- \sum_{\text{products}} n[(\Delta h_f^\circ + h_T - h_{T_0}) - T_0(s_T^\circ - \mathscr{R} \ln x)]$$

where Δh_f° is the standard-state enthalpy of formation of a gas, h_T is its enthalpy at T, h_0 is its enthalpy at T_0, s_T° is its absolute entropy at T and $p_0 = 1$ atm, and x is its mole fraction.

For a chemical-reaction process, there is no actual work done, so the irreversibility is given by

$$I = W_{\text{max}} - \overset{0}{\cancel{W}_{\text{actual}}} = W_{\text{max}}$$

The irreversibility can also be calculated by the equation

$$I = T_0(\Delta S_{\text{system}} + \Delta S_{\text{surr}})$$

$$= T_0 \, S_{\text{reaction}} - Q_{\text{release}}$$

For the special case when each component of the reactants and the products can be assumed to be at the standard condition of $T_0 = 25°C$ and $p_0 = 1$ atm, the maximum work is given by

$$W_{\text{max}} = \sum_{\text{reactants}} n \, \Delta g_f^\circ - \sum_{\text{products}} n \, \Delta g_f^\circ$$

PROBLEMS

16-1. Liquid water at 5000 psia and 500°F is in equilibrium with water vapor through a porous wall that allows only water vapor to pass through. Find the state of the vapor on the other side of the wall.

16-2. It is known that $(\partial p/\partial v)_T = 0$ for a pure substance at the critical point. From the condition of stability imposed on the Helmholtz function, show that

$$(\partial^2 p/\partial v^2)_T = 0 \qquad \text{and} \qquad (\partial^3 p/\partial v^3)_T < 0$$

16-3. For a binary vapor–liquid system at 1 atm, the dew-point (saturated vapor) line and the bubble-point (saturated liquid) line on the temperature–composition diagram are assumed to obey the following equations:

$$T = T_2 - (T_2 - T_1)[x_1^{(\beta)}]^2$$

and

$$T = T_2 - (T_2 - T_1)x_1^{(\alpha)}[2 - x_1^{(\alpha)}]$$

respectively. In these equations, T_1 and T_2 are saturation temperatures of pure components 1 and 2 at 1 atm, and $x_1^{(\beta)}$ and $x_1^{(\alpha)}$ are the mole fractions of component 1 in the vapor and liquid phases, respectively. What will be the composition of the vapor phase when a liquid solution of 40 mole percent of component 1 is heated to its boiling temperature?

16-4. Calculate and draw the temperature–composition diagram for the system carbon tatrachloride (CCl$_4$) and tin tetrachloride (SnCl$_4$) at a total pressure of 760 mm Hg, assuming that Raoult's law, Eq. (16-40), and Dalton's law are obeyed. The following is the vapor-pressure data of the pure components:

Temperature (°C)	77	80	90	100	110	114
p° of CCl$_4$ (mm Hg)	760	836	1112	1450	1880	—
p° of SnCl$_4$ (mm Hg)	—	258	362	490	673	760

16-5. Using the vapor-pressure data for carbon tetrachloride and tin tetrachloride given in Prob. 16-4, calculate and draw the pressure–composition diagram at 100°C. Assume that Raoult's law, Eq. (16-40), and Dalton's law are obeyed.

16-6. Calculate and draw the temperature-versus-mole fraction diagram or the system n-hexane and n-heptane at a total pressure of 40 psia, assuming that Raoult's law, Eq. (16-40), and Dalton's law are obeyed. The following is the vapor-pressure data of the pure components:

Temperature, °C	110	115	120	125	130
n-hexane p_1^0, psia	44.8	50.0	55.8	62.4	70.0
n-heptane p_2^0, psia	20.1	23.1	26.3	29.7	33.4

16-7. A dilute liquid solution will freeze at a lower temperature as compared to the pure solvent, provided that when the solution starts to freeze, it is

the pure solvent that forms the solid phase and the solute is confined to the liquid phase. The depression of freezing temperature ΔT corresponding to the mole fraction x of the solute is given by the equation

$$\Delta T = \frac{\mathcal{R}T_0^2}{h_{fg}}x$$

where $\Delta T = T_0 - T$, T is the freezing temperature of the solution, T_0 is the freezing temperature of the pure solvent, h_{fg} is the latent heat of fusion per mole of pure solvent, all at 1 atmospheric presssure, and $\mathcal{R}$ is the universal gas constant. When 5 liters of ethyl alcohol (C$_2$H$_5$OH) are poured into a 20-liter automobile radiator in winter, how low will the temperature be before the solution freezes? The density of ethyl alcohol is 0.789 grams/milliliter.

16-8. For a dilute liquid solution containing nonvolatile solutes, the vapor pressure is due entirely to the solvent and thus has a lower value as compared to the pure solvent. Consequently, when a solute such as salt is dissolved in a solvent like water, the solution boils at a higher temperature than the pure solvent. The elevation in boiling temperature ΔT corresponding to the mole fraction x of the solute is given by the equation

$$\Delta T = \frac{\mathcal{R}T_0^2}{h_{fg}}x$$

where $\Delta T = T - T_0$, T is the boiling temperature of the solution, T_0 is the boiling temperature of the pure solvent, h_{fg} is the latent heat of vaporization per mole of the pure solvent, all at 1 atmospheric pressure, and $\mathcal{R}$ is the universal gas constant. In a process of evaporative sea-water desalination, the sea water with 1.4 mole percent of salt (NaCl) is heated to its boiling temperature at 1 atm. The evolved vapor, which is essentially pure water, is subsequently condensed to give the required fresh-water supply. Estimate the boiling temperature of the salt solution.

16-9. It is experimentally determined that when 52.5 grams of benzil (C$_{14}$H$_{10}$O$_2$) are dissolved in 1000 grams of toluene (C$_7$H$_8$), the boiling temperature at 1 atm is raised from 110.62 to 111.40°C. By the use of the equation for the elevation in boiling temperature given in Prob. 16-8, estimate the enthalpy of vaporization of pure toluene at 1 atm.

16-10. When 0.8 g of a nonelectrolyte is dissolved in 150 g of benzene (C$_6$H$_6$), the boiling point at 1 atm is raised from 353 K to 353.08 K. If the enthalpy of vaporization of pure benzene is

7350 cal/gmole, determine the molecular weight of the solute, using the equation for the elevation in boiling temperature given in Prob. 16-8.

16-11. In a simple ammonia absorption-refrigeration system, the pressure limits are 215 psia and 55 psia, the generator temperature is 200°F, and the absorber temperature is 80°F. See Fig. P16-11. Determine (a) the amount of aqua-ammonia liquid that must be pumped for each pound of ammonia that enters the condenser, and (b) the amount of heat added to the generator for each pound of ammonia that enters the condenser.

16-12. Figure P16-12 shows an industrial aqua-ammonia absorption-refrigeration system with ammonia as the refrigerant and water as the absorbent. The principal parts of this system are the same as those shown in Fig. 12-12, except that a rectifying column–dephlegmator has been added to achieve efficient elimination of water vapor from the ammonia vapor being sent to the condenser. With the data given in the figure, determine the energy quantities of the components and the tons of refrigeration produced.

16-13. A lithium bromide–water absoption-refrigeration system has a flow diagram as shown schemati-

cally in Fig. 12-12, with water as the refrigerant and lithium bromide as the absorbent. The following conditions are known: condensing temperature, 43°C; evaporation temperature, 3°C; temperature of strong solution entering the generator, 71°C; and temperature of refrigerant vapor and weak solution leaving the generator, 85°C. State 1 is saturated water vapor, state 3 is saturated liquid water, and states 7 and 8 are saturated liquid solution (use data from Fig. A-18). Solar-energy heated hot water enters the gnerator at 96°C and leaves at 89°C. This machine is used to produce chilled water at 7°C, and warm water from the load returns to the machine at a temperature of 13°C and a flow rate of 250 liters/min. Calculate the hot-water mass-flow rate from the collector to the generator.

16-14. A liquid–vapor mixture of ammonia and water is in equilibrium at 50°C. The composition of the vapor phase is 99% NH$_3$ and 1% H$_2$O by mole numbers. Determine the composition of the liquid phase.

16-15. A liquid–vapor mixture of ammonia and water is in equilibrium at 30°C. The composition of the liquid phase is 60% NH$_3$ and 40% H$_2$O by mole

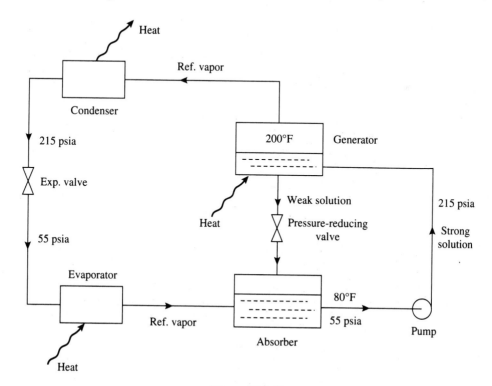

Figure P16-11

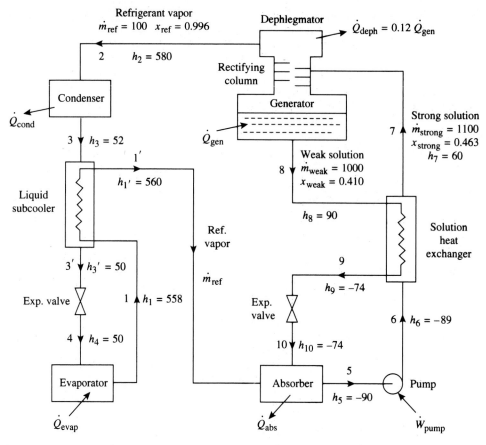

Refrigerant vapor
$\dot{m}_{ref} = 100$ $x_{ref} = 0.996$

Dephlegmator $\dot{Q}_{deph} = 0.12\,\dot{Q}_{gen}$

2   $h_2 = 580$

Rectifying column

Condenser

$\dot{Q}_{cond}$

Generator

$\dot{Q}_{gen}$

Strong solution
7 ↥ $\dot{m}_{strong} = 1100$
$x_{strong} = 0.463$
$h_7 = 60$

3 ↧ $h_3 = 52$

1'

Weak solution
8 ↧ $\dot{m}_{weak} = 1000$
$x_{weak} = 0.410$

$h_{1'} = 560$

Liquid subcooler

$h_8 = 90$

Ref. vapor

Solution heat exchanger

$\dot{m}_{ref}$

3' ↧ $h_{3'} = 50$

9

$h_9 = -74$

Exp. valve   1 ↥ $h_1 = 558$

Exp. valve

6 ↥ $h_6 = -89$

4 ↧ $h_4 = 50$

10 ↧ $h_{10} = -74$

Evaporator

5

Absorber

$h_5 = -90$

Pump

$\dot{Q}_{evap}$

$\dot{Q}_{abs}$

$\dot{W}_{pump}$

Units: $\dot{m}$ in lbm/min, x in lbm NH_3/lbm mix, h in Btu/lbm

Figure P16-12

numbers. Determine the composition of the vapor phase.

16-16. For a two-phase liquid–vapor mixture of water at 100°C, use the steam-table data to show that the specific Gibbs functions of saturated liquid and saturated vapor are equal.

16-17. By using the van't Hoff isobar equation, Eq. (16-67), estimate the enthalpy of reaction at 2000 K in kJ/kgmole for the reaction $CO_2 \rightleftharpoons CO + \frac{1}{2}O_2$. Compare the result with that obtained by using enthalpy data.

16-18. Determine the chemical reaction equilibrium constant at 1000 K for the reaction

$$SO_2 + \tfrac{1}{2}O_2 \;\rightleftharpoons\; SO_3$$

The specific heats at constant pressure for the three substances in the temperature range for this

problem are as follows:

$$c_{p,SO_2} = 4.186(7.116 + 9.512 \times 10^{-3}T$$
$$- 3.511 \times 10^{-6}T^2)$$

$$c_{p,O_2} = 4.186(6.095 + 3.253 \times 10^{-3}T$$
$$- 1.017 \times 10^{-6}T^2)$$

$$c_{p,SO_3} = 4.186(6.077 + 23.537 \times 10^{-3}T$$
$$- 9.687 \times 10^{-6}T^2)$$

where c_p is in kJ/kgmole·K, and T in K. The absolute entropy of SO_2, O_2, and SO_3 at 298 K and 1 atm are 248.65, 206.79, and 256.36 kJ/kgmole·K, respectively. The enthalpy of formation for SO_2 and SO_3 at 298 K and 1 atm are −297,040 and −395,370 kJ/kgmole, respectively.

Phase and Chemical Equilibrium   Chap. 16

16-19. Determine the equilibrium constant K_p at 298 K and 1 atm for the reaction

$$CO_2 + H_2 \rightleftharpoons CO + H_2O$$

by using (a) the Gibbs function of formation data from Table A-28M, and (b) the enthalpy of formation and absolute entropy data from Table A-28M.

16-20. Determine the equilibrium constant K_p at 298 K and 1 atm for the gaseous reaction equation

$$CO_2 + H_2 \rightleftharpoons CO + H_2O$$

by using the K_p data from Table A-29 for the simple reactions

$$CO_2 \rightleftharpoons CO + \tfrac{1}{2}O_2$$

and

$$H_2O \rightleftharpoons H_2 + \tfrac{1}{2}O_2$$

Compare your answer with that read directly from Table A-29 for the original reaction equation.

16-21. Determine the equilibrium constant K_p at 3000 K for the gaseous reaction equation

$$CO_2 + H_2 \rightleftharpoons CO + H_2O$$

by the use of the data from the ideal-gas property tables (Tables A-10M to A-15M). Compare your answer with that given in Table A-29.

16-22. Determine the mole fraction of OH in the product mixture when H_2O is heated to 5000 K at 1 atm. The equilibrium constant for this dissociation reaction at 5000 K is $\log_{10} K_p = 0.731$.

16-23. For the reaction of CO with the stoichiometric amount of O_2 to form CO_2, determine the composition of the equilibrium mixture at 3000 K and (a) 1 atm and (b) 5 atm.

16-24. A mixture consisting initially of 3 moles O_2, 2 moles N_2, and 6 moles Ar reaches equilibrium at

3500 K and 50 atm. Assuming the argon to be inert, determine the equilibrium composition.

16-25. A mixture of 1 mole of CO and 1 mole of O_2 is maintained at 1 atm and 3000 K. Determine the equilibrium composition.

16-26. A mixture of 3 moles of carbon dioxide and $\tfrac{1}{2}$ mole of carbon monoxide is maintained at 1 atm and 2600 K. Determine the equilibrium composition.

16-27. Carbon is burned with 200% theoretical oxygen to form an equilibrium mixture of CO_2, CO, and O_2 at 3000 K and 10 atm. Determine the equilibrium composition when only the CO_2 dissociates as $CO_2 \rightleftharpoons CO + \tfrac{1}{2}O_2$.

16-28. An equilibrium mixture at 1 atm has a molar analysis of 86.53% CO_2, 8.98% CO, and 4.49% O_2. Determine the equilibrium constant for $CO_2 \rightleftharpoons CO + \tfrac{1}{2}O_2$. What is the temperature of the given equilibrium mixture?

16-29. A mixture of 1 mole of H_2 and 1 mole of Ar is heated at a constant pressure of 1 atm until 15% H_2 dissociates into monatomic hydrogen (H). Assuming argon to remain as an inert gas, determine the final temperature of the mixture.

16-30. Water (H_2O) is heated to 3000 K. Assuming the equilibrium dissociated mixture consists of H_2O, OH, H_2, and O_2, determine the equilibrium composition at (a) 1 atm and (b) 10 atm.

16-31. One kgmole of water (H_2O) dissociates to form an equilibrium mixture of H_2O, H_2, and O_2. Determine the equilibrium composition at 3000 K and (a) 1 atm and (b) 10 atm.

16-32. Air consisting of 21% O_2 and 79% N_2 by mole numbers is heated to 3000 K at a pressure of 1 atm. See Fig. P16-32. Determine the equilibrium composition, assuming that only O_2, N_2, O, and NO are present. Is it reasonable to assume

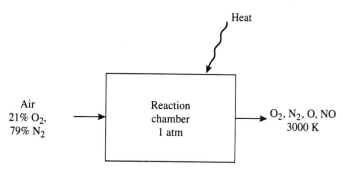

Heat

Air
21% O_2,
79% N_2

Reaction chamber
1 atm

O_2, N_2, O, NO
3000 K

Figure P16-32

that no N will be present in the equilibrium mixture?

16-33. Air consisting of 21% O_2 and 79% N_2 by mole numbers enters a heat exchanger. See Fig. P16-33. An equilibrium mixture of N_2, O_2, and NO exits at 2500 K, 1 atm. Determine the mole fraction of NO in the exiting mixture. Will the amount of NO increase or decrease as temperature increases at fixed pressure?

16-34. Carbon monoxide gas at 37°C and 110 kPa is supplied to a steady-flow combustion chamber at a volume-flow rate of 0.8 m³/min and is burned with oxygen entering the chamber at 25°C, 110 kPa, with a mass-flow rate of 0.7 kg/min. See Fig. P16-34. The combustion products, which consist of CO_2, CO, O_2, and N_2, leave the combustion chamber at 2000 K and 110 kPa. Determine (a) the equilibrium composition of the product gases, and (b) the heat-transfer rate from the combustion chamber.

16-35. A steady-flow combustion chamber is supplied with methane gas (CH_4) at 25°C, 1 atm, and 80% of theoretical air at 25°C, 1 atm. See Fig. P16-35. An equilibrium mixture of CO_2, CO, H_2O

(g), H_2, and N_2 exits at 1700 K, 1 atm. Determine, per kgmole methane, (a) the composition of the exiting mixture, and (b) the heat transfer. Neglect kinetic and potential energies.

16-36. Hydrogen peroxide vapor (H_2O_2) at 10 atm and 25°C enters a heat exchanger, and is heated until it decomposes into water vapor and oxygen, leaving at 10 atm and 600 K. See Fig. P16-36. The mixture of water vapor and oxygen then enters an insulated turbine, and expands to 1 atm and 300 K. Determine, per kgmole of H_2O_2, (a) the heat transfer in the heat exchanger, (b) the work output of the turbine, and (c) the entropy change in the heat exchanger.

16-37. Methane (CH_4) is burned with the theoretical amount of air in a steady-flow process at 1 atm. Both the fuel and the air are supplied at 25°C. The products, which consist of CO_2, CO, H_2O, H_2, O_2, and N_2 in equilibrium, leave the combustion chamber at 2200 K. Determine the composition of the products and the amount of heat transfer in this process per mole of methane.

16-38. Propane gas at 25°C and 1 atm is burned in a steady flow with 40% excess air at 25°C and

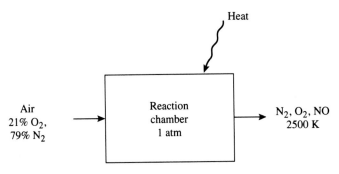

Figure P16-33

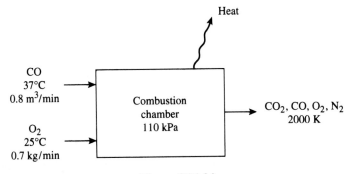

Figure P16-34

Phase and Chemical Equilibrium   Chap. 16

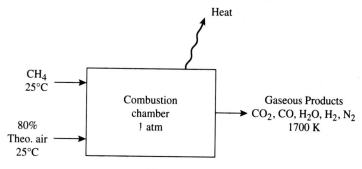

Figure P16-35

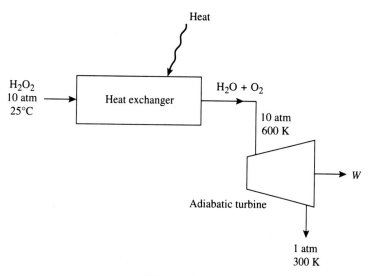

Figure P16-36

1 atm. Assume the reaction is complete. The gaseous products leave as a mixture at 25°C and 1 atm, and the water leaves as a liquid at 25°C and 1 atm. The atmosphere is at 25°C and 1 atm. Determine (a) the heat released by the reaction, (b) the maximum work for the process, and (c) the irreversibility of the reaction, all in kJ/kgmole of fuel. Notice that this is the same problem as Example 16-8, except that the product water leaves separately as a liquid at 25°C and 1 atm.

16-39. Reconsider the same combustion process as in Prob. 16-38, but let it be carried out adiabatically. Determine (a) the maximum combustion temperature, (b) the change of entropy of the adiabatic process, (c) the maximum work for the adiabtic process, (d) the stream availability of the hot products, (e) the stream availability of the

fuel, (f) the ratio of the availability of the hot products to that of the original reactants, and (g) the irreversibility of the reaction.

16-40. Liquid octane (C_8H_{18}) at 25°C and 1 atm enters a steady-flow combustion chamber at a mass-flow rate of 0.2 kg/min. See Fig. P16-40. It is burned with 150% theoretical air, entering at 25°C and 1 atm. The combustion is complete, with water in liquid form in the products. The combustion products are allowed to cool to 25°C before leaving the combustion chamber. Determine (a) the rate of heat transfer from the combustion chamber, (b) the rate of entropy generation, (c) the maximum power, and (d) the irreversibility rate.

16-41. A steady-flow combustion chamber is supplied with ethane gas (C_2H_6) at 1 atm, 25°C, and the stoichiometric amount of air at 1 atm, 25°C. See

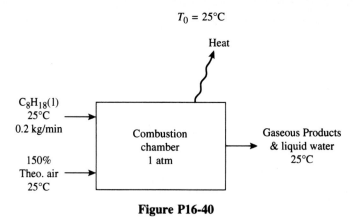

Figure P16-40

Fig. P16-41. The combustion is complete and the products leave at 1 atm, 827°C. Assuming $T_0 = 25°C$, determine (a) the heat transfer, (b) the entropy change, and (c) the irreversibility per kgmole of ethane.

16-42. A steady-flow combustion chamber is supplied with benzene gas (C_6H_6) at 1 atm, 25°C, and 95% theoretical air at 1 atm, 25°C. See Fig. P16-42. All the hydrogen in the fuel burns to H_2O, but part of the carbon burns to CO. Heat is lost to the

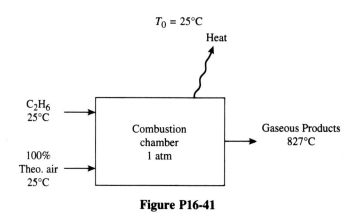

Figure P16-41

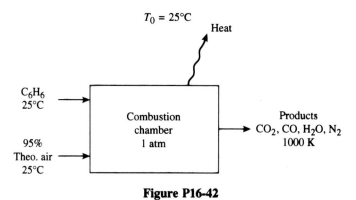

Figure P16-42

surrounding atmosphere at 25°C. The products leave the combustion chamber at 1 atm, 1000 K. Determine, per kgmole of benzene, (a) the heat transfer, (b) the entropy production, and (c) the irreversibility.

16-43. Liquid methyl alcohol (CH_3OH) at 1 atm and 25°C is burned to completion in a steady-flow combustion chamber with the theoretical amount of air entering at 1 atm and 25°C. See Fig. P16-43. The products leave the combustion chamber at 1 atm and 25°C. The water produced is assumed to be a liquid. The environment is at 25°C. Determine, per kgmole of fuel, (a) the heat transfer, (b) the maximum work, and (c) the irreversibility of the process.

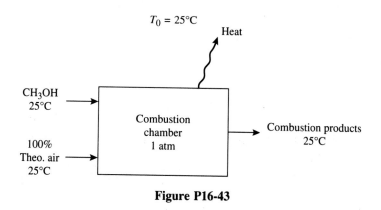

Figure P16-43

Appendix 1

Lagrange's Method of Undetermined Multipliers

Lagrange's method of undetermined multipliers is a general procedure for determining an extremum point such as a maximum or minimum in a continuous function of many variables subject to a number of constraints. Consider a function $f(x_1, x_2, \ldots, x_n)$, where n denotes the number of variables. The variables $x_1, x_2, \ldots, x_n$ are not all independent, but are related according to the following equations of constraint:

$$\phi_j(x_1, x_2, \ldots, x_n) = 0, \qquad j = 1, 2, \cdots, m \qquad \text{(a)}$$

where m is the number of equations of constraint, and $m < n$.

At an extremum point of the function $f(x_1, x_2, \ldots, x_n)$, it is necessary that

$$df = \frac{\partial f}{\partial x_1} dx_1 + \frac{\partial f}{\partial x_2} dx_2 + \cdots + \frac{\partial f}{\partial x_n} dx_n$$

$$= \sum_{i=1}^{n} \frac{\partial f}{\partial x_i} dx_i = 0 \qquad \text{(b)}$$

Of course, if all the variables are independent, the following set of n independent equations

$$\frac{\partial f}{\partial x_i} = 0, \qquad i = 1, 2, \ldots, n \qquad \text{(c)}$$

will give the values of the n independent variables at the extremum condition. But because the variables are not all independent, Eq. (c) is not valid, although an extremum point is still given by Eq. (b). However, we now have the equation of constraint, Eq. (a), in addition to Eq. (b). A straightforward but pedestrian way to find the extremum point is to solve Eq. (a) for m variables in terms of the other $n - m$ variables. These solutions when inserted in $f(x_1, x_2, \ldots, x_n)$ give a function of $n - m$ independent variables, and the extremum can then be found from the $n - m$ equations analogous to Eq. (c).

A more elegant and more useful method to solve this type of problem is to use Lagrange's method of undetermined multipliers. According to this method, one first writes the total differential of each constraining relation to yield

$$d\phi_j = 0 = \sum_{i=1}^{n} \frac{\partial \phi_j}{\partial x_i} dx_i, \qquad j = 1, 2, \ldots, m \tag{d}$$

Next let each $d\phi_j$ of the preceding equations be multiplied by some undetermined multiplier λ_j (the Lagrangian multiplier) and then added to Eq. (b). The result is

$$\sum_{i=1}^{n} \frac{\partial f}{\partial x_i} dx_i + \sum_{j=1}^{m} \left[\lambda_j \sum_{i=1}^{n} \left(\frac{\partial \phi_j}{\partial x_i} dx_i \right) \right] = 0$$

or

$$\left(\frac{\partial f}{\partial x_1} + \lambda_1 \frac{\partial \phi_1}{\partial x_1} + \lambda_2 \frac{\partial \phi_2}{\partial x_1} + \cdots + \lambda_m \frac{\partial \phi_m}{\partial x_1} \right) dx_1$$

$$+ \left(\frac{\partial f}{\partial x_2} + \lambda_1 \frac{\partial \phi_1}{\partial x_2} + \lambda_2 \frac{\partial \phi_2}{\partial x_2} + \cdots + \lambda_m \frac{\partial \phi_m}{\partial x_2} \right) dx_2$$

$$+ \cdots \tag{e}$$

$$+ \left(\frac{\partial f}{\partial x_n} + \lambda_1 \frac{\partial \phi_1}{\partial x_n} + \lambda_2 \frac{\partial \phi_2}{\partial x_n} + \cdots + \lambda_m \frac{\partial \phi_m}{\partial x_n} \right) dx_n = 0$$

Now, if $\lambda_1, \lambda_2, \ldots, \lambda_m$ are so chosen that

$$\frac{\partial f}{\partial x_i} + \lambda_1 \frac{\partial \phi_1}{\partial x_i} + \lambda_2 \frac{\partial \phi_2}{\partial x_i} + \cdots + \lambda_m \frac{\partial \phi_m}{\partial x_i} = 0, \qquad i = 1, 2, \ldots, n \tag{f}$$

then, together with the constraining equations

$$\phi_j(x_1, x_2, \ldots, x_n) = 0, \qquad j = 1, 2, \ldots, m$$

the $n + m$ "unknowns" $x_1, x_2, \ldots, x_n, \lambda_1, \lambda_2, \ldots \lambda_m$ can be determined to satisfy the necessary condition for an extremum of the function $f(x_1, x_2, \ldots, x_n)$.

Appendix 2

Units and Physical Constants

List of Tables

TABLE A-1 SI UNITS

(a) BASIC UNITS

Quantity	Name	Symbol
Length	meter	m
Mass	kilogram	kg
Time	second	s
Electric current	ampere	A
Temperature	kelvin	K
Luminous intensity	candela	cd

(b) SUPPLEMENTARY UNITS

Quantity	Name	Symbol
Plane angle	radian	rad
Solid angle	steradian	sr

(c) DERIVED UNITS WITH SPECIAL NAMES

Quantity	Name	Symbol	Definition
Force	newton	N	$kg \cdot m/s^2$
Energy	joule	J	$kg \cdot m^2/s^2 = N \cdot m$
Power	watt	W	$kg \cdot m^2/s^3 = J/s$
Pressure	pascal	Pa	$kg/m \cdot s^2 = N/m^2$
Electric charge	coulomb	C	$A \cdot s$
Electric potential	volt	V	$kg \cdot m^2/s^3 \cdot A = J/A \cdot s = J/C$
Electric resistance	ohm	Ω	$kg \cdot m^2/s^3 \cdot A^2 = V/A$
Electric capacitance	farad	F	$A^2 \cdot s^4/kg \cdot m^2 = A \cdot s/V = C/V$
Magnetic flux	weber	Wb	$kg \cdot m^2/s^2 \cdot A = V \cdot s$
Magnetic induction (magnetic flux density)	tesla	T	$kg/s^2 \cdot A = V \cdot s/m^2 = Wb/m^2$
Inductance	henry	H	$kg \cdot m^2/s^2 \cdot A^2 = V \cdot s/A$

(d) SOME OTHER DERIVED UNITS

Quantity	Unit	Symbol
Area	square meter	m^2
Volume	cubic meter	m^3
Density	kilogram per cubic meter	kg/m^3
Specific weight	newton per cubic meter	N/m^3
Velocity	meter per second	m/s
Acceleration	meter per second squared	m/s^2
Angular velocity	radian per second	rad/s
Specific heat	joule per kilogram per kelvin	$J/kg \cdot K$
Specific entropy	joule per kilogram per kelvin	$J/kg \cdot K$
Specific enthalpy	joule per kilogram	J/kg
Electric field intensity	volt per meter	V/m
Magnetic field intensity	ampere per meter	A/m

(e) ALTERNATIVE UNITS

Quantity	Name	Symbol	Relation to Basic Units
Temperature	degree Celsius	°C	$T(°C) = T(K) - 273.15$
Volume	liter	liter	$1\ liter = 10^{-3}\ m^3 = 10^3\ cm^3$
Pressure	bar	bar	$1\ bar = 10^5\ N/m^2$

(f) MULTIPLIERS

Multiplier	Prefix	Symbol
10^{12}	tera	T
10^9	giga	G
10^6	mega	M
10^3	kilo	k
10^2	hecto	h
10	deka	da
10^{-1}	deci	d
10^{-2}	centi	c
10^{-3}	milli	m
10^{-6}	micro	μ
10^{-9}	nano	n
10^{-12}	pico	p
10^{-15}	femto	f
10^{-18}	atto	a

TABLE A-2 CONVERSION FACTORS

Length	1 m = 3.28084 ft
	1 ft = 0.3048 m
	1 in. = 2.54 cm
	1 mile = 5280 ft = 1.60934 km
	1 micron (μ) = 10^{-6} m = 3.28084 × 10^{-6} ft
Volume	1 m^3 = 35.31 ft^3 = 1000 liter
	1 $in.^3$ = 16.387 cm^3
	1 liter = 1000 cm^3 = 0.03531 ft^3
	1 gal = 231 $in.^3$
Mass	1 kg = 2.20462 lbm
	1 lbm = 0.453592 kg
	1 slug = 32.174 lbm
Density	1 kg/m^3 = 0.062428 lbm/ft^3
	1 lbm/ft^3 = 16.0185 kg/m^3
Specific volume	1 m^3/kg = 16.0185 ft^3/lbm = 1 liter/g
	1 ft^3/lbm = 0.062428 m^3/kg
	1 liter/gmole = 1 $m^3/kgmole$
Force	1 N = 1 $kg \cdot m/s^2$ = 0.224809 lbf
	1 dyne = 1 $g \cdot cm/s^2$ = 1 × 10^{-5} N
	1 lbf = 1 $slug \cdot ft/s^2$ = 4.44822 N = 4.44822 × 10^5 dynes
Pressure	1 Pa = 1 N/m^2
	1 $lbf/in.^2$ = 6894.76 N/m^2
	1 bar = 10^5 Pa = 0.986923 atm
	1 atm = 14.6959 $lbf/in.^2$ = 1.01325 bars
	= 760 mmHg at 32°F = 29.92 in. Hg at 32°F
Temperature	$T(°R) = 1.8\,T(K)$ or $1K = 1.8°R$
	$T(°F) = 1.8\,T(°C) + 32$
	$T(K) = T(°C) + 273.15$
	$T(°R) = T(°F) + 459.67$
Energy	1 J = 1 N·m = 10^7 ergs
	1 kJ = 0.947817 Btu
	1 Btu = 778.169 ft·lbf = 1.055056 kJ
	1 cal (International Table) = 4.1868 J
	1 erg = 1 dyne·cm = 9.86923 × 10^{-10} atm·cm^3
	1 atm·m^3 = 101.325 kJ
Specific energy	1 J/g = 0.429923 Btu/lbm = 334.553 ft·lbf/lbm
	1 Btu/lbm = 2.326 J/g = 0.555556 cal/g
	1 Btu/lbmole = 2.326 kJ/kgmole
	1 kJ/kgmole = 0.429923 Btu/lbmole
	1 cal/gmole = 4.1868 kJ/kgmole
Power	1 W = 1 J/s
	1 kW = 3412.14 Btu/h
	1 hp = 550 ft·lbf/s = 2544.43 Btu/h
Specific entropy, specific heat, gas constant	1 kJ/kg·K = 0.2388 Btu/lbm·°R
	1 Btu/lbm·°R = 4.1868 kJ/kg·K
	1 kJ/kgmole·K = 0.2388 Btu/lbmole·°R
	1 Btu/lbmole·°R = 4.1868 kJ/kgmole·K
Velocity	1 mph = 1.467 ft/s = 0.4470 m/s
	1 ft/s = 0.3048 m/s

TABLE A-3 PHYSICAL CONSTANTS

Avogadro's number	6.022169×10^{26} (kgmole)$^{-1}$
Boltzmann constant	1.380622×10^{-23} J/K
Planck's constant	6.626196×10^{-34} J·s
Speed of light	2.9979250×10^{8} m/s
Electronic charge	$1.6021917 \times 10^{-19}$ C
Bohr magneton	9.274096×10^{-24} A·m^2
Permeability of free space	4×10^{-7} N/A^2
Standard gravitational acceleration	9.80665 m/s^2 = 32.174 ft/s^2

TABLE A-4 UNIVERSAL GAS CONSTANT

8314.29 J/kgmole·K
8.31429 kJ/kgmole·K
0.0820560 atm·m³/kgmole·K
1.98583 kcal/kgmole·K
1545.31 ft·lbf/lbmole·°R
1.98583 Btu/lbmole·°R
0.730225 atm·ft³/lbmole·°R

Appendix 3

Properties of Substances—Tables

List of Tables

TABLE A-5 CRITICAL CONSTANTS

Substance	Formula	Molecular weight*	T_c, K	p_c, atm	v_c, $\dfrac{m^3}{kgmole}$	Z_c
Acetylene	C_2H_2	26.038	309.5	61.6	0.113	0.274
Ammonia	NH_3	17.031	405.5	111.3	0.0725	0.243
Argon	Ar	39.948	151.	48.0	0.0752	0.291
Benzene	C_6H_6	78.115	562.61	48.6	0.260	0.274
n-Butane	C_4H_{10}	58.124	425.17	37.47	0.255	0.274
i-Butane	C_4H_{10}	58.124	408.14	36.00	0.263	0.283
1-Butene	C_4H_8	56.108	419.6	39.7	0.241	0.278
Carbon dioxide	CO_2	44.010	304.2	72.9	0.0940	0.275
Carbon disulfide	CS_2	76.139	552.	78.	0.170	0.293
Carbon monoxide	CO	28.011	133.	34.5	0.0931	0.294
Carbon tetrachloride	CCl_4	153.823	556.4	45.	0.276	0.272
Chlorine	Cl_2	70.906	417.	76.1	0.124	0.276
n-Decane	$C_{10}H_{22}$	142.287	619.	20.8	0.602	0.247
Ethane	C_2H_6	30.070	305.43	48.20	0.148	0.285
Ethylene	C_2H_4	28.055	283.06	50.50	0.124	0.270
Freon-12	CCl_2F_2	120.914	384.7	39.6	0.218	0.273
Helium	He	4.003	5.20	2.25	0.0575	0.303
n-Heptane	C_7H_{16}	100.206	540.16	27.01	0.426	0.260
n-Hexane	C_6H_{14}	86.178	507.90	29.92	0.368	0.264
Hydrogen	H_2	2.016	33.3	12.80	0.0650	0.304
Hydrogen sulfide	H_2S	34.080	373.6	88.9	0.0977	0.283
Krypton	Kr	83.80	209.4	54.3	0.0922	0.291
Methane	CH_4	16.043	190.7	45.80	0.099	0.290
Methyl chloride	CH_3Cl	50.488	416.3	65.9	0.143	0.276
Neon	Ne	20.183	44.5	26.9	0.0417	0.307
Nitric oxide	NO	30.006	180.	64.	0.058	0.251
Nitrogen	N_2	28.013	126.2	33.5	0.0901	0.291
Nitrous oxide	N_2O	44.013	309.7	71.7	0.0963	0.272
n-Nonane	C_9H_{20}	128.260	595.	22.5	0.543	0.250
n-Octane	C_8H_{18}	114.233	569.4	24.64	0.486	0.256
Oxygen	O_2	31.999	154.6	49.8	0.0734	0.288
n-Pentane	C_5H_{12}	72.151	469.78	33.31	0.311	0.269
i-Pentane	C_5H_{12}	72.151	461.0	32.9	0.308	0.268
neo-Pentane	C_5H_{12}	72.151	433.76	31.57	0.303	0.269
Propane	C_3H_8	44.097	369.97	42.01	0.200	0.277
Propylene	C_3H_6	42.081	365.1	45.4	0.181	0.274
Sulfur dioxide	SO_2	64.063	430.7	77.8	0.122	0.269
Sulfur trioxide	SO_3	80.062	491.4	83.8	0.126	0.262
Toluene	C_7H_8	92.142	594.0	40.	0.320	0.263
Water	H_2O	18.015	647.4	218.3	0.056	0.230
Xenon	Xe	131.30	289.75	58.0	0.1188	0.290

*The values of molecular weight are based on the international atomic mass scale in which $M_{C^{12}} = 12$ units of mass per mole.

Source: Principally from K. A. Kobe and R. E. Lynn, *The Critical Properties of Elements and Compounds, Chem. Rev.*, vol. 52, p. 117 (1953). Reprinted with permission from American Chemical Society.

(a) for six common gases, c_p and c_v in kJ/kg·K

Temp. °K	c_p	c_v	k	c_p	c_v	k	c_p	c_v	k	Temp. °K
	Air			Carbon dioxide, CO_2			Carbon monoxide, CO			
250	1.003	0.716	1.401	0.791	0.602	1.314	1.039	0.743	1.400	250
300	1.005	0.718	1.400	0.846	0.657	1.288	1.040	0.744	1.399	300
350	1.008	0.721	1.398	0.895	0.706	1.268	1.043	0.746	1.398	350
400	1.013	0.726	1.395	0.939	0.750	1.252	1.047	0.751	1.395	400
450	1.020	0.733	1.391	0.978	0.790	1.239	1.054	0.757	1.392	450
500	1.029	0.742	1.387	1.014	0.825	1.229	1.063	0.767	1.387	500
550	1.040	0.753	1.381	1.046	0.857	1.220	1.075	0.778	1.382	550
600	1.051	0.764	1.376	1.075	0.886	1.213	1.087	0.790	1.376	600
650	1.063	0.776	1.370	1.102	0.913	1.207	1.100	0.803	1.370	650
700	1.075	0.788	1.364	1.126	0.937	1.202	1.113	0.816	1.364	700
750	1.087	0.800	1.359	1.148	0.959	1.197	1.126	0.829	1.358	750
800	1.099	0.812	1.354	1.169	0.980	1.193	1.139	0.842	1.353	800
900	1.121	0.834	1.344	1.204	1.015	1.186	1.163	0.866	1.343	900
1000	1.142	0.855	1.336	1.234	1.045	1.181	1.185	0.888	1.335	1000
	Hydrogen, H_2			Nitrogen, N_2			Oxygen, O_2			
250	14.051	9.927	1.416	1.039	0.742	1.400	0.913	0.653	1.398	250
300	14.307	10.183	1.405	1.039	0.743	1.400	0.918	0.658	1.395	300
350	14.427	10.302	1.400	1.041	0.744	1.399	0.928	0.668	1.389	350
400	14.476	10.352	1.398	1.044	0.747	1.397	0.941	0.681	1.382	400
450	14.501	10.377	1.398	1.049	0.752	1.395	0.956	0.696	1.373	450
500	14.513	10.389	1.397	1.056	0.759	1.391	0.972	0.712	1.365	500
550	14.530	10.405	1.396	1.065	0.768	1.387	0.988	0.728	1.358	550
600	14.546	10.422	1.396	1.075	0.778	1.382	1.003	0.743	1.350	600
650	14.571	10.447	1.395	1.086	0.789	1.376	1.017	0.758	1.343	650
700	14.604	10.480	1.394	1.098	0.801	1.371	1.031	0.771	1.337	700
750	14.645	10.521	1.392	1.110	0.813	1.365	1.043	0.783	1.332	750
800	14.695	10.570	1.390	1.121	0.825	1.360	1.054	0.794	1.327	800
900	14.822	10.698	1.385	1.145	0.849	1.349	1.074	0.814	1.319	900
1000	14.983	10.859	1.380	1.167	0.870	1.341	1.090	0.830	1.313	1000

(b) for monatomic gases He, Ar, etc., c_v = 12.5 kJ/kgmole·K and c_p = 20.8 kJ/kgmole·K

Source: Data adapted from *Tables of Thermal Properties of Gases,* NBS Circular 564, 1955.

TABLE A-6E IDEAL-GAS SPECIFIC HEATS FOR SOME COMMON GASES
(ENGLISH UNITS)

(a) for six common gases, c_p and c_v in Btu/lbm·°F

Temp. °F	c_p	c_v	k	c_p	c_v	k	c_p	c_v	k	Temp. °F
	Air			Carbon dioxide, CO_2			Carbon monoxide, CO			
40	0.240	0.171	1.401	0.195	0.150	1.300	0.248	0.177	1.400	40
100	0.240	0.172	1.400	0.205	0.160	1.283	0.249	0.178	1.399	100
200	0.241	0.173	1.397	0.217	0.172	1.262	0.249	0.179	1.397	200
300	0.243	0.174	1.394	0.229	0.184	1.246	0.251	0.180	1.394	300
400	0.245	0.176	1.389	0.239	0.193	1.233	0.253	0.182	1.389	400
500	0.248	0.179	1.383	0.247	0.202	1.223	0.256	0.185	1.384	500
600	0.250	0.182	1.377	0.255	0.210	1.215	0.259	0.188	1.377	600
700	0.254	0.185	1.371	0.262	0.217	1.208	0.262	0.191	1.371	700
800	0.257	0.188	1.365	0.269	0.224	1.202	0.266	0.195	1.364	800
900	0.259	0.191	1.358	0.275	0.230	1.197	0.269	0.198	1.357	900
1000	0.263	0.195	1.353	0.280	0.235	1.192	0.273	0.202	1.351	1000
1500	0.276	0.208	1.330	0.298	0.253	1.178	0.287	0.216	1.328	1500
2000	0.286	0.217	1.312	0.312	0.267	1.169	0.297	0.226	1.314	2000
	Hydrogen, H_2			Nitrogen, N_2			Oxygen, O_2			
40	3.397	2.412	1.409	0.248	0.177	1.400	0.219	0.156	1.397	40
100	3.426	2.441	1.404	0.248	0.178	1.399	0.220	0.158	1.394	100
200	3.451	2.466	1.399	0.249	0.178	1.398	0.223	0.161	1.387	200
300	3.461	2.476	1.398	0.250	0.179	1.396	0.226	0.164	1.378	300
400	3.466	2.480	1.397	0.251	0.180	1.393	0.230	0.168	1.368	400
500	3.469	2.484	1.397	0.254	0.183	1.388	0.235	0.173	1.360	500
600	3.473	2.488	1.396	0.256	0.185	1.383	0.239	0.177	1.352	600
700	3.477	2.492	1.395	0.260	0.189	1.377	0.242	0.181	1.344	700
800	3.494	2.509	1.393	0.262	0.191	1.371	0.246	0.184	1.337	800
900	3.502	2.519	1.392	0.265	0.194	1.364	0.249	0.187	1.331	900
1000	3.513	2.528	1.390	0.269	0.198	1.359	0.252	0.190	1.326	1000
1500	3.618	2.633	1.374	0.283	0.212	1.334	0.263	0.201	1.309	1500
2000	3.758	2.773	1.355	0.293	0.222	1.319	0.270	0.208	1.298	2000

(b) for monatomic gases He, Ar, etc., $c_v = 3.0$ Btu/lbmole·°F and $c_p = 5.0$ Btu/lbmole·°F

Source: Data adapted from *Tables of Thermal Properties of Gases*, NBS Circular 564, 1955.

TABLE A-7M EMPIRICAL c_p EQUATIONS FOR VARIOUS GASES AT ZERO PRESSURE (SI UNITS)

$$c_p = a + bT + cT^2 + dT^3$$
$$T \text{ in K}, \ c_p \text{ in kJ/kgmole·K}$$

Substance	Formula	a	b	c	d	Temperature range K
Nitrogen	N_2	28.90	-0.1571×10^{-2}	0.8081×10^{-5}	-2.873×10^{-9}	273–1800
Oxygen	O_2	25.48	1.520×10^{-2}	-0.7155×10^{-5}	1.312×10^{-9}	273–1800
Air		28.11	0.1967×10^{-2}	0.4802×10^{-5}	-1.966×10^{-9}	273–1800
Hydrogen	H_2	29.11	-0.1916×10^{-2}	0.4003×10^{-5}	-0.8704×10^{-9}	273–1800
Carbon monoxide	CO	28.16	0.1675×10^{-2}	0.5372×10^{-5}	-2.222×10^{-9}	273–1800
Carbon dioxide	CO_2	22.26	5.981×10^{-2}	-3.501×10^{-5}	7.469×10^{-9}	273–1800
Water vapor	H_2O	32.24	0.1923×10^{-2}	1.055×10^{-5}	-3.595×10^{-9}	273–1800
Nitric oxide	NO	29.34	-0.09395×10^{-2}	0.9747×10^{-5}	-4.187×10^{-9}	273–1500
Nitrous oxide	N_2O	24.11	5.8632×10^{-2}	-3.562×10^{-5}	10.58×10^{-9}	273–1500
Nitrogen dioxide	NO_2	22.9	5.715×10^{-2}	-3.52×10^{-5}	7.87×10^{-9}	273–1500
Ammonia	NH_3	27.568	2.5630×10^{-2}	0.99072×10^{-5}	-6.6909×10^{-9}	273–1500
Sulfur	S_2	27.21	2.218×10^{-2}	-1.628×10^{-5}	3.986×10^{-9}	273–1800
Sulfur dioxide	SO_2	25.78	5.795×10^{-2}	-3.812×10^{-5}	8.612×10^{-9}	273–1800
Sulfur trioxide	SO_3	16.40	14.58×10^{-2}	-11.20×10^{-5}	32.42×10^{-9}	273–1300
Acetylene	C_2H_2	21.8	9.2143×10^{-2}	-6.527×10^{-5}	18.21×10^{-9}	273–1500
Benzene	C_6H_6	-36.22	48.475×10^{-2}	-31.57×10^{-5}	77.62×10^{-9}	273–1500
Methanol	CH_4O	19.0	9.152×10^{-2}	-1.22×10^{-5}	-8.039×10^{-9}	273–1000
Ethanol	C_2H_6O	19.9	20.96×10^{-2}	-10.38×10^{-5}	20.05×10^{-9}	273–1500
Hydrogen chloride	HCl	30.33	-0.7620×10^{-2}	1.327×10^{-5}	-4.338×10^{-9}	273–1500
Methane	CH_4	19.89	5.024×10^{-2}	1.269×10^{-5}	-11.01×10^{-9}	273–1500
Ethane	C_2H_6	6.900	17.27×10^{-2}	-6.406×10^{-5}	7.285×10^{-9}	273–1500
Propane	C_3H_8	-4.04	30.48×10^{-2}	-15.72×10^{-5}	31.74×10^{-9}	273–1500
n-Butane	C_4H_{10}	3.96	37.15×10^{-2}	-18.34×10^{-5}	35.00×10^{-9}	273–1500
i-Butane	C_4H_{10}	-7.913	41.60×10^{-2}	-23.01×10^{-5}	49.91×10^{-9}	273–1500
n-Pentane	C_5H_{12}	6.774	45.43×10^{-2}	-22.46×10^{-5}	42.29×10^{-9}	273–1500
n-Hexane	C_6H_{14}	6.938	55.22×10^{-2}	-28.65×10^{-5}	57.69×10^{-9}	273–1500
Ethylene	C_2H_4	3.95	15.64×10^{-2}	-8.344×10^{-5}	17.67×10^{-9}	273–1500
Propylene	C_3H_6	3.15	23.83×10^{-2}	-12.18×10^{-5}	24.62×10^{-9}	273–1500

Source: B. G. Kyle, *Chemical and Process Thermodynamics*, Prentice-Hall, Englewood Cliffs, NJ, 1984. Used with permission.

Properties of Substances-Tables

TABLE A-7E EMPIRICAL c_p EQUATIONS FOR VARIOUS GASES AT ZERO PRESSURE (ENGLISH UNITS)

$$c_p = a + bT + cT^2 + dT^3$$
$$T \text{ in } °R, c_p \text{ in Btu/lbmole·}°R$$

Substance	Formula	a	b	c	d	Temperature range R
Nitrogen	N_2	6.903	$-0.020\,85 \times 10^{-2}$	$0.059\,57 \times 10^{-5}$	-0.1176×10^{-9}	491–3240
Oxygen	O_2	6.085	0.2017×10^{-2}	$-0.052\,75 \times 10^{-5}$	$0.053\,72 \times 10^{-9}$	491–3240
Air	—	6.713	$0.026\,09 \times 10^{-2}$	$0.035\,40 \times 10^{-5}$	$-0.080\,52 \times 10^{-9}$	491–3240
Hydrogen	H_2	6.952	$-0.025\,42 \times 10^{-2}$	$0.029\,52 \times 10^{-5}$	$-0.035\,65 \times 10^{-9}$	491–3240
Carbon monoxide	CO	6.726	$0.022\,22 \times 10^{-2}$	$0.039\,60 \times 10^{-5}$	$-0.091\,00 \times 10^{-9}$	491–3240
Carbon dioxide	CO_2	5.316	$0.793\,61 \times 10^{-2}$	-0.2581×10^{-5}	0.3059×10^{-9}	491–3240
Water vapor	H_2O	7.700	$0.025\,52 \times 10^{-2}$	$0.077\,81 \times 10^{-5}$	-0.1472×10^{-9}	491–3240
Nitric oxide	NO	7.008	$-0.012\,47 \times 10^{-2}$	$0.071\,85 \times 10^{-5}$	-0.1715×10^{-9}	491–2700
Nitrous oxide	N_2O	5.758	0.7780×10^{-2}	-0.2596×10^{-5}	0.4331×10^{-9}	491–2700
Nitrogen dioxide	NO_2	5.48	0.7583×10^{-2}	-0.260×10^{-5}	0.322×10^{-9}	491–2700
Ammonia	NH_3	6.5846	$0.340\,28 \times 10^{-2}$	$0.073\,034 \times 10^{-5}$	$-0.274\,02 \times 10^{-9}$	491–2700
Sulfur	S_2	6.499	0.2943×10^{-2}	-0.1200×10^{-5}	0.1632×10^{-9}	491–3240
Sulfur dioxide	SO_2	6.157	0.7689×10^{-2}	-0.2810×10^{-5}	0.3527×10^{-9}	491–3240
Sulfur trioxide	SO_3	3.918	1.935×10^{-2}	-0.8256×10^{-5}	1.328×10^{-9}	491–2340
Acetylene	C_2H_2	5.21	1.2227×10^{-2}	-0.4812×10^{-5}	0.7457×10^{-9}	491–2700
Benzene	C_6H_6	−8.650	6.4322×10^{-2}	-2.327×10^{-5}	3.179×10^{-9}	491–2700
Methanol	CH_4O	4.55	1.214×10^{-2}	-0.0898×10^{-5}	-0.329×10^{-9}	491–1800
Ethanol	C_2H_6O	4.75	2.781×10^{-2}	-0.7651×10^{-5}	0.821×10^{-9}	491–2700
Hydrogen chloride	HCl	7.244	-0.1011×10^{-2}	$0.097\,83 \times 10^{-5}$	-0.1776×10^{-9}	491–2740
Methane	CH_4	4.750	0.6666×10^{-2}	$0.093\,52 \times 10^{-5}$	-0.4510×10^{-9}	491–2740
Ethane	C_2H_6	1.648	2.291×10^{-2}	-0.4722×10^{-5}	0.2984×10^{-9}	491–2740
Propane	C_3H_8	−0.966	4.044×10^{-2}	-1.159×10^{-5}	1.300×10^{-9}	491–2740
n-Butane	C_4H_{10}	0.945	4.929×10^{-2}	-1.352×10^{-5}	1.433×10^{-9}	491–2740
i-Butane	C_4H_{10}	−1.890	5.520×10^{-2}	-1.696×10^{-5}	2.044×10^{-9}	491–2740
n-Pentane	C_5H_{12}	1.618	6.028×10^{-2}	-1.656×10^{-5}	1.732×10^{-9}	491–2740
n-Hexane	C_6H_{14}	1.657	7.328×10^{-2}	-2.112×10^{-5}	2.363×10^{-9}	491–2740
Ethylene	C_2H_4	0.944	2.075×10^{-2}	-0.6151×10^{-5}	0.7326×10^{-9}	491–2740
Propylene	C_3H_6	0.753	3.162×10^{-2}	-0.8981×10^{-5}	1.008×10^{-9}	491–2740

Source: B. G. Kyle, *Chemical and Process Thermodynamics*. Prentice-Hall, Englewood Cliffs, NJ, 1984. Used with permission.

Appendix 3

c_p in kJ/kg·°C

A. Liquids

Substance	State	c_p	Substance	State	c_p
Water	1 atm, 273°K	4.217	Benzene	1 atm, 15°C	1.80
	1 atm, 280°K	4.198		1 atm, 65°C	1.92
	1 atm, 300°K	4.179	Glycerin	1 atm, 10°C	2.32
	1 atm, 320°K	4.180		1 atm, 50°C	2.58
	1 atm, 340°K	4.188	Mercury	1 atm, 10°C	0.138
	1 atm, 360°K	4.203		1 atm, 315°C	0.134
	1 atm, 373°K	4.218	Sodium	1 atm, 95°C	1.38
Ammonia	sat., −20°C	4.52		1 atm, 540°C	1.26
	sat., 50°C	5.10	Propane	1 atm, 0°C	2.41
Refrigerant 12	sat., −40°C	0.883	Bismuth	1 atm, 425°C	0.144
	sat., −20°C	0.908		1 atm, 760°C	0.164
	sat., 50°C	1.02	Ethyl alcohol	1 atm, 25°C	2.43

B. Solids

Substance	Temp.	c_p	Substance	Temp.	c_p
Ice	200°K	1.56	Silver	20°C	0.233
	220°K	1.71		200°C	0.243
	240°K	1.86	Lead	−173°C	0.118
	260°K	2.01		−50°C	0.126
	270°K	2.08		27°C	0.129
	273°K	2.11		100°C	0.131
Aluminum	200°K	0.797		200°C	0.136
	250°K	0.859	Copper	−173°C	0.254
	300°K	0.902		−100°C	0.342
	350°K	0.929		−50°C	0.367
	400°K	0.949		0°C	0.381
	450°K	0.973		27°C	0.386
	500°K	0.997		100°C	0.393
Iron	20°K	0.448		200°C	0.403

TABLE A-8E SPECIFIC HEATS OF SOME COMMON LIQUIDS AND SOLIDS (ENGLISH UNITS)

$$c_p \text{ in Btu/lbm·°F}$$

A. Liquids

Substance	State	c_p	Substance	State	c_p
Water	1 atm, 32°F	1.007	Glycerin	1 atm, 50°F	0.554
	1 atm, 77°F	0.998		1 atm, 120°F	0.617
	1 atm, 212°F	1.007	Bismuth	1 atm, 800°F	0.0345
Ammonia	sat., 0°F	1.08		1atm, 1400°F	0.0393
	sat., 120°F	1.22	Mercury	1 atm, 50°F	0.033
Refrigerant 12	sat., −40°F	0.211		1 atm, 600°F	0.032
	sat., 0°F	0.217	Sodium	1 atm, 200°F	0.33
	sat., 120°F	0.244		1 atm, 1000°F	0.30
Benzene	1 atm, 60°F	0.43	Propane	1 atm, 32°F	0.576
	1 atm, 150°F	0.46			

B. Solids

Substance	T, °F	c_p	Substance	T, °F	c_p
Ice	−100	0.375	Lead	−455	0.0008
	−50	0.424		−435	0.0073
	0	0.471		−150	0.0283
	20	0.491		32	0.0297
	32	0.502		210	0.0320
Aluminum	−150	0.167		570	0.0356
	−100	0.192	Copper	−240	0.0674
	32	0.212		−150	0.0784
	100	0.218		−60	0.0862
	200	0.224		0	0.0893
	300	0.229		100	0.0925
	400	0.235		200	0.0938
	500	0.240		390	0.0963
Iron	68	0.107	Silver	68	0.0558

T in K, *t* in °C, *h* and *u* in kJ/kg, and ϕ in kJ/kg·K

T	t	h	p_r	u	v_r	φ	T	t	h	p_r	u	v_r	φ
200	−73.15	200.13	.33468	142.72	171.52	5.2950	700	426.85	713.51	28.679	512.59	7.0058	6.5725
210	−63.15	210.15	.39684	149.88	151.89	5.3439	710	436.85	724.27	30.245	520.47	6.7380	6.5877
220	−53.15	220.18	.46684	157.03	135.26	5.3905	720	446.85	735.05	31.876	528.39	6.4832	6.6028
230	−43.15	230.20	.54524	164.18	121.08	5.4350	730	456.85	745.85	33.575	536.31	6.2407	6.6177
240	−33.15	240.22	.63263	171.34	108.89	5.4777	740	466.85	756.68	35.344	544.28	6.0097	6.6324
250	−23.15	250.25	.7296	178.49	98.353	5.5186	750	476.85	767.53	37.184	552.26	5.7894	6.6470
260	−13.15	260.28	.8368	185.65	89.188	5.5580	760	486.85	778.41	39.098	560.27	5.5795	6.6614
270	−3.15	270.31	.9547	192.81	81.173	5.5958	770	496.85	789.31	41.088	568.30	5.3791	6.6757
280	6.85	280.35	1.0842	199.98	74.129	5.6323	780	506.85	800.23	43.156	576.35	5.1878	6.6898
290	16.85	290.39	1.2258	207.15	67.909	5.6676	790	516.85	811.18	45.304	584.43	5.0052	6.7037
300	26.85	300.43	1.3801	214.32	62.393	5.7016	800	526.85	822.15	47.535	592.53	4.8306	6.7175
310	36.85	310.48	1.5480	221.50	57.481	5.7346	810	536.85	833.15	49.852	600.65	4.6637	6.7312
320	46.85	320.53	1.7301	228.68	53.091	5.7665	820	546.85	844.16	52.256	608.80	4.5041	6.7447
330	56.85	330.59	1.9271	235.87	49.152	5.7974	830	556.85	855.20	54.749	616.97	4.3514	6.7581
340	66.85	340.66	2.1398	243.07	45.608	5.8275	840	566.85	866.26	57.336	625.16	4.2052	6.7713
350	76.85	350.73	2.3689	250.27	42.407	5.8567	850	576.85	877.35	60.017	633.37	4.0652	6.7844
360	86.85	360.81	2.6154	257.48	39.509	5.8851	860	586.85	888.45	62.795	641.61	3.9310	6.7974
370	96.85	370.91	2.8799	264.71	36.877	5.9127	870	596.85	899.58	65.674	649.87	3.8024	6.8103
380	106.85	381.01	3.1633	271.94	34.481	5.9397	880	606.85	910.73	68.655	658.15	3.6791	6.8230
390	116.85	391.12	3.4664	279.18	32.293	5.9660	890	616.85	921.90	71.742	666.45	3.5608	6.8356
400	126.85	401.25	3.7902	286.43	30.292	5.9916	900	626.85	933.10	74.937	674.77	3.4473	6.8482
410	136.85	411.38	4.1356	293.70	28.456	6.0166	910	636.85	944.31	78.243	683.11	3.3383	6.8605
420	146.85	421.54	4.5035	300.98	26.769	6.0411	920	646.85	955.55	81.663	691.48	3.2336	6.8728
430	156.85	431.70	4.8949	308.28	25.215	6.0650	930	656.85	966.80	85.200	699.87	3.1331	6.8850
440	166.85	441.88	5.3106	315.59	23.781	6.0884	940	666.85	978.08	88.856	708.27	3.0365	6.8971
450	176.85	452.07	5.7519	322.91	22.456	6.1113	950	676.85	989.38	92.63	716.70	2.9436	6.9090
460	186.85	462.28	6.2197	330.25	21.228	6.1338	960	686.85	1000.69	96.54	725.14	2.8543	6.9209
470	196.85	472.51	6.7150	337.61	20.090	6.1557	970	696.85	1012.03	100.57	733.61	2.7683	6.9326
480	206.85	482.76	7.2391	344.98	19.032	6.1773	980	706.85	1023.39	104.74	742.10	2.6857	6.9443
490	216.85	493.02	7.7930	352.38	18.048	6.1985	990	716.85	1034.76	109.04	750.60	2.6061	6.9558
500	226.85	503.30	8.378	359.79	17.130	6.2193	1000	726.85	1046.16	113.48	759.13	2.5294	6.9673
510	236.85	513.60	8.995	367.22	16.274	6.2396	1010	736.85	1057.57	118.06	767.67	2.4556	6.9786
520	246.85	523.93	9.645	374.67	15.474	6.2597	1020	746.85	1069.01	122.78	776.23	2.3845	6.9899
530	256.85	534.27	10.331	338.14	14.726	6.2794	1030	756.85	1080.46	127.65	784.81	2.3160	7.0010
540	266.85	544.63	11.052	389.63	14.025	6.2988	1040	766.85	1091.93	132.68	793.41	2.2499	7.0121
550	276.85	555.01	11.810	397.15	13.367	6.3178	1050	776.85	1103.41	137.86	802.03	2.1862	7.0231
560	286.85	565.42	12.608	404.68	12.749	6.3366	1060	786.85	1114.92	143.20	810.66	2.1247	7.0340
570	296.85	575.84	13.445	412.24	12.169	6.3550	1070	796.85	1126.44	148.70	819.32	2.0654	7.0448
580	306.85	586.29	14.324	419.82	11.623	6.3732	1080	806.85	1137.98	154.37	827.99	2.0082	7.0556
590	316.85	596.77	15.245	427.42	11.108	6.3911	1090	816.85	1149.54	160.20	836.67	1.9529	7.0661
600	326.85	607.26	16.212	435.04	10.623	6.4087	1100	826.85	1161.11	166.21	845.38	1.8996	7.0768
610	336.85	617.78	17.224	442.69	10.165	6.4261	1110	836.85	1172.70	172.40	854.10	1.8481	7.0873
620	346.85	628.32	18.284	450.36	9.733	6.4433	1120	846.85	1184.31	178.76	862.83	1.7983	7.0977
630	356.85	638.89	19.393	458.06	9.324	6.4602	1130	856.85	1195.93	185.32	871.59	1.7502	7.1080
640	366.85	649.47	20.553	465.77	8.938	6.4768	1140	866.85	1207.57	192.06	880.35	1.7038	7.1183
650	376.85	660.09	21.766	473.52	8.5718	6.4933	1150	876.85	1219.23	198.99	889.14	1.6588	7.1285
660	386.85	670.72	23.033	481.28	8.2249	6.5095	1160	886.85	1230.90	206.12	897.94	1.6154	7.1386
670	396.85	681.38	24.356	489.07	7.8960	6.5256	1170	896.85	1242.58	213.45	906.76	1.5733	7.1486
680	406.85	692.07	25.736	496.89	7.5839	6.5414	1180	906.85	1254.28	220.99	915.59	1.5327	7.1586
690	416.85	702.78	27.177	504.73	7.2875	6.5570	1190	916.85	1266.00	228.73	924.43	1.4933	7.1684

T in K, t in °C, h and u in kJ/kg, and φ in kJ/kg·K

T	t	h	p_r	u	v_r	φ	T	t	h	p_r	u	v_r	φ
1200	926.85	1277.73	236.69	933.29	1.4552	7.1783	1700	1426.85	1879.58	1017.9	1391.63	.47937	7.5970
1210	936.85	1289.48	244.86	942.17	1.4184	7.1880	1725	1451.85	1910.31	1083.6	1415.18	.45694	7.6149
1220	946.85	1301.24	253.26	951.06	1.3827	7.1977	1750	1476.85	1941.09	1152.6	1438.78	.43582	7.6326
1230	956.85	1313.01	261.89	959.96	1.3481	7.2073	1775	1501.85	1971.91	1225.0	1462.43	.41592	7.6501
1240	966.85	1324.80	270.74	968.88	1.3146	7.2168	1800	1526.85	2002.78	1300.9	1486.12	.39714	7.6674
1250	976.85	1336.60	279.83	977.81	1.2822	7.2263	1825	1551.85	2033.69	1380.6	1509.86	.37943	7.6844
1260	986.85	1348.42	289.16	986.76	1.2507	7.2357	1850	1576.85	2064.65	1464.0	1533.65	.36270	7.7013
1270	996.85	1360.25	298.74	995.72	1.2202	7.2451	1875	1601.85	2095.66	1551.5	1557.47	.34689	7.7179
1280	1006.85	1372.09	308.57	1004.69	1.1907	7.2544	1900	1626.85	2126.70	1643.0	1581.34	.33194	7.7344
1290	1016.85	1383.95	318.65	1013.68	1.1620	7.2636	1925	1651.85	2157.79	1738.7	1605.25	.31779	7.7506
1300	1026.85	1395.81	328.98	1022.67	1.1342	7.2728	1950	1676.85	2188.91	1838.8	1629.20	.30439	7.7667
1310	1036.85	1407.70	339.59	1031.69	1.1073	7.2819	1975	1701.85	2220.08	1943.4	1653.19	.29170	7.7826
1320	1046.85	1419.59	350.46	1040.71	1.0811	7.2909	2000	1726.85	2251.28	2052.6	1677.22	.27967	7.7983
1330	1056.85	1431.50	361.61	1049.75	1.0557	7.2999	2025	1751.85	2282.52	2166.7	1701.28	.26826	7.8138
1340	1066.85	1443.42	373.03	1058.79	1.0311	7.3088	2050	1776.85	2313.80	2285.7	1725.38	.25743	7.8292
1350	1076.85	1455.35	384.74	1067.86	1.00716	7.3177	2075	1801.85	2345.11	2409.9	1749.52	.24714	7.8443
1360	1086.85	1467.29	396.74	1076.93	.98393	7.3265	2100	1826.85	2376.46	2539.3	1773.70	.23737	7.8594
1370	1096.85	1479.25	409.03	1086.01	.96138	7.3353	2125	1851.85	2407.85	2674.2	1797.90	.22808	7.8742
1380	1106.85	1491.21	421.62	1095.11	.93947	7.3440	2150	1876.85	2439.26	2814.7	1822.15	.21924	7.8889
1390	1116.85	1503.19	434.52	1104.22	.91819	7.3526	2175	1901.85	2470.71	2961.0	1846.42	.21084	7.9035
1400	1126.85	1515.18	447.73	1113.34	.89752	7.3612	2200	1926.85	2502.20	3113.3	1870.73	.20283	7.9179
1410	1136.85	1527.18	461.25	1122.47	.87742	7.3698	2225	1951.85	2533.71	3271.7	1895.07	.19520	7.9321
1420	1146.85	1539.20	475.10	1131.61	.85789	7.3783	2250	1976.85	2565.26	3436.4	1919.44	.18793	7.9462
1430	1156.85	1551.22	489.27	1140.77	.83891	7.3867	2275	2001.85	2596.84	3607.7	1943.84	.18100	7.9602
1440	1166.85	1563.26	503.78	1149.93	.82045	7.3951	2300	2026.85	2628.45	3785.6	1968.27	.17439	7.9740
1450	1176.85	1575.30	518.63	1159.11	.80250	7.4034	2325	2051.85	2660.08	3970.4	1992.74	.16808	7.9877
1460	1186.85	1587.36	533.81	1168.29	.78504	7.4117	2350	2076.85	2691.75	4162.3	2017.23	.16206	8.0012
1470	1196.85	1599.42	549.35	1177.49	.76806	7.4199	2375	2101.85	2723.45	4361.4	2041.75	.15630	8.0146
1480	1206.85	1611.50	565.25	1186.69	.75154	7.4281	2400	2126.85	2755.17	4568.1	2066.29	.15080	8.0279
1490	1216.85	1623.59	581.51	1195.91	.73546	7.4363	2425	2151.85	2786.92	4782.4	2090.87	.14554	8.0411
1500	1226.85	1635.68	598.14	1205.14	.71982	7.4444	2450	2176.85	2818.70	5004.7	2115.47	.14051	8.0541
1510	1236.85	1647.79	615.14	1214.37	.70459	7.4524	2475	2201.85	2850.50	5235.0	2140.10	.13570	8.0670
1520	1246.85	1659.91	632.52	1223.62	.68976	7.4604	2500	2226.85	2882.34	5473.7	2164.76	.13110	8.0798
1530	1256.85	1672.03	650.29	1232.88	.67533	7.4684	2525	2251.85	2914.19	5720.9	2189.44	.12669	8.0925
1540	1266.85	1684.17	668.45	1242.14	.66128	7.4763	2550	2276.85	2946.07	5976.9	2214.15	.12246	8.1051
1550	1276.85	1696.32	687.01	1251.42	.64759	7.4841	2575	2301.85	2977.98	6241.9	2238.88	.11841	8.1175
1560	1286.85	1708.47	705.98	1260.70	.63425	7.4919	2600	2326.85	3009.91	6516.1	2263.63	.11453	8.1299
1570	1296.85	1720.64	725.36	1270.00	.62127	7.4997	2625	2351.85	3041.87	6799.8	2288.41	.11081	8.1421
1580	1306.85	1732.81	745.16	1279.30	.60861	7.5074	2650	2376.85	3073.85	7093.2	2313.22	.10723	8.1542
1590	1316.85	1744.99	765.38	1288.61	.59628	7.5151	2675	2401.85	3105.85	7396.5	2338.05	.10381	8.1662
1600	1326.85	1757.19	786.04	1297.94	.58426	7.5228	2700	2426.85	3137.88	7710.1	2362.90	.10052	8.1781
1610	1336.85	1769.39	807.13	1307.27	.57255	7.5304	2725	2451.85	3169.93	8034.1	2387.77	.09736	8.1900
1620	1346.85	1781.60	828.67	1316.61	.56113	7.5379	2750	2476.85	3202.00	8368.8	2412.67	.09432	8.2017
1630	1356.85	1793.82	850.67	1325.95	.54999	7.5455	2775	2501.85	3234.09	8714.5	2437.58	.09140	8.2133
1640	1366.85	1806.04	873.12	1335.31	.53913	7.5529	2800	2526.85	3266.21	9071.4	2462.52	.08860	8.2248
1650	1376.85	1818.28	896.05	1344.68	.52855	7.5604	2825	2551.85	3298.35	9439.8	2487.48	.08590	8.2362
1660	1386.85	1830.52	929.44	1354.05	.51822	7.5678	2850	2576.85	3330.50	9820	2512.46	.08330	8.2476
1670	1396.85	1842.78	943.32	1363.43	.50814	7.5751	2875	2601.85	3362.68	10212	2537.47	.08081	8.2588
1680	1406.85	1855.04	967.69	1372.82	.49832	7.5825	2900	2626.85	3394.88	10617	2562.49	.07840	8.2700
1690	1416.85	1867.30	992.55	1382.22	.48873	7.5897	2925	2651.85	3427.10	11034	2587.53	.07609	8.2810

Source: Adapted by permission from J. H. Keenan, J. Chao, and J. Kaye, *Gas Tables*, 2nd ed., SI Units, John Wiley, New York, 1983.

TABLE A-9E IDEAL-GAS PROPERTIES OF AIR (ENGLISH UNITS)

T in °R, t in °F, h and u in Btu/lbm, and ϕ in Btu/lbm·°R

T	t	h	p_r	u	v_r	ϕ	T	t	h	p_r	u	v_r	ϕ
400	−59.67	95.62	.4835	68.19	306.5	.52875	1400	940.33	342.99	42.69	247.02	12.150	.83592
420	−39.67	100.40	.5733	71.61	271.4	.54043	1420	960.33	348.22	45.06	250.87	11.675	.83963
440	−19.67	105.19	.6745	75.03	241.7	.55157	1440	980.33	353.46	47.54	254.74	11.223	.84329
460	.33	109.98	.7878	78.45	216.31	.56222	1460	1000.33	358.71	50.11	258.62	10.793	.84691
480	20.33	114.78	.9142	81.87	194.52	.57241	1480	1020.33	363.98	52.80	262.51	10.384	.85050
500	40.33	119.57	1.0544	85.29	175.68	.58220	1500	1040.33	369.25	55.60	266.42	9.995	.85404
520	60.33	124.36	1.2094	88.71	159.29	.59160	1520	1060.33	374.54	58.52	270.34	9.623	.85754
540	80.33	129.16	1.3801	92.14	144.96	.60065	1540	1080.33	379.84	61.55	274.27	9.270	.86100
560	100.33	133.96	1.5675	95.57	132.35	.60938	1560	1100.33	385.15	64.70	278.21	8.932	.86443
580	120.33	138.76	1.7725	99.00	121.23	.61781	1580	1120.33	390.48	67.98	282.16	8.610	.86782
600	140.33	143.57	1.996	102.44	111.35	.62595	1600	1140.33	395.81	71.39	286.12	8.303	.87118
620	160.33	148.38	2.240	105.88	102.56	.63384	1620	1160.33	401.16	74.94	290.10	8.009	.87450
640	180.33	153.20	2.504	109.32	94.70	.64148	1640	1180.33	406.52	78.62	294.09	7.728	.87778
660	200.33	158.01	2.790	112.77	87.65	.64890	1660	1200.33	411.89	82.44	298.08	7.460	.88104
680	220.33	162.84	3.099	116.22	81.30	.65610	1680	1220.33	417.27	86.41	302.09	7.203	.88426
700	240.33	167.67	3.432	119.68	75.57	.66310	1700	1240.33	422.66	90.52	306.11	6.958	.88745
720	260.33	172.51	3.790	123.14	70.38	.66991	1720	1260.33	428.06	94.79	310.14	6.723	.89061
740	280.33	177.35	4.175	126.62	65.66	.67655	1740	1280.33	433.47	99.21	314.18	6.497	.89374
760	300.33	182.20	4.588	130.10	61.36	.68301	1760	1300.33	438.89	103.80	318.23	6.282	.89684
780	320.33	187.06	5.031	133.58	57.44	.68932	1780	1320.33	444.32	108.29	322.29	6.075	.89990
800	340.33	192.92	5.504	137.08	53.85	.69548	1800	1340.33	449.77	113.48	326.37	5.877	.90294
820	360.33	196.79	6.008	140.58	50.56	.70150	1820	1360.33	455.22	118.57	330.45	5.686	.90596
840	380.33	201.68	6.547	144.09	47.53	.70738	1840	1380.33	460.68	123.85	334.54	5.504	.90894
860	400.33	206.57	7.120	147.61	44.75	.71314	1860	1400.33	466.16	129.31	338.64	5.329	.91190
880	420.33	211.47	7.730	151.14	42.18	.71877	1880	1420.33	471.64	134.96	342.96	5.161	.91483
900	440.33	216.38	8.378	154.68	39.80	.72429	1900	1440.33	477.13	140.81	346.87	4.999	.91774
920	460.33	221.30	9.066	158.23	37.60	.72970	1920	1460.33	482.63	146.85	351.00	4.844	.92062
940	480.33	226.23	9.795	161.79	35.55	.73500	1940	1480.33	488.14	153.09	355.14	4.695	.92347
960	500.33	231.18	10.567	165.36	33.66	.74020	1960	1500.33	493.66	159.54	359.29	4.551	.92630
980	520.33	236.13	11.384	168.95	31.89	.74531	1980	1520.33	499.19	166.21	363.45	4.413	.92911
1000	540.33	241.10	12.248	172.54	30.25	.75032	2000	1540.33	504.72	173.10	367.61	4.281	.93189
1020	560.33	246.07	13.161	176.15	28.71	.75525	2020	1560.33	510.27	180.20	371.79	4.153	.93465
1040	580.33	251.06	14.125	179.76	27.28	.76010	2040	1580.33	515.82	187.54	375.97	4.030	.93739
1060	600.33	256.06	15.141	183.39	25.94	.76486	2060	1600.33	521.39	195.11	380.16	3.911	.94010
1080	620.33	261.08	16.212	187.03	24.68	.76954	2080	1620.33	526.96	202.93	384.36	3.797	.94279
1100	640.33	266.10	17.339	190.69	23.50	.77415	2100	1640.33	532.54	211.0	388.57	3.687	.94546
1120	660.33	271.14	18.526	194.35	22.40	.77869	2120	1660.33	538.13	219.3	392.79	3.582	.94811
1140	680.33	276.19	19.774	198.03	21.36	.78316	2140	1680.33	543.72	227.9	397.01	3.479	.95074
1160	700.33	281.25	21.09	201.72	20.381	.78756	2160	1700.33	549.32	236.7	401.24	3.381	.95334
1180	720.33	286.33	22.46	205.43	19.462	.79190	2180	1720.33	554.94	245.8	405.48	3.286	.95593
1200	740.33	291.41	23.91	209.15	18.595	.79618	2200	1740.33	560.56	255.2	409.73	3.194	.95850
1220	760.33	296.51	25.42	212.88	17.777	.80039	2220	1760.33	566.18	264.8	413.99	3.106	.96104
1240	780.33	301.63	27.01	216.62	17.006	.80455	2240	1780.33	571.82	274.8	418.25	3.020	.96357
1260	800.33	306.75	28.68	220.37	16.276	.80865	2260	1800.33	577.46	285.0	422.52	2.938	.96607
1280	820.33	311.89	30.42	244.14	15.587	.81270	2280	1820.33	583.10	295.5	426.80	2.858	.96856
1300	840.33	317.04	32.25	227.92	14.935	.81669	2300	1840.33	588.76	306.4	431.08	2.781	.97103
1320	860.33	322.21	34.16	231.72	14.317	.82063	2320	1860.33	594.42	317.5	435.37	2.707	.97348
1340	880.33	327.39	36.15	235.52	13.732	.82453	2340	1880.33	600.09	329.0	439.67	2.635	.97592
1360	900.33	332.58	38.24	239.34	13.177	.82837	2360	1900.33	605.77	340.8	443.98	2.566	.97833
1380	920.33	337.78	40.42	243.17	12.650	.83217	2380	1920.33	611.45	352.9	448.29	2.498	.98073

Properties of Substances-Tables

T in °R, t in °F, h and u in Btu/lbm, and ϕ in Btu/lbm·°R

T	t	h	p_r	u	v_r	ϕ	T	t	h	p_r	u	v_r	ϕ
2400	1940.33	617.14	365.4	452.61	2.433	.98311	3400	2940.33	908.38	1601.8	675.29	.7864	1.08443
2420	1960.33	622.84	378.2	456.93	2.371	.98547	3450	2990.33	923.22	1706.3	686.70	.7491	1.08877
2440	1980.33	628.54	391.4	461.26	2.310	.98782	3500	3040.33	938.09	1816.1	698.14	.7140	1.09304
2460	2000.33	634.25	404.9	465.60	2.251	.99015	3550	3090.33	952.97	1931.5	709.60	.6809	1.09727
2480	2020.33	639.96	418.8	469.94	2.194	.99247	3600	3140.33	967.88	2052.6	721.07	.6498	1.10143
2500	2040.33	645.68	433.1	474.29	2.1386	.99476	3650	3190.33	982.80	2179.7	732.57	.6204	1.10555
2520	2060.33	651.41	447.7	478.65	2.0852	.99704	3700	3240.33	997.74	2312.9	744.09	.5927	1.10962
2540	2080.33	657.14	462.8	483.01	2.0334	.99931	3750	3290.33	1012.71	2452.4	755.62	.5665	1.11363
2560	2100.33	662.88	478.2	487.38	1.9832	1.00156	3800	3340.33	1027.69	2598.6	767.18	.5418	1.11760
2580	2120.33	668.63	494.1	491.75	1.9346	1.00380	3850	3390.33	1042.69	2742.6	778.75	.5184	1.12153
2600	2140.33	674.38	510.3	496.13	1.8875	1.00602	3900	3440.33	1057.71	2911.6	790.34	.4962	1.12540
2620	2160.33	680.14	527.0	500.52	1.8418	1.00822	3950	3490.33	1072.74	3078.9	801.94	.4753	1.12923
2640	2180.33	685.90	544.1	504.91	1.7974	1.01041	4000	3540.33	1087.79	3254	813.57	.4554	1.13302
2660	2200.33	691.67	561.7	509.31	1.7545	1.01259	4050	3590.33	1102.86	3436	825.21	.4366	1.13676
2680	2220.33	697.44	579.7	513.71	1.7128	1.01475	4100	3640.33	1117.95	3627	836.87	.4188	1.14046
2700	2240.33	703.22	598.1	518.12	1.6723	1.01690	4150	3690.33	1133.05	3826	848.54	.4018	1.14412
2720	2260.33	709.00	617.0	522.53	1.6331	1.01903	4200	3740.33	1148.17	4034	860.23	.3858	1.14775
2740	2280.33	714.79	636.4	526.95	1.5950	1.02116	4250	3790.33	1163.30	4250	871.93	.3705	1.15133
2760	2300.33	720.58	656.3	531.37	1.5580	1.02326	4300	3840.33	1178.45	4475	883.65	.3560	1.15487
2780	2320.33	726.38	676.6	535.80	1.5221	1.02536	4350	3890.33	1193.61	4710	895.39	.3421	1.15838
2800	2340.33	732.19	697.5	540.23	1.4872	1.02744	4400	3940.33	1208.78	4955	907.14	.3290	1.16185
2820	2360.33	738.00	718.9	544.67	1.4533	1.02950	4450	3990.33	1223.98	5209	918.90	.3165	1.16528
2840	2380.33	743.81	740.7	549.11	1.4204	1.03156	4500	4040.33	1239.18	5474	930.68	.3046	1.16868
2860	2400.33	749.63	763.1	553.56	1.3885	1.03360	4550	4090.33	1254.40	5749	942.47	.2932	1.17204
2880	2420.33	755.45	786.0	558.01	1.3574	1.03563	4600	4140.33	1269.63	6035	954.27	.2824	1.17537
2900	2440.33	761.28	809.5	562.47	1.3272	1.03765	4650	4190.33	1284.88	6332	966.09	.2721	1.17867
2920	2460.33	767.12	833.5	566.93	1.2978	1.03965	4700	4240.33	1300.13	6641	977.92	.2622	1.18193
2940	2480.33	772.95	858.1	571.40	1.2693	1.04164	4750	4290.33	1315.41	6962	989.76	.2528	1.18516
2960	2500.33	778.80	883.3	575.87	1.2416	1.04362	4800	4340.33	1330.69	7294	1001.62	.2438	1.18836
2980	2520.33	784.64	909.0	580.35	1.2146	1.04559	4850	4390.33	1345.98	7640	1013.49	.2352	1.19153
3000	2540.33	790.49	935.3	584.83	1.1883	1.04755	4900	4440.33	1361.29	7998	1025.37	.2270	1.19467
3020	2560.33	796.35	962.2	589.31	1.1628	1.04949	4950	4490.33	1376.61	8369	1037.26	.2191	1.19778
3040	2580.33	802.21	989.8	593.80	1.1379	1.05143	5000	4540.33	1391.94	8754	1049.16	.21161	1.20086
3060	2600.33	808.07	1017.9	598.29	1.1137	1.05335	5050	4590.33	1407.29	9152	1061.08	.20442	1.20392
3080	2620.33	813.94	1046.7	602.79	1.0902	1.05526	5100	4640.33	1422.64	9565	1073.00	.19753	1.20694
3100	2640.33	819.82	1076.1	607.29	1.0672	1.05716	5150	4690.33	1438.01	9993	1084.94	.19093	1.20994
3120	2660.33	825.69	1106.2	611.80	1.0449	1.05905	5200	4740.33	1453.38	10435	1096.89	.18461	1.21291
3140	2680.33	831.57	1136.9	616.31	1.0232	1.06093	5250	4790.33	1468.77	10893	1108.85	.17855	1.21586
3160	2700.33	837.46	1168.3	620.82	1.0020	1.06280	5300	4840.33	1484.17	11367	1120.82	.17273	1.21878
3180	2720.33	843.35	1200.4	625.34	.9814	1.06466	5350	4890.33	1499.57	11857	1132.80	.16716	1.22167
3200	2740.33	849.24	1233.2	629.86	.9613	1.06651	5400	4940.33	1514.99	12364	1144.79	.16181	1.22454
3220	2760.33	855.14	1266.7	634.39	.9417	1.06834	5450	4990.33	1530.42	12888	1156.79	.15667	1.22738
3240	2780.33	862.04	1300.9	638.92	.9227	1.07017	5500	5040.33	1545.86	13429	1168.80	.15173	1.23020
3260	2800.33	866.94	1335.9	643.45	.9041	1.07199	5550	5090.33	1561.31	13988	1180.82	.14700	1.23300
3280	2820.33	872.85	1371.5	647.99	.8860	1.07379	5600	5140.33	1576.76	14565	1192.85	.14244	1.23577
3300	2840.33	878.77	1408.0	652.53	.8683	1.07559	5650	5190.33	1592.23	15161	1204.89	.13806	1.23852
3320	2860.33	884.68	1445.2	657.07	.8511	1.07738	5700	5240.33	1607.71	15776	1216.94	.13385	1.24125
3340	2880.33	890.60	1483.1	661.62	.8343	1.07916	5750	5290.33	1623.19	16411	1228.99	.12980	1.24395
3360	2900.33	896.52	1521.9	666.18	.8179	1.08092	5800	5340.33	1638.68	17066	1241.06	.12591	1.24663
3380	2920.33	902.45	1561.4	670.73	.8020	1.08268	5850	5390.33	1654.19	17742	1253.13	.12216	1.24930

Source: Abridged from Joseph H. Keenan, Jing Chao, and Joseph Kaye, *Gas Tables,* 2nd ed., English units, Wiley, New York, 1980.

TABLE A-10 IDEAL-GAS PROPERTIES OF NITROGEN (SI UNITS)

T in K, h and u in kJ/kgmole, and ϕ in kJ/kgmole·K

T	h	p_r	u	v_r	ϕ	T	h	p_r	u	v_r	ϕ
0	0										
100	2902.2	.0219	2070.8	37.998	159.667						
200	5812.8	.2477	4149.9	6.714	179.842	1600	50571.4	555.53	37268.4	.0239	243.993
298	8665.8	1.0014	6188.1	2.478	191.448	1700	54096.3	718.33	39961.8	.0197	246.129
300	8724.1	1.0243	6229.7	2.435	191.646	1800	57644.5	916.77	42678.6	.0163	248.158
400	11641.3	2.8103	8315.6	1.183	200.037	1900	61213.5	1156.24	45416.1	.0137	250.087
500	14581.0	6.1846	10423.8	.672	206.596	2000	64800.9	1442.65	48172.1	.0115	251.927
600	17564.1	11.894	12575.5	.419	212.033	2100	68404.6	1782.4	50944.4	.00980	253.685
700	20606.8	20.906	14786.7	.278	216.722	2200	72023.1	2182.4	53731.4	.00838	255.369
800	23716.2	34.442	17064.7	.193	220.873	2300	75654.7	2650.0	56531.6	.00722	256.983
900	26892.6	54.009	19409.7	.139	224.614	2400	79298.3	3193.3	59343.7	.00625	258.534
1000	30132.4	81.421	21818.0	.102	228.027	2500	82952.7	3820.9	62166.7	.00544	260.025
1100	33429.7	118.81	24283.9	.0770	231.169	2600	86617.0	4541.9	64999.5	.00476	261.463
1200	36778.3	168.67	26801.0	.0592	234.082	2700	90290.3	5366.0	67841.4	.00418	262.849
1300	40172.0	233.84	29363.3	.0462	236.798	2800	93971.8	6303.5	70691.5	.00369	264.188
1400	43605.3	317.55	31965.1	.0367	239.343	2900	97661.0	7365.5	73549.3	.00327	265.482
1500	47073.2	423.43	34601.6	.0295	241.735	3000	101357.3	8563.6	76414.1	.00291	266.735

Source for Tables A-10 to A-16: Adapted by permission from J. H. Keenan, J. Chao, and J. Kaye, *Gas Tables,* 2nd ed., John Wiley, New York, 1980 (English Units) and 1983 (SI Units).

TABLE A-11 IDEAL-GAS PROPERTIES OF OXYGEN (SI UNITS)

T in K, h and u in kJ/kgmole, and ϕ in kJ/kgmole·K

T	h	p_r	u	v_r	ϕ	T	h	p_r	u	v_r	ϕ
0	0										
100	2904.5	.1108	2073.1	7.506	173.152						
200	5815.7	1.2543	4152.8	1.326	193.330	1600	52946.2	3938.6	39643.2	.00338	260.278
298	8676.9	5.0951	6201.2	.487	204.975	1700	56636.7	5154.7	42502.2	.00274	262.515
300	8737.6	5.2127	6243.3	.479	205.174	1800	60350.6	6654.0	45384.7	.00225	264.638
400	11708.8	14.5639	8383.0	.228	213.717	1900	64087.5	8484.2	48290.1	.00186	266.658
500	14767.7	33.0826	10610.5	.126	220.538	2000	67846.8	10698.7	51218.0	.00155	268.586
600	17927.3	66.124	12938.7	.0754	226.296	2100	71628.4	13357	54168.1	.001307	270.431
700	21182.0	120.872	15361.9	.0482	231.312	2200	75431.8	16524	57140.2	.001107	272.201
800	24518.8	206.541	17867.3	.0322	235.766	2300	79256.9	20274	60133.7	.000943	273.901
900	27924.2	334.563	20441.2	.0224	239.776	2400	83103.1	24685	63148.5	.000808	275.538
1000	30386.2	518.777	23071.8	.0160	243.423	2500	86970.2	29846	66184.2	.000696	277.116
1100	34895.2	775.63	25749.3	.01179	246.768	2600	90857.6	35853	69240.2	.000603	278.641
1200	38443.8	1124.42	28466.5	.00887	249.855	2700	94764.9	42811	72316.1	.000524	280.116
1300	42026.6	1587.50	31217.9	.00681	252.723	2800	98691.6	50833	75411.3	.000458	281.544
1400	45639.6	2190.61	33999.4	.00531	255.400	2900	102637.0	60043	78525.2	.000402	282.928
1500	49280.1	2963.16	36808.5	.00421	257.912	3000	106600.4	70574	81657.2	.000353	284.272

TABLE A-12 IDEAL-GAS PROPERTIES OF WATER VAPOR (SI UNITS)

T in K, h and u in kJ/kgmole, and ϕ in kJ/kgmole·K

T	h	p_r	u	v_r	ϕ	T	h	p_r	u	v_r	ϕ
0	0										
100	3289.6	.0090	2458.1	92.477	152.272						
200	6621.8	.1446	4958.9	11.498	175.369	1600	62811.9	1758.5	49508.9	.00756	253.573
298	9899.0	.7195	7421.3	3.451	188.697	1700	67662.0	2504.5	53527.5	.00564	256.513
300	9966.1	.7384	7471.8	3.378	188.925	1800	72596.8	3515.9	57630.9	.00426	259.334
400	13355.6	2.3846	10029.8	1.395	198.672	1900	77609.5	4870.7	61812.1	.00324	262.044
500	16828.4	6.0530	12671.2	.687	206.417	2000	82693.7	6665.9	66064.9	.00249	264.652
600	20405.3	13.257	15416.6	.3763	212.935	2100	87843.6	9017	70383.3	.001936	267.164
700	24095.7	26.272	18275.6	.2215	218.622	2200	93053.8	12068	74762.1	.001516	269.588
800	27906.2	48.436	21254.7	.1373	223.708	2300	98319.5	15992	79196.4	.001196	271.928
900	31841.2	84.566	24358.3	.0885	228.342	2400	103636.5	20994	83682.0	.000950	274.191
1000	35904.1	141.489	27589.7	.0588	232.621	2500	109001.3	27320	88215.2	.000761	276.381
1100	40094.4	228.71	30948.6	.0400	236.614	2600	114410.3	35261	92792.8	.000613	278.502
1200	44410.1	359.24	34432.8	.0278	240.368	2700	119860.8	45158	97411.9	.000497	280.559
1300	48846.1	550.58	38037.4	.0196	243.918	2800	125349.8	57413	102069.4	.000405	282.556
1400	53396.6	825.94	41756.5	.0141	247.290	2900	130875.5	72493	106763.7	.000333	284.495
1500	58054.5	1215.57	45582.9	.0103	250.503	3000	136436.7	90942	111493.5	.000274	286.380

TABLE A-13 IDEAL-GAS PROPERTIES OF CARBON DIOXIDE (SI UNITS)

T in K, h and u in kJ/kgmole, and ϕ in kJ/kgmole·K

T	h	p_r	u	v_r	ϕ	T	h	p_r	u	v_r	ϕ
0	0										
100	2909.6	.0022	2078.1	376.315	178.893						
200	5952.3	.0275	4289.4	60.453	199.859	1600	76935.6	2846.9	63632.6	.00467	295.868
298	9359.7	.1448	6882.0	17.152	213.657	1700	82847.5	4380.8	68713.0	.00323	299.452
300	9433.7	.1490	6939.4	16.737	213.908	1800	88797.9	6594.4	73832.0	.00227	302.852
400	13368.5	.5795	10042.8	5.739	225.199	1900	94784.7	9739.8	78987.3	.00162	306.095
500	17671.7	1.8362	13514.5	2.264	234.788	2000	100805.5	14114.8	84176.7	.00118	309.179
600	22273.3	5.032	17284.6	.9914	243.170	2100	106853.6	20130	89393.4	.000867	312.131
700	27120.9	12.355	21300.8	.4711	250.639	2200	112929.0	28282	94637.3	.000647	314.958
800	32173.5	27.804	25522.0	.2392	257.382	2300	119024.8	39182	99901.7	.000488	317.668
900	37397.4	58.264	29914.5	.1284	263.533	2400	125143.3	53583	105188.7	.000372	320.271
1000	42764.6	114.994	34450.2	.0723	269.186	2500	131282.3	72474	110496.3	.000287	322.782
1100	48250.8	215.72	39105.0	.0424	274.417	2600	137437.1	96856	115819.7	.000223	325.193
1200	53840.2	386.83	43862.9	.0258	279.272	2700	143610.1	128153	121161.2	.000175	327.521
1300	59515.8	668.46	48707.1	.0162	283.820	2800	149796.8	168044	126516.5	.000139	329.774
1400	65261.8	1116.53	53621.7	.0104	288.086	2900	155999.3	218377	131887.6	.000110	331.953
1500	71071.5	1808.00	58599.9	.0069	292.093	3000	162215.5	281244	137272.3	.000089	334.056

TABLE A-14 IDEAL-GAS PROPERTIES OF HYDROGEN (SI UNITS)

T in K, h and u in kJ/kgmole and ϕ in Kj/kgmole·K

T	h	p_r	u	v_r	ϕ	T	h	p_r	u	v_r	ϕ
0	0										
100	3175.3	.2136	2343.9	3.892	102.034						
200	5697.2	1.7100	4034.3	.972	119.328	1600	48010.1	2791.9	34707.1	.00476	180.838
298	8463.1	6.6072	5985.4	.375	130.558	1700	51303.3	3549.2	37168.8	.00398	182.834
300	8520.7	6.7564	6026.4	.369	130.752	1800	54635.1	4462.6	39669.2	.00335	184.738
400	11426.7	18.4763	8100.9	.180	139.117	1900	58007.8	5555.2	42210.4	.00284	186.559
500	14349.2	40.4671	10192.0	.103	145.635	2000	61419.0	6860.2	44790.2	.00242	188.313
600	17278.3	76.899	12289.7	.0649	150.973	2100	64864.4	8396	47404.1	.002080	189.992
700	20216.1	132.620	14396.0	.0439	155.504	2200	68343.8	10204	50052.1	.001793	191.614
800	23168.1	213.042	16516.6	.0312	159.445	2300	71855.0	12302	52731.9	.001555	193.169
900	26143.3	324.566	18660.4	.0231	162.946	2400	75395.8	14757	55441.2	.001352	194.682
1000	29148.1	474.608	20833.7	.0175	166.105	2500	78966.0	17580	58180.0	.001182	196.137
1100	32186.9	672.83	23041.1	.01359	169.007	2600	82563.6	20837	60946.1	.001037	197.550
1200	35264.3	928.42	25287.1	.01075	171.684	2700	86186.1	24551	63737.2	.000914	198.914
1300	38384.9	1253.24	27576.2	.00862	174.178	2800	89833.6	28810	66553.2	.000808	200.244
1400	41548.6	1663.18	29908.5	.00700	176.531	2900	93506.0	33641	69394.3	.000717	201.533
1500	44757.8	2170.02	32286.2	.00575	178.743	3000	97201.2	39085	72258.0	.000638	202.780

TABLE A-15 IDEAL-GAS PROPERTIES OF CARBON MONOXIDE (SI UNITS)

T in K, h and u in kJ/kgmole, and ϕ in kJ/kgmole·K

T	h	p_r	u	v_r	ϕ	T	h	p_r	u	v_r	ϕ
0	0										
100	2902.5	.0453	2071.1	18.3578	165.716						
200	5813.2	.5126	4150.3	3.2437	185.891	1600	51047.6	1224.2	37744.6	.01087	250.562
298	8666.7	2.0733	6189.1	1.1969	197.499	1700	54605.8	1586.8	40471.3	.00891	252.719
300	8725.0	2.1208	6230.7	1.1761	197.697	1800	58185.4	2029.5	43219.4	.00737	254.765
400	11647.1	5.8282	8321.4	.5706	206.102	1900	61783.7	2564.5	45986.3	.00616	256.710
500	14601.7	12.8769	10444.5	.3228	212.693	2000	65398.6	3205.2	48769.8	.00519	258.565
600	17612.1	24.912	12623.5	.2002	218.180	2100	69028.3	3966	51568.1	.00440	260.336
700	20692.3	44.091	14872.2	.1320	222.927	2200	72671.3	4863	54379.6	.00376	262.030
800	23845.9	73.157	17194.4	.0909	227.137	2300	76326.2	5912	57203.1	.00323	263.655
900	27069.9	115.495	19586.9	.0648	230.933	2400	79991.9	7132	60037.3	.00280	265.215
1000	30357.9	175.181	22043.5	.0475	234.397	2500	83667.3	8543	62881.3	.00243	266.715
1100	33702.6	257.03	24556.8	.0356	237.584	2600	87351.7	10164	65734.2	.00213	268.160
1200	37096.7	366.63	27119.4	.0272	240.537	2700	91044.2	12019	68595.3	.00187	269.554
1300	40533.7	510.40	29724.9	.0212	243.288	2800	94744.3	14131	71464.0	.00165	270.900
1400	44007.6	695.63	32367.5	.0167	245.862	2900	98451.3	16524	74339.6	.00146	272.200
1500	47513.7	930.51	35042.1	.0134	248.281	3000	102164.8	19225	77221.6	.00130	273.459

TABLE A-16 IDEAL-GAS PROPERTIES OF SOME COMMON GASES (ENGLISH UNITS)

T in °R, h and u in Btu/lbmole, and ϕ in Btu/lbmole·°R

$$u = h - 1.985 \bar{8} 3T$$

T	N₂ h	N₂ pr	N₂ vr	N₂ φ	O₂ h	O₂ φ	H₂O h	H₂O φ	CO₂ h	CO₂ φ	H₂ h	H₂ φ	CO h	CO φ
0	0				0		0		0		0		0	
300	2081.9	.13082	24611	41.687	2083.0	44.908	2369.1	40.434	2106.3	46.361	2066.2	27.337	2082.1	43.132
400	2777.1	.35813	11986	43.687	2778.6	46.909	3165.7	42.726	2873.3	48.563	2712.6	29.194	2777.3	45.132
500	3472.5	.78227	6859	45.238	3476.2	48.466	3964.2	44.507	3705.1	50.417	3388.5	30.701	3472.7	46.684
537	3729.8	1.0046	5738	45.735	3735.4	48.966	4260.5	45.079	4029.1	51.041	3642.6	31.192	3730.2	47.180
600	4168.2	1.4817	4346	46.507	4178.9	49.747	4767.4	45.971	4599.8	52.046	4078.0	31.958	4169.0	47.953
700	4865.2	2.5452	2951	47.581	4890.3	50.843	5578.5	47.222	5551.0	53.511	4773.1	33.031	4867.3	49.029
800	5564.8	4.074	2107.3	48.516	5613.1	51.808	6400.4	48.319	6552.1	54.847	5470.6	33.960	5569.5	49.967
900	6268.7	6.185	1561.7	49.344	6349.0	52.675	7234.9	49.302	7597.5	56.078	6169.0	34.784	6277.6	50.801
1000	6978.4	9.012	1190.8	50.092	7098.3	53.464	8083.4	50.196	8682.3	57.221	6868.4	35.521	6993.2	51.555
1100	7695.2	12.712	928.6	50.775	7860.9	54.191	8946.5	51.018	9802.5	58.288	7568.5	36.188	7717.4	52.245
1200	8420.2	17.465	737.3	51.406	8636.0	54.865	9824.9	51.782	10954.6	59.290	8269.7	36.798	8451.2	52.883
1300	9153.9	23.47	594.3	51.993	9422.6	55.495	10718.8	52.498	12135.6	60.236	8972.7	37.360	9194.7	53.478
1400	9896.5	30.97	485.1	52.544	10219.6	56.085	11628.9	53.172	13342.5	61.130	9678.0	37.884	9947.9	54.037
1500	10648.2	40.21	400.3	53.062	11026.2	56.642	12555.3	53.811	14573.2	61.979	10386.2	38.371	10710.7	54.563
1600	11408.6	51.48	333.5	53.553	11841.3	57.168	13498.7	54.420	15825.2	62.787	11097.1	38.830	11482.5	55.061
1700	12177.6	65.10	280.2	54.019	12664.0	57.666	14458.8	55.002	17096.6	63.557	11812.8	39.264	12263.0	55.534
1800	12954.6	81.42	237.24	54.463	13493.6	58.141	15436.0	55.561	18385.1	64.294	12531.4	39.674	13051.6	55.985
1900	13739.3	100.81	202.26	54.887	14329.5	58.593	16430.1	56.098	19689.6	64.999	13254.9	40.065	13847.6	56.415
2000	14531.1	123.69	173.52	55.294	15170.9	59.024	17441.0	56.616	21008.7	65.676	13984.3	40.440	14650.7	56.827
2100	15329.6	150.50	149.74	55.683	16017.6	59.437	18468.5	57.118	22341.5	66.326	14717.6	40.798	15460.1	57.222
2200	16134.4	181.72	129.92	56.058	16868.9	59.833	19512.3	57.603	23686.1	66.951	15457.8	41.141	16275.6	57.601
2300	16945.1	217.88	113.29	56.418	17724.7	60.214	20571.9	58.074	25042.3	67.555	16202.8	41.473	17096.5	57.966
2400	17761.1	259.51	99.25	56.765	18584.6	60.580	21646.8	58.532	26407.4	68.137	16953.6	41.794	17922.5	58.318
2500	18582.1	307.23	87.33	57.100	19448.3	60.932	22736.7	58.977	27781.2	68.699	17710.4	42.104	18753.2	58.657
2600	19407.8	361.64	77.15	57.424	20315.7	61.272	23840.8	59.410	29164.8	69.239	18473.9	42.400	19588.2	58.984
2700	20237.8	423.43	68.43	57.737	21186.6	61.601	24958.9	59.832	30555.2	69.765	19242.4	42.692	20427.2	59.301
2800	21071.9	493.3	60.91	58.041	22060.9	61.919	26090.1	60.243	31953.5	70.272	20017.7	42.974	21269.9	59.607
2900	21909.6	572.0	54.41	58.335	22938.6	62.227	27234.1	60.644	33357.6	70.764	20796.8	43.246	22116.0	59.904
3000	22750.9	660.4	48.75	58.620	23819.4	62.526	28390.1	61.036	34768.5	71.243	21582.9	43.512	22965.3	60.192
3100	23595.4	759.2	43.82	58.897	24703.3	62.815	29557.8	61.419	36185.3	71.707	22372.8	43.772	23817.5	60.472
3200	24442.9	869.3	39.50	59.166	25590.4	63.097	30736.4	61.793	37606.0	72.158	23168.6	44.025	24672.5	60.743
3300	25293.2	991.8	35.71	59.428	26480.5	63.371	31925.7	62.159	39032.5	72.597	23971.2	44.271	25530.0	61.007
3400	26146.2	1127.4	32.36	59.682	27373.6	63.638	33125.0	62.517	40463.0	73.026	24776.7	44.511	26389.9	61.264
3500	27001.6	1277.4	29.40	59.930	28269.8	63.897	34333.9	62.868	41899.3	73.441	25589.1	44.747	27252.1	61.514
3600	27859.3	1442.7	26.78	60.172	29168.9	64.151	35551.9	63.211	43338.6	73.846	26405.4	44.978	28116.3	61.757
3700	28719.3	1624.4	24.44	60.407	30070.9	64.398	36778.6	63.547	44781.7	74.241	27226.6	45.202	28982.5	61.994
3800	29581.2	1824	22.361	60.637	30975.9	64.639	38013.5	63.876	46228.8	74.629	28052.6	45.423	29850.6	62.226
3900	30445.1	2042	20.497	60.862	31883.7	64.875	39256.5	64.199	47678.8	75.004	28882.6	45.639	30720.4	62.452
4000	31310.8	2280	18.826	61.081	32794.5	65.106	40507.0	64.516	49132.7	75.373	29716.5	45.850	31591.8	62.672

TABLE A-17M IDEAL-GAS ENTHALPY OF EIGHT GASES (SI UNITS) Temperature in K, enthalpy in kJ/kgmole

T, K	Methane CH_4	Ethane C_2H_6	Propane C_3H_8	n-Butane C_4H_{10}	n-Octane C_8H_{18}	Methanol CH_3OH	Ammonia NH_3	Hydrazine N_2H_4	T, K
0	0	0	0	0	0	0	0	0	0
200	6644	7259	8414	10744	21581		6665	7070	200
298	10016	11874	14740	19276	37782	11427	10058	11460	298
300	10083	11975	14874	19460	38116	11510	10125	11550	300
400	13887	17874	23276	30648	59580	16263	13836	16930	400
500	18238	25058	33623	44350	85939	21811	17870	23150	500
600	13192	33430	45689	60250	116650	28146	22230	30080	600
700	28727	42844	59287	78115	151084	35192	26907	37610	700
800	34819	53220	74182	97613	188698	42865	31878	45650	800
900	41417	64434	90207	118533	228466	51087	37129	54160	900
1000	48492	76358	107194	140666	270705	59810	42639	63080	1000
1100	55982	88910	125102	163929			48396	72370	1100

Source: TRC "Thermodynamic Tables—Hydrocarbons and Non-Hydrocarbons," Thermodynamics Research Center, The Texas A&M University System: College Station, TX.

TABLE A-17E IDEAL-GAS ENTHALPY OF EIGHT GASES (ENGLISH UNITS) Temperature in °R, enthalpy in Btu/lbmole

T, R	Methane CH_4	Ethane C_2H_6	Propane C_3H_8	n-Butane C_4H_{10}	n-Octane C_8H_{18}	Methanol CH_3OH	Ammonia NH_3	Hydrazine N_2H_4	T, R
0	0	0	0	0	0	0	0	0	0
537	4306	5105	6337	8287	16243	4913	4324	4948	537
600	4853	5929	7499	9841	19215	5589	4866	5733	600
800	6782	9005	11903	15693	30446	8011	6700	8542	800
1000	8999	12717	17261	22769	44087	10853	8708	11776	1000
1200	11525	17018	23466	30927	59839	14085	10884	15367	1200
1400	14359	21850	30412	40028	77374	17671	13219	19257	1400
1600	17482	27148	37993	49932	96340	21564	15705	23402	1600
1800	20857	32837	46105	60501	116354	25712	18332	27767	1800
2000	24427	38831	54646	71594	137009	30050	21088	32332	2000

Source: TRC "Thermodynamic Tables—Hydrocarbons and Non-Hydrocarbons," Thermodynamics Research Center, The Texas A&M University System: College Station, TX.

TABLE A-18M PROPERTIES OF WATER: SATURATED STEAM, TEMPERATURE TABLE (SI UNITS)

Temp C T	Abs. Press. kPa p	Specific Volume, m^3/kg Sat. Liq. v_f	Evap. v_{fg}	Sat. Vapor v_g	Enthalpy, kJ/kg Sat. Liquid h_f	Evap. h_{fg}	Sat. Vapor h_g	Entropy, kJ/kg·K Sat. Liquid s_f	Evap. s_{fg}	Sat. Vapor s_g	Temp C T
0.01	0.6112*	0.0010002	206.16	206.16	0.00	2501.6	2501.6	0.0000	9.1575	9.1575	0.01
1.0	0.6566	0.0010001	192.61	192.61	4.17	2499.2	2503.4	0.0153	9.1158	9.1311	1.0
2.0	0.7055	0.0010001	179.92	179.92	8.39	2496.8	2505.2	0.0306	9.0741	9.1047	2.0
3.0	0.7575	0.0010001	168.17	168.17	12.60	2494.5	2507.1	0.0459	9.0326	9.0785	3.0
4.0	0.8129	0.0010000	157.27	157.27	16.80	2492.1	2508.9	0.0611	8.9915	9.0526	4.0
5.0	0.8718	0.0010000	147.16	147.16	21.01	2489.7	2510.7	0.0762	8.9507	9.0269	5.0
6.0	0.9345	0.0010000	137.78	137.78	25.21	2487.4	2512.6	0.0193	8.9102	9.0015	6.0
7.0	1.0012	0.0010001	129.06	129.06	29.41	2485.0	2514.4	0. 3	8.8699	8.9762	7.0
8.0	1.0720	0.0010001	120.96	120.97	33.60	2482.6	2516.2	0.1213	8.8300	8.9513	8.0
9.0	1.1472	0.0010002	113.43	113.44	37.80	2480.3	2518.1	0.1362	8.7903	.9265	9.0
10.0	1.2270	0.0010003	106.43	106.43	41.99	2477.9	2519.9	0.1510	8.7510	8.9020	10.0
12.0	1.4014	0.0010004	93.83	93.84	50.38	2473.2	2523.6	0.1805	8.6731	8.8536	12.0
14.0	1.5973	0.0010007	82.90	82.90	58.75	2468.5	2527.2	0.2098	8.5963	8.8060	14.0
16.0	1.8168	0.0010010	73.38	73.38	67.13	2463.8	2530.9	0.2388	8.5205	8.7593	16.0
18.0	2.0624	0.0010013	65.09	65.09	75.50	2459.0	2534.5	0.2677	8.4458	8.7135	18.0
20.0	2.337	0.0010017	57.84	57.84	83.86	2454.3	2538.2	0.2963	8.3721	8.6684	20.0
22.0	2.642	0.0010022	51.49	51.49	92.23	2449.6	2541.8	0.3247	8.2994	8.6241	22.0
24.0	2.982	0.0010026	45.92	45.93	100.59	2444.9	2545.5	0.3530	8.2277	8.5806	24.0
26.0	3.360	0.0010032	41.03	41.03	108.95	2440.2	2549.1	0.3810	8.1569	8.5379	26.0
28.0	3.778	0.0010037	36.73	36.73	117.31	2435.4	2552.7	0.4088	8.0870	8.4959	28.0
30.0	4.241	0.0010043	32.93	32.93	125.66	2430.7	2556.4	0.4365	8.0181	8.4546	30.0
32.0	4.753	0.0010049	29.57	29.57	134.02	2425.9	2560.0	0.4640	7.9500	8.4140	32.0
34.0	5.318	0.0010056	26.60	26.60	142.38	2421.2	2563.6	0.4913	7.8828	8.3740	34.0
36.0	5.940	0.0010063	23.97	23.97	150.74	2416.4	2567.2	0.5184	7.8164	8.3348	36.0
38.0	6.624	0.0010070	21.63	21.63	159.06	2411.7	2570.8	0.5453	7.7509	8.2962	38.0
40.0	7.375	0.0010078	19.545	19.546	167.45	2406.9	2574.4	0.5721	7.6861	8.2583	40.0
42.0	8.198	0.0010086	17.691	17.692	175.81	2402.1	2577.9	0.5987	7.6222	8.2209	42.0
44.0	9.100	0.0010094	16.035	16.036	184.17	2397.3	2581.5	0.6252	7.5590	8.1842	44.0
46.0	10.086	0.0010103	14.556	14.557	192.53	2392.5	2585.1	0.6514	7.4966	8.1481	46.0
48.0	11.162	0.0010112	13.232	13.233	200.89	2387.7	2588.6	0.6776	7.4350	8.1125	48.0
50.0	12.335	0.0010121	12.045	12.046	209.26	2382.9	2592.2	0.7035	7.3741	8.0776	50.0
52.0	13.613	0.0010131	10.979	10.980	217.62	2378.1	2595.7	0.7293	7.3138	8.0432	52.0

54.0	8.0093	7.2543	0.7550	2599.2	2373.2	225.99	10.022	10.021	0.0010140	15.002	54.0
56.0	7.9759	7.1955	0.7804	2602.7	2368.4	234.35	9.159	9.158	0.0010150	16.511	56.0
58.0	7.9431	7.1373	0.8058	2606.2	2363.5	242.72	8.381	8.380	0.0010161	18.147	58.0
60.0	7.9108	7.0798	0.8310	2609.7	2358.6	251.09	7.679	7.678	0.0010171	19.920	60.0
62.0	7.8790	7.0230	0.8560	2613.2	2353.7	259.46	7.044	7.043	0.0010182	21.838	62.0
64.0	7.8477	6.9667	0.8809	2616.6	2348.8	267.84	6.469	6.468	0.0010193	23.912	64.0
66.0	7.8168	6.9111	0.9057	2620.1	2343.9	276.21	5.948	5.947	0.0010205	26.150	66.0
68.0	7.7864	6.8561	0.9303	2623.6	2338.9	284.59	5.476	5.475	0.0010217	28.563	68.0
70.0	7.7565	6.8017	0.9548	2626.9	2334.0	292.97	5.046	5.045	0.0010288	31.16	70.0
72.0	7.7270	6.7478	0.9792	2630.3	2329.0	301.36	4.656	4.655	0.0010241	33.96	72.0
74.0	7.6979	6.6945	1.0034	2633.7	2324.0	309.74	4.300	4.299	0.0010253	36.96	74.0
76.0	7.6693	6.6418	1.0275	2637.1	2318.9	318.13	3.976	3.976	0.0010266	40.10	76.0
78.0	7.6410	6.5896	1.0514	2640.4	2313.9	326.52	3.680	3.679	0.0010279	43.65	78.0
80.0	7.6132	6.5380	1.0753	2643.8	2308.8	334.92	3.409	3.408	0.0010292	47.36	80.0
82.0	7.5858	6.4868	1.0990	2647.1	2303.8	343.31	3.162	3.161	0.0010305	51.33	82.0
84.0	7.5588	6.4362	1.1225	2650.4	2298.6	351.71	2.935	2.934	0.0010319	55.57	84.0
86.0	7.5351	6.3861	1.1460	2653.6	2293.5	360.12	2.727	2.726	0.0010333	60.11	86.0
88.0	7.5058	6.3365	1.1693	2656.9	2288.4	368.53	2.536	2.535	0.0010347	64.95	88.0
90.0	7.4799	6.2873	1.1925	2660.1	2283.2	376.94	2.3613	2.3603	0.0010361	70.11	90.0
92.0	7.4543	6.2387	1.2156	2663.4	2278.0	385.36	2.2002	2.1992	0.0010376	75.61	92.0
94.0	7.4291	6.1905	1.2386	2666.6	2272.9	393.78	2.0519	2.0509	0.0010391	81.46	94.0
96.0	7.4042	6.1427	1.2615	2669.7	2267.5	402.20	1.9153	1.9143	0.0010406	87.69	96.0
98.0	7.3796	6.0954	1.2842	2672.9	2262.2	410.63	1.7893	1.7883	0.0010421	94.30	98.0
100.0	7.3554	6.0485	1.3069	2676.0	2256.9	419.06	1.6730	1.6720	0.0010437	101.33	100.0
105.0	7.2962	5.9331	1.3630	2683.7	2243.6	440.17	1.4193	1.4182	0.0010477	120.80	105.0
110.0	7.2388	5.8203	1.4185	2691.3	2230.0	461.32	1.2099	1.2089	0.0010519	143.27	110.0
115.0	7.1832	5.7099	1.4733	2698.7	2216.2	482.50	1.0363	1.0352	0.0010562	169.06	115.0
120.0	7.1293	5.6017	1.5276	2706.0	2202.2	503.72	0.8915	0.8905	0.0010606	198.54	120.0
125.0	7.0769	5.4957	1.5813	2713.0	2188.0	524.99	0.7702	0.7692	0.0010652	232.1	125.0
130.0	7.0261	5.3917	1.6344	2719.9	2173.6	546.31	0.6681	0.6671	0.0010700	270.1	130.0
134.0	6.9766	5.2897	1.6869	2726.6	2158.9	567.68	0.5818	0.5807	0.0010750	313.1	134.0
140.0	6.9284	5.1894	1.7390	2733.1	2144.0	589.10	0.5085	0.5074	0.0010801	361.4	140.0
145.0	6.8815	5.0910	1.7906	2739.3	2128.7	610.59	0.4460	0.4449	0.0010853	415.5	145.0
150.0	6.8358	4.9941	1.8416	2745.4	2113.2	632.15	0.3924	0.3914	0.0010908	476.0	150.0
155.0	6.7911	4.8989	1.8923	2751.2	2097.4	653.77	0.3464	0.3453	0.0010964	543.3	155.0
160.0	6.7475	4.8050	1.9425	2756.7	2081.3	675.47	0.3068	0.3057	0.0011022	618.1	160.0
165.0	6.7048	4.7126	1.9923	2762.0	2064.8	697.25	0.2724	0.2713	0.0011082	700.8	165.0
170.0	6.6630	4.6214	2.0416	2767.1	2047.9	719.12	0.2426	0.2414	0.0011145	792.0	170.0

* Approximate triple point.

TABLE A-18M (Continued)

Temp C T	Abs. Press. kPa p	Specific Volume, m³/kg Sat. Liq. v_f	Evap. v_{fg}	Sat. Vapor v_g	Enthalpy, kJ/kg Sat. Liquid h_f	Evap. h_{fg}	Sat. Vapor h_g	Entropy, kJ/kg·K Sat. Liquid s_f	Evap. s_{fg}	Sat. Vapor s_g	Temp C T
175.0	892.4	0.0011209	0.21542	0.21654	741.07	2030.7	2771.8	2.0906	4.5314	6.6221	175.0
180.0	1002.7	0.0011275	0.19267	0.19380	763.12	2013.2	2776.3	2.1393	4.4426	6.5819	180.0
185.0	1123.3	0.0011344	0.17272	0.17386	785.26	1995.2	2780.4	2.1876	4.3548	6.5424	185.0
190.0	1255.1	0.0011415	0.15517	0.15632	807.52	1976.7	2784.3	2.2356	4.2680	6.5036	190.0
195.0	1398.7	0.0011489	0.13969	0.14084	829.88	1957.9	2787.8	2.2833	4.1821	6.4654	195.0
200.0	1554.9	0.0011565	0.12600	0.12716	852.37	1938.6	2790.9	2.3307	4.0971	6.4278	200.0
205.0	1724.3	0.0011644	0.11386	0.11503	874.99	1918.8	2793.8	2.3778	4.0128	6.3906	205.0
210.0	1907.7	0.0011726	0.10307	0.10424	897.73	1898.5	2796.2	2.4247	3.9293	6.3539	210.0
215.0	2106.0	0.0011811	0.09344	0.09463	920.63	1877.6	2798.3	2.4713	3.8463	6.3176	215.0
220.0	2319.8	0.0011900	0.08485	0.08604	943.67	1856.2	2799.9	2.5178	3.7639	6.2817	220.0
225.0	2550.	0.0011992	0.07715	0.07835	966.88	1834.3	2801.2	2.5641	3.6820	6.2461	225.0
230.0	2798.	0.0012087	0.07024	0.07145	990.27	1811.7	2802.0	2.6102	3.6006	6.2107	230.0
235.0	3063.	0.0012187	0.06403	0.06525	1013.83	1788.5	2802.3	2.6561	3.5194	6.1756	235.0
240.0	3348.	0.0012291	0.05843	0.05965	1037.60	1764.6	2802.2	2.7020	3.4386	6.1406	240.0
245.0	3652.	0.0012399	0.05337	0.05461	1061.58	1740.0	2801.6	2.7478	3.3579	6.1057	245.0
250.0	3978.	0.0012513	0.04879	0.05004	1085.78	1714.7	2800.4	2.7935	3.2773	6.0708	250.0
255.0	4325.	0.0012632	0.04463	0.04590	1110.23	1688.5	2798.7	2.8392	3.1968	6.0359	255.0
260.0	4694.	0.0012756	0.04086	0.04213	1134.94	1661.5	2796.4	2.8848	3.1161	6.0010	260.0
265.0	5088.	0.0012887	0.03742	0.03871	1159.93	1633.5	2793.5	2.9306	3.0353	5.9658	265.0
270.0	5506.	0.0013025	0.03429	0.03559	1185.23	1604.6	2789.9	2.9763	2.9541	5.9304	270.0
275.0	5950.	0.0013170	0.03142	0.03274	1210.86	1574.7	2785.5	3.0222	2.8725	5.8947	275.0
280.0	6420.	0.0013324	0.02879	0.03013	1236.84	1543.6	2780.4	3.0683	2.7903	5.8586	280.0
285.0	6919.	0.0013487	0.02638	0.02773	1263.21	1511.3	2774.5	3.1146	2.7074	5.8220	285.0
290.0	7446.	0.0013659	0.02417	0.02554	1290.01	1477.6	2767.6	3.1611	2.6237	5.7848	290.0
295.0	8004.	0.0013844	0.02213	0.02351	1317.27	1442.6	2759.8	3.2079	2.5389	5.7469	295.0
300.0	8593.	0.0014041	0.020245	0.021649	1345.05	1406.0	2751.0	3.2552	2.4529	5.7081	300.0
305.0	9214.	0.0014252	0.018502	0.019927	1373.40	1367.7	2741.1	3.3029	2.3656	5.6685	305.0
310.0	9870.	0.0014480	0.016886	0.018334	1402.39	1327.6	2730.0	3.3512	2.2766	5.6278	310.0
315.0	10561.	0.0014726	0.015383	0.016856	1432.09	1285.5	2717.6	3.4002	2.1856	5.5858	315.0
320.0	11289.	0.0014995	0.013980	0.015480	1462.60	1241.1	2703.7	3.4500	2.0923	5.5423	320.0
325.0	12056.	0.0015289	0.012666	0.014195	1494.03	1194.0	2688.0	3.5008	1.9961	5.4969	325.0
330.0	12863.	0.0015615	0.011428	0.012989	1526.52	1143.6	2670.2	3.5528	1.8962	5.4490	330.0
335.0	13712.	0.0015978	0.010256	0.011854	1560.25	1089.5	2649.7	3.6963	1.7916	5.3979	335.0

340.0	14605.	0.0016387	0.009142	0.010780·	1595.47	1030.7	2626.2	3.6616	1.6811	5.3427	340.0
345.0	15545.	0.0016858	0.008077	0.009763	1632.52	966.4	2598.9	3.7193	1.5636	5.2828	345.0
350.0	16535.	0.0017411	0.007058	0.008799	1671.94	895.7	2567.7	3.7800	1.4376	5.2177	350.0
355.0	17577.	0.0018085	0.006051	0.007859	1716.63	813.8	2530.4	3.8489	1.2953	5.1442	355.0
360.0	18675.	0.0018959	0.005044	0.006940	1764.17	721.3	2485.4	3.9210	1.1390	5.0600	360.0
365.0	19833.	0.0020160	0.003996	0.006012	1817.96	610.0	2428.0	4.0021	0.9558	4.9479	365.0
370.0	21054.	0.0022136	0.002759	0.004973	1890.21	452.6	2342.8	4.1108	0.7036	4.8144	370.0
371.0	21306.	0.0022778	0.002446	0.004723	1910.50	407.4	2317.9	4.1414	0.6324	4.7738	371.0
372.0	21562.	0.0023636	0.002075	0.004439	1935.57	351.4	2287.0	4.1794	0.5446	4.7240	372.0
373.0	21820.	0.0024963	0.001588	0.004084	1970.50	273.5	2244.0	4.2326	0.4233	4.6559	373.0
374.0	22081.	0.0028427	0.000623	0.003466	2046.72	109.5	2156.2	4.3493	0.1692	4.5185	374.0
374.15	22120.†	0.00317	0.0	0.00317	2107.37	0.0	2107.4	4.4229	0.0	4.4429	374.15

† Approximate critical point.

Source for Tables A-18 to A-20 (M and E): Abridged by permission from ASME, *Steam Tables*, American Society of Mechanical Engineers, N.Y., 1967.

TABLE A-19M PROPERTIES OF WATER: SATURATED STEAM, PRESSURE TABLE (SI UNITS)

Temp. C T	Abs. Press kPa p	Specific Volume, m³/kg Sat. Liquid v_f	Evap. v_{fg}	Sat. Vapor v_g	Enthalpy, kJ/kg Sat. Liquid h_f	Evap. h_{fg}	Sat. Vapor h_g	Entropy, kJ/kg·K Sat. Liquid s_f	Evap. s_{fg}	Sat. Vapor s_g	Energy, kJ/kg Sat. Liquid u_f	Sat. Vapor u_g	Temp. C T
6.983	1.0	0.0010001	129.21	129.21	29.34	2485.0	2514.4	0.1060	8.8706	8.9767	29.33	2385.2	6.983
8.380	1.1	0.0010001	118.04	118.04	35.20	2481.7	2516.9	0.1269	8.8149	8.9418	35.20	2387.1	8.380
9.668	1.2	0.0010002	108.70	108.70	40.60	2478.7	2519.3	0.1461	8.7640	8.9101	40.60	2388.9	9.668
10.866	1.3	0.0010003	100.76	100.76	45.62	2475.9	2521.5	0.1638	8.7171	8.8809	45.62	2390.5	10.866
11.985	1.4	0.0010004	93.92	93.92	50.31	2473.2	2523.5	0.1803	8.6737	8.8539	50.31	2392.0	11.985
13.036	1.5	0.0010006	87.98	87.98	54.71	2470.7	2525.5	0.1957	8.6332	8.8288	54.71	2393.5	13.036
14.026	1.6	0.0010007	82.76	82.77	58.86	2468.3	2527.4	0.2101	8.5952	8.8054	58.86	2394.8	14.026
15.855	1.8	0.0010010	74.03	74.03	66.52	2464.1	2530.6	0.2367	8.5260	8.7627	66.52	2397.4	15.855
17.513	2.0	0.0010012	67.01	67.01	73.46	2460.2	2533.6	0.2607	8.4639	8.7246	73.46	2399.6	17.513
19.031	2.2	0.0010015	61.23	61.23	79.81	2456.6	2536.4	0.2825	8.4077	8.6901	79.81	2401.7	19.031
20.433	2.4	0.0010018	56.39	56.39	85.67	2453.3	2539.0	0.3025	8.3563	8.6587	85.67	2403.6	20.433
21.737	2.6	0.0010021	52.28	52.28	91.12	2450.2	2541.3	0.3210	8.3089	8.6299	91.12	2405.4	21.737
22.955	2.8	0.0010024	48.74	48.74	96.22	2447.3	2543.6	0.3382	8.2650	8.6033	96.21	2407.1	22.955
24.100	3.0	0.0010027	45.67	45.67	101.00	2444.6	2545.6	0.3544	8.2241	8.5785	101.00	2408.6	24.100
26.694	3.5	0.0010033	39.48	39.48	111.85	2438.5	2550.4	0.3907	8.1325	8.5232	111.84	2412.2	26.694
28.983	4.0	0.0010040	34.80	34.80	121.40	2433.1	2554.5	0.4225	8.0530	8.4755	121.41	2415.3	28.983
31.035	4.5	0.0010046	31.14	31.14	129.99	2428.2	2558.2	0.4507	7.9827	8.4335	129.98	2418.1	31.035
32.898	5.0	0.0010052	28.19	28.19	137.77	2423.8	2561.6	0.4763	7.9197	8.3960	137.77	2420.6	32.898
34.605	5.5	0.0010058	25.77	25.77	144.91	2419.8	2564.7	0.4995	7.8626	8.3621	144.90	2422.9	34.605
36.183	6.0	0.0010064	23.74	23.74	151.50	2416.0	2567.5	0.5209	7.8104	8.3312	151.50	2425.1	36.183
37.651	6.5	0.0010069	22.015	22.016	157.64	2412.5	2570.2	0.5407	7.7622	8.3029	157.63	2427.0	37.651
39.025	7.0	0.0010074	20.530	20.531	163.38	2409.2	2572.6	0.5591	7.7176	8.2767	163.37	2428.9	39.025
40.316	7.5	0.0010079	19.238	19.239	168.77	2406.2	2574.9	0.5763	7.6760	8.2523	168.76	2430.6	40.316
41.534	8.0	0.0010084	18.104	18.105	173.86	2403.2	2577.1	0.5925	7.6370	8.2296	173.86	2432.3	41.534
43.787	9.0	0.0010094	16.203	16.204	183.28	2397.9	2581.1	0.6224	7.5657	8.1881	183.27	2435.3	43.787
45.833	10.	0.0010102	14.674	14.675	191.83	2392.9	2584.8	0.6493	7.5018	8.1511	191.82	2438.0	45.833
47.710	11.	0.0010111	13.415	13.416	199.68	2388.4	2588.1	0.6738	7.4439	8.1177	199.67	2440.5	47.710
49.446	12.	0.0010119	12.361	12.362	206.94	2384.3	2591.2	0.6963	7.3909	8.0872	206.93	2442.8	49.446
51.062	13.	0.0010126	11.465	11.466	213.70	2380.3	2594.0	0.7172	7.3420	8.0592	213.68	2445.0	51.062
52.574	14.	0.0010133	10.693	10.694	220.02	2376.7	2596.7	0.7367	7.2967	8.0334	220.01	2447.0	52.574
53.997	15.	0.0010140	10.022	10.023	225.97	2373.2	2599.2	0.7549	7.2544	8.0093	225.96	2448.9	53.997
55.341	16.	0.0010147	9.432	9.433	231.59	2370.0	2601.6	0.7721	7.2148	7.9869	231.58	2450.6	55.341
57.826	18.	0.0010160	8.444	8.445	241.99	2363.9	2605.9	0.8036	7.1424	7.9460	241.98	2453.9	57.826

60.086	20.	0.0010172	7.649	7.650	251.45	2358.4	2609.9	0.8321	7.0774	7.9094	251.43	2456.9	60.086
62.162	22.	0.0010183	6.994	6.995	260.14	2353.3	2613.5	0.8581	7.0184	7.8764	260.12	2459.6	62.162
64.082	24.	0.0010194	6.446	6.447	268.18	2348.6	2616.8	0.8820	6.9644	7.8464	268.16	2462.1	64.082
65.871	26.	0.0010204	5.979	5.980	275.67	2344.2	2619.9	0.9041	6.9147	7.8188	275.65	2464.4	65.871
67.547	28.	0.0010214	5.578	5.579	282.69	2340.0	2622.7	0.9248	6.8685	7.7933	282.66	2466.5	67.547
69.124	30.	0.0010223	5.228	5.229	289.30	2336.1	2625.4	0.9441	6.8254	7.7695	289.27	2468.6	69.124
72.709	35.	0.0010245	4.525	4.526	304.33	2327.2	2631.5	0.9878	6.7288	7.7166	304.29	2473.1	72.709
75.886	40.	0.0010265	3.992	3.993	317.65	2319.2	2636.9	1.0261	6.6448	7.6709	317.61	2477.1	75.886
78.743	45.	0.0010284	3.575	3.576	329.64	2312.0	2641.7	1.0603	6.5704	7.6307	329.59	2480.7	78.743
81.345	50.	0.0010301	3.239	3.240	340.56	2305.4	2646.0	1.0912	6.5035	7.5947	340.51	2484.0	81.345
83.737	55.	0.0010317	2.963	2.964	350.61	2299.3	2649.9	1.1194	6.4428	7.5623	350.56	2486.9	83.737
85.954	60.	0.0010333	2.731	2.732	359.93	2293.6	2653.6	1.1454	6.3873	7.5327	359.86	2489.7	85.954
88.021	65.	0.0010347	2.5335	2.5346	368.62	2288.3	2656.9	1.1696	6.3360	7.5055	368.55	2492.2	88.021
89.959	70.	0.0010361	2.3637	2.3647	376.77	2283.3	2660.1	1.1921	6.2883	7.4804	376.70	2494.5	89.959
91.785	75.	0.0010375	2.2158	2.2169	384.45	2278.6	2663.0	1.2131	6.2439	7.4570	384.37	2496.7	91.785
93.512	80.	0.0010387	2.0859	2.0870	391.72	2274.1	2665.8	1.2330	6.2022	7.4352	391.64	2498.8	93.512
96.713	90.	0.0010412	1.8682	1.8692	405.21	2265.0	2670.9	1.2696	6.1258	7.3954	405.11	2502.6	96.713
99.632	100.	0.0010434	1.6927	1.6937	417.51	2257.9	2675.4	1.3027	6.0571	7.3598	417.41	2506.1	99.632
102.317	110.	0.0010455	1.5482	1.5492	428.84	2250.8	2679.6	1.3330	5.9947	7.3277	428.73	2509.2	102.317
104.808	120.	0.0010476	1.4271	1.4281	439.36	2244.1	2683.4	1.3609	5.9375	7.2984	439.24	2512.1	104.808
107.133	130.	0.0010495	1.3240	1.3251	449.19	2237.8	2687.0	1.3868	5.8847	7.2715	449.05	2514.7	107.133
109.315	140.	0.0010513	1.2353	1.2363	458.42	2231.9	2690.3	1.4109	5.8356	7.2465	458.27	2517.2	109.315
111.37	150.	0.0010530	1.1580	1.1590	467.13	2226.2	2693.4	1.4336	5.7898	7.2234	466.97	2519.5	111.37
113.32	160.	0.0010547	1.0901	1.0911	475.38	2220.9	2696.2	1.4550	5.7467	7.2017	475.25	2521.7	113.32
116.93	180.	0.0010579	0.9762	0.9772	490.70	2210.8	2701.5	1.4944	5.6678	7.1622	490.51	2525.6	116.93
120.23	200.	0.0010608	0.8844	0.8854	504.70	2201.6	2706.3	1.5301	5.5967	7.1268	504.49	2529.2	120.23
123.27	220.	0.0010636	0.8088	0.8098	517.62	2193.0	2710.6	1.5627	5.5321	7.0949	517.39	2532.4	123.27
126.09	240.	0.0010663	0.7454	0.7465	529.6	2184.9	2714.5	1.5929	5.4728	7.0657	529.38	2535.4	126.09
128.73	260.	0.0010688	0.6914	0.6925	540.9	2177.3	2718.2	1.6209	5.4180	7.0389	540.60	2538.1	128.73
131.20	280.	0.0010712	0.6450	0.6460	551.4	2170.1	2721.5	1.6471	5.3670	7.0140	551.14	2540.6	131.20
133.54	300.	0.0010735	0.6045	0.6056	561.4	2163.2	2724.7	1.6716	5.3193	6.9909	561.11	2543.0	133.54
138.89	350.	0.0010789	0.5229	0.5240	584.3	2147.4	2731.6	1.7273	5.2119	6.9392	583.89	2548.2	138.89
143.62	400.	0.0010839	0.4611	0.4622	604.7	2133.0	2737.6	1.7764	5.1179	6.8943	604.24	2552.7	143.62
147.92	450.	0.0010885	0.4127	0.4138	623.2	2119.7	2742.9	1.8204	5.0343	6.8547	622.67	2556.7	147.92
151.84	500.	0.0010928	0.3736	0.3747	640.1	2107.4	2747.5	1.8604	4.9588	6.8192	639.57	2560.2	151.84
155.47	550.	0.0010969	0.3414	0.3425	655.8	2095.9	2751.7	1.8970	4.8900	6.7870	655.20	2563.3	155.47
158.84	600.	0.0011009	0.3144	0.3155	670.4	2085.0	2755.5	1.9308	4.8267	6.7575	669.76	2566.2	158.84

60.086	20.
62.162	22.
64.082	24.
65.871	26.
67.547	28.
69.124	30.
72.709	35.
75.886	40.
78.743	45.
81.345	50.
83.737	55.
85.954	60.
88.021	65.
89.959	70.
91.785	75.
93.512	80.
96.713	90.
99.632	100.
102.317	110.
104.808	120.
107.133	130.
109.315	140.
111.37	150.
113.32	160.
116.93	180.
120.23	200.
123.27	220.
126.09	240.
128.73	260.
131.20	280.
133.54	300.
138.89	350.
143.62	400.
147.92	450.
151.84	500.
155.47	550.
158.84	600.

TABLE A-19M (Continued)

Temp. C T	Abs. Press kPa p	Specific Volume, m³/kg — Sat. Liquid v_f	Evap. v_{fg}	Sat. Vapor v_g	Enthalpy, kJ/kg — Sat. Liquid h_f	Evap. h_{fg}	Sat. Vapor h_g	Entropy, kJ/kg·K — Sat. Liquid s_f	Evap. s_{fg}	Sat. Vapor s_g	Energy, kJ/kg — Sat. Liquid u_f	Sat. Vapor u_g	Temp. C T
161.99	650.	0.0011046	0.29138	0.29249	684.1	2074.7	2758.9	1.9623	4.7681	6.7304	683.42	2568.7	161.99
164.96	700.	0.0011082	0.27157	0.27268	697.1	2064.9	2762.0	1.9918	4.7134	6.7052	696.29	2571.1	164.96
167.76	750.	0.0011116	0.25431	0.25543	709.3	2055.5	2764.8	2.0195	4.6139	6.6596	708.64	2573.3	167.76
170.41	800.	0.0011150	0.23914	0.24026	720.9	2046.5	2767.5	2.0457	4.6139	6.6596	720.04	2575.3	170.41
175.36	900.	0.0011213	0.21369	0.21481	742.6	2029.5	2772.1	2.0941	4.5250	6.6192	741.63	2578.8	175.36
179.88	1000.	0.0011274	0.19317	0.19429	762.6	2013.6	2776.2	2.1382	4.4446	6.5828	761.48	2581.9	179.88
184.07	1100.	0.0011331	0.17625	0.17738	781.1	1998.5	2779.7	2.1786	4.3711	6.5497	779.88	2584.5	184.07
187.96	1200.	0.0011386	0.16206	0.16320	798.4	1984.3	2782.7	2.2161	4.3033	6.5194	797.06	2586.9	187.96
191.61	1300.	0.0011438	0.14998	0.15113	814.7	1970.7	2785.4	2.2510	4.2403	6.4913	813.21	2589.0	191.61
195.04	1400.	0.0011489	0.13957	0.14072	830.1	1957.7	2787.8	2.2837	4.1814	6.4651	828.47	2590.8	195.04
198.29	1500.	0.0011539	0.13050	0.13166	844.7	1945.2	2789.9	2.3145	4.1261	6.4406	842.93	2592.4	198.29
201.37	1600.	0.0011586	0.12253	0.12369	858.6	1933.2	2791.7	2.3436	4.0739	6.4175	856.71	2593.8	201.37
207.11	1800.	0.0011678	0.10915	0.11032	884.6	1910.3	2794.8	2.3976	3.9775	6.3751	882.47	2596.3	207.11
212.37	2000.	0.0011766	0.09836	0.09954	908.6	1888.6	2797.2	2.4469	3.8898	6.3367	906.24	2598.2	212.37
217.24	2200.	0.0011850	0.08947	0.09065	931.0	1868.1	2799.1	2.4922	3.8093	6.3015	928.35	2599.6	217.24
221.78	2400.	0.0011932	0.08201	0.08320	951.9	1848.5	2800.4	2.5343	3.7347	6.2690	949.07	2600.7	221.78
226.04	2600.	0.0012011	0.07565	0.07686	971.7	1829.6	2801.4	2.5736	3.6651	6.2387	968.60	2601.5	226.04
230.05	2800.	0.0012088	0.07018	0.07139	990.5	1811.5	2802.0	2.6106	3.5998	6.2104	987.10	2602.1	230.05
233.84	3000.	0.0012163	0.06541	0.06663	1008.4	1793.9	2802.3	2.6455	3.5382	6.1837	1004.70	2602.1	233.84
242.54	3500.	0.0012345	0.05579	0.05703	1049.8	1752.2	2802.0	2.7253	3.3976	6.1228	1045.44	2602.4	242.54
250.33	4000.	0.0012521	0.04850	0.04975	1087.4	1712.9	2800.3	2.7965	3.2720	6.0685	1082.4	2601.3	250.33
257.41	4500.	0.0012691	0.04277	0.04404	1122.1	1675.6	2797.7	2.8612	3.1579	6.0191	1116.4	2599.5	257.41
263.91	5000.	0.0012858	0.03814	0.03943	1154.5	1639.7	2794.2	2.9206	3.0529	5.9735	1148.0	2597.0	263.91
269.93	5500.	0.0013023	0.03433	0.03563	1184.9	1605.0	2789.9	2.9757	2.9552	5.9309	1177.7	2594.0	269.93
275.55	6000.	0.0013187	0.03112	0.03244	1213.7	1571.3	2785.0	3.0273	2.8635	5.8908	1205.8	2590.4	275.55
280.82	6500.	0.0013350	0.028384	0.029719	1241.1	1538.4	2779.5	3.0759	2.7768	5.8527	1232.5	2586.3	280.82
285.79	7000.	0.0013513	0.026022	0.027373	1267.4	1506.0	2773.5	3.1219	2.6943	5.8162	1258.0	2581.0	285.79
290.50	7500.	0.0013677	0.023959	0.025327	1292.7	1474.2	2766.9	3.1657	2.6153	5.7811	1282.4	2577.0	290.50
294.97	8000.	0.0013842	0.022141	0.023525	1317.1	1442.8	2759.9	3.2076	2.5395	5.7471	1306.0	2571.7	294.97
303.31	9000.	0.0014179	0.019078	0.020495	1363.7	1380.9	2744.6	3.2867	2.3953	5.6820	1351.0	2560.1	303.31
310.96	10000.	0.0014526	0.016589	0.018041	1408.0	1319.7	2727.7	3.3605	2.2593	5.6198	1393.5	2547.3	310.96
318.05	11000.	0.0014887	0.014517	0.016006	1450.6	1258.7	2709.3	3.4304	2.1291	5.5595	1434.2	2533.2	318.05
324.65	12000.	0.0015268	0.012756	0.014283	1491.8	1197.4	2689.2	3.4972	2.0030	5.5002	1473.4	2517.8	324.65

330.83	13000.	0.0015672	0.011230	0.012797	1532.0	1135.0	2667.0	3.5616	1.8792	5.4408	1511.6	2500.6	330.83
336.64	14000.	0.0016106	0.009884	0.011495	1571.6	1070.7	2642.4	3.6242	1.7560	5.3803	1549.1	2481.4	336.64
342.13	15000.	0.0016579	0.008682	0.010340	1611.0	1004.0	2615.0	3.6859	1.6320	5.3178	1586.1	2459.9	342.13
347.33	16000.	0.0017103	0.007597	0.009308	1650.5	934.3	2584.9	3.7471	1.5060	5.2531	1623.2	2436.0	347.33
352.26	17000.	0.0017696	0.006601	0.008371	1691.7	859.9	2551.6	3.8107	1.3748	5.1855	1661.6	2409.3	352.26
356.96	18000.	0.0018399	0.005658	0.007498	1734.8	779.1	2513.9	3.8765	1.2362	5.1128	1701.7	2378.9	356.96
361.43	19000.	0.0019260	0.004751	0.006678	1778.7	692.0	2470.6	3.9429	1.0903	5.0332	1742.1	2343.8	361.43
365.70	20000.	0.0020370	0.003840	0.005877	1826.5	591.9	2418.4	4.0149	0.9263	4.9412	1785.7	2300.8	365.70
369.78	21000.	0.0022015	0.002822	0.005023	1866.3	461.3	2347.6	4.1048	0.7175	4.8223	1840.0	2242.1	369.78
373.69	22000.	0.0026714	0.001056	0.003728	2011.1	184.5	2195.6	4.2947	0.2852	4.5799	1952.4	2113.6	373.69
374.15*	22120.*	0.00317	0.0	0.00317	2107.4	0.0	2107.4	4.4429	0.0	4.4429	2037.3	2037.3	374.15*

* Approximate critical point.

TABLE A-20M PROPERTIES OF WATER: SUPERHEATED STEAM (SI UNITS)

Abs. Press. kPa (Sat. Temp, C)		Sat. Liquid	Sat. Vapor	Temperature—C						
				40.	60.	80.	100.	120.	140.	160.
1.0	v	0.0010	129.2	144.47	153.71	162.95	172.19	181.42	190.66	199.89
(6.983)	h	29.34	2514.4	2575.9	2613.3	2650.9	2688.6	2726.5	2764.6	2802.9
	s	0.1060	9.9767	9.1842	9.3001	9.4096	9.5136	9.6125	9.7070	9.7975
2.0	v	0.0010	67.01	72.211	76.837	81.459	86.080	90.700	95.319	99.936
(17.51)	h	73.46	2533.6	2575.6	2613.1	2650.7	2688.5	2726.4	2764.5	2802.8
	s	0.2607	8.7246	8.8637	8.9797	9.0894	9.1934	9.2924	9.3870	9.4775
3.0	v	0.0010	45.67	48.124	51.211	54.296	57.378	60.460	63.540	66.619
(24.10)	h	101.00	2545.6	2575.4	2612.9	2650.6	2688.4	2726.3	2764.5	2802.8
	s	0.3544	8.5785	8.6760	8.7922	8.9019	9.0060	9.1051	9.1997	9.2902
4.0	v	0.0010	34.80	36.081	38.398	40.714	43.027	45.339	47.650	49.961
(28.98)	h	121.41	2554.5	2575.2	2612.7	2650.4	2688.3	2726.2	2764.4	2802.7
	s	0.4225	8.4755	8.5426	8.6589	8.7688	8.8730	8.9721	9.0668	9.1573
5.0	v	0.0010	28.19	28.854	30.711	32.565	34.417	36.267	38.117	39.966
(32.90)	h	137.77	2561.6	2574.9	2612.6	2650.3	2688.1	2726.1	2764.3	2802.6
	s	0.4763	8.3960	8.4390	8.5555	8.6655	8.7698	8.8690	8.9636	9.0542
6.0	v	0.0010	23.74	24.037	25.586	27.132	28.676	30.219	31.761	33.302
(36.18)	h	151.50	2567.5	2574.7	2612.4	2650.1	2688.0	2726.0	2764.2	2802.6
	s	0.5209	8.3312	8.3543	8.4709	8.5810	8.6854	8.7846	8.8793	8.9700
8.0	v	0.0010	18.105		19.179	20.341	21.501	22.659	23.816	24.973
(41.53)	h	173.86	2577.1		2612.0	2649.8	2687.8	2725.8	2764.1	2802.4
	s	0.5925	8.2296		8.3372	8.4476	8.5521	8.6515	8.7463	8.8370
10.0	v	0.0010	14.675		15.336	16.266	17.195	18.123	19.050	19.975
(45.83)	h	191.85	2584.8		2611.6	2649.5	2687.5	2725.6	2763.9	2802.3
	s	0.6493	8.1511		8.2334	8.3439	8.4486	8.5481	8.6430	8.7338
15.0	v	0.0010	10.023		10.210	10.834	11.455	12.075	12.694	13.312
(54.00)	h	225.97	2599.2		2610.6	2648.8	2686.9	2725.1	2763.5	2802.0
	s	0.7549	8.0093		8.0440	8.1551	8.2601	8.3599	8.4551	8.5460

Abs. Press. kPa (Sat. Temp, C)		Sat. Liquid	Sat. Vapor	80.	100.	120.	140.	160.	180.	200.
20.0	v	0.0010	7.650	8.1172	8.5847	9.0508	9.516	9.980	10.444	10.907
(60.09)	h	251.45	2609.9	2648.0	2686.3	2724.6	2763.1	2801.6	2840.3	2879.2
	s	0.8321	7.9094	8.0206	8.1261	8.2262	8.3215	8.4127	8.5000	8.5839
30.0	v	0.0010	5.229	5.4007	5.7144	6.0267	6.3379	6.6483	6.9582	7.2675
(69.12)	h	239.30	2625.4	2646.5	2685.1	2723.6	2762.3	2801.0	2839.8	2878.7
	s	0.9441	7.7695	7.8300	7.9363	8.0370	8.1329	8.2243	8.3119	8.3960
40.0	v	0.0010	3.993	4.0424	4.2792	4.5146	4.7489	4.9825	5.2154	5.4478
(75.89)	h	317.65	2636.9	2644.9	2683.8	2722.6	2761.4	2800.3	2839.2	2878.2
	s	1.0261	7.6709	7.6937	7.8009	7.9023	7.9985	8.0903	8.1782	8.2625
50.0	v	0.0010	3.240		3.4181	3.6074	3.7955	3.9829	4.1697	4.3560
(81.35)	h	340.56	2646.0		2682.6	2721.6	2760.6	2799.6	2838.6	2877.7
	s	1.0912	7.5947		7.6953	7.7972	7.8940	7.9861	8.0742	8.1587

v = specific volume, m³/kg h = enthalpy, kJ/kg s = entropy, kJ/kg · K

Properties of Substances-Tables

Abs. Press kPa (Sat. Temp, C)		180.	200.	Temperature—C 300.	400.	500.	600.	700.
1.0 (6.983)	v h s	209.12 2841.4 9.8843	218.35 2880.1 9.9679	264.51 3076.8 10.3450	310.66 3279.7 10.6711	356.81 3489.2 10.9612	402.97 3705.6 11.2243	449.12 3928.9 11.4663
2.0 (17.51)	v h s	104.55 2841.3 9.5643	109.17 2880.0 9.6479	132.25 3076.8 10.0251	155.33 3279.7 10.3512	178.41 3489.2 10.6413	201.48 3705.6 10.9044	224.56 3928.8 11.1464
3.0 (24.10)	v h s	69.698 2841.3 9.3771	72.777 2880.0 9.4607	88.165 3076.8 9.8379	103.55 3279.7 10.1641	118.94 3489.2 10.4541	134.32 3705.6 10.7173	149.70 3928.8 10.9593
4.0 (28.98)	v h s	52.270 2841.2 9.2443	54.580 2879.9 9.3279	66.122 3076.8 9.7051	77.662 3279.7 10.0313	89.201 3489.2 10.3214	100.74 3705.6 10.5845	112.28 3928.8 10.8265
5.0 (32.90)	v h s	41.814 2841.2 9.1412	43.661 2879.9 9.2248	52.897 3076.7 9.6021	62.129 3279.7 9.9283	71.360 3489.2 10.2184	80.592 3705.6 10.4815	89.822 3928.8 10.7235
6.0 (36.18)	v h s	34.843 2841.1 9.0569	36.383 2879.8 9.1406	44.079 3076.7 9.5179	51.773 3279.6 9.8441	59.467 3489.2 10.1342	67.159 3705.6 10.3973	74.852 3928.8 10.6394
7.0 (41.53)	v h s	26.129 2841.0 8.9240	27.284 2879.7 9.0077	33.058 3076.7 9.3851	38.829 3279.6 9.7113	44.599 3489.1 10.0014	50.369 3705.5 10.2646	56.138 3928.8 10.5066
10.0 (45.83)	v h s	20.900 2840.9 8.8208	21.825 2879.6 8.9045	26.445 3076.6 9.2820	31.062 3279.6 9.6083	35.679 3489.1 9.8984	40.295 3705.5 10.1616	44.910 3928.8 10.4036
15.0 (54.00)	v h s	13.929 2840.6 8.6332	14.546 2879.4 8.7170	17.628 3076.5 9.0948	20.707 3279.5 9.4211	23.785 3489.1 9.7112	26.863 3705.5 9.9744	29.940 3928.8 10.2164

		240.	280.	300.	400.	500.	600.	700.
20.0 (60.09)	v h s	11.832 2957.4 8.7426	12.295 2996.9 8.8180	13.219 3076.4 8.9618	15.529 3279.4 9.2882	17.838 3489.0 9.5784	20.146 3705.4 9.8416	22.455 3928.7 10.0836
30.0 (69.12)	v h s	7.8854 2957.1 8.5550	8.5024 3036.2 8.7035	8.8108 3076.1 8.7744	10.351 3279.3 9.1010	11.891 3488.9 9.3912	13.430 3705.4 9.6544	14.969 3928.7 9.8965
40.0 (75.89)	v h s	5.9118 2956.7 8.4217	6.3751 3036.0 8.5704	6.6065 3075.9 8.6413	7.7625 3279.1 8.9680	8.9176 3488.8 9.2583	10.072 3705.3 9.5216	11.227 3928.6 9.7636
50.0 (81.35)	v h s	4.7277 2956.4 8.3182	5.0986 3035.7 8.4671	5.2839 3075.7 8.5380	6.2091 3279.0 8.8649	7.1335 3488.7 9.1552	8.0574 3705.2 9.4185	8.9810 3928.6 9.6606

Abs. Press. kPa (Sat. Temp, C)	Sat. Liquid	Sat. Vapor	Temperature—C					

Abs. Press. kPa (Sat. Temp, C)	Sat. Liquid	Sat. Vapor	100.	120.	140.	160.	180.	200.
v	0.0010	2.732	2.8440	3.0025	3.1599	3.3165	3.4726	3.6281
60.0 h	359.93	2653.6	2681.3	2720.6	2759.8	2798.9	2838.1	2877.3
(85.95) s	1.1454	7.5327	7.6085	7.7111	7.8083	7.9008	7.9891	8.0738
v	0.0010	2.0870	2.1262	2.2464	2.3654	2.4836	2.6011	2.7183
80.0 h	391.72	2665.8	2678.8	2718.6	2758.1	2797.5	2836.9	2876.3
(93.51) s	1.2330	7.4352	7.4703	7.5742	7.6723	7.7655	7.8544	7.9395
v	0.0010	1.6937	1.6955	1.7927	1.8886	1.9838	2.0783	2.1723
100.0 h	417.51	2675.4	2676.2	2716.5	2756.4	2796.2	2835.8	2875.4
(99.63) s	1.3027	7.3598	7.3618	7.4670	7.5662	7.6601	7.7495	7.8349
v	0.0011	1.1590		1.1876	1.2529	1.3173	1.3811	1.4444
150.0 h	467.13	2693.4		2711.2	2752.2	2792.7	2832.9	2872.9
(111.4) s	1.4336	7.2234		7.2693	7.3709	7.4667	7.5574	7.6439
v	0.0011	0.8854			0.9349	0.9840	1.0325	1.0804
200.0 h	504.70	2706.3			2747.8	2789.1	2830.0	2870.5
(120.2) s	1.5301	7.1268			7.2298	7.3275	7.4196	7.5072
v	0.0011	0.6056			0.6167	0.6506	0.6837	0.7164
300.0 h	561.4	2724.7			2738.8	2781.8	2824.0	2865.5
(133.5) s	1.6716	6.9909			7.0254	7.1271	7.2222	7.3119
v	0.0011	0.4622				0.4837	0.5093	0.5343
400.0 h	604.7	2737.6				2774.2	2817.8	2860.4
(143.6) s	1.7764	6.8943				6.9805	7.0788	7.1708

	Sat. Liquid	Sat. Vapor	200.	240.	280.	300.	340.	380.	400.
v	0.0011	0.3747	0.4250	0.4647	0.4841	0.5226	0.5606	0.5984	0.6172
500.0 h	640.1	2747.5	2855.1	2940.1	2981.9	3064.8	3147.4	3230.4	3272.1
(151.8) s	1.8604	6.8192	7.0592	7.2317	7.3115	7.4614	7.6008	7.7319	7.7948
v	0.0011	0.3155	0.3520	0.3857	0.4021	0.4344	0.4663	0.4979	0.5136
600.0 h	670.4	2755.5	2849.7	2936.4	2978.7	3062.3	3145.4	3228.7	3270.6
(158.8) s	1.9308	6.7575	6.9662	7.1419	7.2228	7.3740	7.5143	7.6459	7.7090
v	0.0011	0.2403	0.2608	0.2869	0.2995	0.3241	0.3483	0.3723	0.3842
800.0 h	720.9	2767.5	2838.6	2928.6	2972.1	3057.3	3141.4	3225.4	3267.5
(170.4) s	2.0457	6.6596	6.8148	6.9976	7.0807	7.2348	7.3767	7.5094	7.5729
v	0.0011	0.1943	0.2059	0.2276	0.2379	0.2580	0.2776	0.2969	0.3065
1000.0 h	762.6	2776.3	2826.8	2920.6	2965.2	3052.1	3137.4	3222.0	3264.4
(179.9) s	2.1382	6.5828	6.6922	6.8825	6.9680	7.1251	7.2689	7.4027	7.4665
v	0.0012	0.1317	0.1324	0.1483	0.1556	0.1697	0.1832	0.1964	0.2029
1500.0 h	844.7	2789.9	2794.7	2899.2	2947.3	3038.9	3127.0	3213.5	3256.6
(198.3) v	2.3145	6.4406	6.4508	6.6630	6.7550	6.9207	7.0693	7.2060	7.2709
v	0.0012	0.09954		0.1084	0.1144	0.1255	0.1360	0.1461	0.1511
2000.0 h	908.6	2797.2		2875.0	2928.1	3025.0	3116.3	3204.9	3248.7
(212.4) s	2.4469	6.3367		6.4943	6.5941	6.7696	6.9235	7.0635	7.1296

Abs. Press. kPa (Sat. Temp, C)		Temperature—C						
		240.	280.	300.	400.	500.	600.	700.
60.0 (85.95)	v	3.9383	4.2477	4.4022	5.1736	5.9441	6.7141	7.4839
	h	2956.0	3035.4	3075.4	3278.8	3488.6	3705.1	3928.5
	s	8.2336	8.3826	8.4536	8.7806	9.0710	9.3343	9.5764
80.0 (93.51)	v	2.9515	3.1840	3.3000	3.8792	4.4574	5.0351	5.6126
	h	2955.3	3034.9	3075.0	3278.5	3488.4	3705.0	3928.4
	s	8.0998	8.2491	8.3202	8.6475	8.9380	9.2014	9.4436
100.0 (99.63)	v	2.3595	2.5458	2.6387	3.1025	3.5653	4.0277	4.4898
	h	2954.6	3034.4	3074.5	3278.2	3488.1	3704.8	3928.2
	s	7.9958	8.1454	8.2166	8.5442	8.8348	9.0982	9.3405
150.0 (111.4)	v	1.5700	1.6948	1.7570	2.0669	2.3759	2.6845	2.9927
	h	2952.9	3033.0	3073.3	3277.5	3487.6	3704.4	3927.9
	s	7.8061	7.9565	8.0280	8.3562	8.6472	8.9108	9.1531
200.0 (120.2)	v	1.1753	1.2693	1.3162	1.5492	1.7812	2.0129	2.2442
	h	2951.1	3031.7	3072.1	3276.7	3487.0	3704.0	3927.6
	s	7.6707	7.8219	7.8937	8.2226	8.5139	8.7776	9.0201
300.0 (133.5)	v	0.7805	0.8438	0.8753	1.0314	1.1865	1.3412	1.4957
	h	2947.5	3028.9	3069.7	3275.2	3486.0	3703.2	3927.0
	s	7.4783	7.6311	7.7034	8.0338	8.3257	8.5898	8.8325
400.0 (143.6)	v	0.5831	0.6311	0.6549	0.7725	0.8892	1.0054	1.1214
	h	2943.9	3026.2	3067.2	3273.6	3484.9	3702.3	3926.4
	s	7.3402	7.4947	7.5675	7.8994	8.1919	8.4563	8.6992
		440.	480.	500.	600.	650.	700.	800.
500.0 (151.8)	v	0.6547	0.6921	0.7108	0.8039	0.8504	0.8968	0.9896
	h	3356.1	3441.0	3483.8	3701.5	3812.8	3925.8	4156.4
	s	7.9160	8.0318	8.0879	8.3526	8.4766	8.5957	8.8213
600.0 (158.8)	v	0.5450	0.5762	0.5918	0.6696	0.7084	0.7471	0.8245
	h	3354.8	3439.8	3482.7	3700.7	3812.1	3925.1	4155.9
	s	7.8305	7.9465	8.0027	8.2678	8.3919	8.5111	8.7368
800.0 (170.4)	v	0.4078	0.4314	0.4432	0.5017	0.5309	0.5600	0.6181
	h	3352.1	3437.5	3480.5	3699.1	3810.7	3923.9	4155.0
	s	7.6950	7.8115	7.8678	8.1336	8.2579	8.3773	8.6033
1000.0 (179.9)	v	0.3256	0.3445	0.3540	0.4010	0.4244	0.4477	0.4943
	h	3349.5	3435.1	3478.3	3697.4	3809.3	3922.7	4154.1
	s	7.5893	7.7062	7.7627	8.0292	8.1537	8.2734	8.4997
1500.0 (198.3)	v	0.2158	0.2287	0.2350	0.2667	0.2824	0.2980	0.3292
	h	3342.8	3429.3	3472.8	3693.3	3805.7	3919.6	4151.7
	s	7.3953	7.5133	7.5703	7.8385	7.9636	8.0838	8.3108
2000.0 (212.4)	v	0.1610	0.1707	0.1756	0.1995	0.2114	0.2232	0.2467
	h	3336.0	3423.4	3467.3	3689.2	3802.1	3916.5	4149.4
	s	7.2555	7.3748	7.4323	7.7022	7.8279	7.9485	8.1763

Abs. Press. kPa (Sat. Temp. C)		Sat. Liquid	Sat. Vapor	Temperature—C					
				240.	280.	300.	340.	380.	400.
3000.0 (233.8)	v	0.0012	0.0666	0.06816	0.07712	0.08166	0.08871	0.09584	0.09931
	h	1008.4	2802.3	2822.9	2942.0	2995.1	3093.9	3187.0	3232.5
	s	2.6455	6.1837	6.2241	6.4479	6.5422	6.7088	6.8561	6.9246
4000.0 (250.3)	v	0.0013	0.0498		0.05544	0.05883	0.06499	0.07066	0.07338
	h	1087.4	2800.3		2902.0	2962.0	3069.8	3168.4	3215.7
	s	2.7965	6.0685		6.2576	6.3642	6.5461	6.7019	6.7733
5000.0 (263.9)	v	0.0013	0.03943		0.04222	0.04530	0.05070	0.05551	0.05779
	h	1154.5	2794.2		2856.9	2925.5	3044.1	3148.8	3198.3
	s	2.9206	5.9735		6.0886	6.2105	6.4106	6.5762	6.6508
6000.0 (275.5)	v	0.0013	0.0324		0.03317	0.03614	0.04111	0.04539	0.04738
	h	1213.7	2785.0		2804.9	2885.0	3016.5	3128.3	3180.1
	s	3.0273	5.8908		5.9270	6.0692	6.2913	6.4680	6.5462
8000.0 (295.0)	v	0.0014	0.0235			0.02426	0.02896	0.03265	0.03431
	h	1317.1	2759.9			2786.8	2955.3	3084.2	3141.6
	s	3.2076	5.7471			5.7942	6.0790	6.2828	6.3694

Abs. Press. kPa (Sat. Temp. C)		Sat. Liquid	Sat. Vapor	320.	360.	380.	400.	440.	480.	500.
10000.0 (311.0)	v	0.0015	0.01804	0.01926	0.02331	0.02493	0.02641	0.02911	0.03158	0.03276
	h	1408.8	2727.7	2783.5	2964.8	3035.7	3099.9	3216.2	3323.2	3374.6
	s	3.3605	5.6198	5.7145	6.0110	6.1213	6.2182	6.3861	6.5321	6.5994
15000.0 (342.1)	v	0.0017	0.0134		0.01256	0.01428	0.01566	0.01794	0.01989	0.02080
	h	1611.0	2615.0		2770.8	2887.7	2979.1	3126.9	3252.4	3310.6
	s	3.6859	5.3178		5.5677	5.7497	5.8876	6.1010	6.2724	6.3487
20000.0 (365.7)	v	0.0020	0.0059			0.008246	0.009947	0.01224	0.01399	0.01477
	h	1826.5	2418.4			2660.2	2820.5	3023.7	3174.4	3241.1
	s	4.0149	4.9412			5.3165	5.5585	5.8523	6.0581	6.1456
30000.0	v					0.001874	0.002831	0.006227	0.007985	0.008681
	h					1837.7	2161.8	2754.0	2993.9	3085.0
	s					4.0021	4.4896	5.3499	5.6779	5.7972
40000.0	v					0.001682	0.001909	0.003200	0.004941	0.005616
	h					1776.4	1934.1	2399.4	2779.8	2906.8
	s					3.8814	4.1190	4.7893	5.3097	5.4762
50000.0	v					0.001589	0.001729	0.002269	0.003308	0.003882
	h					1746.8	1877.7	2199.7	2564.9	2723.0
	s					3.8110	4.0083	4.4723	4.9709	5.1782
60000.0	v					0.001528	0.001632	0.001962	0.002565	0.002952
	h					1728.4	1847.3	2113.5	2418.8	2570.6
	s					3.7589	3.9383	4.3221	4.7385	4.9374
80000.0	v					0.001445	0.001518	0.001710	0.001999	0.002188
	h					1707.0	1814.2	2036.6	2272.8	2397.4
	s					3.6807	3.8425	4.1633	4.4855	4.6488
100000.0	v					0.001390	0.001446	0.001587	0.001777	0.001893
	h					1696.3	1797.6	2000.3	2207.7	2316.1
	s					3.6211	3.7738	4.0664	4.3492	4.4913

Properties of Substances-Tables

Abs. Press. kPa (Sat. Temp, C)		440.	480.	500.	600.	650.	700.	800.
					Temperature—C			
3000.0 (233.8)	v	0.1061	0.1128	0.1161	0.1323	0.1404	0.1483	0.1641
	h	3322.3	3411.6	3456.2	3681.0	3795.0	3910.3	4144.7
	s	7.0543	7.1760	7.2345	7.5079	7.6349	7.7564	7.9857
4000.0 (250.3)	v	0.07866	0.08381	0.08634	0.09876	0.1049	0.1109	0.1229
	h	3308.3	3399.6	3445.0	3672.8	3787.9	3904.1	4140.0
	s	6.9069	7.0314	7.0909	7.3680	7.4961	7.6187	7.8495
5000.0 (263.9)	v	0.06218	0.06642	0.06849	0.07862	0.08356	0.08845	0.09809
	h	3294.0	3387.4	3433.7	3664.5	3780.7	3897.9	4135.3
	s	6.7890	6.9164	6.9770	7.2578	7.3872	7.5108	7.7431
6000.0 (275.5)	v	0.05118	0.05482	0.05659	0.06518	0.06936	0.07348	0.08159
	h	3279.3	3375.0	3422.2	3656.2	3773.5	3891.7	4130.7
	s	6.6893	6.8199	6.8818	7.1664	7.2971	7.4217	7.6554
8000.0 (295.0)	v	0.03740	0.04030	0.04170	0.04839	0.05161	0.05477	0.06096
	h	3248.7	3349.6	3398.8	3639.5	3759.2	3879.2	4121.3
	s	6.5240	6.6617	6.7262	7.0191	7.1523	7.2790	7.5158
		540.	580.	600.	650.	700.	750.	800.
10000.0 (311.0)	v	0.03504	0.03724	0.03832	0.04096	0.04355	0.04608	0.04858
	h	3475.1	3573.7	3622.7	3744.7	3866.8	3989.1	4112.0
	s	6.7261	6.8446	6.9013	7.0373	7.1660	7.2886	7.4058
15000.0 (342.1)	v	0.02250	0.02411	0.02488	0.02677	0.02859	0.03036	0.03209
	h	3421.4	3527.7	3579.8	3708.3	3835.4	3962.1	4088.6
	s	6.4885	6.6160	6.6764	6.8195	6.9536	7.0806	7.2013
20000.0 (365.7)	v	0.01621	0.01753	0.01816	0.01967	0.02111	0.02250	0.02385
	h	3364.7	3479.9	3535.5	3671.1	3803.8	3935.0	4065.3
	s	6.3015	6.4398	6.5043	6.6554	6.7953	6.9267	7.0511
30000.0	v	0.009890	0.01095	0.01144	0.01258	0.01365	0.01465	0.01562
	h	3241.7	3378.9	3443.0	3595.0	3739.7	3880.3	4018.5
	s	5.9949	6.1597	6.2340	6.4033	6.5560	6.6970	6.8288
40000.0	v	0.006735	0.007667	0.008088	0.009053	0.009930	0.01075	0.01152
	h	3108.0	3272.4	3346.4	3517.0	3674.8	3825.5	3971.7
	s	5.7302	5.9276	6.0135	6.2035	6.3701	6.5210	6.6606
50000.0	v	0.004888	0.005734	0.006111	0.006960	0.007720	0.008420	0.009076
	h	2968.9	3163.2	3248.3	3438.9	3610.2	3770.9	3925.3
	s	5.4886	5.7221	5.8207	6.0331	6.2138	6.3749	6.5222
60000.0	v	0.003755	0.004496	0.004835	0.005596	0.006269	0.006885	0.007460
	h	2838.3	3055.8	3151.6	3362.4	3547.0	3717.4	3879.6
	s	5.2755	5.5367	5.6477	5.8827	6.0775	6.2483	6.4031
80000.0	v	00.02641	0.003132	0.003379	0.003974	0.004519	0.005017	0.005481
	h	2648.2	2874.9	2980.3	3220.3	3428.7	3616.7	3792.8
	s	4.9650	5.2374	5.3595	5.6270	5.8470	6.0354	6.2034
100000.0	v	0.002168	0.002493	0.002668	0.003106	0.003536	0.003952	0.004341
	h	2538.6	2754.5	2857.5	3105.3	3324.4	3526.1	3714.3
	s	4.7719	5.0311	5.1505	5.4267	5.6579	5.8600	6.0397

TABLE A-18E PROPERTIES OF WATER: SATURATED STEAM, TEMPERATURE TABLE (ENGLISH UNITS)

Temp Fahr. T	Abs. Press lbf/in.² p	Specific Volume, ft³/lbm			Enthalpy, Btu/lbm			Entropy, Btu/lbm R			Temp Fahr. T
		Sat. Liquid v_f	Evap. v_{fg}	Sat. Vapor v_g	Sat. Liquid h_f	Evap. h_{fg}	Sat. Vapor h_g	Sat. Liquid s_f	Evap. s_{fg}	Sat. Vapor s_g	
32.0*	0.08859	0.016022	3304.7	3304.7	−0.0179	1075.5	1075.5	0.0000	2.1873	2.1873	32.0*
34.0	0.09600	0.016021	3061.9	3061.9	1.996	1074.4	1076.4	0.0041	2.1762	2.1802	34.0
36.0	0.10395	0.016020	2839.0	2839.0	4.008	1073.2	1077.2	0.0081	2.1651	2.1732	36.0
38.0	0.11249	0.016019	2634.1	2634.2	6.018	1072.1	1078.1	0.0122	2.1541	2.1663	38.0
40.0	0.12163	0.016019	2445.8	2445.8	8.027	1071.0	1079.0	0.0162	2.1432	2.1594	40.0
42.0	0.13143	0.016019	2272.4	2272.4	10.035	1069.8	1079.9	0.0202	2.1325	2.1527	42.0
44.0	0.14192	0.016019	2112.8	2112.8	12.041	1068.7	1080.7	0.0242	2.1217	2.1459	44.0
46.0	0.15314	0.016020	1965.7	1965.7	14.047	1067.6	1081.6	0.0282	2.1111	2.1393	46.0
48.0	0.16514	0.016021	1830.0	1830.0	16.051	1066.4	1082.5	0.0321	2.1006	2.1327	48.0
50.0	0.17796	0.016023	1704.8	1704.8	18.054	1065.3	1083.4	0.0361	2.0901	2.1262	50.0
52.0	0.19165	0.016024	1589.2	1589.2	20.057	1064.2	1084.2	0.0400	2.0798	2.1197	52.0
54.0	0.20625	0.016026	1482.4	1482.4	22.058	1063.1	1085.1	0.0439	2.0695	2.1134	54.0
56.0	0.22183	0.016028	1383.6	1383.6	24.059	1061.9	1086.0	0.0478	2.0593	2.1070	56.0
58.0	0.23843	0.016031	1292.2	1292.2	26.060	1060.8	1086.9	0.0516	2.0491	2.1008	58.0
60.0	0.25611	0.016033	1207.6	1207.6	28.060	1059.7	1087.7	0.0555	2.0391	2.0946	60.0
62.0	0.27494	0.016036	1129.2	1129.2	30.059	1058.5	1088.6	0.0593	2.0291	2.0885	62.0
64.0	0.29497	0.016039	1056.5	1056.5	32.058	1057.4	1089.5	0.0632	2.0192	2.0824	64.0
66.0	0.31626	0.016043	989.0	989.1	34.056	1056.3	1090.4	0.0670	2.0094	2.0764	66.0
68.0	0.33889	0.016046	926.5	926.5	36.054	1055.2	1091.2	0.0708	1.9996	2.0704	68.0
70.0	0.36292	0.016050	868.3	868.4	38.052	1054.0	1092.1	0.0745	1.9900	2.0645	70.0
72.0	0.38844	0.016054	814.3	814.3	40.049	1052.9	1093.0	0.0783	1.9804	2.0587	72.0
74.0	0.41550	0.016058	764.1	764.1	42.046	1051.8	1093.8	0.0821	1.9708	2.0529	74.0
76.0	0.44420	0.016063	717.4	717.4	44.043	1050.7	1094.7	0.0858	1.9614	2.0472	76.0
78.0	0.47461	0.016067	673.8	673.9	46.040	1049.5	1095.6	0.0895	1.9520	2.0415	78.0

* Approximate triple point.

Temp											Temp
80.0	0.50683	0.016072	633.3	633.3	48.037	1048.4	1096.4	0.0932	1.9426	2.0359	80.0
82.0	0.54093	0.016077	595.5	595.5	50.033	1047.3	1097.3	0.0969	1.9334	2.0303	82.0
84.0	0.57702	0.016082	560.3	560.3	52.029	1046.1	1098.2	0.1006	1.9242	2.0248	84.0
86.0	0.61518	0.016087	227.5	527.5	54.026	1045.0	1099.0	0.1043	1.9151	2.0193	86.0
88.0	0.65551	0.016093	496.8	496.8	56.022	1043.9	1099.9	0.1079	1.9060	2.0139	88.0
90.0	0.69813	0.016099	468.1	468.1	58.018	1042.7	1100.8	0.1115	1.8970	2.0086	90.0
92.0	0.74313	0.016105	441.3	441.3	60.014	1041.6	1101.6	0.1152	1.8881	2.0033	92.0
94.0	0.79062	0.016111	416.3	416.3	62.010	1040.5	1102.5	0.1188	1.8792	1.9980	94.0
96.0	0.84072	0.016117	392.8	392.9	64.006	1039.3	1103.3	0.1224	1.8704	1.9928	96.0
98.0	0.89356	0.016123	370.9	370.9	66.003	1038.2	1104.2	0.1260	1.8617	1.9876	98.0
100.0	0.94924	0.016130	350.4	350.4	67.999	1037.1	1105.1	0.1295	1.8530	1.9825	100.0
102.0	1.00789	0.016137	331.1	331.1	69.995	1035.9	1105.9	0.1331	1.8444	1.9775	102.0
104.0	1.06965	0.016144	313.1	313.1	71.992	1034.8	1106.8	0.1366	1.8358	1.9725	104.0
106.0	1.1347	0.016151	296.18	296.16	73.99	1033.6	1107.6	0.1402	1.8273	1.9675	106.0
108.0	1.2030	0.016158	280.30	280.28	75.98	1032.5	1108.5	0.1437	1.8188	1.9626	108.0
110.0	1.2750	0.016165	265.39	265.37	77.98	1031.4	1109.3	0.1472	1.8105	1.9577	110.0
112.0	1.3505	0.016173	251.38	251.37	79.98	1030.2	1110.2	0.1507	1.8021	1.9528	112.0
114.0	1.4299	0.016180	238.22	238.21	81.97	1029.1	1111.0	0.1542	1.7938	1.9480	114.0
116.0	1.5133	0.016188	225.85	225.84	83.97	1027.9	1111.9	0.1577	1.7856	1.9433	116.0
118.0	1.6009	0.016196	214.21	214.20	85.97	1026.8	1112.7	0.1611	1.7774	1.9386	118.0
120.0	1.6927	0.016204	203.26	203.25	87.97	1025.6	1113.6	0.1646	1.7693	1.9339	120.0
122.0	1.7891	0.016213	192.95	192.94	89.96	1024.5	1114.4	0.1680	1.7613	1.9293	122.0
124.0	1.8901	0.016221	183.24	183.23	91.96	1023.3	1115.3	0.1715	1.7533	1.9247	124.0
126.0	1.9959	0.016229	174.09	174.08	93.96	1022.2	1116.1	0.1749	1.7453	1.9202	126.0
128.0	2.1068	0.016238	165.47	165.45	95.96	1021.0	1117.0	0.1783	1.7374	1.9157	128.0
130.0	2.2230	0.016247	157.33	157.32	97.96	1019.8	1117.8	0.1817	1.7295	1.9112	130.0
132.0	2.3445	0.016256	149.66	149.64	99.95	1018.7	1118.6	0.1851	1.7217	1.9068	132.0
134.0	2.4717	0.016265	142.41	142.40	101.95	1017.5	1119.5	0.1884	1.7140	1.9024	134.0
136.0	2.6047	0.016274	135.57	135.55	103.95	1016.4	1120.3	0.1918	1.7063	1.8980	136.0
138.0	2.7438	0.016284	129.11	129.09	105.95	1015.2	1121.1	0.1951	1.6986	1.8937	138.0
140.0	2.8892	0.016293	123.00	122.98	107.95	1014.0	1122.0	0.1985	1.6910	1.8895	140.0
142.0	3.0411	0.016303	117.22	117.21	109.95	1012.9	1122.8	0.2018	1.6834	1.8852	142.0
144.0	3.1997	0.016312	111.76	111.74	111.95	1011.7	1123.6	0.2051	1.6759	1.8810	144.0
146.0	3.3653	0.016322	106.59	106.58	113.95	1010.5	1124.5	0.2084	1.6684	1.8769	146.0
148.0	3.5381	0.016332	101.70	101.68	115.95	1009.3	1125.3	0.2117	1.6610	1.8727	148.0
150.0	3.7184	0.016343	97.07	97.05	117.95	1008.2	1126.1	0.2150	1.6536	1.8686	150.0
152.0	3.9065	0.016353	92.68	92.66	119.95	1007.0	1126.9	0.2183	1.6463	1.8646	152.0
154.0	4.1025	0.016363	88.52	88.50	121.95	1005.8	1127.7	0.2216	1.6390	1.8606	154.0
156.0	4.3068	0.016374	84.57	84.56	123.95	1004.6	1128.6	0.2248	1.6318	1.8566	156.0
158.0	4.5197	0.016384	80.83	80.82	125.96	1003.4	1129.4	0.2281	1.6245	1.8526	158.0
160.0	4.7414	0.016395	77.29	77.27	127.96	1002.2	1130.2	0.2313	1.6174	1.8487	160.0
162.0	4.9722	0.016406	73.92	73.90	129.96	1001.0	1131.0	0.2345	1.6103	1.8448	162.0
164.0	5.2124	0.016417	70.72	70.70	131.96	999.8	1131.8	0.2377	1.6032	1.8409	164.0
166.0	5.4623	0.016428	67.68	67.67	133.97	998.6	1132.6	0.2409	1.5961	1.8371	166.0
168.0	5.7223	0.016440	64.80	64.78	135.97	997.4	1133.4	0.2441	1.5892	1.8333	168.0

Appendix 3

TABLE A-18E (Continued)

Temp Fahr. T	Abs. Press lbf/in.² p	Specific Volume, ft³/lbm Sat. Liquid v_f	Evap. v_{fg}	Sat. Vapor v_g	Enthalpy, Btu/lbm Sat. Liquid h_f	Evap. h_{fg}	Sat. Vapor h_g	Entropy, Btu/lbm R Sat. Liquid s_f	Evap. s_{fg}	Sat. Vapor s_g	Temp Fahr. T
170.0	5.9926	0.016451	62.04	62.06	137.97	996.2	1134.2	0.2473	1.5822	1.8295	170.0
172.0	6.2735	0.016463	59.43	59.45	139.98	995.0	1135.0	0.2505	1.5753	1.8258	172.0
174.0	6.5656	0.016474	56.95	56.97	141.98	993.8	1135.8	0.2537	1.5684	1.8221	174.0
176.0	6.8690	0.016486	54.59	54.61	143.99	992.6	1136.6	0.2568	1.5616	1.8184	176.0
178.0	7.1840	0.016498	52.35	52.36	145.99	991.4	1137.4	0.2600	1.5548	1.8147	178.0
180.0	7.5110	0.016510	50.21	50.22	148.00	990.2	1138.2	0.2631	1.5480	1.8111	180.0
182.0	7.850	0.016522	48.172	48.189	150.01	989.0	1139.0	0.2662	1.5413	1.8075	182.0
184.0	8.203	0.016534	46.232	46.249	152.01	987.8	1139.8	0.2694	1.5346	1.8040	184.0
186.0	8.568	0.016547	44.383	44.400	154.02	986.5	1140.5	0.2725	1.5279	1.8004	186.0
188.0	8.947	0.016559	42.621	42.638	156.03	985.3	1141.3	0.2756	1.5213	1.7969	188.0
190.0	9.340	0.016572	40.941	40.957	158.04	984.1	1142.1	0.2787	1.5148	1.7934	190.0
192.0	9.747	0.016585	39.337	39.354	160.05	982.8	1142.9	0.2818	1.5082	1.7900	192.0
194.0	10.168	0.016598	37.808	37.824	162.05	981.6	1143.7	0.2848	1.5017	1.7865	194.0
196.0	10.605	0.016611	36.348	36.364	164.06	980.4	1144.4	0.2879	1.4952	1.7831	196.0
198.0	11.058	0.016624	34.954	34.970	166.08	979.1	1145.2	0.2910	1.4888	1.7798	198.0
200.0	11.526	0.016637	33.622	33.639	168.09	977.9	1146.0	0.2940	1.4824	1.7764	200.0
208.0	13.568	0.016691	28.862	28.878	176.14	972.8	1149.0	0.3061	1.4571	1.7632	208.0
216.0	15.901	0.016747	24.878	24.894	184.20	967.8	1152.0	0.3181	1.4323	1.7505	216.0
224.0	18.556	0.016805	21.529	21.545	192.27	962.6	1154.9	0.3300	1.4081	1.7380	224.0
232.0	21.567	0.016864	18.701	18.718	200.35	957.4	1157.8	0.3417	1.3842	1.7260	232.0
240.0	24.968	0.016926	16.304	16.321	208.45	952.1	1160.6	0.3533	1.3609	1.7142	240.0
248.0	28.796	0.016990	14.264	14.281	216.56	946.8	1163.4	0.3649	1.3379	1.7028	248.0
256.0	33.091	0.017055	12.520	12.538	224.69	941.4	1166.1	0.3763	1.3154	1.6917	256.0
264.0	37.894	0.017123	11.025	11.042	232.83	935.9	1168.7	0.3876	1.2933	1.6808	264.0
272.0	43.249	0.017193	9.738	9.755	240.99	930.3	1171.3	0.3987	1.2715	1.6702	272.0
280.0	49.200	0.017264	8.627	8.644	249.17	924.6	1173.8	0.4098	1.2501	1.6599	280.0
288.0	55.795	0.01734	7.6634	7.6807	257.4	918.8	1176.2	0.4208	1.2290	1.6498	288.0
296.0	63.084	0.01741	6.8259	6.8433	265.6	913.0	1178.6	0.4317	1.2082	1.6400	296.0
304.0	71.119	0.01749	6.0955	6.1130	273.8	907.0	1180.9	0.4426	1.1877	1.6303	304.0
312.0	79.953	0.01757	5.4566	5.4742	282.1	901.0	1183.1	0.4533	1.1676	1.6209	312.0
320.0	89.643	0.01766	4.8961	4.9138	290.4	894.8	1185.2	0.4640	1.1477	1.6116	320.0
328.0	100.245	0.01774	4.4030	4.4208	298.7	888.5	1187.2	0.4745	1.1280	1.6025	328.0
336.0	111.820	0.01783	3.9681	3.9859	307.1	882.1	1189.1	0.4850	1.1086	1.5936	336.0
344.0	124.430	0.01792	3.5834	3.6013	315.5	875.5	1191.0	0.4954	1.0894	1.5849	344.0
352.0	138.138	0.01801	3.2423	3.2603	323.9	868.9	1192.7	0.5058	1.0705	1.5763	352.0
360.0	153.010	0.01811	2.9392	2.9573	332.3	862.1	1194.4	0.5161	1.0517	1.5678	360.0
368.0	169.113	0.01821	2.6691	2.6873	340.8	855.1	1195.9	0.5263	1.0332	1.5595	368.0
376.0	186.517	0.01831	2.4279	2.4462	349.3	848.1	1197.4	0.5365	1.0148	1.5513	376.0
384.0	205.294	0.01842	2.2120	2.2304	357.9	840.8	1198.7	0.5466	0.9966	1.5432	384.0
392.0	225.516	0.01853	2.0184	2.0369	366.5	833.4	1199.9	0.5567	0.9786	1.5352	392.0

Temp											Temp
400.0	247.259	0.01864	1.8444	1.8630	375.1	825.9	1201.0	0.5667	0.9607	1.5274	**400.0**
408.0	270.600	0.01875	1.6877	1.7064	383.8	818.2	1201.9	0.5766	0.9429	1.5195	**408.0**
416.0	295.617	0.01887	1.5463	1.5651	392.5	810.2	1202.8	0.5866	0.9253	1.5118	**416.0**
424.0	322.391	0.01900	1.4184	1.4374	401.3	802.2	1203.5	0.5964	0.9077	1.5042	**424.0**
432.0	351.00	0.01913	1.30266	1.32179	410.1	793.9	1204.0	0.6063	0.8903	1.4966	**432.0**
440.0	381.54	0.01926	1.19761	1.21687	419.0	785.4	1204.4	0.6161	0.8729	1.4890	**440.0**
448.0	414.09	0.01940	1.10212	1.12152	428.0	776.7	1204.7	0.6259	0.8557	1.4815	**448.0**
456.0	448.73	0.01954	1.01518	1.03472	437.0	767.8	1204.8	0.6356	0.8385	1.4741	**456.0**
464.0	485.56	0.01969	0.93588	0.95557	446.1	758.6	1204.7	0.6454	0.8213	1.4667	**464.0**
472.0	524.67	0.01984	0.86345	0.88329	455.2	749.3	1204.5	0.6551	0.8042	1.4592	**472.0**
480.0	566.15	0.02000	0.79716	0.81717	464.5	739.6	1204.1	0.6648	0.7871	1.4518	**480.0**
488.0	610.10	0.02017	0.73641	0.75658	473.8	729.7	1203.5	0.6745	0.7700	1.4444	**488.0**
496.0	656.61	0.02034	0.68065	0.70100	483.2	719.5	1202.7	0.6842	0.7528	1.4370	**496.0**
504.0	705.78	0.02053	0.62938	0.64991	492.7	709.0	1201.7	0.6939	0.7357	1.4296	**504.0**
512.0	757.72	0.02072	0.58218	0.60289	502.3	698.2	1200.5	0.7036	0.7185	1.4221	**512.0**
520.0	812.53	0.02091	0.53864	0.55956	512.0	687.0	1199.0	0.7133	0.7013	1.4146	**520.0**
528.0	870.31	0.02112	0.49843	0.51955	521.8	675.5	1197.3	0.7231	0.6839	1.4070	**528.0**
536.0	931.17	0.02134	0.46123	0.48257	531.7	663.6	1195.4	0.7329	0.6665	1.3993	**536.0**
544.0	995.22	0.02157	0.42677	0.44834	541.8	651.3	1193.1	0.7427	0.6489	1.3915	**544.0**
552.0	1062.59	0.02182	0.39479	0.41660	552.0	638.5	1190.6	0.7525	0.6311	1.3837	**552.0**
560.0	1133.38	0.02207	0.36507	0.38714	562.4	625.3	1187.7	0.7625	0.6132	1.3757	**560.0**
568.0	1207.72	0.02235	0.33741	0.35975	572.9	611.5	1184.5	0.7725	0.5950	1.3675	**568.0**
576.0	1285.74	0.02264	0.31162	0.33426	583.7	597.2	1180.9	0.7825	0.5766	1.3592	**576.0**
584.0	1367.7	0.02295	0.28753	0.31048	594.6	582.4	1176.9	0.7927	0.5580	1.3507	**584.0**
592.0	1453.3	0.02328	0.26499	0.28827	605.7	566.8	1172.6	0.8030	0.5390	1.3420	**592.0**
600.0	1543.2	0.02364	0.24384	0.26747	617.1	550.6	1167.7	0.8134	0.5196	1.3330	**600.0**
608.0	1637.3	0.02402	0.22394	0.24796	628.8	533.6	1162.4	0.8240	0.4997	1.3238	**608.0**
616.6	1735.9	0.02444	0.20516	0.22960	640.8	515.6	1156.4	0.8348	0.4794	1.3141	**616.0**
624.0	1839.0	0.02489	0.18737	0.21226	653.1	496.6	1149.8	0.8458	0.4583	1.3041	**624.0**
632.0	1947.0	0.02539	0.17044	0.19583	665.9	476.4	1142.2	0.8571	0.4364	1.2934	**632.0**
640.0	2059.9	0.02595	0.15427	0.18021	679.1	454.6	1133.7	0.8686	0.4134	1.2821	**640.0**
648.0	2178.1	0.02657	0.13876	0.16534	692.9	431.1	1124.0	0.8806	0.3893	1.2699	**648.0**
656.0	2301.7	0.02728	0.12387	0.15115	707.4	405.7	1113.1	0.8931	0.3637	1.2567	**656.0**
664.0	2431.1	0.02811	0.10947	0.13757	722.9	377.7	1100.6	0.9064	0.3361	1.2425	**664.0**
672.0	2566.6	0.02911	0.09514	0.12424	740.2	345.7	1085.9	0.9212	0.3054	1.2266	**672.0**
680.0	2708.6	0.03037	0.08080	0.11117	758.5	310.1	1068.5	0.9365	0.2720	1.2086	**680.0**
688.0	2857.4	0.03204	0.06595	0.09799	778.8	268.2	1047.0	0.9535	0.2337	1.1872	**688.0**
696.0	3013.4	0.03455	0.04916	0.08371	804.4	212.8	1017.2	0.9749	0.1841	1.1591	**696.0**
700.0	3094.3	0.03662	0.03857	0.07519	822.4	172.7	995.2	0.9901	0.1490	1.1390	**700.0**
702.0	3135.5	0.03824	0.03173	0.06997	835.0	144.7	979.7	1.0006	0.1246	1.1252	**702.0**
704.0	3177.2	0.04108	0.02192	0.06300	854.2	102.0	956.2	1.0169	0.0876	1.1046	**704.0**
705.0	3198.3	0.04427	0.01304	0.05730	873.0	61.4	934.4	1.0329	0.0527	1.0856	**705.0**
705.47†	3208.2	0.05078	0.00000	0.05078	906.0	0.0	906.0	1.0612	0.0000	1.0612	**705.47†**

† Critical point.

787

TABLE A-19E PROPERTIES OF WATER: SATURATED STEAM, PRESSURE TABLE (ENGLISH UNITS)

Abs. Press. lbf/in.² p	Temp Fahr. T	Sat. Liquid v_f	Evap. v_{fg}	Sat. Vapor v_g	Sat. Liquid h_f	Evap. h_{fg}	Sat. Vapor h_g	Sat. Liquid s_f	Evap. s_{fg}	Sat. Vapor s_g	Abs. Press. lbf/in.² p
		Specific Volume, ft³/lbm			Enthalpy, Btu/lbm			Entropy, Btu/lbm·R			
0.08865	32.018	0.016022	3302.4	3302.4	0.0003	1075.5	1075.5	0.0000	2.1872	2.1872	0.08865
0.50	79.586	0.016071	641.5	641.5	47.623	1048.6	1096.3	0.0925	1.9446	2.0370	0.50
1.0	101.74	0.016136	333.59	333.60	69.73	1036.1	1105.8	0.1326	1.8455	1.9781	1.0
5.0	162.24	0.016407	73.515	73.532	130.20	1000.9	1131.1	0.2349	1.6094	1.8443	5.0
10.0	193.21	0.016592	38.404	38.420	161.26	982.1	1143.3	0.2836	1.5043	1.7879	10.0
14.696	212.00	0.016719	26.782	26.799	180.17	970.3	1150.5	0.3121	1.4447	1.7568	14.696
15.0	213.03	0.016726	26.274	26.290	181.21	969.7	1150.9	0.3137	1.4415	1.7552	15.0
20.0	227.96	0.016834	20.070	20.087	196.27	960.1	1156.3	0.3358	1.3962	1.7320	20.0
30.0	250.34	0.017009	13.7266	13.7436	218.9	945.2	1164.1	0.3682	1.3313	1.6995	30.0
40.0	267.25	0.017151	10.4794	10.4965	236.1	933.6	1169.8	0.3921	1.2844	1.6765	40.0
50.0	281.02	0.017274	8.4967	8.5140	250.2	923.9	1174.1	0.4112	1.2474	1.6586	50.0
60.0	292.71	0.017383	7.1562	7.1736	262.2	915.4	1177.6	0.4273	1.2167	1.6440	60.0
70.0	302.93	0.017482	6.1875	6.2050	272.7	907.8	1180.6	0.4411	1.1905	1.6316	70.0
80.0	312.04	0.017573	5.4536	5.4711	282.1	900.9	1183.1	0.4534	1.1675	1.6208	80.0
90.0	320.28	0.017659	4.8779	4.8953	290.7	894.6	1185.3	0.4643	1.1470	1.6113	90.0
100.0	327.82	0.017740	4.4133	4.4310	298.5	888.6	1187.2	0.4743	1.1284	1.6027	100.0
110.0	334.79	0.01782	4.0306	4.0484	305.8	883.1	1188.9	0.4834	1.1115	1.5950	110.0
120.0	341.27	0.01789	3.7097	3.7275	312.6	877.8	1190.4	0.4919	1.0960	1.5879	120.0
130.0	347.33	0.01796	3.4364	3.4544	319.0	872.8	1191.7	0.4998	1.0815	1.5813	130.0
140.0	353.04	0.01803	3.2010	3.2190	325.0	868.0	1193.0	0.5071	1.0681	1.5752	140.0
150.0	358.43	0.01809	2.9958	3.0139	330.6	863.4	1194.1	0.5141	1.0554	1.5695	150.0
160.0	363.55	0.01815	2.8155	2.8336	336.1	859.0	1195.1	0.5206	1.0435	1.5641	160.0
170.0	368.42	0.01821	2.6556	2.6738	341.2	854.8	1196.0	0.5269	1.0322	1.5591	170.0
180.0	373.08	0.01827	2.5129	2.5312	346.2	850.7	1196.9	0.5328	1.0215	1.5543	180.0
190.0	377.53	0.01833	2.3847	2.4030	350.9	846.7	1197.6	0.5384	1.0113	1.5498	190.0
200.0	381.80	0.01839	2.2689	2.2873	355.5	842.8	1198.3	0.5438	1.0016	1.5454	200.0
210.0	385.91	0.01844	2.16373	2.18217	359.9	839.1	1199.1	0.5490	0.9923	1.5413	210.0
220.0	389.88	0.01850	2.06779	2.08629	364.2	835.4	1199.6	0.5540	0.9834	1.5374	220.0
230.0	393.70	0.01855	1.97991	1.99846	368.3	831.8	1200.1	0.5588	0.9748	1.5336	230.0
240.0	397.39	0.01860	1.89909	1.91769	372.3	828.4	1200.6	0.5634	0.9665	1.5299	240.0
250.0	400.97	0.01865	1.82452	1.84317	376.1	825.0	1201.1	0.5679	0.9585	1.5264	250.0
260.0	404.44	0.01870	1.75548	1.77418	379.9	821.6	1201.5	0.5722	0.9508	1.5230	260.0
270.0	407.80	0.01875	1.69137	1.71013	383.6	818.3	1201.9	0.5764	0.9433	1.5197	270.0
280.0	411.07	0.01880	1.63169	1.65049	387.1	815.1	1202.3	0.5805	0.9361	1.5166	280.0
290.0	414.25	0.01885	1.57597	1.59482	390.6	812.0	1202.6	0.5844	0.9291	1.5135	290.0
300.0	417.35	0.01889	1.52384	1.54274	394.0	808.9	1202.9	0.5882	0.9223	1.5105	300.0

400.0	444.60	0.01934	1.14162	1.16095	424.2	780.4	1204.6	0.6217	0.8630	1.4847	400.0
500.0	467.01	0.01975	0.90787	0.92762	449.5	755.1	1204.7	0.6490	0.8148	1.4639	500.0
600.0	486.20	0.02013	0.74962	0.76975	471.7	732.0	1203.7	0.6723	0.7738	1.4461	600.0
700.0	503.08	0.02050	0.63505	0.65556	491.6	710.2	1201.8	0.6928	0.7377	1.4304	700.0
800.0	518.21	0.02087	0.54809	0.56896	509.8	689.6	1199.4	0.7111	0.7051	1.4163	800.0
900.0	531.95	0.02123	0.47968	0.50091	526.7	669.7	1196.4	0.7279	0.6753	1.4032	900.0
1000.0	544.58	0.02159	0.42436	0.44596	542.6	650.4	1192.9	0.7434	0.6476	1.3910	1000.0
1100.0	556.28	0.02195	0.37863	0.40058	557.5	631.5	1189.1	0.7578	0.6216	1.3794	1100.0
1200.0	567.19	0.02232	0.34013	0.36245	571.9	613.0	1184.8	0.7714	0.5969	1.3683	1200.0
1300.0	577.42	0.02269	0.30722	0.32991	585.6	594.6	1180.2	0.7843	0.5733	1.3577	1300.0
1400.0	587.07	0.02307	0.27871	0.30178	598.8	576.5	1175.3	0.7966	0.5507	1.3474	1400.0
1500.0	596.20	0.02346	0.25372	0.27719	611.7	558.4	1170.1	0.8085	0.5288	1.3373	1500.0
1600.0	604.87	0.02387	0.23159	0.25545	624.2	540.3	1164.5	0.8199	0.5076	1.3274	1600.0
1700.0	613.13	0.02428	0.21178	0.23607	636.5	522.2	1158.6	0.8309	0.4867	1.3176	1700.0
1800.0	621.02	0.02472	0.19390	0.21861	648.5	503.8	1152.3	0.8417	0.4662	1.3079	1800.0
1900.0	628.56	0.02517	0.17761	0.20278	660.4	485.2	1145.6	0.8522	0.4459	1.2981	1900.0
2000.0	635.80	0.02565	0.16266	0.18831	672.1	466.2	1138.3	0.8625	0.4256	1.2881	2000.0
2200.0	649.45	0.02669	0.13603	0.16272	695.5	426.7	1122.2	0.8828	0.3848	1.2676	2200.0
2400.0	662.11	0.02790	0.11287	0.14076	719.0	384.8	1103.7	0.9031	0.3430	1.2460	2400.0
2600.0	673.91	0.02938	0.09172	0.12110	744.5	337.6	1082.0	0.9247	0.2977	1.2225	2600.0
2700.0	679.53	0.03029	0.08165	0.11194	757.3	312.3	1069.7	0.9356	0.2741	1.2097	2700.0
2800.0	684.96	0.03134	0.07171	0.10305	770.7	285.1	1055.8	0.9468	0.2491	1.1958	2800.0
2900.0	690.22	0.03262	0.06158	0.09420	785.1	254.7	1039.8	0.9588	0.2215	1.1803	2900.0
3000.0	695.33	0.03428	0.05073	0.08500	801.8	218.4	1020.3	0.9728	0.1891	1.1619	3000.0
3100.0	700.28	0.03681	0.03771	0.07452	824.0	169.3	993.3	0.9914	0.1460	1.1373	3100.0
3200.0	705.08	0.04472	0.01191	0.05663	875.5	56.1	931.6	1.0351	0.0482	1.0832	3200.0
3208.2*	705.47	0.05078	0.00000	0.05078	906.0	0.0	906.0	1.0612	0.0000	1.0612	3208.2*

* Approximate critical point.

TABLE A-20E PROPERTIES OF WATER: SUPERHEATED STEAM (ENGLISH UNITS)

Abs. Press. lbf/in.² (Sat. Temp. F)		Sat. Liq.	Sat. Vap.	Temperature— F													
				200	250	300	350	400	450	500	600	700	800	900	1000	1100	1200
1 (101.74)	v	0.01614	333.6	392.5	422.4	452.3	482.1	511.9	541.7	571.5	631.1	690.7	750.3	809.8	869.4	929.0	988.6
	h	69.73	1105.8	1150.2	1172.9	1195.7	1218.7	1241.8	1265.1	1288.6	1336.1	1384.5	1433.7	1483.8	1534.9	1586.8	1639.7
	s	0.1326	1.9781	2.0509	2.0841	2.1152	2.1445	2.1722	2.1985	2.2237	2.2708	2.3144	2.3551	2.3934	2.4296	2.4640	2.4969
5 (162.24)	v	0.01641	73.53	78.14	84.21	90.24	96.25	102.24	108.23	114.21	126.15	138.08	150.01	161.94	173.86	185.78	197.70
	h	130.20	1131.1	1148.6	1171.7	1194.8	1218.0	1241.3	1264.7	1288.2	1335.9	1384.3	1433.6	1483.7	1534.7	1586.7	1639.6
	s	0.2349	1.8443	1.8716	1.9054	1.9369	1.9664	1.9943	2.0208	2.0460	2.0932	2.1369	2.1776	2.2159	2.2521	2.2866	2.3194
10 (193.21)	v	0.01659	38.42	38.84	41.93	44.98	48.02	51.03	54.04	57.04	63.03	69.00	74.98	80.94	86.91	92.87	98.84
	h	161.26	1143.3	1146.6	1170.2	1193.7	1217.1	1240.6	1264.1	1287.8	1335.5	1384.0	1433.4	1483.5	1534.6	1586.6	1639.5
	s	0.2836	1.7879	1.7928	1.8273	1.8593	1.8892	1.9173	1.9439	1.9692	2.0166	2.0603	2.1011	2.1394	2.1757	2.2101	2.2430
14.696 (212.00)	v	.0167	26.799		28.42	30.52	32.60	34.67	36.72	38.77	42.86	46.93	51.00	55.06	59.13	63.19	67.25
	h	180.17	1150.5		1168.8	1192.6	1216.3	1239.9	1263.6	1287.4	1335.2	1383.8	1433.2	1483.4	1534.5	1586.5	1639.4
	s	.3121	1.7568		1.7833	1.8158	1.8459	1.8743	1.9010	1.9265	1.9739	2.0177	2.0585	2.0969	2.1332	2.1676	2.2005
15 (213.03)	v	0.01673	26.290		27.837	29.899	31.939	33.963	35.977	37.985	41.986	45.978	49.964	53.946	57.926	61.905	65.882
	h	181.21	1150.9		1168.7	1192.5	1216.2	1239.9	1263.6	1287.3	1335.2	1383.8	1433.2	1483.4	1534.5	1586.5	1639.4
	s	0.3137	1.7552		1.7809	1.8134	1.8437	1.8720	1.8988	1.9242	1.9717	2.0155	2.0563	2.0946	2.1309	2.1653	2.1982
20 (227.96)	v	0.01683	20.087		20.788	22.356	23.900	25.428	26.946	28.457	31.466	34.465	37.458	40.447	43.435	46.420	49.405
	h	196.27	1156.3		1167.1	1191.4	1215.4	1239.2	1263.0	1286.9	1334.9	1383.5	1432.9	1483.2	1534.3	1586.3	1639.3
	s	0.3358	1.7320		1.7475	1.7805	1.8111	1.8397	1.8666	1.8921	1.9397	1.9836	2.0244	2.0628	2.0991	2.1336	2.1665
25 (240.07)	v	0.01693	16.301		16.558	17.829	19.076	20.307	21.527	22.740	25.153	27.557	29.954	32.348	34.740	37.130	39.518
	h	208.52	1160.6		1165.6	1190.2	1214.5	1238.5	1262.5	1286.4	1334.6	1383.3	1432.7	1483.0	1534.2	1586.2	1639.2
	s	0.3535	1.7141		1.7212	1.7547	1.7856	1.8145	1.8415	1.8672	1.9149	1.9588	1.9997	2.0381	2.0744	2.1089	2.1418
30 (250.34)	v	0.01701	13.744			14.810	15.859	16.892	17.914	18.929	20.945	22.951	24.952	26.949	28.943	30.936	32.927
	h	218.93	1164.1			1189.0	1213.6	1237.8	1261.9	1286.0	1334.2	1383.0	1432.5	1482.8	1534.0	1586.1	1639.0
	s	0.3682	1.6995			1.7334	1.7647	1.7937	1.8210	1.8467	1.8946	1.9386	1.9795	2.0179	2.0543	2.0888	2.1217
35 (259.29)	v	0.01708	11.896			12.654	13.562	14.453	15.334	16.207	17.939	19.662	21.379	23.092	24.803	26.512	28.220
	h	228.03	1167.1			1187.8	1212.7	1237.1	1261.3	1285.5	1333.9	1382.8	1432.3	1482.7	1533.9	1586.0	1638.9
	s	0.3809	1.6872			1.7152	1.7468	1.7761	1.8035	1.8294	1.8774	1.9214	1.9624	2.0009	2.0372	2.0717	2.1046

v = specific volume, ft³/lbm h = enthalpy, Btu/lbm s = entropy, Btu/lbm · R

Superheated steam — properties tabulated as v (specific volume), h (enthalpy), s (entropy), by absolute pressure (psia) and saturation temperature (°F).

Temperature—Fahr. (upper table)

Abs. Press. (Sat. Temp.)		Sat. Liq.	Sat. Vap.	300	350	400	450	500	600	700	800	900	1000	1100	1200
40 (267.25)	v	0.01715	10.497	11.036	11.838	12.624	13.398	14.165	15.685	17.195	18.699	20.199	21.697	23.194	24.689
	h	236.14	1169.8	1186.6	1211.7	1236.4	1260.8	1285.0	1333.6	1382.5	1432.1	1482.5	1533.7	1585.8	1638.8
	s	0.3921	1.6765	1.6992	1.7312	1.7608	1.7883	1.8143	1.8624	1.9065	1.9476	1.9860	2.0224	2.0569	2.0899
45 (274.44)	v	0.01721	9.399	9.777	10.497	11.201	11.892	12.577	13.932	15.276	16.614	17.950	19.282	20.613	21.943
	h	243.49	1172.1	1185.4	1210.4	1235.7	1260.2	1284.6	1333.3	1382.3	1431.9	1482.3	1533.6	1585.7	1638.7
	s	0.4021	1.6671	1.6849	1.7173	1.7471	1.7748	1.8010	1.8492	1.8934	1.9345	1.9730	2.0093	2.0439	2.0768
50 (281.02)	v	0.01727	8.514	8.769	9.424	10.062	10.688	11.306	12.529	13.741	14.947	16.150	17.350	18.549	19.746
	h	250.21	1174.1	1184.1	1209.9	1234.9	1259.6	1284.1	1332.9	1382.0	1431.7	1482.2	1533.4	1585.6	1638.6
	s	0.4112	1.6586	1.6720	1.7048	1.7349	1.7628	1.7890	1.8374	1.8816	1.9227	1.9613	1.9977	2.0322	2.0652
55 (287.07)	v	0.01733		7.945	8.546	9.130	9.702	10.267	11.381	12.485	13.583	14.677	15.769	16.859	17.948
	h	256.43		1182.9	1208.9	1234.2	1259.1	1283.6	1332.6	1381.8	1431.5	1482.0	1533.3	1585.5	1638.5
	s	0.4196		1.6601	1.6933	1.7237	1.7518	1.7781	1.8266	1.8710	1.9121	1.9507	1.987	2.022	2.055
60 (292.71)	v	0.01738	7.174	7.257	7.815	8.354	8.881	9.400	10.425	11.438	12.446	13.450	14.452	15.452	16.450
	h	262.21	1177.6	1181.6	1208.0	1233.5	1258.5	1283.2	1332.3	1381.5	1431.3	1481.5	1532.3	1585.3	1638.4
	s	0.4273	1.6440	1.6492	1.6934	1.7134	1.7417	1.7681	1.8168	1.8612	1.9024	1.9410	1.9774	2.0120	2.0450
70 (302.93)	v	0.01748	6.205		6.664	7.133	7.590	8.039	8.922	9.793	10.659	11.522	12.382	13.240	14.097
	h	272.74	1180.6		1206.0	1232.0	1257.3	1282.2	1331.6	1381.0	1430.9	1481.5	1532.9	1585.1	1638.2
	s	0.4411	1.6316		1.6640	1.6951	1.7237	1.7504	1.7993	1.8439	1.8852	1.9238	1.9603	1.9949	2.0279

Temperature—Fahr. (lower table)

Abs. Press. (Sat. Temp.)		Sat. Liq.	Sat. Vap.	350	400	450	500	550	600	700	800	900	1000	1100	1200	1300	1400
80 (312.04)	v	0.01757	5.471	5.801	6.218	6.622	7.018	7.408	7.794	8.560	9.319	10.075	10.829	11.581	12.331	13.081	13.829
	h	282.15	1183.1	1204.0	1230.5	1256.1	1281.3	1306.2	1330.9	1380.5	1430.5	1481.1	1532.6	1584.9	1638.0	1692.0	1746.8
	s	0.4534	1.6208	1.6473	1.6790	1.7080	1.7349	1.7602	1.7842	1.8289	1.8702	1.9089	1.9454	1.9800	2.0131	2.0446	2.0750
90 (320.28)	v	0.01766	4.895	5.128	5.505	5.869	6.223	6.572	6.917	7.600	8.277	8.950	9.621	10.290	10.958	11.625	12.290
	h	290.69	1185.3	1202.0	1228.9	1254.9	1280.3	1305.4	1330.2	1380.0	1430.1	1480.8	1532.3	1584.6	1637.8	1691.8	1746.7
	s	0.4643	1.6113	1.6323	1.6646	1.6940	1.7212	1.7467	1.7707	1.8156	1.8570	1.8957	1.9323	1.9669	2.0000	2.0316	2.0619
100 (327.82)	v	0.01774	4.431	4.590	4.935	5.266	5.588	5.904	6.216	6.833	7.443	8.050	8.655	9.258	9.860	10.460	11.060
	h	298.54	1187.2	1199.9	1227.4	1253.7	1279.3	1304.6	1329.6	1379.5	1429.7	1480.4	1532.0	1584.4	1637.6	1691.6	1746.5
	s	0.4743	1.6027	1.6187	1.6516	1.6814	1.7088	1.7344	1.7586	1.8036	1.8451	1.8839	1.9205	1.9552	1.9883	2.0199	2.0502
110 (334.79)	v	0.01782	4.048	4.149	4.468	4.772	5.068	5.357	5.642	6.205	6.761	7.314	7.865	8.413	8.961	9.507	10.053
	h	305.80	1188.9	1197.7	1225.8	1252.5	1278.3	1303.8	1328.9	1379.0	1429.2	1480.1	1531.7	1584.1	1637.4	1691.4	1746.4
	s	0.4834	1.5950	1.6061	1.6396	1.6698	1.6975	1.7233	1.7476	1.7928	1.8344	1.8732	1.9099	1.9446	1.9777	2.0093	2.0397
120 (341.27)	v	0.01789	3.7275	3.7815	4.0786	4.3610	4.6341	4.9009	5.1637	5.6813	6.1928	6.7006	7.2060	7.7096	8.2119	8.7130	9.2134
	h	312.58	1190.4	1195.6	1224.1	1251.2	1277.4	1302.9	1328.2	1378.4	1428.8	1479.8	1531.4	1583.9	1637.1	1691.3	1746.2
	s	0.4919	1.5879	1.5943	1.6286	1.6592	1.6872	1.7132	1.7376	1.7829	1.8246	1.8635	1.9001	1.9349	1.9680	1.9996	2.0300
140 (353.04)	v	0.01803	3.2190		3.4661	3.7143	3.9526	4.1844	4.4119	4.8588	5.2995	5.7364	6.1709	6.6036	7.0349	7.4652	7.8946
	h	324.96	1193.0		1220.8	1248.7	1275.3	1301.3	1326.8	1377.4	1428.0	1479.1	1530.8	1583.4	1636.7	1690.9	1745.9
	s	0.5071	1.5752		1.6085	1.6400	1.6686	1.6949	1.7196	1.7652	1.8071	1.8461	1.8828	1.9176	1.9508	1.9825	2.0129

TABLE A-20E (Continued)

Temperature—F

Pressures 160–200 lbf/in²

Abs. Press. lbf/in² (Sat. Temp, F)		Sat. Liq.	Sat. Vap.	350	400	450	500	550	600	700	800	900	1000	1100	1200	1300	1400
160 (363.55)	v	0.01815	2.8336		3.0060	3.2288	3.4413	3.6469	3.8480	4.2420	4.6295	5.0132	5.3945	5.7741	6.1522	6.5293	6.9055
	h	336.07	1195.1		1217.4	1246.0	1273.3	1299.6	1325.4	1376.4	1427.2	1478.4	1530.3	1582.9	1636.3	1690.5	1745.6
	s	0.5206	1.5641		1.5906	1.6231	1.6522	1.6790	1.7039	1.7499	1.7919	1.8310	1.8678	1.9027	1.9359	1.9676	1.9980
180 (373.08)	v	0.01827	2.5312		2.6474	2.8508	3.0433	3.2286	3.4093	3.7621	4.1084	4.4508	4.7907	5.1289	5.4657	5.8014	6.1363
	h	346.19	1196.9		1213.8	1243.4	1271.2	1297.9	1324.0	1375.3	1426.3	1477.7	1529.7	1582.4	1635.9	1690.2	1745.3
	s	0.5328	1.5543		1.5743	1.6078	1.6376	1.6647	1.6900	1.7362	1.7784	1.8176	1.8545	1.8894	1.9227	1.9545	1.9849
200 (381.80)	v	0.01839	2.2873		2.3598	2.5480	2.7247	2.8939	3.0583	3.3783	3.6915	4.0008	4.3077	4.6128	4.9165	5.2191	5.5209
	h	355.51	1198.3		1210.1	1240.6	1269.0	1296.2	1322.6	1374.3	1425.5	1477.0	1529.1	1581.9	1635.4	1689.8	1745.0
	s	0.5438	1.5454		1.5593	1.5938	1.6242	1.6518	1.6773	1.7239	1.7663	1.8057	1.8426	1.8776	1.9109	1.9427	1.9732

Pressures 220–360 lbf/in²

Abs. Press. lbf/in² (Sat. Temp, F)		Sat. Liq.	Sat. Vap.	400	450	500	550	600	700	800	900	1000	1100	1200	1300	1400	1500
220 (389.88)	v	0.01850	2.0863	2.1240	2.2999	2.4638	2.6199	2.7710	3.0642	3.3504	3.6327	3.9125	4.1905	4.4671	4.7426	5.0173	5.2913
	h	364.17	1199.6	1206.3	1237.8	1266.9	1294.5	1321.2	1373.2	1424.7	1476.3	1528.5	1581.4	1635.0	1689.2	1744.7	1800.6
	s	0.5540	1.5374	1.5453	1.5808	1.6120	1.6400	1.6658	1.7128	1.7553	1.7948	1.8318	1.8668	1.9002	1.9320	1.9625	1.9919
240 (397.39)	v	0.01860	1.9177	1.9268	2.0928	2.2462	2.3915	2.5316	2.8024	3.0661	3.3259	3.5831	3.8385	4.0926	4.3456	4.5977	4.8492
	h	372.27	1200.6	1202.4	1234.9	1264.6	1292.7	1319.7	1372.1	1423.8	1475.6	1527.9	1580.9	1634.6	1689.1	1744.3	1800.4
	s	0.5634	1.5299	1.5320	1.5687	1.6006	1.6291	1.6552	1.7025	1.7452	1.7848	1.8219	1.8570	1.8904	1.9223	1.9528	1.9822
260 (404.44)	v	0.01870	1.7742		1.9173	2.0619	2.1981	2.3289	2.5808	2.8256	3.0663	3.3044	3.5408	3.7758	4.0097	4.2427	4.4750
	h	379.90	1201.5		1231.9	1262.4	1290.9	1318.2	1371.1	1423.0	1474.9	1527.3	1580.4	1634.2	1688.7	1744.0	1800.1
	s	0.5722	1.5230		1.5573	1.5899	1.6189	1.6453	1.6930	1.7359	1.7756	1.8128	1.8480	1.8814	1.9133	1.9439	1.9732
280 (411.07)	v	0.01880	1.6505		1.7665	1.9037	2.0322	2.1551	2.3909	2.6194	2.8437	3.0655	3.2855	3.5042	3.7217	3.9384	4.1543
	h	387.12	1202.3		1228.8	1260.0	1289.1	1316.8	1370.0	1422.1	1474.2	1526.8	1579.9	1633.8	1688.1	1743.7	1799.8
	s	0.5805	1.5166		1.5464	1.5798	1.6093	1.6361	1.6841	1.7273	1.7671	1.8043	1.8395	1.8730	1.9050	1.9356	1.9649
300 (417.35)	v	0.01889	1.5427		1.6356	1.7665	1.8883	2.0044	2.2263	2.4407	2.6509	2.8585	3.0643	3.2688	3.4721	3.6746	3.8764
	h	393.99	1202.9		1225.7	1257.7	1287.2	1315.2	1368.9	1421.3	1473.6	1526.2	1579.4	1633.3	1688.0	1743.4	1799.6
	s	0.5882	1.5105		1.5361	1.5703	1.6003	1.6274	1.6758	1.7192	1.7591	1.7964	1.8317	1.8652	1.8972	1.9278	1.9572
320 (423.31)	v	0.01899	1.4480		1.5207	1.6462	1.7623	1.8725	2.0823	2.2843	2.4821	2.6774	2.8708	3.0628	3.2538	3.4438	3.6332
	h	400.53	1203.4		1222.5	1255.2	1285.3	1313.7	1367.8	1420.5	1472.9	1525.6	1578.9	1632.9	1687.6	1743.1	1799.3
	s	0.5956	1.5048		1.5261	1.5612	1.5918	1.6192	1.6680	1.7116	1.7516	1.7890	1.8243	1.8579	1.8899	1.9206	1.9500
340 (428.99)	v	0.01908	1.3640		1.4191	1.5399	1.6511	1.7561	1.9552	2.1463	2.3333	2.5175	2.7000	2.8811	3.0611	3.2402	3.4186
	h	406.80	1203.8		1219.2	1252.8	1283.4	1312.2	1366.7	1419.6	1472.2	1525.0	1578.4	1632.5	1687.3	1742.8	1799.0
	s	0.6026	1.4994		1.5165	1.5525	1.5836	1.6114	1.6606	1.7044	1.7445	1.7820	1.8174	1.8510	1.8831	1.9138	1.9432
360 (434.41)	v	0.01917	1.2891		1.3285	1.4454	1.5521	1.6525	1.8421	2.0237	2.2009	2.3755	2.5482	2.7196	2.8898	3.0592	3.2279
	h	412.81	1204.1		1215.8	1250.3	1281.5	1310.6	1365.6	1418.7	1471.5	1524.4	1577.9	1632.1	1686.9	1742.5	1798.8
	s	0.6092	1.4943		1.5073	1.5441	1.5758	1.6040	1.6536	1.6976	1.7379	1.7754	1.8109	1.8445	1.8766	1.9073	1.9368

v = specific volume, ft³/lbm   h = enthalpy, Btu/lbm   s = entropy, Btu/lbm · R

P (T Sat)		Sat. Liq.	Sat. Vap.	450	500	550	600	650	700	800	900	1000	1100	1200	1300	1400	1500
380 (439.61)	v	0.01925	1.2218	1.2472	1.3606	1.4635	1.5598		1.7410	1.9139	2.0825	2.2484	2.4124	2.5750	2.7366	2.8973	3.0572
	h	418.59	1204.4	1212.4	1247.7	1279.5	1309.0		1364.5	1417.9	1470.8	1523.8	1577.4	1631.6	1686.5	1742.2	1798.5
	s	0.6156	1.4894	1.4982	1.5360	1.5683	1.5969		1.6470	1.6911	1.7315	1.7692	1.8047	1.8384	1.8705	1.9012	1.9307
400 (444.60)	v	0.01934	1.1610	1.1738	1.2841	1.3836	1.4763	1.5646	1.6499	1.8151	1.9759	2.1339	2.2901	2.4450	2.5987	2.7515	2.9037
	h	424.17	1204.6	1208.8	1245.1	1277.5	1307.4	1335.9	1363.4	1417.0	1470.1	1523.3	1576.9	1631.2	1686.2	1741.9	1798.2
	s	0.6217	1.4847	1.4894	1.5282	1.5611	1.5901	1.6163	1.6406	1.6850	1.7255	1.7632	1.7988	1.8325	1.8647	1.8955	1.9250
440 (454.03)	v	0.01950	1.0554		1.1517	1.2454	1.3319	1.4138	1.4926	1.6445	1.7918	1.9363	2.0790	2.2203	2.3605	2.4998	2.6384
	h	434.77	1204.8		1239.7	1273.4	1304.2	1333.2	1361.1	1415.3	1468.7	1522.1	1575.9	1630.4	1685.5	1741.2	1797.7
	s	0.6332	1.4759		1.5132	1.5474	1.5772	1.6040	1.6286	1.6734	1.7142	1.7521	1.7878	1.8216	1.8538	1.8847	1.9143
480 (462.82)	v	0.01967	0.9668		1.0409	1.1300	1.2115	1.2881	1.3615	1.5023	1.6384	1.7716	1.9030	2.0330	2.1619	2.2900	2.4173
	h	444.75	1204.8		1234.1	1269.1	1300.8	1330.5	1358.8	1413.6	1467.3	1520.9	1574.9	1629.5	1684.7	1740.6	1797.2
	s	0.6439	1.4677		1.4990	1.5346	1.5652	1.5925	1.6176	1.6628	1.7038	1.7419	1.7777	1.8116	1.8439	1.8748	1.9045
520 (471.07)	v	0.01982	0.8914		0.9466	1.0321	1.1094	1.1816	1.2504	1.3819	1.5085	1.6323	1.7542	1.8746	1.9940	2.1125	2.2302
	h	454.18	1204.5		1228.3	1264.8	1297.4	1327.7	1356.5	1411.8	1465.9	1519.7	1573.9	1628.7	1684.0	1740.0	1796.7
	s	0.6540	1.4601		1.4853	1.5223	1.5539	1.5818	1.6072	1.6530	1.6943	1.7325	1.7684	1.8024	1.8348	1.8657	1.8954
560 (478.84)	v	0.01998	0.8264		0.8653	0.9479	1.0217	1.0902	1.1552	1.2787	1.3972	1.5129	1.6266	1.7388	1.8500	1.9603	2.0699
	h	463.14	1204.2		1222.2	1260.3	1293.9	1324.9	1354.2	1410.0	1464.1	1518.6	1572.9	1627.8	1683.3	1739.4	1796.1
	s	0.6634	1.4529		1.4720	1.5106	1.5431	1.5717	1.5975	1.6438	1.6853	1.7237	1.7598	1.7939	1.8263	1.8573	1.8870
600 (486.20)	v	0.02013	0.7697		0.7944	0.8746	0.9456	1.0109	1.0726	1.1892	1.3008	1.4093	1.5160	1.6211	1.7252	1.8284	1.9309
	h	471.70	1203.7		1215.9	1255.6	1290.3	1322.0	1351.8	1408.3	1463.0	1517.4	1571.9	1627.0	1682.6	1738.8	1795.6
	s	0.6723	1.4461		1.4590	1.4993	1.5329	1.5621	1.5884	1.6351	1.6769	1.7155	1.7517	1.7859	1.8184	1.8494	1.8792
700 (503.08)	v	0.02050	0.6556			0.7271	0.7928	0.8520	0.9072	1.0102	1.1078	1.2023	1.2948	1.3858	1.4757	1.5647	1.6530
	h	491.60	1201.8			1243.4	1281.0	1314.6	1345.6	1403.7	1459.4	1514.4	1569.4	1624.8	1680.7	1737.2	1794.3
	s	0.6928	1.4304			1.4726	1.5090	1.5399	1.5673	1.6154	1.6580	1.6970	1.7335	1.7679	1.8006	1.8318	1.8617
800 (518.21)	v	0.02087	0.5690			0.6151	0.6774	0.7323	0.7828	0.8759	0.9631	1.0470	1.1289	1.2093	1.2885	1.3669	1.4446
	h	509.81	1199.4			1230.1	1271.1	1306.8	1339.3	1399.1	1455.8	1511.4	1566.9	1622.7	1678.9	1735.7	1792.9
	s	0.7111	1.4163			1.4472	1.4869	1.5198	1.5484	1.5980	1.6413	1.6807	1.7175	1.7522	1.7851	1.8164	1.8464
900 (531.95)	v	0.02123	0.5009			0.5263	0.5869	0.6388	0.6858	0.7713	0.8504	0.9262	0.9998	1.0720	1.1430	1.2131	1.2825
	h	526.70	1196.4			1215.5	1260.6	1298.6	1332.7	1394.4	1452.2	1508.5	1564.4	1620.6	1677.1	1734.1	1791.6
	s	0.7279	1.4032			1.4223	1.4659	1.5010	1.5311	1.5822	1.6263	1.6662	1.7033	1.7382	1.7713	1.8028	1.8329

P (T Sat)		Sat. Liq.	Sat. Vap.	550	600	650	700	750	800	850	900	1000	1100	1200	1300	1400	1500
1000 (544.58)	v	0.02159	0.4460	0.4535	0.5137	0.5636	0.6080	0.6489	0.6875	0.7245	0.7603	0.8295	0.8966	0.9622	1.0266	1.0901	1.1529
	h	542.55	1192.9	1199.3	1249.3	1290.1	1325.9	1358.7	1389.6	1419.4	1448.5	1505.4	1561.9	1618.4	1675.3	1732.5	1790.3
	s	0.7434	1.3910	1.3973	1.4457	1.4833	1.5149	1.5426	1.5677	1.5908	1.6126	1.6530	1.6905	1.7256	1.7589	1.7905	1.8207
1200 (567.19)	v	0.02232	0.3624		0.4016	0.4497	0.4905	0.5273	0.5615	0.5939	0.6250	0.6845	0.7418	0.7974	0.8519	0.9055	0.9584
	h	571.85	1184.8		1224.2	1271.8	1311.5	1346.9	1379.7	1410.8	1440.9	1499.4	1556.9	1614.2	1671.6	1729.4	1787.6
	s	0.7714	1.3683		1.4061	1.4501	1.4851	1.5150	1.5415	1.5658	1.5883	1.6298	1.6679	1.7035	1.7371	1.7691	1.7996

v = specific volume, ft³/lbm   h = enthalpy, Btu/lbm   s = entropy, Btu/lbm · R

TABLE A-20E *(Continued)*

Temperature—F

Abs. Press. lbf/in.² (Sat. Temp. F)		Sat. Liq.	Sat. Vap.	550	600	650	700	750	800	850	900	1000	1100	1200	1300	1400	1500
1400 (587.07)	v	0.02307	0.3018		0.3176	0.3667	0.4059	0.4400	0.4712	0.5004	0.5282	0.5809	0.6311	0.6798	0.7272	0.7737	0.8195
	h	598.83	1175.3		1194.1	1251.4	1296.1	1334.5	1369.3	1402.0	1433.2	1493.2	1551.8	1609.9	1668.0	1726.3	1785.0
	s	0.7966	1.3474		1.3652	1.4181	1.4575	1.4900	1.5182	1.5436	1.5670	1.6096	1.6484	1.6845	1.7185	1.7508	1.7815
1600 (604.87)	v	0.02387	0.2555			0.3026	0.3415	0.3741	0.4032	0.4301	0.4555	0.5031	0.5482	0.5915	0.6336	0.6748	0.7153
	h	624.20	1164.5			1228.3	1279.4	1321.4	1358.5	1392.8	1425.2	1486.9	1546.6	1605.6	1664.3	1723.2	1782.3
	s	0.8199	1.3274			1.3861	1.4312	1.4667	1.4968	1.5235	1.5478	1.5916	1.6312	1.6678	1.7022	1.7347	1.7657
1800 (621.02)	v	0.02472	0.2186			0.2505	0.2906	0.3223	0.3500	0.3752	0.3988	0.4426	0.4836	0.5229	0.5609	0.5980	0.6343
	h	648.49	1152.3			1201.2	1261.1	1307.4	1347.2	1383.3	1417.1	1480.6	1541.4	1601.2	1660.7	1720.1	1779.7
	s	0.8417	1.3079			1.3526	1.4054	1.4446	1.4768	1.5049	1.5302	1.5753	1.6156	1.6528	1.6876	1.7204	1.7516
2000 (635.80)	v	0.02565	0.1883			0.2056	0.2488	0.2805	0.3072	0.3312	0.3534	0.3942	0.4320	0.4680	0.5027	0.5365	0.5695
	h	672.11	1138.3			1168.3	1240.9	1292.6	1335.4	1373.5	1408.7	1474.1	1536.2	1596.9	1657.0	1717.0	1771.1
	s	0.8625	1.2881			1.3154	1.3794	1.4231	1.4578	1.4874	1.5138	1.5603	1.6014	1.6391	1.6743	1.7075	1.7389
2200 (649.45)	v	0.02669	0.1627			0.1636	0.2134	0.2458	0.2720	0.2950	0.3161	0.3545	0.3897	0.4231	0.4551	0.4862	0.5165
	h	695.46	1122.2			1123.9	1218.0	1276.8	1323.1	1363.3	1400.0	1467.6	1530.9	1592.5	1653.3	1713.9	1774.4
	s	0.8828	1.2676			1.2691	1.3523	1.4020	1.4395	1.4708	1.4984	1.5463	1.5883	1.6266	1.6622	1.6956	1.7273

Temperature—F

Abs. Press. lbf/in.² (Sat. Temp. F)		Sat. Liq.	Sat. Vap.	700	750	800	850	900	950	1000	1050	1100	1150	1200	1300	1400	1500
2400 (662.11)	v	0.02790	0.1408	0.1824	0.2164	0.2424	0.2648	0.2850	0.3037	0.3214	0.3382	0.3545	0.3703	0.3856	0.4155	0.4443	0.4724
	h	718.95	1103.7	1191.6	1259.6	1310.1	1352.8	1391.2	1426.9	1460.9	1493.7	1525.6	1557.0	1588.1	1649.6	1710.8	1771.8
	s	0.9031	1.2460	1.3232	1.3808	1.4217	1.4549	1.4837	1.5095	1.5332	1.5553	1.5761	1.5959	1.6149	1.6509	1.6847	1.7167
2600 (673.91)	v	0.02938	0.1211	0.1544	0.1909	0.2171	0.2390	0.2585	0.2765	0.2933	0.3093	0.3247	0.3395	0.3540	0.3819	0.4088	0.4350
	h	744.47	1082.0	1160.2	1241.1	1296.5	1341.9	1382.1	1419.2	1454.1	1487.7	1520.2	1552.2	1583.7	1646.0	1707.7	1769.1
	s	0.9247	1.2225	1.2908	1.3592	1.4042	1.4395	1.4696	1.4964	1.5208	1.5434	1.5646	1.5848	1.6040	1.6405	1.6746	1.7068
2800 (684.96)	v	0.03134	0.1030	0.1278	0.1685	0.1952	0.2168	0.2358	0.2531	0.2693	0.2845	0.2991	0.3132	0.3268	0.3532	0.3785	0.4030
	h	770.69	1055.8	1121.2	1220.6	1282.2	1330.7	1372.8	1411.2	1447.2	1481.6	1514.8	1547.5	1579.3	1642.2	1704.5	1766.5
	s	0.9468	1.1958	1.2527	1.3368	1.3867	1.4245	1.4561	1.4838	1.5089	1.5321	1.5537	1.5742	1.5938	1.6306	1.6651	1.6975
3000 (695.33)	v	0.03428	0.0850	0.0982	0.1483	0.1759	0.1975	0.2161	0.2329	0.2484	0.2630	0.2770	0.2904	0.3033	0.3282	0.3522	0.3753
	h	801.84	1020.3	1060.5	1197.9	1267.0	1319.0	1363.2	1403.1	1440.2	1475.4	1509.4	1542.4	1574.8	1638.5	1701.4	1763.8
	s	0.9728	1.1619	1.1966	1.3131	1.3692	1.4097	1.4429	1.4717	1.4976	1.5213	1.5434	1.5642	1.5841	1.6214	1.6561	1.6888
3200 (705.08)	v	0.04472	0.0566		0.1300	0.1588	0.1804	0.1987	0.2151	0.2301	0.2442	0.2576	0.2704	0.2827	0.3065	0.3291	0.3510
	h	875.54	931.6		1172.3	1250.9	1306.9	1353.4	1394.9	1433.1	1469.2	1503.8	1537.1	1570.3	1634.8	1698.3	1761.2
	s	1.0351	1.0832		1.2877	1.3515	1.3951	1.4300	1.4600	1.4866	1.5110	1.5335	1.5547	1.5749	1.6126	1.6477	1.6806
3400	v				0.1129	0.1435	0.1653	0.1834	0.1994	0.2140	0.2276	0.2405	0.2528	0.2646	0.2872	0.3088	0.3296
	h				1143.2	1233.7	1294.3	1343.4	1386.4	1425.9	1462.9	1498.3	1532.4	1565.8	1631.1	1695.1	1758.5
	s				1.2600	1.3334	1.3807	1.4174	1.4486	1.4761	1.5010	1.5240	1.5456	1.5660	1.6042	1.6396	1.6728

3600	v	0.0966	0.1296	0.1517	0.1697	0.1854	0.1996	0.2128	0.2252	0.2371	0.2485	0.2702	0.2908	0.3106
	h	1108.6	1215.3	1281.2	1333.0	1377.9	1418.6	1456.5	1492.6	1527.4	1561.3	1627.3	1692.0	1755.9
	s	1.2281	1.3148	1.3662	1.4050	1.4374	1.4658	1.4914	1.5149	1.5369	1.5576	1.5962	1.6320	1.6654
4000	v	0.0631	0.1052	0.1284	0.1463	0.1616	0.1752	0.1877	0.1994	0.2105	0.2210	0.2411	0.2601	0.2783
	h	1007.4	1174.3	1253.4	1311.6	1360.2	1403.6	1443.6	1481.3	1517.3	1552.2	1619.8	1685.7	1750.6
	s	1.1396	1.2754	1.3371	1.3807	1.4158	1.4461	1.4730	1.4976	1.5203	1.5417	1.5812	1.6177	1.6516
4400	v	0.0421	0.0846	0.1090	0.1270	0.1420	0.1552	0.1671	0.1782	0.1887	0.1986	0.2174	0.2351	0.2519
	h	909.5	1127.3	1223.3	1289.0	1342.0	1388.3	1430.4	1469.7	1507.1	1543.0	1612.3	1679.4	1745.3
	s	1.0556	1.2325	1.3073	1.3566	1.3949	1.4272	1.4556	1.4812	1.5048	1.5268	1.5673	1.6044	1.6389

v = specific volume, ft³/lbm h = enthalpy, Btu/lbm s = entropy, Btu/lbm · R

TABLE A-21M PROPERTIES OF WATER: COMPRESSED LIQUID (SI UNITS)

T °C	v m³/kg	u kJ/kg	h kJ/kg	s kJ/(kg·K)	v m³/kg	u kJ/kg	h kJ/kg	s kJ/(kg·K)	v m³/kg	u kJ/kg	h kJ/kg	s kJ/(kg·K)
	P = 5 MPa (263.99°C)				P = 10 MPa (311.06°C)				P = 15 MPa (342.24°C)			
Sat.	0.001 285 9	1147.8	1154.2	2.9202	0.001 452 4	1393.0	1407.6	3.3596	0.001 658 1	1585.6	1610.5	3.6848
0	0.000 997 7	0.04	5.04	0.0001	0.000 995 2	0.09	10.04	0.0002	0.000 992 8	0.15	15.05	0.0004
20	0.000 999 5	83.65	88.65	0.2956	0.000 997 2	83.36	93.33	0.2945	0.000 995 0	83.06	97.99	0.2934
40	0.001 005 6	166.95	171.97	0.5705	0.001 003 4	165.76	176.38	0.5686	0.001 001 3	165.76	180.78	0.5666
60	0.001 014 9	250.23	255.30	0.8285	0.001 012 7	249.36	259.49	0.8258	0.001 010 5	248.51	263.67	0.8232
80	0.001 026 8	333.72	338.85	1.0720	0.001 024 5	332.59	342.83	1.0688	0.001 022 2	331.48	346.81	1.0656
100	0.001 041 0	417.52	422.72	1.3030	0.001 038 5	416.12	426.50	1.2992	0.001 036 1	414.74	430.28	1.2955
120	0.001 057 6	501.80	507.09	1.5233	0,001 054 9	500.08	510.64	1.5189	0.001 052 2	498.40	514.19	1.5145
140	0.001 076 8	586.76	592.15	1.7343	0.001 073 7	584.68	595.42	1.7292	0.001 070 7	582.66	598.72	1.7242
160	0.001 098 8	672.62	678.12	1.9375	0.001 095 3	670.13	681.08	1.9317	0.001 091 8	667.71	684.09	1.9260
180	0.001 124 0	759.63	765.25	2.1341	0.001 119 9	756.65	767.84	2.1275	0.001 115 9	753.76	770.50	2.1210
200	0.001 153 0	848.1	853.9	2.3255	0.001 148 0	844.5	856.0	2.3178	0.001 143 3	841.0	858.2	2.3104
220	0.001 186 6	938.4	944.4	2.5128	0.001 180 5	934.1	945.9	2.5039	0.001 174 8	929.9	947.5	2.4953
240	0.001 226 4	1031.4	1037.5	2.6979	0.001 218 7	1026.0	1038.1	2.6872	0.001 211 4	1020.8	1039.0	2.6771
260	0.001 274 9	1127.9	1134.3	2.8830	0.001 264 5	1121.1	1133.7	2.8699	0.001 255 0	1114.6	1133.4	2.8576
280					0.001 321 6	1220.9	1234.1	3.0548	0.001 308 4	1212.5	1232.1	3.0393
300					0.001 397 2	1328.4	1342.3	3.2469	0.001 377 0	1316.6	1337.3	3.2260
320									0.001 472 4	1431.1	1453.2	3.4247
340									0.001 631 1	1567.5	1591.9	3.6546
	P = 20 MPa (365.81°C)				P = 30 MPa				P = 50 MPa			
Sat.	0.002 036	1785.6	1826.3	4.0139								
0	0.000 990 4	0.19	20.01	0.0004	0.000 985 6	0.25	29.82	0.0001	0.000 976 6	0.20	49.03	0.0014
20	0.000 992 8	82.77	102.62	0.2923	0.000 988 6	82.17	111.84	0.2899	0.000 980 4	81.00	130.02	0.2848
40	0.000 999 2	165.17	185.16	0.5646	0.000 995 1	164.04	193.89	0.5607	0.000 987 2	161.86	211.21	0.5527
60	0.001 008 4	247.68	267.85	0.8206	0.001 004 2	246.06	276.19	0.8154	0.000 996 2	242.98	292.79	0.8052
80	0.001 019 9	330.40	350.80	1.0624	0.001 015 6	328.30	358.77	1.0561	0.001 007 3	324.34	374.70	1.0440
100	0.001 033 7	413.39	434.06	1.2917	0.001 029 0	410.78	441.66	1.2844	0.001 020 1	405.88	456.89	1.2703
120	0.001 049 6	496.76	517.76	1.5102	0.001 044 5	493.59	524.93	1.5018	0.001 034 8	487.65	539.39	1.4857
140	0.001 067 8	580.69	602.04	1.7193	0.001 062 1	576.88	608.75	1.7098	0.001 051 5	569.77	622.35	1.6915
160	0.001 088 5	665.35	687.12	1.9204	0.001 082 1	660.82	693.28	1.9096	0.001 070 3	652.41	705.92	1.8891
180	0.001 112 0	750.95	773.20	2.1147	0.001 104 7	745.59	778.73	2.1024	0.001 091 2	735.69	790.25	2.0794
200	0.001 138 8	837.7	860.5	2.3031	0.001 130 2	831.4	865.3	2.2893	0.001 114 6	819.7	875.5	2.2634
220	0.001 169 5	925.9	949.3	2.4870	0.001 159 0	918.3	953.1	2.4711	0.001 140 8	904.7	961.7	2.4419
240	0.001 204 6	1016.0	1040.0	2.6674	0.001 192 0	1006.9	1042.6	2.6490	0.001 170 2	990.7	1049.2	2.6158
260	0.001 246 2	1108.6	1133.5	2.8459	0.001 230 3	1097.4	1134.3	2.8243	0.001 203 4	1078.1	1138.2	2.7860
280	0.001 296 5	1204.7	1230.6	3.0248	0.001 275 5	1190.7	1229.0	2.9986	0.001 241 5	1167.2	1229.3	2.9537
300	0.001 359 6	1306.1	1333.3	3.2071	0.001 330 4	1287.9	1327.8	3.1741	0.001 286 0	1258.7	1323.0	3.1200
320	0.001 443 7	1415.7	1444.6	3.3979	0.001 399 7	1390.7	1432.7	3.3539	0.001 338 8	1353.3	1420.2	3.2868
340	0.001 568 4	1539.7	1571.0	3.6075	0.001 492 0	1501.7	1546.5	3.5426	0.001 403 2	1452.0	1522.1	3.4557
360	0.001 822 6	1702.8	1739.3	3.8772	0.001 626 5	1626.6	1675.4	3.7494	0.001 483 8	1556.0	1630.2	3.6291
380					0.001 869 1	1781.4	1837.5	4.0012	0.001 588 4	1667.2	1746.6	3.8101

TABLE A-21E PROPERTIES OF WATER: COMPRESSED LIQUID (ENGLISH UNITS)

T °F	v ft³/lbm	u Btu/lbm	h Btu/lbm	s Btu/(lbm·R)	v ft³/lbm	u Btu/lbm	h Btu/lbm	s Btu/(lbm·R)	v ft³/lbm	u Btu/lbm	h Btu/lbm	s Btu/(lbm·R)
	P = 500 psia (467.13°F)				P = 1000 psia (544.75°F)				P = 1500 psia (596.39°F)			
Sat.	0.019 748	447.70	449.53	0.649 04	0.021 591	538.39	542.38	0.743 20	0.023 461	604.97	611.48	0.808 24
32	0.015 994	0.00	1.49	0.000 00	0.015 967	0.03	2.99	0.000 05	0.015 939	0.05	4.47	0.000 07
50	0.015 998	18.02	19.50	0.035 99	0.015 972	17.99	20.94	0.035 92	0.015 946	17.95	22.38	0.035 84
100	0.016 106	67.87	69.36	0.129 32	0.016 082	67.70	70.68	0.129 01	0.016 058	67.53	71.99	0.128 70
150	0.016 318	117.66	119.17	0.214 57	0.016 293	117.38	120.40	0.214 10	0.016 268	117.10	121.62	0.213 64
200	0.016 608	167.65	169.19	0.293 41	0.016 580	167.26	170.32	0.292 81	0.016 554	166.87	171.46	0.292 21
250	0.016 972	217.99	219.56	0.367 02	0.016 941	217.47	220.61	0.366 28	0.016 910	216.96	221.65	0.365 54
300	0.017 416	268.92	270.53	0.436 41	0.017 379	268.24	271.46	0.435 52	0.017 343	267.58	272.39	0.434 63
350	0.017 954	320.71	322.37	0.502 49	0.017 909	319.83	323.15	0.501 40	0.017 865	318.98	323.94	0.500 34
400	0.018 608	373.68	375.40	0.566 04	0.018 550	372.55	375.98	0.564 72	0.018 493	371.45	376.59	0.563 43
450	0.019 420	428.40	430.19	0.627 98	0.019 340	426.89	430.47	0.626 32	0.019 264	425.44	430.79	0.624 70
500					0.020 36	483.8	487.5	0.6874	0.020 24	481.8	487.4	0.6853
550									0.021 58	542.1	548.1	0.7469
	P = 2000 psia (636.00°F)				P = 3000 psia (695.52°F)				P = 5000 psia			
Sat.	0.025 649	662.40	671.89	0.862 27	0.034 310	783.45	802.50	0.973 20				
32	0.015 912	0.06	5.95	0.000 08	0.015 859	0.09	8.90	0.000 09	0.015 755	0.11	14.70	-0.000 01
50	0.015 920	17.91	23.81	0.035 75	0.015 870	17.84	26.65	0.035 55	0.015 773	17.67	32.26	0.035 08
100	0.016 034	67.37	73.30	0.128 39	0.015 987	67.04	75.91	0.127 77	0.015 897	66.40	81.11	0.126 51
200	0.016 527	166.49	172.60	0.291 62	0.016 476	165.74	174.89	0.290 46	0.016 376	164.32	179.47	0.288 18
300	0.017 308	266.93	273.33	0.433 76	0.017 240	265.66	275.23	0.432 05	0.017 110	263.25	279.08	0.428 75
400	0.018 439	370.38	377.21	0.562 16	0.018 334	368.32	378.50	0.559 70	0.018 141	364.47	381.25	0.555 06
450	0.019 191	424.04	431.14	0.623 13	0.019 053	421.36	431.93	0.620 11	0.018 803	416.44	433.84	0.614 51
500	0.020 14	479.8	487.3	0.6832	0.019 944	476.2	487.3	0.679 4	0.019 603	469.8	487.9	0.6724
560	0.021 72	551.8	559.8	0.7565	0.021 382	546.2	558.0	0.750 8	0.020 835	536.7	556.0	0.7411
600	0.023 30	605.4	614.0	0.8086	0.022 74	597.0	609.6	0.8004	0.021 91	584.0	604.2	0.7876
640					0.024 75	654.3	668.0	0.8545	0.023 34	634.6	656.2	0.8357
680					0.028 79	728.4	744.3	0.9226	0.025 35	690.6	714.1	0.8873
700									0.026 76	721.8	746.6	0.9156

Source of Tables A-21M and A-21E: Adapted by permission from J. H. Keenan, F. G. Keyes, P. G. Hill, and J. G. Moore, "Steam Tables," John Wiley, N.Y. 1969 (English Units) and 1978 (SI Units).

TABLE A-22M PROPERTIES OF REFRIGERANT 12 (CCl$_2$F$_2$)

(a) Saturation Temperature Table (SI Units)

Temp. °C	Abs. Pressure MPa P	Specific Volume m³/kg Sat. Liquid v_l	Evap. v_{lg}	Sat. Vapor v_v	Enthalpy kJ/kg Sat. Liquid h_l	Evap. h_{lg}	Sat. Vapor h_g	Entropy kJ/kg·K Sat. Liquid s_l	Evap. s_{lg}	Sat. Vapor s_g
−90	0.0028	0.000 608	4.414 937	4.415 545	−43.243	189.618	146.375	−0.2084	1.0352	0.8268
−85	0.0042	0.000 612	3.036 704	3.037 316	−38.968	187.608	148.640	−0.1854	0.9970	0.8116
−80	0.0062	0.000 617	2.137 728	2.138 345	−34.688	185.612	150.924	−0.1630	0.9609	0.7979
−75	0.0088	0.000 622	1.537 030	1.537 651	−30.401	183.625	153.224	−0.1411	0.9266	0.7855
−70	0.0123	0.000 627	1.126 654	1.127 280	−26.103	181.640	155.536	−0.1197	0.8940	0.7744
−65	0.0168	0.000 632	0.840 534	0.841 166	−21.793	179.651	157.857	−0.0987	0.8630	0.7643
−60	0.0226	0.000 637	0.637 274	0.637 910	−17.469	177.653	160.184	−0.0782	0.8334	0.7552
−55	0.0300	0.000 642	0.490 358	0.491 000	−13.129	175.641	162.512	−0.0581	0.8051	0.7470
−50	0.0391	0.000 648	0.382 457	0.383 105	−8.772	173.611	164.840	−0.0384	0.7779	0.7396
−45	0.0504	0.000 654	0.302 029	0.302 682	−4.396	171.558	167.163	−0.0190	0.7519	0.7329
−40	0.0642	0.000 659	0.241 251	0.241 910	−0.000	169.479	169.479	−0.0000	0.7269	0.7269
−35	0.0807	0.000 666	0.194 732	0.195 398	4.416	167.368	171.784	0.0187	0.7027	0.7214
−30	0.1004	0.000 672	0.158 703	0.159 375	8.854	165.222	174.076	0.0371	0.6795	0.7165
−25	0.1237	0.000 679	0.130 487	0.131 166	13.315	163.037	176.352	0.0552	0.6570	0.7121
−20	0.1509	0.000 685	0.108 162	0.108 847	17.800	160.810	178.610	0.0730	0.6352	0.7082
−15	0.1826	0.000 693	0.090 326	0.091 018	22.312	158.534	180.846	0.0906	0.6141	0.7046
−10	0.2191	0.000 700	0.075 946	0.076 646	26.851	156.207	183.058	0.1079	0.5936	0.7014
−5	0.2610	0.000 708	0.064 255	0.064 963	31.420	153.823	185.243	0.1250	0.5736	0.6986

TABLE A-22M *(Continued)*

(a) Saturation Temperature Table (SI Units) (continued)

Temp. °C	Abs. Pressure MPa P	Specific Volume m³/kg			Enthalpy kJ/kg			Entropy kJ/kg·K		
		Sat. Liquid v_l	Evap. v_{lg}	Sat. Vapor v_v	Sat. Liquid h_l	Evap. h_{lg}	Sat. Vapor h_g	Sat. Liquid s_f	Evap. s_{lg}	Sat. Vapor s_g
0	0.3086	0.000 716	0.054 673	0.055 389	36.022	151.376	187.397	0.1418	0.5542	0.6960
5	0.3626	0.000 724	0.046 761	0.047 485	40.659	148.859	189.518	0.1585	0.5351	0.6937
10	0.4233	0.000 733	0.040 180	0.040 914	45.337	146.265	191.602	0.1750	0.5165	0.6916
15	0.4914	0.000 743	0.034 671	0.035 413	50.058	143.586	193.644	0.1914	0.4983	0.6897
20	0.5673	0.000 752	0.030 028	0.030 780	54.828	140.812	195.641	0.2076	0.4803	0.6879
25	0.6516	0.000 763	0.026 091	0.026 854	59.653	137.933	197.586	0.2237	0.4626	0.6863
30	0.7449	0.000 774	0.022 734	0.023 508	64.539	134.936	199.475	0.2397	0.4451	0.6848
35	0.8477	0.000 786	0.019 855	0.020 641	69.494	131.805	201.299	0.2557	0.4277	0.6834
40	0.9607	0.000 798	0.017 373	0.018 171	74.527	128.525	203.051	0.2716	0.4104	0.6820
45	1.0843	0.000 811	0.015 220	0.016 032	79.647	125.074	204.722	0.2875	0.3931	0.6806
50	1.2193	0.000 826	0.013 344	0.014 170	84.868	121.430	206.298	0.3034	0.3758	0.6792
55	1.3663	0.000 841	0.011 701	0.012 542	90.201	117.565	207.766	0.3194	0.3582	0.6777
60	1.5259	0.000 858	0.010 253	0.011 111	95.665	113.443	209.109	0.3355	0.3405	0.6760
65	1.6988	0.000 877	0.008 971	0.009 847	101.279	109.024	210.303	0.3518	0.3224	0.6742
70	1.8858	0.000 897	0.007 828	0.008 725	107.067	104.255	211.321	0.3683	0.3038	0.6721
75	2.0874	0.000 920	0.006 802	0.007 723	113.058	99.068	212.126	0.3851	0.2845	0.6697
80	2.3046	0.000 946	0.005 875	0.006 821	119.291	93.373	212.665	0.4023	0.2644	0.6667
85	2.5380	0.000 976	0.005 029	0.006 005	125.818	87.047	212.865	0.4201	0.2430	0.6631
90	2.7885	0.001 012	0.004 246	0.005 258	132.708	79.907	212.614	0.4385	0.2200	0.6585
95	3.0569	0.001 056	0.003 508	0.004 563	140.068	71.658	211.726	0.4579	0.1946	0.6526
100	3.3440	0.001 113	0.002 790	0.003 903	148.076	61.768	209.843	0.4788	0.1655	0.6444
105	3.6509	0.001 197	0.002 045	0.003 242	157.085	49.014	206.099	0.5023	0.1296	0.6319
110	3.9784	0.001 364	0.001 098	0.002 462	168.059	28.425	196.484	0.5322	0.0742	0.6064
112	4.1155	0.001 792	0.000 005	0.001 797	174.920	0.151	175.071	0.5651	0.0004	0.5655

Source for Tables A-22 and A-23: Adapted by permission from E. I. du Pont de Nemours & Company, copyright 1955, 1956, and 1969.

(b) Saturation Pressure Table (SI Units)

Press., MPa, P	Temp., °C T	Specific volume, m³/kg Sat. liquid, v_f	Evap., v_{fg}	Sat. vapor, v_g	Internal energy, kJ/kg Sat. liquid, u_f	Evap., u_{fg}	Sat. vapor, u_g	Enthalpy, kJ/kg Sat. liquid, h_f	Evap., h_{fg}	Sat. vapor, h_g	Entropy, kJ/kg·K Sat. liquid, s_f	Evap., s_{fg}	Sat. vapor, s_g
0.06	−41.42	0.0006578	0.2568	0.2575	−1.29	154.8	153.49	−1.25	170.19	168.94	−0.0054	0.7344	0.7290
0.10	−30.10	0.0006719	0.1593	0.1600	8.71	149.4	158.15	8.78	165.37	174.15	0.0368	0.6803	0.7171
0.12	−25.74	0.0006776	0.1342	0.1349	12.58	147.4	159.95	12.66	163.48	176.14	0.0526	0.6607	0.7133
0.14	−21.91	0.0006828	0.1161	0.1168	15.99	145.5	161.52	16.09	161.78	177.87	0.0663	0.6439	0.7102
0.16	−18.49	0.0006876	0.1024	0.1031	19.07	143.8	162.91	19.18	160.23	179.41	0.0784	0.6292	0.7076
0.18	−15.38	0.0006921	0.09156	0.09225	21.86	142.3	164.19	21.98	158.82	180.80	0.0893	0.6161	0.7054
0.20	−12.53	0.0006962	0.08284	0.08354	24.43	140.9	165.36	24.57	157.50	182.07	0.0992	0.6043	0.7035
0.24	−7.42	0.0007040	0.06963	0.07033	29.06	138.4	167.44	29.23	155.09	184.32	0.1168	0.5836	0.7004
0.28	−2.93	0.0007111	0.06005	0.06076	33.15	136.1	169.26	33.35	152.92	186.27	0.1321	0.5659	0.6980
0.32	1.11	0.0007177	0.05279	0.05351	36.85	134.0	170.88	37.08	150.92	188.00	0.1457	0.5503	0.6960
0.40	8.15	0.0007299	0.04248	0.04321	43.35	130.3	173.69	43.64	147.33	190.97	0.1691	0.5237	0.6928
0.50	15.60	0.0007438	0.03408	0.03482	50.30	126.3	176.61	50.67	143.35	194.02	0.1935	0.4964	0.6899
0.60	22.00	0.0007566	0.02837	0.02913	56.35	122.7	179.09	56.80	139.77	196.57	0.2142	0.4736	0.6878
0.70	27.65	0.0007686	0.02424	0.02501	61.75	119.5	181.23	62.29	136.45	198.74	0.2324	0.4536	0.6860
0.80	32.74	0.0007802	0.02110	0.02188	66.68	116.5	183.13	67.30	133.33	200.63	0.2487	0.4358	0.6845
0.90	37.37	0.0007914	0.01845	0.01942	71.22	113.6	184.81	71.93	130.36	202.29	0.2634	0.4198	0.6832
1.0	41.64	0.0008023	0.01664	0.01744	75.46	110.9	186.32	76.26	127.50	203.76	0.2770	0.4050	0.6820
1.2	49.31	0.0008237	0.01359	0.01441	83.22	105.7	188.95	84.21	122.03	206.24	0.3015	0.3784	0.6799
1.4	56.09	0.0008448	0.01138	0.01222	90.28	100.8	191.11	91.46	116.76	208.22	0.3232	0.3546	0.6778
1.6	62.19	0.0008660	0.00967	0.01054	96.80	96.2	192.95	98.19	111.62	209.81	0.3329	0.3429	0.6758

TABLE A-23M PROPERTIES OF REFRIGERANT 12 (CCl_2F_2): SUPERHEATED VAPOR (SI UNITS)

Temp. °C	v m³/kg	h kJ/kg	s kJ/kg·K	v m³/kg	h kJ/kg	s kJ/kg·K	v m³/kg	h kJ/kg	s kJ/kg·K
	0.05 MPa			0.10 MPa			0.15 MPa		
−20.0	0.341 857	181.042	0.7912	0.167 701	179.861	0.7401			
−10.0	0.356 227	186.757	0.8133	0.175 222	185.707	0.7628	0.114 716	184.619	0.7318
0.0	0.370 508	192.567	0.8350	0.182 647	191.628	0.7849	0.119 866	190.660	0.7543
10.0	0.384 716	198.471	0.8562	0.189 994	197.628	0.8064	0.124 932	196.762	0.7763
20.0	0.398 863	204.469	0.8770	0.197 277	203.707	0.8275	0.129 930	202.927	0.7977
30.0	0.412 959	210.557	0.8974	0.204 506	209.866	0.8482	0.134 873	209.160	0.8186
40.0	0.427 012	216.733	0.9175	0.211 691	216.104	0.8684	0.139 768	215.463	0.8390
50.0	0.441 030	222.997	0.9372	0.218 839	222.421	0.8883	0.144 625	221.835	0.8591
60.0	0.455 017	229.344	0.9565	0.225 955	228.815	0.9078	0.149 450	228.277	0.8787
70.0	0.468 978	235.774	0.9755	0.233 044	235.285	0.9269	0.154 247	234.789	0.8980
80.0	0.482 917	242.282	0.9942	0.240 111	241.829	0.9457	0.159 020	241.371	0.9169
90.0	0.496 838	248.868	1.0126	0.247 159	248.446	0.9642	0.163 774	248.020	0.9354
	0.20 MPa			0.25 MPa			0.30 MPa		
0.0	0.088 608	189.669	0.7320	0.069 752	188.644	0.7139	0.057 150	187.583	0.6984
10.0	0.092 550	195.878	0.7543	0.073 024	194.969	0.7366	0.059 984	194.034	0.7216
20.0	0.096 418	202.135	0.7760	0.076 218	201.322	0.7587	0.062 734	200.490	0.7440
30.0	0.100 228	208.446	0.7972	0.079 350	207.715	0.7801	0.065 418	206.969	0.7658
40.0	0.103 989	214.814	0.8178	0.082 431	214.153	0.8010	0.068 049	213.480	0.7869
50.0	0.107 710	221.243	0.8381	0.085 470	220.642	0.8214	0.070 635	220.030	0.8075
60.0	0.111 397	227.735	0.8578	0.088 474	227.185	0.8413	0.073 185	226.627	0.8276
70.0	0.115 055	234.291	0.8772	0.091 449	233.785	0.8608	0.075 705	233.273	0.8473
80.0	0.118 690	240.910	0.8962	0.094 398	240.443	0.8800	0.078 200	239.971	0.8665
90.0	0.122 304	247.593	0.9149	0.097 327	247.160	0.8987	0.080 673	246.723	0.8853
100.0	0.125 901	254.339	0.9332	0.100 238	253.936	0.9171	0.083 127	253.530	0.9038
110.0	0.129 483	261.147	0.9512	0.103 134	260.770	0.9352	0.085 566	260.391	0.9220

T	0.40 MPa			0.50 MPa			0.60 MPa		
20.0	0.045 836	198.762	0.7199	0.035 646	196.935	0.6999			
30.0	0.047 971	205.428	0.7423	0.037 464	203.814	0.7230	0.030 422	202.116	0.7063
40.0	0.050 046	212.095	0.7639	0.039 214	210.656	0.7452	0.031 966	209.154	0.7291
50.0	0.052 072	218.779	0.7849	0.040 911	217.484	0.7667	0.033 450	216.141	0.7511
60.0	0.054 059	225.488	0.8054	0.042 565	224.315	0.7875	0.034 887	223.104	0.7723
70.0	0.056 014	232.230	0.8253	0.044 184	231.161	0.8077	0.036 285	230.062	0.7929
80.0	0.057 941	239.012	0.8448	0.045 774	238.031	0.8275	0.037 653	237.027	0.8129
90.0	0.059 846	245.837	0.8638	0.047 340	244.932	0.8467	0.038 995	244.009	0.8324
100.0	0.061 731	252.707	0.8825	0.048 886	251.869	0.8656	0.040 316	251.016	0.8514
110.0	0.063 600	259.624	0.9008	0.050 415	258.845	0.8840	0.041 619	258.053	0.8700
120.0	0.065 455	266.590	0.9187	0.051 929	265.862	0.9021	0.042 907	265.124	0.8882
130.0	0.067 298	273.605	0.9364	0.053 430	272.923	0.9198	0.044 181	272.231	0.9061

T	0.70 MPa			0.80 MPa			0.90 MPa		
40.0	0.026 761	207.580	0.7148	0.022 830	205.924	0.7016	0.019 744	204.170	0.6982
50.0	0.028 100	214.745	0.7373	0.024 068	213.290	0.7248	0.020 912	211.765	0.7131
60.0	0.029 387	221.854	0.7590	0.025 247	220.558	0.7469	0.022 012	219.212	0.7358
70.0	0.030 632	228.931	0.7799	0.026 380	227.766	0.7682	0.023 062	226.564	0.7575
80.0	0.031 843	235.997	0.8002	0.027 477	234.941	0.7888	0.024 072	233.856	0.7785
90.0	0.033 027	243.066	0.8199	0.028 545	242.101	0.8088	0.025 051	241.113	0.7987
100.0	0.034 189	250.146	0.8392	0.029 588	249.260	0.8283	0.026 005	248.355	0.8184
110.0	0.035 332	257.247	0.8579	0.030 612	256.428	0.8472	0.026 937	255.593	0.8376
120.0	0.036 458	264.374	0.8763	0.031 619	263.613	0.8657	0.027 851	262.839	0.8562
130.0	0.037 572	271.531	0.8943	0.032 612	270.820	0.8838	0.028 751	270.100	0.8745
140.0	0.038 673	278.720	0.9119	0.033 592	278.055	0.9016	0.029 639	277.381	0.8923
150.0	0.039 764	285.946	0.9292	0.034 563	285.320	0.9189	0.030 515	284.687	0.9098

TABLE A-23M (Continued)

Temp. °C	v m³/kg	h kJ/kg	s kJ/kg·K	v m³/kg	h kJ/kg	s kJ/kg·K	v m³/kg	h kJ/kg	s kJ/kg·K
		1.00 MPa			1.20 MPa			1.40 MPa	
50.0	0.018 366	210.162	0.7021	0.014 483	206.661	0.6812			
60.0	0.019 410	217.810	0.7254	0.015 463	214.805	0.7060	0.012 579	211.457	0.6876
70.0	0.020 397	225.319	0.7476	0.016 368	222.687	0.7293	0.013 448	219.822	0.7123
80.0	0.021 341	232.739	0.7689	0.017 221	230.398	0.7514	0.014 247	227.891	0.7355
90.0	0.022 251	240.101	0.7895	0.018 032	237.995	0.7727	0.014 997	235.766	0.7575
100.0	0.023 133	247.430	0.8094	0.018 812	245.518	0.7931	0.015 710	243.512	0.7785
110.0	0.023 993	254.743	0.8287	0.019 567	252.993	0.8129	0.016 393	251.170	0.7988
120.0	0.024 835	262.053	0.8475	0.020 301	260.441	0.8320	0.017 053	258.770	0.8183
130.0	0.025 661	269.369	0.8659	0.021 018	267.875	0.8507	0.017 695	266.334	0.8373
140.0	0.026 474	276.699	0.8839	0.021 721	275.307	0.8689	0.018 321	273.877	0.8558
150.0	0.027 275	284.047	0.9015	0.022 412	282.745	0.8867	0.018 934	281.411	0.8738
160.0	0.028 068	291.419	0.9187	0.023 093	290.195	0.9041	0.019 535	288.946	0.8914

Temp. °C	v m³/kg	h kJ/kg	s kJ/kg·K	v m³/kg	h kJ/kg	s kJ/kg·K	v m³/kg	h kJ/kg	s kJ/kg·K
		1.60 MPa			1.80 MPa			2.00 MPa	
70.0	0.011 208	216.650	0.6959	0.009 406	213.049	0.6794			
80.0	0.011 984	225.177	0.7204	0.010 187	222.198	0.7057	0.008 704	218.859	0.6909
90.0	0.012 698	233.390	0.7433	0.010 884	230.835	0.7298	0.009 406	228.056	0.7166
100.0	0.013 366	241.397	0.7651	0.011 526	239.155	0.7524	0.010 035	236.760	0.7402
110.0	0.014 000	249.264	0.7859	0.012 126	247.264	0.7739	0.010 615	245.154	0.7624
120.0	0.014 608	257.035	0.8059	0.012 697	255.228	0.7944	0.011 159	253.341	0.7835
130.0	0.015 195	264.742	0.8253	0.013 244	263.094	0.8141	0.011 676	261.384	0.8037
140.0	0.015 765	272.406	0.8440	0.013 772	270.891	0.8332	0.012 172	269.327	0.8232
150.0	0.016 320	280.044	0.8623	0.014 284	278.642	0.8518	0.012 651	277.201	0.8420
160.0	0.016 864	287.669	0.8801	0.014 784	286.364	0.8698	0.013 116	285.027	0.8603
170.9	0.017 398	295.290	0.8975	0.015 272	294.069	0.8874	0.013 570	292.822	0.8781
180.0	0.017 923	302.914	0.9145	0.015 752	301.767	0.9046	0.014 013	300.598	0.8955

°C	2.50 MPa			3.00 MPa			3.50 MPa		
90.0	0.006 595	219.562	0.6823						
100.0	0.007 264	229.852	0.7103	0.005 231	220.529	0.6770			
110.0	0.007 837	239.271	0.7352	0.005 886	232.068	0.7075	0.004 324	222.121	0.6750
120.0	0.008 351	248.192	0.7582	0.006 419	242.208	0.7336	0.004 959	234.875	0.7078
130.0	0.008 827	256.794	0.7798	0.006 887	251.632	0.7573	0.005 456	245.661	0.7349
140.0	0.009 273	265.180	0.8003	0.007 313	260.620	0.7793	0.005 884	255.524	0.7591
150.0	0.009 697	273.414	0.8200	0.007 709	269.319	0.8001	0.006 270	264.846	0.7814
160.0	0.010 104	281.540	0.8390	0.008 083	277.817	0.8200	0.006 626	273.817	0.8023
170.0	0.010 497	289.589	0.8574	0.008 439	286.171	0.8391	0.006 961	282.545	0.8222
180.0	0.010 879	297.583	0.8752	0.008 782	294.422	0.8575	0.007 279	291.100	0.8413
190.0	0.011 250	305.540	0.8926	0.009 114	302.597	0.8753	0.007 584	299.528	0.8597
200.0	0.011 614	313.472	0.9095	0.009 436	310.718	0.8927	0.007 878	307.864	0.8775

°C	4.00 MPa		
120.0	0.003 736	224.863	0.6771
130.0	0.004 325	238.443	0.7111
140.0	0.004 781	249.703	0.7386
150.0	0.005 172	259.904	0.7630
160.0	0.005 522	269.492	0.7854
170.0	0.005 845	278.684	0.8063
180.0	0.006 147	287.602	0.8262
190.0	0.006 434	296.326	0.8453
200.0	0.006 708	304.906	0.8636
210.0	0.006 972	313.380	0.8813
220.0	0.007 228	321.774	0.8985
230.0	0.007 477	330.108	0.9152

TABLE A-22E PROPERTIES OF REFRIGERANT 12 (CCl_2F_2)

(a) Saturation Temperature Table (English Units)

Temp. °F T	Abs. Pressure lbf/in² P	Specific volume, ft³/lbm Sat. Liquid v_f	Evap. v_{fg}	Sat. Vapor v_g	Enthalpy, Btu/lbm Sat. Liquid h_f	Evap. h_{fg}	Sat. Vapor h_g	Entropy, Btu/lbm·°R Sat. Liquid s_f	Evap. s_{fg}	Sat. Vapor s_g
−130	0.41224	0.009736	70.7203	70.730	−18.609	81.577	62.968	−0.04983	0.24743	0.19760
−120	0.64190	0.009816	46.7312	46.741	−16.565	80.617	64.052	−0.04372	0.23731	0.19359
−110	0.97034	0.009899	31.7671	31.777	−14.518	79.663	65.145	−0.03779	0.22780	0.19002
−100	1.4280	0.009985	21.1541	22.164	−12.466	78.714	66.248	−0.03200	0.21883	0.18683
−90	2.0509	0.010073	15.8109	15.821	−10.409	77.764	67.355	−0.02637	0.21034	0.18398
−80	2.8807	0.010164	11.5228	11.533	−8.3451	76.812	68.467	−0.02086	0.20229	0.18143
−70	3.9651	0.010259	8.5584	8.5687	−6.2730	75.853	69.580	−0.01548	0.19464	0.17916
−60	5.3575	0.010357	6.4670	6.4774	−4.1919	74.885	70.693	−0.01021	0.18716	0.17714
−50	7.1168	0.010459	4.9637	4.9742	−2.1011	73.906	71.805	−0.00506	0.18038	0.17533
−40	9.3076	0.010564	3.8644	3.8750	0	72.913	72.913	0	0.17373	0.17373
−30	11.999	0.010674	3.0478	3.0585	2.1120	71.903	74.015	0.00496	0.16733	0.17229
−20	15.267	0.010788	2.4321	2.4429	4.2357	70.874	75.110	0.00983	0.16119	0.17102
−10	19.189	0.010906	1.9628	1.9727	6.3716	69.824	76.196	0.01462	0.15527	0.16989
0	23.849	0.011030	1.5979	1.6089	8.5207	68.750	77.271	0.01932	0.14956	0.16888
10	29.335	0.011160	1.3129	1.3241	10.684	67.651	78.335	0.02395	0.14403	0.16798
20	35.736	0.011296	1.0875	1.0988	12.863	66.522	79.385	0.02852	0.13867	0.16719
30	43.148	0.011438	0.90736	0.91880	15.058	65.361	80.419	0.03301	0.13347	0.16648
40	51.667	0.011588	0.76198	0.77357	17.273	64.163	81.436	0.03745	0.12841	0.16586
50	61.394	0.011746	0.64362	0.65537	19.507	62.926	82.433	0.04184	0.12346	0.16530
60	72.433	0.011913	0.54648	0.55839	21.766	61.643	83.409	0.04618	0.11861	0.16479
70	84.888	0.012089	0.46609	0.47818	24.050	60.309	84.359	0.05048	0.11386	0.16434
80	98.870	0.012277	0.39907	0.41135	26.365	58.917	85.282	0.05475	0.10917	0.16392
90	114.49	0.012478	0.34281	0.35529	28.713	57.461	86.174	0.05900	0.10453	0.16353
100	131.86	0.012693	0.29525	0.30794	31.100	55.929	87.029	0.06323	0.09992	0.16315

110	151.11	0.012924	0.25577	0.26769	33.531	54.313	87.844	0.06745	0.09534	0.16279
120	172.35	0.013174	0.22019	0.23326	36.013	52.597	88.610	0.07168	0.09073	0.16241
130	195.71	0.013447	0.19019	0.20364	38.553	50.768	89.321	0.07583	0.08609	0.16202
140	221.32	0.013746	0.16424	0.17799	41.162	48.805	89.967	0.08021	0.08138	0.16159
150	249.31	0.014078	0.14156	0.15564	43.850	46.684	90.534	0.08453	0.07657	0.16110
160	279.82	0.014449	0.12159	0.13604	46.633	44.373	91.006	0.08893	0.07260	0.16053
170	313.00	0.014871	0.10386	0.11873	49.529	41.830	91.359	0.09342	0.06643	0.15985
180	349.00	0.015360	0.08794	0.10330	52.562	38.999	91.561	0.09804	0.06096	0.15900
190	387.98	0.015942	0.073476	0.089418	55.769	35.792	91.561	0.10284	0.05511	0.15793
200	430.09	0.016659	0.060069	0.076728	59.203	32.075	91.278	0.10789	0.04862	0.15651
210	475.52	0.017601	0.047242	0.064843	62.959	27.599	90.558	0.11332	0.03921	0.15453
220	524.43	0.018986	0.035154	0.053140	67.246	21.790	89.036	0.11943	0.03206	0.15149
230	577.03	0.021854	0.017581	0.039435	72.893	12.229	85.122	0.12739	0.01773	0.14512
233.6 (critical)	596.9	0.02870	0	0.2870	78.86	0	78.86	0.1359	0	0.1359

TABLE A-22E (Continued)

(b) Saturation Pressure Table (English Units)

Press. psia p	Temp. °F T	Specific volume, ft³/lbm		Internal energy, Btu/lbm		Enthalpy, Btu/lbm			Entropy, Btu/lbm·°R	
		Sat. liquid v_f	Sat. vapor v_g	Sat. liquid u_f	Sat. vapor u_g	Sat. liquid h_f	Evap. h_{fg}	Sat. vapor h_g	Sat. liquid s_f	Sat. vapor s_g
5	−62.35	0.0103	6.9069	−4.69	64.04	−4.68	75.11	70.43	−0.0114	0.1776
10	−37.23	0.0106	3.6246	0.56	66.51	0.58	72.64	73.22	0.0014	0.1733
15	−20.75	0.0108	2.4835	4.05	68.13	4.08	70.95	75.03	0.0095	0.1711
20	−8.13	0.0109	1.8977	6.73	69.37	6.77	69.63	76.40	0.0155	0.1697
30	11.11	0.0112	1.2964	10.86	71.25	10.93	67.53	78.45	0.0245	0.1679
40	25.93	0.0114	0.9874	14.08	72.69	14.16	65.84	80.00	0.0312	0.1668
50	38.15	0.0116	0.7982	16.75	73.86	16.86	64.39	81.25	0.0366	0.1660
60	48.64	0.0117	0.6701	19.07	74.86	19.20	63.10	82.30	0.0413	0.1654
70	57.90	0.0119	0.5772	21.13	75.73	21.29	61.92	83.21	0.0453	0.1649
80	66.21	0.0120	0.5068	23.00	76.50	23.18	60.82	84.00	0.0489	0.1645
90	73.79	0.0122	0.4514	24.72	77.20	24.92	59.79	84.71	0.0521	0.1642
100	80.76	0.0123	0.4067	26.31	77.82	26.54	58.81	85.35	0.0551	0.1639
120	93.29	0.0126	0.3389	29.21	78.93	29.49	56.97	86.46	0.0604	0.1634
140	104.35	0.0128	0.2896	31.82	79.89	32.15	55.24	87.39	0.0651	0.1630
160	114.30	0.0130	0.2522	34.21	80.71	34.59	53.59	88.18	0.0693	0.1626
180	123.38	0.0133	0.2228	36.42	81.44	36.86	52.00	88.86	0.0731	0.1623
200	131.74	0.0135	0.1989	38.50	82.08	39.00	50.44	89.44	0.0767	0.1620
220	139.51	0.0137	0.1792	40.48	82.08	41.03	48.90	89.94	0.0816	0.1616
240	146.77	0.0140	0.1625	42.35	83.14	42.97	47.39	90.36	0.0831	0.1613
260	153.60	0.0142	0.1483	44.16	83.58	44.84	45.88	90.72	0.0861	0.1609
280	160.06	0.0145	0.1359	45.90	83.97	46.65	44.36	91.01	0.0890	0.1605
300	166.18	0.0147	0.1251	47.59	84.30	48.41	42.83	91.24	0.0917	0.1601

TABLE A-23E PROPERTIES OF REFRIGERANT 12 (CCl_2F_2): SUPERHEATED VAPOR (ENGLISH UNITS)

Temp. °F	5 lbf/in.²			10 lbf/in.²			15 lbf/in.²		
	v ft³/lbm	h Btu/lbm	s Btu/lbm·°R	v ft³/lbm	h Btu/lbm	s Btu/lbm·°R	v ft³/lbm	h Btu/lbm	s Btu/lbm·°R
0	8.0611	78.582	0.19663	3.9809	78.246	0.18471	2.6201	77.902	0.17751
20	8.4265	81.309	0.20244	4.1691	81.014	0.19061	2.7494	80.712	0.18349
40	8.7903	84.090	0.20812	4.3556	83.828	0.19635	2.8770	83.561	0.18931
60	9.1528	86.922	0.21367	4.5408	86.689	0.20197	3.0031	86.451	0.19498
80	9.5142	89.806	0.21912	4.7248	89.596	0.20746	3.1281	89.383	0.20051
100	9.8747	92.738	0.22445	4.9079	92.548	0.21283	3.2521	92.357	0.20593
120	10.234	95.717	0.22968	5.0903	95.546	0.21809	3.3754	95.373	0.21122
140	10.594	98.743	0.23481	5.2720	98.586	0.22325	3.4981	98.429	0.21640
160	10.952	101.812	0.23985	5.4533	101.669	0.22830	3.6202	101.525	0.22148
180	11.311	104.925	0.24479	5.6341	104.793	0.23326	3.7419	104.661	0.22646
200	11.668	108.079	0.24964	5.8145	107.957	0.23813	3.8632	107.835	0.23135
220	12.026	111.272	0.25441	5.9946	111.159	0.24291	3.9841	111.046	0.23614

Temp. °F	20 lbf/in.²			25 lbf/in.²			30 lbf/in.²		
	v ft³/lbm	h Btu/lbm	s Btu/lbm·°R	v ft³/lbm	h Btu/lbm	s Btu/lbm·°R	v ft³/lbm	h Btu/lbm	s Btu/lbm·°R
20	2.0391	80.403	0.17829	1.6125	80.088	0.17414	1.3278	79.765	0.17065
40	2.1373	83.289	0.18419	1.6932	83.012	0.18012	1.3969	82.730	0.17671
60	2.2340	86.210	0.18992	1.7723	85.965	0.18591	1.4644	85.716	0.18257
80	2.3295	89.168	0.19550	1.8502	88.950	0.19155	1.5306	88.729	0.18826
100	2.4241	92.164	0.20095	1.9271	91.968	0.19704	1.5957	91.770	0.19379
120	2.5179	95.198	0.20628	2.0032	95.021	0.20240	1.6600	94.843	0.19918
140	2.6110	98.270	0.21149	2.0786	98.110	0.20763	1.7237	97.948	0.20445
160	2.7036	101.380	0.21659	2.1535	101.234	0.21276	1.7868	101.086	0.20960
180	2.7957	104.528	0.22159	2.2279	104.393	0.21778	1.8494	104.258	0.21463
200	2.8874	107.712	0.22649	2.3019	107.588	0.22269	1.9116	107.464	0.21957
220	2.9789	110.932	0.23130	2.3756	110.817	0.22752	1.9735	110.702	0.22440
240	3.0700	114.186	0.23602	2.4491	114.080	0.23225	2.0351	113.973	0.22915

TABLE A-23E *(Continued)*

Temp. °F	v ft³/lbm	h Btu/lbm	s Btu/lbm·°R	v ft³/lbm	h Btu/lbm	s Btu/lbm·°R	v ft³/lbm	h Btu/lbm	s Btu/lbm·°R
	35 lbf/in.²			40 lbf/in.²			50 lbf/in.²		
40	1.1850	82.442	0.17375	1.0258	82.148	0.17112	0.80248	81.540	0.16655
60	1.2442	85.463	0.17968	1.0789	85.206	0.17712	0.84713	84.676	0.17271
80	1.3021	88.504	0.18542	1.1306	88.277	0.18292	0.89025	87.811	0.17862
100	1.3589	91.570	0.19100	1.1812	91.367	0.18854	0.93216	90.953	0.18434
120	1.4148	94.663	0.19643	1.2309	94.480	0.19401	0.97313	94.110	0.18988
140	1.4701	97.785	0.20172	1.2798	97.620	0.19933	1.0133	97.286	0.19527
160	1.5248	100.938	0.20689	1.3282	100.788	0.20453	1.0529	100.485	0.20051
180	1.5789	104.122	0.21195	1.3761	103.985	0.20961	1.0920	103.708	0.20563
200	1.6327	107.338	0.21690	1.4236	107.212	0.21457	1.1307	106.958	0.21064
220	1.6862	110.586	0.22175	1.4707	110.469	0.21944	1.1690	110.235	0.21553
240	1.7394	113.865	0.22651	1.5176	113.757	0.22420	1.2070	113.539	0.22032
260	1.7923	117.175	0.23117	1.5642	117.074	0.22888	1.2447	116.871	0.22502

Temp. °F	v ft³/lbm	h Btu/lbm	s Btu/lbm·°R	v ft³/lbm	h Btu/lbm	s Btu/lbm·°R	v ft³/lbm	h Btu/lbm	s Btu/lbm·°R
	60 lbf/in.²			70 lbf/in.²			80 lbf/in.²		
60	0.69210	84.126	0.16892	0.58088	83.552	0.16556	···	···	···
80	0.72964	87.330	0.17497	0.61458	86.832	0.17175	0.52795	86.316	0.16885
100	0.76588	90.528	0.18079	0.64685	90.091	0.17768	0.55734	89.640	0.17489
120	0.80110	93.731	0.18641	0.67803	93.343	0.18339	0.58556	92.945	0.18070
140	0.83551	96.945	0.19186	0.70836	96.597	0.18891	0.61286	96.242	0.18629
160	0.86928	100.776	0.19716	0.73800	99.862	0.19427	0.63943	99.542	0.19170
180	0.90252	103.427	0.20233	0.76708	103.141	0.19948	0.66543	102.851	0.19696
200	0.93531	106.700	0.20736	0.79571	106.439	0.20455	0.69095	106.174	0.20207
220	0.96775	109.997	0.21229	0.82397	109.756	0.20951	0.71609	109.513	0.20706
240	0.99988	113.319	0.21710	0.85191	113.096	0.21435	0.74090	112.872	0.21193
260	1.0318	116.666	0.22182	0.87959	116.459	0.21909	0.76544	116.251	0.21669
280	1.0634	120.039	0.22644	0.90705	119.846	0.22373	0.78975	119.652	0.22135

Temp. °F	v ft³/lbm	h Btu/lbm	s Btu/lbm·°R	v ft³/lbm	h Btu/lbm	s Btu/lbm·°R	v ft³/lbm	h Btu/lbm	s Btu/lbm·°R
	90 lbf/in.²			100 lbf/in.²			125 lbf/in.²		
100	0.48749	89.175	0.17234	0.43138	88.694	0.16996	0.32943	87.407	0.16455
120	0.51346	92.536	0.17824	0.45562	92.116	0.17597	0.35086	91.008	0.17087
140	0.53845	95.879	0.18391	0.47881	95.507	0.18172	0.37098	94.537	0.17686
160	0.56268	99.216	0.18938	0.50118	98.884	0.18726	0.39015	98.023	0.18258
180	0.58629	102.557	0.19469	0.52291	102.257	0.19262	0.40857	101.484	0.18807
200	0.60941	105.905	0.19984	0.54413	105.633	0.19782	0.42642	104.934	0.19338
220	0.63213	109.267	0.20486	0.56492	109.018	0.20287	0.44380	108.380	0.19853
240	0.65451	112.644	0.20976	0.58538	112.415	0.20780	0.46081	111.829	0.20353
260	0.67662	116.040	0.21455	0.60554	115.828	0.21261	0.47750	115.287	0.20840
280	0.69849	119.456	0.21923	0.62546	119.258	0.21731	0.49394	118.756	0.21316
300	0.72016	122.892	0.22381	0.64518	122.707	0.22191	0.51016	122.238	0.21780
320	0.74166	126.349	0.22830	0.66472	126.176	0.22641	0.52619	125.737	0.22235

150 lbf/in.² · 175 lbf/in.² · 200 lbf/in.²

T	150 lbf/in.²			175 lbf/in.²			200 lbf/in.²		
120	0.28007	89.800	0.16629	...	...	...	...	...	...
140	0.29845	93.498	0.17256	0.24595	92.373	0.16859	0.20579	91.137	0.16480
160	0.31566	97.112	0.17849	0.26198	96.142	0.17478	0.22121	95.100	0.17130
180	0.33200	100.675	0.18415	0.27697	99.823	0.18062	0.23535	98.921	0.17737
200	0.34769	104.206	0.18958	0.29120	103.447	0.18620	0.24860	102.652	0.18311
220	0.36285	107.720	0.19483	0.30485	107.036	0.19156	0.26117	106.325	0.18860
240	0.37761	111.226	0.19992	0.31804	110.605	0.19674	0.27323	109.962	0.19387
260	0.39203	114.732	0.20485	0.33087	114.162	0.20175	0.28489	113.576	0.19896
280	0.40617	118.242	0.20967	0.34339	117.717	0.20662	0.29623	117.178	0.20390
300	0.42008	121.761	0.21436	0.35567	121.273	0.21137	0.30730	120.775	0.20870
320	0.43379	125.290	0.21894	0.36773	124.835	0.21599	0.31815	124.373	0.21337
340	0.44733	128.833	0.22343	0.37963	128.407	0.22052	0.32881	127.974	0.21793

250 lbf/in.² · 300 lbf/in.² · 400 lbf/in.²

T	250 lbf/in.²			300 lbf/in.²			400 lbf/in.²		
160	0.15249	92.717	0.16462	...	...	...	...	...	...
180	0.17605	96.925	0.17130	0.13482	94.556	0.16537	...	...	...
200	0.18824	100.930	0.17747	0.14697	98.975	0.17217	0.091005	93.718	0.16092
220	0.19952	104.809	0.18326	0.15774	103.136	0.17838	0.10316	99.046	0.16888
240	0.21014	108.607	0.18877	0.16761	107.140	0.18419	0.11300	103.735	0.17568
260	0.22027	112.351	0.19404	0.17685	111.043	0.18969	0.12163	108.105	0.18183
280	0.23001	116.060	0.19913	0.18562	114.879	0.19495	0.12949	112.286	0.18756
300	0.23944	119.747	0.20405	0.19402	118.670	0.20000	0.13680	116.343	0.19298
320	0.24862	123.420	0.20882	0.20214	122.430	0.20489	0.14372	120.318	0.19814
340	0.25759	127.088	0.21346	0.21002	126.171	0.20963	0.15032	124.235	0.20310
360	0.26639	130.754	0.21799	0.21770	129.900	0.21423	0.15668	128.112	0.20789
380	0.27504	134.423	0.22241	0.22522	133.624	0.21872	0.16285	131.961	0.21258

500 lbf/in.² · 600 lbf/in.²

T	500 lbf/in.²			600 lbf/in.²		
220	0.064207	92.397	0.15683	...	...	...
240	0.077620	99.218	0.16672	0.047488	91.024	0.15335
260	0.087054	104.526	0.17421	0.061922	99.741	0.16566
280	0.094923	109.277	0.18072	0.070859	105.637	0.17374
300	0.10190	113.729	0.18666	0.078059	110.729	0.18053
320	0.10829	117.997	0.19221	0.084333	115.420	0.18663
340	0.11426	122.143	0.19746	0.090017	119.871	0.19227
360	0.11992	126.205	0.20247	0.095289	124.167	0.19757
380	0.12533	130.207	0.20730	0.10025	128.355	0.20262
400	0.13054	134.166	0.21196	0.10498	132.466	0.20746
420	0.13559	138.096	0.21648	0.10952	136.523	0.21213
440	0.14051	142.004	0.22087	0.11391	140.539	0.21664

TABLE A-24M PROPERTIES OF AMMONIA: SATURATION TEMPERATURE TABLE (SI UNITS)

Temp. °C T	Abs. Pressure kPa P	Specific Volume m³/kg			Enthalpy kJ/kg			Entropy kJ/kg·K		
		Sat. Liquid v_f	Evap. v_{fg}	Sat. Vapor v_g	Sat. Liquid h_f	Evap. h_{fg}	Sat. Vapor h_g	Sat. Liquid s_f	Evap. s_{fg}	Sat. Vapor s_g
−50	40.88	0.001 424	2.6239	2.6254	−44.3	1416.7	1372.4	−0.1942	6.3502	6.1561
−48	45.96	0.001 429	2.3518	2.3533	−35.5	1411.3	1375.8	−0.1547	6.2696	6.1149
−46	51.55	0.001 434	2.1126	2.1140	−26.6	1405.8	1379.2	−0.1156	6.1902	6.0746
−44	57.69	0.001 439	1.9018	1.9032	−17.8	1400.3	1382.5	−0.0768	6.1120	6.0352
−42	64.42	0.001 444	1.7155	1.7170	−8.9	1394.7	1385.8	−0.0382	6.0349	5.9967
−40	71.77	0.001 449	1.5506	1.5521	0.0	1389.0	1389.0	0.0000	5.9589	5.9589
−38	79.80	0.001 454	1.4043	1.4058	8.9	1383.3	1392.2	0.0380	5.8840	5.9220
−36	88.54	0.001 460	1.2742	1.2757	17.8	1377.6	1395.4	0.0757	5.8101	5.8858
−34	98.05	0.001 465	1.1582	1.1597	26.8	1371.8	1398.5	0.1132	5.7372	5.8504
−32	108.37	0.001 470	1.0547	1.0562	35.7	1365.9	1401.6	0.1504	5.6652	5.8156
−30	119.55	0.001 476	0.9621	0.9635	44.7	1360.0	1404.6	0.1873	5.5942	5.7815
−28	131.64	0.001 481	0.8790	0.8805	53.6	1354.0	1407.6	0.2240	5.5241	5.7481
−26	144.70	0.001 487	0.8044	0.8059	62.6	1347.9	1410.5	0.2605	5.4548	5.7153
−24	158.78	0.001 492	0.7373	0.7388	71.6	1341.8	1413.4	0.2967	5.3864	5.6831
−22	173.93	0.001 498	0.6768	0.6783	80.7	1335.6	1416.2	0.3327	5.3188	5.6515
−20	190.22	0.001 504	0.6222	0.6237	89.7	1329.3	1419.0	0.3684	5.2520	5.6205
−18	207.71	0.001 510	0.5728	0.5743	98.8	1322.9	1421.7	0.4040	5.1860	5.5900
−16	226.45	0.001 515	0.5280	0.5296	107.8	1316.5	1424.4	0.4393	5.1207	5.5600
−14	246.51	0.001 521	0.4874	0.4889	116.9	1310.0	1427.0	0.4744	5.0561	5.5305
−12	267.95	0.001 528	0.4505	0.4520	126.0	1303.5	1429.5	0.5093	4.9922	5.5015

−10	290.85	0.001 534	0.4169	0.4185	135.2	1296.8	1432.0	0.5440	4.9290	5.4730
−8	315.25	0.001 540	0.3863	0.3878	144.3	1290.1	1434.4	0.5785	4.8664	5.4449
−6	341.25	0.001 546	0.3583	0.3599	153.5	1283.3	1436.8	0.6128	4.8045	5.4173
−4	368.90	0.001 553	0.3328	0.3343	162.7	1276.4	1439.1	0.6469	4.7432	5.3901
−2	398.27	0.001 559	0.3094	0.3109	171.9	1269.4	1441.3	0.6808	4.6825	5.3633
0	429.44	0.001 566	0.2879	0.2895	181.1	1262.4	1443.5	0.7145	4.6223	5.3369
2	462.49	0.001 573	0.2683	0.2698	190.4	1255.2	1445.6	0.7481	4.5627	5.3108
4	497.49	0.001 580	0.2502	0.2517	199.6	1248.0	1447.6	0.7815	4.5037	5.2852
6	534.51	0.001 587	0.2335	0.2351	208.9	1240.6	1449.6	0.8148	4.4451	5.2599
8	573.64	0.001 594	0.2182	0.2198	218.3	1233.2	1451.5	0.8479	4.3871	5.2350
10	614.95	0.001 601	0.2040	0.2056	227.6	1225.7	1453.3	0.8808	4.3295	5.2104
12	658.52	0.001 608	0.1910	0.1926	237.0	1218.1	1455.1	0.9136	4.2725	5.1861
14	704.44	0.001 616	0.1789	0.1805	246.4	1210.4	1456.8	0.9463	4.2159	5.1621
16	752.79	0.001 623	0.1677	0.1693	255.9	1202.6	1458.5	0.9788	4.1597	5.1385
18	803.66	0.001 631	0.1574	0.1590	265.4	1194.7	1460.0	1.0112	4.1039	5.1151
20	857.12	0.001 639	0.1477	0.1494	274.9	1186.7	1461.5	1.0434	4.0486	5.0920
22	913.27	0.001 647	0.1388	0.1405	284.4	1178.5	1462.9	1.0755	3.9937	5.0692
24	972.19	0.001 655	0.1305	0.1322	294.0	1170.3	1464.3	1.1075	3.9392	5.0467
26	1033.97	0.001 663	0.1228	0.1245	303.6	1162.0	1465.6	1.1394	3.8850	5.0244
28	1098.71	0.001 671	0.1156	0.1173	313.2	1153.6	1466.8	1.1711	3.8312	5.0023
30	1166.49	0.001 680	0.1089	0.1106	322.9	1145.0	1467.9	1.2028	3.7777	4.9805
32	1237.41	0.001 689	0.1027	0.1044	332.6	1136.4	1469.0	1.2343	3.7246	4.9589
34	1311.55	0.001 698	0.0969	0.0986	342.3	1127.6	1469.9	1.2656	3.6718	4.9374
36	1389.03	0.001 707	0.0914	0.0931	352.1	1118.7	1470.8	1.2969	3.6192	4.9161
38	1469.92	0.001 716	0.0863	0.0880	361.9	1109.7	1471.5	1.3281	3.5669	4.8950
40	1554.33	0.001 726	0.0815	0.0833	371.7	1100.5	1472.2	1.3591	3.5148	4.8740
42	1642.35	0.001 735	0.0771	0.0788	381.6	1091.2	1472.8	1.3901	3.4630	4.8530
44	1734.09	0.001 745	0.0728	0.0746	391.5	1081.7	1473.2	1.4209	3.4112	4.8322
46	1829.65	0.001 756	0.0689	0.0707	401.5	1072.0	1473.5	1.4518	3.3595	4.8113
48	1929.13	0.001 766	0.0652	0.0669	411.5	1062.2	1473.7	1.4826	3.3079	4.7905
50	2032.62	0.001 777	0.0617	0.0635	421.7	1052.0	1473.7	1.5135	3.2561	4.7696

Source for Tables A-24 and A-25: Adapted from National Bureau of Standards Circular No. 142, *Tables of Thermodynamic Properties of Ammonia.*

811

TABLE A-25M PROPERTIES OF AMMONIA: SUPERHEATED VAPOR (SI UNITS)

v in m³/kg, h in kJ/kg, and s in kJ/kg·K

Abs. Pressure kPa (Sat. Temp. °C)		Temperature, °C											
		−20	−10	0	10	20	30	40	50	60	70	80	100
50 (−46.54)	v	2.4474	2.5481	2.6482	2.7479	2.8473	2.9464	3.0453	3.1441	3.2427	3.3413	3.4397	
	h	1435.8	1457.0	1478.1	1499.2	1520.4	1541.7	1563.0	1584.5	1606.1	1627.8	1649.7	
	s	6.3256	6.4077	6.4865	6.5625	6.6360	6.7073	6.7766	6.8441	6.9099	6.9743	7.0372	
75 (−39.18)	v	1.6233	1.6915	1.7591	1.8263	1.8932	1.9597	2.0261	2.0923	2.1584	2.2244	2.2903	
	h	1433.0	1454.7	1476.1	1497.5	1518.9	1540.3	1561.8	1583.4	1605.1	1626.9	1648.9	
	s	6.1190	6.2028	6.2828	6.3597	6.4339	6.5058	6.5756	6.6434	6.7096	6.7742	6.8373	
100 (−33.61)	v	1.2110	1.2631	1.3145	1.3654	1.4160	1.4664	1.5165	1.5664	1.6163	1.6659	1.7155	1.8145
	h	1430.1	1452.2	1474.1	1495.7	1517.3	1538.9	1560.5	1582.2	1604.1	1626.0	1648.0	1692.6
	s	5.9695	6.0552	6.1366	6.2144	6.2894	6.3618	6.4321	6.5003	6.5668	6.6316	6.6950	6.8177
125 (−29.08)	v	0.9635	1.0059	1.0476	1.0889	1.1297	1.1703	1.2107	1.2509	1.2909	1.3309	1.3707	1.4501
	h	1427.2	1449.8	1472.0	1493.9	1515.7	1537.5	1559.3	1581.1	1603.0	1625.0	1647.2	1691.8
	s	5.8512	5.9389	6.0217	6.1006	6.1763	6.2494	6.3201	6.3887	6.4555	6.5206	6.5842	6.7072
150 (−25.23)	v	0.7984	0.8344	0.8697	0.9045	0.9388	0.9729	1.0068	1.0405	1.0740	1.1074	1.1408	1.2072
	h	1424.1	1447.3	1469.8	1492.1	1514.1	1536.1	1558.0	1580.0	1602.0	1624.1	1646.3	1691.1
	s	5.7526	5.8424	5.9266	6.0066	6.0831	6.1568	6.2280	6.2970	6.3641	6.4295	6.4933	6.6167

Sat. Press. (Temp.)		C1	C2	C3	C4	C5	C6	C7	C8	C9	C10	C11
200 (−18.86)	v	0.6199	0.6471	0.6738	0.7001	0.7261	0.7519	0.7774	0.8029	0.8282	0.8533	0.9035
	h	1442.0	1465.5	1488.4	1510.9	1533.2	1555.5	1577.7	1599.9	1622.2	1644.6	1689.6
	s	5.6863	5.7737	5.8559	5.9342	6.0091	6.0813	6.1512	6.2189	6.2849	6.3491	6.4732
250 (−13.67)	v	0.4910	0.5135	0.5354	0.5568	0.5780	0.5989	0.6196	0.6401	0.6605	0.6809	0.7212
	h	1436.6	1461.0	1484.5	1507.6	1530.3	1552.9	1575.4	1597.8	1620.3	1642.8	1688.2
	s	5.5609	5.6517	5.7365	5.8165	5.8928	5.9661	6.0368	6.1052	6.1717	6.2365	6.3613
300 (−9.23)	v		0.4243	0.4430	0.4613	0.4792	0.4968	0.5143	0.5316	0.5488	0.5658	0.5997
	h		1456.3	1480.6	1504.2	1527.4	1550.3	1573.0	1595.7	1618.4	1641.1	1686.7
	s		5.5493	5.6366	5.7186	5.7963	5.8707	5.9423	6.0114	6.0785	6.1437	6.2693
350 (−5.35)	v		0.3605	0.3770	0.3929	0.4086	0.4239	0.4391	0.4541	0.4689	0.4837	0.5129
	h		1451.5	1476.5	1500.7	1524.4	1547.6	1570.7	1593.6	1616.5	1639.3	1685.2
	s		5.4600	5.5502	5.6342	5.7135	5.7890	5.8615	5.9314	5.9990	6.0647	6.1910
400 (−1.89)	v		0.3125	0.3274	0.3417	0.3556	0.3692	0.3826	0.3959	0.4090	0.4220	0.4478
	h		1446.5	1472.4	1497.2	1521.3	1544.9	1568.3	1591.5	1614.5	1637.6	1683.7
	s		5.3803	5.4735	5.5597	5.6405	5.7173	5.7907	5.8613	5.9296	5.9957	6.1228
450 (1.26)	v		0.2752	0.2887	0.3017	0.3143	0.3266	0.3387	0.3506	0.3624	0.3740	0.3971
	h		1441.3	1468.1	1493.6	1518.2	1542.2	1565.9	1589.3	1612.6	1635.8	1682.2
	s		5.3078	5.4042	5.4926	5.5752	5.6532	5.7275	5.7989	5.8678	5.9345	6.0623

TABLE A-25M *(Continued)*

Abs. Pressure kPa (Sat. Temp. °C)

		20	30	40	50	60	70	80	100	120	140	160	180
500 (4.14)	v	0.2698	0.2813	0.2926	0.3036	0.3144	0.3251	0.3357	0.3565	0.3771	0.3975		
	h	1489.9	1515.0	1539.5	1563.4	1587.1	1610.6	1634.0	1680.7	1727.5	1774.7		
	s	5.4314	5.5157	5.5950	5.6704	5.7425	5.8120	5.8793	6.0079	6.1301	6.2472		
600 (9.29)	v	0.2217	0.2317	0.2414	0.2508	0.2600	0.2691	0.2781	0.2957	0.3130	0.3302		
	h	1482.4	1508.6	1533.8	1558.5	1582.7	1606.6	1630.4	1677.7	1724.9	1772.4		
	s	5.3222	5.4102	5.4923	5.5697	5.6436	5.7144	5.7826	5.9129	6.0363	6.1541		
700 (13.81)	v	0.1874	0.1963	0.2048	0.2131	0.2212	0.2291	0.2369	0.2522	0.2672	0.2821		
	h	1474.5	1501.9	1528.1	1553.4	1578.2	1602.6	1626.8	1674.6	1722.4	1770.2		
	s	5.2259	5.3179	5.4029	5.4826	5.5582	5.6303	5.6997	5.8316	5.9562	6.0749		
800 (17.86)	v	0.1615	0.1696	0.1773	0.1848	0.1920	0.1991	0.2060	0.2196	0.2329	0.2459	0.2589	
	h	1466.3	1495.0	1522.2	1548.3	1573.7	1598.6	1623.1	1671.6	1719.8	1768.0	1816.4	
	s	5.1387	5.2351	5.3232	5.4053	5.4827	5.5562	5.6268	5.7603	5.8861	6.0057	6.1202	
900 (21.54)	v		0.1488	0.1559	0.1627	0.1693	0.1757	0.1820	0.1942	0.2061	0.2178	0.2294	
	h		1488.0	1516.2	1543.0	1569.1	1594.4	1619.4	1668.5	1717.1	1765.7	1814.4	
	s		5.1593	5.2508	5.3354	5.4147	5.4897	5.5614	5.6968	5.8237	5.9442	6.0594	

Temperature, °C

1000 (24.91)	v	0.1321	0.1388	0.1450	0.1511	0.1570	0.1627	0.1739	0.1847	0.1954	0.2058	0.2162
	h	1480.6	1510.0	1537.7	1564.4	1590.3	1615.6	1665.4	1714.5	1763.4	1812.4	1861.7
	s	5.0889	5.1840	5.2713	5.3525	5.4299	5.5021	5.6392	5.7674	5.8888	6.0047	6.1159
1200 (30.96)	v		0.1129	0.1185	0.1238	0.1289	0.1338	0.1434	0.1526	0.1616	0.1705	0.1792
	h		1497.1	1526.6	1554.7	1581.7	1608.0	1659.2	1709.2	1758.9	1808.5	1858.2
	s		5.0629	5.1560	5.2416	5.3215	5.3970	5.5379	5.6687	5.7919	5.9091	6.0214
1400 (36.28)	v		0.0944	0.0995	0.1042	0.1088	0.1132	0.1216	0.1297	0.1376	0.1452	0.1528
	h		1483.4	1515.1	1544.7	1573.0	1600.2	1652.8	1703.9	1754.3	1804.5	1854.7
	s		4.9534	5.0530	5.1434	5.2270	5.3053	5.4501	5.5836	5.7087	5.8273	5.9406
1600 (41.05)	v			0.0851	0.0895	0.0937	0.0977	0.1053	0.1125	0.1195	0.1263	0.1330
	h			1502.9	1534.4	1564.0	1592.3	1646.4	1698.5	1749.7	1800.5	1851.2
	s			4.9584	5.0543	5.1419	5.2232	5.3722	5.5084	5.6355	5.7555	5.8699
1800 (45.39)	v			0.0739	0.0781	0.0820	0.0856	0.0926	0.0992	0.1055	0.1116	0.1177
	h			1490.0	1523.5	1554.6	1584.1	1639.8	1693.1	1745.1	1796.5	1847.7
	s			4.8693	4.9715	5.0635	5.1482	5.3018	5.4409	5.5699	5.6914	5.8069
2000 (49.38)	v			0.0648	0.0688	0.0725	0.0760	0.0824	0.0885	0.0943	0.0999	0.1054
	h			1476.1	1512.0	1544.9	1575.6	1633.2	1687.6	1740.4	1792.4	1844.1
	s			4.7834	4.8930	4.9902	5.0786	5.2371	5.3793	5.5104	5.6333	5.7499

TABLE A-24E PROPERTIES OF AMMONIA: SATURATION TEMPERATURE TABLE (ENGLISH UNITS)

Temp. F	Abs. Press. lbf/in.² P	Specific Volume ft³/lbm Sat. Liquid v_f	Evap. v_{fg}	Sat. Vapor v_g	Enthalpy Btu/lbm Sat. Liquid h_f	Evap. h_{fg}	Sat. Vapor h_g	Entropy Btu/lbm R Sat. Liquid s_f	Evap. s_{fg}	Sat. Vapor s_g
−60	5.55	0.0228	44.707	44.73	−21.2	610.8	589.6	−0.0517	1.5286	1.4769
−55	6.54	0.0229	38.357	38.38	−15.9	607.5	591.6	−0.0386	1.5017	1.4631
−50	7.67	0.0230	33.057	33.08	−10.6	604.3	593.7	−0.0256	1.4753	1.4497
−45	8.95	0.0231	28.597	28.62	−5.3	600.9	595.6	−0.0127	1.4495	1.4368
−40	10.41	0.02322	24.837	24.86	0	597.6	597.6	0.000	1.4242	1.4242
−35	12.05	0.02333	21.657	21.68	5.3	594.2	599.5	0.0126	1.3994	1.4120
−30	13.90	0.0235	18.947	18.97	10.7	590.7	601.4	0.0250	1.3751	1.4001
−25	15.98	0.0236	16.636	16.66	16.0	587.2	603.2	0.0374	1.3512	1.3886
−20	18.30	0.0237	14.656	14.68	21.4	583.6	605.0	0.0497	1.3277	1.3774
−15	20.88	0.02381	12.946	12.97	26.7	580.0	606.7	0.0618	1.3044	1.3664
−10	23.74	0.02393	11.476	11.50	32.1	576.4	608.5	0.0738	1.2820	1.3558
−5	26.92	0.02406	10.206	10.23	37.5	572.6	610.1	0.0857	1.2597	1.3454
0	30.42	0.02419	9.092	9.116	42.9	568.9	611.8	0.0975	1.2377	1.3352
5	34.27	0.02432	8.1257	8.150	48.3	565.0	613.3	0.1092	1.2161	1.3253
10	38.51	0.02446	7.2795	7.304	53.8	561.1	614.9	0.1208	1.1949	1.3157
15	43.14	0.02460	6.5374	6.562	59.2	557.1	616.3	0.1323	1.1739	1.3062
20	48.21	0.02474	5.8853	5.910	64.7	553.1	617.8	0.1437	1.1532	1.2969
25	53.73	0.02488	5.3091	5.334	70.2	548.9	619.1	0.1551	1.1328	1.2879
30	59.74	0.02503	4.8000	4.825	75.7	544.8	620.5	0.1663	1.1127	1.2790
35	66.26	0.02518	4.3478	4.373	81.2	540.5	621.7	0.1775	1.0929	1.2704
40	73.32	0.02533	3.9457	3.971	86.8	536.2	623.0	0.1885	1.0733	1.2618
45	80.96	0.02548	3.5885	3.614	92.3	531.8	624.1	0.1996	1.0539	1.2535
50	89.19	0.02564	3.2684	3.294	97.9	527.3	625.2	0.2105	1.0348	1.2453
55	98.06	0.02581	2.9822	3.008	103.5	522.8	626.3	0.2214	1.0159	1.2373
60	107.6	0.02597	2.7250	2.751	109.2	518.1	627.3	0.2322	0.9972	1.2294
65	117.8	0.02614	2.4939	2.520	114.8	513.4	628.2	0.2430	0.9786	1.2216
70	128.8	0.02632	2.2857	2.312	120.5	508.6	629.1	0.2537	0.9603	1.2140
75	140.5	0.02650	2.0985	2.125	126.2	503.7	629.9	0.2643	0.9422	1.2065
80	153.0	0.02668	1.9283	1.955	132.0	498.7	630.7	0.2749	0.9242	1.1991
85	166.4	0.02687	1.7741	1.801	137.8	493.6	631.4	0.2854	0.9064	1.1918
90	180.6	0.02707	1.6339	1.661	143.5	488.5	632.0	0.2958	0.8888	1.1846
95	195.8	0.02727	1.5067	1.534	149.4	483.2	632.6	0.3062	0.8713	1.1775
100	211.9	0.02747	1.3915	1.419	155.2	477.8	633.0	0.3166	0.8539	1.1705
105	228.9	0.02769	1.2853	1.313	161.1	472.3	633.4	0.3269	0.8366	1.1635
110	247.0	0.02790	1.1891	1.217	167.0	466.7	633.7	0.3372	0.8194	1.1566
115	266.2	0.02813	1.0999	1.128	173.0	460.9	633.9	0.3474	0.8023	1.1497
120	286.4	0.02836	1.0186	1.047	179.0	455.0	634.0	0.3576	0.7851	1.1427
125	307.8	0.02860	0.9444	0.973	185.1	448.9	634.0	0.3679	0.7679	1.1358

TABLE A-25E PROPERTIES OF AMMONIA: SUPERHEATED VAPOR (ENGLISH UNITS)

v in ft³/lbm, h in Btu/lbm, and s in Btu/lbm·°R

Pressure, psia (Saturation temperature in italics)

Temp. °F	5 −63.11°			10 −41.34°			15 −27.29°			20 −16.64°		
	v	h	s	v	h	s	v	h	s	v	h	s
Sat.	*49.31*	*588.3*	*1.4857*	*25.81*	*597.1*	*1.4276*	*17.67*	*602.4*	*1.3938*	*13.50*	*606.2*	*1.3700*
−50	51.05	595.2	1.5025	.....	.....	.....	.....	.....	.....	.....	.....	.....
−40	52.36	600.3	1.5149	25.90	597.8	1.4293	.....	.....	.....	.....	.....	.....
−30	53.67	605.4	1.5269	26.58	603.2	1.4420	.....	.....	.....	.....	.....	.....
−20	54.97	610.4	1.5385	27.26	608.5	1.4542	18.01	606.4	1.4031	.....	.....	.....
−10	56.26	615.4	1.5498	27.92	613.7	1.4659	18.47	611.9	1.4154	13.74	610.0	1.3784
0	57.55	620.4	1.5608	28.58	618.9	1.4773	18.92	617.2	1.4272	14.09	615.5	1.3907
10	58.84	625.4	1.5716	29.24	624.0	1.4884	19.37	622.5	1.4386	14.44	621.0	1.4025
20	60.12	630.4	1.5821	29.90	629.1	1.4992	19.82	627.8	1.4497	14.78	626.4	1.4138
30	61.41	635.4	1.5925	30.55	634.2	1.5097	20.26	633.0	1.4604	15.11	631.7	1.4248
40	62.69	640.4	1.6026	31.20	639.3	1.5200	20.70	638.2	1.4709	15.45	637.0	1.4356
50	63.96	645.5	1.6125	31.85	644.4	1.5301	21.14	643.4	1.4812	15.78	642.3	1.4460
60	65.24	650.5	1.6223	32.49	649.5	1.5400	21.58	648.5	1.4912	16.12	647.5	1.4562
70	66.51	655.5	1.6319	33.14	654.6	1.5497	22.01	653.7	1.5011	16.45	652.8	1.4662
80	67.79	660.6	1.6413	33.78	659.7	1.5593	22.44	658.9	1.5108	16.78	658.0	1.4760
90	69.06	665.6	1.6506	34.42	664.8	1.5687	22.88	664.0	1.5203	17.10	663.2	1.4856
100	70.33	670.7	1.6598	35.07	670.0	1.5779	23.31	669.2	1.5296	17.43	668.5	1.4950
110	71.60	675.8	1.6689	35.71	675.1	1.5870	23.74	674.4	1.5388	17.76	673.7	1.5042
120	72.87	680.9	1.6778	36.35	680.3	1.5960	24.17	679.6	1.5478	18.08	678.9	1.5133
130	74.14	686.1	1.6865	36.99	685.4	1.6049	24.60	684.8	1.5567	18.41	684.2	1.5223
140	75.41	691.2	1.6952	37.62	690.6	1.6136	25.03	690.0	1.5655	18.73	689.4	1.5312
150	76.68	696.4	1.7038	38.26	695.8	1.6222	25.46	695.3	1.5742	19.05	694.7	1.5399
160	77.95	701.6	1.7122	38.90	701.1	1.6307	25.88	700.5	1.5827	19.37	700.0	1.5485
170	79.21	706.8	1.7206	39.54	706.3	1.6391	26.31	705.8	1.5911	19.70	705.3	1.5569
180	80.48	712.1	1.7289	40.17	711.6	1.6474	26.74	711.1	1.5995	20.02	710.6	1.5653
190	.....	.....	.....	40.81	716.9	1.6556	27.16	716.4	1.6077	20.34	715.9	1.5736
200	.....	.....	.....	41.45	722.2	1.6637	27.59	721.7	1.6158	20.66	721.2	1.5817
220	.....	.....	.....	.....	.....	.....	28.44	732.4	1.6318	21.30	732.0	1.5978
240	.....	.....	.....	.....	.....	.....	.....	.....	.....	21.94	742.8	1.6135

TABLE A-25E (Continued)

Pressure, psia (saturation temperature in italics)

Temp. °F	30 −0.57° v	h	s	40 11.66° v	h	s	50 21.67° v	h	s	60 30.21° v	h	s
Sat.	*9.236*	*611.6*	*1.3364*	*7.047*	*615.4*	*1.3125*	*5.710*	*618.2*	*1.2939*	*4.805*	*620.5*	*1.2787*
0	9.250	611.9	1.3371	.....	.....	.....	.....	.....	.....	.....	.....	.....
10	9.492	617.8	1.3497	.....	.....	.....	.....	.....	.....	.....	.....	.....
20	9.731	623.5	1.3618	7.203	620.4	1.3231	.....	.....	.....	.....	.....	.....
30	9.966	629.1	1.3733	7.387	626.3	1.3353	5.838	623.4	1.3046	.....	.....	.....
40	10.20	634.6	1.3845	7.568	632.1	1.3470	5.988	629.5	1.3169	4.933	626.8	1.2913
50	10.43	640.1	1.3953	7.746	637.8	1.3583	6.135	635.4	1.3286	5.060	632.9	1.3035
60	10.65	645.5	1.4059	7.922	643.4	1.3692	6.280	641.2	1.3399	5.184	638.0	1.3152
70	10.88	650.9	1.4161	8.096	648.9	1.3797	6.423	646.9	1.3508	5.307	644.9	1.3265
80	11.10	656.2	1.4261	8.268	654.4	1.3900	6.564	652.6	1.3613	5.428	650.7	1.3373
90	11.33	661.6	1.4359	8.439	659.9	1.4000	6.704	658.2	1.3716	5.547	656.4	1.3479
100	11.55	666.9	1.4456	8.609	665.3	1.4098	6.843	663.7	1.3816	5.665	662.1	1.3581
110	11.77	672.2	1.4550	8.777	670.7	1.4194	6.980	669.2	1.3914	5.781	667.7	1.3681
120	11.99	677.5	1.4642	8.945	676.1	1.4288	7.117	674.7	1.4009	5.897	673.3	1.3778
130	12.21	682.9	1.4733	9.112	681.5	1.4381	7.252	680.2	1.4103	6.012	678.9	1.3873
140	12.43	688.2	1.4823	9.278	686.9	1.4471	7.387	685.7	1.4195	6.126	684.4	1.3966
150	12.65	693.5	1.4911	9.444	692.3	1.4561	7.521	691.1	1.4286	6.239	689.9	1.4058
160	12.87	698.8	1.4998	9.609	697.7	1.4648	7.655	696.6	1.4374	6.352	695.5	1.4148
170	13.08	704.2	1.5083	9.774	703.1	1.4735	7.788	702.1	1.4462	6.464	701.0	1.4236
180	13.30	709.6	1.5168	9.938	708.5	1.4820	7.921	707.5	1.4548	6.576	706.5	1.4323
190	13.52	714.9	1.5251	10.10	714.0	1.4904	8.053	713.0	1.4633	6.687	712.0	1.4409
200	13.73	720.3	1.5334	10.27	719.4	1.4987	8.185	718.5	1.4716	6.798	717.5	1.4493
220	14.16	731.1	1.5495	10.59	730.3	1.5150	8.448	729.4	1.4880	7.019	728.6	1.4658
240	14.59	742.0	1.5653	10.92	741.3	1.5309	8.710	740.5	1.5040	7.238	739.7	1.4819
260	15.02	753.0	1.5808	11.24	752.3	1.5465	8.970	751.6	1.5197	7.457	750.9	1.4976
280	.....	.....	.....	11.56	763.4	1.5617	9.230	762.7	1.5350	7.675	762.1	1.5130
300	.....	.....	.....	11.88	774.6	1.5766	9.489	774.0	1.5500	7.892	773.3	1.5281

Temp. °F	80 44.40° v	80 44.40° h	80 44.40° s	100 56.05° v	100 56.05° h	100 56.05° s	120 66.02° v	120 66.02° h	120 66.02° s	140 74.79° v	140 74.79° h	140 74.79° s
Sat.	3.655	624.0	1.2545	2.952	626.5	1.2356	2.476	628.4	1.2201	2.132	629.9	1.2068
50	3.712	627.7	1.2619	.....	.....	.....	.....	.....	.....	.....	.....	.....
60	3.812	634.3	1.2745	2.985	629.3	1.2409	.....	.....	.....	.....	.....	.....
70	3.909	640.6	1.2866	3.068	636.0	1.2539	2.505	631.3	1.2255	.....	.....	.....
80	4.005	646.7	1.2981	3.149	642.6	1.2661	2.576	638.3	1.2386	2.166	633.8	1.2140
90	4.098	652.8	1.3092	3.227	649.0	1.2778	2.645	645.0	1.2510	2.228	640.9	1.2272
100	4.190	658.7	1.3199	3.304	655.2	1.2891	2.712	651.6	1.2628	2.288	647.8	1.2396
110	4.281	664.6	1.3303	3.380	661.3	1.2999	2.778	658.0	1.2741	2.347	654.5	1.2515
120	4.371	670.4	1.3404	3.454	667.3	1.3104	2.842	664.2	1.2850	2.404	661.1	1.2628
130	4.460	676.1	1.3502	3.527	673.3	1.3206	2.905	670.4	1.2956	2.460	667.4	1.2738
140	4.548	681.8	1.3598	3.600	679.2	1.3305	2.967	676.5	1.3058	2.515	673.7	1.2843
150	4.635	687.5	1.3692	3.672	685.0	1.3401	3.029	682.5	1.3157	2.569	679.9	1.2945
160	4.722	693.2	1.3784	3.743	690.8	1.3495	3.089	688.4	1.3254	2.622	686.0	1.3045
170	4.808	698.8	1.3874	3.813	696.6	1.3588	3.149	694.3	1.3348	2.675	692.0	1.3141
180	4.893	704.4	1.3963	3.883	702.3	1.3678	3.209	700.2	1.3441	2.727	698.0	1.3236
190	4.978	710.0	1.4050	3.952	708.0	1.3767	3.268	706.0	1.3531	2.779	704.0	1.3328
200	5.063	715.6	1.4136	4.021	713.7	1.3854	3.326	711.8	1.3620	2.830	709.9	1.3418
210	5.147	721.3	1.4220	4.090	719.4	1.3940	3.385	717.6	1.3707	2.880	715.8	1.3507
220	5.231	726.9	1.4304	4.158	725.1	1.4024	3.442	723.4	1.3793	2.931	721.6	1.3594
230	5.315	732.5	1.4386	4.226	730.8	1.4108	3.500	729.2	1.3877	2.981	727.5	1.3679
240	5.398	738.1	1.4467	4.294	736.5	1.4190	3.557	734.9	1.3960	3.030	733.3	1.3763
250	5.482	743.8	1.4547	4.361	742.2	1.4271	3.614	740.7	1.4042	3.080	739.2	1.3846
260	5.565	749.4	1.4626	4.428	747.9	1.4350	3.671	746.5	1.4123	3.129	745.0	1.3928
280	5.730	760.7	1.4781	4.562	759.4	1.4507	3.783	758.0	1.4281	3.227	756.7	1.4088
300	5.894	772.1	1.4933	4.695	770.8	1.4660	3.895	769.6	1.4435	3.323	768.3	1.4243

TABLE A-25E (Continued)

Pressure, psia (saturation temperature in italics)

Temp. °F	170 86.29°			200 96.34°			230 105.30°			260 113.42°		
	v	*h*	*s*	*v*	*h*	*s*	*v*	*h*	*s*	*v*	*h*	*s*
Sat.	*1.764*	*631.6*	*1.1900*	*1.502*	*632.7*	*1.1756*	*1.307*	*633.4*	*1.1631*	*1.155*	*633.9*	*1.1518*
90	*1.784*	*634.4*	*1.1952*	.....	.....	.....	.....	.....	.....	.....	.....	.....
100	1.837	641.9	1.2087	1.520	635.6	1.1809	.....	.....	.....	.....	.....	.....
110	1.889	649.1	1.2215	1.567	643.4	1.1947	1.328	637.4	1.1700	.....	.....	.....
120	1.939	656.1	1.2336	1.612	650.9	1.2077	1.370	645.4	1.1840	1.182	639.5	1.1617
130	1.988	662.8	1.2452	1.656	658.1	1.2200	1.410	653.1	1.1971	1.220	647.8	1.1757
140	2.035	669.4	1.2563	1.698	665.0	1.2317	1.449	660.4	1.2095	1.257	655.6	1.1889
150	2.081	675.9	1.2669	1.740	671.8	1.2429	1.487	667.6	1.2213	1.292	663.1	1.2014
160	2.127	682.3	1.2773	1.780	678.4	1.2537	1.524	674.5	1.2325	1.326	670.4	1.2132
170	2.172	688.5	1.2873	1.820	684.9	1.2641	1.559	681.3	1.2434	1.359	677.5	1.2245
180	2.216	694.7	1.2971	1.859	691.3	1.2742	1.594	687.9	1.2538	1.391	684.4	1.2354
190	2.260	700.8	1.3066	1.897	697.7	1.2840	1.629	694.4	1.2640	1.422	691.1	1.2458
200	2.303	706.9	1.3159	1.935	703.9	1.2935	1.663	700.9	1.2738	1.453	697.7	1.2560
210	2.346	713.0	1.3249	1.972	710.1	1.3029	1.696	707.2	1.2834	1.484	704.3	1.2658
220	2.389	719.0	1.3338	2.009	716.3	1.3120	1.729	713.5	1.2927	1.514	710.7	1.2754
230	2.431	724.9	1.3426	2.046	722.4	1.3209	1.762	719.8	1.3018	1.543	717.1	1.2847
240	2.473	730.9	1.3512	2.082	728.4	1.3296	1.794	726.0	1.3107	1.572	723.4	1.2938
250	2.514	736.8	1.3596	2.118	734.5	1.3382	1.826	732.1	1.3195	1.601	729.7	1.3027
260	2.555	742.8	1.3679	2.154	740.5	1.3467	1.857	738.3	1.3281	1.630	736.0	1.3115
270	2.596	748.7	1.3761	2.189	746.5	1.3550	1.889	744.4	1.3365	1.658	742.2	1.3200
280	2.637	754.6	1.3841	2.225	752.5	1.3631	1.920	750.5	1.3448	1.686	748.4	1.3285
290	2.678	760.5	1.3921	2.260	758.5	1.3712	1.951	756.5	1.3530	1.714	754.5	1.3367
300	2.718	766.4	1.3999	2.295	764.5	1.3791	1.982	762.6	1.3610	1.741	760.7	1.3449
320	2.798	778.3	1.4153	2.364	776.5	1.3947	2.043	774.7	1.3767	1.796	772.9	1.3608
340	2.878	790.1	1.4303	2.432	788.5	1.4099	2.103	786.8	1.3921	1.850	785.2	1.3763
360	.....	.....	.....	2.500	800.5	1.4247	2.163	798.9	1.4070	1.904	797.4	1.3914
380	.....	.....	.....	2.568	812.5	1.4392	2.222	811.1	1.4217	1.957	809.6	1.4062

TABLE A-26 PROPERTIES OF REFRIGERANT-113: SATURATION TEMPERATURE TABLE (ENGLISH UNITS)

Temp, °F	Pressure, psia	Volume, ft³/lbm		Enthalpy, Btu/lbm			Entropy, Btu/lbm·°R	
		Liquid	Vapor	Liquid	Latent	Vapor	Liquid	Vapor
−35.00	0.2599	0.009 413	93.380 39	0.993	71.842	72.835	0.002 35	0.171 51
−20.00	0.4473	0.009 515	56.118 36	4.002	71.001	75.003	0.009 31	0.170 79
0.00	0.8657	0.009 565	30.244 49	8.090	69.848	77.937	0.018 40	0.170 35
20.00	1.5727	0.009 803	17.319 90	12.263	68.652	80.915	0.027 29	0.170 41
40.00	2.7032	0.009 956	10.451 00	16.530	67.397	83.928	0.036 00	0.170 88
60.00	4.4268	0.010 117	6.599 86	20.878	66.090	86.968	0.044 52	0.171 69
80.00	6.9474	0.010 285	4.335 76	25.308	64.718	90.026	0.052 88	0.172 79
100.00	10.502	0.010 463	2.947 96	29.817	63.278	93.094	0.061 07	0.174 13
120.00	15.359	0.010 651	2.065 09	34.406	61.760	96.166	0.069 11	0.175 65
140.00	21.812	0.010 851	1.484 85	39.063	60.168	99.231	0.076 99	0.177 32
160.00	30.179	0.011 064	1.092 10	43.787	58.496	102.283	0.084 71	0.179 10
180.00	40.799	0.011 293	0.819 16	48.573	56.740	105.313	0.092 27	0.180 97
200.00	54.028	0.011 540	0.624 82	53.428	54.884	108.312	0.099 70	0.182 90
220.00	70.244	0.011 809	0.483 54	58.339	52.930	111.269	0.106 99	0.184 86
240.00	89.839	0.012 105	0.378 69	63.319	50.851	114.171	0.114 14	0.186 82
260.00	113.23	0.012 434	0.299 52	68.367	48.636	117.003	0.121 18	0.188 76
280.00	140.86	0.012 804	0.238 64	73.503	46.240	119.743	0.128 13	0.190 64
300.00	173.22	0.013 229	0.191 08	78.742	43.621	122.363	0.135 02	0.192 44
320.00	210.82	0.013 728	0.153 31	84.117	40.707	124.824	0.141 88	0.194 09
340.00	254.26	0.014 333	0.122 77	89.681	37.381	127.062	0.148 78	0.195 53
360.00	304.23	0.015 104	0.097 57	95.516	33.460	128.976	0.155 82	0.196 64
380.00	361.53	0.016 170	0.076 15	101.786	28.573	130.360	0.163 18	0.197 20
400.00	427.26	0.017 909	0.056 78	108.902	21.779	130.681	0.171 31	0.196 64
420.00	503.46	0.025 202	0.031 01	120.875	3.977	124.852	0.184 72	0.189 24

Source for Tables A-26 and A-27: From M. J. Mastroianni, R. F. Stahl, and P. N. Sheldon, "Physical and Thermodynamic Properties of 1,1,2-Trifluorotrichloroethane (R-113)," *Journal of Chemical and Engineering Data,* vol. 23, no. 2, (1978). Copyright American Chemical Society.

TABLE A-27 PROPERTIES OF REFRIGERANT-113: SUPERHEATED VAPOR (ENGLISH UNITS)
v in ft³/lbm, h in Btu/lbm, and s in Btu/lbm·°R

P, psia		20°F	60°F	100°F	140°F	180°F	220°F	260°F	300°F	340°F	380°F
0.6	v	45.639	49.485	53.323	57.158	60.988	64.816	68.642	72.467	76.291	80.107
	h	81.011	87.276	93.763	100.453	107.332	114.383	121.593	128.948	136.437	144.048
	s	0.18076	0.19330	0.20533	0.21687	0.22797	0.23866	0.24897	0.25892	0.26852	0.27781
1.0	v	27.324	29.642	31.953	34.260	36.564	38.864	41.163	43.461	45.747	48.053
	h	80.972	87.244	93.737	100.432	107.313	114.367	121.579	128.936	136.426	144.038
	s	0.17529	0.18785	0.19988	0.21143	0.22254	0.23323	0.24354	0.25349	0.26310	0.27239
4.0	v		7.3174	7.9112	8.5003	9.0857	9.6685	10.249	10.829	11.407	11.984
	h		87.002	93.538	100.266	107.174	114.248	121.476	128.846	136.346	143.967
	s		0.17282	0.18493	0.19654	0.20769	0.21841	0.22875	0.23871	0.24833	0.25763
10.0	v			3.1014	3.3470	3.5893	3.829	4.0662	4.3020	4.5368	4.7706
	h			93.129	99.928	106.890	114.006	121.267	128.663	136.185	143.823
	s			0.17469	0.18642	0.19766	0.20845	0.21883	0.22883	0.23848	0.24780

P, psia		260°F	300°F	340°F	380°F	420°F	460°F	500°F	540°F	580°F	620°F
100	v	0.34920	0.38135	0.41145	0.44008	0.46775	0.49466	0.52099	0.54687	0.57240	0.59765
	h	117.646	125.600	133.540	141.503	149.507	157.560	165.668	173.831	182.049	190.321
	s	0.19075	0.20151	0.21170	0.22141	0.23072	0.23968	0.24830	0.25664	0.26470	0.27251
200	v			0.16597	0.19615	0.21360	0.22981	0.24520	0.25997	0.27427	0.28823
	h			125.399	138.424	146.877	155.264	163.630	171.999	180.385	188.797
	s			0.19524	0.21133	0.22117	0.23049	0.23940	0.24794	0.25616	0.26410
800	v						0.02028	0.03077	0.04124	0.04921	0.05584
	h						125.457	142.603	156.347	167.618	177.885
	s						0.18846	0.20670	0.22075	0.23181	0.24150

TABLE A-28M ENTHALPY OF FORMATION, GIBBS FUNCTION OF FORMATION, ABSOLUTE ENTROPY, AND ENTHALPY OF VAPORIZATION AT 25 °C AND 1 atm (SI UNITS)

Δh_f°, Δg_f°, h_{fg} in kJ/kgmole; s° in kJ/kgmole·K

Substance	Formula	Δh_f^0	Δg_f^0	s^0	h_{fg}
Carbon	C(s)	0	0	5.74	
Hydrogen	$H_2(g)$	0	0	130.57	
Nitrogen	$N_2(g)$	0	0	191.50	
Oxygen	$O_2(g)$	0	0	205.04	
Carbon monoxide	CO(g)	−110,530	−137,150	197.56	
Carbon dioxide	$CO_2(g)$	−393,520	−394,380	213.67	
Water	$H_2O(g)$	−241,820	−228,590	188.72	
Water	$H_2O(l)$	−285,830	−237,180	69.95	44,010
Hydrogen peroxide	$H_2O_2(g)$	−136,310	−105,600	232.63	61,090
Ammonia	$NH_3(g)$	−46,190	−16,590	192.33	
Oxygen	O(g)	249,170	231,770	160.95	
Hydrogen	H(g)	218,000	203,290	114.61	
Nitrogen	N(g)	472,680	455,510	153.19	
Hydroxyl	OH(g)	39,040	34,280	183.75	
Methane	$CH_4(g)$	−74,850	−50,790	186.16	
Acetylene (Ethyne)	$C_2H_2(g)$	226,730	209,170	200.85	
Ethylene (Ethene)	$C_2H_4(g)$	52,280	68,120	219.83	
Ethane	$C_2H_6(g)$	−84,680	−32,890	229.49	
Propylene (Propene)	$C_3H_6(g)$	20,410	62,720	266.94	
Propane	$C_3H_8(g)$	−103,850	−23,490	269.91	15,060
n-Butane	$C_4H_{10}(g)$	−126,150	−15,710	310.03	21,060
n-Pentane	$C_5H_{12}(g)$	−146,440	−8,200	348.40	31,410
n-Octane	$C_8H_{18}(g)$	−208,450	17,320	463.67	41,460
n-Octane	$C_8H_{18}(l)$	−249,910	6,610	360.79	
n-Dodecane	$C_{12}H_{26}(g)$	−291,010	50,150	622.83	
Benzene	$C_6H_6(g)$	82,930	129,660	269.20	33,830
Methyl alcohol	$CH_3OH(g)$	−200,890	−162,140	239.70	37,900
Methyl alcohol	$CH_3OH(l)$	−238,810	−166,290	126.80	
Ethyl alcohol	$C_2H_5OH(g)$	−235,310	−168,570	282.59	42,340
Ethyl alcohol	$C_2H_5OH(l)$	−277,690	−174,890	160.70	
Mercury	Hg(l)	0	0	77.24	

Source: K. Wark, *Thermodynamics,* 5th ed., p. 873, McGraw-Hill, New York, 1988.

TABLE A-28E ENTHALPY OF FORMATION, GIBBS FUNCTION OF FORMATION, ABSOLUTE ENTROPY, AND ENTHALPY OF VAPORIZATION AT 77°F and 1 atm (ENGLISH UNITS)

Δh_f°, Δg_f°, h_{fg} in Btu/lbmole; s° in Btu/lbmole·°R

Substance	Formula	Δh_f^0	Δg_f^0	s^0	h_{fg}
Carbon	$C(s)$	0	0	1.36	
Hydrogen	$H_2(g)$	0	0	31.21	
Nitrogen	$N_2(g)$	0	0	45.77	
Oxygen	$O_2(g)$	0	0	49.00	
Carbon monoxide	$CO(g)$	−47,540	−59,010	47.21	
Carbon dioxide	$CO_2(g)$	−169,300	−169,680	51.07	
Water	$H_2O(g)$	−104,040	−98,350	45.11	
Water	$H_2O(l)$	−122,970	−102,040	16.71	
Hydrogen peroxide	$H_2O_2(g)$	−58,640	−45,430	55.60	26,260
Ammonia	$NH_3(g)$	−19,750	−7,140	45.97	
Methane	$CH_4(g)$	−32,210	−21,860	44.49	
Acetylene	$C_2H_2(g)$	+97,540	+87,990	48.00	
Ethylene	$C_2H_4(g)$	+22,490	+29,306	52.54	
Ethane	$C_2H_6(g)$	−36,420	−14,150	54.85	
Propylene	$C_3H_6(g)$	+8,790	+26,980	63.80	
Propane	$C_3H_8(g)$	−44,680	−10,105	64.51	6,480
n-Butane	$C_4H_{10}(g)$	−54,270	−6,760	74.11	9,060
n-Octane	$C_8H_{18}(g)$	−89,680	+7,110	111.55	
n-Octane	$C_8H_{18}(l)$	−107,530	+2,840	86.23	17,835
n-Dodecane	$C_{12}H_{26}(g)$	−125,190	+21,570	148.86	
Benzene	$C_6H_6(g)$	+35,680	+55,780	64.34	14,550
Methyl alcohol	$CH_3OH(g)$	−86,540	−69,700	57.29	
Methyl alcohol	$CH_3OH(l)$	−102,670	−71,570	30.30	16,090
Ethyl alcohol	$C_2H_5OH(g)$	−101,230	−72,520	67.54	
Ethyl alcohol	$C_2H_5OH(l)$	−119,470	−75,240	38.40	18,220
Oxygen	$O(g)$	+107,210	+99,710	38.47	
Hydrogen	$H(g)$	+93,780	+87,460	27.39	
Nitrogen	$N(g)$	+203,340	+195,970	36.61	
Hydroxyl	$OH(g)$	+16,790	+14,750	43.92	

Source: K. Wark, *Thermodynamics*, 5th ed., p. 923, McGraw-Hill, New York, 1988.

TABLE A-29 LOGARITHMS TO THE BASE 10 OF THE EQUILIBRIUM CONSTANT K_p

$$K_p = \frac{x_3^{\nu_3} x_4^{\nu_4}}{x_1^{\nu_1} x_2^{\nu_2}} \left(\frac{p}{p_0}\right)^{\nu_3+\nu_4-\nu_1-\nu_2}$$ and $p_0 = 1$ atm for the reaction $\nu_1 C_1 + \nu_2 C_2 \rightleftharpoons \nu_3 C_3 + \nu_4 C_4$

T, K	$H_2 \rightarrow 2H$	$N_2 \rightarrow 2N$	$O_2 \rightarrow 2O$	$CO_2 \rightarrow$ $CO + \frac{1}{2}O_2$	$H_2O \rightarrow$ $OH + \frac{1}{2}H_2$	$H_2O \rightarrow$ $H_2 + \frac{1}{2}O_2$	$\frac{1}{2}N_2 + \frac{1}{2}O_2$ $\rightarrow NO$	$CO_2 + H_2 \rightarrow$ $CO + H_2O$	T, R
298	−71.23	−159.6	−81.20	−45.07	−46.05	−40.05	−15.17	−5.018	537
500	−40.32	−92.69	−45.88	−25.03	−26.13	−22.89	−8.783	−2.139	900
1000	−17.29	−43.06	−19.61	−10.02	−11.28	−10.06	−4.062	−0.159	1800
1200	−13.41	−34.76	−15.21	−7.764	−8.789	−7.899	−3.275	0.135	2160
1400	−10.63	−28.82	−12.05	−6.014	−7.003	−6.347	−2.712	0.333	2520
1500	−9.514	−26.44	−10.79	−5.316	−6.288	−5.725	−2.487	0.409	2700
1600	−8.534	−24.36	−9.682	−4.706	−5.662	−5.180	−2.290	0.474	2880
1700	−7.668	−22.52	−8.706	−4.169	−5.109	−4.699	−2.116	0.530	3060
1800	−6.896	−20.88	−7.836	−3.693	−4.617	−4.270	−1.962	0.577	3240
1900	−6.206	−19.42	−7.056	−3.267	−4.177	−3.886	−1.823	0.619	3420
2000	−5.582	−18.10	−6.354	−2.884	−3.780	−3.540	−1.699	0.656	3600
2100	−5.018	−16.90	−5.720	−2.539	−3.422	−3.227	−1.586	0.688	3780
2200	−4.504	−15.82	−5.142	−2.226	−3.095	−2.942	−1.484	0.716	3960
2300	−4.032	−14.82	−4.614	−1.940	−2.798	−2.682	−1.391	0.742	4140
2400	−3.602	−13.91	−4.130	−1.679	−2.525	−2.443	−1.305	0.764	4320
2500	−3.204	−13.08	−3.684	−1.440	−2.274	−2.224	−1.227	0.784	4500
2600	−2.836	−12.30	−3.272	−1.219	−2.042	−2.021	−1.154	0.802	4680
2700	−2.496	−11.59	−2.890	−1.015	−1.828	−1.833	−1.087	0.818	4860
2800	−2.178	−10.92	−2.536	−0.825	−1.628	−1.658	−1.025	0.833	5040
2900	−1.884	−10.30	−2.206	−0.649	−1.442	−1.495	−0.967	0.846	5220
3000	−1.608	−9.720	−1.898	−0.485	−1.269	−1.343	−0.913	0.858	5400
3100	−1.350	−9.178	−1.610	−0.332	−1.107	−1.201	−0.863	0.869	5580
3200	−1.108	−8.668	−1.338	−0.189	−0.955	−1.067	−0.815	0.878	5760
3300	−0.880	−8.190	−1.084	−0.054	−0.813	−0.942	−0.771	0.888	5940
3400	−0.664	−7.740	−0.844	0.071	−0.679	−0.824	−0.729	0.895	6120
3500	−0.462	−7.316	−0.620	0.190	−0.552	−0.712	−0.690	0.902	6300

Source: Based on data of JANAF, *Thermochemical Tables*, 2nd ed., National Bureau of Standards, NSRDS-NBS 37, 1971, and revisions published in the *Journal of Physical and Chemical Data* through 1982.

TABLE A-30 CONSTANTS OF THE VAN DER WAALS EQUATION OF STATE*

Units: atm, m³/kgmole, K

Substance	a	b
Acetylene	4.417	0.05154
Ammonia	4.197	0.03737
Argon	1.349	0.03227
Benzene	18.50	0.1187
n-Butane	13.70	0.1164
i-Butane	13.14	0.1163
1-Butene	12.60	0.1084
Carbon dioxide	3.606	0.04280
Carbon disulfide	11.10	0.07259
Carbon monoxide	1.456	0.03954
Carbon tetrachloride	19.54	0.1268
Chlorine	6.491	0.05621
n-Decane	52.33	0.3052
Ethane	5.498	0.06500
Ethylene	4.507	0.05749
Freon-12	10.62	0.09964
Helium	0.03414	0.02371
n-Heptane	30.69	0.2051
n-Hexane	24.49	0.1741
Hydrogen	0.2461	0.02668
Hydrogen sulfide	4.460	0.04311
Krypton	2.294	0.03955
Methane	2.256	0.04271
Methyl chloride	7.470	0.06480
Neon	0.2091	0.01697
Nitric oxide	1.438	0.02885
Nitrogen	1.350	0.03864
Nitrous oxide	3.800	0.04430
n-Nonane	44.70	0.2712
n-Octane	37.38	0.2370
Oxygen	1.363	0.03184
n-Pentane	18.82	0.1447
i-Pentane	18.35	0.1437
neo-Pentane	16.93	0.1409
Propane	9.255	0.09033
Propylene	8.340	0.08249
Sulfur dioxide	6.773	0.05678
Sulfur trioxide	8.185	0.06015
Toluene	25.06	0.1523
Water	5.454	0.03042
Xenon	4.112	0.05124

* Calculated by the use of Eqs. (14-3a) and (14-3b) from the critical data given in Table A-5.

TABLE A-31 CONSTANTS OF THE BEATTIE–BRIDGEMAN EQUATION OF STATE*

Units: atm, m^3/kgmole, K

Gas	A_0	a	B_0	b	$c \times 10^{-4}$
He	0.0216	0.05984	0.01400	0.0	0.0040
Ne	0.2125	0.02196	0.02060	0.0	0.101
Ar	1.2907	0.02328	0.03931	0.0	5.99
Kr	2.4230	0.02865	0.05261	0.0	14.89
Xe	4.6715	0.03311	0.07503	0.0	30.02
H_2	0.1975	−0.00506	0.02096	−0.04359	0.0504
N_2	1.3445	0.02617	0.05046	−0.00691	4.20
O_2	1.4911	0.02562	0.04624	0.004208	4.80
Air	1.3012	0.01931	0.04611	−0.01101	4.34
I_2	17.0	0.0	0.325	0.0	4000.
CO_2	5.0065	0.07132	0.10476	0.07235	66.00
NH_3	2.3930	0.17031	0.03415	0.19112	476.87
CH_4	2.2769	0.01855	0.05587	−0.01587	12.83
C_2H_4	6.1520	0.04964	0.12156	0.03597	22.68
C_2H_6	5.8800	0.05861	0.09400	0.01915	90.00
C_3H_8	11.9200	0.07321	0.18100	0.04293	120.00
$1\text{-}C_4H_8$	16.6979	0.11988	0.24046	0.10690	300.00
$iso\text{-}C_4H_8$	16.9600	0.10860	0.24200	0.08750	250.00
$n\text{-}C_4H_{10}$	17.7940	0.12161	0.24620	0.09423	350.00
$iso\text{-}C_4H_{10}$	16.6037	0.11171	0.23540	0.07697	300.00
$n\text{-}C_5H_{12}$	28.2600	0.15099	0.39400	0.13960	400.00
$neo\text{-}C_5H_{12}$	23.3300	0.15174	0.33560	0.13358	400.00
$n\text{-}C_7H_{16}$	54.520	0.20066	0.70816	0.19179	400.00
CH_3OH	33.309	0.09246	0.60362	0.09929	32.03
$(C_2H_5)_2O$	31.278	0.12426	0.45446	0.11954	33.33

* J. O. Hirschfelder, C. F. Curtiss, and R. B. Bird, *Molecular Theory of Gases and Liquids,* John Wiley, New York, 1964.

TABLE A-32 CONSTANTS OF THE BENEDICT–WEBB–RUBIN EQUATION OF STATE*

Units: atm, m³/kgmole, K

Substance	a	A_0	b	B_0	$c \times 10^{-6}$	$C_0 \times 10^{-6}$	α	γ
CH_4	0.0494000	1.85500	0.00338004	0.0426000	0.00254500	0.0225700	0.000124359	0.0060000
C_2H_6	0.345160	4.15556	0.0111220	0.0627724	0.0327670	0.179592	0.000243389	0.0118000
C_3H_8	0.947700	6.87225	0.0225000	0.0973130	0.129000	0.508256	0.000607175	0.0220000
$n\text{-}C_4H_{10}$	1.88231	10.0847	0.0399983	0.124361	0.316400	0.992830	0.00110132	0.0340000
$i\text{-}C_4H_{10}$	1.93763	10.23264	0.0424352	0.137544	0.286010	0.849943	0.00107408	0.0340000
$n\text{-}C_5H_{12}$	4.07480	12.1794	0.0668120	0.156751	0.824170	2.12121	0.00181000	0.0475000
$i\text{-}C_5H_{12}$	3.75620	12.7959	0.0668120	0.160053	0.695000	1.74632	0.00170000	0.0463000
$neo\text{-}C_5H_{12}$	3.4905	12.9635	0.0668120	0.170530	0.546	1.273	0.002	0.05
$n\text{-}C_6H_{14}$	7.11671	14.4373	0.109131	0.177813	1.51276	3.31935	0.00281086	0.0666849
$n\text{-}C_7H_{16}$	10.36475	17.5206	0.151954	0.199005	2.47000	4.74574	0.00435611	0.0900000
C_2H_4	0.259000	3.33958	0.0086000	0.0556833	0.021120	0.131140	0.000178000	0.00923000
C_3H_6	0.774056	6.11220	0.0187059	0.0850647	0.102611	0.439182	0.000455696	0.0182900
$i\text{-}C_4H_8$	1.69270	8.95325	0.0348156	0.116025	0.274920	0.927280	0.000910889	0.0295945
C_6H_6	5.570	6.509772	0.07663	0.05030055	1.176418	3.42997	0.0007001	0.02930
NH_3	0.10354029	3.7892819	0.00071958516	0.051646121	0.00015753298	0.17857089	0.0000046521779	0.019805156
Ar	0.0288358	0.823417	0.00215289	0.022282597	0.0007982437	0.01314125	0.00003558895	0.0023382711
CO_2	0.136814	2.73742	0.00721045	0.0499101	0.0149180	0.138567	0.0000847	0.005394
CO	0.03665	1.34122	0.00263158	0.0545425	0.001040	0.00856209	0.000135	0.006
He	-0.00057339	0.040962	-0.00000019727	0.023661	-0.000000005521	-0.000000016227	-0.0000072673	0.00077942
N_2	0.025102	1.053642	0.0023277	0.0407426	0.00072841	0.00805900	0.0001272	0.005300
O_2	0.162689940	0.950851963	0.00358834736	0.0000000035328505	0.0128273741	0.0326435918	-3.927058894	0.0301
SO_2	0.84468	2.12044	0.014653	0.026182	0.11335	0.79384	0.000071955	0.0059236

* H. W. Cooper and J. C. Goldfrank, *Hydrocarbon Processing*, vol. 46, no. 12, p. 141 (1967).

Properties of Substances—Figures

List of Figures

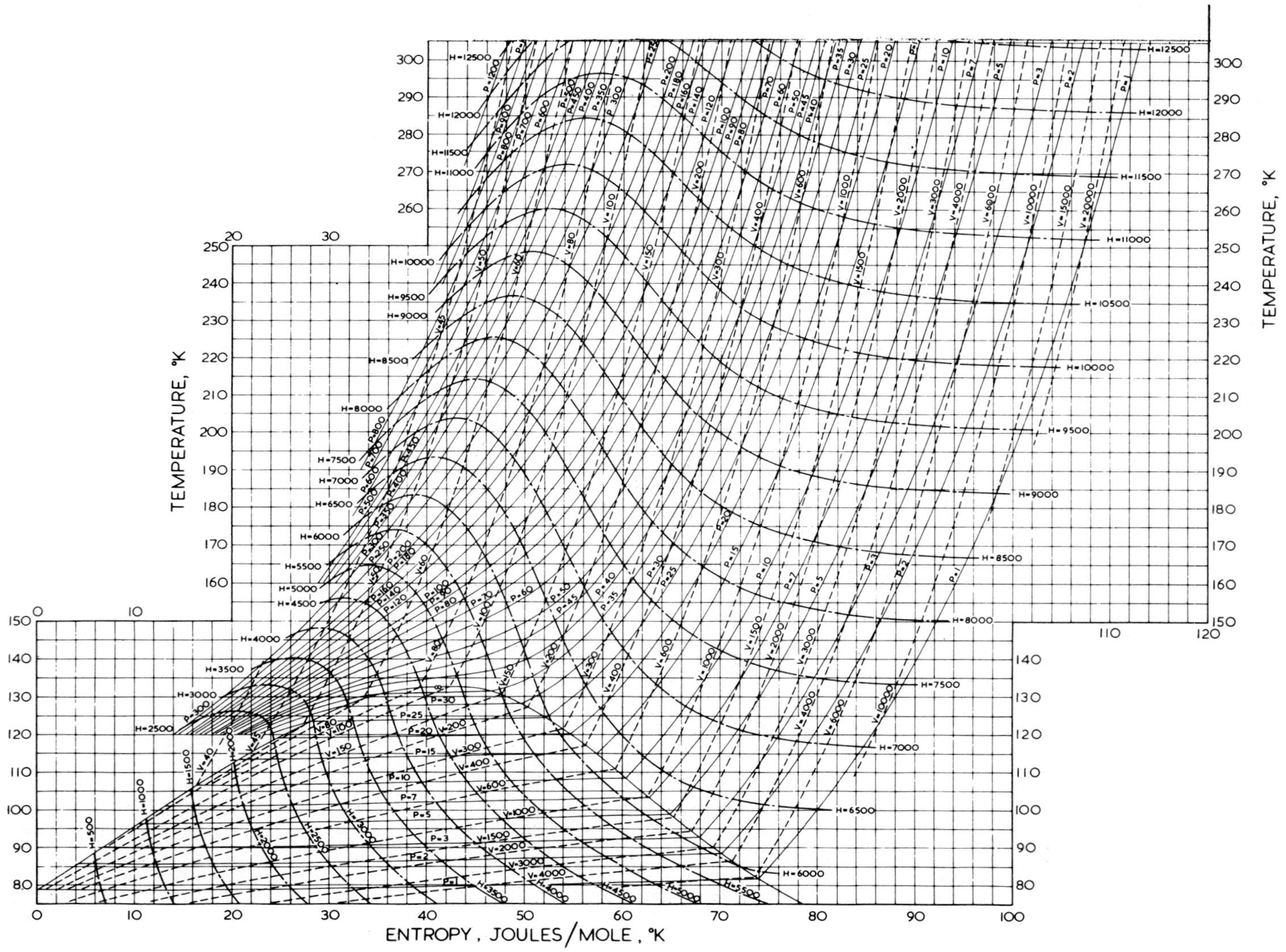

Figure A-1 Temperature–entropy diagram for air. (From F. Din, *Thermodynamic Functions of Gases*, vol. 2, Butterworths, London, 1956.)

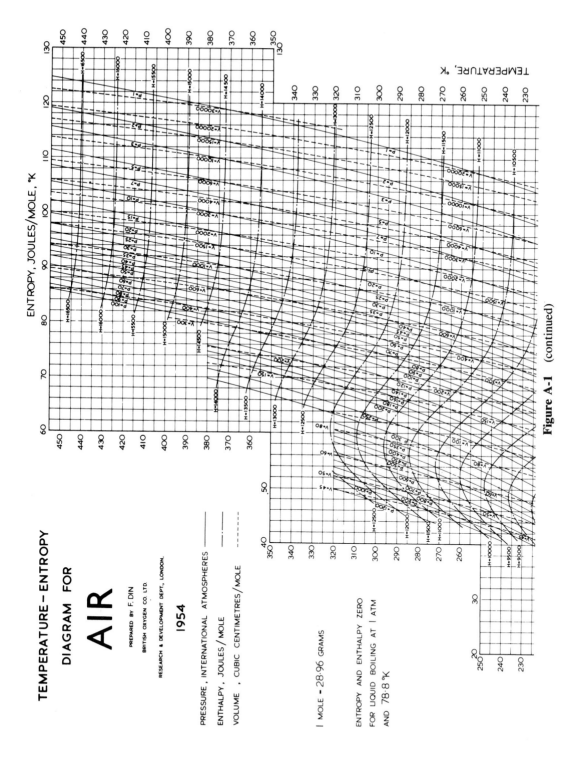

Figure A-1 (continued)

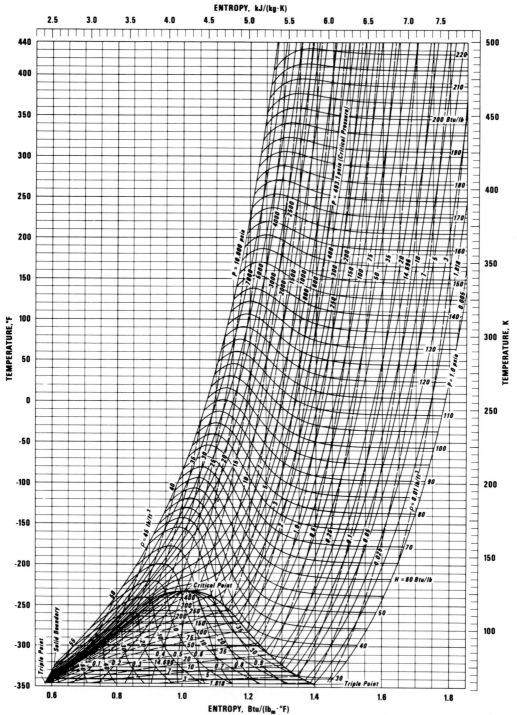

Figure A-2 Temperature–entropy diagram for nitrogen. (From National Institute of Standards and Technology, Boulder, CO.)

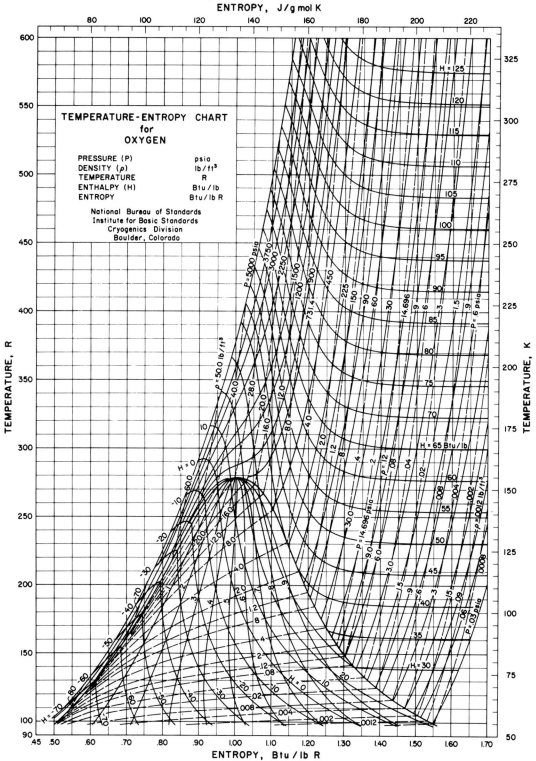

Figure A-3 Temperature–entropy diagram for oxygen. (From National Institute of Standards and Technology, Boulder, CO.)

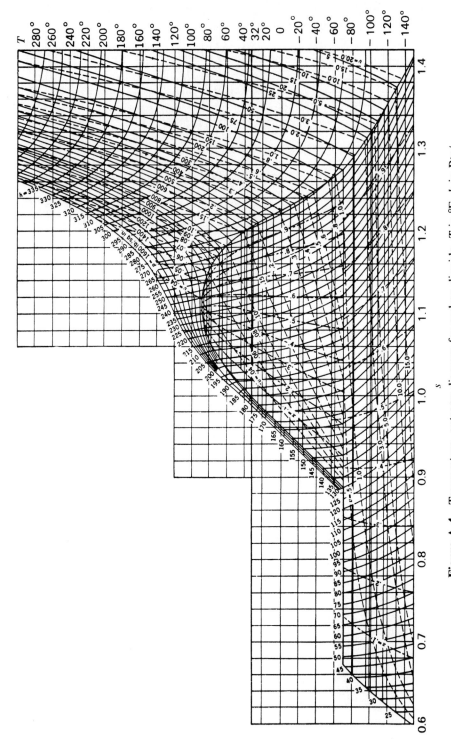

Figure A-4 Temperature–entropy diagram for carbon dioxide. T in °F, h in Btu/lbm, v in ft³/lbm, s in Btu/lbm·°R. (From K. Wark, *Thermodynamics*, McGraw-Hill, New York, 1988.)

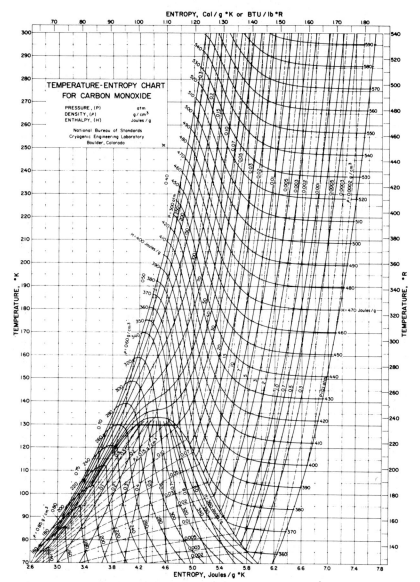

Figure A-5 Temperature–entropy diagram for carbon monoxide. (From National Institute of Standards and Technology, Boulder, CO.)

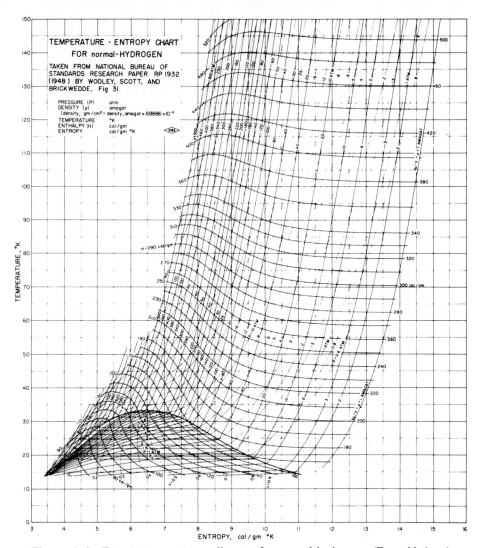

Figure A-6 Temperature–entropy diagram for normal hydrogen. (From National Institute of Standards and Technology, Boulder, CO.)

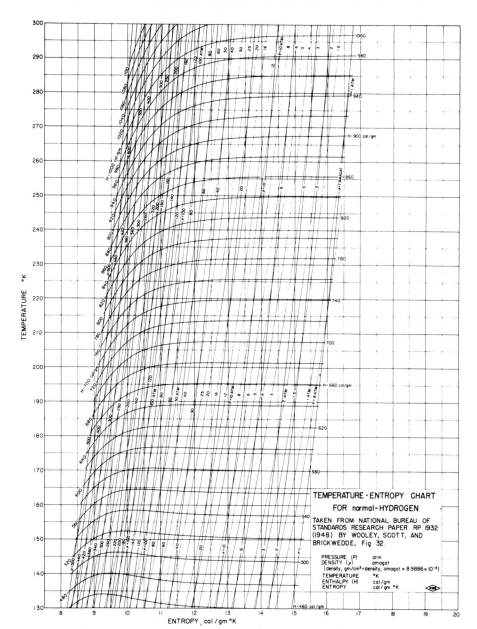

Figure A-6 (continued)

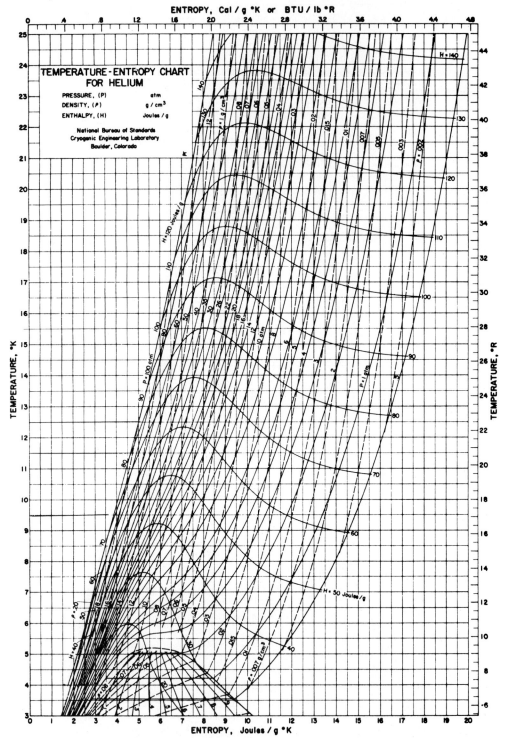

Figure A-7 Temperature–entropy diagram for helium. (From National Institute of Standards and Technology, Boulder, CO.)

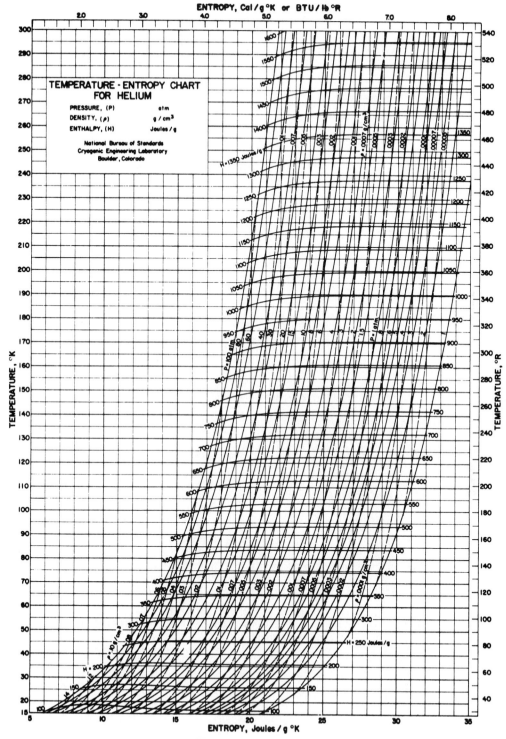

Figure A-7 (continued)

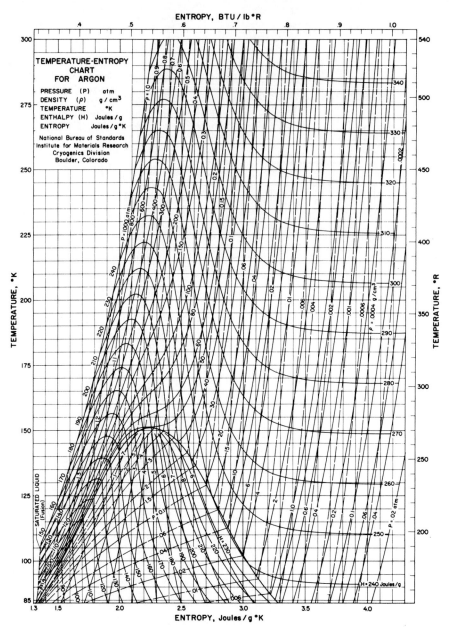

Figure A-8 Temperature–entropy diagram for argon. (From National Institute of Standards and Technology, Boulder, CO.)

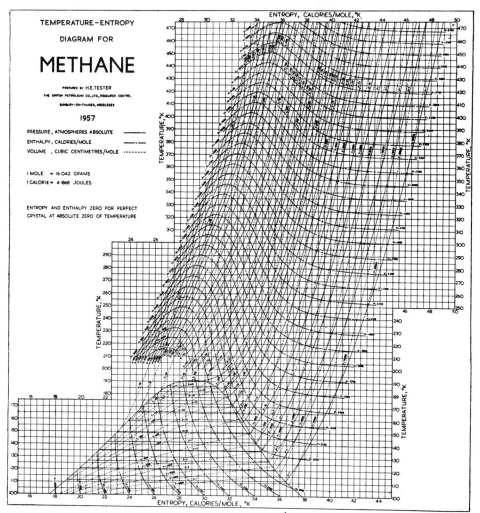

Figure A-9 Temperature–entropy diagram for methane. (From F. Din, *Thermodynamic Functions of Gases,* vol. 3, Butterworths, London, 1961.)

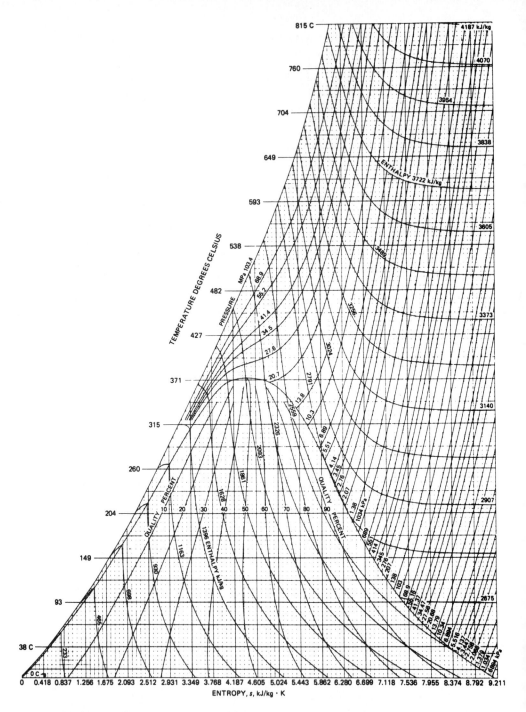

Figure A-10M Temperature–entropy diagram for steam (SI units). (Reprinted by permission from ASME "Steam Tables," American Society of Mechanical Engineers, N.Y. 1967.)

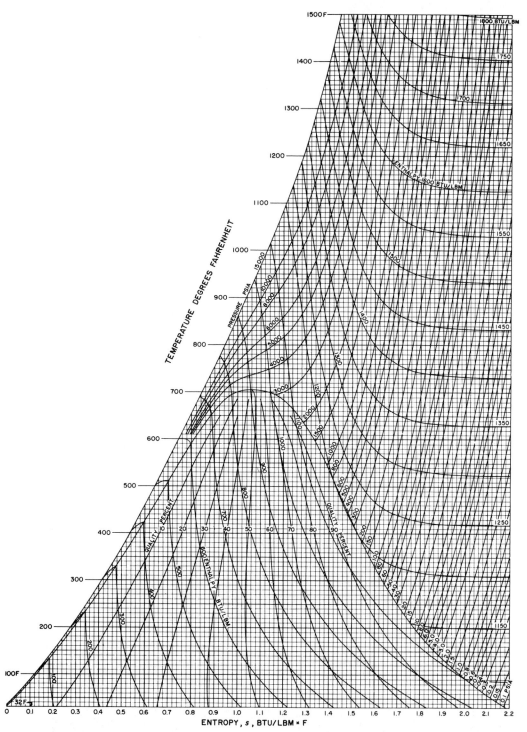

Figure A-10E Temperature–entropy diagram for steam (English units). (Reprinted by permission from ASME "Steam Tables," American Society of Mechanical Engineers, N.Y. 1967.)

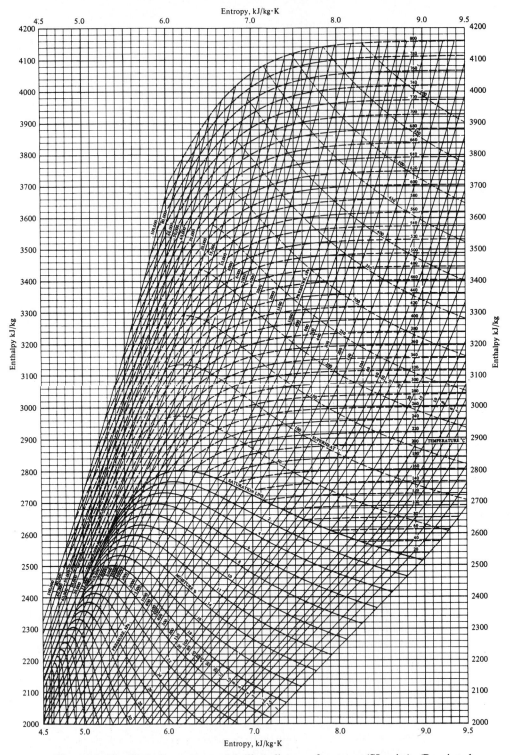

Figure A-11 Enthalpy–entropy (Mollier) diagram for steam (SI units). (Reprinted by permission from ASME "Steam Tables," American Society of Mechanical Engineers, N.Y. 1967.)

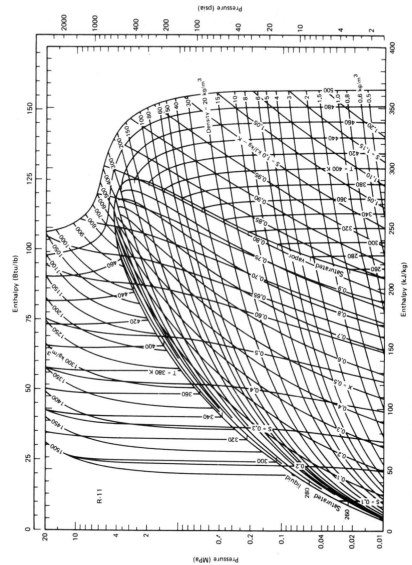

Figure A-12M Pressure–enthalpy diagram for Refrigerant-11 (Freon-11) (SI units). (Reprinted by permission of E. I. Du Pont de Nemours & Company.)

PRESSURE-ENTHALPY DIAGRAM

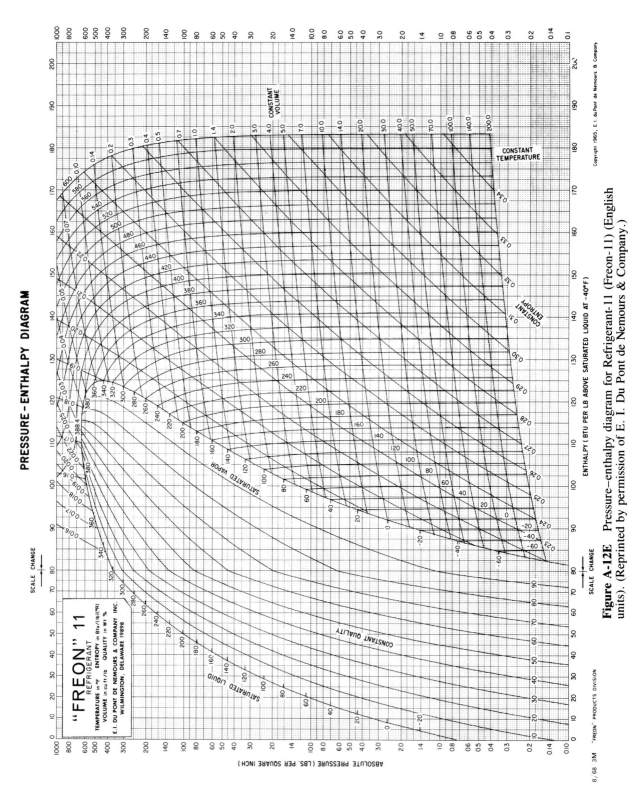

Figure A-12E Pressure–enthalpy diagram for Refrigerant-11 (Freon–11) (English units). (Reprinted by permission of E. I. Du Pont de Nemours & Company.)

PRESSURE-ENTHALPY DIAGRAM

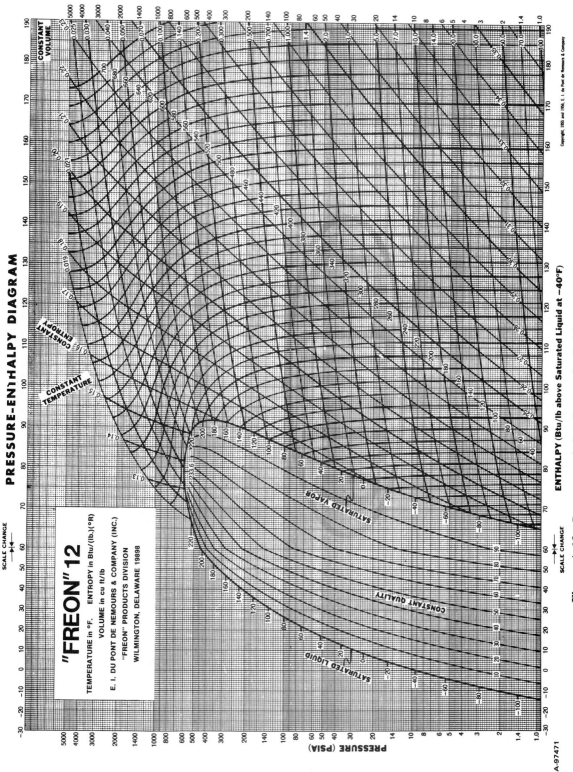

Figure A-13 Pressure–enthalpy diagram for Refrigerant-12 (Freon-12) (English units). (Reprinted by permission of E. I. Du Pont de Nemours & Company.)

847

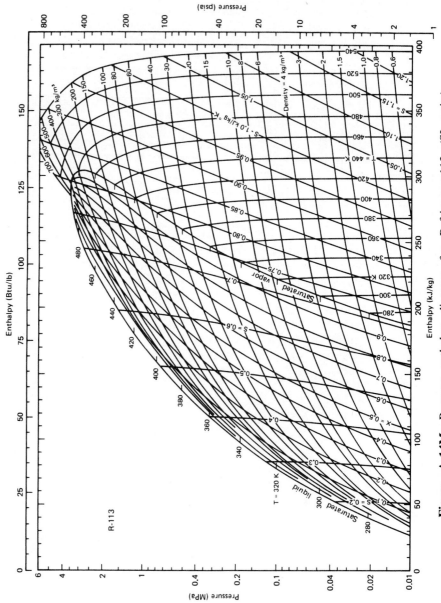

Figure A-14M Pressure–enthalpy diagram for Refrigerant-113 (SI units). (Reprinted by permission from the 1989 ASHRAE Handbook-Fundamentals.)

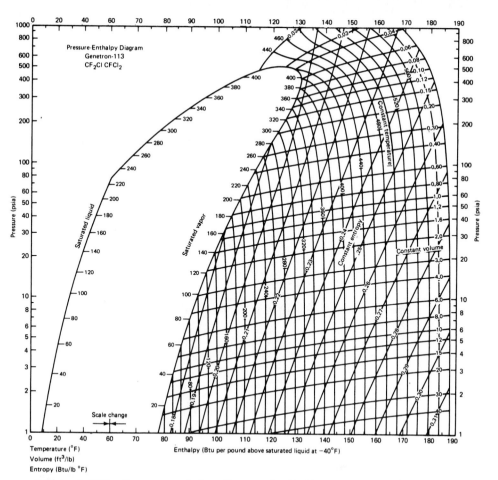

Figure A-14E Pressure–enthalpy diagram for Refrigerant-113 (English units). (Reprinted by permission of Allied-Signal Inc.)

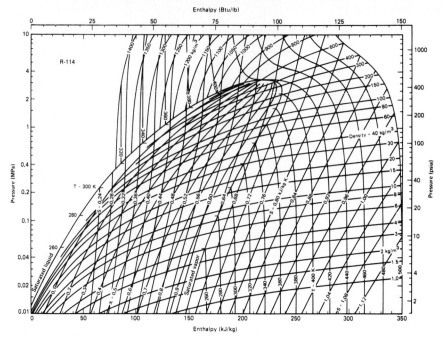

Figure A-15M Pressure–enthalpy diagram for Refrigerant-114 (SI units). (Reprinted by permission from the 1989 ASHRAE Handbook-Fundamentals.)

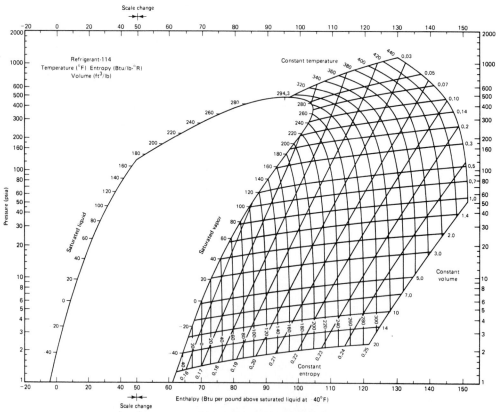

Figure A-15E Pressure–enthalpy diagram for Refrigerant-114 (English units). (Reprinted by permission of E. I. Du Pont de Nemours & Company.)

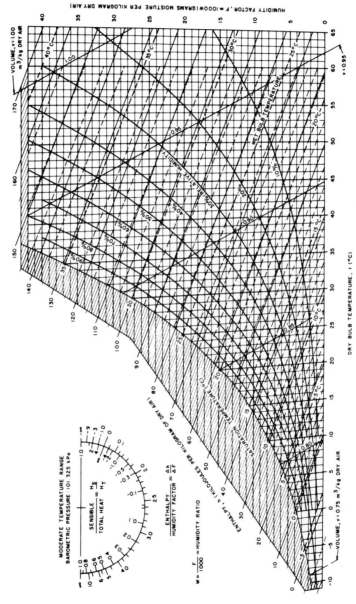

Figure A-16M Psychrometric chart (SI units). (Reprinted by permission from the 1989 ASHRAE Handbook-Fundamentals.)

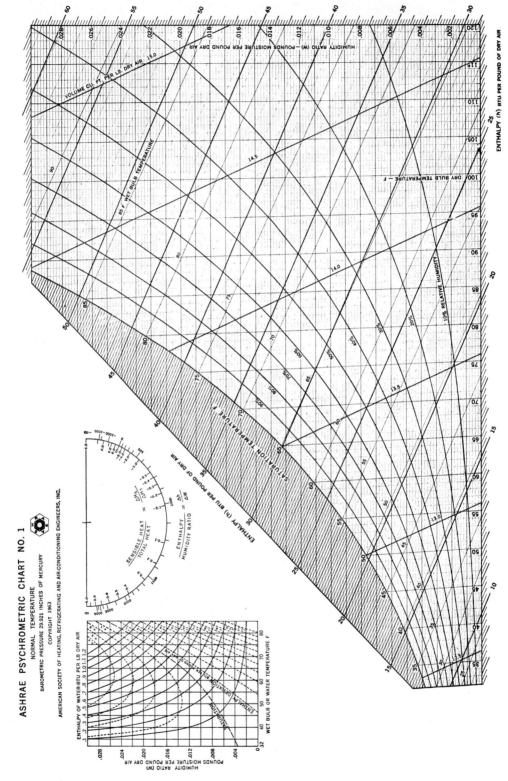

Figure A-16E Psychrometric chart (English units). (Reprinted by permission from the 1989 ASHRAE Handbook-Fundamentals.)

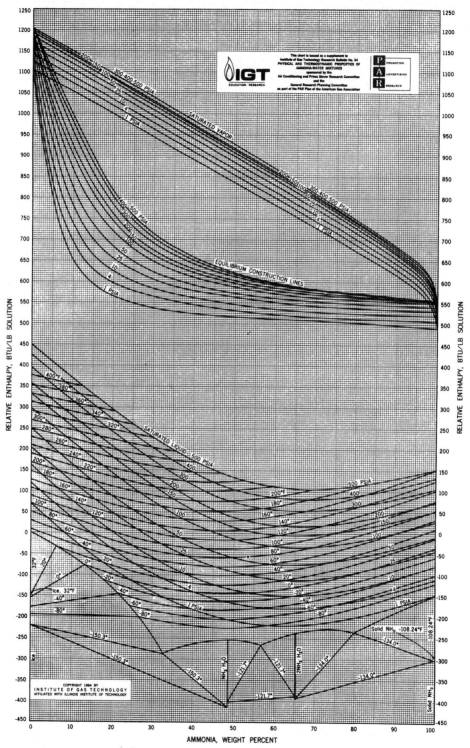

Figure A-17 Enthalpy-concentration diagram for ammonia–water solution. (Reprinted by permission from *Physical and Thermodynamic Properties of Ammonia–Water Mixtures*, Research Bulletin 34, Institute of Gas Technology, Chicago, 1964.)

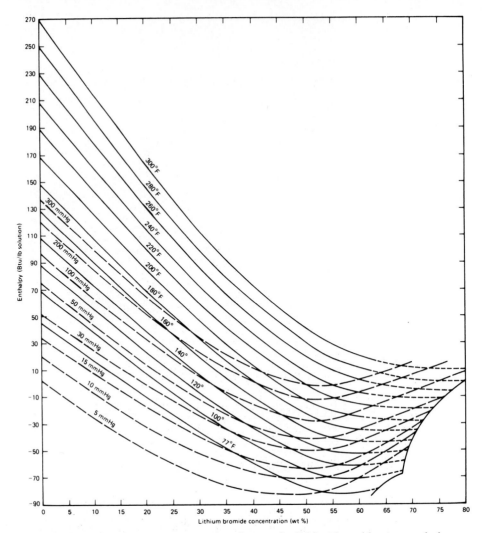

Figure A-18 Enthalpy-concentration diagram for lithium bromide–water solution. (Reprinted by permission from R. T. Ellington, G. Kunst, R. E. Peck, and J. F. Reed, *The Absorption Cooling Process,* Research Bulletin 14, Institute of Gas Technology, Chicago, 1957.)

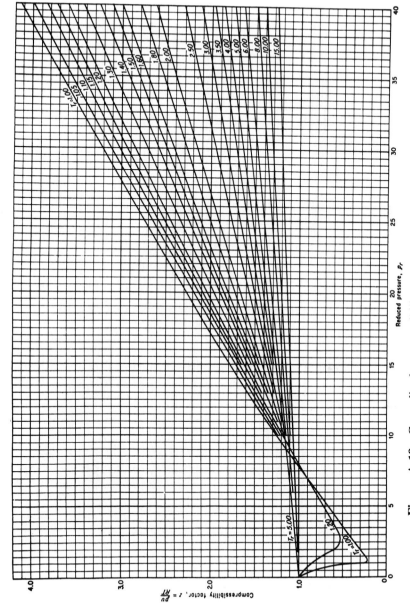

Figure A-19 Generalized compressibility chart up to $p_r = 40$. (After Professor E. F. Obert, University of Wisconsin, *Concepts of Thermodynamics*, McGraw-Hill, New York, 1960.)

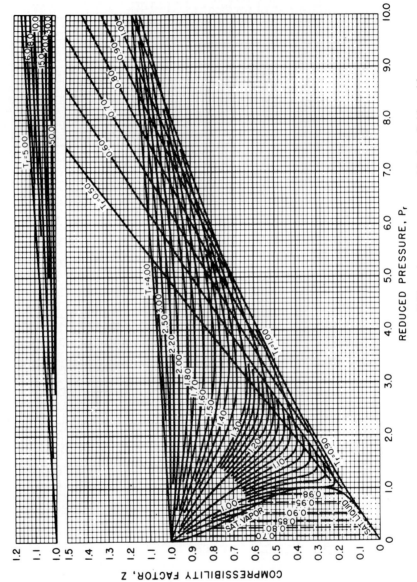

Figure A-20 Generalized compessibility chart up to $p_r = 10$ for fluids with $Z_c = 0.29$. (From J. S. Hsieh, *Journal of Engineering for Industry, Transactions ASME,* vol. 88, 1966, p. 263.)

REDUCED PRESSURE, P_r

COMPRESSIBILITY FACTOR, Z

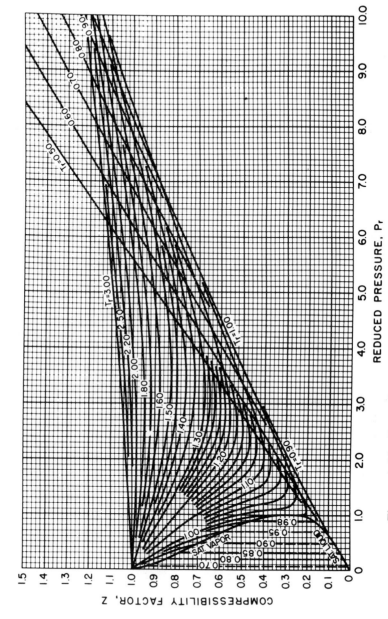

Figure A-21 Generalized compessibility chart up to $p_r = 10$ for fluids with $Z_c = 0.27$. (From J. S. Hsieh, *Journal of Engineering for Industry, Transactions ASME*, vol. 88, 1966, p. 263.)

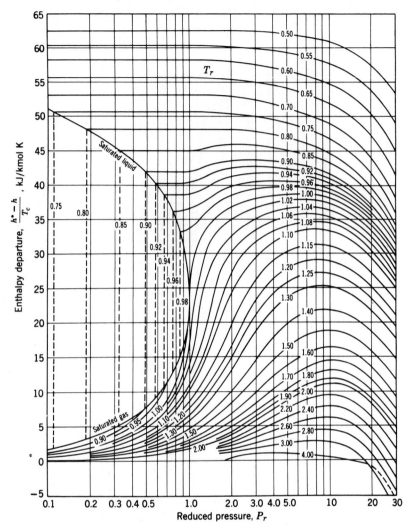

Figure A-22 Generalized enthalpy correction chart. (From G. J. Van Wylen and R. E. Sonntag, *Fundamentals of Classical Thermodynamics,* 3rd ed., John Wiley, New York, 1986.)

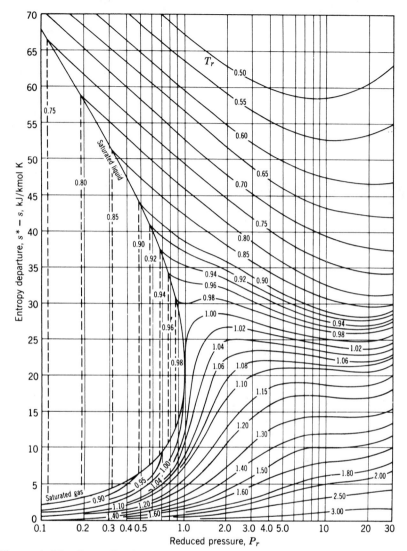

Figure A-23 Generalized entropy correction charg. (From G. J. Van Wylen and R. E. Sonntag, *Fundamentals of Classical Thermodynamics*, 3rd ed., John Wiley, New York, 1986.)

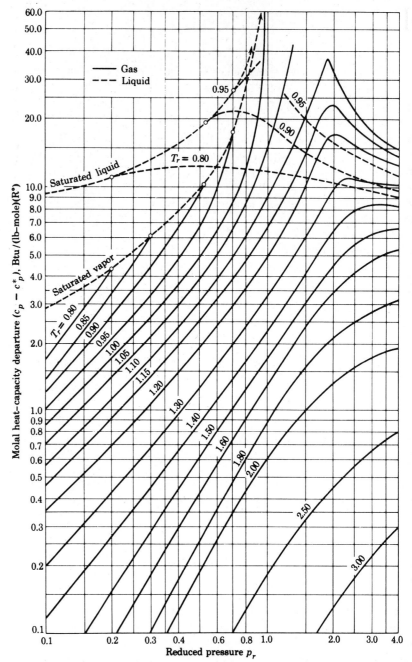

Figure A-24 Generalized specific heat correction chart. (From O. A. Hougen, K. M. Watson, and R. A. Ragatz, *Chemical Process Principles, Part II, Thermodynamics,* John Wiley, New York, 1959.)

Bibliography

1. R. T. BALMER, *Thermodynamics*, West Publishing Co., St. Paul, MN, 1990.

2. R. BARRON, *Cryogenic Systems*, McGraw-Hill, New York, 1966.

3. A. BEJAN, *Advanced Engineering Thermodynamics*, John Wiley, New York, 1988.

4. W. Z. BLACK and J. G. HARTLEY, *Thermodynamics*, Harper Collins, New York, 1991.

5. M. D. BURGHARDT, *Engineering Thermodynamics with Applications*, Harper & Row, New York, 1986.

6. Y. A. CENGAL and M. A. BOLES, *Thermodynamics: An Engineering Approach*, McGraw-Hill, New York, 1989.

7. E. G. CRAVALHO and J. L. SMITH, *Engineering Thermodynamics*, Pitman, Boston, 1981.

8. R. C. FELLINGER and W. J. COOK, *Introduction to Engineering Thermodynamics*, W.C. Brown, Dubuque, IA, 1985.

9. E. P. GYFTOPOULOS and G. P. BERETTA, *Thermodynamics: Foundations and Applications*, Macmillan, New York, 1991.

10. G. N. HATSOPOULOS and J. H. KEENAN, *Principles of General Thermodynamics*, John Wiley, New York, 1965.

11. J. P. HOLMAN, *Thermodynamics*, McGraw-Hill, New York, 1988.

12. O. A. HOUGEN, K. M. WATSON, and R. A. RAGATZ, *Chemical Process Principles, Part II: Thermodynamics*, John Wiley, New York, 1959.

13. J. R. HOWELL and R. O. BUCKIUS, *Fundamentals of Engineering Thermodynamics*, McGraw-Hill, New York, 1987.

14. J. S. HSIEH, *Principles of Thermodynamics*, McGraw-Hill, New York, 1975.

15. J. S. HSIEH, *Solar Energy Engineering*, Prentice-Hall, Englewood Cliffs, NJ, 1986.

16. F. F. HUANG, *Engineering Thermodynamics*, Macmillan, New York, 1988.

17. J. B. JONES and G. A. HAWKINS, *Engineering Thermodynamics*, John Wiley, New York, 1986.

18. R. C. JORDAN and G. B. PRIESTER, *Refrigeration and Air Conditioning*, Prentice-Hall, Englewood Cliffs, NJ, 1956.

19. B. V. KARLEKAR, *Thermodynamics for Engineers*, Prentice-Hall, Englewood Cliffs, NJ, 1983.

20. J. H. KEENAN, J. CHAO, and J. KAYE, *Gas Tables*, John Wiley, New York, English Units, 1980; SI Units, 1983.

21. J. H. KEENAN, F. G. KEYES, P. G. HILL, and J. G. MOORE, *Steam Tables*, John Wiley, New York, English Units, 1969; SI Units, 1978.

22. J. KESTIN, *A Course in Thermodynamics, Volumes I and II*, McGraw-Hill, New York, 1979.

23. D. C. LOOK and H. J. SAUER, *Engineering Thermodynamics*, PWS Publishers, Boston, 1986.

24. M. C. MARTIN, *Elements of Thermodynamics*, Prentice-Hall, Englewood Cliffs, NJ 1986.

25. M. J. MORAN and H. N. SHAPIRO, *Fundamentals of Engineering Thermodynamics*, John Wiley, New York, 1992.

26. G. E. MEYERS, *Engineering Thermodynamics*, Prentice-Hall, Englewood Cliffs, NJ 1989.

27. E. F. OBERT, Concepts of Thermodynamics, McGraw-Hill, New York, 1960.

28. W. C. REYNOLDS, *Thermodynamic Properties in SI: Graphs, Tables, and Computational Equations for 40 Substances*, Department of Mechanical Engineering, Stanford University, Stanford, CA 1979.

29. W. C. REYNOLDS and H. C. PERKINS, *Engineering Thermodynamics*, McGraw-Hill, New York, 1977.

30. A. H. SHAPIRO, *The Dynamics and Thermodynamics of Compressible Fluid Flow*, Ronald Press, New York, 1953.

31. J. L. THRELKELD, *Thermal Environmental Engineering*, Prentice-Hall, Englewood Cliffs, NJ, 1970.

32. G. J. VAN WYLEN and R. E. SONNTAG, *Fundamentals of Classical Thermodynamics*, John Wiley, New York, 1986.

33. K. WARK, *Thermodynamics*, McGraw-Hill, New York, 1988.

34. B. D. WOOD, *Applications of Thermodynamics*, Addison-Wesley, Reading, MA, 1982.

35. M. W. ZEMANSKY and R. H. DITTMAN, *Heat and Thermodynamics*, McGraw-Hill, New York, 1981.

Answers to Selected Problems

1-3. **(a)** 10 lbf, **(b)** 4.445×10^6 dynes, **(c)** 44.45 N

1-6. 4 kg, 39.228 N

1-10. 82.1 kg, 805 N

1-13. 9.93 m/s^2

1-15. $-40°F = -40°C$

1-18. 98.6°F

1-24. 1253.3 Pa

1-28. 793 m

1-31. 2800 kPa

1-34. **(a)** 1361 kg, **(b)** 96.6 km/h, **(c)** 489 kJ

1-38. 1.84×10^{10} kg

1-43. **(a)** 98.1 kW, **(b)** 188.1 kW, **(c)** -21.9 kW

2-3. 11.67×10^5 J

2-7. **(a)** 240 kJ, **(b)** 96.6 kJ, **(c)** 71.2 kJ, **(d)** 216 kJ

2-10. -54 kJ/kg

2-14. 1445 ft·lbf

2-19. -0.0317 J

2-21. -162 J/kg

2-24. -1.1 kW, 1.05 kW, -0.05 kW

2-29. **(a)** 62.26 kJ, **(b)** 1200 W, 162.4 W

2-35. 20.94 kW, 72.55 kW, 426.72 kJ

2-38. 2.58 m/s

2-41. **(a)** -13.0 Btu, **(b)** -16.0 Btu, **(c)** 11.8 Btu

2-48. 12.45 MJ

2-52. 704,742 J

3-7. **(a)** 416.12 kJ/kg, **(b)** 631.68 kJ/kg, **(c)** 182.6 kJ/kg, **(d)** 0.05915 m^3/kg

3-9. 100.03°C

3-12. **(a)** 161.1°C, **(b)** 134.3°C

3-16. 13.97 kg, 0.0614 m^3

3-23. -17.8 kJ

3-27. 671.7 kJ, 85.4 kJ

3-35. 1.03, -30.0 kJ/kg

3-39. 7.65%, 87.2 Btu

3-44. $-44,985$ kJ

3-47. 0.01094 kg, 112.3 kJ

3-52. **(a)** 111°F, **(b)** 111°F

3-56. 52.2 kg

4-4. 5 kPa, 50°C

4-7. -15.31 kJ/kg

4-10. 33.26 kJ

4-15. 311.6 K

4-20. **(a)** 225.6 kJ/kg, **(b)** 226.6 kJ/kg

4-25. **(a)** 2.0608×10^4, **(b)** 2.0629×10^4, **(c)** 2.0489×10^4 (all in kJ/kgmole H$_2$)

4-30. **(a)** 147°C, **(b)** -124 kJ/kg, **(c)** -30 kJ/kg

4-35. **(a)** 240.2°C, **(b)** 1.73, **(c)** -77.15 kJ, **(d)** 63.59 kJ

4-39. 59°C

4-44. $w_{12} = 129.2$, $w_{23} = 0$, $w_{31} = -86.1$, $q_{12} = 775.4$, $q_{23} = -430.8$, $q_{31} = -301.5$, net work $= 43.1$ (all in kJ/kg)

5-2. 1106 m/s

5-7. **(a)** 5.36 hp, **(b)** 1.08 in., **(c)** 61.9 ft/s

5-10. 250 m/s, 10 m/s

5-15. 6.39 kg/s, 16 m/s, 0.012°C

5-20. 109 ft/s, 184 ft

5-26. 3.4 kJ/kg, 5054 kW

5-31. 1.45 kW

5-35. 117.5°C, -61.9 kW

5-39. 272°F

5-44. -6.25°C

5-46. 1.95

5-53. **(a)** 172°C, **(b)** 138×10^6 kg/h

5-55. 6 panels

5-59. 193°C, 105 kg

5-62. 600 kJ

5-67. 184.3 MJ

5-69. 0.282 kg

6-2. 0.109 kJ/kg·K

6-6. 2976 kJ/kgmole, 8.555 kJ/kgmole·K

6-9. 13.2%

6-14. 104 ft^2

6-23. 39.4 kW

6-28. Clain is faulty.

6-31. **(a)** -25.26 kJ, **(b)** -0.08569 kJ/K, **(c)** -0.002324 kJ/K Process is impossible.

6-35. (a) 0.2686 kJ/K, (b) −0.2610 kJ/K, (c) 0.0076 kJ/K

6-41. −44.6 kJ/kg

6-46. 0.0117 Btu/lbm · °R

6-52. 69.8 m/s

6-54. 0.300 W/K for the incandescent light bulb, 0.066 W/K for the fluorescent light bulb. The fluorescent light is more efficient.

6-57. Work: 0, 81.3 Btu/lbm, −62.3 Btu/lbm
Heat: 206.8 Btu/lbm, 83.4 Btu/lbm, −271.2 Btu/lbm
Th. eff: 6.55%

6-62. 0.173 kg

7-1. (a) 49.9°C, (b) 0.823 W/K

7-5. 4055 kW, 870 W/K

7-8. 227°C, 2.56 kJ/K

7-11. .0593 kJ/K

7-16. 7.56 kW

7-20. (a) 9.1 m, (b) 0.000358 kJ/K · s
The process does not violate the second law.

7-26. 160°C, 253°C

7-33. −9.8 J/kg · K

7-38. 43.1 Btu

7-42. 0.615, 0.0867 kJ/kg · K

7-49. (a) Q_{12} = 129.5 Btu, Q_{23} = −18.5 Btu, Q_{31} = −61.1 Btu, (b) η_{th} = 38.5%, (c) For helium, $Q_{12} > 0$, $Q_{23} > 0$, $Q_{31} < 0$

7-54. 155°R, −0.1186 Btu/lbm · °R

7-61. 60.2°F, 0.0000769 Btu/°R · s

7-64. 2505°C, 1.378 J/K

8-1. (a) −86.6 kJ, (b) −7.70 kJ

8-7. (a) 261.4 kJ/kg, (b) 211.5 kJ/kg

8-9. 39.6°C, 58.4 kJ/kg

8-14. 1297 kW

8-20. 89.2 kW, 89.2 kW

8-23. (a) 36.5°C, (b) −47,700 kJ/h, 7,700 kJ/h, −40,000 kJ/h, (c) 40,000 kJ/h

8-25. 48.9 kg, 382 kJ

8-27. 6.97 Btu

8-30. (a) −89,127 kJ, (b) 42,775 kJ

8-36. 30°C, −773 kJ, 773 kJ

8-39. 162 kJ/kg

8-41. (a) −8006 kJ/s, 5669 kJ/s, −2337 kJ/s, (b) 2337 kJ/s

8-50. (a) 2.31 kJ/kg, (b) 1.39 kJ/kg

8-53. (a) 28,840 kJ, (b) 1069 kJ, (c) 1069 kJ

8-56. 0.58 kW

8-59. 2413 kJ, 25.4%

8-61. 37.5%, 91.9%

9-3. (a) 5.00 kg/kgmole, (b) 1.66 kJ/kg · K, (c) 0.03 MPa, 0.27 MPa, (d) 361 K

9-6. 478 K, −5697 kJ/kgmole mixture

9-11. 1647 kJ, 3.757 kJ/K, 1120 kJ

9-19. 3.132 kJ/K · min, 0.9025 kJ/K · min, 1202 kJ/min

9-23. (a) 217.5 kW, (b) 85.3%, (c) 23.9 kW

9-28. (a) 0.0147 lbm W/lbm da, (b) 0.0329

9-32. 36.9 kJ/kg da, 0.0038 kg w/kg da

9-39. 60.7°F

9-41. 53.9°C, 115 kPa

9-45. (a) 1.9°C, (b) 1.27 kW, (c) 0.218 kW, (d) 0.000339 kg/s

9-51. (a) 0.00411 kg w/kg da, (b) 38.2%, (c) 0.889 kJ/kg da

9-53. (a) 36°C, 80%, (b) 18.8 kW

10-2. (a) 14 psia, 209.9 psia, 430.3 psia, 31.6 psia, 560°R, 1200°R, 2460°R, 1263°R, (b) 261.9 Btu/lbm, (c) 134.0 Btu/lbm, (d) 51.2%, (e) 0.148 Btu/lbm · °R, −0.148 Btu/lbm · °R

10-7. 693.1 kJ/kg, 58.8%

10-15. (a) 13,700 ft · lbf, (b) 53.7%, (c) 26.3 Btu, (d) 8.7 Btu

10-25. 149.6 Btu/lbm, 72.8 Btu/lbm, 165.9 Btu/lbm, 89.1 Btu/lbm, 46.3%

10-29. (a) 34.1%, (b) 79,600 ft³/min, (c) 185,000 ft³/min

10-36. (a) 683 N, 837 kJ/s, (b) 663 N, 753 kJ/s

10-44. (a) $\Delta\varphi$: 329.1 kJ/kg, 299.3 kJ/kg, 673.0 kJ/kg, −940.0 kJ/kg, −361.4 kJ/kg
(b) i: 0, 69.1 kJ/kg, 61.7 kJ/kg, 0, 361.4 kJ/kg, 492.2 kJ/kg

10-48. (a) 28,490 kJ/kgmole, 12,300 kJ/kgmole, (b) 16,190 kJ/kgmole, (c) 56.8%, (d) 0, 129 kJ/kgmole, 2821 kJ/kgmole, 0, 4602 kJ/kgmole

11-4. (a) 119,040 kg/min, (b) 17.75 × 10⁶ kJ/min, (c) 17.16 × 10⁶ kJ/min, (d) 3.3%, (e) 1.414 × 10⁶ kg/min, (f) 1.360 × 10⁶ kg/min

11-8. (a) 580.5 Btu/lbm, 8.53 Btu/lbm, 1395.8 Btu/lbm, 823.9 Btu/lbm, (b) 41.0%, (c) 5.97 lbm/kW·h

11-11. 46.0%

11-15. 1349 kJ/kg, 15.2 kJ/kg, 3095 kJ/kg, 1761 kJ/kg, 43.1%

11-23. 2.212 kg/s, 0.375 kg/s, 0.304 kg/s, 5190 kJ/s, 3190 kJ/s, 38.5%

11-27. 28 kJ/kg, 178 kJ/kg, 150 kJ/kg, 15.7%

11-28. $\dot{m}_1$ = 62.5 kg st/s, $\dot{m}_2$ = 12.5 kg st/s, 45.9%, $\dot{m}_s$ = 364 kg Na/s, $\dot{m}_r$ = 222 kg Na/s, $\dot{m}_c$ = 631 kg Na/s

11-32. (a) 6.43 kg air/kg steam, (b) 815.6 kJ/kg steam at turbine inlet, (c) 258.4 kJ/kg air, (d) 1251 kJ/s, (e) 48.0%

11-35. (a) 15,060 kJ/s, (b) 4300 kW, (c) 57.2%

11-43. (a) 156.8 kg/s, 1756.9 kJ/kg, 320.2 kJ/kg, (c) 249.9 kJ/kg, 1.1 kJ/kg

11-45. 34.0%, 72.6%

12-5. (a) 2.37, (b) 11.6 kg/min

12-10. 9.7 tons, 5.0

12-14. (a) 6.13, (b) 11.8 kg/min, (c) 4.52 kW

12-20. (a) 29 kJ/kg, (b) 101 kJ/kg

12-26. (a) 5°C, (b) 1.49, (c) 96.2 kW

12-29. 0.53 hp

12-34. 250 atm, 475 kJ/kg

12-38. 0.209 lbm liquid/lbm compressed 483 Btu/lbm liquefied

12-41. 11.4 kW

12-43. (a) 2.89, (b) 3.48 tons, (c) 1.24 kW, 0.567 kW, (d) −18.82 kJ/kg, −5.41 kJ/kg

12-49. (a) 0.136 kg/s, (b) 15.73 kW, (c) 2.26 kW, (d) 7.96, (e) 0, 0.0684 kJ/s, 0.4008 kJ/s, 0.3157 kJ/s

13-30. 5.12 Btu/lbm

13-34. 103.5 Pa

13-40. 16.9 K/MPa, 56.6°C

13-51. −134.9 J, −134.9 J, −112.4 J/K, 0.032 K

14-3. (a) 306.9 K, (b) 336.6 K, (c) 342 K, (d) 342.4 K

14-4. (a) 11.84 MPa, (b) 10.07 MPa, (c) 9.99 MPa

14-7. (a) 471 K, 90.9 liters, −691 kJ (b) 481 K, 84.2 liters, −671 kJ

14-13. 33.5 kJ/kg

14-19. 146.7 kJ/kg, 460 J/kg·K

14-21. 1.851 Btu/lbm·°R, 1794 Btu/lbm

14-25. −154 kJ/kg, −168 kJ/kg

14-32. 337 K, 94 m/s

14-40. 1.28 kg

14-44. (a) 432°C, (b) −6505 kJ, (c) 4002 kJ

14-46. (a) 678.7 kJ, (b) 123 kJ

14-52. v = 0.94200×10^2 cm³/gmole
h = 0.14380×10^5 J/gmole
u = 0.11517×10^5 J/gmole
s = 0.13946×10^3 J/gmole·K
g = -0.27459×10^5 J/gmole
a = -0.30322×10^5 J/gmole
c_p = 0.36566×10^2 J/gmole·K
c_v = 0.20686×10^2 J/gmole·K

15-5. 22.3 kg air/kg ethane

15-11. (a) carbon 90.39%, hydrogen 9.61%, (b) 22.08 kg air/kg fuel, (c) 162%

15-20. (a) 1.105 lbm water condensed/lbm fuel (b) 1.413 lbm water condensed /lbm fuel

15-23. (a) 14.0 kg air/kg fuel, (b) 12.3 m³/min

15-30. 10.5 kg air/kg fuel

15-36. 31.4 kW

15-38. 2249 K

15-43. 2929 K, 10.4 atm

15-49. 17.4 g/s

15-51. (a) 5369 kJ/s, (b) 2340 kJ/s, (c) 967.9 kJ/s.

15-53. 641.6 m/s

16-7. −7.41°C

16-11. (a) 5.58 lbm, (b) 1371 Btu

16-13. 265.7 kg/min

16-17. 277,861 kJ/kgmole

16-21. $\log_{10} K_p$ = 0.860

16-27. CO, 6.11%; CO_2, 42.37%; O_2, 51.53%

16-29. 3117 K

16-31. (a) $0.852\ H_2O + 0.148\ H_2 + 0.074\ O_2$ (b) $0.928\ H_2O + 0.072\ H_2 + 0.036\ O_2$

16-35. (a) $0.5674\ CO_2 + 0.4326\ CO + 0.3674\ H_2 + 1.6326\ H_2O + 6.016\ N_2$ (b) −146,120 kJ/kgmole CH_4

16-39. 7,844,707 kJ/kgmole $C_{12}H_{26}$

16-42. (a) −2,112,518 kJ/kgmole C_6H_6 (b) 8715 kJ/K·kgmole C_6H_6 (c) 2,597,000 kJ/kgmole C_6H_6

Index